T0181033

Lecture Notes in Computer Science 12356

More information about this subseries at http://www.springer.com/series/7412

Andrea Vedaldi · Horst Bischof ·
Thomas Brox · Jan-Michael Frahm (Eds.)

Computer Vision – ECCV 2020

16th European Conference
Glasgow, UK, August 23–28, 2020
Proceedings, Part XI

 Springer

Editors
Andrea Vedaldi 🆔
University of Oxford
Oxford, UK

Horst Bischof 🆔
Graz University of Technology
Graz, Austria

Thomas Brox 🆔
University of Freiburg
Freiburg im Breisgau, Germany

Jan-Michael Frahm
University of North Carolina at Chapel Hill
Chapel Hill, NC, USA

ISSN 0302-9743 ISSN 1611-3349 (electronic)
Lecture Notes in Computer Science
ISBN 978-3-030-58620-1 ISBN 978-3-030-58621-8 (eBook)
https://doi.org/10.1007/978-3-030-58621-8

LNCS Sublibrary: SL6 – Image Processing, Computer Vision, Pattern Recognition, and Graphics

This Springer imprint is published by the registered company Springer Nature Switzerland AG
The registered company address is: Gewerbestrasse 11, 6330 Cham, Switzerland

Foreword

Hosting the European Conference on Computer Vision (ECCV 2020) was certainly an exciting journey. From the 2016 plan to hold it at the Edinburgh International Conference Centre (hosting 1,800 delegates) to the 2018 plan to hold it at Glasgow's Scottish Exhibition Centre (up to 6,000 delegates), we finally ended with moving online because of the COVID-19 outbreak. While possibly having fewer delegates than expected because of the online format, ECCV 2020 still had over 3,100 registered participants.

Although online, the conference delivered most of the activities expected at a face-to-face conference: peer-reviewed papers, industrial exhibitors, demonstrations, and messaging between delegates. In addition to the main technical sessions, the conference included a strong program of satellite events with 16 tutorials and 44 workshops.

Furthermore, the online conference format enabled new conference features. Every paper had an associated teaser video and a longer full presentation video. Along with the papers and slides from the videos, all these materials were available the week before the conference. This allowed delegates to become familiar with the paper content and be ready for the live interaction with the authors during the conference week. The live event consisted of brief presentations by the oral and spotlight authors and industrial sponsors. Question and answer sessions for all papers were timed to occur twice so delegates from around the world had convenient access to the authors.

As with ECCV 2018, authors' draft versions of the papers appeared online with open access, now on both the Computer Vision Foundation (CVF) and the European Computer Vision Association (ECVA) websites. An archival publication arrangement was put in place with the cooperation of Springer. SpringerLink hosts the final version of the papers with further improvements, such as activating reference links and supplementary materials. These two approaches benefit all potential readers: a version available freely for all researchers, and an authoritative and citable version with additional benefits for SpringerLink subscribers. We thank Alfred Hofmann and Aliaksandr Birukou from Springer for helping to negotiate this agreement, which we expect will continue for future versions of ECCV.

August 2020

Vittorio Ferrari
Bob Fisher
Cordelia Schmid
Emanuele Trucco

Preface

Welcome to the proceedings of the European Conference on Computer Vision (ECCV 2020). This is a unique edition of ECCV in many ways. Due to the COVID-19 pandemic, this is the first time the conference was held online, in a virtual format. This was also the first time the conference relied exclusively on the Open Review platform to manage the review process. Despite these challenges ECCV is thriving. The conference received 5,150 valid paper submissions, of which 1,360 were accepted for publication (27%) and, of those, 160 were presented as spotlights (3%) and 104 as orals (2%). This amounts to more than twice the number of submissions to ECCV 2018 (2,439). Furthermore, CVPR, the largest conference on computer vision, received 5,850 submissions this year, meaning that ECCV is now 87% the size of CVPR in terms of submissions. By comparison, in 2018 the size of ECCV was only 73% of CVPR.

The review model was similar to previous editions of ECCV; in particular, it was double blind in the sense that the authors did not know the name of the reviewers and vice versa. Furthermore, each conference submission was held confidentially, and was only publicly revealed if and once accepted for publication. Each paper received at least three reviews, totalling more than 15,000 reviews. Handling the review process at this scale was a significant challenge. In order to ensure that each submission received as fair and high-quality reviews as possible, we recruited 2,830 reviewers (a 130% increase with reference to 2018) and 207 area chairs (a 60% increase). The area chairs were selected based on their technical expertise and reputation, largely among people that served as area chair in previous top computer vision and machine learning conferences (ECCV, ICCV, CVPR, NeurIPS, etc.). Reviewers were similarly invited from previous conferences. We also encouraged experienced area chairs to suggest additional chairs and reviewers in the initial phase of recruiting.

Despite doubling the number of submissions, the reviewer load was slightly reduced from 2018, from a maximum of 8 papers down to 7 (with some reviewers offering to handle 6 papers plus an emergency review). The area chair load increased slightly, from 18 papers on average to 22 papers on average.

Conflicts of interest between authors, area chairs, and reviewers were handled largely automatically by the Open Review platform via their curated list of user profiles. Many authors submitting to ECCV already had a profile in Open Review. We set a paper registration deadline one week before the paper submission deadline in order to encourage all missing authors to register and create their Open Review profiles well on time (in practice, we allowed authors to create/change papers arbitrarily until the submission deadline). Except for minor issues with users creating duplicate profiles, this allowed us to easily and quickly identify institutional conflicts, and avoid them, while matching papers to area chairs and reviewers.

Papers were matched to area chairs based on: an affinity score computed by the Open Review platform, which is based on paper titles and abstracts, and an affinity score computed by the Toronto Paper Matching System (TPMS), which is based on the paper's full text, the area chair bids for individual papers, load balancing, and conflict avoidance. Open Review provides the program chairs a convenient web interface to experiment with different configurations of the matching algorithm. The chosen configuration resulted in about 50% of the assigned papers to be highly ranked by the area chair bids, and 50% to be ranked in the middle, with very few low bids assigned.

Assignments to reviewers were similar, with two differences. First, there was a maximum of 7 papers assigned to each reviewer. Second, area chairs recommended up to seven reviewers per paper, providing another highly-weighed term to the affinity scores used for matching.

The assignment of papers to area chairs was smooth. However, it was more difficult to find suitable reviewers for all papers. Having a ratio of 5.6 papers per reviewer with a maximum load of 7 (due to emergency reviewer commitment), which did not allow for much wiggle room in order to also satisfy conflict and expertise constraints. We received some complaints from reviewers who did not feel qualified to review specific papers and we reassigned them wherever possible. However, the large scale of the conference, the many constraints, and the fact that a large fraction of such complaints arrived very late in the review process made this process very difficult and not all complaints could be addressed.

Reviewers had six weeks to complete their assignments. Possibly due to COVID-19 or the fact that the NeurIPS deadline was moved closer to the review deadline, a record 30% of the reviews were still missing after the deadline. By comparison, ECCV 2018 experienced only 10% missing reviews at this stage of the process. In the subsequent week, area chairs chased the missing reviews intensely, found replacement reviewers in their own team, and managed to reach 10% missing reviews. Eventually, we could provide almost all reviews (more than 99.9%) with a delay of only a couple of days on the initial schedule by a significant use of emergency reviews. If this trend is confirmed, it might be a major challenge to run a smooth review process in future editions of ECCV. The community must reconsider prioritization of the time spent on paper writing (the number of submissions increased a lot despite COVID-19) and time spent on paper reviewing (the number of reviews delivered in time decreased a lot presumably due to COVID-19 or NeurIPS deadline). With this imbalance the peer-review system that ensures the quality of our top conferences may break soon.

Reviewers submitted their reviews independently. In the reviews, they had the opportunity to ask questions to the authors to be addressed in the rebuttal. However, reviewers were told not to request any significant new experiment. Using the Open Review interface, authors could provide an answer to each individual review, but were also allowed to cross-reference reviews and responses in their answers. Rather than PDF files, we allowed the use of formatted text for the rebuttal. The rebuttal and initial reviews were then made visible to all reviewers and the primary area chair for a given paper. The area chair encouraged and moderated the reviewer discussion. During the discussions, reviewers were invited to reach a consensus and possibly adjust their ratings as a result of the discussion and of the evidence in the rebuttal.

After the discussion period ended, most reviewers entered a final rating and recommendation, although in many cases this did not differ from their initial recommendation. Based on the updated reviews and discussion, the primary area chair then made a preliminary decision to accept or reject the paper and wrote a justification for it (meta-review). Except for cases where the outcome of this process was absolutely clear (as indicated by the three reviewers and primary area chairs all recommending clear rejection), the decision was then examined and potentially challenged by a secondary area chair. This led to further discussion and overturning a small number of preliminary decisions. Needless to say, there was no in-person area chair meeting, which would have been impossible due to COVID-19.

Area chairs were invited to observe the consensus of the reviewers whenever possible and use extreme caution in overturning a clear consensus to accept or reject a paper. If an area chair still decided to do so, she/he was asked to clearly justify it in the meta-review and to explicitly obtain the agreement of the secondary area chair. In practice, very few papers were rejected after being confidently accepted by the reviewers.

This was the first time Open Review was used as the main platform to run ECCV. In 2018, the program chairs used CMT3 for the user-facing interface and Open Review internally, for matching and conflict resolution. Since it is clearly preferable to only use a single platform, this year we switched to using Open Review in full. The experience was largely positive. The platform is highly-configurable, scalable, and open source. Being written in Python, it is easy to write scripts to extract data programmatically. The paper matching and conflict resolution algorithms and interfaces are top-notch, also due to the excellent author profiles in the platform. Naturally, there were a few kinks along the way due to the fact that the ECCV Open Review configuration was created from scratch for this event and it differs in substantial ways from many other Open Review conferences. However, the Open Review development and support team did a fantastic job in helping us to get the configuration right and to address issues in a timely manner as they unavoidably occurred. We cannot thank them enough for the tremendous effort they put into this project.

Finally, we would like to thank everyone involved in making ECCV 2020 possible in these very strange and difficult times. This starts with our authors, followed by the area chairs and reviewers, who ran the review process at an unprecedented scale. The whole Open Review team (and in particular Melisa Bok, Mohit Unyal, Carlos Mondragon Chapa, and Celeste Martinez Gomez) worked incredibly hard for the entire duration of the process. We would also like to thank René Vidal for contributing to the adoption of Open Review. Our thanks also go to Laurent Charling for TPMS and to the program chairs of ICML, ICLR, and NeurIPS for cross checking double submissions. We thank the website chair, Giovanni Farinella, and the CPI team (in particular Ashley Cook, Miriam Verdon, Nicola McGrane, and Sharon Kerr) for promptly adding material to the website as needed in the various phases of the process. Finally, we thank the publication chairs, Albert Ali Salah, Hamdi Dibeklioglu, Metehan Doyran, Henry Howard-Jenkins, Victor Prisacariu, Siyu Tang, and Gul Varol, who managed to compile these substantial proceedings in an exceedingly compressed schedule. We express our thanks to the ECVA team, in particular Kristina Scherbaum for allowing open access of the proceedings. We thank Alfred Hofmann from Springer who again serve as the publisher. Finally, we thank the other chairs of ECCV 2020, including in particular the general chairs for very useful feedback with the handling of the program.

August 2020   Andrea Vedaldi
                                                       Horst Bischof
                                                       Thomas Brox
                                           Jan-Michael Frahm

Organization

General Chairs

Vittorio Ferrari	Google Research, Switzerland
Bob Fisher	University of Edinburgh, UK
Cordelia Schmid	Google and Inria, France
Emanuele Trucco	University of Dundee, UK

Program Chairs

Andrea Vedaldi	University of Oxford, UK
Horst Bischof	Graz University of Technology, Austria
Thomas Brox	University of Freiburg, Germany
Jan-Michael Frahm	University of North Carolina, USA

Industrial Liaison Chairs

Jim Ashe	University of Edinburgh, UK
Helmut Grabner	Zurich University of Applied Sciences, Switzerland
Diane Larlus	NAVER LABS Europe, France
Cristian Novotny	University of Edinburgh, UK

Local Arrangement Chairs

Yvan Petillot	Heriot-Watt University, UK
Paul Siebert	University of Glasgow, UK

Academic Demonstration Chair

Thomas Mensink	Google Research and University of Amsterdam, The Netherlands

Poster Chair

Stephen Mckenna	University of Dundee, UK

Technology Chair

Gerardo Aragon Camarasa	University of Glasgow, UK

Tutorial Chairs

Carlo Colombo   University of Florence, Italy
Sotirios Tsaftaris   University of Edinburgh, UK

Publication Chairs

Albert Ali Salah   Utrecht University, The Netherlands
Hamdi Dibeklioglu   Bilkent University, Turkey
Metehan Doyran   Utrecht University, The Netherlands
Henry Howard-Jenkins   University of Oxford, UK
Victor Adrian Prisacariu University of Oxford, UK
Siyu Tang   ETH Zurich, Switzerland
Gul Varol   University of Oxford, UK

Website Chair

Giovanni Maria Farinella   University of Catania, Italy

Workshops Chairs

Adrien Bartoli   University of Clermont Auvergne, France
Andrea Fusiello   University of Udine, Italy

Area Chairs

Lourdes Agapito   University College London, UK
Zeynep Akata   University of Tübingen, Germany
Karteek Alahari   Inria, France
Antonis Argyros   University of Crete, Greece
Hossein Azizpour   KTH Royal Institute of Technology, Sweden
Joao P. Barreto   Universidade de Coimbra, Portugal
Alexander C. Berg   University of North Carolina at Chapel Hill, USA
Matthew B. Blaschko   KU Leuven, Belgium
Lubomir D. Bourdev   WaveOne, Inc., USA
Edmond Boyer   Inria, France
Yuri Boykov   University of Waterloo, Canada
Gabriel Brostow   University College London, UK
Michael S. Brown   National University of Singapore, Singapore
Jianfei Cai   Monash University, Australia
Barbara Caputo   Politecnico di Torino, Italy
Ayan Chakrabarti   Washington University, St. Louis, USA
Tat-Jen Cham   Nanyang Technological University, Singapore
Manmohan Chandraker   University of California, San Diego, USA
Rama Chellappa   Johns Hopkins University, USA
Liang-Chieh Chen   Google, USA

Yung-Yu Chuang	National Taiwan University, Taiwan
Ondrej Chum	Czech Technical University in Prague, Czech Republic
Brian Clipp	Kitware, USA
John Collomosse	University of Surrey and Adobe Research, UK
Jason J. Corso	University of Michigan, USA
David J. Crandall	Indiana University, USA
Daniel Cremers	University of California, Los Angeles, USA
Fabio Cuzzolin	Oxford Brookes University, UK
Jifeng Dai	SenseTime, SAR China
Kostas Daniilidis	University of Pennsylvania, USA
Andrew Davison	Imperial College London, UK
Alessio Del Bue	Fondazione Istituto Italiano di Tecnologia, Italy
Jia Deng	Princeton University, USA
Alexey Dosovitskiy	Google, Germany
Matthijs Douze	Facebook, France
Enrique Dunn	Stevens Institute of Technology, USA
Irfan Essa	Georgia Institute of Technology and Google, USA
Giovanni Maria Farinella	University of Catania, Italy
Ryan Farrell	Brigham Young University, USA
Paolo Favaro	University of Bern, Switzerland
Rogerio Feris	International Business Machines, USA
Cornelia Fermuller	University of Maryland, College Park, USA
David J. Fleet	Vector Institute, Canada
Friedrich Fraundorfer	DLR, Austria
Mario Fritz	CISPA Helmholtz Center for Information Security, Germany
Pascal Fua	EPFL (Swiss Federal Institute of Technology Lausanne), Switzerland
Yasutaka Furukawa	Simon Fraser University, Canada
Li Fuxin	Oregon State University, USA
Efstratios Gavves	University of Amsterdam, The Netherlands
Peter Vincent Gehler	Amazon, USA
Theo Gevers	University of Amsterdam, The Netherlands
Ross Girshick	Facebook AI Research, USA
Boqing Gong	Google, USA
Stephen Gould	Australian National University, Australia
Jinwei Gu	SenseTime Research, USA
Abhinav Gupta	Facebook, USA
Bohyung Han	Seoul National University, South Korea
Bharath Hariharan	Cornell University, USA
Tal Hassner	Facebook AI Research, USA
Xuming He	Australian National University, Australia
Joao F. Henriques	University of Oxford, UK
Adrian Hilton	University of Surrey, UK
Minh Hoai	Stony Brooks, State University of New York, USA
Derek Hoiem	University of Illinois Urbana-Champaign, USA

Timothy Hospedales	University of Edinburgh and Samsung, UK
Gang Hua	Wormpex AI Research, USA
Slobodan Ilic	Siemens AG, Germany
Hiroshi Ishikawa	Waseda University, Japan
Jiaya Jia	The Chinese University of Hong Kong, SAR China
Hailin Jin	Adobe Research, USA
Justin Johnson	University of Michigan, USA
Frederic Jurie	University of Caen Normandie, France
Fredrik Kahl	Chalmers University, Sweden
Sing Bing Kang	Zillow, USA
Gunhee Kim	Seoul National University, South Korea
Junmo Kim	Korea Advanced Institute of Science and Technology, South Korea
Tae-Kyun Kim	Imperial College London, UK
Ron Kimmel	Technion-Israel Institute of Technology, Israel
Alexander Kirillov	Facebook AI Research, USA
Kris Kitani	Carnegie Mellon University, USA
Iasonas Kokkinos	Ariel AI, UK
Vladlen Koltun	Intel Labs, USA
Nikos Komodakis	Ecole des Ponts ParisTech, France
Piotr Koniusz	Australian National University, Australia
M. Pawan Kumar	University of Oxford, UK
Kyros Kutulakos	University of Toronto, Canada
Christoph Lampert	IST Austria, Austria
Ivan Laptev	Inria, France
Diane Larlus	NAVER LABS Europe, France
Laura Leal-Taixe	Technical University Munich, Germany
Honglak Lee	Google and University of Michigan, USA
Joon-Young Lee	Adobe Research, USA
Kyoung Mu Lee	Seoul National University, South Korea
Seungyong Lee	POSTECH, South Korea
Yong Jae Lee	University of California, Davis, USA
Bastian Leibe	RWTH Aachen University, Germany
Victor Lempitsky	Samsung, Russia
Ales Leonardis	University of Birmingham, UK
Marius Leordeanu	Institute of Mathematics of the Romanian Academy, Romania
Vincent Lepetit	ENPC ParisTech, France
Hongdong Li	The Australian National University, Australia
Xi Li	Zhejiang University, China
Yin Li	University of Wisconsin-Madison, USA
Zicheng Liao	Zhejiang University, China
Jongwoo Lim	Hanyang University, South Korea
Stephen Lin	Microsoft Research Asia, China
Yen-Yu Lin	National Chiao Tung University, Taiwan, China
Zhe Lin	Adobe Research, USA

Haibin Ling	Stony Brooks, State University of New York, USA
Jiaying Liu	Peking University, China
Ming-Yu Liu	NVIDIA, USA
Si Liu	Beihang University, China
Xiaoming Liu	Michigan State University, USA
Huchuan Lu	Dalian University of Technology, China
Simon Lucey	Carnegie Mellon University, USA
Jiebo Luo	University of Rochester, USA
Julien Mairal	Inria, France
Michael Maire	University of Chicago, USA
Subhransu Maji	University of Massachusetts, Amherst, USA
Yasushi Makihara	Osaka University, Japan
Jiri Matas	Czech Technical University in Prague, Czech Republic
Yasuyuki Matsushita	Osaka University, Japan
Philippos Mordohai	Stevens Institute of Technology, USA
Vittorio Murino	University of Verona, Italy
Naila Murray	NAVER LABS Europe, France
Hajime Nagahara	Osaka University, Japan
P. J. Narayanan	International Institute of Information Technology (IIIT), Hyderabad, India
Nassir Navab	Technical University of Munich, Germany
Natalia Neverova	Facebook AI Research, France
Matthias Niessner	Technical University of Munich, Germany
Jean-Marc Odobez	Idiap Research Institute and Swiss Federal Institute of Technology Lausanne, Switzerland
Francesca Odone	Università di Genova, Italy
Takeshi Oishi	The University of Tokyo, Tokyo Institute of Technology, Japan
Vicente Ordonez	University of Virginia, USA
Manohar Paluri	Facebook AI Research, USA
Maja Pantic	Imperial College London, UK
In Kyu Park	Inha University, South Korea
Ioannis Patras	Queen Mary University of London, UK
Patrick Perez	Valeo, France
Bryan A. Plummer	Boston University, USA
Thomas Pock	Graz University of Technology, Austria
Marc Pollefeys	ETH Zurich and Microsoft MR & AI Zurich Lab, Switzerland
Jean Ponce	Inria, France
Gerard Pons-Moll	MPII, Saarland Informatics Campus, Germany
Jordi Pont-Tuset	Google, Switzerland
James Matthew Rehg	Georgia Institute of Technology, USA
Ian Reid	University of Adelaide, Australia
Olaf Ronneberger	DeepMind London, UK
Stefan Roth	TU Darmstadt, Germany
Bryan Russell	Adobe Research, USA

Mathieu Salzmann	EPFL, Switzerland
Dimitris Samaras	Stony Brook University, USA
Imari Sato	National Institute of Informatics (NII), Japan
Yoichi Sato	The University of Tokyo, Japan
Torsten Sattler	Czech Technical University in Prague, Czech Republic
Daniel Scharstein	Middlebury College, USA
Bernt Schiele	MPII, Saarland Informatics Campus, Germany
Julia A. Schnabel	King's College London, UK
Nicu Sebe	University of Trento, Italy
Greg Shakhnarovich	Toyota Technological Institute at Chicago, USA
Humphrey Shi	University of Oregon, USA
Jianbo Shi	University of Pennsylvania, USA
Jianping Shi	SenseTime, China
Leonid Sigal	University of British Columbia, Canada
Cees Snoek	University of Amsterdam, The Netherlands
Richard Souvenir	Temple University, USA
Hao Su	University of California, San Diego, USA
Akihiro Sugimoto	National Institute of Informatics (NII), Japan
Jian Sun	Megvii Technology, China
Jian Sun	Xi'an Jiaotong University, China
Chris Sweeney	Facebook Reality Labs, USA
Yu-wing Tai	Kuaishou Technology, China
Chi-Keung Tang	The Hong Kong University of Science and Technology, SAR China
Radu Timofte	ETH Zurich, Switzerland
Sinisa Todorovic	Oregon State University, USA
Giorgos Tolias	Czech Technical University in Prague, Czech Republic
Carlo Tomasi	Duke University, USA
Tatiana Tommasi	Politecnico di Torino, Italy
Lorenzo Torresani	Facebook AI Research and Dartmouth College, USA
Alexander Toshev	Google, USA
Zhuowen Tu	University of California, San Diego, USA
Tinne Tuytelaars	KU Leuven, Belgium
Jasper Uijlings	Google, Switzerland
Nuno Vasconcelos	University of California, San Diego, USA
Olga Veksler	University of Waterloo, Canada
Rene Vidal	Johns Hopkins University, USA
Gang Wang	Alibaba Group, China
Jingdong Wang	Microsoft Research Asia, China
Yizhou Wang	Peking University, China
Lior Wolf	Facebook AI Research and Tel Aviv University, Israel
Jianxin Wu	Nanjing University, China
Tao Xiang	University of Surrey, UK
Saining Xie	Facebook AI Research, USA
Ming-Hsuan Yang	University of California at Merced and Google, USA
Ruigang Yang	University of Kentucky, USA

Kwang Moo Yi   University of Victoria, Canada
Zhaozheng Yin   Stony Brook, State University of New York, USA
Chang D. Yoo   Korea Advanced Institute of Science and Technology,
                                  South Korea
Shaodi You   University of Amsterdam, The Netherlands
Jingyi Yu   ShanghaiTech University, China
Stella Yu   University of California, Berkeley, and ICSI, USA
Stefanos Zafeiriou   Imperial College London, UK
Hongbin Zha   Peking University, China
Tianzhu Zhang   University of Science and Technology of China, China
Liang Zheng   Australian National University, Australia
Todd E. Zickler   Harvard University, USA
Andrew Zisserman   University of Oxford, UK

Technical Program Committee

Sathyanarayanan   Samuel Albanie   Pablo Arbelaez
  N. Aakur   Shadi Albarqouni   Shervin Ardeshir
Wael Abd Almgaeed   Cenek Albl   Sercan O. Arik
Abdelrahman   Hassan Abu Alhaija   Anil Armagan
  Abdelhamed   Daniel Aliaga   Anurag Arnab
Abdullah Abuolaim   Mohammad   Chetan Arora
Supreeth Achar   S. Aliakbarian   Federica Arrigoni
Hanno Ackermann   Rahaf Aljundi   Mathieu Aubry
Ehsan Adeli   Thiemo Alldieck   Shai Avidan
Triantafyllos Afouras   Jon Almazan   Angelica I. Aviles-Rivero
Sameer Agarwal   Jose M. Alvarez   Yannis Avrithis
Aishwarya Agrawal   Senjian An   Ismail Ben Ayed
Harsh Agrawal   Saket Anand   Shekoofeh Azizi
Pulkit Agrawal   Codruta Ancuti   Ioan Andrei Bârsan
Antonio Agudo   Cosmin Ancuti   Artem Babenko
Eirikur Agustsson   Peter Anderson   Deepak Babu Sam
Karim Ahmed   Juan Andrade-Cetto   Seung-Hwan Baek
Byeongjoo Ahn   Alexander Andreopoulos   Seungryul Baek
Unaiza Ahsan   Misha Andriluka   Andrew D. Bagdanov
Thalaiyasingam Ajanthan   Dragomir Anguelov   Shai Bagon
Kenan E. Ak   Rushil Anirudh   Yuval Bahat
Emre Akbas   Michel Antunes   Junjie Bai
Naveed Akhtar   Oisin Mac Aodha   Song Bai
Derya Akkaynak   Srikar Appalaraju   Xiang Bai
Yagiz Aksoy   Relja Arandjelovic   Yalong Bai
Ziad Al-Halah   Nikita Araslanov   Yancheng Bai
Xavier Alameda-Pineda   Andre Araujo   Peter Bajcsy
Jean-Baptiste Alayrac   Helder Araujo   Slawomir Bak

Mahsa Baktashmotlagh
Kavita Bala
Yogesh Balaji
Guha Balakrishnan
V. N. Balasubramanian
Federico Baldassarre
Vassileios Balntas
Shurjo Banerjee
Aayush Bansal
Ankan Bansal
Jianmin Bao
Linchao Bao
Wenbo Bao
Yingze Bao
Akash Bapat
Md Jawadul Hasan Bappy
Fabien Baradel
Lorenzo Baraldi
Daniel Barath
Adrian Barbu
Kobus Barnard
Nick Barnes
Francisco Barranco
Jonathan T. Barron
Arslan Basharat
Chaim Baskin
Anil S. Baslamisli
Jorge Batista
Kayhan Batmanghelich
Konstantinos Batsos
David Bau
Luis Baumela
Christoph Baur
Eduardo
   Bayro-Corrochano
Paul Beardsley
Jan Bednavr'ik
Oscar Beijbom
Philippe Bekaert
Esube Bekele
Vasileios Belagiannis
Ohad Ben-Shahar
Abhijit Bendale
Róger Bermúdez-Chacón
Maxim Berman
Jesus Bermudez-cameo

Florian Bernard
Stefano Berretti
Marcelo Bertalmio
Gedas Bertasius
Cigdem Beyan
Lucas Beyer
Vijayakumar Bhagavatula
Arjun Nitin Bhagoji
Apratim Bhattacharyya
Binod Bhattarai
Sai Bi
Jia-Wang Bian
Simone Bianco
Adel Bibi
Tolga Birdal
Tom Bishop
Soma Biswas
Mårten Björkman
Volker Blanz
Vishnu Boddeti
Navaneeth Bodla
Simion-Vlad Bogolin
Xavier Boix
Piotr Bojanowski
Timo Bolkart
Guido Borghi
Larbi Boubchir
Guillaume Bourmaud
Adrien Bousseau
Thierry Bouwmans
Richard Bowden
Hakan Boyraz
Mathieu Brédif
Samarth Brahmbhatt
Steve Branson
Nikolas Brasch
Biagio Brattoli
Ernesto Brau
Toby P. Breckon
Francois Bremond
Jesus Briales
Sofia Broomé
Marcus A. Brubaker
Luc Brun
Silvia Bucci
Shyamal Buch

Pradeep Buddharaju
Uta Buechler
Mai Bui
Tu Bui
Adrian Bulat
Giedrius T. Burachas
Elena Burceanu
Xavier P. Burgos-Artizzu
Kaylee Burns
Andrei Bursuc
Benjamin Busam
Wonmin Byeon
Zoya Bylinskii
Sergi Caelles
Jianrui Cai
Minjie Cai
Yujun Cai
Zhaowei Cai
Zhipeng Cai
Juan C. Caicedo
Simone Calderara
Necati Cihan Camgoz
Dylan Campbell
Octavia Camps
Jiale Cao
Kaidi Cao
Liangliang Cao
Xiangyong Cao
Xiaochun Cao
Yang Cao
Yu Cao
Yue Cao
Zhangjie Cao
Luca Carlone
Mathilde Caron
Dan Casas
Thomas J. Cashman
Umberto Castellani
Lluis Castrejon
Jacopo Cavazza
Fabio Cermelli
Hakan Cevikalp
Menglei Chai
Ishani Chakraborty
Rudrasis Chakraborty
Antoni B. Chan

Kwok-Ping Chan
Siddhartha Chandra
Sharat Chandran
Arjun Chandrasekaran
Angel X. Chang
Che-Han Chang
Hong Chang
Hyun Sung Chang
Hyung Jin Chang
Jianlong Chang
Ju Yong Chang
Ming-Ching Chang
Simyung Chang
Xiaojun Chang
Yu-Wei Chao
Devendra S. Chaplot
Arslan Chaudhry
Rizwan A. Chaudhry
Can Chen
Chang Chen
Chao Chen
Chen Chen
Chu-Song Chen
Dapeng Chen
Dong Chen
Dongdong Chen
Guanying Chen
Hongge Chen
Hsin-yi Chen
Huaijin Chen
Hwann-Tzong Chen
Jianbo Chen
Jianhui Chen
Jiansheng Chen
Jiaxin Chen
Jie Chen
Jun-Cheng Chen
Kan Chen
Kevin Chen
Lin Chen
Long Chen
Min-Hung Chen
Qifeng Chen
Shi Chen
Shixing Chen
Tianshui Chen

Weifeng Chen
Weikai Chen
Xi Chen
Xiaohan Chen
Xiaozhi Chen
Xilin Chen
Xingyu Chen
Xinlei Chen
Xinyun Chen
Yi-Ting Chen
Yilun Chen
Ying-Cong Chen
Yinpeng Chen
Yiran Chen
Yu Chen
Yu-Sheng Chen
Yuhua Chen
Yun-Chun Chen
Yunpeng Chen
Yuntao Chen
Zhuoyuan Chen
Zitian Chen
Anchieh Cheng
Bowen Cheng
Erkang Cheng
Gong Cheng
Guangliang Cheng
Jingchun Cheng
Jun Cheng
Li Cheng
Ming-Ming Cheng
Yu Cheng
Ziang Cheng
Anoop Cherian
Dmitry Chetverikov
Ngai-man Cheung
William Cheung
Ajad Chhatkuli
Naoki Chiba
Benjamin Chidester
Han-pang Chiu
Mang Tik Chiu
Wei-Chen Chiu
Donghyeon Cho
Hojin Cho
Minsu Cho

Nam Ik Cho
Tim Cho
Tae Eun Choe
Chiho Choi
Edward Choi
Inchang Choi
Jinsoo Choi
Jonghyun Choi
Jongwon Choi
Yukyung Choi
Hisham Cholakkal
Eunji Chong
Jaegul Choo
Christopher Choy
Hang Chu
Peng Chu
Wen-Sheng Chu
Albert Chung
Joon Son Chung
Hai Ci
Safa Cicek
Ramazan G. Cinbis
Arridhana Ciptadi
Javier Civera
James J. Clark
Ronald Clark
Felipe Codevilla
Michael Cogswell
Andrea Cohen
Maxwell D. Collins
Carlo Colombo
Yang Cong
Adria R. Continente
Marcella Cornia
John Richard Corring
Darren Cosker
Dragos Costea
Garrison W. Cottrell
Florent Couzinie-Devy
Marco Cristani
Ioana Croitoru
James L. Crowley
Jiequan Cui
Zhaopeng Cui
Ross Cutler
Antonio D'Innocente

Rozenn Dahyot
Bo Dai
Dengxin Dai
Hang Dai
Longquan Dai
Shuyang Dai
Xiyang Dai
Yuchao Dai
Adrian V. Dalca
Dima Damen
Bharath B. Damodaran
Kristin Dana
Martin Danelljan
Zheng Dang
Zachary Alan Daniels
Donald G. Dansereau
Abhishek Das
Samyak Datta
Achal Dave
Titas De
Rodrigo de Bem
Teo de Campos
Raoul de Charette
Shalini De Mello
Joseph DeGol
Herve Delingette
Haowen Deng
Jiankang Deng
Weijian Deng
Zhiwei Deng
Joachim Denzler
Konstantinos G. Derpanis
Aditya Deshpande
Frederic Devernay
Somdip Dey
Arturo Deza
Abhinav Dhall
Helisa Dhamo
Vikas Dhiman
Fillipe Dias Moreira
    de Souza
Ali Diba
Ferran Diego
Guiguang Ding
Henghui Ding
Jian Ding

Mingyu Ding
Xinghao Ding
Zhengming Ding
Robert DiPietro
Cosimo Distante
Ajay Divakaran
Mandar Dixit
Abdelaziz Djelouah
Thanh-Toan Do
Jose Dolz
Bo Dong
Chao Dong
Jiangxin Dong
Weiming Dong
Weisheng Dong
Xingping Dong
Xuanyi Dong
Yinpeng Dong
Gianfranco Doretto
Hazel Doughty
Hassen Drira
Bertram Drost
Dawei Du
Ye Duan
Yueqi Duan
Abhimanyu Dubey
Anastasia Dubrovina
Stefan Duffner
Chi Nhan Duong
Thibaut Durand
Zoran Duric
Iulia Duta
Debidatta Dwibedi
Benjamin Eckart
Marc Eder
Marzieh Edraki
Alexei A. Efros
Kiana Ehsani
Hazm Kemal Ekenel
James H. Elder
Mohamed Elgharib
Shireen Elhabian
Ehsan Elhamifar
Mohamed Elhoseiny
Ian Endres
N. Benjamin Erichson

Jan Ernst
Sergio Escalera
Francisco Escolano
Victor Escorcia
Carlos Esteves
Francisco J. Estrada
Bin Fan
Chenyou Fan
Deng-Ping Fan
Haoqi Fan
Hehe Fan
Heng Fan
Kai Fan
Lijie Fan
Linxi Fan
Quanfu Fan
Shaojing Fan
Xiaochuan Fan
Xin Fan
Yuchen Fan
Sean Fanello
Hao-Shu Fang
Haoyang Fang
Kuan Fang
Yi Fang
Yuming Fang
Azade Farshad
Alireza Fathi
Raanan Fattal
Joao Fayad
Xiaohan Fei
Christoph Feichtenhofer
Michael Felsberg
Chen Feng
Jiashi Feng
Junyi Feng
Mengyang Feng
Qianli Feng
Zhenhua Feng
Michele Fenzi
Andras Ferencz
Martin Fergie
Basura Fernando
Ethan Fetaya
Michael Firman
John W. Fisher

Matthew Fisher

Boris Flach

Corneliu Florea

Wolfgang Foerstner

David Fofi

Gian Luca Foresti

Per-Erik Forssen

David Fouhey

Katerina Fragkiadaki

Victor Fragoso

Jean-Sébastien Franco

Ohad Fried

Iuri Frosio

Cheng-Yang Fu

Huazhu Fu

Jianlong Fu

Jingjing Fu

Xueyang Fu

Yanwei Fu

Ying Fu

Yun Fu

Olac Fuentes

Kent Fujiwara

Takuya Funatomi

Christopher Funk

Thomas Funkhouser

Antonino Furnari

Ryo Furukawa

Erik Gärtner

Raghudeep Gadde

Matheus Gadelha

Vandit Gajjar

Trevor Gale

Juergen Gall

Mathias Gallardo

Guillermo Gallego

Orazio Gallo

Chuang Gan

Zhe Gan

Madan Ravi Ganesh

Aditya Ganeshan

Siddha Ganju

Bin-Bin Gao

Changxin Gao

Feng Gao

Hongchang Gao

Jin Gao

Jiyang Gao

Junbin Gao

Katelyn Gao

Lin Gao

Mingfei Gao

Ruiqi Gao

Ruohan Gao

Shenghua Gao

Yuan Gao

Yue Gao

Noa Garcia

Alberto Garcia-Garcia

Guillermo

   Garcia-Hernando

Jacob R. Gardner

Animesh Garg

Kshitiz Garg

Rahul Garg

Ravi Garg

Philip N. Garner

Kirill Gavrilyuk

Paul Gay

Shiming Ge

Weifeng Ge

Baris Gecer

Xin Geng

Kyle Genova

Stamatios Georgoulis

Bernard Ghanem

Michael Gharbi

Kamran Ghasedi

Golnaz Ghiasi

Arnab Ghosh

Partha Ghosh

Silvio Giancola

Andrew Gilbert

Rohit Girdhar

Xavier Giro-i-Nieto

Thomas Gittings

Ioannis Gkioulekas

Clement Godard

Vaibhava Goel

Bastian Goldluecke

Lluis Gomez

Nuno Gonçalves

Dong Gong

Ke Gong

Mingming Gong

Abel Gonzalez-Garcia

Ariel Gordon

Daniel Gordon

Paulo Gotardo

Venu Madhav Govindu

Ankit Goyal

Priya Goyal

Raghav Goyal

Benjamin Graham

Douglas Gray

Brent A. Griffin

Etienne Grossmann

David Gu

Jiayuan Gu

Jiuxiang Gu

Lin Gu

Qiao Gu

Shuhang Gu

Jose J. Guerrero

Paul Guerrero

Jie Gui

Jean-Yves Guillemaut

Riza Alp Guler

Erhan Gundogdu

Fatma Guney

Guodong Guo

Kaiwen Guo

Qi Guo

Sheng Guo

Shi Guo

Tiantong Guo

Xiaojie Guo

Yijie Guo

Yiluan Guo

Yuanfang Guo

Yulan Guo

Agrim Gupta

Ankush Gupta

Mohit Gupta

Saurabh Gupta

Tanmay Gupta

Danna Gurari

Abner Guzman-Rivera

JunYoung Gwak
Michael Gygli
Jung-Woo Ha
Simon Hadfield
Isma Hadji
Bjoern Haefner
Taeyoung Hahn
Levente Hajder
Peter Hall
Emanuela Haller
Stefan Haller
Bumsub Ham
Abdullah Hamdi
Dongyoon Han
Hu Han
Jungong Han
Junwei Han
Kai Han
Tian Han
Xiaoguang Han
Xintong Han
Yahong Han
Ankur Handa
Zekun Hao
Albert Haque
Tatsuya Harada
Mehrtash Harandi
Adam W. Harley
Mahmudul Hasan
Atsushi Hashimoto
Ali Hatamizadeh
Munawar Hayat
Dongliang He
Jingrui He
Junfeng He
Kaiming He
Kun He
Lei He
Pan He
Ran He
Shengfeng He
Tong He
Weipeng He
Xuming He
Yang He
Yihui He

Zhihai He
Chinmay Hegde
Janne Heikkila
Mattias P. Heinrich
Stéphane Herbin
Alexander Hermans
Luis Herranz
John R. Hershey
Aaron Hertzmann
Roei Herzig
Anders Heyden
Steven Hickson
Otmar Hilliges
Tomas Hodan
Judy Hoffman
Michael Hofmann
Yannick Hold-Geoffroy
Namdar Homayounfar
Sina Honari
Richang Hong
Seunghoon Hong
Xiaopeng Hong
Yi Hong
Hidekata Hontani
Anthony Hoogs
Yedid Hoshen
Mir Rayat Imtiaz Hossain
Junhui Hou
Le Hou
Lu Hou
Tingbo Hou
Wei-Lin Hsiao
Cheng-Chun Hsu
Gee-Sern Jison Hsu
Kuang-jui Hsu
Changbo Hu
Di Hu
Guosheng Hu
Han Hu
Hao Hu
Hexiang Hu
Hou-Ning Hu
Jie Hu
Junlin Hu
Nan Hu
Ping Hu

Ronghang Hu
Xiaowei Hu
Yinlin Hu
Yuan-Ting Hu
Zhe Hu
Binh-Son Hua
Yang Hua
Bingyao Huang
Di Huang
Dong Huang
Fay Huang
Haibin Huang
Haozhi Huang
Heng Huang
Huaibo Huang
Jia-Bin Huang
Jing Huang
Jingwei Huang
Kaizhu Huang
Lei Huang
Qiangui Huang
Qiaoying Huang
Qingqiu Huang
Qixing Huang
Shaoli Huang
Sheng Huang
Siyuan Huang
Weilin Huang
Wenbing Huang
Xiangru Huang
Xun Huang
Yan Huang
Yifei Huang
Yue Huang
Zhiwu Huang
Zilong Huang
Minyoung Huh
Zhuo Hui
Matthias B. Hullin
Martin Humenberger
Wei-Chih Hung
Zhouyuan Huo
Junhwa Hur
Noureldien Hussein
Jyh-Jing Hwang
Seong Jae Hwang

Hansung Kim
Heewon Kim
Hyo Jin Kim
Hyunwoo J. Kim
Jinkyu Kim
Jiwon Kim
Jongmin Kim
Junsik Kim
Junyeong Kim
Min H. Kim
Namil Kim
Pyojin Kim
Seon Joo Kim
Seong Tae Kim
Seungryong Kim
Sungwoong Kim
Tae Hyun Kim
Vladimir Kim
Won Hwa Kim
Yonghyun Kim
Benjamin Kimia
Akisato Kimura
Pieter-Jan Kindermans
Zsolt Kira
Itaru Kitahara
Hedvig Kjellstrom
Jan Knopp
Takumi Kobayashi
Erich Kobler
Parker Koch
Reinhard Koch
Elyor Kodirov
Amir Kolaman
Nicholas Kolkin
Dimitrios Kollias
Stefanos Kollias
Soheil Kolouri
Adams Wai-Kin Kong
Naejin Kong
Shu Kong
Tao Kong
Yu Kong
Yoshinori Konishi
Daniil Kononenko
Theodora Kontogianni
Simon Korman

Adam Kortylewski
Jana Kosecka
Jean Kossaifi
Satwik Kottur
Rigas Kouskouridas
Adriana Kovashka
Rama Kovvuri
Adarsh Kowdle
Jedrzej Kozerawski
Mateusz Kozinski
Philipp Kraehenbuehl
Gregory Kramida
Josip Krapac
Dmitry Kravchenko
Ranjay Krishna
Pavel Krsek
Alexander Krull
Jakob Kruse
Hiroyuki Kubo
Hilde Kuehne
Jason Kuen
Andreas Kuhn
Arjan Kuijper
Zuzana Kukelova
Ajay Kumar
Amit Kumar
Avinash Kumar
Suryansh Kumar
Vijay Kumar
Kaustav Kundu
Weicheng Kuo
Nojun Kwak
Suha Kwak
Junseok Kwon
Nikolaos Kyriazis
Zorah Lähner
Ankit Laddha
Florent Lafarge
Jean Lahoud
Kevin Lai
Shang-Hong Lai
Wei-Sheng Lai
Yu-Kun Lai
Iro Laina
Antony Lam
John Wheatley Lambert

Xiangyuan lan
Xu Lan
Charis Lanaras
Georg Langs
Oswald Lanz
Dong Lao
Yizhen Lao
Agata Lapedriza
Gustav Larsson
Viktor Larsson
Katrin Lasinger
Christoph Lassner
Longin Jan Latecki
Stéphane Lathuilière
Rynson Lau
Hei Law
Justin Lazarow
Svetlana Lazebnik
Hieu Le
Huu Le
Ngan Hoang Le
Trung-Nghia Le
Vuong Le
Colin Lea
Erik Learned-Miller
Chen-Yu Lee
Gim Hee Lee
Hsin-Ying Lee
Hyungtae Lee
Jae-Han Lee
Jimmy Addison Lee
Joonseok Lee
Kibok Lee
Kuang-Huei Lee
Kwonjoon Lee
Minsik Lee
Sang-chul Lee
Seungkyu Lee
Soochan Lee
Stefan Lee
Taehee Lee
Andreas Lehrmann
Jie Lei
Peng Lei
Matthew Joseph Leotta
Wee Kheng Leow

Gil Levi
Evgeny Levinkov
Aviad Levis
Jose Lezama
Ang Li
Bin Li
Bing Li
Boyi Li
Changsheng Li
Chao Li
Chen Li
Cheng Li
Chenglong Li
Chi Li
Chun-Guang Li
Chun-Liang Li
Chunyuan Li
Dong Li
Guanbin Li
Hao Li
Haoxiang Li
Hongsheng Li
Hongyang Li
Houqiang Li
Huibin Li
Jia Li
Jianan Li
Jianguo Li
Junnan Li
Junxuan Li
Kai Li
Ke Li
Kejie Li
Kunpeng Li
Lerenhan Li
Li Erran Li
Mengtian Li
Mu Li
Peihua Li
Peiyi Li
Ping Li
Qi Li
Qing Li
Ruiyu Li
Ruoteng Li
Shaozi Li

Sheng Li
Shiwei Li
Shuang Li
Siyang Li
Stan Z. Li
Tianye Li
Wei Li
Weixin Li
Wen Li
Wenbo Li
Xiaomeng Li
Xin Li
Xiu Li
Xuelong Li
Xueting Li
Yan Li
Yandong Li
Yanghao Li
Yehao Li
Yi Li
Yijun Li
Yikang LI
Yining Li
Yongjie Li
Yu Li
Yu-Jhe Li
Yunpeng Li
Yunsheng Li
Yunzhu Li
Zhe Li
Zhen Li
Zhengqi Li
Zhenyang Li
Zhuwen Li
Dongze Lian
Xiaochen Lian
Zhouhui Lian
Chen Liang
Jie Liang
Ming Liang
Paul Pu Liang
Pengpeng Liang
Shu Liang
Wei Liang
Jing Liao
Minghui Liao

Renjie Liao
Shengcai Liao
Shuai Liao
Yiyi Liao
Ser-Nam Lim
Chen-Hsuan Lin
Chung-Ching Lin
Dahua Lin
Ji Lin
Kevin Lin
Tianwei Lin
Tsung-Yi Lin
Tsung-Yu Lin
Wei-An Lin
Weiyao Lin
Yen-Chen Lin
Yuewei Lin
David B. Lindell
Drew Linsley
Krzysztof Lis
Roee Litman
Jim Little
An-An Liu
Bo Liu
Buyu Liu
Chao Liu
Chen Liu
Cheng-lin Liu
Chenxi Liu
Dong Liu
Feng Liu
Guilin Liu
Haomiao Liu
Heshan Liu
Hong Liu
Ji Liu
Jingen Liu
Jun Liu
Lanlan Liu
Li Liu
Liu Liu
Mengyuan Liu
Miaomiao Liu
Nian Liu
Ping Liu
Risheng Liu

Sheng Liu
Shu Liu
Shuaicheng Liu
Sifei Liu
Siqi Liu
Siying Liu
Songtao Liu
Ting Liu
Tongliang Liu
Tyng-Luh Liu
Wanquan Liu
Wei Liu
Weiyang Liu
Weizhe Liu
Wenyu Liu
Wu Liu
Xialei Liu
Xianglong Liu
Xiaodong Liu
Xiaofeng Liu
Xihui Liu
Xingyu Liu
Xinwang Liu
Xuanqing Liu
Xuebo Liu
Yang Liu
Yaojie Liu
Yebin Liu
Yen-Cheng Liu
Yiming Liu
Yu Liu
Yu-Shen Liu
Yufan Liu
Yun Liu
Zheng Liu
Zhijian Liu
Zhuang Liu
Zichuan Liu
Ziwei Liu
Zongyi Liu
Stephan Liwicki
Liliana Lo Presti
Chengjiang Long
Fuchen Long
Mingsheng Long
Xiang Long

Yang Long
Charles T. Loop
Antonio Lopez
Roberto J. Lopez-Sastre
Javier Lorenzo-Navarro
Manolis Lourakis
Boyu Lu
Canyi Lu
Feng Lu
Guoyu Lu
Hongtao Lu
Jiajun Lu
Jiasen Lu
Jiwen Lu
Kaiyue Lu
Le Lu
Shao-Ping Lu
Shijian Lu
Xiankai Lu
Xin Lu
Yao Lu
Yiping Lu
Yongxi Lu
Yongyi Lu
Zhiwu Lu
Fujun Luan
Benjamin E. Lundell
Hao Luo
Jian-Hao Luo
Ruotian Luo
Weixin Luo
Wenhan Luo
Wenjie Luo
Yan Luo
Zelun Luo
Zixin Luo
Khoa Luu
Zhaoyang Lv
Pengyuan Lyu
Thomas Möllenhoff
Matthias Müller
Bingpeng Ma
Chih-Yao Ma
Chongyang Ma
Huimin Ma
Jiayi Ma

K. T. Ma
Ke Ma
Lin Ma
Liqian Ma
Shugao Ma
Wei-Chiu Ma
Xiaojian Ma
Xingjun Ma
Zhanyu Ma
Zheng Ma
Radek Jakob Mackowiak
Ludovic Magerand
Shweta Mahajan
Siddharth Mahendran
Long Mai
Ameesh Makadia
Oscar Mendez Maldonado
Mateusz Malinowski
Yury Malkov
Arun Mallya
Dipu Manandhar
Massimiliano Mancini
Fabian Manhardt
Kevis-kokitsi Maninis
Varun Manjunatha
Junhua Mao
Xudong Mao
Alina Marcu
Edgar Margffoy-Tuay
Dmitrii Marin
Manuel J. Marin-Jimenez
Kenneth Marino
Niki Martinel
Julieta Martinez
Jonathan Masci
Tomohiro Mashita
Iacopo Masi
David Masip
Daniela Massiceti
Stefan Mathe
Yusuke Matsui
Tetsu Matsukawa
Iain A. Matthews
Kevin James Matzen
Bruce Allen Maxwell
Stephen Maybank

Helmut Mayer
Amir Mazaheri
David McAllester
Steven McDonagh
Stephen J. Mckenna
Roey Mechrez
Prakhar Mehrotra
Christopher Mei
Xue Mei
Paulo R. S. Mendonca
Lili Meng
Zibo Meng
Thomas Mensink
Bjoern Menze
Michele Merler
Kourosh Meshgi
Pascal Mettes
Christopher Metzler
Liang Mi
Qiguang Miao
Xin Miao
Tomer Michaeli
Frank Michel
Antoine Miech
Krystian Mikolajczyk
Peyman Milanfar
Ben Mildenhall
Gregor Miller
Fausto Milletari
Dongbo Min
Kyle Min
Pedro Miraldo
Dmytro Mishkin
Anand Mishra
Ashish Mishra
Ishan Misra
Niluthpol C. Mithun
Kaushik Mitra
Niloy Mitra
Anton Mitrokhin
Ikuhisa Mitsugami
Anurag Mittal
Kaichun Mo
Zhipeng Mo
Davide Modolo
Michael Moeller

Pritish Mohapatra
Pavlo Molchanov
Davide Moltisanti
Pascal Monasse
Mathew Monfort
Aron Monszpart
Sean Moran
Vlad I. Morariu
Francesc Moreno-Noguer
Pietro Morerio
Stylianos Moschoglou
Yael Moses
Roozbeh Mottaghi
Pierre Moulon
Arsalan Mousavian
Yadong Mu
Yasuhiro Mukaigawa
Lopamudra Mukherjee
Yusuke Mukuta
Ravi Teja Mullapudi
Mario Enrique Munich
Zachary Murez
Ana C. Murillo
J. Krishna Murthy
Damien Muselet
Armin Mustafa
Siva Karthik Mustikovela
Carlo Dal Mutto
Moin Nabi
Varun K. Nagaraja
Tushar Nagarajan
Arsha Nagrani
Seungjun Nah
Nikhil Naik
Yoshikatsu Nakajima
Yuta Nakashima
Atsushi Nakazawa
Seonghyeon Nam
Vinay P. Namboodiri
Medhini Narasimhan
Srinivasa Narasimhan
Sanath Narayan
Erickson Rangel
    Nascimento
Jacinto Nascimento
Tayyab Naseer

Lakshmanan Nataraj
Neda Nategh
Nelson Isao Nauata
Fernando Navarro
Shah Nawaz
Lukas Neumann
Ram Nevatia
Alejandro Newell
Shawn Newsam
Joe Yue-Hei Ng
Trung Thanh Ngo
Duc Thanh Nguyen
Lam M. Nguyen
Phuc Xuan Nguyen
Thuong Nguyen Canh
Mihalis Nicolaou
Andrei Liviu Nicolicioiu
Xuecheng Nie
Michael Niemeyer
Simon Niklaus
Christophoros Nikou
David Nilsson
Jifeng Ning
Yuval Nirkin
Li Niu
Yuzhen Niu
Zhenxing Niu
Shohei Nobuhara
Nicoletta Noceti
Hyeonwoo Noh
Junhyug Noh
Mehdi Noroozi
Sotiris Nousias
Valsamis Ntouskos
Matthew O'Toole
Peter Ochs
Ferda Ofli
Seong Joon Oh
Seoung Wug Oh
Iason Oikonomidis
Utkarsh Ojha
Takahiro Okabe
Takayuki Okatani
Fumio Okura
Aude Oliva
Kyle Olszewski

Björn Ommer
Mohamed Omran
Elisabeta Oneata
Michael Opitz
Jose Oramas
Tribhuvanesh Orekondy
Shaul Oron
Sergio Orts-Escolano
Ivan Oseledets
Aljosa Osep
Magnus Oskarsson
Anton Osokin
Martin R. Oswald
Wanli Ouyang
Andrew Owens
Mete Ozay
Mustafa Ozuysal
Eduardo Pérez-Pellitero
Gautam Pai
Dipan Kumar Pal
P. H. Pamplona Savarese
Jinshan Pan
Junting Pan
Xingang Pan
Yingwei Pan
Yannis Panagakis
Rameswar Panda
Guan Pang
Jiahao Pang
Jiangmiao Pang
Tianyu Pang
Sharath Pankanti
Nicolas Papadakis
Dim Papadopoulos
George Papandreou
Toufiq Parag
Shaifali Parashar
Sarah Parisot
Eunhyeok Park
Hyun Soo Park
Jaesik Park
Min-Gyu Park
Taesung Park
Alvaro Parra
C. Alejandro Parraga
Despoina Paschalidou

Nikolaos Passalis
Vishal Patel
Viorica Patraucean
Badri Narayana Patro
Danda Pani Paudel
Sujoy Paul
Georgios Pavlakos
Ioannis Pavlidis
Vladimir Pavlovic
Nick Pears
Kim Steenstrup Pedersen
Selen Pehlivan
Shmuel Peleg
Chao Peng
Houwen Peng
Wen-Hsiao Peng
Xi Peng
Xiaojiang Peng
Xingchao Peng
Yuxin Peng
Federico Perazzi
Juan Camilo Perez
Vishwanath Peri
Federico Pernici
Luca Del Pero
Florent Perronnin
Stavros Petridis
Henning Petzka
Patrick Peursum
Michael Pfeiffer
Hanspeter Pfister
Roman Pflugfelder
Minh Tri Pham
Yongri Piao
David Picard
Tomasz Pieciak
A. J. Piergiovanni
Andrea Pilzer
Pedro O. Pinheiro
Silvia Laura Pintea
Lerrel Pinto
Axel Pinz
Robinson Piramuthu
Fiora Pirri
Leonid Pishchulin
Francesco Pittaluga

Daniel Pizarro
Tobias Plötz
Mirco Planamente
Matteo Poggi
Moacir A. Ponti
Parita Pooj
Fatih Porikli
Horst Possegger
Omid Poursaeed
Ameya Prabhu
Viraj Uday Prabhu
Dilip Prasad
Brian L. Price
True Price
Maria Priisalu
Veronique Prinet
Victor Adrian Prisacariu
Jan Prokaj
Sergey Prokudin
Nicolas Pugeault
Xavier Puig
Albert Pumarola
Pulak Purkait
Senthil Purushwalkam
Charles R. Qi
Hang Qi
Haozhi Qi
Lu Qi
Mengshi Qi
Siyuan Qi
Xiaojuan Qi
Yuankai Qi
Shengju Qian
Xuelin Qian
Siyuan Qiao
Yu Qiao
Jie Qin
Qiang Qiu
Weichao Qiu
Zhaofan Qiu
Kha Gia Quach
Yuhui Quan
Yvain Queau
Julian Quiroga
Faisal Qureshi
Mahdi Rad

Filip Radenovic
Petia Radeva
Venkatesh
  B. Radhakrishnan
Ilija Radosavovic
Noha Radwan
Rahul Raguram
Tanzila Rahman
Amit Raj
Ajit Rajwade
Kandan Ramakrishnan
Santhosh
  K. Ramakrishnan
Srikumar Ramalingam
Ravi Ramamoorthi
Vasili Ramanishka
Ramprasaath R. Selvaraju
Francois Rameau
Visvanathan Ramesh
Santu Rana
Rene Ranftl
Anand Rangarajan
Anurag Ranjan
Viresh Ranjan
Yongming Rao
Carolina Raposo
Vivek Rathod
Sathya N. Ravi
Avinash Ravichandran
Tammy Riklin Raviv
Daniel Rebain
Sylvestre-Alvise Rebuffi
N. Dinesh Reddy
Timo Rehfeld
Paolo Remagnino
Konstantinos Rematas
Edoardo Remelli
Dongwei Ren
Haibing Ren
Jian Ren
Jimmy Ren
Mengye Ren
Weihong Ren
Wenqi Ren
Zhile Ren
Zhongzheng Ren

Zhou Ren
Vijay Rengarajan
Md A. Reza
Farzaneh Rezaeianaran
Hamed R. Tavakoli
Nicholas Rhinehart
Helge Rhodin
Elisa Ricci
Alexander Richard
Eitan Richardson
Elad Richardson
Christian Richardt
Stephan Richter
Gernot Riegler
Daniel Ritchie
Tobias Ritschel
Samuel Rivera
Yong Man Ro
Richard Roberts
Joseph Robinson
Ignacio Rocco
Mrigank Rochan
Emanuele Rodolà
Mikel D. Rodriguez
Giorgio Roffo
Grégory Rogez
Gemma Roig
Javier Romero
Xuejian Rong
Yu Rong
Amir Rosenfeld
Bodo Rosenhahn
Guy Rosman
Arun Ross
Paolo Rota
Peter M. Roth
Anastasios Roussos
Anirban Roy
Sebastien Roy
Aruni RoyChowdhury
Artem Rozantsev
Ognjen Rudovic
Daniel Rueckert
Adria Ruiz
Javier Ruiz-del-solar
Christian Rupprecht

Chris Russell
Dan Ruta
Jongbin Ryu
Ömer Sümer
Alexandre Sablayrolles
Faraz Saeedan
Ryusuke Sagawa
Christos Sagonas
Tonmoy Saikia
Hideo Saito
Kuniaki Saito
Shunsuke Saito
Shunta Saito
Ken Sakurada
Joaquin Salas
Fatemeh Sadat Saleh
Mahdi Saleh
Pouya Samangouei
Leo Sampaio
  Ferraz Ribeiro
Artsiom Olegovich
  Sanakoyeu
Enrique Sanchez
Patsorn Sangkloy
Anush Sankaran
Aswin Sankaranarayanan
Swami Sankaranarayanan
Rodrigo Santa Cruz
Amartya Sanyal
Archana Sapkota
Nikolaos Sarafianos
Jun Sato
Shin'ichi Satoh
Hosnieh Sattar
Arman Savran
Manolis Savva
Alexander Sax
Hanno Scharr
Simone Schaub-Meyer
Konrad Schindler
Dmitrij Schlesinger
Uwe Schmidt
Dirk Schnieders
Björn Schuller
Samuel Schulter
Idan Schwartz

William Robson Schwartz
Alex Schwing
Sinisa Segvic
Lorenzo Seidenari
Pradeep Sen
Ozan Sener
Soumyadip Sengupta
Arda Senocak
Mojtaba Seyedhosseini
Shishir Shah
Shital Shah
Sohil Atul Shah
Tamar Rott Shaham
Huasong Shan
Qi Shan
Shiguang Shan
Jing Shao
Roman Shapovalov
Gaurav Sharma
Vivek Sharma
Viktoriia Sharmanska
Dongyu She
Sumit Shekhar
Evan Shelhamer
Chengyao Shen
Chunhua Shen
Falong Shen
Jie Shen
Li Shen
Liyue Shen
Shuhan Shen
Tianwei Shen
Wei Shen
William B. Shen
Yantao Shen
Ying Shen
Yiru Shen
Yujun Shen
Yuming Shen
Zhiqiang Shen
Ziyi Shen
Lu Sheng
Yu Sheng
Rakshith Shetty
Baoguang Shi
Guangming Shi

Hailin Shi
Miaojing Shi
Yemin Shi
Zhenmei Shi
Zhiyuan Shi
Kevin Jonathan Shih
Shiliang Shiliang
Hyunjung Shim
Atsushi Shimada
Nobutaka Shimada
Daeyun Shin
Young Min Shin
Koichi Shinoda
Konstantin Shmelkov
Michael Zheng Shou
Abhinav Shrivastava
Tianmin Shu
Zhixin Shu
Hong-Han Shuai
Pushkar Shukla
Christian Siagian
Mennatullah M. Siam
Kaleem Siddiqi
Karan Sikka
Jae-Young Sim
Christian Simon
Martin Simonovsky
Dheeraj Singaraju
Bharat Singh
Gurkirt Singh
Krishna Kumar Singh
Maneesh Kumar Singh
Richa Singh
Saurabh Singh
Suriya Singh
Vikas Singh
Sudipta N. Sinha
Vincent Sitzmann
Josef Sivic
Gregory Slabaugh
Miroslava Slavcheva
Ron Slossberg
Brandon Smith
Kevin Smith
Vladimir Smutny
Noah Snavely

Roger
   D. Soberanis-Mukul
Kihyuk Sohn
Francesco Solera
Eric Sommerlade
Sanghyun Son
Byung Cheol Song
Chunfeng Song
Dongjin Song
Jiaming Song
Jie Song
Jifei Song
Jingkuan Song
Mingli Song
Shiyu Song
Shuran Song
Xiao Song
Yafei Song
Yale Song
Yang Song
Yi-Zhe Song
Yibing Song
Humberto Sossa
Cesar de Souza
Adrian Spurr
Srinath Sridhar
Suraj Srinivas
Pratul P. Srinivasan
Anuj Srivastava
Tania Stathaki
Christopher Stauffer
Simon Stent
Rainer Stiefelhagen
Pierre Stock
Julian Straub
Jonathan C. Stroud
Joerg Stueckler
Jan Stuehmer
David Stutz
Chi Su
Hang Su
Jong-Chyi Su
Shuochen Su
Yu-Chuan Su
Ramanathan Subramanian
Yusuke Sugano

Masanori Suganuma
Yumin Suh
Mohammed Suhail
Yao Sui
Heung-Il Suk
Josephine Sullivan
Baochen Sun
Chen Sun
Chong Sun
Deqing Sun
Jin Sun
Liang Sun
Lin Sun
Qianru Sun
Shao-Hua Sun
Shuyang Sun
Weiwei Sun
Wenxiu Sun
Xiaoshuai Sun
Xiaoxiao Sun
Xingyuan Sun
Yifan Sun
Zhun Sun
Sabine Susstrunk
David Suter
Supasorn Suwajanakorn
Tomas Svoboda
Eran Swears
Paul Swoboda
Attila Szabo
Richard Szeliski
Duy-Nguyen Ta
Andrea Tagliasacchi
Yuichi Taguchi
Ying Tai
Keita Takahashi
Kouske Takahashi
Jun Takamatsu
Hugues Talbot
Toru Tamaki
Chaowei Tan
Fuwen Tan
Mingkui Tan
Mingxing Tan
Qingyang Tan
Robby T. Tan

Xiaoyang Tan
Kenichiro Tanaka
Masayuki Tanaka
Chang Tang
Chengzhou Tang
Danhang Tang
Ming Tang
Peng Tang
Qingming Tang
Wei Tang
Xu Tang
Yansong Tang
Youbao Tang
Yuxing Tang
Zhiqiang Tang
Tatsunori Taniai
Junli Tao
Xin Tao
Makarand Tapaswi
Jean-Philippe Tarel
Lyne Tchapmi
Zachary Teed
Bugra Tekin
Damien Teney
Ayush Tewari
Christian Theobalt
Christopher Thomas
Diego Thomas
Jim Thomas
Rajat Mani Thomas
Xinmei Tian
Yapeng Tian
Yingli Tian
Yonglong Tian
Zhi Tian
Zhuotao Tian
Kinh Tieu
Joseph Tighe
Massimo Tistarelli
Matthew Toews
Carl Toft
Pavel Tokmakov
Federico Tombari
Chetan Tonde
Yan Tong
Alessio Tonioni

Andrea Torsello
Fabio Tosi
Du Tran
Luan Tran
Ngoc-Trung Tran
Quan Hung Tran
Truyen Tran
Rudolph Triebel
Martin Trimmel
Shashank Tripathi
Subarna Tripathi
Leonardo Trujillo
Eduard Trulls
Tomasz Trzcinski
Sam Tsai
Yi-Hsuan Tsai
Hung-Yu Tseng
Stavros Tsogkas
Aggeliki Tsoli
Devis Tuia
Shubham Tulsiani
Sergey Tulyakov
Frederick Tung
Tony Tung
Daniyar Turmukhambetov
Ambrish Tyagi
Radim Tylecek
Christos Tzelepis
Georgios Tzimiropoulos
Dimitrios Tzionas
Seiichi Uchida
Norimichi Ukita
Dmitry Ulyanov
Martin Urschler
Yoshitaka Ushiku
Ben Usman
Alexander Vakhitov
Julien P. C. Valentin
Jack Valmadre
Ernest Valveny
Joost van de Weijer
Jan van Gemert
Koen Van Leemput
Gul Varol
Sebastiano Vascon
M. Alex O. Vasilescu

Subeesh Vasu
Mayank Vatsa
David Vazquez
Javier Vazquez-Corral
Ashok Veeraraghavan
Erik Velasco-Salido
Raviteja Vemulapalli
Jonathan Ventura
Manisha Verma
Roberto Vezzani
Ruben Villegas
Minh Vo
MinhDuc Vo
Nam Vo
Michele Volpi
Riccardo Volpi
Carl Vondrick
Konstantinos Vougioukas
Tuan-Hung Vu
Sven Wachsmuth
Neal Wadhwa
Catherine Wah
Jacob C. Walker
Thomas S. A. Wallis
Chengde Wan
Jun Wan
Liang Wan
Renjie Wan
Baoyuan Wang
Boyu Wang
Cheng Wang
Chu Wang
Chuan Wang
Chunyu Wang
Dequan Wang
Di Wang
Dilin Wang
Dong Wang
Fang Wang
Guanzhi Wang
Guoyin Wang
Hanzi Wang
Hao Wang
He Wang
Heng Wang
Hongcheng Wang

Hongxing Wang
Hua Wang
Jian Wang
Jingbo Wang
Jinglu Wang
Jingya Wang
Jinjun Wang
Jinqiao Wang
Jue Wang
Ke Wang
Keze Wang
Le Wang
Lei Wang
Lezi Wang
Li Wang
Liang Wang
Lijun Wang
Limin Wang
Linwei Wang
Lizhi Wang
Mengjiao Wang
Mingzhe Wang
Minsi Wang
Naiyan Wang
Nannan Wang
Ning Wang
Oliver Wang
Pei Wang
Peng Wang
Pichao Wang
Qi Wang
Qian Wang
Qiaosong Wang
Qifei Wang
Qilong Wang
Qing Wang
Qingzhong Wang
Quan Wang
Rui Wang
Ruiping Wang
Ruixing Wang
Shangfei Wang
Shenlong Wang
Shiyao Wang
Shuhui Wang
Song Wang

Tao Wang
Tianlu Wang
Tiantian Wang
Ting-chun Wang
Tingwu Wang
Wei Wang
Weiyue Wang
Wenguan Wang
Wenlin Wang
Wenqi Wang
Xiang Wang
Xiaobo Wang
Xiaofang Wang
Xiaoling Wang
Xiaolong Wang
Xiaosong Wang
Xiaoyu Wang
Xin Eric Wang
Xinchao Wang
Xinggang Wang
Xintao Wang
Yali Wang
Yan Wang
Yang Wang
Yangang Wang
Yaxing Wang
Yi Wang
Yida Wang
Yilin Wang
Yiming Wang
Yisen Wang
Yongtao Wang
Yu-Xiong Wang
Yue Wang
Yujiang Wang
Yunbo Wang
Yunhe Wang
Zengmao Wang
Zhangyang Wang
Zhaowen Wang
Zhe Wang
Zhecan Wang
Zheng Wang
Zhixiang Wang
Zilei Wang
Jianqiao Wangni

Anne S. Wannenwetsch
Jan Dirk Wegner
Scott Wehrwein
Donglai Wei
Kaixuan Wei
Longhui Wei
Pengxu Wei
Ping Wei
Qi Wei
Shih-En Wei
Xing Wei
Yunchao Wei
Zijun Wei
Jerod Weinman
Michael Weinmann
Philippe Weinzaepfel
Yair Weiss
Bihan Wen
Longyin Wen
Wei Wen
Junwu Weng
Tsui-Wei Weng
Xinshuo Weng
Eric Wengrowski
Tomas Werner
Gordon Wetzstein
Tobias Weyand
Patrick Wieschollek
Maggie Wigness
Erik Wijmans
Richard Wildes
Olivia Wiles
Chris Williams
Williem Williem
Kyle Wilson
Calden Wloka
Nicolai Wojke
Christian Wolf
Yongkang Wong
Sanghyun Woo
Scott Workman
Baoyuan Wu
Bichen Wu
Chao-Yuan Wu
Huikai Wu
Jiajun Wu

Jialin Wu
Jiaxiang Wu
Jiqing Wu
Jonathan Wu
Lifang Wu
Qi Wu
Qiang Wu
Ruizheng Wu
Shangzhe Wu
Shun-Cheng Wu
Tianfu Wu
Wayne Wu
Wenxuan Wu
Xiao Wu
Xiaohe Wu
Xinxiao Wu
Yang Wu
Yi Wu
Yiming Wu
Ying Nian Wu
Yue Wu
Zheng Wu
Zhenyu Wu
Zhirong Wu
Zuxuan Wu
Stefanie Wuhrer
Jonas Wulff
Changqun Xia
Fangting Xia
Fei Xia
Gui-Song Xia
Lu Xia
Xide Xia
Yin Xia
Yingce Xia
Yongqin Xian
Lei Xiang
Shiming Xiang
Bin Xiao
Fanyi Xiao
Guobao Xiao
Huaxin Xiao
Taihong Xiao
Tete Xiao
Tong Xiao
Wang Xiao

Yang Xiao
Cihang Xie
Guosen Xie
Jianwen Xie
Lingxi Xie
Sirui Xie
Weidi Xie
Wenxuan Xie
Xiaohua Xie
Fuyong Xing
Jun Xing
Junliang Xing
Bo Xiong
Peixi Xiong
Yu Xiong
Yuanjun Xiong
Zhiwei Xiong
Chang Xu
Chenliang Xu
Dan Xu
Danfei Xu
Hang Xu
Hongteng Xu
Huijuan Xu
Jingwei Xu
Jun Xu
Kai Xu
Mengmeng Xu
Mingze Xu
Qianqian Xu
Ran Xu
Weijian Xu
Xiangyu Xu
Xiaogang Xu
Xing Xu
Xun Xu
Yanyu Xu
Yichao Xu
Yong Xu
Yongchao Xu
Yuanlu Xu
Zenglin Xu
Zheng Xu
Chuhui Xue
Jia Xue
Nan Xue

Tianfan Xue
Xiangyang Xue
Abhay Yadav
Yasushi Yagi
I. Zeki Yalniz
Kota Yamaguchi
Toshihiko Yamasaki
Takayoshi Yamashita
Junchi Yan
Ke Yan
Qingan Yan
Sijie Yan
Xinchen Yan
Yan Yan
Yichao Yan
Zhicheng Yan
Keiji Yanai
Bin Yang
Ceyuan Yang
Dawei Yang
Dong Yang
Fan Yang
Guandao Yang
Guorun Yang
Haichuan Yang
Hao Yang
Jianwei Yang
Jiaolong Yang
Jie Yang
Jing Yang
Kaiyu Yang
Linjie Yang
Meng Yang
Michael Ying Yang
Nan Yang
Shuai Yang
Shuo Yang
Tianyu Yang
Tien-Ju Yang
Tsun-Yi Yang
Wei Yang
Wenhan Yang
Xiao Yang
Xiaodong Yang
Xin Yang
Yan Yang

Yanchao Yang
Yee Hong Yang
Yezhou Yang
Zhenheng Yang
Anbang Yao
Angela Yao
Cong Yao
Jian Yao
Li Yao
Ting Yao
Yao Yao
Zhewei Yao
Chengxi Ye
Jianbo Ye
Keren Ye
Linwei Ye
Mang Ye
Mao Ye
Qi Ye
Qixiang Ye
Mei-Chen Yeh
Raymond Yeh
Yu-Ying Yeh
Sai-Kit Yeung
Serena Yeung
Kwang Moo Yi
Li Yi
Renjiao Yi
Alper Yilmaz
Junho Yim
Lijun Yin
Weidong Yin
Xi Yin
Zhichao Yin
Tatsuya Yokota
Ryo Yonetani
Donggeun Yoo
Jae Shin Yoon
Ju Hong Yoon
Sung-eui Yoon
Laurent Younes
Changqian Yu
Fisher Yu
Gang Yu
Jiahui Yu
Kaicheng Yu

Ke Yu
Lequan Yu
Ning Yu
Qian Yu
Ronald Yu
Ruichi Yu
Shoou-I Yu
Tao Yu
Tianshu Yu
Xiang Yu
Xin Yu
Xiyu Yu
Youngjae Yu
Yu Yu
Zhiding Yu
Chunfeng Yuan
Ganzhao Yuan
Jinwei Yuan
Lu Yuan
Quan Yuan
Shanxin Yuan
Tongtong Yuan
Wenjia Yuan
Ye Yuan
Yuan Yuan
Yuhui Yuan
Huanjing Yue
Xiangyu Yue
Ersin Yumer
Sergey Zagoruyko
Egor Zakharov
Amir Zamir
Andrei Zanfir
Mihai Zanfir
Pablo Zegers
Bernhard Zeisl
John S. Zelek
Niclas Zeller
Huayi Zeng
Jiabei Zeng
Wenjun Zeng
Yu Zeng
Xiaohua Zhai
Fangneng Zhan
Huangying Zhan
Kun Zhan

Xiaohang Zhan
Baochang Zhang
Bowen Zhang
Cecilia Zhang
Changqing Zhang
Chao Zhang
Chengquan Zhang
Chi Zhang
Chongyang Zhang
Dingwen Zhang
Dong Zhang
Feihu Zhang
Hang Zhang
Hanwang Zhang
Hao Zhang
He Zhang
Hongguang Zhang
Hua Zhang
Ji Zhang
Jianguo Zhang
Jianming Zhang
Jiawei Zhang
Jie Zhang
Jing Zhang
Juyong Zhang
Kai Zhang
Kaipeng Zhang
Ke Zhang
Le Zhang
Lei Zhang
Li Zhang
Lihe Zhang
Linguang Zhang
Lu Zhang
Mi Zhang
Mingda Zhang
Peng Zhang
Pingping Zhang
Qian Zhang
Qilin Zhang
Quanshi Zhang
Richard Zhang
Rui Zhang
Runze Zhang
Shengping Zhang
Shifeng Zhang

Shuai Zhang
Songyang Zhang
Tao Zhang
Ting Zhang
Tong Zhang
Wayne Zhang
Wei Zhang
Weizhong Zhang
Wenwei Zhang
Xiangyu Zhang
Xiaolin Zhang
Xiaopeng Zhang
Xiaoqin Zhang
Xiuming Zhang
Ya Zhang
Yang Zhang
Yimin Zhang
Yinda Zhang
Ying Zhang
Yongfei Zhang
Yu Zhang
Yulun Zhang
Yunhua Zhang
Yuting Zhang
Zhanpeng Zhang
Zhao Zhang
Zhaoxiang Zhang
Zhen Zhang
Zheng Zhang
Zhifei Zhang
Zhijin Zhang
Zhishuai Zhang
Ziming Zhang
Bo Zhao
Chen Zhao
Fang Zhao
Haiyu Zhao
Han Zhao
Hang Zhao
Hengshuang Zhao
Jian Zhao
Kai Zhao
Liang Zhao
Long Zhao
Qian Zhao
Qibin Zhao

Qijun Zhao
Rui Zhao
Shenglin Zhao
Sicheng Zhao
Tianyi Zhao
Wenda Zhao
Xiangyun Zhao
Xin Zhao
Yang Zhao
Yue Zhao
Zhichen Zhao
Zijing Zhao
Xiantong Zhen
Chuanxia Zheng
Feng Zheng
Haiyong Zheng
Jia Zheng
Kang Zheng
Shuai Kyle Zheng
Wei-Shi Zheng
Yinqiang Zheng
Zerong Zheng
Zhedong Zheng
Zilong Zheng
Bineng Zhong
Fangwei Zhong
Guangyu Zhong
Yiran Zhong
Yujie Zhong
Zhun Zhong
Chunluan Zhou
Huiyu Zhou
Jiahuan Zhou
Jun Zhou
Lei Zhou
Luowei Zhou
Luping Zhou
Mo Zhou
Ning Zhou
Pan Zhou
Peng Zhou
Qianyi Zhou
S. Kevin Zhou
Sanping Zhou
Wengang Zhou
Xingyi Zhou

Yanzhao Zhou
Yi Zhou
Yin Zhou
Yipin Zhou
Yuyin Zhou
Zihan Zhou
Alex Zihao Zhu
Chenchen Zhu
Feng Zhu
Guangming Zhu
Ji Zhu
Jun-Yan Zhu
Lei Zhu
Linchao Zhu
Rui Zhu
Shizhan Zhu
Tyler Lixuan Zhu

Wei Zhu
Xiangyu Zhu
Xinge Zhu
Xizhou Zhu
Yanjun Zhu
Yi Zhu
Yixin Zhu
Yizhe Zhu
Yousong Zhu
Zhe Zhu
Zhen Zhu
Zheng Zhu
Zhenyao Zhu
Zhihui Zhu
Zhuotun Zhu
Bingbing Zhuang
Wei Zhuo

Christian Zimmermann
Karel Zimmermann
Larry Zitnick
Mohammadreza
   Zolfaghari
Maria Zontak
Daniel Zoran
Changqing Zou
Chuhang Zou
Danping Zou
Qi Zou
Yang Zou
Yuliang Zou
Georgios Zoumpourlis
Wangmeng Zuo
Xinxin Zuo

Additional Reviewers

Victoria Fernandez
   Abrevaya
Maya Aghaei
Allam Allam
Christine
   Allen-Blanchette
Nicolas Aziere
Assia Benbihi
Neha Bhargava
Bharat Lal Bhatnagar
Joanna Bitton
Judy Borowski
Amine Bourki
Romain Brégier
Tali Brayer
Sebastian Bujwid
Andrea Burns
Yun-Hao Cao
Yuning Chai
Xiaojun Chang
Bo Chen
Shuo Chen
Zhixiang Chen
Junsuk Choe
Hung-Kuo Chu

Jonathan P. Crall
Kenan Dai
Lucas Deecke
Karan Desai
Prithviraj Dhar
Jing Dong
Wei Dong
Turan Kaan Elgin
Francis Engelmann
Erik Englesson
Fartash Faghri
Zicong Fan
Yang Fu
Risheek Garrepalli
Yifan Ge
Marco Godi
Helmut Grabner
Shuxuan Guo
Jianfeng He
Zhezhi He
Samitha Herath
Chih-Hui Ho
Yicong Hong
Vincent Tao Hu
Julio Hurtado

Jaedong Hwang
Andrey Ignatov
Muhammad
   Abdullah Jamal
Saumya Jetley
Meiguang Jin
Jeff Johnson
Minsoo Kang
Saeed Khorram
Mohammad Rami Koujan
Nilesh Kulkarni
Sudhakar Kumawat
Abdelhak Lemkhenter
Alexander Levine
Jiachen Li
Jing Li
Jun Li
Yi Li
Liang Liao
Ruochen Liao
Tzu-Heng Lin
Phillip Lippe
Bao-di Liu
Bo Liu
Fangchen Liu

Hanxiao Liu
Hongyu Liu
Huidong Liu
Miao Liu
Xinxin Liu
Yongfei Liu
Yu-Lun Liu
Amir Livne
Tiange Luo
Wei Ma
Xiaoxuan Ma
Ioannis Marras
Georg Martius
Effrosyni Mavroudi
Tim Meinhardt
Givi Meishvili
Meng Meng
Zihang Meng
Zhongqi Miao
Gyeongsik Moon
Khoi Nguyen
Yung-Kyun Noh
Antonio Norelli
Jaeyoo Park
Alexander Pashevich
Mandela Patrick
Mary Phuong
Bingqiao Qian
Yu Qiao
Zhen Qiao
Sai Saketh Rambhatla
Aniket Roy
Amelie Royer
Parikshit Vishwas
    Sakurikar
Mark Sandler
Mert Bülent Sarıyıldız
Tanner Schmidt
Anshul B. Shah

Ketul Shah
Rajvi Shah
Hengcan Shi
Xiangxi Shi
Yujiao Shi
William A. P. Smith
Guoxian Song
Robin Strudel
Abby Stylianou
Xinwei Sun
Reuben Tan
Qingyi Tao
Kedar S. Tatwawadi
Anh Tuan Tran
Son Dinh Tran
Eleni Triantafillou
Aristeidis Tsitiridis
Md Zasim Uddin
Andrea Vedaldi
Evangelos Ververas
Vidit Vidit
Paul Voigtlaender
Bo Wan
Huanyu Wang
Huiyu Wang
Junqiu Wang
Pengxiao Wang
Tai Wang
Xinyao Wang
Tomoki Watanabe
Mark Weber
Xi Wei
Botong Wu
James Wu
Jiamin Wu
Rujie Wu
Yu Wu
Rongchang Xie
Wei Xiong

Yunyang Xiong
An Xu
Chi Xu
Yinghao Xu
Fei Xue
Tingyun Yan
Zike Yan
Chao Yang
Heran Yang
Ren Yang
Wenfei Yang
Xu Yang
Rajeev Yasarla
Shaokai Ye
Yufei Ye
Kun Yi
Haichao Yu
Hanchao Yu
Ruixuan Yu
Liangzhe Yuan
Chen-Lin Zhang
Fandong Zhang
Tianyi Zhang
Yang Zhang
Yiyi Zhang
Yongshun Zhang
Yu Zhang
Zhiwei Zhang
Jiaojiao Zhao
Yipu Zhao
Xingjian Zhen
Haizhong Zheng
Tiancheng Zhi
Chengju Zhou
Hao Zhou
Hao Zhu
Alexander Zimin

Contents – Part XI

MessyTable: Instance Association in Multiple Camera Views

Zhongang Cai[1], Junzhe Zhang[1,2], Daxuan Ren[1,2], Cunjun Yu[1], Haiyu Zhao[1], Shuai Yi[1], Chai Kiat Yeo[2], and Chen Change Loy[2(✉)]

[1] SenseTime Research, Tai Po, Hong Kong
{caizhongang,yucunjun,zhaohaiyu,yishuai}@sensetime.com
[2] Nanyang Technological University, Singapore, Singapore
{junzhe001,daxuan001}@e.ntu.edu.sg,
{asckyeo,ccloy}@ntu.edu.sg

Abstract. We present an interesting and challenging dataset that features a large number of scenes with messy tables captured from multiple camera views. Each scene in this dataset is highly complex, containing multiple object instances that could be identical, stacked and occluded by other instances. The key challenge is to associate all instances given the RGB image of all views. The seemingly simple task surprisingly fails many popular methods or heuristics that we assume good performance in object association. The dataset challenges existing methods in mining subtle appearance differences, reasoning based on contexts, and fusing appearance with geometric cues for establishing an association. We report interesting findings with some popular baselines, and discuss how this dataset could help inspire new problems and catalyse more robust formulations to tackle real-world instance association problems. (Project page: https://caizhongang.github.io/projects/MessyTable/.)

1 Introduction

We introduce a new and interesting dataset, MessyTable. It contains over 5,000 scenes, each of which captured by nine cameras in one of the 600 configurations of camera poses. Each scene is shot with a random cluttered background and different lighting conditions with about 30 general objects on average. The objects are chosen arbitrarily from 120 classes of possible instances. Figure 1 depicts some scene examples. The goal is to associate the different objects in a scene, i.e., finding the right match of the same instance across views.

The seemingly easy task on this dataset is surprisingly challenging. The relative pose between two cameras can be large, and therefore, an object may

---

Z. Cai and J. Zhang—Indicates equal contribution.

Electronic supplementary material The online version of this chapter (https://doi.org/10.1007/978-3-030-58621-8_1) contains supplementary material, which is available to authorized users.

© Springer Nature Switzerland AG 2020
A. Vedaldi et al. (Eds.): ECCV 2020, LNCS 12356, pp. 1–16, 2020.
https://doi.org/10.1007/978-3-030-58621-8_1

Fig. 1. MessyTable is a large-scale multi-camera general object dataset with cross-camera association labels. One of the 5,579 scenes is shown in this figure (with only 4 out of 9 cameras depicted). MessyTable is challenging due to severe partial occlusion (instance 2 in camera 9 and instance 31 in camera 2), large camera angle differences, similar or identical-looking instances (instances 31 and 27) and more. Objects are arbitrarily chosen from 120 classes that are commonly found on table. The brand names are pixelated in all illustrations

appear to be very different when viewed from different angles. Objects are heavily occluded while some of them can be elevated by other objects in a cluttered scene. Hence, appearance viewed from different cameras is always partial. The problem is further complicated with similar-looking or even identical instances.

Solving the aforementioned problems is non-trivial. The geometric constraint is hard to use right away. Multi-view epipolar geometry is ambiguous when a pixel can correspond to all points on the epipolar line in the other view. Homographic projection assumes a reference plane, which is not always available. To associate an object across camera views, a method needs to distinguish subtle differences between similar-looking objects. Fine-grained recognition is still a challenging problem in computer vision. When a scene contains identical-looking objects, the method is required to search for neighbouring cues in the vicinity of the objects to differentiate them. The neighbouring configuration, however, can be occluded and highly non-rigid with changing relative position due to different camera views.

While the method developed from MessyTable can be applied to some immediate applications such as automatic check-out [29] in supermarkets, e.g., leveraging multiple views to prevent counting error on the merchandise, the instance association problem found in this dataset is reminiscent of many real-world problems such as person re-identification or object tracking across views. Both examples of real-world problems require some sort of association between objects, either through appearances, group configurations, or temporal cues. Solving these real-world problems requires one to train a model using domain-specific data. Nonetheless, they still share similar challenges and concerns as to the setting in MessyTable.

MessyTable is not expected to replace the functionalities of domain-specific datasets. It aims to be a general dataset offering fundamental challenges to existing vision algorithms, with the hope of inspiring new problems and encouraging novel solutions. In this paper, apart from describing the details of MessyTable, we also present the results of applying some baselines and heuristics to address the instance association problem. We also show how a deep learning-based method developed from MessyTable can be migrated to other real-world multi-camera domains and achieve good results.

2   Related Work

Related Problems. Various computer vision tasks, such as re-identification and tracking, can be viewed as some forms of instance association. Despite differences in the problem settings, they share common challenges as featured in MessyTable, including subtle appearance difference, heavy occlusion and viewpoint variation. We take inspirations from methods in these fields for developing a potential solution for multi-camera instance association.

Re-identification (ReID) aims to associate the query instance (e.g., a person) and the instances in the gallery [9]. Association performance suffers from drastic appearance differences caused by viewpoint variation, and heavy occlusion in crowded scenes. Therefore, the appearance feature alone can be insufficient for satisfactory results. In this regard, instead of distinguishing individual persons in isolation (e.g.,[26,38,40]), an alternative solution proposed by [39] exploits contextual cues: as people often walk in groups in crowded scenes, it associates the same group of people over space and time.

Multi-object Tracking (MOT) is the task to associate instances across sequential frames, leveraging the availability of both appearance features and temporal cues [19,22,24]. It suffers from ID switches and fragmentation primarily caused by occlusion [14,17]. MOT in a multi-camera setting is formally referred to as Multi-Target Multi-Camera Tracking (MTMCT) [20], which also suffers from viewpoint variation in cross-camera tracklet association [8,12,37]. In addition, MTMCT with overlapping field of view [4,32] is similar to MessyTable's multi-camera setting. Thus, studies conducted on MessyTable might be inspiring for a better cross-camera association performance in MTMCT.

Related Datasets. Many datasets for ReID and MOT offer prominent challenges [10] that are common in real life. For instance, CUHK03 [13], MSMT17 [28], and MOT16 [18] feature occlusion and viewpoint variation, and many other datasets [31,34] also feature illumination variations.

There are several multi-camera datasets. Though originally designed for different purposes, they can be used for evaluating instance association. MPII Multi-Kinect (MPII MK) [27] is designed for object instance detection and collected on a flat kitchen countertop with nine classes of kitchenwares captured in four fixed views. The dataset features some level of occlusion, but the scenes are relatively simple for evaluating general object association. EPFL Multi-View Multi-Class (EPFL MVMC) [21] contains only people, cars, and buses and is built from video sequences of six static cameras taken at the university campus (with the road, bus stop, and parking slots). WILDTRACK [4], captured with seven static cameras in a public open area, is the latest challenging dataset for multi-view people detection.

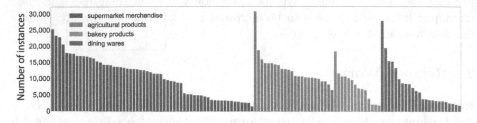

Fig. 2. Number of instances per class in MessyTable. MessyTable contains 120 classes: 60 supermarket merchandise, 23 agricultural products, 13 bakery products, and 24 dinning wares. A full list of classes is included in the Supplementary Materials

Compared to existing datasets, MessyTable aims to offer fundamental challenges that are not limited to specific classes. MessyTable also contains a large number of camera setup configurations for a larger variety of camera poses and an abundance of identical instances that are absent in other datasets.

3   MessyTable Dataset

MessyTable is a large-scale multi-camera general object dataset designed for instance association tasks. It comprises 120 classes of common on-table objects (Fig. 2), encompassing a wide variety of sizes, colors, textures, and materials. Nine cameras are arranged in 567 different setup configurations, giving rise to 20,412 pairs of relative camera poses (Sect. 3.1). A total of 5,579 scenes, each containing 6 to 73 randomly selected instances, are divided into three levels of difficulties. Harder scenes have more occlusions, more similar- or identical-looking instances, and proportionally fewer instances in the overlapping areas (Sect. 3.2). The total 50,211 images in MessyTable are densely annotated with 1,219,240 bounding boxes. Each bounding box has an instance ID for instance association across cameras (Sect. 3.3). To make MessyTable more representative, varying light conditions and backgrounds are added. Details of the data collection can be found in the Supplementary Materials.

3.1   Variations in View Angles

For a camera pair with a large angle difference, the same instance may appear to be very different (e.g., instance ID 5 in Fig. 1) in the two views. Existing multi-camera datasets typically have their cameras installed on static structures [21,27], even at similar heights [4]. This significantly limits the variation as the data essentially collapses to a very limited set of modes. In contrast, MessyTable

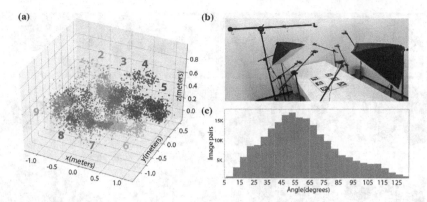

Fig. 3. MessyTable has a rich variety of camera poses. (a) The diverse distribution of camera positions covers the entire space. The positions of cameras #2-9 are projected to the coordinate system of camera #1, visualized by eight clusters of different colors. Camera #1 is fixed providing the bird's eye view. (b) The studio for data collection. The checkerboard is used for camera extrinsic calibration (see the Supplementary Materials for details); (c) Distribution of the angle differences of image pairs. The angle is computed between line-of-sights of cameras in each image pair. A significant portion of image pairs have larger angles than that in wide-baseline stereo matching, which is typically capped at 45°[3] (Color figure online)

has not only a high number of cameras but also a large variation in cameras' poses. The camera stands are arbitrarily adjusted between scenes, resulting in an extremely diverse distribution of camera poses and large variance of angle difference between cameras, as shown in Fig. 3.

3.2   Variations in Scenes

Partial and Full Occlusions. As shown in Fig. 4(a) and (b), partial occlusion results in loss of appearance features [1], making matching more difficult across cameras; full occlusion completely removes the object from one's view despite its existence in the scene. To effectively benchmark algorithms, in addition to dense clutter, artificial obstacles (such as cardboard boxes) are inserted into the scene.

Similar- or Identical-Looking Instances. It is common to have similar and identical objects placed in the vicinity as illustrated in Fig. 4(c) and (d). Similar or identical instances are challenging for association. MessyTable has multiple duplicates of the same appearance included in each scene, a unique feature that is not present in other datasets such as [27].

Variations in Elevation. Many works simplify the matching problem by assuming all objects are in contact with the same plane [1,5,7,15]. However, this assumption often does not hold as the scene gets more complicated. To mimic the most general and realistic scenarios in real life, object instances in MessyTable are allowed to be stacked or placed on elevated surfaces as shown in Fig. 4(e) and (f).

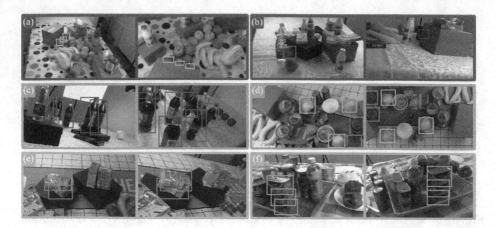

Fig. 4. Visualization of some of the design features that impose challenges for association: (a) Partial occlusion. (b) Full occlusion. (c) Similar-looking instances. (d) Identical-looking instances. (e) and (f): Object stacking and placing objects on elevated surfaces

Data Splits. The dataset is collected with three different complexity levels: Easy, Medium, and Hard, with each subset accounting for 30%, 50%, and 20% of the total scenes. For each complexity level, we randomly partition data equally (1:1:1) into the training, validation, and test sets.

The camera angle differences are similar among the three levels of complexity. But the number of instances, the fraction of overlapped instances and the fraction of identical instances are significantly different as shown in Fig. 5. Furthermore, as shown in the example scenes, the larger number of instances in a harder scene significantly increases the chance of heavy occlusion. We empirically show that these challenges undermine the association performance of various methods.

3.3   Data Annotation

We use OpenCV [2] for calibrating intrinsic and extrinsic parameters. As for the instance association annotation, we gather a team of 40 professional annotators and design a three-stage annotation scheme to obtain reliable annotations in a timely manner. The annotators first annotate bounding boxes to enclose all foreground objects (localization stage), followed by assigning class labels to the bounding boxes (classification stage). In the last stage, we develop an interactive tool for the annotators to group bounding boxes of the same object in all nine views and assign them the same instance ID (association stage). Bounding boxes in different views with the same instance ID are associated (corresponding to the same object instance in the scene) and the ID is unique in each view. For each stage, the annotators are split into two groups in the ratio of 4:1 for annotation and quality inspection.

It is worth mentioning that the interactive tool has two desirable features to boost efficiency and minimize errors: first, the class labels are used to filter out

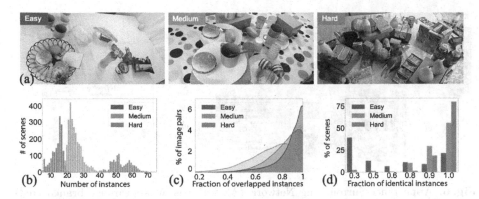

Fig. 5. The comparison amongst Easy, Medium and Hard subsets. (a) Example scenes by difficulty: harder scenes have more severe dense clutter and heavy occlusion. (b) Distributions of scenes by the number of instances per scene: harder scenes have more instances. (c) Distribution of image pairs by the extent of instance overlap: the graph peak of harder scenes is shifted leftward, indicating they have less fraction of instances visible in both cameras. (d) Distribution of scenes by the fraction of identical instances in a scene: harder scenes have more identical-looking instances

irrelevant bounding boxes during the association stage; second, the association results correct errors in the classification stage as the disagreement of classification labels from different views triggers reannotation. The details of the data annotation can be found in the Supplementary Materials. In short, MessyTable provides the following annotations:

- intrinsic and extrinsic parameters for the cameras;
- regular bounding boxes (with class labels) for all 120 foreground objects;
- instance ID for each bounding box

4   Baselines

In this section, we describe a few popular methods and heuristics that leverage appearance and geometric cues, and a new baseline that additionally exploits contextual information. We adopt a multiple instance association framework, in which pairwise distances between two sets of instances are computed first. After that, the association is formulated as a maximum bipartite matching problem and solved by the Kuhn-Munkres (Hungarian) algorithm.

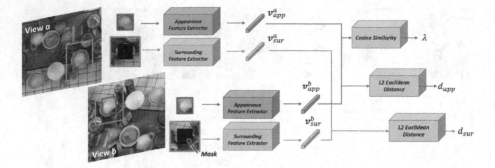

Fig. 6. Appearance-Surrounding Network (ASNet) has an appearance branch (red) and a surrounding branch (blue). Each instance and its immediate surrounding are cropped as the input to their respective branches (Color figure online)

4.1 Appearance Information

The appearance feature of the instance itself is the most fundamental information for instance association. As the instances are defined by bounding boxes, which are essentially image patches, we find instance association largely resembles the patch matching (local feature matching) problem.

Local feature matching is one of the key steps for low-level multiple camera tasks [3,30,36]. Various hand-crafted feature descriptors, e.g., SIFT [16], have been widely used in this task. We implement a classical matching pipeline including SIFT keypoint description of the instances and K-means clustering for the formation of a visual bag of words (VBoW). The distance between two VBoW representations is computed via chi-square (χ^2).

The application of deep learning has led to significant progress in patch matching [6,11,35]. The recent works highlight the use of CNN-based discriminative feature extractors such as DeepDesc [25] to directly output feature vectors, and the distance between two vectors can be computed using L2 distance; MatchNet [11] uses metric networks instead of L2 distance for better performance; DeepCompare [35] proposes to use both multi-resolution patches and metric networks. We use these three works as baselines.

We also implement a standard triplet network architecture with a feature extractor supervised by the triplet loss during training, which has been proven effective to capture subtle appearance difference in face recognition [23]. It is referred to as TripletNet in the experiments and L2 distance is used to measure the feature dissimilarity.

4.2 Surrounding Information

Inspired by Zheng et al. [39] in ReID, which addresses occlusion and view variation by associating a group of people instead of the individuals, we propose to look outside of the tight bounding box and involve the neighboring information, hoping to tackle not only occlusion and viewpoint variation, but also the existence of similar-looking instances.

The most intuitive idea is to expand the receptive field by a linear scaling ratio, i.e., cropping an area larger than the actual bounding box. This modification on the network input is referred to in the experiments as **zoom-out**, and the ratio as **zoom-out ratio**.

We take further inspiration from the human behavior: one may look at the surroundings for more visual cues *only if* the appearance of the instances themselves are not informative. Hence, we design a simple network (named Appearance-Surrounding Network, Fig. 6), which has two branches for appearance and surrounding feature extraction, fused as follows:

$$d^{ab} = (1 - \lambda) \times D_{l_2}(\mathbf{v}_{app}^a, \mathbf{v}_{app}^b) + \lambda \times D_{l_2}(\mathbf{v}_{sur}^a, \mathbf{v}_{sur}^b) \tag{1}$$

$$\lambda = S_c(\mathbf{v}_{app}^a, \mathbf{v}_{app}^b) \tag{2}$$

where a and b are superscripts for two patches, $\mathbf{v}_{app}$ and $\mathbf{v}_{sur}$ are appearance and surrounding feature vectors, respectively. D_{l_2} is L2 distance, and S_c is cosine similarity. λ is the weighting factor to fuse the appearance and surrounding branches. The fusion is designed such that, if the appearance features are similar, the network will place more weight on the surrounding than the appearance. Note that λ is jointly optimized in an end-to-end network; it is not a hyperparameter to be set manually.

Table 1. Baseline performances on MessyTable shows a combination of appearance, surrounding and geometric cues is most effective for instance association. [†]: with metrics learning; ★: upgraded backbone for a fair comparison; ZO: zoom-out; ESC: epipolar soft constraint; [‡]: triplet architecture. ↑: the higher the value, the better; ↓: the lower the value, the better; the same applies for the rest of the tables and figures

Model/Method	AP↑	FPR-95↓	IPAA-100↑	IPAA-90↑	IPAA-80↑
Homography	0.049	0.944	0	0	0
SIFT	0.063	0.866	0	0	0
MatchNet[†]	0.193	0.458	0.010	0.012	0.033
MatchNet[†]★	0.138	0.410	0.002	0.003	0.010
DeepCompare[†]	0.202	0.412	0.023	0.025	0.063
DeepCompare[†]★	0.129	0.402	0.005	0.005	0.01
DeepDesc	0.090	0.906	0.011	0.011	0.018
DeepDesc★	0.171	0.804	0.027	0.032	0.058
TripletNet[‡]	0.467	0.206	0.168	0.220	0.376
TripletNet[‡]+ZO	0.430	0.269	0.047	0.062	0.161
ASNet[‡]	0.524	0.209	0.170	0.241	0.418
ASNet[‡]+ESC	**0.577**	**0.157**	**0.219**	**0.306**	**0.499**

4.3   Geometric Methods

Homographic projection-based methods are very popular and used extensively in past works on Multi-Target Multi-Camera Tracking (MTMCT)[32,33] and Multi-View People Detection [1,5,7,15]. The mid-points of the bottom edges of the bounding boxes [15] are typically projected to a common coordinate system. The instances can thus be associated based on L2 distance between two sets of projected points. It assumes that all instances are placed on one reference 2D plane (e.g., the ground) and this simplification allows for an unambiguous pixel-to-pixel projection across cameras.

We also make use of epipolar geometry, which does not assume a reference plane. For a pair of bounding boxes in two views, the bounding box center in the first view is used to compute an epipolar line in the second view using the calibrated camera parameters. The distance between the bounding box center in second view and the epipolar line is added to the overall distance between the two bounding boxes. It is a soft constraint, since it does not accept or reject the matches, but penalizes unlikely matches by a large distance.

5   Experiments

Unless specified otherwise, we choose ResNet-18 as a light-weight backbone for all models, zoom-out ratio of 2 for models with zoom-out, a mixture of Easy, Medium, and Hard sets are used for training and evaluation.

5.1   Evaluation Metrics

AP: Class-agnostic Average Precision is used to evaluate the algorithm's ability to differentiate positive and negative matches, independent of the choice of the threshold value. All distances are scaled into a range of 0 and 1, and the confidence score is obtained by 1 - x, where x is the scaled distance.

FPR-95: False Positive Rate at 95% recall [11] is commonly used in patch-based matching tasks and is adopted as a supplement to AP. However, it is worth noting that in the patch matching problem, the positive and negative examples are balanced in the evaluation, which is not the case in our task where the negative examples largely outnumber the positive ones.

IPAA: We introduce a new metric, Image Pair Association Accuracy (IPAA), that evaluates the image-pair level association results instead of the instance-pair level confidence scores. IPAA is computed as the fraction of image pairs with no less than X% of the objects associated correctly (written as IPAA-X). In our experiments, we observed that IPAA is more stringent than AP, making it ideal for showing differences between models with reasonably high AP values. Details can be found in the Materials.

5.2   Benchmarking Baselines on MessyTable

In this section, we analyze and provide explanations for the performances of baselines on MessyTable, collated in Table 1.

Homographic projection performs poorly on MessyTable. The result is not surprising as objects in MessyTable can be placed on different elevated surfaces, violating the 2D reference plane assumption that is critical to accurate projection.

The SIFT-based classical method gives a poor performance as the hand-crafted key points tend to cluster around edges and texture-rich areas, leading to an unbalanced distribution. Hence, texture-less instances have very scarce key points, resulting in ineffective feature representation.

Deep learning-based patch matching SOTAs such as MatchNet [11], Deep-Compare [35], and DeepDesc [25] give sub-optimal results as they struggle in distinguishing identical objects, which are abundant in MessyTable. Interestingly, our experiments show that a deeper backbone does not improve performance for MatchNet and DeepCompare, as their performances may be bottlenecked by their simple metric network designs. TripletNet with a triplet architecture outperforms these three models with a Siamese architecture by a clear margin (around a 0.25 increment in AP).

We compare TripletNet and ASNet on surrounding information extraction. Naive inclusion of surrounding information (TripletNet+ZO) worsens the association results, as a larger receptive field may introduce noises. In contrast, ASNet trains a specialized branch for the surrounding information to extract meaningful features. Figure 7 visualizes the feature map activations, showing that ASNet effectively learns to use surrounding information whereas TripletNet+ZO tends to focus on the instance itself. However, we highlight that despite a considerable improvement, the ASNet only achieves a moderate AP of 0.524. This leaves a great potential for improvements.

We also show that adding soft geometric constraints to ASNet gives further improvement (around 0.05 improvement in AP), indicating that the geometric information is complementary to appearance and surrounding information. However, the performance, especially in terms of the stringent metric IPAA-100, is still unsatisfactory.

5.3   Effects of View Variation and Scene Complexity

We ablate the challenges featured in MessyTable and their effects on instance association.

We compare the performances of several relatively strong baseline methods at various angle differences in Fig. 8. It is observed that the performance by all three metrics deteriorate rapidly with an increase in the angle differences. As shown in Fig. 1, large angle difference leads to differences in not only the appearance of an instance itself, but also its relative position within its context.

In addition, we test the same trained model on Easy, Medium, and Hard test sets. The three test sets have the same distribution of angle differences, but different scene complexity in terms of the number of instances, percentage of identical objects, and the extent of overlapping (Fig. 5). The performance drops significantly in harder scenes, as shown in Table 2. We offer explanations: first, with more instances on the table, harder scenes contain more occlusion, as shown in Fig. 5(a). Second, it is more common to have identical objects closely placed or stacked together, leading to similar surrounding features and geometric distances (Fig. 9), making such instances indistinguishable. Third, harder scenes have a smaller fraction of instances in the overlapping area, this may lead to more false positive matches between non-overlapped similar or identical objects, which contributes to higher FPR-95 values.

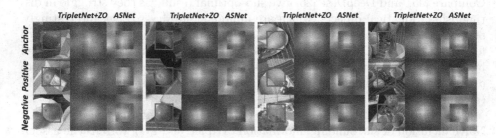

Fig. 7. Visualization of feature map activations after conv_5 of ResNet-18 for Triplet-Net+ZO (zoom-out) and ASNet. We normalize feature maps to the same scale, then map the values to colors. The higher activation in the surroundings for ASNet indicates that it effectively learns to use surrounding features compared to TripleNet with naive zoom-out

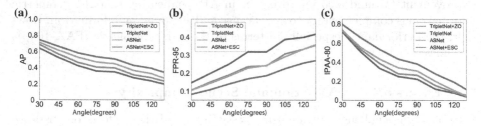

Fig. 8. Effects of angle differences. As the angle difference increases, cross-camera association performance deteriorates for all metrics. (a) AP↑. (b) FPR-95↓. (c) IPAA-80↑

The above challenges demand a powerful feature extractor that is invariant to viewpoint changes, robust under occlusion, and able to learn the surrounding feature effectively, yet, all baselines have limited capabilities.

Table 2. Results on Easy, Medium, and Hard test sets separately. As the scene get more complicated, the performances of all models worsen. Models are trained on a mixture of Easy, Medium and Hard train sets

Subsets	Model	AP↑	FPR-95↓	IPAA-100↑	IPAA-90↑	IPAA-80↑
Easy	TripletNet	0.618	0.156	0.266	0.342	0.561
	ASNet	0.667	0.151	0.408	0.427	0.660
	ASNet+ESC	0.709	0.122	0.497	0.500	0.734
Medium	TripletNet	0.494	0.211	0.063	0.150	0.290
	ASNet	0.547	0.207	0.101	0.226	0.391
	ASNet+ESC	0.594	0.163	0.151	0.296	0.476
Hard	TripletNet	0.341	0.259	0.003	0.023	0.078
	ASNet	0.396	0.255	0.003	0.026	0.100
	ASNet+ESC	0.457	0.185	0.007	0.048	0.155

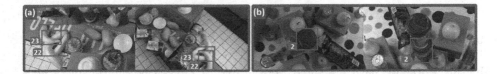

Fig. 9. Failure cases: wrong associations are visualized with bounding boxes and instance IDs. Most failure cases occur when identical objects are close to one another such as being (a) placed in a heavy clutter or (b) stacked, leading to indistinguishable surrounding features and similar penalty distances by the soft epipolar constraint. This remains as a challenge worth future research

5.4 MessyTable as a Benchmark and a Training Source

We further validate the usefulness of MessyTable by conducting experiments on three public multi-camera datasets (Table 3), which gives the following insights:

First, methods that saturate MPII MK and EPFL MVMC are far from saturating MessyTable (Table 1). Note that both datasets have a limited number of classes and instances. Hence, this result highlights the need for MessyTable, a more realistic and challenging dataset for research of instance association.

Second, it is observed that algorithms show consistent trends on MessyTable and other datasets, that is, an algorithm that performs better on MessyTable also performs better on all other datasets. This shows MessyTable can serve as a highly indicative benchmark for multi-camera instance association.

Third, models pretrained on MessyTable consistently perform better than those pretrained on ImageNet, showing MessyTable is a better training source for instance association tasks. Note that EPFL MVMC has three classes (people, cars, and buses) and WILDTRACK is a challenging people dataset. It shows that a model trained on the general objects in MessyTable, learns feature extraction that is readily transferable across domains without feature engineering.

Table 3. Experiments on three other multi-camera datasets. MessyTable is highly indicative as algorithms benchmarked on MessyTable show consistent trends on all other datasets. Moreover, models pretrained on MessyTable consistently give better performances, showing MessyTable's potential as a training source

Dataset	Model	Pretraining	AP↑	FPR-95↓	IPAA-100↑
MPII MK	TripletNet	ImageNet	0.847	0.196	0.333
	ASNet	ImageNet	0.881	0.167	0.696
	ASNet	MessyTable	0.905	0.119	0.765
EPFL MVMC	TripletNet	ImageNet	0.921	0.056	0.529
	ASNet	ImageNet	0.950	0.038	0.559
	ASNet	MessyTable	0.969	0.031	0.575
WILDTRACK	TripletNet	ImageNet	0.616	0.207	0.095
	ASNet	ImageNet	0.718	0.094	0.304
	ASNet	MessyTable	0.734	0.083	0.321

6   Discussion

In this work, we have presented MessyTable, a large-scale multi-camera general object dataset for instance association. MessyTable features the prominent challenges for instance association such as appearance inconsistency due to view angle differences, partial and full occlusion, similar and identical-looking objects, difference in elevation, and limited usefulness of geometric constraints. We show in the experiments that it is useful in two more ways. First, MessyTable is a highly indicative benchmark for instance association algorithms. Second, it can be used as a training source for domain-specific instance association tasks.

By benchmarking baselines on MessyTable, we obtain important insights for instance association: appearance feature is insufficient especially in the presence of identical objects; our proposed simple baseline, ASNet, incorporates the surrounding information into association and effectively improves the association performance. In addition, we show that epipolar geometry as a soft constraint is complementary to ASNet.

Although the combined use of appearance features, context information, and geometric cues achieves reasonably good performance, ASNet is still inadequate to tackle all challenges. Therefore, we ask three important questions: (1) how to extract stronger appearance, neighbouring and geometric cues? (2) is there a smarter way to fuse these cues? (3) is there more information that we can leverage to tackle instance association?

The experiment results on MessyTable set many directions worth exploring. First, increasing view angle difference leads to a sharp deterioration of instance association performance of all baselines, highlighting the need for research on methods that capture view-invariant features and non-rigid contextual information. Second, methods give poorer performances as the scenes get more complicated; failure cases show that identical instances placed close to each other are extremely difficult to address despite that the strongest baseline already leverages appearance, surrounding and geometric cues. Hence, more in-depth object relationship reasoning may be helpful to distinguish such instances.

Acknowledgements. This research was supported by SenseTime-NTU Collaboration Project, Singapore MOE AcRF Tier 1 (2018-T1-002-056), NTU SUG, and NTU NAP.

References

1. Baqué, P., Fleuret, F., Fua, P.: Deep occlusion reasoning for multi-camera multi-target detection. In: ICCV (2017)
2. Bradski, G.: The OpenCV library. Dr. Dobb's J. Softw. Tools **25**, 120–125 (2000)
3. Caliskan, A., Mustafa, A., Imre, E., Hilton, A.: Learning dense wide baseline stereo matching for people. In: ICCVW (2019)
4. Chavdarova, T., et al.: WILDTRACK: a multi-camera HD dataset for dense unscripted pedestrian detection. In: CVPR (2018)
5. Chavdarova, T., et al.: Deep multi-camera people detection. In: ICMLA (2017)
6. Csurka, G., Humenberger, M.: From handcrafted to deep local features for computer vision applications. CoRR abs/1807.10254 (2018)
7. Fleuret, F., Berclaz, J., Lengagne, R., Fua, P.: Multicamera people tracking with a probabilistic occupancy map. PAMI **30**, 267–282 (2007)
8. Gao, J., Nevatia, R.: Revisiting temporal modeling for video-based person ReID. CoRR abs/1805.02104 (2018)
9. Raja, Y., Gong, S.: Scalable multi-camera tracking in a metropolis. In: Gong, S., Cristani, M., Yan, S., Loy, C.C. (eds.) Person Re-Identification. ACVPR, pp. 413–438. Springer, London (2014). https://doi.org/10.1007/978-1-4471-6296-4_20
10. Gou, M., et al.: A systematic evaluation and benchmark for person re-identification: features, metrics, and datasets. PAMI (2018)
11. Han, X., Leung, T., Jia, Y., Sukthankar, R., Berg, A.C.: MatchNet: unifying feature and metric learning for patch-based matching. In: CVPR (2015)
12. Hsu, H.M., Huang, T.W., Wang, G., Cai, J., Lei, Z., Hwang, J.N.: Multi-camera tracking of vehicles based on deep features Re-ID and trajectory-based camera link models. In: CVPRW (2019)
13. Li, W., Zhao, R., Xiao, T., Wang, X.: DeepReID: deep filter pairing neural network for person re-identification. In: CVPR (2014)
14. Li, W., Mu, J., Liu, G.: Multiple object tracking with motion and appearance cues. In: ICCVW (2019)
15. López-Cifuentes, A., Escudero-Viñolo, M., Bescós, J., Carballeira, P.: Semantic driven multi-camera pedestrian detection. CoRR abs/1812.10779 (2018)
16. Lowe, D.G.: Distinctive image features from scale-invariant keypoints. IJCV **60**, 91–110 (2004)
17. Luo, W., et al.: Multiple object tracking: a literature review. CoRR abs/1409.7618 (2014)
18. Milan, A., Leal-Taixé, L., Reid, I., Roth, S., Schindler, K.: MOT16: a benchmark for multi-object tracking. CoRR abs/1603.00831 (2016)
19. Milan, A., Rezatofighi, S.H., Dick, A., Reid, I., Schindler, K.: Online multi-target tracking using recurrent neural networks. In: AAAI (2017)

20. Ristani, E., Solera, F., Zou, R., Cucchiara, R., Tomasi, C.: Performance measures and a data set for multi-target, multi-camera tracking. In: Hua, G., Jégou, H. (eds.) ECCV 2016. LNCS, vol. 9914, pp. 17–35. Springer, Cham (2016). https://doi.org/ 10.1007/978-3-319-48881-3_2
21. Roig, G., Boix, X., Shitrit, H.B., Fua, P.: Conditional random fields for multi-camera object detection. In: ICCV (2011)
22. Sadeghian, A., Alahi, A., Savarese, S.: Tracking the untrackable: learning to track multiple cues with long-term dependencies. In: ICCV (2017)
23. Schroff, F., Kalenichenko, D., Philbin, J.: FaceNet: a unified embedding for face recognition and clustering. In: CVPR (2015)
24. Schulter, S., Vernaza, P., Choi, W., Chandraker, M.: Deep network flow for multi-object tracking. In: CVPR (2017)
25. Simo-Serra, E., Trulls, E., Ferraz, L., Kokkinos, I., Fua, P., Moreno-Noguer, F.: Discriminative learning of deep convolutional feature point descriptors. In: ICCV (2015)
26. Sun, Y., Zheng, L., Yang, Y., Tian, Q., Wang, S.: Beyond part models: person retrieval with refined part pooling (and a strong convolutional baseline). In: Ferrari, V., Hebert, M., Sminchisescu, C., Weiss, Y. (eds.) ECCV 2018. LNCS, vol. 11208, pp. 501–518. Springer, Cham (2018). https://doi.org/10.1007/978-3-030-01225-0_30
27. Susanto, W., Rohrbach, M., Schiele, B.: 3D object detection with multiple kinects. In: Fusiello, A., Murino, V., Cucchiara, R. (eds.) ECCV 2012. LNCS, vol. 7584, pp. 93–102. Springer, Heidelberg (2012). https://doi.org/10.1007/978-3-642-33868-7_10
28. Wei, L., Zhang, S., Gao, W., Tian, Q.: Person transfer GAN to bridge domain gap for person re-identification. In: CVPR (2018)
29. Wei, X.S., Cui, Q., Yang, L., Wang, P., Liu, L.: RPC: a large-scale retail product checkout dataset. CoRR abs/1901.07249 (2019)
30. Winder, S., Hua, G., Brown, M.: Picking the best DAISY. In: CVPR (2009)
31. Xu, Y., Zhou, X., Chen, S., Li, F.: Deep learning for multiple object tracking: a survey. IET Comput. Vis. 13, 355–368 (2019)
32. Xu, Y., Liu, X., Liu, Y., Zhu, S.C.: Multi-view people tracking via hierarchical trajectory composition. In: CVPR (2016)
33. Xu, Y., Liu, X., Qin, L., Zhu, S.C.: Cross-view people tracking by scene-centered spatio-temporal parsing. In: AAAI (2017)
34. Ye, M., Shen, J., Lin, G., Xiang, T., Shao, L., Hoi, S.C.: Deep learning for person re-identification: a survey and outlook. CoRR abs/2001.04193 (2020)
35. Zagoruyko, S., Komodakis, N.: Learning to compare image patches via convolutional neural networks. In: CVPR (2015)
36. Zbontar, J., LeCun, Y.: Computing the stereo matching cost with a convolutional neural network. In: CVPR (2015)
37. Zhang, Z., Wu, J., Zhang, X., Zhang, C.: Multi-target, multi-camera tracking by hierarchical clustering: recent progress on DukeMTMC project. CoRR abs/1712.09531 (2017)
38. Zhao, H., et al.: Spindle net: person re-identification with human body region guided feature decomposition and fusion. In: CVPR (2017)
39. Zheng, W.S., Gong, S., Xiang, T.: Associating groups of people. In: BMVC (2009)
40. Zhou, Y., Shao, L.: Aware attentive multi-view inference for vehicle re-identification. In: CVPR (2018)

A Unified Framework for Shot Type Classification Based on Subject Centric Lens

Anyi Rao[1]([✉]), Jiaze Wang[1], Linning Xu[1], Xuekun Jiang[2], Qingqiu Huang[1], Bolei Zhou[1], and Dahua Lin[1]

[1] CUHK - SenseTime Joint Lab, The Chinese University of Hong Kong, Sha Tin, Hong Kong
{anyirao,hq016,bzhou,dhlin}@ie.cuhk.edu.hk, jzwang.cuhk@gmail.com, linningxu@link.cuhk.edu.cn
[2] Communication University of China, Beijing, China
xkjiang@cuc.edu.cn

Abstract. *Shots* are key narrative elements of various videos, *e.g.* movies, TV series, and user-generated videos that are thriving over the Internet. The types of shots greatly influence how the underlying ideas, emotions, and messages are expressed. The technique to analyze shot types is important to the understanding of videos, which has seen increasing demand in real-world applications in this era. Classifying shot type is challenging due to the additional information required beyond the video content, such as the spatial composition of a frame and camera movement. To address these issues, we propose a learning framework Subject Guidance Network (SGNet) for shot type recognition. SGNet separates the subject and background of a shot into two streams, serving as separate guidance maps for scale and movement type classification respectively. To facilitate shot type analysis and model evaluations, we build a large-scale dataset *MovieShots*, which contains $46K$ shots from $7K$ movie trailers with annotations of their scale and movement types. Experiments show that our framework is able to recognize these two attributes of shot accurately, outperforming all the previous methods.

1 Introduction

In 1900, film pioneer George Albert Smith firstly introduced shot type transitions into videos, which revolutionized traditional narrative thought and this technique remains widely used in today's video editing [45]. *Shot*, a series of visual continuous frames, plays an important role in presenting the story.It can

Electronic supplementary material The online version of this chapter (https:// doi.org/10.1007/978-3-030-58621-8_2) contains supplementary material, which is available to authorized users.

be recognized from multiple attributes, such as *scale* and *movement*. As illustrated in Fig. 1, five scale types and four movement types of shots are widely adopted in video editing, serving for different scenes and emotional expressions.

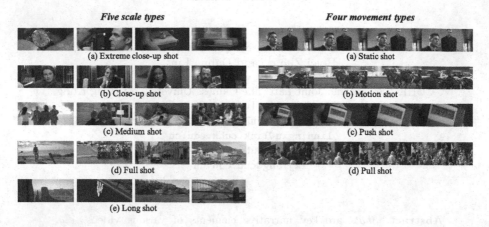

Fig. 1. Demonstrations of five *scale* types and four *movement* types of video shots sampled from *MovieShots* dataset. It is noticed that shot scales can reveal information of a story from different aspects. For example, *long shots* usually indicate the location information, while *close-up shots* are widely used for emphasizing the identities of the characters. *Medium shots* and *full shots* are good at depicting an event, while *extreme close-up shots* are used for symbolic expressions or intensifying the story emotion. For movement types, we notice that *static shots* are mainly used for narrative purposes and *motion shots* try to track moving objects. *Push shots* aim to emphasize the content information of the main subject while pull shots shrink the figure of the main subject and gradually reveal its surrounding environment

We are in the era of web 2.0 where user-generated videos proliferate, the techniques to analyze shot types have seen increasing demand in real-world applications: 1) With the capability of recognizing shot types, the videos shared online can be automatically classified or organized not only by their content, *e.g.* object categories, but also by shot types. Thus the system will be able to respond to queries like "long shots over a city" etc. 2) By analyzing the sequence of shots in movies, we may provide a data-driven view of how professional movies are constructed. Such insight can help ordinary users to make videos that look more professional – one can even build softwares to guide video production for amateurs.

Despite the potential value of shot type analysis, it is true that most previous works in computer vision primarily focus on objective content understanding. For example in video analysis, we focus on classifying and localizing the actions [16, 31, 41], while shot type analysis has been rarely investigated and lacks appropriate benchmark. Existing datasets on shot type classification are either too small or not publicly available. However, we believe that the analysis of cinematic techniques (e.g. shot types) are also equally important.

To facilitate researches along this direction, we construct a new dataset *MovieShots*, which composes of over $46K$ shots collected from public movie trailers, with annotations on five scale types and four movement types. We select out scale and movement from many other shot attributes, as they are the two most common and distinguishable attributes that can uniquely characterize a shot in video, where the *scale* type is decided by the amount of subjects within the frame, and the *movement* type is determined by the camera motion [18].

We further propose a novel subject centric framework, namely Subject Guidance Network (SGNet), to classify the scale and movement type of a shot. The key point here is to find out the dominant subject in a given shot, then we can decide its scale according to the portion it takes, and differentiate between the camera movement and subject movement to determine the movement type. SGNet successfully separates the subject and background in a shot and takes them to guide the full images to predict the labels for scale and movement type.

The contributions of this work are as follows: 1) We construct *MovieShots*, a large-scale $46K$ shot dataset with professionally annotated scale and movement attributes for each shot. 2) SGNet is proposed to classify scale and movement type simultaneously based on the subject centric lens. Our experiments show that this framework greatly improves the classification performance comparing to traditional methods and conventional deep networks TSN [41] and I3D [7].

2   Related Work

Shot Type Classification Datasets. Traditional shot type classifications mainly focus on sports videos [14,15,26]. Sports video is a special kind of video that contains many clips such as video replays or comments, which is hard to transfer to general video scenarios. Previous movie shot type researches [6,40] are limited on their evaluation benchmarks. They collect no more than twenty films with about one-thousand shots. There is no public available dataset to test the functionality of these methods. It is noticed that these datasets annotate either *scale* or *movement* attribute only, lacking a comprehensive description for a shot. In order to solve these limitations, we collect a $10\times$–$100\times$ larger dataset, with $46K$ video shots annotated with both *scale* and *movement* attributes from more than $7K$ public movie trailers.

Shot Type Classification Methods. Conventional methods for shot *scale* classification use SVM with dominate color region [28], low-level texture features [1,52], or optical flow [26]. Decision tree method [47] sets up fixed rules to classify the scale type of a shot. Scene depth [3] is applied to infer the scale but is limited to the depth approximation accuracy and lacks generalization ability. For *movement* type classification, traditional approaches rely on the manual design of a motion descriptor. *e.g.* [21,32] design *motion vectors* CAMHID and 2DMH to capture the camera movement. [40] leverage optical flow to find an alternative of the motion vectors. However, all these methods heavily depend on hand-crafted features that are not applicable to general cases. Our SGNet separates the subject from image and considers both the spatial and temporal

Table 1. Comparisons with other datasets

	#Shot	#Video	Scale	Move
Lie 2014 [4]	327	327		✓
Unified 2005 [14]	430	1	✓	
Sports 2007 [55]	1,364	8	✓	
Soccer 2009 [28]	1,838	1	✓	
Cinema 2013 [6]	3,000	12	✓	
Context 2011 [52]	3,206	4	✓	
Taxon 2009 [40]	5,054	7		✓
MovieShots	46,857	7,858	✓	✓

Table 2. Statistics of *MovieShots*

	Train	Val	Test	Total
Number of Movies	4,843	1,062	1,953	7,858
Number of Shots	32,720	4,610	9,527	46,857
Avg. Dur. of Shot (s)	3.84	5.31	3.78	3.95

configurations of a given shot, achieving much improved performance with better generalization ability.

Video Analysis and Understanding in One Shot. Most previous single shot video understanding tasks [16,31] are about action recognition [17,41,43] and temporal action localization [8,50,59]. Video object detection [12,20], video object segmentation [51,57], video person recognition [46,54], video-text retrieval [48,56] and some low-level vision tasks, *e.g.* video inpainting [53] video super-resolution [29,58] are also applied in single shot videos. However, research on video shot type is rarely explored, despite of its huge potential for video understanding. We set up a benchmark with our *MovieShots* dataset and conduct a detailed study on it.

3   *MovieShots* Dataset

To facilitate the shot type analysis in videos, we collect *MovieShots*, a large-scale shot type annotation set that contains $46K$ shots from 7858 movies. The details of this dataset are specified as follows.

3.1   Shot Categories

Following previous definition on shot type [18,25,27,37,40,52], shot *scale* is defined by the amount of subject figure that is included within the frame, while shot *movement* is determined by the camera movement or the lens change.

Shot *scale* has five categories: 1) *long shot* (LS) is taken from a long distance, sometimes as far as a quarter of a mile away; 2) *full shot* (FS) barely includes the human body in full; 3) *medium shot* (MS) contains a figure from the knees or waist up; 4) *close-up shot* (CS) concentrates on a relatively small object, showing the face or the hand of a person; (5) *extreme close-up shot* (ECS) shows even smaller parts such as the image of an eye or a mouth.

Shot *movement* has four categories: 1) in *static shot*, the camera is fixed but the subject is flexible to move; 2) for *motion shot*, the camera moves or rotates; 3) the camera zooms in for *push shot*, and 4) zooms out for *pull shot*. While all the four movement types are widely used in movies, the use of *push* and *pull* shots only takes a very small portion. The usage of different shots usually depends on the movie genres and the preferences of the filmmakers.

3.2   Dataset Statistics

MovieShots consists of 46,857 shots from 7,858 movie trailers, covering a wide variety of movie genres to ensure the inclusion of all scale and movement types of shot. Table 1 compares *MovieShots* with existing private shot type datasets, noting that none of them are publicly available. *MovieShots* is significantly larger than others in terms of the shot number and the video coverage, with a more comprehensive annotation covering both the *scale* type and the *movement* type for each shot.

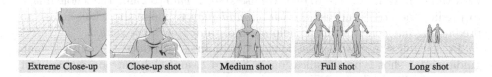

Fig. 2. Prototypes of annotation corresponding to extreme close-up shot, close-up shot, medium shot, full shot and long shot

3.3   Annotation Procedure

Building a large-scale shot dataset is challenging in two aspects: appropriate data collections and accurate data annotations. We firstly crawled more than $10K$ movie trailers online. Because movie trailers usually contain advertisements and big subtitles displaying the actor/director information, we firstly cleaned them with auto advertisement/big text detection and went through a second round of manual check. Noting that shot detection is a well solved problem and we used an off-the-shelf approach [33,39] to cut shots in these trailers and filtered out failure cases with manual check. All our annotators are cinematic professionals in film industries or cinematic arts majors, who provide high quality labels. We also set up annotation prototypes for these well defined criterion of shot types, as illustrated in Fig. 2. Additionally, three rounds annotation procedures have been done to ensure the high consistency. We finally achieve 95% annotation consistency, with those inconsistent shots being filtered out in our experiments.

4   SGNet: Subject Guidance Network

In this section, we introduce our Subject Guidance Network (SGNet) for *scale* and *movement* type classification. The overall framework is shown in Fig. 3.

We firstly divide a shot into N clips to capture the contextual variations along the temporal dimension. Each clip passes through a two-branch classification network and outputs a feature vector. The feature vectors coming from the N clips are pooled together and pass through a fully-connected layer to get the final prediction.

It is noticed that, the separation of subject and background information is critical for both two tasks. While the *scale* type depends on the portion of the subject in the shot, the *movement* type relies on the background motion rather than the subject motion, as the changes of the background information are closely related to the camera motion. To reduce the burden of the whole pipeline, a light-weight *subject map generator* is designed to separate the subject and background[1] in both image and flow in an effective way. The *subject map* here is a saliency map sharing the same width and height as the original image with values in range of zero to one. We use *subject map* to guide the whole image to predict the *scale*, and use the *background map* to guide the whole image to predict the *movement* with the help of an obtained *variance map*.

In the following two subsections, we will first introduce the subject map guidance for shot type classification in Sect. 4.1, and elaborate on our subject map generation in Sect. 4.2.

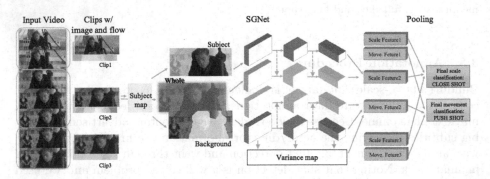

Fig. 3. Pipeline of Subject Guidance Network (SGNet). A subject map is generated from each clip's image. With this subject map, we take subject and background to guide scale and movement prediction respectively

4.1   Subject Map Guidance Classification

As discussed before, the separation of subject and background is crucial to shot type classification. In this section, we firstly explain how the obtained subject map guides scale and movement classification respectively.

For *scale* type classification, the subject map $[1 \times W \times H]$ element-wise multiplies with the whole image $[3 \times W \times H]$ to get an subject image $[3 \times W \times H]$. Then we take the whole image and the obtained subject image as two input pathways.

---

[1] Background image is equal to the whole image minus the subject part.

Each of them is sent into a ResNet50 [22] network. To apply subject guidance, we fues the subject feature map from different stages of feature representations into the whole image pathway. Specifically, these fusion connections are applied right after $pool_1$, res_2, res_3, and res_4 in the ResNet50 [22] backbone. The fusion is conducted by lateral connections [11].

For *movement* type classification, the background guidance is applied in a similar way to the scale type prediction. Additionally, a *variance map* module is further introduced, inspired by the fact that the changes of appearance along time is a cue for movement classification. For example, in a *static* shot, the appearances of background among different clips are almost the same, while the background appearances changes significantly as time changes in *motion* shots, *push* shots and *pull* shots. We calculate one variance map $\mathbf{V}_m \in \mathcal{R}^{N \times N}$ for each shot among its different clips ($N = 8$ clips in our experiments) at different stages of the backbone ResNet50. Specifically, these stages include those after $pool_1$, res_2, res_3, and res_4, the same as those used in the previous fusions. We apply inner product between the two normalized feature maps $\mathbf{F}_{m,i}, \mathbf{F}_{m,j} \in \mathcal{R}^{h_m \times w_m \times c_m}$ from two clips i, j at stage m to get $\mathbf{V}_{m,(i,j)}$. The inner product here is equivalent to calculating the cosine similarity, which captures the similarities between different clips. $\mathbf{V}_m$ is achieved by concatenating all possible clips pair $\mathbf{V}_{m,(i,j)}$. Finally, all $\mathbf{V} = \{\mathbf{V}_m\}$ among M stages are concatenated along the channel-wise dimension and are fed into a two-layer FC for classification. The classification results using variance maps are fused with the image classification results for the final prediction.

4.2   Subject Map Generation

Now we elaborate on how we separate the subject from the background with our light-weight *subject map generator*.

Conventional saliency/attention map methods employ hand-crafted visual features or heuristic priors [9,60], which are incapable of capturing high-level semantic knowledge, making the predicted map unsatisfactory. Pre-trained state-of-the-art deep networks [13,30] are usually very large with more than 50–100 layers and are not easy to be taken as a submodule in the designed networks to fine tune, considering the high computational costs. From another perspective, to train a randomly initialized subject network from scratch with only shot type label is impractical, since the supervision signal is too weak and the network is unable to converge, considering the subject map is a pixel-wise prediction but the annotation is a video-level label.

To strike a balance between the performance and the computational effi-ciency, we resort to knowledge distillation (KD) [23,49], considering that it is easy and flexible to learn, and achieves state-of-the-art performance on classifica-tion problems [10]. A light-weight *student generator* (a 6-layer CNN) learns from its *teacher network* (a MSRA10K [24] pre-trained R^3Net [13]). Note that a naive KD using $\mathcal{L}_2$ loss is usually suboptimal because it is difficult to learn the true data distribution from the teacher and may result in missing generation details.

Therefore, an additional adversarial loss with the help of Generative Adversarial Networks (GAN) [2,19,38,42,44] is adopted. In all, the student generator is trained by minimizing the following three-term loss,

$$\mathcal{L} = \alpha\mathcal{L}_2 + \beta\mathcal{L}_{\mathrm{adv}} + \mathcal{L}_{\mathrm{cross}}.$$

The first loss term L_2 is the least square error between the generated subject map and its corresponding pseudo subject map, which aims to mimic the output of teacher network. L_2 loss alone is not able to teach the student network to generate fine grained details since it does not consider the constraints from the whole data distribution. The second term $\mathcal{L}_{\mathrm{adv}}$ is given by a learned discriminator, which is trained to compete with the student generator to learn the true data distribution. The discriminator takes the subject map from the teacher network as *real* and the output from student generator as *fake*. Finally, we take cross-entropy loss L_{cross} as our classification loss and be jointly trained with the whole pipeline to encourage right predictions.

5 Experiments

5.1 Experiments Setup

Data. All the experiments are conducted on *MovieShots*. The whole dataset is split into *Train*, *Val*, and *Test* sets with a ratio 7:1:2, as shown in Table 2.

Implementation Details. We take cross-entropy loss for the classification result. The shot is evenly split into 3 clips in training and 25 clips in testing. The fusing function from the subject/background map to whole image is implemented by concatenating the output from the two branches. Image and flow are set up as two inputs and their classification score are fused to get the final results. We train these models for 60 epochs with mini-batch SGD, where the batch size is set to 128 and the momentum is set to 0.9. The initial learning rate is 0.001 and the learning rate will be divided by 10 at the 20*th* and 40*th* epoch.

Evaluation Metrics. We take the commonly used Top-1 accuracy as the evaluation metric. Specifically, in our experiment, we denote Acc_S for scale classification performance and Acc_M for movement classification performance.

5.2 Overall Results

We reproduce DCR [28], CAMHID [21] and 2DMH [32] according to their papers. DCR [28] clusters dominant color sets and predicts shot type based on the ratio of different color sets. CAMHID [21], 2DMH [32] are based on motion vectors. CAMHID [21] takes SVD to get the dominant components. 2DMH [32] disentangles the magnitude and orientation of motion vectors. TSN [41] and I3D [7] are experimented using authors' code repositories. SGNet adopts ResNet50 [22]

Table 3. The overall results on shot scale and movement type classification

Models	Acc_S (↑)	Acc_M (↑)
DCR, Li *et al.* [28]	51.53	33.20
CAMHID, Wang *et al.* [21]	52.37	40.19
2DMH, Prasertsakul *et al.* [32]	52.35	40.34
I3D-ResNet50 [7]	76.79	78.45
I3D-ResNet50-Kinetics [7]	77.11	83.25
TSN-ResNet50 (img) [41]	84.08	70.46
TSN-ResNet50-Kinetics (img) [41]	84.18	71.61
TSN-ResNet50 (img + flow) [41]	84.10	77.13
TSN-ResNet152 (img + flow) [41]	84.95	78.02
SGNet (img)	87.21	71.30
SGNet (img + flow)	87.50	80.65
SGNet w/ Var (img)	87.42	80.57
SGNet w/ Var (img + flow)	**87.57**	**81.86**
SGNet w/ Var-Kinetics (img + flow)	**87.77**	**83.72**

as the backbone. All the network weights are initialized with pre-trained models from ImageNet [36] unless specially stated.

Overall Results Analysis. 1) *Traditional Methods.* The overall results are shown in Table 3. The performances of DCR [28], CAMHID [21] and 2DMH [32] are restricted by their poor representations.

2) *3D Networks.* For movement classification, I3D-ResNet50 achieves better result than TSN-ResNet50 (img + flow) since it captures more temporal relationships. With Kinetics400 [7] pre-trained, I3D-ResNet50 gets 4.8 boost on Acc_M. But in scale classification, I3D-ResNet50 performs worse than TSN-ResNet50 (img). The reason might be that I3D-ResNet50 is not good at capturing the spatial configuration of frames in predicting the shot scale. The performance of I3D-ResNet101 is similar to I3D-ResNet50 since deeper 3D networks needs more data to improve the performance.

From another perspective, 3D CNNs are much more computational expensive and need dense samples from videos, which causes the low speed for training and inference. We choose 2D TSN-ResNet50 as our backbone. The results prove that this 2D network can achieve better results than 3D networks with our careful designs. Deep 2D network TSN [41] using image (TSN img) achieves ∼30% raise on Acc_S and Acc_M than traditional methods, as it captures high-level semantic information such as the subject contours and the temporal relationship in a shot.

3) *Deeper Backbones.* To show that the improvement does not come from the increase of model parameters, we compare SGNet w/ Var (img + flow) (use ResNet50 backbone) with TSN-ResNet152 (img + flow). SGNet w/ Var (img

+ flow) outperforms TSN-ResNet152 by a margin of 2.62 on Acc_S and 3.84 on Acc_M, with 15% fewer parameters and 19% fewer GFLOPs.

4) *2D Networks and Kinetics Pre-training.* Our full model SGNet w/ Var (img + flow) which includes subject map guidance, motion information flow, and variance map, improves 3.49 (relatively 4.12%) on Acc_S and 11.40 (relatively 16.11%) on Acc_M compared to TSN (img), and 3.47 (relatively 4.15%) on Acc_S and 4.73 (relatively 6.13%) on Acc_M compared to TSN (img + flow). The full model get further improvements by 0.2 on Acc_S and 1.8 Acc_M with Kinetics [7] pre-trained. This result shows that action recognition dataset can bring more help to shot movement predictions.

Table 4. Comparison of different subject or/and background map guidance.

#	Settings	Acc_S (↑)	Acc_M (↑)
1	Base (TSN w/ Var img+flow)	84.15	77.25
2	Subject only	79.97	74.65
3	Back only	79.60	75.80
4	Base+Subj	**87.57**	80.86
5	Base+Back	87.10	**81.86**
6	Base+Subj+Back	87.31	81.54

Analysis of Our Framework. Based on TSN (img), SGNet (img) takes the advantage of subject map guidance and improves the scale and movement results by 3.13 and 0.96 respectively, which shows the usefulness of subject guidance especially for scale type prediction. With the help of variance map, SGNet w/ Var (img) raise the movement classification performance from 70.46 to 80.57 (relatively 14.35%). Similarly, flow (SGNet img+ flow) helps the model to improve the movement results from 70.46 to 80.65 (relatively 14.46%). These results show that variance map and flow both capture the movement information and contribute to the great performance on movement type classification. As for scale type classification, variance map and flow bring slight improvements (0.2–0.3), which shows that the movement information captured by variance map and flow provide a weak assistance to the scale type classification. Finally, combining variance map and flow, SGNet w/Var (img + flow) further gains improvement on scale (87.57) and on movement (81.86) classification and achieves the best performance among all (without Kinetics pre-trained).

5.3   Ablation Studies

We conduct ablation studies on the following designs to verify their effectiveness: 1) subject map guidance, 2) subject map generation, and 3) joint training.

The Effects of Different Subject and Background Map Guidances. In the first block of Table 4, we take TSN model using image and flow with variance map (TSN w/ Var img+flow) as our baseline to test the effects of different subject map guidances. It takes a single-branch ResNet50 as backbone and two models for image and flow respectively, and fuses their scores at the end. We observe that using only subject or background information is inferior to the performance of using the whole image and flow, with ∼5/∼2 drop on Acc_S/Acc_M.

Setting 4, 5 in Table 4 are two branches setting with either subject or background guidance. In these experiments, we take a two-branch ResNet50 as backbone for image and flow model, one branch for subject/background and the other one for the whole image/flow. The output obtained from the first branch is concatenated with the output of second branch (+Subj and +Back) as guidance, and send to following networks. Generally, subject guidance achieves 0.4–0.8 performance gain on scale classification and background guidance outperforms subject guidance on movement prediction with 0.8–1.0 better results.

Setting 6 (Base+Subj+Back) in Table 4 use a four-branch ResNet50. Two branches are for subject guidance, and the rest two are for background guidance. With more information, the performance drops a little since the subject and background information might be mutually exclusive to each other.

The Influence of Different Subject Map Generations. As discussed above, a subject map generation module is needed to guide the network prediction. This module has many alternatives. Table 5 shows the comparisons between our approaches and self-attention SBS (Saliency-Based Sampling Layer) [34], and fine-tuned/fixed models R^3Net-ResNet18/50 [13]. We take TSN model using image and flow with variance map (TSN w/ Var img+flow) as our baseline to test the influence of different subject map generations.

Table 5. Comparison of different subject map generation modules.

#	Settings	Acc_S (↑)	Acc_M (↑)
1	Base (TSN w/ Var img+flow)	84.15	77.25
2	SBS-ResNet50 [34]	83.82	76.36
3	R^3Net-ResNet18-fixed [13]	84.55	78.14
4	R^3Net-ResNet50-fixed [13]	85.10	79.56
5	R^3Net-ResNet18-finetuned [13]	86.15	81.24
6	R^3Net-ResNet50-finetuned [13]	**88.10**	**82.58**
7	Student generator w/ $\mathcal{L}_2 + \mathcal{L}_{cross}$	85.34	79.11
8	Student generator w/ $\mathcal{L}_2 + \mathcal{L}_{adv} + \mathcal{L}_{cross}$	**87.08**	**81.13**

Self-attention generation method SBS (Saliency-Based Sampling Layer) [34] does not bring improvement compared with the baseline. The reason might be that self-attention is hard to learn from these weak labels, *i.e.*shot types. The pre-trained fixed networks (settings 3,4) bring gains to the performance, and the performance increases as the network becomes deeper. Moreover, when we

Table 6. Parameters and computational complexity of different networks

Network Architecture	Params(M)	GFLOPs
Student generator	0.04	2.38
R^3Net-ResNet18 [13]	23.66	19.95
R^3Net-ResNet50 [13]	37.53	22.54

Table 7. Comparison of the performance of joint training sharing different modules.

Settings	Acc_S (↑)	Acc_M (↑)
Separate	87.57	81.86
Joint-training (Share SMG)	**88.12**	**82.19**
Joint-training (Share till res_1)	87.24	81.10
Joint-training (Share till res_4)	86.17	80.29

fine tune these networks on our tasks (settings 5,6), the performance improves further with ~2 gains on both Acc_S and Acc_M.

Our light-weight subject map generation module is driven by two losses besides the classification cross entropy loss. The performance of using $\mathcal{L}_2$ loss (setting 7) is worse than fine-tuned R^3Net-ResNet50. With the help of both $\mathcal{L}_2$ loss and adversarial loss $\mathcal{L}_{adv}$ (setting 8), student generator is on par with R^3Net-ResNet50. However, compared with R^3Net-ResNet50, our light-weight subject map student generator has 99.8% fewer parameters and 89.4% fewer GFLOPs (shown in Table 6), which largely speeds up the training and inference processes.

Two-Task Joint Training. To investigate the relationship between the scale and movement classification, we conduct the joint training experiments on these two tasks, as shown in Table 7. We take our full model SGNet w/ Var (img + flow) as the baseline. We testify the coupling of scale and movement type by sharing the same modules from bottom to top gradually. As lined out in the second row, sharing the subject map generation (SMG) module is helpful to the performance, where Acc_S and Acc_M raises 0.55 and 0.33 respectively. However, when we further share these two tasks classification till ResNet50's res_1 and res_4 modules, we observe that joint training is harmful to the performance when they share more branches. These prove that both scale and movement benefit from the subject guidance. The spatial layout learnt from scale and the camera motion learnt from the movement contribute complementaily to the subject map generation. While the subject guidance is shared by both tasks, the distinct goals of the two tasks still require task-specific designs in the later part to learn better representations.

5.4 Qualitative Results

In this section, we show the qualitative results of subject map generation and the variance map computation.

Subject Map. Figure 4(a) compares our generated subject map with those generated by R^3Net-ResNet50 in fixed setting. Our generated subject map achieves much better generation result that are consistent with our human judgment. The first row in the figure is an over-the-shoulder static close-up shot, where both methods successfully predict the subject woman rather than the man with back head. But our method outputs much less noise. The second row is a full shot. Our method successfully detects both two people and does not include the background stone into the subject map. In the third and fourth row cases, R^3Net-ResNet50-fixed outputs blurred area around the contours of two people while our method obtains a sharp shape of the subject.

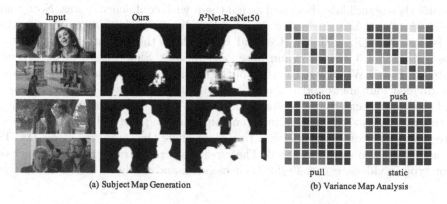

(a) Subject Map Generation   (b) Variance Map Analysis

Fig. 4. (a) Comparison of our generated subject map and R^3Net-ResNet50-fixed generated map. (b) Variance map visualization of different movement types using gray-scale colors to indicate the similarity. The lighter the color, the lower the similarity score

Variance Map. The variance map is important for predicting shot movement. We divide a shot video into 8 clips, plot the variance map for each movement type in the test set and average these variance maps in Fig. 4(b). The variance map is of size 8×8, and these gray scale blocks show the similarity among clips in the variance map. As noted from the plot, the variance map of the static shot is nearly an all-one matrix, meaning that there is no significant change between the eight clips. The near identity matrix shape of motion shots reveal that it has the least similarities between consecutive clips.

6 Application

Shot type analysis has a wide range of potential applications. In this section, we illustrate one such application of realizing automatic video editing with the help of shot type classification.

Automatic Video Editing: Shot Type Changing. Video editors usually try out different shot types to convey emotions and stories, which consumes a lot of time and resources. In many cases, the change of shot type changes the semantic of a movie and effects audience's emotions, revealing the intent of the directors.

The model we propose in this paper could classify the shot type of any given video shot. Figure 5 shows a shot clip[2] from the famous film *Titanic*, demonstrating how our model can be applied to changing shot scales to achieve a

---

[2] [39] is adopted here to cut shots from the film.

desired artistic expression. The original shot is a *medium shot*. Suppose we want to emphasize the role of speaker in this *dialogue scene*, we may want to use some *close-up shot* to emphasize the speaker. Firstly, we propose many cropping regions randomly depend on the position of the speaker and generate the corresponding candidate shots. Secondly, we use our shot classification model to classify these candidate shots and assign them with confidence scores. Note that the original shot is a single shot. We divide it into four shots depending on the active-speaker [35]. In Shot 1, Rose and Jack walk on the deck of the ship; Shot 2, Rose talks; Shot 3, they stop and Rose looks at Jack; Shot 4, Jack talks, as illustrated in Fig. 5. We change the style of the original shot by selecting parts from the divided four shots and replace them with the candidates with high scores and the desired scale types, as shown in Fig. 6. After these changes, the emotion of this clip turns to be more intense and the speaking cast is being emphasized after changing from a middle shot to a close shot. One more result on DAVIS dataset [5] is also shown in Fig. 6. These results demonstrate the importance of shot type in videos, especially for their emotion and aesthetic analysis.[3]

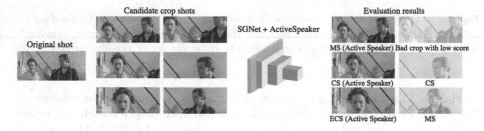

Fig. 5. A sample shot from a dialogue scene in *Titanic*, showing how we use our proposed shot type classification framework to aid the shot type changing

Fig. 6. Editing results on a medium shot clip from *Titanic* to emphasize the speaker and on a full shot clip *Parkour* from the DAVIS dataset [5] to emphasize the action

7   Conclusion

In this work, we construct a large-scale dataset *MovieShots* for shot analysis, which containing $46K$ shots from $7K$ movie trailers with professionally annotated scale and movement attributes. We propose a Subject Guidance Network

---

[3] More results and their corresponding videos are shown in the supplementary videos.

(SGNet) to capture the contextual information and the spatial and temporal configuration of a shot for our shot type classification task. Experiments show that this network is very effective and achieves better results than existing methods. All the studies in this paper together show that shot type analysis is a promising direction for edited video analysis which deserves further research efforts.

Acknowledgement. This work is partially supported by the SenseTime Collaborative Grant on Large-scale Multi-modality Analysis (CUHK Agreement No. TS1610626 & No. TS1712093), the General Research Fund (GRF) of Hong Kong (No. 14203518 & No. 14205719), and Innovation and Technology Support Program (ITSP) Tier 2, ITS/431/18F.

References

1. Bagheri-Khaligh, A., Raziperchikolaei, R., Moghaddam, M.E.: A new method for shot classification in soccer sports video based on SVM classifier. In: 2012 IEEE Southwest Symposium on Image Analysis and Interpretation, pp. 109–112. IEEE (2012)
2. Belagiannis, V., Farshad, A., Galasso, F.: Adversarial network compression. In: Leal-Taixé, L., Roth, S. (eds.) ECCV 2018. LNCS, vol. 11132, pp. 431–449. Springer, Cham (2019). https://doi.org/10.1007/978-3-030-11018-5_37
3. Benini, S., Canini, L., Leonardi, R.: Estimating cinematographic scene depth in movie shots. In: 2010 IEEE International Conference on Multimedia and Expo, pp. 855–860. IEEE (2010)
4. Bhattacharya, S., Mehran, R., Sukthankar, R., Shah, M.: Classification of cinematographic shots using lie algebra and its application to complex event recognition. IEEE Trans. Multimed. **16**(3), 686–696 (2014)
5. Caelles, S., Pont-Tuset, J., Perazzi, F., Montes, A., Maninis, K.K., Van Gool, L.: The 2019 DAVIS challenge on VOS: unsupervised multi-object segmentation. arXiv:1905.00737 (2019)
6. Canini, L., Benini, S., Leonardi, R.: Classifying cinematographic shot types. Multimed. Tools Appl. **62**(1), 51–73 (2013)
7. Carreira, J., Zisserman, A.: Quo vadis, action recognition? a new model and the kinetics dataset. In: Proceedings of the IEEE Conference on Computer Vision and Pattern Recognition, pp. 6299–6308 (2017)
8. Chao, Y.W., Vijayanarasimhan, S., Seybold, B., Ross, D.A., Deng, J., Sukthankar, R.: Rethinking the faster R-CNN architecture for temporal action localization. In: Proceedings of the IEEE Conference on Computer Vision and Pattern Recognition, pp. 1130–1139 (2018)
9. Cheng, M.M., Mitra, N.J., Huang, X., Torr, P.H., Hu, S.M.: Global contrast based salient region detection. IEEE Trans. Pattern Anal. Mach. Intell. **37**(3), 569–582 (2014)
10. Cheng, Y., Wang, D., Zhou, P., Zhang, T.: A survey of model compression and acceleration for deep neural networks. arXiv preprint arXiv:1710.09282 (2017)
11. Christoph, R., Pinz, F.A.: Spatiotemporal residual networks for video action recognition. In: Advances in Neural Information Processing Systems, pp. 3468–3476 (2016)
12. Deng, J., Pan, Y., Yao, T., Zhou, W., Li, H., Mei, T.: Relation distillation networks for video object detection. In: Proceedings of the IEEE International Conference on Computer Vision, pp. 7023–7032 (2019)

13. Deng, Z., et al.: R3Net: recurrent residual refinement network for saliency detection. In: Proceedings of the 27th International Joint Conference on Artificial Intelligence, pp. 684–690. AAAI Press (2018)
14. Duan, L.Y., Xu, M., Tian, Q., Xu, C.S., Jin, J.S.: A unified framework for semantic shot classification in sports video. IEEE Trans. Multimed. **7**(6), 1066–1083 (2005)
15. Ekin, A., Tekalp, A.M.: Shot type classification by dominant color for sports video segmentation and summarization. In: Proceedings of the 2003 IEEE International Conference on Acoustics, Speech, and Signal Processing, 2003, ICASSP 2003, vol. 3, pp. III-173. IEEE (2003)
16. Heilbron, F.C., Escorcia, V., Ghanem, B., Niebles, J.C.: ActivityNet: a large-scale video benchmark for human activity understanding. In: Proceedings of the IEEE Conference on Computer Vision and Pattern Recognition, pp. 961–970 (2015)
17. Feichtenhofer, C., Fan, H., Malik, J., He, K.: Slowfast networks for video recognition. arXiv preprint arXiv:1812.03982 (2018)
18. Giannetti, L.D., Leach, J.: Understanding Movies, vol. 1. Prentice Hall, Upper Saddle River (1999)
19. Goldblum, M., Fowl, L., Feizi, S., Goldstein, T.: Adversarially robust distillation. In: Thirty-Fourth AAAI Conference on Artificial Intelligence (2020)
20. Guo, C., et al.: Progressive sparse local attention for video object detection. In: Proceedings of the IEEE International Conference on Computer Vision, pp. 3909–3918 (2019)
21. Hasan, M.A., Xu, M., He, X., Xu, C.: CAMHID: camera motion histogram descriptor and its application to cinematographic shot classification. IEEE Trans. Circuits Syst. Video Technol. **24**(10), 1682–1695 (2014)
22. He, K., Zhang, X., Ren, S., Sun, J.: Deep residual learning for image recognition. In: Proceedings of the IEEE Conference on Computer Vision and Pattern Recognition, pp. 770–778 (2016)
23. Hinton, G., Vinyals, O., Dean, J.: Distilling the knowledge in a neural network. In: NIPS Deep Learning and Representation Learning Workshop (2015)
24. Hou, Q., Cheng, M.M., Hu, X., Borji, A., Tu, Z., Torr, P.: Deeply supervised salient object detection with short connections. IEEE TPAMI **41**(4), 815–828 (2019)
25. Huang, Q., Xiong, Y., Rao, A., Wang, J., Lin, D.: Movienet: a holistic dataset for movie understanding. In: The European Conference on Computer Vision (ECCV). Springer, Cham (2020)
26. Jiang, H., Zhang, M.: Tennis video shot classification based on support vector machine. In: 2011 IEEE International Conference on Computer Science and Automation Engineering, vol. 2, pp. 757–761. IEEE (2011)
27. Kowdle, A., Chen, T.: Learning to segment a video to clips based on scene and camera motion. In: Fitzgibbon, A., Lazebnik, S., Perona, P., Sato, Y., Schmid, C. (eds.) ECCV 2012. LNCS, vol. 7574, pp. 272–286. Springer, Heidelberg (2012). https://doi.org/10.1007/978-3-642-33712-3_20
28. Li, L., Zhang, X., Hu, W., Li, W., Zhu, P.: Soccer video shot classification based on color characterization using dominant sets clustering. In: Muneesawang, P., Wu, F., Kumazawa, I., Roeksabutr, A., Liao, M., Tang, X. (eds.) PCM 2009. LNCS, vol. 5879, pp. 923–929. Springer, Heidelberg (2009). https://doi.org/10.1007/978-3-642-10467-1_83
29. Li, S., He, F., Du, B., Zhang, L., Xu, Y., Tao, D.: Fast spatio-temporal residual network for video super-resolution. In: The IEEE Conference on Computer Vision and Pattern Recognition (CVPR) (2019)
30. Li, X., et al.: DeepSaliency: multi-task deep neural network model for salient object detection. IEEE Trans. Image Process. **25**(8), 3919–3930 (2016)

31. Monfort, M., et al.: Moments in time dataset: one million videos for event under-standing. IEEE Trans. Pattern Anal. Mach. Intell. **42**, 502–508 (2019)
32. Prasertsakul, P., Kondo, T., Iida, H.: Video shot classification using 2D motion histogram. In: 2017 14th International Conference on Electrical Engineering/Electronics, Computer, Telecommunications and Information Technology (ECTI-CON), pp. 202–205. IEEE (2017)
33. Rao, A., et al.: A local-to-global approach to multi-modal movie scene segmentation. In: Proceedings of the IEEE/CVF Conference on Computer Vision and Pattern Recognition, pp. 10146–10155 (2020)
34. Recasens, A., Kellnhofer, P., Stent, S., Matusik, W., Torralba, A.: Learning to zoom: a saliency-based sampling layer for neural networks. In: Ferrari, V., Hebert, M., Sminchisescu, C., Weiss, Y. (eds.) ECCV 2018. LNCS, vol. 11213, pp. 52–67. Springer, Cham (2018). https://doi.org/10.1007/978-3-030-01240-3_4
35. Roth, J., et al.: AVA-ActiveSpeaker: an audio-visual dataset for active speaker detection. arXiv preprint arXiv:1901.01342 (2019)
36. Russakovsky, O., et al.: Imagenet large scale visual recognition challenge. Int. J. Comput. Vis. **115**(3), 211–252 (2015)
37. Savardi, M., Signoroni, A., Migliorati, P., Benini, S.: Shot scale analysis in movies by convolutional neural networks. In: 2018 25th IEEE International Conference on Image Processing (ICIP), pp. 2620–2624. IEEE (2018)
38. Shou, Z., et al.: DMC-Net: generating discriminative motion cues for fast compressed video action recognition. In: Proceedings of the IEEE Conference on Computer Vision and Pattern Recognition, pp. 1268–1277 (2019)
39. Sidiropoulos, P., Mezaris, V., Kompatsiaris, I., Meinedo, H., Bugalho, M., Trancoso, I.: Temporal video segmentation to scenes using high-level audiovisual features. IEEE Trans. Circuits Syst. Video Technol. **21**(8), 1163–1177 (2011)
40. Wang, H.L., Cheong, L.F.: Taxonomy of directing semantics for film shot classification. IEEE Trans. Circuits Syst. Video Technol. **19**(10), 1529–1542 (2009)
41. Wang, L., et al.: Temporal segment networks: towards good practices for deep action recognition. In: Leibe, B., Matas, J., Sebe, N., Welling, M. (eds.) ECCV 2016. LNCS, vol. 9912, pp. 20–36. Springer, Cham (2016). https://doi.org/10.1007/978-3-319-46484-8_2
42. Wang, X., Zhang, R., Sun, Y., Qi, J.: KDGAN: knowledge distillation with generative adversarial networks. In: Bengio, S., Wallach, H., Larochelle, H., Grauman, K., Cesa-Bianchi, N., Garnett, R. (eds.) Advances in Neural Information Processing Systems 31, pp. 775–786. Curran Associates, Inc. (2018)
43. Wang, X., Girshick, R., Gupta, A., He, K.: Non-local neural networks. In: The IEEE Conference on Computer Vision and Pattern Recognition (CVPR) (2018)
44. Wang, Y., Xu, C., Xu, C., Tao, D.: Adversarial learning of portable student networks. In: Thirty-Second AAAI Conference on Artificial Intelligence (2018)
45. Wikipedia: As seen through a telescope. https://en.wikipedia.org/. Accessed 18 Feb 2020
46. Xia, J., Rao, A., Huang, Q., Xu, L., Wen, J., Lin, D.: Online multi-modal person search in videos. In: Vedaldi, A., Bischof, H., Brox, T., Frahm, J.-M. (eds.) ECCV 2020. LNCS, vol. 12357, pp. 174–190. Springer, Cham (2020). https://doi.org/10.1007/978-3-030-58610-2_11
47. Tong, X.-F., Liu, Q.-S., Lu, H.-Q., Jin, H.-L.: Shot classification in sports video. In: Proceedings 7th International Conference on Signal Processing, ICSP 2004, vol. 2, pp. 1364–1367 (2004)

48. Xiong, Y., Huang, Q., Guo, L., Zhou, H., Zhou, B., Lin, D.: A graph-based framework to bridge movies and synopses. In: The IEEE International Conference on Computer Vision (ICCV) (2019)
49. Xu, G., Liu, Z., Li, X., Loy, C.C.: Knowledge distillation meets self-supervision. In: European Conference on Computer Vision (ECCV). Springer, Cham (2020)
50. Xu, H., Das, A., Saenko, K.: R-C3D: region convolutional 3D network for temporal activity detection. In: Proceedings of the IEEE International Conference on Computer Vision, pp. 5783–5792 (2017)
51. Xu, K., Wen, L., Li, G., Bo, L., Huang, Q.: Spatiotemporal CNN for video object segmentation. In: Proceedings of the IEEE Conference on Computer Vision and Pattern Recognition, pp. 1379–1388 (2019)
52. Xu, M., et al.: Using context saliency for movie shot classification. In: 2011 18th IEEE International Conference on Image Processing, pp. 3653–3656. IEEE (2011)
53. Xu, R., Li, X., Zhou, B., Loy, C.C.: Deep flow-guided video inpainting. In: The IEEE Conference on Computer Vision and Pattern Recognition (CVPR) (2019)
54. Yang, J., Zheng, W.S., Yang, Q., Chen, Y.C., Tian, Q.: Spatial-temporal graph convolutional network for video-based person re-identification. In: Proceedings of the IEEE/CVF Conference on Computer Vision and Pattern Recognition, pp. 3289–3299 (2020)
55. Laradji, I.H., Rostamzadeh, N., Pinheiro, P.O., Vazquez, D., Schmidt, M.: Where are the blobs: counting by localization with point supervision. In: Ferrari, V., Hebert, M., Sminchisescu, C., Weiss, Y. (eds.) ECCV 2018. LNCS, vol. 11206, pp. 560–576. Springer, Cham (2018). https://doi.org/10.1007/978-3-030-01216-8_34
56. Yuan, L., et al.: Central similarity quantization for efficient image and video retrieval. In: Proceedings of the IEEE/CVF Conference on Computer Vision and Pattern Recognition, pp. 3083–3092 (2020)
57. Zeng, X., Liao, R., Gu, L., Xiong, Y., Fidler, S., Urtasun, R.: DMM-Net: differentiable mask-matching network for video object segmentation. arXiv preprint arXiv:1909.12471 (2019)
58. Zhang, H., Liu, D., Xiong, Z.: Two-stream oriented video super-resolution for action recognition. arXiv preprint arXiv:1903.05577 (2019)
59. Zhao, Y., Xiong, Y., Wang, L., Wu, Z., Tang, X., Lin, D.: Temporal action detection with structured segment networks. In: Proceedings of the IEEE International Conference on Computer Vision, pp. 2914–2923 (2017)
60. Zhu, W., Liang, S., Wei, Y., Sun, J.: Saliency optimization from robust background detection. In: Proceedings of the IEEE Conference on Computer Vision and Pattern Recognition, pp. 2814–2821 (2014)

BSL-1K: Scaling Up Co-articulated Sign Language Recognition Using Mouthing Cues

Samuel Albanie[1], Gül Varol[1(✉)], Liliane Momeni[1], Triantafyllos Afouras[1], Joon Son Chung[1,2], Neil Fox[3], and Andrew Zisserman[1]

[1] Visual Geometry Group, University of Oxford, Oxford, UK
{albanie,gul,liliane,afourast,joon,az}@robots.ox.ac.uk
[2] Naver Corporation, Seoul, South Korea
[3] Deafness, Cognition and Language Research Centre, University College London, London, UK
neil.fox@ucl.ac.uk

Abstract. Recent progress in fine-grained gesture and action classification, and machine translation, point to the possibility of automated sign language recognition becoming a reality. A key stumbling block in making progress towards this goal is a lack of appropriate training data, stemming from the high complexity of sign annotation and a limited supply of qualified annotators. In this work, we introduce a new scalable approach to data collection for sign recognition in continuous videos. We make use of weakly-aligned subtitles for broadcast footage together with a keyword spotting method to automatically localise sign-instances for a vocabulary of 1,000 signs in 1,000 h of video. We make the following contributions: (1) We show how to use mouthing cues from signers to obtain high-quality annotations from video data—the result is the BSL-1K dataset, a collection of British Sign Language (BSL) signs of unprecedented scale; (2) We show that we can use BSL-1K to train strong sign recognition models for co-articulated signs in BSL and that these models additionally form excellent pretraining for other sign languages and benchmarks—we exceed the state of the art on both the MSASL and WLASL benchmarks. Finally, (3) we propose new large-scale evaluation sets for the tasks of *sign recognition* and *sign spotting* and provide baselines which we hope will serve to stimulate research in this area.

Keywords: Sign language recognition · Visual keyword spotting

S. Albanie and G. Varol—Equal contribution

Electronic supplementary material The online version of this chapter (https://doi.org/10.1007/978-3-030-58621-8_3) contains supplementary material, which is available to authorized users.

A. Vedaldi et al. (Eds.): ECCV 2020, LNCS 12356, pp. 35–53, 2020.
https://doi.org/10.1007/978-3-030-58621-8_3

1   Introduction

With the continual increase in the performance of human action recognition there has been a renewed interest in the challenge of recognising sign languages such as American Sign Language (ASL), British Sign Language (BSL), and Chinese Sign Language (CSL). Although in the past isolated sign recognition has seen some progress, recognition of continuous sign language remains extremely challenging [10]. Isolated signs, as in dictionary examples, do not suffer from the *naturally* occurring complication of co-articulation (i.e. transition motions) between preceding and subsequent signs, making them visually very different from continuous signing. If we are to recognise ASL and BSL performed *naturally* by signers, then we need to recognise co-articulated signs.

Similar problems were faced by Automatic Speech Recognition (ASR) and the solution, as always, was to learn from very large scale datasets, using a parallel corpus of speech and text. In the vision community, a related path was taken with the modern development of automatic lip reading: first isolated words were recognised [16], and later sentences were recognised [15]—in both cases tied to the release of large datasets. The objective of this paper is to design a scalable *method* to generate large-scale datasets of continuous signing, for training and testing sign language recognition, and we demonstrate this for BSL. We start from the perhaps counter-intuitive observation that signers often mouth the word they sign simultaneously, as an additional signal [5,53,54], performing similar lip movements as for the spoken word. This differs from mouth gestures which are not derived from the spoken language [21]. The mouthing helps disambiguate between different meanings of the same manual sign [60] or in some cases simply provides redundancy. In this way, a sign is not only defined by the hand movements and hand shapes, but also by facial expressions and mouth movements [20].

We harness word mouthings to provide a method of automatically annotating continuous signing. The key idea is to exploit the readily available and abundant supply of sign-language translated TV broadcasts that consist of an overlaid interpreter performing signs and subtitles that correspond to the audio content. The availability of subtitles means that the annotation task is in essence one of alignment between the words in the subtitle and the mouthings of the overlaid signer. Nevertheless, this is a *very* challenging task: a continuous sign may last for only a fraction (e.g. 0.5) of a second, whilst the subtitles may last for several seconds and are not synchronised with the signs produced by the signer; the word order of the English need not be the same as the word order of the sign language; the sign may not be mouthed; and furthermore, words may not be signed or may be signed in different ways depending on the context. For example, the word "fish" has a different visual sign depending on referring to the animal or the food, introducing additional challenges when associating subtitle words to signs.

To detect the mouthings we use *visual keyword spotting*—the task of determining *whether* and *when* a keyword of interest is uttered by a talking face using *only* visual information—to address the alignment problem described above. Two

Table 1. Summary of previous public sign language datasets: The BSL-1K dataset contains, to the best of our knowledge, the largest source of annotated sign data in any dataset. It comprises of co-articulated signs outside a lab setting.

Dataset	Lang	Co-articulated	#signs	#annos (avg. per sign)	#signers	Source
ASLLVD [4]	ASL	✗	2742	9K (3)	6	lab
Devisign [14]	CSL	✗	2000	24K (12)	8	lab
MSASL [33]	ASL	✗	1000	25K (25)	222	lexicons, web
WLASL [39]	ASL	✗	2000	21K (11)	119	lexicons, web
S-pot [57]	FinSL	✓	1211	4K (3)	5	lab
Purdue RVL-SLLL [59]	ASL	✓	104	2K (19)	14	lab
Video-based CSL [31]	CSL	✓	178	25K (140)	50	lab
SIGNUM [58]	DGS	✓	455	3K (7)	25	lab
RWTH-Phoenix [10,34]	DGS	✓	1081	65K (60)	9	TV
BSL Corpus [50]	BSL	✓	5K	50K (10)	249	lab
BSL-1K	BSL	✓	1064	273K (257)	40	TV

factors motivate its use: (1) direct lip reading of arbitrary isolated mouthings is a fundamentally difficult task, but searching for a particular known word within a short temporal window is considerably less challenging; (2) the recent availability of large scale video datasets with aligned audio transcriptions [1,17] now allows for the training of powerful visual keyword spotting models [32,51,62] that, as we show in the experiments, work well for this application.

We make the following contributions: (1) we show how to use visual keyword spotting to recognise the mouthing cues from signers to obtain high-quality annotations from video data—the result is the BSL-1K dataset, a large-scale collection of BSL (British Sign Language) signs with a 1K sign vocabulary; (2) We show the value of BSL-1K by using it to train strong sign recognition models for co-articulated signs in BSL and demonstrate that these models additionally form excellent pretraining for other sign languages and benchmarks—we exceed the state of the art on both the MSASL and WLASL benchmarks with this approach; (3) We propose new evaluation datasets for *sign recognition* and *sign spotting* and provide baselines for each of these tasks to provide a foundation for future research[1].

2 Related Work

Sign Language Datasets. We begin by briefly reviewing public benchmarks for studying automatic sign language recognition. Several benchmarks have been proposed for American [4,33,39,59], German [34,58], Chinese [14,31], and Finnish [57] sign languages. BSL datasets, on the other hand, are scarce. One exception is the ongoing development of the linguistic corpus [49,50] which provides fine-grained annotations for the atomic elements of sign production. Whilst

---

[1] The project page is at: https://www.robots.ox.ac.uk/~vgg/research/bsl1k/.

its high annotation quality provides an excellent resource for sign linguists, the annotations span only a fraction of the source videos so it is less appropriate for training current state-of-the-art data-hungry computer vision pipelines.

Table 1 presents an overview of publicly available datasets, grouped according to their provision of *isolated* signs or *co-articulated* signs. Earlier datasets have been limited in the size of their video instances, vocabularies, and signers. Within the isolated sign datasets, Purdue RVL-SLLL [59] has a limited vocabulary of 104 signs (ASL comprises more than 3K signs in total [56]). ASLLVD [4] has only 6 signers. Recently, MSASL [33] and WLASL [39] large-vocabulary isolated sign datasets have been released with 1K and 2K signs, respectively. The videos are collected from lexicon databases and other instructional videos on the web.

Due to the difficulty of annotating co-articulated signs in long videos, continuous datasets have been limited in their vocabulary, and most of them have been recorded in lab settings [31,58,59]. RWTH-Phoenix [34] is one of the few realistic datasets that supports training complex models based on deep neural networks. A recent extension also allows studying sign language translation [10]. However, the videos in [10,34] are only from weather broadcasts, restricting the domain of discourse. In summary, the main constraints of the previous datasets are one or more of the following: (i) they are limited in size, (ii) they have a large total vocabulary but only of isolated signs, or (iii) they consist of natural co-articulated signs but cover a limited domain of discourse. The BSL-1K dataset provides a considerably greater number of annotations than all previous public sign language datasets, and it does so in the co-articulated setting for a large domain of discourse.

Sign Language Recognition. Early work on sign language recognition focused on hand-crafted features computed for hand shape and motion [24,25,52,55]. Upper body and hand pose have then been widely used as part of the recognition pipelines [7,9,19,46,48]. Non-manual features such as face [24,34,45], and mouth [3,35,37] shapes are relatively less considered. For sequence modelling of signs, HMMs [2,23,27,52], and more recently LSTMs [9,31,63,64], have been utilised. Koller et al. [38] present a hybrid approach based on CNN-RNN-HMM to iteratively re-align sign language videos to the sequence of sign annotations. More recently 3D CNNs have been adopted due to their representation capacity for spatio-temporal data [6,8,30,33,39]. Two recent concurrent works [33,39] showed that I3D models [13] significantly outperform their pose-based counterparts. In this paper, we confirm the success of I3D models, while also showing improvements using pose distillation as pretraining. There have been efforts to use sequence-to-sequence translation models for sign language translation [10], though this has been limited to the weather discourse of RWTH-Phoenix, and the method is limited by the size of the training set. The recent work of [40] localises signs in continuous news footage to improve an isolated sign classifier.

In this work, we utilise mouthings to localise signs in weakly-supervised videos. Previous work [7,17,18,48] has used weakly aligned subtitles as a source of training data, and both one-shot [48] (from a visual dictionary) and zero-shot [6] (from a textual description) have also been used. Though no previous

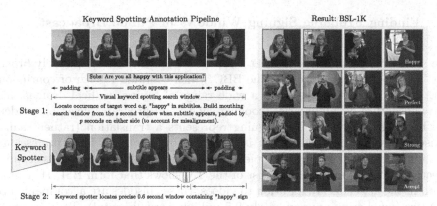

Fig. 1. Keyword-driven sign annotation: (Left, the annotation pipeline): Stage 1: for a given target sign (e.g. "happy") each occurrence of the word in the subtitles provides a candidate temporal window when the sign may occur (this is further padded by several seconds on either side to account for misalignment of subtitles and signs); Stage 2: a keyword spotter uses the mouthing of the signer to perform precise localisation of the sign within this window. (Right): Examples from the BSL-1K dataset—produced by applying keyword spotting for a vocabulary of 1K words.

work, to our knowledge, has put these ideas together. The sign spotting problem was formulated in [22,57].

Using the Mouth Patterns. The mouth has several roles in sign language that can be grouped into spoken components (mouthings) and oral components (mouth gestures) [60]. Several works focus on recognising mouth shapes [3,37] to recover mouth gestures. Few works [35,36] attempt to recognise mouthings in sign language data by focusing on a few categories of visemes, i.e., visual correspondences of phonemes in the lip region [26]. Most closely related to our work, [47] similarly searches subtitles of broadcast footage and uses the mouth as a cue to improve alignment between the subtitles and the signing. Two key differences between our work and theirs are: (1) we achieve precise localisation through keyword spotting, whereas they only use an open/closed mouth classifier to reduce the number of candidates for a given sign; (2) scale—we gather signs over 1,000 h of signing (in contrast to the 30 h considered in [47]).

3   Learning Sign Recognition with Automatic Labels

In this section, we describe the process used to collect BSL-1K, a large-scale dataset of BSL signs. An overview of the approach is provided in Fig. 1. In Sect. 3.1, we describe how large numbers of video clips that are likely to contain a given sign are sourced from public broadcast footage using subtitles; in Sect. 3.2, we show how automatic keyword spotting can be used to precisely localise specific signs to within a fraction of a second; in Sect. 3.3, we apply this technique to efficiently annotate a large-scale dataset with a vocabulary of 1K signs.

3.1   Finding Probable Signing Windows in Public Broadcast Footage

The source material for the dataset comprises 1,412 episodes of publicly broadcast TV programs produced by the BBC which contains 1,060 h of continuous BSL signing. The episodes cover a wide range of topics: medical dramas, history and nature documentaries, cooking shows and programs covering gardening, business and travel. The signing represents a translation (rather than a transcription) of the content and is produced by a total of forty professional BSL interpreters. The signer occupies a fixed region of the screen and is cropped directly from the footage. A full list of the TV shows that form BSL-1K can be found in the appendix. In addition to videos, these episodes are accompanied by subtitles (numbering approximately 9.5 million words in total). To locate temporal windows in which instances of signs are likely to occur within the source footage, we first identify a candidate list of words that: (i) are present in the subtitles; (ii) have entries in both BSL signbank[2] and sign BSL[3], two online dictionaries of isolated signs (to ensure that we query words that have valid mappings to signs). The result is an initial vocabulary of 1,350 words, which are used as queries for the keyword spotting model to perform sign localisation—this process is described next.

3.2   Precise Sign Localisation Through Visual Keyword Spotting

By searching the content of the subtitle tracks for instances of words in the initial vocabulary, we obtain a set of candidate temporal windows in which instances of signs may occur. However, two factors render these temporal proposals extremely noisy: (1) the presence of a word in the subtitles does not guarantee its presence in the signing; (2) even for subtitled words that are signed, we find through inspection that their appearance in the subtitles can be misaligned with the sign itself by several seconds.

To address this challenge, we turn to *visual keyword spotting*. Our goal is to detect and precisely localise the presence of a sign by identifying its "spoken components" [54] within a temporal sequence of mouthing patterns. Two hypotheses underpin this approach: (a) that mouthing provides a strong localisation signal for signs as they are produced; (b) that this mouthing occurs with sufficient frequency to form a useful localisation cue. Our method is motivated by studies in the Sign Linguistics literature which find that spoken components frequently serve to identify signs—this occurs most prominently when the mouth pattern is used to distinguish between manual homonyms[4] (see [54] for a detailed discussion). However, even if these hypotheses hold, the task remains extremely challenging—signers typically do not mouth continuously and the mouthings that are produced may only correspond to a portion of the word [54]. For this

---

[2] https://bslsignbank.ucl.ac.uk/.

[3] https://www.signbsl.com/.

[4] These are signs that use identical hand movements (e.g. "king" and "queen") whose meanings are distinguished by mouthings.

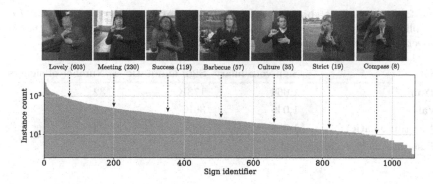

Fig. 2. BSL-1K sign frequencies: Log-histogram of instance counts for the 1,064 words constituting the BSL-1K vocabulary, together with example signs. The long-tail distribution reflects the *real* setting in which some signs are more frequent than others.

reason, existing lip reading approaches cannot be used directly (indeed, an initial exploratory experiment we conducted with the state-of-the-art lip reading model of [1] achieved zero recall on five-hundred randomly sampled sentences of signer mouthings from the BBC source footage).

The key to the effectiveness of visual keyword spotting is that rather than solving the general problem of lip reading, it solves the much easier problem of identifying a single token from a small collection of candidates within a short temporal window. In this work, we use the subtitles to construct such windows. The pipeline for automatic sign annotations therefore consists of two stages (Fig. 1, left): (1) For a given target sign e.g. "happy", determine the times of all occurrences of this sign in the subtitles accompanying the video footage. The subtitle time provides a short window during which the word was spoken, but not necessarily when its corresponding sign is produced in the translation. We therefore extend this candidate window by several seconds to increase the likelihood that the sign is present in the sequence. We include ablations to assess the influence of this padding process in Sect. 5 and determine empirically that padding by four seconds on each side of the subtitle represents a good choice. (2) The resulting temporal window is then provided, together with the target word, to a keyword spotting model (described in detail in Sect. 4.1) which estimates the probability that the sign was mouthed at each time step (we apply the keyword spotter with a stride of 0.04 s—this choice is motivated by the fact that the source footage has a frame rate of 25 fps). When the keyword spotter asserts with high confidence that it has located a sign, we take the location of the peak posterior probability as an anchoring point for one endpoint of a 0.6 s window (this value was determined by visual inspection to be sufficient for capturing individual signs). The peak probability is then converted into a decision about whether a sign is present using a threshold parameter. To build the BSL-1K dataset, we select a value of 0.5 for this parameter after conducting

Table 2. Statistics of the proposed BSL-1K dataset: The *Test-(manually veri-fied)* split represents a sample from the Test-(automatic) split annotations that have been verified by human annotators (see Sect. 3.3 for details).

Set	Sign vocabulary	Sign annotations	Number of signers
Train	1,064	173K	32
Val	1,049	36K	4
Test-(automatic)	1,059	63K	4
Test-(manually verified)	334	2103	4

experiments (reported in Table 3) to assess its influence on the downstream task of sign recognition performance.

3.3   BSL-1K Dataset Construction and Validation

Following the sign localisation process described above, we obtain approximately 280k localised signs from a set of 2.4 million candidate subtitles. To ensure that the dataset supports study of signer-independent sign recognition, we then compute face embeddings (using an SENet-50 [29] architecture trained for verification on the VGGFace2 dataset [11]) to group the episodes according to which of the forty signers they were translated by. We partition the data into three splits, assigning thirty-two signers for training, four signers for validation and four signers for testing. We further sought to include an equal number of hearing and non-hearing signers (the validation and test sets both contain an equal number of each, the training set is approximately balanced with 13 hearing, 17 non-hearing and 2 signers whose deafness is unknown). We then perform a further filtering step on the vocabulary to ensure that each word included in the dataset is represented with high confidence (at least one instance with confidence 0.8) in the training partition, which produces a final dataset vocabulary of 1,064 words (see Fig. 2 for the distribution and the appendix for the full word list).

Validating the Automatic Annotation Pipeline. One of the key hypotheses underpinning this work is that keyword spotting is capable of correctly locating signs. We first verify this hypothesis by presenting a randomly sampled subset of the test partition to a native BSL signer, who was asked to assess whether the short temporal windows produced by the keyword spotting model with high confidence (each 0.6 s in duration) contained correct instances of the target sign. A screenshot of the annotation tool developed for this task is provided in the appendix. A total of 1k signs were included in this initial assessment, of which 70% were marked as correct, 28% were marked as incorrect and 2% were marked as uncertain, validating the key idea behind the annotation pipeline. Possible reasons for incorrect marks include: BSL mouthing patterns are not always identical to spoken English and mouthings many times do not represent the full word (e.g., "fsh" for "finish") [54].

Constructing a Manually Verified Test Set. To construct a high quality, human verified test set and to maximise yield from the annotators, we started from a collection of sign predictions where the keyword model was highly confident (assigning a peak probability of greater than 0.9) yielding 5,826 sign predictions. Then, in addition to the validated 980 signs (corrections were provided as labels for the signs marked as incorrect and uncertain signs were removed), we further expanded the verified test set with non-native (BSL level 2 or above) signers who annotated a further 2k signs. We found that signers with lower levels of fluency were able to confidently assert that a sign was correct for a portion of the signs (at a rate of around 60%), but also annotated a large number of signs as "unsure", making it challenging to use these annotations as part of the validation test for the effectiveness of the pipeline. Only signs marked as correct were included into the final verified test set, which ultimately comprised 2,103 annotations covering 334 signs from the 1,064 sign vocabulary. The statistics of each partition of the dataset are provided in Table 2. All experimental test set results in this paper refer to performance on the verified test set (but we retain the full automatic test set, which we found to be useful for development).

In addition to the keyword spotting approach described above, we explore techniques for further dataset expansion based on other cues in the appendix.

4   Models and Implementation Details

In this section, we first describe the visual keyword spotting model used to collect signs from mouthings (Sect. 4.1). Next, we provide details of the model architecture for sign recognition and spotting (Sect. 4.2). Lastly, we describe a method for obtaining a good initialisation for the sign recognition model (Sect. 4.3).

4.1   Visual Keyword Spotting Model

We use the improved visual-only keyword spotting model of Stafylakis et al. [51] from [44] (referred to in their paper as "P2G [51] baseline"), provided by the authors. The model of [51] combines visual features with a fixed-length keyword embedding to determine whether a user-defined keyword is present in an input video clip. The performance of [51] is improved in [44] by switching the keyword encoder-decoder from grapheme-to-phoneme (G2P) to phoneme-to-grapheme (P2G).

In more detail, the model consists of four stages: (i) visual features are first extracted from the sequence of face-cropped image frames from a clip (this is performed using a 512×512 SSD architecture [42] trained for face detection on WIDER faces [61]), (ii) a fixed-length keyword representation is built using a P2G encoder-decoder, (iii) the visual and keyword embeddings are concatenated and passed through BiLSTMs, (iv) finally, a sigmoid activation is applied on the output to approximate the posterior probability that the keyword occurs in the video clip for each input frame. If the maximum posterior over all frames is greater than a threshold, the clip is predicted to contain the keyword. The predicted location of the keyword is the position of the maximum posterior. Finally, non-maximum suppression is run with a temporal window of 0.6 s over the untrimmed source videos to remove duplicates.

4.2 Sign Recognition Model

We employ a spatio-temporal convolutional neural network architecture that takes a multiple-frame video as input, and outputs class probabilities over sign categories. Specifically, we follow the I3D architecture [13] due to its success on action recognition benchmarks, as well as its recently observed success on sign recognition datasets [33,39]. To retain computational efficiency, we only use an RGB stream. The model is trained on 16-frame consecutive frames (i.e., 0.64 sec for 25 fps), as [7,47,57] observed that co-articulated signs last roughly for 13 frames. We resize our videos to have a spatial resolution of 224×224. For training, we randomly subsample a fixed-size, temporally contiguous input from the spatio-temporal volume to have $16 \times 224 \times 224$ resolution in terms of number of frames, width, and height, respectively. We minimise the cross-entropy loss using SGD with momentum (0.9) with mini-batches of size 4, and an initial learning rate of 10^{-2} with a fixed schedule. The learning rate is decreased twice with a factor of 10^{-1} at epochs 20 and 40. We train for 50 epochs. Colour, scale, and horizontal flip augmentations are applied on the input video. When pretraining is used (e.g. on Kinetics-400 [13] or on other data where specified), we replace the last linear layer with the dimensionality of our classes, and fine-tune all network parameters (we observed that freezing part of the model is suboptimal). Finally, we apply dropout on the classification layer with a probability of 0.5.

At test time, we perform centre-cropping and apply a sliding window with a stride of 8 frames before averaging the classification scores to obtain a video-level prediction.

4.3 Video Pose Distillation

Given the significant focus on pose estimation in the sign language recognition literature, we investigate how explicit pose modelling can be used to improve the I3D model. To this end, we define a *pose distillation* network that takes in a sequence of 16 consecutive frames, but rather than predicting sign categories, the 1024-dimensional (following average pooling) embedding produced by the network is used to regress the poses of individuals appearing in each of the frames of its input. In more detail, we assume a single individual per-frame (as is the case in cropped sign translation footage) and task the network with predicting 130 human pose keypoints (18 body, 21 per hand, and 70 facial) produced by an OpenPose [12] model (trained on COCO [41]) that is evaluated per-frame. The key idea is that, in order to effectively predict pose across multiple frames from a single video embedding, the model is encouraged to encode information not only about pose, but also descriptions of relevant dynamic gestures. The model is trained on a portion of the BSL-1K training set (due to space constraints,

Table 3. Trade-off between training noise vs. size: Training (with Kinetics initialisation) on the full training set BSL-1K$_{m.5}$ versus the subset BSL-1K$_{m.8}$, which correspond to a mouthing score threshold of 0.5 and 0.8, respectively. Even when noisy, with the 0.5 threshold, mouthings provide automatic annotations that allow supervised training at scale, resulting in 70.61% accuracy on the manually validated test set.

Training data	#videos	Per-instance		Per-class	
		Top-1	Top-5	Top-1	Top-5
BSL-1K$_{m.8}$ (mouthing $\geq$ 0.8)	39K	69.00	83.79	45.86	64.42
BSL-1K$_{m.5}$ (mouthing $\geq$ 0.5)	173K	**70.61**	**85.26**	**47.47**	**68.13**

further details of the model architecture and training procedure are provided in the appendix).

5 Experiments

We first provide several ablations on our sign recognition model to answer questions such as which cues are important, and how to best use human pose. Then, we present baseline results for sign recognition and sign spotting, with our best model. Finally, we compare to the state of the art on ASL benchmarks to illustrate the benefits of pretraining on our data.

5.1 Ablations for the Sign Recognition Model

In this section, we evaluate our sign language recognition approach and investigate (i) the effect of mouthing score threshold, (ii) the comparison to pose-based approaches, (iii) the contribution of multi-modal cues, and (iv) the video pose distillation. Additional ablations about the influence of the temporal extent of the automatic annotations and the search window size for the keyword spotting can be found in the appendix.

Evaluation Metrics. Following [33,39], we report both top-1 and top-5 classification accuracy, mainly due to ambiguities in signs which can be resolved in context. Furthermore, we adopt both per-instance and per-class accuracy metrics. Per-instance accuracy is computed over all test instances. Per-class accuracy refers to the average over the sign categories present in the test set. We use this metric due to the unbalanced nature of the datasets.

The Effect of the Mouthing Score Threshold. The keyword spotting method, being a binary classification model, provides a confidence score, which we threshold to obtain our automatically annotated video clips. Reducing this threshold yields an increased number of sign instances at the cost of a potentially noisier set of annotations. We denote the training set defined by a mouthing threshold 0.8 as BSL-1K$_{m.8}$. In Table 3, we show the effect of changing this hyper-parameter between a low- and high-confidence model with 0.5 and 0.8

Table 4. Contribution of individual cues: We compare I3D (pretrained on Kinetics) with a keypoint-based baseline both trained and evaluated on a subset of BSL-1K$_{m.8}$, where we have the pose estimates. We also quantify the contribution of the body&hands and the face regions. We see that significant information can be attributed to both types of cues, and the combination performs the best.

	Body&hands	Face	Per-instance		Per-class	
			Top-1	Top-5	Top-1	Top-5
Pose→Sign (70 points)	✗	✓	24.41	47.59	9.74	25.99
Pose→Sign (60 points)	✓	✗	40.47	59.45	20.24	39.27
Pose→Sign (130 points)	✓	✓	**49.66**	**68.02**	**29.91**	**49.21**
I3D (face-crop)	✗	✓	42.23	69.70	21.66	50.51
I3D (mouth-masked)	✓	✗	46.75	66.34	25.85	48.02
I3D (full-frame)	✓	✓	**65.57**	**81.33**	**44.90**	**64.91**

Table 5. Effect of pretraining the I3D model on various tasks before fine-tuning for sign recognition on BSL-1K$_{m.8}$. Our dynamic pose features learned on 16-frame videos provide body-motion-aware cues and outperform other pretraining strategies.

Pretraining		Per-instance		Per-class	
Task	Data	Top-1	Top-5	Top-1	Top-5
Random init.	–	39.80	61.01	15.76	29.87
Gesture recognition	Jester [43]	46.93	65.95	19.59	36.44
Sign recognition	WLASL [39]	69.90	83.45	44.97	62.73
Action recognition	Kinetics [13]	69.00	83.79	45.86	64.42
Video pose distillation	Signers	**70.38**	**84.50**	**46.24**	**65.31**

mouthing thresholds, respectively. The larger set of training samples obtained with a threshold of 0.5 provide the best performance. For the remaining ablations, we use the smaller BSL-1K$_{m.8}$ training set for faster iterations, and return to the larger BSL-1K$_{m.5}$ set for training the final model.

Pose-Based Model Versus I3D. We next verify that I3D is a suitable model for sign language recognition by comparing it to a pose-based approach. We implement Pose→Sign, which follows a 2D ResNet architecture [28] that operates on $3 \times 16 \times P$ dimensional dynamic pose images, where P is the number of keypoints. In our experiments, we use OpenPose [12] (pretrained on COCO [41]) to extract 18 body, 21 per hand, and 70 facial keypoints. We use 16-frame inputs to make it comparable to the I3D counterpart. We concatenate the estimated normalised xy coordinates of a keypoint with its confidence score to obtain the 3 channels. In Table 4, we see that I3D significantly outperforms the explicit 2D pose-based method (65.57% vs 49.66% per-instance accuracy). This conclusion is in accordance with the recent findings of [33,39].

Fig. 3. Qualitative analysis: We present results of our sign recognition model on BSL-1K for success (top) and failure (bottom) cases, together with their confidence scores in parentheses. To the right of each example, we show a random training sample for the predicted sign (in small). We observe that failure modes are commonly due to high visual similarity in the gesture (bottom-left) and mouthing (bottom-right).

Table 6. Benchmarking: We benchmark our best sign recognition model (trained on BSL-1K$_{m.5}$, initialised with pose distillation, with 4-frame temporal offsets) for sign recognition and sign spotting task to establish strong baselines on BSL-1K.

	Per-instance		Per-class			mAP
	Top-1	Top-5	Top-1	Top-5		(334 sign classes)
Sign recognition	75.51	88.83	52.76	72.14	Sign spotting	0.159

Contribution of Individual Cues. We carry out two set of experiments to determine how much our sign recognition model relies on signals from the mouth and face region versus the manual features from the body and hands: (i) using Pose→Sign, which takes as input the 2D keypoint locations over several frames, (ii) using I3D, which takes as input raw video frames. For the pose-based model, we train with only 70 facial keypoints, 60 body&hand keypoints, or with the combination. For I3D, we use the pose estimations to mask the pixels outside of the face bounding box, to mask the mouth region, or use all the pixels from the videos. The results are summarised in Table 4. We observe that using only the face provides a strong baseline, suggesting that mouthing is a strong cue for recognising signs, e.g., 42.23% for I3D. However, using all the cues, including body and hands (65.57%), significantly outperforms using individual modalities.

Pretraining for Sign Recognition. Next we investigate different forms of pretraining for the I3D model. In Table 5, we compare the performance of a model trained with random initialisation (39.80%), fine-tuning from gesture recognition (46.93%), sign recognition (69.90%), and action recognition (69.00%). Video pose distillation provides a small boost over the other pretraining strategies (70.38%), suggesting that it is an effective way to force the I3D model to pay attention to the dynamics of the human keypoints, which is relevant for sign recognition.

5.2   Benchmarking Sign Recognition and Sign Spotting

Next, we combine the parameter choices suggested by each of our ablations to establish baseline performances on the BSL-1K dataset for two tasks: (i) sign recognition, (ii) sign spotting. Specifically, the model comprises an I3D architecture trained on BSL-1K$_{m.5}$ with pose-distillation as initialisation and random temporal offsets of up to 4 frames around the sign during training (the ablation studies for this temporal augmentation parameter are included in the appendix). The sign recognition evaluation protocol follows the experiments conducted in the ablations, the sign spotting protocol is described next.

Sign Spotting. Differently from sign recognition, in which the objective is to classify a pre-defined temporal segment into a category from a given vocabulary, *sign spotting* aims to locate all instances of a particular sign within long sequences of untrimmed footage, enabling applications such as content-based search and efficient indexing of signing videos for which subtitles are not available. The evaluation protocol for assessing sign spotting on BSL-1K is defined as follows: for each sign category present amongst the human-verified test set annotations (334 in total), windows of 0.6-s centred on each verified instance are marked as positive and all other times within the subset of episodes that contain at least one instance of the sign are marked as negative. To avoid false penalties at signs that were not discovered by the automatic annotation process, we exclude windows of 8 s of footage centred at each location in the original footage at which the target keyword appears in the subtitles, but was not detected by the visual keyword spotting pipeline. In aggregate this corresponds to locating approximately one positive instance of a sign in every 1.5 h of continuous signing negatives. A sign is considered to have been correctly spotted if its temporal overlap with the model prediction exceeds an IoU (intersection-over-union) of 0.5, and we report mean Average Precision (mAP) over the 334 sign classes as the metric for performance.

Table 7. Transfer to ASL: Performance on American Sign Language (ASL) datasets with and without pretraining on our data. I3D results are reported from the original papers for MSASL [33] and WLASL [39]. I3D† denotes our implementation and training, adopting the hyper-parameters from [33]. We show that our features provide good initialisation, even if it is trained on BSL.

	Pretraining	WLASL [39]				MSASL [33]			
		Per-instance		Per-class		Per-instance		Per-class	
		Top-1	Top-5	Top-1	Top-5	Top-1	Top-5	Top-1	Top-5
I3D [33]	Kinetics	–	–	–	–	–	–	57.69	81.08
I3D [39]	Kinetics	32.48	57.31	–	–	–	–	–	–
I3D†	Kinetics	40.85	74.10	39.06	73.33	60.45	82.05	57.17	80.02
I3D	BSL-1K	**46.82**	**79.36**	**44.72**	**78.47**	**64.71**	**85.59**	**61.55**	**84.43**

We report the performance of our strongest model for both the sign recognition and sign spotting benchmarks in Table 6. In Fig. 3, we provide some qualitative results from our sign recognition method and observe some modes of failure which are driven by strong visual similarity in sign production.

5.3 Comparison with the State of the Art on ASL Benchmarks

BSL-1K, being significantly larger than the recent WLASL [39] and MSASL [33] benchmarks, can be used for pretraining I3D models to provide strong initialisation for other datasets. Here, we transfer the features from BSL to ASL, which are two distinct sign languages.

As the models from [33,39] were not available at the time writing, we first reproduce the I3D Kinetics pretraining baseline with our implementation to achieve fair comparisons. We use 64-frame inputs as isolated signs in these datasets are significantly slower than co-articulated signs. We then train I3D from BSL-1K pretrained features. Table 7 compares pretraining on Kinetics versus our BSL-1K data. BSL-1K provides a significant boost in the performance, outperforming the state-of-the-art results (46.82% and 64.71% top-1 accuracy). Find additional details, as well as similar experiments on co-articulated datasets in the appendix.

6   Conclusion

We have demonstrated the advantages of using visual keyword spotting to automatically annotate continuous sign language videos with weakly-aligned subtitles. We have presented BSL-1K, a large-scale dataset of co-articulated signs that, coupled with a 3D CNN training, allows high-performance recognition of signs from a large-vocabulary. Our model has further shown beneficial as initialisation for ASL benchmarks. Finally, we have provided ablations and baselines for sign recognition and sign spotting tasks. A potential future direction is leveraging our automatic annotations and recognition model for sign language translation.

Acknowledgements. This work was supported by EPSRC grant ExTol. We also thank T. Stafylakis, A. Brown, A. Dutta, L. Dunbar, A. Thandavan, C. Camgoz, O. Koller, H. V. Joze, O. Kopuklu for their help.

References

1. Afouras, T., Chung, J.S., Senior, A., Vinyals, O., Zisserman, A.: Deep audio-visual speech recognition. IEEE Trans. Pattern Anal. Mach. Intell. (2019)
2. Agris, U., Zieren, J., Canzler, U., Bauer, B., Kraiss, K.F.: Recent developments in visual sign language recognition. Univ. Access Inf. Soc. **6**, 323–362 (2008). https://doi.org/10.1007/s10209-007-0104-x
3. Antonakos, E., Roussos, A., Zafeiriou, S.: A survey on mouth modeling and analysis for sign language recognition. In: IEEE International Conference and Workshops on Automatic Face and Gesture Recognition (2015)

4. Athitsos, V., et al.: The American sign language lexicon video dataset. In: CVPRW (2008)
5. Bank, R., Crasborn, O., Hout, R.: Variation in mouth actions with manual signs in sign language of the Netherlands (NGT). Sign Lang. Linguist. **14**, 248–270 (2011)
6. Bilge, Y.C., Ikizler, N., Cinbis, R.: Zero-shot sign language recognition: can textual data uncover sign languages? In: BMVC (2019)
7. Buehler, P., Zisserman, A., Everingham, M.: Learning sign language by watching TV (using weakly aligned subtitles). In: CVPR (2009)
8. Camgoz, N.C., Hadfield, S., Koller, O., Bowden, R.: Using convolutional 3D neural networks for user-independent continuous gesture recognition. In: IEEE International Conference of Pattern Recognition, ChaLearn Workshop (2016)
9. Camgoz, N.C., Hadfield, S., Koller, O., Bowden, R.: SubUNets: end-to-end hand shape and continuous sign language recognition. In: ICCV (2017)
10. Camgoz, N.C., Hadfield, S., Koller, O., Ney, H., Bowden, R.: Neural sign language translation. In: CVPR (2018)
11. Cao, Q., Shen, L., Xie, W., Parkhi, O.M., Zisserman, A.: VGGFace2: a dataset for recognising faces across pose and age. In: International Conference on Automatic Face and Gesture Recognition (2018)
12. Cao, Z., Hidalgo, G., Simon, T., Wei, S.E., Sheikh, Y.: OpenPose: realtime multi-person 2D pose estimation using Part Affinity Fields. arXiv preprint arXiv:1812.08008 (2018)
13. Carreira, J., Zisserman, A.: Quo vadis, action recognition? A new model and the Kinetics dataset. In: CVPR (2017)
14. Chai, X., Wang, H., Chen, X.: The devisign large vocabulary of Chinese sign language database and baseline evaluations. Technical report VIPL-TR-14-SLR-001. Key Lab of Intelligent Information Processing of Chinese Academy of Sciences (CAS), Institute of Computing Technology, CAS (2014)
15. Chung, J.S., Senior, A., Vinyals, O., Zisserman, A.: Lip reading sentences in the wild. In: CVPR (2017)
16. Chung, J.S., Zisserman, A.: Lip reading in the wild. In: ACCV (2016)
17. Chung, J.S., Zisserman, A.: Signs in time: encoding human motion as a temporal image. In: Workshop on Brave New Ideas for Motion Representations, ECCV (2016)
18. Cooper, H., Bowden, R.: Learning signs from subtitles: a weakly supervised approach to sign language recognition. In: CVPR (2009)
19. Cooper, H., Pugeault, N., Bowden, R.: Reading the signs: a video based sign dictionary. In: ICCVW (2011)
20. Cooper, H., Holt, B., Bowden, R.: Sign language recognition. In: Moeslund, T., Hilton, A., Krüger, V., Sigal, L. (eds.) Visual Analysis of Humans, pp. 539–562. Springer, London (2011). https://doi.org/10.1007/978-0-85729-997-0_27
21. Crasborn, O.A., Van Der Kooij, E., Waters, D., Woll, B., Mesch, J.: Frequency distribution and spreading behavior of different types of mouth actions in three sign languages. Sign Lang. Linguist. **11**, 45–67 (2008)
22. Ong, E.-J., Koller, O., Pugeault, N., Bowden, R.: Sign spotting using hierarchical sequential patterns with temporal intervals. In: CVPR (2014)
23. Farhadi, A., Forsyth, D.: Aligning ASL for statistical translation using a discriminative word model. In: CVPR (2006)
24. Farhadi, A., Forsyth, D.A., White, R.: Transfer learning in sign language. In: CVPR (2007)

25. Fillbrandt, H., Akyol, S., Kraiss, K.: Extraction of 3D hand shape and posture from image sequences for sign language recognition. In: IEEE International SOI Conference (2003)
26. Fisher, C.G.: Confusions among visually perceived consonants. J. Speech Hear. Res. **11**(4), 796–804 (1968)
27. Forster, J., Oberdörfer, C., Koller, O., Ney, H.: Modality combination techniques for continuous sign language recognition. In: Sanches, J.M., Micó, L., Cardoso, J.S. (eds.) IbPRIA 2013. LNCS, vol. 7887, pp. 89–99. Springer, Heidelberg (2013). https://doi.org/10.1007/978-3-642-38628-2_10
28. He, K., Zhang, X., Ren, S., Sun, J.: Deep residual learning for image recognition. In: CVPR (2016)
29. Hu, J., Shen, L., Albanie, S., Sun, G., Wu, E.: Squeeze-and-excitation networks. IEEE Trans. Pattern Anal. Mach. Intell. **42**, 2011–2023 (2019)
30. Huang, J., Zhou, W., Li, H., Li, W.: Sign language recognition using 3D convolutional neural networks. In: International Conference on Multimedia and Expo (ICME) (2015)
31. Huang, J., Zhou, W., Zhang, Q., Li, H., Li, W.: Video-based sign language recognition without temporal segmentation. In: AAAI (2018)
32. Jha, A., Namboodiri, V.P., Jawahar, C.V.: Word spotting in silent lip videos. In: WACV (2018)
33. Joze, H.R.V., Koller, O.: MS-ASL: a large-scale data set and benchmark for understanding American sign language. In: BMVC (2019)
34. Koller, O., Forster, J., Ney, H.: Continuous sign language recognition: towards large vocabulary statistical recognition systems handling multiple signers. Comput. Vis. Image Underst. **141**, 108–125 (2015)
35. Koller, O., Ney, H., Bowden, R.: Read my lips: continuous signer independent weakly supervised viseme recognition. In: Fleet, D., Pajdla, T., Schiele, B., Tuytelaars, T. (eds.) ECCV 2014. LNCS, vol. 8689, pp. 281–296. Springer, Cham (2014). https://doi.org/10.1007/978-3-319-10590-1_19
36. Koller, O., Ney, H., Bowden, R.: Weakly supervised automatic transcription of mouthings for gloss-based sign language corpora. In: LREC Workshop on the Representation and Processing of Sign Languages: Beyond the Manual Channel (2014)
37. Koller, O., Ney, H., Bowden, R.: Deep learning of mouth shapes for sign language. In: 3rd Workshop on Assistive Computer Vision and Robotics, ICCV (2015)
38. Koller, O., Zargaran, S., Ney, H.: Re-sign: re-aligned end-to-end sequence modelling with deep recurrent CNN-HMMs. In: CVPR (2017)
39. Li, D., Opazo, C.R., Yu, X., Li, H.: Word-level deep sign language recognition from video: a new large-scale dataset and methods comparison. In: WACV (2019)
40. Li, D., Yu, X., Xu, C., Petersson, L., Li, H.: Transferring cross-domain knowledge for video sign language recognition. In: CVPR (2020)
41. Lin, T.-Y., et al.: Microsoft COCO: common objects in context. In: Fleet, D., Pajdla, T., Schiele, B., Tuytelaars, T. (eds.) ECCV 2014. LNCS, vol. 8693, pp. 740–755. Springer, Cham (2014). https://doi.org/10.1007/978-3-319-10602-1_48
42. Liu, W., et al.: SSD: single shot multibox detector. In: Leibe, B., Matas, J., Sebe, N., Welling, M. (eds.) ECCV 2016. LNCS, vol. 9905, pp. 21–37. Springer, Cham (2016). https://doi.org/10.1007/978-3-319-46448-0_2
43. Materzynska, J., Berger, G., Bax, I., Memisevic, R.: The Jester dataset: A large-scale video dataset of human gestures. In: ICCVW (2019)
44. Momeni, L., Afouras, T., Stafylakis, T., Albanie, S., Zisserman, A.: Seeing wake words: Audio-visual keyword spotting. BMVC (2020)

45. Nguyen, T.D., Ranganath, S.: Tracking facial features under occlusions and recognizing facial expressions in sign language. In: International Conference on Automatic Face and Gesture Recognition (2008)
46. Ong, E., Cooper, H., Pugeault, N., Bowden, R.: Sign language recognition using sequential pattern trees. In: CVPR (2012)
47. Pfister, T., Charles, J., Zisserman, A.: Large-scale learning of sign language by watching TV (using co-occurrences). In: BMVC (2013)
48. Pfister, T., Charles, J., Zisserman, A.: Domain-adaptive discriminative one-shot learning of gestures. In: Fleet, D., Pajdla, T., Schiele, B., Tuytelaars, T. (eds.) ECCV 2014. LNCS, vol. 8694, pp. 814–829. Springer, Cham (2014). https://doi.org/10.1007/978-3-319-10599-4_52
49. Schembri, A., Fenlon, J., Rentelis, R., Cormier, K.: British Sign Language Corpus Project: A corpus of digital video data and annotations of British Sign Language 2008–2017, 3rd edn. (2017). http://www.bslcorpusproject.org
50. Schembri, A., Fenlon, J., Rentelis, R., Reynolds, S., Cormier, K.: Building the British sign language corpus. Lang. Doc. Conserv. 7, 136–154 (2013)
51. Stafylakis, T., Tzimiropoulos, G.: Zero-shot keyword spotting for visual speech recognition in-the-wild. In: Ferrari, V., Hebert, M., Sminchisescu, C., Weiss, Y. (eds.) ECCV 2018. LNCS, vol. 11208, pp. 536–552. Springer, Cham (2018). https://doi.org/10.1007/978-3-030-01225-0_32
52. Starner, T.: Visual Recognition of American Sign Language Using Hidden Markov Models. Master's thesis, Massachusetts Institute of Technology (1995)
53. Sutton-Spence, R.: Mouthings and simultaneity in British sign language. In: Simultaneity in Signed Languages: Form and Function, pp. 147–162. John Benjamins (2007)
54. Sutton-Spence, R., Woll, B.: The Linguistics of British Sign Language: An Introduction. Cambridge University Press, Cambridge (1999)
55. Tamura, S., Kawasaki, S.: Recognition of sign language motion images. Pattern Recogn. 21(4), 343–353 (1988)
56. Valli, C., University, G.: The Gallaudet Dictionary of American Sign Language. Gallaudet University Press, Washington, D.C. (2005)
57. Viitaniemi, V., Jantunen, T., Savolainen, L., Karppa, M., Laaksonen, J.: S-pot - a benchmark in spotting signs within continuous signing. In: LREC (2014)
58. von Agris, U., Knorr, M., Kraiss, K.: The significance of facial features for automatic sign language recognition. In: 2008 8th IEEE International Conference on Automatic Face Gesture Recognition (2008)
59. Wilbur, R.B., Kak, A.C.: Purdue RVL-SLLL American sign language database. School of Electrical and Computer Engineering Technical Report, TR-06-12, Purdue University, W. Lafayette, IN 47906 (2006)
60. Woll, B.: The sign that dares to speak its name: echo phonology in British sign language (BSL). In: Boyes-Braem, P., Sutton-Spence, R. (eds.) The Hands are the Head of the Mouth: The Mouth as Articulator in Sign Languages, pp. 87–98. Signum Press, Hamburg (2001)
61. Yang, S., Luo, P., Loy, C.C., Tang, X.: Wider face: a face detection benchmark. In: CVPR (2016)
62. Yao, Y., Wang, T., Du, H., Zheng, L., Gedeon, T.: Spotting visual keywords from temporal sliding windows. In: Mandarin Audio-Visual Speech Recognition Challenge (2019)

63. Ye, Y., Tian, Y., Huenerfauth, M., Liu, J.: Recognizing American sign language gestures from within continuous videos. In: CVPRW (2018)
64. Zhou, H., Zhou, W., Zhou, Y., Li, H.: Spatial-temporal multi-cue network for continuous sign language recognition. CoRR abs/2002.03187 (2020)

HTML: A Parametric Hand Texture Model for 3D Hand Reconstruction and Personalization

Neng Qian[1,2], Jiayi Wang[1], Franziska Mueller[1(✉)], Florian Bernard[1,3], Vladislav Golyanik[1], and Christian Theobalt[1]

[1] Max Planck Institute for Informatics, Saarland Informatics Campus, Saarbrücken, Germany
{nqian,jwang,frmueller,fbernard,golyanik,theobalt}@mpi-inf.mpg.de
[2] RWTH Aachen University, Aachen, Germany
[3] Technical University of Munich, München, Germany

Abstract. 3D hand reconstruction from images is a widely-studied problem in computer vision and graphics, and has a particularly high relevance for virtual and augmented reality. Although several 3D hand reconstruction approaches leverage hand models as a strong prior to resolve ambiguities and achieve more robust results, most existing models account only for the hand shape and poses and do not model the texture. To fill this gap, in this work we present HTML, the first parametric texture model of human hands. Our model spans several dimensions of hand appearance variability (*e.g.*, related to gender, ethnicity, or age) and only requires a commodity camera for data acquisition. Experimentally, we demonstrate that our appearance model can be used to tackle a range of challenging problems such as 3D hand reconstruction from a single monocular image. Furthermore, our appearance model can be used to define a neural rendering layer that enables training with a self-supervised photometric loss. We make our model publicly available.

Keywords: Hand texture model · Appearance modeling · Hand tracking · 3D hand reconstruction

1 Introduction

Hands are one of the most natural ways for humans to interact with their environment. As interest in virtual and augmented reality grows, so does the need for reconstructing a user's hands to enable intuitive and immersive interactions with the virtual environment. Ideally, this reconstruction contains accurate hand

https://handtracker.mpi-inf.mpg.de/projects/HandTextureModel/.

Electronic supplementary material The online version of this chapter (https://doi.org/10.1007/978-3-030-58621-8_4) contains supplementary material, which is available to authorized users.

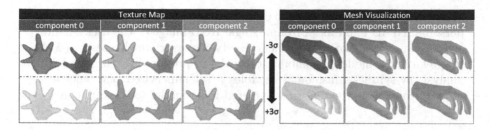

Fig. 1. We present the first parametric hand texture model. Our model successfully captures appearance variations from different gender, age, and ethnicity.

shape, pose, and appearance. However, it is a challenging task to capture a user's hands from just images due to the complexity of hand interactions and self-occlusion. In recent years, there has been significant progress in hand pose estimation from monocular depth [1,8,12,25,30,53,54] and RGB [7,22,46,55,57] images. Although most of these works estimate only joint positions, a few recent works attempt to reconstruct the hand geometry as well [2,6,26,27,56].

Despite these recent advances, there is little work that addresses the reconstruction of hand appearance. However, hand appearance personalization is important for increasing immersion and the sense of "body-ownership" in VR applications [19], and for improved tracking and pose estimation through analysis-by-synthesis approaches. Without a personalized appearance model, existing pose estimation methods must use much coarser hand silhouettes [2,6,56] as an approximation of appearance. One approach to obtain a personalized hand texture is to project the tracked geometry to the RGB image and copy the observed color to the texture map [23]. However, only a partial appearance of the observed hand parts can be recovered with this method and tracking errors can lead to unnatural appearances. In addition, without explicit lighting estimation, lighting effects will be baked into the results of these projection-based methods.

To address this gap, we present *HTML*, the first data-driven parametric **H**and Texture Mode**L** (see Fig. 1). We captured a large variety of hands and aligned the scans in order to enable principal component analysis (PCA) and build a textured parametric hand model. PCA compresses the variations of natural hand appearances to a low dimensional appearance basis, thus enabling a more robust appearance fitting. Our model can additionally produce plausible appearance of the entire hand from fitting to partial observations from a single RGB image. Our main **contributions** can be summarized as follows:

- We introduce a novel parametric model of hand texture, HTML, that we make publicly available. Our model is based on a dataset of high-resolution hand scans of 51 subjects with variety in gender, age, and ethnicity.
- We register our scans to the popular MANO hand model [39] in order to create a statistical hand appearance model that is also compatible with MANO.

- We demonstrate that our new parametric texture model allows to obtain a personalized 3D hand mesh from a single RGB image of the user's hand.
- We present a proof-of-concept neural network layer which uses the MANO shape and pose model in combination with our proposed texture model in an analysis-by-synthesis fashion. It enables a self-supervised photometric loss, directly comparing the textured rendered hand model to the input image.

2   Related Work

The use of detailed, yet computationally efficient, hand models for hand tracking applications is well studied [31,35,41,48]. Nevertheless, many such methods require time-consuming expert adjustments to personalize the model to a user's hand, making them difficult to deploy to the end-user. Therefore, we focus our review to methods that can automatically generate personalized articulated hand models from images. However, we will see that almost all these methods exclusively consider shape personalization and do not include texture or appearance.

Modeling Hand Geometry. Two types of personalizable hand models exist in the literature, *i.e.*, heuristic parameterizations that directly move and scale the geometric primitives of the models [23,37,47,51,52], and data-driven statistical parameterizations that model the covariance of hand geometry [20,39]. Although heuristic approaches are expressive, infeasible hand-shape configurations can arise when fitting such models to single images due to ambiguities between shape and pose. Thus, existing approaches must perform the personalization offline over a set of depth images [37,47,51], or design additional heuristic constraints [52] to resolve these ambiguities. On the other hand, data-driven parameterizations [20] provide a low-dimensional shape representation and natural priors on hand poses. The recent MANO model [39] additionally provides learned data-driven pose-dependent shape corrections to the geometry to avoid artifacts in posing a hand model through *linear blend skinning* (LBS). This model has been applied in many recent hand pose estimation methods [2,6,16,56] and has been used to annotate hand pose estimation benchmarks [15,16,58].

Nonetheless, and despite the popularity of the MANO model of hand geometry, there exists no data-driven parametric texture model for providing realistic appearance. As such, in this work we present for the first time a hand appearance model that is fully compatible with MANO. Although MANO has a rather low-resolution mesh (778 vertices), our appearance model is defined in the texture space so that a much higher texture resolution is available.

Modeling Appearance. With a few exceptions [23,24], the previously mentioned works do not model hand texture. The works of de La Gorce *et al.* [23,24] incorporate heuristic texture personalization for hand-tracking using an analysis-by-synthesis approach. Their approach obtains only a partial estimate of the hand texture using the current pose estimate, and relies on a smoothness prior to transfer color to unobserved parts by a diffusion process on a per-frame basis.

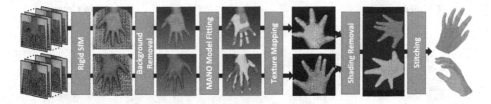

Fig. 2. Overview of our hand texture acquisition pipeline. We run rigid structure from motion (SfM) on a set of input images to obtain a scanned mesh for back and palm side of the hand, respectively. After removing background vertices, we fit the MANO template mesh to extract the texture from the scan. We remove lighting effects and seamlessly stitch the front and back texture, resulting in a complete texture for the captured hand (visualized on the 3D hand mesh from 2 virtual views on the right).

Romero *et al.* [39] provide the raw RGB scans used to register the MANO model, but they contain strong lighting effects like shadows and over-exposed regions. Hence, it is not possible to recover accurate appearance from these scans as we show in the supplementary document. Despite the lack of a parametric hand texture model, the benefits of having such a model can be readily seen in face modeling literature. For example, 3D morphable face models (3DMM) [5, 9, 13, 18, 33] provide parametric geometry and appearance models for faces that have been used to drive research in many recent works in diverse applications [10]. For example, these 3DMMs were used within analysis-by-synthesis frameworks for RGB tracking [38, 50], and as unsupervised loss for learning-based methods [49]. Our proposed parametric hand appearance model HTML has the potential to drive similar advances in the hand pose estimation and modeling community.

3   Textured Parametric Hand Model

Our hand texture acquisition pipeline is summarized in Fig. 2. First, we record two image sequences observing the palm side and the back side of the hand, respectively. Subsequently, we run rigid structure from motion (SfM) [3, 40] to obtain a 3D reconstruction of the observed hand side (Sect. 3.1). Next, we remove the scene background, and register both (partial) hand scans to the MANO model [39] based on nonlinear optimization. Afterwards, the texture of the partial hand scans is mapped to the registered mesh. We then remove shading effects from the textures and stitch them to obtain a complete hand texture (Sect. 3.2). The parametric texture model in subsequently generated using PCA (Sect. 3.3).

3.1   Data Acquisition

In total, we captured 51 subjects with varying gender, age, and ethnicity (see Fig. 3). To minimize hand motion during scanning, we record the palm side and backside of the hand separately, so that the subjects can rest their hand on a

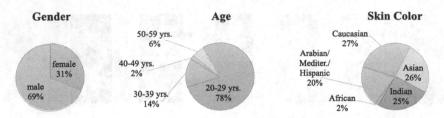

Fig. 3. Distribution of age, gender, and skin color for our 51 captured subjects. We use the Goldman world classification scale [42] for classifying skin color.

flat surface. As such, for each subject we obtain four scans, *i.e.*, back and palm sides for both left and right hands. The scanning takes ∼90 s for one hand side, so that the total scanning time of ∼6 min is required per person.

To obtain 3D hand scans, we use SONY's 3DCreator App [44]. The 3D reconstruction pipeline includes three stages, *i.e.*, initial anchor point extraction, simultaneous localization and mapping (SLAM) with sparse points [21], and online dense 3D reconstruction (sculpting) [45]. The output is a textured high-resolution surface mesh (of one hand side as well as the background), which contains ∼6.2k vertices and ∼11k triangles in the hand area on average. By design, our hand texture model is built for the right hand. For model creation, we mirror the left hand meshes, so that we use a total of 102 "right" hands for modeling. We note that by mirroring we can also use the texture model of "right" hand for the left hand. In the following, we will abstract away this technical detail and describe our texture modeling approach for a single hand.

3.2   Data Canonicalization

To learn the texture variations in a data-driven manner it is crucial that the acquired 3D scans are brought into a common representation. Due to the popularity and the wide use of the MANO model of hand geometry, we decided to build the hand texture in the MANO space. This has the advantage that existing hand reconstruction and tracking frameworks that are based on MANO, such as [6,16,29], can be directly extended to also incorporate hand texture. We point out that our texture model can also be used with other models by defining the respective UV mapping. Our data canonicalization comprises several consecutive stages, *i.e.*, *background removal*, *MANO model fitting*, *texture mapping*, *shading removal*, and *seamless stitching*, which we describe next.

Background Removal. For each hand we have reconstructed two textured meshes, one that shows the hand palm-down on a flat surface, and one that shows the hand palm-up on a flat surface (cf. Sect. 3.1). In both cases, the background, *i.e.*, the flat surface that the hand is resting on, is also reconstructed as part of the mesh. Hence, in order to remove the background, we perform a robust plane fitting based on RANSAC [11], where a plane is fitted to the flat background surface. To this end, we sample 100 random configurations of three vertices, fit a plane to the sampled points, and then count the number of inliers. Any point that has a distance to the fitted plane that is smaller than the median edge length of the input scanned mesh is considered as inlier. Eventually, the plane that leads to the largest inlier count is considered the background plane. We have empirically found that this approach is robust and able to reliably identify the flat surface in all cases. Eventually, we use a combination of distance-based and color-based thresholding to discard background vertices in the scanned mesh. In particular, we discard a vertex if its distance from the background plane is less than 1cm and the difference between the red and green channel of the vertex color is smaller than 30 (RGB $\in [0, 255]^3$). This yields better preservation of hand vertices that are close to the background plane.

MANO Model Fitting. Subsequently, we fit the MANO hand model to the filtered hand scan mesh (*i.e.,* the one without background). To this end, we first obtain the MANO shape and pose parameters based on the hand tracking approach of Mueller *et al.* [29]. The approach uses a Gauss-Newton optimization scheme that makes use of additional information based on trained machine learning predictors (*e.g.,* for correspondence estimation). Since their method was developed for 3D reconstruction and tracking of hands in *depth images*, we render synthetic depth images from our partial hand scan meshes. Note that the approach [29] was partially trained on synthetic depth images and thus we have found that it is able to produce sufficiently good fits of the MANO geometry to our data.

However, since the MANO model is relatively coarse (778 vertices), and more importantly, it has a limited expressivity of hand shape (it only spans the variations of their training set of 31 subjects), we have found that there are still some misalignments. To also allow for deformations outside the shape space of the MANO model, we hence use a complementary non-rigid refinement of the previously fitted MANO mesh to the hand scan. To this end, we use a variant of non-rigid *iterative closet point* (ICP) [4] that optimizes for individual vertex displacements that further refine the template, which in our case is the fitted MANO model. As our objective function, we use 3D point-to-point and point-to-plane distances together with a spatial smoothness regularizer [14]. An accurate alignment is especially important at salient points, like fingertips, to ensure high perceptual quality. Hence, we add prior correspondences for the fingertips and the wrist to the non-rigid ICP fitting. We automatically obtain these correspondences in the input scanned mesh using OpenPose [43]. The influence of the prior correspondences is shown in our evaluation (see Sect. 5.1).

Texture Mapping. After having obtained an accurate alignment of the hand template, *i.e.,* the fitted MANO model plus non-rigid deformation for refinement, to our textured high-resolution hand scan, we transfer the scan texture to a texture map. To this end, we have manually defined UV coordinates for the MANO model template by unwrapping the mesh to a plane (see texture mapping step in Fig. 2). We project each vertex in the high-resolution hand scan to the closest point on the surface of the fitted MANO hand template. Using the barycentric coordinates of this projected point together with the UV coordinates of the template mesh, we transfer the color to the texture map. After performing this procedure for all vertices of our high-resolution hand scan, there can still be some texels (pixels in the texture map) that are not set (we have found that about 6.5% of the hand interior does not have a defined texture). To deal with that, holes are filled based on inpainting with neighboring texels.

Shading Removal. We ensured that our scans have low-frequency shading by using controlled lighting. Thus, we implicitly made the assumptions of having a mostly Lambertian surface and no casted shadows. Since the smooth shading effects have low frequency (see Fig. 4a), they can be separated and removed using a frequency-based method like the Laplacian image pyramid. To this end, we first build a Laplacian pyramid with five levels from the texture map that we obtained in the previous step. We observe that the deepest level separates the (almost) constant skin color as well as the smooth shading from the texture details that are kept on earlier levels of the pyramid. We replace this deepest level with a constant skin color for palm and back side, respectively, effectively removing the smooth shading. We obtain this constant skin color by averaging in the well-lit area (see blue rectangles in Fig. 4). Note how the texture details from higher levels are preserved in the modified texture map (see Fig. 4b).

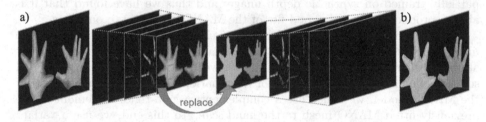

Fig. 4. Shading removal. (a) Original texture and its Laplacian pyramid decomposition. (b) The shading effects are removed by modifying the deepest level. (Color figure online)

Seamless Texture Stitching. Since so far this texture mapping is performed both for the palm-up and palm-down facing meshes, we eventually blend both partial texture maps to obtain a complete texture map of the hand. To this end, we use a recent gradient-domain texture stitching approach that directly operates in the texture atlas domain while preserving continuity induced by the 3D mesh topology across atlas chart boundaries [34].

3.3   Texture Model Creation

Let $\{T_i\}_{i=1}^n$ be the collection of 2D texture maps that we obtain after data canonicalization as described in Sect. 3.2. In order to create a parametric texture model we employ PCA. We vectorize each T_i to obtain the vector $t_i \in \mathbb{R}^{618,990}$ that stacks the red, green and blue channels of all hand texels. PCA first computes the data covariance matrix

$$C = \frac{1}{n-1} \sum_{i=1}^{n} (t_i - \bar{t})(t_i - \bar{t})^\top ,\tag{1}$$

for $\bar{t} = \frac{1}{n}\sum_{i=1}^{n} t_i$ being the average texture. Subsequently, eigenvalue decomposition of $C = \Phi\Lambda\Phi^T$ is used to obtain the principal components Φ and the diagonal matrix of eigenvalues Λ. With that we obtain the parametric texture model for the parameter vector $\alpha \in \mathbb{R}^k, k = 101$ as

$$t(\alpha) = \bar{t} + \Phi\alpha .\tag{2}$$

Fig. 5. 3D hand personalization from a single image. Starting from a single RGB input image (left), we first initialize the mesh using the method by Boukhayma *et al.* [6]. Next, we refine the fit non-rigidly and extract the partial hand texture. By fitting our parametric texture model, we are able to obtain a complete texture which minimizes the error to the input texture (right).

4   Applications

To demonstrate possible use cases of our parametric hand appearance model, we present two applications. First, we consider 3D hand reconstruction and personalization from a single monocular RGB image. Subsequently, we show the usage as a neural network layer enabling a self-supervised photometric loss.

4.1   3D Hand Personalization from a Single Image

Given a single monocular RGB image of a hand, we aim to reconstruct a 3D hand mesh that is personalized to the user's shape and appearance. This application consists of four steps: (1) initialization of shape and pose parameters of the MANO model, (2) non-rigid shape and pose refinement, (3) partial texture extraction, and (4) estimation of appearance parameters of our model.

Shape and Pose Initialization. We use the method of Boukhayma *et al.* [6] to obtain an initial pose and shape estimate of the MANO template mesh from a single RGB image. As discussed before, the MANO shape space is not always expressive enough to perfectly fit the user's hand shape. In addition, the results from the method by Boukhayma *et al.* do not yield sufficiently accurate reprojection of the mesh onto the image plane as shown in Fig. 5 (second from the left). Hence, this initial mesh is further refined.

Non-Rigid Refinement of the Initial Mesh. We non-rigidly refine the initial mesh estimate to better fit the hand silhouette in the image. Therefore, we optimize the 3D displacement of each vertex using ICP constraints on the boundary vertices. We define the set of boundary vertices of the hand mesh $\bar{\mathcal{V}} \subset \mathcal{V}$, *i.e.*, the set of vertices on the silhouette. Let $\Pi : \mathbb{R}^3 \rightarrow \Omega$ be the camera projection converting from 3D world coordinates to 2D pixel locations. For each boundary vertex $\bar{\boldsymbol{v}}_i$, we first find the closest hand silhouette pixel $\bar{\boldsymbol{p}}_i$ in the image domain Ω as

$$\bar{\boldsymbol{p}}_i = \underset{\boldsymbol{p} \in \Omega}{arg\,min}\, ||\Pi(\bar{\boldsymbol{v}}_i) - \boldsymbol{p}||_2 \quad \text{s.t.} \quad n(\boldsymbol{p})^\top \Pi(n(\bar{\boldsymbol{v}}_i)) > \eta. \tag{3}$$

Here, $n(\boldsymbol{p})$ is the 2D boundary normal at pixel $\boldsymbol{p}$ (calculated by Sobel filtering), and $\Pi(n(\bar{\boldsymbol{v}}_i))$ is the 2D image-plane projection of the 3D vertex normal at $\bar{\boldsymbol{v}}_i$. The threshold $\eta = 0.8$ discards unsuitable pixels based on normal dissimilarity. We then use this closest hand silhouette pixel $\bar{\boldsymbol{p}}_i$ as correspondence for boundary vertex $\bar{\boldsymbol{v}}_i$ if it is closer than δ (= 4% of the image size):

$$\bar{\boldsymbol{c}}_i = \begin{cases} \bar{\boldsymbol{p}}_i, & \text{if } ||\Pi(\bar{\boldsymbol{v}}_i) - \bar{\boldsymbol{p}}_i||_2 < \delta \\ \emptyset, & \text{otherwise} \end{cases}. \tag{4}$$

We can then optimize for the refined 3D vertex positions using the computed correspondences in the following objective function:

$$E(\mathcal{V}) = \frac{1}{|\bar{\mathcal{V}}|} \sum_{\bar{\boldsymbol{v}}_i \in \bar{\mathcal{V}}} ||\Pi(\bar{\boldsymbol{v}}_i) - \bar{\boldsymbol{p}}_i||_2^2 + w_{\text{smth}} \sum_{\boldsymbol{v}_j \in \mathcal{V}} \sum_{\boldsymbol{v}_k \in \mathcal{N}_j} \frac{1}{|\mathcal{N}_j|} ||(\boldsymbol{v}_j - \boldsymbol{v}_k) - (\boldsymbol{v}_j^0 - \boldsymbol{v}_k^0)||_2^2, \tag{5}$$

where $\mathcal{N}_j$ is the set of neighboring vertices of $\boldsymbol{v}_j$, and $\mathcal{V}^0 = \{\boldsymbol{v}_\bullet^0\}$ are the vertex positions from the previous ICP iteration. In total, we use 20 ICP iterations and initialize $\mathcal{V}, \mathcal{V}^0$ from the shape and pose initialization step as described above.

Partial Texture Extraction. For each fully visible triangle, *i.e.*, when all its 3 vertices are visible, we extract the color from the input image and copy it to the texture map. This yields a partial texture map where usually at most half the texels have a value assigned and all other texels are set to $\emptyset$. We then obtain the vectorized target texture map t^{trgt} (as for model creation in Sect. 3.3).

Estimation of Appearance Parameters. Subsequently, we find the appearance parameters of our model that best fit the user's hand by solving the least-squares problem with Tikhonov regularization:

$$\underset{\alpha \in \mathbb{R}^k}{arg\,min} \sum_{t_i^{\text{trgt}} \neq \emptyset} (t_i^{\text{trgt}} - t(\alpha)_i)^2 + w_{\text{reg}}||\alpha||_2^2 \,. \qquad (6)$$

Note that our proposed parametric appearance model enables us to obtain a complete texture. In contrast to the extracted partial texture, the result is free of lighting effects and artifacts caused by small misalignments of the hand model.

4.2 Self-supervised Photometric Loss

Previous works have trained neural networks to regress joint positions or MANO model parameters from RGB images [7,28,46,55,57,58]. The most common loss is the Euclidean distance between the regressed and ground truth joint positions. Some works have also explored a silhouette loss between the mesh and the hand region in the image [2,6,56]. Our HTML enables the use of a *self-supervised* photometric loss, which complements the existing fully supervised losses. With that, when training a network to predict shape and pose with such an approach, we additionally obtain a hand texture estimate. To this end, we introduce a *textured hand model layer*, which we explain now.

Textured Hand Model Layer. Given a pair of MANO shape and pose parameters (β, θ), as well as the texture parameters α, our model layer computes the textured 3D hand mesh $\mathcal{M}(\beta, \theta, \alpha)$. An image of this mesh is then rendered using a scaled orthographic projection. As such, this rendered image can directly be compared to the input image $\mathcal{I}$ using a photometric loss in an analysis-by-synthesis manner. We formulate the photometric loss as

$$\mathcal{L}_{\text{photo}}(\beta, \theta, \alpha) = \frac{1}{|\Gamma|} \sum_{(u,v) \in \Gamma} ||\text{render}(\mathcal{M}(\beta, \theta, \alpha))(u, v) - \mathcal{I}(u, v)||_2 \,, \qquad (7)$$

where Γ is the set of pixels which the estimated hand mesh projects to. The use of a differentiable renderer makes the photometric loss $\mathcal{L}_{\text{photo}}$ fully differentiable and enables backpropagation for neural network training.

Network Training. We train a residual network with the architecture of ResNet-34 [17] to regress the shape β, pose θ, and texture parameters α from a given input image. In addition to the self-supervised photometric loss, we employ losses on 2D joint positions, 3D joint positions, and L2-regularizers on the magnitude of the shape, pose, and texture parameters. The network is trained in PyTorch [32], using the differentiable renderer provided in PyTorch3D [36]. We assume a single fixed illumination condition for training. We leave the joint estimation of additional lighting and material properties to future work.

5 Experiments

In this section, we evaluate our proposed parametric hand texture model, explore different design choices in our texture acquisition pipeline, and present results of our two example applications.

5.1   Texture Model Evaluation

Compactness. Figure 6 (left) shows the compactness of our texture model. The plot describes how much the explained variance in the training dataset increases with the number of used principal components. The first few components already explain a significant amount of variation since they account for more global changes in the texture, *e.g.*, skin tone. However, adding more components continuously increases the explained variance.

Generalization. For evaluating generalization, we use a leave-one-subject-out protocol. We remove the data of one subject, *i.e.*, the two texture samples from left and right hand, and rebuild the PCA model. Then, we reconstruct the left-out textures using the built model and measure the reconstruction error as the mean absolute distance (MAD) of the vectorized textures. As shown in Fig. 6 (middle), the reconstruction error decreases monotonically for an increasing number of components for both of the two models.

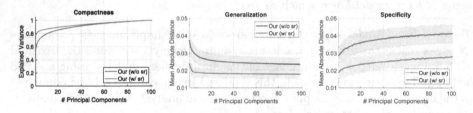

Fig. 6.   Evaluation of compactness, generalization and specificity. Using shading removal ("w/ sr") substantially outperforms not using shading removal ("w/o sr").

Specificity. We also report the specificity, which quantifies the similarity between random samples from the model and the training data. To this end, we first sample a texture instance from our model based on a multivariate standard Normal distribution (due to the Gaussian assumption of PCA). Then, we find the nearest texture in our training dataset in terms of the MAD. We repeat this procedure 200 times, and report the statistics of the MAD in Fig. 6 (right).

Influence of Shading Removal. Figure 6 also shows compactness, generalization, and specificity for a version of the texture model that was built without shading removal ("w/o sr"). It can be seen that the version without shading removal performs worse compared to the one with shading removal ("w/ sr") in all metrics. When the lighting effects are not removed, they increase the variance in the training dataset. Hence, more principal components are necessary to explain variation and the reconstruction of unseen test samples has a higher error. In the supplemental material, we also show visually that the principal components for the model without shading removal have to account for strong lighting variation.

Influence of Prior Correspondences. To ensure a good alignment of the hand template mesh and the scanned mesh, as explained in Sect. 3.2, for the non-rigid ICP-based refinement step in our model building stage we make use of prior correspondences for the fingertips and the wrist. Figure 7 compares the textures obtained by running the non-rigid ICP fitting with and without them. Especially for the thumb, the tip is often not well-aligned, resulting in a missing finger nail in the texture. Using explicit prior correspondences alleviates this issue.

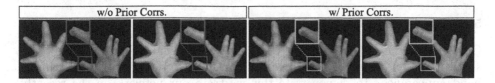

Fig. 7. Using non-rigid ICP-based refinement with prior correspondences for fingertips and the wrist improves the alignment of the hand template mesh to the scanned mesh, yielding better textures (right). (Textures shown before shading removal.)

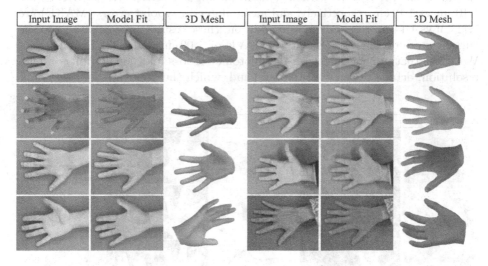

Fig. 8. Hand personlization from a single RGB input image for different subjects. (Color figure online)

5.2 Application Results: 3D Hand Personalization

Here, we show results for obtaining a personalized 3D hand model from a single RGB image (see Sect. 4.1). As previously discussed, since the output meshes of state-of-the-art regression approaches [6] do not have a low reprojection error, we use non-rigid refinement based on silhouettes. To simplify segmentation in our example application, we captured the images of the users in front of a green screen. In future work, this could be replaced by a dedicated hand segmentation method. Figure 8 shows hand model fits and complete recovered textures from a single RGB image for several subjects. Since we use a low-dimensional PCA space to model hand texture variation, we can robustly estimate a plausible and complete texture from noisy or partially corrupted input (see Fig. 9). In contrast, a texture that is directly obtained by projecting the input image onto a mesh obtained by the method of Boukhayma et al. [6] contains large misalignments and a significant amount of background pixels, and thus is severely corrupted.

5.3   Application Results: Photometric Neural Network Loss

Our self-supervised photometric loss (see Sect. 4.2) enables to not only obtain shape and pose estimates as in previous work, but in addition to also estimate hand appearance. To demonstrate this we train our network on the recently proposed FreiHAND dataset [58]. For details of the experimental setup, please see the document. In Fig. 10, we show hand model fits predicted by a neural network trained with and without our photometric loss (cf. Sect. 4.2). We note that the pose and shape prediction with the photometric loss are quantitatively similar to the predictions without (the mean aligned vertex errors (MAVE) are 1.10 cm vs 1.14 cm respectively, and mean aligned keypoint errors (MAKE) are 1.11 cm vs 1.14 cm respectively). In addition, these results are comparable to the current state of the art [58] with a MAVE of 1.09 cm and MAKE of 1.10 cm. We stress that our method with the photometric loss additionally infers a high resolution, detailed texture of the full hand, which the other methods do not.

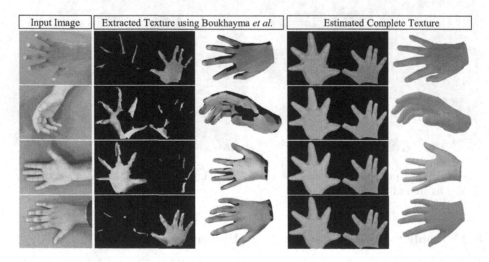

Fig. 9. Fitting to noisy or corrupted input textures is robust and yields a realistic and complete texture estimate due to the low-dimensional PCA space built by our model.

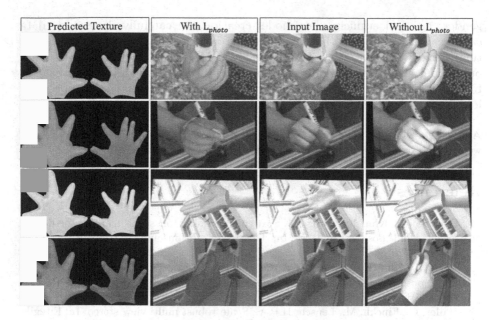

| Predicted Texture | With L_{photo} | Input Image | Without L_{photo} |

Fig. 10. We show the predicted pose and texture from a neural network trained using a photometric loss L_{photo} enabled by our parametric hand texture model.

6   Limitations and Discussion

Our experiments have shown that HTML can be used to recover personalized 3D hand shape and appearance. Although our model provides detailed texture, the underlying geometry of the MANO mesh is coarse (778 vertices). This could be improved by using a higher-resolution mesh and extending the MANO shape space with more detailed geometry. Non-linear models, *e.g.*, an autoencoder neural network, can be explored for capturing variations that a linear PCA model cannot. As hand appearance varies during articulation, modeling pose-dependent texture changes can increase the realism. This would need a more complicated capture and registration setup and a significantly larger dataset to capture the whole pose space and diverse users. In terms of applications, estimating lighting in addition to or jointly with the texture parameters can better reconstruct input observations. Correctly modeling lighting for hands, where shadow casting often occurs, is a challenge that would need to be addressed. Other applications of our model, such as exploring how self-supervision can alleviate the need for keypoint annotations or improve pose estimation, can be directions for future research.

7   Conclusion

In this work, we introduced HTML—the first parametric texture model of hands. The model is based on data that captures 102 hands of people with varying gender, age and ethnicity. For model creation, we carefully designed a data canonicalization pipeline that entails background removal, geometric model fitting, texture mapping, and shading removal. Moreover, we demonstrated that our model enables two highly relevant applications: 3D hand personalization from a single RGB image, and learning texture estimation using a self-supervised loss. We make our model publicly available to encourage future work in the area.

Acknowledgments. The authors thank all participants of the data recordings. This work was supported by the ERC Consolidator Grant 4DRepLy (770784).

References

1. Baek, S., In Kim, K., Kim, T.K.: Augmented skeleton space transfer for depth-based hand pose estimation. In: Computer Vision and Pattern Recognition (CVPR) (2018)
2. Baek, S., Kim, K.I., Kim, T.K.: Pushing the envelope for RGB-based dense 3D hand pose estimation via neural rendering. In: Computer Vision and Pattern Recognition (CVPR) (2019)
3. Bailer, C., Finckh, M., Lensch, H.P.A.: Scale robust multi view stereo. In: Fitzgibbon, A., Lazebnik, S., Perona, P., Sato, Y., Schmid, C. (eds.) ECCV 2012. LNCS, vol. 7574, pp. 398–411. Springer, Heidelberg (2012). https://doi.org/10.1007/978-3-642-33712-3_29
4. Besl, P.J., McKay, N.D.: A method for registration of 3-D shapes. Trans. Pattern Anal. Mach. Intell. (TPAMI) **14**(2), 239–256 (1992)
5. Blanz, V., Vetter, T.: A morphable model for the synthesis of 3D faces. In: SIGGRAPH, pp. 187–194 (1999)
6. Boukhayma, A., de Bem, R., Torr, P.H.: 3D hand shape and pose from images in the wild. In: Computer Vision and Pattern Recognition (CVPR) (2019)
7. Cai, Y., Ge, L., Cai, J., Yuan, J.: Weakly-supervised 3D hand pose estimation from monocular RGB images. In: Ferrari, V., Hebert, M., Sminchisescu, C., Weiss, Y. (eds.) ECCV 2018. LNCS, vol. 11210, pp. 678–694. Springer, Cham (2018). https://doi.org/10.1007/978-3-030-01231-1_41
8. Chen, Y., Tu, Z., Ge, L., Zhang, D., Chen, R., Yuan, J.: SO-HandNet: self-organizing network for 3D hand pose estimation with semi-supervised learning. In: International Conference on Computer Vision (ICCV) (2019)
9. Dai, H., Pears, N., Smith, W.A., Duncan, C.: A 3D morphable model of craniofacial shape and texture variation. In: International Conference on Computer Vision (ICCV), pp. 3085–3093 (2017)
10. Egger, B., et al.: 3D morphable face models - past, present and future. ACM Trans. Graph. **39**, 157:1–157:38 (2020)
11. Fischler, M.A., Bolles, R.C.: Random sample consensus: a paradigm for model fitting with applications to image analysis and automated cartography. Commun. ACM **24**(6), 381–395 (1981)
12. Ge, L., Cai, Y., Weng, J., Yuan, J.: Hand PointNet: 3D hand pose estimation using point sets. In: Computer Vision and Pattern Recognition (CVPR) (2018)
13. Gerig, T., et al.: Morphable face models-an open framework. In: International Conference on Automatic Face & Gesture Recognition (FG), pp. 75–82 (2018)

14. Habermann, M., Xu, W., Zollhoefer, M., Pons-Moll, G., Theobalt, C.: LiveCap: real-time human performance capture from monocular video. ACM Trans. Graph. (TOG) **38**(2), 1–17 (2019)

15. Hampali, S., Rad, M., Oberweger, M., Lepetit, V.: HOnnotate: a method for 3D annotation of hand and object poses. In: Conference on Computer Vision and Pattern Recognition (CVPR), pp. 3196–3206 (2020)

16. Hasson, Y., et al.: Learning joint reconstruction of hands and manipulated objects. In: Computer Vision and Pattern Recognition (CVPR) (2019)

17. He, K., Zhang, X., Ren, S., Sun, J.: Deep residual learning for image recognition. In: Computer Vision and Pattern Recognition (CVPR), pp. 770–778 (2016)

18. Huber, P., et al.: A multiresolution 3D morphable face model and fitting framework. In: International Joint Conference on Computer Vision, Imaging and Computer Graphics Theory and Applications (VISIGRAPP) (2016)

19. Jung, S., Hughes, C.: Body ownership in virtual reality. In: International Conference on Collaboration Technologies and Systems (CTS), pp. 597–600 (10 2016)

20. Khamis, S., Taylor, J., Shotton, J., Keskin, C., Izadi, S., Fitzgibbon, A.: Learning an efficient model of hand shape variation from depth images. In: Computer Vision and Pattern Recognition (CVPR) (2015)

21. Klein, G., Murray, D.: Parallel tracking and mapping for small AR workspaces. In: International Symposium on Mixed and Augmented Reality (ISMAR) (2007)

22. Kovalenko, O., Golyanik, V., Malik, J., Elhayek, A., Stricker, D.: Structure from articulated motion: accurate and stable monocular 3D reconstruction without training data. Sensors **19**(20), 4603 (2019)

23. de La Gorce, M., Fleet, D.J., Paragios, N.: Model-based 3D hand pose estimation from monocular video. Trans. Pattern Anal. Mach. Intell. (PAMI) **33**(9), 1793–1805 (2011)

24. de La Gorce, M., Paragios, N., Fleet, D.J.: Model-based hand tracking with texture, shading and self-occlusions. In: Computer Vision and Pattern Recognition (CVPR) (2008)

25. Li, S., Lee, D.: Point-to-pose voting based hand pose estimation using residual permutation equivariant layer. In: Computer Vision and Pattern Recognition (CVPR) (2019)

26. Malik, J., et al.: HandVoxNet: deep voxel-based network for 3D hand shape and pose estimation from a single depth map. In: Computer Vision and Pattern Recognition (CVPR) (2020)

27. Malik, J., et al.: DeepHPS: end-to-end estimation of 3D hand pose and shape by learning from synthetic depth. In: International Conference on 3D Vision (3DV) (2018)

28. Mueller, F., et al.: GANerated hands for real-time 3D hand tracking from monocular RGB. In: Computer Vision and Pattern Recognition (CVPR) (2018)

29. Mueller, F., et al.: Real-time pose and shape reconstruction of two interacting hands with a single depth camera. ACM Trans. Graph. (TOG) **38**(4), 1–13 (2019)

30. Oberweger, M., Wohlhart, P., Lepetit, V.: Training a feedback loop for hand pose estimation. In: International Conference on Computer Vision (ICCV) (2015)

31. Oikonomidis, I., Kyriazis, N., Argyros, A.A.: Efficient model-based 3D tracking of hand articulations using Kinect. In: British Machine Vision Conference (BMVC) (2011)

32. Paszke, A., et al.: PyTorch: an imperative style, high-performance deep learning library. In: Advances in Neural Information Processing Systems (NeurIPS), pp. 8024–8035 (2019)

33. Paysan, P., Knothe, R., Amberg, B., Romdhani, S., Vetter, T.: A 3D face model for pose and illumination invariant face recognition. In: International Conference on Advanced Video and Signal Based Surveillance (AVSS), pp. 296–301 (2009)
34. Prada, F., Kazhdan, M., Chuang, M., Hoppe, H.: Gradient-domain processing within a texture atlas. ACM Trans. Graph. (TOG) 37(4), 154:1–154:14 (2018)
35. Qian, C., Sun, X., Wei, Y., Tang, X., Sun, J.: Realtime and robust hand tracking from depth. In: Computer Vision and Pattern Recognition (CVPR) (2014)
36. Ravi, N., et al.: PyTorch3D (2020). https://github.com/facebookresearch/pytorch3d
37. Remelli, E., Tkach, A., Tagliasacchi, A., Pauly, M.: Low-dimensionality calibration through local anisotropic scaling for robust hand model personalization. In: International Conference on Computer Vision (ICCV) (2017)
38. Romdhani, S., Vetter, T.: Estimating 3D shape and texture using pixel intensity, edges, specular highlights, texture constraints and a prior. Comput. Vis. Pattern Recogn. (CVPR) 2, 986–993 (2005)
39. Romero, J., Tzionas, D., Black, M.J.: Embodied hands: modeling and capturing hands and bodies together. ACM Trans. Graph. (Proc. SIGGRAPH Asia) 36(6), 245:1–245:17 (2017)
40. Schönberger, J.L., Frahm, J.M.: Structure-from-motion revisited. In: Computer Vision and Pattern Recognition (CVPR) (2016)
41. Sharp, T., et al.: Accurate, robust, and flexible real-time hand tracking. In: ACM Conference on Human Factors in Computing Systems (CHI) (2015)
42. Shiffman, M.A., Mirrafati, S., Lam, S.M., Cueteaux, C.G.: Simplified Facial Rejuvenation. Springer, Heidelberg (2007). https://doi.org/10.1007/978-3-540-71097-4
43. Simon, T., Joo, H., Matthews, I., Sheikh, Y.: Hand keypoint detection in single images using multiview bootstrapping. In: Computer Vision and Pattern Recognition (CVPR) (2017)
44. SONY 3D Creator. https://3d-creator.sonymobile.com/
45. Sony Corporation: 3D Creator App (White Paper), version 3: August 2018 (2018). https://dyshcs8wkvd5y.cloudfront.net/docs/3D-Creator-Whitepaper.pdf
46. Spurr, A., Song, J., Park, S., Hilliges, O.: Cross-modal deep variational hand pose estimation. In: Computer Vision and Pattern Recognition (CVPR) (2018)
47. Sridhar, S., Oulasvirta, A., Theobalt, C.: Interactive markerless articulated hand motion tracking using RGB and depth data. In: International Conference on Computer Vision (ICCV) (2013)
48. Taylor, J., et al.: Articulated distance fields for ultra-fast tracking of hands interacting. ACM Trans. Graph. (TOG) 36(6), 1–12 (2017)
49. Tewari, A., et al.: MoFA: model-based deep convolutional face autoencoder for unsupervised monocular reconstruction. In: International Conference on Computer Vision (ICCV) (2017)
50. Thies, J., Zollhöfer, M., Stamminger, M., Theobalt, C., Nießner, M.: Face2Face: real-time face capture and reenactment of RGB videos. In: Computer Vision and Pattern Recognition (CVPR) (2016)
51. Tkach, A., Pauly, M., Tagliasacchi, A.: Sphere-meshes for real-time hand modeling and tracking. ACM Trans. Graph. (ToG) 35(6), 1–11 (2016)
52. Tkach, A., Tagliasacchi, A., Remelli, E., Pauly, M., Fitzgibbon, A.: Online generative model personalization for hand tracking. ACM Trans. Graph. (ToG) 36(6), 1–11 (2017)
53. Tompson, J., Stein, M., Lecun, Y., Perlin, K.: Real-time continuous pose recovery of human hands using convolutional networks. ACM Trans. Graph. 33, 169:1–169:10 (2014)

54. Wan, C., Probst, T., Van Gool, L., Yao, A.: Crossing nets: combining GANs and VAEs with a shared latent space for hand pose estimation. In: Computer Vision and Pattern Recognition (CVPR) (2017)

55. Yang, L., Li, S., Lee, D., Yao, A.: Aligning latent spaces for 3D hand pose estimation. In: International Conference on Computer Vision (ICCV) (2019)

56. Zhang, X., Li, Q., Mo, H., Zhang, W., Zheng, W.: End-to-end hand mesh recovery from a monocular RGB image. In: International Conference on Computer Vision (ICCV) (2019)

57. Zimmermann, C., Brox, T.: Learning to estimate 3D hand pose from single RGB images. In: International Conference on Computer Vision (ICCV) (2017)

58. Zimmermann, C., Ceylan, D., Yang, J., Russell, B., Argus, M., Brox, T.: Frei-HAND: a dataset for markerless capture of hand pose and shape from single RGB images. In: International Conference on Computer Vision (ICCV) (2019)

CycAs: Self-supervised Cycle Association for Learning Re-identifiable Descriptions

Zhongdao Wang[1], Jingwei Zhang[1], Liang Zheng[2], Yixuan Liu[1], Yifan Sun[3],
Yali Li[1], and Shengjin Wang[1(✉)]

[1] Department of Electronic Engineering, Tsinghua University, Beijing, China
{zjw18,wcd17}@mails.tsinghua.edu.cn, {liyali13,wgsgj}@tsinghua.edu.cn
[2] Australian National University, Canberra, Australia
liang.zheng@anu.edu.au
[3] MEGVII Technology, Beijing, China
peter@megvii.com

Abstract. This paper proposes a self-supervised learning method for the person re-identification (re-ID) problem, where existing unsupervised methods usually rely on pseudo labels, such as those from video tracklets or clustering. A potential drawback of using pseudo labels is that errors may accumulate and it is challenging to estimate the number of pseudo IDs. We introduce a different unsupervised method that allows us to learn pedestrian embeddings from raw videos, without resorting to pseudo labels. The goal is to construct a self-supervised pretext task that matches the person re-ID objective. Inspired by the *data association* concept in multi-object tracking, we propose the **Cyc**le **As**sociation (**CycAs**) task: after performing data association between a pair of video frames forward and then backward, a pedestrian instance is supposed to be associated to itself. To fulfill this goal, the model must learn a meaningful representation that can well describe correspondences between instances in frame pairs. We adapt the discrete association process to a differentiable form, such that end-to-end training becomes feasible. Experiments are conducted in two aspects: We first compare our method with existing unsupervised re-ID methods on seven benchmarks and demonstrate CycAs' superiority. Then, to further validate the practical value of CycAs in real-world applications, we perform training on self-collected videos and report promising performance on standard test sets.

Keywords: Self-supervised · Cycle consistency · person re-ID

1 Introduction

Self-supervised learning is a recent solution to the lack of labeled data in various computer vision areas like optical flow estimation [17,45], disparity/depth

Electronic supplementary material The online version of this chapter (https://doi.org/10.1007/978-3-030-58621-8_5) contains supplementary material, which is available to authorized users.

© Springer Nature Switzerland AG 2020
A. Vedaldi et al. (Eds.): ECCV 2020, LNCS 12356, pp. 72–88, 2020.
https://doi.org/10.1007/978-3-030-58621-8_5

estimation [7, 20, 37], pixel/object tracking [10, 26, 29] and universal representation learning [1, 4, 6, 19, 28, 40]. As a branch of unsupervised learning, the idea of self-supervised learning is to construct a *pretext task*. It is supposed that free supervision signals of the pretext task can be generated directly from the data, and the challenges lie in the design of the pretext task, so that the learned representation matches the task objective.

Self-supervised/unsupervised learning finds critical significance in the person re-identification (re-ID) area, because of the high annotation cost. The goal of the re-ID task is to search for cross-camera bounding boxes containing the same person with a query: the cross-camera requirement increases the burden of data annotation. Existing unsupervised methods usually rely on pseudo labels that can be obtained from video tracklets [13, 34] or clustering [5, 16]. This strategy achieves descent accuracy, but its potential drawback consists of the error accumulation and the challenge in estimating the number of pseudo identities.

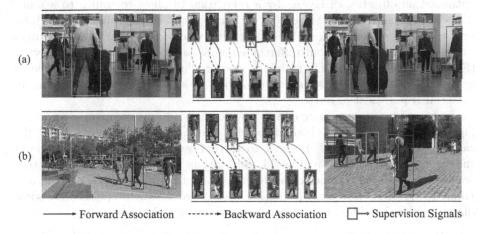

Fig. 1. Cycle association between (a). Temporal consecutive frames in a single video; (b). Temporal-aligned frames from two cameras that shares an overlapped visual field. In these examples, (a) shows a symmetric case (two frames contain the same group of identities) and (b) shows an asymmetric case.

We are interested in finding a pretext task for re-ID, such that pedestrian descriptors can be learned in a self-supervised way. We are motivated by the cycle association between a pair of video frames that contain multiple persons. Considering two temporal-consecutive frames from such a video, because of the short time interval between them, they usually share the same group of identities (Fig. 1 (a)). With perfect person representations, if we apply data association[1] between the two frames, we can find accurate correspondences between the two

---

[1] In Multi-Object Tracking (MOT) [11], data association means matching observations in a new frame to a set of tracked trajectories. In our case, we simplify the concept to matching observations between a frame pair.

sets of identities. Further, if we perform forward data association and then backward, an instance is supposed to be associated to itself. Based on this motivation, we construct the **Cy**cle **A**ssociation (CycAs) pretext task: apply data association in a cycle, *i.e.,* forward then backward, and use the inconsistency in the cycle association matrix as supervision signals. To maximize the cycle consistency or minimize inconsistency, the model inclines to learn a meaningful representation that can well-describe correspondences between instances [10,29].

In the above pretext task, we face a dilemma, *i.e.,* appearance diversity vs. accurate association. To learn robust descriptors, we need a pedestrian to exhibit sufficient diversity between two frames, *e.g.,* people from frames with long temporal interval. It means that persons from the two frames do not come from the exact same group of identities, creating asymmetry and leading to inaccurate association. On the other hand, if we use consecutive frames from a single video to ensure accurate association, the appearance variation would be small, compromising the descriptor robustness. To address this dilemma, we modify the optimization objective of CycAs by a relaxation to allow tolerance to a moderate level of asymmetry. Moreover, we adopt a two-stage training procedure: first train the pretext task using consecutive video frames, and then for fine-tuning add frame pairs from different cameras with overlapping field of view (FOV) (see Fig. 1 (b)). This training process allows learning correspondences across cameras and benefits feature robustness to large appearance variation, an essential requirement of re-ID.

Experiments are conducted in two aspects: First, we compare with existing unsupervised re-ID methods on a wide range of public datasets under the same train / test protocol. Second, to further validate the practical value of the proposed method in real-world applications, we train CycAs with self-collected videos and conduct cross-domain evaluation on Market-1501 [42] and DukeMTMC-ReID [21]. Very promising results are shown compared with some direct transfer baselines. Our strengths are summarized below,

(1) We propose CycAs, a self-supervised pretext task for person re-ID. Strong features can be learned from associating persons between videos frames.
(2) We design a relaxed optimization objective for CycAs, allowing leveraging frames with large appearance variation. It significantly improves the discriminative ability of learned representations.
(3) We showcase the strength of CycAs on public benchmarks. We further validate the practical value of CycAs using self-collected videos as training data.

2   Related Work

Unsupervised Person re-ID. Most existing deep learning based unsupervised person re-ID approaches in literature can be categories into there paradigms:

(1). Domain adaptation methods [3,25,43,44]. These methods start with a supervised learned model which is pre-trained using the source domain data, and then transfer knowledge from the unlabeled target domain data.

(2). Clustering-based methods [5,16,33]. These methods usually adopt the iterative clustering-and-training strategy. Unlabeled data are grouped with clustering algorithms and assigned pseudo labels, then the model is fine-tuned
with these pseudo labels. Such procedure repeats until convergence.

(3). Tracklet-based methods [13,14,34]. These methods label different trackelts,
from a specific camera, as different identities, and train multiple classification tasks for multiple cameras in a parallel manner. Cross-camera matching
is usually modeled as metric learning with pseudo labels.

Domain adaptation methods need supervised learned pre-train models so
they are not totally unsupervised in essence. Clustering/Tracklet-based methods
all rely on pseudo labels. In contrast, the proposed CycAs is unsupervised and
does not require pseudo labels.

Self-supervised Learning. As a form of unsupervised learning, self-supervised
learning seeks to learn from unlabeled data by constructing pretext tasks. For
instance, image-level pretext tasks such as predicting context [4], rotation [6]
and color [40] are useful for learning universal visual representations.

Video-level pretext tasks [10,28,29] are recently prevalent due to the large
amount of available web videos and the fertile information they contain. Our
work is closely related to a line of video-based self-supervised methods that
utilize cycle consistency as free supervision [10,26,29]. While these methods
usually focus on learning fine-grained correspondences between pixels [10,29],
thus mainly tackle the tracking problem, ours focus more on learning high-level
semantic correspondences and is more adaptive to the re-ID problem. To the
best of our knowledge, we are the first to provide a self-supervised solution to
learn re-identifiable object descriptions.

3 Proposed Cycle Association Task

3.1 Overview

Our goal is to learn a discriminative pedestrian embedding function Φ by learning
correspondences between two sets of person images I_1 and I_2. Specifically, I_1
and I_2 are detected pedestrian bounding boxes in a pair of frames. In this paper,
we design two strategies to sample the frame pairs as follows (Fig. 2).

- **Intra-sampling.** Frame pairs are sampled from the same video within a
  short temporal interval, $i.e.$, 2 s.
- **Inter-sampling.** Each frame pair is sampled at the same timestamp from
  two different cameras that capture an overlapped FOV.

In both strategies, with proper selection of temporal interval or deployment
of cameras, which should not be too difficult to control, a reasonable identity
overlap between I_1 and I_2 can be guaranteed. We define $\tau = \frac{\# \ overlapped \ IDs}{max\{|I_1|,|I_2|\}}$ as
the *symmetry* between I_1 and I_2, and its impact on our system is investigated
in Sect. 4.1. For ease of illustration, let us begin with the absolute symmetric

case $\tau = 1$, *i.e.*, for any instance in I_1 there is a correspondence in I_2, and vice versa. Then we introduce how we deal with asymmetry in Sect. 3.3. The training procedure is presented in Fig. 3.

3.2   Association Between Symmetric Pairs

Consider all the images in $I_1 \cup I_2$ forming a minibatch. Suppose the size of the two sets $|I_1| = |I_2| = K$ ($\tau = 1$, *i.e.*, absolute symmetry). The bounding boxes are mapped to the embedding space by Φ, such that $X_1 = \Phi(I_1)$ and $X_2 = \Phi(I_2)$, where $X_1 = [x_1^1, x_1^2, ..., x_1^K] \in \mathbb{R}^{D \times K}$ and $X_2 = [x_2^1, x_2^2, ..., x_2^K] \in \mathbb{R}^{D \times K}$ are embedding matrices composed of K embedding vectors of dimension D. All the embedding vectors are ℓ_2-normalized. To capture similarity between instances, we compute an affinity matrix between all instances in X_1 and X_2 by calculating the pairwise cosine similarities,

$$S = X_1^\top X_2 \in \mathbb{R}^{K \times K}. \tag{1}$$

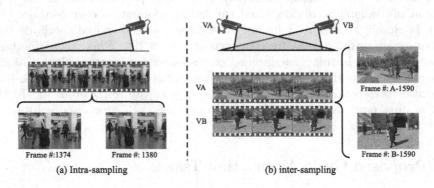

Frame #:1374	Frame #: 1380
(a) Intra-sampling	(b) inter-sampling

Fig. 2. Illustration on the two frame pair sampling methods. (a) Intra-sampling, frame pairs are drawn from a single video within a short temporal interval. (b) Inter-sampling, frame pairs are drawn from cameras capturing an overlapped FOV, at the same time.

We take S as input to perform data association, which aims to predict correspondences in X_2 for each instance in X_1. Formally, the goal is to obtain an assignment matrix,

$$A = \psi(S) \in \{0, 1\}^{K \times K}, \tag{2}$$

where 1 indicates correspondence. In MOT, it is usually modeled as a linear assignment problem, and the solution ψ can be found by the Hungarian algorithm (examples can be found in many MOT algorithms [30,32,38]).

Suppose the embedding function Φ is perfect, *i.e.*, the cosine similarity between vectors of the same identity equals 1, while the cosine similarity between vectors from different identities equals -1. The Hungarian algorithm can output the optimal assignment $A^* = \frac{S+1}{2}$ for the forward association process $X_1 \to X_2$.

The backward association process $X_2 \rightarrow X_1$ is similar, and the optimal assignment matrix $A'^* = \frac{S^\top + 1}{2}$. A Cycle Association pass is then defined as a forward association pass plus a backward association pass,

$$A^{\texttt{cycle}} = AA'. \tag{3}$$

Intuitively, if $A = A^*$ and $A' = A'^*$, an instance will be associated to itself. In other words, the cycle association matrix underpinning perfect association $A^{\texttt{cycle}}$ should equal the identity matrix I. Accordingly, the difference between $A^{\texttt{cycle}}$ and I can be used as signals to implicitly supervise the model to learn correspondences between X_1 and X_2.

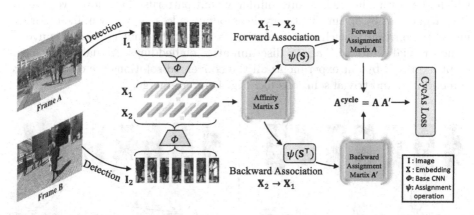

Fig. 3. An overview of the proposed cycle association task. First, two sets of detected pedestrians are mapped to embeddings via the base CNN Φ. Pairwise affinity matrix S is computed from the two sets of embeedings, then forward/backward assignment matrix A and A' if computed from S. Finally the inconsistency in the cycle association matrix $A^{\texttt{cycle}} = AA'$ is used as supervision signals.

The whole process needs to be differentiable for end-to-end training. However, the assignment operation ψ (Hungarian algorithm) is not differentiable. This motivates us to design a differentiable ψ. We notice that if the one-to-one correspondence constraint is removed, ψ can be approximated by the row-wise `argmax` function. Considering `argmax` is not differentiable either, we further soften this operation by the row-wise softmax function. Now, the assignment matrix is computed as,

$$A_{i,j} = \psi_{i,j}(S) = \frac{e^{TS_{i,j}}}{\sum_{j'}^{K} e^{TS_{i,j'}}}, \tag{4}$$

where $A_{i,j}$ is the element of A in the i-th row and the j-th colomn, and T is the temperature of the softmax operation. The backward association pass has a different temperature T'. T and T' are designed to be adaptive to the size of A and A', and more details will be described in Sect. 3.4.

Combing Eq. 1, Eq. 4 and Eq. 3, a cycle association matrix A^{cycle} can be computed with all operations therein being differentiable. Finally, the loss function is defined as the mean ℓ_1 error between A^{cycle} and I,

$$\mathcal{L}_{\text{symmetric}} = \frac{1}{K^2} \| A^{\text{cycle}} - I \|_1. \tag{5}$$

Discussion. Theoretically, cycle consistency is a *necessary* but not *sufficient* requirement for discriminative embeddings. In Fig. 4 we present a trivial solution: what the embedding model has learned is matching an identity to the next identity, and the last identity is matched to the first. Such a trivial solution requires the model to learn strong correlations between random identity pairs, which share very limited, if any, similar visual patterns. Therefore, we reasonably argue that by optimizing the cycle association loss, it is very unlikely for the model to converge to such trivial solutions and that it is much easier to converge to non-trivial solutions, *i.e.*, the discriminative embeddings. Actually, this argument is proved by our experiment: all the converged solutions are discriminative embeddings, and trivial solutions like Fig. 4 never emerge.

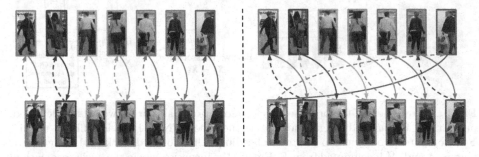

Fig. 4. A non-trivial solution (**Left**) *v.s.* a trivial solution (**Right**) of Eq. 5. Solid and dotted lines means forward and backward correspondences, respectively. We argue the trivial solutions are very unlikely to be learned.

3.3   Relaxation on Asymmetric Pairs

In practice, asymmetric always arises along with large appearance diversity. The main reasons are multi-folds: First, pedestrians enter and leave the visual field of the camera; second, real-world videos usually contain high-velocity motions, severe occlusions, and low-quality persons, so the detector may fail sometimes; for inter-sampled data, it's also impossible to ensure FOVs overlap exactly. For better leveraging such appearance-diverse data, we make the following efforts to reduce the negative impact of the asymmetry.

First, consider descriptor matrices X_1 and X_2 with the number of descriptors being K_1 and K_2, respectively. According to Eq. 3, the resulting A^{cycle} is of size $K_1 \times K_1$. If $K_1 > K_2$, there will exist at least $K_1 - K_2$ instances that cannot be associated back to themselves. This will introduce ambiguity. Therefore, we swap X_1 and X_2 in such cases to ensure $K_1 \leq K_2$ always holds.

Second, the learning objective is modified. In the asymmetric scenario, the loss function $\mathcal{L}_{\text{symmetric}}$ is sub-optimal, because some instances may lose correspondences in cycle association, and thus the corresponding diagonal elements in A^{cycle} are not supposed to equal 1. Simply changing the supervision of these lost instances from 1 to 0 is not feasible, because there are no annotations and we do not know which instances are lost. To address this, our solution is to relax the learning objective. More specifically, we expect a diagonal element $A^{\text{cycle}}_{i,i}$ to be greater than all the other elements along the same row and column, by a given margin m. The loss function is formulated as,

$$\mathcal{L}_{\text{asymmetric}} = \frac{1}{K_1} \sum_{i=1}^{K_1} \left[\left(\max_{j \neq i} A^{\text{cycle}}_{i,j} - A^{\text{cycle}}_{i,i} + m \right)_+ + \left(\max_{k \neq i} A^{\text{cycle}}_{k,i} - A^{\text{cycle}}_{i,i} + m \right)_+ \right], \quad (6)$$

which has a similar form as the triplet loss [22]. The margin m is a hyper-parameter ranging in $(0, 1)$ with smaller values indicating softer constrains. We set $m = 0.5$ in all the experiment if not specified.

We will show through experiment that the relaxation of the loss function benefits learning in both the asymmetric and symmetric cases. In the experiment, we use $\mathcal{L}_{\text{asymmetric}}$ by default unless specified.

3.4 Adapt Softmax Temperature to Varying Sizes

Cosider two vectors with different sizes, $v = (1, 0.5)^\top$ and $u = (1, 0.5, 0.5)^\top$. Let σ be the softmax operation, then $\sigma(v) = (0.62, 0.38)^\top$ and $\sigma(u) = (0.45, 0.27, 0.27)^\top$. The *Soft-Max* operation, as we observe, has different levels of softening ability on inputs with different sizes. The Max value in a longer vector is less highlighted, or maxed, and vice versa. To alleviate this problem and stabelize the training, we let the softmax temperature be adaptive to the varying input size, so that for input vectors of different sizes, the max values in them are equally highlighted.

To fulfill this goal, we let the temprature $T = \frac{1}{\epsilon} \log \left[\frac{\delta(K-1)+1}{1-\delta} \right]$, where ϵ and δ are two hyper-parameters ranging from 0 to 1. In fact, the only hyper-parameter that matters is ϵ, and δ can be simply set to 0.5. This leads to the final form $T = \frac{1}{\epsilon} \log(K+1)$. Detailed derivation and discussion can be found in the supplementary material.

3.5 Two-Stage Training

The two types of sampling strategies have their respective advantages and drawbacks. To get the best of both worlds, we design a two-stage training procedure that initializes from a model pretrained on ImageNet [2].

In Stage I, we train the model with intra-sampled data only, and the temporal interval for sampling is set to rather small, *e.g.*, 2 s, so that the appearance variation between two person sets is small, or the symmetric τ is high (>0.9).

After convergence of Stage I, we start Stage II. In this stage, we train the model using both the inter-sampled data and intra-sampled data in a multi-task manner with 1 : 1 loss weights. The inter-sampled data have much higher appearance variations but a lower τ (around 0.6).

Discussions. This training strategy is carefully designed so as to converge well. Directly starting from Stage II converges with a slower speed *w.r.t.* our progressive training strategy, while starting from inter-sampled data only fails in converging. We will quantitatively demonstrate the effectiveness of this training strategy in Sect. 4.2.

Alternatively, we give an intuitive illustration by visualizations on the embedding space at each training state, shown in Fig. 5. In the initial embedding space before training, , most embeddings from different identities are not separable Fig. 5 (a). In Stage I training, the model learns to find correspondence between persons within the same camera, so in the resulting embedding space embeddings of the same identity from the same cameras are grouped together, while embeddings of the same identity from different cameras are still separate (see red rectangles in Fig. 5 (b)). Finally, in Stage II training, the model learns to associate across different cameras, in which case the appearance variations are large. Therefore, the resulting embedding can handle large appearance variation, thus is camera-invariant(Fig. 5 (c)). To summarize, training Stage I functions as a "warm-up" process for Stage II, while in Stage II the model learns meaningful camera-invariant representation for the re-ID task.

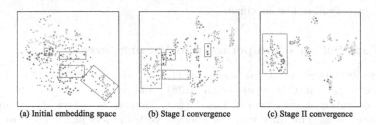

(a) Initial embedding space   (b) Stage I convergence   (c) Stage II convergence

Fig. 5. Illustration on the evolution of the embedding space from (a) Initial model to (b) Stage I convergence then to (c) Stage II convergence. Different colors indicate different identities. Different markers indicate different cameras. Visualized via Barnes-Hut t-SNE [24]. (Color figure online)

4   Experiment

The experiments are organized as follows. First, we adopt standard train / test protocols on seven video- / image-based person re-ID datasets and compare our method with existing unsupervised methods (Sect. 4.1). Then we investigate the impact of different components and hyper-parameters (Sect. 4.2). Finally, to be more practical, we perform experiment using self-collected pedestrian videos as training data and compare with direct transfer (supervised) models (Sect. 4.3). We use ResNet-50 [8] as the backbone network in all experiments.

4.1   Experiment with Standard Datasets

Setup. We test the proposed CycAs on both video-based (MARS [41], iLIDS [27], PRID2011 [9]) and image-based (Market-1501 [42], CUHK03 [15], DukeMTMC-ReID [21], MSMT17 [31]) person re-id datasets. All the datasets provide camera annotations and the video-based datasets additionally provide tracklets. Following existing practice [13,14,34], for image-based datasets, we assume all images per ID per camera are drawn from a single tracklet. Consider a mini-batch with batch size B, to mimic the intra- / inter-sampling, we first randomly sample $B/2$ identities. For intra-sampling, a tracklet is sampled for each of these identities; then, two bounding boxes are sampled within each tracklet. For inter-sampling, two tracklets from different cameras are sampled for each of the $B/2$ identities; then, one image is is sampled from each tracklet. Results on image- / video-based re-ID datasets are shown in Table 1 and Table 2, respectively.

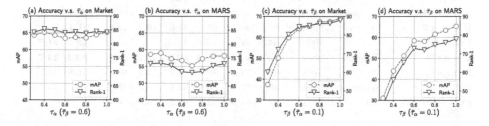

Fig. 6. Robustness against different levels of (a-b) intra-sampled data symmetry τ_α and (c-d) inter-sampled data symmetry τ_β. For evaluating τ_α, in each mini-batch, we fix τ_β and draw a random τ_α from $\mathcal{N}(\bar{\tau}_\alpha, 0.01)$, and plot the performance curve $w.r.t.$ different mean $\bar{\tau}_\alpha$. The impact of τ_β is evaluated in a similar way.

Performance Upper Bound Analysis. According to above sampling strategy, the data are absolutely symmetric, $i.e.$, $\tau = 1$. The performance under this setting can be seen as an upper bound of the proposed method, denoted as CycAs^{sym}. For comparison, we implement a supervised baseline (IDE [42]). We observe that the performance of CycAs^{sym} is consistently competitive on all the datasets. Compared with IDE, the rank-1 accuracy of CycAs^{sym} is lower by only 1.1%, 0.3% and 1.6% on Market-1501, DukeMTMC-ReID and MSMT17, respectively. On CUHK03, CycAs^{sym} even surpasses IDE by +2.2%. This result partially prove the good alignment between the CycAs task and the objective of re-ID. We also list state-of-the-art supervised methods [12,23] for comparison and observe that the performance gap between CycAs^{sym} and these methods is not too large. These results suggest the potential of CycAs in the re-ID task.

Robustness Against Different Levels of Data Symmetry τ. To investigate the robustness of CycAs against different levels of symmetry, we introduce asymmetry to the sampled data and observe how the ReID accuracy changes. Specifically, we control intra-sampling symmmetry τ_α and inter-sampling asymmetry τ_β by replacing a portion of images in I_2 with randomly sampled images from irrelevant identities. To evaluate the impact of τ_α, in each mini-batch, we fix τ_β and draw τ_α from a gaussian distribution

Table 1. Comparison with state-of-the-art methods on image-based re-ID datasets. Note all the methods starts from a ImageNet pretrained model. The requirement *Pretain* and *None* refers to whether pretraining on labeled re-ID datasets is needed. CycAs^{asy} is our method, and CycAs^{sym} refers to an upper bound of our method.

Method	Category	Require	Market [42]		Duke [21]		CUHK03 [15]		MSMT17 [31]	
			R1	mAP	R1	mAP	R1	mAP	R1	mAP
SPGAN [3]	UDA	Pretrain	51.5	22.8	41.1	22.3	–	–	–	–
SPGAN+LMP [3]	UDA	Pretrain	57.7	26.7	46.4	26.2	–	–	–	–
TJ-AIDL [25]	UDA	Pretrain	58.2	26.5	44.3	23.0	–	–	–	–
HHL [43]	UDA	Pretrain	62.2	31.4	46.9	27.2	–	–	–	–
ECN [44]	UDA	Pretrain	75.1	43.0	63.3	40.4	–	–	30.2	10.2
PUL [5]	Clustering	Pretrain	44.7	20.1	30.4	16.4	–	–	–	–
CAMEL [39]	Clustering	Pretrain	54.5	26.3	–	–	39.4	–	–	–
BUC [16]	Clustering	None	66.2	38.3	47.4	27.5	–	–	–	–
CDS [33]	Clustering	Pretrain	71.6	39.9	67.2	42.7	–	–	–	–
TAUDL [13]	Tracklet	MOT	63.7	41.2	61.7	43.5	44.7	31.2	28.4	12.5
UTAL [14]	Tracklet	MOT	69.2	46.2	62.3	44.6	**56.3**	**42.3**	31.4	13.1
UGA [34]	Tracklet	MOT	**87.2**	**70.3**	*75.0*	*53.3*	–	–	*49.5*	*21.7*
CycAs^{asy}	Self-Sup	None	*84.8*	*64.8*	**77.9**	**60.1**	*47.4*	*41.0*	**50.1**	**26.7**
CycAs^{sym}	–	None	88.1	71.8	79.7	62.7	56.4	49.6	61.8	36.2
IDE	Supervised	Label	89.2	73.9	80.0	63.1	54.2	47.2	60.2	33.4
PCB+RPP [23]	Supervised	Label	93.8	81.6	83.3	69.2	63.7	57.5	–	–

Table 2. Comparison with state-of-the-art methods on video-based re-ID datasets.

Method	Category	Require	MARS [41]		PRID [9]	iLIDS [15]
			R1	mAP	R1	R1
DGM+IDE [36]	Clustering	Pretrain	36.8	21.3	56.4	36.2
RACE [35]	Clustering	MOT	43.2	24.5	50.6	19.3
BUC [16]	Clustering	None	61.1	38.0	–	–
SMP [18]	Tracklet	MOT	23.9	10.5	80.9	41.7
TAUDL [13]	Tracklet	MOT	43.8	29.1	49.4	26.7
UTAL [14]	Tracklet	MOT	49.9	35.2	54.7	35.1
UGA [34]	Tracklet	MOT	58.1	39.3	80.9	57.3
CycAs^{asy}	Self-Sup	None	**72.8**	**58.4**	**86.5**	**73.3**
CycAs^{sym}	–	None	79.2	67.5	85.4	77.3
IDE	Supervised	Label	81.7	67.8	90.5	78.4
GLTR [12]	Supervised	Label	87.0	78.5	95.5	86.0

$\mathcal{N}(\bar{\tau}_\alpha, 0.01)$, truncate the value in range $(0, 1)$, and plot the model performance against the mean $\bar{\tau}_\alpha$. The impact of τ_β is evaluated in a similar way. We report results in Fig. 6.

Two observations can be made from the curves. First, with a moderate fixed value of τ_β, *i.e.*, 0.6 in our case, the model accuracy is robust to a wide range of $\bar{\tau}_\alpha$. For example, in Fig. 6 (a), the rank-1 accuracy is both 84.2% when $\bar{\tau}_\alpha$ is set to 1 and 0.3, respectively. Second, we observe that the accuracy improves when $\bar{\tau}_\beta$ becomes larger. The main reason is explained in Sect. 3.5 and Fig. 5: Training Stage I (Training with intra-sampled data) only functions as a "warm-up" process , to provide a meaningful initial point for Stage II. The knowledge learned from intra-sampled data contributes less on the overall performance. Therefore the final accuracy is less sensitive to $\bar{\tau}_\alpha$. In contrast, learning from inter-sampled data aligns with the objective of re-ID task, therefore the final accuracy is more sensitive to $\bar{\tau}_\beta$.

Remarks. Note that in Fig. 6 (c-d), the curves drop very slowly when $\tau_\beta = 1$ decreases from 1 to 0.6. This suggests that CycAs has a good ability to handle data with reasonably asymmetry. Such a property is valuable, because in practice we can control $\bar{\tau}_\beta$ in a reasonable range (say from 0.6 to 0.9), by carefully placing the cameras. Comparing with manually annotating data, this requires less effort.

Comparison with the State of the Art. For fair comparisons, we train CycAs under a practically reasonable asymmetric assumption. We fix $\tau_\alpha = 0.9$ and $\tau_\beta = 0.6$, and compare the results (denoted as CycAs^{asy}) with existing unsupervised re-ID approaches. Three categories of existing methods are compared, i.e., unsupervised domain adaptation (UDA) [3,25,43,44], clustering-based methods [5,16,33,35,36,39], and tracklet-based methods [13,14,34]. Beside re-ID accuracy, we also compare another dimension, i.e., ease of use, by listing the requirements of each method in Table 1 and Table 2.

Under image-based unsupervised learning, CycAs^{asy} achieves state-of-the-art accuracy on two larger datasets, i.e., DukeMTMC and MSMT17. The mAP improvement over the second best method [34] is +6.8% and +5.0% on DukeMTMC and MSMT17, respectively. On Market-1501 and CUHK03, CycAs^{asy} is very competitive to the best performing methods [34].

Under video-based unsupervised learning, CycAs^{asy} achieves state-of-the-art results on three datasets. The rank-1 accuracy improvement over the second best method is +14.7%, +5.6% and +16.0% on MARS, PRID and iLIDS, respectively.

Comparing with other unsupervised strategies, CycAs requires less external supervision. For example, UDA methods use a labeld source re-ID dataset, and most clustering-based methods need a pre-trained model for initialization, which also uses external labeled re-ID datasets. The tracklet-based methods do not require re-ID labels, but require a good tracker to provide good supervision signals.Training such a good tracker also requires external pedestrian labels. Note that ImageNet pretraining is needed by all the methods. CycAs learns person representations directly from videos and does not require any external annotation. Its requiring less supervision and competitive accuracy making it potentially a more practical solution for unsupervised re-ID.

4.2 Ablation Study

$\mathcal{L}_{symmetric}$ v.s. $\mathcal{L}_{asymmetric}$. To prove the proposed relaxed loss $\mathcal{L}_{asymmetric}$ is superior to the original loss $\mathcal{L}_{symmetric}$, we compare the two losses with both symmetric and asymmetric training data. Results on Market dataset is shown in Table 3. We observe that with both symmetric/asymmetric training data, $\mathcal{L}_{asymmetric}$ consistently outperforms $\mathcal{L}_{symmetric}$. These results reveal a misunderstanding that $\mathcal{L}_{symmetric}$ is better for symmetric data and $\mathcal{L}_{asymmetric}$ is better for asymmetric data. In contrast, the results suggest $\mathcal{L}_{asymmetric}$ is essentially a more reasonable objective for the CycAs task, and it results in a more desirable embedding space. We speculate $\mathcal{L}_{symmetric}$ is too rigorous for the

Table 3. Comparison between two losses under different data symmetry settings on Market-1501. We see the asymmetric loss always outperforms the symmetric loss no matter with asymmetric or symmetric data.

Data	Loss	Rank-1	mAP
symmetric			
$\tau_\alpha, \tau_\beta =$	$\mathcal{L}_{\text{symmetric}}$	78.9	59.1
$(1.0, 1.0)$	$\mathcal{L}_{\text{asymmetric}}$	**88.1**	**71.8**
asymmetric			
$\tau_\alpha, \tau_\beta =$	$\mathcal{L}_{\text{symmetric}}$	67.1	47.1
$(0.9, 0.6)$	$\mathcal{L}_{\text{asymmetric}}$	**84.8**	**64.8**

Table 4. Impact of the intra-sampling temporal interval and different training strategy. Evaluated on Market-1501.

Training		R-1	mAP
Stage I Only	*Interval:*		
	2 s	**29.4**	**11.9**
	4 s	25.8	9.7
	8 s	27.2	10.4
Stage II Only	*Data:*		
	intra + inter	84.6	64.7
	inter	–	–
Stage I + Stage II	*Interval:*		
	2 s	**84.8**	**64.8**
	4 s	84.2	64.1
	8 s	84.0	53.3

task, and the large magnitudes of losses from those well-associated cycles hamper the training.

Impact of the Temporal Interval for Intra-sampling. We investigate the impact of the temporal interval for intra-sampling and show comparisons in Table 4. When only intra-sampled data are used for training (Stage I only), the sampling interval slightly affects the final performance. However, longer temporal interval, *e.g.*, 8 s, does not bring performance gain. We speculate the reason could be the large asymmetry brought by long temporal interval .

Two-Stage Training Strategy. The effectiveness of the proposed two-stage training strategy is also investigated, and results are shown in Table 4. For comparison, we first train with Stage I only, *i.e.*, only intra-sampled data are used. It can be observed the final accuracy is quite poor. As discussed in previous section, the reason is training with intra-sampled data does not align with the objective of the re-ID task, *i.e.*, cross-camera retrieval. We also compare with training with Stage II only. If both intra- and inter-sampled data are used, the model converges to a decent solution. However, if we remove intra-sampled data from training, the model fails in converging. This suggest training with intra-sampled data is necessary for converging to a meaningful solution. Finally, if we use the proposed two-stage training, the results are as good as training with Stage II only, and the benefit is that the model converges with a faster speed.

4.3 Experiment Using Self-collected Videos as Training Data

To our knowledge, prior works in unsupervised re-ID usually evaluate their systems on standard benchmarks by simulating real-world scenarios. In this paper, to further assess the practical potential of CycAs, we report experimental results obtained by training with real-world videos. The videos are captured by hand-hold cameras in several scenes with high pedestrian density, such as the airport, shopping mall and campus. The total length of the videos is about 6 h. Among these videos, about 5 h are captured from a single view, which can only be used for intra-sampling; The rest 1 h videos are captured from two different views, and thus can be used for inter-sampling.

We employ an open-source pedestrian detector [30] to detect persons in every 7 frames and crop the detected persons. For intra-sampling, we set the maximum temporal interval between two frames to 2 s. In every mini-batch, we sample 8 frame pairs to enlarge the batch size for high training efficiency. Training lasts for 10k iterations for Stage I and another 35k iterations for Stage II.

Table 5. Results with real-world videos as training data. Comparisons are made with supervised baselines and existing unsupervised methods.

	Training data	Market-1501		DukeMTMC	
		mAP	R1	mAP	R1
Supervised.M	Market-1501 [42]	73.9	89.2	_16.6_	_33.4_
Supervised.D	DukeMTMC [21]	_21.4_	_48.1_	63.1	80.0
Supervised.C	CUHK03 [15]	19.8	43.2	13.1	26.3
BUC [16]	6-h unlabelled videos	14.2	29.8	11.2	21.5
UGA [34]	6-h unlabelled videos	17.8	37.2	15.4	25.6
CycAs	6-h unlabelled videos	**23.3**	**50.8**	19.2	**34.6**

We evaluate the performance on Market-1501 [42] and DukeMTMC-ReID [21] test sets. Note that we do not use the training sets of Market-1501 and DukeMTMC-ReID. Since we use self-collected videos that are under completely different environments from Market-1501 and Duke-ReID, there is a large domain gap between our training data (self-collected video) and the test data. Results are presented in Table 5. We make two observations.

First, our method is significantly superior to unsupervised methods BUC [16] and UGA [34]. Both models are trained on our self-collected videos for fair comparison. For BUC we use the public code; For UGA we use our own implementation, and employ the JDE [30] tracker to generate the tracklets. CycAs outperforms UGA and BUC by +13.6%, and +21.0% in rank-1 accuracy on Market. It shows the promising potential of CycAs in real-world applications.

Second, our method is very competitive or slightly superior to supervised models. For example, when trained on Market-1501 and tested on DukeMTMC, the IDE model obtains an mAP of 16.6%. In comparison, we achieve 19.2% mAP on DukeMTMC, which is +2.6% higher. Similarly, our test performance on Market-1501 is 1.9% higher than IDE trained on DukeMTMC. Moreover, our results on Market-1501 and DukeMTMC are consistently higher than IDE trained on CUHK03. In this experiment, the strength of CycAs lies in two aspects: 1) We utilize more training data. 2) We utilize unlabeled data in a more effective way. We also note that IDE is significantly higher than our method when IDE is trained and tested on the same domain. We reasonably think that if our system can be trained in a similar environment to Market-1501 or DukeMTMC (with properly deployed cameras), we would have a much better accuracy with a smaller domain gap.

5 Conclusion

This paper presents CycAs, a self-supervised person ReID approach. We carefully design the pretext task—cycle association—to closely match the objective of re-ID. Objective function relaxations are made to allow end-to-end learning and introduce higher appearance variations in the training data. We train CycAs using public data and self-collected videos, and both settings validate the competitive performance of CycAs.

References

1. Caron, M., Bojanowski, P., Joulin, A., Douze, M.: Deep clustering for unsupervised learning of visual features. In: Ferrari, V., Hebert, M., Sminchisescu, C., Weiss, Y. (eds.) Computer Vision – ECCV 2018. LNCS, vol. 11218, pp. 139–156. Springer, Cham (2018). https://doi.org/10.1007/978-3-030-01264-9_9
2. Deng, J., Dong, W., Socher, R., Li, L.J., Li, K., Fei-Fei, L.: Imagenet: a large-scale hierarchical image database. In: CVPR (2009)
3. Deng, W., Zheng, L., Ye, Q., Kang, G., Yang, Y., Jiao, J.: Image-image domain adaptation with preserved self-similarity and domain-dissimilarity for person re-identification. In: CVPR (2018)
4. Doersch, C., Gupta, A., Efros, A.A.: Unsupervised visual representation learning by context prediction. In: ICCV (2015)
5. Fan, H., Zheng, L., Yan, C., Yang, Y.: Unsupervised person re-identification: Clustering and fine-tuning. ACM Trans. Multimed. Comput. Commun. Appl. (TOMM) 14(4), 83 (2018)
6. Gidaris, S., Singh, P., Komodakis, N.: Unsupervised representation learning by predicting image rotations. In: ICLR (2018)
7. Godard, C., Mac Aodha, O., Firman, M., Brostow, G.J.: Digging into self-supervised monocular depth estimation. In: ICCV (2019)
8. He, K., Zhang, X., Ren, S., Sun, J.: Deep residual learning for image recognition. In: CVPR (2016)
9. Hirzer, M., Beleznai, C., Roth, P.M., Bischof, H.: Person re-identification by descriptive and discriminative classification. In: Heyden, A., Kahl, F. (eds.) SCIA 2011. LNCS, vol. 6688, pp. 91–102. Springer, Heidelberg (2011). https://doi.org/10.1007/978-3-642-21227-7_9
10. Lai, Z., Xie, W.: Self-supervised learning for video correspondence flow. In: BMVC (2019)
11. Leal-Taixé, L., Milan, A., Reid, I., Roth, S., Schindler, K.: Motchallenge 2015: towards a benchmark for multi-target tracking. arXiv preprint arXiv:1504.01942 (2015)
12. Li, J., Wang, J., Tian, Q., Gao, W., Zhang, S.: Global-local temporal representations for video person re-identification. In: ICCV (2019)
13. Li, M., Zhu, X., Gong, S.: Unsupervised person re-identification by deep learning tracklet association. In: Ferrari, V., Hebert, M., Sminchisescu, C., Weiss, Y. (eds.) ECCV 2018. LNCS, vol. 11208, pp. 772–788. Springer, Cham (2018). https://doi.org/10.1007/978-3-030-01225-0_45
14. Li, M., Zhu, X., Gong, S.: Unsupervised tracklet person re-identification. IEEE Trans. Pattern Anal. Mach. Intell. (2019)

15. Li, W., Zhao, R., Xiao, T., Wang, X.: Deepreid: deep filter pairing neural network for person re-identification. In: CVPR (2014)
16. Lin, Y., Dong, X., Zheng, L., Yan, Y., Yang, Y.: A bottom-up clustering approach to unsupervised person re-identification. In: AAAI (2019)
17. Liu, P., Lyu, M., King, I., Xu, J.: Selflow: self-supervised learning of optical flow. In: CVPR (2019)
18. Liu, Z., Wang, D., Lu, H.: Stepwise metric promotion for unsupervised video person re-identification. In: ICCV (2017)
19. Noroozi, M., Favaro, P.: Unsupervised learning of visual representations by solving Jigsaw puzzles. In: Leibe, B., Matas, J., Sebe, N., Welling, M. (eds.) ECCV 2016. LNCS, vol. 9910, pp. 69–84. Springer, Cham (2016). https://doi.org/10.1007/978-3-319-46466-4_5
20. Pillai, S., Ambruş, R., Gaidon, A.: Superdepth: self-supervised, super-resolved monocular depth estimation. In: ICRA (2019)
21. Ristani, E., Solera, F., Zou, R., Cucchiara, R., Tomasi, C.: Performance measures and a data set for multi-target, multi-camera tracking. In: Hua, G., Jégou, H. (eds.) ECCV 2016. LNCS, vol. 9914, pp. 17–35. Springer, Cham (2016). https://doi.org/10.1007/978-3-319-48881-3_2
22. Schroff, F., Kalenichenko, D., Philbin, J.: Facenet: a unified embedding for face recognition and clustering. In: CVPR (2015)
23. Sun, Y., Zheng, L., Yang, Y., Tian, Q., Wang, S.: Beyond part models: person retrieval with refined part pooling (and a strong convolutional baseline). In: Ferrari, V., Hebert, M., Sminchisescu, C., Weiss, Y. (eds.) ECCV 2018. LNCS, vol. 11208, pp. 501–518. Springer, Cham (2018). https://doi.org/10.1007/978-3-030-01225-0_30
24. Van Der Maaten, L.: Accelerating t-SNE using tree-based algorithms. J. Mach. Learn. Res. 15(1), 3221–3245 (2014)
25. Wang, J., Zhu, X., Gong, S., Li, W.: Transferable joint attribute-identity deep learning for unsupervised person re-identification. In: CVPR (2018)
26. Wang, N., Song, Y., Ma, C., Zhou, W., Liu, W., Li, H.: Unsupervised deep tracking. In: CVPR (2019)
27. Wang, T., Gong, S., Zhu, X., Wang, S.: Person re-identification by video ranking. In: Fleet, D., Pajdla, T., Schiele, B., Tuytelaars, T. (eds.) ECCV 2014. LNCS, vol. 8692, pp. 688–703. Springer, Cham (2014). https://doi.org/10.1007/978-3-319-10593-2_45
28. Wang, X., He, K., Gupta, A.: Transitive invariance for self-supervised visual representation learning. In: ICCV (2017)
29. Wang, X., Jabri, A., Efros, A.A.: Learning correspondence from the cycle-consistency of time. In: CVPR (2019)
30. Wang, Z., Zheng, L., Liu, Y., Wang, S.: Towards real-time multi-object tracking. arXiv preprint arXiv:1909.12605 (2019)
31. Wei, L., Zhang, S., Gao, W., Tian, Q.: Person transfer GAN to bridge domain gap for person re-identification. In: CVPR (2018)
32. Wojke, N., Bewley, A., Paulus, D.: Simple online and realtime tracking with a deep association metric. In: ICIP (2017)
33. Wu, J., Liao, S., Wang, X., Yang, Y., Li, S.Z., et al.: Clustering and dynamic sampling based unsupervised domain adaptation for person re-identification. In: ICME (2019)
34. Wu, J., Yang, Y., Liu, H., Liao, S., Lei, Z., Li, S.Z.: Unsupervised graph association for person re-identification. In: ICCV (2019)

35. Ye, M., Lan, X., Yuen, P.C.: Robust anchor embedding for unsupervised video person re-identification in the wild. In: Ferrari, V., Hebert, M., Sminchisescu, C., Weiss, Y. (eds.) ECCV 2018. LNCS, vol. 11211, pp. 176–193. Springer, Cham (2018). https://doi.org/10.1007/978-3-030-01234-2_11
36. Ye, M., Ma, A.J., Zheng, L., Li, J., Yuen, P.C.: Dynamic label graph matching for unsupervised video re-identification. In: ICCV (2017)
37. Ye, M., Johns, E., Handa, A., Zhang, L., Pratt, P., Yang, G.Z.: Self-supervised siamese learning on stereo image pairs for depth estimation in robotic surgery. arXiv preprint arXiv:1705.08260 (2017)
38. Yu, F., Li, W., Li, Q., Liu, Yu., Shi, X., Yan, J.: POI: multiple object tracking with high performance detection and appearance feature. In: Hua, G., Jégou, H. (eds.) ECCV 2016. LNCS, vol. 9914, pp. 36–42. Springer, Cham (2016). https://doi.org/10.1007/978-3-319-48881-3_3
39. Yu, H.X., Wu, A., Zheng, W.S.: Cross-view asymmetric metric learning for unsupervised person re-identification. In: ICCV, pp. 994–1002 (2017)
40. Zhang, R., Isola, P., Efros, A.A.: Colorful image colorization. In: Leibe, B., Matas, J., Sebe, N., Welling, M. (eds.) ECCV 2016. LNCS, vol. 9907, pp. 649–666. Springer, Cham (2016). https://doi.org/10.1007/978-3-319-46487-9_40
41. Zheng, L., et al.: MARS: a video benchmark for large-scale person re-identification. In: Leibe, B., Matas, J., Sebe, N., Welling, M. (eds.) ECCV 2016. LNCS, vol. 9910, pp. 868–884. Springer, Cham (2016). https://doi.org/10.1007/978-3-319-46466-4_52
42. Zheng, L., Shen, L., Tian, L., Wang, S., Wang, J., Tian, Q.: Scalable person re-identification: a benchmark. In: CVPR (2015)
43. Zhong, Z., Zheng, L., Li, S., Yang, Y.: Generalizing a person retrieval model hetero- and homogeneously. In: Ferrari, V., Hebert, M., Sminchisescu, C., Weiss, Y. (eds.) ECCV 2018. LNCS, vol. 11217, pp. 176–192. Springer, Cham (2018). https://doi.org/10.1007/978-3-030-01261-8_11
44. Zhong, Z., Zheng, L., Luo, Z., Li, S., Yang, Y.: Invariance matters: exemplar memory for domain adaptive person re-identification. In: CVPR (2019)
45. Zhu, A.Z., Yuan, L., Chaney, K., Daniilidis, K.: EV-FlowNet: self-supervised optical flow estimation for event-based cameras. arXiv preprint arXiv:1802.06898 (2018)

Open-Edit: Open-Domain Image Manipulation with Open-Vocabulary Instructions

Xihui Liu[1]([✉])[iD], Zhe Lin[2][iD], Jianming Zhang[2], Handong Zhao[2], Quan Tran[2], Xiaogang Wang[1], and Hongsheng Li[1][iD]

[1] The Chinese University of Hong Kong, Hong Kong, China
{xihuiliu,xgwang,hsli}@ee.cuhk.edu.hk
[2] Adobe Research, San Jose, USA
{zlin,jianmzha,hazhao,qtran}@adobe.com

Abstract. We propose a novel algorithm, named Open-Edit, which is the first attempt on open-domain image manipulation with open-vocabulary instructions. It is a challenging task considering the large variation of image domains and the lack of training supervision. Our approach takes advantage of the unified visual-semantic embedding space pretrained on a general image-caption dataset, and manipulates the embedded visual features by applying text-guided vector arithmetic on the image feature maps. A structure-preserving image decoder then generates the manipulated images from the manipulated feature maps. We further propose an on-the-fly sample-specific optimization approach with cycle-consistency constraints to regularize the manipulated images and force them to preserve details of the source images. Our approach shows promising results in manipulating open-vocabulary color, texture, and high-level attributes for various scenarios of open-domain images (Code is released at https://github.com/xh-liu/Open-Edit).

1   Introduction

Automatic image editing, aiming at manipulating images based on the user instructions, is a challenging problem with extensive applications. It helps users to edit photographs and create art works with higher efficiency.

Several directions have been explored towards image editing with generative models. Image-to-image translation [12,19,49] translates an image from a source domain to a target domain. But it is restricted to the predefined domains, and cannot be generalized to manipulating images with arbitrary instructions. GAN Dissection [5] and GANPaint [4] are able to add or remove certain objects by manipulating related units in the latent space. However, they are limited to editing a small number of pre-defined objects and stuff that can be identified by semantic segmentation and can be disentangled in the latent space.

X. Liu—This work was done during Xihui Liu's internship at Adobe.

A. Vedaldi et al. (Eds.): ECCV 2020, LNCS 12356, pp. 89–106, 2020.
https://doi.org/10.1007/978-3-030-58621-8_6

Fig. 1. Examples of open-edit. The editing instruction (top), source image (left), and manipulated image (right) are shown for each example. Our approach edits open-vocabulary color, texture, and semantic attributes of open-domain images.

Most relevant to our problem setting is language-based image editing [13, 16,28,31,51]. Some previous work [8,11,14] annotates the manipulation instructions and ground-truth manipulated images for limited images and scenarios. But it is infeasible to obtain such annotations for large-scale datasets. To avoid using ground-truth manipulated images, other work [13,16,28,31,51] only use images and caption annotations as training data. Given an image A and a mismatched caption B, the model is required to edit A to match B. The manipulated images are encouraged to be realistic and to match the manipulation instructions, without requiring ground-truth manipulated images as training supervision. However, it is assumed that any randomly sampled caption is a feasible manipulation instruction for the image. For example, given an image of a red flower, we can use "a yellow flower" as the manipulation instruction. But it is meaningless to use "a blue bird" as the manipulation instruction for the image of a red flower. So this approach is restricted to datasets from a specific domain (*e.g.*, flowers or birds in previous work [30]) with human-annotated fine-grained descriptions for each image, and cannot generalize to open-domain images.

In this work, we aim to manipulate open-domain images by open-vocabulary instructions with minimal supervision, which is a challenging task and has not been explored in previous work. We propose *Open-Edit*, which manipulates the visual feature maps of source images based on the open-vocabulary instructions, and generates the manipulated images from the manipulated visual feature maps. It takes advantages of the universal visual-semantic embedding pretrained on a large-scale image-caption dataset, Conceptual Captions [35]. The visual-semantic embedding model encodes any open-domain images and open-vocabulary instructions into a joint embedding space. Features within the joint embedding space can be used for localizing instruction-related regions of the input images and for manipulating the related visual features. The manipulations are performed by vector arithmetic operations between the visual feature maps and the textual features, *e.g.*, visual embedding of green apple = visual embedding of red apple - textual embedding of "*red apple*" + textual embedding of "*green apple*". Then a structure-preserving image decoder generates the manipulated images based on the manipulated visual feature maps. The image generator is trained with image reconstruction supervision and does not require any paired manipulation instruction for training. So our approach naturally handles open-vocabulary open-domain image manipulations with minimal supervision.

Moreover, to better preserve details and regularize the manipulated images, we introduce *sample-specific optimization* to optimize the image decoder with the specific input image and manipulation instruction. Since we cannot apply direct supervisions on the manipulated images, we adopt reconstruction and cycle-consistency constraints to optimize the small perturbations added to the inter-mediate decoder layers. The reconstruction constraint forces the image decoder to reconstruct the source images from their visual feature maps; The cycle-consistency constraint performs a cycle manipulation (*e.g.*, red apple → green apple → red apple) and forces the final image to be similar to the original ones.

Our proposed framework, Open-Edit, is the first attempt for open-domain image manipulation with open-vocabulary instructions, with several unique advantages: (1) Unlike previous approaches that require single-domain images and fine-grained human-annotated descriptions, we only use noisy image-captions pairs harvested from the web for training. Results in Fig. 1 demonstrates that our model is able to manipulate open-vocabulary colors, textures, and semantic attributes of open-domain images. (2) By controlling the coefficients of the vector arithmetic operation, we can smoothly control the manipulation strength and achieve visual appearances with interpolated attributes. (3) The sample-specific optimization with cycle-consistency constraints further regularizes the manipulated images and preserves details of the source images. Our results achieve better visual quality than previous language-based image editing approaches.

2 Related Work

Image Manipulation with Generative Models. Zhu *et al.* [48] to defines coloring, sketching, and warping brush as editing operations and used constrained optimization to update images. Similarly, Andrew *et al.* [6] proposes Introspective Adversarial Network (IAN) which optimizes the latent space to generate manipulated images according to the input images and user brush inputs. GAN-Paint [4] manipulates the latent space of the input image guided by GAN Dissection [5], which relies on a segmentation model to identify latent units related to specific objects. This approach therefore is mainly suitable for adding or removing specific types of objects from images. Another line of work focuses on face or fashion attribute manipulation with predefined attributes and labeled images on face or fashion datasets [1,10,33,36,37,39,43]. In contrast, our approach aims to handle open-vocabulary image manipulation on arbitrary colors, textures, and high-level attributes without attribute annotations for training.

Language-Based Image Editing. The interaction between language and vision has been studied for various applications [7,25,26,40,42]. Language-based image editing enables user-friendly control for image editing by free-form sentences or phrases as the manipulation instructions. [8,11,14] collects paired data (*i.e.*original images, manipulation queries, and images after manipulation) for training. However, collecting such data is time-consuming and infeasible for most editing scenarios. Other works [13,16,28,31,45,51] only require image-caption pairs for training, but those methods are restricted to specific image domains with fine-grained descriptions such as flowers or birds. Our work extends the problem setting to open-domain images. Moreover, our approach does not rely on fine-grained accurate captions. Instead, we use Conceptual Captions dataset, where the images and captions are harvested from the web. Concurrent work [23] conducts language-based image editing on COCO dataset. But it takes a trade-off between reconstructing and editing, restricting the model from achieving both high-quality images and effective editing at the same time.

Image-to-Image Translation. Supervised image-to-image translation [9,19, 27,32,41] translates images between different domains with paired training data. [21,24,34,38,44,49] focus on unsupervised translation with unpaired training data. Consequent works focus on multi-domain [12] or multi-modal [2,18,22,50]. However, one have to define domains and collect domain-specific images for image-to-image translation, which is not able to tackle arbitrary manipulation instructions. On the contrary, our approach performs open-vocabulary image manipulation without defining domains and collecting domain-specific images.

3   Method

Our goal of open-vocabulary open-domain image manipulation is to edit an arbitrary image based on an open-vocabulary manipulation instruction. The manipulation instructions should indicate the source objects or attributes to be edited as well as the target objects or attributes to be added. For example, the manipulation instruction could be "red apple → green apple".

There are several challenges for open-vocabulary open-domain image manipulation: (1) It is difficult to obtain a plausible manipulation instruction for each training image. And it is infeasible to collect large-scale ground-truth manipulated images for fully supervised training. (2) The open-domain images are of high variations, compared with previous work which only consider single-domain images like flowers or birds. (3) The manipulated images may fail to preserve all details of the source images. Previous work on language-guided image editing uses other images' captions as the manipulation instruction for an image to train the model. However, it assumes that all images are from the same domain, while cannot handle open-domain images, *e.g.*, a caption for a flower image cannot be used as the manipulation instruction for a bird image.

To achieve open-vocabulary open-domain image manipulation, we propose a simple but effective pipeline, named Open-Edit, as shown in Fig. 2. It exploits the visual-semantic joint embedding space to manipulate visual features by textual features, and then decodes the images from the manipulated feature maps. It is composed of visual-semantic embedding, text-guided visual feature manipulation, structure-preserving image decoding, and sample-specific optimization.

There are two stages for training. In the first stage, we pretrain the *visual-semantic embedding* (VSE) model on a large-scale image-caption dataset to embed any images and texts into latent codes in the visual-semantic embedding space (Fig. 2(a)). Once trained, the VSE model is fixed to provide image

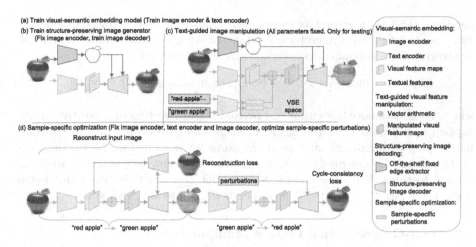

Fig. 2. The pipeline of our Open-Edit framework. (a) and (b) show the training process. (c) and (d) illustrate the testing process. To simplify the demonstration, edge extractor and text encoder are omitted in (d).

and text embeddings. In the second stage, the *structure-preserving image decoder* is trained to reconstruct the images from the visual feature maps encoded by the VSE model, as shown in Fig. 2(b). The whole training process only requires the images and noisy captions harvested from the web, and does not need any human annotated manipulation instructions or ground-truth manipulated images.

During inference (Fig. 2(c)), the visual-semantic embedding model encodes the input images and manipulation instructions into visual feature maps and textual features in the joint embedding space. Then *text-guided visual feature manipulation* is performed to ground the manipulation instructions on the visual feature maps and manipulate the corresponding regions of the visual feature maps with the provided textual features. Next, the structure-preserving image decoder generates the manipulated images from the manipulated feature maps.

Furthermore, in order to regularize the manipulated images and preserve details of the source images, we introduce small sample-specific perturbations added to the intermediate layers of the image decoder, and propose a *sample-specific optimization* approach to optimize the perturbations based on the input image and instruction, shown in Fig. 2(d). For a specific image and manipulation instruction, we put constraint on both the reconstructed images and the images generated by a pair of cycle manipulations (*e.g.*, red apple → green apple → red apple). In this way, we adapt the image generator to the specific input image and instruction and achieve higher quality image manipulations.

3.1 A Revisit of Visual-Semantic Embedding

To handle open-vocabulary instructions and open-domain images, we use a large-scale image-caption dataset to learn a universal visual-semantic embedding space. Convolutional neural networks (CNN) and long short-term memory networks (LSTM) are used as encoders to transform images and captions into visual and textual feature vectors. A triplet ranking loss with hardest negatives, as shown below, is applied to train the visual and textual encoders [15].

$$\mathcal{L}(\mathbf{v}, \mathbf{t}) = \max_{\hat{\mathbf{t}}}[m + \langle \mathbf{v}, \hat{\mathbf{t}} \rangle - \langle \mathbf{v}, \mathbf{t} \rangle]_+ + \max_{\hat{\mathbf{v}}}[m + \langle \hat{\mathbf{v}}, \mathbf{t} \rangle - \langle \mathbf{v}, \mathbf{t} \rangle]_+ \qquad (1)$$

where $\mathbf{v}$ and $\mathbf{t}$ denote the visual and textual feature vectors of a positive image-caption pair. $\hat{\mathbf{v}}$ and $\hat{\mathbf{t}}$ are the negative image and caption features in the mini-batch. $[x]_+ = \max(0, x)$, and m is the constant margin for the ranking loss. $\langle \mathbf{v}, \mathbf{t} \rangle$ denotes the dot product to measure the similarity between the visual and textual features. With the trained VSE model, the visual feature maps before average pooling $\mathbf{V} \in \mathbb{R}^{1024 \times 7 \times 7}$ is also embedded into the VSE space.

3.2 Text-Guided Visual Feature Manipulation

The universal visual-semantic embedding space enables us to manipulate the visual feature maps with the text instructions by vector arithmetic operations, similar to that of word embeddings (*e.g.*, "king" - "man" + "woman" = "queen") [29]. When manipulating certain objects or attributes, we would like to only modify specific regions while keeping other regions unchanged. So instead of editing the global visual feature vector, we conduct vector arithmetic operations between the visual feature maps $\mathbf{V} \in \mathbb{R}^{1024 \times 7 \times 7}$ and textual feature vectors.

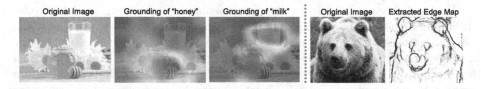

Fig. 3. Left: an example of grounding results by visual-semantic embedding. Right: an example of edge map extracted by off-the-shelf edge detector.

We first identify which regions in the feature map to manipulate, *i.e.*, ground the manipulation instructions on the spatial feature map. The VSE model provides us a soft grounding for textual queries by a weighted summation of the image feature maps, similar to class activation maps (CAM) [47]. We use the textual feature vector $\mathbf{t} \in \mathbb{R}^{1024 \times 1}$ as weights to compute the weighted summation of the image feature maps $\mathbf{g} = \mathbf{t}^\top \mathbf{V}$. This scheme gives us a soft grounding map $\mathbf{g} \in \mathbb{R}^{7 \times 7}$, which is able to roughly localize corresponding regions in the visual feature maps related to the textual instruction. Examples of the text-guided soft grounding results are shown in Fig. 3 (left). We adopt the grounding map as location-adaptive coefficients to control the manipulation strength at different locations. We further adopt a coefficient α to control the global manipulation strength, which enables continuous transitions between source images and the manipulated ones. The visual feature vector at spatial location (i, j) (where $i, j \in \{0, 1, ...6\}$) in the visual feature map $\mathbf{V} \in \mathbb{R}^{1024 \times 7 \times 7}$, is denoted as $\mathbf{v}^{i,j} \in \mathbb{R}^{1024}$. We define the following types of manipulations by vector arithmetics weighted by the soft grounding map and the coefficient α.

Changing Attributes. Changing object attributes or global attributes is one of the most common manipulations. The textual feature embeddings of the source and target concepts are denoted as $\mathbf{t}_1$ and $\mathbf{t}_2$. respectively. For example, if we want to change a "red apple" into a "green apple", $\mathbf{t}_1$ would be the textual embedding of phrase "red apple" and $\mathbf{t}_2$ would be the embedding of phrase "green apple". The manipulation of image feature vector $\mathbf{v}^{i,j}$ at location (i, j) is,

$$\mathbf{v}_m^{i,j} = \mathbf{v}^{i,j} - \alpha \langle \mathbf{v}^{i,j}, \mathbf{t}_1 \rangle \mathbf{t}_1 + \alpha \langle \mathbf{v}^{i,j}, \mathbf{t}_1 \rangle \mathbf{t}_2, \tag{2}$$

where $i, j \in \{0, 1, ...6\}$, and $\mathbf{v}_m^{i,j}$ is the manipulated visual feature vector at location (i, j) of the 7×7 feature map. We remove the source features $\mathbf{t}_1$ and add the target features $\mathbf{t}_2$ to each visual feature vector $\mathbf{v}^{i,j}$. $\langle \mathbf{v}^{i,j}, \mathbf{t}_1 \rangle$ is the value of the soft grounding map at location (i, j), calculated as the dot product of the image feature vector and the source textual features. We can also interpret the dot product as the projection of the visual embedding $\mathbf{v}^{i,j}$ onto the direction of the textual embedding $\mathbf{t}_1$. It serves as a location-adaptive manipulation strength to control which regions in the image should be edited. α is a hyper-parameter that controls the image-level manipulation strength. By smoothly increasing α, we can achieve smooth transitions from source to target attributes.

Removing Concepts. In certain scenarios, objects, stuff or attributes need to be removed, *e.g.*, remove the beard from a face. Denote the semantic embedding of the concept we would like to remove as $\mathbf{t}$. The removing operation is

$$\mathbf{v}_m^{i,j} = \mathbf{v}^{i,j} - \alpha \langle \mathbf{v}^{i,j}, \mathbf{t} \rangle \mathbf{t}. \tag{3}$$

Relative Attributes. Our framework also handles relative attribute manipulation, such as making a red apple less red or tuning the image to be brighter. Denote the semantic embedding of the relative attribute as $\mathbf{t}$. The strength of the relative attribute is controlled by the hyper-parameter α. By smoothly adjusting α, we can gradually strengthen or weaken the relative attribute as

$$\mathbf{v}_m^{i,j} = \mathbf{v}^{i,j} \pm \alpha \langle \mathbf{v}^{i,j}, \mathbf{t} \rangle \mathbf{t}. \tag{4}$$

3.3   Structure-Preserving Image Decoding

After deriving the manipulated feature map $\mathbf{V}_m \in \mathbb{R}^{1024 \times 7 \times 7}$, an image decoder takes $\mathbf{V}_m$ as input and generates the manipulated images.

Since we do not have paired data for training and it is difficult to generate plausible manipulation instructions for each image, we train the image decoder with only the reconstruction supervisions, as shown in Fig. 2(b). Specifically, we fix the VSE model to transform an image $\mathbf{I}$ into the feature maps $\mathbf{V}$ in the joint embedding space, and train a generative adversarial network to reconstruct the input image from $\mathbf{V}$. The generator is trained with the hinge-based adversarial loss, discriminator feature matching loss, and perceptual loss.

$$\mathcal{L}_G = -\mathbb{E}[D(G(\mathbf{V}))] + \lambda_{VGG}\mathbb{E}[\sum_{k=1}^{N} \frac{1}{n_k}||F_k(G(\mathbf{V})) - F_k(\mathbf{I})||_1]$$

$$+ \lambda_{FM}\mathbb{E}[\sum_{k=1}^{N} \frac{1}{m_k}||D_k(G(\mathbf{V})) - D_k(\mathbf{I})||_1],$$

$$\mathcal{L}_D = -\mathbb{E}[\min(0, -1 + D(\mathbf{I}))] - \mathbb{E}[min(0, -1 - D(G(\mathbf{V})))], \quad (5)$$

where $\mathcal{L}_D$ and the first term of $\mathcal{L}_G$ are the hinge-based adversarial loss. The second term of $\mathcal{L}_G$ is the perceptual loss, calculated as the VGG feature distance between the reconstructed image and the input image. The third term of $\mathcal{L}_G$ is the discriminator feature matching loss, which matches the intermediate features of the discriminator between the reconstructed image and the input image. n_k and m_k are the number of elements in the k-th layer of the VGG network and discriminator, respectively. λ_{VGG} and λ_{FM} are the loss weights. Although not being trained on manipulated feature maps, the image decoder learns a general image prior. So during inference, the decoder is able to generate manipulated images when given the manipulated feature maps as input.

Furthermore, we incorporate edge constraints into the image decoder to preserve the structure information when editing image appearances. We adopt an off-the-shelf CNN edge detector [17] to extract edges from the input images. The extracted edges, as shown in Fig. 3 (right), are fed into intermediate layers of the image decoder by spatially-adaptive normalization [32]. Specifically, we use the edge maps to predict the spatially-adaptive scale and bias parameters of batch-normalization layers. We denote the edge map as $\mathcal{E}$. Denote the feature map value of the n-th image in the mini-batch at channel c and location (h, w) as $f_{n,c,h,w}$. Denote the mean and standard deviation of the feature maps at channel c as μ_c and σ_c, respectively. The spatially-adaptive normalization is

$$\gamma_{c,h,w}(\mathcal{E})\frac{f_{n,c,h,w} - \mu_c}{\sigma_c} + \beta_{c,h,w}(\mathcal{E}), \quad (6)$$

where γ and β are two-layer convolutions to predict spatially-adaptive scale and bias for BN layers. With the edge constraints, the decoder is able to preserve the structures and edges of the source images when editing the image appearances.

3.4   Sample-Specific Optimization with Cycle-Consistency Constraints

The vector arithmetic manipulation operations may not be precise enough, because some attributes might be entangled and the visual-semantic embedding space may not be strictly linear. Moreover, the image decoder trained with only reconstruction supervision is not perfect and might not be able to reconstruct all details of the source image. To mitigate those problems, we adopt a sample-specific optimization approach to adapt the decoder to the specific input image and manipulation instruction.

For each image and manipulation instruction (*e.g.*, "red apple" → "green apple"), we apply a pair of cycle manipulations to exchange the attributes forth and back (*e.g.*, "red apple" → "green apple" → "red apple"). The corresponding source and manipulated images are denoted as $\mathbf{I} \rightarrow \mathbf{I}_m \rightarrow \mathbf{I}_c$. We incorporate a cycle-consistency loss to optimize the decoder to adapt to the specific image and manipulation instruction. In this way, we can regularize the manipulated image and complete the details missed during encoding and generating. We also adopt a reconstruction loss to force the optimized decoder to reconstruct the source image without manipulating the latent visual features. The reconstruction loss $\mathcal{L}_{rec}$ and cycle-consistency loss $\mathcal{L}_{cyc}$ are the summation of L_1 loss and perceptual loss, computed between the source image $\mathbf{I}$ and the reconstructed $\mathbf{I}_r$ or the cycle manipulated image $\mathbf{I}_c$,

$$\mathcal{L}_{cyc} = ||\mathbf{I}_c - \mathbf{I}||_1 + \lambda \sum_{k=1}^{N} \frac{1}{n_k} ||F_k(\mathbf{I}_c) - F_k(\mathbf{I})||_1, \tag{7}$$

$$\mathcal{L}_{rec} = ||\mathbf{I}_r - \mathbf{I}||_1 + \lambda \sum_{k=1}^{N} \frac{1}{n_k} ||F_k(\mathbf{I}_r) - F_k(\mathbf{I})||_1, \tag{8}$$

where λ is the loss weight for perceptual loss and F_k is the k-th layer of the VGG network with n_k features.

However, directly finetuning the decoder parameters for a specific image and manipulation instruction would cause severe overfitting to the source image, and the finetuned decoder would not be able to generate the high-quality manipulated image. Alternatively, we fix the decoder parameters and only optimize a series of additive perturbations of the decoder network, as shown in Fig. 2(d). For each specific image and manipulation, the sample-specific perturbations are initialized as zeros and added to the intermediate layers of the decoder. The perturbation parameters are optimized with the manipulation cycle-consistency loss and reconstruction loss on that specific image and manipulation instruction. So when generating the manipulated images, the optimized perturbations can complete the high-frequency details of the source images, and regularize the manipulated images. Specifically, the image decoder is divided into several decoder blocks $G_1, G_2, \cdots, G_n$ ($n = 4$ in our implementation), and the perturbations are added to the decoder between the n blocks,

$$G'(\mathbf{V}) = G_n(G_{n-1}(\cdots (G_1(\mathbf{V}) + \mathbf{P_1}) \cdots) + \mathbf{P_{n-1}}), \tag{9}$$

where $\mathbf{P_1}, \cdots, \mathbf{P_{n-1}}$ are the introduced perturbations. We optimize the perturbations by the summation of reconstructions loss, manipulation cycle-consistency loss, and a regularization loss $\mathcal{L}_{reg} = \sum_{i=1}^{n-1} ||\mathbf{P}_i||_2^2$.

Those optimization steps are conducted only during testing. We adapt the perturbations to the input image and manipulation instruction by the introduced optimization process. Therefore, the learned sample-specific perturbations models high-frequency details of the source images, and regularizes the manipulated images. In this way, the generator with optimized perturbations is able to generate photo-realistic and detail-preserving manipulated images.

4   Experiments

4.1   Datasets and Implementation Details

Our visual-semantic embedding model (including image encoder and text encoder) and image decoder are trained on Conceptual Captions dataset [35] with 3 million image-caption pairs harvested from the web. The images are from various domains and of various styles, including portrait, objects, scenes, and others. Instead of human-annotated fine-grained descriptions in other image captioning datasets, the captions of Conceptual Captions dataset are harvested from the Alt-text HTML attribute associated with web images. Although the images are of high variations and the captions are noisy, results show that with large datasets, the VSE model is able to learn an effective visual-semantic embedding space for image manipulation. The image decoder trained with images from Conceptual Captions dataset learns a general image prior for open-domain images.

The model structure and training process of the VSE model follow that of VSE++ [15]. The image decoder takes $1024 \times 7 \times 7$ feature maps as input, and is composed of 7 ResNet Blocks with upsampling layers in between, which generates 256×256 images. The discriminator is a Multi-scale Patch-based discriminator following [32]. The decoder is trained with GAN loss, perceptual loss, and discriminator feature matching loss. The edge extractor is an off-the-shelf bi-directional cascade network [17] trained on BSDS500 dataset [3].

4.2   Applications and Results

Our approach can achieve open-domain image manipulation with open-vocabulary instructions, which has various applications. We demonstrate several examples in Fig. 4, including changing color, texture, and global or local high-level attributes.

Results in the first row demonstrate that our model is able to change *color* for objects and stuff while preserving other details of the image. Moreover, it preserves the lighting conditions and relative color strengths very well when changing colors. Our model is also able to change *textures* of the images with language instructions, for example, editing object materials or changing sea to grass, as shown in the second row. Results indicate that the VSE model learns effective texture features in the joint embedding space, and that the generator is able to generate reasonable textures based on the manipulated features. Besides low-level attributes, our model is also able to handle *high-level semantic attributes*, such as removing lights, changing photos to paintings, sunny to cloudy, and transferring seasons, in the third and fourth rows.

Quantitative Evaluation. Since ground-truth manipulated images are not available, we conduct evaluations by user study, L2 error, and LPIPS.

Fig. 4. Applications and manipulation results of our method.

The user study is conducted to evaluate human perceptions of our approach. For each experiment, we randomly pick 60 images and manually choose the appropriate manipulation instructions for them. The images cover a wide variety of styles and the instructions range from color and texture manipulation to high-level attribute manipulation. 10 users are asked to score the 60 images for each experiment by three criteria, (1) visual quality, (2) how well the manipulated images preserve the details of the source image, and (3) how well the manipulation results match the instruction. The scores range from 1 (worst) to 5 (best), and we will analyze the results shown in Table 1 in the following.

To evaluate the visual quality and content preservation, we calculate the L2 error and Perceptual similarity (LPIPS) [46] between the reconstructed images and input images, as shown in Table 2 and analyzed in the following.

Comparison with Previous Work. Since this is the first work to explore open-domain image manipulation with open-vocabulary instructions, our problem setting is much more challenging than previous approaches. Nonetheless, we compare with two representative approaches, CycleGAN [49] and TAGAN [31].

CycleGAN is designed for image-to-image translation, but we have to define domains and collect domain-specific images for training. So it is not able to tackle open-vocabulary instructions. To compare with CycleGAN, we train three Cycle-GAN models for translating between blue and red objects, translating between red and green objects, and translating between beach and sea, respectively. The images for training CycleGAN are retrieved from Conceptual Captions with our visual-semantic embedding model. Qualitative comparison are shown in Fig. 5, and user study are shown in Table 1. Results indicate that both our approach

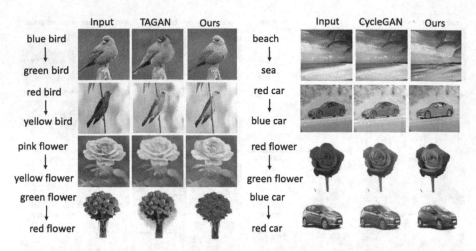

Fig. 5. Comparison between with previous language-based image editing method TAGAN [31] (left) and image-to-image translation method CycleGAN [49] (right).

and CycleGAN is able to preserve details of the input images very well, but CycleGAN worse at transferring desired attributes in some scenarios.

State-of-the-art language-based image manipulation method TAGAN [31] uses mismatched image-caption pairs as training samples for manipulation. It is able to handle language instructions, but is limited to only one specific domain such as flowers (Oxford-102) or birds (CUB). It also requires fine-grained human-annotated descriptions of each image in the dataset. While our approach handles open-domain images with noisy captions harvested from the web for training. Since TAGAN only has the models for manipulating bird or flower images, we compare our results with theirs on flower and bird image manipulation in Fig. 5. We also compare user evaluation in Table 1. Quantitative evaluation of L2 error and perceptual metric (LPIPS) between reconstructed images and original images are shown in Table 2. Our model is not trained on the Oxford-102 or CUB datasets, but still performs better than the TAGAN models specifically trained on those datasets. Moreover, the L2 reconstruction error also shows that our model has the potential to preserve the detailed contents of the source images.

4.3   Component Analysis

The Effectiveness of Instruction Grounding. Our text-guided visual feature manipulation module uses the soft instruction grounding maps as location-adaptive manipulation coefficients to control the local manipulation strength at different locations. The instruction grounding map is very important when the manipulation instruction is related to local areas or objects in the image. Figure 6(a) demonstrates the effectiveness of adopting grounding for manipulation, where we aim to change the green apple into a yellow apple and keep the

Table 1. User study results on visual quality, how well the manipulated images preserve details, and how well the manipulated images match the instruction. In the table, "edge" represents edge constraints in the image decoder, and "opt" represents the sample-specific optimization.

	CycleGAN [49]	TAGAN [31]	w/o edge, w/o opt	w/ edge, w/o opt	**w/ edge, w/ opt**
Visual quality	4.0	3.1	1.3	4.1	**4.4**
Preserve details	4.2	2.7	1.2	3.7	**4.3**
Match instruction	1.9	4.2	4.0	4.5	**4.5**

Table 2. L2 error and LPIPS between reconstructed and original images of TAGAN and ablations of our approach. Lower L2 reconstruction error and LPIPS metric indicates that the reconstructed images preserve details of the source images better.

	TAGAN [31]	w/o edge, w/o opt	w/ edge, w/o opt	**w/ edge, w/ opt**
L2 error on Oxford-102 test set	0.11	0.19	0.10	**0.05**
L2 on Conceptual Captions val	N/A	0.20	0.12	**0.07**
LPIPS on Conceptual Captions val	N/A	0.33	0.17	**0.06**

red apple unchanged. The grounding map is able to roughly locate the green apple, and with the help of the soft grounding map, the model is able to change the color of the green apple while keeping the red apple and the background unchanged. On the contrary, the model without grounding changes not only the green apple, but also the red apple and the background.

Edge Constraints. Our structure-aware image decoder exploits edge constraints to preserve the structure information when generating the reconstructed and manipulated images. In Fig. 6(b), we show an example of image reconstruction and manipulation with and without edges. The image decoder is able to reconstruct and generate higher-quality images with clear structures and edges with edge constraints. User study results in Table 1 and quantitative evaluation in Table 2 also indicate that the generated images are of better visual quality and preserve details better with the structure-aware image decoder.

Adjusting Coefficient α for Smooth Attribute Transition. The hyperparameter α controls the global attribute manipulation strength, which can be adjusted according to user requirements. By gradually increasing α, we obtain a smooth transition from the source images to the manipulated images with different manipulation strengths. Figure 6(c)(d) illustrates the smooth transition of an image from dark to bright, and from red apple to green apple, respectively.

The Effectiveness of Sample-Specific Optimization and Cycle-Consistency Constraints. Figure 6(e)[1] demonstrates the effectiveness of sample-

---

[1] The decoder for Fig. 6(e) is trained on FFHQ [20] to learn the face image prior.

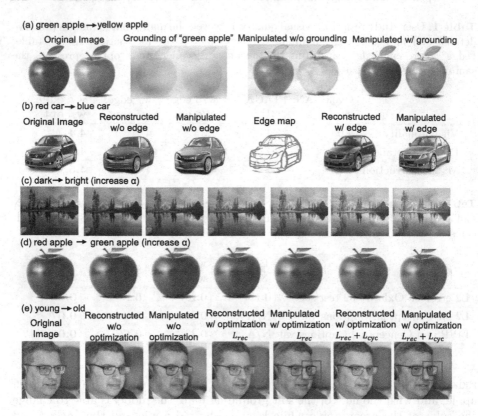

Fig. 6. Component analysis of our approach. The examples from top to bottom show analysis on grounding, edge constraints, adjusting coefficient, and the sample-specific optimization and cycle-consistency constraints, respectively.

specific optimization and cycle-consistency constraints. The reconstructed image and manipulated image without sample-specific optimization miss some details such as the shape of the glasses. With the perturbation optimization by reconstruction loss, our model is able to generate better-quality reconstructed and manipulated images. Optimizing the perturbations with both reconstruction loss and manipulation cycle-consistency loss further improves the quality of the generated images, *e.g.*, the glasses are more realistic and the person identity appearance is better preserved. User study in Table 1 and quantitative evaluation in Table 2 indicate that the sample-specific optimization has the potential of enhancing details of the generated images.

5   Conclusion and Discussions

We propose Open-Edit, the first framework for open-vocabulary open-domain image manipulation with minimal training supervision. It takes advantage of the pretrained visual-semantic embedding, and manipulates visual features by vector arithmetic with textual embeddings in the joint embedding space. The sample-specific optimization further regularizes the manipulated images and encourages realistic and detail-preserving results. Impressive color, texture, and semantic attribute manipulation are shown on various types of images.

We believe that this is a challenging and promising direction towards more general and practical image editing, and that our attempt would inspire future work to enhance editing qualities and extend the application scenario. In this work we focus on editing appearance-related attributes without changing the structure of images. Further work can be done on more challenging structure-related editing and image editing with more complex sentence instructions.

References

1. Ak, K.E., Lim, J.H., Tham, J.Y., Kassim, A.A.: Attribute manipulation generative adversarial networks for fashion images. In: Proceedings of the IEEE International Conference on Computer Vision, pp. 10541–10550 (2019)
2. Almahairi, A., Rajeswar, S., Sordoni, A., Bachman, P., Courville, A.: Augmented cyclegan: learning many-to-many mappings from unpaired data. arXiv preprint arXiv:1802.10151 (2018)
3. Arbelaez, P., Maire, M., Fowlkes, C., Malik, J.: Contour detection and hierarchical image segmentation. IEEE Trans. Pattern Anal. Mach. Intell. **33**(5), 898–916 (2010)
4. Bau, D., et al.: Semantic photo manipulation with a generative image prior. ACM Trans. Graph. (TOG) **38**(4), 59 (2019)
5. Bau, D., et al.: GAN dissection: visualizing and understanding generative adversarial networks. arXiv preprint arXiv:1811.10597 (2018)
6. Brock, A., Lim, T., Ritchie, J.M., Weston, N.: Neural photo editing with introspective adversarial networks. arXiv preprint arXiv:1609.07093 (2016)
7. Chen, D., et al.: Improving deep visual representation for person re-identification by global and local image-language association. In: Ferrari, V., Hebert, M., Sminchisescu, C., Weiss, Y. (eds.) ECCV 2018. LNCS, vol. 11220, pp. 56–73. Springer, Cham (2018). https://doi.org/10.1007/978-3-030-01270-0_4
8. Chen, J., Shen, Y., Gao, J., Liu, J., Liu, X.: Language-based image editing with recurrent attentive models. In: Proceedings of the IEEE Conference on Computer Vision and Pattern Recognition, pp. 8721–8729 (2018)
9. Chen, Q., Koltun, V.: Photographic image synthesis with cascaded refinement networks. In: Proceedings of the IEEE International Conference on Computer Vision, pp. 1511–1520 (2017)
10. Chen, Y.C., et al.: Semantic component decomposition for face attribute manipulation. In: Proceedings of the IEEE Conference on Computer Vision and Pattern Recognition, pp. 9859–9867 (2019)
11. Cheng, Y., Gan, Z., Li, Y., Liu, J., Gao, J.: Sequential attention gan for interactive image editing via dialogue. arXiv preprint arXiv:1812.08352 (2018)
12. Choi, Y., Choi, M., Kim, M., Ha, J.W., Kim, S., Choo, J.: StarGAN: unified generative adversarial networks for multi-domain image-to-image translation. In: Proceedings of the IEEE Conference on Computer Vision and Pattern Recognition, pp. 8789–8797 (2018)

13. Dong, H., Yu, S., Wu, C., Guo, Y.: Semantic image synthesis via adversarial learning. In: Proceedings of the IEEE International Conference on Computer Vision, pp. 5706–5714 (2017)
14. El-Nouby, A., et al.: Keep drawing it: iterative language-based image generation and editing. arXiv preprint arXiv:1811.09845 (2018)
15. Faghri, F., Fleet, D.J., Kiros, J.R., Fidler, S.: VSE++: improving visual-semantic embeddings with hard negatives. arXiv preprint arXiv:1707.05612 (2017)
16. Günel, M., Erdem, E., Erdem, A.: Language guided fashion image manipulation with feature-wise transformations. arXiv preprint arXiv:1808.04000 (2018)
17. He, J., Zhang, S., Yang, M., Shan, Y., Huang, T.: Bi-directional cascade network for perceptual edge detection. In: Proceedings of the IEEE Conference on Computer Vision and Pattern Recognition, pp. 3828–3837 (2019)
18. Huang, X., Liu, M.-Y., Belongie, S., Kautz, J.: Multimodal unsupervised image-to-image translation. In: Ferrari, V., Hebert, M., Sminchisescu, C., Weiss, Y. (eds.) ECCV 2018. LNCS, vol. 11207, pp. 179–196. Springer, Cham (2018). https://doi.org/10.1007/978-3-030-01219-9_11
19. Isola, P., Zhu, J.Y., Zhou, T., Efros, A.A.: Image-to-image translation with conditional adversarial networks. In: Proceedings of the IEEE Conference on Computer Vision and Pattern Recognition, pp. 1125–1134 (2017)
20. Karras, T., Laine, S., Aila, T.: A style-based generator architecture for generative adversarial networks. arXiv preprint arXiv:1812.04948 (2018)
21. Kim, T., Cha, M., Kim, H., Lee, J.K., Kim, J.: Learning to discover cross-domain relations with generative adversarial networks. In: Proceedings of the 34th International Conference on Machine Learning, vol. 70, pp. 1857–1865. JMLR. org (2017)
22. Lee, H.-Y., Tseng, H.-Y., Huang, J.-B., Singh, M., Yang, M.-H.: Diverse image-to-image translation via disentangled representations. In: Ferrari, V., Hebert, M., Sminchisescu, C., Weiss, Y. (eds.) ECCV 2018. LNCS, vol. 11205, pp. 36–52. Springer, Cham (2018). https://doi.org/10.1007/978-3-030-01246-5_3
23. Li, B., Qi, X., Lukasiewicz, T., Torr, P.H.: Manigan: text-guided image manipulation. In: Proceedings of the IEEE/CVF Conference on Computer Vision and Pattern Recognition, pp. 7880–7889 (2020)
24. Liu, M.Y., Tuzel, O.: Coupled generative adversarial networks. In: Advances in Neural Information Processing Systems, pp. 469–477 (2016)
25. Liu, X., Li, H., Shao, J., Chen, D., Wang, X.: Show, tell and discriminate: image captioning by self-retrieval with partially labeled data. In: Ferrari, V., Hebert, M., Sminchisescu, C., Weiss, Y. (eds.) ECCV 2018. LNCS, vol. 11219, pp. 353–369. Springer, Cham (2018). https://doi.org/10.1007/978-3-030-01267-0_21
26. Liu, X., Wang, Z., Shao, J., Wang, X., Li, H.: Improving referring expression grounding with cross-modal attention-guided erasing. In: Proceedings of the IEEE Conference on Computer Vision and Pattern Recognition, pp. 1950–1959 (2019)
27. Liu, X., Yin, G., Shao, J., Wang, X., Li, H.: Learning to predict layout-to-image conditional convolutions for semantic image synthesis. In: Advances in Neural Information Processing Systems, pp. 570–580 (2019)
28. Mao, X., Chen, Y., Li, Y., Xiong, T., He, Y., Xue, H.: Bilinear representation for language-based image editing using conditional generative adversarial networks. In: ICASSP 2019–2019 IEEE International Conference on Acoustics, Speech and Signal Processing (ICASSP), pp. 2047–2051. IEEE (2019)
29. Mikolov, T., Sutskever, I., Chen, K., Corrado, G.S., Dean, J.: Distributed representations of words and phrases and their compositionality. In: Advances in Neural Information Processing Systems, pp. 3111–3119 (2013)

30. Mo, S., Cho, M., Shin, J.: Instagan: instance-aware image-to-image translation. arXiv preprint arXiv:1812.10889 (2018)
31. Nam, S., Kim, Y., Kim, S.J.: Text-adaptive generative adversarial networks: manipulating images with natural language. In: Advances in Neural Information Processing Systems, pp. 42–51 (2018)
32. Park, T., Liu, M.Y., Wang, T.C., Zhu, J.Y.: Semantic image synthesis with spatially-adaptive normalization. arXiv preprint arXiv:1903.07291 (2019)
33. Perarnau, G., Van De Weijer, J., Raducanu, B., Álvarez, J.M.: Invertible conditional GANs for image editing. arXiv preprint arXiv:1611.06355 (2016)
34. Royer, A., et al.: Xgan: unsupervised image-to-image translation for many-to-many mappings. arXiv preprint arXiv:1711.05139 (2017)
35. Sharma, P., Ding, N., Goodman, S., Soricut, R.: Conceptual captions: a cleaned, hypernymed, image alt-text dataset for automatic image captioning. In: Proceedings of ACL (2018)
36. Shen, Y., Gu, J., Tang, X., Zhou, B.: Interpreting the latent space of GANs for semantic face editing. arXiv preprint arXiv:1907.10786 (2019)
37. Shu, Z., Yumer, E., Hadap, S., Sunkavalli, K., Shechtman, E., Samaras, D.: Neural face editing with intrinsic image disentangling. In: Proceedings of the IEEE Conference on Computer Vision and Pattern Recognition, pp. 5541–5550 (2017)
38. Taigman, Y., Polyak, A., Wolf, L.: Unsupervised cross-domain image generation. arXiv preprint arXiv:1611.02200 (2016)
39. Usman, B., Dufour, N., Saenko, K., Bregler, C.: PuppetGAN: cross-domain image manipulation by demonstration. In: Proceedings of the IEEE International Conference on Computer Vision, pp. 9450–9458 (2019)
40. Vinyals, O., Toshev, A., Bengio, S., Erhan, D.: Show and tell: a neural image caption generator. In: Proceedings of the IEEE Conference on Computer Vision and Pattern Recognition, pp. 3156–3164 (2015)
41. Wang, T.C., Liu, M.Y., Zhu, J.Y., Tao, A., Kautz, J., Catanzaro, B.: High-resolution image synthesis and semantic manipulation with conditional GANs. In: Proceedings of the IEEE Conference on Computer Vision and Pattern Recognition, pp. 8798–8807 (2018)
42. Wang, Z., et al.: Camp: cross-modal adaptive message passing for text-image retrieval. In: Proceedings of the IEEE International Conference on Computer Vision, pp. 5764–5773 (2019)
43. Xiao, T., Hong, J., Ma, J.: ELEGANT: exchanging latent encodings with GAN for transferring multiple face attributes. In: Ferrari, V., Hebert, M., Sminchisescu, C., Weiss, Y. (eds.) ECCV 2018. LNCS, vol. 11214, pp. 172–187. Springer, Cham (2018). https://doi.org/10.1007/978-3-030-01249-6_11
44. Yi, Z., Zhang, H., Tan, P., Gong, M.: DualGAN: unsupervised dual learning for image-to-image translation. In: Proceedings of the IEEE International Conference on Computer Vision, pp. 2849–2857 (2017)
45. Yu, X., Chen, Y., Liu, S., Li, T., Li, G.: Multi-mapping image-to-image translation via learning disentanglement. In: Advances in Neural Information Processing Systems, pp. 2994–3004 (2019)
46. Zhang, R., Isola, P., Efros, A.A., Shechtman, E., Wang, O.: The unreasonable effectiveness of deep features as a perceptual metric. In: Proceedings of the IEEE Conference on Computer Vision and Pattern Recognition, pp. 586–595 (2018)
47. Zhou, B., Khosla, A., Lapedriza, A., Oliva, A., Torralba, A.: Learning deep features for discriminative localization. In: Proceedings of the IEEE Conference on Computer Vision and Pattern Recognition, pp. 2921–2929 (2016)

48. Zhu, J.-Y., Krähenbühl, P., Shechtman, E., Efros, A.A.: Generative visual manipulation on the natural image manifold. In: Leibe, B., Matas, J., Sebe, N., Welling, M. (eds.) ECCV 2016. LNCS, vol. 9909, pp. 597–613. Springer, Cham (2016). https://doi.org/10.1007/978-3-319-46454-1_36
49. Zhu, J.Y., Park, T., Isola, P., Efros, A.A.: Unpaired image-to-image translation using cycle-consistent adversarial networks. In: Proceedings of the IEEE International Conference on Computer Vision, pp. 2223–2232 (2017)
50. Zhu, J.Y., et al.: Toward multimodal image-to-image translation. In: Advances in Neural Information Processing Systems, pp. 465–476 (2017)
51. Zhu, S., Urtasun, R., Fidler, S., Lin, D., Change Loy, C.: Be your own prada: fashion synthesis with structural coherence. In: Proceedings of the IEEE International Conference on Computer Vision, pp. 1680–1688 (2017)

Towards Real-Time Multi-Object Tracking

Zhongdao Wang[1], Liang Zheng[2], Yixuan Liu[1], Yali Li[1], and Shengjin Wang[1]([⊠])

[1] Department of Electronic Engineering, Tsinghua University, Beijing, China
`wcd17@mails.tsinghua.edu.cn`, {`liyali13,wgsgj`}`@tsinghua.edu.cn`
[2] Australian National University, Canberra, Australia
`liang.zheng@anu.edu.au`

Abstract. Modern multiple object tracking (MOT) systems usually follow the *tracking-by-detection* paradigm. It has 1) a detection model for target localization and 2) an appearance embedding model for data association. Having the two models separately executed might lead to efficiency problems, as the running time is simply a sum of the two steps without investigating potential structures that can be shared between them. Existing research efforts on real-time MOT usually focus on the association step, so they are essentially real-time association methods but not real-time MOT system. In this paper, we propose an MOT system that allows target detection and appearance embedding to be learned in a shared model. Specifically, we incorporate the appearance embedding model into a single-shot detector, such that the model can simultaneously output detections and the corresponding embeddings. We further propose a simple and fast association method that works in conjunction with the joint model. In both components the computation cost is significantly reduced compared with former MOT systems, resulting in a neat and fast baseline for future follow-ups on real-time MOT algorithm design. To our knowledge, this work reports the first (near) real-time MOT system, with a running speed of 22 to 40 FPS depending on the input resolution. Meanwhile, its tracking accuracy is comparable to the state-of-the-art trackers embodying separate detection and embedding (SDE) learning (64.4% MOTA *v.s.* 66.1% MOTA on MOT-16 challenge). Code and models are available at https://github.com/Zhongdao/Towards-Realtime-MOT.

Keyword: Multi-Object Tracking

1 Introduction

Multiple object tracking (MOT), which aims at predicting trajectories of multiple targets in video sequences, underpins critical application significance ranging from autonomous driving to smart video analysis.

The dominant strategy to this problem, *i.e.*, *tracking-by-detection* [6,24,40] paradigm, breaks MOT down to two steps: 1) the detection step, in which targets in single video frames are localized; and 2) the association step, where

© Springer Nature Switzerland AG 2020
A. Vedaldi et al. (Eds.): ECCV 2020, LNCS 12356, pp. 107–122, 2020.
https://doi.org/10.1007/978-3-030-58621-8_7

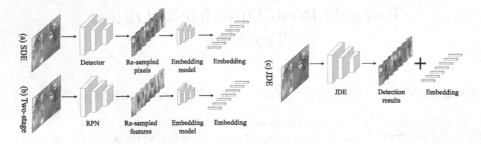

Fig. 1. Comparison between (a) the Separate Detection and Embedding (SDE) model, (b) the two-stage model and (c) the proposed Joint Detection and Embedding (JDE).

detected targets are assigned and connected to existing trajectories. It means the system requires at least two compute-intensive components: a detector and an embedding (re-ID) model. We term those methods as the Separate Detection and Embedding (SDE) methods for convenience. The overall inference time, therefore, is roughly the summation of the two components, and will increase as the target number increases. The characteristics of SDE methods bring critical challenges in building a real-time MOT system, an essential demand in practice.

In order to save computation, a feasible idea is to integrate the detector and the embedding model into a single network. The two tasks thus can share the same set of low-level features, and re-computation is avoided. One choice for joint detector and embedding learning is to adopt the Faster R-CNN framework [28], a type of two-stage detectors. Specifically, the first stage, the region proposal network (RPN), remains the same with Faster R-CNN and outputs detected bounding boxes; the second stage, Fast R-CNN [11], can be converted to an embedding model by replacing the classification supervision with the metric learning supervision [36,39]. In spite of saving some computation, this method is still limited in speed due to its two-stage design and usually runs at fewer than 10 frames per second (FPS), far from real-time. Moreover, the runtime of the second stage also increases as target number increases like SDE methods.

This paper is dedicated to the improving efficiency of an MOT system. We introduce an early attempt that Jointly learns the Detector and Embedding model (JDE) in a *single-shot* deep network. In other words, the proposed JDE employs a single network to *simultaneously* output detection results and the corresponding appearance embeddings of the detected boxes. In comparison, SDE methods and two-stage methods are characterized by re-sampled pixels (bounding boxes) and feature maps, respectively. Both the bounding boxes and feature maps are fed into a separate re-ID model for appearance feature extraction. Figure 1 briefly illustrates the difference between the SDE methods, the two-stage methods and the proposed JDE. Our method is near real-time while being almost as accurate as the SDE methods. For example, we obtain a running time of 20.2 FPS with MOTA = 64.4% on the MOT-16 test set. In comparison, Faster R-CNN + QAN embedding [40] only runs at <6 FPS with MOTA = 66.1% on the MOT-16 test set.

To build a joint learning framework with high efficiency and accuracy, we explore and deliberately design the following fundamental aspects: training data, network architecture, learning objectives, optimization strategies, and validation metrics. First, we collect six publicly available datasets on pedestrian detection and person search to form a unified large-scale multi-label dataset. In this unified dataset, all the pedestrian bounding boxes are labeled, and a portion of the pedestrian identities are labeled. Second, we choose the Feature Pyramid Network (FPN) [21] as our base architecture and discuss with which type of loss functions the network learns the best embeddings. Then, we model the training process as a multi-task learning problem with anchor classification, box regression, and embedding learning. To balance the importance of each individual task, we employ task-dependent uncertainty [16] to dynamically weight the heterogenous losses. A simple and fast association algorithm is proposed to further improve efficiency. Finally, we employ the following evaluation metrics. The average precision (AP) is employed to evaluate the performance of the detector. The retrieval metric True Accept Rate (TAR) at certain False Alarm Rate (FAR) is adopted to evaluate the quality of the embedding. The overall MOT accuracy is evaluated by the CLEAR metrics [2], especially the MOTA score. This paper also provides new settings and baselines for joint detection and embedding learning, which we believe will facilitate research towards real-time MOT.

The contributions of our work are summarized as follows,

- We introduce JDE, a single-shot framework for joint detection and embedding learning. It runs in (near) real-time and is comparably accurate to the separate detection + embedding (SDE) state-of-the-art methods.
- We conduct thorough analysis and experiments on how to build such a joint learning framework from multiple aspects including training data, network architecture, learning objectives and optimization strategy.
- Experiments with the same training data show the JDE performs as well as a range of strong SDE model combinations and achieves the fastest speed.
- Experiments on MOT-16 demonstrate the advantage of our method over state-of-the-art MOT systems considering the amount of training data, accuracy and speed.

2 Related Work

Recent progresses on multiple object tracking can be primarily categorized into the following aspects:

1) Ones that model the association problem as certain form of optimization problem on graphs [17, 37, 42].
2) Ones that make efforts to model the association process by an end-to-end neural network [32, 49].
3) Ones that seek novel tracking paradigm other than tracking-by-detection [1].

Among them, the first two categories have been the prevailing solution to MOT in the past decade. In these methods, detection results and appearance embeddings are given as input, and the only problem to be solved is data association. A standard formulation is using a graph, where nodes represent a detected targets, and edges indicate the possibility of linkages among nodes. Data association thus can be solved by minimizing some fixed [15,26,43] or learned [19] cost, or by more complex optimization such as multi-cuts [35] and minimum cliques [42]. Some recent works attempt to model the association problem using graph networks [4,20], so that end-to-end association can be achieved. Graph-based association shows good tracking accuracy especially in hard cases such as large occlusions, but their efficiency is always a problem. Although some methods [6] claim to be able to attain real-time speed, the runtime of the detector is excluded, such that the overall system still has some distance from the claim. In contrast, in this work, we consider the runtime of *the entire MOT system* rather than the association step only. Achieving efficiency on the entire system is more practically significant.

The third category attempts to explore novel MOT paradigms, for instance, incorporating single object trackers into the detector by predicting the spatial offsets [1]. These methods are appealing owning to their simplicity, but tracking accuracy is not satisfying unless an additional embedding model is introduced. As such, the trade-off between performance and speed still needs improvement.

The spirit of our approach, that learning auxiliary associative embeddings simultaneously with the main task, also shows good performance in many other vision tasks, such as person search [39], human pose estimation [25], and point-based object detection [18].

3   Joint Learning of Detection and Embedding

3.1   Problem Settings

The objective of JDE is to simultaneously output the location and appearance embeddings of targets in a single forward pass. Formally, suppose we have a training dataset $\{\mathbf{I}, \mathbf{B}, \mathbf{y}\}_{i=1}^{N}$. Here, $\mathbf{I} \in \mathbb{R}^{c \times h \times w}$ indicates an image frame, and $\mathbf{B} \in \mathbb{R}^{k \times 4}$ represents the bounding box annotations for the k targets in this frame. $\mathbf{y} \in \mathbb{Z}^{k}$ denotes the partially annotated identity labels, where -1 indicates targets without an identity label. JDE aims to output predicted bounding boxes $\hat{\mathbf{B}} \in \mathbb{R}^{\hat{k} \times 4}$ and appearance embeddings $\hat{\mathbf{F}} \in \mathbb{R}^{\hat{k} \times D}$, where D is the dimension of the embedding. The following objectives should be satisfied.

- $\mathbf{B}^{*}$ is as close to $\mathbf{B}$ as possible.
- Given a distance metric $d(\cdot)$, $\forall (k_t, k_{t+\Delta t}, k'_{t+\Delta t})$ that satisfy $\mathbf{y}_{k_{t+\Delta t}} = \mathbf{y}_{k_t}$ and $\mathbf{y}_{k'_{t+\Delta t}} \neq \mathbf{y}_{k_t}$, we have $d(f_{k_t}, f_{k_{t+\Delta t}}) < d(f_{k_t}, f_{k'_{t+\Delta t}})$, where f_{k_t} is a row vector from $\hat{\mathbf{F}}_t$ and $f_{k_{t+\Delta t}}, f_{k'_{t+\Delta t}}$ are row vectors from $\hat{\mathbf{F}}_{t+\Delta t}$, *i.e.*, embeddings of targets in frame t and $t + \Delta t$, respectively,

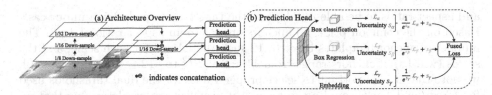

Fig. 2. Illustration of (a) the network architecture and (b) the prediction head. Prediction heads are added upon multiple FPN scales. In each prediction head the learning of JDE is modeled as a multi-task learning problem. We automatically weight the heterogeneous losses by learning a set of auxiliary parameters, *i.e.*, the task-dependent uncertainty.

The first objective requires the model to detect targets accurately. The second objective requires the appearance embedding to have the following property. The distance between observations of the same identity in consecutive frames should be smaller than the distance between different identities. The distance metric $d(\cdot)$ can be the Euclidean distance or the cosine distance. Technically, if the two objectives are both satisfied, even a simple association strategy, *e.g.*, the Hungarian algorithm, would produce good tracking results.

3.2 Architecture Overview

We employ the architecture of Feature Pyramid Network (FPN) [21]. FPN makes predictions from multiple scales, thus bringing improvement in pedestrian detection where the scale of targets varies a lot. Figure 2 briefly shows the neural architecture used in JDE. An input video frame first undergoes a forward pass through a backbone network to obtain feature maps at three scales, namely, scales with $\frac{1}{32}$, $\frac{1}{16}$ and $\frac{1}{8}$ down-sampling rate, respectively. Then, the feature map with the smallest size (also the semantically strongest features) is up-sampled and fused with the feature map from the second smallest scale by skip connection, and the same goes for the other scales. Finally, prediction heads are added upon fused feature maps at all the three scales. A prediction head consists of several stacked convolutional layers and outputs a dense prediction map of size $(6A + D) \times H \times W$, where A is the number of anchor templates assigned to this scale, and D is the dimension of the embedding. The dense prediction map is divided into three parts (tasks):

1) the box classification results of size $2A \times H \times W$;
2) the box regression coefficients of size $4A \times H \times W$;
3) the dense embedding map of size $D \times H \times W$.

In the following sections, we will detail how these tasks are trained.

3.3 Learning to Detect

In general the detection branch is similar to the standard RPN [28], but with two modifications. First, we redesign the anchors in terms of numbers, scales, and aspect ratios to be able to adapt to the targets, *i.e.*, pedestrian in our case. Based on the common prior, all anchors are set to an aspect ratio of $1 : 3$. The number of anchor templates is set to 12 such that $A = 4$ for each scale, and the scales (widths) of anchors range from $11 \approx 8 \times 2^{1/2}$ to $512 = 8 \times 2^{12/2}$. Second, we note that it is important to select proper values for the dual thresholds used for foreground/background assignment. By visualization we determine that an $IOU > 0.5$ *w.r.t.* the ground truth approximately ensures a foreground, which is consistent with the common setting in generic object detection. On the other hand, those boxes that have an $IOU < 0.4$ *w.r.t.* the ground truth should be regarded as background in our case rather than 0.3 used in generic scenarios. Our preliminary experiment indicates that these thresholds effectively suppress false alarms, which usually happens under heavy occlusions.

The learning objective of detection has two loss functions, namely the foreground/background classification loss $\mathcal{L}_\alpha$, and the bounding box regression loss $\mathcal{L}_\beta$. $\mathcal{L}_\alpha$ is formulated as a cross-entropy loss and $\mathcal{L}_\beta$ as a smooth-L1 loss. The regression targets are encoded in the same manner as [28].

3.4   Learning Appearance Embeddings

The second objective is a *metric learning* problem, *i.e.*, learning a embedding space where instances of the same identity are close to each other while instances of different identities are far apart. To achieve this goal, an effective solution is to use the triplet loss [29]. The triplet loss has also been used in previous MOT works [36]. Formally, we use triplet loss $\mathcal{L}_{triplet} = \max(0, f^\top f^- - f^\top f^+)$, where $f^\top$ is an instance in a mini-batch selected as an anchor, f^+ represents a positive sample *w.r.t.* $f^\top$, and f^- is a negative sample. The margin term is neglected for convenience. This naive formulation of the triplet loss has several challenges. The first is the huge sampling space in the training set. In this work we address this problem by looking at a mini-batch and mining all the negative samples and the hardest positive sample in this mini-batch, such that,

$$\mathcal{L}_{triplet} = \sum_i \max \left(0, f^\top f_i^- - f^\top f^+\right), \tag{1}$$

where f^+ is the hardest positive sample in a mini-batch.

The second challenge is that training with the triplet loss can be unstable and the convergence might be slow. To stabilize the training process and speed up convergence, it is proposed in [31] to optimize over a smooth upper bound of the triplet loss,

$$\mathcal{L}_{upper} = \log \left(1 + \sum_i \exp \left(f^\top f_i^- - f^\top f^+\right)\right). \tag{2}$$

Note that this smooth upper bound of triplet loss can be also written as,

$$\mathcal{L}_{upper} = -\log \frac{\exp(f^\top f^+)}{\exp(f^\top f^+) + \sum_i \exp(f^\top f_i^-)}. \tag{3}$$

It is similar to the formulation of the cross-entropy loss,

$$\mathcal{L}_{CE} = -\log \frac{\exp(f^\top g^+)}{\exp(f^\top g^+) + \sum_i \exp(f^\top g_i^-)}, \tag{4}$$

where we denote the class-wise weight of the positive class (to which the anchor instance belongs) as g^+ and weights of negative classes as g_i^-. The major ditinctions between $\mathcal{L}_{upper}$ and $\mathcal{L}_{CE}$ are two-fold. First, the cross-entropy loss employs learnable class-wise weights as proxies of class instances rather than using the embeddings of instances directly. Second, all the negative classes participate in the loss computation in $\mathcal{L}_{CE}$ such that the anchor instance is pulled away from all the negative classes in the embedding space. In contrast, in $\mathcal{L}_{upper}$, the anchor instance is only pulled away from the sampled negative instances.

In light of the above analysis, we speculate the performance of the three losses under our case should be $\mathcal{L}_{CE} > \mathcal{L}_{upper} > \mathcal{L}_{triplet}$. Experimental result in the experiment section confirms this. As such, we select the cross-entropy loss as the objective for embedding learning (hereinafter referred to as $\mathcal{L}_\gamma$).

Specifically, if an anchor box is labeled as the foreground, the corresponding embedding vector is extracted from the dense embedding map. Extracted embeddings are fed into a *shared* fully-connected layer to output the class-wise logits, and then the cross-entropy loss is applied upon the logits. In this manner, embeddings from multiple scales shares the same space, and association across scales is feasible. Embeddings with label -1, *i.e.*, foregrounds with box annotations but without identity annotations, are ignored when computing the embedding loss.

3.5   Automatic Loss Balancing

The learning objective of each prediction head in JDE can be modeled as a multi-task learning problem. The joint objective can be written as a weighted linear sum of losses from every scale and every component,

$$\mathcal{L}_{total} = \sum_i^M \sum_{j=\alpha,\beta,\gamma} w_j^i \mathcal{L}_j^i, \tag{5}$$

where M is the number of prediction heads and $w_j^i, i = 1, ..., M, j = \alpha, \beta, \gamma$ are loss weights. A simple way to determine loss weights are described below.

1. Let $w_\alpha^i = w_\beta^i$, as suggested in [28]
2. Let $w_{\alpha/\gamma/\beta}^1 = ... = w_{\alpha/\gamma/\beta}^M$.
3. Search for the remaining two independent loss weights for the best performance.

Searching loss weights with this strategy can yield decent results within several attempts. However, the reduction of searching space also brings strong restrictions on the loss weights, such that the resulting loss weights might be far from optimal. Instead, we adopt an automatic learning scheme for loss weights proposed in [16] by using the concept of task-independent uncertainty. Formally, the learning objective with automatic loss balancing is written as,

$$\mathcal{L}_{total} = \sum_i \sum_{j=\alpha,\beta,\gamma} \frac{1}{2}\left(\frac{1}{e^{s_j^i}}\mathcal{L}_j^i + s_j^i\right), \tag{6}$$

where s_j^i is the task-dependent uncertainty for each individual loss and is modeled as learnable parameters. We refer readers to [16] for more detailed derivation and discussion.

3.6 Online Association

Although the association algorithm is not the focus of this work, here we introduce a simple and fast online association strategy to work in conjunction with JDE. Specifically, a tracklet is described with an appearance state e_i and a motion state $m_i = (x, y, \gamma, h, \dot{x}, \dot{y}, \dot{\gamma}, \dot{h})$, where x, y indicate the bounding box center position, h indicates the bounding box height

Table 1. Comparison between our association method and SORT. Inputs are the same.

Method	Density	FPS	MOTA	IDF-1
SORT [3]	Low	44.1	66.9	55.8
Ours	Low	**46.2**	**67.5**	**67.6**
SORT [3]	High	26.4	35.0	32.4
Ours	High	**33.9**	**35.4**	**35.5**

and γ indicates the aspect ratio, and $\dot{x}$ indicates the velocity along x direction. The tracklet appearance e_i is initialized with the appearance embedding of the first observation f_i^0. We maintain a tracklet pool containing all the reference tracklets that observations are probable to be associated with. For an incoming frame, we compute the pair-wise motion affinity matrix A_m and appearance affinity matrix A_e between all the observations and the traklets from the pool. The appearance affinity is computed using cosine similarity, and the motion affinity is computed using Mahalanobis distance. Then we solve the linear assignment problem by Hungarian algorithm with cost matrix $C = \lambda A_e + (1 - \lambda)A_m$. The motion state m_i of all matched tracklets are updated by the Kalman filter, and the appearance state e_i is updated by

$$e_i^t = \alpha e_i^{t-1} + (1 - \alpha)f_i^t \tag{7}$$

Where f_i^t is the appearance embedding of the current matched observation, $\alpha = 0.9$ is a momentum term. Finally observations that are not assigned to any tracklets are initialized as new tracklets if they consecutively appear in 2 frames. A tracklet is terminated if it is not updated in the most current 30 frames.

Note this association method is simpler than the cascade matching strategy proposed in SORT [3], since we only apply association once for one frame and resort to a buffer pool to deal with those shortly lost tracklets. Moreover, we also implement a vectorized version of the Kalman filter and find it critical for high FPS, especially when the model is already fast. A comparison between SORT and our association method, based on the same JDE model, is shown in Table 1. We use MOT-15 [24] for testing the low density scenario and CVPR-19-01 [7] for high density. It can be observed that our method outperforms SORT in both accuracy and speed, especially under the high-density case.

4 Experiments

4.1 Datasets and Evaluation Metrics

Performing experiments on small datasets may lead to biased results and conclusions may not hold when applying the same algorithm to large-scale datasets. Therefore, we build a large-scale training set by

Table 2. Statistics of the joint training set.

Dataset	ETH	CP	CT	M16	CS	PRW	Total
# img	2K	3K	27K	53K	11K	6K	54K
# box	17K	21K	46K	112K	55K	18K	270K
# ID	–	–	0.6K	0.5K	7K	0.5K	8.7K

putting together six publicly available datasets on pedestrian detection, MOT and person search. These datasets can be categorized into two types: ones that only contain bounding box annotations, and ones that have both bounding box and identity annotations. The first category includes the ETH dataset [9] and the CityPersons (CP) dataset [44]. The second category includes the CalTech (CT) dataset [8], MOT-16 (M16) dataset [24], CUHK-SYSU (CS) dataset [39] and PRW dataset [47]. Training subsets of all these datasets are gathered to form the joint training set, and videos in the ETH dataset that overlap with the MOT-16 **test** set are excluded for fair evaluation. Table 2 shows the statistics of the joint training set.

For validation/evaluation, three aspects of performance need to be evaluated: the detection accuracy, the discriminative ability of the embedding, and the tracking performance of the entire MOT system. To evaluate detection accuracy, we compute average precision (AP) at IOU threshold of 0.5 over the Caltech validation set. To evaluate the appearance embedding, we extract embeddings of all ground truth boxes over the validation sets of the Caltech dataset, the CUHK-SYSU dataset and the PRW dataset, apply $1 : N$ retrieval among these instances and report the true positive rate at false accept rate 0.1 (TPR@FAR=0.1). To evaluate the tracking accuracy of the entire MOT system, we employ the CLEAR metric [2], particularly the MOTA metric that aligns best with human perception. In validation, we use the MOT-15 **training** set with duplicated sequences with the training set removed. During testing, we use the MOT-16 **test** set to compare with existing methods.

4.2 Implementation Details

We employ DarkNet-53 [27] as the backbone network in JDE. The network is trained with standard SGD for 30 epochs. The learning rate is initialized as 10^{-2} and is decreased by 0.1 at the 15th and the 23th epoch. Several data augmentation techniques, such as random rotation, random scale and color jittering, are applied to reduce overfitting. Finally, the augmented images are adjusted to a fixed resolution. The input resolution is 1088×608 if not specified.

4.3   Experimental Results

Comparison of the three loss functions for appearance embedding learning. We first compare the discriminative ability of appearance embeddings trained with the cross-entropy loss, the triplet loss and its upper bound variant, described in the previous section. For models trained with $\mathcal{L}_{triplet}$ and $\mathcal{L}_{upper}$, $B/2$ pairs of temporal consecutive frames are sampled to form a mini-batch with size B. This ensures that there always exist positive samples. For models

Table 3. Comparing different embedding losses and loss weighting strategies. TPR is short for TPR@FAR=0.1 on the embedding validation set, and IDs means times of ID switches on the tracking validation set. ↓ means the smaller the better; ↑ means the larger the better. In each column, the best result is in **bold**, and the second best is underlined.

Embed. Loss	Weighting strategy	Det	Emb	MOT	
		AP↑	TPR↑	MOTA↑	IDs↓
$\mathcal{L}_{triplet}$	App.Opt	81.6	42.2	59.5	375
$\mathcal{L}_{upper}$	App.Opt	81.7	44.3	59.8	346
$\mathcal{L}_{CE}$	App.Opt	<u>82.0</u>	88.2	<u>64.3</u>	223
$\mathcal{L}_{CE}$	Uniform	6.8	**94.8**	36.9	366
$\mathcal{L}_{CE}$	MGDA-UB	8.3	<u>93.5</u>	38.3	357
$\mathcal{L}_{CE}$	Loss.Norm	80.6	82.1	57.9	321
$\mathcal{L}_{CE}$	Uncertainty	**83.0**	90.4	**65.8**	**207**

trained with $\mathcal{L}_{CE}$, images are randomly sampled to form a mini-batch. Table 3 presents comparisons of the three loss functions.

As expected, $\mathcal{L}_{CE}$ outperforms both $\mathcal{L}_{triplet}$ and $\mathcal{L}_{upper}$. Surprisingly, the performance gap is large (+46.0/+43.9 TAR@FAR=0.1). A possible reason for the large gap is that the cross-entropy loss requires the similarity between one instance and its positive class be higher than the similarities between this instance and *all* negative classes. This objective is more rigorous than the triplet loss family, which exerts constraints merely in a sampled mini-batch. Considering its effectiveness and simplicity, we adopt the cross-entropy loss in JDE.

Comparison of Different Loss Weighting Strategies. The loss weighting strategy is crucial to learn good joint representation for JDE. In this paper, three loss weighting strategies are implemented. The first is a loss normalization method (named "Loss.Norm"), where the losses are weighted by the reciprocal of their moving average magnitude. The second is the "MGDA-UB" algorithm proposed in [30] and the last is the weight-by-uncertainty strategy described in Sect. 3.5. Moreover, we have two baselines. The first trains all the tasks with identical loss weights, named as "Uniform". The second, referred to as "App.Opt", uses a set of approximate optimal loss weights by searching under the two-independent-variable assumption as described in Sect. 3.5. Table 3 summarizes the comparisons of these strategies. Two observations are made.

First, the Uniform baseline produces poor detection results, and thus the tracking accuracy is not good. This is because the scale of the embedding loss is much larger than the other two losses and dominates the training process. Once we set proper loss weights to let al.l tasks learn at a similar rate, as in the "App.Opt" baseline, both the detection and embedding tasks yield good performance.

Second, results indicate that the "Loss.Norm" strategy outperforms the "Uniform" baseline but is inferior to the "App.Opt" baseline. The MGDA-UB algorithm, despite being the most theoretically sound method, fails in our case because it assign too large weights to the embedding loss, such that its performance is similar to the Uniform baseline. The only method that outperforms the App.Opt baseline is the weight-by-uncertainty strategy.

Comparison with SDE Methods. To demonstrate the superiority of JDE to the Separate Detection and Embedding (SDE) methods, we implemented several state-of-the-art detectors and person re-id models and compare their combinations with JDE in terms of both tracking accuracy (MOTA) and runtime (FPS). The detectors include JDE with ResNet-50 and ResNet-101 [13] as backbone, Faster R-CNN [28] with ResNet-50 and ResNet-101 as backbone, and Cascade R-CNN [5] with ResNet-50 and ResNet-101 as backbone. The person re-id models include IDE [46], Triplet [14] and PCB [33]. In the association step, we use the same online association approach described in Sect. 3.6 for all the SDE models. For fair comparison, all the training data are the same as used in JDE.

In Fig. 3, we plot the MOTA metric and the IDF-1 score against the runtime for SDE combinations of the above detectors and person re-id models. Runtime of all models are tested on a single Nvidia Titan xp GPU. Figure 3 (a) and (c) show comparisons on the MOT-15 train set, in which the pedestrian density is low, e.g., less than 20. In contrast, Fig. 3 (b) and (d) show comparisons on a video sequence that contains crowd in high-density (CVPR19-01 from the CVPR19 MOT challenge datast, with density 61.1). Several observations can be made.

First, cosidering the MOTA metric, the proposed JDE produces competitive tracking accuracy meanwhile runs much faster than strong SDE combinations, reaching the best trade-off between accuracy and speed in both low-density and high-density cases. Specifically, JDE with DarkNet-53 (JDE-DN53) runs at 22 FPS and produces tracking accuracy nearly as good as the combination of the Cascade RCNN detector with ResNet-101 (Cascade-R101) + PCB embedding (65.8% v.s. 66.2%), while the latter only runs at ∼6 FPS. In the other hand, Considering the IDF-1 score which reflects the association performance, our JDE is also competitive with strong SDE combinations in the low-density case. Specifically, JDE with DarkNet-53 presents 66.2% IDF-1 score at 22 FPS, while Cascade RCNN with ResNet-101 + PCB presents 69.6% IDF-1 score at 7.6 FPS. In the high-density crowd case, performance of all methods rapidly degrades, and we observe that IDF-1 score of JDE degrades slightly more than strong SDE combinations. We find the major reason is that, in the crowd case, pedestrian often overlap with each other, and since JDE employs a single-stage detector the detected boxes often drift in such case. The misalignment of boxes brings ambiguity in the embedding, so that ID switches increase and IDF-1 score drops. Figure 3 shows an example of such failure case.

Second, the tracking accuracy of JDE is very close to the combinations of JDE+IDE, JDE+Triplet and JDE+PCB (see the cross markers in Fig. 3). With other components fixed, JDE even outperforms the JDE+ IDE combination. This strongly suggests the jointly learned embedding is almost as discriminative as the separately learned embedding.

Finally, comparing the runtime of a same model between Fig. 3 (a) and (b), it can be observed that all the SDE models suffer a significant speed drop under the crowded case. This is because the runtime of the embedding model increases with the number of detected targets. This drawback does not exist in JDE because the embedding is computed together with the detection results. As such, the runtime difference between JDE under

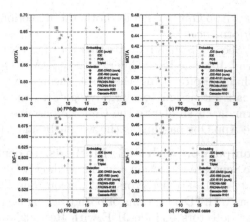

Fig. 3. Comparing JDE and various SDE combinations in terms of tracking accuracy (MOTA/IDF-1) and speed (FPS). (a) and (c) show comparisons under the case where the pedestrian density is low (MOT-15 **train** set), (b) and (d) show comparisons under the crowded scenario (MOT-CVPR19-01). Different colors represent different embedding models, and different shapes denote different detectors. We clearly observe that the proposed JDE method (JDE Embedding + JDE-DN53) has the best time-accuracy trade-off. Best viewed in color. (Color figure online)

the usual case and the crowded case is much smaller (see the red markers). In fact, the speed drop of JDE is due to the increased time in the association step, which is positively related to the target number.

Comparison with the State-of-the-Art MOT Systems. Since we train JDE using additional data instead of the MOT-16 **train** set, we compare JDE under the "private data" protocol of the MOT-16 benchmark. State-of-the-art online MOT methods under the private protocol are compared, including Deep-SORT_2 [38], RAR16wVGG [10], TAP [48], CNNMTT [23] and POI [40]. All these methods employ the same detector, *i.e.*, Faster-RCNN with VGG-16 as backbone, which is trained on a large private pedestrian detection dataset. The main differences among these methods reside in their embedding models and the association strategies. For instance, DeepSORT_2 employs Wide Residual Network (WRN) [41] as the embedding model and uses the MARS [45] dataset to train the appearance embedding. RAR16withVGG, TAP, CNNMTT and POI use Inception [34], Mask-RCNN [12], a 5-layer CNN, and QAN [22] as their embedding models, respectively. Training data of these embedding models also differ from each other. For clear comparison, we list the number of training data for all these methods in Table 4. Accuracy and speed metrics are also presented.

Table 4. Comparison with the state-of-the-art online MOT systems under the private data protocol on the MOT-16 benchmark. The performance is evaluated with the CLEAR metrics, and runtime is evaluated with three metrics: frames per second of the detector (FPSD), frame per second of the association step (FPSA), and frame per second of the overall system (FPS). * indicates estimated timing. We clearly observe our method has the best efficiency and a comparable accuracy.

Method	#box	#id	MOTA↑	IDF1↑	MT↑	ML ↓	IDs ↓	FPSD ↑	FPSA ↑	FPS ↑
DeepSORT	429K	1.2k	61.4	62.2	32.8	18.2	781	<15*	17.4	<8.1
RAR16	429K	-	63.0	63.8	39.9	22.1	**482**	<15*	1.6	<1.5
TAP	429K	-	64.8	**73.5**	**40.6**	22.0	794	<15*	18.2	<8.2
CNNMTT	429K	0.2K	65.2	62.2	32.4	21.3	946	<15*	11.2	<6.4
POI	429K	16K	**66.1**	65.1	34.0	21.3	805	<15*	9.9	<6
JDE864	270K	8.7K	62.1	56.9	34.4	16.7	1,608	**34.3**	**259.8**	**30.3**
JDE1088	270K	8.7K	64.4	55.8	35.4	20.0	1,544	24.5	236.5	22.2

Considering the overall tracking accuracy, *e.g.*, the MOTA metric, JDE is generally comparable. Our result is higher than DeepSORT_2 by +3.0% and is lower than POI by 1.7%. In terms of running speed, it is not feasible to directly compare these methods because their runtimes are not all reported. Therefore, we re-implemented the VGG-16 based Faster R-CNN detector and benchmark its running speed, and then estimate the running speed upper bounds of the entire MOT system for these methods. Note that for some methods the runtime of the embedding model is not taken into account, so the speed upper bounds are far from being tight. Even with such relaxed upper bound, the proposed JDE runs at least 2–3× faster than existing methods, reaching a near real-time speed, *i.e.*, 22.2 FPS at an image resolution of as high as 1088 × 608. When we down-sample the input frames to a lower resolution of 864 × 408, the runtime of JDE can be further sped up to 30.3 FPS with only a minor performance drop ($\Delta = -2.6\%$ MOTA).

Visualization. To show the discrimination of the joint learned embedding intuitively, we perform a simple retrieval experiment and visualize the results in Fig. 4. We extract the feature of a pedestrian in one video frame as a query and compute pixel-wise cosine similarity with the feature map of another frame. We compare the retrieval results between using detection feature map as the feature and using the dense embedding as the feature, and it is clearly observed the dense embedding results in better correspondence between the query and the target.

Analysis and Discussions. One may notice that JDE has a lower IDF1 score and more ID switches than existing methods. At first we suspect the reason is that the jointly learned embedding might be weaker than a separately learned embedding. However, when we replace the jointly learned embedding with the separately learned embedding, the IDF1 score and the number of ID switches remain almost the same. Finally we find that the major reason lies in the inac-

Fig. 4. Visualization of the retrieval performance of the detection feature map and the dense embedding. Similarity maps are computed as the cosine similarity between the query feature and the target feature map. The joint learned dense embedding presents good correspondence between the query and the target.

curate detection when multiple pedestrians have large overlaps with each other. Such inaccurate boxes introduce lots of ID switches, and unfortunately, such ID switches often occur in the middle of a trajectory, hence the IDF1 score is lower. In our future work, it remains to be solved how to improve JDE to make more accurate boxes predictions when pedestrian overlaps are significant.

5   Conclusion

In this paper, we introduce JDE, an MOT system that allows target detection and appearance features to be learned in a shared model. Our design significantly reduces the runtime of an MOT system, making it possible to run at a (near) real-time speed. Meanwhile, the tracking accuracy of our system is comparable with the state-of-the-art online MOT methods. Moreover, we have provided thorough analysis, discussions and experiments about good practices and insights in building such a joint learning framework. In the future, we will investigate deeper into the time-accuracy trade-off issue.

References

1. Bergmann, P., Meinhardt, T., Leal-Taixe, L.: Tracking without bells and whistles. arXiv preprint arXiv:1903.05625 (2019)
2. Bernardin, K., Stiefelhagen, R.: Evaluating multiple object tracking performance: the clear mot metrics. J. Image Video Process. **2008**, 1 (2008)
3. Bewley, A., Ge, Z., Ott, L., Ramos, F., Upcroft, B.: Simple online and realtime tracking. In: ICIP (2016)
4. Brasó, G., Leal-Taixé, L.: Learning a neural solver for multiple object tracking. In: CVPR (2020)
5. Cai, Z., Vasconcelos, N.: Cascade R-CNN: delving into high quality object detection. In: CVPR (2018)
6. Choi, W.: Near-online multi-target tracking with aggregated local flow descriptor. In: ICCV (2015)

7. Dendorfer, P., et al.: CVPR19 tracking and detection challenge: how crowded can it get? arXiv preprint arXiv:1906.04567 (2019)
8. Dollár, P., Wojek, C., Schiele, B., Perona, P.: Pedestrian detection: a benchmark. In: CVPR (2009)
9. Ess, A., Leibe, B., Schindler, K., van Gool, L.: A mobile vision system for robust multi-person tracking. In: CVPR (2008)
10. Fang, K., Xiang, Y., Li, X., Savarese, S.: Recurrent autoregressive networks for online multi-object tracking. In: WACV (2018)
11. Girshick, R.: Fast R-CNN. In: ICCV (2015)
12. He, K., Gkioxari, G., Dollár, P., Girshick, R.: Mask R-CNN. In: ICCV (2017)
13. He, K., Zhang, X., Ren, S., Sun, J.: Deep residual learning for image recognition. In: CVPR (2016)
14. Hermans, A., Beyer, L., Leibe, B.: In defense of the triplet loss for person re-identification. arXiv preprint arXiv:1703.07737 (2017)
15. Jiang, H., Fels, S., Little, J.J.: A linear programming approach for multiple object tracking. In: CVPR (2007)
16. Kendall, A., Gal, Y., Cipolla, R.: Multi-task learning using uncertainty to weigh losses for scene geometry and semantics. In: CVPR (2018)
17. Kim, C., Li, F., Ciptadi, A., Rehg, J.M.: Multiple hypothesis tracking revisited. In: ICCV (2015)
18. Law, H., Deng, J.: CornerNet: detecting objects as paired keypoints. In: Ferrari, V., Hebert, M., Sminchisescu, C., Weiss, Y. (eds.) Computer Vision – ECCV 2018. LNCS, vol. 11218, pp. 765–781. Springer, Cham (2018). https://doi.org/10.1007/978-3-030-01264-9_45
19. Leal-Taixé, L., Fenzi, M., Kuznetsova, A., Rosenhahn, B., Savarese, S.: Learning an image-based motion context for multiple people tracking. In: CVPR (2014)
20. Li, J., Gao, X., Jiang, T.: Graph networks for multiple object tracking. In: CVPR (2020)
21. Lin, T.Y., Dollár, P., Girshick, R., He, K., Hariharan, B., Belongie, S.: Feature pyramid networks for object detection. In: CVPR (2017)
22. Liu, Y., Yan, J., Ouyang, W.: Quality aware network for set to set recognition. In: CVPR (2017)
23. Mahmoudi, N., Ahadi, S.M., Rahmati, M.: Multi-target tracking using CNN-based features: CNNMTT. Multimed. Tools Appl. **78**(6), 7077–7096 (2019)
24. Milan, A., Leal-Taixé, L., Reid, I., Roth, S., Schindler, K.: MOT16: a benchmark for multi-object tracking. arXiv preprint arXiv:1603.00831 (2016)
25. Newell, A., Huang, Z., Deng, J.: Associative embedding: end-to-end learning for joint detection and grouping. In: NIPS (2017)
26. Pirsiavash, H., Ramanan, D., Fowlkes, C.C.: Globally-optimal greedy algorithms for tracking a variable number of objects. In: CVPR (2011)
27. Redmon, J., Farhadi, A.: YOLOV3: an incremental improvement. arXiv preprint arXiv:1804.02767 (2018)
28. Ren, S., He, K., Girshick, R., Sun, J.: Faster R-CNN: towards real-time object detection with region proposal networks. In: NIPS (2015)
29. Schroff, F., Kalenichenko, D., Philbin, J.: Facenet: a unified embedding for face recognition and clustering. In: CVPR (2015)
30. Sener, O., Koltun, V.: Multi-task learning as multi-objective optimization. In: NIPS (2018)
31. Sohn, K.: Improved deep metric learning with multi-class n-pair loss objective. In: NIPS (2016)

32. Sun, S., Akhtar, N., Song, H., Mian, A.S., Shah, M.: Deep affinity network for multiple object tracking. IEEE Trans. Pattern Anal. Mach. Intell. (2019)
33. Sun, Y., Zheng, L., Yang, Y., Tian, Q., Wang, S.: Beyond part models: person retrieval with refined part pooling (and a strong convolutional baseline). In: Ferrari, V., Hebert, M., Sminchisescu, C., Weiss, Y. (eds.) ECCV 2018. LNCS, vol. 11208, pp. 501–518. Springer, Cham (2018). https://doi.org/10.1007/978-3-030-01225-0_30
34. Szegedy, C., et al.: Going deeper with convolutions. In: CVPR (2015)
35. Tang, S., Andriluka, M., Andres, B., Schiele, B.: Multiple people tracking by lifted multicut and person re-identification. In: CVPR (2017)
36. Voigtlaender, P., et al.: Mots: Multi-object tracking and segmentation. In: CVPR (2019)
37. Wen, L., Li, W., Yan, J., Lei, Z., Yi, D., Li, S.Z.: Multiple target tracking based on undirected hierarchical relation hypergraph. In: CVPR (2014)
38. Wojke, N., Bewley, A., Paulus, D.: Simple online and realtime tracking with a deep association metric. In: ICIP (2017)
39. Xiao, T., Li, S., Wang, B., Lin, L., Wang, X.: Joint detection and identification feature learning for person search. In: CVPR (2017)
40. Yu, F., Li, W., Li, Q., Liu, Yu., Shi, X., Yan, J.: POI: multiple object tracking with high performance detection and appearance feature. In: Hua, G., Jégou, H. (eds.) ECCV 2016. LNCS, vol. 9914, pp. 36–42. Springer, Cham (2016). https://doi.org/10.1007/978-3-319-48881-3_3
41. Zagoruyko, S., Komodakis, N.: Wide residual networks. arXiv preprint arXiv:1605.07146 (2016)
42. Roshan Zamir, A., Dehghan, A., Shah, M.: GMCP-tracker: global multi-object tracking using generalized minimum clique graphs. In: Fitzgibbon, A., Lazebnik, S., Perona, P., Sato, Y., Schmid, C. (eds.) ECCV 2012. LNCS, vol. 7573, pp. 343–356. Springer, Heidelberg (2012). https://doi.org/10.1007/978-3-642-33709-3_25
43. Zhang, L., Li, Y., Nevatia, R.: Global data association for multi-object tracking using network flows. In: CVPR (2008)
44. Zhang, S., Benenson, R., Schiele, B.: Citypersons: a diverse dataset for pedestrian detection. In: CVPR (2017)
45. Zheng, L., et al.: MARS: a video benchmark for large-scale person re-identification. In: Leibe, B., Matas, J., Sebe, N., Welling, M. (eds.) ECCV 2016. LNCS, vol. 9910, pp. 868–884. Springer, Cham (2016). https://doi.org/10.1007/978-3-319-46466-4_52
46. Zheng, L., Yang, Y., Hauptmann, A.G.: Person re-identification: Past, present and future. arXiv preprint arXiv:1610.02984 (2016)
47. Zheng, L., Zhang, H., Sun, S., Chandraker, M., Yang, Y., Tian, Q.: Person re-identification in the wild. In: CVPR (2017)
48. Zhou, Z., Xing, J., Zhang, M., Hu, W.: Online multi-target tracking with tensor-based high-order graph matching. In: ICPR (2018)
49. Zhu, J., Yang, H., Liu, N., Kim, M., Zhang, W., Yang, M.-H.: Online multi-object tracking with dual matching attention networks. In: Ferrari, V., Hebert, M., Sminchisescu, C., Weiss, Y. (eds.) ECCV 2018. LNCS, vol. 11209, pp. 379–396. Springer, Cham (2018). https://doi.org/10.1007/978-3-030-01228-1_23

A Balanced and Uncertainty-Aware Approach for Partial Domain Adaptation

Jian Liang[1]([✉])[iD], Yunbo Wang[2][iD], Dapeng Hu[1], Ran He[3][iD], and Jiashi Feng[1][iD]

[1] Department of ECE, National University of Singapore (NUS), Singapore, Singapore
liangjian92@gmail.com, dapeng.hu@u.nus.edu, elefjia@nus.edu.sg
[2] Peking University, Beijing, China
wangyunbo09@gmail.com
[3] Institute of Automation, Chinese Academy of Sciences, Beijing, China
rhe@nlpr.ia.ac.cn

Abstract. This work addresses the unsupervised domain adaptation problem, especially in the case of class labels in the target domain being only a subset of those in the source domain. Such a partial transfer setting is realistic but challenging and existing methods always suffer from two key problems, negative transfer and uncertainty propagation. In this paper, we build on domain adversarial learning and propose a novel domain adaptation method BA^{3}US with two new techniques termed Balanced Adversarial Alignment (BAA) and Adaptive Uncertainty Suppression (AUS), respectively. On one hand, negative transfer results in misclassification of target samples to the classes only present in the source domain. To address this issue, BAA pursues the balance between label distributions across domains in a fairly simple manner. Specifically, it randomly leverages a few source samples to augment the smaller target domain during domain alignment so that classes in different domains are symmetric. On the other hand, a source sample would be denoted as uncertain if there is an incorrect class that has a relatively high prediction score, and such uncertainty easily propagates to unlabeled target data around it during alignment, which severely deteriorates adaptation performance. Thus we present AUS that emphasizes uncertain samples and exploits an adaptive weighted complement entropy objective to encourage incorrect classes to have uniform and low prediction scores. Experimental results on multiple benchmarks demonstrate our BA^{3}US surpasses state-of-the-arts for partial domain adaptation tasks. Code is available at https://github.com/tim-learn/BA3US.

Keywords: Partial transfer learning · Domain adaptation · Adversarial alignment · Uncertainty propagation · Object recognition

1 Introduction

Over the past two decades, many research efforts have been devoted to unsupervised domain adaptation (UDA), which aims to leverage labeled source domain

© Springer Nature Switzerland AG 2020
A. Vedaldi et al. (Eds.): ECCV 2020, LNCS 12356, pp. 123–140, 2020.
https://doi.org/10.1007/978-3-030-58621-8_8

data to learn to classify unlabeled target domain data. Typically, existing UDA methods minimize the discrepancy between two domains by matching their statistical distribution moments [18,37,41,47] or by domain adversarial learning [10,26,39,40]. Once the domain shift is mitigated, source classifiers can be easily transferred to the target domain even with no labeled target domain data available. However, both UDA strategies always assume that different domains share the same label space. Such an assumption may not hold in practice and target domain labels may be only a subset of source domain labels. This introduces an unsupervised partial domain adaptation (PDA) problem that receives increasing research attention recently [4,5,30,49].

The PDA problem is challenging since source-only classes may occur in the target domain during distribution alignment, which is well-known as class mismatch that potentially causes *negative transfer*. Several previous PDA approaches [4,5] mitigate negative transfer by jointly filtering out source-only classes and promote positive transfer by matching the data distribution in the shared classes. Samples from source-only classes are expected to have lower weights in the adaptation module such that the marginal distributions of two domains can be aligned well. However, it is rather risky to rule out the source-only classes, especially when the estimation of label distribution in the target domain is inaccurate.

To match two non-identical label spaces, we view the PDA problem from a new perspective and propose to augment the target label space to be the same as the source label space. Specifically, we develop a simple balanced alignment solution, termed Balanced Adversarial Alignment (BAA), that borrows fewer and fewer samples from the source domain to the target domain within an iterative adversarial learning framework. We expect that the augmented target domain looks much more similar to the source domain w.r.t. the label distribution, and the challenging PDA problem can be transformed to a well-studied UDA task. To focus on the originally shared classes, we propose to filter out the source-only classes via a class-level weighting scheme meanwhile, making the large UDA task more compact.

Besides, existing domain adaptation methods always employ a conventional cross-entropy loss to merely promote the prediction score of ground-truth classes but neglect to suppress those of incorrect classes, which may result in a new problem termed *uncertainty propagation*. Intuitively, if incorrect classes have relatively high prediction scores for source data, some wrong classes would possibly have the largest prediction scores for the aligned target data around them. This problem is quite critical but has been always ignored in the domain adaptation field. To circumvent the issue, we develop an uncertainty suppression solution termed Adaptive Uncertainty Suppression (AUS) that exploits complement entropy [7] in the labeled source domain to prohibit possibly high prediction scores from incorrect classes. Specifically, we emphasize more the uncertain samples corresponding to smaller cross-entropy loss (confidence) and propose a confidence-weighted complement entropy objective in addition to the primary cross-entropy objective.

Generally, our baseline is built on the seminal domain adversarial networks [9,10] and exploits conditional entropy minimization and an entropy-aware weight strategy [26]. In this paper, we equip the baseline model with two proposed techniques mentioned above and finally formulate a unified framework BA^{3}US which well addresses the negative transfer and uncertainty propagation problems in partial domain adaptation. We also empirically discover that the uncertainty suppression technique works well for vanilla closed-set domain adaptation.

To sum up, we make the following contributions. To our best knowledge, this is the first work that tackles partial domain adaptation by augmenting the target domain and transforming it into a UDA-like problem. The proposed balanced augmentation technique is fairly simple and works very well for PDA tasks. Besides, we address an overlooked issue in this field named uncertainty propagation by designing an adaptive weighted complement entropy for the source domain. Extensive results demonstrate that our approach yields new state-of-the-art results on several visual benchmark datasets, including Office31 [34], Office-Home [42], and ImageNet-Caltech [6].

2   Related Work

The past two decades have witnessed remarkable progress in domain adaptation. Interested readers can refer to [8,19,50] for taxonomy and survey.

Unsupervised Domain Adaptation (UDA). Compared with its supervised counterpart, UDA is more practical and challenging since no labeled data in the target domain are available. Recently, deep convolutional neural networks have achieved great success for visual recognition tasks, and we focus on deep UDA methods in this work. They can be categorized into three main groups. The first group aims to minimize the domain discrepancy by matching different statistic moments like maximum mean discrepancy (MMD) [22,25,27,41] and higher-order moment matching [18,37,47]. The second group that is widely used introduces a domain discriminator and exploits the idea of adversarial learning [12] to encourage domain confusion so that the discriminator can not decide which domain the data come from. Some typical examples are [2,10,39,40]. A third group is a reconstruction-based approach, assuming the reconstruction of both source and target domain samples to be important and helpful. Among them, [11,51] utilize encoder-decoder reconstruction and adversarial reconstruction, respectively. Despite their success for vanilla UDA, they are easily stuck by negative transfer for PDA due to the mismatched marginal label distributions.

Partial Domain Adaptation (PDA). In reality, PDA can be considered as a special case of imbalanced domain adaptation, where the target label distribution is quite dissimilar to that of the source domain. Until recent years, [31] first introduces an imbalanced scenario where label numbers of the source and target domains are not the same, which draws the attention of many researchers [4–6,17,30,38,49]. Different from shallow methods [31,38], deep methods [4,5,30,49] are mainly based on the domain adversarial learning framework and achieve promising recognition accuracy. Selective adversarial network (SAN) [4] exploits a multi-discriminator domain adversarial network and tries to select source-only classes by imposing different localized weights on different discriminators. Importance weighted adversarial nets (IWAN) [49] apply only one domain discriminator and weigh each source sample with the probability of being a target sample. Partial adversarial domain adaptation (PADA) [5] estimates the target label distribution and then feeds the class-wise weights to both the source discriminator and the domain discriminator, while two weighted inconsistency-reduced networks (TWINs) [30] leverage two independent networks to estimate the target label distribution and minimize the domain difference measured by the classifiers' inconsistency on the target samples. Deep Residual Correction Network (DRCN) [21] proposes a weighted class-wise matching strategy to explicitly align target data with the most relevant source subclasses. Recently, Example Transfer Network (ETN) [6] jointly learns domain-invariant representations across domains and a progressive weighting scheme to quantify the transferability of source examples, which achieves state-of-the-art results on several benchmark datasets. Generally, all the PDA methods above attempt to filter out the large source domain to match the small target domain. Comparatively, our method tries to augment the small target domain to match the source domain from a different perspective.

Data Synthesis and Augmentation. Recently, synthesis and augmentation techniques like CycleGAN [51] and *mixup* [48] are favored by UDA and semi-supervised learning methods for improving performance. For example, [16] directly exploits CycleGAN to generate target-like images from source samples to narrow the domain shift for adaptive semantic segmentation. [29,43] extend *mixup* to domain adaptation and generate pseudo training samples via interpolating between certain source samples and uncertain target samples. To some degree, target augmentation in this paper is like a special case of *mixup* where the mixup coefficient is always binary. However, the motivations are totally different. Our method considers neither the interpolated semantic label nor the interpolated domain label for a classification loss.

3 Proposed Method

We elaborate on the proposed framework for partial domain adaptation (PDA) in this section. First, we give definitions and notations. We follow the protocol of unsupervised PDA where we have a labeled source domain dataset $\mathcal{D}_s = \{(x_i^s, y_i^s)\}_{i=1}^{n_s}, x_i^s \in \mathbb{R}^d$ and an unlabeled target domain dataset $\mathcal{D}_t = \{(x_i^t)\}_{i=1}^{n_t}, x_i^t \in \mathbb{R}^d$ during the training stage. These two domains have different feature distributions: $p_s(x_s) \neq p_t(x_t)$ due to the domain shift. Notably, different from vanilla UDA, the target labels are a subset of the source labels for PDA: $\mathcal{Y}_t \subseteq \mathcal{Y}_s$, and C denotes the total number of classes in $\mathcal{Y}_s$.

We aim to learn a deep neural network $h : \mathcal{X} \to \mathcal{Y}$ that consists of two components: $h = g \circ f$. Here $f : \mathcal{X} \to \mathcal{Z}$ denotes the feature extractor and $g : \mathcal{Z} \to \mathcal{Y}$

denotes a class predictor. Since we target at learning domain-invariant features, the prediction function is assumed identical, i.e., $g = g_s = g_t$. For simplicity, we also share the feature extractor f for different domains. We introduce an adversarial classifier $D : \mathcal{Z} \rightarrow \{0,1\}$ to mitigate the distribution discrepancy across domains as explained later.

3.1 Domain Adversarial Learning Revisited

Generative adversarial network (GAN) [12] learns two competing components: the discriminator D and the generator F which play a minimax two-player game, where F tries to fool D by generating examples that are as realistic as possible and D tries to classify the generated samples as fake. Inspired by the idea of GAN, domain adversarial neural networks (DANN) [9,10] develops a two-player game for UDA, where one player $D(\cdot; \theta_d)$ (i.e., domain discriminator) tries to distinguish the source domain datum from that of the target domain and the other player $F(\cdot; \theta_f)$ (i.e., feature extractor) is trained to confuse the domain discriminator $D(\cdot; \theta_d)$. Generally, the minimax game of DANN is formulated as

$$\min_{\theta_f, \theta_g} \max_{\theta_d} \mathcal{L}_{cls}(\theta_f, \theta_g) + \lambda \, \mathcal{L}_{adv}(\theta_f, \theta_d),$$

$$\mathcal{L}_{adv}(\theta_f, \theta_d) = \frac{1}{n_s} \sum_{i=1}^{n_s} \log[D(F(x_i^s))] + \frac{1}{n_t} \sum_{j=1}^{n_t} \log[1 - D(F(x_j^t))], \quad (1)$$

$$\mathcal{L}_{cls}(\theta_f, \theta_g) = \frac{1}{n_s} \sum_{i=1}^{n_s} l_{ce}(G(F(x_i^s)), y_i^s),$$

where $l_{ce}(\cdot, \cdot)$ represents the softmax cross-entropy loss, $G(\cdot; \theta_g)$ denotes the source classifier, $F(\cdot; \theta_f)$ represents the domain-shared feature extractor, and λ is a hyper-parameter to trade-off the source risk and domain adversary. Different from a label flipping step in GAN, a gradient reversal layer (GRL) is further defined in [9] to optimize the objective in Eq. (1). Due to its simplicity and effectiveness, the idea of domain adversarial learning has been adopted in many previous domain adaptation works [24,26,40].

As shown in Eq. (1), each sample from both source and target domains is equally involved in the adversarial loss $\mathcal{L}_{adv}$, which seems not reasonable. If we only have samples distributed in the margin of the classifier (called 'hard' samples) from different domains and pursue domain alignment via them, it may perform badly for those 'easy' samples across domains. As such, we expect those 'hard' samples and 'easy' samples to own lower and higher weights during domain adversarial alignment. Specifically, we quantify the difficulty via the entropy criterion $H(h) = -\sum_{c=1}^{C} h_c \log(h_c)$ and adopt the same weighting strategy as [26] to impose an entropy-aware weight $w(x) = 1 + e^{-H(h(x))}$ on each sample,

$$\mathcal{L}_{adv}^e(\theta_f, \theta_d) = \frac{1}{n_s} \sum_{i=1}^{n_s} w(x_i^s) \log[D(F(x_i^s))] + \frac{1}{n_t} \sum_{j=1}^{n_t} w(x_j^t) \log[1 - D(F(x_j^t))]. \quad (2)$$

Obviously, if there is no domain shift, UDA (including PDA) degenerates to a typical semi-supervised learning problem. On one hand, we aim to mitigate the domain shift; on the other hand, this motivates us to adopt the popular entropy

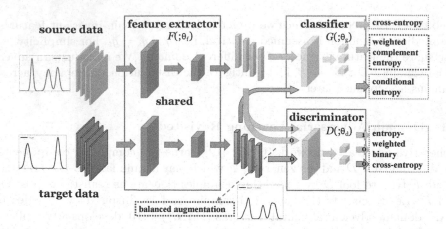

Fig. 1. Architecture of our domain adaptation method. There are three modules: a shared feature extractor F, a classifier G and a domain discriminator D. Different from domain adversarial learning [9], it contains two extra components with marked red border, i.e., balanced augmentation and weighted complement entropy. (Color figure online)

minimization principle [13] in semi-supervised learning to the PDA task. It is desirable that all the unlabeled target samples have highly-confident predictions. This is encouraged via the following conditional entropy term,

$$\mathcal{L}_{ent}(\theta_f, \theta_y) = \frac{1}{n_t} \sum\nolimits_{j=1}^{n_t} H(G(F(x_j^t))). \tag{3}$$

Inspired by the observation in [4] that redundant information is not beneficial for adaptation, we adopt the following optimization objective of Entropy-regularized DANN (**E-DANN**) as an initial model of our method,

$$\min_{\theta_f, \theta_g} \max_{\theta_d} \mathcal{L}_{cls}^w(\theta_f, \theta_g) + \alpha \mathcal{L}_{ent}(\theta_f, \theta_g) + \lambda \mathcal{L}_{adv}^e(\theta_f, \theta_d), \tag{4}$$

where $\mathcal{L}_{cls}^w(\theta_f, \theta_g) = \frac{1}{n_s} \sum_{i=1}^{n_s} m(y_i^s) l_{ce}(G(F(x_i^s)), y_i^s)$, m denotes the normalized estimated class-level weight vector via the target domain, and α, λ are two empirical trade-off parameters.

3.2   Balanced Adversarial Alignment (BAA)

The joint distribution shift is the actual root to negative transfer [45]. For example, the marginal label distributions are not symmetric in PDA, and thus source-only classes are prone to be matched with target classes, resulting in a negative transfer problem. Generally, a class-level source re-weighting scheme sounds a natural choice for PDA, since it is expected to filter out source-only classes and promote positive transfer between the shared classes across domains. Previous methods [5,30] resort to target predictions to generate class-level weights and effectively avoid negative transfer to some degree. Ideally, PDA with a weighting scheme behaves like a small UDA problem, but it heavily relies on accurate target predictions to calculate suitable weights.

In contrast, we propose an extremely simple scheme as shown in Fig. 1 for the challenging distribution alignment in PDA. Intuitively, we pursue the balance between different label distributions across domains with quite an opposite solution, i.e., augmenting the target domain using original source samples instead of weighting the source domain. This idea looks weird but is actually reasonable because we readily turn the PDA task into a large UDA-like task and the negative transfer effects caused by source-only classes can be well alleviated. The detailed formulation of balanced augmentation (alignment) is shown below,

$$\mathcal{L}_{adv}^{ba}(\theta_f, \theta_d) = \frac{1}{n_s} \sum\nolimits_{i=1}^{n_s} w(x_i^s) m(y_i^s) \log[D(F(x_i^s))]$$

$$+ \frac{1}{n_t} \sum\nolimits_{j=1}^{n_t} w(x_j^t) \log[1 - D(F(x_j^t))] + \frac{\rho}{n_s} \sum\nolimits_{i=1}^{n_s} w(x_i^s) m(y_i^s) \log[1 - D(F(x_i^s))].$$

$$(5)$$

Specifically, we adopt a progressive strategy for the target augmentation scheme. We borrow the source samples with different ratios, and the ratio ρ gradually decreases to 0 as the number of iterations increases. This is because the learned feature representations in early iterations are not quite transferable and we need more source samples to avoid class mismatch. When the learned features are desirably discriminative and transferable, the estimation of class-level weights becomes more accurate, making the augmentation trivial.

Note that we borrow original samples from the source domain rather than exploiting a generative model like CycleGAN [51] to synthesize target-like source images. The reason is that obtaining data-dependent translation models between such heterogeneous domains is quite time-consuming and we find translation does not even improve the adaptation results.

3.3   Adaptive Uncertainty Suppression (AUS)

Previous DA methods focus on strengthening the feature transferability by developing various domain alignment strategies, but they mostly ignore the feature discriminability and simply use the conventional cross-entropy loss in the labeled source domain to learn the features. In that case, even though the domain shift is mitigated, the

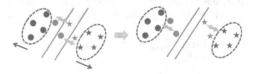

Fig. 2. Mitigating the effects of uncertainty propagation from source. [blue: source, red: target, gray: adversarial alignment.] (Color figure online)

classifier may perform worse on target data. This is because source classes are not equally separated from each other, which may lead to the propagation of the confusion (uncertainty) to target predictions and thus confusing features in the target domain, which is termed as the uncertainty propagation problem. Take a

3-way classification problem as an example. There is no class overlap in the source domain, but class 1 and 2 may be close to each other. A toy example of how class 1 and class 2 behave is shown in Fig. 2 where a few samples lie close to the decision boundary between class 1 and class 2. During domain alignment, some unlabeled target data are enforced to match these source data, which would be easily misclassified. However, such a critical problem has always been overlooked in prior literature. Looking back at the cross-entropy loss $l_{ce}(\hat{y}, y) = -\sum_i y_i \log(\hat{y}_i)$, it only exploits the information from the ground-truth class while ignoring that from other incorrect classes. For example, the source output [**0.6, 0.3, 0.1**] is more uncertain than [**0.6, 0.2, 0.2**], but both have the same cross-entropy loss.

Though several previous methods [20,35] incorporate virtual adversarial training [32] in the classification term to implicitly increase the margin between different classes, more computation complexity is required and several parameters need to be tuned. Inspired by [7], we exploit a complement entropy that expects uniform and low prediction scores for incorrect classes for labeled source samples. To accurately suppress uncertainty, we further place more emphasis on the uncertain samples that own smaller cross-entropy loss (confidence) and propose a confidence-weighted complement entropy objective below,

$$\mathcal{L}_{wce}^{w}(\theta_f, \theta_g) = \frac{1}{n_s \log(K-1)} \sum_{i=1}^{n_s} m(y_i^s) l_{wce}(G(F(x_i^s)), y_i^s),$$

$$\text{where } l_{wce}(\hat{y}, y) = (1 - \hat{y}_a)^{\xi} \sum_{j \neq a} \frac{\hat{y}_j}{1 - \hat{y}_a} \log(\frac{\hat{y}_j}{1 - \hat{y}_a}),$$

(6)

where ξ is a hyper-parameter and a is the index of ground-truth class in y. Different from the complementary training strategy in [7], we exploit the adaptive weighted complement entropy $\mathcal{L}_{wce}$ as a regularizer, which is more efficient. We also assign a class-level weight for each sample in Eq. (6) like that in $\mathcal{L}_{cls}^{w}$.

3.4   Unified Minimax Optimization Problem of BA^{3}US

Finally, we integrate all the terms in Eqs. (4, 5, 6) on both source and target samples to avoid negative transfer and uncertainty prorogation and derive a unified framework for PDA. The overall min-max objective is formulated as

$$\min_{\theta_f, \theta_g} \max_{\theta_d} \mathcal{L}_{cls}^{w}(\theta_f, \theta_g) + \alpha \mathcal{L}_{ent}(\theta_f, \theta_g) + \beta \mathcal{L}_{wce}^{w}(\theta_f, \theta_g) + \lambda \mathcal{L}_{adv}^{ba}(\theta_f, \theta_d), \quad (7)$$

where β is another trade-off hyper-parameter to balance the complement entropy and the cross-entropy term. Again, $m \propto \sum_{j=1}^{n_t} G(F(x_i^t))$ is the normalized estimated class-level weight vector. To optimize the objective above, we follow [9] to introduce a gradient reversal layer and adopt the same progressive strategy for parameter λ, i.e., increasing λ from 0 to 1 as the number of iterations grows.

Our method is closely related to DANN [9], sharing similar formalism of the domain adaptation theory [1] that the expected target risk $\epsilon_T(H)$ on the target examples is bounded by the source risk $\epsilon_S(H)$ and other two terms below,

$$\epsilon_T(H) \leq \epsilon_S(H) + |\epsilon_S(H, H^*) - \epsilon_T(H, H^*)| + [\epsilon_S(H^*) + \epsilon_T(H^*)], \quad (8)$$

where $H^* = \arg\min_{x \in \mathcal{H}}[\epsilon_S(x) + \epsilon_T(x)]$ is the ideal joint hypothesis for the combined risk. DANN further discovers the second term $\mathcal{H}\Delta\mathcal{H}$-distance can be upper bounded by error of the domain adversarial classifier. As we do not have labels of the target domain, we expect the entropy minimization term on the target domain to help reduce the last term. Besides, the proposed complement entropy objective alleviates uncertainty propagation to make the weight estimation more accurate, and thus our method would turn PDA into a small UDA task. In this way, we can expect that our method minimizes the empirical target risk $\mathbf{E}_{(x_t, y_t) \sim p_t}[f(x_t) \neq y_t]$. The detailed optimization is summarized in Algorithm 1.

4 Experiments

4.1 Setup

Datasets. Office31 dataset [34] includes images of 31 object classes from three different domains, i.e., Amazon, DSLR, and Webcam. We follow the standard protocol used in [5] and pick up images of 10 categories shared by Office31 and Caltech256 [14] as target domains. **Office-Home** dataset [42] consists of 4 different domains with each containing 65 kinds of everyday objects, i.e., Artistic, Clipart, Product images, and Real World images. Likewise, we follow [5] to select the first 25 categories (in alphabetical order) in each domain as a partial target domain. **ImageNet-Caltech** is a large-scale object recognition dataset that consists of two subsets, ImageNet-1K [33] and Caltech256 [14]. Here we use images from the public validation set of ImageNet-1K for the target domain. In reality, each source domain contains 1,000 and 256 classes, and each target domain contains 84 classes.

Baseline Methods. We utilize all the source and target samples and report the average classification accuracy and standard deviation over 3 random trials. **A → B** means **A** is the source domain and **B** is the partial target domain. For comprehensive comparison, we provide the recognition results of our methods including E-DANN, Ours (w/ BAA) and Ours (BA³US) on each dataset, and compare them with some popular UDA methods [26,40] and existing PDA methods, including SAN [4], IWAN [49], PADA [5], SSPDA [3], MWPDA [17], DRCN [21], ETN [6], and SAFN [46].

Implementation Details. If not specified, all the methods adopt **ResNet-50** [15] as backbone. We fine-tune the pre-trained ImageNet model in **PyTorch** using a NVIDIA Titan X (12 GB memory). The adversarial layer and classifier layer are trained through back-propagation, and the learning rate of the classifier layer is 10 times that of lower layers. We adopt mini-batch SGD with momentum of 0.9 and the learning rate annealing strategy as [10,26]: the learning rate is adjusted by $\eta_p = \eta_0(1 + \hat{\alpha}p)^{-\hat{\beta}}$, where p denotes the training progress changing from 0 to 1, and $\eta_0 = 0.01$, $\hat{\alpha} = 10$, $\hat{\beta} = 0.75$ as suggested by [26]. Based on the number of classes and balancing analysis in [7], we use $\beta = 5$ for Office31 and

Table 1. Accuracy (%) on **Office-Home** dataset for *partial domain adaptation* via ResNet-50 [15]. The best in bold red; the second best in italic blue

Method	Ar→Cl	Ar→Pr	Ar→Rw	Cl→Ar	Cl→Pr	Cl→Rw	Pr→Ar	Pr→Cl	Pr→Rw	Rw→Ar	Rw→Cl	Rw→Pr	Avg.
ResNet-50 [15]	46.33	67.51	75.87	59.14	59.94	62.73	58.22	41.79	74.88	67.40	48.18	74.17	61.35
ADDA [40]	45.23	68.79	79.21	64.56	60.01	68.29	57.56	38.89	77.45	70.28	45.23	78.32	62.82
CDAN+E [26]	47.52	65.91	75.65	57.07	54.12	63.42	59.60	44.30	72.39	66.02	49.91	72.80	60.73
IWAN [49]	53.94	54.45	78.12	61.31	47.95	63.32	54.17	52.02	81.28	76.46	56.75	82.90	63.56
SAN [4]	44.42	68.68	74.60	67.49	64.99	77.80	59.78	44.72	80.07	72.18	50.21	78.66	65.30
PADA [5]	51.95	67.00	78.74	52.16	53.78	59.03	52.61	43.22	78.79	73.73	56.60	77.09	62.06
SSPDA [3]	52.02	63.64	77.95	65.66	59.31	73.48	70.49	51.54	84.89	76.25	60.74	80.86	68.07
MWPDA [17]	55.39	77.53	81.27	57.08	61.03	62.33	68.74	56.42	86.67	76.70	57.67	80.06	68.41
ETN [6]	59.24	77.03	79.54	62.92	65.73	75.01	68.29	55.37	84.37	75.72	57.66	84.54	70.45
DRCN [21]	54.00	76.40	83.00	62.10	64.50	71.00	70.80	49.80	80.50	77.50	59.10	79.90	69.00
SAFN [46]	58.93	76.25	81.42	70.43	72.97	77.78	72.36	55.34	80.40	75.81	60.42	79.92	71.83
	±0.50	±0.33	±0.27	±0.46	±1.39	±0.52	±0.31	±0.46	±0.78	±0.37	±0.83	±0.20	
E-DANN	54.05	74.12	84.06	67.06	64.95	75.15	71.29	53.09	83.42	76.00	58.17	81.53	70.24
	±0.45	±0.12	±0.19	±0.19	±0.59	±0.19	±0.37	±0.22	±0.65	±0.51	±0.26		
Ours (w/ BAA)	56.20	79.55	86.21	70.86	69.94	81.06	72.51	57.91	86.47	77.10	59.34	83.64	73.40
	±0.28	±0.45	±0.32	±0.67	±2.52	±0.30	±0.32	±0.30	±0.32	±0.62	±0.46	±0.39	
Ours (BA³US)	60.62	83.16	88.39	71.75	72.79	83.40	75.45	61.59	86.53	79.25	62.80	86.05	75.98
	± 0.45	±0.12	±0.19	±0.19	±1.10	±0.59	±0.19	±0.37	±0.22	±0.65	±0.51	±0.26	

ImageNet-Caltech, and $\beta = 1$ for Office-Home. Besides, we set the number of intervals N/N_u to 10, and N_u is set to 200, 500, and 4,000 for Office31, Office-Home, and ImageNet-Caltech, respectively. Note that our method does not use the ten-crop technique [5,26] at the evaluation phase for better performance.

4.2   Quantitative Results for Partial Domain Adaptation

The results on three object recognition datasets including **Office-Home**, **Office31** and **ImageNet-Caltech** for PDA are shown in Table 1 and 2, with some baseline results directly reported from ETN [6] with the same protocol. Obviously, BA³US achieves the best or second-best results on 10 out of 12 transfer tasks on the **Office-Home** dataset, and Ours (w/ BAA) and SAFN obtain the second and third best results, respectively. Regarding the average accuracy, BA³US advances the state-of-the-art result on **Office-Home** in SAFN [46] by 5.78%, from 71.83% to 75.98%. For two specific tasks $Cl \rightarrow Ar$ and $Pr \rightarrow Rw$, BA³US merely performs slightly worse than the best method. Besides, a state-of-the-art UDA approach CDAN+E [26] performs worse than ResNet-50, which implies the difficulty of the partial transfer task. Further compared with UDA methods, even PDA methods like IWAN [49] and PADA [5] do not work well, which again indicates the partial transfer is quite challenging.

On the small-scale **Office31** dataset, BA³US again obtains the best average accuracy and performs the best in 3 out of 6 transfer tasks. For transfer tasks from a large source domain A to small target domains (D, W), BA³US remarkably outperforms other PDA methods. BA³US performs slightly worse for the $D \rightarrow A$ task because the target domain D is very small, making the proposed target augmentation and adaptive uncertainty suppression techniques inefficient. On the large-scale **ImageNet-Caltech** dataset, BA³US performs the best for both transfer tasks and still holds the best average accuracy with significant improvements. Moreover, UDA methods do not always perform better than ResNet-50, which implies they may suffer from the negative transfer problem.

Besides the ResNet-50 backbone network, we further investigate the effectiveness of BA^{3}US with another backbone network VGG-16 [36] on **Office31** and compare it with state-of-the-art methods in Table 3. It can be clearly observed that BA^{3}US achieves the best or second-best results for all the tasks, significantly advancing the average accuracy from 93.88% to 95.84%. Compared with the results in Table 2, we find BA^{3}US (97.81%→95.84%) is also robust than ETN (96.73%→93.88%) w.r.t. the change of the backbone network.

Table 2. Accuracy (%) on **Office31** and **ImageNet-Caltech** for *partial domain adaptation* via ResNet-50 [15]. The best in bold red; the second best in italic blue.

Method	Office31							ImageNet-Caltech		
	A → D	A → W	D → A	D → W	W → A	W → D	Avg.	I → C	C → I	Avg.
ResNet-50[15]	$83.44_{\pm1.12}$	$75.59_{\pm1.09}$	$83.92_{\pm0.95}$	$96.27_{\pm0.85}$	$84.97_{\pm0.86}$	$98.09_{\pm0.74}$	87.05	$69.69_{\pm0.78}$	$71.29_{\pm0.74}$	70.49
ADDA[40]	$83.41_{\pm0.17}$	$75.67_{\pm0.17}$	$83.62_{\pm0.14}$	$95.38_{\pm0.23}$	$84.25_{\pm0.13}$	$99.85_{\pm0.12}$	87.03	$71.82_{\pm0.45}$	$69.32_{\pm0.41}$	70.57
CDAN+E [26]	$77.07_{\pm0.90}$	$80.51_{\pm1.20}$	$93.58_{\pm0.07}$	$98.98_{\pm0.00}$	$91.65_{\pm0.00}$	$98.09_{\pm0.00}$	89.98	$72.45_{\pm0.07}$	$72.02_{\pm0.13}$	72.24
IWAN[49]	$90.45_{\pm0.36}$	$89.15_{\pm0.37}$	$95.62_{\pm0.29}$	$99.32_{\pm0.32}$	$94.26_{\pm0.25}$	$99.36_{\pm0.24}$	94.69	$78.06_{\pm0.40}$	$73.33_{\pm0.46}$	75.70
SAN[4]	$94.27_{\pm0.28}$	$93.90_{\pm0.45}$	$94.15_{\pm0.36}$	$99.32_{\pm0.52}$	$88.73_{\pm0.44}$	$99.36_{\pm0.12}$	94.96	$77.75_{\pm0.36}$	$75.26_{\pm0.42}$	76.51
PADA[5]	$82.17_{\pm0.37}$	$86.54_{\pm0.31}$	$92.69_{\pm0.29}$	$99.32_{\pm0.45}$	$95.41_{\pm0.33}$	$100.0_{\pm0.00}$	92.69	$75.03_{\pm0.36}$	$70.48_{\pm0.44}$	72.76
SSPDA[3]	90.87	91.52	90.61	92.88	94.36	98.94	93.20	-	-	-
MWPDA[17]	*95.12*	*96.61*	95.02	100.0	95.51	100.0	97.05	-	-	-
DRCN[21]	86.00	88.05	95.60	100.0	95.80	100.0	94.30	75.30	78.90	77.10
ETN[6]	$95.03_{\pm0.22}$	$94.52_{\pm0.20}$	$96.21_{\pm0.27}$	$100.0_{\pm0.00}$	$94.64_{\pm0.24}$	$100.0_{\pm0.00}$	96.73	$83.23_{\pm0.24}$	$74.93_{\pm0.44}$	79.08
E-DANN	$92.36_{\pm0.00}$	$93.22_{\pm0.00}$	$94.61_{\pm0.05}$	$100.0_{\pm0.00}$	$94.71_{\pm0.05}$	$98.73_{\pm0.00}$	95.60	$78.31_{\pm0.81}$	$77.69_{\pm0.25}$	78.00
Ours(w/ BAA)	$93.63_{\pm0.00}$	$93.90_{\pm0.00}$	$94.89_{\pm0.09}$	$100.0_{\pm0.00}$	$94.78_{\pm0.00}$	$100.0_{\pm0.00}$	*96.20*	$82.97_{\pm0.49}$	*$79.34_{\pm0.08}$*	*81.16*
Ours(BA^{3}US)	$99.36_{\pm0.00}$	$98.98_{\pm0.28}$	$94.82_{\pm0.05}$	$100.0_{\pm0.00}$	$94.99_{\pm0.06}$	$98.73_{\pm0.00}$	97.81	$84.00_{\pm0.15}$	$83.35_{\pm0.28}$	83.68

Table 3. Accuracy (%) on **Office31** for *partial domain adaptation* via VGG-16 [36]. The best in bold red; the second best in italic blue.

Method	A → D	A → W	D → A	D → W	W → A	W → D	Avg.
VGG-16 [36]	76.43 ± 0.48	60.34 ± 0.84	72.96 ± 0.56	97.97 ± 0.63	79.12 ± 0.54	*99.36 ± 0.36*	81.03
IWAN [49]	*90.95 ± 0.33*	82.90 ± 0.31	89.57 ± 0.24	79.75 ± 0.26	93.36 ± 0.22	88.53 ± 0.16	87.51
SAN [4]	90.70 ± 0.20	83.39 ± 0.36	87.16 ± 0.23	*99.32 ± 0.45*	91.85 ± 0.35	100.0 ± 0.00	92.07
PADA [5]	81.73 ± 0.34	*86.05 ± 0.35*	93.00 ± 0.24	100.0 ± 0.00	*95.26 ± 0.27*	100.0 ± 0.00	92.54
ETN [6]	89.43 ± 0.17	85.66 ± 0.16	95.93 ± 0.23	100.0 ± 0.00	92.28 ± 0.20	100.0 ± 0.00	*93.88*
Ours (BA^{3}US)	95.54 ± 0.00	89.83 ± 0.00	*94.92 ± 0.05*	99.32 ± 0.00	95.41 ± 0.00	100.0 ± 0.00	95.84

Ablation Study. As shown in Tables 1 and 2, we provide the results of E-DANN and Ours (w/ BAA) along with BA^{3}US on three datasets. Ours (w/ BAA) extends E-DANN by using the proposed balanced adversarial alignment technique in Sect. 3.2 instead, while Ours (BA^{3}US) extends Ours (w/ BAA) by considering the proposed adaptive complement entropy objective in Eq. (6). Firstly, the baseline method E-DANN always performs well for partial domain adaptation since it removes the irrelevant classes in the source classification term like [5]. Secondly, results on all three datasets demonstrate that Ours (BA³US) performs better than Ours (w/ BAA) and Ours (w/ BAA) performs better than E-DANN, which verify the effectiveness of two proposed techniques in Sect. 3.2 and Sect. 3.3.

4.3 Quantitative Results for Closed-Set Domain Adaptation

This section further investigates the effectiveness of the proposed uncertainty suppression technique for vanilla closed-set domain adaptation. Here we consider integrating them with DANN and CDAN [26] respectively, and compare our methods with state-of-the-art UDA approaches including [44,46] on the most-favored **Office-Home** dataset. As shown in Table 4, the proposed method BA³US built on CDAN obtains the best average accuracy and ranks the top two in 9 out of 12 different transfer tasks. It is obvious that the adaptive uncertainty suppression technique works well for closed-set domain adaptation, advancing the average accuracy from 67.6% to 68.7% and from 68.0% to 69.2%. In fact, we also study the effectiveness of balanced adversarial alignment but find it hardly improve the performance since the label distributions in UDA have already been symmetric.

Table 4. Accuracy (%) on **Office-Home** dataset for *vanilla unsupervised domain adaptation* via ResNet-50 [15]. Methods* utilize augmentation during evaluation.

Method	Ar→Cl	Ar→Pr	Ar→Rw	Cl→Ar	Cl→Pr	Cl→Rw	Pr→Ar	Pr→Cl	Pr→Rw	Rw→Ar	Rw→Cl	Rw→Pr	Avg.
CDAN+E [26]*	50.7	70.6	76.0	57.6	70.0	70.0	57.4	50.9	77.3	70.9	56.7	81.6	65.8
MCS [23]	55.9	73.8	79.0	57.5	69.9	71.3	58.4	50.3	78.2	65.9	53.2	82.2	66.3
DRCN [21]	50.6	72.4	76.8	61.9	69.5	71.3	60.4	48.6	76.8	72.9	56.1	81.4	66.6
CDAN+TransNorm [44]*	50.2	71.4	77.4	59.3	72.7	73.1	61.0	53.1	79.5	71.9	59.0	82.9	67.6
SAFN [46]*	54.4	73.3	77.9	65.2	71.5	73.2	63.6	52.6	78.2	72.3	58.0	82.1	68.5
SAFN [46]	52.0	71.7	76.3	64.2	69.9	71.9	63.7	51.4	77.1	70.9	57.1	81.5	67.3
Ours (w/ BAA)	50.9	72.0	77.5	61.2	72.6	72.7	62.8	52.7	79.9	70.8	56.6	82.7	67.6
Ours (BA³US)	51.2	73.8	78.1	63.3	73.4	73.6	63.3	54.5	80.4	72.6	56.7	83.7	68.7
Ours (w/ BAA)+CDAN	52.2	73.1	77.9	61.4	72.7	73.2	61.0	51.8	80.0	72.0	57.8	83.3	68.0
Ours (BA³US)+CDAN	54.1	74.2	77.7	62.9	73.6	74.6	63.4	54.9	80.4	73.1	58.2	83.6	69.2

4.4 Qualitative Results for Partial Domain Adaptation

We study our methods with different numbers of target classes in Fig. 3(a). The performance decreases when the number is larger than 15, and BA³US always obtains the best results. As expected, the proposed augmentation technique becomes important when the number of target classes is small.

Weight Visualization. As stated before, the weighting scheme plays an important role in PDA. Thus we investigate the quality of the estimated class-level weight m in Algorithm 1. As shown in Fig. 3(b–c), we plot the estimated weights of BA³US for the two specific tasks. Since the Cl domain and Ar domain are quite different, making the estimation challenging. This is also evidenced in other PDA methods in Table 1 since the accuracy is around 55%. For the relatively easy

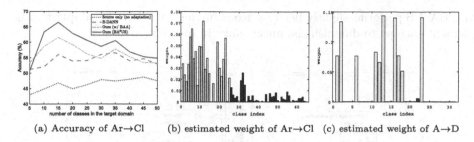

(a) Accuracy of Ar→Cl   (b) estimated weight of Ar→Cl   (c) estimated weight of A→D

Fig. 3. (a) Accuracy with varying number of target classes. (b–c) Estimated class-level weights. Yellow bins denote ground-truth classes. Best viewed in color. (Color figure online)

task A→D, the weight estimation seems accurate, resulting in high classification accuracy.

Parameter Sensitivity. We study the sensitivity of parameters ξ and β of BA^{3}US on the **Office-Home** and **Office31** datasets. The mean accuracy is reported in Table 5 and Table 6, respectively. Note that the complement entropy [7] can be considered as a special case of the proposed adaptive one where $\xi = 0$. The results indicate using an adaptive objective is much better, and the performance is relatively stable. Regarding the parameter β, the accuracy around 1.0 and 5.0 is also stable, implying our method is not sensitive.

Table 5. Sensitivity of parameter ξ.

Avg. (%)	0.0 [7]	0.1	0.3	0.5	0.7	0.9	1.0
Office-Home	75.32	75.88	**76.28**	*76.10*	*76.10*	75.81	75.98
Office31	97.68	97.71	97.65	97.67	97.64	**97.84**	*97.81*

Table 6. Sensitivity of parameter β.

Avg. (%)	0.0	0.1	0.5	1.0	5.0	10.0
Office-Home	73.40	73.58	75.09	**75.98**	*75.69*	73.25
Office31	96.20	96.13	96.50	96.63	*97.81*	**97.83**

Convergence Performance. As shown in Fig. 4, we study the convergence performance of the proposed methods for Ar→Cl and A→D. Obviously, the 'source only' method works worse without the domain alignment module, and E-DANN performs much better than it and quickly converges after 1,000 iterations. Besides, both BA^{3}US and Ours (w/ BAA) obtain similar promising results, and BA³US performs slightly better since it further considers the adaptive complement entropy to diminish the uncertainty in source predictions.

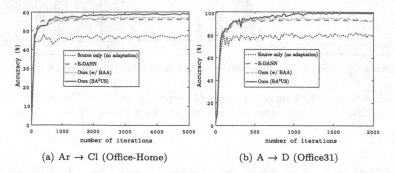

(a) Ar → Cl (Office-Home)   (b) A → D (Office31)

Fig. 4. Convergence analysis of proposed methods on two different transfer tasks. Accuracy (%) is given w.r.t. number of iterations. Best viewed in color. (Color figure online)

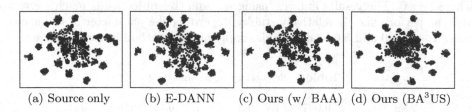

(a) Source only   (b) E-DANN   (c) Ours (w/ BAA)   (d) Ours (BA³US)

Fig. 5. t-SNE visualizations for the transfer task Ar→Cl. Blue: source; red: target. (Color figure online)

Feature Visualization. We plot in Fig. 5 the t-SNE embeddings [28] of the features learned by 'source only', E-DANN and BA³US for the transfer task Ar→Cl. It is easy to discover that the features of target data are rather confusing in 'no adaptation' while the balanced alignment module helps mitigate the domain gap and the adaptive uncertainty suppression module helps increase the discrimination of the features.

Algorithm 1: Pseudo code of our method termed BA^{3}US.

Input: Labeled source domain $\mathcal{D}_s$, unlabeled target domain $\mathcal{D}_t$;
Parameters: Total training iterations N, updating interval N_u, batch size $B_s = 36$, $\rho_0 = 1/4$, $\xi = 1$, $\alpha = 0.1$, $\beta \in \{1, 5\}$, λ;
Initialize the model parameters $\theta_f, \theta_g, \theta_d$;
Initialize the class-level weight vector m, $m_i = 1/C$;
for $i = 1$ to N do
  Obtain B_s samples from $\mathcal{D}_s$ and $\mathcal{D}_t$, respectively;
  Obtain ρB_s random samples from $\mathcal{D}_s$;
  Update $\theta_f, \theta_g, \theta_d$ by optimizing Eq. (7) and gradient reversaral layer;
  if $i \% N_u == 0$ then
    update the class-level weight vector m;
    gray% note that this step is ignored in closed-set domain adaptation
    calcuate $\mathcal{L}_{ent}$ in Eq. (3) for model selection;
    $\rho \leftarrow \rho_0 (1 - N_u/N)$;
  end
end
Output: Target outputs corresponding to the minimal value of $\mathcal{L}_{ent}$.

---

5 Conclusion

We develop a novel adversarial learning-based method BA^{3}US for partial domain adaptation, which well addresses two key problems, negative transfer and uncertainty propagation. To tackle the asymmetric label distributions, BA^{3}US offers a very simple solution by augmenting the target domain with samples from the source domain. Then, it uncovers an overlooked issue in the field termed uncertainty propagation and designs an adaptive complement entropy objective to well suppress the uncertainty in source predictions. Empirical results show it also works well for vanilla closed-set domain adaptation. Further experiments have validated that BA^{3}US improves existing methods with substantial gains, establishing new state-of-the-art.

Acknowledgment. J. Feng was partially supported by NUS ECRA FY17 P08, AISG-100E2019-035, and MOE Tier 2 MOE2017-T2-2-151. The authors also thank Quanhong Fu for her help to improve the technical writing aspect of this paper.

References

1. Ben-David, S., Blitzer, J., Crammer, K., Kulesza, A., Pereira, F., Vaughan, J.W.: A theory of learning from different domains. Mach. Learn. **79**(1–2), 151–175 (2010)
2. Bousmalis, K., Silberman, N., Dohan, D., Erhan, D., Krishnan, D.: Unsupervised pixel-level domain adaptation with generative adversarial networks. In: Proceedings of the CVPR, pp. 3722–3731 (2017)
3. Bucci, S., D'Innocente, A., Tommasi, T.: Tackling partial domain adaptation with self-supervision. arXiv preprint arXiv:1906.05199 (2019)
4. Cao, Z., Long, M., Wang, J., Jordan, M.I.: Partial transfer learning with selective adversarial networks. In: Proceedings of the CVPR, pp. 2724–2732 (2018)

5. Cao, Z., Ma, L., Long, M., Wang, J.: Partial adversarial domain adaptation. In: Ferrari, V., Hebert, M., Sminchisescu, C., Weiss, Y. (eds.) ECCV 2018. LNCS, vol. 11212, pp. 139–155. Springer, Cham (2018). https://doi.org/10.1007/978-3-030-01237-3_9
6. Cao, Z., You, K., Long, M., Wang, J., Yang, Q.: Learning to transfer examples for partial domain adaptation. In: Proceedings of the CVPR, pp. 2985–2994 (2019)
7. Chen, H.Y., et al.: Complement objective training. In: Proceedings of the ICLR (2019)
8. Csurka, G.: Domain adaptation for visual applications: A comprehensive survey. arXiv preprint arXiv:1702.05374 (2017)
9. Ganin, Y., Lempitsky, V.: Unsupervised domain adaptation by backpropagation. In: Proceedings of the ICML, pp. 1180–1189 (2015)
10. Ganin, Y., et al.: Domain-adversarial training of neural networks. J. Mach. Learn. Res. **17**(1), 1–35 (2016)
11. Ghifary, M., Kleijn, W.B., Zhang, M., Balduzzi, D., Li, W.: Deep reconstruction-classification networks for unsupervised domain adaptation. In: Leibe, B., Matas, J., Sebe, N., Welling, M. (eds.) ECCV 2016. LNCS, vol. 9908, pp. 597–613. Springer, Cham (2016). https://doi.org/10.1007/978-3-319-46493-0_36
12. Goodfellow, I., et al.: Generative adversarial nets. In: Proceedings of the NeurIPS, pp. 2672–2680 (2014)
13. Grandvalet, Y., Bengio, Y.: Semi-supervised learning by entropy minimization. In: Proceedings of the NeurIPS, pp. 529–536 (2005)
14. Griffin, G., Holub, A., Perona, P.: Caltech-256 object category dataset (2007)
15. He, K., Zhang, X., Ren, S., Sun, J.: Deep residual learning for image recognition. In: Proceedings of the CVPR, pp. 770–778 (2016)
16. Hoffman, J., et al.: CyCADA: cycle-consistent adversarial domain adaptation. In: Proceedings of the ICML, pp. 1989–1998 (2018)
17. Hu, J., Wang, C., Qiao, L., Zhong, H., Jing, Z.: Multi-weight partial domain adaptation. In: Proceedings of the BMVC (2019)
18. Koniusz, P., Tas, Y., Porikli, F.: Domain adaptation by mixture of alignments of second-or higher-order scatter tensors. In: Proceedings of the CVPR, pp. 7139–7148 (2017)
19. Kouw, W.M., Loog, M.: A review of single-source unsupervised domain adaptation. IEEE Trans. Pattern Anal. Mach. Intell. (2019, in press)
20. Kumar, A., et al.: Co-regularized alignment for unsupervised domain adaptation. In: Proceedings of the NeurIPS, pp. 9345–9356 (2018)
21. Li, S., et al.: Deep residual correction network for partial domain adaptation. IEEE Trans. Pattern Anal. Mach. Intell. (2020, in press)
22. Liang, J., He, R., Sun, Z., Tan, T.: Aggregating randomized clustering-promoting invariant projections for domain adaptation. IEEE Trans. Pattern Anal. Mach. Intell. **41**(5), 1027–1042 (2019)
23. Liang, J., He, R., Sun, Z., Tan, T.: Distant supervised centroid shift: a simple and efficient approach to visual domain adaptation. In: Proceedings of the CVPR, pp. 2975–2984 (2019)
24. Liu, M.Y., Tuzel, O.: Coupled generative adversarial networks. In: Proceedings of the NeurIPS, pp. 469–477 (2016)
25. Long, M., Cao, Y., Wang, J., Jordan, M.: Learning transferable features with deep adaptation networks. In: Proceedings of the ICML, pp. 97–105 (2015)
26. Long, M., Cao, Z., Wang, J., Jordan, M.I.: Conditional adversarial domain adaptation. In: Proceedings of the NeurIPS, pp. 1647–1657 (2018)

27. Long, M., Zhu, H., Wang, J., Jordan, M.I.: Deep transfer learning with joint adaptation networks. In: Proceedings of the ICML, pp. 2208–2217 (2017)
28. van der Maaten, L., Hinton, G.: Visualizing data using t-SNE. J. Mach. Learn. Res. **9**, 2579–2605 (2008)
29. Mao, X., Ma, Y., Yang, Z., Chen, Y., Li, Q.: Virtual mixup training for unsupervised domain adaptation. arXiv preprint arXiv:1905.04215 (2019)
30. Matsuura, T., Saito, K., Harada, T.: TWINs: Two weighted inconsistency-reduced networks for partial domain adaptation. arXiv preprint arXiv:1812.07405 (2018)
31. Ming Harry Hsu, T., Yu Chen, W., Hou, C.A., Hubert Tsai, Y.H., Yeh, Y.R., Frank Wang, Y.C.: Unsupervised domain adaptation with imbalanced cross-domain data. In: Proceedings of the ICCV, pp. 4121–4129 (2015)
32. Miyato, T., Maeda, S., Koyama, M., Ishii, S.: Virtual adversarial training: a regularization method for supervised and semi-supervised learning. IEEE Trans. Pattern Anal. Mach. Intell. **41**(8), 1979–1993 (2018)
33. Russakovsky, O., et al.: ImageNet large scale visual recognition challenge. Int. J. Comput. Vis. **115**(3), 211–252 (2015)
34. Saenko, K., Kulis, B., Fritz, M., Darrell, T.: Adapting visual category models to new domains. In: Daniilidis, K., Maragos, P., Paragios, N. (eds.) ECCV 2010. LNCS, vol. 6314, pp. 213–226. Springer, Heidelberg (2010). https://doi.org/10.1007/978-3-642-15561-1_16
35. Shu, R., Bui, H.H., Narui, H., Ermon, S.: A DIRT-T approach to unsupervised domain adaptation. In: Proceedings of the ICLR (2018)
36. Simonyan, K., Zisserman, A.: Very deep convolutional networks for large-scale image recognition. arXiv preprint arXiv:1409.1556 (2014)
37. Sun, B., Saenko, K.: Deep CORAL: correlation alignment for deep domain adaptation. In: Hua, G., Jégou, H. (eds.) ECCV 2016. LNCS, vol. 9915, pp. 443–450. Springer, Cham (2016). https://doi.org/10.1007/978-3-319-49409-8_35
38. Tsai, Y.H.H., Hou, C.A., Chen, W.Y., Yeh, Y.R., Wang, Y.C.F.: Domain-constraint transfer coding for imbalanced unsupervised domain adaptation. In: Proceedings of the AAAI, pp. 3597–3603 (2016)
39. Tzeng, E., Hoffman, J., Darrell, T., Saenko, K.: Simultaneous deep transfer across domains and tasks. In: Proceedings of the ICCV, pp. 4068–4076 (2015)
40. Tzeng, E., Hoffman, J., Saenko, K., Darrell, T.: Adversarial discriminative domain adaptation. In: Proceedings of the CVPR, pp. 2962–2971 (2017)
41. Tzeng, E., Hoffman, J., Zhang, N., Saenko, K., Darrell, T.: Deep domain confusion: Maximizing for domain invariance. arXiv preprint arXiv:1412.3474 (2014)
42. Venkateswara, H., Eusebio, J., Chakraborty, S., Panchanathan, S.: Deep hashing network for unsupervised domain adaptation. In: Proceedings of the CVPR, pp. 5018–5027 (2017)
43. Wang, Q., Li, W., Van Gool, L.: Semi-supervised learning by augmented distribution alignment. In: Proceedings of the ICCV, pp. 1466–1475 (2019)
44. Wang, X., Jin, Y., Long, M., Wang, J., Jordan, M.I.: Transferable normalization: towards improving transferability of deep neural networks. In: Proceedings of the NeurIPS, pp. 1951–1961 (2019)
45. Wang, Z., Dai, Z., Póczos, B., Carbonell, J.: Characterizing and avoiding negative transfer. In: Proceedings of the CVPR, pp. 11293–11302 (2019)
46. Xu, R., Li, G., Yang, J., Lin, L.: Larger norm more transferable: an adaptive feature norm approach for unsupervised domain adaptation. In: Proceedings of the ICCV, pp. 1426–1435 (2019)

47. Zellinger, W., Grubinger, T., Lughofer, E., Natschläger, T., Saminger-Platz, S.: Central moment discrepancy (CMD) for domain-invariant representation learning. In: Proceedings of the ICLR (2016)
48. Zhang, H., Cisse, M., Dauphin, Y.N., Lopez-Paz, D.: mixup: beyond empirical risk minimization. In: Proceedings of the ICLR (2018)
49. Zhang, J., Ding, Z., Li, W., Ogunbona, P.: Importance weighted adversarial nets for partial domain adaptation. In: Proceedings of the CVPR, pp. 8156–8164 (2018)
50. Zhang, L.: Transfer adaptation learning: A decade survey. arXiv preprint arXiv:1903.04687 (2019)
51. Zhu, J.Y., Park, T., Isola, P., Efros, A.A.: Unpaired image-to-image translation using cycle-consistent adversarial networks. In: Proceedings of the ICCV, pp. 2223–2232 (2017)

Unsupervised Deep Metric Learning with Transformed Attention Consistency and Contrastive Clustering Loss

Yang Li[1]([✉])(iD), Shichao Kan[2](iD), and Zhihai He[1](iD)

[1] University of Missouri, Columbia, MO, USA
yltb5@mail.missouri.edu, hezhi@missouri.edu
[2] Beijing Jiaotong University, Beijing, China
161120620@bjtu.edu.cn

Abstract. Existing approaches for unsupervised metric learning focus on exploring self-supervision information within the input image itself. We observe that, when analyzing images, human eyes often compare images against each other instead of examining images individually. In addition, they often pay attention to certain keypoints, image regions, or objects which are discriminative between image classes but highly consistent within classes. Even if the image is being transformed, the attention pattern will be consistent. Motivated by this observation, we develop a new approach to unsupervised deep metric learning where the network is learned based on self-supervision information across images instead of within one single image. To characterize the consistent pattern of human attention during image comparisons, we introduce the idea of transformed attention consistency. It assumes that visually similar images, even undergoing different image transforms, should share the same consistent visual attention map. This consistency leads to a pairwise self-supervision loss, allowing us to learn a Siamese deep neural network to encode and compare images against their transformed or matched pairs. To further enhance the inter-class discriminative power of the feature generated by this network, we adapt the concept of triplet loss from supervised metric learning to our unsupervised case and introduce the contrastive clustering loss. Our extensive experimental results on benchmark datasets demonstrate that our proposed method outperforms current state-of-the-art methods for unsupervised metric learning by a large margin.

Keywords: Unsupervised metric learning · Attention map · Consistency loss · Contrastive loss

Electronic supplementary material The online version of this chapter (https://doi.org/10.1007/978-3-030-58621-8_9) contains supplementary material, which is available to authorized users.

1   Introduction

Deep metric learning aims to learn discriminative features that can aggregate visually similar images into compact clusters in the high-dimensional feature space while separating images of different classes from each other. It has many important applications, including image retrieval [11,14,38], face recognition [37], visual tracking [33] and person re-identification [15,43]. In supervised deep metric learning, we assume that the labels for training data are available. In this paper, we consider unsupervised deep metric learning where the image labels are not available. Learning directly and automatically from images in an unsupervised manner without human supervision represents a very important yet challenging task in computer vision and machine learning.

Clustering is one of the earliest methods developed for unsupervised learning. Recently, motivated by the remarkable success of deep learning, researchers have started to develop unsupervised learning methods using deep neural networks [2]. Auto-encoder trains an encoder deep neural network to output feature representations with sufficient information to reconstruct input images by a paired decoder [44]. As we know, during deep neural network training, the network model is updated and learned in an iterative and progressive manner so that the network output can match the target. In other works, deep neural networks need human supervision to provide ground-truth labels. However, in unsupervised learning, there are no labels available. To address this issue, researchers have exploited the unique characteristics of images and videos to create various self-supervised labels, objective functions, or loss functions, which essentially convert the unsupervised learning into a supervised one so that the deep neural networks can be successfully trained. For example, in DeepCluster [2], clustering is used to generate pseudo labels for images. Various supervised learning methods have been developed to train networks to predict the relative position of two randomly sampled patches [6], solve Jigsaw image puzzles [26], predict pixel values of missing image patches [5], classify image rotations of four discrete angles [10], reconstruct image transforms [44], etc. Once successfully trained by these pretext tasks, the baseline network should be able to generate discriminative features for subsequent tasks, such as image retrieval, classification, matching, etc. [13].

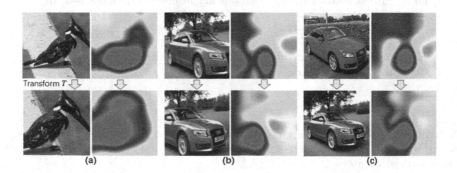

(a)   (b)   (c)

Fig. 1. Consistency of visual attention across images under transforms.

In this work, we propose to explore a new approach to unsupervised deep metric learning. We observe that existing methods for unsupervised metric learning focus on learning a network to analyze the input image itself. As we know, when examining and classifying images, human eyes compare images back and forth in order to identify discriminative features [9]. In other words, comparison plays an important role in human visual learning. When comparing images, they often pay attention to certain keypoints, image regions, or objects which are discriminative between image classes but highly consistent across image within classes. Even when the image is being transformed, the attention areas will be consistent. To further illustrate this, we provide three examples in Fig. 1. In (a), human eyes can easily tell the top image A of the first column and the bottom image B are the same bird since they have the same visual characteristics. The attention will be on the feather texture and head shape. In the pixel domain, A and B are up to a spatial transform, specifically, cropping plus resizing. When the human eyes moves from image A to its transformed version B, the attention will be also transformed so that it can be still focused on the head and feather. If we represent this attention using the attention map in deep neural networks, the attention map $\mathcal{M}(A)$ for image A and the attention map $\mathcal{M}(B)$ for image B should also follow the same transform, as shown in the second column of Fig. 1(a). We can also see this consistency of attention across image under different transforms in other examples in Figs. 1(b) and (c).

This lead to our idea of transformed attention consistency. Based on this idea, we develop a new approach to unsupervised deep metric learning based on image comparison. Specifically, using this consistency, we can define a pairwise self-supervision loss, allowing us to learn a Siamese deep neural network to encode and compare images against their transformed or matched pairs. To further enhance the inter-class discriminative power of the feature generated by this network, we adapt the concept of triplet loss from supervised metric learning to our unsupervised case and introduce the contrastive clustering loss. Our extensive experimental results on benchmark datasets demonstrate that our proposed method outperforms current state-of-the-art methods by a large margin.

2   Related Work and Major Contributions

This work is related to deep metric learning, self-supervised representation learning, unsupervised metric learning, and attention mechanisms.

(1) Deep Metric Learning. The main objective of deep metric learning is to learn a non-linear transformation of an input image by deep neural networks. In a common practice [35,42], the backbone in deep metric learning can be pre-trained on 1000 classes ImageNet [29] classification, and is then jointly trained on the metric learning task with an additional linear embedding layer. Many recent deep metric learning methods are built on pair-based [4,12,27] and triplet relationships [22,30,40]. Triplet loss [30] defines a positive pair and a negative pair based on the same anchor point. It encourages the embedding distance of positive pair to be smaller than the distance of negative pair by a given margin. Multi-similarity loss [35] considers multiple similarities and provides a more powerful approach for mining and weighting informative pairs by considering multiple similarities. The ability of mining informative pairs in existing methods is limited by the size of mini-batch. Cross-batch memory (XBM) [36] provides a memory bank for the feature embeddings of past iterations. In this way, the informative pairs can be identified across the dataset instead of a mini-batch.

(2) Self-supervised Representation Learning. Self-supervised representation learning directly derives information from unlabeled data itself by formulating predictive tasks to learn informative feature representations. DeepCluster [2] uses k-means clustering to assign pseudo-labels to the features generated by the deep neural network and introduces a discriminative loss to train the network. Gidaris *et al.* [10] explore the geometric transformation and propose to predict the angle ($0°$, $90°$, $180°$, and $270°$) of image rotation as a four-way classification. Zhang *et al.* [44] propose to predict the randomly sampled transformation from the encoded features by Auto-encoding transformation (AET). The encoder is forced to extract the features with visual structure information, which are informative enough for the decoder to decode the transformation. Self-supervision has been widely used to initialize and pre-train backbone on unlabeled data, and is then fine-tuned on a labeled training data for evaluating different tasks.

(3) Unsupervised Metric Learning. Unsupervised metric learning is a relatively new research topic. It is a more challenging task since the training classes have no labels and it does not overlap with the testing classes. Iscen *et al.* [17] propose an unsupervised method to mine hard positive and negative samples based on manifold-aware sampling. The feature embedding can be trained with standard contrastive and triplet loss. Ye *et al.* [42] propose to utilize the instance-wise relationship instead of class information in the learning process. It optimizes the instance feature embedding directly based on the positive augmentation invariant and negative separated properties.

(4) Attention Mechanism. The goal of the attention mechanism is to capture the most informative feature in the image. It explores important parts of features and suppress unnecessary parts [1,18,23]. Convolutional block attention module (CBAM) [39] is an effective attention method with channel and spatial attention module which can be integrated into existing convolutional neural network architectures. Fu *et al.* [8] propose to produce the attention proposals and train the attention module and embedding module in an iterative two-stage manner. Chen *et al.* [3] propose the hybrid-attention system by random walk graph propagation for object attention and the adversary constraint for channel attention.

Compared to existing methods, the ***unique contributions*** of this paper can be summarized as follows. (1) Unlike existing methods which focus on information analysis of the input image only, we explore a new approach for unsupervised deep metric learning based on image comparison and cross-image consistency. (2) Motivated by the human visual experience, we introduce the new approach of transformed attention consistency to effectively learn a deep neural network which can focus on discriminative features. (3) We extend the existing triplet loss developed for supervised metric learning to unsupervised learning using k-mean clustering to assign pseudo labels and memory bank to allow its access to all training samples, instead of samples in the current mini-batch. (4) Our experimental results demonstrate that our proposed approach has improved the state-of-the-art performance by a large margin.

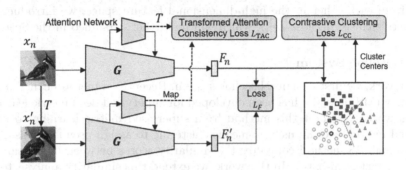

Fig. 2. Overview of the proposed approach for unsupervised deep metric learning with transformed attention consistency and contrastive clustering loss.

3 Method

3.1 Overview

Suppose that we have a set of unlabeled images $\mathcal{X} = \{x_1, x_2, \ldots, x_N\}$. Our goal is to learn a deep neural network to extract their features $\mathbf{G}(x_n) \in \mathbb{R}^d$, where d is the feature dimension. Figure 2 shows the overall design of our proposed method for unsupervised deep metric learning based on transformed attention consistency and contrastive clustering loss (TAC-CCL). Given an input image x_n, we apply a transform T, which is randomly sampled from a set of image transforms $\mathcal{T}$, to x_n, to obtain its transformed version $x'_n = T(x_n)$. In our experiments, we mainly consider spatial transforms, including cropping (sub-image), rotation, zooming, and perspective transform. Each transform is controlled by a set of transform parameters. For example, the cropping is controlled by its bounding box. The perspective transform is controlled by its 6 parameters. Image pairs (x_n, x'_n) are inputs to the Siamese deep neural network. These two identical networks will be trained to extract features F_n and F'_n for these two images. As illustrated in Fig. 2, each network is equipped with an attention network to learn the attention map which will modulate the output feature map. The attention map can enforce the network to focus on discriminative local features for the specific learning tasks or datasets. Let M_n and M'_n be the attention maps for images x_n and x'_n, respectively. According to the transformed attention consistency, we shall have

$$M'_n = T(M_n). \tag{1}$$

Based on this constraint, we introduce the transformed attention consistency loss L_{TAC} to train the feature embedding network $\mathbf{G}$, which will be further explained in Sect. 3.3. Besides this attention consistency, we also require that the output features F_n and F_n' should be similar to each other since the corresponding input images x_n and x_n' are visually the same. To enforce this constraint, we introduce the feature similarity loss $L_F = ||F_n - F_n'||_2$ which is the L_2-normal between these two features. To ensure that image features from the same class aggregate into compact clusters while image features from different classes are pushed away from each other in the high-dimensional feature space, we introduce the contrastive clustering loss L_{CC}, which will be further explained in the Sect. 3.3.

3.2 Baseline System

In this work, we first design a baseline system. Recently, a method called multi-similarity (MS) loss [35] has been developed for supervised deep metric learning. In this work, we adapt this method from supervised metric learning to unsupervised metric learning using k-means clustering to assign pseudo labels. Also, the original MS method computes the similarity scores between image samples in the current mini-batch. In this work, we extend this similarity analysis to the whole training set using the approach of memory bank [36]. The features of all training samples generated by the network are stored in the memory bank by the enqueue method. When the memory bank is full, the features and corresponding labels of the oldest mini-batch are removed by the dequeue method. Using this approach, the current mini-batch has access to the whole training set. We can then compute the similarity scores between all samples in the mini-batch and all samples in the training set. Our experiments demonstrate that this enhanced similarity matrix results in significantly improved performance in unsupervised metric learning. In this work, we use this network as the baseline system, denoted by TAC-CLL (baseline).

3.3 Loss Functions

To further improve the performance of the baseline system, we introduce the ideas of transformed attention consistency and contrastive clustering loss, which are explained in the following.

The transformed attention consistency aims to enforce the feature embedding network $\mathbf{G}$ to focus visually important features instead of other background noise. Let $M_n(u, v)$ and $M_n'(u, v)$ be the attention maps for input image pair x_n and x_n', where (u, v) represents a point location in the attention map. Under the transform T, this point is mapped to a new location denoted by $(T_u(u, v), T_v(u, v))$. According to (1), if we transform the attention map $M_n(u, v)$ for the original image x_n by T, it should match the attention map $M_n'(u, v)$ for the transformed image $x_n' = T(x_n)$. Based on this, the proposed transformed attention consistency loss L_{TAC} is defined as follows

$$L_{TAC} = \sum_{(u,v)} |M_n(u, v) - M_n'(T_u(u, v), T_v(u, v))|^2, \tag{2}$$

where $u' = T_u(u,v)$ and $v' = T_v(u,v)$ are the mapped location of (u,v) in image x'_n.

The constrastive clustering loss extends the triplet loss [30] developed in supervised deep metric learning, where an anchor sample x is associated with a positive sample x_+ and a negative sample x_-. The triplet loss aims to maximize the ratio $S(x,x_+)/S(x,x_-)$, where $S(\cdot,\cdot)$ represents the cosine similarity between two features. It should be noted that this triplet loss requires the knowledge of image labels, which however are not available in our unsupervised case. To extend this triplet loss to unsupervised metric learning, we propose to cluster the image features into K clusters. In the high-dimensional feature space, we wish these clusters are compact and are well separated from each other by large margins. Let $\{C_k\}$, $1 \leq k \leq K$, be the cluster centers. Let $C_+(F_n)$ be the nearest center which has the minimum distance to the input image feature F_n and the corresponding distance is denoted by $d_+(F_n) = ||F_n - C_+(F_n)||_2$. Let $C_-(F_n)$ be the cluster center which has the second minimum distance to F_n and the corresponding distance is denoted by $d_-(F_n) = ||F_n - C_-(F_n)||_2$. If the contrastive ratio of $d_+(F_n)/d_+(F_n)$ is small, then this feature has more discriminative power. We define the following contrastive clustering loss

$$L_{CC} = \mathcal{E}_{F_n} \left\{ \frac{||F_n - C_+(F_n)||_2}{||F_n - C_-(F_n)||_2} \right\}, \tag{3}$$

which is the average contrastive ratio of all input image features. During the training process, the network $\mathbf{G}$, as well as the feature for each input, is progressively updated. For example, the clustering is performed and the cluster centers are updated for every 20 epochs.

3.4 Transformed Attention Consistency with Cross-Images Supervision

Note that, in our proposed approach, we transform or augment the input image x_n to create its pair x'_n. These two are from the same image source. We also notice that most of existing self-supervision methods, such as predicting locations of image patches and classifying the rotation of an image [10], and reconstructing the transform of the image [44], all focus on self-supervision information within the image itself. The reason behind this is that image patches from the same image will automatically have the same class label. This provides an important self-supervision constraint to train the network. However, this one-image approach will limit the learning capability of the network since the network is not able to compare multiple images. As we know, when human eyes are examining images to determine which features are discriminative, they need to compare multiple images to determine which set of features are consistent across images and which set of features are background noise [9]. Therefore, in unsupervised learning, it is highly desirable to utilize the information across images.

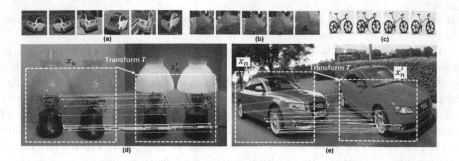

Fig. 3. Sub-image matching for cross-image supervision.

Figures 3(a)–(c) show image samples from the Cars and SOP benchmark datasets. We can see that images from the same class exhibit strong similarity between images, especially in the object regions. The question is how to utilize these unlabeled images to create reliable self-supervision information for unsupervised learning? In this work, we propose to perform keypoint or sub-image matching across images. Specifically, as illustrated in Fig. 3(d) and (e), for a given image sample I_n, in the pre-processing stage, we perform affine-SIFT [24] keypoint matching between I_n and other images in the dataset and find the top matches with confidence scores about a very high threshold. We then crop out the sub-images containing high-confidence keypoints as x_n and x'_n which are related by a transform T. This high-confidence constraint aims to ensure that x_n and x'_n are having the same object class or semantic label. In this way, for each image in the k-means cluster, we can find multiple high-confidence matched sub-images. For example, on the CUB dataset, for top-2 matching in each cluster, the label accuracy is 77.0%, which is much higher than the true positive pair rate obtained by k-means clustering (39.1%). This will significantly augment the training set, establish cross-image self-supervision, and provide significantly enhanced visual diversity for the network to learn more robust and discriminative features. In this work, we combine this cross-image supervision with the transformed attention consistency. Let $\{(u_i, v_i)\}$ and $\{(u'_i, v'_i)\}$, $1 \leq i \leq N$, be the set of matched keypoints in x_n and x'_n. We wish that, within the small neighborhoods of these matched keypoints, the attention maps M_n and M'_n are consistent. To define a small neighborhood around a keypoint (u_i, v_i), we use the following 2-D Gaussian kernel,

$$\phi(u - u_i, v - v_i) = e^{-\frac{(u-u_i)^2}{2\sigma_u^2} - \frac{(v-v_i)^2}{2\sigma_v^2}}. \tag{4}$$

Let

$$\Gamma(u, v) = \sum_{i=1}^{M} \phi(u - u_i, v - v_i), \quad \Gamma'(u, v) = \sum_{i=1}^{M} \phi(u - u'_i, v - v'_i), \tag{5}$$

which define two masks to indicate the neighborhood areas around these matched keypoints in these two attention maps. The extended transformed attention consistency becomes

$$L_{TAC} = \sum_{(u,v)} |M_n(u,v) \cdot \Gamma(u,v) - M'_n(u,v) \cdot \Gamma'(u,v)|^2, \tag{6}$$

which compares the difference between these two attention maps around these matched keypoints. Compared to the label propagation method developed for semi-supervised learning [45,46], our cross-image supervision method is unique in the following aspects: (1) it discovers sub-images of the same label (with very high probability) from unlabeled images. (2) It establishes the transform between these two sub-images and combines with the transformed attention consistency to achieve efficient unsupervised deep metric learning.

4   Experimental Results

In this section, we conduct extensive experiments on benchmark datasets in image retrieval settings to evaluate the proposed TAC-CCL method for unsupervised deep metric learning.

4.1   Datasets

We follow existing papers on unsupervised deep metric learning [42] to evaluate our proposed methods on the following three benchmark datasets. **(1) CUB-200-2011 (CUB)** [34] is composed of 11,788 images of birds from 200 classes. The first 100 classes (5864 images) are used for training, with the rest 100 classes (5924 images) for testing. **(2) Cars-196 (Cars)** [21] contains 16,185 images of 196 classes of car models. We use the first 98 classes with 8054 images for training, and remaining 98 classes (8131 images) for testing. **(3) Stanford Online Products (SOP)** [27] has 22,634 classes (120,053 images) of online products. We use the first 11,318 products (59,551 images) for training and the remaining 11,316 products (60,502 images) for testing. The training classes are separated from the test classes. We use the standard image retrieval performance metric (Recall@K), for performance evaluations and comparisons.

4.2   Implementation Details

We implement our proposed method by PyTorch and follow the standard experimental settings in existing papers [27,35,42] for performance comparison. We use the same GoogLeNet [32] pre-trained on ImageNet as the backbone network [25,27,31] and a CBAM [39] attention module is placed after the *inception_5b* layer. A fully connected layer is then added on the top of the network as the embedding layer. The default dimension of embedding is set as 512. For the clustering, we set the number of clusters K to be 100 for the CUB and Cars datasets, and $K = 10000$ for the SOP dataset. For each batch, we follow the data sampling strategy in multi-similarity loss [35] to sample 5 images per class.

For data augmentation, images in the training set are randomly cropped at size 227×227 with random horizontal flipping, while the images in testing set is center cropped. Adam optimizer [20] is used in all experiments and the weigh decay is set as $5e^{-4}$.

Table 1. Recall@K (%) performance on CUB and Cars datasets in comparison with other methods.

Methods	Backbone	CUB				Cars			
		R@1	R@2	R@4	R@8	R@1	R@2	R@4	R@8
Supervised methods									
ABIER [28]	GoogLeNet	57.5	68.7	78.3	86.2	82.0	89.0	93.2	96.1
ABE [19]	GoogLeNet	60.6	71.5	79.8	87.4	85.2	90.5	94.0	96.1
Multi-similarity [35]	BN-Inception	65.7	77.0	86.3	91.2	84.1	90.4	94.0	96.5
Unsupervised methods									
Examplar [7]	GoogLeNet	38.2	50.3	62.8	75.0	36.5	48.1	59.2	71.0
NCE [41]	GoogLeNet	39.2	51.4	63.7	75.8	37.5	48.7	59.8	71.5
DeepCluster [2]	GoogLeNet	42.9	54.1	65.6	76.2	32.6	43.8	57.0	69.5
MOM [17]	GoogLeNet	45.3	57.8	68.6	78.4	35.5	48.2	60.6	72.4
Instance [42]	GoogLeNet	46.2	59.0	70.1	80.2	41.3	52.3	63.6	74.9
TAC-CCL (baseline)	GoogLeNet	**53.9**	**66.2**	**76.9**	**85.8**	**43.0**	**53.8**	**65.3**	**76.0**
TAC-CCL	GoogLeNet	**57.5**	**68.8**	**78.8**	**87.2**	**46.1**	**56.9**	**67.5**	**76.7**
Gain		+11.3	+9.8	+8.7	+7.0	+4.8	+4.6	+3.9	+1.8

4.3   Performance Comparisons with State-of-the-Art Methods

We compare the performance of our proposed methods with the state-of-the-art unsupervised methods on image retrieval tasks. The mining on manifolds (MOM) [17] and the invariant and spreading instance feature method (denoted by Instance) [42] are current state-of-the-art methods for unsupervised metric learning. They both use the GoogLeNet [32] as the backbone encoder. In the Instance paper [42], the authors have also implemented three other state-of-the-art methods originally developed for feature learning and adapted them to unsupervised metric learning tasks: Examplar [7], NCE (Noise-Contrastive Estimation) [41], and DeepCluster [2]. We include the results of these methods for comparisons. We have also included the performance of recent supervised deep metric learning methods for comparison so that we can see the performance difference between unsupervised metric learning and supervised one. These methods include: ABIER [28], and ABE [19], and MS (Multi-Similarity) [35]. Both ABIER and ABE methods are using the GoogLeNet as the backbone encoder. The MS method is using the BN-Inception network [16] as the backbone encoder.

The results for the CUB, Cars, and SOP datasets are summarized in Tables 1 and 2, respectively. We can see that our proposed TAC-CCL method achieves new state-of-the-art performance in unsupervised metric learning on both fine-grained CUB and Cars datasets and the large-scale SOP dataset. On the CUB dataset, our TAC-CCL improves the Recall@1 by 11.3% and is even competitive to some supervised metric learning methods, e.g., ABIER [28]. On the Cars dataset, our TAC-CCL outperforms the current state-of-the-art Instance method [42] by 4.8%. On SOP, our method achieves 63.9% and outperforms existing methods by a large margin of 15%. For other Recall@K rates with large values of k, the amount of improvement is also very significant. Note that our baseline system achieves a large improvement over existing methods. The proposed TAC-CCL approach further improves upon this baseline system by another 1.4–3.6%.

Figure 4 shows examples of retrieval results from the CUB, Cars, and SOP datasets. In each row, the first image highlighted with a blue box is the query image. The rest images are the top 15 retrieval results. Images highlighted with red boxes are from different classes. It should be noted that some classes have very small number of samples. We can see that our TAC-CCL can learn discriminative features to achieve satisfying retrieval results, even for these challenging tasks. For example, at the first row of the SOP dataset, our model is able to learn the glass decoration feature under the lampshade, which is a unique feature of the query images. In addition, the negative retrieved results are also visually closer to the query images.

Table 2. Recall@K (%) performance on SOP dataset in comparison with other methods.

Methods	Backbone	SOP		
		R@1	R@10	R@100
Supervised methods				
ABIER [28]	GoogLeNet	74.2	86.9	94.0
ABE [19]	GoogLeNet	76.3	88.4	94.8
Multi-similarity [35]	BN-Inception	78.2	90.5	96.0
Unsupervised methods				
Examplar [7]	GoogLeNet	45.0	60.3	75.2
NCE [41]	GoogLeNet	46.6	62.3	76.8
DeepCluster [2]	GoogLeNet	34.6	52.6	66.8
MOM [17]	GoogLeNet	43.3	57.2	73.2
Instance [42]	GoogLeNet	48.9	64.0	78.0
TAC-CCL (baseline)	GoogLeNet	**62.5**	**76.5**	**87.2**
TAC-CCL	GoogLeNet	**63.9**	**77.6**	**87.8**
Gain		+15.0	+13.6	+9.8

4.4   Ablation Studies

In this section, we conduct ablation studies to perform in-depth analysis of our proposed method and its different components.

(1) Impact of the Number of Clusters. The proposed contrastive clustering loss is based on clustering in the feature space. The number of clusters K is a

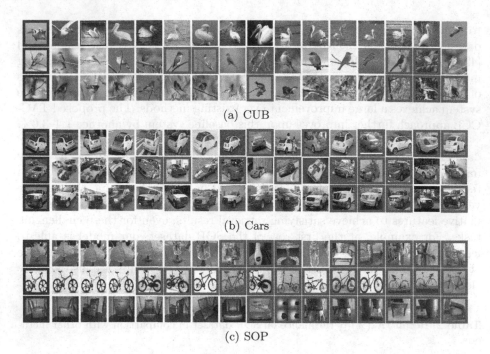

(a) CUB

(b) Cars

(c) SOP

Fig. 4. Retrieval results of some example queries on CUB, Cars, and SOP datasets. The query images and the negative retrieved images are highlighted with blue and red. (Color figure online)

critical parameter for the proposed method since it determines the number of pseudo labels. We conduct the following ablation study experiment on the CUB data to study the impact of K. The first plot in Fig. 5a shows the Recall@1 results with different values of K: 50, 100, 200, 500, and 1000. The other three plots show the results for Recall@2, 4, and 8. We can see that, on this dataset, the best value of K is 100, which is the number of test classes in the CUB dataset. The performance drops when K increases. This study suggests that the best value of K is close to the truth number test classes of the dataset.

(2) Impact of Different Embedding Sizes. In this ablation study, we follow existing supervised metric learning methods [28,35] to study the impact of different embedding sizes, or the size of the embedded feature. For example, the feature size ranges from 64, 128, 256, 512, to 1024. The first plot of Fig. 5b shows the Recall@1 results for different embedding size. The results for Recall@2, 4, 8 are shown in the other three plots. We can see that unsupervised metric learning performance increases with the embedding size since it contains more feature information with enhanced discriminative power.

(3) Impact of the Pre-trained Model. We follow the recent state-of-the-art unsupervised metric learning Instance method [42] and evaluate the performance of our proposed method on the large-scale SOP dataset by using the Resnet-18

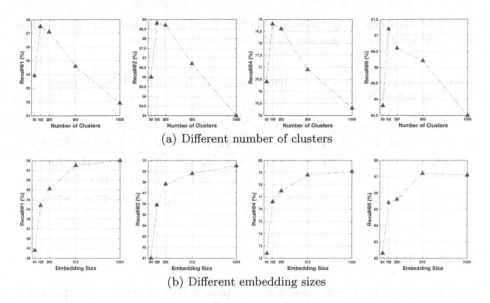

(a) Different number of clusters

(b) Different embedding sizes

Fig. 5. Recall@K (%) performance on CUB dataset in comparison with different number of clusters and different embedding size.

network without pre-trained parameters. From Table 3, we can see our proposed method outperforms Instance method [42] by more than 7%.

(4) Performance Contributions of Different Algorithm Components. Our proposed system has three major components: the baseline system for unsupervised deep metric learning, transformed attention consistency (TAC), and contrastive clustering loss (CCL). In this ablation study, we aim to identify the contribution of each algorithm component on different datasets. Table 4 summarizes the performance results on the CUB, Cars, and SOP datasets using three different method configurations: (1) the baseline system, (2) baseline with CCL, and (3) baseline with CCL + TAC. We can see that both the CCL and TAC approaches significantly improve the performance.

Table 3. Recall@K (%) performance on SOP dataset using Resnet-18 network without pre-trained parameters.

Methods	SOP		
	R@1	R@10	R@100
Random	18.4	29.4	46.0
Examplar [7]	31.5	46.7	64.2
NCE [41]	34.4	49.0	65.2
MOM [17]	16.3	27.6	44.5
Instance [42]	39.7	54.9	71.0
Ours	**47.0**	**62.6**	**77.5**

In our Supplementary Material, we will provide additional method implementation details, experimental results, and ablation studies.

Table 4. The performance of different components from our TAC-CCL method on CUB, Cars, and SOP datasets.

	CUB				Cars				SOP		
	R@1	R@2	R@4	R@8	R@1	R@2	R@4	R@8	R@1	R@10	R@100
Baseline	53.9	66.2	76.9	85.8	43.0	53.8	65.3	76.0	62.5	76.5	87.2
+CCL	55.7	67.8	77.5	86.2	44.7	55.6	65.9	75.7	63.0	76.8	87.2
+TAC	**57.5**	**68.8**	**78.8**	**87.2**	**46.1**	**56.9**	**67.5**	**76.7**	**63.9**	**77.6**	**87.8**

5   Conclusions

In this work, we have developed a new approach to unsupervised deep metric learning based on image comparisons, transformed attention consistency, and constrastive clustering loss. This transformed attention consistency leads to a pairwise self-supervision loss, allowing us to learn a Siamese deep neural network to encode and compare images against their transformed or matched pairs. To further enhance the inter-class discriminative power of the feature generated by this network, we have adapted the concept of triplet loss from supervised metric learning to our unsupervised case and introduce the contrastive clustering loss. Our extensive experimental results on benchmark datasets demonstrate that our proposed method outperforms current state-of-the-art methods by a large margin.

Acknowledgement. This work was supported in part by National Science Foundation under grants 1647213 and 1646065. Any opinions, findings, and conclusions or recommendations expressed in this material are those of the authors and do not necessarily reflect the views of the National Science Foundation.

References

1. Ba, J., Mnih, V., Kavukcuoglu, K.: Multiple object recognition with visual attention. arXiv preprint arXiv:1412.7755 (2014)
2. Caron, M., Bojanowski, P., Joulin, A., Douze, M.: Deep clustering for unsupervised learning of visual features. In: Ferrari, V., Hebert, M., Sminchisescu, C., Weiss, Y. (eds.) Computer Vision – ECCV 2018. LNCS, vol. 11218, pp. 139–156. Springer, Cham (2018). https://doi.org/10.1007/978-3-030-01264-9_9
3. Chen, B., Deng, W.: Hybrid-attention based decoupled metric learning for zero-shot image retrieval. In: Proceedings of the IEEE Conference on Computer Vision and Pattern Recognition, pp. 2750–2759 (2019)

4. Chopra, S., Hadsell, R., LeCun, Y.: Learning a similarity metric discriminatively, with application to face verification. In: 2005 IEEE Computer Society Conference on Computer Vision and Pattern Recognition, CVPR 2005, vol. 1, pp. 539–546. IEEE (2005)
5. DeVries, T., Taylor, G.W.: Improved regularization of convolutional neural networks with cutout. arXiv preprint arXiv:1708.04552 (2017)
6. Doersch, C., Gupta, A., Efros, A.A.: Unsupervised visual representation learning by context prediction. In: Proceedings of the IEEE International Conference on Computer Vision, pp. 1422–1430 (2015)
7. Dosovitskiy, A., Fischer, P., Springenberg, J.T., Riedmiller, M., Brox, T.: Discriminative unsupervised feature learning with exemplar convolutional neural networks. IEEE Trans. Pattern Anal. Mach. Intell. **38**(9), 1734–1747 (2015)
8. Fu, J., Zheng, H., Mei, T.: Look closer to see better: Recurrent attention convolutional neural network for fine-grained image recognition. In: Proceedings of the IEEE Conference on Computer Vision and Pattern Recognition, pp. 4438–4446 (2017)
9. Gazzaniga, M.S.: The Cognitive Neurosciences. MIT Press, Cambridge (2009)
10. Gidaris, S., Singh, P., Komodakis, N.: Unsupervised representation learning by predicting image rotations. arXiv preprint arXiv:1803.07728 (2018)
11. Grabner, A., Roth, P.M., Lepetit, V.: 3D pose estimation and 3D model retrieval for objects in the wild. In: Proceedings of the IEEE Conference on Computer Vision and Pattern Recognition, pp. 3022–3031 (2018)
12. Hadsell, R., Chopra, S., LeCun, Y.: Dimensionality reduction by learning an invariant mapping. In: 2006 IEEE Computer Society Conference on Computer Vision and Pattern Recognition, CVPR 2006, vol. 2, pp. 1735–1742. IEEE (2006)
13. He, K., Fan, H., Wu, Y., Xie, S., Girshick, R.: Momentum contrast for unsupervised visual representation learning. arXiv preprint arXiv:1911.05722 (2019)
14. He, X., Zhou, Y., Zhou, Z., Bai, S., Bai, X.: Triplet-center loss for multi-view 3D object retrieval. In: Proceedings of the IEEE Conference on Computer Vision and Pattern Recognition, pp. 1945–1954 (2018)
15. Hermans, A., Beyer, L., Leibe, B.: In defense of the triplet loss for person re-identification. arXiv preprint arXiv:1703.07737 (2017)
16. Ioffe, S., Szegedy, C.: Batch normalization: Accelerating deep network training by reducing internal covariate shift. arXiv preprint arXiv:1502.03167 (2015)
17. Iscen, A., Tolias, G., Avrithis, Y., Chum, O.: Mining on manifolds: metric learning without labels. In: Proceedings of the IEEE Conference on Computer Vision and Pattern Recognition, pp. 7642–7651 (2018)
18. Jaderberg, M., Simonyan, K., Zisserman, A., et al.: Spatial transformer networks. In: Advances in Neural Information Processing Systems, pp. 2017–2025 (2015)
19. Kim, W., Goyal, B., Chawla, K., Lee, J., Kwon, K.: Attention-based ensemble for deep metric learning. In: Ferrari, V., Hebert, M., Sminchisescu, C., Weiss, Y. (eds.) ECCV 2018. LNCS, vol. 11205, pp. 760–777. Springer, Cham (2018). https://doi.org/10.1007/978-3-030-01246-5_45
20. Kingma, D.P., Ba, J.: Adam: A method for stochastic optimization. arXiv preprint arXiv:1412.6980 (2014)
21. Krause, J., Stark, M., Deng, J., Fei-Fei, L.: 3D object representations for fine-grained categorization. In: 4th International IEEE Workshop on 3D Representation and Recognition, 3dRR-13, Sydney, Australia (2013)
22. Kumar, V., Harwood, B., Carneiro, G., Reid, I., Drummond, T.: Smart mining for deep metric learning. arXiv preprint arXiv:1704.01285 (2017)

23. Mnih, V., Heess, N., Graves, A., et al.: Recurrent models of visual attention. In: Advances in Neural Information Processing Systems, pp. 2204–2212 (2014)
24. Morel, J.M., Yu, G.: ASIFT: a new framework for fully affine invariant image comparison. SIAM J. Imaging Sci. **2**(2), 438–469 (2009)
25. Movshovitz-Attias, Y., Toshev, A., Leung, T.K., Ioffe, S., Singh, S.: No fuss distance metric learning using proxies. In: Proceedings of the IEEE International Conference on Computer Vision, pp. 360–368 (2017)
26. Noroozi, M., Favaro, P.: Unsupervised learning of visual representations by solving jigsaw puzzles. In: Leibe, B., Matas, J., Sebe, N., Welling, M. (eds.) ECCV 2016. LNCS, vol. 9910, pp. 69–84. Springer, Cham (2016). https://doi.org/10.1007/978-3-319-46466-4_5
27. Oh Song, H., Xiang, Y., Jegelka, S., Savarese, S.: Deep metric learning via lifted structured feature embedding. In: Proceedings of the IEEE conference on computer vision and pattern recognition, pp. 4004–4012 (2016)
28. Opitz, M., Waltner, G., Possegger, H., Bischof, H.: Deep metric learning with bier: boosting independent embeddings robustly. IEEE Trans. Pattern Anal. Mach. Intell. **42**, 276–290 (2018)
29. Russakovsky, O., et al.: ImageNet large scale visual recognition challenge. Int. J. Comput. Vis. **115**(3), 211–252 (2015)
30. Schroff, F., Kalenichenko, D., Philbin, J.: FaceNet: a unified embedding for face recognition and clustering. In: Proceedings of the IEEE Conference on Computer Vision and Pattern Recognition, pp. 815–823 (2015)
31. Sohn, K.: Improved deep metric learning with multi-class N-pair loss objective. In: Advances in Neural Information Processing Systems, pp. 1857–1865 (2016)
32. Szegedy, C., et al.: Going deeper with convolutions. In: Proceedings of the IEEE Conference on Computer Vision and Pattern Recognition, pp. 1–9 (2015)
33. Tao, R., Gavves, E., Smeulders, A.W.: Siamese instance search for tracking. In: Proceedings of the IEEE Conference on Computer Vision and Pattern Recognition, pp. 1420–1429 (2016)
34. Wah, C., Branson, S., Welinder, P., Perona, P., Belongie, S.: The Caltech-UCSD Birds-200-2011 (2011)
35. Wang, X., Han, X., Huang, W., Dong, D., Scott, M.R.: Multi-similarity loss with general pair weighting for deep metric learning. In: Proceedings of the IEEE Conference on Computer Vision and Pattern Recognition, pp. 5022–5030 (2019)
36. Wang, X., Zhang, H., Huang, W., Scott, M.R.: Cross-batch memory for embedding learning. arXiv preprint arXiv:1912.06798 (2019)
37. Wen, Y., Zhang, K., Li, Z., Qiao, Yu.: A discriminative feature learning approach for deep face recognition. In: Leibe, B., Matas, J., Sebe, N., Welling, M. (eds.) ECCV 2016. LNCS, vol. 9911, pp. 499–515. Springer, Cham (2016). https://doi.org/10.1007/978-3-319-46478-7_31
38. Wohlhart, P., Lepetit, V.: Learning descriptors for object recognition and 3D pose estimation. In: Proceedings of the IEEE Conference on Computer Vision and Pattern Recognition, pp. 3109–3118 (2015)
39. Woo, S., Park, J., Lee, J.-Y., Kweon, I.S.: CBAM: convolutional block attention module. In: Ferrari, V., Hebert, M., Sminchisescu, C., Weiss, Y. (eds.) ECCV 2018. LNCS, vol. 11211, pp. 3–19. Springer, Cham (2018). https://doi.org/10.1007/978-3-030-01234-2_1
40. Wu, C.Y., Manmatha, R., Smola, A.J., Krahenbuhl, P.: Sampling matters in deep embedding learning. In: Proceedings of the IEEE International Conference on Computer Vision, pp. 2840–2848 (2017)

41. Wu, Z., Xiong, Y., Yu, S.X., Lin, D.: Unsupervised feature learning via non-parametric instance discrimination. In: Proceedings of the IEEE Conference on Computer Vision and Pattern Recognition, pp. 3733–3742 (2018)
42. Ye, M., Zhang, X., Yuen, P.C., Chang, S.F.: Unsupervised embedding learning via invariant and spreading instance feature. In: Proceedings of the IEEE Conference on Computer Vision and Pattern Recognition, pp. 6210–6219 (2019)
43. Yu, R., Dou, Z., Bai, S., Zhang, Z., Xu, Y., Bai, X.: Hard-aware point-to-set deep metric for person re-identification. In: Ferrari, V., Hebert, M., Sminchisescu, C., Weiss, Y. (eds.) ECCV 2018. LNCS, vol. 11220, pp. 196–212. Springer, Cham (2018). https://doi.org/10.1007/978-3-030-01270-0_12
44. Zhang, L., Qi, G.J., Wang, L., Luo, J.: AET vs. AED: unsupervised representation learning by auto-encoding transformations rather than data. In: Proceedings of the IEEE Conference on Computer Vision and Pattern Recognition, pp. 2547–2555 (2019)
45. Zhu, X., Ghahramani, Z.: Learning from labeled and unlabeled data with label propagation. Technical report. Citeseer (2002)
46. Zhu, X.J.: Semi-supervised learning literature survey. University of Wisconsin-Madison Department of Computer Sciences, Technical report (2005)

STEm-Seg: Spatio-Temporal Embeddings for Instance Segmentation in Videos

Ali Athar[1(✉)], Sabarinath Mahadevan[1(✉)], Aljoša Ošep[2(✉)],
Laura Leal-Taixé[2(✉)], and Bastian Leibe[1(✉)]

[1] RWTH Aachen University, Aachen, Germany
{athar,mahadevan,leibe}@vision.rwth-aachen.de
[2] Technical University of Munich, Munich, Germany
{aljosa.osep,leal.taixe}@tum.de

Abstract. Existing methods for instance segmentation in videos typically involve multi-stage pipelines that follow the tracking-by-detection paradigm and model a video clip as a sequence of images. Multiple networks are used to detect objects in individual frames, and then associate these detections over time. Hence, these methods are often non-end-to-end trainable and highly tailored to specific tasks. In this paper, we propose a different approach that is well-suited to a variety of tasks involving instance segmentation in videos. In particular, we model a video clip as a single 3D spatio-temporal volume, and propose a novel approach that segments and tracks instances across space and time in a single stage. Our problem formulation is centered around the idea of spatio-temporal embeddings which are trained to cluster pixels belonging to a specific object instance over an entire video clip. To this end, we introduce (i) novel mixing functions that enhance the feature representation of spatio-temporal embeddings, and (ii) a single-stage, proposal-free network that can reason about temporal context. Our network is trained end-to-end to learn spatio-temporal embeddings as well as parameters required to cluster these embeddings, thus simplifying inference. Our method achieves state-of-the-art results across multiple datasets and tasks. Code and models are available at https://github.com/sabarim/STEm-Seg.

1 Introduction

The task of segmenting and tracking multiple objects in videos is becoming increasingly popular due to a surge in development of autonomous vehicles and robots that are able to perceive and accurately track surrounding objects. These advances are driven by the recent emergence of new datasets [11,80,94] containing videos with dense, per-pixel annotations of object instances. The underlying

---

A. Athar and S. Mahadevan—Equal contribution.

Electronic supplementary material The online version of this chapter (https://doi.org/10.1007/978-3-030-58621-8_10) contains supplementary material, which is available to authorized users.

A. Vedaldi et al. (Eds.): ECCV 2020, LNCS 12356, pp. 158–177, 2020.
https://doi.org/10.1007/978-3-030-58621-8_10

task tackled in these datasets can be summarized as follows: given an input video containing multiple objects, each pixel has to be uniquely assigned to a specific object instance or to the background.

Fig. 1. Our method is applicable to multi-object segmentation tasks such as VOS *(top)*, VIS *(middle)* and MOTS *(bottom)*.

State-of-the-art methods tackling these tasks [80,83,94] usually operate in a *top-down* fashion and follow the *tracking-by-detection* paradigm which is well-established in multi-object tracking (MOT) [8,37,47,62]. Such methods usually employ multiple networks to detect objects in individual images [32], associate them over consecutive frames, and resolve occlusions using learned appearance models. Though these approaches yield high-quality results, they involve multiple networks, are computationally demanding, and not end-to-end trainable.

Inspired by the Perceptual Grouping Theory [66], we learn to segment object instances in videos in a *bottom-up* fashion by leveraging *spatio-temporal* embeddings. To this end, we propose an efficient, single-stage network that operates directly on a 3D spatio-temporal volume. We train the embeddings in a category-agnostic setting, such that pixels belonging to the same object instance across the spatio-temporal volume are mapped to a single cluster in the embedding space. This way, we can infer object instances by simply assigning pixels to their respective clusters. Our method outperforms proposal-based methods for tasks involving pixel-precise tracking such as Unsupervised Video Object Segmentation (UVOS) [9,11], Video Instance Segmentation (VIS) [94], and Multi-Object Tracking and Segmentation (MOTS) [80].

To summarize, our contributions are the following: (i) We propose a unified approach for tasks involving instance segmentation in videos [11,80,94]. Our method performs consistently well under highly varied settings (see Fig. 1) such as automotive driving scenes [80], semantically diverse YouTube videos [94] and scenarios where object classes are not pre-defined [11]. (ii) We propose using spatio-temporal embeddings for the aforementioned set of tasks. To this end, we propose a set of mixing functions (Sect. 3.2) that improve performance by modifying the feature representation of these embeddings. Our method enables a simple inference procedure based on clustering within a 3D spatio-temporal volume, thus alleviating the need for external components for temporal association. (iii) We propose a single-stage network architecture which is able to effectively incorporate temporal context and learn the spatio-temporal embeddings.

2   Related Work

Image-Level Instance Segmentation: Several existing methods for image-level instance segmentation, which is closely related to our task, operate in a *top-down* fashion by using a Region Proposal Network to generate object proposals [13,69,70,73], which are then classified and segmented [32]. Other methods operate in a *bottom-up* fashion by grouping pixels belonging to the same object instance [18,19,42,57–59]. Novotny *et al.* [59] introduce an embedding mixing function to overcome appearance ambiguity, and predict a displacement field from the instance-specific center, which is similar to Hough voting based methods for object detection [46,81]. Recent methods for 3D object detection and instance segmentation also follow this trend [21,22,40,72,100]. Neven *et al.* [57] extend [18] by training a network to predict object centers and clustering bandwidths, thus alleviating the need for density-based clustering algorithms [16,23,54]. This serves as a basis for our work. However, in contrast to our approach, the aforementioned methods are only suitable for image-level segmentation.

Video Segmentation: Temporally consistent object segmentation in videos can benefit several other tasks such as action/activity recognition [29,34], video object detection [25,41] and object discovery [17,44,64,82,90,91]. Several *bottom-up* methods segment moving regions by grouping based on optical flow [2,92] or point trajectories [7,60,67,91]. By contrast, our method segments and tracks both moving and static objects. Other methods obtain video object proposals [25,29,34,41] based on cues such as actions [29,34] or image-level objects of interest [41,65]. Feature representations are then learned for these proposals in order to perform classification. Instead, we propose a single-stage approach that localizes objects with pixel-level precision.

Pixel-Precise Tracking of Multiple Objects: Multi object tracking (MOT) has its roots in robotic perception [38,68,76,86]. Although some works extend MOT with pixel-precise masks [56,63], a much larger set of works can be found in the domain of Video Object Segmentation (VOS), which encompasses multiple tasks related to pixel-precise tracking. In the *semi-supervised* variant of VOS [10], the ground-truth masks of the objects which are meant to be tracked are given for the first frame. State-of-the-art approaches to this task [9,36,61,77–79,89, 93,95,99] usually involve online fine-tuning on the first frame masks and/or follow the *object proposal generation and mask propagation* approach [53,99].

Chen *et al.* [14] tackle this problem by learning embeddings from the first-frame masks, and then associating pixels in the remaining frames.

More relevant to our work is the task of *unsupervised* VOS [11,71]. Here, no ground truth information is provided at test-time, and the goal is to segment and track all dominant "foreground" objects in the video clip. Current state-of-the-art methods for this task are either proposal-based [99,101] or focus on foreground-background segmentation [35,39,74,75, 77,78,96,97]. Li *et al.* [48] propose an approach that groups pixel embeddings based on *objectness* and optical flow cues. In contrast to ours, this method processes each frame separately, employs K-means clustering for object localization, and cannot separate different object instances. Wang *et al.* [84] employ an attentive graph neural network [30] and use differentiable message passing to propagate information across time; we compare our results to theirs in Sect. 4.5.

Recently, the task of Multi-Object Tracking and Segmentation (MOTS) [80] and Video Instance Segmentation (VIS) [94] were introduced. Voigtlaender *et al.* [80] extended the KITTI [28] and MOTChallenge [45,55] datasets with pixel-precise annotations, and proposed a baseline method that adapts Mask R-CNN [32] to associate object proposals across time. Hu *et al.* [1] use a mask network to filter foreground embeddings and perform mean-shift clustering to generate object masks, which are then associated over time using a distance threshold unlike our single-stage end-to-end method.

The YouTube-VIS [94] contains a large number of YouTube videos with per-pixel annotations for a variety of object classes. Compared to MOTS, these videos contain fewer instances, but are significantly more diverse. In addition to adapting several existing methods [6,31,79,85,95] for this task, Yang *et al.* [94] also proposed their own method (MaskTrack-RCNN) which extends Mask R-CNN with additional cues for data association. Methods such as [20,26,51] also rely on object proposals and/or heuristic post-processing to associate objects over time, unlike our end-to-end bottom-up approach.

3   Our Method

We tackle the task of instance segmentation in videos by modeling the video clip as a 3D spatio-temporal volume and using a network to learn an embedding for each pixel in that volume. This network is trained to push pixels belonging to different object instances towards different, non-overlapping clusters in the embedding space. This differs from most existing approaches, which first generate object detections per-frame, and then associate them over time. The following sections explain our problem formulation, the network architecture and loss functions employed, and the inference process.

3.1   Problem Formulation

As input, we assume a video clip with T frames of resolution $H \times W$. Let us denote the set of RGB pixels in this clip with $\mathcal{X} \in \mathbb{R}^{N \times 3}$ where $N = T \times H \times W$. Assuming there are K object instances in this clip (K is unknown), our aim is to produce a segmentation that assigns each pixel in the clip to either the background, or to exactly one of these K instances. We design a network that predicts video instance tubes by clustering pixels simultaneously across space and time based on a learned embedding function. Instead of learning just the embedding function and then using standard density-based clustering algorithms [27,52,54] to obtain instance tubes, we take inspiration from Neven et al. [57] and design a network which estimates the cluster centers as well as their corresponding variances, thus enabling efficient inference. Formally, our network can be viewed as a function that maps the set of pixels $\mathcal{X}$ to three outputs; (1) $\mathcal{E} \in \mathbb{R}^{N \times E}$: an E-dimensional embedding for each pixel, (2) $\mathcal{V} \in \mathbb{R}_{+}^{N \times E}$: a positive variance for each pixel and each embedding dimension, and (3) $\mathcal{H} \in [0,1]^N$: an object instance center heat-map.

Instance Representation: The network is trained such that the embeddings belonging to the jth instance in the video clip are modeled by a multivariate Gaussian distribution $\mathcal{N}(\boldsymbol{\mu}_j, \boldsymbol{\Sigma}_j)$. Assuming that this instance comprises the set of pixel coordinates $\mathcal{C}_j$ with cardinality N_j, we denote the embeddings and variances output by the network at these coordinates using $\mathcal{E}_j \subset \mathcal{E}$, $\mathcal{E}_j \in \mathbb{R}^{N_j \times E}$ and $\mathcal{V}_j \subset \mathcal{V}$, $\mathcal{V}_j \in \mathbb{R}_{+}^{N_j \times E}$, respectively. During training, $\mathcal{C}_j$ (i.e., the ground truth mask tube for instance j) is known. Using it, the mean $\boldsymbol{\mu}_j$ and covariance $\boldsymbol{\Sigma}_j$ of the distribution are computed by averaging over the per-pixel outputs:

$$\boldsymbol{\mu}_j = \frac{1}{N_j} \sum_{e \in \mathcal{E}_j} e \in \mathbb{R}^E, \qquad \boldsymbol{\Sigma}_j = \frac{1}{N_j} \operatorname{diag}\left(\sum_{v \in \mathcal{V}_j} v\right) \in \mathbb{R}^{E \times E}. \qquad (1)$$

This single distribution models all embeddings belonging to instance j across the entire clip (not individually for each frame). We can now use the distribution $\mathcal{N}(\boldsymbol{\mu}_j, \boldsymbol{\Sigma}_j)$ to compute the probability p_{ij} of any embedding $e_i \in \mathcal{E}$, anywhere in the input clip, of belonging to instance j:

$$p_{ij} = \frac{1}{(2\pi)^{\frac{E}{2}} |\boldsymbol{\Sigma}_j|^{\frac{1}{2}}} \exp\left(-\frac{1}{2}(e_i - \boldsymbol{\mu}_j)^T \boldsymbol{\Sigma}_j^{-1}(e_i - \boldsymbol{\mu}_j)\right). \qquad (2)$$

Using Eq. 2 to compute p_{ij}, $\forall\, i \in \{1, .., N\}$, we can obtain the set of pixels $\widehat{\mathcal{C}}_j$ comprising the predicted mask tube for instance j by thresholding the probabilities at 0.5:

$$\widehat{\mathcal{C}}_j = \{(x_i, y_i, t_i) \,|\, i \in \{1, .., N\},\ p_{ij} > 0.5\}. \qquad (3)$$

Training: This way, the training objective can be formulated as one of learning the optimal parameters $\boldsymbol{\mu}_j^{\text{opt}}$ and $\boldsymbol{\Sigma}_j^{\text{opt}}$ which maximize the intersection-over-union (IoU) between the predicted and ground-truth mask tubes in the clip:

$$\boldsymbol{\mu}_j^{\text{opt}}, \boldsymbol{\Sigma}_j^{\text{opt}} = \underset{\boldsymbol{\mu}_j, \boldsymbol{\Sigma}_j}{\text{argmax}} \; \frac{\widehat{\mathcal{C}}_j \cap \mathcal{C}_j}{\widehat{\mathcal{C}}_j \cup \mathcal{C}_j}, \tag{4}$$

that is, all pixels in the ground truth mask tube $\mathcal{C}_j$ should have probability larger than 0.5, and vice versa. A classification loss such as cross-entropy could be used to optimize the pixel probabilities. However, we employ the Lovàsz hinge loss [3,4,57,98] (details in supplementary material), which is a differentiable, convex surrogate of the Jaccard index that directly optimizes the IoU. This formulation allows $\boldsymbol{\mu}_j$ and $\boldsymbol{\Sigma}_j$ to be implicitly learned by the network.

Using Eqs. 1–4, we can define and optimize the distribution for every instance ($i.e. \forall j \in \{1, \ldots, K\}$). Note that only a single forward pass of the network is required regardless of the number of instances in the clip. This is in contrast to common approaches for Video Object Segmentation, which only process one instance at a time [61,89].

We further remark that ours is a *bottom-up approach* which detects and tracks objects in a single step, thus mitigating the inherent drawback of *top-down* approaches that often require different networks/cues for single-image object detection and temporal association. A further advantage of our approach is that it can implicitly resolve occlusions, insofar as they occur within the clip.

3.2   Embedding Representation

Under the formulation described in Sect. 3.1, the network can learn arbitrary representations for the embedding dimensions. However, it is also possible to fix the representation by using a mixing function $\phi : \mathbb{R}^E \to \mathbb{R}^E$ that modifies the embeddings $\mathcal{E}$ as follows: $\mathcal{E} \leftarrow \{\phi(e), \,|\, e \in \mathcal{E}\}$.

In [57], for the task of single-image instance segmentation, 2D embeddings were used ($E = 2$) in conjunction with a spatial coordinate mixing function $\phi_{\text{xy}}(e_i) = e_i + [x_i, y_i]^1$. With this setting, the embeddings could be interpreted as offsets to the (x, y) coordinates of their respective locations. The network thus learned to cluster the embeddings belonging to a given object towards some object-specific point on the image. It follows that the predicted variances could be interpreted as the network's estimate of the size of the object along the x and y axes. We postulate that the reason for this formulation yielding good results is that the (x, y) coordinates already serve as a good initial feature for instance separation; the network can then enhance this representation by producing offsets which further improve the clustering behavior. Furthermore, this can be done in an end-to-end trainable fashion which allows the network to adjust the clustering parameters, *i.e.*, the Gaussian distribution parameters, for each instance. In general, it has been shown that imparting spatial coordinate information to CNNs can improve performance for a variety of tasks [50,59].

---

[1] This notation denotes element-wise addition between e_i and $[x_i, y_i]$.

Compared to single-image segmentation, the task of associating pixels across space and time in videos poses additional challenges, *e.g.*, camera ego-motion, occlusions, appearance/pose changes. To tackle these challenges we propose (and experimentally validate in Sect. 4.3) the following extensions:

Spatio-Temporal Coordinating Mixing: Since we operate on video clips instead of single images, a logical extension is to use 3D embeddings ($E = 3$) with a spatio-temporal coordinate mixing function $\phi_{\text{xyt}}(e_i) = e_i + [x_i, y_i, t_i]$.

Free Dimensions: In addition to the spatial (and temporal) coordinate dimensions, it can be beneficial to include extra dimensions whose representation is left for the network to decide. The motivation here is to improve instance clustering quality by allowing additional degrees of freedom in the embedding space. From here on, we shall refer to these extra embedding dimensions as *free dimensions*. For example, if $E = 4$ with 2 spatial coordinate dimensions and 2 free dimensions, the mixing function is denoted as $\phi_{\text{xyff}}(e_i) = e_i + [x_i, y_i, 0, 0]$.

There is, however, a caveat with free dimensions: since the spatial (and temporal) dimensions already achieve reasonable instance separation, the network may converge to a poor local minimum by producing very large variances for the free dimensions instead of learning a discriminative feature representation. Consequently, the free dimensions may end up offering no useful instance separation during inference. We circumvent this problem at the cost of introducing one extra hyper-parameter by setting the variances for the free dimensions to a fixed value v_{free}. We justify our formulation quantitatively using multiple datasets and different variants of the mixing function $\phi(\cdot)$ in Sect. 4.3.

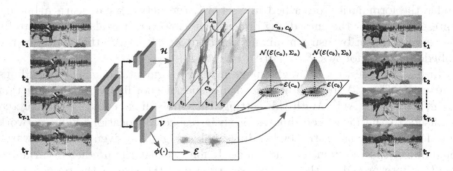

Fig. 2. For an input video clip, our network produces embeddings ($\mathcal{E}$), variances ($\mathcal{V}$), and instance center heat map ($\mathcal{H}$). $\mathcal{H}$ contains one peak per object for the entire spatio-temporal volume (c_a for the rider, c_b for the horse). $\mathcal{E}(c_a)$, $\mathcal{E}(c_b)$ and $\mathcal{V}(c_a)$, $\mathcal{V}(c_b)$ are the corresponding embeddings and variances at c_a and c_b, respectively. These quantities are then used to define the Gaussian distribution for each object.

3.3   Inference

Since the ground truth mask tube is not known during inference, it is not possible to obtain μ_j and Σ_j using Eq. 1. This is where the instance center heat map $\mathcal{H}$ comes into play. For each pixel $c_i = (x_i, y_i, t_i)$, the value $\mathcal{H}(c_i) \in [0, 1]$ in the heat map at this location gives us the probability of the embedding vector $\mathcal{E}(c_i)$ at this location being an instance center. The sequential process of inferring object instances in a video clip is described in the following algorithm:

1. Identify the coordinates of the most likely instance center $c_j = \text{argmax}_i \mathcal{H}(c_i)$.
2. Find the corresponding embedding vector $\mathcal{E}(c_j)$ and variances $\mathcal{V}(c_j)$.
3. Using $\mu_j \leftarrow \mathcal{E}(c_j)$ and $\Sigma_j \leftarrow \text{diag}(\mathcal{V}(c_j))$, generate the 3D mask tube $\widehat{\mathcal{C}}_j$ for this instance by computing per-pixel probabilities using Eq. 2, and then thresholding them as in Eq. 3.
4. Since the pixels in $\widehat{\mathcal{C}}_j$ have now been assigned to an instance, the embeddings, variances and heat map probabilities at these pixel locations are masked out and removed from further consideration:

$$\mathcal{E} \leftarrow \mathcal{E} \setminus \widehat{\mathcal{E}}_j, \qquad \mathcal{V} \leftarrow \mathcal{V} \setminus \widehat{\mathcal{V}}_j, \qquad \mathcal{H} \leftarrow \mathcal{H} \setminus \widehat{\mathcal{H}}_j. \tag{5}$$

5. Repeat steps 1–4 until either $\mathcal{E} = \mathcal{V} = \mathcal{H} = \emptyset$, or the next highest probability in the heat map falls below some threshold.

Even though this final clustering step (Fig. 2) depends on the number of instances in a clip, in practice, the application of Eqs. 2 and 3 carries little computational overhead and its run-time is negligible compared to a forward pass.

Video Clip Stitching: Due to memory constraints, the clip length that can be input to the network is limited. In order to apply our framework to videos of arbitrary length, we split the input video into clips of length T with an overlap of T_c frames between consecutive clips. Linear assignment [43] is then used to associate the predicted tracklets in consecutive clips. The cost metric for this assignment is the IoU between tracks in overlapping frames. Our approach is currently *near online* because, given a new frame, the delay until its output becomes available is at most $T - T_c - 1$.

3.4   Losses

Our model's loss function is a linear combination of three terms:

$$L_{\text{total}} = L_{\text{emb}} + L_{\text{smooth}} + L_{\text{center}} \tag{6}$$

Embedding Loss L_{emb}: As mentioned in Sect. 3.1, we use the Lovàsz hinge loss to optimize the IoU between the predicted and ground truth masks for a given instance. The embedding loss for the entire input clip is calculated as the mean of the Lovàsz hinge loss for all object instances in that clip.

Variance Smoothness Loss L_{smooth}: To ensure that the variance values at every pixel belonging to an object are consistent, we employ a smoothness loss L_{smooth} similar to [57]. This regresses the variances $\mathcal{V}_j$ for instance j to be close to the average value of all the variances for that instance, $i.e.\,\text{Var}[\mathcal{V}_j]$.

Instance Center Heat Map Loss L_{center}: For all pixels belonging to instance j, the corresponding outputs in the sigmoid activated heat map $\mathcal{H}_j$ are trained with an L_2 regression loss to match the output of Eq. 2. The outputs for background pixels are regressed to 0. During inference, this enables us to sample the highest values from the heat map which corresponds to the peak of the learned Gaussian distributions for the object instances in an input volume, as explained in Sect. 3.3.

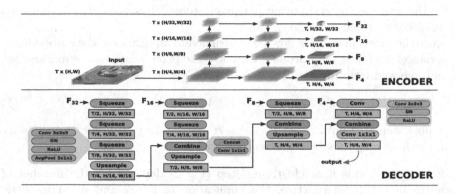

Fig. 3. The network has an encoder-decoder structure. GN: Group Normalization [87].

3.5   Network Architecture

The network (Fig. 3) consists of an encoder with two decoders. The first decoder outputs the embeddings $\mathcal{E}$ and variances $\mathcal{V}$, while the second outputs the instance center heat map $\mathcal{H}$. The encoder comprises a backbone with Feature Pyramid Network (FPN) that produces feature maps at 4 different scales for each image in the input clip. The feature maps at each scale are then stacked along the temporal dimension before being input to each of the decoders.

Our decoder consists of 3D convolutions and pooling layers which first compress the feature maps along the temporal dimension before gradually expanding them back to the input size. The underlying idea here is to allow the network to learn temporal context in order to enable pixel association across space and time. To reduce the decoder's memory footprint and run-time, the large sized feature maps undergo a lower degree of temporal pooling (*i.e.*, fewer convolution/normalization/pooling layers). We call our decoder a *temporal squeeze-expand decoder* (abbreviated as TSE decoder). In Sect. 4.5, we experimentally validate our network's stand-alone ability to learn spatio-temporal context.

3.6   Category Prediction

Our task formulation is inherently category-agnostic. However, some datasets [80,94] require a category label for each predicted object track. For such cases we add an additional TSE decoder to the network that performs semantic segmentation for all pixels in the input clip, and that is trained using a standard cross-entropy loss. During inference, the logits for all pixels belonging to a given object instance are averaged, and the highest scoring category label is assigned. Note that this is merely a post-processing step; the instance clustering still happens in a category-agnostic manner.

4   Experimental Evaluation

To demonstrate our method's effectiveness and generalization capability, we apply it to three different tasks and datasets involving pixel-precise segmentation and tracking of multiple objects in videos.

4.1   Training

For all experiments, we use a ResNet-101 [33] backbone initialized with weights from Mask R-CNN [32] trained for image instance segmentation on COCO [49]. The temporal squeeze-expand decoders are initialized with random weights. The network is optimized end-to-end using SGD with momentum 0.9 and an initial learning rate of 10^{-3} which decays exponentially.

Augmented Images. Since the amount of publicly available video data with dense per-pixel annotations is limited, we utilize image instance segmentation datasets by synthesizing training clips from single images. We apply on-the-fly random affine transformations and motion blur and show that this technique is effective (Sect. 4.3) even though such augmented image sequences have little visual resemblance to video clips.

4.2   Benchmarks

DAVIS Unsupervised: DAVIS [11] is a popular video object segmentation dataset with 90 videos (60 for training and 30 for validation) containing multiple moving objects of diverse categories. Several DAVIS benchmarks have been introduced over the years; we evaluate our method on the 2019 *Unsupervised Video Object Segmentation* (UVOS) benchmark in which the salient "foreground objects" in each video have to be segmented and tracked. The evaluation measures employed are (i) $\mathcal{J}$-score (the IoU between predicted and ground truth mask tubes), and (ii) $\mathcal{F}$-score (accuracy of predicted mask boundaries against ground truth). The mean of those measures, $\mathcal{J}\&\mathcal{F}$, serves as the final score.

YouTube-VIS: The YouTube Video Instance Segmentation (YT-VIS) [94] dataset contains 2,883 high quality YouTube videos with ~131k object instances spanning 40 known categories. The task requires all objects belonging to the known category set to be segmented, tracked and assigned a category label. The evaluation measures used for this task are an adaptation of the Average Precision ($\mathcal{AP}$) and Average Recall ($\mathcal{AR}$) metrics used for image instance segmentation.

KITTI-MOTS: Multi-Object Tracking and Segmentation (MOTS) [80] extends the KITTI multi-object tracking dataset [28] with pixel-precise instance masks. It contains 21 videos (12 for training and 9 for validation) of driving scenes captured from a moving vehicle. The task here is to segment and track all *car* and *pedestrian* instances in the videos. The evaluation measures used are an extension of the CLEAR MOT measures [5] to account for pixel-precise tracking (details in [80]). We report these measures separately for each object class.

4.3 Ablation Studies

Embedding Formulation: We first ablate the impact of using different mixing functions for the embeddings (Sect. 3.2) on the DAVIS'19 `val` dataset with clip length $T = 8$, and summarize the results in Table 1a. Compared to the identity function baseline ϕ_{identity}, imparting a spatial coordinate offset (ϕ_{xy}) improves the $\mathcal{J}\&\mathcal{F}$ from 57.3% to 61.6%. Adding another embedding dimension with a temporal coordinate offset, as in ϕ_{xyt}, yields a further improvement to 62.6%.

Comparing the mixing function pairs (ϕ_{xyf}, ϕ_{xyt}) where $E = 3$, and (ϕ_{xyff}, ϕ_{xytf}) where $E = 4$, we note that having a free dimension is slightly better than having a temporal dimension since there is a difference of 0.2% $\mathcal{J}\&\mathcal{F}$ for both pairs of functions. This is the case for both DAVIS'19 and YT-VIS,[2] where ϕ_{xyff} yields the best results. For KITTI-MOTS however, (see footnote 2) temporal and free dimensions yield roughly the same performance. Our intuitive explanation is that for DAVIS and YT-VIS, object instances normally persist throughout the video clip. Therefore, in contrast to the spatial coordinates which serve as a useful feature for instance separation, the temporal coordinate provides no useful separation cue, thus rendering the temporal embedding dimension less effective. On the other hand, using a free dimension enables the network to learn a more discriminative feature representation that can better aid in separating instances. By contrast, objects in KITTI-MOTS driving scenes undergo fast motion and often enter/exit the scene midway through a clip. Thus, the temporal dimension becomes useful for instance separation.

Having additional embedding dimensions is beneficial, but only up to a certain point (*i.e.*, $E = 4$). Beyond that, test-time performance drops, as can be seen by comparing ϕ_{xyff} and ϕ_{xyfff}. We conclude by noting that our proposed formulations for $\phi(\cdot)$ improve performance on video-related tasks compared to existing formulations for single-image tasks [57,59]. For further discussion and results we refer to the supplementary material.

---

[2] Results for various $\phi(\cdot)$ on YT-VIS and KITTI-MOTS are given in supplementary.

Table 1. Ablation studies on DAVIS'19 `val`: (a) Impact of different embedding mixing functions; (b) Effect of temporal context; (c) Analysis of training data; (d) Impact of Semantic head on a custom validation split of YT-VIS.

Mixing Function	$\mathcal{J}\&\mathcal{F}$	E
ϕ_{identity}	57.3	2
ϕ_{xy}	61.6	2
ϕ_{xyt}	62.6	3
ϕ_{xyf}	62.8	3
ϕ_{xytf}	64.2	4
ϕ_{xyff}	64.4	4
ϕ_{xyfff}	62.4	5

(a)

Clip Length (T)	$\mathcal{J}\&\mathcal{F}$
4	62.2
8	64.4
16	64.7
24	63.1

(b)

(c)

Training Data	$\mathcal{J}\&\mathcal{F}$
Images	57.1
Video	60.7
Images + Video	64.4

Object Category	$\mathcal{AP}$	$\mathcal{AR}$
Oracle	33.0	34.5
Predicted	24.7	31.8

(d)

Temporal Window Size: Next, we investigate the effect of varying the input clip length on DAVIS'19 `val` (ϕ_{xyff} is used throughout). As shown in Table 1b, larger temporal length helps the TSE decoder to predict better quality mask tubes. Increasing the input clip length from $T = 4$ to $T = 16$ improves the $\mathcal{J}\&\mathcal{F}$ from 62.2% to 64.7%, respectively. Above $T = 24$, the performance decreases.

Training Data: For the DAVIS Unsupervised task, we train on image datasets (COCO [49] and Pascal VOC [24]) in conjunction with video datasets (YT-VIS [94] and DAVIS [11]). As shown in Table 1c, this combination yields 64.4% $\mathcal{J}\&\mathcal{F}$ compared to 60.7% when using only video datasets, and 57.1% when using only image datasets. This highlights the benefit of using a combination of image-augmented and video data. For this ablation, $T = 8$ and ϕ_{xyff} were used.

Semantic Head: Since our network does not produce semantic labels for objects, we adapt it to tasks requiring such labels by adding a semantic segmentation decoder as explained in Sect. 3.6. Here, we compare the quality of our semantic output to an *oracle* by training our network on a custom train-validation split of YT-VIS. This was done because ground truth annotations for the official validation set are not publicly available. The results presented in Table 1d show that using *oracle* category labels improves $\mathcal{AP}$ performance by 8.3 from 24.7 to 33.0. This suggests that our results on the official validation set could be further improved by using a better semantic classifier. We leave this for future work.

4.4 Comparison with State of the Art

Video Object Segmentation: Table 2 summarizes our results on the DAVIS'19 unsupervised validation set. `OF-Tracker` and `RI-Tracker` are our own optical flow and re-id baselines which use proposals from a Mask-RCNN [32] network trained with the same backbone and data as our method (see supplementary for details). Our method (64.7% $\mathcal{J}\&\mathcal{F}$) outperforms these baselines and the other published methods by a significant margin, even though we use a single, proposal-free network. AGNN [84], with the second best score of 61.1%, uses object proposals from Mask R-CNN [32] on the salient regions detected by their network, and associates them over time using optical flow. We also list the top entries of the DAVIS'19 Challenge Workshop in gray. UnOVOST [101] achieves a higher score (67.0%), but (i) uses several networks along with heuristic-based post-processing, (ii) is an order of magnitude slower (1 vs. 7FPS), and (iii) is highly tailored to this benchmark. To validate this, we adapted UnOVOST to KITTI-MOTS by re-training its networks and optimizing the post-processing parameters with grid search (for further details and analysis of this experiment, we refer to the supplementary). As can be seen in Table 4, it does not generalize well to the task of Multi-object Tracking and Segmentation.

Table 2. DAVIS'19 validation results for the unsupervised track. P/D: Proposals/Detections, OF: optical flow, RI: Re-Id, *: uses heuristic post-processing.

DAVIS 2019 unsupervised											
Method	P/D	OF	RI	$\mathcal{J}\&\mathcal{F}$	$\mathcal{J}$ Mean	$\mathcal{J}$ Recall	$\mathcal{J}$ Decay	$\mathcal{F}$ Mean	$\mathcal{F}$ Recall	$\mathcal{F}$ Decay	fps
KIS* [15]	✓		✓	59.9	–	–	–	–	–	–	–
UnOVOST* [101]	✓	✓	✓	67.0	65.6	75.5	0.3	68.4	75.9	3.7	<1
RVOS [78]				41.2	36.8	40.2	0.5	45.7	46.4	1.7	20+
AGNN [84]	✓	✓		61.1	58.9	65.7	11.7	63.2	67.1	14.3	–
OF-Tracker	✓	✓		54.6	53.4	60.9	−1.3	55.9	63	1.1	~1
RI-Tracker	✓		✓	56.9	55.5	63.3	2.7	58.2	64.4	6.4	<1
Ours				**64.7**	**61.5**	**70.4**	**−4**	**67.8**	**75.5**	**1.2**	7

Video Instance Segmentation (VIS): This task requires object instances to be segmented, tracked and also assigned a category label. We therefore adapt our network to this setting as explained in Sect. 3.6. We train jointly on YT-VIS [94] and augmented images from COCO [49] (for COCO, we only use the 20 object classes which overlap with the YT-VIS class set). Since this is a new task with few published works, we compare our method to various baselines and adaptions of existing works from [94] in Table 3. As can be seen, our method performs best with respect to all evaluation metrics. Compared to MaskTrack-RCNN [94], we improve the $\mathcal{AP}$ from 30.3 to 34.6, even though MaskTrack-RCNN uses a two stage object detector and incorporates additional cues during post-processing. Since MaskTrack-RCNN uses a ResNet-50 backbone, we also applied this backbone to our network and still improve the $\mathcal{AP}$ from 30.3 to 30.6.

Multi-object Tracking and Segmentation (MOTS): Here we again adapt the network for category prediction as required by the task formulation. We outline the results of comparing our method with TrackR-CNN [80] and UnOVOST [101] in Table 4. Current top-performing method on MOTS is the two-stage network that extends Mask-RCNN with a re-id head, trained to learn

Table 3. YouTube-VIS validation results. P/D: Proposals/Detections, FF: First Frame Proposals, OF: Optical Flow.

YouTube video instance segmentation								
Method	FF	P/D	OF	$\Updownarrow \mathcal{AP}$	$\mathcal{AP}$@50	$\mathcal{AP}$@75	$\mathcal{AR}$@1	$\mathcal{AR}$@10
OSMN MaskProp [95]	✓			23.4	36.5	25.7	28.9	31.1
FEELVOS [79]	✓			26.9	42.0	29.7	29.9	33.4
IoUTracker+ [94]		✓		23.6	39.2	25.5	26.2	30.9
OSMN [95]		✓		27.5	45.1	29.1	28.6	33.1
DeepSORT [85]		✓		26.1	42.9	26.1	27.8	31.3
MaskTrack R-CNN [94]			✓	30.3	51.1	32.6	31.0	35.5
SeqTracker [94]				27.5	45.7	28.7	29.7	32.5
Ours (ResNet-50)				30.6	50.7	33.5	31.6	37.1
Ours				**34.6**	**55.8**	**37.9**	**34.4**	**41.6**

Table 4. KITTI MOTS validation set results for Car and Pedestrian class. P/D: Proposals/Detections, OF: optical flow, RI: Re-Id.

KITTI MOTS											
Method	P/D	OF	RI	Car				Pedestrian			
				sMOTSA	MOTSA	MOTSP	IDS	sMOTSA	MOTSA	MOTSP	IDS
UnOVOST [101]	✓	✓	✓	50.7	60.2	85.6	151	33.4	47.7	76.0	68
MaskRCNN+maskprop [80]	✓	✓		75.1	86.6	87.1	–	45.0	63.5	75.6	–
TrackRCNN [80]	✓		✓	**76.2**	**87.8**	**87.2**	93	46.8	65.1	75.7	78
Ours				72.7	83.8	**87.2**	76	**50.4**	**66.1**	**77.7**	14

an appearance embedding vector for each object detection, used for data association. Our method achieves the highest sMOTSA score (50.4) on the *pedestrian* class, but TrackR-CNN performs better on the *car* class. However, for both classes, STEm-Seg suffers significantly fewer ID switches (IDS) compared to TrackR-CNN (76 vs. 93 for the *car* class and 14 vs. 78 for the *pedestrian* class). This measure is of particular interest to the tracking community since it directly reflects temporal association accuracy.

Similar to UnOVOST [101], TrackR-CNN does not generalize well to other tasks. To validate this, we retrained TrackR-CNN on the YT-VIS dataset. However, the resulting $\mathcal{AP}$ was less than 10. We assume this is due to TrackR-CNN that is forced to use a 14×14 ROI-Align layer [32] due to memory constraints. This results in coarse segmentation masks which are heavily penalized by the $\mathcal{AP}$ measure. Furthermore, the ReID-based embeddings can only learn an appearance model, which is a limitation in YT-VIS where similar looking objects often occur in the same scene.

4.5   Segmentation of Salient Regions in Videos

Finally, we apply our method to the DAVIS'16 unsupervised benchmark [71], where the task is to produce a binary segmentation for the salient regions in a given video clip. Since separating object instances is not required, we simplify our network to one decoder with two output channels trained for binary segmentation using a bootstrapped cross-entropy loss [88] on randomly selected clips from the YT-VIS and DAVIS datasets.

Table 5. Results on DAVIS'16 val. for the unsupervised track. OF: optical flow, CRF: post-processing using CRF.

DAVIS 2016 unsupervised					
Method	OF	CRF	$\mathcal{J}$ & $\mathcal{F}$	$\mathcal{J}$-mean	$\mathcal{F}$-mean
LVO [75]	✓	✓	–	75.9	72.1
PDB [75]		✓	–	77.2	74.5
MotAdapt [74]			–	77.2	77.4
3D-CNN [35]			–	78.3	77.2
FusionSeg [39]	✓		–	70.7	65.3
LVO [77]	✓	✓	–	75.9	72.1
ARP [97]	✓		–	76.2	70.6
PDB [75]	✓		–	77.2	74.5
AD-Net [96]			78.8	79.4	78.2
AGNN [84]		✓	79.9	**80.7**	79.1
Ours			**80.6**	80.6	**80.6**

Although competing methods are specifically engineered for this task, our simple setup obtains state-of-the-art results (Table 5). We note that while AGNN [84] and AD-Net [96] use a stronger DeepLabv3 [12] backbone, we use additional video data for training. This is needed as we work with 3D input volumes. Since we do not perform post-processing, we report results from AD-Net [96] without their DAVIS-specific post-processing for a fair comparison.

5   Conclusion

We have proposed a novel *bottom-up* approach for instance segmentation in videos which models video clips as 3D space-time volumes and then separates object instances by clustering learned embeddings. We enhanced the feature representation of these embeddings using novel mixing functions which yield considerable performance improvements over existing formulations. We applied our method to multiple, diverse datasets and achieved state-of-the-art results under both category-aware and category-agnostic settings. We further showed that, compared to existing dataset-specific state-of-the-arts, our approach generalizes much better across different datasets. Finally, we validated our network's temporal context learning ability by performing a separate experiment on video saliency detection and showed that our good results also generalize there.

Acknowledgements. This project was funded, in parts, by ERC Consolidator Grant DeeVise (ERC-2017-COG-773161), EU project CROWDBOT (H2020-ICT-2017-779942) and the Humboldt Foundation through the Sofja Kovalevskaja Award. Computing resources for several experiments were granted by RWTH Aachen University under project 'rwth0519'. We thank Sebastian Hennen for help with experiments and Francis Engelmann, Theodora Kontogianni, Paul Voigtlaender, Gulliem Brasó and Aysim Toker for helpful discussions.

References

1. Hu, A., Kendall, A., Cipolla, R.: Learning a spatio-temporal embedding for video instance segmentation. arxiv preprint arXiv:1912:08969v (2019)
2. Van den Bergh, M., Roig, G., Boix, X., Manen, S., Van Gool, L.: Online video seeds for temporal window objectness. In: ICCV (2013)
3. Berman, M., Blaschko, M.B.: Optimization of the Jaccard index for image segmentation with the Lovász hinge. In: CVPR (2018)
4. Berman, M., Rannen Triki, A., Blaschko, M.B.: The Lovász-Softmax loss: a tractable surrogate for the optimization of the intersection-over-union measure in neural networks. In: CVPR (2018)
5. Bernardin, K., Stiefelhagen, R.: Evaluating multiple object tracking performance: the CLEAR MOT metrics. JIVP **2008**, 1:1–1:10 (2008)
6. Bochinski, E., Eiselein, V., Sikora, T.: High-speed tracking-by-detection without using image information. In: AVSS (2017)
7. Brox, T., Malik, J.: Object segmentation by long term analysis of point trajectories. In: Daniilidis, K., Maragos, P., Paragios, N. (eds.) ECCV 2010. LNCS, vol. 6315, pp. 282–295. Springer, Heidelberg (2010). https://doi.org/10.1007/978-3-642-15555-0_21
8. Butt, A.A., Collins, R.T.: Multi-target tracking by Lagrangian relaxation to min-cost network flow. In: CVPR (2013)
9. Caelles, S., Maninis, K.K., Pont-Tuset, J., Leal-Taixé, L., Cremers, D., Van Gool, L.: One-shot video object segmentation. In: CVPR (2017)
10. Caelles, S., et al.: The 2018 DAVIS challenge on video object segmentation. arXiv preprint arXiv:1803.00557 (2018)
11. Caelles, S., Pont-Tuset, J., Perazzi, F., Montes, A., Maninis, K., Gool, L.V.: The 2019 DAVIS challenge on VOS: unsupervised multi-object segmentation. arXiv arXiv:1905.00737 (2019)
12. Chen, L., Papandreou, G., Schroff, F., Adam, H.: Rethinking atrous convolution for semantic image segmentation. arXiv preprint arXiv:1706.05587 (2017)
13. Chen, X., Girshick, R., He, K., Dollár, P.: TensorMask: a foundation for dense object segmentation. In: ICCV (2019)
14. Chen, Y., Pont-Tuset, J., Montes, A., Van Gool, L.: Blazingly fast video object segmentation with pixel-wise metric learning. In: CVPR (2018)
15. Cho, D., Hong, S., Kim, J., Kang, S.: Key instance selection for unsupervised video object segmentation. In: The 2019 DAVIS Challenge on Video Object Segmentation - CVPR Workshops (2019)
16. Comaniciu, D., Meer, P.: Mean shift: a robust approach toward feature space analysis. PAMI **24**(5), 603–619 (2002)
17. Dave, A., Tokmakov, P., Ramanan, D.: Towards segmenting everything that moves. arXiv preprint arXiv:1902.03715 (2019)
18. De Brabandere, B., Neven, D., Van Gool, L.: Semantic instance segmentation for autonomous driving. In: CVPR Workshops (2017)
19. De Brabandere, B., Neven, D., Van Gool, L.: Semantic instance segmentation with a discriminative loss function. arXiv preprint arXiv:1708.02551 (2017)
20. Dong, M., et al.: Temporal feature augmented network for video instance segmentation. In: ICCV Workshops (2019)
21. Elich, C., Engelmann, F., Schult, J., Kontogianni, T., Leibe, B.: 3D-BEVIS: birds-eye-view instance segmentation. In: German Conference on Pattern Recognition (GCPR) (2019)

22. Engelmann, F., Bokeloh, M., Fathi, A., Leibe, B., Nießner, M.: 3D-MPA: multi proposal aggregation for 3D semantic instance segmentation. In: IEEE Conference on Computer Vision and Pattern Recognition (CVPR) (2020)
23. Ester, M., Kriegel, H.P., Sander, J., Xu, X., et al.: A density-based algorithm for discovering clusters in large spatial databases with noise. In: ACM Conference on Knowledge Discovery and Data Mining (KDD) (1996)
24. Everingham, M., Van Gool, L., Williams, C., Winn, J., Zisserman, A.: The pascal visual object classes (VOC) challenge. IJCV **88**(2), 303–338 (2010)
25. Feichtenhofer, C., Pinz, A., Zisserman, A.: Detect to track and track to detect. In: ICCV (2017)
26. Feng, Q., Yang, Z., Li, P., Wei, Y., Yang, Y.: Dual embedding learning for video instance segmentation. In: ICCV Workshops (2019)
27. Fukunaga, K., Hostetler, L.: The estimation of the gradient of a density function, with applications in pattern recognition. IEEE Trans. Inf. Theory **21**(1), 32–40 (1975)
28. Geiger, A., Lenz, P., Urtasun, R.: Are we ready for autonomous driving? The KITTI vision benchmark suite. In: CVPR (2012)
29. Gkioxari, G., Malik, J.: Finding action tubes. In: CVPR (2015)
30. Gori, M., Monfardini, G., Scarselli, F.: A new model for learning in graph domains. In: IJCNN (2005)
31. Han, W., et al.: Seq-NMS for video object detection. arXiv preprint arXiv:1602.08465 (2016)
32. He, K., Gkioxari, G., Dollár, P., Girshick, R.: Mask R-CNN. In: ICCV (2017)
33. He, K., Zhang, X., Ren, S., Sun, J.: Deep residual learning for image recognition. In: CVPR (2016)
34. Hou, R., Chen, C., Shah, M.: Tube convolutional neural network (T-CNN) for action detection in videos. In: ICCV (2017)
35. Hou, R., Chen, C., Sukthankar, R., Shah, M.: An efficient 3D CNN for action/object segmentation in video. In: BMVC (2019)
36. Hu, Y., Huang, J., Schwing, A.: MaskRNN: instance level video object segmentation. In: NIPS (2017)
37. Huang, C., Wu, B., Nevatia, R.: Robust object tracking by hierarchical association of detection responses. In: Forsyth, D., Torr, P., Zisserman, A. (eds.) ECCV 2008. LNCS, vol. 5303, pp. 788–801. Springer, Heidelberg (2008). https://doi.org/10.1007/978-3-540-88688-4_58
38. Jain, R., Nagel, H.H.: On the analysis of accumulative difference pictures from image sequences of real world scenes. PAMI **1**, 206–214 (1979)
39. Jain, S., Xiong, B., Grauman, K.: FusionSeg: learning to combine motion and appearance for fully automatic segmentation of generic objects in videos. In: CVPR (2017)
40. Jiang, L., Zhao, H., Shi, S., Liu, S., Fu, C.W., Jia, J.: PointGroup: dual-set point grouping for 3D instance segmentation. In: CVPR (2020)
41. Kang, K., et al.: Object detection in videos with tubelet proposal networks. In: CVPR (2017)
42. Kong, S., Fowlkes, C.C.: Recurrent pixel embedding for instance grouping. In: CVPR (2018)
43. Kuhn, H.W., Yaw, B.: The Hungarian method for the assignment problem. Naval Res. Logist. Q. **2**, 83–97 (1955)
44. Kwak, S., Cho, M., Laptev, I., Ponce, J., Schmid, C.: Unsupervised object discovery and tracking in video collections. In: ICCV (2015)

45. Leal-Taixé, L., Milan, A., Reid, I., Roth, S., Schindler, K.: MOTChallenge 2015: towards a benchmark for multi-target tracking. arXiv preprint arXiv:1504.01942 (2015)

46. Leibe, B., Leonardis, A., Schiele, B.: Robust object detection with interleaved categorization and segmentation. IJCV **77**(1–3), 259–289 (2008)

47. Leibe, B., Schindler, K., Cornelis, N., Gool, L.V.: Coupled object detection and tracking from static cameras and moving vehicles. PAMI **30**(10), 1683–1698 (2008)

48. Li, S., Seybold, B., Vorobyov, A., Fathi, A., Huang, Q., Kuo, C.C.J.: Instance embedding transfer to unsupervised video object segmentation. In: CVPR (2018)

49. Lin, T.-Y.: Microsoft COCO: common objects in context. In: Fleet, D., Pajdla, T., Schiele, B., Tuytelaars, T. (eds.) ECCV 2014. LNCS, vol. 8693, pp. 740–755. Springer, Cham (2014). https://doi.org/10.1007/978-3-319-10602-1_48

50. Liu, R., et al.: An intriguing failing of convolutional neural networks and the CoordConv solution. In: NIPS (2018)

51. Liu, X., Ye, T.: Spatio-temporal attention network for video instance segmentation. In: ICCV Workshops (2019)

52. Lloyd, S.: Least squares quantization in PCM. IEEE Trans. Inf. Theory **28**(2), 129–137 (1982)

53. Luiten, J., Voigtlaender, P., Leibe, B.: PReMVOS: proposal-generation, refinement and merging for video object segmentation. In: Jawahar, C.V., Li, H., Mori, G., Schindler, K. (eds.) ACCV 2018. LNCS, vol. 11364, pp. 565–580. Springer, Cham (2019). https://doi.org/10.1007/978-3-030-20870-7_35

54. McInnes, L., Healy, J., Astels, S.: HDBSCAN: hierarchical density based clustering. J. Open Source Softw. **2**(11), 205 (2017)

55. Milan, A., Leal-Taixé, L., Reid, I., Roth, S., Schindler, K.: MOT16: a benchmark for multi-object tracking. arXiv preprint arXiv:1603.00831 (2016)

56. Milan, A., Leal-Taixé, L., Schindler, K., Reid, I.: Joint tracking and segmentation of multiple targets. In: CVPR (2015)

57. Neven, D., Brabandere, B.D., Proesmans, M., Gool, L.V.: Instance segmentation by jointly optimizing spatial embeddings and clustering bandwidth. In: CVPR (2019)

58. Newell, A., Huang, Z., Deng, J.: Associative embedding: end-to-end learning for joint detection and grouping. In: NIPS (2017)

59. Novotny, D., Albanie, S., Larlus, D., Vedaldi, A.: Semi-convolutional operators for instance segmentation. In: Ferrari, V., Hebert, M., Sminchisescu, C., Weiss, Y. (eds.) ECCV 2018. LNCS, vol. 11205, pp. 89–105. Springer, Cham (2018). https://doi.org/10.1007/978-3-030-01246-5_6

60. Ochs, P., Brox, T.: Higher order motion models and spectral clustering. In: CVPR (2012)

61. Oh, S.W., Lee, J.Y., Xu, N., Kim, S.J.: Video object segmentation using space-time memory networks. In: ICCV (2019)

62. Okuma, K., Taleghani, A., de Freitas, N., Little, J.J., Lowe, D.G.: A boosted particle filter: multitarget detection and tracking. In: Pajdla, T., Matas, J. (eds.) ECCV 2004. LNCS, vol. 3021, pp. 28–39. Springer, Heidelberg (2004). https://doi.org/10.1007/978-3-540-24670-1_3

63. Ošep, A., Mehner, W., Voigtlaender, P., Leibe, B.: Track, then decide: category-agnostic vision-based multi-object tracking. In: ICRA (2018)

64. Ošep, A., Voigtlaender, P., Luiten, J., Breuers, S., Leibe, B.: Large-scale object mining for object discovery from unlabeled video (2019)

65. Ošep, A., Voigtlaender, P., Weber, M., Luiten, J., Leibe, B.: 4D generic video object proposals. In: ICRA (2020)
66. Palmer, S.E.: Organizing objects and scenes. In: Foundations of Cognitive Psychology: Core Readings, pp. 189–211 (2002)
67. Palou, G., Salembier, P.: Hierarchical video representation with trajectory binary partition tree. In: CVPR (2013)
68. Paragios, N., Deriche, R.: Geodesic active contours and level sets for the detection and tracking of moving objects. PAMI **22**, 266–280 (2000)
69. Pinheiro, P.O., Lin, T.-Y., Collobert, R., Dollár, P.: Learning to refine object segments. In: Leibe, B., Matas, J., Sebe, N., Welling, M. (eds.) ECCV 2016. LNCS, vol. 9905, pp. 75–91. Springer, Cham (2016). https://doi.org/10.1007/978-3-319-46448-0_5
70. Pinheiro, P., Collobert, R., Dollár, P.: Learning to segment object candidates. In: NIPS (2015)
71. Pont-Tuset, J., Perazzi, F., Caelles, S., Arbeláez, P., Sorkine-Hornung, A., Gool, L.V.: A benchmark dataset and evaluation methodology for video object segmentation. In: CVPR (2016)
72. Qi, C.R., Litany, O., He, K., Guibas, L.J.: Deep Hough voting for 3D object detection in point clouds. In: CVPR (2019)
73. Ren, S., He, K., Girshick, R., Sun, J.: Faster R-CNN: towards real-time object detection with region proposal networks. In: NIPS (2015)
74. Siam, M., et al.: Video segmentation using teacher-student adaptation in a human robot interaction (HRI) setting. In: ICRA (2018)
75. Song, H., Wang, W., Zhao, S., Shen, J., Lam, K.-M.: Pyramid dilated deeper ConvLSTM for video salient object detection. In: Ferrari, V., Hebert, M., Sminchisescu, C., Weiss, Y. (eds.) ECCV 2018. LNCS, vol. 11215, pp. 744–760. Springer, Cham (2018). https://doi.org/10.1007/978-3-030-01252-6_44
76. Teichman, A., Levinson, J., Thrun, S.: Towards 3D object recognition via classification of arbitrary object tracks. In: ICRA (2011)
77. Tokmakov, P., Alahari, K., Schmid, C.: Learning video object segmentation with visual memory. In: ICCV (2017)
78. Ventura, C., Bellver, M., Girbau, A., Salvador, A., Marqués, F., Gir'o i Nieto, X.: RVOS: end-to-end recurrent network for video object segmentation. CVPR (2019)
79. Voigtlaender, P., Chai, Y., Schroff, F., Adam, H., Leibe, B., Chen., L.C.: FEELVOS: fast end-to-end embedding learning for video object segmentation. In: CVPR (2019)
80. Voigtlaender, P., et al.: MOTS: multi-object tracking and segmentation. In: CVPR (2019)
81. Wang, H., Luo, R., Maire, M., Shakhnarovich, G.: Pixel consensus voting for panoptic segmentation. In: CVPR (2020)
82. Wang, L., Hua, G., Sukthankar, R., Xue, J., Zheng, N.: Video object discovery and co-segmentation with extremely weak supervision. In: Fleet, D., Pajdla, T., Schiele, B., Tuytelaars, T. (eds.) ECCV 2014. LNCS, vol. 8692, pp. 640–655. Springer, Cham (2014). https://doi.org/10.1007/978-3-319-10593-2_42
83. Wang, Q., He, Y., Yang, X., Yang, Z., Torr, P.: An empirical study of detection-based video instance segmentation. In: ICCV Workshops (2019)
84. Wang, W., Lu, X., Shen, J., Crandall, D.J., Shao, L.: Zero-shot video object segmentation via attentive graph neural networks. In: The IEEE International Conference on Computer Vision (ICCV) (2019)

85. Wojke, N., Bewley, A., Paulus., D.: Onboard contextual classification of 3D point clouds with learned high-order Markov random fields. In: ICIP (2017)

86. Wren, C.R., Azarbayejani, A., Darrell, T., Pentland, A.: Pfinder: real-time tracking of the human body. PAMI **19**, 780–785 (1997)

87. Wu, Y., He, K.: Group normalization. In: Ferrari, V., Hebert, M., Sminchisescu, C., Weiss, Y. (eds.) ECCV 2018. LNCS, vol. 11217, pp. 3–19. Springer, Cham (2018). https://doi.org/10.1007/978-3-030-01261-8_1

88. Wu, Z., Shen, C., van den Hengel, A.: Wider or deeper: revisiting the ResNet model for visual recognition. arXiv preprint arXiv:1611.10080 (2016)

89. Wug Oh, S., Lee, J.Y., Sunkavalli, K., Joo Kim, S.: Fast video object segmentation by reference-guided mask propagation. In: CVPR (2018)

90. Xiao, F., Jae Lee, Y.: Track and segment: an iterative unsupervised approach for video object proposals. In: CVPR (2016)

91. Xie, C., Xiang, Y., Harchaoui, Z., Fox, D.: Object discovery in videos as foreground motion clustering. In: CVPR (2019)

92. Xu, C.: Evaluation of super-voxel methods for early video processing. In: CVPR (2012)

93. Xu, N., et al.: YouTube-VOS: sequence-to-sequence video object segmentation. In: Ferrari, V., Hebert, M., Sminchisescu, C., Weiss, Y. (eds.) ECCV 2018. LNCS, vol. 11209, pp. 603–619. Springer, Cham (2018). https://doi.org/10.1007/978-3-030-01228-1_36

94. Yang, L., Fan, Y., Xu, N.: Video instance segmentation. In: ICCV (2019)

95. Yang, L., Wang, Y., Xiong, X., Yang, J., Katsaggelos, A.K.: Efficient video object segmentation via network modulation. In: CVPR (2018)

96. Yang, Z., Wang, Q., Bertinetto, L., Hu, W., Bai, S., Torr, P.H.S.: Anchor diffusion for unsupervised video object segmentation. In: ICCV (2019)

97. Jun Koh, Y., Kim, C.S.: Primary object segmentation in videos based on region augmentation and reduction. In: CVPR (2017)

98. Yu, J., Blaschko, M.: Learning submodular losses with the Lovász hinge. In: International Conference on Machine Learning (ICML) (2015)

99. Zeng, X., Liao, R., Gu, L., Xiong, Y., Fidler, S., Urtasun, R.: DMM-Net: differentiable mask-matching network for video object segmentation. In: ICCV (2019)

100. Zhang, D., Chun, J., Cha, S.K., Kim, Y.M.: Spatial semantic embedding network: fast 3D instance segmentation with deep metric learning. arXiv preprint arXiv:2007.03169 (2020)

101. Zulfikar, I.E., Luiten, J., Leibe, B.: UnOVOST: unsupervised offline video object segmentation and tracking for the 2019 unsupervised DAVIS challenge. In: The 2019 DAVIS Challenge on Video Object Segmentation - CVPR Workshops (2019)

Hierarchical Style-Based Networks
for Motion Synthesis

Jingwei Xu[1], Huazhe Xu[2], Bingbing Ni[1(✉)], Xiaokang Yang[1], Xiaolong Wang[3], and Trevor Darrell[2]

[1] Shanghai Jiao Tong University, Shanghai, China
nibingbing@sjtu.edu.cn
[2] University of California, Berkeley, USA
[3] University of California, San Diego, USA

Abstract. Generating diverse and natural human motion is one of the long-standing goals for creating intelligent characters in the animated world. In this paper, we propose an unsupervised method for generating long-range, diverse and plausible behaviors to achieve a specific goal location. Our proposed method learns to model the motion of human by decomposing a long-range generation task in a hierarchical manner. Given the starting and ending states, a memory bank is used to retrieve motion references as source material for short-range clip generation. We first propose to explicitly disentangle the provided motion material into style and content counterparts via bi-linear transformation modelling, where diverse synthesis is achieved by free-form combination of these two components. The short-range clips are then connected to form a long-range motion sequence. Without ground truth annotation, we propose a parameterized bi-directional interpolation scheme to guarantee the physical validity and visual naturalness of generated results. On large-scale skeleton dataset, we show that the proposed method is able to synthesise long-range, diverse and plausible motion, which is also generalizable to unseen motion data during testing. Moreover, we demonstrate the generated sequences are useful as subgoals for actual physical execution in the animated world. Please refer to our project page for more synthesised results (https://sites.google.com/view/hsnms/home).

Keywords: Long-range motion generation · Motion style transfer

1 Introduction

Human motion is naturally continuous in time and diverse between different individuals. The same action performed by different people can look very different,

---

J. Xu and H. Xu—Equal contribution.

Electronic supplementary material The online version of this chapter (https://doi.org/10.1007/978-3-030-58621-8_11) contains supplementary material, which is available to authorized users.

A. Vedaldi et al. (Eds.): ECCV 2020, LNCS 12356, pp. 178–194, 2020.
https://doi.org/10.1007/978-3-030-58621-8_11

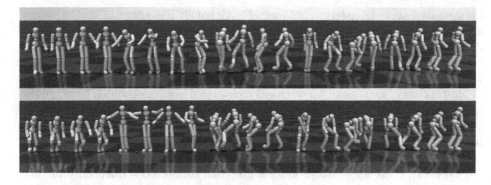

Fig. 1. Visualization of motion sequences generated by proposed hierarchical style-based networks. The generation is achieved by transferring the "style" of multiple subsequences to new ones and then connect these clips smoothly in a temporal sequential manner. We present two synthesized motion sequences, which consist of the same motion "content", but capture different motion "style". The specific definition of "content" and "style" is described in the method part. The long-range and diverse motion generation is thus achieved by free-form composition of "content" and "style" and connecting short-range clips to a long-range one.

and even performed by the same actor twice could hardly be identical (Fig. 1). Capturing this diverse and stylized motion has been a long-standing demand in animation production and video games [10]. By generating this natural motion automatically, it provides useful tools for player customization of action skills in video games [10,36]. However, synthesizing long-range and diverse motion is an extremely challenging task. On one hand, given the nature of multi-modal human behaviours, it is difficult to generate diverse motion without access to the distribution of motion states. On the other hand, it is common in sequential generative models that the error will accumulate through time, which restricts the maximum length of the generated motion sequence [6].

To generate natural and diverse motion, researchers have proposed statistical models based on optimization [2,17,34]. For example, Style Machine is a probabilistic generative model introduce in [2]. It is capable of generating motion with different styles (e.g., motion of novice ballet or modern dance of an expert) by optimizing with a cross-entropy framework. Motivated by this work, Motion texture [17] further proposed a two-level statistical model designed to capture diverse motion transitions, which achieves visually appealing state switching within seen sequence clips. However, these optimization-based methods are commonly restricted by the optimization complexity and can hardly be applied with large-scale dataset. Meanwhile, it is hard for these approaches to generalize to unseen distribution of data.

To adapt and generalize to large-scale data, deep neural networks are utilized in several recent works [10,11,36]. For example, Holden et al. [11] introduced to synthesize the motion sequence using deep networks by first taking the control signal as inputs, the outputs of the deep networks are then furthered edited via image-based style transfer techniques [7]. Although the adopted style transfer approach works well for transferring texture and color in the image domain, it does not necessarily generalize to motion stylization. Without carefully designed training strategies for post-processing, unnatural pose and other artifacts, e.g., foot sliding, is commonly observed during motion synthesis [9].

In this work, we present hierarchical style-based networks, which leverage the large spectrum of human behaviours from unlabeled and unsegmented data to generate long-range, diverse and visually plausible motion sequences (as shown in Fig. 1). Our framework is in a 2-level hierarchical structure: (i) Locally, our network first generates multiple short-range motion sequences independently; (ii) Globally, these short-range motion are then connected sequentially in time to a long-range motion sequence.

For obtaining diverse long-range motion sequences, we will first need to generate diverse short-range motion clips. To achieve this, we propose to disentangle the feature representation for short-range motion to two parts, representing the content and the style of the motion. The *content* refers to the specific action performed at each step, e.g., walking, running and dancing. The *style* refers to some pattern/property existing throughout the whole sequence, which keeps constant along with time. Take the walking motion for example: senior individuals and child walking sequences demonstrate two different styles for the same "walking" action. With this disentangled representation, we can obtain diverse motion from the free-form combination of motion content vectors and style vectors. However, labeling content and style is expensive and most of the time it is even hard to define the style semantically. Thus, instead of defining the style, we propose an unsupervised learning approach to automatically discover the disentanglement between content and style. Specifically, we utilize a bi-linear transformation to explicitly decompose the motion feature into two components, and the two features are then combined together to reconstruct the input motion in an auto-encoder structure. The key for disentanglement is that we enforce the intermediate style feature to be consistent in time.

Given the synthesised short-range clips, we propose to connect them sequentially in time with motion interpolation between each two clips. Our network takes the starting and ending states from two different clips as inputs. To interpolate between these two states, we propose a parameterized representation which maintains the component scale of the generated sequences (e.g., bone lengths are fixed along time). Specifically, the representation is parameterized by a bi-directional LSTM model, which concurrently leverages the motion information of starting and ending states. In this way, we can obtain more plausible and visually appealing interpolation results.

Extensive experiments are conducted on large-scale human motion datasets [11]. We show that the proposed method outperforms existing motion generation baselines in terms of synthesis length, diversity and plausibility, e.g., being useful as sub-goals for actual physical execution in the animated world. We also demonstrate the proposed model is capable of synthesising novel motion based on unseen data without additional fine-tuning procedure, which indicates the generalization ability of our method.

2   Related Work

Motion Interpolation. Given start and end states, this task aims to synthesize intermediate states which smoothly translate between them [31]. For video interpolation [18–20,22], similar as prediction task, where start and end states are two consecutive frames, the final result is expected to increase frame rate of original video to a higher value. Previous researches often utilize phase dynamics [20], flow based feature [19] and other motion information [22] to facilitate this task. Our work is different from this branch of work because there exists large motion gap between start and end states in our settings. Another branch of work is video completion [4,18,32]. It receives two *nonconsecutive* frames as input and aims to fill the motion gap between start and end states. [4] firstly attempts to solve this task and more specifically, propose to select out a rational path in the latent space with BFGS [3] algorithm. [18] incorporates the 3D convolution layers and LSTM network into a unified model, which tries to automatically find the optimal results for intermediate frames. Despite much progress has been made in this filed, the high dimensional data (i.e., video frames) severely restricts video completion within *simple* and *seen* motion categories. However, we do not limit the start and end states belonging to the same sequence. Meanwhile, we expect the interpolated sequence as diverse as possible meanwhile with natural transition between synthesised states. This has not been deeply addressed in previous motion completion works [15,33]. As a potential downstream application, our model could be used to construct motion planning [21] algorithm. Compared to goal-driven RL [14,15], our model gets rid of requirements hard to achieve, i.e., known dynamics of agent, which is more general and applicable to more motion planning scenarios.

Motion Synthesis in Computer Graphics. In the context of computer graphics, there is a branch of researches [11,13,16,23,28] which also concentrate on motion generation, i.e., obtaining a continuous trajectory from a discrete set of poses. Our work shares similar target with this branch of work. However, we would like to emphasise that these works [13,23,28] are in parallel with ours and have completely different research routine on this task. More specifically, graphics methods [2,17] focus on formulating an optimal and explicit statistic modelling framework, which is not easy scale-up to large datasets and unseen motion. For example, [2] introduce a fully data-driven method for articulated motion generation, which needs online optimization if given new demonstration data as inputs. The involved learning procedure is relatively more complex compared to ours and needs careful hyperparameter tuning. On the contrary, our model combines both advantages of deep model and purely data-driven method. Another related work [17] proposes a statistical model for approximation of original motion distribution, which is represented by a transition matrix indicating

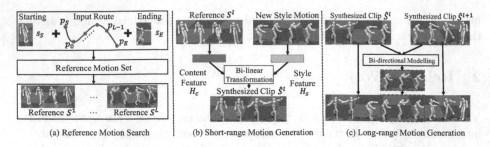

(a) Reference Motion Search   (b) Short-range Motion Generation   (c) Long-range Motion Generation

Fig. 2. Hierarchical style-based motion synthesis framework. (a) *Reference Motion Search*: Given starting and ending states, we search for reference subsequences in the training dataset for generation; (b) *Short-range Motion Generation*: given each reference subsequence, we generate a novel subsequence with motion style transfer; (c) *Long-range Motion Generation*: all synthesized subsequences are connected together in time with bi-directional modelling.

the possibility of state change. This method requires all motion data are available for generation, which is in turn restricted to seen data. Differently, our model could generalize to unseen data, which implies more practical value for downstream application.

3   Method

We aim to synthesize visually natural, diverse and long-range motion sequence constrained by a pair of starting/ending states in a hierarchical manner. We denote a motion sequence with length of M as $\mathbf{S} = [\mathbf{s}_1, \ldots, \mathbf{s}_i, \ldots, \mathbf{s}_M]$, where $\mathbf{s}_i$ is the state at time stamp i. The starting and ending states are denoted as $\mathbf{s}_S$ and $\mathbf{s}_E$ respectively. Note that $\mathbf{s} \in \mathbb{R}^{J \times 3}$, where J refers to the number of joints, which is represented with the x-y-z Cartesian coordinate.

Our motion synthesis framework contains 3 steps as shown in Fig. 2: (i) *Reference Motion Search*. Given the input starting/ending states, we first divide the route to L segments by adding $L - 1$ sub-goals in between on the ground; then, for a segment l, sampling 1 reference subsequence $\mathbf{S}^l \in \mathbb{R}^{M \times J \times 3}$ from the training dataset for motion generation, where M represents the subsequence length. Each sub-goal is represented by a spatial point $\mathbf{p}^l \in \mathbb{R}^2, l = 1, \ldots, L - 1$. We denote $\mathbf{p}_S$ and $\mathbf{p}_E$ as the projected locations of root joint of $\mathbf{s}_S$ and $\mathbf{s}_E$. (ii) *Short-range Motion Generation*. Toward diverse motion generation, when generating each subsequence l, we advocate a synthesis paradigm in the motion style transfer manner, which keeps the content identical to that of $\mathbf{S}^l$ while changing the style based on another subsequence that is randomly sampled from the dataset. We denote the output short-range subsequence as $\hat{\mathbf{S}}^l$. (iii) *Long-range Motion Generation*. Given the short-range motion subsequences, we connect each two consecutive subsequences by adding a transitional motion in between. We denote the transitional motion between $\hat{\mathbf{S}}^l$ and $\hat{\mathbf{S}}^{l+1}$ as $\hat{\mathbf{S}}^{l,l+1} \in \mathbb{R}^{N \times J \times 3}$, where N is

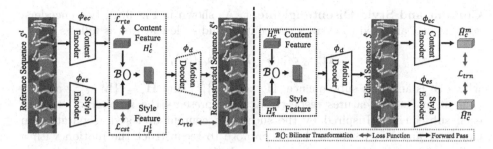

Fig. 3. Short-range motion generation. (a) We train an auto-encoder for motion sequence $\mathbf{S}^l$ reconstruction. The bi-linear transformation is utilized for content and style disentanglement. (b) Training with content and style features from different sources of motion subsequences. Given two features as inputs, we generate the motion which can be used to reconstruct the feature inputs. The training objective is defined on the content and style feature reconstruction error $\mathcal{L}_{trn}$.

the length of the transitional sequence. The final long-range generated sequence is presented as: $[\mathbf{s}_S, \hat{\mathbf{S}}^{0,1}, \hat{\mathbf{S}}^1, \ldots, \hat{\mathbf{S}}^l, \hat{\mathbf{S}}^{l,l+1}, \hat{\mathbf{S}}^{l+1}, \ldots, \hat{\mathbf{S}}^L, \hat{\mathbf{S}}^{L,L+1}, \mathbf{s}_E]$.

Reference Motion Search. The searching procedure is conducted as follows: (i) *Reference length calculation*: For a reference subsequence $\mathbf{S}$ in the dataset, we calculate the distance (denoted as d) between the first and last states of $\mathbf{S}$. The minimum and maximum values are denoted as d_{min} and d_{max} respectively. (ii) *Sub-goal sampling*: We randomly sample $L-1$ sub-goals on the ground, $[\mathbf{p}_1, \ldots, \mathbf{p}_{L-1}]$. The distance between two consecutive sub-goals (denoted as d_l) is sampled from $\mathcal{U}(d_{min}, d_{max})$. $\mathcal{U}()$ is the uniform distribution. The direction specified by the vector, $\overrightarrow{\mathbf{p}_l\mathbf{p}_{l+1}} = \mathbf{p}_{l+1} - \mathbf{p}_l$, is sampled from $\mathcal{U}(-\frac{\pi}{2}, \frac{\pi}{2})$. $\mathbf{s}_S$ and $\mathbf{s}_E$ are translated with an offset respectively, to make sure that the length of $\overrightarrow{\mathbf{p}_S\mathbf{p}_1}$ and $\overrightarrow{\mathbf{p}_{L-1}\mathbf{p}_E}$ (denoted as d_S and d_E) satisfy $d_{min} < d_S, d_E < d_{max}$. (iii) *Subsequence match*: For each sub-goal pair $(\mathbf{p}_l, \mathbf{p}_{l+1})$, we select the reference subsequence whose length matches d_l within a tolerance σ, i.e., $d_l - \sigma < d < d_l$ and $\sigma = 0.05$. Finally, the selected subsequence $\mathbf{S}^l$ is rotated to match $\overrightarrow{\mathbf{p}_l\mathbf{p}_{l+1}}$.

3.1 Short-Range Motion Generation

In this section, we present the details of generating short-range motion clips. This part is formulated as a motion style transfer task, i.e., new motion clips are synthesised via altering the style of reference subsequence $\mathbf{S}^l$ while keeping their motion contents unchanged. Recall that the *content* refers to the specific action performed at each step, e.g., walking, running and dancing, while the *style* refers to some pattern/property existing throughout the whole sequence, which keeps constant along with time. The free-form combination of content and style information is used for diverse synthesis of subsequences. We denote the generated subsequence as $\hat{\mathbf{S}}^l = \{\hat{\mathbf{s}}_i^l\}_{i=1}^M$. Without annotation of style, we first propose to learn corresponding representations in a disentangling manner.

Content and Style Disentanglement. As shown in Fig. 3(a), two encoders, ϕ_{ec} and ϕ_{es}, are used to extract the content and style features as follows,

$$\mathbf{H}_c^l = \phi_{ec}(\mathbf{S}^l), \mathbf{H}_s^l = \phi_{es}(\mathbf{S}^l), \tag{1}$$

where $\mathbf{S}^l$ is an input subsequence, $\mathbf{H}_c^l = \{\mathbf{h}_{c,i}^l\}_{i=1}^M$ and $\mathbf{H}_s^l = \{\mathbf{h}_{s,i}^l\}_{i=1}^M$ are the content and style features respectively, M represents the number of steps in each subsequence. Inspired by the success of content and style separation in character and image [29] fields, we propose to reconstruct the motion with a bi-linear transformation scheme, where each time step i is represented as:

$$\hat{\mathbf{s}}_i^l = \phi_d(\mathcal{B}(\mathbf{h}_{c,i}^l, \mathbf{h}_{s,i}^l)), \tag{2}$$

where $\mathcal{B}(\cdot)$ is the bi-linear transformation [29], i.e., $\mathcal{B}(\mathbf{h}_{c,i}^l, \mathbf{h}_{s,i}^l) = \mathbf{h}_{c,i}^l \mathcal{W}(\mathbf{h}_{s,i}^l)^\mathsf{T}$ and $\mathcal{W} \in \mathbb{R}^{C_{ctn} \times C_{out} \times C_{sty}}$ is a bi-linear transformation weight. ϕ_d is the motion decoder. As a two-factor method, the bi-linear transformation possesses an elegant mathematical property, i.e., separability: their outputs are linear in either component when the other is kept unchanged. Facilitated by bi-linear transformation, the contribution of two components can be effectively disentangled and fused into a representative latent feature that is generalizable to unseen factor modes with new contents.

Training Objective. Encoder and decoder are jointly trained by a L2 reconstruction loss, i.e., $\mathcal{L}_{rec} = \|\mathbf{S}^l - \hat{\mathbf{S}}^l\|_2^2$. To guarantee the style consistency of the reconstructed subsequence, an explicit L2 loss is incorporated into training procedure, i.e., $\mathcal{L}_{cst} = \|\hat{\mathbf{h}}_{s,i}^l - \hat{\mathbf{h}}_{s,j}^l\|_2^2$, where i, j are randomly sampled two style features corresponding to time step i and j respectively. Meanwhile, the moving route $\mathbf{c}^l$ (including root position and velocity, foot position and velocity [11]) is extracted as a part of content feature and trained with a L2 loss to prevent foot sliding, i.e., $\mathcal{L}_{rte} = \|\mathbf{c}^l - \hat{\mathbf{c}}^l\|_2^2$, where $\mathbf{c}^l$ is the input control signal. Note that the moving route is treated as a part of content feature $\mathbf{h}_c^l$, i.e., $\mathbf{c}^l \in \mathbf{h}_c^l$. The training pipeline of this motion auto-encoder is shown in Fig. 3(a).

At each iteration we sample a new batch of data for training, to further guarantee the consistency at feature level. Suppose a subsequence (denoted as $\hat{\mathbf{S}}$) is generated with the content $\mathbf{H}_c^m$ of $\mathbf{S}^m$ and style $\mathbf{H}_s^n$ of $\mathbf{S}^n$. $\hat{\mathbf{S}}$ is subsequently fed into ϕ_{ec} and ϕ_{es} to obtain reconstructed features, i.e., $\hat{\mathbf{H}}_c^m$ and $\hat{\mathbf{H}}_s^n$, as shown in Fig. 3(b). This part is trained with $\mathcal{L}_{trn} = \|\mathbf{H}_c^m - \hat{\mathbf{H}}_c^m\|_2^2 + \|\mathbf{H}_s^n(\mathbf{H}_s^n)^\mathsf{T} - \hat{\mathbf{H}}_s^n(\hat{\mathbf{H}}_s^n)^\mathsf{T}\|_2^2$. The latter part for style consistency is inspired from Gram Matrix utilized in image translation [7]. The high-order statistics of style feature has been proven to be more critical for translation in image domain. The final training objective is presented as follows,

$$\mathcal{L} = \mathcal{L}_{rec} + 0.01\mathcal{L}_{cst} + 0.5\mathcal{L}_{rte} + \mathcal{L}_{trn}. \tag{3}$$

Note that both the encoder and decoder are jointly trained under the supervision of $\mathcal{L}_{trn}$. We present the model architecture details in the material.

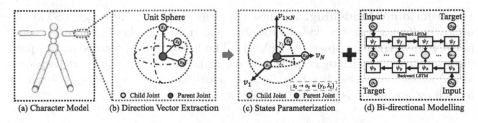

(a) Character Model   (b) Direction Vector Extraction   (c) States Parameterization   (d) Bi-directional Modelling

Fig. 4. Proposed model for long-range motion generation. Starting with skeleton model of human subject (leftmost), we extract the corresponding direction vector from 3D coordinate (second part), which are parameterized by the starting/ending states of one subsequence (third part). This leads to more compact solution space with learnable parameters (γ, λ), which are generated by the proposed bi-directional model.

3.2   Long-Range Motion Generation

This part refers to connecting edited short-range subsequences $(\hat{\mathbf{S}}^l, \hat{\mathbf{S}}^{l+1})$ into a long-range one with interpolated subsequence $\hat{\mathbf{S}}^{l,l+1}$. To achieve smooth and natural transition, we propose to Parameterize the original 3D coordinate space as shown in Fig. 4, which leads to a more compact output space.

Parameterized Representation. Given the state $\mathbf{s}_t$ at time step t we first obtain the direction vector $\mathbf{v}_t(p,q) = \frac{\mathbf{s}_t(p)-\mathbf{s}_t(q)}{||\mathbf{s}_t(p)-\mathbf{s}_t(q)||_2^2}$, where (p,q) corresponds to one child-parent pair of character joint according to the skeleton topology (Fig. 4(b)). The transition subsequence is generated in the direction vector space. Given two vectors $(\mathbf{v}_1, \mathbf{v}_N)$ as bases (the length of transition is assumed to be N), interpolation is essentially the combination of $\mathbf{v}_1$ and $\mathbf{v}_N$. The interpolation procedure is inspired from Quaternion Slerp [27] but generalized to non-linear situation. As shown in Fig. 4(c), supposing $\mathbf{v}_1$ and $\mathbf{v}_N$ are two bases in 3D space, the third basis is obtained by outer product: $\mathbf{v}_{1 \times N} = \mathbf{v}_1 \times \mathbf{v}_N$. An arbitrary direction vector $\mathbf{v}_t$ could be represented in the parameterization manner,

$$\mathbf{v}_t = \frac{\sin(1-\gamma_t)\Lambda}{\sin \Lambda}\left(\frac{\sin(1-\lambda_t)\Omega}{\sin \Omega}\mathbf{v}_1 + \frac{\sin(\lambda_t \Omega)}{\sin \Omega}\mathbf{v}_N\right) + \frac{\sin(\gamma_t \Lambda)}{\sin \Lambda}\mathbf{v}_{1 \times N}, \qquad (4)$$

where Ω is directed angle between $\mathbf{v}_1$ and $\mathbf{v}_N$, $\Lambda = \pi/2$ because of outer product. $\mathbf{v}_t$ is thus parameterized by $\mathbf{v}_1$ and $\mathbf{v}_N$ with (γ_t, λ_t). Note that γ_t, λ_t are two learnable parameters in our work, which is modelled by a bi-directional LSTM [8] introduced in following paragraph.

Intuitively, the interpolation defined by Eq. 4 is analogous to locating on the earth with the longitude and latitude. The Eq. 4 conducts quaternion slerp twice, where the first one, inside the brackets, decides the "longitude" (λ) and the second one, outside the brackets, decides the "latitude" (γ). One one hand, the output dimension at each time stamp is reduced from 4J (quaternion) to 2J (γ, λ), leading to a more compact solution space. On the other hand, outputs are naturally valid, i.e., unit vector required by direction vector, which avoids additional normalization procedure.

Bi-directional Modelling. As shown in Fig. 4(d), we utilize two LSTM [8] networks (forward and backward) to achieve natural and plausible interpolation in the parametrization space ($o = \{\lambda, \gamma\}$). The forward one (denoted as ψ_f) predicts future states with o_1 as origin and o_N as target. The counterpart (denoted as ψ_b) takes the reverse direction, i.e., o_N as origin and o_1 as target. The whole modelling procedure is executed as follows,

$$\hat{o}_{t+1,f} = \psi_f(\hat{o}_{t,f}, o_N), \hat{o}_{t-1,b} = \psi_b(\hat{o}_{t,b}, o_1). \tag{5}$$

The complementary pair of forward/backward outputs, i.e., $(\hat{o}_{t,f}, \hat{o}_{N-t,b})$ are firstly converted back to the x-y-z coordinate space $(\hat{s}_{t,f}, \hat{s}_{N-t,b})$ with forward kinematics [25], and then fused as follows,

$$\hat{s}_t = \psi_{fse}(\hat{s}_{t,f}, \hat{s}_{N-t,b}), \tag{6}$$

$$\mathcal{L}_{pos} = ||s_t - \hat{s}_t||_2^2. \tag{7}$$

With supervision signal in the position space, the topology information of character skeleton is better utilized and motion generation artifacts, e.g, foot sliding, could be explicitly punished during training procedure.

Adversarial Training: Generalizing beyond Training Sequence. Above interpolation is trained with starting/ending pair belonging to the same sequence, i.e., with ground truth. However, proposed model should be tested with arbitrary state pairs for practical usage. It lacks ground truth signal for training under such condition. We thus utilize adversarial training to facilitate generalization ability of motion interpolation. More specifically, with s_1, s_N sampled from different sequences, $\psi_{f,b}$ give prediction of intermediate states as described above, which are regarded as fake sequences. A motion discriminator D is further proposed for adversarial training as follows,

$$\mathcal{L}_D = \frac{1}{2}((1 - D(S))^2 + D^2(G(s_1, s_N))), \tag{8}$$

$$\mathcal{L}_G = (1 - D(G(s_1, s_N)))^2, \tag{9}$$

where G refers to proposed interpolation model in this section. The discriminator architecture follows the work of Chen et al. [5], i.e., built based on residual block and designed for synthesising more realistic 3D poses.

Implementation Details. We train the models for short-range and long-range motion generation in a 2-step procedure. For the long-range motion generation, the adversarial training scheme is alternating the supervision between $\mathcal{L}_{pos} + 0.01\mathcal{L}_G$ and $0.01\mathcal{L}_D$. We adopt Adam [12] as the optimizer, where learning rate, learning decay and weight decay are set to $5e^{-4}, 0.97, 1e^{-5}$ respectively. Both short-range and long-range models are trained with 200 epochs.

4   Experiments

4.1   Evaluation Settings

Datasets. We use the CMU Mocap dataset[1] for training. Reference subsequences are obtained in a sliding window manner, where $M = 120$ and $N = 40$. The dimension of state at a single time step $\mathbf{s}_t$ is 63, which is $J = 21$ joints with 3D coordinates. To demonstrate that our model could generate novel behaviour never seen during training, we keep a held-out reference set (denoted as $\mathcal{D}_R$) from training data (denoted as $\mathcal{D}_T$) for further testing.

Baselines. Considering the hierarchical structure of our model, we conduct comparison experiments at both two levels, i.e., short-range and long-range. For short-range generation, we compare our model with two strong baselines: the work of Holden et al. [11] (termed as DeepSyn in this paper) and QuaterNet [25]. Both baselines are retrained with the same data used in this work. The input dimension is adjusted to match with our data. For long-range generation, we compare our model with temporal prediction baselines: HP-GAN [1] and MT-VAE [35]. For these two baselines, the ending states are feed as input with the default hyper-parameter setting.

4.2   Evaluation of Short-Range Generation

Generation Diversity Evaluation. We compare our model with DeepSyn [11] and QuaterNet [25] to evaluate the motion diversity, under different percentages of data used for training. As illustrated in Fig. 5(a), 10%, 30%, 50% and 100% training data are used respectively. We calculate the averaged standard deviation of all joints with a higher value indicating higher diversity. QuaterNet [25] achieves lowest diversity under all settings. DeepSyn [11] keeps relative constant motion diversity which is comparable with 10% training data. On the contrary, the motion diversity of our model constantly increases if more data used for training, which is mainly facilitated by the hierarchical synthesis framework.

Short-range Generation Visualization. Figure 6 shows the short-range results (from top to bottom: our model, DeepSyn [11] and QuaterNet [25]). We can observe that DeepSyn [11] synthesises a abnormal walking sequence with unnatural behavior (middle row in Fig. 6, fall-down pose during synthesis). QuaterNet [25] is able to generate a visually natural walking sequence but struggles to produce diverse motion behaviour (last row in Fig. 6, restricted within simple locomotion synthesis). Different from all these models, our hierarchical style-based model achieves natural state transition throughout the whole sequence (first row in Fig. 6), meanwhile provides natural and diverse motion (i.e., walking-turning-dancing) behaviours during generation.

---

[1] http://mocap.cs.cmu.edu/.

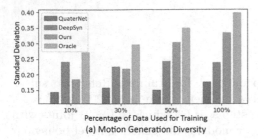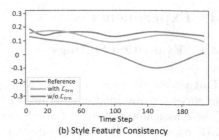

(a) Motion Generation Diversity   (b) Style Feature Consistency

Fig. 5. Short-range motion evaluation: (a) Motion diversity evaluation. X-axis is the proportion of data used for training. Y-axis is the deviation of joint coordinates. Our model is able to synthesize more diverse sequences than baseline models. (b) Style feature visualization. We can see that facilitated by the loss $\mathcal{L}_{trn}$, the style feature of synthesized sequence is closer to the input one (reference) and kept relatively constant. (Color figure online)

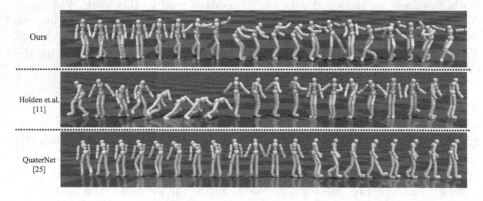

Fig. 6. Comparison of motion naturalness with DeepSyn [11] and QuaterNet [25].

Style Feature Consistency Evaluation. As shown in Fig. 5(b), we plot one representative dimension of the learned style feature from a complete sequence. The purple/green/gray curves refer to the style feature of reference sequence, synthesised sequence trained with and without $\mathcal{L}_{trn}$ respectively. We can observe that training without the supervision of $\mathcal{L}_{trn}$ leads to the transferred style feature (gray curve) drifts away from the reference one (purple curve). On the contrary, training with the supervision of $\mathcal{L}_{trn}$ (green curve) effectively facilitates transferring the style pattern from reference sequence to the synthesised one, i.e., the green curve is close to the purple curve.

Style-based Generation Evaluation. As shown in Fig. 7, we provide two synthesized examples (middle part) which possess the general motion style from one subsequence (right part), but detailed motion pattern from another one (left part). For the first sequence (top row in Fig. 7), the target style motion shows a walking sequence with back bent down (style), while the reference motion

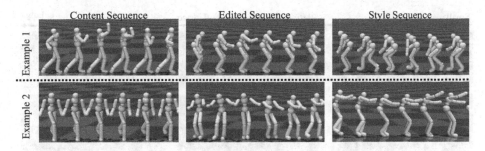

Fig. 7. Visualization of generated short-range subsequences. The synthesized short-range sequence (middle column) well preserves the detailed motion from content sequence (e.g., the arm moving in the first row) and constant pattern from style sequence (e.g., the arm lifting in the second row).

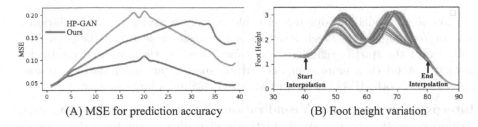

(A) MSE for prediction accuracy   (B) Foot height variation

Fig. 8. Evaluation of long-range motion generation in terms of MSE (a) and foot height variation (b). (a) shows that our model (blue line) achieves lower interpolation error than baseline models. (b) demonstrates the non-linearity of interpolation results. (Color figure online)

content (sampled from held-out set $\mathcal{D}_R$) is a sequence with arm waving. We can notice that the synthesised motion (second column, top row) preserves the arm motion and learns the style pattern (back bent down) successfully. For the second example (bottom row in Fig. 7), the target style motion shows a walking sequence with arm lifted horizontally (style), while the reference motion content (sampled from training set $\mathcal{D}_T$) is a regular walking sequence. The synthesised motion (second column, bottom row) fully captures the style of upper body meanwhile preserves the walking motion from reference sequence.

4.3   Evaluation of Long-Range Motion Generation

Generation Accuracy Evaluation. We compare with two goal-conditioned prediction models (HP-GAN [1] and MT-VAE [35]) evaluated by prediction accuracy. The training data is a subset chosen from original one, consisting of locomotion, punching, kicking and dancing sequences. The test data belongs to the same motion category of no overlap with training data. We calculate the MSE value (the lower the better) of predicted clips given starting/ending states from the same sequences. As shown in Fig. 8(A), we can see that facilitated by the

Fig. 9. States transition visualization for evaluation of smoothness. From left to right, the pose difference between starting and ending states becomes larger. Our model is able to generate smooth and natural transition under all situations.

bi-directional modelling, our model could equally consider the contribution of starting/ending states, which leads to high transition accuracy (i.e., low MSE value) near the starting/ending states. Facilitated by the proposed parameterization method (i.e., more compact output space), the interpolation accuracy outperforms both HP-GAN [1] and MT-VAE [35] by a large margin.

Interpolation Sequence Visualization. Figure 9 demonstrates long-range motion generation results given starting and ending states from different reference subsequences, respectively. Note that here the starting and ending states are sampled from held-out set $(\mathcal{D}_R)$ for evaluation. Note that for all sequences shown in Fig. 9, the starting as well as ending states are from held-out reference set. We can observe that our model is able to generate smooth and natural transition when starting and ending states are similar (left part in Fig. 9). Moreover, when encountered large motion difference (right part in Fig. 9), e.g., from walking to dancing, turning back with a relatively large degree, our model still makes it to generate visually natural transition sequence.

Non-trivial Interpolation Verification. To evaluate whether our model learns non-linear interpolation between two sequences, we report the height variation of the right foot in a interpolation motion sequence. Meanwhile, we randomly rotate the second sequence to show that our model is robust to a wide range of direction difference between two sequences. As shown in Fig. 8(B), we present multiple curves which correspond to different rotation angles. All recorded curves are highly non-linear but smooth between starting and ending points. Moreover, our model adaptively changes foot height with different rotation configurations, which leads to smooth and natural motion.

Visualization of Final Synthesized Sequences. With both short-range and long-range motion generation, we are able to synthesize final sequences. Recall that our model is constrained by given starting and ending states for motion interpolation. To this end, we present three synthesized sequences in Fig. 10, of which each result leverages three reference subsequences for motion generation. As shown in left part of Fig. 10, starting from the same state, we are able to

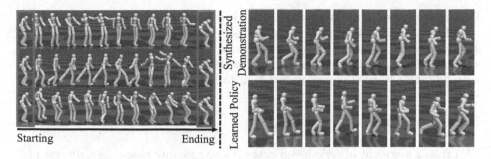

Fig. 10. Left: Diverse motion generation given the same starting and ending states. Right: Expert demonstration guidance for imitation learning.

synthesize long-range and visually natural motion boosted by the short-range and long-range motion generation. We are also able to generate diverse behaviour (complex hand and arm motion) facilitated hierarchical modeling.

Reality Evaluation. We train three classifiers with the ground truth as positive samples. Negative samples are generated by three models (i.e., DeepSyn [11], QuaterNet [25], and Ours) respectively. As shown in Table 1, we report the proportion of generated sequences classified as positive ones. Row refers to the model used to train the classifier. Column refers to the model evaluated by the classifiers. Our model generates more realistic sequences than these two baselines (first two rows). The proportion of the other two models' results classified as positive by the our model (third row) is lower.

Table 1. Classification Accuracy for evaluating the reality of synthesised sequences.

	DeepSyn [11]	QuaterNet [25]	Ours
DeepSyn [11]	—	37.6%	**66.3%**
QuaterNet [25]	41.9%	—	**75.3%**
Ours	28.3%	50.4%	—

Distribution Similarity Evaluation. Inspired by FVD score [30] we measure the distribution similarity between the original and generated sequences with motion features as inputs. The scores (the lower the better) are 281.5 (DeepSyn [11]), 341.1 (QuaterNet [25]), and 179.8 (ours), i.e., the distribution of our results is closer to that of the original data. Meanwhile, we compare the motion diversity of short-range sequences with those generated by the full model. Motion deviation (the higher the better) is reported, i.e., 0.226 (short-range sequence) and 0.347 (full sequence) respectively. We can see that the long-range generation model facilitates increasing the motion diversity by a large margin.

Demonstration Guidance for Imitation Learning. To further show our model produces realistic motion, generated results are used for demonstration guidance of imitation learning. As shown in Fig. 10, the right part is demonstration synthesised by our model (top) and the bottom one is learned policy with [26]. Our learned motion succeeds in following the synthesised one.

5 Conclusion

We present hierarchical style-based networks to generate long-range, diverse and visually plausible motion sequences. Our model trained with large-scale skeleton dataset is also able to generalize to unseen motion much better than previous baselines. We believe our method not only will facilitate graphics applications, but also can be used to generate demonstration for imitation learning.

Acknowledgement. This work was supported in part by BAIR and BDD. This work was also supported in part by grant No. 18DZ1112300, No. 61976137, No. 61527804, No. U1611461, No. U19B2035, and No. 2016YFB1001003.

References

1. Barsoum, E., Kender, J., Liu, Z.: HP-GAN: probabilistic 3D human motion prediction via GAN. In: CVPR Workshops, pp. 1418–1427 (2018)
2. Brand, M., Hertzmann, A.: Style machines. In: SIGGRAPH, pp. 183–192 (2000)
3. Byrd, R.H., Lu, P., Nocedal, J., Zhu, C.: A limited memory algorithm for bound constrained optimization. SIAM J. Sci. Comput. **16**(5), 1190–1208 (1995)
4. Cai, H., Bai, C., Tai, Y.-W., Tang, C.-K.: Deep video generation, prediction and completion of human action sequences. In: Ferrari, V., Hebert, M., Sminchisescu, C., Weiss, Y. (eds.) ECCV 2018. LNCS, vol. 11206, pp. 374–390. Springer, Cham (2018). https://doi.org/10.1007/978-3-030-01216-8_23
5. Chen, C., et al.: Unsupervised 3D pose estimation with geometric self-supervision. In: CVPR, pp. 5714–5724 (2019)
6. Finn, C., Goodfellow, I.J., Levine, S.: Unsupervised learning for physical interaction through video prediction. In: NIPS, pp. 64–72 (2016)
7. Gatys, L.A., Ecker, A.S., Bethge, M.: Image style transfer using convolutional neural networks. In: CVPR, pp. 2414–2423 (2016)
8. Hochreiter, S., Schmidhuber, J.: Long short-term memory. Neural Comput. **9**(8), 1735–1780 (1997)
9. Holden, D., Habibie, I., Kusajima, I., Komura, T.: Fast neural style transfer for motion data. IEEE Comput. Graph. Appl. **37**(4), 42–49 (2017)
10. Holden, D., Komura, T., Saito, J.: Phase-functioned neural networks for character control. ACM Trans. Graph. **36**(4), 42:1–42:13 (2017)
11. Holden, D., Saito, J., Komura, T.: A deep learning framework for character motion synthesis and editing. ACM Trans. Graph. **35**(4), 138:1–138:11 (2016)
12. Kingma, D.P., Ba, J.: Adam: a method for stochastic optimization. In: ICLR (2015)
13. Kovar, L., Gleicher, M.: Flexible automatic motion blending with registration curves. In: SIGGRAPH/Eurographics Symposium, pp. 214–224 (2003)

14. Kulkarni, T.D., Narasimhan, K., Saeedi, A., Tenenbaum, J.: Hierarchical deep reinforcement learning: integrating temporal abstraction and intrinsic motivation. In: NIPS, pp. 3675–3683 (2016)
15. Lee, Y., Sun, S., Somasundaram, S., Hu, E.S., Lim, J.J.: Composing complex skills by learning transition policies. In: ICLR (2019)
16. Levine, S., Wang, J.M., Haraux, A., Popovic, Z., Koltun, V.: Continuous character control with low-dimensional embeddings. ACM Trans. Graph. **31**(4), 28:1–28:10 (2012)
17. Li, Y., Wang, T., Shum, H.: Motion texture: a two-level statistical model for character motion synthesis. ACM Trans. Graph. **21**(3), 465–472 (2002)
18. Li, Y., Roblek, D., Tagliasacchi, M.: From here to there: video inbetweening using direct 3D convolutions. CoRR abs/1905.10240 (2019)
19. Liu, Z., Yeh, R.A., Tang, X., Liu, Y., Agarwala, A.: Video frame synthesis using deep voxel flow. In: ICCV, pp. 4473–4481 (2017)
20. Meyer, S., Wang, O., Zimmer, H., Grosse, M., Sorkine-Hornung, A.: Phase-based frame interpolation for video. In: CVPR, pp. 1410–1418 (2015)
21. Myers, D.R.: Robot Motion: Planning and Control edited by Michael Brady M.I.T. Press, Cambridge MA, USA, 1983 (£33.95). Robotica **1**(2), 109 (1983)
22. Niklaus, S., Mai, L., Liu, F.: Video frame interpolation via adaptive separable convolution. In: ICCV, pp. 261–270 (2017)
23. Park, S.I., Shin, H.J., Shin, S.Y.: On-line locomotion generation based on motion blending. In: SIGGRAPH/Eurographics Symposium, pp. 105–111 (2002)
24. Paszke, A., et al.: Pytorch: an imperative style, high-performance deep learning library. In: NeurIPS, pp. 8024–8035 (2019)
25. Pavllo, D., Feichtenhofer, C., Auli, M.: Modeling human motion with quaternion-based neural networks. In: IJCV (2019)
26. Peng, X.B., Abbeel, P., Levine, S., van de Panne, M.: DeepMimic: example-guided deep reinforcement learning of physics-based character skills. ACM Trans. Graph. (TOG) **37**(4), 143 (2018)
27. Shoemake, K.: Animating rotation with quaternion curves. In: SIGGRAPH, pp. 245–254 (1985)
28. Tan, C.I., Tai, W.: Characteristics preserving racer animation: a data-driven race path synthesis in formation space. J. Vis. Comput. Anim. **23**(3–4), 215–223 (2012)
29. Tenenbaum, J.B., Freeman, W.T.: Separating style and content. In: NeurIPS, pp. 662–668 (1996)
30. Unterthiner, T., van Steenkiste, S., Kurach, K., Marinier, R., Michalski, M., Gelly, S.: Towards accurate generative models of video: a new metric & challenges. CoRR abs/1812.01717 (2018). http://arxiv.org/abs/1812.01717
31. Urtasun, R., Fleet, D.J., Geiger, A., Popovic, J., Darrell, T., Lawrence, N.D.: Topologically-constrained latent variable models. In: ICML, pp. 1080–1087 (2008)
32. Wexler, Y., Shechtman, E., Irani, M.: Space-time completion of video. IEEE Trans. Pattern Anal. Mach. Intell. **29**(3), 463–476 (2007)
33. Xia, G., Sun, H., Liu, Q., Hang, R.: Learning-based sphere nonlinear interpolation for motion synthesis. IEEE Trans. Ind. Inform. **15**(5), 2927–2937 (2019). https://doi.org/10.1109/TII.2019.2894113
34. Xia, S., Wang, C., Chai, J., Hodgins, J.K.: Realtime style transfer for unlabeled heterogeneous human motion. ACM Trans. Graph. **34**(4), 119:1–119:10 (2015)

35. Yan, X., et al.: MT-VAE: learning motion transformations to generate multimodal human dynamics. In: Ferrari, V., Hebert, M., Sminchisescu, C., Weiss, Y. (eds.) ECCV 2018. LNCS, vol. 11209, pp. 276–293. Springer, Cham (2018). https://doi.org/10.1007/978-3-030-01228-1_17
36. Zhang, H., Starke, S., Komura, T., Saito, J.: Mode-adaptive neural networks for quadruped motion control. ACM Trans. Graph. **37**(4)

Who Left the Dogs Out? 3D Animal Reconstruction with Expectation Maximization in the Loop

Benjamin Biggs[1]([⊠]), Oliver Boyne[1], James Charles[1], Andrew Fitzgibbon[2], and Roberto Cipolla[1]

[1] Department of Engineering, University of Cambridge, Cambridge, UK
{bjb56,ob312,jjc75,rc10001}@cam.ac.uk
[2] Microsoft, Cambridge, UK
awf@microsoft.com

Abstract. We introduce an automatic, end-to-end method for recovering the 3D pose and shape of dogs from monocular internet images. The large variation in shape between dog breeds, significant occlusion and low quality of internet images makes this a challenging problem. We learn a richer prior over shapes than previous work, which helps regularize parameter estimation. We demonstrate results on the Stanford Dog Dataset, an 'in the wild' dataset of 20,580 dog images for which we have collected 2D joint and silhouette annotations to split for training and evaluation. In order to capture the large shape variety of dogs, we show that the natural variation in the 2D dataset is enough to learn a detailed 3D prior through expectation maximization (EM). As a by-product of training, we generate a new parameterized model (including limb scaling) SMBLD which we release alongside our new annotation dataset *StanfordExtra* to the research community.

1 Introduction

Animals contribute greatly to our society, in numerous ways both economic and otherwise (there are more than 63 million pet dogs in the US alone [3]). Consequently, there has been considerable attention in the computer vision research community to the interpretation of animal imagery. Although these techniques share similarities to those used for understanding images of humans, a key difference is that obtaining labelled training data for animals is more difficult than for humans. This is due to the wide range of shapes and species of animals, and the difficulty of educating manual labellers in animal physiology.

Dogs are a particular species of interest, however it is noticeable that existing work has not yet demonstrated effective 3D reconstruction of dogs over large test sets. We postulate that this is partially because dog breeds are remarkably

Electronic supplementary material The online version of this chapter (https://doi.org/10.1007/978-3-030-58621-8_12) contains supplementary material, which is available to authorized users.

© Springer Nature Switzerland AG 2020
A. Vedaldi et al. (Eds.): ECCV 2020, LNCS 12356, pp. 195–211, 2020.
https://doi.org/10.1007/978-3-030-58621-8_12

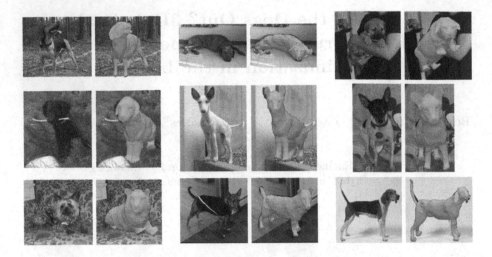

Fig. 1. End-to-end 3D dog reconstruction from monocular images. We propose a novel method that, given a monocular input image of a dog, directly predicts a set of SMBLD parameters to generate an accurate 3D dog model consistent in terms of shape and pose with the input. We regularize learning using a multi-modal shape prior, which is tuned during training with an expectation maximization scheme.

dissimilar in shape and texture, presenting a challenge to the current state of the art. The methods we propose extend the state of the art in several ways. While each of these qualities exist in some existing works, we believe ours is the first to exhibit this combination, leading to a new state of the art in terms of scale and object diversity (Fig. 1).

1. We reconstruct pose and shape on a test set of 1703 low-quality internet images of a complex 3D object class (dogs).
2. We directly regress to object pose and shape from a single image without a model fitting stage.
3. We use easily obtained 2D annotations in training, and none at test time.
4. We incorporate fitting of a new multi-modal prior into the training phase (via EM update steps), rather than fitting it to 3D data as in previous work.
5. We introduce new degrees of freedom to the SMAL model, allowing explicit scaling of subparts.

1.1   Related Work

The closest work in terms of scale is the category-specific mesh reconstruction of Kanazawa et al. [15], where 2850 images of birds were reconstructed. However, doing so for the complex pose and shape variations of dogs required the advances described in this paper.

Table 1 summarizes previous work on animal reconstruction. It is interesting to note that while several papers demonstrate reconstruction across species,

Table 1. Literature summary: Our paper extends large-scale 'in the wild' reconstruction to the difficult class of diverse breeds of dogs. MLQ: Medium-to-large quadrupeds. J2: 2D Joints. S2: 2D Silhouettes. T3: 3D Template. P3: 3D Priors. M3: 3D Model.

Paper	Animal class	Training requirements	Template model	Video required	Test time annotation	Model fitting	Test size
This paper	Dogs	J2, S2, T3, P3	SMAL	No	None	No	1703
3D-Safari [32]	Zebras, horses	M3 (albeit synthetic), J2, S2, P3	SMAL	3–7 frames/animal	None	Yes	200
Lions, Tigers and Bears (SMALR) [33]	MLQ	Not trained	SMAL	3–7 frames/animal	J2, S2	Yes	14
3D Menagerie (SMAL) [34]	MLQ	Not trained	SMAL	No	J2, S2	Yes	48
Creatures Great and SMAL [5]	MLQ	Not trained	SMAL	Yes	S2 (for best results shown)	Yes	9
Category Specific Mesh Reconstructions [15]	Birds	J2, S2	Bird convex hull	No	None	No	2850
What Shape are Dolphins [7]	Dolphins, Pigeons	Not trained	Dolphin Template	25 frames/category	J2, S2	Yes	25
Animated 3D Creatures [29]	MLQ	Not trained	Generalized Cylinders	Yes	J2, S2	Yes	15

which *prima facie* is a richer class than just dogs, the test-time requirements (e.g. manually-clicked keypoints/silhouette segmentations, input image quality etc.) are considerably higher for those systems. Thus we claim that the achievement of reconstructing a full range of dog breeds, with variable fur length, varying shape and pose of ears, and with considerable occlusion, is a significant contribution.

Monocular 3D Reconstruction of Human Bodies. The majority of recent work in 3D pose and shape recovery from monocular images tackles the special case of 3D *human* reconstruction. As a result, the research community has collected a multitude of open source human datasets which provide strong supervisory signals for training deep neural networks. These include accurate 3D deformable template models [23] generated from real human scans, 3D motion capture datasets [11,24] and large 2D datasets [4,12,22] which provide keypoint and silhouette annotations.

The abundance of available human data has supported the development of successful monocular 3D reconstruction pipelines [13,21]. Such approaches rely on accurate 3D data to build detailed priors over the distribution of human shapes and poses, and use large 2D keypoint datasets to promote generalization to 'in the wild' scenarios. Silhouette data has also been shown to assist in accurate reconstruction of clothes, hair and other appearance detail [2,30]. While the dominant paradigm in human reconstruction is now end-to-end deep learning methods, SPIN [20] shows impressive improvement by incorporating an energy minimization process within their training loop to further minimize a 2D reprojection loss subject to fixed pose & shape priors. Inspired by this innovation, we learn an iteratively-improving shape prior by applying expectation maximization during the training process.

Monocular 3D Reconstruction of Animal Categories. While animals are often featured in computer vision literature, there are still relatively few works that focus on accurate 3D animal reconstruction.

A primary reason for this is absence of large scale 3D datasets[1] stemming from the practical challenges associated with 3D motion capture, as well as a lack of 2D data which captures a wide variety of animals. The recent Animal Pose dataset [6] is one such 2D alternative, but contains significantly fewer labelled images than in our new StanfordDogs dataset (4,000 compared to 20,580). On the other hand, animal silhouette data is plentiful [9,18,22].

Zuffi et al. [34] made a significant contribution to 3D animal reconstruction research by releasing SMAL, a deformable 3D quadruped model (analogous to SMPL [23] for human reconstruction) from 41 scans of artist-designed toy figurines. The authors also released shape and pose priors generated from artist data. In this work we develop *SMBLD*, an extension of SMAL that better represents the diverse dog category by adding scale parameters and refining the shape prior using our large image dataset.

While there have been various 'model-free' approaches which do not rely on an initial template model to generate the 3D animal reconstruction, these techniques often do not produce a mesh [1,26] or rely heavily on input 2D keypoints or video at test-time [28,31]. An exception is the end-to-end network of Kanazawa et al. [15], although we argue that the bird category exhibits more limited articulation than our dog category.

We instead focus on model-based approaches. The SMAL authors [34] demonstrate fitting their deformable 3D model to quadruped species using user-provided keypoint and silhouette dataset. SMALR [33] then demonstrated fitting to broader animal categories by incorporating multi-view constraints from video sequences. Biggs et al. [5] overcame the need for hand-clicked keypoints by training a joint predictor on synthetic data. 3D-Safari [32] further improve by training a deep network on synthetic data (built using SMALR [33]) to recover detailed zebra shapes 'in the wild'.

A drawback of these approaches is their reliance on a test-time energy-based optimization procedure, which is susceptible to failure with poor quality keypoint/silhouette predictions and increases the computational burden. By contrast our method requires no additional energy-based refinement, and is trained purely from single 'in the wild' images. The experimental section of this paper contains a robust comparison between our end-to-end method and relevant optimization-based approaches.

---

[1] Released after the submission of this paper, RGBD-Dog dataset [17] is the first open-source 3D motion capture dataset for dogs.

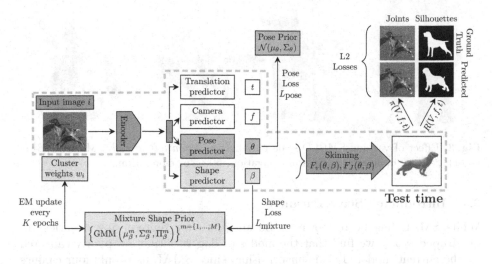

Fig. 2. Our method consists of (1) a deep CNN encoder which condenses the input image into a feature vector (2) a set of prediction heads which generate SMBLD parameters for shape β, pose θ, camera focal length f and translation t (3) skinning functions F_v and F_J which construct the mesh from a set of parameters, and (4) loss functions which minimise the error between projected and ground truth joints and silhouettes. Finally, we incorporate a mixture shape prior (5) which regularises the predicted 3D shape and is iteratively updated during training using expectation maximisation. At test time, our system (1) condenses the input image, (2) generates the SMBLD parameters and (3) constructs the mesh.

A major impediment to research in 3D animal reconstruction has been the lack of a strong evaluation benchmark, with most of the above methods showing only qualitative evaluations or providing quantitative results on fewer than 50 examples. To remedy this, we introduce *StanfordExtra*, a new large-scale dataset which we hope will drive further progress in the field.

2   Parametric Animal Model

At the heart of our method is a parametric representation of a 3D animal mesh, which is based on the Skinned Multi-Animal Linear (SMAL) model proposed by [34]. SMAL is a deformable 3D animal mesh parameterized by shape and pose. The *shape* $\beta \in \mathbb{R}^B$ parameters are PCA coefficients of an undeformed template mesh with limbs in default position. The *pose* $\theta \in \mathbb{R}^P$ parameters meanwhile govern the joint angle rotations (35 × 3 Rodrigues parameters) which effect the articulated limb movement. The model consists of a linear blend skinning function $F_v : (\theta, \beta) \mapsto V$, which generates a set of vertex positions $V \in \mathbb{R}^{3889 \times 3}$, and a joint function $F_J : (\theta, \beta) \mapsto J$, which generates a set of joint positions $J \in \mathbb{R}^{35 \times 3}$.

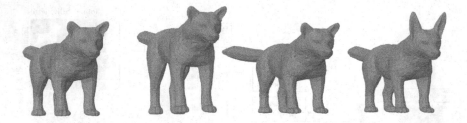

Fig. 3. Effect of varying SMBLD scale parameters. *From left to right*: Mean SMBLD model, 25% leg elongation, 50% tail elongation, 50% ear elongation.

2.1 Introducing Scale Parameters

While SMAL has been shown to be adequate for representing a variety of quadruped types, we find that the modes of dog variation are poorly captured by the current model. This is unsurprising, since SMAL used only four canines in its construction.

We therefore introduce a simple but effective way to improve the model's representational power over this particularly diverse animal category. We augment the set of shape parameters β with an additional set κ which independently scale parts of the mesh. For each model joint, we define parameters $\kappa_x, \kappa_y, \kappa_z$ which apply a local scaling of the mesh along the local coordinate x, y, z axes, before pose is applied. Allowing each joint to scale entirely independently can however lead to unrealistic deformations, so we share scale parameters between multiple joints, e.g. leg lengths. The new Skinned Multi-Breed Linear Model for Dogs (SMBLD) is therefore adapted from SMAL by adding 6 scale parameters to the existing set of shape parameters. Figure 3 shows how introducing scale parameters increases the flexibility of the SMAL model. We also extend the provided SMAL shape prior (which later initializes our EM procedure) to cover the new scale parameters by fitting SMBLD to a set of 13 artist-designed 3D dog meshes. Further details left to the supplementary.

3 End-to-End Dog Reconstruction from Monocular Images

We now consider the task of reconstructing a 3D dog mesh from a monocular image. We achieve this by training an end-to-end convolutional network that predicts a set of SMBLD model and perspective camera parameters. In particular, we train our network to predict pose θ and shape β SMBLD parameters together with translation t and focal length f for a perspective camera. A complete overview of the proposed system is shown in Fig. 2.

3.1 Model Architecture

Our network architecture is inspired by the model of 3D-Safari [32]. Given an input image cropped to (224, 224), we apply a Resnet-50 [10] backbone network to encode a 1024-dimensional feature map. These features are passed through various linear prediction heads to produce the required parameters. The pose, translation and camera prediction modules follow the design of 3D-Safari, but we describe the differences in our shape module.

Pose, Translation and Camera Prediction. These modules are independent multi-layer perceptrons which map the above features to the various parameter types. As with 3D-Safari we use two linear layers to map to a set of 35 × 3 3D pose parameters (three parameters for each joint in the SMBLD kinematic tree) given in Rodrigues form. We use independent heads to predict camera frame translation $t_{x,y}$ and depth t_z independently. We also predict the focal length of the perspective camera similarly to 3D-Safari.

Shape and Scale Prediction. Unlike 3D-Safari, we design our network to predict the set of shape parameters (including scale) rather than vertex offsets. We observe improvement by handling the standard 20 blend-shape parameters and our new scale parameters in separate linear prediction heads. We retrieve the scale parameters by $\kappa = \exp x$ where x are the network predictions, as we find predicting log scale helps stabilise early training.

3.2 Training Losses

A common approach for training such an end-to-end system would be to supervise the prediction of (θ, β, t, f) with 3D ground truth annotations [14,20,27]. However, building a suitable 3D annotation dataset would require an experienced graphics artist to design an accurate ground truth mesh for each of 20,520 StanfordExtra dog images, a prohibitive expense.

We instead develop a method that instead relies on *weak 2D supervision* to guide network training. In particular, we rely on only 2D keypoints and silhouette segmentations, which are significantly cheaper to obtain.

The rest of this section describes the set of losses used to supervise the network at train time.

Joint Reprojection. The most important loss to promote accurate limb positioning is the joint reprojection loss L_{joints}, which compares the projected model joints $\pi(F_J(\theta, \beta), t, f)$ to the ground truth annotations $\hat{X}$. Given the parameters predicted by the network, we apply the SMBLD model to transform the pose and shape parameters into a set of 3D joint positions $J \in \mathbb{R}^{35 \times 3}$, and project them to the image plane using translation and camera parameters. The joint loss L_{joints} is given by the ℓ_2 error between the ground truth and projected joints:

$$L_{\text{joints}}(\theta, \beta, t, f; \hat{X}) = \|\hat{X} - \pi(F_J(\theta, \beta), t, f)\|_2 \tag{1}$$

Note that many of our training images exhibit significant occlusion, so $\hat{X}$ contains many invisible joints. We handle this by masking L_{joints} to prevent invisible joints contributing to the loss.

Silhouette Loss. The silhouette loss L_{sil} is used to promote shape alignment between the SMBLD dog mesh and the input dog. In order to compute the silhouette loss, we define a rendering function $R : (\nu, t, f) \mapsto S$ which projects the SMBLD mesh to produce a binary segmentation mask. In order to allow derivatives to be propagated through R, we implement R using the differentiable Neural Mesh Renderer [16]. The loss is computed as the ℓ_2 difference between a projected silhouette and the ground truth mask $\hat{S}$:

$$L_{\text{sil}}(\theta, \beta, t, f; \hat{S}) = \|\hat{S} - R(F_V(\theta, \beta), t, f)\|_2 \tag{2}$$

Priors. In the absence of 3D ground truth training data, we rely on priors obtained from artist graphics models to encourage realism in the network predictions. We model both pose and shape using a multivariate Gaussian prior, consisting of means μ_θ, μ_β and covariance matrices $\Sigma_\theta, \Sigma_\beta$. The loss is given as the log likelihood of a given shape or pose vector under these distributions, which corresponds to the Mahalanobis distance between the predicted parameters and their corresponding means:

$$L_{\text{pose}}(\theta; \mu_\theta, \Sigma_\theta) = (\theta - \mu_\theta)^T \Sigma_\theta^{-1} (\theta - \mu_\theta) \tag{3}$$

$$L_{\text{shape}}(\beta; \mu_\beta, \Sigma_\beta) = (\beta - \mu_\beta)^T \Sigma_\beta^{-1} (\beta - \mu_\beta) \tag{4}$$

Unlike previous work, we find there is no need to use a loss to penalize pose parameters if they exceed manually specified joint angle limits. We suspect our network learns this regularization naturally because of our large dataset.

3.3   Learning a Multi-modal Shape Prior

The previous section introduced a unimodal, multivariate Gaussian shape prior, based on mean μ_β and covariance matrix Σ_β. However, we find enforcing this prior throughout training tends to result in predictions which appear similar in 3D shape, even when tested on dog images of different breeds. We propose to improve diversity among predicted 3D dog shapes by extending the above formulation to a mixture of M Gaussians prior. The mixture shape loss is then given as:

$$L_{\text{mixture}}(\beta_i; \mu_\beta, \Sigma_\beta, \Pi_\beta) = \sum_{m=1}^{M} \Pi_\beta^m (\beta_i - \mu_\beta^m)^T \Sigma_\beta^{m-1} (\beta_i - \mu_\beta^m) \tag{5}$$

$$= \sum_{m=1}^{M} \Pi_\beta^m L_{\text{shape}}(\beta_i; \mu_\beta^m, \Sigma_\beta^m) \tag{6}$$

Where μ_β^m, Σ_β^m and Π_β^m are the mean, covariance and mixture weight respectively for Gaussian component m. For each component the mean is sampled from our existing unimodal prior, and the covariance is set equal to the unimodal prior i.e. $\Sigma_\beta^m := \Sigma_\beta$. All mixture weights are initially set to $\frac{1}{M}$.

Each training image i is assigned a set of latent variables $\{w_i^1, \ldots w_i^M\}$ encoding the probability of the dog shape in image i being generated by component m.

3.4 Expectation Maximization in the Loop

As previously discussed, our initial shape prior is obtained from artist data which we find is unrepresentative of the diverse shapes present in our real dog dataset. We address this by proposing to recover the latent variables w_i^m and parameters $(\mu_\beta^m, \Sigma_\beta^m$ and $\Pi_\beta^m)$ of our 3D shape prior by learning from monocular images of 'in the wild' dogs and their 2D training labels in our training dataset.

We achieve this using Expectation Maximization (EM), which regularly updates the means and variances for each mixture component and per-image mixture weights based on the observed shapes in the training set. While training our 3D reconstruction network, we progressively update our shape mixture model with an alternating 'E' step and 'M' step described below:

The 'E' Step. The 'E' step computes the expected value of the latent variables w_i^m assuming fixed $(\mu_\beta^m, \Sigma_\beta^m, \Pi_\beta^m)$ for all $i \in \{1, \dots, N\}, m \in \{1, \dots, M\}$.

The update equation for an image i with latest shape prediction β_i and cluster m with parameters $(\mu_\beta^m, \Sigma_\beta^m, \Pi_\beta^m)$ is given as:

$$w_i^m := \frac{\mathcal{N}(\beta_i|\mu_\beta^m, \Sigma_\beta^m)\Pi_\beta^m}{\sum_{m'}^{M} \mathcal{N}(\beta_i|\mu_\beta^{m'}, \Sigma_\beta^{m'})\Pi_\beta^{m'}} \tag{7}$$

The 'M' Step. The 'M' step computes new values for $(\mu_\beta^m, \Sigma_\beta^m, \Pi_\beta^m)$, assuming fixed w_i^m for all $i \in \{1, \dots, N\}, m \in \{1, \dots, M\}$.

The update equations are given as follows:

$$\mu_\beta^m := \frac{\sum_i w_i^m \beta_i}{\sum_i w_i^m} \quad \Sigma_\beta^m := \frac{\sum_i w_i^m (\beta_i - \Sigma_\beta^m)(\beta_i - \Sigma_\beta^m)^T}{\sum_i w_i^m} \quad \Pi_\beta^m := \frac{1}{N} \sum_i w_i^m \tag{8}$$

4 Experiments

In this section we compare our method to competitive baselines. We begin by describing our new large-scale dataset of annotated dog images, followed by a quantitative and qualitative evaluation.

4.1 StanfordExtra: A New Large-Scale Dog Dataset with 2D Keypoint and Silhouette Annotations

In order to evaluate our method, we introduce StanfordExtra: a new large-scale dataset with annotated 2D keypoints and binary segmentation masks for dogs. We opted to take source images from the existing Stanford Dog Dataset [19], which consists of 20,580 dog images taken 'in the wild' and covers 120 dog breeds. The dataset contains vast shape and pose variation between dogs, as well as nuisance factors such as self/environmental occlusion, interaction with

Fig. 4. StanfordExtra example images. *Left*: outlined segmentations and labelled keypoints for a selection of StanfordExtra images. *Right*: heatmap showing annotator uncertainty in per-keypoint clicking, computed over the entire dataset.

humans/other animals and partial views. Figure 4 (left) shows samples from the new dataset.

We used Amazon Mechanical Turk to collect a binary silhouette mask and 20 keypoints per image: 3 per leg (knee, ankle, toe), 2 per ear (base, tip), 2 per tail (base, tip), 2 per face (nose and jaw). We can approximate the difficulty of the dataset by analysing the variance between 3 annotators at both the joint labelling and silhouette task. Figure 4 (right) illustrates typical per-joint variance in joint labelling. Further details of the data curation procedure are left to the supplementary materials.

4.2   Evaluation Protocol

Our evaluation is based on our new StanfordExtra dataset. In line with other methods which tackle 'in the wild' 3D reconstruction of articulated subjects [20, 21], we filter images from the original dataset of 20,580 for which the majority of dog keypoints are invisible. We consider these images unsuitable for our full-body dog reconstruction task. We also remove images for which the consistency in keypoint/silhouette segmentations between the 3 annotators is below a set threshold. This leaves us with 8,476 images which we divide per-breed into an 80%/20% train and test split.

We consider two primary evaluation metrics. IoU is the intersection-over-union of the projected model silhouette compared to the ground truth annotation and indicates the quality of the reconstructed 3D shape. Percentage of Correct Keypoints (PCK) computes the percentage of joints which are within a normalized distance (based on square root of 2D silhouette area) to the ground truth locations, and evaluates the quality of reconstructed 3D pose. We also produce PCK results on various joint groups (legs, tail, ears, face) to compare the reconstruction accuracy for different parts of the dog model.

4.3   Training Procedure

We train our model in two stages. The first omits the silhouette loss which we find can lead the network to unsatisfactory local minima if applied too early. With the silhouette loss turned off, we find it satisfactory to use the simple unimodal prior (and without EM) for this preliminary stage since there is no loss to specifically encourage a strong shape alignment. After this, we introduce the silhouette loss, the mixture prior and begin applying the expectation maximization updates over $M = 10$ clusters. We train the first stage for 250 epochs, the second stage for 150 and apply the EM step every 50 epochs. All losses are weighted, as described in the supplementary. The entire training procedure takes 96 h on a single P100 GPU.

4.4   Comparison to Baselines

We first compare our method to various baseline methods. SMAL [34] is an approach which fits the 3D SMAL model using per-image energy minimization. Creatures Great and SMAL (CGAS) [5] is a three-stage method, which employs a joint predictor on silhouette renderings from synthetic 3D dogs, applies a genetic algorithm to clean predictions, and finally applies the SMAL optimizer to produce the 3D mesh.

At test-time both SMAL and CGAS rely on manually-provided segmentation masks, and SMAL also relies on hand-clicked keypoints. In order to produce a fair comparison, we produce a set of *predicted* keypoints for StanfordExtra by training the Stacked Hourglass Network [25] with 8 stacks and 1 block, and *predicted* segmentation masks using DeepLab v3+ [8]. The Stacked Hourglass Network achieves 71.4% PCK score, DeepLab v3+ achieves 83.4% IoU score and the CGAS joint predictor achieves 41.8% PCK score.

Table 2 and Fig. 5 show the comparison against competitive methods. For full examination, we additionally provide results for SMAL and CGAS in the scenario that ground-truth keypoints and/or segmentations are available at test time. The results show our end-to-end method outperforms the competitors when they are provided with predicted keypoints/segmentations (white rows). Our method therefore achieves a new state-of-the-art on this 3D reconstruction task. In addition, we show our method achieves improved average IoU/PCK scores than competitive methods, even when they are provided ground truth annotations at test time (grey rows). We also demonstrate wider applicability of two contributions from our work (scale parameters and improved prior) by showing improved performance of the SMAL method when these are incorporated. Finally, our model's test-time speed is significantly faster than the competitors as it does not require an optimizer.

4.5   Generalization to Unseen Dataset

Table 3 shows an experiment to compare how well our model generalizes to a new data domain. We test our model against the SMAL [34] method (using predicted

Table 2. Quantitative comparison to baselines. PCK and silhouette IoU metrics are shown against competitive methods in various conditions. *White rows*: Experiments which use no test-time ground truth annotations, *Grey rows*: Experiments which either use test-time ground truth annotations (GT) or incorporate components introduced in this paper (SMBLD scale parameters or mixture shape prior).

Method	Kps	Seg	IoU	PCK Avg	Legs	Tail	Ears	Face
SMAL [34]	Pred	Pred	67.9	67.1	65.7	79.5	54.9	87.4
SMAL	GT	GT	69.2	72.6	69.9	**92.0**	58.6	**96.9**
SMAL	GT	Pred	68.6	72.6	70.2	91.5	58.1	**96.9**
SMAL	Pred	GT	68.5	67.4	66.0	79.9	55.0	88.2
CGAS [5]	CGAS	Pred	62.4	43.7	46.5	64.1	36.5	21.4
CGAS	CGAS	GT	63.1	43.6	46.3	64.2	36.3	21.6
SMAL + scaling	Pred	Pred	69.3	69.6	69.4	79.3	56.5	87.6
SMAL + scaling + new prior	Pred	Pred	70.7	71.6	71.5	80.7	59.3	88.0
Ours	—	—	**73.6**	**75.7**	**75.0**	77.6	**69.9**	90.0

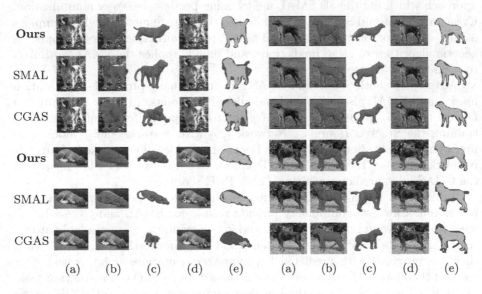

Ours   SMAL   CGAS   Ours   SMAL   CGAS

(a)   (b)   (c)   (d)   (e)   (a)   (b)   (c)   (d)   (e)

Fig. 5. Qualitative comparison to baselines. Comparison between our method, SMAL [34] and CGAS [5]. (a) Input image, (b) predicted 3D mesh, (c) canonical view 3D mesh, (d) reprojected model joints and (e) silhouette reprojection error.

keypoints and segmentations as above for fairness) on the recent Animal Pose dataset [6]. The data preparation process is the same as for StanfordExtra and no fine-tuning was used for either method. We achieve strong results in this unseen domain and still improve over the SMAL optimizer.

Table 3. Animal Pose dataset [6]. Evaluation on recent Animal Pose dataset with no fine-tuning to our method nor joint/silhouette predictors used for SMAL.

Method	IoU	PCK				
		Avg	Legs	Tail	Ears	Face
SMAL [34]	63.6	69.1	60.9	83.5	75.0	93.0
Ours	**66.9**	**73.8**	**65.1**	**85.6**	**84.0**	**93.6**

Table 4. Ablation study. Evaluation with the following contributions removed: (a) EM updates, (b) Mixture Shape Prior, (c) SMBLD scale parameters.

Method	IoU	PCK				
		Avg	Legs	Tail	Ears	Face
Ours	**73.6**	**75.7**	**75.0**	**77.6**	69.9	90.0
−EM	67.7	74.6	72.9	75.2	**72.5**	88.3
−MoG	68.0	74.9	74.3	73.3	70.0	**90.2**
−Scale	67.3	72.6	72.9	75.3	62.3	89.1

4.6 Ablation Study

We also produce a study in which we ablate individual components of our method and examine the effect on the PCK/IoU performance. We evaluate three variants: (1) **Ours w/o EM** that omits EM updates, (2) **Ours w/o MoG** which replaces our mixture shape prior with a unimodal prior, (3) **Ours w/o Scale** which removes the scale parameters.

The results in Table 4 indicate that each individual component has a positive impact on the overall method performance. In particular, it can be seen that the inclusion of the EM and mixture of Gaussians prior leads to an improvement in IoU, suggesting that the shape prior refinements steps help the model accurately fit the exact dog shape. Interestingly, we notice that adding the mixture of Gaussians prior but omitting EM steps slightly hinders performance, perhaps due to a sub-optimal initialization for the M clusters. However, we find adding EM updates to the mixture of Gaussian model improves all metrics except the ear keypoint accuracy. We observe the error here is caused by the our shape prior learning slightly imprecise shapes for dogs with extremely 'floppy' ears. Although there is good silhouette coverage for these regions, the fact our model has only a single articulation point per ear causes a lack of flexibility that results in occasionally misplaced ear tips for these instances. This could be improved in future work by adding additional model joints to the ear. Finally, our results show the increased model flexibility afforded by the SMBLD scale parameters has a positive effect on IoU/PCK scores.

4.7 Qualitative Evaluation

Figure 5 shows a range of predicted 3D reconstructions from our method when tested on a wide range of StanfordExtra and Animal Pose [6] dogs. The examples include challenging poses, large diversity in dog shape and size, interaction with humans, various environmental conditions and handling occluded keypoints. Note that only StanfordExtra is used for training and we use no fine-tuning to produce results on the Animal Pose dataset (Fig. 6).

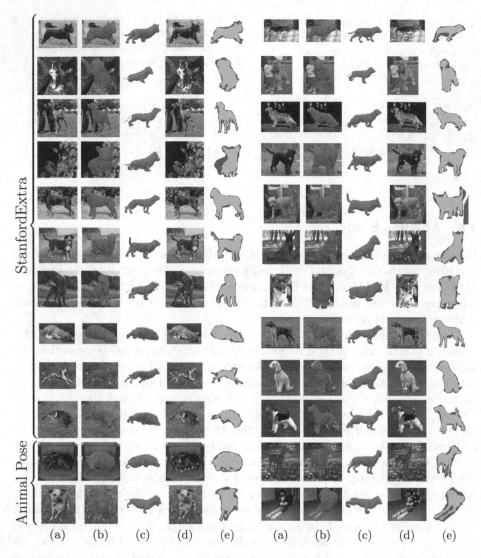

Fig. 6. Qualitative results on StanfordExtra and Animal Pose [6]. For each sample we show: (a) input image, (b) predicted 3D mesh, (c) canonical view 3D mesh, (d) reprojected model joints and (e) silhouette reprojection error.

5   Conclusions

This paper presents an end-to-end method for automatic, monocular 3D dog reconstruction. We achieve this using only weak 2D supervision, provided by our novel StanfordExtra dataset. Furthermore, we show we can learn a more detailed shape prior by tuning a Gaussian mixture during model training and this leads to improved reconstructions. We also show our method improves over competitive baselines, even when they are given access to ground truth data at test time.

Future work should involve tackling some failure cases of our system, for example handling multiple overlapping dogs or dealing with heavy motion blur. Other areas for research include extending our EM formulation to handle video input to take advantage of multi-view shape constraints, and transferring knowledge accumulated through training on StanfordExtra dogs to other species.

Acknowlegements. The authors would like to thank the GSK AI team for providing access to their GPU cluster, Michael Sutcliffe, Matthew Allen, Thomas Roddick and Peter Fisher for useful technical discussions, and the GSK TDI team for project sponsorship.

References

1. Agudo, A., Pijoan, M., Moreno-Noguer, F.: Image collection pop-up: 3D reconstruction and clustering of rigid and non-rigid categories. In: Proceedings of the IEEE Conference on Computer Vision and Pattern Recognition (CVPR) (2018)
2. Alldieck, T., Magnor, M., Bhatnagar, B.L., Theobalt, C., Pons-Moll, G.: Learning to reconstruct people in clothing from a single RGB camera. In: Proceedings of the IEEE Conference on Computer Vision and Pattern Recognition, pp. 1175–1186 (2019)
3. American Pet Products Association: 2019–2020 APPA National Pet Owners Survey (2020). http://www.americanpetproducts.org
4. Andriluka, M., Pishchulin, L., Gehler, P., Schiele, B.: 2D human pose estimation: new benchmark and state of the art analysis. In: IEEE Conference on Computer Vision and Pattern Recognition (CVPR) (2014)
5. Biggs, B., Roddick, T., Fitzgibbon, A., Cipolla, R.: Creatures great and SMAL: recovering the shape and motion of animals from video. In: Jawahar, C.V., Li, H., Mori, G., Schindler, K. (eds.) ACCV 2018. LNCS, vol. 11365, pp. 3–19. Springer, Cham (2019). https://doi.org/10.1007/978-3-030-20873-8_1
6. Cao, J., Tang, H., Fang, H., Shen, X., Tai, Y., Lu, C.: Cross-domain adaptation for animal pose estimation. In: 2019 IEEE/CVF International Conference on Computer Vision (ICCV), pp. 9497–9506 (2019)
7. Cashman, T.J., Fitzgibbon, A.W.: What shape are dolphins? Building 3D morphable models from 2D images. IEEE Trans. Pattern Anal. Mach. Intell. **35**(1), 232–244 (2013)
8. Chen, L.C., Papandreou, G., Kokkinos, I., Murphy, K., Yuille, A.L.: DeepLab: semantic image segmentation with deep convolutional nets, atrous convolution, and fully connected CRFs. CoRR abs/1606.00915 (2016)
9. Everingham, M., Van Gool, L., Williams, C.K., Winn, J., Zisserman, A.: The pascal visual object classes (VOC) challenge. Int. J. Comput. Vis. **88**(2), 303–338 (2010)
10. He, K., Zhang, X., Ren, S., Sun, J.: Deep residual learning for image recognition. In: Proceedings of the IEEE Conference on Computer Vision and Pattern Recognition, pp. 770–778 (2016)
11. Ionescu, C., Papava, D., Olaru, V., Sminchisescu, C.: Human3.6M: large scale datasets and predictive methods for 3D human sensing in natural environments. IEEE Trans. Pattern Anal. Mach. Intell. **36**(7), 1325–1339 (2013)

12. Johnson, S., Everingham, M.: Clustered pose and nonlinear appearance models for human pose estimation. In: Proceedings of the British Machine Vision Conference (2010). https://doi.org/10.5244/C.24.12

13. Kanazawa, A., Black, M.J., Jacobs, D.W., Malik, J.: End-to-end recovery of human shape and pose. In: Computer Vision and Pattern Recognition (CVPR) (2018)

14. Kanazawa, A., Black, M.J., Jacobs, D.W., Malik, J.: End-to-end recovery of human shape and pose. In: Proceedings of the CVPR (2018)

15. Kanazawa, A., Tulsiani, S., Efros, A.A., Malik, J.: Learning category-specific mesh reconstruction from image collections. In: Ferrari, V., Hebert, M., Sminchisescu, C., Weiss, Y. (eds.) ECCV 2018. LNCS, vol. 11219, pp. 386–402. Springer, Cham (2018). https://doi.org/10.1007/978-3-030-01267-0_23

16. Kato, H., Ushiku, Y., Harada, T.: Neural 3D mesh renderer. In: The IEEE Conference on Computer Vision and Pattern Recognition (CVPR) (2018)

17. Kearney, S., Li, W., Parsons, M., Kim, K.I., Cosker, D.: RGBD-dog: predicting canine pose from RGBD sensors. In: IEEE/CVF Conference on Computer Vision and Pattern Recognition (CVPR) (2020)

18. Khoreva, A., Benenson, R., Ilg, E., Brox, T., Schiele, B.: Lucid data dreaming for object tracking. In: The 2017 DAVIS Challenge on Video Object Segmentation - CVPR Workshops (2017)

19. Khosla, A., Jayadevaprakash, N., Yao, B., Fei-Fei, L.: Novel dataset for fine-grained image categorization. In: First Workshop on Fine-Grained Visual Categorization, IEEE Conference on Computer Vision and Pattern Recognition, Colorado Springs, CO (2011)

20. Kolotouros, N., Pavlakos, G., Black, M.J., Daniilidis, K.: Learning to reconstruct 3D human pose and shape via model-fitting in the loop. In: Proceedings of the IEEE International Conference on Computer Vision, pp. 2252–2261 (2019)

21. Kolotouros, N., Pavlakos, G., Daniilidis, K.: Convolutional mesh regression for single-image human shape reconstruction. In: Proceedings of the CVPR (2019)

22. Lin, T.-Y., et al.: Microsoft COCO: common objects in context. In: Fleet, D., Pajdla, T., Schiele, B., Tuytelaars, T. (eds.) ECCV 2014. LNCS, vol. 8693, pp. 740–755. Springer, Cham (2014). https://doi.org/10.1007/978-3-319-10602-1_48. https://www.microsoft.com/en-us/research/publication/microsoft-coco-common-objects-in-context/

23. Loper, M., Mahmood, N., Romero, J., Pons-Moll, G., Black, M.J.: SMPL: a skinned multi-person linear model. ACM Trans. Graph. (TOG) **34**(6), 248 (2015)

24. von Marcard, T., Henschel, R., Black, M.J., Rosenhahn, B., Pons-Moll, G.: Recovering accurate 3D human pose in the wild using IMUs and a moving camera. In: Ferrari, V., Hebert, M., Sminchisescu, C., Weiss, Y. (eds.) ECCV 2018. LNCS, vol. 11214, pp. 614–631. Springer, Cham (2018). https://doi.org/10.1007/978-3-030-01249-6_37

25. Newell, A., Yang, K., Deng, J.: Stacked hourglass networks for human pose estimation. In: Leibe, B., Matas, J., Sebe, N., Welling, M. (eds.) ECCV 2016. LNCS, vol. 9912, pp. 483–499. Springer, Cham (2016). https://doi.org/10.1007/978-3-319-46484-8_29

26. Novotny, D., Ravi, N., Graham, B., Neverova, N., Vedaldi, A.: C3DPO: canonical 3D pose networks for non-rigid structure from motion. In: Proceedings of the ICCV (2019)

27. Pavlakos, G., Zhu, L., Zhou, X., Daniilidis, K.: Learning to estimate 3D human pose and shape from a single color image. In: Proceedings of the CVPR (2018)

28. Probst, T., Paudel, D.P., Chhatkuli, A., Van Gool, L.: Incremental non-rigid structure-from-motion with unknown focal length. In: Ferrari, V., Hebert, M., Sminchisescu, C., Weiss, Y. (eds.) ECCV 2018. LNCS, vol. 11217, pp. 776–793. Springer, Cham (2018). https://doi.org/10.1007/978-3-030-01261-8_46

29. Reinert, B., Ritschel, T., Seidel, H.P.: Animated 3D creatures from single-view video by skeletal sketching. In: Proceedings of the Graphics Interface (2016)

30. Saito, S., Huang, Z., Natsume, R., Morishima, S., Kanazawa, A., Li, H.: PIFu: pixel-aligned implicit function for high-resolution clothed human digitization. arXiv preprint arXiv:1905.05172 (2019)

31. Vicente, S., Agapito, L.: Balloon shapes: reconstructing and deforming objects with volume from images. In: 2013 International Conference on 3D Vision - 3DV 2013, pp. 223–230 (2013)

32. Zuffi, S., Kanazawa, A., Berger-Wolf, T., Black, M.J.: Three-D Safari: learning to estimate zebra pose, shape, and texture from images "in the wild". In: The IEEE International Conference on Computer Vision (ICCV) (2019)

33. Zuffi, S., Kanazawa, A., Black, M.J.: Lions and tigers and bears: capturing non-rigid, 3D, articulated shape from images. In: IEEE Conference on Computer Vision and Pattern Recognition (CVPR). IEEE Computer Society (2018)

34. Zuffi, S., Kanazawa, A., Jacobs, D., Black, M.J.: 3D menagerie: modeling the 3D shape and pose of animals. In: IEEE Conference on Computer Vision and Pattern Recognition (CVPR) (2017)

Learning to Count in the Crowd
from Limited Labeled Data

Vishwanath A. Sindagi[1]([✉]), Rajeev Yasarla[1], Deepak Sam Babu[2],
R. Venkatesh Babu[2], and Vishal M. Patel[1]

[1] Johns Hopkins University, Baltimore, MD 21218, USA
{vishwanathsindagi,ryasarl1,vpatel36}@jhu.edu
[2] Indian Institute of Science, Bangalore 560012, India
{deepaksam,venky}@iisc.ac.in

Abstract. Recent crowd counting approaches have achieved excellent performance. However, they are essentially based on fully supervised paradigm and require large number of annotated samples. Obtaining annotations is an expensive and labour-intensive process. In this work, we focus on reducing the annotation efforts by learning to count in the crowd from limited number of labeled samples while leveraging a large pool of unlabeled data. Specifically, we propose a Gaussian Process-based iterative learning mechanism that involves estimation of pseudo-ground truth for the unlabeled data, which is then used as supervision for training the network. The proposed method is shown to be effective under the reduced data (semi-supervised) settings for several datasets like Shang-haiTech, UCF-QNRF, WorldExpo, UCSD, *etc.* Furthermore, we demonstrate that the proposed method can be leveraged to enable the network in learning to count from synthetic dataset while being able to generalize better to real-world datasets (synthetic-to-real transfer).

Keywords: Crowd counting · Semi-supervised learning ·
Pseudo-labeling · Domain adaptation · Synthetic to real transfer

1   Introduction

Due to its significance in several applications (like video surveillance [12,44,50], public safety monitoring [58], microscopic cell counting [15], environmental studies [23], *etc.*), crowd counting has attracted a lot of interest from the deep learning research community. Several convolutional neural network (CNN) based approaches have been developed that address various issues in counting like scale variations, occlusion, background clutter [2,3,17–19,22,28,33,34,36,37,39,42, 43,59], *etc.* While these methods have achieved excellent improvements in terms of the overall error rate, they follow a fully-supervised paradigm and require

Electronic supplementary material The online version of this chapter (https://doi.org/10.1007/978-3-030-58621-8_13) contains supplementary material, which is available to authorized users.

© Springer Nature Switzerland AG 2020
A. Vedaldi et al. (Eds.): ECCV 2020, LNCS 12356, pp. 212–229, 2020.
https://doi.org/10.1007/978-3-030-58621-8_13

several labeled data samples. There is a wide variety of scenes and crowded scenarios that these networks need to handle to in the real world. Due to a distribution gap between the training and testing environments, these networks have limited generalization abilities and hence, procuring annotations becomes especially important. However, annotating data for crowd counting typically involves obtaining point-wise annotations at head locations, and this is a labour intensive and expensive process. Hence, it is infeasible to procure annotations for all possible scenarios. Considering this, it is crucial to reduce the annotation efforts, especially for crowd counting methods which get deployed in a wide variety of scenarios.

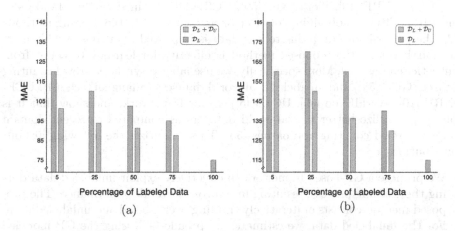

(a) (b)

Fig. 1. Results of semi-supervised learning experiments. (a) ShanghaiTech A (b) UCF-QNRF. For both datasets, the error increases with reduction in the %-age of labeled data. By leveraging the unlabeled dataset using the proposed GP-based framework, we are able to reduce the error considerably. Note that $\mathcal{D}_{\mathcal{L}}$ and $\mathcal{D}_{\mathcal{U}}$ indicate labeled and unlabeled dataset, respectively.

With the exception of a few works [6, 22, 55], reducing annotation efforts while maintaining good performance is relatively less explored for the task of crowd counting. Hence, in this work, we focus on learning to count using limited labeled data while leveraging unlabeled data to improve the performance. Specifically, we propose a Gaussian Process (GP) based iterative learning framework where we augment the existing networks with capabilities to leverage unlabeled data, thereby resulting in overall improvement in the performance. Inspired by [57], the proposed framework follows a pseudo-labeling approach, where we estimate the pseudo-ground truth (pseudo-GT) for the unlabeled data, which is then used to supervise the network. The network is trained iteratively on labeled and unlabeled data. In the labeled stage, the network weights are updated by minimizing the L_2 error between predictions and the ground-truth (GT) for the labeled data. In addition, we save the latent space vectors of the labeled data along with the ground-truths. In the unlabeled stage, we first model the relationship between the latent space vectors of the labeled images along with the corresponding ground-truth and unlabeled latent space vectors jointly using GP. Next, we estimate the pseudo-GT for the unlabeled inputs using the GP modeled earlier. This pseudo-GT is then used to supervise the network for the unlabeled data. Minimizing the error between the unlabeled data predictions and the pseudo-GT results in improved performance. Figure 1 illustrates the effectiveness of the proposed GP-based framework in exploiting unlabeled data on two datasets (ShanghaiTech-A [61] and UCF-QNRF[10]) in the reduced data setting. It can be observed that the proposed method is able to leverage unlabeled data effectively resulting in lower error across various settings.

The proposed method is evaluated on different datasets like ShanghaiTech [61], UCF-QNRF [10], WorldExpo [59], UCSD [4], *etc.* in the reduced data settings. In addition to obtaining lower error as compared to the existing methods [22], the performance drop due to less data is improved by a considerable margin. Furthermore, the proposed method is effective for learning to count from synthetic data as well. More specifically, we use labeled synthetic crowd counting dataset (GCC [55]) and unlabeled real-world datasets (ShanghaiTech [61], UCF-QNRF [10], WorldExpo [59], UCSD [5]) in our framework, and show that it is able to generalize better to real-world datasets as compared to recent domain adaptive crowd counting approaches [55]. To summarize, the following are our contributions:

- We propose a GP-based framework to effectively exploit unlabeled data during the training process, resulting in improved overall performance. The proposed method consists of iteratively training over labeled and unlabeled data. For the unlabeled data, we estimate the pseudo-GT using the GP modeled during labeled phase.
- We demonstrate that the proposed framework is effective in semi-supervised and synthetic-to-real transfer settings. Through various ablation studies, we show that the proposed method is generalizable to different network architectures and various reduced data settings.

2   Related Work

Crowd Counting. Traditional approaches in crowd counting ([7,9,15,16,27, 31,56]) typically involved feature extraction techniques and training regression algorithms. Recently, CNN-based approaches like [1,26,36,42,51,54,59,61] have surpassed the traditional approaches by a large margin in terms of the overall error rate. Most of these methods focus on addressing the issue of large variations in scales. Approaches like [36,42,61] focus on improving the receptive field. Different from these, approaches like [28,32,41,47] focus on effective ways of fusing multi-scale information from deep networks. In addition to scale variation, recent approaches have addressed other issues in crowd counting like improving the quality of predicted density maps using adversarial regularization [37,42], use of deep negative correlation-based learning for obtaining more generalizable features, and scale-based feature aggregation [3]. Most recently, several methods have employed additional information like segmentation and semantic priors [53,62], attention [20,45,46], perspective [38], context information [21], multiple-views [60] and multi-scale features [11], adaptive density maps [52] into the network. In other efforts, researchers have made important contributions by creating large-scale datasets for counting like UCF-QNRF [10], GCC [55] and JHU-CROWD [48,49]. For a more detailed discussion on these methods, the reader is referred to recent comprehensive surveys [8,43].

Learning from Limited Data. Recent research in crowd counting has been largely focused on improving the counting performance in the fully-supervised paradigm. Very few works like [6,22,55] have made efforts on minimizing annotation efforts. Loy *et al.*[6] proposed a semi-supervised regression framework that exploit underlying geometric structures of crowd patterns to assimilate the count estimation of two nearby crowd pattern points in the manifold. However, this approach is specifically designed for video-based crowd counting.

Recently, Liu *et al.*[22] proposed to leverage additional unlabeled data for counting by introducing a learning to rank framework. They assume that any sub-image of a crowded scene image is guaranteed to contain the same number or fewer persons than the super-image. They employ pairwise ranking hinge loss to enforce this ranking constraint for unlabeled data in addition to the L_2 error to train the network. In our experiments we observed that this constraint is almost always satisfied, and it provides relatively less supervision over unlabeled data.

Babu *et al.*[35] focus on a different approach, where they train 99.9% of their parameters from unlabeled data using a novel unsupervised learning framework based on winner-takes-all (WTA) strategy. However, they still train the remaining set of parameters using labeled data.

Wang *et al.*[55] take a totally different approach to minimize annotation efforts by creating a new synthetic crowd counting dataset (GCC). Additionally, they propose a Cycle-GAN based domain adaptive approach for generalizing the network trained on synthetic dataset to real-world dataset. However, there is a large gap in terms of the style and also the crowd count between the synthetic and real-world scenarios. Domain adaptive approaches have limited abilities in handling such scenarios. In order to obtain successful adaptation, the authors in [55] manually select the samples from the synthetic dataset that are closer to the real-world scenario in terms of crowd count for training the network. This selection is possible when one has information about the count from the real-world datasets, which violates the assumption of lack of unlabeled data in the target domain for unsupervised domain adaptation.

Considering the drawbacks of existing approaches, we propose a new GP-based iterative training framework to exploit unlabeled data.

3   Preliminaries

In this section, we briefly review the concepts (crowd counting, semi-supervised learning and Gaussian Process) that are used in this work.

Crowd Counting. Following recent works [59,61], we employ the approach of density estimation technique. That is, an input crowd image is forwarded through the network, and the network outputs a density map. This density map indicates the per-pixel count of people in the image. The count in the image is obtained by integrating over the density map. For training the network using labeled data, the ground-truth density maps are obtained by imposing 2D Gaussians at head location x_g using $D(x) = \sum_{x_g \in S} \mathcal{N}(x - x_g, \sigma)$. Here, σ is the Gaussian kernel's scale and S is the list of all locations of people.

Problem Formulation. We are given a set of labeled dataset of input-GT pairs ($\{x, y\} \in \mathcal{D}_\mathcal{L}$) and a set of unlabeled input data samples $x \in \mathcal{D}_\mathcal{U}$. The objective is to fit a mapping-function $f(x|\phi)$ (with parameters defined by ϕ) that accurately estimates target label y for unobserved samples. Note that this definition applies to both semi-supervised setting and synthetic-to-real transfer setting. In the case of synthetic-to-real transfer, the synthetic dataset is labeled and hence, can be used as the labeled dataset ($\mathcal{D}_\mathcal{L}$). Similarly, the real-world dataset is unlabeled and can be used as the unlabeled dataset ($\mathcal{D}_\mathcal{U}$).

In order to learn the parameters, both labeled and unlabeled datasets are exploited. Typically, loss functions such as L_1, L_2 or cross entropy error are used for labeled data. For exploiting unlabeled data $\mathcal{D}_\mathcal{U}$, existing approaches augment $f(x|\phi)$ with information like shape of the data manifold [25] via different techniques such as enforcing consistent regularization [13], virtual adversarial training [24] or pseudo-labeling [14]. In this work, we employ pseudo-labeling based approach where we estimate pseudo-GT for unlabeled data, and then use them for supervising the network using traditional supervised loss functions.

Gaussian Process. A Gaussian process (GP) $f(v)$ is an infinite collection of random variables, any finite subset of which have a joint Gaussian distribution. A GP is fully specified by its mean function ($m(v)$) and covariance function $K(v, v')$. These are defined below:

$$m(v) = \mathbb{E}[f(v)], \tag{1}$$

$$K(v, v') = \mathbb{E}\left[(f(v) - m(v))(f(v') - m(v'))\right], \tag{2}$$

where $v, v' \in \mathcal{V}$ denote the possible inputs that index the GP. The covariance matrix is computed from the covariance function K which expresses the notion of smoothness of the underlying function. GP can then be formulated as follows:

$$f(v) \sim \mathcal{GP}(m(v), K(v, v') + \sigma_\epsilon^2 I), \tag{3}$$

where I is identity matrix and σ_ϵ^2 is the variance of the additive noise. Any collection of function values is then jointly Gaussian as follows

$$f(V) = [f(v_1), \dots, f(v_n)]^T \sim \mathcal{N}\left(\mu, K(V, V') + \sigma_\epsilon^2 I\right), \tag{4}$$

with mean vector and covariance matrix defined by the GP as mentioned earlier. To make predictions at unlabeled points, one can compute a Gaussian posterior distribution in closed form by conditioning on the observed data. For more details, we refer the reader to [29].

4   GP-Based Iterative Learning

Figure 2 gives an overview of the proposed method. The network is constructed using an encoder $f_e(x, \phi_e)$ and a decoder $f_d(z, \phi_d)$, that are parameterized by ϕ_e and ϕ_d, respectively. The proposed framework is agnostic to the encoder network, and we show in the experiments section that it generalizes well to architectures such as VGG16 [40], ResNet-50 and ResNet-101 [30]. The decoder consists of a set of 2 conv-relu layers (see supplementary material for more details). Typically, an input crowd image x is forwarded through the encoder network to obtain the corresponding latent space vector z. This vector is then forwarded through the decoder network to obtain the crowd density output y, $i.e$, $y = f_d(f_e(x, \phi_e), \phi_d)$.

We are given a training dataset, $\mathcal{D} = \mathcal{D}_\mathcal{L} \cup \mathcal{D}_\mathcal{U}$, where $\mathcal{D}_\mathcal{L} = \{x_l^i, y_l^i\}_{i=1}^{N_l}$ is a labeled dataset containing N_l training samples and $\mathcal{D}_\mathcal{U} = \{x_u^i\}_{i=1}^{N_u}$ is an unlabeled dataset containing N_u training samples. The proposed framework effectively leverages both the datasets by iterating the training process over labeled $\mathcal{D}_\mathcal{L}$ and unlabeled datasets $\mathcal{D}_\mathcal{U}$. More specifically, the training process consists of two stages: (i) Labeled training stage: In this stage, we employ supervised loss function $\mathcal{L}_s$ to learn the network parameters using labeled dataset, and (ii) Unlabeled training stage: We generate pseudo GTs for the unlabeled data points using the GP formulation, which is then used for supervising the network on the unlabeled dataset. In what follows, we describe these stages in detail.

4.1   Labeled Stage

Since the labeled dataset $\mathcal{D}_\mathcal{L}$ comes with annotations, we employ L_2 error between the predictions and the GTs as supervision loss for training the network. This loss objective is defined as follows:

$$\mathcal{L}_s = \mathcal{L}_2 = \|y_l^{pred} - y_l\|_2, \tag{5}$$

where $y_l^{pred} = g(z_l, \phi_d)$ is the predicted output, y_l is the ground-truth, $z = h(x, \phi_e)$ is the intermediate latent space vector. Note that, the subscript l in the above quantities indicate that these are defined for labeled data.

Along with performing supervision on the labeled data, we additionally save feature vectors z_l^i's from the intermediate latent space in a matrix F_{z_l}. Specifically, $F_{z_l} = \{z_l^i\}_{i=1}^{N_l}$. This matrix is used for computing the pseudo-GTs for unlabeled data at a later stage. The dimension of F_{z_l} matrix is $N_l \times M$. Here, M is the dimension of the latent space vector z_l. In our case, the latent space vector dimension is $64 \times 32 \times 32$ (see supplementary material for more details), which is reshaped to $1 \times 65,536$. Hence, $M = 65,536$.

4.2   Unlabeled Stage

Since the unlabeled data $\mathcal{D}_\mathcal{U}$ does not come with any GT annotations, we estimate pseudo-GTs which are then used as supervision for training the network on unlabeled data. For this purpose, we model the relationship between the latent

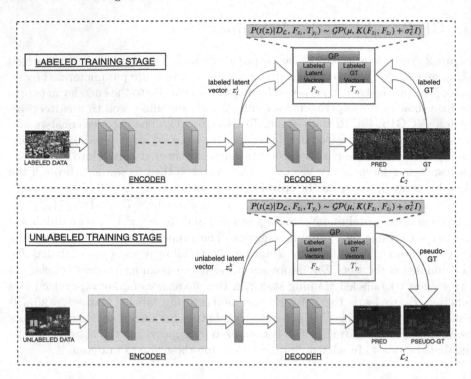

Fig. 2. Illustration of the proposed framework. Training is performed iteratively over labeled and unlabeled data. For labeled data, we minimize the L_2 error between the predictions and GT. For unlabeled data, we minimize the L_2 error between the predictions and pseudo-GT.

space vectors of the labeled images F_{z_l} along with the corresponding GT T_{y_l} and unlabeled latent space vectors z_u^{pred} jointly using GP.

Estimation of Pseudo-GT: As discussed earlier, the training process iterates over labeled $\mathcal{D}_\mathcal{L}$ and unlabeled data $\mathcal{D}_\mathcal{U}$. After the labeled stage, the labeled latent space vectors F_{z_l} and their corresponding GT density maps T_{y_l} are used to model the function t which maps the relationship between the latent vectors and the output density maps as, $y = t(z)$. Using GP, we model this function $t(.)$ as an infinite collection of functions of which any finite subset is jointly Gaussian. More specifically, we jointly model the distribution of the function values $t(.)$ of the latent space vectors of the labeled and the unlabeled samples using GP as follows:

$$P(t(z)|\mathcal{D}_\mathcal{L}, F_{z_l}, T_{y_l}) \sim \mathcal{GP}(\mu, K(F_{z_l}, F_{z_l}) + \sigma_\epsilon^2 I), \qquad (6)$$

where μ is the function value computed using GP, σ_ϵ^2 is set equal to 1, and K is the kernel function. Based on this, the conditional joint distribution for the latent space vector z_u^k of the kth unlabeled sample x_u^k can be expressed as the following Gaussian distribution:

$$P(t(z_u^k)|\mathcal{D}_\mathcal{L}, F_{z_l}, T_{z_l}) = \mathcal{N}(\mu_u^k, \Sigma_u^k), \tag{7}$$

where

$$\mu_u^k = K(z_u^k, F_{z_l})[K(F_{z_l}, F_{z_l}) + \sigma_\epsilon^2 I]^{-1} T_{y_l}, \tag{8}$$

$$\Sigma_u^k = K(z_u^k, z_u^k) - K(z_u^k, F_{z_l})[K(F_{z_l}, F_{z_l}) + \sigma_\epsilon^2 I]^{-1} K(F_{z_l}, z_u^k) + \sigma_\epsilon^2 \tag{9}$$

where σ_ϵ^2 is set equal to 1 and K is a kernel function with the following definition:

$$K(Z, Z)_{k,i} = \kappa(z_u^k, z_l^i) = \frac{\langle z_u^k, z_l^i \rangle}{|z_u^k| \cdot |z_l^i|}. \tag{10}$$

Considering the large dimensionality of the latent space vector, $K(F_{z_l}, F_{z_l})$ can grow quickly in size especially if the number of labeled data samples N_l is high. In such cases, the computational and memory requirements become prohibitively high. Additionally, all the latent vectors may not be necessarily effective since these vectors correspond to different regions of images in terms of content and size/density of the crowd. In order to overcome these issues, we use only those labeled vectors that are similar to the unlabeled latent vector. Specifically, we consider only N_n nearest labeled vectors corresponding to an unlabeled vector. That is, we replace F_{z_l} by $F_{z_l,n}$ in Eq. (7)–(9). Here $F_{z_l,n} = \{z_l^j : z_l^j \in nearest(z_u^k, F_{z_l}, N_n)\}$, and $T_{y_l,n} = \{y_l^j : z_l^j \in nearest(z_u^k, F_{z_l}, N_n)\}$ with $nearest(p, Q, N_n)$ being a function that finds top N_n nearest neighbors of p in Q.

The pseudo-GT for unlabeled data sample is given by the mean predicted in Eq. (8), $i.e$, $y_{u,pseudo}^k = \mu_u^k$. The L_2 distance between the predictions $y_{u,pred}^k = g(z_u^k, \phi_e)$ and the pseudo-GT $y_{u,pseudo}^k$ is used as supervision for updating the parameters of the encoder $f_e(\cdot, \phi_e)$ and the decoder $f_d(., \phi_d)$.

Furthermore, the pseudo-GT estimated using Eq. (8) may not be necessarily perfect. Errors in pseudo-GT will limit the performance of the network. To overcome this, we explicitly exploit the variance modeled by the GP. Specifically, we minimize the predictive variance by considering Eq. (9) in the loss function. As discussed earlier, using all the latent space vectors of labeled data may not be necessarily effective. Hence, we minimize the variance $\Sigma_{u,n}^k$ computed between z_u^k and the N_n nearest neighbors in the latent space vectors using GP. Thus, the loss function during the unlabeled stage is defined as:

$$\mathcal{L}_{un} = \frac{1}{|\Sigma_{u,n}^k|}\|y_{u,pred}^k - y_{u,pseudo}^k\|_2 + \log \Sigma_{u,n}^k, \tag{11}$$

where $y_{u,pred}^k$ is the crowd density map prediction obtained by forwarding an unlabeled input image x_u^k through the network, $y_{u,pseudo}^k = \mu_u^k$ is the pseudo-GT (see Eq. (8)), and $\Sigma_{u,n}^k$ is the predictive variance obtained by replacing F_{z_l} in Eq. (9) with $F_{z_l,n}$. Note that the prediction error (the first term) is scaled by loss by inverse of the variance. This ensures that the loss from uncertain pseudo-gts are down-weighted and hence, only accurate pseudo-gts are used for training.

4.3   Final Objective Function

We combine the supervised loss Eq. (5) and unsupervised loss Eq. (11) to obtain the final objective function as follows:

$$\mathcal{L}_f = \mathcal{L}_s + \lambda_{un}\mathcal{L}_{un}, \tag{12}$$

where λ_{un} is a hyper-parameter that weighs the unsupervised loss.

5   Experiments and Results

In this section, we discuss the details of the various experiments conducted to demonstrate the effectiveness of the proposed method. Since the proposed method is able to leverage unlabeled data to improve the overall performance, we performed evaluation in two settings: (i) *Semi-supervised settings*: In this setting, we varied the percentage of labeled samples from 5% to 75%. We first show that with the base network, there is performance drop due to the reduced data. Later, we show that the proposed method is able to recover a major percentage of the performance drop. (ii) *Synthetic-to-real transfer settings*: In this setting, the goal is to train on synthetic dataset (labeled), while adapting to real-world dataset. Unlabeled images from the real-world are available during training. In both settings, the proposed method is able to achieve better results as compared to recent methods. Details of the datasets are provided in the supplementary material.

5.1   Semi-supervised Settings

In this section, we conduct experiments in the semi-supervised settings by reducing the amount of labeled data available during training. The rest of the samples in the dataset are considered as unlabeled samples wherever applicable. In the following sub-sections, we present comparison of the proposed method in the 5% setting with other recent methods. For comparison, we used 4 datasets: ShanghaiTech (SH-A/B)[61], UCF-QNRF [10], WorldExpo [59] and UCSD [4]. This is followed by a detailed ablation study involving different architectures and various percentages of labeled data used during training. For ablation, we chose ShanghaiTech-A and UCF-QNRF datasets since they contain a wide diversity of scenes and large variation in count and scales.

Implementation Details. We train the network using Adam optimizer with a learning rate of $10e - 5$ and a momentum of 0.9 on an NVIDIA Titan Xp GPU. We use batch size of 24. During training, random crops of size 256×256 are used. During inference, the entire image is forwarded through the network. For evaluation, we use mean absolute error (MAE) and mean squared error (MSE) metrics, which are defined as: $MAE = \frac{1}{N}\sum_{i=1}^{N}|y_i - y_i'|$ and $MSE = \sqrt{\frac{1}{N}\sum_{i=1}^{N}|y_i - y_i'|^2}$, respectively. Here, N is the total number of test images, y_i

is the ground-truth/target count of people in the image and y_i' is the predicted count of people in to the ith image. We set aside 10% of the training set for the purpose of validation. The hyper-parameter λ_{un} was chosen based on the validation performance. More details are provided in the supplementary.

Table 1. Comparison of results in SSL settings. Reducing labeled data to 5% results in performance drop by a big margin as compared to 100% data. ResNet-50 was used as the encoder network for all the methods. RL: Ranking-Loss. GP: Gaussian-Process. AG: Average Gain %. (see footnote 1)

Method	$\mathcal{D}_\mathcal{L}$	$\mathcal{D}_\mathcal{U}$	SH-A			SH-B			UCF-QNRF			WExpo		UCSD		
			MAE	MSE	AG	MAE	MSE	AG	MAE	MSE	AG	MAE	AG	MAE	MSE	AG
ResNet-50 (Oracle)	100%	–	76	126	–	8.4	14.5	–	114	195	–	10.1	–	1.7	2.1	–
ResNet-50 ($\mathcal{D}_\mathcal{L}$-only)	5%	–	118	211	–	21.2	34.2	–	186	295	–	14.2	–	2.2	2.8	–
ResNet-50+RL	5%	95%	115	208	2.0	20.1	32.9	4.0	182	291	1.7	14.0	0.01	2.2	2.8	0
ResNet-50+GP(Ours)	5%	95%	102	172	16	15.7	27.9	22	160	275	10	12.8	10	2.0	2.4	12

Comparison with Recent Approaches. Here, we compare the effectiveness of the proposed method with a recent method by Liu et al.[22] on 4 different datasets. In order to get a better understanding of the overall improvements, we also provide the results of the base network with (i) 100% labeled data supervision that is the oracle performance, and (ii) 5% labeled data supervision.

For all the methods (except oracle), we limited the labeled data used during training to 5% of the training dataset. Rest of the samples were used as unlabeled samples. We used ResNet-50 as the encoder network. The results of the experiments are shown in Table 1. For all the experiments that we conducted, we report the average of the results for 5 trials. The standard deviations are reported in the supplementary. We make the following observations for all the datasets: (i) Compared to using the entire dataset, reducing the labeled data during training (to 5%) leads to significant increase in error. (ii) The proposed GP-based framework is able to reduce the performance drop by a large margin. Further, the proposed method achieves an average gain $(AG)^1$ of anywhere between 10%–22% over the $\mathcal{D}_\mathcal{L}$-only baseline across all datasets. (iii) The proposed method is able to leverage the unlabeled data more effectively as compared to Liu et al.[22]. This is because the authors in [22] using a ranking loss on the unlabeled data which is based on the assumption that sub-image of a crowded scene is guaranteed to contain the same or fewer number of people compared to the entire image. We observed that this constraint is satisfied naturally for most of the unlabeled images, and hence it provides less supervision (see supplementary material for a detailed analysis).

Ablation of Labeled Data Percentage. We conducted an ablation study where we varied the percentage of labeled data used during the training process. More specifically, we used 4 different settings: 5%, 25%, 50% and 75%.

---

[1] $AG = \frac{G_{mae}+G_{mse}}{2}$, $G_{mae} = \frac{mae(\mathcal{D}_\mathcal{U}+\mathcal{D}_\mathcal{L})-mae(\mathcal{D}_\mathcal{L})}{mae(\mathcal{D}_\mathcal{L})}$, $G_{mse} = \frac{mse(\mathcal{D}_\mathcal{U}+\mathcal{D}_\mathcal{L})-mse(\mathcal{D}_\mathcal{L})}{mse(\mathcal{D}_\mathcal{L})}$.

Table 2. Results of ablation study with different %-ages of labeled data. The proposed method achieves significant gains across different percentages of labeled data. We used ResNet-50 as the encoder network for all the experiments. AG: Average Gain %. (see footnote 1)

$\mathcal{D_L}$ %	SH-A					UCF-QNRF				
	No-GP ($\mathcal{D_L}$-only)		GP ($\mathcal{D_L} + \mathcal{D_U}$)		AG %	No-GP ($\mathcal{D_L}$-only)		GP ($\mathcal{D_L} + \mathcal{D_U}$)		AG %
	MAE	MSE	MAE	MSE		MAE	MSE	MAE	MSE	
5	118	211	102	172	16	186	295	160	275	10
25	110	160	91	149	12	178	252	147	226	14
50	102	149	89	148	6.1	158	250	136	218	13
75	93	146	88	139	4.7	139	240	129	210	9.8
100	76	126	–	–	–	114	195	–	–	–

The remaining data were used as unlabeled samples. We used ResNet-50 as the network encoder for all the settings. This ablation study was conducted on 2 datasets: ShanghaiTech-A (SH-A) and UCF-QNRF. The results of this ablation study are shown in Table 2. It can be observed for both datasets that as the percentage of labeled data is reduced, the performance of the baseline network drops significantly. However, the proposed GP-based framework is able to leverage unlabeled data in all the cases to reduce this performance drop by a considerable margin. Figure 3 show sample qualitative results on ShanghaiTech-A dataset for the semi-supervised protocol with 5% labeled data setting. It can be observed that the proposed method is able to predict the density maps more accurately as compared to the baseline method that does not consider unlabeled data.

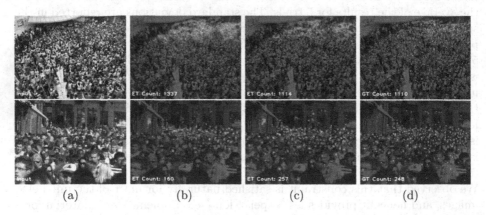

(a)   (b)   (c)   (d)

Fig. 3. Results of SSL experiments on the ShanghaiTech-A [61] dataset using the 5% labeled data setting. *(a):* Input. *(b)* No-GP *(c)* Proposed Method *(d)* Ground-truth.

Architecture Ablation. We conducted an ablation study where we evaluated the proposed method using different architectures. More specifically, we used dif-

Table 3. Results of ablation study with different networks. The proposed method is able to exploit unlabeled data irrespective of different architectures. We used 5% of the training data as labeled set, and the rest as unlabeled samples. AG: Average Gain %. (see footnote 1)

Net	$\mathcal{D_L}$%	SH-A					AG %	UCF-QNRF					AG %
		No-GP($\mathcal{D_L}$-only)		GP($\mathcal{D_L} + \mathcal{D_U}$)				No-GP ($\mathcal{D_L}$-only)		GP ($\mathcal{D_L} + \mathcal{D_U}$)		
		MAE	MSE	MAE	MSE			MAE	MSE	MAE	MSE		
ResNet-50	100	76	126	–	–		–	114	195	–	–		–
	5	118	211	102	172		16	186	295	160	275		10
ResNet-101	100	76	117	–	–		–	116	197	–	–		–
	5	131	200	110	162		18	196	324	174	288		11
VGG16	100	74	118	–	–		–	120	197	–	–		–
	5	121	205	112	163		14	188	316	175	291		7.4

ferent networks like ResNet-50, ResNet-101 and VGG16 as encoder network. The ablation was performed on 2 datasets: ShanghaiTech-A (SH-A) and UCF-QNRF. For all the experiments, we used 5% of the training dataset as labeled dataset, and the rest were used as unlabeled samples. The results of this experiment are shown in Table 3. Based on these results, we make the following observations: (i) Since networks like VGG16 and ResNet-101 have higher number of parameters, they tend to overfit more in the reduced-data setting as compared to ResNet-50. (ii) The proposed GP-based method obtains consistent gains by leveraging unlabeled dataset across different architectures.

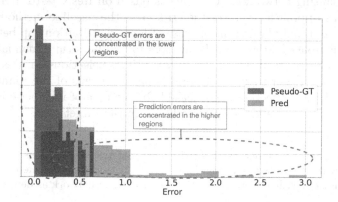

Fig. 4. Histogram for pseudo-GT errors (err^u_{pseudo}) and prediction errors (err^u_{pred}) on unlabeled data during training. Note that pseudo-GT errors are concentrated on the lower end, implying that they are more closer to the ground truth as compared to the predictions. Hence, pseudo-GTs provide meaningful supervision.

Pseudo-GT Analysis. In order to gain a deeper understanding about the effectiveness of the proposed approach, we plot the histogram of normalized errors with respect to the predictions y^u_{pred} of the network and the pseudo-GT y^u_{pseudo} for the unlabeled data during the training process. Specifically, we plot histograms of err^u_{pred} and err^u_{pseudo}, where $err^u_{pred} = \frac{|y^u_{pred} - y^u_{gt}|}{y^u_{gt}}$ and $err^u_{pseudo} = \frac{|y^u_{pseudo} - y^u_{gt}|}{y^u_{gt}}$. Here, y^u_{gt} is the actual GT corresponding to the unlabeled data sample. The plot is shown in Fig. 4. It can be observed that the pseudo-GT errors are concentrated in the lower end of the error region as compared to the prediction errors. This implies that the pseudo-GTs are more closer to the GTs than the predictions. Hence, the pseudo-GTs obtained using the proposed method are able to provide good quality supervision on the unlabeled data.

5.2 Synthetic-to-Real Transfer Setting

Recently, Wang et al.[55] proposed a synthetic crowd counting dataset (GCC) that consists of 15,212 images with a total of 7,625,843 annotations. The primary purpose of this dataset is to reduce the annotation efforts by training the networks on the synthetic dataset, thereby eliminating the need for labeling. However, due to a gap between the synthetic and real-world data distributions, the networks trained on synthetic dataset perform poorly on real-world images. In order to overcome this issue, the authors in [55] proposed a Cycle-GAN based domain adaptive approach that additionally enforces SSIM consistency. More specifically, they first learn to translate from synthetic crowd images to real-world images using SSIM-based Cycle-GAN. This transfers the style in the synthetic image to more real-world style. The translated synthetic images are then used to train a counting network (SFCN) that is based on ResNet-101 architecture.

While this approach improves the error over the baseline methods, its performance is essentially limited in the case of large distribution gap between real and synthetic images. Moreover, the authors in [55] perform a manual selection of synthetic samples for training the network. This selections ensures that only samples that are closer to the real-world images in terms of the count are used for training. Such a selection is not feasible in the case of unsupervised domain adaptation where we have no access to labels in the target dataset.

Table 4. Comparison of results in synthetic-to-real transfer settings. We train the network on synthetic crowd counting dataset (GCC), and leverage the training set of real-world datasets without any labels. We used the same network as described in [55].

Method	SH-A		SH-B		UCF-QNRF		UCF-CC-50		WExpo
	MAE	MSE	MAE	MSE	MAE	MSE	MAE	MSE	MAE
No Adapt	160	217	22.8	30.6	276	459	487	689	42.8
Cycle GAN [63]	143	204	24.4	39.7	257	401	405	548	32.4
SE Cycle GAN [55]	123	193	19.9	28.3	230	384	373	529	26.3
Proposed Method	**121**	**181**	**12.8**	**19.2**	**210**	**351**	**355**	**505**	**20.4**

The proposed GP-based framework overcomes these drawbacks easily and can be extended to the synthetic-to-real transfer setting as well. We consider the synthetic data as labeled training set and real-world training set as unlabeled dataset, and train the network to leverage the unlabeled dataset. The results of this experiment are reported in Table 4. We used the same network (SFCN) and training process as described in [55]. As it can be observed, the proposed method achieves considerable improvements compared to the recent approach. Since we estimate the pseudo-GT for unlabeled real-world images and use it as supervision directly, the distribution gap that the network needs to handle is much lesser. This results in better performance compared to the domain adaptive approach [55]. Unlike [55], we train the network on the unlabeled data and hence, we do not need to perform any synthetic sample selection. Figure 5 show sample qualitative results on the ShanghaiTech-A dataset for the synthetic-to-real transfer protocol. The proposed method is able to predict the density maps more accurately as compared to the baseline.

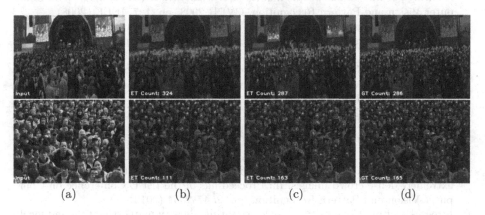

(a)   (b)   (c)   (d)

Fig. 5. Results of Synthetic-to-Real transfer experiments on ShanghaiTech-A dataset. *(a):* Input. *(b)* No Adapt *(c)* Proposed Method *(d)* Ground-truth.

6   Conclusions

In this work, we focused on learning to count in the crowd from limited labeled data. Specifically, we proposed a GP-based iterative learning framework that involves estimation of pseudo-GT for unlabeled data using Gaussian Processes, which is then used as supervision for training the network. Through various experiments, we show that the proposed method can be effectively used in a variety of scenarios that involve unlabeled data like learning with less data or synthetic to real-world transfer. In addition, we conducted detailed ablation studies to demonstrate that the proposed method generalizes well to different network architectures and is able to achieve consistent gains for different amounts of labeled data.

Acknowledgement. This work was supported by the NSF grant 1910141.

References

1. Arteta, C., Lempitsky, V., Zisserman, A.: Counting in the wild. In: Leibe, B., Matas, J., Sebe, N., Welling, M. (eds.) ECCV 2016. LNCS, vol. 9911, pp. 483–498. Springer, Cham (2016). https://doi.org/10.1007/978-3-319-46478-7_30
2. Babu Sam, D., Sajjan, N.N., Venkatesh Babu, R., Srinivasan, M.: Divide and grow: capturing huge diversity in crowd images with incrementally growing CNN. In: Proceedings of the IEEE Conference on Computer Vision and Pattern Recognition, pp. 3618–3626 (2018)
3. Cao, X., Wang, Z., Zhao, Y., Su, F.: Scale aggregation network for accurate and efficient crowd counting. In: Ferrari, V., Hebert, M., Sminchisescu, C., Weiss, Y. (eds.) ECCV 2018. LNCS, vol. 11209, pp. 757–773. Springer, Cham (2018). https://doi.org/10.1007/978-3-030-01228-1_45
4. Chan, A.B., Liang, Z.S.J., Vasconcelos, N.: Privacy preserving crowd monitoring: counting people without people models or tracking. In: IEEE Conference on Computer Vision and Pattern Recognition, CVPR 2008, pp. 1–7. IEEE (2008)
5. Chan, A.B., Vasconcelos, N.: Bayesian Poisson regression for crowd counting. In: 2009 IEEE 12th International Conference on Computer Vision, pp. 545–551. IEEE (2009)
6. Change Loy, C., Gong, S., Xiang, T.: From semi-supervised to transfer counting of crowds. In: Proceedings of the IEEE International Conference on Computer Vision, pp. 2256–2263 (2013)
7. Chen, K., Loy, C.C., Gong, S., Xiang, T.: Feature mining for localised crowd counting. In: European Conference on Computer Vision (2012)
8. Gao, G., Gao, J., Liu, Q., Wang, Q., Wang, Y.: CNN-based density estimation and crowd counting: a survey. arXiv preprint arXiv:2003.12783 (2020)
9. Idrees, H., Saleemi, I., Seibert, C., Shah, M.: Multi-source multi-scale counting in extremely dense crowd images. In: Proceedings of the IEEE Conference on Computer Vision and Pattern Recognition, pp. 2547–2554 (2013)
10. Idrees, H., et al.: Composition loss for counting, density map estimation and localization in dense crowds. In: Ferrari, V., Hebert, M., Sminchisescu, C., Weiss, Y. (eds.) ECCV 2018. LNCS, vol. 11206, pp. 544–559. Springer, Cham (2018). https://doi.org/10.1007/978-3-030-01216-8_33
11. Jiang, X., et al.: Crowd counting and density estimation by trellis encoder-decoder network. arXiv preprint arXiv:1903.00853 (2019)
12. Kang, D., Ma, Z., Chan, A.B.: Beyond counting: comparisons of density maps for crowd analysis tasks-counting, detection, and tracking. arXiv preprint arXiv:1705.10118 (2017)
13. Laine, S., Aila, T.: Temporal ensembling for semi-supervised learning. arXiv preprint arXiv:1610.02242 (2016)
14. Lee, D.H.: Pseudo-label: the simple and efficient semi-supervised learning method for deep neural networks (2013)
15. Lempitsky, V., Zisserman, A.: Learning to count objects in images. In: Advances in Neural Information Processing Systems, pp. 1324–1332 (2010)
16. Li, M., Zhang, Z., Huang, K., Tan, T.: Estimating the number of people in crowded scenes by mid based foreground segmentation and head-shoulder detection. In: 19th International Conference on Pattern Recognition, ICPR 2008, pp. 1–4. IEEE (2008)

17. Li, T., Chang, H., Wang, M., Ni, B., Hong, R., Yan, S.: Crowded scene analysis: a survey. IEEE Trans. Circ. Syst. Video Technol. **25**(3), 367–386 (2015)
18. Li, W., Mahadevan, V., Vasconcelos, N.: Anomaly detection and localization in crowded scenes. IEEE Trans. Pattern Anal. Mach. Intell. **36**(1), 18–32 (2014)
19. Li, Y., Zhang, X., Chen, D.: CSRNet: dilated convolutional neural networks for understanding the highly congested scenes. In: Proceedings of the IEEE Conference on Computer Vision and Pattern Recognition, pp. 1091–1100 (2018)
20. Liu, N., Long, Y., Zou, C., Niu, Q., Pan, L., Wu, H.: ADCrowdNet: an attention-injective deformable convolutional network for crowd understanding. arXiv preprint arXiv:1811.11968 (2018)
21. Liu, W., Salzmann, M., Fua, P.: Context-aware crowd counting. In: Proceedings of the IEEE Conference on Computer Vision and Pattern Recognition, pp. 5099–5108 (2019)
22. Liu, X., van de Weijer, J., Bagdanov, A.D.: Leveraging unlabeled data for crowd counting by learning to rank. In: The IEEE Conference on Computer Vision and Pattern Recognition (CVPR), June 2018
23. Lu, H., Cao, Z., Xiao, Y., Zhuang, B., Shen, C.: TasselNet: counting maize tassels in the wild via local counts regression network. Plant Methods **13**(1), 79 (2017). https://doi.org/10.1186/s13007-017-0224-0
24. Miyato, T., Maeda, S.I., Koyama, M., Ishii, S.: Virtual adversarial training: a regularization method for supervised and semi-supervised learning. IEEE Trans. Pattern Anal. Mach. Intell. **41**(8), 1979–1993 (2018)
25. Oliver, A., Odena, A., Raffel, C.A., Cubuk, E.D., Goodfellow, I.: Realistic evaluation of deep semi-supervised learning algorithms. In: Advances in Neural Information Processing Systems, pp. 3235–3246 (2018)
26. Oñoro-Rubio, D., López-Sastre, R.J.: Towards perspective-free object counting with deep learning. In: Leibe, B., Matas, J., Sebe, N., Welling, M. (eds.) ECCV 2016. LNCS, vol. 9911, pp. 615–629. Springer, Cham (2016). https://doi.org/10.1007/978-3-319-46478-7_38
27. Pham, V.Q., Kozakaya, T., Yamaguchi, O., Okada, R.: Count forest: co-voting uncertain number of targets using random forest for crowd density estimation. In: Proceedings of the IEEE International Conference on Computer Vision, pp. 3253–3261 (2015)
28. Ranjan, V., Le, H., Hoai, M.: Iterative crowd counting. In: Ferrari, V., Hebert, M., Sminchisescu, C., Weiss, Y. (eds.) ECCV 2018. LNCS, vol. 11211, pp. 278–293. Springer, Cham (2018). https://doi.org/10.1007/978-3-030-01234-2_17
29. Rasmussen, C.E.: Gaussian processes in machine learning. In: Bousquet, O., von Luxburg, U., Rätsch, G. (eds.) ML -2003. LNCS (LNAI), vol. 3176, pp. 63–71. Springer, Heidelberg (2004). https://doi.org/10.1007/978-3-540-28650-9_4
30. Ren, S., He, K., Girshick, R., Sun, J.: Faster R-CNN: towards real-time object detection with region proposal networks. In: Advances in Neural Information Processing Systems, pp. 91–99 (2015)
31. Ryan, D., Denman, S., Fookes, C., Sridharan, S.: Crowd counting using multiple local features. In: Digital Image Computing: Techniques and Applications, DICTA 2009, pp. 81–88. IEEE (2009)
32. Sam, D.B., Babu, R.V.: Top-down feedback for crowd counting convolutional neural network. In: Thirty-Second AAAI Conference on Artificial Intelligence (2018)
33. Sam, D.B., et al.: Locate, size and count: accurately resolving people in dense crowds via detection. arXiv preprint arXiv:1906.07538 (2019)

34. Sam, D.B., Peri, S.V., Mukuntha, N., Babu, R.V.: Going beyond the regression paradigm with accurate dot prediction for dense crowds. In: 2020 IEEE Winter Conference on Applications of Computer Vision (WACV), pp. 2853–2861. IEEE (2020)
35. Sam, D.B., Sajjan, N.N., Maurya, H., Babu, R.V.: Almost unsupervised learning for dense crowd counting. In: Thirty-Third AAAI Conference on Artificial Intelligence (2019)
36. Sam, D.B., Surya, S., Babu, R.V.: Switching convolutional neural network for crowd counting. In: Proceedings of the IEEE Conference on Computer Vision and Pattern Recognition (2017)
37. Shen, Z., Xu, Y., Ni, B., Wang, M., Hu, J., Yang, X.: Crowd counting via adversarial cross-scale consistency pursuit. In: The IEEE Conference on Computer Vision and Pattern Recognition (CVPR), June 2018
38. Shi, M., Yang, Z., Xu, C., Chen, Q.: Revisiting perspective information for efficient crowd counting. In: Proceedings of the IEEE Conference on Computer Vision and Pattern Recognition, pp. 7279–7288 (2019)
39. Shi, Z., et al.: Crowd counting with deep negative correlation learning. In: The IEEE Conference on Computer Vision and Pattern Recognition (CVPR), June 2018
40. Simonyan, K., Zisserman, A.: Very deep convolutional networks for large-scale image recognition. In: International Conference on Learning Representations (2015)
41. Sindagi, V.A., Patel, V.M.: CNN-based cascaded multi-task learning of high-level prior and density estimation for crowd counting. In: 2017 IEEE International Conference on Advanced Video and Signal Based Surveillance (AVSS). IEEE (2017)
42. Sindagi, V.A., Patel, V.M.: Generating high-quality crowd density maps using contextual pyramid CNNs. In: The IEEE International Conference on Computer Vision (ICCV), October 2017
43. Sindagi, V.A., Patel, V.M.: A survey of recent advances in CNN-based single image crowd counting and density estimation. Pattern Recogn. Lett. **107**, 3–16 (2017)
44. Sindagi, V.A., Patel, V.M.: DAFE-FD: Density aware feature enrichment for face detection. In: 2019 IEEE Winter Conference on Applications of Computer Vision (WACV), pp. 2185–2195. IEEE (2019)
45. Sindagi, V.A., Patel, V.M.: HA-CCN: Hierarchical attention-based crowd counting network. arXiv preprint arXiv:1907.10255 (2019)
46. Sindagi, V.A., Patel, V.M.: Inverse attention guided deep crowd counting network. arXiv preprint (2019)
47. Sindagi, V.A., Patel, V.M.: Multi-level bottom-top and top-bottom feature fusion for crowd counting. In: Proceedings of the IEEE International Conference on Computer Vision, pp. 1002–1012 (2019)
48. Sindagi, V.A., Yasarla, R., Patel, V.M.: Pushing the frontiers of unconstrained crowd counting: new dataset and benchmark method. In: Proceedings of the IEEE International Conference on Computer Vision, pp. 1221–1231 (2019)
49. Sindagi, V.A., Yasarla, R., Patel, V.M.: JHU-CROWD++: large-scale crowd counting dataset and a benchmark method. arXiv preprint arXiv:2004.03597 (2020)
50. Toropov, E., Gui, L., Zhang, S., Kottur, S., Moura, J.M.: Traffic flow from a low frame rate city camera. In: 2015 IEEE International Conference on Image Processing (ICIP), pp. 3802–3806. IEEE (2015)
51. Walach, E., Wolf, L.: Learning to count with CNN boosting. In: Leibe, B., Matas, J., Sebe, N., Welling, M. (eds.) ECCV 2016. LNCS, vol. 9906, pp. 660–676. Springer, Cham (2016). https://doi.org/10.1007/978-3-319-46475-6_41

52. Wan, J., Chan, A.: Adaptive density map generation for crowd counting. In: Proceedings of the IEEE International Conference on Computer Vision, pp. 1130–1139 (2019)
53. Wan, J., Luo, W., Wu, B., Chan, A.B., Liu, W.: Residual regression with semantic prior for crowd counting. In: Proceedings of the IEEE Conference on Computer Vision and Pattern Recognition, pp. 4036–4045 (2019)
54. Wang, C., Zhang, H., Yang, L., Liu, S., Cao, X.: Deep people counting in extremely dense crowds. In: Proceedings of the 23rd ACM international conference on Multimedia, pp. 1299–1302. ACM (2015)
55. Wang, Q., Gao, J., Lin, W., Yuan, Y.: Learning from synthetic data for crowd counting in the wild. arXiv preprint arXiv:1903.03303 (2019)
56. Xu, B., Qiu, G.: Crowd density estimation based on rich features and random projection forest. In: 2016 IEEE Winter Conference on Applications of Computer Vision (WACV), pp. 1–8. IEEE (2016)
57. Yasarla, R., Sindagi, V.A., Patel, V.M.: Syn2Real transfer learning for image deraining using Gaussian processes. In: Proceedings of the IEEE/CVF Conference on Computer Vision and Pattern Recognition, pp. 2726–2736 (2020)
58. Zhan, B., Monekosso, D.N., Remagnino, P., Velastin, S.A., Xu, L.Q.: Crowd analysis: a survey. Mach. Vis. Appl. **19**(5–6), 345–357 (2008). https://doi.org/10.1007/s00138-008-0132-4
59. Zhang, C., Li, H., Wang, X., Yang, X.: Cross-scene crowd counting via deep convolutional neural networks. In: Proceedings of the IEEE Conference on Computer Vision and Pattern Recognition, pp. 833–841 (2015)
60. Zhang, Q., Chan, A.B.: Wide-area crowd counting via ground-plane density maps and multi-view fusion CNNs. In: Proceedings of the IEEE Conference on Computer Vision and Pattern Recognition, pp. 8297–8306 (2019)
61. Zhang, Y., Zhou, D., Chen, S., Gao, S., Ma, Y.: Single-image crowd counting via multi-column convolutional neural network. In: Proceedings of the IEEE Conference on Computer Vision and Pattern Recognition, pp. 589–597 (2016)
62. Zhao, M., Zhang, J., Zhang, C., Zhang, W.: Leveraging heterogeneous auxiliary tasks to assist crowd counting. In: Proceedings of the IEEE Conference on Computer Vision and Pattern Recognition, pp. 12736–12745 (2019)
63. Zhu, J.Y., Park, T., Isola, P., Efros, A.A.: Unpaired image-to-image translation using cycle-consistent adversarial networks. In: Proceedings of the IEEE International Conference on Computer Vision, pp. 2223–2232 (2017)

SPOT: Selective Point Cloud Voting for Better Proposal in Point Cloud Object Detection

Hongyuan Du[✉], Linjun Li, Bo Liu, and Nuno Vasconcelos

Department of Electrical and Computer Engineering,
University of California, San Diego, USA
{hdu,lili,boliu,nvasconcelos}@ucsd.edu

Abstract. The sparsity of point clouds limits deep learning models on capturing long-range dependencies, which makes features extracted by the models ambiguous. In point cloud object detection, ambiguous features make it hard for detectors to locate object centers (Fig. 1) and finally lead to bad detection results. In this work, we propose Selective Point clOud voTing (SPOT) module, a simple effective component that can be easily trained end-to-end in point cloud object detectors to solve this problem. Inspired by probabilistic Hough voting, SPOT incorporates an attention mechanism that helps detectors focus on less ambiguous features and preserves their diversity of mapping to multiple object centers. For evaluating our module, we implement SPOT on advanced baseline detectors and test on two benchmark datasets of clutter indoor scenes, ScanNet and SUN RGB-D. Baselines enhanced by our module can stably improve results in agreement by a large margin and achieve new state-or-the-art detection, especially under more strict evaluation metric that adopts larger IoU threshold, implying our module is the key leading to high-quality object detection in point clouds.

1 Introduction

3D object detection is important for many applications, such as indoor robot navigation, augmented reality, and autonomous driving. While it can be performed using data from many sensing modalities, there has recently been interest in point clouds, due to their ability to accurately represent geometric information, their lightweight nature, and the popularity of LIDAR sensors. It is, however, challenging to implement object detection on point clouds, for two main reasons. First their non-Euclidean structure [5] makes them poorly suited for classic deep-learning architectures. Second, their sparsity increases the challenges of feature extraction. The first problem has received substantial interest in the recent computer vision literature, with the introduction of many deep architectures tailored for point clouds [13,17,18,22,25,27,29,37]. However, considerably less progress has been observed on the second.

H. Du and L. Li—Equal contribution.

© Springer Nature Switzerland AG 2020
A. Vedaldi et al. (Eds.): ECCV 2020, LNCS 12356, pp. 230–247, 2020.
https://doi.org/10.1007/978-3-030-58621-8_14

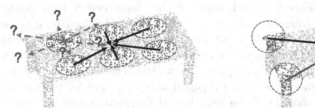
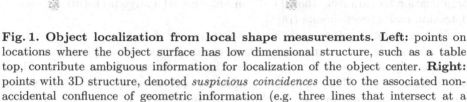

Fig. 1. Object localization from local shape measurements. Left: points on locations where the object surface has low dimensional structure, such as a table top, contribute ambiguous information for localization of the object center. **Right:** points with 3D structure, denoted *suspicious coincidences* due to the associated non-accidental confluence of geometric information (e.g. three lines that intersect at a point), are much more informative for this localization.

Modern point cloud architectures for object detection, attempt to mitigate the sparseness problem by aggregating information from multiple points [20, 26, 32–34, 43, 46]. An object is usually defined in terms of its center or a bounding box, which are detected by aggregating local shape information from the points on the object surface. This can be seen as a voting mechanism, where each point contributes information for both the localization and identification of the object. For example, the aggregation of geometric information from all points in the surface of each of the tables of Fig. 1 is what allows the perception of these point clouds as tables. However, the consolidation of the *local* measurements into a *global* object percept is a difficult problem, because not all points on an object are equally informative of object identity and location.

Consider, for example, the localization of the table of known dimensions of Fig. 1, from local shape measurements derived from sets of points on the surface of the object, such as those shown of the figure. As shown on the left, a neighborhood on the surface of the table is consistent with many object centers. This can be seen from the fact that any 2D translation along the tabletop leaves the neighborhood unchanged. Any amount of noise in the point cloud can originate a vote to an incorrect center or bounding box. Hence, such points are not reliable indicators of the object location. This is not the case for the neighborhood shown on the right, which is centered on a corner of the table. In this case, the neighborhood is only consistent with a center vote. Hence, the point is a reliable indicator of the object location.

For object class detection, the situation is obviously more complex, since the table can have any height and length. Nevertheless, it remains true that points where the object surface has 3D structure (e.g. table corners) are much more informative than points of 2D (table edges) or 1D structure (table top). This is similar to the *aperture problem* in optic flow estimation, where object corners are known to be more informative of object motion than other image points. In fact, the importance of these informative points for object recognition and localization has long been pointed out in the vision literature. This dates back to at least the work Attneave [1] which equated the visual cortex to a detective that makes inductive inferences about the environment by looking out for "suspicious coincidences", such as the confluence of three 3D edges into a single point. This is what enables the recognition of a table from a hand-drawn sketch depicting some lines and corners. In computer vision, the detection of suspicious, or non-accidental coincidences has been proposed as a principle for perceptual organization by various authors [4,23] and motivated a large literature on corner detection and interest points [14,31].

Non-accidental coincidences are important for detection exactly because their non-accidental nature makes them *rare*. Hence, when objects are sampled sparsely, they are likely to either be missed or immersed in an ocean of less informative points. This increases the difficulty of recovering object identity and location. In this work, we seek to address this problem by focusing the attention of the object detector in points of suspicious coincidences. For this we introduce a *Selective Point clOud voTing* (SPOT) module, which seeks to increase the attention of the point cloud around points of suspicious coincidences and reduce it everywhere else. SPOT consists of a combination of two operations: 1) detection of locations of suspicious coincidences, and 2) voting synthesis in the neighborhood of these locations. The two operations are performed on the 3D interest points produced by popular detection architectures in the literature. The first is implemented by a softmax network and the second by a set of non-linear regressions. This allows the implementation of both operations with a simple module that can be easily integrated into most existing point cloud detectors, to enable end-to-end training. We demonstrate this by implementing SPOT on three point cloud object detectors, VoteNet [26], PointRCNN [33], and a self-implemented version of PointRCNN that uses the Sparse Convolutional Network [13] as backbone. Evaluation on two large datasets of indoor scenes, ScanNet [8] and SUN RGB-D [35], shows that a simple implementation of SPOT without any bells and whistles can enhance all the baseline models by a large margin. In particular, it is shown that SPOT improves performance under more strict evaluation metrics, using higher IoU thresholds. This suggests that selective voting is important for high quality point cloud object detection.

2   Related Work

Feature Learning for Point Cloud Analysis. To deal with the irregular format of point cloud, one popular direction is to convert points into voxels in regular 3D grids and then utilize 3D CNNs for feature learning [25,28,42,46]. Recent works adopt Sparse Convolutional Networks [13] to reduce the computation cost of 3D convolution so that much larger point cloud input can be processed for vision tasks like semantic segmentation [13] and object detection [20,43,47]. Another trend is to use neural networks specially formulated for point cloud data. PointNet [27] and PointNet++ [29] are pioneers in this area that take point coordinates as input and learn permutation invariant features by multi-layer perceptrons and MaxPooling, showing strong performance on modeling point cloud geometry. In this work, we evaluate SPOT on both kinds of feature learning schemes, showing that our module is an universally useful structure for enhancing point cloud object detectors.

Point Cloud Object Detection. Due to the growing applications of high-resolution lidar sensors and the challenge of 2D-3D sensor fusion, recent methods are proposed to directly detect objects in 3D using point clouds. Some of these convert point clouds into voxels and use 3D CNNs to form backbones [20,43,46, 47]. PointRCNN [33] and VoteNet [26] utilize PointNet [27] and PointNet++ [29] to do detection on raw point clouds. More recently, several works [6,32,34,44] explore the hybrid of voxel and point representation to take the advantages from both. Our work investigates the impact of suspicious coincidence on point cloud object detectors and proposes a method to make them more robust.

Hough Voting in 2D/3D Object Detection. Hough transform/voting is a good paradigm for bottom-up detection. Origin Hough transform [16] lets edge points vote in parameter space for detecting simple shapes like lines and circles. Generalized Hough transform [2] can detect arbitrary shapes, by recording a matching table of the mappings from an edge orientation to possible positions of a reference point on the shape. Leibe et al. [21] further extend this idea to general object detection and segmentation in images, by using more discriminative features and probabilistic voting that learns the likelihood of a vote being an object center in a data-driven manner. Improved methods also show success in 3D recognition problems [19,24,38,41]. Recent works attempt embedding Hough voting in deep learning models for 3D object detection. [9,39] cast votes according to the weights of convolutional kernels. VoteNet [26] includes a voting module to cast one-to-one votes, with each local feature voting for one object center. Similar schemes are implemented in PointRCNN [33], where one foreground point is used to predict a single proposal. In contrast, our work inherits and extends the idea of probabilistic Hough voting [21] that selectively allows a local feature to cast multiple votes with probability weighting and implements it in an end-to-end trainable style, showing strong performance on high quality object detection in point clouds.

3   Selective Point Cloud Voting

3.1   Overview

While the idea of selective voting can be of interest for many operations on point clouds, in this work we consider its deployment in the context of the two-stage object detection architecture show at the top of Fig. 2. This is a general architecture, implemented by several popular detectors in the literature. The first stage generates object proposals. Given an input of N points with XYZ coordinates, a backbone network is used to abstract the point cloud and learn deep features. It outputs a subset of the input containing M *interest points* $\mathbf{q}^i =$

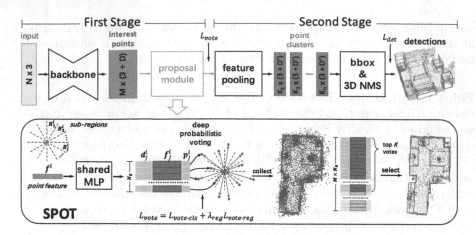

Fig. 2. Detection pipeline and Selective Point clOud voTing (SPOT). The proposal module of existing point cloud detectors is replaced by the proposed SPOT for better localization of object centers.

$(\mathbf{z}^i, \mathbf{f}^i)$, each composed by a vector $\mathbf{z}^i$ of 3D coordinates and a D-dimensional descriptor $\mathbf{f}^i$ of the local object geometry. Interest points are all the information retained for object proposal generation. Proposals are generated by a *proposal module*, which maps the interest point descriptors $\mathbf{f}^i$ into a preliminary prediction of the locations of scene objects. The second stage performs a pooling or NMS of proposals to infer a refined set of descriptors. Finally, those are processed by a detection head that includes classification, bounding box regression, and NMS modules to output the final detection.

SPOT works on the proposal module in the first stage. Commonly used proposal module has slightly different implementations on different detectors. For example, PointRCNN [33] predicts 3D bounding boxes as proposals that is similar to region proposal [30] in image object detection; in VoteNet [26], object centers are regressed as proposals instead of whole bounding boxes, and the local shape descriptors $\mathbf{f}^i$ are propagated to the proposals for its second stage. Though implemented differently, a common behavior is that all interest points uniformly generate proposals, which gives no preference to the points of suspicious coincidences, such as the corner on the right of Fig. 1. Since these points are rare, they can be missed altogether, or have small contribution to the set of proposals considered in the subsequent stages of the detector. Instead, the large majority of the proposals available to the later stages originate from points that are much less informative of the object identity and location, such as the tabletop points on the left of Fig. 1. These proposals are likely to be less accurate than those rooted at locations of suspicious coincidences.

The two-stage detector is generally supervised by a combination of a proposal loss on the first stage and a detection loss on the second stage. The detection loss L_{det} consists of objectness, bounding box regression and semantic classification.

In our case, the first stage is supervised by a voting loss L_{vote} and the whole system is trained with the loss

$$L = L_{\text{vote}} + \lambda_{\text{det}} L_{\text{det}} \qquad (1)$$

In this section, we will discuss the novel SPOT model trained with the addition of the voting loss L_{vote}.

3.2   SPOT

SPOT can be seen as an attention mechanism that aims to focus the proposal stage of Fig. 2 on interest points indicative of suspicious coincidences, such as the formation of a 3D corner by the confluence of three 3D lines on the same point[1]. Given a set of interest points $\mathbf{q}^i = (\mathbf{z}^i, \mathbf{f}^i)$, it produces a set of *votes* $\mathbf{v}_j^i = (\mathbf{d}_j^i, \mathbf{f}_j^i, p_j^i)$. Each interest point can contribute none or multiple votes, depending on how suspicious it is. Vote $\mathbf{v}_j^i$ is composed by a 3D coordinate $\mathbf{d}_j^i$, a D-dimensional descriptor $\mathbf{f}_j^i$ and a probability p_j^i. The 3D coordinate is the prediction of the object center, the descriptor is a refined version of the descriptor $\mathbf{f}^i$ provided by the input interest point, and the probability is the posterior of this vote being predicted as a valid object center. The goal of SPOT is to increase the attention of the point cloud around points of suspicious coincidences and reduce it everywhere else. This is performed by a sequence of two operations: 1) detection of locations of suspicious coincidences, and 2) selectively voting for object centers in the neighborhood of these locations.

Suspicious Coincidences. The central operation for the detection of suspicious coincidences is the estimation of the certainty with which the object center can be determined from the shape information contained in each interest point. This is inspired by Fig. 1. Interest points located at points of the object surface rich in 3D structure (such as corners) provide stronger constraints for localization of the object center than interest points at locations where the object surface has lower dimensional structure (such as table tops). Since the amount of 3D structure in the vicinity of the interest point can in principle be derived from the local shape descriptor $\mathbf{f}^i$, it should be possible to detect suspicious interest points by analyzing the shape descriptors.

   SPOT implements this intuition as follows. Given interest point $\mathbf{q}^i = (\mathbf{z}^i, \mathbf{f}^i)$, a neighborhood of $\mathbf{z}^i$, composed by a series of pre-specified sub-regions $\mathcal{R}_j^i, j \in \{1, \ldots, N_R\}$, is defined as shown in Fig. 2. It is assumed that the object center is within one of the regions $\mathcal{R}_j^i$, identified by a label $y^i \in \{1, \ldots, N_R\}$. The posterior probability of this label is then predicted by a classifier

$$[p_1^i, \ldots, p_{N_R}^i]^T = [P(y^i = 1|\mathbf{f}^i), \ldots, P(y^i = N_R|\mathbf{f}^i)]^T = g(\mathbf{f}^i; \theta^g), \qquad (2)$$

---

[1] While the description provided in this section is tailored to VoteNet, SPOT can be deployed on other detectors with minor modifications. Some variants are discussed in the experiments section.

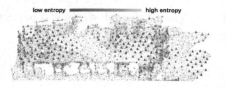

Fig. 3. Detection of suspicious coincidences. Left: A 3D scene composed of a large table. **Right:** Corresponding point cloud. For suspicious coincidences, the entropy of the distribution of center locations has low-entropy. These tends to happen around object regions informative for object identification and localization, such as table corners and edges.

where $g(; \theta^g)$ is a NN of parameters θ^g with softmax activation. During detector training, this classifier is optimized on a set of object interest points $\mathcal{Q} = \{(\mathbf{z}^i, \mathbf{f}^i, \mathbf{c}^i)\}$, of coordinate $\mathbf{z}^i$, descriptor $\mathbf{f}^i$, and ground-truth object center $\mathbf{c}^i$. For each interest point, a label y^i is set to the index of the region $\mathcal{R}_j^i$ that contains the object center, i.e. $y^i = j | \mathbf{c}^i \in \mathcal{R}_j^i$. The center location classifier is then optimized by minimizing the cross-entropy loss

$$L_{\text{vote-cls}} = -\sum_i \log g_{y^i}(\mathbf{f}^i; \theta^g). \tag{3}$$

During inference, given interest point $\mathbf{q}^i = (\mathbf{z}^i, \mathbf{f}^i)$, the classifier $g(.; \theta^g)$ is used to estimate the probabilities of (2). The detection of suspicious coincidences then follows from the intuition of Fig. 1. While interest points in the neighborhood of these coincidences (e.g. the centers of the patches on the right of the figure) should have distributions of low uncertainty, those consistent with many object centers (e.g. on the left of the figure) should generate high uncertainty. This is confirmed by Fig. 3, which shows a typical example of the information entropy of the posterior distribution

$$\mathbf{H}(\mathbf{f}^i) = -\sum_{j=1}^{N_R} g_j(\mathbf{f}^i; \theta^g) \log g_j(\mathbf{f}^i; \theta^g) \tag{4}$$

for many interest points on an object surface. Points of lower dimensional structure, such as the center of a tabletop, the ground, etc., generate larger entropies than points on the vicinity of the 3D edges and corners that demarcate the table boundaries.

This suggests that the detection of suspicious coincidences can be framed as an instance of the problem of assessing classification confidence, which has received recent interest in the literature on calibration of deep classifiers [7, 11, 12, 40]. Methods based on the thresholding of the entropy of (4) or the largest probability at the classifier output

$$\mathbf{C}(\mathbf{f}^i) = \max_j g_j(\mathbf{f}^i; \theta^g) \tag{5}$$

are commonly used in this literature to assess whether the classifier is confident in the classification of a given example. However, the calibration of the network probabilities is known to be difficult. For example, some methods require the training of a secondary network just for this purpose [7, 12, 40], while others rely on computational expensive Monte-Carlo dropout [10] procedures.

In this work, we rely on an alternative approach, which aims to increase the robustness of suspicious coincidence detection. This is implemented as follows. Given a point cloud $\mathcal{P}$ of M interest points $\mathbf{q}^i$ of shape descriptor $\mathbf{f}^i$, the posterior probabilities from all interest points are stored in an array $\mathbf{G} \in [0,1]^{M \times N_R}$ such that $\mathbf{G}_{ij} = g_j(\mathbf{f}^i; \theta^g)$. The entries of $\mathbf{G}$ are then ranked in decreasing order and the *minimum confidence* threshold for suspicious coincidence detection set to the value of the Kth largest entry, i.e.

$$T(\mathcal{P}; K) = rank_K(\mathbf{G}). \tag{6}$$

The set of suspicious coincidences is then defined as the set of interest points for which the confidence measure of (5) is above this threshold, i.e.

$$\mathcal{S}(K) = \{\mathbf{q}^i \in \mathcal{P} \mid \mathbf{C}(\mathbf{f}^i) \geq T(\mathcal{P}; K)\}. \tag{7}$$

where $\mathbf{C}$ is the confidence measure of (5). Note that the computations above can be easily performed by standard neural network operations. $\mathbf{C}$ is a simple max-pooling operation and (6) is implemented by sorting the outputs of g_j in each iteration.

The parameter K controls the number $|\mathcal{S}(K)|$ of detected suspicious coincidences. $|\mathcal{S}(K)|$ is usually smaller than K, because suspicious coincidences can assign strong probability to more than one object center location. This is typically the case for complex scenes, where an interest point located at a suspicious coincidence, e.g. the corner of a table, may be located near another object, e.g. a chair. In this case, because the receptive field centered on the interest point overlaps with both objects, the interest point may confidently vote for both of them. In our experience, it is not uncommon for suspicious coincidences to vote for two or three different center locations.

The overall procedure can be seen as an attention mechanism that focuses the detector on the object locations most informative for object identification and detection. An additional benefit is that, in 3D object detection, the background is usually composed of planar structures, such as the ground or the walls of a room. Since, as shown in Fig. 3, interest points located on these structures are unlikely to be declared suspicious, SPOT tends to suppress background clutter. Hence, in addition to being an attention mechanism that highlights informative object features, it also acts as an object-level attention mechanism, which declares objects as overall salient from background walls and ground.

Selective Voting. Selective voting attempts to aggregate local features from the locations of suspicious coincidences while rejects those from elsewhere. This is done as follows. During training, given an interest point $(\mathbf{z}^i, \mathbf{f}^i)$ whose corresponding object centered at $\mathbf{c}^i$, the region $\mathcal{R}_j^i$ that contains the center is first identified, which gives the label $y_i = j$. A ground truth offset $\Delta \mathbf{z}_*^i = \mathbf{c}^i - \mathbf{z}^i$ is then computed. The set of offsets associated with the same region label $y_i = j$ are then assembled into an offset training set $\mathcal{O}_j = \{\Delta \mathbf{z}_*^i | \mathbf{z}^i + \Delta \mathbf{z}_*^i \in \mathcal{R}_j^i\}$. This set is then used to train a *center location* regression function $\Delta \mathbf{z}_j^i = \phi_j(\mathbf{f}^i; \theta_j^\phi)$ for the region $\mathcal{R}_j^i$. The center location regression functions ϕ_j are learned by minimizing the regression loss

$$L_{\text{vote-reg}} = \sum_{j=1}^{N_R} \sum_{i|y_i=j} ||\phi_j(\mathbf{f}^i; \theta_j^\phi) - \Delta \mathbf{z}_*^i|| \tag{8}$$

Combined with (3), the whole voting loss is then implemented as

$$L_{\text{vote}} = L_{\text{vote-cls}} + \lambda_{\text{reg}} L_{\text{vote-reg}}. \tag{9}$$

During inference, given interest point $(\mathbf{z}^i, \mathbf{f}^i)$, the center location regressors ϕ_j are used to predict an estimate of the location of the center for each of the regions $\mathcal{R}_j^i$

$$\mathbf{d}_j^i = \mathbf{z}^i + \phi_j(\mathbf{f}^i; \theta_j^\phi). \tag{10}$$

Finally, a descriptor

$$\mathbf{f}_j^i = \mathbf{f}^i + \varphi_j(\mathbf{f}^i; \theta_j^\varphi) \tag{11}$$

is synthesized for each new center location $\mathbf{d}_j^i$. This is a refinement of the descriptor $\mathbf{f}^i$ of the interest point $(\mathbf{z}^i, \mathbf{f}^i)$, which accommodates variations of local geometry between $\mathbf{z}^i$ and $\mathbf{d}_j^i$. Since ground truth descriptors are not available, φ_j functions are learned end-to-end, using supervision from the second stage loss L_{det} of (1).

Overall, a single interest point produces multiple object center votes $\mathbf{v}_j^i = (\mathbf{d}_j^i, \mathbf{f}_j^i, p_j^i)$, where $\mathbf{d}_j^i$ is the center predicted by (10), $\mathbf{f}_j^i$ the shape descriptor predicted by (11) and p_j^i the probability $P(y^i = j | \mathbf{f}^i)$ of (2). $\mathcal{S}(K)$ of (7) then takes effect as a selection mechanism that only the votes of interest points in $\mathcal{S}(K)$ are passed to the subsequent stages of the network. Furthermore, the votes of these interest points are pruned by considering only those whose confidence is larger than the threshold $T(\mathcal{P}; K)$ of (6). This leads to a final set of votes

$$\mathcal{V} = \{(\mathbf{d}_j^i, \mathbf{f}_j^i, p_j^i) | p_j^i \geq T(\mathcal{P}; K)\}. \tag{12}$$

The functions g of (2), ϕ_j of (10), and φ_j of (8) are implemented with a shared MLP. The network implementation of SPOT is summarized at the bottom of Fig. 2.

4   Experiments

In this section, we discuss several experiments performed to evaluate the performance of SPOT.

4.1   Dataset and Evaluation Metrics

SUN RGB-D [35] is a dataset of RGB-D images for scene understanding. It contains 10K RGB-D images densely annotated with 58,657 3D bounding boxes with orientations for 47 object categories. For fair comparison, we follow the evaluation protocol of [26], which prunes the dataset to ~5K samples and the 10 most common categories. RGB-D images are converted to clouds of 20K points per image. ScanNetV2 [8] is an indoor scene dataset of 3D meshes reconstructed from RGB-D images. It contain 1,201 scans from hundreds of rooms, annotated with instance segmentation for 18 object categories. Following [26], we convert the meshes to clouds of 40K points per scene by sampling mesh vertices, and evaluate object detection performance on aligned circumscribed bounding boxes of instance segmentations.

4.2   Implementation Details

The impact of SPOT is evaluated on three detectors: VoteNet [26], PointR-CNN [33] and a variant of PointRCNN based on Sparse Convolution [13].

VoteNet. The original voting module is replaced by SPOT. This is implemented with interest point neighborhoods of 24 sub-regions $\mathcal{R}_1$ to $\mathcal{R}_{24}$. These are defined by 12 radial partitions, as illustrated in the bottom of Fig. 2, and one partition along the Z-axis. The rank parameter K of (6) is chosen so that the number of votes $|\mathcal{V}|$ of (12) is equal to $1,024$.

PointRCNN. To minimize the changes to the original PointRCNN model, we modify its bin-based localization to SPOT. After the point cloud is segmented into background and foreground, each foreground point is considered as an interest point, predicting the object center coordinates along the X and Y axes. These axes are binned into 6 segments, forming a set of square regions $\mathcal{R}_1, \ldots, \mathcal{R}_{36}$. The probability of the object center being located $\mathcal{R}_j$ is then computed as $p_j = p_{xj} \cdot p_{yj}$, where p_{xj} and p_{yj} are the probabilities computed by the original network for the X and Y bins corresponding to $\mathcal{R}_j$.

PointRCNN-SC. To investigate the effectiveness of SPOT on different backbone networks, we have also implemented a variant of PointRCNN using the submanifold sparse U-Net [13] as backbone to extract pointwise features. The remaining components are unaltered. This is denoted as PointRCNN with sparse convolutions (PointRCNN-SC).

4.3   Ablation Studies

We start with a series of experiments using VoteNet and ScanNet V2 to ablate several parameters of SPOT.

Definition of Suspicious Coincidences. SPOT defines suspicious coincidences as in (7), from which the set of center votes $\mathcal{V}$ of (12) is extracted. Table 1 compares this strategy to other possibilities for proposal selection. These are the simple regression of the baseline VoteNet, and three alternative approaches that keep the best 1, 2, and 3 votes of largest probability from each interest point. Selecting the best vote from each interest point outperforms the baseline, confirming the advantages of sub-region center votes as compared to a single center regression. On the other hand, selecting the best 2 or 3 votes per interest point degrades performance. This is because these votes are not reliable for interest points that are not co-located with suspicious coincidences. In fact, as discussed in Fig. 1, even the top vote is usually unreliable when this is the case. The detection of suspicious coincidences by SPOT eliminates such ambiguous interest points, allowing the detector to focus attention on the ones that most informative of object center locations. Hence, for the same number of votes as the best 1 strategy, and significantly less than the others, SPOT enables the best detection performance. When compared to the baseline, it enables significant gains of more than 3 points under both AR@0.5 and mAP@0.5 metrics.

Table 1. Ablation study of vote selecting method. I.P. means interest point. Our method outperforms others by dynamically selecting votes based on the spatial probability.

	total votes	min votes/I.P	max votes/I.P	AR@0.5	mAP@0.5
VoteNet (regression)	1024	1	1	53.3	33.5
best 1 per I.P	1024	1	1	56.0	38.7
best 2 per I.P	2048	2	2	55.0	38.4
best 3 per I.P	3072	3	3	54.3	37.1
SPOT	1024	0	3	**57.8**	**40.4**

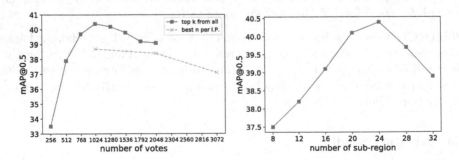

Fig. 4. Analysis of number of votes and sub-regions. Left: Number of votes v.s. mAP performance, under the case of 24 sub-regions. **Right:** Number of sub-regions v.s. mAP performance, under the case of 1024 votes.

Number of Votes. We next investigated the impact of the number of votes $|\mathcal{V}|$ on overall detector performance. This is shown in Fig. 4 a) for mAP@0.5. Performance increases until $|\mathcal{V}| = 1,024$ and decreases after that. Also shown is the performance of the three best n per I.P. strategies. For $|\mathcal{V}| = 512$ the performance of SPOT is already superior to the baseline detector and for $|\mathcal{V}| > 768$ SPOT is always superior to all other strategies. While the choice of the threshold of (6) has some impact on the mAP@0.5 performance, SPOT has the best performance for a large range of thresholds. These results illustrate its robustness.

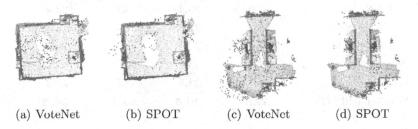

(a) VoteNet   (b) SPOT   (c) VoteNet   (d) SPOT

Fig. 5. Votes from baseline model and SPOT. Red dots are center votes.(Color figure online)

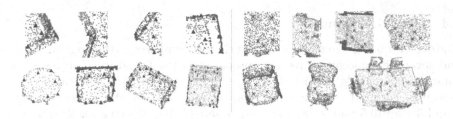

Fig. 6. Local geometry of different interest points. Left (blue): Interest points that cast deterministic votes. **Right (red):** Interest points that cast multiple votes or filtered out.(Color figure online)

Number of Sub-regions. The impact of the number of sub-regions N_R on detection performance was also investigated, by varying the number of spatial sectors $\mathcal{R}_j^i$ under a bird's eye view. The results are shown in Fig. 4 b). When the number of sub-region is too small detection performance degrades. This is not surprising, because the sub-regions $\mathcal{R}_j^i$ become less selective for center location. On the other hand, too many sub-regions can also lead to a drop in performance, because classifier labels become noisier and there are fewer examples per region, increasing the difficulty of learning all the classification and regression functions

Table 2. 3D object detection results on SUN RGB-D and ScanNet V2 validation set with 3D IoU threshold 0.25 and 0.50. DSS, F-PointNet and 3D-SIS results are from [15], GSPN are from [45], VoteNet are from [26], PointRCNN is implemented base on [33] and Sparse Conv backbone is implemented base on [13].

	Scan @0.25	Scan @0.5	SUN @0.25	SUN @0.5
DSS [36]	15.2	6.8	42.1	–
F-PointNet [6]	19.8	10.8	54.0	–
GSPN [45]	30.6	17.7	–	–
3D-SIS [15]	40.2	22.5	40.2	22.5
PRCNN [33]	53.0	25.4	53.7	23.4
PRCNN + SPOT	55.2	27.8	57.6	25.3
PointRCNN-SC	57.0	31.8	53.0	24.5
PointRCNN-SC + SPOT	57.4	33.1	59.5	27.7
VoteNet [26]	58.7	33.5	57.7	32.3
VoteNet + SPOT	**59.8**	**40.4**	**60.4**	**36.3**

of SPOT. While the careful section of the number of sub-regions can make a significant difference, in the example of the figure a gain of almost 3 points for $N_R = 24$, the curve is not very sharp and there is some flexibility in this parameter.

4.4   Detection Results

The detector baselines enhanced by SPOT were compared to the original versions and several enhancements recently proposed in the literature. The results are summarized in Table 2, which shows that SPOT improves the performance of all detectors. For most combinations of dataset and detector, the gains are between 2 and 3 map@0.5 points, with the larger gains being observed for the strongest baseline, which is VoteNet. This and the improvements of the PointRCNN with different backbone designs suggest that SPOT should improve the quality of other point cloud detectors.

Table 3. AP gains v.s. object sizes. The best 6 gains (top) and the worst 6 gains (bottom) out of 17 object categories of ScanNet V2 along with objects' mean sizes.

	bed	table	desk	refrigerator	bathtub	counter
AP gain	+7.3	+7.4	+7.2	+7.6	+6.2	+4.8
mean size	(1.9, 1.8, 1.2)	(1.0, 1.2, 0.6)	(1.0, 1.4, 0.9)	(0.7, 0.7, 1.3)	(1.2, 1.1, 0.5)	(1.4, 1.9, 0.3)
	chair	toilet	sink	garbagebin	picture	cabinet
AP gain	+1.7	+1.4	+1.3	+1.1	+0.2	−1.5
mean size	(0.6, 0.6, 0.7)	(0.6, 0.6, 0.7)	(0.5, 0.5, 0.3)	(0.5, 0.5, 0.6)	(0.2, 0.4, 0.5)	(0.8, 0.8, 0.9)

Qualitative Results. To understand how SPOT improves detection accuracy, we visualize the votes produced by it and compare with those generated by the baseline model. Figure 5 shows how SPOT focus the attention of the detector on suspicious coincidences indicative of object presence. Note the much smaller number of votes on the ground or table tops and the concentration of votes on object surfaces. Figure 6 shows the local geometry of different kinds of interest points. Interest points that cast deterministic votes, i.e., one vote with high probability score, tend to gather around 3D structures of objects like corners and edges. On the other hand interest points that cast multiple votes or are prunned by SPOT tend to gather on low dimensional structures, such as flat surfaces, or locations where multiple object intersect. Table 3 summarizes AP gains per object class, showing that SPOT benefits the most the detection of large objects, such as beds, tables, or desks. This is because these objects contain larger surface areas of low dimensional structure than small objects and are more likely to produce interest points unaligned with suspicious coincidences. Figure 7 and Figure 8 show how the addition of SPOT affects the object detection of VoteNet on ScanNetV2 and SUN RGB-D. As shown in these figures, the addition of SPOT leads to a reduction of false positives. For example, in Fig. 7, the vote clustering of VoteNet produces a set of detections around each object, even after NMS. With SPOT the results are much more accurate, due to the attention of votes around suspicious coincidences.

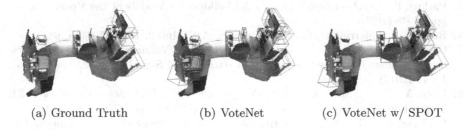

(a) Ground Truth   (b) VoteNet   (c) VoteNet w/ SPOT

Fig. 7. Qualitative results for ScanNetV2

(a) Ground Truth   (b) VoteNet   (c) VoteNet w/ SPOT

Fig. 8. Qualitative results for SUN RGB-D

5 Conclusion

In this work, we considered point cloud object detection, and proposed a procedure for selective point cloud voting (SPOT). This can be seen as an attention mechanism, which increases the attention of the point cloud in the neighborhood of suspicious coincidences, i.e. features that are most informative of object identity and location. SPOT was shown to be a valuable addition to several state of the art detectors based on different architectures, achieving state-of-the-art results on both ScanNet and SUN-RGBD. All of these observations confirm long-standing arguments for the importance of suspicious coincidences in object recognition [1,3], and suggest that selective point cloud voting should be useful for future object detector designs.

Acknowledgment:. This work was partially funded by NSF awards IIS-1637941, IIS-1924937, and NVIDIA GPU donations.

References

1. Attneave, F.: Information aspects of visual perception. Psychol. Rev. **61**, 183–193 (1954)
2. Ballard, D.H.: Generalizing the Hough transform to detect arbitrary shapes. Pattern Recogn. **13**(2), 111–122 (1981)
3. Barlow, H.: Cerebral cortex as a model builder. In: Models of the Visual Cortex, pp. 37–46 (1985)
4. Binford, T.: Inferring surfaces from images. Artif. Intell. **17**, 205–244 (1981)
5. Bronstein, M.M., Bruna, J., LeCun, Y., Szlam, A., Vandergheynst, P.: Geometric deep learning: going beyond euclidean data. IEEE Signal Process. Mag. **34**(4), 18–42 (2017)
6. Chen, Y., Liu, S., Shen, X., Jia, J.: Fast point R-CNN. In: Proceedings of the IEEE International Conference on Computer Vision (ICCV), pp. 9775–9784 (2019)
7. Corbière, C., Thome, N., Bar-Hen, A., Cord, M., Pérez, P.: Addressing failure prediction by learning model confidence. In: Advances in Neural Information Processing Systems (NeurIPS), pp. 2902–2913 (2019)
8. Dai, A., Chang, A.X., Savva, M., Halber, M., Funkhouser, T., Nießner, M.: ScanNet: richly-annotated 3D reconstructions of indoor scenes. In: Proceedings of the IEEE Conference on Computer Vision and Pattern Recognition (CVPR), pp. 5828–5839 (2017)
9. Engelcke, M., Rao, D., Wang, D.Z., Tong, C.H., Posner, I.: Vote3Deep: fast object detection in 3D point clouds using efficient convolutional neural networks. In: 2017 IEEE International Conference on Robotics and Automation (ICRA), pp. 1355–1361 (2017)
10. Gal, Y., Ghahramani, Z.: Dropout as a Bayesian approximation: representing model uncertainty in deep learning. In: International Conference on Machine Learning (ICML), pp. 1050–1059 (2016)
11. Geifman, Y., El-Yaniv, R.: Selective classification for deep neural networks. In: Advances in Neural Information Processing Systems (NeurIPS), pp. 4878–4887 (2017)

12. Geifman, Y., El-Yaniv, R.: SelectiveNet: a deep neural network with an integrated reject option. In: International Conference on Machine Learning (ICML), pp. 2151–2159 (2019)
13. Graham, B., Engelcke, M., Van Der Maaten, L.: 3D semantic segmentation with submanifold sparse convolutional networks. In: Proceedings of the IEEE Conference on Computer Vision and Pattern Recognition (CVPR), pp. 9224–9232 (2018)
14. Harris, C.G., et al.: A combined corner and edge detector. In: Alvey Vision Conference, vol. 15, pp. 10–5244 (1988)
15. Hou, J., Dai, A., Nießner, M.: 3D-SIS: 3D semantic instance segmentation of RGB-D scans. In: Proceedings of the IEEE Conference on Computer Vision and Pattern Recognition (CVPR), pp. 4421–4430 (2019)
16. Hough, P.V.: Machine analysis of bubble chamber pictures. In: Proceedings of the International Conference on High Energy Accelerators and Instrumentation, pp. 554–556 (1959)
17. Huang, Q., Wang, W., Neumann, U.: Recurrent slice networks for 3D segmentation of point clouds. In: Proceedings of the IEEE Conference on Computer Vision and Pattern Recognition (CVPR), pp. 2626–2635 (2018)
18. Klokov, R., Lempitsky, V.: Escape from cells: deep Kd-networks for the recognition of 3D point cloud models. In: Proceedings of the IEEE International Conference on Computer Vision (ICCV), pp. 863–872 (2017)
19. Knopp, J., Prasad, M., Willems, G., Timofte, R., Van Gool, L.: Hough transform and 3D SURF for robust three dimensional classification. In: Daniilidis, K., Maragos, P., Paragios, N. (eds.) ECCV 2010. LNCS, vol. 6316, pp. 589–602. Springer, Heidelberg (2010). https://doi.org/10.1007/978-3-642-15567-3_43
20. Lang, A.H., Vora, S., Caesar, H., Zhou, L., Yang, J., Beijbom, O.: PointPillars: fast encoders for object detection from point clouds. In: Proceedings of the IEEE Conference on Computer Vision and Pattern Recognition (CVPR), pp. 12697–12705 (2019)
21. Leibe, B., Leonardis, A., Schiele, B.: Robust object detection with interleaved categorization and segmentation. Int. J. Comput. Vis. (IJCV) 77(1–3), 259–289 (2008)
22. Liu, Y., Fan, B., Xiang, S., Pan, C.: Relation-shape convolutional neural network for point cloud analysis. In: Proceedings of the IEEE Conference on Computer Vision and Pattern Recognition, pp. 8895–8904 (2019)
23. Lowe, D.: Three-dimensional object recognition from single two-dimensional images. Artif. Intell. 31, 355–395 (1987)
24. Maji, S., Malik, J.: Object detection using a max-margin Hough transform. In: Proceedings of the IEEE Conference on Computer Vision and Pattern Recognition (CVPR), pp. 1038–1045 (2009)
25. Maturana, D., Scherer, S.: VoxNet: a 3D convolutional neural network for real-time object recognition. In: 2015 IEEE/RSJ International Conference on Intelligent Robots and Systems (IROS), pp. 922–928 (2015)
26. Qi, C.R., Litany, O., He, K., Guibas, L.J.: Deep Hough voting for 3D object detection in point clouds. In: Proceedings of the IEEE International Conference on Computer Vision (ICCV), pp. 9277–9286 (2019)
27. Qi, C.R., Su, H., Mo, K., Guibas, L.J.: PointNet: deep learning on point sets for 3D classification and segmentation. In: Proceedings of the IEEE Conference on Computer Vision and Pattern Recognition (CVPR), pp. 652–660 (2017)

28. Qi, C.R., Su, H., Nießner, M., Dai, A., Yan, M., Guibas, L.J.: Volumetric and multi-view CNNs for object classification on 3D data. In: Proceedings of the IEEE Conference on Computer Vision and Pattern Recognition (CVPR), pp. 5648–5656 (2016)
29. Qi, C.R., Yi, L., Su, H., Guibas, L.J.: PointNet++: deep hierarchical feature learning on point sets in a metric space. In: Advances in Neural Information Processing Systems (NeurIPS), pp. 5099–5108 (2017)
30. Ren, S., He, K., Girshick, R., Sun, J.: Faster R-CNN: towards real-time object detection with region proposal networks. In: Advances in Neural Information Processing Systems (NeurIPS), pp. 91–99 (2015)
31. Schmid, C., Mohr, R., Bauckhage, C.: Evaluation of interest point detectors. Int. J. Comput. Vis. (IJCV) **37**, 151–172 (2000)
32. Shi, S., et al.: PV-RCNN: Point-Voxel feature set abstraction for 3D object detection. In: Proceedings of the IEEE Conference on Computer Vision and Pattern Recognition (CVPR), pp. 10529–10538 (2020)
33. Shi, S., Wang, X., Li, H.: PointRCNN: 3D object proposal generation and detection from point cloud. In: Proceedings of the IEEE Conference on Computer Vision and Pattern Recognition (CVPR), pp. 770–779 (2019)
34. Shi, S., Wang, Z., Shi, J., Wang, X., Li, H.: From points to parts: 3D object detection from point cloud with part-aware and part-aggregation network. IEEE Trans. Pattern Anal. Mach. Intell. (TPAMI) (2020)
35. Song, S., Lichtenberg, S.P., Xiao, J.: SUN RGB-D: a RGB-D scene understanding benchmark suite. In: Proceedings of the IEEE Conference on Computer Vision and Pattern Recognition (CVPR), pp. 567–576 (2015)
36. Song, S., Xiao, J.: Deep sliding shapes for amodal 3D object detection in RGB-D images. In: Proceedings of the IEEE conference on Computer Vision and Pattern Recognition (CVPR), pp. 808–816 (2016)
37. Su, H., et al.: SPLATNet: sparse lattice networks for point cloud processing. In: Proceedings of the IEEE Conference on Computer Vision and Pattern Recognition (CVPR), pp. 2530–2539 (2018)
38. Velizhev, A., Shapovalov, R., Schindler, K.: Implicit shape models for object detection in 3D point clouds. In: International Society of Photogrammetry and Remote Sensing Congress, vol. 2, p. 2 (2012)
39. Wang, D.Z., Posner, I.: Voting for voting in online point cloud object detection. In: Robotics: Science and Systems, vol. 1, pp. 10–15607 (2015)
40. Wang, P., Vasconcelos, N.: Towards realistic predictors. In: Proceedings of the European Conference on Computer Vision (ECCV), pp. 36–51 (2018)
41. Woodford, O.J., Pham, M.T., Maki, A., Perbet, F., Stenger, B.: Demisting the hough transform for 3D shape recognition and registration. Int. J. Comput. Vis. (IJCV) **106**, 332–341 (2014)
42. Wu, Z., et al.: 3D ShapeNets: a deep representation for volumetric shapes. In: Proceedings of the IEEE Conference on Computer Vision and Pattern Recognition (CVPR), pp. 1912–1920 (2015)
43. Yan, Y., Mao, Y., Li, B.: Second: sparsely embedded convolutional detection. Sensors **18**(10), 3337 (2018)
44. Yang, Z., Sun, Y., Liu, S., Shen, X., Jia, J.: STD: Sparse-to-dense 3D object detector for point cloud. In: Proceedings of the IEEE International Conference on Computer Vision (ICCV), pp. 1951–1960 (2019)

45. Yi, L., Zhao, W., Wang, H., Sung, M., Guibas, L.J.: GSPN: Generative shape proposal network for 3D instance segmentation in point cloud. In: Proceedings of the IEEE Conference on Computer Vision and Pattern Recognition (CVPR), pp. 3947–3956 (2019)
46. Zhou, Y., Tuzel, O.: VoxelNet: end-to-end learning for point cloud based 3D object detection. In: Proceedings of the IEEE Conference on Computer Vision and Pattern Recognition (CVPR), pp. 4490–4499 (2018)
47. Zhu, B., Jiang, Z., Zhou, X., Li, Z., Yu, G.: Class-balanced grouping and sampling for point cloud 3D object detection. arXiv preprint arXiv:1908.09492 (2019)

Explainable Face Recognition

Jonathan R. Williford[1], Brandon B. May[1(⊠)], and Jeffrey Byrne[1,2]

[1] Systems and Technology Research, Woburn, MA 01801, USA
{jonathan.williford,brandon.may}@stresearch.com
[2] Visym Labs, Cambridge, MA 02140, USA
jeff@visym.com
https://www.stresearch.com

Abstract. Explainable face recognition (XFR) is the problem of explaining the matches returned by a facial matcher, in order to provide insight into why a probe was matched with one identity over another. In this paper, we provide the first comprehensive benchmark and baseline evaluation for XFR. We define a new evaluation protocol called the "inpainting game", which is a curated set of 3648 triplets (probe, mate, nonmate) of 95 subjects, which differ by synthetically inpainting a chosen facial characteristic like the nose, eyebrows or mouth creating an inpainted nonmate. An XFR algorithm is tasked with generating a network attention map which best explains which regions in a probe image match with a mated image, and not with an inpainted nonmate for each triplet. This provides ground truth for quantifying what image regions contribute to face matching. Finally, we provide a comprehensive benchmark on this dataset comparing five state-of-the-art XFR algorithms on three facial matchers. This benchmark includes two new algorithms called subtree EBP and Density-based Input Sampling for Explanation (DISE) which outperform the state-of-the-art XFR by a wide margin.

1 Introduction

Explainable AI [29] is the problem of interpreting, understanding and visualizing machine learning models. Deep convolutional network trained at large scales are traditionally considered blackbox systems, where designers have an understanding of the dataset and loss functions for training, but limited understanding of the learned model. Furthermore, predictions generated by the system are often not explainable as to why the system generated this output for that input. An explainable AI system would enable interpretation of what the ML model has learned [2,26], enable transparency to understand and identify biases or failure modes in the system [3,13,15,25] and provide user friendly visualizations to build user trust in critical applications [31,33,42].

Explainable face recognition (XFR) is the problem of explaining why a face matching system matches faces. Human adjudicators have a long history in

Electronic supplementary material The online version of this chapter (https://doi.org/10.1007/978-3-030-58621-8_15) contains supplementary material, which is available to authorized users.

© Springer Nature Switzerland AG 2020
A. Vedaldi et al. (Eds.): ECCV 2020, LNCS 12356, pp. 248–263, 2020.
https://doi.org/10.1007/978-3-030-58621-8_15

Fig. 1. Explainable Face Recognition (XFR). Given a image triplet of (*probe, mate 1, nonmate*), an explainable face recognition algorithm is tasked with estimating which pixels belong in a region that is discriminative for the mate - i.e. a region is more similar to the mate than the non-mate. These estimations are given as a saliency map. The nonmate has been synthesized by inpainting a given region (e.g. eyebrows) that changes the identity according to the given network. This provides ground truth for a quantitative evaluation of XFR algorithms using the "inpainting game" protocol.

explaining face recognition in the field of forensic face matching. Professional facial analysts follow the FISWG standards [11] which leverage comparing facial morphology, measuring facial landmarks and matching scars, marks and blemishes. These features are used to match a controlled mugshot of a proposed candidate to an uncontrolled probe, such as a security camera image. However, these approaches require a candidate list for human adjudication, and a candidate list in a modern workflow is returned from a facial matching system [24]. Why did the face matching system return that candidate list for this probe? What facial features did the face matching system use, and are they the same as the FISWG standards? Is the face matcher biased or noisy? The goal of XFR is to explore such questions, and answer why a system matched a pair of faces. A successful explainable system would increase confidence in a face matching system for professional examiners, enable intepretation of the internal face representations by machine learning researchers and generate trust by the user community.

What is an "explanation" in face recognition? Explainable AI has explored various forms of explanation for machine learning systems in the form of: activation maximization [32], synthesizing optimal images [21], network attention [12,23,39], network dissection [2] or synthesizing linguistic explanations [16]. However, a key challenge in explainable AI is the lack of ground truth to compare and quantify explainable results across networks. XFR is especially challenging because the difference between near-mates or *doppelgangers* is subtle, the explanations are non-obvious, and differences are rarely well localized in a compact facial feature [6].

In this paper, we provide the first comprehensive benchmark for explainable face recognition (XFR). Figure 1 shows the structure of this problem. An XFR system is given a triplet of (*probe, mate, nonmate*) images. The XFR system is tasked with generating a saliency map that best captures the regions of the probe image that increase similarity to the mate and decrease similarity to the nonmate. This provides an explanation for why the matcher provides a high verification score for the pair (*probe*, *mate*) and a low verification score for (*probe*, *nonmate*). This explanation can be quantitatively evaluated by synthesizing non-mates that differ from the mate only in specific regions (e.g. nose, eyes, mouth), such that if the saliency algorithm selects these regions, then it performs well on this metric. This paper makes the following contributions:

1. **XFR baseline**. We provide a baseline for XFR based on five algorithms for network attention evaluated on three publicly available convolutional networks trained for face recognition: LightCNN [35], VGGFace2 [5] and ResNet-101. These baselines include two new algorithms for network attention called subtree EBP (Sect. 3.1) and DISE (Sect. 3.2).
2. **Inpainting game protocol and dataset**. We provide a standardized evaluation protocol (Sect. 4.1) and dataset (Sect. 4.2) for fine grained discriminative visualization of faces. This provides a quantitative metric for objectively comparing XFR systems.
3. **XFR evaluation**. We provide the first comprehensive evaluation of XFR using the baseline algorithms on the inpainting game protocol to provide a benchmark for future research (Sect. 5.1). Furthermore, in the supplemental material, we show a qualitative evaluation on novel (non inpainted) images to draw conclusions about the utility of the methods for explanation on real images.

2   Related Work

The related work most relevant for our proposed approach to XFR can be broadly categorized into two areas: network attention models for convolutional networks and interpretable face recognition.

Network attention is the problem of generating an image based saliency map which visualizes the input regions that best explains a class activation output of a network. Gradient-based methods [31,33,42] attempt to compute the derivative of the class signal with respect to the input image, while other approaches [4] modify network architectures to capture these signals or localize attribution [17]. Excitation backprop [39], contrastive EBP [39] or truncated contrastive EBP [6] formulate the saliency map as marginal probabilities in a probabilistic absorbing Markov chain. Layerwise relevance propagation [1,18,28] provides network attention through a set of layerwise propagation rules theoretically justified by deep Taylor decomposition. Latent attention networks learn an auxiliary network to map input to attention, rather than exploring the network directly [14]. Inversion methods [19] seek to recover natural images that have the same feature representation as a given image. However, the same insights have not yet been applied to fine grained categorization for face recognition. Finally, black box methods have explored network attention for systems that do not have an exposed convolutional network [4,8,12,23]. The approaches to XFR explored in this paper are most closely related to EBP [39], RISE [23], and methods for network attention for pairwise similarity [34].

Recently, there has been emerging research on the interpretation of face recognition systems [6, 9, 27, 36–38, 41]. Visual psychophysics [27] have provided a set of tools for the controlled manipulation of input stimuli and metrics for the output responses evoked in a face matching system. This approach was inspired by Cambridge Face Memory Test [10], which involves progressively perturbing face images using a chosen transformation function (e.g. adding noise) to investigate controlled degradation of matching performance [27]. This approach enables detailed studies of the failure modes of a face matcher or exploring how facial attributes are expressed in a network [9, 41]. In contrast, our approach generates controlled degradations using inpainting, to provide localized ground truth for evaluation of network attention models. In [37], the authors propose a novel loss function to encourage part separability during network dissection of parts in a convolutional network for face matching. This approach is primarily concerned with training new networks to maximize interpretability, rather than studying existing networks. In [38], the authors study pairwise matching of faces, to visualize features that lead towards classification decisions. This is similar in spirit to our proposed approach, however we provide a performance metric for evaluating a saliency approach as well as extending visualizations to mated and nonmated triplets. Finally, in [36], the authors visualize the features of shape and texture that underlie subject identity decisions. This approach uses 3D modeling to generate a controlled dataset, rather than inpainting. However, given the authors conclusions that texture has a much larger effect of matching than morphology, having a ground truth dataset that includes texture variation would be an appropriate metric for explainable face recognition.

3 Explainable Face Recognition (XFR)

XFR is the problem of explaining why a face matcher matches faces. Figure 1 shows the structure of this problem. Given a triplet of (*probe, mate, nonmate*), the XFR algorithm is tasked with generating a saliency map that explains the regions of the probe image that maximize the similarity to the mate and minimize the similarity to the nonmate. This provides an explanation for why the matcher returns this image for the mated identity.

Why triplets? Previous work has shown that pairwise similarity between faces is heavily dominated by the periocular region and nose [6], as confirmed by the qualitative visualization study performed in the supplementary material The periocular region and nose is almost always used for facial classification, but this level of XFR is not very helpful in explaining finer levels of discrimination. Our goal is to highlight those regions for a probe that are more similar to a presumptive mate and *simultaneously* less similar to a nonmate. This triplet of (*probe, mate, nonmate*) provides a deeper explanation beyond facial class activation maps for the *relative importance* of facial regions.

In this section, we describe five approaches for network attention in XFR. These approaches are all whitebox methods, which assume access to the underlying convolutional network used for facial matching. The objective of XFR is

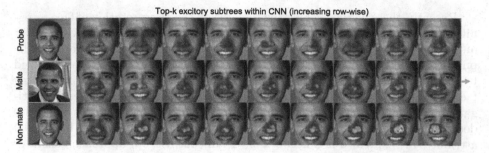

Fig. 2. Subtree EBP. Given a triplet of (*probe, mate, nonmate*) subtree EBP explores the activations of individual nodes in a convolutional network that minimizes a triplet loss, which maximizes similarity to the mate and minimizes similarity to the nonmate. The excitory regions for each node are visualized independently, sorted by loss and combined into a saliency map that best explains how to discriminate the probe.

to generate a non-negative saliency map, that captures the underlying image regions of the probe that are most similar to the mate and least similar to the nonmate. The XFR algorithm can use any property of the convolutional network to generate this saliency map. For our benchmark evaluation, we selected three state-of-the-art approaches for network attention (excitation backprop, contrastive excitation backprop and truncated contrastive excitation backprop) following the survey and evaluation results in [6]. In this section, we introduce two new methods to improve upon these published approaches: subtree EBP (Sect. 3.1) and DISE (Sect. 3.2).

Excitation Backprop (EBP). Excitation backprop (EBP) [39] models network attention as a probabilistic winner-take-all (WTA) process. EBP calculates the probability of traversing to a given node in the convolutional network, with the probabilities being defined by the positive weights and non-negative activations. The output of EBP is a saliency map that localizes regions in the image that are excitory for a given class.

In our approach, we replace the cross-entropy loss for EBP with a triplet loss [30]. The original formulation of EBP considers a cross-entropy loss to optimize softmax classification of a set of classes in the training set. In this new formulation, given three embeddings for a mate (m), nonmate (n) and probe (p), the triplet loss function is a max-margin hinge loss

$$L(p, n, m) = \mathtt{max}(0, ||p - m||^2 - ||p - n||^2 + \alpha). \tag{1}$$

This uses the squared Euclidean distance between embeddings to capture similarity, such that the loss is minimized when the distance from the probe and mate is small (similarity is high) and the distance form the probe to the nonmate is large (similarity is low), with margin term α. This loss function extends EBP to cases where a new subject is observed at test time that was not present in the training set, as is commonly the case with face matching systems.

Contrastive EBP (cEBP). Contrastive EBP was introduced [39] to handle fine-grained network attention for closely related classes. This approach discards activations common to a pair of classes, to provide network attention specific to one class and not another. In our approach, contrastive EBP [39] is combined with a triplet loss (Eq. 1).

Truncated Contrastive EBP (tcEBP). Truncated contrastive EBP was introduced [6] as an extension of cEBP that considers the contrastive EBP attention map only within the kth percentile of the EBP saliency map. This addresses an observed instability of cEBP [6] resulting noisy attention maps for faces.

3.1   Subtree EBP

In this section, we introduce *Subtree EBP*, a novel method for whitebox XFR. This approach uses the triplet loss function (Eq. 1), with the following extension. Given a triplet (*probe, mate, nonmate*) images, we compute the gradient of the triplet loss function ($\frac{\partial L}{\partial x_i}$) with respect to every node x_i in the network. This approach uses the standard triplet-based learning, where the mate and nonmate embeddings are assumed constant and the gradient is computed relative to the probe image. Next, we sort the gradients at every node x_j in decreasing order, and select the top-k nodes with the largest positive gradients. These are the top-k nodes in the network that most affect the triplet loss, to increase the similarity to the mate and simultaneously decrease the similarity to the nonmate. Finally, we construct k EBP saliency maps S_i starting from each of the selected interior nodes, then the S_i are combined in a weighted convex combination with weights $w_i = \frac{\partial L}{\partial x_i}$ and

$$S = \frac{1}{\sum_j w_j} \sum_i \frac{\partial L}{\partial x_i} S_i \tag{2}$$

where the weights are given by the loss gradient (w_i), normalized to sum to one. This forms the final subtree EBP saliency map S.

Figure 2 shows an example of the subtree EBP method. This montage shows the top 27 nodes with the largest triplet loss gradient for the shown triplet. Each node results in a saliency map corresponding to the excitory subtree rooted at this node. The weight of the saliency map is proportional to the gradient sorted rowwise, so that the nodes in the bottom right affect the loss more than the nodes in the upper left. Each of these saliency maps are combined into a convex combination (Eq. 2) forming the final network attention map. In this example, the nonmate was synthesized to differ with the mate only in the nose region, and our method is able to correctly localize this region. The supplementary material shows a more detailed example of this selection process starting from the largest excitation node at each layer of a ResNet-101 network. This result shows that nodes selected close to the image will be well localized, nodes in the middle of the network correspond to parts and nodes selected close to the embedding correspond to the whole nose and eyes of the face.

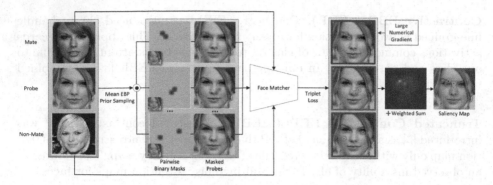

Fig. 3. Density-based Input Sampling for Explanation (DISE). Our approach is an extention of RISE [23] for XFR. This approach occludes small regions in the probe image with grey (i.e. masked pixels), sampled according to a prior density derived from excitation backprop, and computes a numerical gradient for the triplet loss for (*probe*, *mate*, *nonmate*) given these masked probes. Masks with a large numerical gradient are more heavily weighted in the accumulated saliency map.

3.2   Density-Based Input Sampling for Explanation (DISE)

Density-based Input Sampling for Explanation (DISE) is a second novel approach for whitebox XFR introduced in this paper. DISE is an extension of Randomized Input Sampling for Explanation (RISE) [23] using a prior density to aid in sampling. Previous work [12,23] has constructed a saliency map associated with a particular class by randomly perturbing the input image by masking selected pixels, evaluating it using a blackbox system, and accumulating those perturbations based on how confident the system is that the modified input image corresponds to the target class. However, these approaches generate masks to occlude the input image uniformly at random. This sampling process is inefficient, and can be improved by introducing a prior distribution to guide the sampling. In this section, we describe the extension to RISE [23] where the prior density for input sampling is derived from a whitebox EBP with triplet loss.

Figure 3 shows an overview of this approach. Our approach extends RISE [23] for XFR as follows:

1. Using a non-uniform prior for generating the random binary masks
2. Restricting the masks to use a sparse, fixed number of mask elements
3. Defining a numerical gradient of the triplet loss to weight each mask

Non-uniform Prior. Prior research on discriminative features for facial recognition showed that the most important regions of the face were generally located in and around the eyes and nose (Sect. 3). Figure 3 shows an example of this saliency map computed for a probe image of Taylor Swift using the VGG-16 [22] network as the whitebox face matcher. Using this saliency map as our prior probability for generating random masks allows us to sample the space of most salient masks that will affect the loss more efficiently than assuming a uniform probability across the entire image. Further limiting this prior to the upper 50th percentile of the mean EBP effectively eliminates the possibility of masking out unimportant background elements.

Sparse Masks. Next, we restrict the number of masked elements to be sparse. RISE considered random binary masks covering the entire input image. In contrast, we use a sparse mask to highlight the affect of a small localized region of the face on the loss. We used two mask elements per mask, upsampled by a factor of 12 (to avoid pixel level adversarial effects). We found that filling the masks with a blurred version of the image performed quantitatively better on the inpainting game than using grey masks.

Numerical Gradient. Finally, given the probe image which has been masked with the sparse mask sampled from the non-uniform prior, we can compute a numerical gradient of the triplet loss. Let p be an embedding of the probe, m the mated image embedding, n the nonmated image embedding, and $\hat{p}$ the masked probe embedding. Then, the numerical gradient of the triplet loss (Eq. 1) can be approximated as:

$$\frac{\partial L_{dise}}{\partial p} \approx \mathtt{max}(0, \ L(p, m, n) - L(\hat{p}, m, n)) \tag{3}$$

The numerical gradient is an approximation to the true loss gradient computed by perturbing the input by occluding the probe with a pixel mask, and computing the corresponding change in the triplet loss. In other words, when the probe masks out a region that increases for the similarity between the probe and mate and decreases for the probe and nonmate, the numerical gradient should be large. This allows for a loss weighted accumulation of these masks into a saliency map. The final saliency is accumulated following (Eq. 2), where saliency maps S_i are the pairwise binary masks, with non-negative gradient weights (Eq. 3).

4 Experimental Protocol

Recent explainable AI research has focused on class activation maps [4,12,23, 31,33,39,42], which visualize salient regions used for classification. For facial recognition, prior work has shown this is almost always the eyes, nose, and upper lip of the face [6]. In facial identification, a probe image is given to a face matching system, which returns the top K identities from a gallery. A natural question is why the matching system picked the top match instead of the second top match (or remaining top K matches). One way to give an answer to this question is to highlight the region(s) that match a given identity more than the second identity or other identities. This saliency map should be larger for the regions that contribute the most to the identity and not others. In this paper,

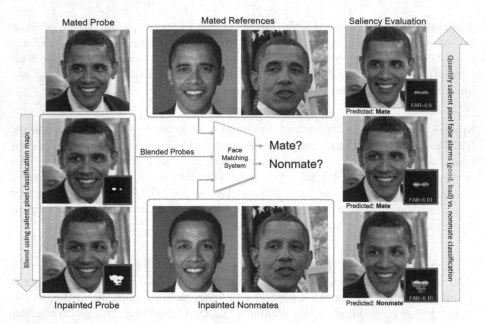

Fig. 4. Inpainting game overview. The XFR algorithm is given triplets (probe, mate, nonmate) labeled in the figure as (mated probe, mated references, inpainted non-mates), and is tasked with estimating a discriminative saliency map that estimates the likelihood that a pixel belongs to a region that is discriminative for the mate. A threshold is applied to the saliency map to classify each pixel as being discriminatively salient (inset, blue squares, left). A high performing XFR algorithm will correctly classify the discriminative pixels within the inpainted region (green, right) while avoiding classifying the pixels that are identical between the mated references and the probes as being discriminative (red, right). See Sect. 4.1 for more details. (Color figure online)

our goal is to highlight the regions that are responsible for matching a given image to one identity versus a similar identity.

A key challenge for evaluating the performance of an XFR approach is generating ground truth. For XFR, ground truth not only depends on the selection of probes, mates, and nonmates, but can also depend on a target network for evaluation. We address this issue by synthesizing inpainted nonmates or *doppelgangers*, where a select region of the face is changed from the original identity. Only the inpainted region differs between the two images and therefore only the inpainted region can be used to discriminate between them. Furthermore, we synthesize doppelgangers based on their ability to reduce the match score for a target network. We call our overall strategy for quantitative evaluation the *inpainting game*.

4.1   The Inpainting Game

An overview of the inpainting game evaluation is given in Fig. 4. The inpainting game uses four (or more) images for each evaluation: a probe image, mate image(s), an inpainted probe and inpainted nonmate(s). The inpainted probe or *probe doppelganger* is subtly different from the probe in a fixed region of the face, such as the eyes, nose or mouth. Similarly, the inpainted nonmate or *mate doppelganger* is subtly different from the mate image, such that the doppelgangers are a different identity. The inpainted probe and inpainted nonmate are constrained to be the same new identity. Sect. 4.2 discusses the construction of this dataset.

The XFR algorithm is given triplets of probes, mates and nonmates, labeled in Fig. 4 as ("mated probe", "mated references" and "inpainted non-mates"). For each triplet, the XFR algorithm is tasked with estimating the likelihood that each pixel belongs to a region that is discriminative for matching the probe to the mated identity over the nonmated/inpainted identity. These discriminative pixel estimations form a saliency map. Each pixel is classified as being discriminatively salient by applying a threshold, which forms a binary saliency map. For each binary saliency map, pixels in the probe are replaced with the pixels from an inpainted probe forming a blended probe. The inpainted probe is generating by inpainting the same facial region as the inpainted nonmates and is not provided to the XFR algorithm, which is sequestered and used for evaluation only. The saliency map is evaluated by how quickly it can flip the identity of the blended probe from the mate to the non-mate, while maximizing saliency (green) in ground truth (grey) while minimizing false alarms (red). See Sect. 4.3 for additional details, including the metrics for the inpainting game.

4.2   Inpainting Dataset for Facial Recognition

The inpainting dataset for face recognition is based on the images from the IJB-C dataset [20]. The inpainting dataset contains 561 images of 95 subjects selected from IJB-C, for an average of 5.9 images per subject. We defined eight facial regions for evaluation: 1) cheeks and jaw, 2) mouth, 3) nose, 4) left eye, 5) right eye, 6) eyebrows, and 7) left face, 8) right face. Each image is inpainted for each of the eight regions forming a total of 4488 inpainted doppelgangers. From this set, we define 3648 triplets, such that each triplet is a combination of (probe, mate and inpainted doppelganger nonmate). The XFR algorithms should not be evaluated on triplets for networks that cannot distinguish the original and inpainted identities. Hence, the only the triplets that contain discriminable identities are included for the network the algorithm is being evaluated on.

The inpainted doppelgangers are generated as follows. In order to systematically mask the regions, we use the pix2face algorithm [7] to fit a 3d face mesh onto each facial image. We then projected the facial region masks onto the images. We use pluralistic inpainting [40] to synthesize an image completion in that masked region. Figure 5 shown examples of these inpainted doppelgangers.

Fig. 5. Example facial inpainted images. This montage shows the first four of eight inpainting regions (cheeks, mouth, nose, eyebrows, left face, right face, left eye, right eye), and synthesized inpainted doppelgangers images. The first seven columns show seven subjects, with the same image repeated along the column. The middle column shows a binary inpainting mask that defines the inpainting region. The last seven columns show the inpainted doppelganger image using the mask region for that row, such that the inpainted image differs from the original image only in the mask region. Observe that the identity may change subtly while looking down a column.

A key challenge of constructing the inpainting dataset is to enforce that the inpainted nonmate is in fact a different identity. Most of our inpainted images are not sufficiently different in similarity from the original mated identity for a specific network. A given triplet of (*probe*, *mate*, and *inpainted nonmate*) is only included in the dataset if a given target network can distinguish the two identities for the mate/mate doppelganger and the probe/probe doppelganger. They are required to be able to distinguish these identities both using a nearest match protocol and an verification protocol, such that the verification match threshold for a target network is calibrated to a false alarm rate of 1e−4. Specifically, each triplet has to fulfill the following criteria in order to be included in the dataset for a given network:

1. The original probe must be: (i) more similar to the original/mated identity than the corresponding inpainted/nonmated identity and (ii) correctly verified as the original/mated identity at the calibrated verification threshold.
2. The inpainted probe must be: (i) more similar to the corresponding inpainted/nonmated identity than the original identity and (ii) correctly verified as the same identity as the inpainted/nonmated identity at the calibrated verification threshold.

The inpainting dataset is filtered for each target network according to the above criteria, resulting in a dataset specific to that target network. For example, for the ResNet-101 based system, the final filtered dataset includes 84 identities and 543 triplets, filtered down from 95 identities and 3648 triplets. Lower performing networks will generally have fewer triplets satisfying the selection criteria than higher performing networks, because they will not be able to discriminate as many of the subtle changes in the inpainted probes.

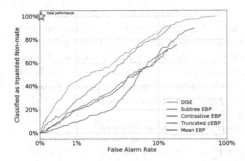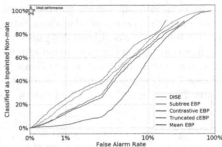

Fig. 6. (left) Inpainting game analysis using VGGFace2 ResNet-50. (right) Inpainting game analysis using Light-CNN. Refer to Table 1 for a summary of performance at fixed operating points on these curves.

4.3 Evaluation Metrics

The XFR algorithms estimate the likelihood that each pixel belongs to a region that is discriminative for matching the probe to the mated identity over the non-mated/inpainted identity. These discriminative pixel estimations form a saliency map, where the brightest pixels are estimated to be most likely to belong to the discriminative region. Figure 4 shows an example and saliency predictions at two thresholds, where the saliency prediction is shown at different thresholds as a binary mask.

In order to motivate our proposed metric, let's first consider using a classic receiver operating characteristic (ROC) curve for evaluation of the inpainting game rather than our proposed metric. A ROC curve can be generated by sweeping a threshold for the pixel saliency estimations, and computing true accept rate and false alarm rate by using the inpainted region as the positive/salient region and the non-inpainted region as the negative/non-salient region (i.e. middle column in Fig. 5). However, not all pixels within the inpainted region contribute equally to the identity, and the saliency algorithm should not be either penalized or credited with this selection.

To address this key challenge, we use mean nonmate classification rate instead of true positive rate for saliency classification. We play a game where the pixels classified as being salient by sweeping the saliency threshold are replaced with the pixels from the "inpainted probe", which is not provided to the saliency algorithm. These "blended probes" can then be classified as original identity or inpainted nonmate identity by the network being tested. High performing XFR algorithms will correctly assign more saliency for the inpainted regions that will change the identity of the blended probes without increasing the false alarm rate of the pixel salience classification. This is the key idea behind our evaluation metric. The false positive rate is calculated from salient pixel classification across all triplets, using the ground truth masks for the blended probe. The mean nonmate classification rate is weighted by the number of triplets within each facial region for a filtered dataset, to avoid bias for subprotocols with more examples. Example of the output curves for this metric is shown in Fig. 6.

Table 1. Inpainting game evaluation results. This table summarizes the performance at two operating points of false alarm rate (1E−2, 5E−2) for the performance curves in Fig. 6 (ResNet-50 and LightCNN) and in the supplementary material. Mean nonmate classification rate is the proportion of triplets where the identity of the blended image was correctly "flipped" to the doppelganger. Results show that our new methods (DISE, Subtree EBP) outperform the state of the art by a wide margin on three matchers. Detailed subprotocol results and curves are provided in the supplemental material.

| System | Saliency | Inpainting Game (Mean Nonmate Classification Rate) | | | | | | | |
| | | All | | Nose | | Eyes | | Eyebrows | |
		FAR=1E-2	FAR=5E-2	FAR=1E-2	FAR=5E-2	FAR=1E-2	FAR=5E-2	FAR=1E-2	FAR=5E-2
ResNet-101	DISE	**0.438**	**0.816**	0.691	0.932	**0.627**	**0.931**	**0.723**	**0.981**
	Subtree EBP	0.274	0.792	**0.740**	**0.973**	0.503	**0.931**	0.065	0.942
	Mean EBP	0.143	0.626	0.208	0.904	0.565	**0.931**	0.010	0.781
	Contrastive EBP	0.132	0.454	0.178	0.589	0.310	0.517	0.040	0.543
	Truncated cEBP	0.167	0.582	0.247	0.699	0.276	0.573	0.066	0.642
VGGFace2 ResNet-50	DISE	**0.443**	**0.761**	**0.730**	0.902	**0.609**	**0.957**	**0.891**	**0.976**
	Subtree EBP	0.285	0.735	0.705	**0.984**	0.418	0.878	0.332	0.928
	Mean EBP	0.092	0.499	0.148	0.705	0.388	0.831	0.108	0.821
	Contrastive EBP	0.195	0.520	0.343	0.705	0.484	0.652	0.323	0.850
	Truncated cEBP	0.205	0.536	0.361	0.754	0.539	0.696	0.329	0.868
LightCNN NiN+ MFM-28	DISE	0.202	0.587	**0.729**	**0.961**	**0.409**	**0.847**	**0.641**	**0.927**
	Subtree EBP	**0.250**	**0.643**	0.699	0.914	0.294	0.778	0.489	0.921
	Mean EBP	0.027	0.307	0.048	0.477	0.085	0.600	0.026	0.558
	Contrastive EBP	0.121	0.526	0.286	0.821	0.211	0.533	0.287	0.842
	Truncated cEBP	0.135	0.557	0.294	0.820	0.209	0.576	0.298	0.892

5   Experimental Results

5.1   Inpainting Game Quantitative Evaluation

We ran the inpainting game evaluation protocol on the inpainting dataset using three target networks: LightCNN [35], VGGFace2 ResNet-50 [5] and a custom trained ResNet-101. We considered the five XFR algorithms described in Sect. 3 forming the benchmark for XFR evaluation.

The evaluation results are summarized in Table 1 and plotted in Fig. 6 and in the supplementary material. The summary table shows for each combination of network and XFR algorithm, at two false alarm rates (1E−2, 5E−2) for the full protocol and three subprotocols: eyes, nose and eyebrows only. Additional results in the supplementary material show the results for the individual facial region subprotocols.

Overall, results show that for deeper networks (ResNet-101, ResNet-50), the top performing XFR algorithm is DISE. However, for shallower networks (LightCNN) then top performing algorithm is Subtree EBP. Both of these new approaches outperform the state of the art (EBP, cEBP, tcEBP) by a wide margin. We believe that DISE outperforms Subtree EBP since subtree EBP cannot localize image regions any better than the underlying network represents faces. For example, consider the eyebrows subprotocol result in the supplementary material, which shows that subtree EBP cannot represent eyebrows independently from the eyes. DISE can mask image regions independently from the underlying target network and correctly localize eyebrow effects.

6 Conclusions

In this paper, we introduced the first comprehensive benchmark for XFR. We motivated the need for XFR and describe a new quantitative method for comparing XFR algorithms using the inpainting game. The results show that the DISE and subtree EBP methods provide a significant performance improvement over the state of the art, which provides a new baseline for visualizing discriminative features for face recognition. This evaluation protocol provides a means to compare different approaches to network saliency, and we believe this form of quantitative evaluation will help encourage research in this emerging area of explainable AI for face recognition. All software and datasets for reproducible research are available for download at http://stresearch.github.io/xfr.

Acknowledgement. This research is based upon work supported by the Office of the Director of National Intelligence (ODNI), Intelligence Advanced Research Projects Activity (IARPA) under contract number 2019-19022600003. The views and conclusions contained herein are those of the authors and should not be interpreted as necessarily representing the official policies or endorsements, either expressed or implied, of ODNI, IARPA, or the U.S. Government. The U.S. Government is authorized to reproduce and distribute reprints for Governmental purpose notwithstanding any copyright annotation thereon.

References

1. Bach, S., Binder, A., Montavon, G., Klauschen, F., Muller, K.-R., Samek, W.: On pixel-wise explanations for non-linear classifier decisions by layer-wise relevance propagation. PLoS ONE **10**(7), e0130140 (2015)
2. Bau, D., Zhou, B., Khosla, A., Oliva, A., Torralba, A.: Network dissection: quantifying interpretability of deep visual representations. In: 2017 IEEE Conference on Computer Vision and Pattern Recognition (CVPR), pp. 3319–3327 (2017)
3. Buolamwini, J., Gebru, T.: Gender shades: intersectional accuracy disparities in commercial gender classification. In: Friedler, S.A., Wilson, C., (eds) Proceedings of the 1st Conference on Fairness, Accountability and Transparency, Proceedings of Machine Learning Research, New York, NY, USA, 23–24 February 2018, vol. 81, pp. 77–91. PMLR (2018)
4. Cao, C., et al.: Look and think twice: capturing top-down visual attention with feedback convolutional neural networks. In: Proceedings of the IEEE International Conference on Computer Vision, pp. 2956–2964 (2015)
5. Cao, Q., Shen, L., Xie, W., Parkhi, O.M., Zisserman, A.: VGGFace2: a dataset for recognising faces across pose and age. In: International Conference on Automatic Face and Gesture Recognition (2018)
6. Castanon, G., Byrne., J.: Visualizing and quantifying discriminative features for face recognition. In: International Conference on Automatic Face and Gesture Recognition (2018)

7. Crispell, D., Bazik, M.: Pix2Face: direct 3D face model estimation. In: 2017 IEEE International Conference on Computer Vision Workshop (ICCVW), pp. 2512–2518, October 2017

8. Dabkowski, P., Gal, Y.: Real time image saliency for black box classifiers. In: Advances in Neural Information Processing Systems, pp. 6967–6976 (2017)

9. Dhar, P., Bansal, A., Castillo, C.D., Gleason, J., Phillips, P.J., Chellappa, R.: How are attributes expressed in face DCNNs? ArXiv, abs/1910.05657 (2019)

10. Duchaine, B., Nakayama, K.: The Cambridge face memory test: results for neurologically intact individuals and an investigation of its validity using inverted face stimuli and prosopagnosic participants. Neuropsychologia **44**, 576–585 (2006)

11. Facial Identification Scientific Working Group. FISWG Guidelines for Facial Comparison Methods. In: FISWG Standards Version 1.0 - 2012–02-02 (2012)

12. Fong, R., Vedaldi, A.: Interpretable Explanations of Black Boxes by Meaningful Perturbation. arXiv preprint (2017)

13. Garvie, C., Bedoya, A., Frankle, J.: The perpetual line-up: unregulated police face recognition in America. Technical Report, Georgetown University Law School (2018)

14. Grimm, C., Arumugam, D., Karamcheti, S., Abel, D., Wong, L.L., Littman, M.L.: Latent attention networks. arXiv:1706.00536v1 (2017)

15. Grother, P., Ngan, M., Hanaoka, K.: Face recognition vendor test (FRVT) Part 3: demographic effects. In: NISTIR 8280 (2019)

16. Hu, R., Andreas, J., Darrell, T., Saenko, K.: Explainable neural computation via stack neural module networks. In: ECCV (2018)

17. Kindermans, P.-J., et al.: Learning how to explain neural networks: patternnet and pattern attribution. arXiv preprint arXiv:1705.05598 (2017)

18. Li, H., Mueller, K., Chen, X.: Beyond saliency: understanding convolutional neural networks from saliency prediction on layer-wise relevance propagation. Image Vis. Comput. **83–84**, 70–86 (2017)

19. Mahendran, A., Vedaldi, A.: Understanding deep image representations by inverting them. In: Proceedings of the IEEE Conference on Computer Vision and Pattern Recognition (CVPR) (2015)

20. Maze, B., et al.: IARPA Janus benchmark-c: face dataset and protocol. In: 2018 International Conference on Biometrics (ICB), pp. 158–165. IEEE (2018)

21. Nguyen, A.M., Dosovitskiy, A., Yosinski, J., Brox, T., Clune, J.: Synthesizing the preferred inputs for neurons in neural networks via deep generator networks. In: NIPS (2016)

22. Parkhi, O., Vedaldi, A., Zisserman, A.: Deep face recognition. In: BMVC (2015)

23. Petsiuk, V., Das, A., Saenko, K.: Rise: randomized input sampling for explanation of black-box models. In: British Machine Vision Conference (BMVC) (2018)

24. Phillips, P.J., et al.: Face recognition accuracy of forensic examiners, superrecognizers, and face recognition algorithms. In: Proceedings of the National Academy of Sciences of the United States of America (2018)

25. Raji, I.D., Buolamwini, J.: Actionable auditing: investigating the impact of publicly naming biased performance results of commercial AI products. In: AIES 2019 (2019)

26. Ribeiro, M.T., Singh, S., Guestrin, C.: Why should i trust you?: Explaining the predictions of any classifier. In: KDD 2016 (2016)

27. RichardWebster, B., Kwon, S.Y., Clarizio, C., Anthony, S.E., Scheirer, W.J.: Visual psychophysics for making face recognition algorithms more explainable. In: European Conference on Computer Vision (ECCV) (2018)

28. Samek, W., Binder, A., Montavon, G., Lapuschkin, S., Müller, K.-R.: Evaluating the visualization of what a deep neural network has learned. IEEE Trans. Neural Netw. Learn. Syst. **28**, 2660–2673 (2015)
29. Samek, W., Montavon, G., Vedaldi, A., Hansen, L.K., Müller, K. (eds.): Explaining and Visualizing Deep Learning Explainable AI: Interpreting. Springer, Heidelberg (2019). https://doi.org/10.1007/978-3-030-28954-6
30. Schroff, F., Kalenichenko, D., Philbin, J.: FaceNet: a unified embedding for face recognition and clustering. In CVPR (2015)
31. Selvaraju, R.R., Cogswell, M., Das, A., Vedantam, R., Parikh, D., Batra, D.: Grad-CAM: visual explanations from deep networks via gradient-based localization. In: 2017 IEEE International Conference on Computer Vision (ICCV), pp. 618–626 (2016)
32. Simonyan, K., Vedaldi, A., Zisserman, A.: Deep inside convolutional networks: visualising image classification models and saliency maps. CoRR, abs/1312.6034 (2013)
33. Simonyan, K., Vedaldi, A., Zisserman, A.: Deep inside convolutional networks: visualising image classification models and saliency maps. In: ICLR, p. 1 (2014)
34. Stylianou, A., Souvenir, R., Pless, R.: Visualizing deep similarity networks. In: 2019 IEEE Winter Conference on Applications of Computer Vision (WACV), pp. 2029–2037. IEEE (2019)
35. Wu, X., He, R., Sun, Z., Tan, T.: A light CNN for deep face representation with noisy labels. IEEE Trans. Inf. Forensics Secur. **13**(11), 2884–2896 (2018)
36. Xu, T., et al.: Deeper interpretability of deep networks. ArXiv, abs/1811.07807 (2018)
37. Yin, B., Tran, L., Li, H., Shen, X., Liu, X.: Towards interpretable face recognition. In: Proceeding of International Conference on Computer Vision, Seoul, South Korea, October 2019
38. Zee, T., Gali, G., Nwogu, I.: Enhancing human face recognition with an interpretable neural network. In: The IEEE International Conference on Computer Vision (ICCV) Workshops, October 2019
39. Zhang, J., Lin, Z., Brandt, J., Shen, X., Sclaroff, S.: Top-down neural attention by excitation backprop. In: Lecture Notes in Computer Science (including subseries Lecture Notes in Artificial Intelligence and Lecture Notes in Bioinformatics), vol. 9908, pp. 543–559 (2016)
40. Zheng, C., Cham, T.-J., Cai, J.: Pluralistic image completion. In: Proceedings of the IEEE Conference on Computer Vision and Pattern Recognition, pp. 1438–1447 (2019)
41. Zhong, Y., Deng, W.: Exploring features and attributes in deep face recognition using visualization techniques. In: IEEE International Conference on Automatic Face and Gesture Recognition (FG 2019) (2019)
42. Zhou, B., Khosla, A., Lapedriza, A., Oliva, A., Torralba, A.: Learning deep features for discriminative localization. In: CVPR (2016)

From Shadow Segmentation to Shadow Removal

Hieu Le[(⊠)] and Dimitris Samaras

Stony Brook University, Stony Brook, NY 11794, USA
hle@cs.stonybrook.edu

Abstract. The requirement for paired shadow and shadow-free images limits the size and diversity of shadow removal datasets and hinders the possibility of training large-scale, robust shadow removal algorithms. We propose a shadow removal method that can be trained using only shadow and non-shadow patches cropped from the shadow images themselves. Our method is trained via an adversarial framework, following a physical model of shadow formation. Our central contribution is a set of physics-based constraints that enables this adversarial training. Our method achieves competitive shadow removal results compared to state-of-the-art methods that are trained with fully paired shadow and shadow-free images. The advantages of our training regime are even more pronounced in shadow removal for videos. Our method can be fine-tuned on a testing video with only the shadow masks generated by a pre-trained shadow detector and outperforms state-of-the-art methods on this challenging test. We illustrate the advantages of our method on our proposed video shadow removal dataset.

Keywords: Shadow removal · GAN · Weakly-supervised · Illumination model · Unpaired · Image-to-image

1 Introduction

Shadows are present in most natural images. Shadow effects make objects harder to detect or segment [23], and scenes with shadows are harder to process and analyze [20]. Realistic shadow removal is an integral part of image editing [3] and can greatly improve performance on various computer vision tasks [21,24, 32,41,56], getting increased attention in recent years [11,13,37]. Data-driven approaches using deep learning models have achieved remarkable performance on shadow removal [5,15,17,22,47,55] thanks to recent large-scale datasets [45,47].

Most of the current deep-learning shadow removal approaches are end-to-end mapping functions trained in a fully supervised manner. Such systems require pairs of shadow images and their shadow-free counter-parts as training signals.

Electronic supplementary material The online version of this chapter (https://doi.org/10.1007/978-3-030-58621-8_16) contains supplementary material, which is available to authorized users.

© Springer Nature Switzerland AG 2020
A. Vedaldi et al. (Eds.): ECCV 2020, LNCS 12356, pp. 264–281, 2020.
https://doi.org/10.1007/978-3-030-58621-8_16

However, this type of data is cumbersome to obtain, lacks diversity, and is error-prone: all current shadow removal datasets exhibit color mismatches between the shadow images and their shadow-free ground truth (see Fig. 1-left panel). Moreover, there are no images with self-cast shadows because the occluders are never visible in the image in the current data acquisition setups [15,37, 47]. This dependency on paired data significantly hinders building large-scale, robust shadow-removal systems. A recent method trying to overcome this issue is MaskShadow-GAN [15], which learns shadow removal from unpaired shadow and shadow-free images. However, such cycle-GAN [58] based systems usually require enough statistical similarity between the two sets of images [2,25]. This requirement can be hard to satisfy when capturing shadow-free images is tricky, such as shadow-free images of urban areas [4] or moving objects [18,36].

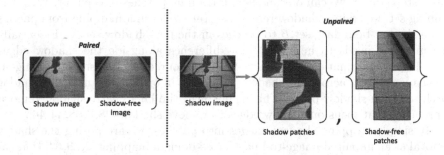

Fig. 1. Paired training data (left) consists of training examples {shadow, shadow-free} images which are expensive to collect, lack diversity, and are sensitive to errors due to possible color mismatches between the two images. Note the slightly different color tone between the two images. In this paper, we propose to learn shadow removal from unpaired shadow and non-shadow patches cropped from the same shadow image (right). This eliminates the need for shadow free images.

In this paper, we propose an alternative solution to the data dependency issue. We first observe that image patches alongside the shadow boundary contain critical information for shadow removal, including non-shadow, umbra and penumbra areas. They sufficiently reflect the characteristics of the shadowing effects, including the color differences between shadow and non-shadow areas as well as the gradual changes of the shadow effects across the shadow boundary [14,33,34]. If we further assume that the shadow effects are fairly consistent in the umbra areas, a patch-based shadow removal can be used to remove shadows in the whole image. Based on this observation, we propose training a patch-based shadow removal system for which we use unpaired shadow and non-shadow patches directly cropped from the shadow images themselves as training data. This approach eliminates the dependency on paired training data and opens up the possibility of handling different types of shadows, since it can be trained with any kind of shadow image. Compared to MaskShadow-GAN, shadow and non-shadow patches cropped from the same image naturally ensure significant statistical similarity. The only supervision required in this data processing scheme are the shadow masks, which are relatively easy to obtain, either manually, semi-interactively [11,45], or automatically using shadow detection methods [5,23,57,59]. Automatic shadow detection is improving, with the main challenge being generalization across datasets. At some point, one can expect to get very good shadow masks automatically, which would allows training our shadow removal method with very little annotation cost.

In particular, to obtain shadow and shadow-free patches, we crop the shadow images into small overlapping patches of size $n \times n$ with a step size of m. Based on the shadow masks, we group these patches into three sets: a non-shadow set ($\mathcal{N}$) containing patches having no shadow pixels, a shadow-boundary set ($\mathcal{B}$) containing patches lying on the shadow boundaries, and a full-shadow set ($\mathcal{F}$) containing patches where all pixels are in shadow. With small enough patch size n and step size m, we can obtain enough training patches in each set. With this training set, we train a shadow removal system to learn a mapping from patches in the shadow-boundary set $\mathcal{B}$ to patches in the non-shadow set $\mathcal{N}$. Essentially, this mapping needs to infer the color difference alongside the shadow edges, including the chromatic attributes of the light source and the smooth change of the shadow effects across the shadow boundary, in order to transform a shadow patch to a non-shadow patch. This is, in spirit, similar to early shadow removal approaches that focus on shadow edges to remove shadows [8,9,38,44,46].

By simply cropping shadow images into patches, we are posing the shadow removal as an unpaired image-to-image cross-domain mapping [2,29,54] that can be estimated via an adversarial framework. In particular, we seek a mapping function G that takes as input a shadow-boundary patch x from the set $\mathcal{B}$, and outputs an image patch $\hat{x}$, such that a critic function D cannot distinguish whether $\hat{x}$ was drawn from the non-shadow set $\mathcal{N}$ or generated by G. Note that one potential solution here is to use Cycle-GAN or MaskShadow-GAN to estimate this transformation. However, the mapping functions learned by these methods are not able to remove shadows from patches in the full-shadow set $\mathcal{F}$.

Training such an unpaired image-to-image mapping for shadow removal is challenging. The mapping is under-constrained and training can collapse easily. [12,27,28,30,31,42]. Here, we propose to systematically constrain the shadow removal process by a physical model of shadow formation [39] and incorporate a number of physical properties of shadows into the framework. We show that these physics-based priors define a transformation closely modelling shadow removal. Driven by an adversarial signal, our framework effectively learns physically-plausible shadow removal without any direct supervision from paired data. Specifically, we constrain the shadow removal process to a shadow image decomposition model [22] that extracts a set of parameters and a matting layer from the shadow image. This set of shadow parameters is responsible for removing shadows on the umbra areas of the shadows via a linear function. Thus, once we estimate these shadow parameters from shadow boundary patches, we can use them to remove shadows for patches fully covered by the same shadow under the assumption that they share the same set of shadow parameters. Based on the physical properties of shadows, we apply the following constraints to the model:

- We limit the search space of the shadow parameters and shadow matte to the appropriate value ranges that correspond to shadow removal.
- Our matting and smoothness losses ensure that shadow removal only happens in the shadow areas and transitions smoothly across shadow boundaries.
- Our boundary loss on the generated shadow-free image enforces color similarity between the inner and outer areas alongside shadow boundaries.

With these constraints and the adversarial signal, our method achieves shadow removal results that are competitive with state-of-the-art methods that were trained in a fully supervised manner with paired shadow and non-shadow images [22,37,47]. We further compare our method to state-of-the-art methods on a novel and challenging video shadow removal dataset including static videos with various scenes and shadow conditions. This test exposes the weaknesses of data-driven methods trained on datasets lacking diversity. Our patch-based method seems to generalize better than other methods when evaluated on this video shadow removal test. Most importantly, we can easily fine-tune our pre-trained model on a single testing video to further improve shadow removal results, showcasing this advantage of our training scheme.

In short, our contributions are:

- We propose the use of an adversarial critic to train a shadow remover from unpaired shadow and non-shadow patches, providing an alternative solution to the paired data dependency issue.
- We propose a set of physics-based constraints that define a transformation closely modelling shadow removal, which enables shadow remover training with only an adversarial training signal.
- Our system trained without any shadow-free images has competitive results compared to fully-supervised state-of-the-art methods on the ISTD dataset.
- We collect a novel video shadow removal dataset. Our shadow removal system can be fine-tuned for free to better remove shadows on testing videos.

2 Related Work

Shadows are physical phenomena. Early shadow removal works, without much training data, usually focused on studying different physical shadow properties [1,6–10,26,53]. Many works look for cues to remove shadows starting from shadow edges. Finlayson et al.[9] used shadow edges to estimate a scaling factor that differentiates shadow areas from their non-shadow counterparts. Wu & Tang [51] imposed a smoothness constraint alongside the shadow boundaries to handle penumbra areas. Wu et al.[50] detected strong shadow-edges to remove shadows on the whole image. Shor & Lischinki [39] defined an affine relationship between shadow and non-shadow pixels where they used the areas surrounding the shadow edges to estimate the parameters of such affine transforms.

Shadow boundary effects can also be modeled via image matting [14]. Wu *et al.*[52] estimated a matte layer representing the pixel-wise shadow probability to estimate a color transfer function to remove shadows. Chuang *et al.* [3] computed a shadow matte from video for shadow editing. They computed the lit and shadow images by finding min-max values at each pixel location throughout all frames of a video captured by a static camera. We use this technique to create a video dataset for testing shadow removal methods in Sect. 4.4.

Current shadow removal methods [5,17,22,47,55] use deep-learning models trained with full supervision on large-scale datasets [37,47] of paired shadow and shadow-free images. Pairs are obtained by taking a photo with shadows, then removing the occluders from the scene to take the photo without shadows. Deshadow-Net [37] extracted multi-context features to predict a matte layer that removes shadows. Some works use adversarial frameworks to train their shadow removal. In [47] a unified adversarial framework predicted shadow masks and removed shadows. Similarly, Ding *et al.*[5] used an adversarial signal to improve shadow removal in an iterative manner. Note that these methods use the shadow-free image as the main training signal while our method is trained only through an adversarial loss. In prior work [22] we constrained shadow removal by a physical model of shadow formation. We trained networks to extract shadow parameters and a matte layer to remove shadows. We adapt this model to patch-based shadow removal. Note that in [22], all shadow parameters and matting layers were pre-computed using paired training images and the network was trained to simply regress those values, whereas our model automatically estimates them through adversarial training. MaskShadow-GAN [17] is the only deep-learning method that learns shadow removal from just unpaired training data.

3   Method

We describe our patch-based shadow removal in Sect. 3.1. Our whole image pipeline for shadow removal is described in Sect. 3.2. For both image-level and patch-level shadow removal, we use shadow matting [3,35,40,49] to express a shadow-free image $I^{\text{shadow-free}}$ by:

$$I^{\text{shadow-free}} = I^{\text{relit}} \cdot \alpha + I^{\text{shadow}} \cdot (1 - \alpha) \tag{1}$$

with I^{shadow} the shadow image, α the matting layer, and I^{relit} the relit image. The relit image contains shadow pixels relit to their non-shadow values, computed via a linear function following a physical shadow formation model [22,39]:

$$I_i^{\text{relit}} = w \cdot I_i^{\text{shadow}} + b \tag{2}$$

The unknown factors in this shadow matting formula are the set of shadow parameters (w, b) which define the linear function that removes the shadow effects in the umbra areas of the shadow, and the matte layer α that models the shadow effects on the shadow boundaries. We train a system of three networks to estimate these unknown factors via adversarial training. We use the annotated shadow segmentation masks for training. For testing, we obtain a segmentation mask for the image using the shadow detector proposed by Zhu *et al.* [59].

3.1 Patch-Based Shadow Removal

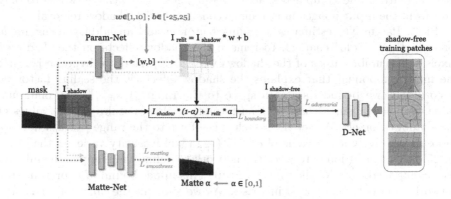

Fig. 2. Weakly-supervised shadow decomposition. Our framework consists of three networks: Param-Net, Matte-Net, and D-Net. Param-Net and Matte-Net predict the shadow parameters (w, b) and the matte layer α respectively to jointly remove the shadow. Param-Net takes as input the input image patch and its shadow mask to predict three sets of shadow parameters (w, b) for the three color channels, which is used to obtain a relit image. The input image patch, shadow mask, and relit image are input into Matte-Net to predict a matte layer. D-Net is the critic function distinguishing between the generated image patches and the real shadow-free patches. The only supervision signal is the set of shadow-free patches. The four losses guiding this training are the matting loss, smoothness loss, boundary loss, and adversarial loss. (Color figure online)

Figure 2 summarizes our framework to remove shadows from a single image patch, which consists of three networks: Param-Net, Matte-Net, and D-Net. Param-Net and Matte-Net predict the shadow parameters (w, b) and the matte layer α respectively to jointly remove shadows. D-Net is the critic distinguishing between the generated image patches and the real shadow-free patches. With Param-Net and Matte-Net being the generators and D-Net being the discriminator, the three networks form an adversarial training framework where the main source of training signal is the set of shadow-free patches.

In theory, as D-Net is trained to distinguish patches containing shadow boundaries from patches without any shadows, a natural solution to fool D-Net is to remove the shadows in the input shadow patches to make them indistinguishable from shadow-free patches. However, such an adversarial signal from D-Net alone often cannot guide the generators, (Param-Net and Matte-Net) to actually remove shadows. The parameter search space is very large and the mapping is extremely under-constrained. In practice, we observe that without any constraints, Param-Net tends to output consistently high values of (w, b) as they would directly increase the overall brightness of the image patches, and Matte-Net tends to introduce artifacts similar to visual patterns frequently appearing in the non-shadow areas. Thus, our main idea is to constrain this framework with physical shadow properties. Constraining the output shadow parameters, shadow mattes, and combined shadow-free images, forces the networks to only transform the input images in a manner consistent with shadow removal.

First, Param-Net estimates a scaling factor w and an additive constant b, for each R,G,B color channel, to remove the shadow effects on the shadowed pixels in the umbra areas of the shadows via Eq. (2). Here we hypothesize that the main component that explains the shadow effects is the scaling factor w. Accordingly, we bound its search space to the range $[1; s_{max}]$. The minimum value of $w = 1$ ensures that the transformation always scales up the values of the shadowed pixels. We set the search space for b to the range $[-c, c]$ where we choose a relatively small value of $c = 25$ (the pixel intensity varies in the range $[0, 255]$). Our intuition is to force the network to define the mapping mainly via the scaling factor w. We choose $s_{max} = 10$. This upper bound of w prevents the network from collapsing as w increases. As we show in the ablation study, the network fails to learn a shadow removal without proper search space limitation.

Matte-Net estimates a blending layer α that combines the shadow image patch and the relit image patch into a shadow-free image patch via Eq. 1. The value of a pixel i in the output image patch, I_i^{output}, is computed as:

$$I_i^{output} = I_i^{relit} \cdot \alpha_i + I_i^{shadow} \cdot (1 - \alpha_i) \qquad (3)$$

We map the output of Matte-Net to [0,1] as α is being used as a matting layer and constrain the value of α_i as follows:

- If i indicates a non-shadow pixel, we enforce $\alpha_i = 0$ so that the value of the output pixel I_i^{output} equals its value in the input image I_i^{shadow}.
- If i indicates a pixel in the umbra areas of the shadows, we enforce $\alpha_i = 1$ so that the value of the output pixel I_i^{output} equals its relit value I_i^{relit}.
- We do not control the value of α in the penumbra areas of the shadows and rely on the training of the network to estimate these values.

where the umbra, non-shadow or penumbra areas can be roughly specified using the shadow mask. We define two areas alongside the shadow boundary, denoted as $\mathcal{M}_{in}$ and $\mathcal{M}_{out}$ - see Fig. 3. $\mathcal{M}_{out}$ is the area right outside the boundary, computed by subtracting the shadow mask, $\mathcal{M}$, from its dilated version $\mathcal{M}_{dilated}$. The inside area $\mathcal{M}_{in}$ is computed similarly by subtracting an eroded shadow mask from the shadow mask. These two areas $\mathcal{M}_{in}$ and $\mathcal{M}_{out}$ roughly define a small area surrounding the shadow boundary, which can be considered as the penumbra area of the shadow. Then the above constraints are implemented as the matting loss $\mathcal{L}_{mat-\alpha}$ computed by the following formula for every pixel i:

$$\mathcal{L}_{mat-\alpha} = \sum_{i \in (\mathcal{M} - \mathcal{M}_{in})} |\alpha_i - 1| + \sum_{i \notin \mathcal{M}_{dilated}} |\alpha_i| \qquad (4)$$

Moreover, since the shadow effects are assumed to vary smoothly across the shadow boundaries, we enforce an $L1$ smoothness loss on the spatial gradients

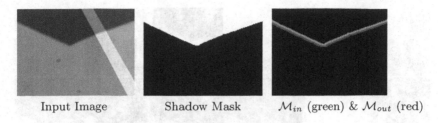

| Input Image | Shadow Mask | $\mathcal{M}_{in}$ (green) & $\mathcal{M}_{out}$ (red) |

Fig. 3. The penumbra area of the shadow. We define two areas alongside the shadow boundary, denoted as $\mathcal{M}_{in}$ (shown in green) and $\mathcal{M}_{out}$ (shown in red). These two areas roughly define a small region surrounding the shadow boundary, which can be considered as the penumbra area of the shadow. (Color figure online)

of the matte layer, α. This smoothness loss $\mathcal{L}_{sm}$ also prevents Matte-Net from producing undesired artifacts since it enforces local uniformity. This loss is:

$$\mathcal{L}_{sm-\alpha} = |\nabla \alpha| \tag{5}$$

Then, given a set of estimated parameters (w, b) and a matte layer α, we obtain an output image I^{output} via the image decomposition formula (1). We penalize the $L1$ difference between the average intensity of pixels lying right outside and inside the shadow boundary, which are the two areas $\mathcal{M}_{in}$ and $\mathcal{M}_{out}$. This shadow boundary loss $\mathcal{L}_{bd}$ is computed by:

$$\mathcal{L}_{bd} = \left| \frac{\sum_{i \in \mathcal{M}_{in}} I_i^{output}}{\sum_{i \in \mathcal{M}_{in}}} - \frac{\sum_{i \in \mathcal{M}_{out}} I_i^{output}}{\sum_{i \in \mathcal{M}_{out}}} \right| \tag{6}$$

Last, we compute the adversarial loss with the feedback from D-Net:

$$\mathcal{L}_{GAN} = \log(1 - D(I^{output})) \tag{7}$$

where $D(\cdot)$ denotes the output of D-Net.

The final objective function to train Param-Net and Matte-Net is to minimize a weighted sum of the above losses:

$$\mathcal{L}_{final} = \lambda_{sm}\mathcal{L}_{sm-\alpha} + \lambda_{mat}\mathcal{L}_{mat-\alpha} + \lambda_{bd}\mathcal{L}_{bd} + \lambda_{adv}\mathcal{L}_{GAN} \tag{8}$$

All these losses are essential for training our networks, as shown in our ablation study in Sect. 4.3. By using all the proposed losses together, our method is able to automatically extract a set of shadow parameters and an α layer from an input image patch. Figure 4 visualizes the components extracted from our framework for two challenging input patches. In the first row, a dark shadow area is lit correctly to its non-shadow value. In the second row, the matte layer α is not affected by the dark material of the surface.

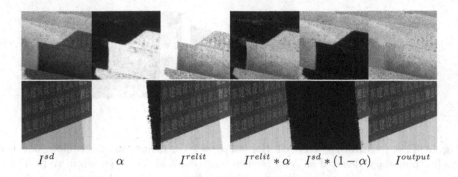

I^{sd} α I^{relit} $I^{relit} * \alpha$ $I^{sd} * (1 - \alpha)$ I^{output}

Fig. 4. Weakly-supervised shadow image decomposition. With only shadow mask supervision, our method automatically learns to decompose the shadow effect in the input image patch I^{sd} into a matte layer α and a relit image I^{relit}. The matte layer α combines I^{sd} and I^{relit} to obtain a shadow-free image patch I^{output} via Eq. (1).

3.2   Image Shadow Removal Using a Patch-Based Model

We estimate a set of shadow parameters and a matte layer for the input image to remove shadows via Eq. (1). First, we obtain a shadow mask using the shadow detector of Zhu *et al.* [59]. We crop the input shadow image into overlapping patches. All patches containing the shadow boundaries are then input into the three networks. We approximate the whole image shadow parameters from the patch shadow parameters, under the assumption that they share the same or very similar parameters. We simply compute the image shadow parameters as a linear combination of the patch shadow parameters. Similarly, we compute the values of each pixel in the matte layer by combining the overlapping matte patches. We set the matte layer pixels in the non-shadow area to 0 and those in the umbra area to 1. We observe that the classification scores obtained from the critic function D-Net correlate with the quality of the generated image patches. Thus, we normalize these scores to sum to 1 and use them as coefficients for the linear combinations that form the image shadow parameters and matte layer.

4   Experiments

4.1   Network Architectures and Implementation Details

We use a VGG-19 architecture for Param-Net and a U-Net architecture for Matte-Net. D-Net is a simple 5-layer convolutional network. To map the outputs of the networks to a certain range, we use Tanh functions together with scaling and additive constants. We use stochastic gradient descent with the Adam solver [19] to train our model. The initial learning rate for Matte-Net and D-Net is 0.0002 and for Param-Net is 0.00002. All networks were trained from scratch.

We experimentally set our training parameters (λ_{bd}, $\lambda_{mat-\alpha}$, $\lambda_{sm-\alpha}$, λ_{adv}) to (0.5, 100, 10, 0.5). We train our network with batch size 96 for 150 epochs[1].

We use the ISTD dataset [47] for training. Each original training image of size 640×480 is cropped into patches of size 128×128 with a step size of 32. This creates 311,220 image patches from 1,330 training shadow images. This training set includes 151,327 non-shadow patches, 147,312 shadow-boundary patches, and 12,581 full-shadow patches.

4.2 Shadow Removal Evaluation

We first evaluate our method on the adjusted testing set of the ISTD dataset [22,47]. Following previous work [14,22,37,47], we compute the root-mean-square-error (RMSE) in the LAB color space on the shadow area, non-shadow area, and the whole image, where all shadow removal results are re-sized to 256×256. Note that our method can take any size image as input. We used the Zhu et al. [59] shadow detector, pre-trained on the SBU dataset and fine-tuned on the ISTD dataset, to obtain the shadow masks for our testing, as in [22].

Table 1. Shadow removal results of our networks compared to state-of-the-art shadow removal methods on the adjusted ISTD testing set [22,47]. The metric is RMSE (the lower, the better). Best results are in bold.

Methods	Training data	Shadow	Non-shadow	All
Input Image	–	40.2	2.6	8.5
Yang et al. [53]	–	24.7	14.4	16.0
Guo et al. [14]	Shd. Free + Shd. Mask	22.0	3.1	6.1
Gong et al. [11]	–	13.3	–	–
ST-CGAN [47]	Shd. Free + Shd. Mask	13.4	7.7	8.7
DeshadowNet [37]	Shd. Free	15.9	6.0	7.6
MaskShadow-GAN [15]	Shd. Free (Unpaired)	12.4	4.0	5.3
SP+M-Net [22]	Shd. Free + Shd.Mask	**7.9**	3.1	**3.9**
Ours	Shd. Mask	9.7	**3.0**	4.0

In Table 1, we compare our weakly-supervised methods with the recent state-of-the-art methods of Guo et al. [14], Gong et al. [11], Yang et al. [53], ST-CGAN [47], DeshadowNet [37], MaskShadow-GAN [15], and SP+M-Net [22]. The second column shows the training data of each method. All other deep-learning methods require paired shadow-free images as training signal except MaskShadow-GAN, which is trained on unpaired shadow and shadow-free images from the ISTD dataset. ST-CGAN and SP+M-Net also require the training

---

[1] All code, trained models, and data are available at: https://www3.cs.stonybrook.edu/~cvl/projects/FSS2SR/index.html.

shadow masks. Our method, trained without any shadow-free image, got 9.7 RMSE on the shadow areas, which is competitive with SP+M-Net. However, SP+M-Net requires full supervision.

Our method outperforms MaskShadow-GAN by 22%, reducing the RMSE in the shadow area from 12.4 to 9.7 while also achieving lower RMSE on the non-shadow area. We outperform DeshadowNet and ST-CGAN, two methods that were trained with paired shadow and shadow-free images, reducing the RMSE by 38% and 26% respectively.

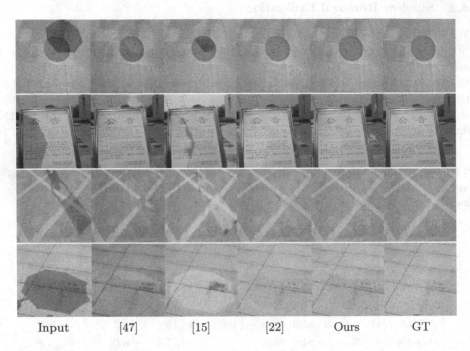

Input   [47]   [15]   [22]   Ours   GT

Fig. 5. Comparison of shadow removal on ISTD dataset. Qualitative comparison between our method and the state-of-the-art methods: ST-CGAN [47], MaskShadow-GAN [15], SP+M-Net [22]. Our method, trained without any shadow-free images, produces clean shadow-free images with very few artifacts.

Figure 5 compares qualitative shadow removal results from our method with other state-of-the-art methods on the ISTD dataset. Our method, trained with just an adversarial signal, produces clean shadow-free images with very few artifacts. On the other hand, ST-CGAN and MaskShadow-GAN tend to produce blurry images, introduce artifacts, and often relight the wrong image parts. Our method generates images which are visually similar to that of SP+M-Net. While SP+M-Net is less affected by the error in the shadow masks (shown in the 2nd row), our method generates images with more consistent colors between areas inside and outside the shadow boundaries (3rd and 4th rows). In all cases, our method preserves almost perfectly the textures beneath the shadows (last row).

4.3 Ablation Studies

We conduct ablation studies to better understand the effects of each proposed component in our framework. Starting from the original model with all the proposed features and losses, we train new models removing the proposed components one at a time. Table 2 summarizes these experiments. The first row shows the results of our model when we set the search space of the scaling factor w to $[-10, 10]$ and the search space of the additive constant b to $[-255, 255]$. In this case, the model collapses and consistently outputs uniformly dark images. Similarly, the model collapses when we omit the boundary loss $\mathcal{L}_{bd}$. We observe that this loss is essential in stabilizing the training as it prevents the Param-Net from outputting consistently high values.

The matting loss $\mathcal{L}_{mat-\alpha}$ and $\mathcal{L}_{GAN}$ loss are critical for learning proper shadow removal. We observe that without the matting loss $\mathcal{L}_{mat-\alpha}$, the model behaves similarly to an image inpainting model where it tends to modify all parts of the images to fool the discriminator. Last, dropping the smoothness loss $\mathcal{L}_{sm}$ only results in a slight drop in shadow removal performance, from 9.7 to 10.2 RMSE on the shadow areas. However, we observe more visible boundary artifacts on the output images without this loss.

Table 2. Ablation Studies. We train our network without a certain loss or feature and report the shadow removal performances on the ISTD dataset [47]. The metric is RMSE (the lower, the better). The table shows that all the proposed features in our model are essential in training for shadow removal.

Methods	Shadow	Non-shadow	All
Input Image	40.2	2.6	8.5
Ours w/o limiting search space	47.5	2.9	9.9
Ours w/o $\mathcal{L}_{bd}$	41.7	3.9	9.8
Ours w/o $\mathcal{L}_{mat-\alpha}$	38.7	3.1	9.0
Ours w/o $\mathcal{L}_{sm-\alpha}$	10.2	2.8	4.0
Ours w/o $\mathcal{L}_{GAN}$	26.9	2.9	6.8
Ours	9.7	3.0	4.0

4.4 Video Shadow Removal

Video Shadow Removal is challenging for shadow removal methods. A video sequence has hundreds of frames with changing shadows. It is even harder for videos with a moving camera, moving objects, and illumination changes.

To better evaluate the performance of shadow removal methods in videos, we collected a set of 8 videos, each containing a static scene without visible moving objects. We cropped those videos to obtain clips with the only dominant motions caused by the shadows (either by direct light motion or motion of the unseen occluders). As can be seen from the top row of Fig. 6, the dataset includes videos containing shadows cast by close-up occluders, far distance occluders, videos with simple-to-complex shadows, and shadows on various types of backgrounds and materials. Inspired by [3], we propose a "max-min" technique to obtain a single pseudo shadow-free frame for each video: since the camera is static and there is no visible moving object in the frames, the changes in the video are caused by the moving shadows. We first obtain two images V_{max} and V_{min} by taking the maximum and minimum intensity values at each pixel location across the whole video. V_{max} is then the image that contains the shadow-free values of pixels if they ever go out of the shadows. Similarly, their shadowed values, if they ever go into the shadows, are captured in V_{min}. Figure 6 shows these two images for a video named "plant". From these two images, we can trivially obtain a mask, namely moving-shadow $\mathcal{M}$, marking the pixels appearing in both the shadow and non-shadow areas in the video:

$$\mathcal{M}_i = \begin{cases} 1 & \text{if } V_{max,i} > V_{min,i} + \epsilon \\ 0 & \text{otherwise,} \end{cases} \qquad (9)$$

where we set a small threshold of $\epsilon = 40$. This method allows us to obtain pairs of shadow and non-shadow pixel values in the moving-shadow mask, $\mathcal{M}$, for free.

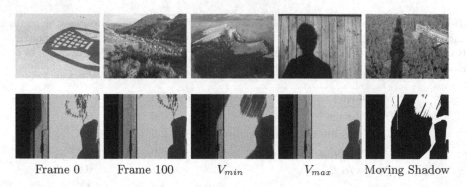

| Frame 0 | Frame 100 | V_{min} | V_{max} | Moving Shadow |

Fig. 6. Examples of Video Shadow Removal dataset. The dataset consists of videos where both the scene and the visible objects remaining static. The top row shows frames of different videos in our dataset. The second row visualizes our method to obtain the shadow-free frames for evaluating shadow removal.

To measure shadow removal performance, we input the frames of these videos into the shadow removal algorithm and measure the RMSE on the LAB color channel between the output frame and the image V_{max} on the moving-shadow area $\mathcal{M}$. We compute RMSE on each video and take their average to measure the shadow removal performance on the whole dataset. Table 3 summarizes the performance of our methods compared to MaskShadow-GAN [15] and SP+M-Net [22] on these videos. Our method outperforms SP+M-Net and MaskShadow-GAN, reducing the RMSE by 5% and 11% respectively. As our method only needs shadow segmentation masks for training, we use a pre-trained shadow detection model [59] to obtain a set of shadow masks for each video. While these shadow mask sets are imperfect, fine-tuning our model using this free supervision results in a 10% error reduction, showing the advantage of our training scheme. Figure 7 visualizes two example shadow removal results for different methods. We show a single input frame of each video. From left to right are the input frame, the shadow removal results of MaskShadow-GAN [15], the results of SP+M-Net [22], the results of our model trained on the ISTD dataset, and the result of our model fine-tuned with each testing video for 1 epoch. The top row shows an example where all methods perform relatively well. Our method seems to have better color balance between the relit pixels and the non-shadow pixels, although there is a visible boundary artifact due to imperfect shadow masks. After 1 epoch of fine-tuning, these artifacts are greatly suppressed. The bottom row shows a challenging case where all methods fail to remove shadows properly.

Table 3. Shadow removal results on our proposed Video Shadow Removal dataset. The metric is RMSE (the lower, the better), compared to the pseudo shadow-free frame on the moving shadow mask. All methods were pre-trained on the ISTD dataset. Ours+ denotes our model fine-tuned for one epoch on each video using the shadow masks generated by a shadow detector [59] pre-trained on the SBU dataset [43]

Methods	Input frame	[15]	[22]	Ours	Ours+
RMSE	32.9	23.5	22.2	20.9	18.0

Input Frame   MaskShadow-GAN   SP+M-Net   Ours   Ours+

Fig. 7. Shadow Removal on Videos. We visualize the shadow removal results of different methods on two frames extracted from our video dataset. "Ours+" denotes the results of our model fine-tuned with each testing video for 1 epoch. Top row shows an example where all methods perform relatively well. Bottom row shows a challenging case where all methods fail to remove shadow properly.

5   Conclusion

We presented a novel patch-based deep-learning model to remove shadows from images. This method can be trained on patches cropped directly from the shadow images, using the shadow segmentation mask as the only supervision signal. This obviates the dependency on paired training data and allows us to train this system on any kind of shadow image. The main contribution of this paper is a set of physics-based constrains that enable the training of this mapping. We have illustrated the effectiveness of our method on the standard ISTD dataset [47] and on our novel Video Shadow Removal dataset. As shadow detection methods mature with the aid of recently proposed shadow detection datasets [16,48], our method can be trained to remove shadows for a very low annotation cost.

Acknowledgements. This work was partially supported by the Partner University Fund, the SUNY2020 ITSC, and a gift from Adobe. Computational support provided by IACS and a GPU donation from NVIDIA. We thank Kumara Kahatapitiya and Cristina Mata for assistance with the manuscript.

References

1. Arbel, E., Hel-Or, H.: Shadow removal using intensity surfaces and texture anchor points. IEEE Trans. Pattern Anal. Mach. Intell. **33**, 1202–1216 (2011)
2. Choi, Y., Choi, M.J., Kim, M., Ha, J.W., Kim, S., Choo, J.: StarGAN: unified generative adversarial networks for multi-domain image-to-image translation. In: 2018 IEEE/CVF Conference on Computer Vision and Pattern Recognition, pp. 8789–8797 (2017)
3. Chuang, Y.Y., Goldman, D.B., Curless, B., Salesin, D.H., Szeliski, R.: Shadow matting and compositing. ACM Trans. Graph. **22**(3), 494–500 (2003). Special Issue of SIGGRAPH 2003 Proceeding
4. Dare, P.: Shadow analysis in high-resolution satellite imagery of urban areas. Photogram. Eng. Remote Sens. **71**, 169–177 (2005). https://doi.org/10.14358/PERS.71.2.169
5. Ding, B., Long, C., Zhang, L., Xiao, C.: ARGAN: attentive recurrent generative adversarial network for shadow detection and removal. In: 2019 IEEE/CVF International Conference on Computer Vision (ICCV), pp. 10212–10221 (2019)
6. Drew, M.S.: Recovery of chromaticity image free from shadows via illumination invariance. In: IEEE Workshop on Color and Photometric Methods in Computer Vision, ICCV'03, pp. 32–39 (2003)
7. Finlayson, G., Drew, M.S.: 4-sensor camera calibration for image representation invariant to shading, shadows, lighting, and specularities. In: Proceedings of the International Conference on Computer Vision, vol. 2, pp. 473–480 (2001). https://doi.org/10.1109/ICCV.2001.937663
8. Finlayson, G., Hordley, S., Lu, C., Drew, M.: On the removal of shadows from images. IEEE Trans. Pattern Anal. Mach. Intell. **28**(1), 59–68 (2006)
9. Finlayson, C., Hordley, S.D., Drew, M.S.: Removing shadows from images. In: Heyden, A., Sparr, G., Nielsen, M., Johansen, P. (eds.) Computer Vision – ECCV 2002. Lecture Notes in Computer Science, vol. 2353, pp. 823–836. Springer, Berlin, Heidelberg (2002). https://doi.org/10.1007/3-540-47979-1_55

10. Fredembach, C., Finlayson, G.D.: Hamiltonian path based shadow removal. In: BMVC (2005)
11. Gong, H., Cosker, D.: Interactive removal and ground truth for difficult shadow scenes. J. Opt. Soc. Am. A **33**(9), 1798–1811 (2016). https://doi.org/10.1364/JOSAA.33.001798. http://josaa.osa.org/abstract.cfm?URI=josaa-33-9-1798
12. Gulrajani, I., Ahmed, F., Arjovsky, M., Dumoulin, V., Courville, A.C.: Improved training of wasserstein GANs. In: Advances in Neural Information Processing Systems, pp. 5767–5777 (2017)
13. Guo, R., Dai, Q., Hoiem, D.: Single-image shadow detection and removal using paired regions. In: Proceedings of the IEEE Conference on Computer Vision and Pattern Recognition (2011)
14. Guo, R., Dai, Q., Hoiem, D.: Paired regions for shadow detection and removal. IEEE Trans. Pattern Anal. Mach. Intell. **35**(12), 2956–2967 (2012)
15. Hu, X., Jiang, Y., Fu, C.W., Heng, P.A.: Mask-ShadowGAN: learning to remove shadows from unpaired data. In: ICCV (2019)
16. Hu, X., Wang, T., Fu, C.W., Jiang, Y., Wang, Q., Heng, P.A.: Revisiting shadow detection: a new benchmark dataset for complex world. arXiv:abs/1911.06998 (2019)
17. Hu, X., Zhu, L., Fu, C.W., Qin, J., Heng, P.A.: Direction-aware spatial context features for shadow detection. In: Proceedings of the IEEE Conference on Computer Vision and Pattern Recognition (2018)
18. KaewTrakulPong, P., Bowden, R.: An improved adaptive background mixture model for real- time tracking with shadow detection (2002)
19. Kingma, D.P., Ba, J.: Adam: a method for stochastic optimization. In: Proceedings of the International Conference on Learning Representations (2015)
20. Le, H., Goncalves, B., Samaras, D., Lynch, H.: Weakly labeling the antarctic: the penguin colony case. In: The IEEE Conference on Computer Vision and Pattern Recognition (CVPR) Workshops (2019)
21. Le, H., Nguyen, V., Yu, C.P., Samaras, D.: Geodesic distance histogram feature for video segmentation. In: ACCV (2016)
22. Le, H., Samaras, D.: Shadow removal via shadow image decomposition. In: Proceedings of the International Conference on Computer Vision (2019)
23. Le, H., Vicente, T.F.Y., Nguyen, V., Hoai, M., Samaras, D.: A+D Net: training a shadow detector with adversarial shadow attenuation. In: Proceedings of the European Conference on Computer Vision (2018)
24. Le, H., Yu, C.P., Zelinsky, G., Samaras, D.: Co-localization with category-consistent features and geodesic distance propagation. In: ICCV 2017 Workshop on CEFRL: Compact and Efficient Feature Representation and Learning in Computer Vision (2017)
25. Li, Y., Tang, S., Zhang, R., Zhang, Y., Li, J., Yan, S.: Asymmetric gan for unpaired image-to-image translation. IEEE Trans. Image Process. **28**, 5881–5896 (2019)
26. Liu, F., Gleicher, M.: Texture-consistent shadow removal. In: Forsyth, D., Torr, P., Zisserman, A. (eds.) Computer Vision – ECCV 2008. Lecture Notes in Computer Science, vol. 5305, pp. 437–450. Springer, Berlin, Heidelberg (2008). https://doi.org/10.1007/978-3-540-88693-8_32
27. Liu, H., Gu, X., Samaras, D.: Wasserstein GAN with quadratic transport cost. In: The IEEE International Conference on Computer Vision (ICCV) (2019)
28. Liu, H., Xianfeng, G., Samaras, D.: A two-step computation of the exact GAN Wasserstein distance. In: International Conference on Machine Learning, pp. 3165–3174 (2018)

29. Liu, M.Y., Breuel, T., Kautz, J.: Unsupervised image-to-image translation networks. arXiv:abs/1703.00848 (2017)
30. Mescheder, L., Nowozin, S., Geiger, A.: Which training methods for GANs do actually converge? In: International Conference on Machine Learning (2018)
31. Miyato, T., Kataoka, T., Koyama, M., Yoshida, Y.: Spectral normalization for generative adversarial networks. In: International Conference on Machine Learning (2018)
32. Müller, T., Erdnüeß, B.: Brightness correction and shadow removal for video change detection with UAVs. In: Defense + Commercial Sensing (2019)
33. Panagopoulos, A., Wang, C., Samaras, D., Paragios, N.: Estimating shadows with the bright channel cue. In: Kutulakos, K.N. (ed.) Trends and Topics in Computer Vision – ECCV 2010. Lecture Notes in Computer Science, vol. 6554, pp. 1–12. Springer, Berlin, Heidelberg (2010). https://doi.org/10.1007/978-3-642-35740-4_1
34. Panagopoulos, A., Wang, C., Samaras, D., Paragios, N.: Simultaneous cast shadows, illumination and geometry inference using hypergraphs. IEEE Trans. Pattern Anal. Mach. Intell. **35**(2), 437–449 (2013). https://doi.org/10.1109/TPAMI.2012.110
35. Porter, T., Duff, T.: Compositing digital images. Proc. ACM SIGGRAPH Conf. Comput. Graph. **18**(3), 1–12 (1984)
36. Prati, A., Mikic, I., Trivedi, M.M., Cucchiara, R.: Detecting moving shadows: algorithms and evaluation. IEEE Trans. Pattern Anal. Mach. Intell. **25**, 918–923 (2003)
37. Qu, L., Tian, J., He, S., Tang, Y., Lau, R.W.H.: DeshadowNet: a multi-context embedding deep network for shadow removal. In: Proceedings of the IEEE Conference on Computer Vision and Pattern Recognition (2017)
38. Shiting, W., Hong, Z.: Clustering-based shadow edge detection in a single color image. In: International Conference on Mechatronic Sciences, Electric Engineering and Computer, pp. 1038–1041 (2013). https://doi.org/10.1109/MEC.2013.6885215
39. Shor, Y., Lischinski, D.: The shadow meets the mask: pyramid-based shadow removal. Comput. Graph. Forum **27**(2), 577–586 (2008)
40. Smith, A.R., Blinn, J.F.: Blue screen matting. In: Proceedings of the ACM SIGGRAPH Conference on Computer Graphics (1996)
41. Su, N., Zhang, Y., Tian, S., Yan, Y., Miao, X.: Shadow detection and removal for occluded object information recovery in urban high-resolution panchromatic satellite images. IEEE J. Sel. Top. Appl. Earth Obs. Remote Sens. **9**, 2568–2582 (2016)
42. Thanh-Tung, H., Tran, T., Venkatesh, S.: Improving generalization and stability of generative adversarial networks. In: International Conference on Learning Representations (2019)
43. Vicente, T.F.Y., Hoai, M., Samaras, D.: Noisy label recovery for shadow detection in unfamiliar domains. In: Proceedings of the IEEE Conference on Computer Vision and Pattern Recognition (2016)
44. Vicente, T.F.Y., Hoai, M., Samaras, D.: Leave-one-out kernel optimization for shadow detection and removal. IEEE Transactions on Pattern Analysis and Machine Intelligence **40**(3), 682–695 (2018)
45. Vicente, T.F.Y., Hou, L., Yu, C.P., Hoai, M., Samaras, D.: Large-scale training of shadow detectors with noisily-annotated shadow examples. In: Proceedings of the European Conference on Computer Vision (2016)
46. Vicente, T.F.Y., Samaras, D.: Single image shadow removal via neighbor-based region relighting. In: Proceedings of the European Conference on Computer Vision Workshops (2014)

47. Wang, J., Li, X., Yang, J.: Stacked conditional generative adversarial networks for jointly learning shadow detection and shadow removal. In: Proceedings of the IEEE Conference on Computer Vision and Pattern Recognition (2018)
48. Wang, T., Hu, X., Wang, Q., Heng, P.A., Fu, C.W.: Instance shadow detection. In: CVPR (2020)
49. Wright, S.: Digital Compositing for Film and Video. Focal Press (2001)
50. Wu, Q., Zhang, W.,Vijay Kumar, B.V.K.: Strong shadow removal via patch-based shadow edge detection. In: 2012 IEEE International Conference on Robotics and Automation, pp. 2177–2182 (2012)
51. Wu, T.P., Tang, C.K.: A Bayesian approach for shadow extraction from a single image. In: Tenth IEEE International Conference on Computer Vision (ICCV'05), vol. 1 1, pp. 480–487 (2005)
52. Wu, T.P., Tang, C.K., Brown, M.S., Shum, H.Y.: Natural shadow matting. ACM Trans. Graph. **26**, 2 (2007). https://doi.org/10.1145/1243980.1243982. http://doi.acm.org/10.1145/1243980.1243982
53. Yang, Q., Tan, K., Ahuja, N.: Shadow removal using bilateral filtering. IEEE Trans. Image Process. **21**, 4361–4368 (2012)
54. Yi, Z., Zhang, H., Tan, P., Gong, M.: DualGAN: unsupervised dual learning for image-to-image translation. In: 2017 IEEE International Conference on Computer Vision (ICCV), pp. 2868–2876 (2017)
55. Zhang, L., Long, C., Zhang, X., Xiao, C.: RIS-GAN: explore residual and illumination with generative adversarial networks for shadow removal. In: AAAI Conference on Artificial Intelligence (AAAI) (2020)
56. Zhang, W., Zhao, X., Morvan, J.M., Chen, L.: Improving shadow suppression for illumination robust face recognition. IEEE Trans. Pattern Anal. Mach. Intell. **41**, 611–624 (2019)
57. Zheng, Q., Qiao, X., Cao, Y., Lau, R.W.H.: Distraction-aware shadow detection. In: 2019 IEEE/CVF Conference on Computer Vision and Pattern Recognition (CVPR), pp. 5162–5171 (2019)
58. Zhu, J.Y., Park, T., Isola, P., Efros, A.A.: Unpaired image-to-image translation using cycle-consistent adversarial networks. In: 2017 IEEE International Conference on Computer Vision (ICCV) (2017)
59. Zhu, L., et al.: Bidirectional feature pyramid network with recurrent attention residual modules for shadow detection. In: Proceedings of the European Conference on Computer Vision (2018)

Diverse and Admissible Trajectory Forecasting Through Multimodal Context Understanding

Seong Hyeon Park[1](✉), Gyubok Lee[2], Jimin Seo[3], Manoj Bhat[4],
Minseok Kang[5], Jonathan Francis[4,6], Ashwin Jadhav[4], Paul Pu Liang[4],
and Louis-Philippe Morency[4]

[1] Hanyang University, Seoul, Korea
shpark@spa.hanyang.ac.kr
[2] Yonsei University, Seoul, Korea
glee48@yonsei.ac.kr
[3] Korea University, Seoul, Korea
jmseo0607@korea.ac.kr
[4] Carnegie Mellon University, Pittsburgh, PA, USA
{mbhat,jmf1,arjadhav,pliang,morency}@cs.cmu.edu
[5] Sogang University, Seoul, Korea
ahstarwab@sogang.ac.kr
[6] Bosch Research Pittsburgh, Pittsburgh, PA, USA

Abstract. Multi-agent trajectory forecasting in autonomous driving requires an agent to accurately anticipate the behaviors of the surrounding vehicles and pedestrians, for safe and reliable decision-making. Due to partial observability in these dynamical scenes, directly obtaining the posterior distribution over future agent trajectories remains a challenging problem. In realistic embodied environments, each agent's future trajectories should be both *diverse* since multiple plausible sequences of actions can be used to reach its intended goals, and *admissible* since they must obey physical constraints and stay in drivable areas. In this paper, we propose a model that synthesizes multiple input signals from the multimodal world—the environment's scene context and interactions between multiple surrounding agents—to best model all diverse and admissible trajectories. We compare our model with strong baselines and ablations across two public datasets and show a significant performance improvement over previous state-of-the-art methods. Lastly, we offer new metrics incorporating admissibility criteria to further study and evaluate the diversity of predictions. Codes are at: https://github.com/kami93/CMU-DATF.

J. Seo and M. Bhat—Authors contributed equally.

Electronic supplementary material The online version of this chapter (https://doi.org/10.1007/978-3-030-58621-8_17) contains supplementary material, which is available to authorized users.

A. Vedaldi et al. (Eds.): ECCV 2020, LNCS 12356, pp. 282–298, 2020.
https://doi.org/10.1007/978-3-030-58621-8_17

Keywords: Trajectory forecasting · Diversity · Admissibility · Generative modeling · Autonomous driving

1   Introduction

Trajectory forecasting is an important problem in autonomous driving scenarios, where an autonomous vehicle must anticipate the behavior of other surrounding agents (e.g., vehicles and pedestrians), within a dynamically-changing environment, in order to plan its own actions accordingly. However, since the contexts of agents' behavior such as intentions, social interactions, or environmental constraints are not directly observed, predicting future trajectories is a challenging problem [22,26,33]. It requires an estimation of most likely agent actions based on key observable environmental features (e.g., road structures, agent interactions) as well as the simulation of agents' hypothetical future trajectories toward their intended goals. In realistic embodied environments, there are multiple plausible sequences of actions that an agent can take to reach its intended goals. However, each trajectory must obey physical constraints (e.g., Newton's laws) and stay in the statistically plausible locations in the environment (i.e., the drivable areas). In this paper, we refer to these attributes as *diverse* and *admissible* trajectories, and illustrate some examples in Fig. 1. Achieving diverse and admissible trajectory forecasting for autonomous driving allows each agent to make the best predictions, by taking into account all valid actions that other agents could take.

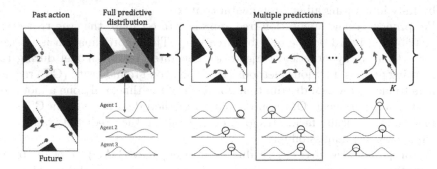

Fig. 1. Diverse and admissible trajectory forecasting. Based on the existing context, there can be multiple valid hypothetical future trajectories. Therefore, the predictive distribution of the trajectories should reflect various modes, representing different plausible goals (*diversity*) while penalizing implausible trajectories that either conflict with the other agents or are outside valid drivable areas (*admissibility*).

To predict a diverse set of admissible trajectories, each agent must understand its *multimodal* environment, consisting of the scene context as well as interactions between other surrounding agents. Scene context refers to the typical and spatial activity of surrounding objects, presence of traversable area, etc.

which contribute to forecasting next maneuvers. These help to understand some semantic constraints on the agent's motion (e.g., traffic laws, road etiquette) and can be inferred from the present and corresponding mutlimodal data i.e spatial as well as social, temporal motion behavior data. Therefore, the model's ability to extract and meaningfully represent multimodal cues is crucial.

Concurrently, another challenging aspect of trajectory forecasting lies in encouraging models to make diverse predictions about future trajectories. Due to high-costs in data collection, diversity is not explicitly present in most public datasets, but only one annotated future trajectories. [7,16,18]. Vanilla predictive models that fit future trajectories based only on the existing annotations would severely underestimate the diversity of all possible trajectories. In addition, measuring the quality of predictions using existing annotation-based measures (e.g., displacement errors [23]) does not faithfully score diverse and admissible trajectory predictions.

As a step towards multimodal understanding for diverse and admissible trajectory forecasting, our contributions are *three*-fold:

1. We propose a model that addresses the lack of diversity and admissibility for trajectory forecasting through the understanding of the multimodal environmental context. As illustrated in Fig. 2, our approach explicitly models agent-to-agent and agent-to-scene interactions through "self-attention" [27] among multiple agent trajectory encodings, and a conditional trajectory-aware "visual attention" [31] over the map, respectively. Together with a constrained flow-based decoding, trained with symmetric cross-entropy [21], this allows our model to generate diverse and admissible trajectory candidates by fully integrating all environmental contexts.
2. We propose a new annotation-free approach to estimate the true trajectory distribution based on the drivable-area map. This approximation is utilized for evaluating hypothetical trajectories generated by our model during the learning process. Previous methods [21] depend on ground-truth (GT) recordings to model the real distribution; for most of the time, only one annotation is available per agent. Our approximation method does not rely on GT samples and empirically facilitates greater diversity in the predicted trajectories while ensuring admissibility.
3. We propose a new metric, Drivable Area Occupancy (DAO), to evaluate the diversity of the trajectory predictions while ensuring admissibility. This new metric makes another use of the drivable-area map, without requiring multiple annotations of trajectory futures. We couple this new metric with standard metrics from prior art, such as Average Displacement Error (ADE) and Final Displacement Error (FDE), to compare our model with existing baselines.

Additionally, we publish tools to replicate our data and results which we hope will advance the study of diverse trajectory forecasting. Our codes are available at: https://github.com/kami93/CMU-DATF.

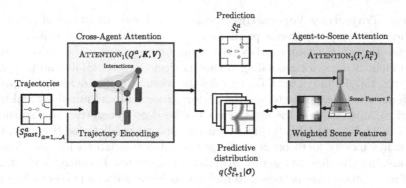

Fig. 2. Overview of our multimodal attention approach. Best viewed in color. The cross-agent attention module (left) generates an attention map over the encoded trajectories of neighboring agents. The agent-to-scene attention model (right) generates an attention map over the scene, based on the drivable-area map. (Color figure online)

2   Related Work

Multimodal Trajectory Forecasting requires a detailed understanding of the agent's environment. Many works integrate information from multiple sensory cues, such as LiDAR point-cloud information to model the surrounding environment [17,21,22], high dimensional map data to model vehicle lane segmentation [3,6,33], and RGB image to capture the environmental context [17,18,24]. Other methods additionally fuse different combinations of interactions with the intention of jointly capturing all interactions between the agents [1,11,18,24]. Without mechanisms to jointly and explicitly model such agent-to-scene and agent-to-agent relations, we hypothesize that these models are unable to capture complex interactions in the high-dimensional input space and propose methods to explicitly model these interactions via sophisticated attention mechanisms.

Multi-agent Modeling aims to learn representations that summarize the behavior of one agent given its surrounding agents. These interactions are often modeled through either spatial-oriented methods or neural attention-based methods. Some of the spatial-oriented methods simply take into account agent-wise distances through a relative coordinate system [3,13,19,22], while others utilize sophisticated pooling approaches across individual agent representations [8,11,17,33]. On the other hand, the attention-based methods use the attention architecture to model multi-agent interaction in a variety of domains including pedestrians [9,24,28] and vehicles [18,26]. In this paper, we employ the attention based cross-agent module to capture explicit agent-to-agent interactions. Even with the increasing number of agents around the ego-vehicle, our cross-agent module can successfully model the interactions between agents, as supported in one of our experiments.

Diverse Trajectory Forecasting involves stochastic modeling of future trajectories and sampling diverse predictions based on the model distribution. The Dynamic Bayesian network (DBN) is a common approach without deep generative models, utilized for modelling vehicle trajectories [10,25,30] and pedestrian actions [2,15]. Although the DBN enables the models to consider physical process that generates agent trajectories, the performance is often limited for real traffic scenarios. Most state-of-the-art models utilize deep generative models such as GAN [11,24,33] and VAE [17] to encourage diverse predictions. However, these approaches mostly focus on generating multiple candidates while less focusing on analyzing the diversity across distributional modes. Recently, sophisticated sampling functions are proposed to tackle this issue, such as Diversity Sampling Function [32] and Latent Semantic Sampling [12]. Despite some promising empirical results, it remains difficult to evaluate both the diversity and admissibility of predictions. In this paper, we tackle the task of diverse trajectory forecasting with a special emphasis on admissibility in dynamic scenes and propose a new task metric that specifically assess models on the basis of these attributes.

3   Problem Formulation

We define the terminology that constitutes our problem. An *agent* is a dynamic on-road object that is represented as a sequence of 2D coordinates, i.e., a spatial position over time. We denote the position for agent a at time t as $S_t^a \in \mathbb{R}^2$, sequence of such positions from t_1 to t_2 as $S_{t_1:t_2}^a$, and the full sequence as (bold) $\boldsymbol{S}^a$. We set $t = 0$ as *present*, $t \leq 0$ as *past*, and $t > 0$ as *prediction* or simply, *pred*. We often split the sequence into two parts, with respect to the *past* and *pred* subsequences: we denote these as $\boldsymbol{S}_{\text{past}}^a$ and $\boldsymbol{S}_{\text{pred}}^a$. In order to clearly distinguish the predicted values from these variables, we use 'hats' such as $\hat{S}_t^a$ and $\hat{\boldsymbol{S}}_{\text{pred}}$. A *scene* is a high-dimensional structured data that describes the present environmental context around the agent. For this, we utilize a bird's eye view array, denoted as $\boldsymbol{\Phi} \in \mathbb{R}^{H \times W \times C}$, where H and W are the sizes of field around the agent and C is the channel size of the scene, where each channel consists of distinct information such as the drivable area, position, and distance encodings.

Combining the *scene* and all *agent* trajectories yields an *episode*. In the combined setting, there are a variable number of agents which we denote using bold $\boldsymbol{S} \equiv \{\boldsymbol{S}^1, \boldsymbol{S}^2, ..., \boldsymbol{S}^A\}$ and as similarly to other variables, we may split it into two subsets, $\boldsymbol{S}_{\text{past}}$ and $\boldsymbol{S}_{\text{pred}}$ to represent the past and prediction segments. Since $\boldsymbol{S}_{\text{past}}$ and $\boldsymbol{\Phi}$ serve as the observed information cue used for the prediction, they are often called *observation* simply being denoted as $\mathcal{O} \equiv \{\boldsymbol{S}_{\text{past}}, \boldsymbol{\Phi}\}$.

We define *diversity* to be the level of coverage in a model's predictions, across modes in a distribution representing all possible future trajectories. We denote the model distribution as $q(\boldsymbol{S}_{\text{pred}}^a | \mathcal{O})$ and want the model to generate multiple samples interpreting each sample as an independent *hypothesis* that might have happened, given the same observation. We also acknowledge that encouraging a model's predictions to be diverse, alone, is not sufficient for accurate and safe output; the model predictions should lie in the support of the true future trajectory distribution $p(\boldsymbol{S}_{\text{pred}}|\mathcal{O})$, i.e., predictions should be *admissible*. Given the observation $\mathcal{O}$, it is futile to predict samples around regions that are physically and statistically implausible to reach.

To summarize, this paper addresses the task of *diverse and admissible multi-agent trajectory forecasting*, by modeling multiple modes in the posterior distribution over the *prediction* trajectories, given the observation.

4   Proposed Approach

We hypothesize that future trajectories of human drivers should follow distributions of multiple modes, conditioned on the scene context and social behaviors of agents. Therefore, we design our model to explicitly capture both agent-to-scene interactions and agent-to-agent interactions with respect to each agent of interest. Through our objective function, we explicitly encourage the model to learn a distribution with multiple modes by taking into account past trajectories and attended scene context.

4.1   Model Architecture

As illustrated in Fig. 3, our model consists of an encoder-decoder architecture. The encoder includes the cross-agent interaction module. The decoder, on the other hand, comprises the agent-to-scene interaction module to capture the scene interactions. Please refer to Fig. 4 for a detailed illustration of our main proposed modules.

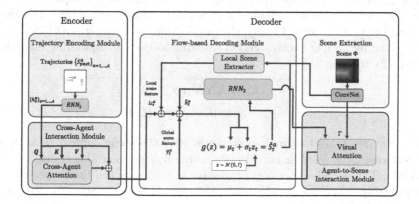

Fig. 3. Model Architecture. The model consists of an encoder-decoder architecture: the encoder takes as past agent trajectories and calculates cross-agent attention, and the flow-based decoder predicts future trajectories by attending scene contexts for each decoding step.

The *encoder* extracts past trajectory encoding for each agent, then calculates and fuses the interaction features among the agents. Given an observation, we

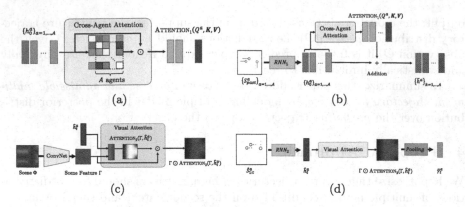

Fig. 4. (a) Cross-agent attention. Interaction between each agent is modeled using attention, (b) Cross-agent interaction module. Agent trajectory encodings are corrected via cross-agent attention. (c) Visual attention. Agent-specific scene features are calculated using attention. (d) Agent-to-scene interaction module. Pooled vectors are retrieved from pooling layer after visual attention.

encode each agent's past trajectory S_{past}^a by feeding it to the trajectory encoding module. The module has the LSTM-based layer RNN_1 to summarize the past trajectory. It iterates through the past trajectory with Eq. (1) and its final output h_0^a (at *present* $t = 0$) is utilized as the agent embedding. The collection of the embeddings for all agents is then passed to the cross-agent interaction module, depicted in Fig. 4a, which uses *self-attention* [27] to generate a cross-agent representation. We linearly transform each agent embedding to get a query-key-value triple, (Q^a, K^a, V^a). Next, we calculate the interaction features through the self-attention layer $Attention_1$. Finally, the fused agent encoding $\tilde{\boldsymbol{h}} \equiv \{\tilde{h}^1, \tilde{h}^2, ..., \tilde{h}^A\}$ is calculated by adding each attended features to the corresponding embedding (see Eq. (2) and Fig. 4b).

$$h_t^a = RNN_1(S_{t-1}^a, h_{t-1}^a) \tag{1}$$

$$\tilde{h}^a = h_0^a + Attention_1(Q^a, \boldsymbol{K}, \boldsymbol{V}) \tag{2}$$

The *decoder* takes the final encoding $\tilde{h}^a$ and the scene context $\boldsymbol{\Phi}$ as inputs. We first extract the scene feature through convolutional neural networks i.e., $\boldsymbol{\Gamma} = CNN(\boldsymbol{\Phi})$. The decoder then autoregressively generates the future position $\hat{S}_t^a$, while referring to both local and global scene context from the agent-to-scene interaction module. The local scene feature is gathered using bilinear interpolation at the local region of $\boldsymbol{\Gamma}$, corresponding to the physical position $\hat{S}_{t-1}^a$. Then, the feature is concatenated with $\tilde{h}^a$ and processed thorough FC layers to form the local context lc_t^a. We call this part the local scene extractor. The global scene feature is calculated using *visual-attention* [31] to generate weighted scene features, as shown in Fig. 4c. To calculate the attention, we first encode previous outputs $\hat{S}_{1:t-1}^a$, using a GRU-based RNN_2 in Eq. (3), whose output is then used to calculate the pixel-wise attention $\tilde{\gamma}_t^a$ at each decoding step, for each agent in Eq. (4).

$$\hat{h}_t^a = RNN_2(\hat{S}_{1:t-1}^a, \hat{h}_{t-1}^a) \tag{3}$$

$$\tilde{\gamma}_t^a = Pool(\boldsymbol{\Gamma} \odot Attention_2(\hat{h}_t^a, \boldsymbol{\Gamma})) \tag{4}$$

The flow-based decoding module generates the future position $\hat{S}_t^a$. The module utilizes *Normalizing Flow* [20], a generative modeling method using bijective and differentiable functions; in particular, we choose an autoregressive design [14,21,22]. We concatenate $\tilde{\gamma}_t^a$, $\hat{h}_t^a$, and lc_t^a then project them down to 6-dimensional vector using FC layers. We split this vector to $\hat{\mu}_t \in \mathbb{R}^2$ and $\hat{\sigma}_t \in \mathbb{R}^{2 \times 2}$. Next, we transform a standard Gaussian sample $z_t \sim \mathcal{N}(\mathbf{0}, \boldsymbol{I}) \in \mathbb{R}^2$, by the mapping $g_\theta(z_t; \mu_t, \sigma_t) = \sigma_t \cdot z_t + \mu_t = \hat{S}_t^a$, where θ is the set of model parameters. 'hats' over $\hat{\mu}_t$ and $\hat{\sigma}_t$ are removed, in order to note that they went through the following details. To ensure the positive definiteness, we apply matrix exponential $\sigma_t = expm(\hat{\sigma}_t)$ using the formula in [4]. Also, to improve the the physical *admissibility* of the prediction, we apply the constraint $\mu_t = \hat{S}_{t-1}^a + \alpha(\hat{S}_{t-1}^a - \hat{S}_{t-2}^a) + \hat{\mu}_t$, where α is a model degradation coefficient. When $\alpha = 1$, the constraint is equivalent to *Verlet integration* [29], used in some previous works [21,22], which gives the a perfect constant velocity (CV) prior to the model. However, we found empirically that, the model easily overfits to datasets when the the perfect CV prior is used, and perturbing the model with α prevents overfitting; we choose $\alpha = 0.5$.

Iterating through time, we get the predictive trajectory $\hat{S}_{\text{pred}}^a$ for each agent. By sampling multiple instances of z_{pred} and mapping them to trajectories, we get various hypotheses of future. Further details of the our network architecture and experiments for the degradation coefficient α are given in the supplementary.

4.2 Learning

Our model learns to predict the distribution over the future trajectories of agents present in a given episode. In detail, we focus on predicting the conditional distribution $p(\boldsymbol{S}_{\text{pred}}|\mathcal{O})$ where the future trajectory $\boldsymbol{S}_{\text{pred}}$ depends on the observation. As described in the previous sections, our model incorporates a bijective and differentiable mapping between standard Gaussian prior q_0 and the future trajectory distribution q_θ. Such technique, commonly aliased 'normalizing flow', enables our model not only to generate multiple candidate samples of future, but also to evaluate the ground-truth trajectory with respect to the predicted distribution q_θ by using the inverse and the change-of-variable formula in Eq. (5).

$$q_\theta(S_{\text{pred}}^a|\mathcal{O}) = q_0(g^{-1}(S_{\text{pred}}^a)) \left| \det(\partial S_{\text{pred}}^a / \partial(g^{-1}(S_{\text{pred}}^a))) \right|^{-1} \tag{5}$$

As a result, our model can simply close the discrepancy between the predictive distribution q_θ and the real world distribution p by optimizing our model parameter θ. In particular, we choose to minimize the combination of forward and reverse cross-entropies, also known as 'symmetric cross-entropy', between the two distributions in Eq. (6). Minimizing the symmetric cross-entropy allows model to learn generating diverse and plausible trajectory, which is mainly used in [21].

$$\min_{\theta} H(p, q_\theta) + \beta H(q_\theta, p) \tag{6}$$

To realize this, we gather the ground-truth trajectories S and scene context Φ from the dataset that we assume to well reflect the real distribution p, then optimize the model parameter θ such that 1) the density of the ground-truth future trajectories on top of the predicted distribution q_θ is maximized and 2) the density of the predicted samples on top of the real distribution p is also maximized as described in Eq. (7). Since this objective is fully differentiable with respect to the model parameter θ, we train our model using the stochastic gradient descent.

$$\min_{\theta} \mathbb{E}_{S_{\mathrm{pred}}, \mathcal{O} \sim p} \left[\mathbb{E}_{S_t^a \in S_{\mathrm{pred}}} - \log q_\theta(S_t^a | \mathcal{O}) + \beta \mathbb{E}_{\hat{S}_t^a \sim q_\theta} - \log p(\hat{S}_t^a | \mathcal{O}) \right] \tag{7}$$

Such symmetric combination of the two cross-entropies guides our model to predict q_θ that covers all plausible modes in the future trajectory while penalizing the bad samples that are less likely under the real distribution p. However, one major problem inherent in this setting is that we cannot actually evaluate p in practice. In this paper, we propose a new method which approximates p using the drivable-area map and we discuss it in the following Subsect. 4.3. Other optimization details are included in the supplementary.

4.3   The Drivable-Area Map and Approximating p

We generate a binary mask that denote the drivable-area around the agents. We refer to this feature as the *drivable-area map* and utilize it for three different purposes: 1) deriving the approximated true trajectory distribution $\tilde{p}$, 2) calculating the diversity and admissibility metrics, and 3) building the scene input Φ. Particularly, $\tilde{p}$ is a key component in our training objective, Eq. (6). Since the reverse cross-entropy penalizes the predicted trajectories with respect to the real distribution, the approximation should not underestimate some region of the real distribution, or diversity in the prediction could be discouraged. Previous works on $\tilde{p}$ utilize the ground-truth (GT) trajectories to model it [21]. However, there is often only one GT annotation available thus deriving $\tilde{p}$ based on the GT could assign awkwardly low density around certain region. To cope with such problem, our method assumes that every drivable locations are equally probable for future trajectories to appear in and that the non-drivable locations are increasingly less probable, proportional to the distance from the drivable-area. To derive it, we encode the distance on each non-drivable location using the distance transform on the drivable-area maps, then apply softmax over the transformed map to constitute it as a probability distribution. The visualizations of the $\tilde{p}$ are available in Fig. 6. Further details on deriving $\tilde{p}$ and the scene context input Φ, as well as additional visualizations and qualitative results are given in the supplementary.

5   Experimental Setup

The primary goal in the following experiments is to evaluate our model, baselines, and ablations on the following criteria—(i) Leveraging mechanisms that explicitly model agent-to-agent and agent-to-scene interactions (experiments 1 and 2). (ii) Producing diverse trajectory predictions, while obeying admissibility constraints on the trajectory candidates, given different approximation methods for the true trajectory distribution p (experiment 3). (iii) Remaining robust to an increasing number of agents in the scene (agent complexity; experiment 4). (iv) Generalizing to other domains (experiment 5).

5.1   Dataset

We utilize two real world datasets to evaluate our model and the baselines: NUSCENES tracking [5] and ARGOVERSE motion forecasting [7]. NUSCENES contains 850 different real-world driving scenarios, where each spanning 20 s of frames and 3D box annotations for the surrounding objects. It also provides drivable-area maps. Based on this setting, we generate trajectories by associating the box annotations of the same agents. While NUSCENES provides trajectories for realistic autonomous driving scenarios, the number of episodes is limited around 25K. On the other hand, ARGOVERSE provides around 320 K episodes from the real world driving along with the drivable-area maps. ARGOVERSE presents independent episodes spanning only 5 s (2 s for the past and 3 s for the prediction), rather than providing long, continuing scenarios as in NUSCENES. However, this setting suffices to test our method and we evaluate baselines and our models on these two real-world datasets to provide complementary validations of each model's diversity, admissibility, and generalizing performance across domains. Further details in the data processing are available in the supplementary.

5.2   Baseline Models

Deterministic Baselines: We compare three deterministic models with our approach, to examine our model's ability to capture agent-to-agent interaction: *LSTM-based encoder-decoder* (LSTM), *LSTM with convolutional social pooling* [8] (CSP), and a deterministic version of *multi-agent tensor fusion* (MATF-D) [33]. For our deterministic model, we use an LSTM with our cross-agent attention module in the encoder, which we refer to as the *cross-agent attention model* (CAM). Because each model is predicated on an LSTM component, we set the capacity to be the same in all cases, to ensure fair comparison.
Stochastic Baselines: We experiment three stochastic baselines. Our first stochastic baseline is a model based on a Variational Autoencoder structure, (DESIRE) [17], which utilizes scene contexts and an iterative refinement process. The second baseline model is a Generative Adversarial Network version of *multi-agent tensor fusion* (MATF-GAN) [33]. Our third baseline is the Reparameterized Pushforward Policy (R2P2-MA) [22] which is a modified version of

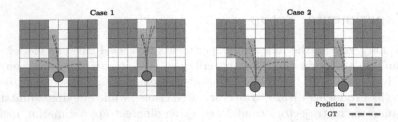

Fig. 5. We motivate the need for multiple metrics, to assess diversity and admissibility. Case 1: DAO measures are equal, even though predictions have differing regard for the modes in the posterior distribution. Case 2: RF measures are equal, despite differing regard for the cost of leaving the drivable area. In both cases, it is important to distinguish between conditions—we do this by using DAO, RF, and DAC together.

R2P2 [21] for multi-agent prediction. To validate our model's ability to extract scene information and generate diverse trajectories, multiple versions of our models are tested. While these models can be used as stand-alone models to predict diverse trajectories, comparison amongst these new models is equivalent to an ablation study of our final model. CAM-NF is a CAM model with a flow-based decoder. LOCAL-CAM-NF is CAM-NF with local scene features. GLOBAL-CAM-NF is LOCAL-CAM-NF with global scene features. Finally, ATTGLOBAL-CAM-NF is GLOBAL-CAM-NF with agent-to-scene attention, which is our main proposed model.

5.3 Metrics

We define multiple metrics that provide a thorough interpretation about the behavior of each model in terms of precision, diversity, and admissibility. To evaluate precision, we calculate Euclidean errors: ADE (*average displacement error*) and FDE (*final displacement error*), or $Error$ to denote both. To evaluate multiple hypotheses, we use the average and the minimum $Error$ among K hypotheses: $avgError = \frac{1}{k}\sum_{i=1}^{k} Error^{(i)}$ and $minError = \min\{Error^{(1)}, ..., Error^{(k)}\}$. A large $avgError$ implies that predictions are spread out, and a small $minError$ implies at least one of predictions has high precision. From this observation, we define a new evaluation metric that capture diversity in predictions: the ratio of $avgFDE$ to $minFDE$, namely rF. rF is robust to the variability of magnitude in velocity in predictions hence provides a handy tool that can distinguish between predictions with multiple modes (diversity) and predictions with a single mode (perturbation).

$$\text{Ratio of } avgFDE \text{ to } minFDE \ (rF) = \frac{avgFDE}{minFDE} \tag{8}$$

$$\text{Drivable Area Occupancy (DAO)} = \frac{\text{count}(\text{traj}_{\text{pix}})}{\text{count}(\text{driv}_{\text{pix}})} \tag{9}$$

We also report performance on additional metrics that are designed to capture diversity and admissibility in predictions. While RF measures the spread of predictions in Euclidean distance, DAO measures diversity in predictions that are only admissible, and DAC measures extreme off-road predictions that defy admissibility. We follow [7] in the use of *Drivable Area Count* (DAC), $DAC = \frac{k-m}{k}$, where m is the number of predictions that go out of the drivable area and k is the number of hypotheses per agent. Next, we propose a new metric, *Drivable Area Occupancy* (DAO), which measures the proportion of pixels that predicted trajectories occupy in the drivable-area. Shown in Eq. (9), count(traj$_\text{pix}$) is the number of pixels occupied by predictions and count(driv$_\text{pix}$) is the total number of pixels of the drivable area, both within a pre-defined grid around the ego-vehicle.

DAO may seem a reasonable standalone measure of capturing both diversity and admissibility, as it considers diversity in a reasonable region of interest. However, DAO itself cannot distinguish between *diversity* (Sect. 3) and arbitrary stochasticity in predictions, as illustrated by Case 1 in Fig. 5: although DAO measures of both predictions are equal, the causality behind each prediction is different and we must distinguish the two. RF and DAO work in a complementary way and we, therefore, use both for measuring diversity. To assure the *admissibility* of predictions, we use DAC which explicitly counts off-road predictions, as shown by Case 2 in Fig. 5. As a result, assessing predictions using DAO along with RF and DAC provides a holistic view of the quantity and the quality of diversity in predictions.

For our experiments, we use MINADE and MINFDE to measure *precision*, and use RF, DAC, and DAO to measure both *diversity* and *admissibility*. Due to the nature of DAO, where the denominator in our case is the number of overlapping pixels in a 224×224 grid, we normalize it by multiplying by $10,000$ when reporting results. For the multi-agent experiment (shown in Table 4), relative improvement (RI) is calculated as we are interested in the relative improvement as the number of agents increases. For all metrics, the number of trajectory hypotheses should be set equally for fair comparison of models. If not specified, the number of hypotheses k is set to 12 when reporting the performance metrics.

6   Results and Discussion

In this section, we show experimental results on numerous settings including the comparison with the baseline, and ablation studies of our model. We first show the effect of our cross-agent interaction module and agent-to-scene interaction module on the model performance, then we analyze the performance with respect to different numbers of agents, and other datasets. All experiments are measured with MINADE, MINFDE, RF, DAC, and DAO for holistic interpretation.

Effectiveness of Cross-Agent Interaction Module: We show the performance of one of our proposed models CAM, which utilizes our cross-agent attention module, along with three deterministic baselines as shown in Table 1. CSP models the interaction through layers of convolutional networks, and the interaction is implicitly calculated within the receptive field of convolutional layers.

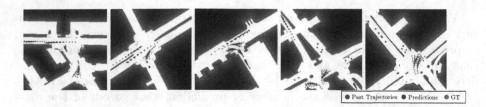

Fig. 6. Our map loss and corresponding model predictions. Each pixel on our map loss denotes probability of future trajectories; higher probability values are represented by brighter pixels. Our approach generates diverse and admissible future trajectories. More visualizations of qualitative results are provided in the supplementary.

Table 1. Deterministic models on NUSCENES. Our proposed model outperforms the existing baselines.

Model	MINADE ($\downarrow$)	MINFDE ($\downarrow$)
LSTM	1.186	2.408
CSP	1.390	2.676
MATF-D [33]	1.261	2.538
CAM (OURS)	**1.124**	**2.318**

$MATF - D$ is an extension of convolutional social pooling with scene information. CAM explicitly defines the interaction between each agent by using attention. The result shows that CAM outperforms other baselines in both MINADE and MINFDE, indicating that the explicit way of modeling agent-to-agent interaction performs better in terms of precision than an implicit way of modeling interaction using convolutional networks used in CSP and MATF-D. Interestingly, CAM outperforms MATF-D that utilizes scene information. This infers that the cross-agent interaction module has ability to learn the structure of roads and permissible region given the trajectories of surrounding agents.

Table 2. Stochastic models on NUSCENES. Our models outperform the existing baselines, achieving the best precisions, diversity, and admissibility. Improvements indicated by arrows.

Model	MINADE ($\downarrow$)	MINFDE ($\downarrow$)	RF ($\uparrow$)	DAO ($\uparrow$)	DAC ($\uparrow$)
DESIRE [17]	1.079	1.844	1.717	16.29	0.776
MATF-GAN [33]	1.053	2.126	1.194	11.64	0.910
R2P2-MA [21]	1.179	2.194	1.636	**25.65**	0.893
CAM-NF (OURS)	0.756	1.381	2.123	23.15	0.914
LOCAL-CAM-NF (OURS)	0.774	1.408	2.063	22.58	**0.921**
GLOBAL-CAM-NF (OURS)	0.743	1.357	2.106	22.65	**0.921**
ATTGLOBAL-CAM-NF (OURS)	**0.639**	**1.171**	**2.558**	22.62	0.918

Table 3. Training ATTGLOBAL-CAM-NF using the proposed annotation-free $\tilde{p}$ outperforms the annotation-dependent counterpart (MSE) in NUSCENES.

Model	MINADE ($\downarrow$)	MINFDE ($\downarrow$)	RF ($\uparrow$)	DAO ($\uparrow$)	DAC ($\uparrow$)
ATTGLOBAL-CAM-NF (MSE)	0.735	1.379	1.918	21.48	**0.924**
ATTGLOBAL-CAM-NF	**0.638**	**1.171**	**2.558**	**22.62**	0.918

Table 4. Multi-agent experiments on NUSCENES (MINFDE). RI denotes relative improvements of MINFDE between 10 and 1 agent. Our approach best models multi-agent interactions.

Model	1 agent	3 agents	5 agents	10 agents	RI(1–10)
LSTM	2.736	2.477	2.442	2.268	17.1%
CSP	2.871	2.679	2.671	2.569	10.5%
DESIRE [17]	2.150	1.846	1.878	1.784	17.0%
MATF GAN [33]	2.377	2.168	2.150	2.011	15.4%
R2P2-MA [21]	2.227	2.135	2.142	2.048	8.0%
ATTGLOBAL-CAM-NF (OURS)	**1.278**	**1.158**	**1.100**	**0.964**	**24.6%**

Effectiveness of Agent-to-Scene Interaction Module: The performance of stochastic models is compared in Table 2. We experiment with removing scene processing operations in the decoder to validate the importance of our proposed agent-to-scene interaction module. As mentioned previously, generating multiple modes of sample requires a strong scene processing module and a diversity-oriented decoder. Our models achieves the best precision. MATF-GAN has a small rF showing that predictions are unimodal, while other models such as VAE-based model (DESIRE) and flow-based models (R2P2-MA and OURS) show spread in their predictions. R2P2-MA shows the highest DAO. We note that OURS has a comparable DAO while keeping the highest RF and DAC, indicating that our models exhibit diverse and admissible predictions by accurately utilizing scene context.

Effectiveness of New $\tilde{p}$: We compare two different versions of our ATTGLOBAL-CAM-NF. One is trained using mean squared error (MSE) between S_{pred} and $\hat{S}_{pred}$ as an example of annotation-based approximation for p, while the other is trained with our drivable area-based (annotation-free) approximation of p in Table 3. Using our new approximation in training shows superior results in most of the reported metrics. In particular, the precision and the diversity of predictions increases drastically as reflected in MINERROR, DAO, and RF while DAC remains comparable. Thus our $\tilde{p}$ considers admissibility while improving precision and diversity i.e drivable-area related approximate enhances the estimate on additional trajectories over the most probable one.

Complexity from Number of Agents: We experiment with varying number of surrounding agents as shown in Table 4. Throughout all models, the performance increases as the number of agents increases even though we observe that many agents in the surrounding do not move significantly. In terms of relative improvement RI, as calculated between 1 agent and 10 agents, our model has

Table 5. Results of baseline models (upper partition) and our proposed models (lower partition). ATTGLOBAL-CAM-NF is our full proposed model and others in the lower partition are the ablations. The metrics are abbreviated as follows: MINADE(**A**), MINFDE(**B**), RF(**C**), DAO(**D**), DAC(**E**). Improvements indicated by arrows.

Model	ARGOVERSE					NUSCENES				
	A (↓)	B (↓)	C (↑)*	D (↑)*	E (↑)*	A (↓)	B (↓)	C (↑)*	D (↑)*	E (↑)*
LSTM	1.441	2.780	1.000	3.435	0.959	1.186	2.408	1.000	3.187	0.912
CSP	1.385	2.567	1.000	3.453	0.963	1.390	2.676	1.000	3.228	0.900
MATF-D [33]	1.344	2.484	1.000	3.372	0.965	1.261	2.538	1.000	3.191	0.906
DESIRE [17]	0.896	1.453	3.188	15.17	0.457	1.079	1.844	1.717	16.29	0.776
MATF-GAN [33]	1.261	2.313	1.175	11.47	0.960	1.053	2.126	1.194	11.64	0.910
R2P2-MA [22]	1.108	1.771	3.054	**37.18**	0.955	1.179	2.194	1.636	**25.65**	0.893
CAM	1.131	2.504	1.000	3.244	**0.973**	1.124	2.318	1.000	3.121	**0.924**
CAM-NF	0.851	1.349	2.915	32.89	0.951	0.756	1.381	2.123	23.15	0.914
LOCAL-CAM-NF	0.808	1.253	3.025	31.80	0.965	0.774	1.408	2.063	22.58	0.921
GLOBAL-CAM-NF	0.806	1.252	3.040	31.59	0.965	0.743	1.357	2.106	22.65	0.921
ATTGLOBAL-CAM-NF	**0.730**	**1.124**	**3.282**	28.64	0.968	**0.639**	**1.171**	**2.558**	22.62	0.918

the most improvement, indicating that our model makes the most use of the fine-grained trajectories of surrounding agents to generate future trajectories.

Generalizability Across Datasets: We compare our model with baselines extensively across another real world dataset ARGOVERSE to test generalization to different environments. We show results in Table 5 where we outperform or achieve comparable results as compared to the baselines. For ARGOVERSE, we additionally outperform MFP3 [26] in MINFDE with 6 hypotheses: our full model shows a MINFDE of 0.915, while MFP3 achieves 1.399.

7 Conclusion

In this paper, we tackled the problem of generating diverse and admissible trajectory predictions by understanding each agent's multimodal context. We proposed a model that learns agent-to-agent interactions and agent-to-scene interactions using attention mechanisms, resulting in better prediction in terms of precision, diversity, and admissibility. We also developed a new approximation method that provides richer information about the true trajectory distribution and allows more accurate training of flow-based generative models. Finally, we present new metrics that provide a holistic view of the quantity and the quality of diversity in prediction, and a NUSCENES trajectory extraction code to support future research in diverse and admissible trajectory forecasting.

Acknowledgements. This work was supported in part by the Technology Innovation Program under Grant 10083646 (Development of Deep Learning-Based Future Prediction and Risk Assessment Technology considering Inter-vehicular Interaction in Cut-in Scenario), funded by the Ministry of Trade, Industry, and Energy, South Korea. We also acknowledge the anonymous reviewers for their constructive comments.

References

1. Alahi, A., Goel, K., Ramanathan, V., Robicquet, A., Fei-Fei, L., Savarese, S.: Social lSTM: human trajectory prediction in crowded spaces. In: Proceedings of the IEEE Conference on Computer Vision and Pattern Recognition, pp. 961–971 (2016)
2. Ballan, L., Castaldo, F., Alahi, A., Palmieri, F., Savarese, S.: Knowledge transfer for scene-specific motion prediction. In: Leibe, B., Matas, J., Sebe, N., Welling, M. (eds.) Computer Vision – ECCV 2016. Lecture Notes in Computer Science, vol. 9905, pp. 697–713. Springer, Cham (2016). https://doi.org/10.1007/978-3-319-46448-0_42
3. Bansal, M., Krizhevsky, A., Ogale, A.: Chauffeurnet: learning to drive by imitating the best and synthesizing the worst. arXiv preprint arXiv:1812.03079 (2018)
4. Bernstein, D.S., So, W.: Some explicit formulas for the matrix exponential. IEEE Trans. Autom. Control **38**(8), 1228–1232 (1993)
5. Caesar, H., et al.: nuScenes: a multimodal dataset for autonomous driving. In: Proceedings of the IEEE/CVF Conference on Computer Vision and Pattern Recognition, pp. 11621–11631 (2020)
6. Casas, S., Luo, W., Urtasun, R.: IntentNet: learning to predict intention from raw sensor data. In: Conference on Robot Learning, pp. 947–956 (2018)
7. Chang, M.F., et al.: Argoverse: 3D tracking and forecasting with rich maps. In: Proceedings of the IEEE Conference on Computer Vision and Pattern Recognition, pp. 8748–8757 (2019)
8. Deo, N., Trivedi, M.M.: Convolutional social pooling for vehicle trajectory prediction. In: Proceedings of the IEEE Conference on Computer Vision and Pattern Recognition Workshops, pp. 1468–1476 (2018)
9. Fernando, T., Denman, S., Sridharan, S., Fookes, C.: Soft + hardwired attention: an lstm framework for human trajectory prediction and abnormal event detection. Neural networks **108**, 466–478 (2018)
10. Gindele, T., Brechtel, S., Dillmann, R.: Learning driver behavior models from traffic observations for decision making and planning. IEEE Intell. Transp. Syst. Mag. **7**(1), 69–79 (2015)
11. Gupta, A., Johnson, J., Fei-Fei, L., Savarese, S., Alahi, A.: Social GAN: socially acceptable trajectories with generative adversarial networks. In: Proceedings of the IEEE Conference on Computer Vision and Pattern Recognition, pp. 2255–2264 (2018)
12. Huang, X., et al.: DiversityGAN: diversity-aware vehicle motion prediction via latent semantic sampling. IEEE Robot. Autom. Lett. (2020)
13. Kim, B., Kang, C.M., Kim, J., Lee, S.H., Chung, C.C., Choi, J.W.: Probabilistic vehicle trajectory prediction over occupancy grid map via recurrent neural network. In: 2017 IEEE 20th International Conference on Intelligent Transportation Systems (ITSC), pp. 399–404. IEEE (2017)
14. Kingma, D.P., Salimans, T., Jozefowicz, R., Chen, X., Sutskever, I., Welling, M.: Improved variational inference with inverse autoregressive flow. In: Advances in Neural Information Processing systems. pp. 4743–4751 (2016)
15. Kooij, J.F.P.: Context-based pedestrian path prediction. In: Fleet, D., Pajdla, T., Tuytelaars, T. (eds.) European Conference on Computer Vision. Lecture Notes in Computer Science, vol. 8694, pp. 618–633. Springer, Cham (2014)
16. Krajewski, R., Bock, J., Kloeker, L., Eckstein, L.: The highd dataset: a drone dataset of naturalistic vehicle trajectories on German highways for validation of highly automated driving systems. In: 2018 21st International Conference on Intelligent Transportation Systems (ITSC), pp. 2118–2125. IEEE (2018)

17. Lee, N., Choi, W., Vernaza, P., Choy, C.B., Torr, P.H., Chandraker, M.: Desire: distant future prediction in dynamic scenes with interacting agents. In: Proceedings of the IEEE Conference on Computer Vision and Pattern Recognition, pp. 336–345 (2017)
18. Ma, Y., Zhu, X., Zhang, S., Yang, R., Wang, W., Manocha, D.: Trafficpredict: Trajectory prediction for heterogeneous traffic-agents. In: Proceedings of the AAAI Conference on Artificial Intelligence, vol. 33, pp. 6120–6127 (2019)
19. Park, S.H., Kim, B., Kang, C.M., Chung, C.C., Choi, J.W.: Sequence-to-sequence prediction of vehicle trajectory via LSTM encoder-decoder architecture. In: 2018 IEEE Intelligent Vehicles Symposium (IV), pp. 1672–1678. IEEE (2018)
20. Rezende, D.J., Mohamed, S.: Variational inference with normalizing flows. arXiv preprint arXiv:1505.05770 (2015)
21. Rhinehart, N., Kitani, K.M., Vernaza, P.: R2p2: A reparameterized pushforward policy for diverse, precise generative path forecasting. In: Proceedings of the European Conference on Computer Vision (ECCV). pp. 772–788 (2018)
22. Rhinehart, N., McAllister, R., Kitani, K., Levine, S.: Precog: Prediction conditioned on goals in visual multi-agent settings. In: Proceedings of the IEEE International Conference on Computer Vision. pp. 2821–2830 (2019)
23. Rudenko, A., Palmieri, L., Herman, M., Kitani, K.M., Gavrila, D.M., Arras, K.O.: Human motion trajectory prediction: A survey. The International Journal of Robotics Research p. 0278364920917446 (2019)
24. Sadeghian, A., Kosaraju, V., Sadeghian, A., Hirose, N., Rezatofighi, H., Savarese, S.: Sophie: An attentive gan for predicting paths compliant to social and physical constraints. In: Proceedings of the IEEE Conference on Computer Vision and Pattern Recognition. pp. 1349–1358 (2019)
25. Schulz, J., Hubmann, C., Löchner, J., Burschka, D.: Interaction-aware probabilistic behavior prediction in urban environments. In: 2018 IEEE/RSJ International Conference on Intelligent Robots and Systems (IROS). pp. 3999–4006. IEEE (2018)
26. Tang, C., Salakhutdinov, R.R.: Multiple futures prediction. In: Advances in Neural Information Processing Systems. pp. 15398–15408 (2019)
27. Vaswani, A., Shazeer, N., Parmar, N., Uszkoreit, J., Jones, L., Gomez, A.N., Kaiser, Ł., Polosukhin, I.: Attention is all you need. In: Advances in neural information processing systems. pp. 5998–6008 (2017)
28. Vemula, A., Muelling, K., Oh, J.: Social attention: Modeling attention in human crowds. In: 2018 IEEE international Conference on Robotics and Automation (ICRA). pp. 1–7. IEEE (2018)
29. Verlet, L.: Computer" experiments" on classical fluids. i. thermodynamical properties of lennard-jones molecules. Physical review 159(1), 98 (1967)
30. Xie, G., Gao, H., Qian, L., Huang, B., Li, K., Wang, J.: Vehicle trajectory prediction by integrating physics-and maneuver-based approaches using interactive multiple models. IEEE Transactions on Industrial Electronics 65(7), 5999–6008 (2017)
31. Xu, K., Ba, J., Kiros, R., Cho, K., Courville, A., Salakhudinov, R., Zemel, R., Bengio, Y.: Show, attend and tell: Neural image caption generation with visual attention. In: International conference on machine learning. pp. 2048–2057 (2015)
32. Yuan, Y., Kitani, K.M.: Diverse trajectory forecasting with determinantal point processes. In: International Conference on Learning Representations (2020)
33. Zhao, T., Xu, Y., Monfort, M., Choi, W., Baker, C., Zhao, Y., Wang, Y., Wu, Y.N.: Multi-agent tensor fusion for contextual trajectory prediction. In: Proceedings of the IEEE Conference on Computer Vision and Pattern Recognition. pp. 12126–12134 (2019)

CONFIG: Controllable Neural Face Image Generation

Marek Kowalski$^{(\boxtimes)}$, Stephan J. Garbin, Virginia Estellers, Tadas Baltrušaitis, Matthew Johnson, and Jamie Shotton

Microsoft, Cambridge, UK

Abstract. Our ability to sample realistic natural images, particularly faces, has advanced by leaps and bounds in recent years, yet our ability to exert fine-tuned control over the generative process has lagged behind. If this new technology is to find practical uses, we need to achieve a level of control over generative networks which, without sacrificing realism, is on par with that seen in computer graphics and character animation. To this end we propose ConfigNet, a neural face model that allows for controlling individual aspects of output images in semantically meaningful ways and that is a significant step on the path towards finely-controllable neural rendering. ConfigNet is trained on real face images as well as synthetic face renders. Our novel method uses synthetic data to factorize the latent space into elements that correspond to the inputs of a traditional rendering pipeline, separating aspects such as head pose, facial expression, hair style, illumination, and many others which are very hard to annotate in real data. The real images, which are presented to the network without labels, extend the variety of the generated images and encourage realism. Finally, we propose an evaluation criterion using an attribute detection network combined with a user study and demonstrate state-of-the-art individual control over attributes in the output images.

Keywords: Neural rendering · Face image manipulation · GAN

1 Introduction

Recent advances in generative adversarial networks (GANs) [5,17,18] have enabled the production of realistic high resolution images of smooth organic objects such as faces. Generating photorealistic human bodies, and faces in particular, with traditional rendering pipelines is notoriously difficult [26], requiring hand-crafted 3D assets. However, once these assets have been generated we can render the face from any direction and in any pose. In contrast, GANs can be used to easily generate realistic head and face images without the need to author expensive 3D assets, by training on curated datasets of 2D images of real human

Electronic supplementary material The online version of this chapter (https://doi.org/10.1007/978-3-030-58621-8_18) contains supplementary material, which is available to authorized users.

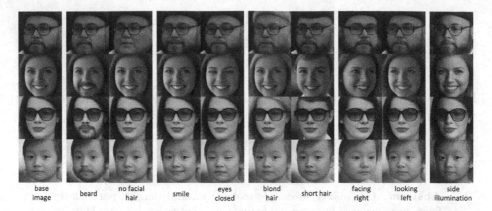

base image	beard	no facial hair	smile	eyes closed	blond hair	short hair	facing right	looking left	side illumination

Fig. 1. ConfigNet learns a factorized latent space, where each part corresponds to a different facial attribute. The first column shows images produced by ConfigNet for certain points in the latent space. The remaining columns show changes to various parts of the latent space vectors, where we can generate attribute combinations outside the distribution of the training set like children or women with facial hair.

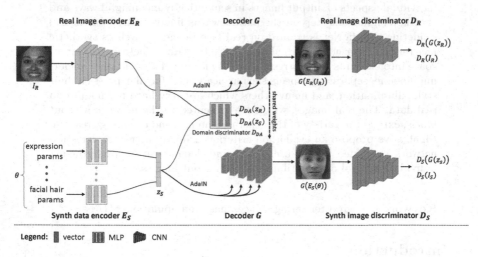

Fig. 2. ConfigNet has two encoders E_R and E_S that encode real face images I_R and the parameters θ of synthetic face images I_S. The encoders output latent space vectors z_R, z_S. The shared decoder, G, generates both real and synthetic images. A domain discriminator D_{DA} ensures the latent distributions generated by E_R and E_S are similar.

faces. However, it is difficult to enable meaningful control over this generation without detailed hand labelling of the dataset. Even when conditional models are trained with detailed labels, they struggle to generalize to out-of-distribution combinations of control parameters such as children with extensive facial hair or young people with gray hair. In order for GAN based rendering techniques to replace traditional rendering pipelines they must enable a greater level of control.

In this paper we present ConfigNet, one of the first methods to enable control of GAN outputs using the same methods as traditional graphics pipelines. The key idea behind ConfigNet is to train the generative model on both real and synthetically generated face images. Since the synthetic images were generated with a traditional graphics pipeline, the renderer parameters for those images are readily available. We use those known correspondences to train a generative model that uses the same input parametrization as the graphics pipeline used to generate the synthetic data. This allows for independent control of various face aspects including: head pose, hair style, facial hair style, expression and illumination. By simultaneously training the model on unlabelled face images, it learns to generate photorealistic looking faces, while enabling full control over these outputs. Figure 1 shows example results produced by ConfigNet.

ConfigNet can be used to both sample novel images and to embed existing ones, which can then be manipulated. The ability to embed face images sets ConfigNet apart from traditional graphics pipelines, which would require person-specific 3D assets to achieve similar results. The use of a parametrization derived from a traditional graphics pipeline makes ConfigNet easy to use for people familiar with digital character animation. For example, facial expressions are controlled with blendshapes with values in $(0, 1)$, head pose is controlled with Euler angles and illumination can be set using an environment map.

Our main contributions are:

1. ConfigNet, a novel method for placing real and synthetic data into a single factorized and disentangled latent space that is parametrized based on a computer graphics pipeline.
2. A method for using ConfigNet to modify existing face images in a fine-grained way that allows for changing parts of the latent space factors meaningfully.
3. Experiments showing our method generating realistic face images with attribute combinations that are not present in the real images of the training set. For example, a face of a child with extensive facial hair.

2   Related Work

Image Generation Driven by 3D Models. One of the uses of synthetic data in image generation is the "synthetic to real" scenario, where the goal is to generate realistic images that belong to a target domain based on synthetic images, effectively increasing their realism. The methods that tackle this problem [8,34,44] usually use a neural network with an adversarial and semantic loss to push a synthetic image closer to the real domain. While those type of methods can generate realistic images that are controllable through synthetic data, the editing of existing images is difficult as it would require fitting the underlying 3D model to an existing image.

3D model parameters can also be a supervision signal for generative models as shown in [20,33] and most recently StyleRig [36]. This group of methods shares the challenges of the synthetic to real scenario as they require fitting the 3D model to existing images.

PuppetGAN [40], is a method designed to edit existing images using synthetic data of the same class of objects. It uses two encoder-decoder pairs, one for real and one for synthetic images, which have a common latent space, part of which is designated for an attribute of interest that can be controlled. An image can be edited by encoding it with the real-data encoder, then swapping the attribute of interest part of the latent space with one encoded from a synthetic image and finally decoding with the real-data decoder. Due to the use of separate decoders for real and synthetic data PuppetGAN struggles to decode images where the attribute of interest is outside of the range seen in real data. The method performs well for a single attribute. In contrast, ConfigNet demonstrates disentanglement of multiple face attributes as well as generation of attribute combinations that do not exist in the real training data.

Disentangled Representation Learning. Supervised disentanglement methods try to learn a factorised representation, parts of which correspond to some semantically meaningful aspects of the generated images, based on labelled data in the target domain (as opposed to the synthetic data domain). The major limitation of these methods [7,24,29,43] is that they are only able to disentangle factors of variations that are labelled in the training set. For human faces, labels are easily obtainable for some attributes, such as identity, but the task becomes more difficult with attributes like illumination and almost impossible with attributes like hair style. This labelling problem also becomes more difficult as the required fidelity of the labels increases (e.g. smile intensity).

Unsupervised disentanglement methods share the above goal but do not require labelled data. Most methods in this family, such as β-VAE [12], Info-GAN [6], ID-GAN [21], place constraints on the latent space that lead to disentanglement. The fundamental problem with those approaches is that there is little control over what factors get disentangled and which part of the latent space corresponds to a given factor of variation. HoloGAN [27] separates the 3D rotation of the object in the image from variation in its shape and appearance. ConfigNet borrows the generator architecture of HoloGAN, while disentangling many additional factors of variation and allowing existing images to be edited.

Detach and Adapt [22] is trained in a semi-supervised way on images from two related domains, only one of which has labels. The resulting model allows generating images in both domains with some control over the labelled attribute.

Face Video Re-enactment. Face video re-enactment methods aim to produce a video of a certain person's articulated face that is driven by a second video of the same or a different person. The methods in [37,41,43] achieve some of the most impressive face manipulation results seen to date. Face2Face [37] fits a 3D face model and illumination parameters to a video of a person and then re-renders the sequence with modified expression parameters that are obtained from a different sequence. This approach potentially allows for modifying any aspects of the rendered face that can be modelled, rendered and fitted to the input video. In practice, due to limitations of existing 3D face models and fitting methods, this approach cannot modify complex face attributes like hair style or attributes that require modelling of the whole head, like head pose.

Zakharov et al. [43] propose a video re-enactment method where the images are generated by a neural network driven by face landmarks from a different video sequence. The method produces impressive results given only a small number of target frames. X2Face [41] uses one neural network to resample the source image into a standard reference frame and a second network that resamples this standardized image into a different head pose or facial expression, which can be driven by images or audio signal. While these two methods produce convincing results, the controllability is limited to head pose and expression.

3   Method

The key concept behind the proposed method is to factorize the latent space into parts that correspond to separate and clearly-defined aspects of face images. These factors can be individually swapped or modified (Sect. 3.6) to modify the corresponding aspect of the output image. The factorization needs labels that fully explain the image content, which would require laborious annotation for real data, but are easily obtained for synthetic data. We thus propose a generative model trained in a semi-supervised way, with labels that are known for synthetic data only. Figure 2 outlines the proposed architecture.

Overview. Our approach is to treat the synthetic images $\mathcal{I}_S$ and real images $\mathcal{I}_R$ as two different subsets of a larger set of all possible face images. Hence, the proposed method consists of a decoder G and two encoders E_R and E_S that embed real and synthetic data into a common factorized latent space z (Sect. 3.1). We will refer to z predicted by E_R and E_S as z_R and z_S respectively. The real data is supplied to its encoder as images $I_R \in \mathcal{I}_R$, while the synthetic data is supplied as vectors $\theta \in \mathbb{R}^m$ that fully describe the content of the corresponding image $I_S \in \mathcal{I}_S$. To increase the realism of the generated images we employ two discriminator networks D_R and D_S for real and synthetic data respectively.

We assume that the synthetic data is a reasonable approximation of the real data so that $\mathcal{I}_S \cap \mathcal{I}_R \neq \emptyset$. Hence, it is desirable for $E_S(\Theta)$ and $E_R(\mathcal{I}_R)$, where Θ is the space of all θ, to also be overlapping. To do so, we introduce a domain discriminator network D_{DA} and train it with a domain adversarial loss [38] on z, that forces z_R and z_S to be close. In Sect. 4.2 we show that this loss is crucial for the method's ability to control the attributes of the output images.

To accurately reproduce and modify existing images we employ one-shot learning (Sect. 3.4) that improves reconstruction accuracy compared to embedding using E_R. To enable the sampling of novel images we train a latent GAN that generates samples of z (Sect. 3.5). Finally, we propose a method for modifying attributes of existing images in a fine-grained way that allows for changing parts of individual factors of z meaningfully (Sect. 3.6).

3.1 Factorized Latent Space

Each synthetic data sample θ is factorised into k parts θ_1 to θ_k, such that:

$$\theta \in \mathbb{R}^m = \mathbb{R}^{m_1} \times \mathbb{R}^{m_2} \times \ldots \times \mathbb{R}^{m_k}. \tag{1}$$

Each θ_i corresponds to semantically meaningful input of the graphics pipeline used to generate $\mathcal{I}_S$. Examples of such inputs are: facial expression, facial hair parameters, head shape, environment map, etc. The synthetic data encoder E_S maps each θ_i to z_i, a part of z, which thus factorizes z into k parts.

The factorized latent space is a key feature of ConfigNet that allows for easy modification of various aspects of the generated images. For example, one might encode a real image into z using E_R and then change the illumination by swapping out the part of z that corresponds to illumination. Note that the part of z that is swapped in might come from θ_i (encoded by E_S), which is semantically meaningful, or it may come from a different real face image encoded by E_R.

3.2 Loss Functions

To ensure that the output image $G(z)$ is close to the ground truth image I_{GT}, we use the perceptual loss $\mathcal{L}_{perc}$ [15], which is the MSE between the activations of a pre-trained neural network computed on $G(z)$ and I_{GT}. We use VGG-19 [35] trained on ImageNet [30] as the pre-trained network. We experimented with using VGGFace [28] as base for the perceptual loss, but didn't see improvement.

While the perceptual loss retains the overall content of the image well, it struggles to preserve some small scale features. Because of that, we use an additional loss with the goal of preserving the eye gaze direction:

$$\mathcal{L}_{eye} = w_M \sum M \circ (I_{GT} - G(z_s)) \quad \text{with} \quad w_M = (1 + |M|_1)^{-1}, \tag{2}$$

where M is a pixel-wise binary mask that denotes the iris, only available for $\mathcal{I}_S$. Thanks to the accurate ground truth segmentation that comes with the synthetic data, similar losses could be added for any part of the face if necessary.

We train the adversarial blocks with the non-saturating GAN loss [9]:

$$\mathcal{L}_{GAN_D}(D, x, y) = \log D(x) + \log(1 - D(y)), \tag{3}$$

$$\mathcal{L}_{GAN_G}(D, y) = \log(D(y)), \tag{4}$$

where $\mathcal{L}_{GAN_D}$ is used for the discriminator and $\mathcal{L}_{GAN_G}$ is used for the generator, D is the discriminator, x is a real sample and y is a network output.

3.3 Two-Stage Training Procedure

First Stage: We train all the sub-networks except E_R, sampling $z_R \sim \mathcal{N}(0, \boldsymbol{I})$ as there is no encoder for real data at this stage. At this stage E_S and G are trained with the following loss:

$$\begin{aligned} \mathcal{L}_1 = \mathcal{L}_{GAN_G}(D_R, G(z_R)) + \mathcal{L}_{GAN_G}(D_{DA}, z_S) \\ + \mathcal{L}_{GAN_G}(D_S, G(z_S)) + \lambda_{eye}\mathcal{L}_{eye} + \lambda_{perc}\mathcal{L}_{perc}(G(z_S), I_S), \end{aligned} \tag{5}$$

where $z_S = E_S(\theta)$ and λ are the loss weights. The domain discriminator D_{DA} acts on E_S to bring the distribution of its outputs closer to $\mathcal{N}(0, \boldsymbol{I})$ and so E_S effectively maps the distribution of each θ_i to $\mathcal{N}(0, \boldsymbol{I})$.

Second Stage: We add the real data encoder E_R so that $z_R = E_R(I_R)$. The loss used for training E_S and G is then:

$$\mathcal{L}_2 = \mathcal{L}_1 + \lambda_{perc}\mathcal{L}_{perc}\left(G(z_R), I_R\right) + \log(1 - D_{DA}(z_R)), \qquad (6)$$

where the goal of $\log(1 - D_{DA}(z_R))$ is to bring the output distribution of E_R closer to that of E_S. In the second stage we increase the weight of λ_{perc}, in the first stage it is set to a lower value as otherwise the total loss for synthetic data would overpower that for real data. Our experiments show that this two-stage training improves controllability and image quality.

3.4   One-Shot Learning by Fine-Tuning

Our architecture allows for embedding face images into z using the real data encoder E_R, individual factors z_i can then be modified to modify the corresponding output image as explained in Sects. 3.1 and 3.6. We have found that while $G(E_R(I_R))$ is usually similar to I_R as a whole image, there is often an identity gap between the face in I_R and in the generated image. A similar finding was made in [43], where the authors proposed to decrease the identity gap by fine-tuning the generator on the images of a given person.

Similarly, we fine-tune our generator on I_R by minimizing the following loss:

$$\mathcal{L}_{ft} = \mathcal{L}_{GAN_G}(D_R, G(\hat{z_R})) + \log(1 - D_{DA}(\hat{z_R}))$$
$$+ \lambda_{perc}[\mathcal{L}_{perc}\left(G(\hat{z_R}), I_R\right) + \mathcal{L}_{face}\left(G(\hat{z_R}), I_R\right)], \qquad (7)$$

where $\mathcal{L}_{face}$ is a perceptual loss with VGGFace [28] as the pre-trained network. We optimize over the weights of G as well as $\hat{z_R}$ which is initialized with $E_R(I_R)$. The addition of a $\mathcal{L}_{face}$ improves the perceptual quality of the generated face images. We believe that this improvement is visible here, but not in the main training phase, as fine-tuning lacks the regularization provided by training on a large number of images and can easily "fool" the single perceptual loss.

3.5   Sampling of z

While the proposed method allows for embedding existing face images into the latent space, sometimes it might be desirable to sample the latent space itself. Samples of the latent space can be used to generate novel images or to sample individual factor z_i. The sampled z_i can then be used to generate additional variations of an existing image that was embedded in z.

To do this, we use a latent GAN [1]. The latent GAN is trained to map between its input $w \sim \mathcal{N}(0, \boldsymbol{I})$ and the latent space z. This simple approach allows for sampling the latent space without the constraints on z imposed by VAEs that lead to reduced quality. The latent GAN is trained with the GAN losses described above, both the discriminator and generator G_{lat} are 3-layer MLPs. Figure 18 in suppl. shows an outline of the method when used with G_{lat}.

3.6   Fine-Grained Control

Given an existing face image embedded into z, we can easily swap any part, z_i, of its embedding with one that is obtained from E_S or E_R. However, sometimes we might want a finer level of control and only modify a single aspect of z_i while leaving the rest the same. If z_i is a face expression, its single aspect might be the intensity of smile, if z_i is illumination, the brightness might be one aspect. These aspects are controlled by individual elements of the corresponding θ_i vector. However θ_i is unknown if z was generated by E_R or G_{lat}.

For this reason, we use an approximation $\tilde{\theta}_i$ obtained by solving the minimization problem $\min_{\tilde{\theta}_i} |z_i - E_{S_i}(\tilde{\theta}_i)|^2$ with gradient descent, where E_{S_i} is the part of E_S that corresponds to θ_i. We incorporate constraints on θ_i into the optimization algorithm. For example, our expression parameters lie in the convex set $[0, 1]$ and we use projected gradient descent to incorporate the constraint into the minimization algorithm. Given $\tilde{\theta}_i$, e.g. a face expression vector, we can modify the part of the vector responsible for an individual expression and use E_S to obtain a new latent code z_i that generates images where only this individual expression is modified. We use this approach to manipulate individual expressions in Figs. 1 and 7 and combinations in Fig. 12 (supplementary). The method described above is also outlined in Algorithm 1 in supplementary.

3.7   Implementation

The architecture of the decoder G is based on the generator used in HoloGAN [27], explained in supplementary. We choose this particular architecture as it decouples object rotation from z and it allows for specifying the rotation with any parametrization. This lets us obtain the poses of the heads in $\mathcal{I}_S$ in ConfigNet parametrization and supply head pose directly, without an encoder.

The remaining $k - 1$ parts of θ are encoded with separate multi layer perceptrons (MLPs) E_{S_i}, each of which consists of 2 layers with number of hidden units equal to the dimensionality of the corresponding θ_i. The real image encoder E_R is a ResNet-50 [10] pre-trained on ImageNet [30]. The domain discriminator D_{DA} is a 4-layer MLP. The two image discriminators D_R and D_S share the same basic convolutional architecture. The supplementary material contains all network details, source code is available at http://aka.ms/confignet.

4   Experiments

Datasets. We use the FFHQ [17] (60k images, 1Mpix each), and SynthFace (30k images, 1Mpix each) datasets as a source of real and synthetic training images. We align the face images from all datasets to a standard reference frame using landmarks from OpenFace [2,4,42] and reduce the resolution to 256×256 pixels.

Our experiments use the 10k images in the validation set of FFHQ to evaluate ConfigNet. The SynthFace dataset was generated using the method of [3] and

Fig. 3. Images from SynthFace dataset, note the domain gap to real images.

Fig. 4. Left: $G(z)$ trained using the two-stage method, where z is sampled from latent GAN. Right: $G(z)$ trained using the first stage only, where z is sampled from prior. Note the large improvement in quality when second stage is added.

setting rotation limits for yaw and pitch to $\pm 30°$ and $\pm 10°$ to cover the typical range of poses in face images. For SynthFace, θ has $m = 304$ dimensions, while z has $n = 145$ dimensions, and is divided into $k = 12$ factors. Table 6 in the supplementary provides the dimensionality of each factor in θ and z, Fig. 3 shows sample SynthFace images.

4.1 Evaluation of ConfigNet

Our experiments evaluate ConfigNet key features: photorealism and control.

Photorealism. Figure 4 shows samples generated by the latent GAN (where E_R, G were trained using the two stage-procedure of Sect. 3.3) and a standard GAN model trained only with the first-stage procedure. We observe a large improvement in photorealism when the second stage of training is added. We believe that the low-quality images produced by the standard GAN are caused by the constraint $z \sim \mathcal{N}(0, I)$, which is relaxed in our second-stage training thus allowing real and synthetic data to co-exist in the same space.

We quantitatively measure the photorealism and coverage of the generated images using the Frechet Inception Distance (FID) [11] in Table 1. The latent GAN achieves scores that are close to those produced by sampling z through E_R, which is the upper limit of its performance. Training only the first stage and sampling $z \sim \mathcal{N}(0, I)$ results in poorer metrics. As expected, the raw synthetic images give the worst result. To further evaluate how much of the photorealism of the generated data is lost due to training on both real and synthetic data, we train ConfigNet without synthetic data and the losses that require its presence. We find that the resulting FID is very close to those produced by our standard

Table 1. FID score for FFHQ, SynthFace, and images obtained with our decoder G and latent codes from the real-image encoder E_R and latent GAN G_{lat}.

Method	FID↓
$G(E_R(I_R))$	33.41
Synthetic data $\mathcal{I}_S$	52.19
$G(E_R(I_R))$ trained without $\mathcal{I}_S$	33.49
$G(G_{lat}(w))$, $w \sim \mathcal{N}(0,(I))$	39.76
$G(z)$, $z \sim \mathcal{N}(0,(I))$ no 2nd stage	43.05

training. This suggests that the photorealism of the results might be limited by our network architecture rather then by the use of synthetic data. We speculate that using a more powerful G and D_R, for example the ones used in StyleGAN, may lead to improved results.

Controllability. We evaluate ConfigNet's controllability analysing how changing a specific attribute (e.g., hair colour) changes the output image: with perfect control, the output image should only change with respect to that attribute.

Figures 1 and 6 show controllability qualitatively. Figure 1 shows that the generator is able to modify individual attributes of faces embedded in its latent space, while Fig. 6 shows that each attribute can take many different values while only influencing certain aspects of the produced image. The second column of Fig. 1 shows that we are able to set facial hair to faces of children and women, demonstrating that the generator is not constrained by the distribution of the real training data. Fine-grained control over individual expressions is shown in Fig. 7 as well as Fig. 12 in the supplementary. The supplementary also includes additional results of face attribute manipulation and interpolation, including a video.

To evaluate if ConfigNet offers this ideal level of control quantitatively, we propose the following experiment: We take a random image I_R from the FFHQ validation set, encode it into latent space $z = E_R(I_R)$ and then swap the latent factor z_i that corresponds to a given attribute v (for example hair colour) with a latent factor obtained with E_S. For each attribute v we output two images: I_+ where the attribute is set to a certain value v_+ (e.g. blond hair) and another I_- with the attribute takes a semantically opposite value v_- (e.g., black hair)[1]. This gives us image pairs (I_+, I_-) that should be identical except for the chosen attribute v, where they should differ. We measure how and where these images differ with an attribute predictor and a user study.

We train an attribute predictor C_{pred} on CelebA [23] to predict 38 face attributes and use it with 1000 FFHQ validation images to estimate (1) if v_+ is present in each set of images pairs (I_+, I_-) and (2) if the other face attributes change. Ideally, $C_{pred}(I_+) = 1$, $C_{pred}(I_-) = 0$ and the Mean Absolute Difference (MD) for other face attributes should be 0. Figure 5a shows how

---

[1] We choose the values of Θ_i for v_+ and v_- by manual inspection, details in suppl.

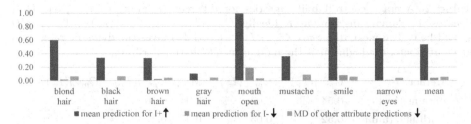

(a) Evaluation of controllability and disentanglement with an attribute predictor.

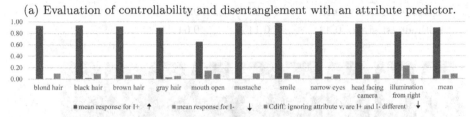

(b) Evaluation of controllability and disentanglement with a user study.

Fig. 5. Evaluation of control and disentanglement of ConfigNet. Blue and orange bars show the predicted values of given attribute for images with that attribute (I_+, higher better) and images with an opposite attribute (I_-, lower better). The gray bars measure differences of other attributes (MD and C_{diff}, lower better). (Color figure online)

$C_{pred}(I_+) \ggg C_{pred}(I_-)$ while the MD of other attributes is close to 0. The best controllability is achieved for the mouth opening and smile attributes, with $C_{pred}(I_+)$ approaching the ideal value of 1, while the poorest results are achieved for the gray hair attribute. We believe those large differences are caused by bias in CelebA, where certain attributes are not distributed evenly across age (for example gray hair) or gender (for example moustache).

Our user study C_{user} follows a similar evaluation protocol: 59 users evaluated the presence of v_+ in a total of 1771 images pairs I_+ and I_- on a 5-level scale and gave a score C_{diff} that measures whether, ignoring v, the images depict the same person. Figure 5b shows the results of the controllability and disentanglement metrics for the user study: users evaluate the controllability of the given attribute higher than the feature predictor C_{pred}, with $C_{user}(I_+) > C_{pred}(I_+)$ and $C_{user}(I_+) - C_{user}(I_-) > C_{pred}(I_+) - C_{pred}(I_-)$ for all features except *mouth open*, while the score C_{diff} measuring whether I_+ and I_- show the same person has low values indicating that features other than v_+ remain close to constant. This results support the result of the feature predictor and show a similar performance for different attributes because user judgements do not suffer from the bias of the attribute predictor trained on CelebA.

4.2 Ablation Study

We evaluate the importance of two stage training and the domain discriminator D_{DA} by training the neural network without them. Table 2 shows how each

Table 2. Average controllability metrics for different variants of ConfigNet. D_{DA} denotes the domain discriminator. Ideally, $C_{pred}(I_+) = 1$, $C_{pred}(I_-) = 0$ and MD should be 0. The mean difference $C_{pred}(I_+) - C_{pred}(I_-)$ gives the dynamic range of a given attribute, the higher it is the more controllable the attribute.

Method	$C_{pred}(I_+)$ ↑	$C_{pred}(I_-)$ ↓	MD↓	$C_{pred}(I_+) - C_{pred}(I_-)$ ↑
Base method	0.54	0.04	0.06	0.50
With fine-tuning	0.52	0.05	0.05	0.47
Without D_{DA}	0.39	0.19	0.03	0.20
Without 1st stage	0.43	0.14	0.04	0.29

Fig. 6. Effects of fine tuning and attribute variety. The first 3 columns show the input image, the results of the encoder embedding and fine tuning. The other columns show different facial attributes controllable modifying $E_{S_i}(\theta_i)$.

of those procedures contributes to controllability of ConfigNet. Compared to the base method, $C_{pred}(I_+) - C_{pred}(I_-)$ decreases by 60% when the domain adversarial loss is removed and by 42% when the first stage training is removed. Quantitatively, the mean absolute difference of the non-altered attributes, MD, is slightly larger for the base method. While this might seem a degradation caused by two stage training and the domain discriminator, we attribute the lower MD to the reduced capability of the network to modify the output image.

One worry with fine-tuning[2] on a single image is that it will change the decoder in a way that negatively affects controllability of the output image. Our experiments show that fine tuning leads to a 6% reduction in $C_{pred}(I_+) - C_{pred}(I_-)$ and no increase of MD, which leads us to believe that the controllability of the fine-tuned generator is not significantly affected. Figure 6 qualitatively shows the effects of fine-tuning compared to embedding using E_R.

An additional ablation study showing the influence of the eye gaze preserving loss $\mathcal{L}_{eye}$ is shown in Fig. 16 in the supplementary.

---

[2] In all fine-tuning experiments we ran the fine-tuning procedure for 50 iterations.

4.3   Comparison to State of the Art

In this section we compare to PuppetGAN [40], which is the most closely related method, additional comparisons to CycleGAN [44] and Face-ID GAN [33] are in the supplementary. For comparison to PuppetGAN we use a figure from [40] that shows control over the degree of mouth opening on frames from several videos from the 300-VW dataset [31]. To generate the figure, the authors of PuppetGAN trained separate models on each of the videos and then demonstrated the ability to change the degree of mouth opening in the frames of the same video.

Fig. 7. Comparison between ConfigNet (left) and PuppetGAN (right). Top row shows the input, left column the desired level of mouth opening for each row. To facilitate comparison, ConfigNet results are cropped to match PuppetGAN.

To generate similar results we use a model trained on FFHQ and fine-tune it on the input frame using the method described in Sect. 3.4. We then use the fine-grained control method (Sect. 3.6) to change only the degree of mouth opening. The results of this comparison are shown in Fig. 7. At a certain level of mouth opening PuppetGAN saturates and is not able to open the mouth more widely, ConfigNet does so, while retaining a similar level of quality and disentanglement. Both methods fail to close the mouth fully for some of the input images. We believe that in case of ConfigNet this is an issue with the disentanglement of the synthetic training set itself, we give further details and describe a solution in supplementary materials. It is also worth noting that PuppetGAN uses hundreds of training images of a specific person, while ConfigNet requires only a single frame and it is able to modify many additional attributes.

4.4 Failure Modes

One of the key issues we have identified is that the z_i that corresponds to head shape is often separated for real and synthetic data. For example, changing the head shape of a real image embedded into z using $E_S(\theta_i)$ results in the face appearing closer to the synthetic image space and some of its features being lost, see Fig. 8a for an example. This separation is placed in the head shape space very consistently, we believe this is because head shape affects the whole image in a significant way, so it's easy for the generator to "hide" the difference between real and synthetic images there.

Fig. 8. Failure modes. Left image pair: changing head shape to one obtained from θ moves the appearance of the image closer to synthetic data. Central image pair: change of z_i corresponding to texture changes style of glasses. Right: frontal image generated from an image I_R with pose outside the supported range.

Another issue is that SynthFace does not model glasses, which leads to ConfigNet hiding the representation of glasses in unrelated face attributes, most commonly texture, head and eyebrow shape, as shown in Fig. 8b. Lastly, we have found that when I_R has a head pose that is out of the rotation range of SynthFace, the encoder E_R hides the rotation in other parts of z, as shown in Fig. 8c. We believe this is a result of constraining the rotation output of E_R to the range seen in SynthFace (details in supplementary). Generating a synthetic dataset with a wider rotation range would likely alleviate this issue.

5 Conclusions

We have presented ConfigNet, a novel face image synthesis method that allows for controlling the output images to an unprecedented degree. Crucially, we show the ability to generate realistic face images with attribute combinations that are outside the distribution of the real training set. This unique ability brings neural rendering closer to traditional rendering pipelines in terms of flexibility.

An open question is how to handle aspects of real face images not present in synthetic data. Adding additional variables in the latent space to model these aspects only for real data is an investigation that we leave to future work.

In the short term, we believe that ConfigNet could be used to enrich existing datasets with samples that are outside of their data distribution or be applied to character animation. In the long term, we hope that similar methods will replace traditional rendering pipelines and allow for controllable, realistic and person-specific face rendering.

Acknowdledgments. The authors would like to thank Nate Kushman for helpful discussions and suggestions.

References

1. Achlioptas, P., Diamanti, O., Mitliagkas, I., Guibas, L.: Learning representations and generative models for 3D point clouds. arXiv preprint arXiv:1707.02392 (2017)
2. Baltrusaitis, T., Robinson, P., Morency, L.P.: Constrained local neural fields for robust facial landmark detection in the wild. In: Proceedings of the IEEE International Conference on Computer Vision Workshops, pp. 354–361 (2013)
3. Baltrusaitis, T., et al.: A high fidelity synthetic face framework for computer vision. Technical report, MSR-TR-2020-24, Microsoft (2020). https://www.microsoft.com/en-us/research/publication/high-fidelity-face-synthetics/
4. Baltrusaitis, T., Zadeh, A., Lim, Y.C., Morency, L.P.: OpenFace 2.0: facial behavior analysis toolkit. In: 2018 13th IEEE International Conference on Automatic Face & Gesture Recognition (FG 2018), pp. 59–66. IEEE (2018)
5. Brock, A., Donahue, J., Simonyan, K.: Large scale GAN training for high fidelity natural image synthesis. arXiv preprint arXiv:1809.11096 (2018)
6. Chen, X., Duan, Y., Houthooft, R., Schulman, J., Sutskever, I., Abbeel, P.: InfoGAN: interpretable representation learning by information maximizing generative adversarial nets. In: Advances in Neural Information Processing Systems, pp. 2172–2180 (2016)
7. Choi, Y., Choi, M., Kim, M., Ha, J.W., Kim, S., Choo, J.: StarGAN: unified generative adversarial networks for multi-domain image-to-image translation. In: Proceedings of the IEEE Conference on Computer Vision and Pattern Recognition, pp. 8789–8797 (2018)
8. Gecer, B., Bhattarai, B., Kittler, J., Kim, T.-K.: Semi-supervised adversarial learning to generate photorealistic face images of new identities from 3D morphable model. In: Ferrari, V., Hebert, M., Sminchisescu, C., Weiss, Y. (eds.) ECCV 2018. LNCS, vol. 11215, pp. 230–248. Springer, Cham (2018). https://doi.org/10.1007/978-3-030-01252-6_14
9. Goodfellow, I., et al.: Generative adversarial nets. In: Advances in Neural Information Processing Systems, pp. 2672–2680 (2014)
10. He, K., Zhang, X., Ren, S., Sun, J.: Deep residual learning for image recognition. In: Proceedings of the IEEE Conference on Computer Vision and Pattern Recognition, pp. 770–778 (2016)
11. Heusel, M., Ramsauer, H., Unterthiner, T., Nessler, B., Hochreiter, S.: GANs trained by a two time-scale update rule converge to a local Nash equilibrium. In: Advances in Neural Information Processing Systems, pp. 6626–6637 (2017)
12. Higgins, I., et al.: Beta-VAE: learning basic visual concepts with a constrained variational framework. ICLR **2**(5), 6 (2017)
13. Huang, G.B., Ramesh, M., Berg, T., Learned-Miller, E.: Labeled faces in the wild: a database for studying face recognition in unconstrained environments. Technical report, 07–49, University of Massachusetts, Amherst (2007)
14. Huang, X., Belongie, S.: Arbitrary style transfer in real-time with adaptive instance normalization. In: Proceedings of the IEEE International Conference on Computer Vision, pp. 1501–1510 (2017)

15. Johnson, J., Alahi, A., Fei-Fei, L.: Perceptual losses for real-time style transfer and super-resolution. In: Leibe, B., Matas, J., Sebe, N., Welling, M. (eds.) ECCV 2016. LNCS, vol. 9906, pp. 694–711. Springer, Cham (2016). https://doi.org/10.1007/978-3-319-46475-6_43

16. Karras, T., Aila, T., Laine, S., Lehtinen, J.: Progressive growing of GANs for improved quality, stability, and variation. arXiv preprint arXiv:1710.10196 (2017)

17. Karras, T., Laine, S., Aila, T.: A style-based generator architecture for generative adversarial networks. In: Proceedings of the IEEE Conference on Computer Vision and Pattern Recognition, pp. 4401–4410 (2019)

18. Karras, T., Laine, S., Aittala, M., Hellsten, J., Lehtinen, J., Aila, T.: Analyzing and improving the image quality of StyleGAN. arXiv preprint arXiv:1912.04958 (2019)

19. Kingma, D.P., Ba, J.: Adam: a method for stochastic optimization. arXiv preprint arXiv:1412.6980 (2014)

20. Kulkarni, T.D., Whitney, W.F., Kohli, P., Tenenbaum, J.: Deep convolutional inverse graphics network. In: Advances in Neural Information Processing Systems, pp. 2539–2547 (2015)

21. Lee, W., Kim, D., Hong, S., Lee, H.: High-fidelity synthesis with disentangled representation. arXiv preprint arXiv:2001.04296 (2020)

22. Liu, Y.C., Yeh, Y.Y., Fu, T.C., Wang, S.D., Chiu, W.C., Frank Wang, Y.C.: Detach and adapt: learning cross-domain disentangled deep representation. In: Proceedings of the IEEE Conference on Computer Vision and Pattern Recognition, pp. 8867–8876 (2018)

23. Liu, Z., Luo, P., Wang, X., Tang, X.: Deep learning face attributes in the wild. In: Proceedings of International Conference on Computer Vision (ICCV) (2015)

24. Mathieu, M.F., Zhao, J.J., Zhao, J., Ramesh, A., Sprechmann, P., LeCun, Y.: Disentangling factors of variation in deep representation using adversarial training. In: Advances in Neural Information Processing Systems, pp. 5040–5048 (2016)

25. Mescheder, L., Geiger, A., Nowozin, S.: Which training methods for GANs do actually converge? arXiv preprint arXiv:1801.04406 (2018)

26. Mori, M., et al.: The uncanny valley. Energy **7**(4), 33–35 (1970)

27. Nguyen-Phuoc, T., Li, C., Theis, L., Richardt, C., Yang, Y.L.: HoloGAN: unsupervised learning of 3D representations from natural images. In: Proceedings of the IEEE International Conference on Computer Vision, pp. 7588–7597 (2019)

28. Parkhi, O.M., Vedaldi, A., Zisserman, A.: Deep face recognition (2015)

29. Qian, S., et al.: Make a face: towards arbitrary high fidelity face manipulation. In: Proceedings of the IEEE International Conference on Computer Vision, pp. 10033–10042 (2019)

30. Russakovsky, O., et al.: ImageNet large scale visual recognition challenge. Int. J. Comput. Vis. **115**(3), 211–252 (2015)

31. Sagonas, C., Tzimiropoulos, G., Zafeiriou, S., Pantic, M.: 300 faces in-the-wild challenge: the first facial landmark localization challenge. In: Proceedings of the IEEE International Conference on Computer Vision Workshops, pp. 397–403 (2013)

32. Sandler, M., Howard, A., Zhu, M., Zhmoginov, A., Chen, L.C.: MobileNetV2: inverted residuals and linear bottlenecks. In: Proceedings of the IEEE Conference on Computer Vision and Pattern Recognition, pp. 4510–4520 (2018)

33. Shen, Y., Luo, P., Yan, J., Wang, X., Tang, X.: FaceID-GAN: learning a symmetry three-player GAN for identity-preserving face synthesis. In: Proceedings of the IEEE Conference on Computer Vision and Pattern Recognition, pp. 821–830 (2018)

34. Shrivastava, A., Pfister, T., Tuzel, O., Susskind, J., Wang, W., Webb, R.: Learning from simulated and unsupervised images through adversarial training. In: Proceedings of the IEEE Conference on Computer Vision and Pattern Recognition, pp. 2107–2116 (2017)
35. Simonyan, K., Zisserman, A.: Very deep convolutional networks for large-scale image recognition. arXiv preprint arXiv:1409.1556 (2014)
36. Tewari, A., et al.: StyleRig: rigging StyleGAN for 3D control over portrait images. In: Proceedings of the IEEE/CVF Conference on Computer Vision and Pattern Recognition, pp. 6142–6151 (2020)
37. Thies, J., Zollhofer, M., Stamminger, M., Theobalt, C., Nießner, M.: Face2Face: real-time face capture and reenactment of RGB videos. In: Proceedings of the IEEE Conference on Computer Vision and Pattern Recognition, pp. 2387–2395 (2016)
38. Tzeng, E., Hoffman, J., Saenko, K., Darrell, T.: Adversarial discriminative domain adaptation. In: Proceedings of the IEEE Conference on Computer Vision and Pattern Recognition, pp. 7167–7176 (2017)
39. Ulyanov, D., Vedaldi, A., Lempitsky, V.: Instance normalization: the missing ingredient for fast stylization. arXiv preprint arXiv:1607.08022 (2016)
40. Usman, B., Dufour, N., Saenko, K., Bregler, C.: PuppetGAN: cross-domain image manipulation by demonstration. In: Proceedings of the IEEE International Conference on Computer Vision, pp. 9450–9458 (2019)
41. Wiles, O., Koepke, A.S., Zisserman, A.: X2Face: a network for controlling face generation using images, audio, and pose codes. In: Ferrari, V., Hebert, M., Sminchisescu, C., Weiss, Y. (eds.) ECCV 2018. LNCS, vol. 11217, pp. 690–706. Springer, Cham (2018). https://doi.org/10.1007/978-3-030-01261-8_41
42. Zadeh, A., Chong Lim, Y., Baltrusaitis, T., Morency, L.P.: Convolutional experts constrained local model for 3D facial landmark detection. In: Proceedings of the IEEE International Conference on Computer Vision Workshops, pp. 2519–2528 (2017)
43. Zakharov, E., Shysheya, A., Burkov, E., Lempitsky, V.: Few-shot adversarial learning of realistic neural talking head models. In: Proceedings of the IEEE International Conference on Computer Vision, pp. 9459–9468 (2019)
44. Zhu, J.Y., Park, T., Isola, P., Efros, A.A.: Unpaired image-to-image translation using cycle-consistent adversarial networks. In: Proceedings of the IEEE International Conference on Computer Vision, pp. 2223–2232 (2017)

Single View Metrology in the Wild

Rui Zhu[1]([✉]) [iD], Xingyi Yang[1] [iD], Yannick Hold-Geoffroy[2] [iD],
Federico Perazzi[2] [iD], Jonathan Eisenmann[2] [iD], Kalyan Sunkavalli[2] [iD],
and Manmohan Chandraker[1] [iD]

[1] University of California San Diego, La Jolla, CA 92093, USA
{rzhu,x3yang,mkchandraker}@eng.ucsd.edu
[2] Adobe Research, San Jose, CA 95110, USA
{holdgeof,perazzi,eisenman,sunkaval}@adobe.com

Abstract. Most 3D reconstruction methods may only recover scene properties up to a global scale ambiguity. We present a novel approach to single view metrology that can recover the *absolute* scale of a scene represented by 3D heights of objects or camera height above the ground as well as camera parameters of orientation and field of view, using just a monocular image acquired in unconstrained condition. Our method relies on data-driven priors learned by a deep network specifically designed to imbibe weakly supervised constraints from the interplay of the unknown camera with 3D entities such as object heights, through estimation of bounding box projections. We leverage categorical priors for objects such as humans or cars that commonly occur in natural images, as references for scale estimation. We demonstrate state-of-the-art qualitative and quantitative results on several datasets as well as applications including virtual object insertion. Furthermore, the perceptual quality of our outputs is validated by a user study.

Keywords: Single view metrology · Absolute scale estimation · Camera calibration · Virtual object insertion

1 Introduction

Reconstructing a 3D scene from images is a fundamental problem in computer vision. Despite many successes on this task, most previous works only reconstruct scenes up to an unknown scale. This is true for many problems including structure-from-motion (SfM) from uncalibrated cameras [13], monocular camera calibration in the wild [16,33,34] and single image depth estimation [9,23]. This ambiguity is inherent to the projective nature of image formation and resolving it requires additional information. For example, the seminal work "Single View Metrology" of Criminisi et al. [6] relies on the size of reference objects in the scene.

Electronic supplementary material The online version of this chapter (https://doi.org/10.1007/978-3-030-58621-8_19) contains supplementary material, which is available to authorized users.

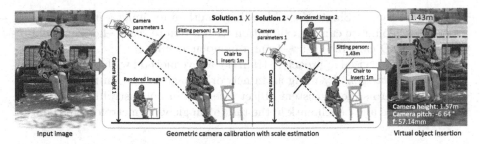

Fig. 1. Given the image on the left, single view metrology can recover the scene and the camera parameters in 3D only up to a global scale factor (for example, the two solutions in the middle). Our method accurately estimates absolute 3D camera parameters and object heights (middle, left) to produce realistic object insertion results (right).

In this work, we consider the problem of single view metrology "in the wild", where only a single image is available for an unconstrained scene composed of objects with unknown sizes. In particular, we plan to achieve this via geometric camera calibration with absolute scale estimation, *i.e.* recovering camera orientation (alternatively, the horizon in the image), field-of-view, and the *absolute* 3D height of the camera from the ground. Given these parameters, it is possible to convert any 2D measurement in image space to 3D measurements.

Our goal is to leverage modern deep networks to build a robust, automatic single view metrology method that is applicable to a broad variety of images. One approach to this problem could be to train a deep neural network to predict the scale of a scene using a database of images with known absolute 3D camera parameters. Unfortunately, no such large-scale dataset currently exists. Instead, our insight is to leverage large-scale datasets with 2D object annotations [9,11, 24,35]. In particular, we make the observation that objects of certain categories such as humans and cars are ubiquitous in images in the wild [24,35] and would make good "reference objects" to infer the 3D scale.

While the idea of using objects of known classes as references to reconstruct camera and scene 3D properties has been used in previous work [15,17], we significantly extend this work by making fewer approximations in our image formation model (*e.g.* full perspective camera vs. zero camera pitch angle, infinite focal length in [15]), leading to better modeling of images in the wild. Moreover, our method learns to predict all camera and scene properties (object and camera height estimation, camera calibration) in an end-to-end fashion; in contrast, previous work relies on individual components that address each sub-task. We demonstrate that this holistic approach leads to state-of-the-art results across all these tasks on a variety of datasets (SUN360, KITTI, IMDB-23K). We also demonstrate the use of our method for applications such as virtual object insertion, where we automatically create semantically meaningful renderings of a 3D object with known dimensions (see Fig. 1). We summarize our contributions as following:

– A state-of-the-art Single View Metrology method for images in the wild that performs geometric camera calibration with absolute scale—horizon, field-of-view, and 3D camera height—from a monocular image.
– A weakly supervised approach to train the above method with only 2D bounding box annotations by using an in-network image formation model.
– Application to scale-consistent object insertion in unconstrained images.
– New datasets and benchmark in the task of single view metrology in the wild.

2   Related Work

Camera Calibration To estimate the camera parameters, numerous efforts have been made for estimating camera intrinsics [4,8,16,33] by explicit reasoning or learning in a data-driven fashion. In addition, to estimate camera extrinsics, *e.g.* camera rotation angles or in the form of horizon estimation, classical methods [3,7,21,37] look for low-level features such as line segments. More recently, methods are proposed to directly regress the horizon from the input image [19,25,34] by learning from large-scale datasets annotated with ground truth horizons. The human sensitivity to calibration errors is studied in [16].

Depth Prediction in the Wild As we discussed in Sect. 1, the problem of scene scale estimation will be solved if we are able to predict pixel-wise depth for the scene. There has been a line of work in this topic. For domains which we can acquire the ground truth absolute depth with depth sensors we may learn to predict depth in a supervised fashion [9,18,31]. However given the limitation of the range or mobility of the sensors, these datasets are more or less limited to specific scenes. In other cases, people are able to acquire ground truth from stereo matching [10,11,32] but a large-scale stereo depth dataset for images in the wild is still absent. Other people have turn to proxy methods for collecting depth via structure-from-motion (SfM) [22,23], or in the form of relative depth [5], or from synthetic images [2,26]. However, these methods either produce depth without absolute scale, or pose a domain gap to natural images.

Single View Metrology Another line of work that seeks to estimate 3D scene parameters from images is Single View Metrology [6], which recovers scene structure in 3D from purely 2D measurements. These methods look for 2D properties such as vanishing lines and vanishing points as well as object locations, to establish relations among 3D sizes of objects in the image based on 2D measurements. Some works have been done to embed Single View Metrology in a framework to estimate the size of an unknown object in the scene or the camera height itself [1,15,17,20], given at least one reference object with known size.

3   Method

3.1   Recovering 3D Parameters from 2D Annotations

We start by describing the image formation model that allows us to associate 3D camera parameters, 3D object sizes (*i.e.* heights) and 2D bounding boxes. This is also illustrated in Fig. 2.

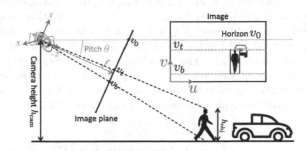

Fig. 2. Camera model of the scene (bottom) and measurements in image space (top).

We assume the world is composed of a dominant ground plane on which all objects are situated, and a camera that observes the scene. We adopt a perspective camera model similar to [15,16], which is parameterized by camera angles (yaw φ, pitch θ and roll ψ), focal length f and camera height h_{cam} to the ground (see Fig. 2). For the measurements in the vertical axis of image frame, the location of the horizon is v_0, while the vertical image center is at v_c. Each object bounding box have a top v_t and bottom v_b location in the image. We assume all images were taken with zero roll, or were rectified beforehand [21]. We further assume, without loss of generality, a null yaw and zero distortion from rectification. Camera pitch θ can be computed from v_c, v_0 and f using

$$\theta = \arctan \frac{v_c - v_0}{f}. \tag{1}$$

Consider a thin object of height h_{obj} with its bottom located at $[x, 0, z]^T$ in 3D and its top at $[x, h_{\mathrm{obj}}, z]^T$. These points project to $[u, v_b]^T$ and $[u, v_t]^T$ respectively in the camera coordinates. Based on the perspective camera model (see Eq. 3 of the supplementary material), we have

$$v_t = \frac{(f\cos\theta + v_c\sin\theta)h_{\mathrm{obj}}\,\|\cos\theta\| + (-f\sin\theta + v_c\cos\theta)z - fh_{\mathrm{cam}}}{h_{\mathrm{cam}}\,\|\cos\theta\|\tan\theta + z\cos\theta}, \tag{2}$$

$$v_b = \frac{-f\sin\theta z + v_c\cos\theta z + f^2 h_{\mathrm{cam}}}{z\cos\theta}, \tag{3}$$

where $[u_c, v_c] \in \mathbb{R}^2$ is the camera optical center which we assume is known. Substituting z from Eq. 3 into Eq. 2, we may derive v_t from camera focal length

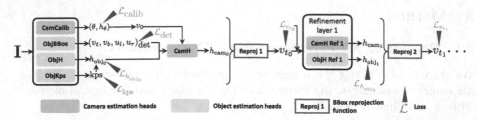

Fig. 3. Overview of our method. From input image **I**, the camera calibration module estimates pitch θ and field of view h_θ, and the object estimation modules estimate keypoints for person, object heights h_{obj} and bounding boxes. The estimated horizon v_0, bounding boxes and object heights are fed into the camera height estimation module to give an initial estimation h_{cam}. The bounding box reprojection errors $\mathcal{L}_{v_t}$ are then computed from the reprojection module (see Eqs. 2 and 6), and together with other variables are fed to the refinement network to estimate updates on the camera and object heights. Several layers of refinement are made to produce the final estimation.

f, pitch θ, camera height h_{cam} and object height h_{obj}. Hoeim et al. [15] make a number of approximations including $\cos\theta \approx 1$, $\sin\theta \approx \theta$ and $(v_c - v_0) \times (v_c - v_0)/f^2 \approx 0$, to linearly solve for the object height:

$$h_{obj} = h_{cam} \frac{v_t - v_b}{v_0 - v_b}. \tag{4}$$

In contrast, we model the full expression accounting for all the camera parameters.

Equations 2 and 3 establish a relationship between the camera parameters, including its 3D height, the 3D heights of objects in the scene, and the 2D projections of these objects into the image. Moreover, note that once we can estimate these parameters, we can directly infer the 3D size of *any* object from its 2D bounding box (Eq. 4), thus resolving the scale ambiguity in monocular reconstruction. In the following section, we introduce ScaleNet, a deep network that leverages this constraint to create weak supervision to learn to predict 3D camera height.

3.2   ScaleNet: Single View Metrology Network with Absolute Scale Estimation

Previous work [15,17] has shown that when scene parameters (*e.g.* camera parameters, object sizes) are reasonable, reprojected 2D bounding boxes should ideally fit the detected ones in the image frame. We follow a similar path in our weakly-supervised learning framework and specifically focus on humans and cars, given that they are the most commonly occurring object categories in datasets of images in the wild (*e.g.* COCO dataset [24]).

Our end-to-end method, referred to as **ScaleNet (SN)**, is split into two parts, which we describe in Fig. 3. First, all the object bounding boxes and camera parameters except camera height are jointly estimated by a geometric camera calibration network. These parameters are directly supervised during training. Second, a cascade of PointNet-like networks [30] estimates and refines the camera height (scene scale) based on the previous outputs. This second part is weakly supervised at each stage using a bounding box reprojection loss.

Camera Calibration Module and Object Heads The camera calibration module is inspired by [16], where we replace their backbone with Mask R-CNN [14,27], to which we add heads to estimate the camera parameters. To train the camera calibration module, we follow the representation of [16], including bins and training loss. However, instead of predicting the focal length f (in pixels), we find it easier to predict the vertical field of view h_θ, which can be converted to the focal length and the horizon midpoint using:

$$f = \frac{\frac{1}{2}h_{\mathrm{im}}}{\tan\left(\frac{1}{2}h_\theta\right)}, \qquad v_0 = \frac{\frac{1}{2}h_{\mathrm{im}}\tan\theta}{\tan\left(\frac{1}{2}h_\theta\right)} + \frac{h_{\mathrm{im}}}{2}, \tag{5}$$

where h_{im} is the height of the image in pixels. We also use additional heads to estimate the object bounding box, height and person keypoints (since we find that a person's 3D height in an image is closely related to their pose) from ROI features which share the same backbone as the camera estimation module. Please refer to the supplementary material for the full architecture of this network. In total, we enforce three losses on this part of the model, i.e. the camera calibration loss $\mathcal{L}_{\mathrm{calib}}(\theta, h_\theta)$ and the detection losses $\mathcal{L}_{\mathrm{det}}$ and $\mathcal{L}_{\mathrm{kps}}$. The i_{th} object with detected 2D top position $v^i_{t_{\det}}$ is reprojected to the image by Eq. 2 to v^i_t with estimated object height, camera height and camera parameters, and we define the bounding reprojection error as

$$\mathcal{L}_{v_t}\left(\{v^i_t\}^N_{i=1}\right) = \frac{1}{N}\sum_{i=1}^{N}\left\|v^i_{t_{\det}} - v^i_t\right\|. \tag{6}$$

Object Height Prior The above bounding box supervision has the same scale ambiguity as previous work. However, explicitly modeling 3D object heights allows us to use a prior on size to regularize the network to produce a meaningful object height estimation. We follow [15] and use a Gaussian prior fit from statistics (i.e. for 1.70 ± 0.09 m for people and 1.59 ± 0.21 m for cars). For an object i of height h^i_{obj} and prior Gaussian distribution $\mathcal{P}(x; \mu, \sigma)$, we define the height prior loss as

$$\mathcal{L}_{h_{\mathrm{obj}}}\left(\{h^i_{\mathrm{obj}}\}^N_{i=1}\right) = -\frac{1}{N}\sum_{i=1}^{N}\mathcal{P}(h^i_{\mathrm{obj}}; \mu, \sigma). \tag{7}$$

Camera Height Estimation Module Directly predicting camera height from images would require the network to learn to be robust to a wide variety of

Fig. 4. Example of images with different camera heights exhibiting similar bounding boxes.

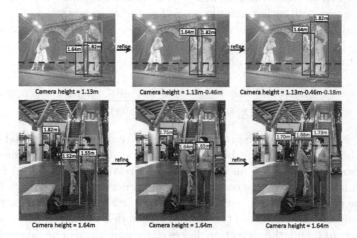

Fig. 5. Example of cascade refinements of camera height (top) and person heights (bottom). The refined parameters are labelled in red. (Color figure online)

appearance properties (object, layout, illumination, *etc.*). Instead, we design a camera height estimation module that leverages the strong *geometric* relationship between camera height, 2D bounding boxes and horizon. As exemplified in Fig. 4, both images are composed of a group of standing people while the horizons are not fully visible in the back. At first glance, both images seem to have the same camera orientation, since the people take roughly the same space in the image. However camera heights are quite different between both images. Based on this observation, instead of estimating camera height from image appearance, we take advantage of middle level representations of the scene (*e.g.* object bounding boxes and estimated horizon line) and feed those to the camera height estimation module which is derived from PointNet [29]. Its input is the concatenation of all object bounding box coordinates and the offset between the bounding box and horizon, *i.e.* $\gamma_0 = [v_0, u_{l_{\text{det}}}, u_{r_{\text{det}}}, v_{t_{\text{det}}}, v_{b_{\text{det}}}, v_{t_{\text{det}}} - v_0, v_{b_{\text{det}}} - v_0, h_{\text{obj}}]^T \in \mathbb{R}^8$ where $u_{l_{\text{det}}}$ and $u_{r_{\text{det}}}$ are the left and right coordinates of the detected bounding box. The network outputs the camera height as a discrete probability distribution. Finally, a weighted sum after `softmax` is applied to obtain the camera height.

Cascade Refinement Layers We observe that we can iteratively refine the camera height by considering all scene parameters jointly. Inspired by the cascade refinement scheme from [30], we propose to look at the error residual— in our case the object bounding box reprojection error—and predict a *difference* to the estimated parameters. The whole process is highlighted in Fig. 5, where the reprojected object bounding boxes are shorter than the detected ones at first. After a first step of refinement, the network reduces the camera height to reduce the object bounding boxes error, and so on. To this end, we design layers of refinement, where in layer $j \in \{1, 2, \ldots, M\}$ a camera height and object height refinement module takes as input the object bounding box reprojection residuals and the other camera parameters, formally as $\gamma_j = [v_0, (u_l^i, u_r^i, v_{t_{j-1}}^i, v_b^i)_{\mathrm{det}}, v_{t_{j-1}}^i - v_{t_{\mathrm{det}}}^i, h_{\mathrm{obj}_{j-1}}^i, h_{\mathrm{cam}_{j-1}}]^T \in \mathbb{R}^8$ where $i \in \{1, 2, \ldots N\}$ is the object index. Each refinement layer j predicts updates $\Delta h_{\mathrm{cam}_j}$ and $\Delta h_{\mathrm{obj}_j}^i$ so that $h_{\mathrm{obj}_j}^i = \Delta h_{\mathrm{obj}_j}^i + h_{\mathrm{obj}_{j-1}}^i$ and $h_{\mathrm{cam}_j} = \Delta h_{\mathrm{cam}_j} + h_{\mathrm{cam}_{j-1}}$. An object bounding box reprojection loss $\mathcal{L}_{v_{t_j}}$ and object height prior $\mathcal{L}_{h_{\mathrm{obj}_j}}$ are enforced for each layer.

The final training loss is a weighted combination of the losses, written as

$$\mathcal{L}_{\mathrm{all}} = \alpha_1 \sum_{j=1}^{M} \mathcal{L}_{v_{t_j}} + \alpha_2 \sum_{j=0}^{M} \mathcal{L}_{h_{\mathrm{obj}_j}} + \alpha_3 \mathcal{L}_{\mathrm{calib}} + \alpha_4 \mathcal{L}_{\mathrm{det}} + \alpha_5 \mathcal{L}_{\mathrm{kps}}, \qquad (8)$$

where $\alpha_{1\ldots5}$ are weighting constants to balance the losses during training.

3.3 Datasets

In the following, we describe the datasets and their preprocessing used for training and evaluation. Data generation details, statistics, sampled visualization of all datasets can be found in the supplemental material.

Calib: Camera Calibration Dataset To train the camera calibration module to estimate camera pitch and field of view from a single image, we follow the data generation pipeline from [16], where data are cropped from the SUN360 database [36] of 360° panoramas with sampled camera parameters. We split the resulting camera calibration dataset into 397,987 images for training and 2,000 images for validation. For simplicity, we refer to this dataset as the Calib dataset.

COCO-Scale: Scale Estimation Dataset from COCO While the Calib dataset provides a large and diversified dataset of images and many ground truth camera pitch and field of view parameters, it does not provide camera height. To complement this dataset, we use the COCO dataset [24], which allows us to train our method in a weakly-supervised way. This dataset features 2D annotations of object bounding boxes, keypoints of person, and stuff annotation. We further extend these annotations by using Mask R-CNN [14] to infer objects of certain categories, *e.g.* person and car. These additional annotations complement the ones provided in the dataset, which together form our candidate object set. We refer to this dataset as the COCO-Scale dataset.

We filter out invalid objects which do not satisfy our scene model, using the stuff annotation (*e.g.* ground, grass, water) to infer the support relationship of an object with its surrounding pixels; we only keep objects that are most likely situated on a plane (*e.g.* people standing on grassland, cars on a street). For the *person* category, we use the detected keypoints from Mask R-CNN [27] to detect people with both head and ankles visible to ensure the obtained bounding box is amodal as in [17]. We further filter the images based on aspect ratio, object size and number of objects to keep bounding boxes of certain shape.

This pruning step yields 10,547 training images, 2,648 validation images, taken from COCO's train2017 and val2017 splits respectively. We further obtain test images from val2017 and ensure no overlap exists between the splits. We call this person-only subset COCO-Scale-person. For the multi-category setting (COCO-Scale-mulCat), we look for images including both cars and people, which provides us with 12,794 images for training, 3,189 for validation, and 584 for testing.

KITTI We use the KITTI [11] dataset to evaluate our camera and object height estimations. We apply the same filtering rules as used on COCO, yielding 298 images for person-only setting (KITTI-person), and 234 images for the multi-category setting (KITTI-mulCat).

IMDB-23K Celebrity Dataset for Person Height Evaluation IMDB-23K [12] is a collection of online images of celebrities, with annotations of body height from the IMDB website. We use this dataset to evaluate our object height prediction. However, these height annotations may not be exact and we treat them as *pseudo* ground truth to draw comparisons. We apply the same filtering rules as on COCO, and the filtered dataset consists of 2,550 test images with one celebrity labelled with height in each image. An image from this dataset is shown in Fig. 6.

Fig. 6. (Left) Annotated person bounding box (red) with ground truth height and detected person (green) with keypoints (colored). (Right) Calculation of upright ratio. (Color figure online)

4 Experiments

4.1 Baseline Methods

We use Hoiem *et al.* [15] as the baseline method. For fair comparison, we employ our object proposals or top predictions from Mask R-CNN [27] as input to this method to replace the original detector [28], which enhances the original method. We set up 2 baseline models: (1) *PGM*: the original model based on a Probabilistic Graphical Model, which takes in object proposals and surface geometry, and predicts camera height and horizon. Object heights can be computed from Eq. 4 by directly minimizing the reprojection error; (2) *PGM-fixedH*: same as PGM but assumes all objects have canonical height. For people, we use 1.7 m which is the mean of the person height prior used in [15]). For cars we use 1.59 m.

4.2 Training

We train our model in two stages. Firstly, we train the camera calibration network with full supervision from the Calib dataset using camera calibration losses and the detection & keypoint estimation heads with full supervision from COCO ground truth with losses following [27]. The backbone, camera calibration head, and detection & keypoint heads are initialized with a pre-trained Mask R-CNN model [27].

For the second stage, the object height estimation module is trained together while other modules are finetuned, in a weakly-supervised fashion, with the full loss in Eq. 8. Training details can be found in the supplemental material.

Variants of ScaleNet (SN) We evaluate several variations of ScaleNet (*SN*): (1) *SN-L1*: one layer architecture with direct prediction of object height and camera height. (2) *SN-L3*: one layer for initial prediction with 2 additional refinement layers. (3) *SN-L3-mulCat*: same as SN-L3 but with objects of multiple categories as input and trained on COCO-Scale-mulCat. (4) *SN-L3-kps-can*: same as SN-L3 but training keypoints prediction and predicting each person in upright height instead of actual height. An upright ratio computed as $l_{actual}/l_{upright}$ in Fig. 6 is an approximation of the actual ration in 3D that takes into account the person's pose. It is multiplied to the predicted upright height to obtain actual height, and the height prior is applied to the predicted upright height.

Training Results on COCO-Scale We calculate the bounding reprojection error from Eq. 2 on the test split which is an indication of 2D bounding box fits. The results are shown in Table 1 and Fig. 7.

Table 1. $\mathcal{L}_{v_t}$ on COCO-Scale-person

		PGM-fixedH	SN-L1	SN-L3	SN-L3-mulCat	SN-L3-kps-can
$\mathcal{L}_{v_t}$	Mean	0.1727	0.1502	0.0717	0.0613	**0.0540**
	Std	0.3598	0.2579	0.1875	0.1874	0.1677
	Med	0.0793	0.0693	0.0116	**0.0074**	0.0094

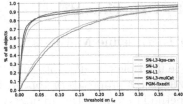

Fig. 7. $\mathcal{L}_{v_t}$ of all objects on COCO-Scale-person under varying thresholds.

4.3   Evaluation on COCO-Scale-Person

Since we do not have ground truth 3D annotations on the COCO-Scale-person dataset, we evaluate performance using a user study on virtual object insertion. We evaluate the plausibility of our estimates on the resulting scale and perspective effects of an inserted object via an A/B test on COCO-Scale-person. In our evaluation, we render a 1 m height chair alongside each object. For each of the 4 pairs of models, we insert the chair in 50 random test images. 10 users for each pair of models were asked to choose which image of the pair is more realistic w.r.t. the scales of all chairs.

As can be seen in Table 2, our results improve when adding multiple categories (*SN-L3-mulCat*) or regress keypoint and account for the pose while computing person height (*SN-L3-kps-can*). The best variant *SN-L3-kps-can* outperforms other methods significantly. This is consistent with the bounding box reprojection error in Table 1 where *SN-L3-kps-can* is the best performing method.

Figure 8 shows a qualitative evaluation of our method. The first row shows the benefit of using the upright ratio, leading to better estimations in *SN-L3-kps-can*. The second row shows our method displays better behavior in cases of multiple objects with diverse heights. In the third row, we demonstrate robustness to outliers (the person on the bus).

Fig. 8. Scene parameters estimation and virtual object insertion results on COCO-Scale. The detected boxes are shown in green and reprojected ones in blue. The horizon is shown as dashed blue line. Camera parameters are overlaid on the top (camera height as y_c, focal length in millimeters as f_{mm} assuming 35 mm full-frame sensor, and pitch as θ). A chair of 1 m tall is inserted alongside each person with the estimated parameters. (Color figure online)

Table 2. A/B test results for scale on COCO-Scale-person

	SN-L3	SN-L1	SN-L3	SN-L3-kps-can	SN-L3	SN-L3-mulCat	SN-L3-kps-can	PGM
preference	**54.6%**	43.8%	42.8%	**57.2%**	**50.7%**	49.3%	**59.5%**	40.5%

Table 3. Evaluation errors on all *pedestrian* objects of KITTI-person. PGM-fixedH assumes a canonical object height, while PGM explicitly solves for the object height that minimizes bounding box reprojection error $\mathcal{L}_{v_t}$ (ideally to zero), as a result of which the $\mathcal{L}_{v_t}$ errors of PGM are grayed out as they are not directly comparable to others. Best results in bold (lower is better).

		PGM-fixedH	PGM	SN-L1	SN-L3	SN-L3-kps-can	SN-L3-mulCat
Input	Car						
	Person	✓	✓	✓	✓	✓	✓
$\mathcal{E}_{h_{obj}}$	mean	0.0863	0.1358	**0.0837**	0.0956	0.1014	0.0849
	Std.	0.0570	0.1406	0.0610	0.0751	0.0864	0.0714
	Med.	0.0800	0.0916	0.0727	0.0770	0.0864	**0.0685**
$\mathcal{L}_{v_t}$	mean	0.0767	0.0016	0.0980	0.0331	0.0585	**0.0283**
	Std.	0.0638	0.0009	0.1000	0.0644	0.0815	0.0415
	Med.	0.0618	0.0003	0.0724	0.0128	0.0815	**0.0127**
$\mathcal{E}_{h_{cam}}$	mean	0.1408		0.2356	0.1988	0.2649	**0.1264**
	Std.	0.1585		0.2860	0.3162	0.3207	0.1147
	Med.	0.1096		0.1821	0.1160	0.1666	**0.0878**

4.4 Quantitative Evaluation on KITTI

Since KITTI provides ground truth for all of the parameters our method estimate, we can directly evaluate the errors in bounding box reprojection $\mathcal{L}_{v_t}$, camera height estimation $\mathcal{E}_{h_{cam}}$ and object height estimation $\mathcal{E}_{h_{obj}}$ as shown in Tables 3 and 4 on KITTI-person and KITTI-mulCat respectively, where

$$\mathcal{E}_{h_{cam}} = \left\| h_{cam} - h_{cam_{gt}} \right\|, \mathcal{E}_{h_{obj}} = \frac{1}{N} \sum_{i=1}^{N} \left\| h_{obj} - h_{obj_{gt}} \right\| \tag{9}$$

h_{cam} and h_{obj}^i are the final estimated camera height and height of object i, and $h_{cam_{gt}}$ and $h_{obj_{gt}}^i$ are their ground truth values respectively.

Our method SN-L3-mulCat outperforms previous work on both KITTI-person and KITTI-mulCat. This method takes into account the cues from multiple categories to perform inference, giving it an advantage on scenes with high-diversity content (see Table 4). Qualitative results are shown in Fig. 9. Please refer to the supplementary material for more results.

Fig. 9. Scene parameters estimation results with SN-L3-mulCat on KITTI-mulCat. Reprojected pedestrians are in blue, while cars are in magenta.

Table 4. Evaluation errors on all *pedestrian* objects of KITTI-mulCat following specifications of Table 3

		PGM	PGM	SN-L3	SN-L3-mulCat	SN-L3-mulCat
Input	Car		✓			✓
	Person	✓	✓	✓	✓	✓
$\mathcal{E}_{h_{obj}}$	mean	0.1198	0.1177	0.1266	0.1092	**0.0956**
	Std.	0.1007	0.0932	0.0994	0.0883	0.0811
	Med.	0.0896	0.0939	0.0968	0.0876	**0.0780**
$\mathcal{L}_{v_t}$	mean	0.0008	0.0008	0.0647	**0.0303**	0.0712
	Std.	0.0013	0.0011	0.1124	0.0465	0.1153
	Med.	0.0003	0.0004	0.0166	**0.0123**	0.0297
$\mathcal{E}_{h_{cam}}$	mean	0.1379	0.1519	0.3464	0.1547	**0.1222**
	Std.	0.1735	0.1676	0.3693	0.1687	0.1235
	Med.	**0.0703**	0.1096	0.2278	0.0991	0.0904

4.5 Quantitative Evaluation on IMDB-23K

IMDB-23K provides annotation for the registered height of a person, which we assume is the height of the person standing straight (the upright height). Since all of our models except SN-L3-kps-can predict the actual person height (influenced by the specific pose, viewpoint, bounding box drawing, etc.), we use the upright ratio (see Fig. 6) computed from detected keypoints to convert the actual height back to upright height. This upright ratio allows us to compute upright height from all methods,and compare against the pseudo ground truth, as included in Table 5 and Fig. 10. Since the ground truth annotations are only valid for

Table 5. $\mathcal{E}_{h_{obj}}$ and $\mathcal{L}_{v_t}$ on IMDB-23K following specifications of Table 3

		PGM-fixedH	PGM	SN-L1	SN-L3	SN-L3-kps-can	SN-L3-mulCat
$\mathcal{E}_{h_{obj}}$	mean	0.0843	0.2234	**0.0832**	0.0891	0.1003	0.0990
	Std.	0.0638	0.2246	0.0688	0.0818	0.0920	0.0915
	Med.	0.0700	0.1644	0.0706	**0.0695**	0.0920	0.0777
$\mathcal{L}_{v_t}$	mean	0.0983	0.0157	0.1011	0.0431	0.0920	**0.0416**
	Std.	0.0546	0.0185	0.1706	0.1324	0.1257	0.1537
	Med.	0.0689	0.0105	0.0441	0.0071	0.1257	**0.0056**

standing people, we further get a subset of the test set where the estimated upright ratio from keypoint prediction is less than 0.90, which typically denotes a non-standing person. Table 6 evaluates the methods on this subset and shows that SN-L3-kps-can, which directly accounts for the upright ratio in training and inference, performs better in getting upright heights compared to other models; visual comparisons are shown in Fig. 10 and the material.

Table 6. $\mathcal{E}_{h_{obj}}$ on IMDB-23K (non-standing person).Best results in bold (lower is better)

		PGM	SN-L3	SN-L3-kps-can
$\mathcal{E}_{h_{obj}}$	Mean	0.1552	0.1591	**0.1212**
	Std.	0.1379	0.1788	0.1072
	Med.	0.1177	0.1013	**0.0909**

Fig. 10. Scene parameters estimation results with SN-L3-kps-can on IMDB-23K.

5   Conclusion and Future Work

We present a learning-based method that performs geometric camera calibration with absolute scale from images in the wild. We demonstrate that our method provides state-of-the-art results on multiple datasets. Despite this advance, our method is hindered by some limitations.

Our single dominant ground plane assumption does not always hold in the wild. Urban scene may provide multiple supporting surfaces at different heights (tables, balconies), so objects may not be laying on the assumed ground plane. Also, the ground may be non-flat in nature environments.

Our method is highly biased on *appearance*. Adding amodal reasoning, as proposed in [17], would be an interesting way forward to perform holistic scene reasoning. We would like to tackle these limitations as future work.

References

1. Andaló, F.A., Taubin, G., Goldenstein, S.: Efficient height measurements in single images based on the detection of vanishing points. Comput. Vis. Image Underst. **138**, 51–60 (2015)
2. Atapour-Abarghouei, A., Breckon, T.P.: Real-time monocular depth estimation using synthetic data with domain adaptation via image style transfer. In: Proceedings of the IEEE Conference on Computer Vision and Pattern Recognition, pp. 2800–2810 (2018)
3. Barinova, O., Lempitsky, V., Tretiak, E., Kohli, P.: Geometric image parsing in man-made environments. In: Daniilidis, K., Maragos, P., Paragios, N. (eds.) Computer Vision – ECCV 2010. Lecture Notes in Computer Science, vol. 6312, pp. 57–70. Springer, Berlin, Heidelberg (2010). https://doi.org/10.1007/978-3-642-15552-9_5
4. Chen, Q., Wu, H., Wada, T.: Camera calibration with two arbitrary coplanar circles. In: Pajdla, T., Matas, J. (eds.) Computer Vision – ECCV 2004. Lecture Notes in Computer Science, vol. 3023, pp. 521–532. Springer, Berlin, Heidelberg (2004). https://doi.org/10.1007/978-3-540-24672-5_41
5. Chen, W., Fu, Z., Yang, D., Deng, J.: Single-image depth perception in the wild. In: Advances in Neural Information Processing Systems, pp. 730–738 (2016)
6. Criminisi, A., Reid, I., Zisserman, A.: Single view metrology. Int. J. Comput. Vis. **40**(2), 123–148 (2000)
7. Denis, P., Elder, J.H., Estrada, F.J.: Efficient edge-based methods for estimating Manhattan frames in urban imagery. In: Forsyth, D., Torr, P., Zisserman, A. (eds.) Computer Vision – ECCV 2008. Lecture Notes in Computer Science, vol. 5303, pp. 197–210. Springer, Berlin, Heidelberg (2008). https://doi.org/10.1007/978-3-540-88688-4_15
8. Deutscher, J., Isard, M., MacCormick, J.: Automatic camera calibration from a single Manhattan image. In: Heyden, A., Sparr, G., Nielsen, M., Johansen, P. (eds.) Computer Vision – ECCV 2002. Lecture Notes in Computer Science, vol. 2353, pp. 175–188. Springer, Berlin, Heidelberg (2002). https://doi.org/10.1007/3-540-47979-1_12
9. Eigen, D., Puhrsch, C., Fergus, R.: Depth map prediction from a single image using a multi-scale deep network. In: Proceedings of the 27th International Conference on Neural Information Processing Systems, vol. 2, pp. 2366–2374 (2014)

10. Garg, R., Wadhwa, N., Ansari, S., Barron, J.T.: Learning single camera depth estimation using dual-pixels. In: Proceedings of the IEEE International Conference on Computer Vision, pp. 7628–7637 (2019)
11. Geiger, A., Lenz, P., Stiller, C., Urtasun, R.: Vision meets robotics: the kitti dataset. Int. J. Robot. Res. **32**(11), 1231–1237 (2013)
12. Gunel, S., Rhodin, H., Fua, P.: What face and body shapes can tell us about height. In: Proceedings of the IEEE International Conference on Computer Vision Workshops, pp. 0–0 (2019)
13. Hartley, R., Zisserman, A.: Multiple View Geometry in Computer Vision (2003)
14. He, K., Gkioxari, G., Dollár, P., Girshick, R.: Mask R-CNN. In: Proceedings of the IEEE International Conference on Computer Vision, pp. 2961–2969 (2017)
15. Hoiem, D., Efros, A.A., Hebert, M.: Putting objects in perspective. Int. J. Comput. Vis. **80**(1), 3–15 (2008)
16. Hold-Geoffroy, Y., et al.: A perceptual measure for deep single image camera calibration. In: CVPR, pp. 2354–2363 (2018)
17. Kar, A., Tulsiani, S., Carreira, J., Malik, J.: Amodal completion and size constancy in natural scenes. In: Proceedings of the IEEE International Conference on Computer Vision, pp. 127–135 (2015)
18. Kim, W., Ramanagopal, M.S., Barto, C., Yu, M.Y., Rosaen, K., Goumas, N., Vasudevan, R., Johnson-Roberson, M.: PedX: benchmark dataset for metric 3-D pose estimation of pedestrians in complex urban intersections. IEEE Robot. Autom. Lett. **4**(2), 1940–1947 (2019)
19. Kluger, F., Ackermann, H., Yang, M.Y., Rosenhahn, B.: Temporally consistent horizon lines. In: 2020 International Conference on Robotics and Automation (ICRA) (2020)
20. Lalonde, J.F., Hoiem, D., Efros, A.A., Rother, C., Winn, J., Criminisi, A.: Photo clip art. ACM Trans. Graph. (TOG) **26**(3), 3 (2007)
21. Lee, H., Shechtman, E., Wang, J., Lee, S.: Automatic upright adjustment of photographs with robust camera calibration. IEEE Trans. Pattern Anal. Mach. Intell. **36**(5), 833–844 (2013)
22. Li, Z., et al.: Learning the depths of moving people by watching frozen people. In: Proceedings of the IEEE Conference on Computer Vision and Pattern Recognition, pp. 4521–4530 (2019)
23. Li, Z., Snavely, N.: MegaDepth: learning single-view depth prediction from internet photos. In: CVPR (2018)
24. Lin, T.Y., et al.: Microsoft coco: common objects in context. In: Fleet, D., Pajdla, T., Schiele, B., Tuytelaars, T. (eds.) Computer Vision – ECCV 2014. Lecture Notes in Computer Science, vol. 8693, pp. 740–755. Springer, Cham (2014). https://doi.org/10.1007/978-3-319-10602-1_48
25. Man, Y., Weng, X., Li, X., Kitani, K.: GroundNet: monocular ground plane estimation with geometric consistency. In: Computer Vision and Pattern Recognition (CVPR) (2018)
26. Martinez II, M.A.: Beyond Grand Theft Auto V for Training, Testing and Enhancing Deep Learning in Self Driving Cars. Ph.D. thesis, Princeton University (2018)
27. Massa, F., Girshick, R.: Maskrcnn-benchmark: fast, modular reference implementation of instance segmentation and object detection algorithms in PyTorch (2018). https://github.com/facebookresearch/maskrcnn-benchmark. Accessed 16 Oct 2019
28. Murphy, K.P., Torralba, A., Freeman, W.T.: Graphical model for recognizing scenes and objects. In: NIPS, pp. 1499–1506 (2003)

29. Qi, C.R., Su, H., Mo, K., Guibas, L.J.: PointNet: deep learning on point sets for 3D classification and segmentation. In: Proceedings of the IEEE Conference on Computer Vision and Pattern Recognition, pp. 652–660 (2017)

30. Ranftl, R., Koltun, V.: Deep fundamental matrix estimation. In: Ferrari, V., Hebert, M., Sminchisescu, C., Weiss, Y. (eds.) Computer Vision – ECCV 2018. Lecture Notes in Computer Science, vol. 11205, pp. 292–309. Springer, Cham (2018). https://doi.org/10.1007/978-3-030-01246-5_18

31. Wang, C., Miguel Buenaposada, J., Zhu, R., Lucey, S.: Learning depth from monocular videos using direct methods. In: Proceedings of the IEEE Conference on Computer Vision and Pattern Recognition, pp. 2022–2030 (2018)

32. Wang, L., et al.: DeepLens: shallow depth of field from a single image. arXiv preprint arXiv:1810.08100 (2018)

33. Workman, S., Greenwell, C., Zhai, M., Baltenberger, R., Jacobs, N.: DEEPFOCAL: a method for direct focal length estimation. In: 2015 IEEE International Conference on Image Processing (ICIP), pp. 1369–1373. IEEE (2015)

34. Workman, S., Zhai, M., Jacobs, N.: Horizon lines in the wild. arXiv preprint arXiv:1604.02129 (2016)

35. Xiang, Y., Mottaghi, R., Savarese, S.: Beyond PASCAL: a benchmark for 3D object detection in the wild. In: IEEE Winter Conference on Applications of Computer Vision, pp. 75–82. IEEE (2014)

36. Xiao, J., Ehinger, K.A., Oliva, A., Torralba, A.: Recognizing scene viewpoint using panoramic place representation. In: 2012 IEEE Conference on Computer Vision and Pattern Recognition, pp. 2695–2702. IEEE (2012)

37. Zhai, M., Workman, S., Jacobs, N.: Detecting vanishing points using global image context in a non-Manhattan world. In: Proceedings of the IEEE Conference on Computer Vision and Pattern Recognition, pp. 5657–5665 (2016)

Procedure Planning in Instructional Videos

Chien-Yi Chang$^{(\boxtimes)}$ [iD], De-An Huang[iD], Danfei Xu[iD], Ehsan Adeli[iD], Li Fei-Fei[iD], and Juan Carlos Niebles[iD]

Stanford University, Stanford, USA
cy3@stanford.edu

Abstract. In this paper, we study the problem of procedure planning in instructional videos, which can be seen as a step towards enabling autonomous agents to plan for complex tasks in everyday settings such as cooking. Given the current visual observation of the world and a visual goal, we ask the question "What actions need to be taken in order to achieve the goal?". The key technical challenge is to learn structured and plannable state and action spaces directly from unstructured videos. We address this challenge by proposing Dual Dynamics Networks (DDN), a framework that explicitly leverages the structured priors imposed by the conjugate relationships between states and actions in a learned plannable latent space. We evaluate our method on real-world instructional videos. Our experiments show that DDN learns plannable representations that lead to better planning performance compared to existing planning approaches and neural network policies.

Keywords: Latent space planning · Task planning · Video understanding · Representation for action and skill

1 Introduction

What does it take for an autonomous agent to perform complex tasks in everyday settings, such as cooking in a kitchen? The most crucial ability is to know what actions should be taken in order to achieve the goal. In other words, the agent needs to make goal-conditioned decisions sequentially based on its perception of the environment. This sequential decision making process is often referred to as *planning* in the robotics literature. While recent works have shown great promise in learning to plan for simple tasks in structured environments such as pushing objects or stacking blocks on a table [7,8,31], it is unclear how these approaches will scale to visually complex, unstructured environments as seen in instructional videos [37,38] (recordings of human performing everyday tasks).

Electronic supplementary material The online version of this chapter (https://doi.org/10.1007/978-3-030-58621-8_20) contains supplementary material, which is available to authorized users.

© Springer Nature Switzerland AG 2020
A. Vedaldi et al. (Eds.): ECCV 2020, LNCS 12356, pp. 334–350, 2020.
https://doi.org/10.1007/978-3-030-58621-8_20

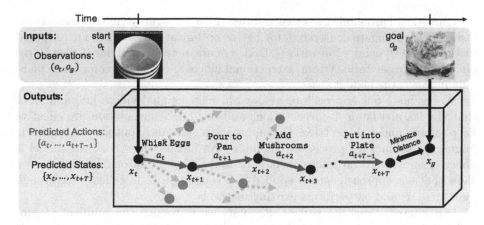

Fig. 1. We study the problem of *procedure planning* in instructional videos. The goal is to generate a sequence of actions towards a desired goal such as making a mushroom omelette. Given a start observation o_t and a visual goal o_g, our model plans a sequence of actions $\{a_i\}$ (red arrows) that can bring the start towards the goal. In addition, our model also predicts a sequence of intermediate states $\{x_i\}$ (black dots). (Color figure online)

In this paper, we study procedure planning in instructional videos, which can be seen as a step towards enabling autonomous agents to plan for complex tasks in everyday settings. Given the current visual observation of the world and a visual goal, we ask the question "What actions need to be taken in order to achieve the goal?". As illustrated in Fig. 1, we define the *procedure planning* problem as: given a current visual observation o_t and a visual goal o_g that indicates the desired final configuration, the model should plan a sequence of high-level actions $\{a_t, \cdots, a_{t+T-1}\}$ that can bring the underlying state of o_t to that of o_g. Here, T is the horizon of planning, which defines how many steps of task-level actions we allow the model to take.

The key technical challenge of the proposed procedure planning problem is how to learn structured and plannable state and action spaces directly from unstructured real videos. Since instructional videos are visually complex and the tasks are often described at high levels of abstractions, one can imagine indefinitely growing semantic state and action spaces from the visually complex scenes and high-level descriptions, which prevents the application of classical symbolic planning approaches [10, 15] as they require a given set of predicates for a well-defined state space.

We address this challenge by explicitly leveraging the conjugate relationships between states and actions to impose structured priors on the learned latent representations. Our key insight is that in addition to modeling the action as a transformation between states, we can also treat the state as the precondition for the next action because the state contains the history of previous actions. Following this intuition, we jointly train two modules: a forward dynamics model that captures the transitional probabilities between states and a conjugate dynamics model that utilizes the conjugate constraints for better optimization. We call our proposed method Dual Dynamics Networks (DDN), a neural network framework that learns plannable representations from instructional videos and performs procedure planning in the learned latent space.

We evaluate our approach on real-world instructional videos [38] and show that the learned latent representations and our DDN formulation are effective for planning under this setting. Our approach significantly outperforms existing planning baselines for procedure planning. We further show that our model is able to generalize to variations of the start and goal observations. In addition, we show that our approach can be applied to a related problem called walkthrough planning [17] and outperforms existing methods.

Our contributions are three-fold. First, we introduce the problem of procedure planning in instructional videos, which can be seen as a step towards enabling autonomous agents to plan for complex tasks in everyday settings such as cooking in a kitchen. Second, we propose Dual Dynamics Network (DDN), a framework that explicitly leverages structured priors imposed by the conjugate relationships between states and actions for latent space planning. Third, we evaluate our approach extensively and show that it significantly outperforms existing planning and video understanding methods for the proposed problem of procedure planning.

2 Related Work

Task Planning. *Procedure planning* is closely related to task planning widely studied in classical AI and robotics. Task planning is the problem of finding a sequence of task-level actions to reach the desired goal state from the current state. In the task planning literature, most studies rely on a pre-defined planning problem domain for the task [10,15,21]. Our work diverges from those because our proposed model can perform task-level, long-horizon planning in the visual and semantic space without requiring a hand-defined symbolic planning domain.

Planning from Pixels. Recent works have shown that deep networks can learn to plan directly from pixel observations in domains such as table-top manipulation [17,30,31], navigation in VizDoom [25], and locomotion in joint space [5]. However, learning to plan from unstructured high-dimensional observations is still challenging [7,11], especially for long-horizon, complex tasks that we want to address in procedure planning. A closely related method is Universal Planning Networks (UPN) [31], which uses a gradient descent planner to learn representations from expert demonstrations. However, it assumes the action space to be differentiable. Alternatively, one can also learn the forward dynamics by optimizing the data log-likelihood from the actions [11]. We use a similar formulation and further propose the conjugate dynamics model to expedite the latent space learning. Without using explicit action supervision, causal InfoGAN [17] extracts state representations by learning salient features that describe the causal structure of toy data. In contrast to [17], our model operates directly on real-world videos and handle the semantics of actions with sequential learning.

Understanding Instructional Videos. There has been a growing interest in analyzing instructional videos [16,23,33,37,38] by studying a variety of challenging tasks. Some of the tasks ask the question "What is happening?", such as action recognition and temporal action segmentation [4,13,28], state understanding [2], video summarization/captioning [29,32,37], retrieval [32] etc. The others ask the question "What is going to happen?", such as early action recognition [9,35], action label prediction [1,6,9,14,22,27,29,36], video prediction [18–20,24,26,34], etc. However, due to the large uncertainty in human activities, the correct answer to this question is often not unique. In this paper, we take a different angle and instead ask the question: "What actions need to be taken in order to achieve a given visual goal?" Rather than requiring the model to predict potentially unbounded future actions conditioned only on history, we ask the model to infer future actions conditioned on both history and goal. Such goal-conditioned formulation immediately resolves the ambiguity when the history does not encode adequate information to bound future actions.

3 Method

We are interested in planning in real-world instructional videos. The key technical challenge is how to learn structured and plannable state and action spaces directly from unstructured real videos. We take a latent space approach by learning plannable representations of the visual observations and actions, along with the forward and conjugate dynamics models in the latent space. We will first define the procedure planning problem setup and how to address it using a latent space planning approach. We will then discuss how we learn the latent space and leverage the conjugate relationships between states and actions to avoid trivial solutions to our optimization. Finally, we will present the algorithms for procedure planning and walkthrough planning [17] in the learned plannable space.

3.1 Problem Formulation

As illustrated in Fig. 1, given a current visual observation o_t and a visual goal o_g that indicates the desired final configuration, we aim to plan a sequence of actions $\pi = \{a_t, \cdots, a_{t+T-1}\}$ that can bring the underlying state of o_t to that of o_g. T is the horizon of planning, which defines how many steps of task-level actions we allow the model to take. We formulate the problem as latent space planning [17,31]. Concretely, we learn a model that can plan in some latent space spanned by the mapping functions f and g that encode the visual observation o and action a to a semantic state $f(o) = x$ (*e.g.*, from the observed frames to cooked eggs with mushrooms) and a latent action state $g(a) = \bar{a}$ respectively.

Inspired by the classical Markovian Decision Process (MDP) [3], we assume that in this latent space there exists a forward dynamics $\mathcal{T}(x_{t+1}|x_t, \bar{a}_t)$ that predicts the future state x_{t+1} given the current state x_t and the applicable action $\bar{a}_t$. Using the forward dynamics, we can perform sampling based planning [10] by applying different actions and search for the desired goal state x_g. Specifically, given $f(\cdot)$, $g(\cdot)$ and $\mathcal{T}$, we can find a plan $\pi = \{a_t, \cdots, a_{t+T-1}\}$ by (i) mapping from the visual space to the latent space $x_t = f(o_t)$, $x_g = f(o_g)$ and (ii) search in the latent space using $\mathcal{T}(x_{t+1}|x_t, \bar{a}_t)$ to find the sequence of actions that can bring x_t to x_g. We will discuss details of this procedure in Sect. 3.3.

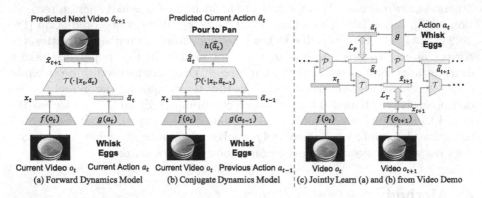

Fig. 2. (a) Our forward dynamics model $\mathcal{T}$ predicts the next state based on the current state and action. (b) We learn the conjugate dynamics model $\mathcal{P}$ jointly with $\mathcal{T}$ to restrict the possible state mapping f and action embedding g. (c) At training time, $\mathcal{T}$ takes as inputs the current state x_t and predicted current action $\hat{\bar{a}}_t$ to predict the next state $\hat{x}_{t+1}$, and the conjugate dynamics model $\mathcal{P}$ takes as inputs the predicted next state $\hat{x}_{t+1}$ and the current action $\bar{a}_t$ to predict the next applicable action $\hat{\bar{a}}_{t+1}$.

3.2 Learning Plannable Representations

In this section, we discuss how to learn the embedding functions $f(\cdot)$, $g(\cdot)$, and forward dynamics $\mathcal{T}$ from data. One possible approach is to directly optimize $f(\cdot)$ with some surrogate loss function such as mutual information [17], without explicitly modeling $\mathcal{T}(\cdot|x, \bar{a})$ and $g(\cdot)$. This approach is limited because it assumes a strong correspondence between the visual observation o and the semantic state $x = f(o)$. In real-world videos, however, a small change in visual space Δo can induce a large variation in the semantic space Δx and vice versa. An alternative approach is to formulate a differentiable objective jointly with $f(\cdot)$, $g(\cdot)$, and $\mathcal{T}(\cdot|x, \bar{a})$ [31]. This approach also falls short when applied to procedure planning, because it often requires the action space to be continuous and differentiable, while we have an unstructured and discrete action space in real-world instructional videos.

Forward Dynamics. To address the above challenges, we propose to learn $f(\cdot)$, $g(\cdot)$, and $\mathcal{T}(\cdot|x, \bar{a})$, jointly with a conjugate dynamics model $\mathcal{P}$ which provides further constraints on the learned latent spaces. We consider training data $\mathcal{D} = \{(o_t^i, a_t^i, o_{t+1}^i)\}_{i=1}^n$, which corresponds to the triplets of current visual observation, the action to be taken, and the next visual observation. The triplets can be curated from existing instructional video datasets and the details will be discussed in Sect. 4. The learning problem is to find $\mathcal{T}$ that minimizes:

$$\mathcal{L}_T(\mathcal{T}, f, g; \mathcal{D}) = -\frac{1}{n} \sum_{i=1}^{n} \log \Pr(x_{t+1}^i | x_t^i, \bar{a}_t^i; \mathcal{T})$$

$$= -\frac{1}{n} \sum_{i=1}^{n} \log \Pr(f(o_{t+1}^i) | f(o_t^i), g(a_t^i); \mathcal{T}). \tag{1}$$

The architectures are shown in Fig. 2(a).

Conjugate Dynamics. There exists a large number of $\mathcal{T}(\cdot | x, \bar{a})$ that can minimize $\mathcal{L}_T$ and our model can easily overfit to the training data in $\mathcal{D}$. In this case, it is hard for the learned model to generalize to unseen visual observations. It is thus crucial to impose structured priors on the model. Our insight is to leverage the conjugate relationship between states and actions [12] to provide further constraints on $f(\cdot)$ and $g(\cdot)$ and improve the optimization of $\mathcal{L}_T$. Please note that this is also how our proposed method diverges from the classical MDP setting and previous works [17,31]. We can treat the loss in Eq. (1) as leveraging the standard relationship between states and actions to learn $f(\cdot)$: Given the current semantic state x_t, applying a latent action $\bar{a}_t$ would bring it to a new state x_{t+1}. $\mathcal{L}_T$ is encouraging $f(\cdot)$ to fit $x_{t+1} = f(o_{t+1})$ and to be consistent with the prediction of $\mathcal{T}(\cdot | x, \bar{a})$. On the other hand, the conjugate relationship between states and actions also implies the following: Given the previous action $\bar{a}_{t-1}$, the current state x_t constraints the possible actions $\bar{a}_t$ because x_t needs to satisfy the precondition of $\bar{a}_t$. For example, if $\bar{a}_{t-1}$ representing *'pour eggs to pan'* is followed by $\bar{a}_t$ representing *'cook it'*, then the state x_t in between must satisfy the precondition of $\bar{a}_t$, that is: to cook the eggs, the eggs should be in the pan. Based on this intuition, we further propose to learn a conjugate dynamics model $\mathcal{P}(\bar{a}_t | x_t, \bar{a}_{t-1})$, which is illustrated in Fig. 2(b). The conjugate dynamics model $\mathcal{P}$ takes as inputs the current state and the previous action to predict the current applicable action. Formally, the learning problem is to find $\mathcal{P}$ that minimizes the following function:

$$\mathcal{L}_P(\mathcal{P}, f, g, h; \mathcal{D}) = -\frac{1}{n} \sum_{i=1}^{n} \left[\log \Pr(\bar{a}_t^i | x_t^i, \bar{a}_{t-1}^i; \mathcal{P}) + \phi(h(g(a_t^i)), a_t^i) \right]$$

$$= -\frac{1}{n} \sum_{i=1}^{n} \left[\log \Pr\left(g(a_t^i) | f(o_t^i), g(a_{t-1}^i); \mathcal{P} \right) + \phi(h(g(a_t^i)), a_t^i) \right]. \tag{2}$$

where the mapping functions f and g have shared parameters across $\mathcal{T}$ and $\mathcal{P}$. h is an inverse mapping function that decodes the action a from the latent action representation $\bar{a}$. $\phi(\cdot, \cdot)$ measures the distance in the action space.

Now we have discussed all the components for Dual Dynamics Network (DDN), a framework for latent space task planning in instructional videos. In our framework, we encode the visual observations and the actions into latent space states and actions with $f(\cdot)$ and $g(\cdot)$, and jointly learn from video demonstrations a forward dynamics model $\mathcal{T}$ that captures the transitional probabilities and a

---

Algorithm 1. Procedure Planning

1: **Inputs:** Current and goal observations o_t, o_g; learned models $f(\cdot)$, $h(\cdot)$, $\mathcal{T}$, $\mathcal{P}$; max
iteration β, threshold ϵ, beam size η
2: $x \leftarrow f(o_t)$, $x_g \leftarrow f(o_g)$
3: $x^* \leftarrow x$, $\bar{a} \leftarrow \emptyset$
4: $q \leftarrow PriorityQueue()$, $q \leftarrow q \cup \{(x, \bar{a})\}$
5: **while** iteration $< \beta$ and $||x^* - x_g||_2^2 > \epsilon$ **do**
6:     $(x, \bar{a}) \leftarrow Pop(q)$
7:     $\bar{a}' \sim \mathcal{P}(\cdot | x, \bar{a})$
8:     **for** $\bar{a}_i \in \bar{a}'$ **do**
9:         $x \leftarrow \mathcal{T}(\cdot | x, \bar{a}_i)$
10:        $q \leftarrow q \cup \{(x, \bar{a}_i)\}$
11:        **if** $||x - x_g||_2^2 < ||x^* - x_g||_2^2$ **then**
12:            $x^* \leftarrow x$
13:    $q \leftarrow Sort(q)$
14:    $q \leftarrow q[: \eta]$
15: $\{\bar{a}_i^*\} \leftarrow Backtracking(q, x^*)$
16: **return** $\{h(\bar{a}_i^*)\}$

---

conjugate dynamics model $\mathcal{P}$ that utilizes the conjugate relationship between states and actions by minimizing the following combined loss function:

$$\mathcal{L}(\mathcal{T}, \mathcal{P}, f, g, g^{-1}; \mathcal{D}) = \alpha \cdot \mathcal{L}_T + \mathcal{L}_P. \tag{3}$$

α is a weighting coefficient for the combined loss. The learning process is illustrated in Fig. 2(c). During training, the forward dynamics model $\mathcal{T}$ takes the current state x_t and predicted current action $\hat{\bar{a}}_t$ to predict the next state $\hat{x}_{t+1}$, and the conjugate dynamics model $\mathcal{P}$ takes the predicted next state $\hat{x}_{t+1}$ and the current action $\bar{a}_t$ to predict the next applicable action $\hat{\bar{a}}_{t+1}$.

3.3   Planning in Latent Space

In this section, we discuss how to use DDN for planning in instructional videos. The general paradigm is illustrated in Fig. 3. At inference time, our full model rollouts by sampling the action from the conjugate dynamics $\mathcal{P}$ and the next state from the forward dynamics $\mathcal{T}$. $\mathcal{P}$ captures the applicable actions from the current state and improves planning performance. In the following, we first discuss how we can leverage the learned models in Sect. 3.2 to perform procedure planning in latent space. We then describe how our models can be applied to walkthrough planning [17], where the objective is to output the visual waypoints/subgoals between the current observation and the goal observation.

Procedure Planning. Using the learned models, we perform sampling-based forward planning [10] to plan a sequence of actions to achieve the goal. The process is shown in Algorithm 1. Given the current and the goal observations o_t and o_g, we first map them to the latent space with $f(\cdot)$: $x_t = f(o_t)$, $x_g = f(o_g)$.

In contrast to symbolic planning, we do not have a list of *applicable* actions to apply in the search process. One additional advantage of having jointly learned the conjugate dynamics model $\mathcal{P}$ is that we can efficiently sample the actions to apply using $\mathcal{P}(\cdot|x_t, \bar{a}_{t-1})$. Based on $\bar{a}_t$ and x_t, we can obtain x_{t+1} using $\mathcal{T}(\cdot|x, \bar{a})$. The search process continues for a max iteration of β and threshold ϵ while maintaining a priority queue of size η.

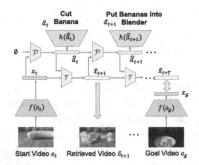

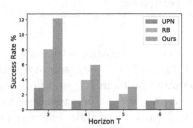

Fig. 3. At inference time, our full model rollouts by sampling the action from the conjugate dynamics $\mathcal{P}$ and the next state from the forward dynamics $\mathcal{T}$. $\mathcal{P}$ essentially captures the applicable actions and improves planning performance.

Fig. 4. Our model consistently outperforms other baselines as the horizon of planning T increases. When $T \geq 6$, the start observation and the visual goal are separated by 6 or more than 6 high-level actions. In this case the success rates of all baselines are less than 1%, suggesting there are more variations in the ways to reach the goal.

Walkthrough Planning. Kurutach *et al.* proposed walkthrough planning in [17]. The outputs of walkthrough planning can serve as visual signals of the subgoals to guide task executions. In addition, it is also helpful for interpretation by visualizing what the model has learned. The details of the process are shown in Algorithm 2. Given the pool of visual observations $\{o_i\}$, we can first construct the score matrix $R_{i,j}$ to capture the transition probability between two video clips o_i and o_j using our learned model $\mathcal{T}$ and $\mathcal{P}$. We can then perform walkthrough planning by finding the path of length T that starts at o_t and ends at o_g, while maximizing the total score. If the pool of video clips is all the clips in the same instructional video, then the problem is equivalent to finding a permutation function $b : \{1, 2, ..., T\} \rightarrow \{1, 2, ..., T\}$ that maximizes the total score along the permutation path, under the constraints that $b(1) = 1, b(T) = T$.

4   Experiments

We aim to answer the following questions in our experiments: (i) Can we learn plannable representations from real-world instructional videos? How does DDN

Algorithm 2. Walkthrough Planning

1: **Inputs:** All observations $\{o_i\}|_{i=1}^T$, learned models $f(\cdot)$, $g(\cdot)$, $\mathcal{T}$, $\mathcal{P}$, horizon T
2: $b \leftarrow \emptyset$
3: **for** i in $\{1 \ldots T\}$ **do**
4:     $x_i \leftarrow f(o_i)$
5: **for** i in $\{1 \ldots T\}$ **do**
6:     **for** j in $\{1 \ldots T\}$ **do**
7:         $R_{i,j} \leftarrow \sum_a \mathcal{T}(x_j | x_i, g(a)) \mathcal{P}(g(a) | x_i)$
8: $\{o_{b(i)}\} \leftarrow \arg\max_{b \in Perm(T)} \sum_{i=1}^T R_{b(i), b(i+1)}$
9: **return** $\{o_{b(i)}\}$

---

compare to existing latent space planning methods? (ii) How important are the forward and conjugate dynamics? (iii) Can we apply DDN to walkthrough planning problem [17] and how does it compare to existing methods?

We answer the first two questions with ablation studies on real-world instructional videos. We show how our model can generalize to various start and goal observations. For the last question, we apply our model to walkthrough planning [17] and show that our approach outperforms existing approaches.

Dataset. To train and test our proposed model, we curate a video dataset of size N that takes the form: $\{(o_t^i, a_t^i, ..., a_{t+T-1}^i, o_{t+T}^i)\}_{i=1}^N$. Each example contains T high-level actions, and each action can last from tens of seconds to 6 min. Our dataset is adapted from the 2750 labeled videos in CrossTask [38], averaging 4.57 min in duration, for a total of 212 h. Each video depicts one of the 18 long-horizon tasks like *Grill Steak*, *Make Pancakes*, or *Change a Tire*. The videos have manually annotated temporal segmentation boundaries and action labels. Please note that since our method requires full supervision, we will leave how to utilize unlabeled instructional videos [23,38] as future work. We encourage the reader refer to the supplementary material for data curation details. We divide the dataset into 70%/30% splits for training and testing. We use the precomputed features provided in CrossTask in all baselines and ablations for fair comparison. Each 3200-dimensional feature vector is a concatenation of the I3D, Resnet-152, and audio VGG features. The action space in CrossTask is given by enumerating all 105 combinations of the predicates and objects, which is shared across all 18 tasks. The distance function ϕ in CrossTask is the cross-entropy.

Implementation Details. The state mapping f and action embedding g are 128-dimensional. $\mathcal{T}$ is a 2-layer MLPs with 128 units that takes the outputs of f and g as inputs. The network $\mathcal{P}$ shares the state and the action embedding with $\mathcal{T}$ and we introduce recurrence to $\mathcal{P}$ by replacing the MLPs with a two-layer RNN of 128 hidden size, and concatenate the goal embedding as input. The action decoder h takes the outputs of $\mathcal{P}$ as inputs and outputs actions as defined by the dataset. We set hyperparameters $\alpha = 0.001$, $\beta = 20T$, $\eta = 20$, and $\epsilon = 1e - 5$. We use a 5-fold cross validation on the training split to set the hyper-parameters, and report performance on the test set. We train our model for 200 epochs with batch size of 256 on a single GTX 1080 Ti GPU. We use

Adam optimizer with learning rate of $1e-4$ and schedule the learning rate to decay with a decay factor 0.5 and a patience epoch of 5.

4.1 Evaluating Procedure Planning

In procedure planning, the inputs are the start video clip o_t and the goal video clip o_g, and the output is a sequence of T actions that brings o_t to o_g.

Baselines. We evaluate our model in learning plannable representations from real-world videos and compare with the following baselines and ablations:

- *Random Policy.* This baseline randomly selects an action from all actions. We include this baseline to show the empirical lower bound of performance.

- *Retrieval-Based (RB).* The procedure planning problem is formulated as a goal-conditioned decision making process. In contrast, one might approach this problem from a more static view which is analogous to [32]: Given the unseen start and goal visual observations, the retrieval-based baseline finds the nearest neighbor of the start and goal pair in the training set by minimizing the distance in the learned feature space, and then directly output the associated ground truth action labels. We use the same features as in our full model for fair comparison.

- *WLTDO* [5]. Ehsani *et al.* proposed an action planning model for egocentric dog videos. This baseline is a recurrent model for the planning task. Given two non-consecutive observations, it predicts a sequence of actions taking the state from the first observation to the second observation. We modify the author's implementation to use the same features as our full model, and add softmax layer to output discrete actions.

- *Uncertainty-Aware Anticipation of Activities (UAAA)* [6]. This baseline has a two-step approach that uses RNN-HMM to infer the action labels in the observed frames, and then use an autoregressive model to predict the future action labels. We re-implement the model and modify it to condition on both the observed frames (start observations) and the visual goal.

Table 1. Procedure planning results. Our model significantly outperforms baselines. With ~10% improvement of accuracy, our model is able to improve success rate by 8 times compared to *Ours w/o T*. This shows the importance of reasoning over the full sequence. * indicates re-implementations.

	$T = 3$			$T = 4$		
	Success Rate	Accuracy	mIoU	Success Rate	Accuracy	mIoU
Random	<0.01%	0.94%	1.66%	<0.01%	0.83%	1.66%
RB [32]*	8.05%	23.3%	32.06%	3.95%	22.22%	36.97%
WLTDO [5]*	1.87%	21.64%	31.70%	0.77%	17.92 %	26.43%
UAAA [6]*	2.15%	20.21%	30.87%	0.98%	19.86%	27.09%
UPN [31]*	2.89%	24.39%	31.56%	1.19%	21.59%	27.85%
Ours w/o $\mathcal{P}$	<0.01%	2.61%	0.86%	<0.01%	2.51%	1.14%
Ours w/o $\mathcal{T}$	1.55%	18.66%	28.81%	0.65%	15.97%	26.54%
Ours	**12.18%**	**31.29%**	**47.48%**	**5.97%**	**27.10%**	**48.46%**

Input Start   **Planning Outputs**   **Input Goal**

GT: Dip Bread in Mixture→ Put Bread in Pan→ Dip Bread in Mixture → Put Bread in Pan
Ours: Dip Bread in Mixture→ Put Bread in Pan→ Dip Bread in Mixture → Put Bread in Pan
RB: Melt Butter → Dip Bread in Mixture → Put Bread in Pan → Flip Bread

GT: Start Loose → Start Loose → Jack Up → Unscrew Wheel
Ours: Get Things Out → Start Loose → Jack Up → Unscrew Wheel
RB: Start Loose → Jack Up → Unscrew Wheel → Withdraw Wheel

Fig. 5. Examples of planned action sequences by DDN (ours) and the RB baseline. In the second example, DDN is not able to capture the subtle visual cues that the tool is already in the man's hand, so there is no need to get the tools out.

- *Universal Planning Networks (UPN)* [31]. UPN is the closest to ours among existing works. Similar to our approach, UPN also aims to learn a plannable representation using supervision from the imitation loss function at training. However, it assumes a continuous and differentiable action space to enable gradient-based planning, which might not be applicable to the discrete action space in CrossTask. We re-implement UPN and adapt it to output discrete actions by adding a softmax layer.

- *Ours w/o $\mathcal{T}$.* We compare to the ablation of our model without learning the forward dynamics $\mathcal{T}$, where we only $\mathcal{P}$ directly outputs the actions based on the previous action and the current state. We implement $\mathcal{P}$ with an RNN and concatenate the goal and start as input. In this case, this ablation is equivalent to a goal-conditional RNN policy directly trained with expert actions. This ablation can also be seen as a re-implementation of the RNN-based model in [1].

- *Ours w/o $\mathcal{P}$.* We compare to the ablation without learning the conjugate dynamics. In this case, the joint optimization of the forward dynamics $\mathcal{T}$ and the mapping $f(\cdot)$ to the latent space can easily overfit to the training sequences.

Metrics. We use three metrics for comparison. The first is *success rate*. Although we do not have access to the underlying environment to evaluate the policies by executing them in simulation, we consider a plan as a success if all the actions in the plan are the same as those in the ground truth. This is a reasonable approximation because we consider a fixed number of steps, and there is less variation in the ways to complete the task. The second metric we consider is the *accuracy* of the actions at each step, which does not require the whole sequence to match the ground truth as in the success rate metric, but only looks at each individual time step. We take the mean over the actions to balance the effect of repeating actions. The third metric we use is *mean Intersection over Union* (mIoU), which is the least strict of all the metrics we use. We compare the IoU by $\frac{|\{a_t\} \cap \{a_t^*\}|}{|\{a_t\} \cup \{a_t^*\}|}$, where $\{a_t^*\}$ is the set of ground truth actions, and $\{a_t\}$ is the set of predicted actions. We use IoU to capture the cases where the model understands what steps are required, but fails to discern the order of actions.

Results. The results are shown in Table 1. UPN is able to learn representations that perform reasonably well compared to the random baseline. However,

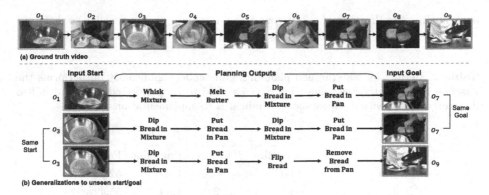

Fig. 6. Examples of generalizations to unseen start/goal. (a) The 9 snapshot observations in making French toasts. (b) Our proposed model is robust to changes of start and goal observations among different stages in the video, and is able to output reasonable plans to achieve the goal and reason about the essential steps. For example, in the third row, the start o_3 and goal o_9 are separated by 6 actions. While the model has never seen such data during training, it successfully plans the 4 essential actions by merging repeated actions and further adding the pivotal actions of flipping and plating.

as the action space in instructional videos is not continuous, the gradient-based planner is not able to work well. Both WLTDO and UAAA perform similar to Ours w/o $\mathcal{T}$, which can be seen an RNN goal-conditional policy directly trained with imitation objectives. We also note that Ours w/o $\mathcal{P}$ cannot learn reasonable plannable representations without the conjugate dynamics. Our full model combines the strengths of planning and action imitation objective as conjugate constraints, which enables us to learn plannable representations from real-world videos to outperform all the baseline approaches on all metrics. In Fig. 4, we further show our model's performance as the planning horizon increases. Our model consistently outperforms the RB baseline for all metrics because DDN explicitly imposes the structured priors using $\mathcal{T}$ and $\mathcal{P}$ to find the sequence of actions that reaches the goal.

Figure 5 shows some qualitative results. We note that it is difficult for the predicted sequence of actions to be exactly the same as the ground-truth, which explains why the success rates in Table 1 and Fig. 4 are low in absolute value. The results are nevertheless semantically reasonable. Furthermore, as shown in Fig. 6, it allows our model to generalize to different start observations when fixing the goal and vice versa. Figure 6(a) shows the 9 snapshot observations in making French toast. In the second row of Fig. 6(b), we pick o_3 and o_7 as the start and goal observations and ask the model to plan for 4 actions to reach the goal. The model successfully recognizes that there are two French toasts and dip and pan fry the bread twice. In the third row, we change the goal to o_9. In this case, the start and goal are separated by 6 actions. While the model has never seen such data during training, it successfully plans the 4 essential actions by merging repeated actions and further adding the pivotal actions of flipping and plating. Similarly, in the first row, the model can also generalize to different start observations when fixing the goal.

Table 2. Results for walkthrough planning. Our model significantly outperforms the baseline by explicitly reasoning what actions need to be performed first, and is less distracted by the visual appearances. * indicates re-implementations.

	$T = 3$		$T = 4$	
	Hamming	Pair Acc.	Hamming	Pair Acc.
Random	1.06	46.85%	1.95	52.23%
RB [32]*	0.88	56.23%	1.80	55.42%
VO [36] *	1.02	49.06%	1.99	50.31%
Causal InfoGAN [17]	0.57	71.55%	1.36	68.41%
Ours w/o $\mathcal{T}$	0.99	50.45%	2.01	47.39%
Ours w/o $\mathcal{P}$	0.33	83.33%	1.08	77.11%
Ours	**0.26**	**86.81%**	**0.88**	**81.21%**

4.2   Evaluating Intermediate States with Walkthrough Planning

Given the start and goal video clips, we have shown that our model is able to plan a sequence of actions that brings the start to the goal. At the same time, our model also predicts a sequence of intermediate states. In this section, we evaluate and visualize these predicted intermediate states. Specifically, we show how our proposed method can be apply to *walkthrough planning* [17], where the objective is to output the visual waypoints between the start and goal observations.

Experimental Setup. We evaluate the predicted intermediate state representations by using them to retrieve visual subgoals for task completion. In this way, the model only needs to predict lower-dimensional representations that can be used to retrieve the correct video clips from a pool of candidates. Specifically, we use all the video clips in the original video as the video clip pool. In this case, the task is equivalent to sorting the intermediate video clips while fixing the first and the last video clips.

Additional Baselines. We further compare to the following approaches for walkthrough planning:

- *Visual Ordering (VO).* As the task is reduced to sorting a pool of video clips [36], one baseline is to directly learn a model $V(o_1, o_2)$ to see if o_1 and o_2 are consecutive video clips in the same video. Given $V(o_1, o_2)$, we can find the order of the candidate video clips by maximizing the total score given by V using a greedy-based search. We learned $V(o_1, o_2)$ using the same setup as our $\mathcal{T}$, only that the actions a are not used as the input.

- *Causal InfoGAN (CIGAN)* [17]. CIGAN learns plannable representations by maximizing the mutual information between the representations and the visual observations. Additionally, the latent space is assumed to follow the forward dynamics of a certain class of actions. CIGAN is able to perform walkthrough planning using minimal supervision. We modify the author's implementation to use the same features as our full model for fair comparison.

Metrics. Let $Y = (y_1, ..., y_T)$ be a sequence of the ground truth order, and $b : \{1, 2, ..., T\} \rightarrow \{1, 2, ..., T\}$ be the permutation function such that the prediction is $\hat{Y} = (y_{b(1)}, ..., y_{b(T)})$. We use the following two metrics to evaluate the walkthrough planning outputs order:

   - *Hamming Distance:* counts the number of $\{i | i \neq b(i)\}$.
   - *Pairwise Accuracy:* calculates if the order between a pair i and j is respected by $b(i)$ and $b(j)$. It is given by $\frac{2}{T(T-1)} \sum_{i<j,i\neq j}^{T} \{b(i) < b(j)\}$.

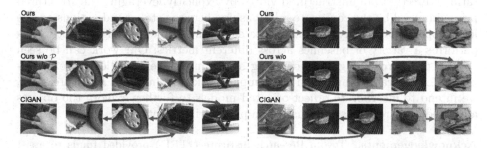

Fig. 7. Examples of visualized predicted intermediate states by DDN (ours) and two baselines. Green/red arrows indicate correct/incorrect transitions between two visual steps. Our model successfully predicts sequence of state representations with correct ordering, while the baselines are more or less distracted by the visual appearances. (Color figure online)

Results. The results are shown in Table 2. VO is unable to improve much over chance without modeling the actions. RB also struggles to perform well because it cannot handle the subsequence level variations that are not seen in the training data. By maximizing the mutual information with an adversarial loss, CIGAN[17] is able to learn reasonable models beyond Random without using action supervision. However, the complexity of the instructional videos requires explicit modeling of the forward dynamics conditioned on the semantic actions. Our full model learned for procedure planning successfully transfers to walkthrough planning and significantly outperforms all baselines on all metrics. It is interesting to see that Ours w/o $\mathcal{P}$ actually outperforms Ours w/o $\mathcal{T}$ in this case. Ours w/o $\mathcal{P}$ learns the forward dynamics $\mathcal{T}$ to directly anticipate the visual effect of an action a, which is better suited for walkthrough planning. By just using $\mathcal{P}$ the model is more accurate at predicting the actions, suggesting a trade-off between action and state space modeling. Our full model successfully combines the strength of the two and imposes the structured priors through conjugate constraints to both procedure planning and walkthrough planning.

In Fig. 7, we visualize some examples of the predicted intermediates states, where the task is to change tires (left) and grill a steak (right). Our full model is able to predict a sequence of intermediate states with correct ordering. Specifically, the most challenging step in changing tire (left) is the second step, where the person goes to take the tools, and the video is visually different from the rest of the steps. Neither of the baseline models is able to understand that to perform the rest of the steps, the person needs to get the tools first.

5   Conclusion

We presented Dual Dynamics Networks (DDN), a framework for procedure planning in real-world instructional videos. DDN is able to learn plannable representations directly from unstructured videos by explicitly leveraging the structured priors imposed by the conjugate relationships between states and actions on the latent space. Our experimental results show our framework significantly outperforms a variety of baselines across different metrics. Our work can be seen as a step towards the goal of enabling autonomous agents to learn from real-world demonstrations and plan for complex tasks like humans. In future work, we intend to incorporate object-oriented models to further explore the objects and predicates relations in visually complex environments.

Acknowledgements.. Toyota Research Institute ("TRI") provided funds to assist the authors with their research but this article solely reflects the opinions and conclusions of its authors and not TRI or any other Toyota entity.

References

1. Abu Farha, Y., Richard, A., Gall, J.: When will you do what?-anticipating temporal occurrences of activities. In: CVPR (2018)
2. Alayrac, J.B., Laptev, I., Sivic, J., Lacoste-Julien, S.: Joint discovery of object states and manipulation actions. In: ICCV (2017)
3. Bellman, R.: A Markovian decision process. J. Math. Mech. **6**, 679–684 (1957)
4. Chang, C.Y., Huang, D.A., Sui, Y., Fei-Fei, L., Niebles, J.C.: D3TW: discriminative differentiable dynamic time warping for weakly supervised action alignment and segmentation. In: CVPR (2019)
5. Ehsani, K., Bagherinezhad, H., Redmon, J., Mottaghi, R., Farhadi, A.: Who let the dogs out? modeling dog behavior from visual data. In: CVPR (2018)
6. Farha, Y.A., Gall, J.: Uncertainty-aware anticipation of activities. arXiv preprint arXiv:1908.09540 (2019)
7. Finn, C., Levine, S.: Deep visual foresight for planning robot motion. In: ICRA (2017)
8. Finn, C., Tan, X.Y., Duan, Y., Darrell, T., Levine, S., Abbeel, P.: Deep spatial autoencoders for visuomotor learning. In: ICRA (2016)

9. Furnari, A., Farinella, G.M.: What would you expect? anticipating egocentric actions with rolling-unrolling LSTMs and modality attention. In: ICCV, pp. 6252–6261 (2019)
10. Ghallab, M., Nau, D., Traverso, P.: Automated Planning: Theory and Practice. Elsevier, Amsterdam (2004)
11. Hafner, D., et al.: Learning latent dynamics for planning from pixels. In: ICML (2019)
12. Hayes, B., Scassellati, B.: Autonomously constructing hierarchical task networks for planning and human-robot collaboration. In: ICRA (2016)
13. Huang, D.-A., Fei-Fei, L., Niebles, J.C.: Connectionist temporal modeling for weakly supervised action labeling. In: Leibe, B., Matas, J., Sebe, N., Welling, M. (eds.) ECCV 2016. LNCS, vol. 9908, pp. 137–153. Springer, Cham (2016). https://doi.org/10.1007/978-3-319-46493-0_9
14. Ke, Q., Fritz, M., Schiele, B.: Time-conditioned action anticipation in one shot. In: CVPR (2019)
15. Konidaris, G., Kaelbling, L.P., Lozano-Perez, T.: From skills to symbols: learning symbolic representations for abstract high-level planning. J. Artif. Intell. Res. **61**, 215–289 (2018)
16. Kuehne, H., Arslan, A., Serre, T.: The language of actions: recovering the syntax and semantics of goal-directed human activities. In: CVPR (2014)
17. Kurutach, T., Tamar, A., Yang, G., Russell, S.J., Abbeel, P.: Learning plannable representations with causal infogan. In: NeurIPS (2018)
18. Lan, T., Chen, T.-C., Savarese, S.: A hierarchical representation for future action prediction. In: Fleet, D., Pajdla, T., Schiele, B., Tuytelaars, T. (eds.) ECCV 2014. LNCS, vol. 8691, pp. 689–704. Springer, Cham (2014). https://doi.org/10.1007/978-3-319-10578-9_45
19. Lee, A.X., Zhang, R., Ebert, F., Abbeel, P., Finn, C., Levine, S.: Stochastic adversarial video prediction. arXiv preprint arXiv:1804.01523 (2018)
20. Lu, C., Hirsch, M., Scholkopf, B.: Flexible spatio-temporal networks for video prediction. In: CVPR (2017)
21. McDermott, D., et al.: PDDL-the planning domain definition language (1998)
22. Mehrasa, N., Jyothi, A.A., Durand, T., He, J., Sigal, L., Mori, G.: A variational auto-encoder model for stochastic point processes. In: CVPR (2019)
23. Miech, A., Zhukov, D., Alayrac, J.B., Tapaswi, M., Laptev, I., Sivic, J.: Howto100m: Learning a text-video embedding by watching hundred million narrated video clips. In: ICCV (2019)
24. Oh, J., Guo, X., Lee, H., Lewis, R.L., Singh, S.: Action-conditional video prediction using deep networks in atari games. In: NeurIPS (2015)
25. Pathak, D., Agrawal, P., Efros, A.A., Darrell, T.: Curiosity-driven exploration by self-supervised prediction. In: CVPR Workshops (2017)
26. Ranzato, M., Szlam, A., Bruna, J., Mathieu, M., Collobert, R., Chopra, S.: Video (language) modeling: a baseline for generative models of natural videos. arXiv preprint arXiv:1412.6604 (2014)
27. Rhinehart, N., Kitani, K.M.: First-person activity forecasting with online inverse reinforcement learning. In: ICCV (2017)
28. Richard, A., Kuehne, H., Iqbal, A., Gall, J.: Neuralnetwork-viterbi: a framework for weakly supervised video learning. In: CVPR (2018)
29. Sener, F., Yao, A.: Zero-shot anticipation for instructional activities. In: ICCV (2019)
30. Sermanet, P., et al.: Time-contrastive networks: self-supervised learning from video. In: ICRA (2018)

31. Srinivas, A., Jabri, A., Abbeel, P., Levine, S., Finn, C.: Universal planning networks. In: ICML (2018)
32. Sun, C., Myers, A., Vondrick, C., Murphy, K., Schmid, C.: Videobert: a joint model for video and language representation learning. In: ICCV (2019)
33. Tang, Y., et al.: Coin: a large-scale dataset for comprehensive instructional video analysis. In: CVPR (2019)
34. Vondrick, C., Pirsiavash, H., Torralba, A.: Anticipating visual representations from unlabeled video. In: CVPR (2016)
35. Wang, X., Hu, J.F., Lai, J.H., Zhang, J., Zheng, W.S.: Progressive teacher-student learning for early action prediction. In: CVPR (2019)
36. Zeng, K.H., Shen, W.B., Huang, D.A., Sun, M., Carlos Niebles, J.: Visual forecasting by imitating dynamics in natural sequences. In: ICCV (2017)
37. Zhou, L., Xu, C., Corso, J.J.: Towards automatic learning of procedures from web instructional videos. In: AAAI (2018)
38. Zhukov, D., Alayrac, J.B., Cinbis, R.G., Fouhey, D., Laptev, I., Sivic, J.: Cross-task weakly supervised learning from instructional videos. In: CVPR (2019)

Funnel Activation for Visual Recognition

Ningning Ma[1], Xiangyu Zhang[2][✉], and Jian Sun[2]

[1] Hong Kong University of Science and Technology, Clear Water Bay, Hong Kong
nmaac@cse.ust.hk
[2] MEGVII Technology, Beijing, China
{zhangxiangyu,sunjian}@megvii.com

Abstract. We present a conceptually simple but effective funnel activation for image recognition tasks, called *Funnel activation (FReLU)*, that extends ReLU and PReLU to a 2D activation by adding a negligible overhead of spatial condition. The forms of ReLU and PReLU are $y = max(x, 0)$ and $y = max(x, px)$, respectively, while FReLU is in the form of $y = max(x, \mathbb{T}(x))$, where $\mathbb{T}(\cdot)$ is the 2D spatial condition. Moreover, the spatial condition achieves a pixel-wise modeling capacity in a simple way, capturing complicated visual layouts with regular convolutions. We conduct experiments on ImageNet, COCO detection, and semantic segmentation tasks, showing great improvements and robustness of FReLU in the visual recognition tasks. Code is available at https://github.com/megvii-model/FunnelAct.

Keywords: Funnel activation · Visual recognition · CNN

1 Introduction

Convolutional neural networks (CNNs) have achieved state-of-the-art performance in many visual recognition tasks, such as image classification, object detection, and semantic segmentation. As popularized in the CNN framework, one major kind of layer is the convolution layer, another is the non-linear activation layer.

First in the convolution layers, capturing the spatial dependency adaptively is challenging, many advances in more complex and effective convolutions have been proposed to grasp the local context adaptively in images [7,18]. The advances achieve great success especially on dense prediction tasks (e.g., semantic segmentation, object detection). Driven by the advances in more complex convolutions and their less efficient implementations, a question arises: *Could regular convolutions achieve similar accuracy, to grasp the challenging complex images?*

Second, usually right after capturing spatial dependency in a convolution layer *linearly*, then an activation layer acts as a scalar non-linear transformation. Many insightful activations have been proposed [5,14,25,31], but improving the performance on visual tasks is challenging, therefore currently the most widely

© Springer Nature Switzerland AG 2020
A. Vedaldi et al. (Eds.): ECCV 2020, LNCS 12356, pp. 351–368, 2020.
https://doi.org/10.1007/978-3-030-58621-8_21

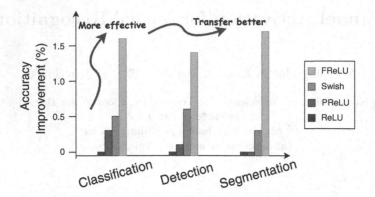

Fig. 1. Effectiveness and **generalization** performance. We set the ReLU network as the baseline, and show the *relative improvement* of accuracy on the three basic tasks in computer vision: image classification (Top-1 accuracy), object detection (mAP), and semantic segmentation (mean_IU). We use the ResNet-50 [15] as the backbone pre-trained on the ImageNet dataset, to evaluate the generalization performance on COCO and CityScape datasets. FReLU is more effective, and transfer better on all of the three tasks.

used activation is still the Rectified Linear Unit (ReLU) [32]. Driven by the distinct roles of the convolution layers and activation layers, another question arises: *Could we design an activation specifically for visual tasks?*

To answer both questions raised above, we show that the simple but effective visual activation, together with the regular convolutions, can also achieve significant improvements on both dense and sparse predictions (e.g. image classification, see Fig. 1). To achieve the results, we identify spatially insensitiveness in activations as the main obstacle impeding visual tasks from achieving significant improvements and propose a new visual activation that eliminates this barrier. In this work, we present a simple but effective visual activation that extends ReLU and PReLU to a 2D visual activation.

Spatially insensitiveness is addressed in modern activations for visual tasks. As popularized in the ReLU activation, non-linearity is performed using a $max(\cdot)$ function, the condition is the hand-designed *zero*, thus in the scalar form: $y = max(x, 0)$. The ReLU activation consistently achieves top accuracy on many challenging tasks. Through a sequence of advances [5,14,25,31], many variants of ReLU modify the condition in various ways and relatively improve the accuracy. However, further improvement is challenging for visual tasks.

Our method, called **Funnel activation (FReLU)**, extends the spirit of ReLU/PReLU by adding a spatial condition (see Fig. 2) which is simple to implement and only adds a negligible computational overhead. Formally, the form of our proposed method is $y = max(x, \mathbb{T}(x))$, where $\mathbb{T}(x)$ represents the simple and efficient spatial contextual feature extractor. By using the spatial condition in activations, it simply extends ReLU and PReLU to a visual parametric ReLU with a pixel-wise modeling capacity.

Our proposed visual activation acts as an efficient but much more effective alternative to previous activation approaches. To demonstrate the effectiveness of the proposed visual activation, we replace the normal ReLU in classification networks, and we use the pre-trained backbone to show its generality on the other two basic vision tasks: object detection and semantic segmentation. The results show that FReLU not only improves performance on a single task but also transfers well to other visual tasks.

2 Related Work

Scalar Activations. Scalar activations are activations with single input and single output, in the form of $y = f(x)$. The Rectified Linear Unit (ReLU) [13, 23, 32] is the most widely used scalar activation on various tasks [26, 38], in the form of $y = max(x, 0)$. It is simple and effective for various tasks and datasets. To modify the negative part, many variants have been proposed, such as Leaky ReLU [31], PReLU [14], ELU [5]. They keep the positive part identity and make the negative part dependent on the sample adaptively.

Other scalar methods such as the sigmoid non-linearity has the form $\sigma(x) = 1/(1+e^{-x})$, and the Tanh non-linearity has the form $tanh(x) = 2\sigma(2x)-1$. These activations are not widely used in deep CNNs mainly because they saturate and kill gradients, also involve expensive operations (exponentials, etc.).

Many advances followed [1, 10, 16, 25, 35, 39, 46], and recent searching technique contributes to a new searched scalar activation called Swish [36] by combing a comprehensive set of unary functions and binary functions. The form is $y = x * Sigmoid(x)$, outperforms other scalar activations on some structures and datasets, and many searched results show great potential.

Contextual Conditional Activations. Besides the scalar activation which only depends on the neuron itself, conditional activation is a many-to-one function, which activates the neurons conditioned on contextual information. A representative method is Maxout [12], it extends the layer to a multi-branch and selects the maximum. Most activations apply a non-linearity on the linear dot product between the weights and the data, which is: $f(w^T x + b)$. Maxout computes the $max(w_1^T x + b_1, w_2^T x + b_2)$, and generalizes ReLU and Leaky ReLU into the same framework. With dropout [17], the Maxout network shows improvement. However, it increases the complexity too much, the numbers of parameters and multiply-adds has doubled and redoubled.

Contextual gating methods [8, 44] use contextual information to enhance the efficacy, especially on RNN based methods, because the feature dimension is relatively smaller. There are also on CNN based methods [34], since 2D feature size has a large dimension, the method is used after a feature reduction.

The contextually conditioned activations are usually channel-wise methods. However, in this paper, we find the spatial dependency is also important in the non-linear activation functions. We use light-weight CNN technique depth-wise separable convolution to help with the reduction of additional complexity.

Spatial Dependency Modeling. Learning better spatial dependency is challenging, Some approaches use different shapes of convolution kernels [40–42] to aggregate the different ranges of spatial dependences. However, it requires a multi-branch that decreases efficiency. Advances in convolution kernels such as atrous convolution [18] and dilated convolution [47] also lead to better performance by increasing the receptive field.

Another type of methods learn the spatial dependency adaptively, such as STN [22], active convolution [24], deformable convolution [7]. These methods adaptively use the spatial transformations to refine the short-range dependencies, especially for dense vision tasks (e.g. object detection, semantic segmentation). Our simple FReLU even outperforms them without complex convolutions.

Moreover, the non-local network provides the methods to capture long-range dependencies to address this problem. GCNet [3] provides a spatial attention mechanism to better use the spatial global context. Long-range modeling methods achieve better performance but still require additional blocks into the origin network structure, which decreases efficiency. Our method address this issue in the non-linear activations, solve this issue better and more efficiently.

Receptive Field. The region and size of receptive field are essential in vision recognition tasks [33,50]. The work on effective receptive field [11,29] finds that different pixels contribute unequally and the center pixels have a larger impact. Therefore, many methods have been proposed to implement the adaptive receptive field [7,49,51]. The methods achieve the adaptive receptive field and improve the performance, by involving additional branches in the architecture, such as developing more complex convolutions or utilizing the attention mechanism. Our method also achieves the same goal, but in a more simple and efficient manner by introducing the receptive field into the non-linear activations. By using the more adaptive receptive field, we can approximate the layouts in common complex shapes, thus achieve even better results than the complex convolutions, by using the efficient regular convolutions.

3   Funnel Activation

FReLU is designed specifically for visual tasks and is conceptually simple: the condition is a hand-designed zero for ReLU and a parametric px for PReLU, to this we modify it to a 2D funnel-like condition dependent on the spatial context. The visual condition helps extract the fine spatial layout of an object. Next, we introduce the key elements of FReLU, including the funnel condition and the pixel-wise modeling capacity, which are the main missing parts in ReLU and its variants.

ReLU. We begin by briefly reviewing the ReLU activation. ReLU, in the form $max(x,0)$, uses the $max(\cdot)$ to serve as non-linearity and uses a hand-designed $zero$ as the condition. The non-linear transformation acts as a supplement of the linear transformation such as convolution and fully-connected layers.

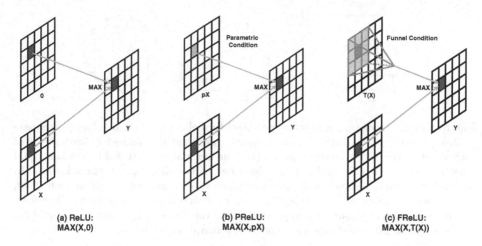

Fig. 2. Funnel activation. We propose a novel activation for visual recognition we call *FReLU* that follows the spirit of ReLU/PReLU and extends them to 2D by adding a visual funnel condition $\mathbb{T}(x)$. (a) ReLU with a condition zero; (b) PReLU with a parametric condition; (c) FReLU with a visual parametric condition.

PReLU. As an advanced variant of ReLU, PReLU has an original form $max(x,0) + p \cdot min(x,0)$, where p is a learnable parameter and initialized as 0.25. However, in most case $p < 1$, under this assumption, we rewrite it to the form: $max(x, px)$, $(p < 1)$. Since p is a channel-wise parameter, it can be interpreted as a 1×1 depth-wise convolution regardless of the bias terms.

Funnel Condition. FReLU adopts the same $max(\cdot)$ as the simple non-linear function. For the condition part, FReLU extends it to be a 2D condition dependent on the spatial context for each pixel (see Fig. 2). This is in contrast to most recent methods whose condition depends on the pixel itself (e.g. [14,31]) or the channel context (e.g. [12]). Our approach follows the spirit of ReLU that uses a $max(\cdot)$ to obtain the maximum between x and a condition.

Formally, we define the funnel condition as $\mathbb{T}(x)$. To implement the spatial condition, we use a **Parametric Pooling Window** to create the spatial dependency, specifically, we define the activation function:

$$f(x_{c,i,j}) = max(x_{c,i,j}, \mathbb{T}(x_{c,i,j})) \tag{1}$$

$$\mathbb{T}(x_{c,i,j}) = x_{c,i,j}^{\omega} \cdot p_c^{\omega} \tag{2}$$

Here, $x_{c,i,j}$ is the input pixel of the non-linear activation $f(\cdot)$ on the c-th channel, at the 2-D spatial position (i, j); function $\mathbb{T}(\cdot)$ denotes the funnel condition, $x_{c,i,j}^{\omega}$ denotes a $k_h \times k_w$ **Parametric Pooling Window** centered on $x_{c,i,j}$, p_c^{ω} denotes the coefficient on this window which is shared in the same channel, and $(\cdot)$ denotes dot multiply.

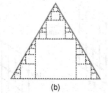

Fig. 3. Graphic depiction of how the per-pixel funnel condition can achieve *pixel-wise modeling capacity*. The distinct sizes of squares represent the distinct *activate fields* of each pixel in the top activation layers. (a) The normal activate field that has equal sizes of squares per-pixel, and can only describe the horizontal and vertical layouts. In contrast, the $max(\cdot)$ allows each pixel to choose *looking around or not* in each layer, after enough number of layers, they have many different sizes of squares. Therefore, the different sizes of squares can approximate (b) the shape of the oblique line, and (c) the shape of an arc, which are more common natural object layouts.

Pixel-Wise Modeling Capacity. Our definition of funnel condition allows the network to generate spatial conditions in the non-linear activations for every pixel. The network conducts non-linear transformations and creates spatial dependencies *simultaneously*. This is different from common practice which creates spatial dependency in the convolution layer and conducts non-linear transformations separately. In that case, the activations do not depend on spatial conditions explicitly; in our case, with the funnel condition, they do.

As a result, the pixel-wise condition makes the network has a pixel-wise modeling capacity, the function $max(\cdot)$ gives per-pixel a choice between *looking at the spatial context or not*. Formally, consider a network $\{F_1, F_2, ..., F_n\}$ with n FReLU layers, each FReLU layer F_i has a $k \times k$ parametric window. For brevity, we only analyze the FReLU layers regardless of the convolution layers. Because the max selection between 1×1 and $k \times k$, each pixel after F_1 has a *activate filed* set $\{1, 1+r\}$ ($r = k - 1$). After the F_n layer, the set becomes $\{1, 1+r, 1+2r, ..., 1+nr\}$, which gives more choices to each pixel and can approximate any layouts if n is sufficiently large. With many distinct sizes of the activate field, the distinct sizes of squares can approximate the shape of the oblique line and arc (see Fig. 3). As we know, the layout of the objects in the images are usually not horizontal or vertical, they are usually in the shape of the oblique line or arc, therefore extracting the spatial structure of objects can be addressed naturally by the pixel-wise modeling capacity provided by the spatial condition. We show by experiments that it captures **irregular and detailed object layouts** better in complex tasks (see Fig. 4).

3.1   Implementation Details

Our proposed change is simple: we avoid the hand-designed condition in activations, we use a simple and effective spatial 2D condition to replace it. The visual activation leads to significant improvements as shown in Fig. 1. We first change the ReLU activations in the classification task on the ImageNet dataset.

We use ResNet [15] as the classification network and use the pre-trained network as backbones for other tasks: object detection and semantic segmentation.

All the regions $x^\omega_{c,i,j}$ in the same channel share the same coefficient p^ω_c, therefore, it only adds a slight additional number of parameters. The region represented by $x^\omega_{c,i,j}$ is a sliding window, the size is default set to a 3×3 square, and we set the 2-D padding to be 1, in this case,

$$x^\omega_{c,i,j} \cdot p^\omega_c = \sum_{i-1\leq h\leq i+1, j-1\leq w\leq j+1} x_{c,h,w} \cdot p_{c,h,w} \qquad (3)$$

Parameter Initialization. We use the gaussian initialization to initialize the hyper-parameters. Therefore we get the condition values close to zero, which does not change the origin network's property too much. We also investigate the cases without parameters, (e.g. max pooling, average pooling), which do not show improvement. That shows the importance of the additional parameters.

Parameter Computation. We assume there is a $K'_h \times K'_w$ convolution with the input feature size of $C \times H \times W$ input, and the output size of $C \times H' \times W'$, then we compute the number of parameters to be $CCK'_h K'_w$, and the FLOPs (floating point operations) to be $CCK'_h K'_w HW$. To this we add our funnel condition with window $K_h \times K_w$, the additional number of parameters is $CK_h K_w$, and the additional number of FLOPs is $CK_h K_w HW$. We assume $K = K_h = K_w$, $K' = K'_h = K'_w$ for simplification.

Therefore the original complexity of parameters is $O(C^2 K'^2)$, after adopting FReLU, it becomes $O(C^2 K'^2 + CK^2)$); and the original complexity of FLOPs is $O(C^2 K'^2 HW)$, after adopting the visual activation, it becomes $O(C^2 K'^2 HW + CK^2 HW)$. Usually, C is much larger than K and K', therefore the additional complexity can be negligible. Actually in practice the additional part is negligible (more details in Table 1). Moreover, the funnel condition is a $k_h \times k_w$ sliding window, and we implement it using the highly optimized depth-wise separable convolution operator followed with a BN [21] layer.

4   Experiments

4.1   Image Classification

To evaluate the effectiveness of our visual activation, first, we conduct our experiments on ImageNet 2012 classification dataset[9,37], which comprises 1.28 million training images and 50K validation images.

Our visual activation is easy to adopt on the network structures, by simply changing the ReLU in the original CNN structure. First, we evaluate the activation on different sizes of ResNet [15]. For the network structure, we use the original implementation. Spatial dependency is important especially in the shallow layers, for the small 224×224 input size, we replace the ReLUs in all the stages except the last stage, which has a small 7×7 feature map size. For the

Table 1. Comparisons with other effective activations [14,36] on ResNets [15] in ImageNet 2012. Image size 224 × 224. Single crop. We evaluate the Top-1 error rate on the
test set.

Model	Activation	#Params	FLOPs	Top-1 Err
ResNet-50	ReLU	25.5M	3.86G	24.0
	PReLU	25.5M	3.86G	23.7
	Swish	25.5M	3.86G	23.5
	FReLU	25.5M	3.87G	**22.4**
ResNet-101	ReLU	44.4M	7.6G	22.8
	PReLU	44.4M	7.6G	22.7
	Swish	44.4M	7.6G	22.7
	FReLU	44.5M	7.6G	**22.1**

training settings, we use a batch size of 256, 600k iterations, a learning rate of
0.1 with linear decay schedule, a weight decay of 1e-4, and a dropout [17] rate of
0.1. We present the Top-1 error rate on the validation set. For a fair comparison,
we run all the results on the same code base.

Comparisons with Scalar Activations. We conduct a comprehensive comparison on ResNets [15] with different depths (e.g. ResNet-50, ResNet-101). We
take ReLU as the baseline and take one of its variants PReLU for comparison.
Further, we compare our visual activation with the activation Swish [36] searched
by the NAS [52,53] technique. Swish has shown its positive influence on various
model structures, comparing with many scalar activations.

Table 1 shows the comparison, our visual activation still outperforms all of
them with a negligible additional complexity. Our visual activation improves
1.6% and 0.7% top-1 accuracy rates on ResNet-50 and ResNet-101. It's remarkable that with the increase of model size and model depth, other scalar activations show limited improvement, while visual activation still has significant
improvement. For example, Swish and PReLU improve the accuracy of 0.1% on
ResNet-101, while visual activation increases still significantly on ResNet-101
with an improvement of 0.7%.

Comparison on Light-Weight CNNs. Besides deep CNNs, we compare the
visual activation with other effective activations on recent light-weight CNNs
such as MobileNets [19] and ShuffleNets [30]. We use the same training settings
in [30]. The model sizes are extremely small, we use a window size of $1 \times 3 + 3 \times 1$
to reduce the additional parameters. Moreover, for MobileNet we slightly refine
the width multiplier from 0.75 to 0.73 to maintain the model complexity. Table 2
shows the comparison results on ImageNet dataset. Our visual activation also
boosts accuracy on light-weight CNNs. ShuffleNetV2 0.5× can improve 2.5%
top-1 accuracy by only adding a slight additional FLOPs.

Table 2. Comparisons among other effective activations [14,36] on light-weight CNNs (MobileNet [19], ShuffleNetV2 [30]) in ImageNet 2012. Image size 224 × 224. Single crop. We evaluate the Top-1 error rate on the test set.

Model	Activation	#Params	FLOPs	Top-1 Err
MobileNet 0.75	ReLU	2.5M	325M	29.8
	PReLU	2.5M	325M	29.6
	Swish	2.5M	325M	28.9
	FReLU	2.5M	328M	**28.5**
ShuffleNetV2	ReLU	1.4M	41M	39.6
	PReLU	1.4M	41M	39.1
	Swish	1.4M	41M	38.7
	FReLU	1.4M	45M	**37.1**

4.2   Object Detection

To evaluate the generalization performance of visual activation on different tasks, we conduct object detection experiments on COCO dataset [28]. The COCO dataset has 80 object categories. We use the *trainval35k* set for training and use the *minival* set for testing.

We present the result on RetinaNet [27] detector. For a fair comparison, we train all the models in the same code base with the same settings. We use a batch size of 2, a weight decay of 1e-4 and a momentum of 0.9. We use anchors for 3 scales and 3 aspect ratios and use a 600-pixel train and test image scale. For the backbone, we use the pre-trained model in Sect. 4.1 as a feature extractor, and compare the generality among different activations.

Table 3 shows the comparison among different activations. The comparison shows that our visual activation increases 1.4% mAP comparing to the ReLU backbone, and increases 0.8% mAP comparing to the Swish backbone. It is worth mentioning that, on all the small, medium, and large objects, FReLU outperforms all the other counterparts significantly.

Table 3. Comparisons of different activations in COCO **object detection**. We use ResNet-50 [15] and ShuffleNetV2 (1.5×) [30] with different activations as the pre-trained backbones. We use the RetinaNet [27] detector.

Model	Activation	#Params	FLOPs	mAP	AP_{50}	AP_{75}	AP_s	AP_m	AP_l
ResNet-50	ReLU	25.5M	3.86G	35.2	53.7	37.5	18.8	39.7	48.8
	Swish	25.5M	3.86G	35.8	54.1	38.7	18.6	40.0	49.4
	FReLU	25.5M	3.87G	**36.6**	**55.2**	**39.0**	**19.2**	**40.8**	**51.9**
ShuffleNetV2	ReLU	3.5M	299M	31.7	49.4	33.7	15.3	35.1	45.2
	Swish	3.5M	299M	32.0	49.9	34.0	16.2	35.2	45.2
	FReLU	3.7M	318M	**32.8**	**50.9**	**34.8**	**17.0**	**36.2**	**46.8**

Table 4. Comparisons on the **semantic segmentation** task in CityScape dataset. We use the PSPNet [48] as the framework and use the ResNet-50 [15] as backbone. The pre-trained backbones are from Table 1.

	ReLU	Swish[36]	FReLU		ReLU	Swish	FReLU
mean_IU	77.2	77.5	**78.9**	terrain	65.0	64.0	64.5
road	98.0	98.1	98.1	sky	94.7	94.9	94.8
sidewalk	84.2	85.0	84.7	person	82.1	83.1	83.2
building	92.3	92.5	92.7	rider	62.3	65.5	64.7
wall	55.0	56.3	59.5	car	95.1	94.8	95.3
fence	59.0	59.6	60.9	truck	77.7	70.1	79.8
pole	63.3	63.6	64.3	bus	84.9	84.0	87.8
traffic light	71.4	72.1	72.2	train	63.3	68.8	74.6
traffic sign	79.0	80.0	79.9	motorcycle	68.3	69.4	69.8
vegetation	92.4	92.7	92.8	bicycle	78.2	78.4	78.7

We also show the comparison on the light-weight CNNs. As the comparison of ResNet-50, we use pre-trained ShuffleNetV2 backbones adopted with different activations. We mainly compare FReLU with ReLU and the effective activation Swish [36]. Table 3 shows visual activation also outperforms much better than ReLU and Swish backbones, to which it improves 1.1% mAP and 0.8% mAP respectively. Moreover, it increases the performance of all the sizes of objects.

4.3   Semantic Segmentation

We further present the semantic segmentation results on CityScape dataset [6]. The dataset is a semantic urban scene understanding dataset, contains 19 categories. It has 5,000 finely annotated images, 2,975 for training, 500 for validation and 1525 for testing.

We use the PSPNet [48] as the segmentation framework, for the training settings we use the poly learning rate policy [4] where the base is 0.01 and the power is 0.9, we use a weight decay of 1e-4, and 8 GPUs with a batch size of 2 on each GPU.

To evaluate the generality of the previous pre-trained models in Sect. 4.1, we use the pre-trained ResNet-50 [15] backbone models with different activations, we compare FReLU with Swish and ReLU respectively.

In Table 4, we show the comparison with scalar activations. From the result, we observe that our visual activation outperforms the ReLU and the searched Swish 1.7% and 1.4% mean_IU, respectively. Moreover, our visual activation has significant improvements in both large and small objects, especially on categories such as 'train', 'bus', 'wall', etc.

For better visualization of the improved performance, Fig. 4 shows the predict results on the testing dataset. It shows that by only changing the backbone activations, the results have obvious improvement. The boundaries of both the large and the small objects are well-segmented because the pixel-wise modeling

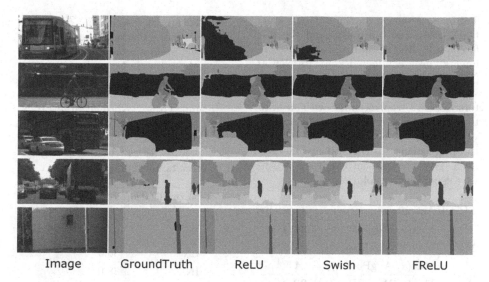

| Image | GroundTruth | ReLU | Swish | FReLU |

Fig. 4. Visualization of semantic segmentation on ResNet-50 [15]-PSPNet [48] with different activations in backbone. We clip the CityScape images to make the differences more clear (better view enlarge images). FReLU has better long-range (large or slender objects) and short-range (small objects) understandings due to its better context capturing capacity. It captures irregular and detailed object layouts in complex cases much better. We note that modern frameworks are finely optimized with ReLU, however, it has obvious improvements by only changing the backbones, thus having the potential for further gains if redesign the frameworks for the visual activation.

capacity can handle both global and detailed regions (see Fig. 3). We note that the modern recognition frameworks are finely designed with the ReLU activation, therefore the visual activation still has great potential for further improving the results, which is beyond the focus of this work.

5   Discussion

The previous sections demonstrate the optimum performance comparing with other effective activations. To further investigate our visual activation, we conduct ablation studies. We first discuss the properties of the visual activation, then we discuss the compatibility with existing methods.

5.1   Properties

Our funnel activation mainly has two components: 1) funnel condition, and 2) $max(\cdot)$ non-linearity. Separately, we investigate the effect of each component.

Table 5. Ablation on the different **spatial condition manners**, and the different **non-linear manners**. The experiments are conducted on ResNet-50 [15]. Model A, B, C compare different visual conditions with/without parameters. Model D replaces *max* with *sum*, to this we add a ReLU, or it will not converge. Model E separates and evaluates the performance of the spatial condition itself. DW(x) represent the 3×3 depthwise separable convolution.

Model	Activation	Top-1 Err
A	Max(x, ParamPool(x))	**22.4**
B	Max(x, MaxPool(x))	24.4
C	Max(x, AvgPool(x))	24.5
A	Max(x, ParamPool(x))	**22.4**
D	Sum(x, ParamPool(x))	23.6
E	Max(DW(x), 0)	23.7

Table 6. Ablation on different normalization methods after the spatial condition layer. We adopt Batch Normalization (BN) [21], Layer Normalization (LN) [2], Instance Normalization (IN) [43] and Group Normalization (GN) [45] after the spatial condition layer which is implemented by depth-wise convolution. ImageNet results on ShuffleNetV2 0.5×.

Normalization	Top-1 Err
-	37.6
BN	37.1
LN	36.5
IN	38.0
GN	36.5

Ablation on the Spatial Condition. First, we compare the different manners of the spatial condition. Besides the manner of parametric pooling that we used, to investigate the importance of the additional parameters, we compare other pooling manners without additional parameters, they are max pooling and average pooling. We simply replace the parametric pooling with the other two non-parametric manners and evaluate the results on the ImageNet dataset.

Table 5 (A, B, C) shows the importance of the parametric pooling. Without additional parameter, the results decrease more than 2% top-1 accuracy, even perform worse than the baseline that does not use spatial condition. Table 6 shows the comparison of different normalization after the spatial condition.

Ablation on the Non-linearity. Second, we also compare the use of non-linearity. In our method, we use the $max(\cdot)$ function to perform the non-linearity, *simultaneously* capturing visual dependency. In contrast, we compare with the manners that *separately* capture visual dependency and non-linearity.

For the spatial context capturing, we use two manners: 1) use the parametric pooling as before, then linearly add up with the original feature, 2) simply add a depth-wise separable convolution layer. For the non-linear transformation, we use the ReLU function. Table 5 (A, D, E) show the results. Comparing with the baseline, the spatial context itself improves about 0.3% accuracy, but together as the non-linear condition in our method, it further increases more than 1%. Therefore, performing the spatial dependency and non-linearity *separately* has not an ideal effect as doing them *simultaneously*.

Table 7. Ablation on the **window size.** We simply change the window size in the funnel condition. We evaluate the top-1 error rate on ImageNet dataset using the ResNet-50 [15] structure.

Model	Window size	Top-1 Err
A	1 × 1	23.7
B	3 × 3	**22.4**
C	5 × 5	22.9
D	7 × 7	23.0
E	Sum(1 × 3, 3 × 1)	22.6
F	Max(1 × 3, 3 × 1)	**22.4**

Table 8. Ablation on **different layers.** We replace the ReLU with FReLU after the 1 × 1 convolution and the 3 × 3 convolution. Results are performed on ResNet-50 [15] and MobileNet [19].

	1 × 1 conv	3 × 3 conv	Top-1 Err
ResNet-50	✓		22.9
		✓	23.0
	✓	✓	**22.4**
MobileNet	✓		29.2
		✓	29.0
	✓	✓	**28.5**

Ablation on the Window Size. In the parametric pooling window, the size of the window decides the size of the area each pixel *looks*. We simply change the window size in the funnel condition and compare different sizes among {1 × 1, 3 × 3, 5 × 5, 7 × 7}. The case of 1 × 1 does not have the spatial condition and it is the case of PReLU since the parameter value is smaller than 1. Table 7 shows the comparison results. We conclude that 3 × 3 is the best choice. The larger window sizes also show benefits but do not outperform 3 × 3.

Further, we consider the case using an irregular window instead of squares. We use multiple windows of sizes 1 × 3 and 3 × 1, we consider to use the sum and max of them as the condition. Table 7 {B, E, F} show the comparison. The results show that irregular window sizes also have the optimum performance since they have a more flexible pixel-wise modeling capacity (Fig. 3).

5.2   Compatibility with Existing Methods

To adopt the new activation into the convolutional networks, we have to choose which layers, and which stages to adopt. Moreover, we also investigate the compatibility with existing effective approaches such as SENet.

Compatibility with Different Convolution Layers. First, we compare the positions after different convolution layers. That is, we investigate the effect of FReLU in different positions after 1 × 1 and 3 × 3 convolutions. We conduct experiments on ResNet-50 [15] and ShuffleNetV2 [30]. We replace the ReLU after the 1 × 1 convolution and the 3 × 3 convolution and observe the improvement. Table 8 shows the results, in the bottleneck of the above two networks. From the results, we can see that the improvements on different layers are comparable, and it has the optimum performance when adopting both of them.

Table 9. Ablation of visual activation on different stages (Stage {2–4} in ResNet-50 [15]). In each stage we replace each ReLU with our visual activation. The results are the top-1 error rates on ImageNet. Image size 224 × 224.

Stage 2	Stage 3	Stage 4	Top-1 Err
✓			23.1
	✓		23.0
		✓	23.3
✓	✓		22.8
	✓	✓	23.0
✓	✓	✓	**22.4**

Table 10. Ablation comparisons of the compatibility between FReLU and SENet [20] on ResNet-50 [15]. The results are the top-1 error rates on ImageNet. Image size 224 × 224. Single crop.

Model	#Params	FLOPs	Top-1
ReLU	25.5M	3.9G	24.0
FReLU	25.5M	3.9G	22.4
ReLU+SE	26.7M	3.9G	22.8
FReLU+SE	26.7M	3.9G	**22.1**

Compatibility with Different Stages. Secondly, we investigate the compatibility with different stages in the CNN structures. The visual activations are important especially on the layer with high spatial dimensions. For the classification network whose shallow layers have larger spatial dimensions and deeper layers have large channel dimensions, there may be differences when we apply visual activations on different stages. For Stage 5 of ResNet-50 with 224 × 224 input, it has a relatively small 7 × 7 feature size, which mainly contains channel dependency instead of spatial dependency. Therefore, we adopt visual activations on Stage {2–4} on ResNet-50, as Table 9 shows. The results reveal that adopting the shallow layers has a larger effect, while a deeper layer has a smaller effect. Moreover, adopting FReLU on all of them has the optimum top-1 accuracy.

Compatibility with SENet. At last, we compare the performance with SENet [20] and show the compatibility with it. Without the complex advances in CNN architecture, it achieves significant improvements on all the three vision tasks, simply together with the regular convolution layers. We further compare visual activation with recent effective attention module SENet, since SENet is one of the most effective attention modules recently.

Table 10 shows the result, although SENet uses an additional block to enhance the model capacity, it is remarkable that the simple visual activation even outperforms SENet. We also wish the visual activation we proposed can co-exist with other techniques, such as the SE module. We adopt the SE module on the last stage in ResNet-50 to avoid overfitting. Table 10 also shows the co-existence between FReLU and SE module. Together with SENet, funnel activation improves 0.3% accuracy further.

6   Conclusions

In this work, we present a funnel activation specifically designed visual tasks, which easily captures complex layouts using the pixel-wise modeling capacity. Our approach is simple, effective, and finely compatible with other techniques, that provides a new alternative activation for image recognition tasks. We note that ReLU has been so influential that many state-of-the-art architectures have been designed for it, however, their settings may not be optimal for the funnel activation. Therefore, it still has a large potential for further improvements.

Acknowledgements. This work is supported by The National Key Research and Development Program of China (No. 2017YFA0700800) and Beijing Academy of Artificial Intelligence (BAAI).

References

1. Agostinelli, F., Hoffman, M., Sadowski, P., Baldi, P.: Learning activation functions to improve deep neural networks. arXiv preprint arXiv:1412.6830 (2014)
2. Ba, J.L., Kiros, J.R., Hinton, G.E.: Layer normalization. arXiv preprint arXiv:1607.06450 (2016)
3. Cao, Y., Xu, J., Lin, S., Wei, F., Hu, H.: Gcnet: Non-local networks meet squeeze-excitation networks and beyond. In: Proceedings of the IEEE International Conference on Computer Vision Workshops (2019)
4. Chen, L.C., Papandreou, G., Kokkinos, I., Murphy, K., Yuille, A.L.: DeepLab: Semantic image segmentation with deep convolutional nets, atrous convolution, and fully connected CRFs. IEEE transactions on pattern analysis and machine intelligence **40**(4), 834–848 (2017)
5. Clevert, D.A., Unterthiner, T., Hochreiter, S.: Fast and accurate deep network learning by exponential linear units (elus). arXiv preprint arXiv:1511.07289 (2015)
6. Cordts, M., et al.: The cityscapes dataset for semantic urban scene understanding. In: Proceedings of the IEEE Conference on Computer Vision and Pattern Recognition, pp. 3213–3223 (2016)
7. Dai, J., Qi, H., Xiong, Y., Li, Y., Zhang, G., Hu, H., Wei, Y.: Deformable convolutional networks. In: Proceedings of the IEEE International Conference on Computer Vision, pp. 764–773 (2017)
8. Dauphin, Y.N., Fan, A., Auli, M., Grangier, D.: Language modeling with gated convolutional networks. In: Proceedings of the 34th International Conference on Machine Learning-Volume 70, pp. 933–941. JMLR. org (2017)
9. Deng, J., et al.: Imagenet: A large-scale hierarchical image database. In: 2009 IEEE Conference on Computer Vision and Pattern Recognition, pp. 248–255. IEEE (2009)
10. Elfwing, S., Uchibe, E., Doya, K.: Sigmoid-weighted linear units for neural network function approximation in reinforcement learning. Neural Netw. **107**, 3–11 (2018)
11. Glorot, X., Bengio, Y.: Understanding the difficulty of training deep feedforward neural networks. In: Proceedings of the Thirteenth International Conference on Artificial Intelligence and Statistics, pp. 249–256 (2010)
12. Goodfellow, I.J., Warde-Farley, D., Mirza, M., Courville, A., Bengio, Y.: Maxout networks. arXiv preprint arXiv:1302.4389 (2013)

13. Hahnloser, R.H., Sarpeshkar, R., Mahowald, M.A., Douglas, R.J., Seung, H.S.: Digital selection and analogue amplification coexist in a cortex-inspired silicon circuit. Nature **405**(6789), 947–951 (2000)
14. He, K., Zhang, X., Ren, S., Sun, J.: Delving deep into rectifiers: Surpassing human-level performance on ImageNet classification. In: Proceedings of the IEEE International Conference on Computer Vision, pp. 1026–1034 (2015)
15. He, K., Zhang, X., Ren, S., Sun, J.: Deep residual learning for image recognition. In: Proceedings of the IEEE Conference on Computer Vision and Pattern Recognition, pp. 770–778 (2016)
16. Hendrycks, D., Gimpel, K.: Bridging nonlinearities and stochastic regularizers with gaussian error linear units (2016)
17. Hinton, G.E., Srivastava, N., Krizhevsky, A., Sutskever, I., Salakhutdinov, R.R.: Improving neural networks by preventing co-adaptation of feature detectors. arXiv preprint arXiv:1207.0580 (2012)
18. Holschneider, M., Kronland-Martinet, R., Morlet, J., Tchamitchian, P.: A real-time algorithm for signal analysis with the help of the wavelet transform. In: Combes, J.M., Grossmann, A., Tchamitchian, P. (eds.) Wavelets. Inverse Problems and Theoretical Imaging, pp. 286–297. Springer, Berlin (1990). https://doi.org/10.1007/978-3-642-75988-8_28
19. Howard, A.G., et al.: Mobilenets: Efficient convolutional neural networks for mobile vision applications. arXiv preprint arXiv:1704.04861 (2017)
20. Hu, J., Shen, L., Sun, G.: Squeeze-and-excitation networks. In: Proceedings of the IEEE Conference on Computer Vision and Pattern Recognition, pp. 7132–7141 (2018)
21. Ioffe, S., Szegedy, C.: Batch normalization: accelerating deep network training by reducing internal covariate shift. arXiv preprint arXiv:1502.03167 (2015)
22. Jaderberg, M., Simonyan, K., Zisserman, A., et al.: Spatial transformer networks. In: Advances in Neural Information Processing Systems, pp. 2017–2025 (2015)
23. Jarrett, K., Kavukcuoglu, K., Ranzato, M., LeCun, Y.: What is the best multi-stage architecture for object recognition? In: 2009 IEEE 12th International Conference on Computer Vision, pp. 2146–2153. IEEE (2009)
24. Jeon, Y., Kim, J.: Active convolution: Learning the shape of convolution for image classification. In: Proceedings of the IEEE Conference on Computer Vision and Pattern Recognition, pp. 4201–4209 (2017)
25. Klambauer, G., Unterthiner, T., Mayr, A., Hochreiter, S.: Self-normalizing neural networks. In: Advances in Neural Information Processing Systems, pp. 971–980 (2017)
26. Krizhevsky, A., Sutskever, I., Hinton, G.E.: ImageNet classification with deep convolutional neural networks. In: Advances in Neural Information Processing Systems, pp. 1097–1105 (2012)
27. Lin, T.Y., Goyal, P., Girshick, R., He, K., Dollár, P.: Focal loss for dense object detection. In: Proceedings of the IEEE International Conference on Computer Vision, pp. 2980–2988 (2017)
28. Lin, Tsung-Yi., et al.: Microsoft COCO: common objects in context. In: Fleet, David, Pajdla, Tomas, Schiele, Bernt, Tuytelaars, Tinne (eds.) ECCV 2014. LNCS, vol. 8693, pp. 740–755. Springer, Cham (2014). https://doi.org/10.1007/978-3-319-10602-1_48
29. Luo, W., Li, Y., Urtasun, R., Zemel, R.: Understanding the effective receptive field in deep convolutional neural networks. In: Advances in Neural Information Processing System, pp. 4898–4906 (2016)

30. Ma, N., Zhang, X., Zheng, H.T., Sun, J.: Shufflenet v2: practical guidelines for efficient CNN architecture design. In: Proceedings of the European Conference on Computer Vision (ECCV), pp. 116–131 (2018)
31. Maas, A.L., Hannun, A.Y., Ng, A.Y.: Rectifier nonlinearities improve neural network acoustic models. In: Processing ICML. vol. 30, p. 3 (2013)
32. Nair, V., Hinton, G.E.: Rectified linear units improve restricted boltzmann machines. In: Proceedings of the 27th International Conference on Machine Learning (ICML-10), pp. 807–814 (2010)
33. Noh, H., Araujo, A., Sim, J., Weyand, T., Han, B.: Large-scale image retrieval with attentive deep local features. In: Proceedings of the IEEE International Conference on Computer Vision, pp. 3456–3465 (2017)
34. Van den Oord, A., Kalchbrenner, N., Espeholt, L., Vinyals, O., Graves, A., et al.: Conditional image generation with PixelCNN decoders. In: Advances in Neural Information Processing Systems, pp. 4790–4798 (2016)
35. Qiu, S., Xu, X., Cai, B.: FReLU: flexible rectified linear units for improving convolutional neural networks. In: 2018 24th International Conference on Pattern Recognition (ICPR), pp. 1223–1228. IEEE (2018)
36. Ramachandran, P., Zoph, B., Le, Q.V.: Searching for activation functions. arXiv preprint arXiv:1710.05941 (2017)
37. Russakovsky, O., et al.: Imagenet large scale visual recognition challenge. Int. J. Comput. Vis. **115**(3), 211–252 (2015). https://doi.org/10.1007/s11263-015-0816-y
38. Simonyan, K., Zisserman, A.: Very deep convolutional networks for large-scale image recognition. arXiv preprint arXiv:1409.1556 (2014)
39. Singh, S., Krishnan, S.: Filter response normalization layer: Eliminating batch dependence in the training of deep neural networks. arXiv preprint arXiv:1911.09737 (2019)
40. Szegedy, C., Ioffe, S., Vanhoucke, V., Alemi, A.A.: Inception-v4, inception-ResNet and the impact of residual connections on learning. In: Thirty-first AAAI Conference on Artificial intelligence (2017)
41. Szegedy, C., et al.: Going deeper with convolutions. In: Proceedings of the IEEE Conference on Computer Vision and Pattern Recognition, pp. 1–9 (2015)
42. Szegedy, C., Vanhoucke, V., Ioffe, S., Shlens, J., Wojna, Z.: Rethinking the inception architecture for computer vision. In: Proceedings of the IEEE Conference on Computer Vision and Pattern Recognition, pp. 2818–2826 (2016)
43. Ulyanov, D., Vedaldi, A., Lempitsky, V.: Instance normalization: the missing ingredient for fast stylization. arXiv preprint arXiv:1607.08022 (2016)
44. Wu, Y., Zhang, S., Zhang, Y., Bengio, Y., Salakhutdinov, R.R.: On multiplicative integration with recurrent neural networks. In: Advances in Neural Information Processing Systems, pp. 2856–2864 (2016)
45. Wu, Y., He, K.: Group normalization. In: Proceedings of the European Conference on Computer Vision (ECCV), pp. 3–19 (2018)
46. Xu, B., Wang, N., Chen, T., Li, M.: Empirical evaluation of rectified activations in convolutional network. arXiv preprint arXiv:1505.00853 (2015)
47. Yu, F., Koltun, V.: Multi-scale context aggregation by dilated convolutions. arXiv preprint arXiv:1511.07122 (2015)
48. Zhao, H., Shi, J., Qi, X., Wang, X., Jia, J.: Pyramid scene parsing network. In: Proceedings of the IEEE Conference on Computer Vision and Pattern Recognition, pp. 2881–2890 (2017)
49. Zhao, H., et al.: PSANet: Point-wise spatial attention network for scene parsing. In: Proceedings of the European Conference on Computer Vision (ECCV), pp. 267–283 (2018)

50. Zhou, B., Khosla, A., Lapedriza, A., Oliva, A., Torralba, A.: Object detectors emerge in deep scene CNNs. arXiv preprint arXiv:1412.6856 (2014)
51. Zhu, X., Hu, H., Lin, S., Dai, J.: Deformable convnets v2: More deformable, better results. In: Proceedings of the IEEE Conference on Computer Vision and Pattern Recognition, pp. 9308–9316 (2019)
52. Zoph, B., Le, Q.V.: Neural architecture search with reinforcement learning. arXiv preprint arXiv:1611.01578 (2016)
53. Zoph, B., Vasudevan, V., Shlens, J., Le, Q.V.: Learning transferable architectures for scalable image recognition. In: Proceedings of the IEEE Conference on Computer Vision and Pattern Recognition, pp. 8697–8710 (2018)

GIQA: Generated Image Quality Assessment

Shuyang Gu[1], Jianmin Bao[2(✉)], Dong Chen[2], and Fang Wen[2]

[1] University of Science and Technology of China, Hefei, China
gsy777@mail.ustc.edu.cn
[2] Microsoft Research, Beijing, China
{jianbao,doch,fangwen}@microsoft.com

Abstract. Generative adversarial networks (GANs) achieve impressive results today, but not all generated images are perfect. A number of quantitative criteria have recently emerged for generative models, but none of them are designed for a single generated image. In this paper, we propose a new research topic, Generated Image Quality Assessment (GIQA), which quantitatively evaluates the quality of each generated image. We introduce three GIQA algorithms from two perspectives: learning-based and data-based. We evaluate a number of images generated by various recent GAN models on different datasets and demonstrate that they are consistent with human assessments. Furthermore, GIQA is available for many applications, like separately evaluating the realism and diversity of generative models, and enabling online hard negative mining (OHEM) in the training of GANs to improve the results.

Keywords: Generative model · Generative adversarial networks · Image quality assessment

1 Introduction

Recent studies have shown remarkable success in generative models for their wide applications like high quality image generation [2,18], image-to-image translation [10,11,15,35], data augmentation [6,7], and so on. However, due to the large variance in quality of generated images, not all generated images are satisfactory for real-world applications. Relying on a manual quality assessment of generated images takes a lot of time and effort. This work proposes a new research topic: Generated Image Quality Assessment (GIQA). The goal of GIQA is to automatically and objectively assess the quality of each image generated by the various generative models.

GIQA is related to Blind/No-Reference Image Quality Assessment (NR-IQA) [8,25,30,34,38]. However, NR-IQA mainly focuses on quality assessment of natural images instead of the generated image. Most of them are distortion-specific; they are capable of performing NR-IQA only if the distortion that afflicts

Please refer to our arxiv version for more paper details.

© Springer Nature Switzerland AG 2020
A. Vedaldi et al. (Eds.): ECCV 2020, LNCS 12356, pp. 369–385, 2020.
https://doi.org/10.1007/978-3-030-58621-8_22

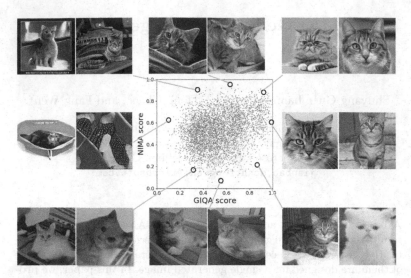

Fig. 1. Score distribution of a NR-IQA method NIMA [34] and our GIQA methods. Score of NIMA is normalized to [0,1] for better comparison, higher score denotes higher image quality. Our GIQA score is more consistent with human observation.

the image is known beforehand, *e.g.*, blur or noise or compression and so on. While the generated images may contain many uncertain model specific artifacts like checkboards [26], droplet-like [20], and unreasonable structure [40]. Unlike low-level degradations, these artifacts are difficult to simulate at different levels for training. Therefore, traditional natural image quality assessment methods are not suitable for generated images as shown in Fig. 1. On the other hand, previous quantitative metrics, like Inception Score [12] and FID [13], focus on the assessment of the generative models, which also cannot be used to assess the quality of each single generated image.

In this paper, we introduce three GIQA algorithms from two perspectives: learning-based and data-based perspectives. For the learning-based method, we apply a CNN model to regress the quality score of a generated image. The difficulty is that it is hard to obtain the labelled quality score for a generated image. To address this problem, we propose a novel semi-supervised learning procedure. We observe that the quality of the generated images gets better and better during the training process of generative models. Based on this, we use images generated by models with different iterations, and use the number of iterations as the pseudo label of the quality score. To eliminate the label noise, we propose a new algorithm that uses multiple binary classifiers as a regressor to implement regression. Our learning-based algorithm can be applied to a variety of different models and databases without any manual annotation.

For data-based methods, the essence is that the similarity between the generated image and the real image could indicate quality, so we convert the GIQA problem into density estimation problem of real images. This problem can be broadly categorized as a parametric and non-parametric method. For the parametric method, we directly adopt the Gaussian Mixture Model (GMM) to capture the probability distribution of real data, then we estimate the probability of a generated image as the quality score. Although this model is very simple, we find it works quite well for most situations. A limitation of the parametric method is that the chosen density might not capture complex distribution, so we propose another non-parametric method by computing the distance between generated image and its K nearest neighbours (KNN), the smaller distance indicates larger probability.

The learning-based method and the data-based method each have their own advantages and disadvantages. The GMM based method is easy to use and can be trained without any generated images, but it can only be applied to relatively simple distributed databases. The KNN based method has a great merit that there is no training phase, but its memory cost is large since it requires the whole training set to be stored. The learning-based method can handle a variety of complex data distributions, but it is also very time-consuming to collect the images generated by various models at different iterations. Considering both effectiveness and efficiency, we recommend GMM-GIQA mostly. We evaluate these 3 methods in detail in the experiments part.

The proposed GIQA methods can be applied in many applications. 1) We can apply it for generative model assessment. Current generative model assessment algorithms like Inception Score [12] and FID [13] evaluate the performance of the generative model in a score which represents the summation of two aspects: realism and diversity. Our proposed GIQA model can evaluate these two aspects separately. 2) By using our GIQA method, we can assess the quality of each generated image for a specific iteration of generator, and rank the quality of these samples, we suggest that the generator pay more attention to the samples with low quality. To achieve this, we adopt online hard negative mining (OHEM) [32] in the discriminator to put larger loss of weight to the lower quality generated samples. Thus the performance of the generator is improved by this strategy. 3) We can leverage GIQA as an image picker to obtain a subset of generated images with higher quality.

Evaluating the GIQA algorithm is an open and challenging problem. It is difficult to get the precise quality annotation for the generated images. In order to evaluate the performance of our methods, we propose a labeled generated image for the quality assessment (LGIQA) dataset. To be specific, we present a series of pairs which consist of two generated images for different observers to choose which has a better quality. We keep the pairs which are annotated with the consistent opinions for evaluating. We will release the data and encourage more research to explore the problem of GIQA.

To summarise, our main contribution are as follows, (1) To our knowledge, we are the first to propose the topic of GIQA. We proposed three straightforward methods from two perspectives to encourage further research. (2) GIQA is general and available to many applications, such as separately evaluating the quality and diversity of generative models and improving the results of generative model through OHEM. (3)We introduce the LGIQA dataset for evaluating different GIQA methods.

2   Related Work

In this section, we briefly review prior natural image quality assessment methods and generative model assessment methods that are most related to our work.

Image Quality Assessment: Traditional Image Quality Assessment(IQA) aims to assess the quality of natural images regarding low-level degradations like noise, blur, compression artifacts, *etc.* It is a traditional technique that is widely used in many applications. Formally, it can be divided into three main categories: Full-reference IQA (FR-IQA), Reduced-reference IQA (RR-IQA) and No-reference IQA (NR-IQA). FR-IQA is a relatively simple problem since we have the reference for the image to be assessed, the most widely used metrics are PSNR [14] and SSIM [36]. RR-IQA [37] aims to evaluate the perceptual quality of a distorted image through partial information of the corresponding reference image. NR-IQA is a more common real-world scenario which needs to estimate the quality of natural image without any reference images. Many NR-IQA approaches [8,25,30,38] focus on some specific distortion. Recently advances in convolution neural networks (CNNs) have spawned many CNNs based methods [1,16,17] for natural image quality assessment. More recent works [22,28] leverage the generative model in their framework to regress the final quality score.

Generative Model Assessment: Recent studies have shown remarkable success in generative models. Many generative models like VAEs [5], GANs [9], and Pixel CNNs [27] have been proposed, so the assessment of generative models has received extensive attention. Many works try to evaluate the generative model by conducting the user study, users are often required to score the generated images. While this will cost a large amount of time and effort. Therefore early work [31] propose a new metric Inception Score (IS) to measure the performance of generative model, the Inception Score evaluates the generative model in two aspects: realism and diversity of the generated images which are synthesized using the generative model. More recent work [13] proposes the Fréchet Inception Distance (FID) score for the assessment of generative models. It takes the real data distribution into consideration and calculates the statistics between the generated samples distribution and real data distribution. [21] proposed precision and recall to measure generative model from quality and diversity separately.

3   Methods

Given a generated image $\mathcal{I}_g$, the target of GIQA is to quantitatively and objectively evaluate its quality score $S(\mathcal{I}_g)$ which should be consistent with human assessment. We propose solving this problem from two different perspectives. The first one is a learning-based method, in which we apply a CNN model to regress the quality score of a generated image. The second one is a data-based method, for which we directly model the probability distribution of real data. Thus we can estimate the quality of a generated image by the estimated probability from the model. We'll describe them in detail in the following sections.

3.1 Learning-Based Methods

For learning-based methods, we aim to apply a CNN model to learn the quality of the generated images. Previous supervised learning method often require large amounts of labeled data for training. However, the quality annotation for the generated images is difficult to obtain since it is impossible for human observers to give the precise score to each generated image. Therefore, we propose a novel semi-supervised learning procedure.

Semi-supervised Learning: We find an important observation that the quality of generated images from most generative models, *e.g.*, PGGAN [18] and StyleGAN [19], is becoming better and better as the training iteration increases. Based on this, we collect images generated by models with different iterations, and use the number of iterations as the pseudo label of the quality score. Note that there is still a gap between the quality of the image generated by the last iteration and the real image. So we suppose that the quality of the generated images ranges from 0 to S_g, where $S_g \in (0,1)$, and the quality of the real images is 1. Formally, the pseudo label of quality score $S_\mathrm{p}(\mathcal{I})$ for image $\mathcal{I}$ is

$$S_\mathrm{p}(\mathcal{I}) = \begin{cases} \frac{S_g \cdot iter}{max\_iter} & if\ \mathcal{I}\ is\ generated \\ 1 & otherwise \end{cases}, \tag{1}$$

where *iter* presents the iteration number, *max_iter* presents the maximum iteration number, S_g defines the maximum quality score for the generated image, we set it to 0.9 in our experiment. Then we are able to build a training dataset $\mathcal{D} = \{\mathcal{I}, S_\mathrm{p}(\mathcal{I})\}$ for semi-supervised learning, where $\mathcal{I}$ represents the generated images or the real images, $\mathcal{S}_\mathrm{p}(\mathcal{I})$ denotes the corresponding quality score (Fig. 2).

Multiple Binary Classifiers as Regressor: A basic solution is to directly adopt a CNN based framework to regress the quality score from the input image. However, we find that this naive regression method is sub-optimal, since the pseudo label contains a lot of noise. Although statistically the longer the training is, the better the quality is, but there is also a large gap in image quality within the same iteration. To solve this problem, inspired by previous work [23], we propose employing multiple binary classifiers to learn the GIQA, which we call MBC-GIQA. To be specific, N binary classifiers are trained. For the i-th classifier, the training data is divided into positive or negative samples according to a threshold T^i, given an image $\mathcal{I} \in \mathcal{D}$, its label c^i for the i-th classifier is:

$$c^i = \begin{cases} 0 & if\ S_p(\mathcal{I}) < T^i \\ 1 & otherwise \end{cases}, \tag{2}$$

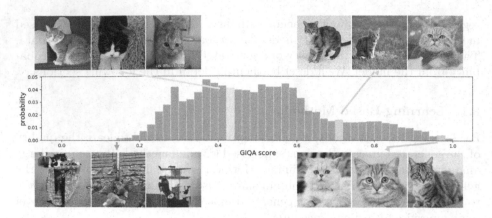

Fig. 2. Generated images from StyleGAN pretrained on LSUN-cat dataset, sorted by GMM-GIQA method. We randomly sample images from different score for better visualization.

where $i = 1, 2, \ldots, N$ and $0 < T^1 < T^2 < \cdots < T^N = 1$. So a quality score $S_p(\mathcal{I})$ can be converted to a set of binary labels $\{c^1, c^2, \ldots, c^N\}$. Each binary classifier learns to distinguish whether the quality value is larger than T^i. Suppose the predicted score for i-th binary classifier is $\hat{c}^i$, $i = 1, 2, \ldots, N$. So the training loss for the framework is:

$$L = - \sum_{I \in \mathcal{D}} \sum_{i=1}^{N} (c^i \log(\hat{c}^i) + (1 - c^i) \log(1 - \hat{c}^i)). \tag{3}$$

Using classification instead of regression in this way can be more robust to noise. Although both positive and negative training samples contain noise, T^i is still statistically the decision boundary of i-th classifier. During the inference time, suppose we get all the predicted scores $\hat{c}^i, i = 1, 2, \ldots, N$ for a generated image $\mathcal{I}_g$. Then the final predicted quality score for $\mathcal{I}_g$ is the average of all predicted scores:

$$S_{\mathrm{MBC}}(\mathcal{I}_g) = \frac{1}{N} \sum_{i=1}^{N} \hat{c}^i. \tag{4}$$

3.2 Data-Based Methods

Data-based methods aims to solve the quality estimation in a probability distribution perspective. We directly model the probability distribution of the real data, then we can estimate the quality of a generated image by the estimated probability from the model. We propose adopting two density estimation methods: Gaussian Mixture Model (GMM) and K Nearest Neighbour (KNN).

Gaussian Mixture Model: We propose adopting the Gaussian Mixture Model (GMM) to capture the real data distribution for GIQA. We call this method

GMM-GIQA. A Gaussian mixture model is a weighted sum of M component Gaussian densities. Suppose the mean vector and covariance matrix for i-th Gaussian component are $\boldsymbol{\mu}^i$ and $\boldsymbol{\Sigma}^i$, respectively. The probability of an image $\mathcal{I}$ is given by:

$$p(\mathbf{x}|\lambda) = \sum_{i=1}^{M} \mathbf{w}^i g(\mathbf{x}|\boldsymbol{\mu}^i, \boldsymbol{\Sigma}^i), \tag{5}$$

where $\mathbf{x}$ is the extracted feature of $\mathcal{I}$. Suppose the feature extractor function is $f(\cdot)$, so $\mathbf{x} = f(\mathcal{I})$. $\mathbf{w}^i$ is the mixture weights, which satisfies the constraint that $\sum_{i=1}^{M} \mathbf{w}^i = 1$. And $g(\mathbf{x}|\boldsymbol{\mu}^i, \boldsymbol{\Sigma}^i)$ is the component Gaussian densities.

The complete Gaussian mixture model is parameterized by the mean vectors, covariance matrices, and mixture weights from all component densities. These parameters are collectively represented by the notation, $\lambda = \{\mathbf{w}^i, \boldsymbol{\mu}^i, \boldsymbol{\Sigma}^i\}$. To estimate these parameters, we adopt the expectation-maximization (EM) algorithm [4] to iteratively update them. Since the probability of a generated image represents its quality score, the quality score of $\mathcal{I}_g$ is given by:

$$S_{\text{GMM}}(\mathcal{I}_g) = p(f(\mathcal{I}_g)|\lambda). \tag{6}$$

K **Nearest Neighbour**: When the real data distribution becomes complicated, it would be difficult to capture the distribution with GMM well. In this situation, we introduce a non-parametric method based on K Nearest Neighbor (KNN). We think the Euclidean distance between generated images and nearby real images in feature space could also represent the probability of generated image, suppose the feature of a generated sample is $\mathbf{x}$. Its k-th nearest real sample's feature is $\mathbf{x}^k$. We can calculate the probability of generated image as:

$$p(\mathbf{x}) = \frac{1}{K} \sum_{k=1}^{K} \frac{1}{||\mathbf{x} - \mathbf{x}^k||^2}. \tag{7}$$

Suppose the feature extractor function is also $f(\cdot)$. The quality score of $\mathcal{I}_g$ is given by:

$$S_{\text{KNN}}(\mathcal{I}_g) = p(f(\mathcal{I}_g)). \tag{8}$$

Above all, we introduce three approaches to get three forms of quality score function $S(\mathcal{I}_g)$: $S_{\text{MBC}}(\mathcal{I}_g)$, $S_{\text{GMM}}(\mathcal{I}_g)$, and $S_{\text{KNN}}(\mathcal{I}_g)$. We believe these methods will serve as baselines for further research. In terms of recommendation, we recommend the GMM-based method, since this method outperforms other methods(Table 1) and is highly efficient.

4   Applications

The proposed GIQA framework is simple and general. In this section, we will show how GIQA can be applied in many applications, such as generative model evaluation, improving the performance of GANs.

4.1   Generative Model Evaluation

Generative model evaluation is an important research topic in the vision community. Recently, a lot of quantitative metrics have been developed to assess the performance of a GAN model based on the realism and diversity of generated images, such as Inception Score [31] and FID [13]. However, both of them summarise these two aspects. Our GIQA model can separately assess the realism and diversity of generated images. Specifically, we employ the mean quality score from our methods to indicate the realistic performance of the generative model. Supposing the generative model is G, the generated samples are $\mathcal{I}_g^i, i = 1, 2, \ldots, N_g$. So the quality score of generator G is calculated with the mean quality of N_g generated samples:

$$QS(G) = \frac{1}{N_g} \sum_i^{N_g} S(\mathcal{I}_g^i), \tag{9}$$

On the other hand, we can also evaluate the diversity of the generative model G. Note that the diversity represents the relative diversity compared to real data distribution. We exchange the positions of real and generated images in data-based GIQA method. We use generated images to build the model and then evaluate the quality of the real images.

Considering if the generated samples have similar distribution with real samples, then the quality of the real samples is high. Otherwise, if the generated samples have the problem of "mode collapse", which means a low diversity, then the probability of the real samples become low. This shows by exchanging the position, the quality of real samples is consistent with the diversity of generative models. Supposing the real samples are $\mathcal{I}_r^i, i = 1, 2, \ldots, N_r$, the score function built with generative model G is $S'(\cdot)$. The diversity score of the generative model is calculated with mean quality of N_r real samples:

$$DS(G) = \frac{1}{N_r} \sum_i^{N_r} S'(\mathcal{I}_r^i), \tag{10}$$

In summary, we have the quality score (QS) and diversity score (DS) to measure the quality and diversity of the generative model separately.

4.2   Improve the Performance of GANs

Another important application of GIQA is to help the generative model achieve better performance. In general, the quality of generated images from a specific iteration of the generator have large variance, we can assess the quality of these generated samples by using our GIQA method, then we force the generator to pay more attention to these samples with low quality. To achieve this, we employ online hard negative mining (OHEM) [32] in the discriminator to apply a higher loss weight to the lower quality samples. To be specific, we set a quality threshold T_q. Samples with quality lower than the threshold T_q will be given a large loss weight $w_l > 1$.

4.3   Image Picker Based on Quality

Another important application of GIQA is to leverage it as an image picker based on quality. For the wide applications of generative models, picking high quality generated images is of great importance and makes these applications more practical. On the other hand, for a generative model to be evaluated, we can take full advantage of the image picker to discard these images with low quality to further improve performance.

5   Experiments

In this section, we first introduce the overall experiment setups and then present extensive experiment results to demonstrate the superiority of our approach.

Datasets and Training Details. We conduct experiments on a variety of generative models trained on different datasets. For unconditional generative models, we choose WGAN-GP [12], PGGAN [18], and StyleGAN [19] trained on FFHQ [19], and LSUN [39] datasets. For conditional generative models, we choose pix2pix [15], pix2pixHD [35], SPADE [29] trained on Cityscapes [3] datasets. FFHQ is a large dataset which contains 70000 high-resolution face images. LSUN contains 10 scene categories and 20 object categories, each category contains a large number of images. The Cityscapes dataset is widely used in conditional generative models. In our experiments, we use all the officially released models of these methods for testing.

For learning-based methods, we need the generated images at different iterations of a generative model for training. Specifically, for unconditional generative models, we collect the generated images in the training process of StyleGAN for training, and test the resulting model on the generated images from PGGAN, StyleGAN, and real images. For the conditional generative model, we use the generated images at different iterations of pix2pixHD for training, and test it on the generated images from pix2pix, pix2pixHD, SPADE and real images. To get these training images, we use the official training code, and collect 200,000 generated images, which consist of images from 4000 iterations, 50 images per iteration. We adopt 8 binary classifiers for the MBC-GIQA approach. For the GMM-GIQA method, we set the number of Gaussian components to 7 for LSUN and Cityscapes datasets, and 70 for FFHQ. For the KNN-GIQA method, we set K to 1 for FFHQ and Cityscapes datasets, 3500 for LSUN. All features are extracted from the inception model [33] which is trained on ImageNet. More details please refer to the supplementary material.

Evaluation Metrics. Evaluating GIQA algorithms is an open and challenging problem. To quantitatively evaluate the performance of these algorithms, we collect a dataset which is annotated by multiple human observers. To be specific, we first use the generated images from PGGAN, StyleGAN, and real images to build 1500 image pairs [1], then we demonstrate these pairs to 3 human observers

---

[1] We not only collect images from pretrained models, but also some low quality images from the training procedure.

Table 1. Comparison of the accuracy on LGIQA dataset for different methods.

Methods	FFHQ	LSUN-cat	Cityscapes
NIMA[34]	0.598	0.583	0.827
DeepIQA[1]	0.581	0.550	0.763
RankIQA[24]	0.573	0.557	0.780
BC-GIQA	0.663	0.710	0.768
IR-GIQA	0.678	0.784	0.837
SGM-GIQA	0.620	0.829	0.847
MBC-GIQA(our)	0.731	0.831	0.886
GMM-GIQA(our)	**0.764**	**0.846**	0.895
KNN-GIQA(our)	0.761	0.843	**0.898**

to choose which image has a better quality. Finally, we discard the pairs which have inconsistent human opinions. The number of remaining pairs are 974, 1206, and 1102 for FFHQ, LSUN-cat, and Cityscapes dataset, respectively. We refer to this dataset as the Labeled Generated Image Quality Assessment(LGIQA) dataset. To evaluate a GIQA algorithm, we employ the algorithm to rank the image quality in each pairs and check if it is consistent with the annotation. Thus we can calculate the accuracy of each algorithm.

5.1   Comparison with Recent Works

Since no previous approach aims to solve the problem of GIQA, we design several baselines and compare our approach with them to validate our approach.

The first baselines are the methods for natural image quality assessment, we choose recent works like DeepIQA [1], NIMA [34], RankIQA [24] for comparison. For DeepIQA and NIMA, we directly apply their released model for testing. For RankIQA, we use their degradation strategy and follow their setting to train a model on our datasets. The second baselines are related to the learning-based method. We adopt the simple idea of directly employing a CNN network to regress pseudo label of quality score $S_p(\mathcal{I})$, which is called IR-GIQA. Another idea is instead of using multiple binary classifiers, we use only 1 classifier to determine whether the image is real or not, we call this BC-GIQA. The third baseline is to capture the real data probability distribution is to use a single Gaussian model, we call this SGM-GIQA.

We present the results in Table 1. We observe that our proposed GIQA methods perform better than those natural image assessment methods. Meanwhile, the MBC-GIQA gets higher accuracy than the baseline IR-GIQA and BC-GIQA, and GMM-GIQA is also better than the SGM-GIQA model. Which demonstrates the effectiveness of our proposed method. Overall, GMM-GIQA achieves the best results, so we use GMM-GIQA for the following experiments.

We qualitatively compare the generated image quality ranking results for our proposed GMM-GIQA and NIMA in Fig. 3, we observe that GMM-GIQA

Top-4 high quality images   Top-4 low quality images

Fig. 3. Generated image quality assessment results for NIMA(the top 3 rows) and our proposed GMM-GIQA(the bottom 3 rows) on LSUN-cat, LSUN-bedroom and LSUN-car datasets. The left are the top-4 high quality images and the right are top-4 low quality images.

achieves a better generated image quality ranking results that is more consistent with human assessment. More results can be found in supplemental material.

5.2   Generative Model Assessment

Quality Distribution Evaluation. The proposed GIQA methods are able to assess the quality for every generated sample. Therefore we first employ our proposed GMM-GIQA to validate the quality distribution of generated samples from several generative models. For unconditional generative models, we choose WGAN-GP, PGGAN, StyleGAN trained on FFHQ, LSUN-cat and LSUN-car datasets. For conditional generative models, we choose pix2pix, pix2pixHD, and SPADE trained on Cityscapes dataset. Each generative model generates 5000 test images, and then apply our GMM-GIQA method to calculate the quality score, the quality score distributions are shown in Fig. 4. Note that all the quality scores are normalized to $[0, 1]$. We find that PGGAN and StyleGAN are much better than WGAN-GP, and StyleGAN is better than PGGAN. SPADE

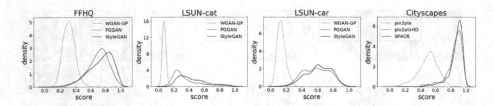

Fig. 4. Quality score distribution of generated images from different generative models.

Table 2. Comparison of FID, Precision [21], Recall [21], QS, and DS metric for the generative model WGAN-GP [12], PGGAN [18], and StyleGAN [19] on three different datasets: FFHQ, LSUN-cat, and LSUN-car.

	FFHQ			LSUN-cat			LSUN-car		
	[12]	[18]	[19]	[12]	[18]	[19]	[12]	[18]	[19]
FID	107.6	14.66	10.54	192.2	49.87	18.67	146.5	14.73	12.70
Prec	0.006	0.640	0.704	0.012	0.487	0.608	0.022	0.608	0.680
Rec	0	0.452	0.555	0	0.356	0.467	0.002	0.487	0.531
QS	0.312	0.694	0.731	0.072	0.347	0.441	0.138	0.583	0.617
DS	0.355	0.815	0.806	0.236	0.789	0.796	0.281	0.801	0.791

and pix2pixHD are much better than pix2pix, SPADE is slightly better than pix2pixHD. All these observations are consistent with human observation.

QS and DS for Generative Models. As we introduced in Sect. 4.2, we propose two new metrics the quality score (QS) and diversity score (DS) to assess the performance of generative models. So we compare these two metrics with the FID and [21]. Table 2 reports the results on WGAN-GP, PGGAN, and StyleGAN. Our QS and DS metrics are consistent with other metrics. Also to validate our metric can be applied for conditional models, we report the QS score 0.498, 0.851, and 0.879 for conditional generative models: pix2pix, pix2pixHD and SPADE, respectively. The results are consistent with human observation.

5.3   Improving the Performance of GANs

One important application of GIQA is to improve the performance of GANs. We find that we can achieve this in two ways, one is to adopt the GIQA to discard low quality images from all the generated images for evaluation. The other one is to take full advantage of the GIQA to achieve OHEM in the training process of GANs, then the performance gets improved.

Image Picker Trick. We conduct this experiment on the StyleGAN model trained on LSUN-cat. We first generate 10000 images, then we use the GMM-GIQA method to rank the quality of these images and retain different percentages of high quality images, finally we randomly sample 5000 remaining images

Table 3. Comparison of truncation trick and image picker trick using StyleGAN on LSUN-cat dataset.

Methods	Metrics	1	0.9	0.8	0.7	0.6	0.5
Truncation rate	FID	18.67	18.05	19.64	23.46	30.48	41.68
	QS	0.441	0.463	0.486	0.510	0.537	0.567
	DS	0.796	0.771	0.756	0.731	0.699	0.686
Remaining rate	FID	18.67	16.65	17.63	20.73	25.84	33.19
	QS	0.441	0.466	0.495	0.520	0.551	0.587
	DS	0.796	0.792	0.780	0.766	0.746	0.712

Table 4. Performance comparison of various settings for StyleGAN.

Datasets	Methods	FID	QS	DS
FFHQ	StyleGAN	17.35	0.697	0.753
	StyleGAN + OHEM	16.89	0.711	0.755
	StyleGAN + OHEM+Picker	16.68	0.723	0.749
LSUN-cat	StyleGAN	18.67	0.441	0.796
	StyleGAN + OHEM	18.12	0.462	0.790
	StyleGAN + OHEM + Picker	16.25	0.482	0.785

for evaluation. We test the generated images with different remaining rates. For comparison, we notice that StyleGAN adopt a "truncation trick" on the latent space which also discards low quality images. With a smaller truncation rate, the quality becomes higher, the diversity becomes lower. We report the FID, QS, DS results of different truncation rate and remaining rate in Table 3. We notice that the FID improves when the truncation rate and remaining rate are set to 0.9, and the remaining rate works better than the truncation rate. Which also perfectly validates the superiority of the QS and DS metric.

OHEM for GANs. To validate whether OHEM improves the performance of GANs, we train two different settings of StyleGAN on FFHQ and LSUN-cat datasets at 256×256 resolution. One follows the original training setting (denoted as StyleGAN), the other applies the OHEM in the training process and puts a large loss weight w_l on low quality images whose quality score is lower than threshold T_q, which we called StyleGAN+OHEM. We set the w_l, T_q to 2, 0.2 in our experiments. After finishing the training process, we evaluate the FID, QS, and DS metric. Table 4 reports the results. We find that OHEM improves the performance of GANs. Besides, based on this model, by using our image picker trick (denoted as StyleGAN+OHEM+Picker), it can further achieve better performance.

5.4   Analysis of the Proposed Methods

In this subsection, we conduct experiments to investigate the sensitiveness of hyper parameters in the proposed three approaches. All the results are evaluated on our LGIQA dataset.

Hyper Parameters for MBC-GIQA. For MBC-GIQA, what we mainly want to explore is how the number of binary classifiers influence the results, we train the model using different numbers of binary classifiers. The number of classifiers is set to 1, 4, 6, 8, 10, 12. Table 5 reports the results. We find that as the number of binary classifiers increases from 1 to 8, the performance becomes better and better, and as the number continues to increase to 12 the performance degrades.

Table 5. Results of different number of binary classifiers N for MBC-GIQA.

N	1	4	6	8	10	12
Accuracy	0.663	0.682	0.722	**0.731**	0.718	0.717

Table 6. Results of different number of Gaussian components M for GMM-GIQA.

M	5	10	20	30	50	70	100
Accuracy	0.648	0.663	0.738	0.733	0.752	**0.764**	0.753

Table 7. Results of different number of nearest neighbors K for KNN-GIQA.

K	1	30	100	500	1000	2000	3500	5000	7000
Accuracy	0.823	0.828	0.833	0.837	0.840	0.842	**0.843**	0.841	0.840

Hyper Parameters of KNN-GIQA. To explore how the number of nearest neighbours K affects the results, we apply different K in the KNN-GIQA. Specifically, we set K to 1, 30, 100, 500, 1000, 2000, 3500, 5000, 7000. The results on LGIQA-LSUN-cat dataset are shown in Table 7, as K increases from 1 to 1000, we get better and better results, and the number continues to increase to 7000, the performance is comparable.

Hyper Parameters of GMM-GIQA. The key factor for GMM is the number of Gaussian components M, therefore we explore how M affects the results of GIQA. We set M to 5, 10, 20, 30, 50, 70, 100 and test the results on our LGIQA-FFHQ dataset. We show the results in Table 6, as the number of Gaussian components M increases from 5 to 70, we get better and better results, and the number continues to increase to 100, the performance degrades.

6   Conclusions

In this paper, we aim to solve the problem of quality evaluation of a single generated image and propose the new research topic: GIQA. To tackle this problem, we propose three novel approaches from two perspectives: learning-based and data-based. Extensive experiments show that our proposed methods can perform quite well on this new topic, also we demonstrate that GIQA can be applied in a wide range of applications.

We are also aware that there exist some limitations of our methods. For the learning-based method MBC-GIQA, it requires the generated images at different iterations for training, while these images may not be easily obtained in some situations. For the data-based method GMM-GIQA, there is a chance of failure when the real data distribution is too complicated. We also notice that our current results are far from solving this problem completely. We hope our approach will serve as a solid baseline and help support future research in GIQA.

References

1. Bosse, S., Maniry, D., Wiegand, T., Samek, W.: A deep neural network for image quality assessment. In: 2016 IEEE International Conference on Image Processing (ICIP), pp. 3773–3777. IEEE (2016)
2. Brock, A., Donahue, J., Simonyan, K.: Large scale GAN training for high fidelity natural image synthesis. arXiv preprint arXiv:1809.11096 (2018)
3. Cordts, M., et al.: The cityscapes dataset for semantic urban scene understanding. In: Proceedings of the IEEE conference on computer vision and pattern recognition, pp. 3213–3223 (2016)
4. Dempster, A.P., Laird, N.M., Rubin, D.B.: Maximum likelihood from incomplete data via the EM algorithm. J. R. Stat. Soc. Ser. B (Methodological) **39**(1), 1–22 (1977)
5. Doersch, C.: Tutorial on variational autoencoders. arXiv preprint arXiv:1606.05908 (2016)
6. Frid-Adar, M., Diamant, I., Klang, E., Amitai, M., Goldberger, J., Greenspan, H.: Gan-based synthetic medical image augmentation for increased CNN performance in liver lesion classification. Neurocomputing **321**, 321–331 (2018)
7. Frid-Adar, M., Klang, E., Amitai, M., Goldberger, J., Greenspan, H.: Synthetic data augmentation using GAN for improved liver lesion classification. In: 2018 IEEE 15th international symposium on biomedical imaging (ISBI 2018), pp. 289–293. IEEE (2018)
8. Golestaneh, S.A., Chandler, D.M.: No-reference quality assessment of jpeg images via a quality relevance map. IEEE Signal Process. Lett. **21**(2), 155–158 (2013)
9. Goodfellow, I., et al: Generative adversarial nets. In: Advances in neural information processing systems, pp. 2672–2680 (2014)
10. Gu, S., Bao, J., Yang, H., Chen, D., Wen, F., Yuan, L.: Mask-guided portrait editing with conditional GANs. In: Proceedings of the IEEE Conference on Computer Vision and Pattern Recognition, pp. 3436–3445 (2019)
11. Gu, S., Chen, C., Liao, J., Yuan, L.: Arbitrary style transfer with deep feature reshuffle. In: Proceedings of the IEEE Conference on Computer Vision and Pattern Recognition, pp. 8222–8231 (2018)

12. Gulrajani, I., Ahmed, F., Arjovsky, M., Dumoulin, V., Courville, A.C.: Improved training of Wasserstein GANs. In: Advances in Neural Information Processing Systems, pp. 5767–5777 (2017)
13. Heusel, M., Ramsauer, H., Unterthiner, T., Nessler, B., Hochreiter, S.: GANs trained by a two time-scale update rule converge to a local nash equilibrium. In: Advances in Neural Information Processing Systems, pp. 6626–6637 (2017)
14. Huynh-Thu, Q., Ghanbari, M.: Scope of validity of PSNR in image/video quality assessment. Electron. Lett. **44**(13), 800–801 (2008)
15. Isola, P., Zhu, J.Y., Zhou, T., Efros, A.A.: Image-to-image translation with conditional adversarial networks. In: Proceedings of the IEEE Conference on Computer Vision and Pattern Recognition, pp. 1125–1134 (2017)
16. Kang, L., Ye, P., Li, Y., Doermann, D.: Convolutional neural networks for no-reference image quality assessment. In: Proceedings of the IEEE Conference on Computer Vision and Pattern Recognition, pp. 1733–1740 (2014)
17. Kang, L., Ye, P., Li, Y., Doermann, D.: Simultaneous estimation of image quality and distortion via multi-task convolutional neural networks. In: 2015 IEEE International Conference on Image Processing (ICIP), pp. 2791–2795. IEEE (2015)
18. Karras, T., Aila, T., Laine, S., Lehtinen, J.: Progressive growing of GANs for improved quality, stability, and variation. arXiv preprint arXiv:1710.10196 (2017)
19. Karras, T., Laine, S., Aila, T.: A style-based generator architecture for generative adversarial networks. In: Proceedings of the IEEE Conference on Computer Vision and Pattern Recognition, pp. 4401–4410 (2019)
20. Karras, T., Laine, S., Aittala, M., Hellsten, J., Lehtinen, J., Aila, T.: Analyzing and improving the image quality of StyleGAN. arXiv preprint arXiv:1912.04958 (2019)
21. Kynkäänniemi, T., Karras, T., Laine, S., Lehtinen, J., Aila, T.: Improved precision and recall metric for assessing generative models. In: Advances in Neural Information Processing Systems, pp. 3927–3936 (2019)
22. Lin, K.Y., Wang, G.: Hallucinated-IQA: no-reference image quality assessment via adversarial learning. In: Proceedings of the IEEE Conference on Computer Vision and Pattern Recognition, pp. 732–741 (2018)
23. Liu, T.J., Liu, K.H., Liu, H.H., Pei, S.C.: Age estimation via fusion of multiple binary age grouping systems. In: 2016 IEEE International Conference on Image Processing (ICIP), pp. 609–613. IEEE (2016)
24. Liu, X., van de Weijer, J., Bagdanov, A.D.: Rankiqa: learning from rankings for no-reference image quality assessment. In: Proceedings of the IEEE International Conference on Computer Vision, pp. 1040–1049 (2017)
25. Moorthy, A.K., Bovik, A.C.: A two-step framework for constructing blind image quality indices. IEEE Signal Process. Lett. **17**(5), 513–516 (2010)
26. Odena, A., Dumoulin, V., Olah, C.: Deconvolution and checkerboard artifacts. Distill (2016). 10.23915/distill.00003, http://distill.pub/2016/deconv-checkerboard
27. Van den Oord, A., Kalchbrenner, N., Espeholt, L., Vinyals, O., Graves, A., et al.: Conditional image generation with PixelCNN decoders. In: Advances in Neural Information Processing Systems, pp. 4790–4798 (2016)
28. Pan, D., Shi, P., Hou, M., Ying, Z., Fu, S., Zhang, Y.: Blind predicting similar quality map for image quality assessment. In: Proceedings of the IEEE Conference on Computer Vision and Pattern Recognition, pp. 6373–6382 (2018)
29. Park, T., Liu, M.Y., Wang, T.C., Zhu, J.Y.: Semantic image synthesis with spatially-adaptive normalization. In: Proceedings of the IEEE Conference on Computer Vision and Pattern Recognition, pp. 2337–2346 (2019)

30. Saad, M.A., Bovik, A.C., Charrier, C.: Blind image quality assessment: a natural scene statistics approach in the DCT domain. IEEE Trans. Image Process. **21**(8), 3339–3352 (2012)
31. Salimans, T., Goodfellow, I., Zaremba, W., Cheung, V., Radford, A., Chen, X.: Improved techniques for training GANs. In: Advances in neural information processing systems, pp. 2234–2242 (2016)
32. Shrivastava, A., Gupta, A., Girshick, R.: Training region-based object detectors with online hard example mining. In: Proceedings of the IEEE Conference on Computer Vision and Pattern Recognition, pp. 761–769 (2016)
33. Szegedy, C., et al.: Going deeper with convolutions. In: Proceedings of the IEEE Conference on Computer Vision and Pattern Recognition, pp. 1–9 (2015)
34. Talebi, H., Milanfar, P.: NIMA: neural image assessment. IEEE Trans. Image Process. **27**(8), 3998–4011 (2018)
35. Wang, T.C., Liu, M.Y., Zhu, J.Y., Tao, A., Kautz, J., Catanzaro, B.: High-resolution image synthesis and semantic manipulation with conditional GANs. In: Proceedings of the IEEE Conference on Computer Vision and Pattern Recognition, pp. 8798–8807 (2018)
36. Wang, Z., Bovik, A.C., Sheikh, H.R., Simoncelli, E.P.: Image quality assessment: from error visibility to structural similarity. IEEE Trans. Image process. **13**(4), 600–612 (2004)
37. Wang, Z., Simoncelli, E.P.: Reduced-reference image quality assessment using a wavelet-domain natural image statistic model. In: Human Vision and Electronic Imaging X. vol. 5666, pp. 149–159. International Society for Optics and Photonics (2005)
38. Yan, Q., Xu, Y., Yang, X.: No-reference image blur assessment based on gradient profile sharpness. In: 2013 IEEE International Symposium on Broadband Multimedia Systems and Broadcasting (BMSB), pp. 1–4. IEEE (2013)
39. Yu, F., Seff, A., Zhang, Y., Song, S., Funkhouser, T., Xiao, J.: LSUN: Construction of a large-scale image dataset using deep learning with humans in the loop. arXiv preprint arXiv:1506.03365 (2015)
40. Zhang, H., Goodfellow, I., Metaxas, D., Odena, A.: Self-attention generative adversarial networks. arXiv preprint arXiv:1805.08318 (2018)

Adversarial Continual Learning

Sayna Ebrahimi[1,2(✉)], Franziska Meier[1], Roberto Calandra[1], Trevor Darrell[2], and Marcus Rohrbach[1]

[1] Facebook AI Research, New York, USA
{fmeier,rcalandra,mrf}@fb.com
[2] UC Berkeley EECS, Berkeley, CA, USA
{sayna,trevor}@eecs.berkeley.edu

Abstract. Continual learning aims to learn new tasks without forgetting previously learned ones. We hypothesize that representations learned to solve each task in a sequence have a shared structure while containing some task-specific properties. We show that shared features are significantly less prone to forgetting and propose a novel hybrid continual learning framework that learns a disjoint representation for task-invariant and task-specific features required to solve a sequence of tasks. Our model combines architecture growth to prevent forgetting of task-specific skills and an experience replay approach to preserve shared skills. We demonstrate our hybrid approach is effective in avoiding forgetting and show it is superior to both architecture-based and memory-based approaches on class incrementally learning of a single dataset as well as a sequence of multiple datasets in image classification. Our code is available at https://github.com/facebookresearch/Adversarial-Continual-Learning.

1 Introduction

Humans can learn novel tasks by augmenting core capabilities with new skills learned based on information for a specific novel task. We conjecture that they can leverage a lifetime of previous task experiences in the form of fundamental skills that are robust to different task contexts. When a new task is encountered, these generic strategies form a base set of skills upon which task-specific learning can occur. We would like artificial learning agents to have the ability to solve many tasks sequentially under different conditions by developing task-specific and task-invariant skills that enable them to quickly adapt while avoiding *catastrophic forgetting* [24] using their memory.

One line of continual learning approaches learns a single representation with a fixed capacity in which they detect important weight parameters for each task and minimize their further alteration in favor of learning new tasks. In contrast, structure-based approaches increase the capacity of the network to accommodate

Electronic supplementary material The online version of this chapter (https://doi.org/10.1007/978-3-030-58621-8_23) contains supplementary material, which is available to authorized users.

new tasks. However, these approaches do not scale well to a large number of tasks if they require a large amount of memory for each task. Another stream of approaches in continual learning relies on explicit or implicit experience replay by storing raw samples or training generative models, respectively.

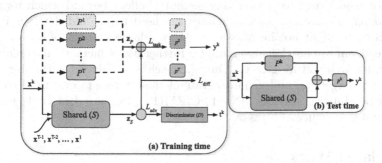

Fig. 1. Factorizing task-specific and task-invariant features in our method (ACL) while learning T sequential tasks at a time. *Left:* Shows ACL at training time where the Shared module is adversarially trained with the discriminator to generate *task-invariant* features ($\mathbf{z}_S$) while the discriminator attempts to predict task labels. Architecture growth occurs at the arrival of the kth task by adding a *task-specific* modules denoted as P^k and p^k, optimized to generate orthogonal representation $\mathbf{z}_P$ to $\mathbf{z}_S$. To prevent forgetting, 1) Private modules are stored for each task and 2) A shared module which is less prone to forgetting, yet is also retrained with experience reply with a limited number of exemplars *Right:* At test time, the discriminator is removed and ACL uses the P^k module for the specific task it is evaluated on.

In this paper, we propose a novel adversarial continual learning (ACL) method in which a disjoint latent space representation composed of *task-specific* or *private* latent space is learned for each task and a *task-invariant* or *shared* feature space is learned for all tasks to enhance better knowledge transfer as well as better recall of the previous tasks. The intuition behind our method is that tasks in a sequence share a part of the feature representation but also have a part of the feature representation which is task-specific. The shared features are notably less prone to forgetting and the tasks-specific features are important to retain to avoid forgetting the corresponding task. Therefore, factorizing these features separates the part of the representation that forgets from that which does not forget. To disentangle the features associated with each task, we propose a novel adversarial learning approach to enforce the shared features to be task-invariant and employ orthogonality constraints [30] to enforce the shared features to not appear in the task-specific space (Fig. 1).

Once factorization is complete, minimizing forgetting in each space can be handled differently. In the task-specific latent space, due to the importance of these features in recalling the task, we freeze the private module and add a new one upon finishing learning a task. The shared module, however, is significantly less susceptible to forgetting and we only use the replay buffer mechanism in this space to the extend that factorization is not perfect, i.e., when tasks have little overlap or have high domain shift in between, using a tiny memory containing samples stored from prior tasks will help with better factorization and hence higher performance. We empirically found that unlike other memory-based methods in which performance increases by increasing the samples from prior tasks, our model requires a very tiny memory budget beyond which its performance remains constant. This alleviates the need to use old data, as in some applications it might not be possible to store a large amount of data if any at all. Instead, our approach leaves room for further use of memory, if available and need be, for architecture growth. Our approach is simple yet surprisingly powerful in not forgetting and achieves state-of-the-art results on visual continual learning benchmarks such as MNIST, CIFAR100, Permuted MNIST, miniImageNet, and a sequence of 5 tasks.

2 Related Work

2.1 Continual Learning

The existing continual learning approaches can be broadly divided into three categories: memory-based, structure-based, and regularization-based methods.

Memory-Based Methods: Methods in this category mitigate forgetting by relying on storing previous experience explicitly or implicitly wherein the former raw samples [6,21,26–28] are saved into the memory for *rehearsal* whereas in the latter a generative model such as a GAN [32] or an autoencoder [16] synthesizes them to perform *pseudo-rehearsal*. These methods allow for simultaneous multi-task learning on i.i.d. data which can significantly reduce forgetting. A recent study on tiny episodic memories in CL [7] compared methods such as GEM [21], A-GEM [6], MER [27], and ER-RES [7]. Similar to [27], for ER-RES they used reservoir sampling using a single pass through the data. Reservoir sampling [39] is a better sampling strategy for long input data compared to random selection. In this work, we explicitly store raw samples into a very tiny memory used for replay buffer and we differ from prior work by how these stored examples are used by specific parts of our model (discriminator and shared module) to prevent forgetting in the features found to be shared across tasks.

Structure-Based Methods: These methods exploit modularity and attempt to localize inference to a subset of the network such as columns [29], neurons [11,41], a mask over parameters [23,31]. The performance on previous tasks is preserved by storing the learned module while accommodating new tasks by augmenting the network with new modules. For instance, Progressive Neural Nets (PNNs) [29] statically grow the architecture while retaining lateral connections to previously frozen modules resulting in guaranteed zero forgetting at the price of quadratic scale in the number of parameters. [41] proposed dynamically expandable networks (DEN) in which, network capacity grows according to tasks *relatedness* by splitting/duplicating the most important neurons while time-stamping them so that they remain accessible and re-trainable at all time.

This strategy despite introducing computational cost is inevitable in continual learning scenarios where a large number of tasks are to be learned and a fixed capacity cannot be assumed.

Regularization Methods: In these methods [1,10,18,25,42], the learning capacity is assumed fixed and continual learning is performed such that the change in parameters is controlled and reduced or prevented if it causes performance downgrade on prior tasks. Therefore, for parameter selection, there has to be defined a *weight importance* measurement concept to prioritize parameter usage. For instance, inspired by Bayesian learning, in elastic weight consolidation (EWC) method [18] important parameters are those to have the highest in terms of the Fisher information matrix. HAT [31] learns an attention mask over *important* parameters. Authors in [10] used per-weight uncertainty defined in Bayesian neural networks to control the change in parameters. Despite the success gained by these methods in maximizing the usage of a fixed capacity, they are often limited by the number of tasks.

2.2   Adversarial Learning

Adversarial learning has been used for different problems such as generative models [13], object composition [2], representation learning [22], domain adaptation [36], active learning [34], etc. The use of an adversarial network enables the model to train in a fully-differentiable manner by adjusting to solve the *minimax* optimization problem [13]. Adversarial learning of the latent space has been extensively researched in domain adaptation [14], active learning [34], and representation learning [17,22]. While previous literature is concerned with the case of modeling single or multiple tasks at once, here we extend this literature by considering the case of continuous learning where multiple tasks need to be learned in a sequential manner.

2.3   Latent Space Factorization

In the machine learning literature, *multi-view* learning, aims at constructing and/or using different views or modalities for better learning performances [3,40]. The approaches to tackle multi-view learning aim at either maximizing the mutual agreement on distinct views of the data or focus on obtaining a latent subspace shared by multiple views by assuming that the input views are generated from this latent subspace using Canonical correlation analysis and clustering [8], Gaussian processes [33], etc. Therefore, the concept of factorizing the latent space into *shared* and *private* parts has been extensively explored for different data modalities. Inspired by the practicality of factorized representation in handling different modalities, here we factorize the latent space learned for different tasks using adversarial learning and orthogonality constraints [30].

3   Adversarial Continual Learning (ACL)

We consider the problem of learning a sequence of T data distributions denoted as $\mathcal{D}^{tr} = \{\mathcal{D}_1^{tr}, \cdots, \mathcal{D}_T^{tr}\}$, where $\mathcal{D}_k^{tr} = \{(\mathbf{X}_i^k, \mathbf{Y}_i^k, \mathbf{T}_i^k)_{i=1}^{n_k}\}$ is the data distribution for task k with n sample tuples of input ($\mathbf{X}^k \in \mathcal{X}$), output label ($\mathbf{Y}^k \in \mathcal{Y}$), and task label ($\mathbf{T}^k \in \mathcal{T}$). The goal is to sequentially learn the model $f_\theta : \mathcal{X} \to \mathcal{Y}$ for each task that can map each task input to its target output while maintaining its performance on all prior tasks. We aim to achieve this by learning a disjoint latent space representation composed of a *task-specific* latent space for each task and a *task-invariant* feature space for all tasks to enhance better knowledge transfer as well as better catastrophic forgetting avoidance of prior knowledge. We mitigate catastrophic forgetting in each space differently. For the *task-invariant* feature space, we assume a limited memory budget of $\mathcal{M}^k$ which stores m samples $x_{i=1\cdots m} \sim \mathcal{D}_{j=1\cdots k-1}^{tr}$ from every single task prior to k.

We begin by learning f_θ^k as a mapping from $\mathbf{X}^k$ to $\mathbf{Y}^k$. For C-way classification task with a cross-entropy loss, this corresponds to

$$\mathcal{L}_{\text{task}}(f_\theta^k, \mathbf{X}^k, \mathbf{Y}^k, \mathcal{M}^k) = -\mathbb{E}_{(x^k, y^k) \sim (\mathbf{X}^k, \mathbf{Y}^k) \cup \mathcal{M}^k} \sum_{c=1}^{C} \mathbb{1}_{[c=y^k]} \log(\sigma(f_\theta^k(x^k))) \quad (1)$$

where σ is the softmax function and the subscript $i = \{1, \cdots, n_t\}$ is dropped for simplicity. In the process of learning a sequence of tasks, an ideal f^k is a model that maps the inputs to two independent latent spaces where one contains the shared features among all tasks and the other remains private to each task. In particular, we would like to disentangle the latent space into the information shared across all tasks ($\mathbf{z}_S$) and the independent or private information of each task ($\mathbf{z}_P$) which are as distinct as possible while their concatenation followed by a task-specific head outputs the desired targets.

To this end, we introduce a mapping called Shared ($S_{\theta_S} : \mathcal{X} \to \mathbf{z}_S$) and train it to generate features that fool an adversarial discriminator D. Conversely, the adversarial discriminator ($D_{\theta_D} : \mathbf{z}_S \to \mathcal{T}$) attempts to classify the generated features by their task labels ($\mathbf{T}^{k \in \{0, \cdots, T\}}$). This is achieved when the discriminator is trained to maximize the probability of assigning the correct task label to generated features while simultaneously S is trained to confuse the discriminator by minimizing $\log(D(S(x^k)))$. This corresponds to the following T-way classification cross-entropy adversarial loss for this minimax game

$$\mathcal{L}_{\text{adv}}(D, S, \mathbf{X}^k, \mathbf{T}^k, \mathcal{M}^k) = \min_S \max_D \sum_{k=0}^{T} \mathbb{1}_{[k=t^k]} \log \left(D \left(S \left(x^k \right) \right) \right). \quad (2)$$

Note that the extra label zero is associated with the 'fake' task label paired with randomly generated noise features of $\mathbf{z}_S' \sim \mathcal{N}(\mu, \Sigma)$. In particular, we use adversarial learning in a different regime that appears in most works related to generative adversarial networks [13] such that the generative modeling of input data distributions is not utilized here because the ultimate task is to learn a discriminative representation.

To facilitate training S, we use the Gradient Reversal layer [12] that optimizes the mapping to maximize the discriminator loss directly ($\mathcal{L}_{\text{task}_S} = -\mathcal{L}_D$). In fact, it acts as an identity function during forward propagation but negates its inputs and reverses the gradients during back propagation. The training for S and D is complete when S is able to generate features that D can no longer predict the correct task label for leading $\mathbf{z}_S$ to become as task-invariant as possible. The private module ($P_{\theta_P} : \mathcal{X} \to \mathbf{z}_P$), however, attempts to accommodate the task-invariant features by learning merely the features that are specific to the task in hand and do not exist in $\mathbf{z}_S$. We further factorize $\mathbf{z}_S$ and $\mathbf{z}_P$ by using orthogonality constraints introduced in [30], also known as "difference" loss in the domain adaptation literature [4], to prevent the shared features between all tasks from appearing in the private encoded features. This corresponds to

$$\mathcal{L}_{\text{diff}}(S, P, \mathbf{X}^k, \mathcal{M}^k) = \sum_{k=1}^{T} ||(S(x^k))^{\text{T}} P^k(x^k)||_F^2, \tag{3}$$

where $|| \cdot ||_F$ is the Frobenius norm and it is summed over the encoded features of all P modules encoding samples for the current tasks and the memory.

Final output predictions for each task are then predicted using a task-specific multi-layer perceptron head which takes $\mathbf{z}_P$ concatenated with $\mathbf{z}_S$ ($\mathbf{z}_P \oplus \mathbf{z}_S$) as an input. Taken together, these loss form the complete objective for ACL as

$$\mathcal{L}_{\text{ACL}} = \lambda_1 \mathcal{L}_{\text{adv}} + \lambda_2 \mathcal{L}_{\text{task}} + \lambda_3 \mathcal{L}_{\text{diff}}, \tag{4}$$

where λ_1, λ_2, and λ_3 are regularizers to control the effect of each component. The full algorithm for ACL is given in Algorithm 1 in the appendix.

3.1   Avoiding Forgetting in ACL

Catastrophic forgetting occurs when a representation learned through a sequence of tasks changes in favor of learning the current task resulting in performance downgrade on previous tasks. The main insight to our approach is decoupling the conventional *single* representation learned for a sequence of tasks into two parts: a part that *must not change* because it contains task-specific features without which complete performance retrieval is not possible, and a part that is *less prone to change* as it contains the core structure of all tasks.

To *fully prevent* catastrophic forgetting in the first part (private features), we use *compact* modules that can be stored into memory. If factorization is successfully performed, the second part remains highly immune to forgetting. However, we empirically found that when disentanglement cannot be fully accomplished either because of the little overlap or large domain shift between the tasks, using a tiny replay buffer containing few samples for old data can be beneficial to retain high ACC values as well as mitigating forgetting.

3.2   Evaluation Metrics

After training for each new task, we evaluate the resulting model on all prior tasks. Similar to [10,21], to measure ACL performance we use ACC as the average test classification accuracy across all tasks. To measure forgetting we report backward transfer, BWT, which indicates how much learning new tasks has influenced the performance on previous tasks. While BWT < 0 directly reports *catastrophic forgetting*, BWT > 0 indicates that learning new tasks has helped with the preceding tasks.

$$\text{BWT} = \frac{1}{T-1} \sum_{i=1}^{T-1} R_{T,i} - R_{i,i}, \quad \text{ACC} = \frac{1}{T} \sum_{i=1}^{T} R_{T,i} \tag{5}$$

where $R_{n,i}$ is the test classification accuracy on task i after sequentially finishing learning the nth task. We also compare methods based on the memory used either in the network architecture growth or replay buffer. Therefore, we convert them into memory size assuming numbers are 32-bit floating point which is equivalent to 4 bytes.

4   Experiments

In this section, we review the benchmark datasets and baselines used in our evaluation as well as the implementation details.

4.1   ACL on Vision Benchmarks

Datasets: We evaluate our approach on the commonly used benchmark datasets for T-split class-incremental learning where the entire dataset is divided into T disjoint susbsets or tasks. We use common image classification datasets including **5-Split MNIST** and **Permuted MNIST** [20], previously used in [10,25,42], **20-Split CIFAR100** [19] used in [6,21,42], and **20-Split miniImageNet** [38] used in [7,43]. We also benchmark ACL on a sequence of **5-Datasets** including **SVHN, CIFAR10, not-MNIST, Fashion-MNIST** and, **MNIST** and report average performance over multiple random task orderings. Dataset statistics are given in Table 5a in the appendix. No data augmentation of any kind has been used in our analysis.

Baselines: From the prior work, we compare with state-of-the-art approaches in all the three categories described in Sect. 2 including Elastic Weight Consolidation (EWC) [18], Progressive neural networks (PNNs) [29], and Hard Attention Mask (HAT) [31] using implementations provided by [31] unless otherwise stated. For memory-based methods including A-GEM, GEM, and ER-RES, for Permuted MNIST, 20-Split CIFAR100, and 20-Split miniImageNet, we relied on the implementation provided by [7], but changed the experimental setting from single to multi-epoch and without using 3 Tasks for cross-validation for a more fair comparison against ACL and other baselines. On Permuted MNIST results for SI [42] are reported from [31], for VCL [25] those are obtained using their original provided code, and for Uncertainty-based CL in Bayesian framework (UCB) [10] are directly reported from the paper. We also perform fine-tuning, and joint training. In fine-tuning (ORD-FT), an ordinary single module network without the discriminator is continuously trained without any forgetting avoidance strategy in the form of experience replay or architecture growth. In joint training with an ordinary network (ORD-JT) and our ACL setup (ACL-JT) we learn all the tasks jointly in a multitask learning fashion using the entire dataset at once which serves as the upper bound for average accuracy on all tasks, as it does not adhere to the continual learning scenario.

Implementation Details: For all ACL experiments except for Permuted MNIST and 5-Split MNIST we used a reduced AlexNet [15] architecture as the backbone for S and P modules for a fair comparison with the majority of our baselines. However, ACL can be also used with more sophisticated architectures (see our code repository for implementation of ACL with reduced ResNet18 backbone). However, throughout this paper, we only report our results using AlexNet. The architecture in S is composed of 3 convolutional and 4 fully-connected (FC) layers whereas P is only a convolutional neural network (CNN) with similar number of layers and half-sized kernels compared to those used in S. The private head modules (p) and the discriminator are all composed of a small 3-layer perceptron. Due to the differences between the structure of our setup and a regular network with a single module, we used a similar CNN structure to S followed by larger hidden FC layers to match the total number of parameters throughout our experiments with our baselines for fair comparisons. For 5-Split MNIST and Permuted MNIST where baselines use a two-layer perceptron with 256 units in each and ReLU nonlinearity, we used a two-layer perceptron of size 784×175 and 175×128 with ReLU activation in between in the shared module and a single-layer of size 784×128 and ReLU for each P. In each head, we also used an MLP with layers of size 256 and 28, ReLU activations, and a 14-unit softmax layer. In all our experiments, no pre-trained model is used. We used stochastic gradient descent in a multi-epoch setting for ACL and all the baselines.

5 Results and Discussion

In the first set of experiments, we measure ACC, BWT, and the memory used by ACL and compare it against state-of-the-art methods with or without memory constraints on 20-Split miniImageNet. Next, we provide more insight and discussion on ACL and its component by performing an ablation study and visualizations on this dataset. In Sect. 6, we evaluate ACL on a more difficult continual learning setting where we sequentially train on 5 different datasets. Finally, in Sect. 7, we demonstrate the experiments on sequentially learning single datasets such as 20-Split CIFAR100 and MNIST variants.

5.1   ACL Performance on 20-Split miniImageNet

Starting with 20-Split miniImageNet, we split it in 20 tasks with 5 classes at a time. Table 1a shows our results obtained for ACL compared to several baselines. We compare ACL with HAT as a regularization based method with no experience replay memory dependency that achieves ACC $= 59.45 \pm 0.05$ with BWT $= -0.04 \pm 0.03\%$. Results for the memory-based methods of ER-RES and A-GEM

Table 1. CL results on 20-Split miniImageNet measuring ACC (%), BWT (%), and Memory (MB). (**) denotes that methods do not adhere to the continual learning setup: ACL-JT and ORD-JT serve as the upper bound for ACC for ACL/ORD networks, respectively. (*) denotes result is re(produced) by us using the original provided code. (†) denotes result is obtained using the re-implementation setup by [31]. BWT of Zero indicates the method is zero-forgetting guaranteed. (b) Cumulative ablation study of ACL on miniImageNet where P: private modules, S: shared module, D: discriminator, $\mathcal{L}_{\mathrm{diff}}$: orthogonality constraint, and RB: replay buffer memory of one sample per class. All results are averaged over 3 runs and standard deviation is given in parentheses

(a)

Method	ACC %	BWT %	Arch (MB)	Replay Buffer (MB)
HAT* [31]	59.45(0.05)	−0.04(0.03)	123.6	–
PNN † [29]	58.96(3.50)	Zero	588	–
ER-RES* [7]	57.32(2.56)	−11.34(2.32)	102.6	110.1
A-GEM* [6]	52.43(3.10)	−15.23(1.45)	102.6	110.1
ORD-FT	28.76(4.56)	−64.23(3.32)	37.6	–
ORD-JT**	69.56(0.78)	–	5100	–
ACL-JT**	66.89(0.32)	–	5100	–
ACL (Ours)	**62.07(0.51)**	**0.00(0.00)**	113.1	8.5

(b)

#	S	P	D	$\mathcal{L}_{\mathrm{diff}}$	RB	ACC %	BWT %
1	x					21.19(4.43)	−60.10(4.14)
2		x				29.09(5.67)	Zero
3	x		x			32.82(2.71)	−28.67(3.61
4	x	x		x		49.13(3.45)	−3.99(0.42)
5	x	x				50.15(1.41)	−14.32(2.34)
6	x	x			x	51.19(1.98)	−9.12(2.98)
7	x	x		x	x	52.07(2.49)	−0.01(0.01)
8	x	x	x			55.72(1.42)	−0.12(0.34)
9	x	x	x	x		57.66(1.44)	−3.71(1.31)
10	x	x	x		x	60.28(0.52)	0.00(0.00)
11	x	x	x	x	x	62.07(0.51)	0.00(0.00)

are re(produced) by us using the implementation provided in [7] by applying modifications to the network architecture to match with ACL in the backbone structure as well as the number of parameters. We only include A-GEM in Table 1a which is only a faster algorithm compared to its precedent GEM with identical performance.

A-GEM and ER-RES use an architecture with 25.6M parameters (102.6 MB) along with storing 13 images of size ($84 \times 84 \times 3$) per class (110.1 MB) resulting in total memory size of 212.7 MB. ACL is able to outperform all baselines in ACC = **62.07 ± 0.51**, BWT=**0.00 ± 0.00**, using total memory of 121.6 MB for architecture growth (113.1 MB) and storing 1 sample per class for replay buffer (8.5 MB). In our ablation study in Sect. 5.2, we will show our performance without using replay buffer for this dataset is ACC = 57.66 ± 1.44. However, ACL is able to overcome the gap by using only one image per class (5 per task) to achieve ACC = **62.07 ± 0.51** without the need to have a large buffer for old data in learning datasets like miniImagenet with diverse sets of classes.

Table 2. Comparison of the effect of the replay buffer size between ACL and other baselines including A-GEM [6], and ER-RES [7] on 20-Split miniImageNet where unlike the baselines, ACL's performance remains unaffected by the increase in number of samples stored per class as discussed in Sect. 5.2. The results from this table are used to generate Fig. 2 in the appendix.

Samples per class		1	3	5	13
A-GEM [6]		45.14(3.42)	49.12(4.69)	50.24(4.56)	52.43(3.10)
ER-RES [7]		40.21(2.68)	46.87(4.51)	53.45(3.45)	57.32(2.56)
ACL (ours)	ACC	62.07(0.51)	61.80(0.50)	61.69(0.61)	61.33(0.40)
	BWT	0.00(0.00)	0.01(0.00)	0.01(0.00)	−0.01(0.02)

5.2   Ablation Studies on 20-Split miniImageNet

We now analyze the major building blocks of our proposed framework including the discriminator, the shared module, the private modules, replay buffer, and the *difference* loss on the miniImagenet dataset. We have performed a complete cumulative ablation study for which the results are summarized in Table 1a and are described as follows:

Shared and Private Modules: Using only a shared module without any other ACL component (ablation #1 in Table 1b) yields the lowest ACC of 21.19 ± 4.43 as well as the lowest BWT performance of −60.10 ± 4.14 while using merely private modules (ablation #2) obtains a slightly better ACC of 29.05 ± 5.67 and a zero-guaranteed forgetting by definition. However, in both scenarios the ACC achieved is too low considering the random chance being 20% which is due to the small size of networks used in S and P.

Discriminator and Orthogonality Constraint ($\mathcal{L}_{\text{diff}}$): The role of adversarial training or presence of D on top of S and P can be seen by comparing the ablations #8 and #5 where in the latter D, as the only disentanglement mechanism, is eliminated. We observe that ACC is improved from 50.15 ± 1.41 to $55.72 \pm 1.42\%$ and BWT is increased from -14.32 ± 2.34 to $-0.12 \pm 0.34\%$. On the other hand, the effect of orthogonality constraint as the only factorization mechanism is shown in ablation #4 where the $\mathcal{L}_{\text{diff}}$ can not improve the ACC performance, but it increases BWT form -14.32 ± 2.34 to -3.99 ± 0.42. Comparing ablations #8 and #4 shows the importance of adversarial training in factorizing the latent spaces versus orthogonality constraint if they were to be used individually. To compare the role of adversarial and diff losses in the presence of replay buffer (RB), we can compare #7 and #10 in which the D and $\mathcal{L}_{\text{diff}}$ are ablated, respectively. It appears again that D improves ACC more than $\mathcal{L}_{\text{diff}}$ by reaching ACC $= 60.28 \pm 0.52$ whereas $\mathcal{L}_{\text{diff}}$ can only achieve ACC $= 52.07 \pm 2.49$. However, the effect of D and $\mathcal{L}_{\text{diff}}$ on BWT is nearly the same.

Replay Buffer: Here we explore the effect of adding the smallest possible memory replay to ACL, i. e., storing one sample per class for each task. Comparing ablation #9 and the most complete version of ACL (#11) shows that adding this memory improves both the ACC and BWT by 4.41% and 3.71%, respectively. We also evaluated ACL using more samples in the memory. Table 2 shows that unlike A-GEM and ER-RES approaches in which performance increases with more episodic memory, in ACL, ACC remains nearly similar to its highest performance. Being insensitive to the amount of old data is a remarkable feature of ACL, not because of the small memory it consumes, but mainly due to the fact that access to the old data might be prohibited or very limited in some real-world applications. Therefore, for a fixed allowed memory size, a method that can effectively use it for architecture growth can be considered as more practical for such applications.

6   ACL Performance on a Sequence of 5-Datasets

In this section, we present our results for continual learning of 5 tasks using ACL in Table 3b. Similar to the previous experiment we look at both ACC and BWT obtained for ACL, finetuning as well as UCB as our baseline. Results for this sequence are averaged over 5 random permutations of tasks and standard deviations are given in parenthesis. CL on a sequence of datasets has been previously performed by two regularization based approaches of UCB and HAT where UCB was shown to be superior [10]. With this given sequence, ACL is able to outperform UCB by reaching ACC $= 78.55(\pm 0.29)$ and BWT $= -0.01$ using only half of the memory size and also no replay buffer. In Bayesian neural networks such as UCB, there exists double number of parameters compared to a regular model representing mean and variance of network weights. It is very encouraging to see that ACL is not only able to continually learn on a single dataset, but also across diverse datasets.

7   Additional Experiments

20-Split CIFAR100: In this experiment we incrementally learn CIFAR100 in 5 classes at a time in 20 tasks. As shown in Table 3, HAT is the most competitive baseline, although it does not depend on memory and uses 27.2 MB to store its architecture in which it learns task-based attention maps reaching ACC = $76.96 \pm 1.23\%$. PNN uses 74.7 MB to store the lateral modules to the memory and guarantees zero forgetting. Results for A-GEM, and ER-Reservoir are re(produced) by us using a CNN similar to our shared module architecture. We use fully connected layers with more number of neurons to compensate for the remaining number of parameters reaching 25.4 MB of memory. We also stored 13 images per class (1300 images of size $(32 \times 32 \times 3)$ in total) which requires 16.0 MB of memory. However, ACL achieves ACC = $(\mathbf{78.08 \pm 1.25})\%$ with BWT=$\mathbf{0.00 \pm 0.01})\%$ using only 25.1 MB to grow private modules with 167.2K parameters (0.6 MB) without using memory for replay buffer which is mainly due to the overuse of parameters for CIFAR100 which is considered as a relevantly 'easy' dataset with all tasks (classes) sharing the same data distribution. Disentangling shared and private latent spaces, prevents ACL from using redundant parameters by only storing task-specific parameters in P and p modules. In fact, as opposed to other memory-based methods, instead of starting from a large network and using memory to store samples, which might not be available in practice due to confidentiality issues (*e.g.* medical data), ACL uses memory to gradually add small modules to accommodate new tasks and relies on knowledge transfer through the learned shared module. The latter is what makes ACL to be different than architecture-based methods such as PNN where the network grows by the entire *column* which results in using a highly disproportionate memory to what is needed to learn a new task with.

Permuted MNIST: Another popular variant of the MNIST dataset in CL literature is Permuted MNIST where each task is composed of randomly permuting pixels of the entire MNIST dataset. To compare against values reported in prior work, we particularly report on a sequence of $T = 10$ and $T = 20$ tasks with ACC, BWT, and memory for ACL and baselines. To further evaluate ACL's ability in handling more tasks, we continually learned up to 40 tasks. As shown in Table 4 in the appendix, among the regularization-based methods, HAT achieves the highest performance of 91.6% [31] using an architecture of size 1.1 MB. Vanilla VCL improves by 7% in ACC and 6.5% in BWT using a K-means core-set memory size of 200 samples per task (6.3 MB) and an architecture size similar to HAT. PNN appears as a strong baseline achieving ACC = 93.5% with guaranteed zero forgetting. Finetuning (ORD-FT) and joint training (ORD-JT) results for an ordinary network, similar to EWC and HAT (a two-layer MLP with 256 units and ReLU activations), are also reported as reference values for lowest BWT and highest achievable ACC, respectively. ACL achieves the highest accuracy among all baselines for both sequences of 10 and 20 equal to ACC = 98.03 ± 0.01 and ACC = 97.81 ± 0.03, and BWT = -0.01% BWT = 0%, respectively which shows that performance of ACL drops only by

0.2% as the number of tasks doubles. ACL also remains efficient in using memory to grow the architecture compactly by adding only 55K parameters (0.2MB) for each task resulting in using a total of 2.4 MB and 5.0 MB when $T = 10$ and $T = 20$, respectively for the entire network including the shared module and the discriminator. We also observed that the performance of our model does not change as the number of tasks increases to 30 and 40 if each new task is accommodated with a new private module. We did not store old data and used memory only to grow the architecture by 55K parameters (0.2 MB).

5-Split MNIST: As the last experiment in this section, we continually learn $0-9$ MNIST digits by following the conventional pattern of learning 2 classes over 5 sequential tasks [10, 25, 42]. As shown in Table 6 in the appendix, we compare ACL with regularization-based methods with no memory dependency (EWC, HAT, UCB, Vanilla VCL) and methods relying on memory only (GEM), and VCL with K-means Core-set (VCL-C) where 40 samples are stored per task. ACL reaches ACC = (**99.76 ± 0.03**)% with zero forgetting outperforming UCB with ACC = 99.63% which uses nearly 40% more memory size. In this task, we only use architecture growth (no experience replay) where 54.3K private parameters are added for each task resulting in memory requirement of 1.6 MB to store

Table 3. CL results on 20-Split CIFAR100 measuring ACC (%), BWT (%), and Memory (MB). (**) denotes that methods do not adhere to the continual learning setup: ACL-JT and ORD-JT serve as the upper bound for ACC for ACL/ORD networks, respectively. (*) denotes result is obtained by using the original provided code. (†) denotes result is obtained using the re-implementation setup by [31]. (°) denotes result is reported by [7]. BWT of Zero indicates the method is zero-forgetting guaranteed. All results are averaged over 3 runs and standard deviation is given in parentheses.

(a) 20-Split CIFAR100

Method	ACC %	BWT %	Arch (MB)	Replay buffer (MB)
HAT * [31]	76.96(1.23)	0.01(0.02)	27.2	–
PNN† [29]	75.25(0.04)	Zero	93.51	–
A-GEM° [6]	54.38(3.84)	−21.99(4.05)	25.4	16
ER-RES° [7]	66.78(0.48)	−15.01(1.11)	25.4	16
ORD-FT	34.71(3.36)	−48.56(3.17)	27.2	–
ORD-JT**	78.67(0.34)	–	764.5	–
ACL-JT**	79.91(0.05)	–	762.6	–
ACL (Ours)	**78.08(1.25)**	**0.00(0.01)**	25.1	–

(b) Sequence of 5 Datasets

Method	ACC %	BWT %	Arch (MB)	Replay buffer (MB)
UCB * [10]	76.34(0.12)	−1.34(0.04)	32.8	–
ORD-FT	27.32(2.41)	−42.12(2.57)	16.5	–
ACL (Ours)	**78.55(0.29)**	**−0.01(0.15)**	16.5	–

all private modules. Our core architecture has a total number of parameters of 420.1K. We also provide naive finetuning results for ACL and a regular single-module network with (268K) parameters (1.1 MB). Joint training results for the regular network (ORD-JT) is computed as ACC = 99.89 ± 0.01 for ACL which requires 189.3 MB for the entire dataset as well as the architecture (Table 4).

Table 4. CL results on Permuted MNIST. measuring ACC (%), BWT (%), and Memory (MB). (**) denotes that methods do not adhere to the continual learning setup: ACL-JT and ORD-JT serve as the upper bound for ACC for ACL/ORD networks, respectively. (*) denotes result is obtained by using the original provided code. (‡) denotes result reported from original work. (°) denotes the results reported by [7] and (°°) denotes results are reported by [31]; T shows the number of tasks. Note the difference between BWT of Zero and 0.00 where the former indicates the method is zero-forgetting guaranteed by definition and the latter is computed using Eq. 5. All results are averaged over 3 runs, the standard deviation is provided in parenthesis.

Method	ACC %	BWT %	Arch (MB)	Replay buffer (MB)
EWC°° [18] (T = 10)	88.2	–	1.1	–
HAT‡ [31] (T = 10)	97.4	–	2.8	–
UCB‡ [10] (T = 10)	91.44(0.04)	−0.38(0.02)	2.2	–
VCL*[25] (T = 10)	88.80(0.23)	−7.90(0.23)	1.1	–
VCL-C* [25] (T = 10)	95.79(0.10)	−1.38(0.12)	1.1	6.3
PNN° [29] (T = 20)	93.5(0.07)	Zero	N/A	–
ORD-FT (T = 10)	44.91(6.61)	−53.69(1.91)	1.1	–
ORD-JT** (T = 10)	96.03(0.02)	–	189.3	–
ACL-JT** (T = 10)	98.45(0.02)	–	194.4	–
ACL (Ours) (T = 10)	**98.03(0.01)**	−0.01(0.01)	2.4	–
ACL (Ours) (T = 20)	**97.81(0.03)**	0.00(0.00)	5.0	–
ACL (Ours) (T = 30)	**97.81(0.03)**	0.00(0.00)	7.2	–
ACL (Ours) (T = 40)	**97.80(0.02)**	0.00(0.00)	9.4	–

8 Conclusion

In this work, we proposed a novel hybrid continual learning algorithm that factorizes the representation learned for a sequence of tasks into *task-specific* and *task-invariant* features where the former is important to be fully preserved to avoid forgetting and the latter is empirically found to be remarkably less prone to forgetting. The novelty of our work is that we use adversarial learning along with orthogonality constraints to disentangle the shared and private latent representations which results in compact private modules that can be stored into memory and hence, efficiently preventing forgetting. A tiny replay buffer, although not critical, can be also integrated into our approach if forgetting occurs in the shared module. We established a new state of the art on CL benchmark datasets.

References

1. Aljundi, R., Babiloni, F., Elhoseiny, M., Rohrbach, M., Tuytelaars, T.: Memory aware synapses: learning what (not) to forget. In: Ferrari, V., Hebert, M., Sminchisescu, C., Weiss, Y. (eds.) ECCV 2018. LNCS, vol. 11207, pp. 144–161. Springer, Cham (2018). https://doi.org/10.1007/978-3-030-01219-9_9
2. Azadi, S., Pathak, D., Ebrahimi, S., Darrell, T.: Compositional GAN: learning image-conditional binary composition. Int. J. Comput. Vis. **128**, 2570–2585 (2020). https://doi.org/10.1007/s11263-020-01336-9
3. Blum, A., Mitchell, T.: Combining labeled and unlabeled data with co-training. In: Proceedings of the Eleventh Annual Conference on Computational Learning Theory, pp. 92–100. ACM (1998)
4. Bousmalis, K., Trigeorgis, G., Silberman, N., Krishnan, D., Erhan, D.: Domain separation networks. In: Advances in Neural Information Processing Systems, pp. 343–351 (2016)
5. Chaudhry, A., Dokania, P.K., Ajanthan, T., Torr, P.H.S.: Riemannian walk for incremental learning: understanding forgetting and intransigence. In: Ferrari, V., Hebert, M., Sminchisescu, C., Weiss, Y. (eds.) ECCV 2018. LNCS, vol. 11215, pp. 556–572. Springer, Cham (2018). https://doi.org/10.1007/978-3-030-01252-6_33
6. Chaudhry, A., Ranzato, M., Rohrbach, M., Elhoseiny, M.: Efficient lifelong learning with A-GEM. In: International Conference on Learning Representations (2019)
7. Chaudhry, A., et al.: Continual learning with tiny episodic memories. arXiv preprint arXiv:1902.10486 (2019)
8. Chaudhuri, K., Kakade, S.M., Livescu, K., Sridharan, K.: Multi-view clustering via canonical correlation analysis. In: Proceedings of the 26th annual international conference on machine learning, pp. 129–136. ACM (2009)
9. Deng, J., Dong, W., Socher, R., Li, L.J., Li, K., Fei-Fei, L.: ImageNet: a large-scale hierarchical image database. In: CVPR 2009 (2009)
10. Ebrahimi, S., Elhoseiny, M., Darrell, T., Rohrbach, M.: Uncertainty-guided continual learning with Bayesian neural networks. In: International Conference on Learning Representations (2020)
11. Fernando, C., et al.: PathNet: evolution channels gradient descent in super neural networks. arXiv preprint arXiv:1701.08734 (2017)
12. Ganin, Y., et al.: Domain-adversarial training of neural networks. J. Mach. Learn. Res. **17**(1), 2030–2096 (2016)
13. Goodfellow, I., et al.: Generative adversarial nets. In: Advances in Neural Information Processing Systems, pp. 2672–2680 (2014)
14. Hoffman, J., et al.: CyCADA: cycle-consistent adversarial domain adaptation. In: International Conference on Machine Learning, pp. 1989–1998 (2018)
15. Iandola, F.N., Han, S., Moskewicz, M.W., Ashraf, K., Dally, W.J., Keutzer, K.: SqueezeNet: AlexNet-level accuracy with 50x fewer parameters and <0.5 mb model size. arXiv preprint arXiv:1602.07360 (2016)
16. Kemker, R., Kanan, C.: FearNet: brain-inspired model for incremental learning. In: International Conference on Learning Representations (2018)
17. Kim, H., Mnih, A.: Disentangling by factorising. arXiv preprint arXiv:1802.05983 (2018)
18. Kirkpatrick, J., et al.: Overcoming catastrophic forgetting in neural networks. In: Proceedings of the National Academy of Sciences, p. 201611835 (2017)
19. Krizhevsky, A., Hinton, G.: Learning multiple layers of features from tiny images. Technical report, Citeseer (2009)

20. LeCun, Y., Bottou, L., Bengio, Y., Haffner, P.: Gradient-based learning applied to document recognition. Proc. IEEE **86**(11), 2278–2324 (1998)
21. Lopez-Paz, D., et al.: Gradient episodic memory for continual learning. In: Advances in Neural Information Processing Systems, pp. 6467–6476 (2017)
22. Makhzani, A., Shlens, J., Jaitly, N., Goodfellow, I., Frey, B.: Adversarial autoencoders. arXiv preprint arXiv:1511.05644 (2015)
23. Mallya, A., Lazebnik, S.: PackNet: adding multiple tasks to a single network by iterative pruning. In: IEEE Conference on Computer Vision and Pattern Recognition (CVPR) (2018)
24. McCloskey, M., Cohen, N.J.: Catastrophic interference in connectionist networks: the sequential learning problem. In: Psychology of Learning and Motivation, vol. 24, pp. 109–165. Elsevier (1989)
25. Nguyen, C.V., Li, Y., Bui, T.D., Turner, R.E.: Variational continual learning. In: ICLR (2018)
26. Rebuffi, S.A., Kolesnikov, A., Sperl, G., Lampert, C.H.: iCaRL: incremental classifier and representation learning. In: CVPR (2017)
27. Riemer, M., et al.: Learning to learn without forgetting by maximizing transfer and minimizing interference. In: International Conference on Learning Representations (2019)
28. Robins, A.: Catastrophic forgetting, rehearsal and pseudorehearsal. Connect. Sci. **7**(2), 123–146 (1995)
29. Rusu, A.A., et al.: Progressive neural networks. arXiv preprint arXiv:1606.04671 (2016)
30. Salzmann, M., Ek, C.H., Urtasun, R., Darrell, T.: Factorized orthogonal latent spaces. In: Proceedings of the Thirteenth International Conference on Artificial Intelligence and Statistics, pp. 701–708 (2010)
31. Serra, J., Suris, D., Miron, M., Karatzoglou, A.: Overcoming catastrophic forgetting with hard attention to the task. In: Dy, J., Krause, A. (eds.) Proceedings of the 35th International Conference on Machine Learning. Proceedings of Machine Learning Research, vol. 80, pp. 4548–4557. PMLR (2018)
32. Shin, H., Lee, J.K., Kim, J., Kim, J.: Continual learning with deep generative replay. In: Advances in Neural Information Processing Systems, pp. 2990–2999 (2017)
33. Shon, A., Grochow, K., Hertzmann, A., Rao, R.P.: Learning shared latent structure for image synthesis and robotic imitation. In: Advances in Neural Information Processing Systems, pp. 1233–1240 (2006)
34. Sinha, S., Ebrahimi, S., Darrell, T.: Variational adversarial active learning. In: Proceedings of the IEEE International Conference on Computer Vision, pp. 5972–5981 (2019)
35. Srivastava, R.K., Masci, J., Kazerounian, S., Gomez, F., Schmidhuber, J.: Compete to compute. In: Advances in Neural Information Processing Systems, pp. 2310–2318 (2013)
36. Tzeng, E., Hoffman, J., Saenko, K., Darrell, T.: Adversarial discriminative domain adaptation. In: Proceedings of the IEEE Conference on Computer Vision and Pattern Recognition, pp. 7167–7176 (2017)
37. Van Der Maaten, L.: Accelerating t-SNE using tree-based algorithms. J. Mach. Learn. Res. **15**(1), 3221–3245 (2014)
38. Vinyals, O., Blundell, C., Lillicrap, T., Wierstra, D., et al.: Matching networks for one shot learning. In: Advances in neural information processing systems. pp. 3630–3638 (2016)

39. Vitter, J.S.: Random sampling with a reservoir. ACM Trans. Math. Softw. (TOMS) **11**(1), 37–57 (1985)
40. Xu, C., Tao, D., Xu, C.: A survey on multi-view learning. arXiv preprint arXiv:1304.5634 (2013)
41. Yoon, J., Yang, E., Lee, J., Hwang, S.J.: Lifelong learning with dynamically expandable networks. In: International Conference on Learning Representations (2018)
42. Zenke, F., Poole, B., Ganguli, S.: Continual learning through synaptic intelligence. In: Precup, D., Teh, Y.W. (eds.) Proceedings of the 34th International Conference on Machine Learning. Proceedings of Machine Learning Research, vol. 70, pp. 3987–3995. PMLR (2017)
43. Zhang, M., Wang, T., Lim, J.H., Feng, J.: Prototype reminding for continual learning. arXiv preprint arXiv:1905.09447 (2019)

Adapting Object Detectors with Conditional Domain Normalization

Peng Su[1,2(✉)], Kun Wang[2], Xingyu Zeng[2], Shixiang Tang[2], Dapeng Chen[2], Di Qiu[2], and Xiaogang Wang[1]

[1] The Chinese University of Hong Kong, Sha Tin, Hong Kong
{psu,xgwang}@ee.cuhk.edu.hk
[2] SenseTime Research, Sha Tin, Hong Kong

Abstract. Real-world object detectors are often challenged by the domain gaps between different datasets. In this work, we present the Conditional Domain Normalization (CDN) to bridge the domain distribution gap. CDN is designed to encode different domain inputs into a shared latent space, where the features from different domains carry the same domain attribute. To achieve this, we first disentangle the domain-specific attribute out of the semantic features from source domain via a domain embedding module, which learns a domain-vector to characterize the domain attribute information. Then this domain-vector is used to encode the features from target domain through a conditional normalization, resulting in different domains' features carrying the same domain attribute. We incorporate CDN into various convolution stages of an object detector to adaptively address the domain shifts of different level's representation. In contrast to existing adaptation works that conduct domain confusion learning on semantic features to remove domain-specific factors, CDN aligns different domain distributions by modulating the semantic features of target domains conditioned on the learned domain-vector of the source domain. Extensive experiments show that CDN outperforms existing methods remarkably on both real-to-real and synthetic-to-real adaptation benchmarks, including 2D image detection and 3D point cloud detection.

1 Introduction

Deep neural networks have achieved remarkable success on visual recognition tasks. However, it is still very challenging for deep networks to generalize on a different domain, whose data distribution is not identical with original training data. Such a problem is known as dataset bias or domain shift [31]. For example, to guarantee safety in autonomous driving, the perception model is required to perform well under all conditions, like sunny, night, rainy, snowy, etc. However, even top-grade object detectors still face significant challenges when deployed in

Electronic supplementary material The online version of this chapter (https://doi.org/10.1007/978-3-030-58621-8_24) contains supplementary material, which is available to authorized users.

© Springer Nature Switzerland AG 2020
A. Vedaldi et al. (Eds.): ECCV 2020, LNCS 12356, pp. 403–419, 2020.
https://doi.org/10.1007/978-3-030-58621-8_24

such varying real-world settings. Although collecting and annotating more data from unseen domains can help, it is prohibitively expensive, laborious and time-consuming. Another appealing application is to adapt from synthetic data to real data, as it can save the amount of cost and time. However, current objector detectors trained with synthetic data can rarely generalize on real data due to a significant domain distribution gap [36,38].

Adversarial domain adaptation emerges as a hopeful method to learn transferable representations across domains. It has achieved noticeable progress in various machine learning tasks, from image classification [24,27], semantic segmentation [36,39,47], object detection [33,46] to reinforcement learning [20,28,38]. According to Ben-David's theory [1], the empirical risk on the target domain is bounded by the source domain risk and the $\mathcal{H}$ domain divergence. Adversarial adaptation dedicates to learn domain invariant representation to reduce the $\mathcal{H}$ divergence, which eventually decreases the upper bound of the empirical error on the target domain.

However, existing adversarial adaptation methods still suffer from several problems. First, previous methods [4,8,38] directly feed semantic features into a domain discriminator to conduct domain confusion learning. But the semantic features contain both image contents and domain attribute information. It's difficult to make the discriminator only focusing on removing domain-specific information without inducing undesirable influence on the images contents. Second, existing adversarial adaptation methods [4,8,38] use domain confusion learning at one or few convolution stages to handle the distribution mismatch, which ignores the differences of domain shifts at various representation levels. For example, the first few convolution layers' features mainly convey low-level information of local patterns, while the higher convolution layers' features include more abstract global patterns with semantics [43]. Such differences born within deep convolution neural networks naturally exhibit different types of domain shift at various convolution stages.

Motivated by this, we propose the Conditional Domain Normalization (CDN) to embed different domain inputs into a shared latent space, where the features of all different domains inputs carry the same domain attribute information. Specifically, CDN utilizes a domain embedding module to learn a domain-vector to characterize the domain attribute information, through disentangling the domain attribute out of the semantic features of domain inputs. We use this domain-vector to encode the semantic features of another domain's inputs via a conditional normalization. Thus different domain features carry the same domain attributes information. The experiment on both real-to-real and synthetic-to-real adaptation benchmarks demonstrate that our method outperforms the-state-of-the-art adaptation methods. To summarize, our contributions are three folds: (1) We propose the Conditional Domain Normalization (CDN) to bridge the domain distribution gap, through embedding different domain inputs into a shared latent space, where the features from different domains carry the same domain attribute. (2) CDN achieves state-of-the-art unsupervised domain adaptation performance on both real-to-real and synthetic-to-real benchmarks, including

2D image and 3D point cloud detection tasks. And we conduct both quantitative and qualitative comparisons to analyze the features learned by CDN. (3) We construct a large-scale synthetic-to-real driving benchmark for 2D object detection, including a variety of public datasets.

2 Related Work

Object Detection is the center topic in computer vision, which is crucial for many real-world applications, such as autonomous driving. In 2D detection, following the pioneering work of RCNN [11], a number of object detection frameworks based on convolutional networks have been developed like Fast R-CNN [10], Faster R-CNN [32], and Mask R-CNN [12], which significantly push forward the state of the art. In 3D detection, spanning from detecting 3d objects from 2d images [3], to directly generate 3D box from point cloud [29,37], abundant works has been successfully explored. All these 2D and 3D objectors have achieved remarkable success on one or few specific public datasets. However, even top-grade object detectors still face significant challenges when deployed in real-world settings. The difficulties usually arise from the changes in environmental conditions.

Domain Adaptation generalizes a model across different domains, and it has been extensively explored in various tasks, spanning from image classification [2,23,24,27,40], semantic segmentation [15,36,39] to reinforcement learning [20,28,38]. For 2D detection, domain confusion learning via a domain discriminator has achieved noticeable progress in cross-domain detection. [4] incorporated a gradient reversal layer [8] into a Faster R-CNN model. [33,46] adopt domain confusion learning on both global and local levels to align source and target distributions. In contrast to existing methods conducting domain confusion learning directly on semantic features, we explicitly disentangle the domain attribute out of semantic features. And this domain attribute is used to encode other domains' features, thus different domain inputs share the same domain attribute in the feature space. For 3D detection, only a few works [17,45] has been explored to adapt object detectors across different point cloud dataset. Different from existing works [17,45] are specifically designed for point cloud data, our proposed CDN is a general adaptation framework that adapts both 2D image and 3D point cloud object detector through the conditional domain normalization.

Conditional Normalization is a technique to modulate the neural activation using a transformation that depends on external data. It has been successfully used in the generative models and style transfer, like conditional batch normalization [6], adaptive instance normalization (AdaIN) [16] and spatial adaptive batch normalization [25]. [16] proposes AdaIN to control the global style of the synthesized image. [41] modulates the features conditioned on semantic masks for image super-resolution. [25] adopts a spatially-varying transformation, making it suitable for image synthesis from semantic masks. Inspired by these works, we propose Conditional Domain Normalization (CDN) to modulate one domain's inputs condition on another domain's attributes information. But our method exhibits significant difference with style transfer works: Style transfer works modify a content image conditioned on another style image, which is a conditional instance normalization by nature; but CDN modulates one domain's features conditioned on the domain embedding learned from another domains' inputs (a group of images), which is like a domain-to-domain translation. Hence we use different types of conditional normalization to achieve different goals.

3   Method

We first introduce the general unsupervised domain adaptation approach in Sect. 3.1. Then we present the proposed Conditional Domain Normalization (CDN) in Sect. 3.2. Last we adapt object detectors with the CDN in Sect. 3.3.

3.1   General Adversarial Adaptation Framework

Given source images and labels $\{(x_i^S, y_i^S)\}_{i=1}^{N_S}$ drawn from P_s, and target images $\{x_i^T\}_{i=1}^{N_T}$ from target domain P_t, the goal of unsupervised domain adaptation is to find a function $f : x \rightarrow y$ that minimize the empirical error on target data. For object detection task, the f can be decomposed as $f = G(\cdot; \theta_g) \circ H(\cdot; \theta_h)$, where $G(\cdot; \theta_g)$ represents a feature extractor network and $H(\cdot; \theta_h)$ denotes a bounding box head network. The adversarial domain adaptation introduces a discriminator network $D(\cdot; \theta_d)$ that tries to determine the domain labels of feature maps generated by $G(\cdot; \theta_g)$.

$$\min_{\theta_g, \theta_h} \mathcal{L}_{det} = \mathcal{L}_{cls}(G(x; \theta_g), H(x; \theta_h)) + \mathcal{L}_{reg}(G(x; \theta_g), H(x; \theta_h))$$

$$\min_{\theta_d} \max_{\theta_g} \mathcal{L}_{adv} = \mathbb{E}_{x \sim P_s}[\log(D(G(x; \theta_g); \theta_d))] + \mathbb{E}_{x \sim P_t}[\log(1 - D(G(x; \theta_g); \theta_d)] \tag{1}$$

As illustrated in Eq. 1, $G(\cdot; \theta_g)$ and $H(\cdot; \theta_h)$ are jointly trained to minimize the detection loss $\mathcal{L}_{det}$ by supervised training on the labeled source domain. At the same time, the backbone $G(\cdot; \theta_g)$ is optimized to maximize the probability of $D(\cdot; \theta_d)$ to make mistakes. Through this two-player min-max game, the final $G(\cdot; \theta_g)$ will converge to extract features that are indistinguishable for $D(\cdot; \theta_d)$, thus domain invariant representations are learned.

3.2   Conditional Domain Normalization

Conditional Domain Normalization (CDN) is designed to embed source and target domain inputs into a shared latent space, where the semantic features from different domains carry the same domain attribute information. Formally, let $v^s \in \mathbb{R}^{N \times C \times H \times W}$ and $v^t \in \mathbb{R}^{N \times C \times H \times W}$ represent feature maps of source and target inputs, respectively. C is the channel dimension and N denotes the mini-batch size. We first learn a domain embedding vector $e_{domain}^s \in \mathbb{R}^{1 \times C \times 1}$ to characterize the domain attribute of source inputs. It is accomplished by a

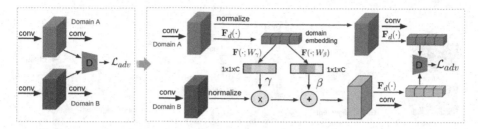

Fig. 1. (Left) Traditional domain adversarial approach. (Right) Conditional Domain Normalization (CDN). The green and blue cubes represent the feature maps of domain A and domain B respectively. (Color figure online)

domain embedding network $\mathbf{F}_d(\cdot; W)$ parameterized by two fully-connect layers with ReLU non-linearity δ as

$$e^s_{domain} = \mathbf{F}_d(v^s_{avg}; W) = \delta(W_2\delta(W_1 v^s_{avg})). \tag{2}$$

And $v^s_{avg} \in \mathbb{R}^{N \times C \times 1}$ represents the channel-wise statistics of source feature v^s generated by global average pooling

$$v^s_{avg} = \frac{1}{HW} \sum_{h=1}^{H} \sum_{w=1}^{W} v^s(h, w). \tag{3}$$

To embed both source and target domain inputs into a shared latent space, where source and target features carry the same domain attributes while preserving individual image contents. We encode the target features v^t with the source domain embedding via an affine transformation as

$$\hat{v}^t = \mathbf{F}(e^s_{domain}; W_\gamma, b_\gamma) \cdot \left(\frac{v^t - \mu^t}{\sigma^t} \right) + \mathbf{F}(e^s_{domain}; W_\beta, b_\beta), \tag{4}$$

where μ^t and σ^t denote the mean and variance of target feature v^t. The affine parameters are learned by function $F(\cdot; W_\gamma, b_\gamma)$ and $F(\cdot; W_\beta, b_\beta)$ conditioned on the source domain embedding vector e^s_{domain},

$$F(e^s_{domain}; W_\gamma, b_\gamma) = W_\gamma e^s_{domain} + b_\gamma, \quad F(e^s_{domain}; W_\beta, b_\beta) = W_\beta e^s_{domain} + b_\beta. \tag{5}$$

For the target feature mean $\mu^t \in \mathbb{R}^{1 \times C \times 1}$ and variance $\sigma^t \in \mathbb{R}^{1 \times C \times 1}$, we calculate it with a standard batch normalization [19]

$$\mu^t_c = \frac{1}{NHW} \sum_{n=1}^{N} \sum_{h=1}^{H} \sum_{w=1}^{W} v^t_{nchw}, \quad \sigma^t_c = \sqrt{\frac{1}{NHW} \sum_{n=1}^{N} \sum_{h=1}^{H} \sum_{w=1}^{W} (v^t_{nchw} - \mu^t_c)^2 + \epsilon}, \tag{6}$$

where μ^t_c and σ^t_c denotes c-th channel of μ^t and σ^t. Finally, we have a discriminator to supervise the encoding process of domain attribute as

$$\min_{\theta_d} \max_{\theta_g} \mathcal{L}_{adv} = \mathbb{E}[\log(D(\mathbf{F}_d(v^s)); \theta_d)] + \mathbb{E}[\log(1 - D(\mathbf{F}_d(\hat{v}^t); \theta_d))], \tag{7}$$

where v^s and v^t are generated by $G(\cdot; \theta_g)$.

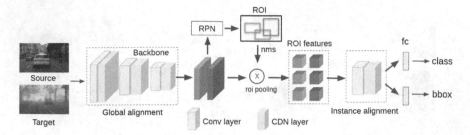

Fig. 2. Faster R-CNN network incorporates with CDN. The CDN is adopted in both backbone network and bounding box head network to adaptively address the domain shift at different representation levels.

Discussion. CDN exhibits a significant difference compared with existing adversarial adaptation works. As shown in Fig. 1, previous methods conduct domain confusion learning directly on semantic features to remove domain-specific factors. However, the semantic features contain both domain attribute and image contents. It is not easy to enforce the domain discriminator only regularizing the domain-specific factors without inducing any undesirable influence on image contents. In contrast, we disentangle the domain attribute out of the semantic features via conditional domain normalization. And this domain attribute is used to encode other domains' features, thus different domain features carry the same domain attribute information.

3.3   Adapting Detector with Conditional Domain Normalization

Convolution neural network's (CNN) success in pattern recognition has been largely attributed to its great capability of learning hierarchical representations [43]. More specifically, the first few layers of CNN focus on low-level features of local pattern, while higher layers capture semantic representations. Given this observation, CNN based object detectors naturally exhibit different types of domain shift at various levels' representations. Hence we incorporate CDN into different convolution stages in object detectors to address the domain mismatch adaptively, as shown in Fig. 2.

Coincident to our analysis, some recent works [33,46] empirically demonstrate that global and local region alignments have different influences on detection performance. For easy comparison, we refer to the CDN located at the backbone network as global alignment, and CDN in the bounding box head networks as local or instance alignment.

As shown in Fig. 2, taking faster-RCNN model [32] with ResNet [13] backbone as an example, we incorporate CDN in the last residual block at each stage. Thus the global alignment loss can be computed by

$$L_{adv}^{global} = \sum_{l=1}^{L} \mathbb{E}[\log(D_l(\mathbf{F}_d^l(v_l^s); \theta_d^l)] + \mathbb{E}[\log(1 - D_l(\mathbf{F}_d^l(\hat{v}_l^t); \theta_d^l))], \qquad (8)$$

where v_l^s and v_l^t denote l-th layer's source feature and the encoded target feature, and D_l represents the corresponding domain discriminator parameterized by θ_d^l.

As for bounding box head network, we adopt CDN on the fixed-size region of interest (ROI) features generated by ROI pooling [32]. Because the original ROIs are often noisy and the quantity of source and target ROIs are not equal, we randomly select $\min(N_{roi}^S, N_{roi}^T)$ ROIs from each domain. N_{roi}^S and N_{roi}^T represent the quantity of source and target ROIs after non-maximum suppression (NMS). Hence we have instance alignment regularization for ROI features as

$$L_{adv}^{instance} = \mathbb{E}[\log(D_{roi}(\mathbf{F}_d^{roi}(v_{roi}^s); \theta_d^{roi})] + \mathbb{E}[\log(1 - D_{roi}(\mathbf{F}_d^{roi}(\hat{v}_{roi}^t); \theta_d^{roi}))]. \quad (9)$$

The overall training objective is to minimize the detection loss $\mathcal{L}_{det}$ (of the labeled source domain) that consists of a classification loss $\mathcal{L}_{cls}$ and a regression loss $\mathcal{L}_{reg}$, and min-max a adversarial loss $\mathcal{L}_{adv}$ of discriminator network

$$\min_{\theta_d} \max_{\theta_g} \ \mathcal{L}_{adv} = \lambda L_{adv}^{global} + L_{adv}^{instance}$$
$$\min_{\theta_g, \theta_h} \ \mathcal{L}_{det} = \mathcal{L}_{cls}(G(x; \theta_g), H(x; \theta_h)) + \mathcal{L}_{reg}(G(x; \theta_g), H(x; \theta_h)), \quad (10)$$

where λ is a weight to balance the global and local alignment regularization.

4 Experiments

We evaluate CDN on various real-to-real (KITTI to Cityscapes) and synthetic-to-real (Virtual KITTI/Synscapes/SIM10K to BDD100K, PreSIL to KITTI) adaptation benchmarks. We also report results on cross-weather adaptation, Cityscapes to Foggy Cityscapes. Mean average precision (mAP) with an intersection-over-union (IOU) threshold of 0.5 is reported for 2D detection experiments. We use Source and Target to represent the results of supervised training on source and target domain, respectively. For 3D point cloud object detection, PointRCNN [37] with backbone of PointNet++ [29] is adopted as our baseline model. Following standard metric on KITTI benchmark [37], we use Average Precision(AP) with IOU threshold 0.7 for car and 0.5 for pedestrian/cyclist.

4.1 Dataset

Cityscapes [5] is a European traffic scene dataset, which contains $2,975$ images for training and 500 images for testing.
Foggy Cityscapes derives from Cityscapes with a fog simulation proposed by [34]. It also includes $2,975$ images for training, 500 images for testing.
KITTI [9] contains $21,260$ images collected from different urban scenes, which includes 2D RGB images and 3D point cloud data.
Virtual KITTI is derived from KITTI with a real-to-virtual cloning technique proposed by [7]. It has the same number of images and categories as KITTI.
Synscapes [42] is a synthetic dataset of street scene, which consists of $25,000$ images created with a photo-realistic rendering technique.

Table 1. Cityscapes to Foggy Cityscapes adaptation.

Method	Person	Rider	Car	Truck	Bus	Train	Motorcycle	Bicycle	mAP
Source	29.3	31.9	43.5	15.8	27.4	9.0	20.3	29.9	26.1
DA-Faster [4]	25.0	31.0	40.5	22.1	35.3	20.2	20.0	27.1	27.9
DT [18]	25.4	39.3	42.4	24.9	40.4	23.1	25.9	30.4	31.5
SCDA [46]	33.5	38.0	48.5	26.5	39.0	23.3	28.0	33.6	33.8
DDMRL [22]	30.8	40.5	44.3	27.2	38.4	**34.5**	28.4	32.2	34.6
SWDA [33]	30.3	42.5	44.6	24.5	36.7	31.6	30.2	35.8	34.8
CDN (ours)	**35.8**	**45.7**	**50.9**	**30.1**	**42.5**	29.8	**30.8**	**36.5**	**36.6**

SIM10K [21] is a street view dataset generated from the realistic computer game GTA-V. It has 10,000 training images and the same categories as in Cityscapes. **PreSIL** [17] is synthetic point cloud dataset derived from GTA-V, which consists of 50,000 frames of high-definition images and point clouds.
BDD100K [44] is a large-scale dataset (contains 100k images) that covers diverse driving scenes. It is a good representative of real data in the wild.

4.2   Implementation Details

We train the Faster R-CNN [32] model for 12 epochs on all experiments. The model is optimized by SGD with multi-step learning rate decay. SGD uses the learning rate of 0.00625 multiplied by the batchsize, and momentum of 0.9. All experiments use sync BN [26] with a batchsize of 32. λ is set as 0.4 by default in all experiments. On synthetic-to-real adaptation, for a fair comparison, we randomly select 7000 images for training and 3000 for testing, for all synthetic datasets and BDD100K dataset. For 3D point cloud detection, we use PointR-CNN [37] model with same setting as [37]. We incorporated the CDN layer in the point-wise feature generation stage (global alignment) and 3D ROIs proposal stage (instance alignment).

5   Experimental Results and Analysis

5.1   Results on Cityscapes to Foggy Cityscapes

We compare CDN with the state-of-the-art methods in Table 1. Following [33, 46], we also report results using Faster R-CNN model with VGG16 backbone. As shown in Table 1, CDN outperforms previous state-of-the-art methods by a large margin of 1.8% mAP. The results demonstrate the effectiveness of CDN on reducing domain gaps. A detailed comparison of different CDN settings can be found at the ablation study Sect. 7. As shown in Fig. 3, our method exhibits good generalization capability under foggy weather conditions.

5.2   Results on KITTI to Cityscapes

Different camera settings may influence the detector performance in real-world applications. We conduct the cross-camera adaptation from KITTI to Cityscapes. Table 2 shows the adaptation results on *car* category produced by Faster R-CNN with VGG16. Global and Instance represent global and local alignment respectively. The results demonstrate that CDN achieves 1.7% mAP improvements over the state-of-the-art methods. We can also find that instance feature alignment contributes to a larger performance boost than global counterpart, which is consistent with previous discovery [33,46].

5.3   Results on SIM10K to Cityscapes

Following the setting of [33], we evaluate the detection performance on *car* on SIM10K-to-Cityscapes benchmark. The results in Table 3 demonstrate CDN constantly performs better than the baseline methods. CDN with both global and instance alignment achieves 49.3% mAP on validation set of Cityscapes, which outperforms the previous state-of-the-art method by 1.6% mAP.

5.4   Results on Synthetic to Real Data

To thoroughly evaluate the performance of the state-of-the-art methods on synthetic-to-real adaptation, we construct a large-scale synthetic-to-real adaptation benchmark on various public synthetic datasets, including Virtual KITTI, Synscapes and SIM10K. "All" represents using the combination of 3 synthetic datasets. Compared with SIM10K-to-Cityscapes, the proposed benchmark is more challenging in terms of much larger image diversity in both real and synthetic domains. We compare CDN with the state-of-the-art method SWDA[33] in Table 4. CDN consistently outperforms SWDA under different backbones, which achieves average 2.2% mAP and 2.1% mAP improvements on Faster-R18 and Faster-R50 respectively. Using the same adaptation method, the detection performance strongly depends on the quality of synthetic data. For instance, the

Table 2. KITTI to Cityscapes.

Method	Global	Instance	mAP (%)
Source only			37.1
DA-Faster [4]	✓	✓	38.3
SWDA [33]	✓	✓	43.2
SCDA [46]	✓	✓	42.9
CDN	✓		40.2
		✓	43.1
	✓	✓	**44.9**

Table 3. SIM10K to Cityscapes.

Method	Global	Instance	mAP (%)
Source only			34.3
DA-Faster [4]	✓	✓	38.3
SWDA [33]	✓	✓	47.7
SCDA [46]	✓	✓	44.1
CDN	✓		41.2
		✓	45.8
	✓	✓	**49.3**

Table 4. Adaptation from different synthetic data to real data. mAP on car is reported on BDD100K validation. The results of supervised training on BDD100K are highlighted in gray.

Model	Method	Virtual KITTI	Synscapes	SIM10K	All
Faster-R18	Source	9.8	24.5	37.7	38.2
	SWDA[33]	15.6	27.0	40.2	41.3
	CDN	**17.5**	**29.1**	**42.7**	**43.6**
	Target	70.5			
Faster-R50	Source	13.9	29.1	41.6	42.8
	SWDA[33]	19.7	31.5	42.9	44.3
	CDN	**21.8**	**33.4**	**45.3**	**47.2**
	Target	75.6			

Fig. 3. Example results on Foggy Cityscapes/Synscapes/SIM10K/BDD100K (from top to bottom). The results are produced by a Faster R-CNN model incorporated with CDN. The class and score predictions are at the top left corner of the bounding box. Zoom in to visualize the details.

adaptation performance of SIM10K is much better than Virtual KITTI. Some example predictions produced by our method are visualized in Fig. 3.

5.5   Adaptation on 3D Point Cloud Detection

We evaluate CDN on adapting 3D object detector from synthetic point cloud (PreSIL) to real point cloud data (KITTI). Table 5 shows that CDN constantly outperforms the state-of-the-art method PointDAN [30] across all categories, with an average improvement of 1.9% AP. We notice that instance alignment contributes to a larger performance boost than global alignment. It can be attributed by the fact that point cloud data spread over a huge 3D space but most information is stored in the local foreground points (see Fig. 4).

Fig. 4. Top:PreSIL; Bottom:KITTI.

Table 5. Adapting from synthetic (PreSIL) to real (KITTI) pint cloud. AP of moderate level on KITTI test is reported.

Model	Global	Instance	Car	Pedestrian	Cyclist
Source			15.7	9.6	5.6
CycleGAN [35]	✓	✓	16.5	10.3	5.9
PointDAN[30]	✓	✓	17.1	10.9	7.5
CDN	✓		17.3	10.5	6.0
		✓	18.5	12.8	8.7
	✓	✓	**19.0**	**13.2**	**9.1**
Target			75.7	41.7	59.6

6   Analysis

6.1   Visualize and Analyze the Feature Maps

Despite the general efficiency on various benchmarks, we are also interested in the underlying principle of CDN. We interpret the learned domain embedding via appending a decoder network after the backbone to reconstruct the RGB images from the feature maps. As shown in Fig. 5, the top row shows the original inputs from Foggy Cityscapes, SIM10K and Synscapes (left to right), and the bottom row shows the reconstructed images from the corresponding features encoded with the domain embedding of another domain. The reconstructed images carry the same domain style of another domain, suggesting the learned domain embedding captures the domain attribute information and CDN can effectively transform the domain style of different domains.

Furthermore, we compute Fréchet Inception Distance (FID) [14] score to quantitatively investigate the difference between source and target features. FID has been a popular metric to evaluate the style similarity between two groups of images in GANs. Lower FID score indicates a smaller style difference. For easy comparison, we normalize the FID score to [0, 1] by dividing the maximum score. As shown in Table 6, the feature learned with CDN achieves significantly smaller FID score compared with feature learned on source domain only, suggesting CDN effectively reduces the domain gap in the feature space. Obviously, supervised joint training on source and target data gets the smallest FID score, which is verified by the best detection performance achieved by joint training. As shown in Fig. 6, synthetic-to-real has larger FID score than real-to-real dataset, since the former owns larger domain gaps.

Fig. 5. Top row: Original inputs from Foggy Cityscapes, SIM10K and Synscapes (left to right); Bottom row: Reconstructed images from features encoded with the learned domain embedding of another domain.

Table 6. FID score and mAP.

Method	SIM to BDD		City to Foggy	
	FID	mAP	FID	mAP
Source	0.94	37.7	0.83	26.1
Joint training	0.67	79.3	0.41	49.5
SWDA [33]	0.83	40.2	0.76	34.8
CDN	0.71	42.7	0.60	36.6

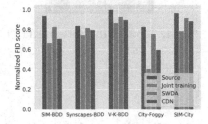

Fig. 6. FID scores on all datasets.

6.2 Analysis on Domain Discrepancy

We adopt symmetric Kullback-Leibler divergence to investigate the discrepancy between source and target domain in feature space. To simplify the analysis, we assume source and target features are drawn from the multivariate normal distribution. The divergence is calculated with the Res5-3 features and plotted in log scale. Figure 7 (a) and (c) show that the domain divergence continues decreasing during training, indicating the Conditional Domain Normalization keeps reducing domain shift in feature space. Benefiting from the reduction of domain divergence, the adaptation performance on the target domain keeps increasing. Comparing with SWDA, CDN achieves lower domain discrepancy and higher adaptation performance.

Figure 7 (b)(d) shows the t-SNE plot of instance features extracted by a Faster R-CNN model incorporated with CDN. The same category features from two domains group in tight clusters, suggesting source and target domain distributions are well aligned in feature space. Besides, features of different categories own clear decision boundaries, indicating discriminative features are learned by our method. These two factors contribute to the target performance boost.

7   Ablation Study

For the ablation study, we use a Faster R-CNN model with ResNet-18 on SIM10K-to-BDD100K adaptation benchmark, and a Faster R-CNN model with VGG16 on Cityscapes-to-Foggy Cityscapes adaptation benchmark. G and I denote adopting CDN in the backbone and bounding box head network, respectively.

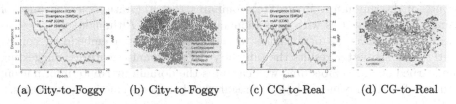

| (a) City-to-Foggy | (b) City-to-Foggy | (c) CG-to-Real | (d) CG-to-Real |

Fig. 7. (a)(c): Divergence and adaptation performance. (c)(d): t-SNE plot of instance features.

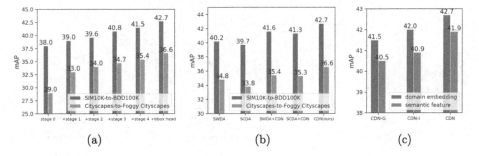

| (a) | (b) | (c) |

Fig. 8. (a) Adopt CDN at different convolution stages of ResNet; (b) Adopt CDN in existing adaptation frameworks; (c) Domain embedding vs. semantic features.

Adopting CDN at Different Convolution Stages. Figure 8(a) compares the results of Faster R-CNN models adopting CDN at different convolution stages. We follow [13] to divide ResNet into 5 stages. Bbox head denotes the bounding box head network. From left to right, adding more CDN layers keeps boosting the adaptation performance on both benchmarks, benefiting from adaptive distribution alignments across different levels' representation. It suggests that adopting CDN in each convolution stage is a better choice than only aligning domain distributions at one or two specific convolution stages.

Comparing with Existing Domain Adaptation Frameworks Adopting CDN. Figure 8(b) shows the results of adopting CDN layer in existing adaptation methods like SWDA [33] and SCDA [46]. Directly adopting CDN in SWDA and SCDA can bring average 1.3% mAP improvements on two adaptation benchmarks, suggesting CDN is more effective to address domain shifts than traditional domain confusion learning. It can be attributed to that CDN disentangle the domain-specific factors out of the semantic features via learning a domain-vector. Leveraging the domain-vector to align the different domain distributions can be more efficient.

Compare Domain Embedding with Semantic Features. In Eq. 7, we can either use semantic features $(v^s, \hat{v}^t)$ or domain embedding $(\mathbf{F}_d(v^s), \mathbf{F}_d(\hat{v}^t))$ as inputs of discriminator. Figure 8(c) compares the adaptation performance of using semantic features with using domain embedding. Although semantic features can improve the performance over baseline, domain embedding consistently achieves better results than directly using semantic features. Suggesting the learned domain embedding well captures the domain attribute information, and it is free from some undesirable regularization on specific image contents.

Value of. λ In Eq. 10, we use λ controls the balance between global and local regularization. Figure 9 (left) shows the influence on adaptation performance by different λ. Because object detectors naturally focus more on local regions, we can see stronger instance regularization largely contributes to detection performance. In our experiments, λ between 0.4 and 0.5 gives the best performance.

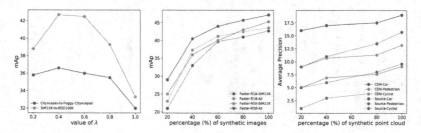

Fig. 9. Left: mAP vs. Value of λ; Middle: mAP vs. Percentage (%) of synthetic image data; Right: AP vs. Percentage (%) of synthetic point cloud.

Scale of Target Domain Dataset. Figure 9 middle/right quantitatively investigate the relation between real data detection performance and percentage of synthetic data used for training. "All" means to use the combination of 3 different synthetic datasets. The larger synthetic dataset provides better adaptation performance, on both 2D image and 3D point cloud detection.

8   Conclusion

We present the Conditional Domain Normalization (CDN) to adapt object detectors across different domains. CDN aims to embed different domain inputs into a shared latent space, where the features from different domains carry the same domain attribute. Extensive experiments demonstrate the effectiveness of CDN on adapting object detectors, including 2D image and 3D point cloud detection tasks. And both quantitative and qualitative comparisons are conducted to analyze the features learned by our method.

References

1. Ben-David, S., Blitzer, J., Crammer, K., Kulesza, A., Pereira, F., Vaughan, J.W.: A theory of learning from different domains. Mach. Learn. **79**, 151–175 (2010). https://doi.org/10.1007/s10994-009-5152-4
2. Bousmalis, K., Silberman, N., Dohan, D., Erhan, D., Krishnan, D.: Unsupervised pixel-level domain adaptation with generative adversarial networks. In: CVPR (2017)
3. Chen, X., et al.: 3D object proposals for accurate object class detection. In: Advances in Neural Information Processing Systems (2015)
4. Chen, Y., Li, W., Sakaridis, C., Dai, D., Van Gool, L.: Domain adaptive faster R-CNN for object detection in the wild. In: CVPR (2018)
5. Cordts, M., et al.: The cityscapes dataset for semantic urban scene understanding. In: CVPR (2016)
6. Dumoulin, V., Shlens, J., Kudlur, M.: A learned representation for artistic style. arXiv preprint arXiv:1610.07629 (2016)
7. Gaidon, A., Wang, Q., Cabon, Y., Vig, E.: Virtual worlds as proxy for multi-object tracking analysis. In: CVPR (2016)
8. Ganin, Y., Lempitsky, V.: Unsupervised domain adaptation by backpropagation. In: ICML (2015)
9. Geiger, A., Lenz, P., Urtasun, R.: Are we ready for autonomous driving? The KITTI vision benchmark suite. In: CVPR (2012)
10. Girshick, R.: Fast R-CNN. In: ICCV (2015)
11. Girshick, R., Donahue, J., Darrell, T., Malik, J.: Rich feature hierarchies for accurate object detection and semantic segmentation. In: CVPR (2014)
12. He, K., Gkioxari, G., Dollár, P., Girshick, R.: Mask R-CNN. In: ICCV (2017)
13. He, K., Zhang, X., Ren, S., Sun, J.: Deep residual learning for image recognition. In: CVPR (2016)
14. Heusel, M., Ramsauer, H., Unterthiner, T., Nessler, B., Hochreiter, S.: GANs trained by a two time-scale update rule converge to a local Nash equilibrium. In: Advances in Neural Information Processing Systems (2017)
15. Hoffman, J., et al.: CyCADA: cycle-consistent adversarial domain adaptation. arXiv preprint arXiv:1711.03213 (2017)
16. Huang, X., Belongie, S.: Arbitrary style transfer in real-time with adaptive instance normalization. In: ICCV (2017)
17. Hurl, B., Czarnecki, K., Waslander, S.: Precise synthetic image and LiDAR (PreSIL) dataset for autonomous vehicle perception. In: 2019 IEEE Intelligent Vehicles Symposium (IV) (2019)
18. Inoue, N., Furuta, R., Yamasaki, T., Aizawa, K.: Cross-domain weakly-supervised object detection through progressive domain adaptation. In: CVPR (2018)
19. Ioffe, S., Szegedy, C.: Batch normalization: accelerating deep network training by reducing internal covariate shift. arXiv preprint arXiv:1502.03167 (2015)
20. James, S., et al.: Sim-to-real via sim-to-sim: data-efficient robotic grasping via randomized-to-canonical adaptation networks. In: CVPR (2019)

21. Johnson-Roberson, M., Barto, C., Mehta, R., Sridhar, S.N., Rosaen, K., Vasudevan, R.: Driving in the matrix: can virtual worlds replace human-generated annotations for real world tasks? arXiv preprint arXiv:1610.01983 (2016)

22. Kim, T., Jeong, M., Kim, S., Choi, S., Kim, C.: Diversify and match: a domain adaptive representation learning paradigm for object detection. In: CVPR (2019)

23. Liu, Z., et al.: Open compound domain adaptation. In: CVPR (2020)

24. Long, M., Cao, Z., Wang, J., Jordan, M.I.: Conditional adversarial domain adaptation. In: Advances in Neural Information Processing Systems (2018)

25. Park, T., Liu, M.Y., Wang, T.C., Zhu, J.Y.: Semantic image synthesis with spatially-adaptive normalization. arXiv preprint arXiv:1903.07291 (2019)

26. Peng, C., et al.: MegDet: a large mini-batch object detector. In: CVPR (2018)

27. Peng, X., Bai, Q., Xia, X., Huang, Z., Saenko, K., Wang, B.: Moment matching for multi-source domain adaptation. In: ICCV (2019)

28. Peng, X.B., Andrychowicz, M., Zaremba, W., Abbeel, P.: Sim-to-real transfer of robotic control with dynamics randomization. In: ICRA. IEEE (2018)

29. Qi, C.R., Yi, L., Su, H., Guibas, L.J.: PointNet++: deep hierarchical feature learning on point sets in a metric space. In: Advances in Neural Information Processing Systems (2017)

30. Qin, C., You, H., Wang, L., Kuo, C.C.J., Fu, Y.: PointDAN: a multi-scale 3D domain adaption network for point cloud representation. In: Advances in Neural Information Processing Systems (2019)

31. Quionero-Candela, J., Sugiyama, M., Schwaighofer, A., Lawrence, N.D.: Dataset Shift in Machine Learning. The MIT Press, Cambridge (2009)

32. Ren, S., He, K., Girshick, R., Sun, J.: Faster R-CNN: towards real-time object detection with region proposal networks. In: Advances in Neural Information Processing Systems (2015)

33. Saito, K., Ushiku, Y., Harada, T., Saenko, K.: Strong-weak distribution alignment for adaptive object detection. In: CVPR (2019)

34. Sakaridis, C., Dai, D., Van Gool, L.: Semantic foggy scene understanding with synthetic data. Int. J. Comput. Vis. **126**, 973–992 (2018). https://doi.org/10.1007/s11263-018-1072-8

35. Saleh, K., et al.: Domain adaptation for vehicle detection from bird's eye view lidar point cloud data. In: ICCV Workshops (2019)

36. Sankaranarayanan, S., Balaji, Y., Jain, A., Nam Lim, S., Chellappa, R.: Learning from synthetic data: addressing domain shift for semantic segmentation. In: CVPR (2018)

37. Shi, S., Wang, X., Li, H.: PointRCNN: 3D object proposal generation and detection from point cloud. In: CVPR (2019)

38. Tobin, J., Fong, R., Ray, A., Schneider, J., Zaremba, W., Abbeel, P.: Domain randomization for transferring deep neural networks from simulation to the real world. In: IROS. IEEE (2017)

39. Tsai, Y.H., Sohn, K., Schulter, S., Chandraker, M.: Domain adaptation for structured output via discriminative representations. arXiv preprint arXiv:1901.05427 (2019)

40. Tzeng, E., Hoffman, J., Saenko, K., Darrell, T.: Adversarial discriminative domain adaptation. In: CVPR (2017)

41. Wang, X., Yu, K., Dong, C., Change Loy, C.: Recovering realistic texture in image super-resolution by deep spatial feature transform. In: CVPR (2018)

42. Wrenninge, M., Unger, J.: Synscapes: a photorealistic synthetic dataset for street scene parsing. arXiv preprint arXiv:1810.08705 (2018)

43. Yosinski, J., Clune, J., Nguyen, A., Fuchs, T., Lipson, H.: Understanding neural networks through deep visualization. arXiv preprint arXiv:1506.06579 (2015)
44. Yu, F., et al.: BDD100K: a diverse driving video database with scalable annotation tooling. arXiv preprint arXiv:1805.04687 (2018)
45. Yue, X., Wu, B., Seshia, S.A., Keutzer, K., Sangiovanni-Vincentelli, A.L.: A LiDAR point cloud generator: from a virtual world to autonomous driving. In: Proceedings of the 2018 ACM on International Conference on Multimedia Retrieval (2018)
46. Zhu, X., Pang, J., Yang, C., Shi, J., Lin, D.: Adapting object detectors via selective cross-domain alignment. In: CVPR (2019)
47. Zou, Y., Yu, Z., Vijaya Kumar, B.V.K., Wang, J.: Unsupervised domain adaptation for semantic segmentation via class-balanced self-training. In: Ferrari, V., Hebert, M., Sminchisescu, C., Weiss, Y. (eds.) ECCV 2018. LNCS, vol. 11207, pp. 297–313. Springer, Cham (2018). https://doi.org/10.1007/978-3-030-01219-9_18

HARD-Net: Hardness-AwaRe Discrimination Network for 3D Early Activity Prediction

Tianjiao Li[1,2], Jun Liu[2(✉)], Wei Zhang[1(✉)], and Lingyu Duan[3]

[1] School of CSE, Shandong University, Jinan, China
tianjiao.lee@mail.sdu.edu.cn, davidzhang@sdu.edu.cn
[2] ISTD Pillar, Singapore University of Technology and Design, Singapore, Singapore
jun_liu@sutd.edu.sg
[3] School of EE and CS, Peking University, Beijing, China
lingyu@pku.edu.cn

Abstract. Predicting the class label from the partially observed activity sequence is a very *hard* task, as the observed early segments of different activities can be very similar. In this paper, we propose a novel Hardness-AwaRe Discrimination Network (HARD-Net) to specifically investigate the relationships between the similar activity pairs that are hard to be discriminated. Specifically, a Hard Instance-Interference Class (HI-IC) bank is designed, which dynamically records the hard similar pairs. Based on the HI-IC bank, a novel adversarial learning scheme is proposed to train our HARD-Net, which thus grants our network with the strong capability in mining subtle discrimination information for 3D early activity prediction. We evaluate our proposed HARD-Net on two public activity datasets and achieve state-of-the-art performance.

Keywords: Early activity prediction · Action/gesture understanding · 3D skeleton data · Hardness-aware learning

1 Introduction

Early human activity prediction (predicting the class label of an *action* or *gesture* before it is completely performed) is an important and hot research problem in the human behavior analysis domain, thanks to its relevance to many real-world applications, such as online human-robot interactions, self-driving vehicles, and security surveillance [10,40,42]. Existing works [32,39] show that the 3D skeleton data [2–5,25,29,35,45], which can be conveniently acquired with low-cost depth cameras, is a concise yet informative and powerful representation for human behavior analysis. Therefore, in this paper, we focus on the task of early human activity prediction from the 3D skeleton data, namely, 3D early activity prediction.

Unlike 3D activity recognition where the full-length skeleton sequences can be used, which often contain sufficient discrimination information, in 3D early

© Springer Nature Switzerland AG 2020
A. Vedaldi et al. (Eds.): ECCV 2020, LNCS 12356, pp. 420–436, 2020.
https://doi.org/10.1007/978-3-030-58621-8_25

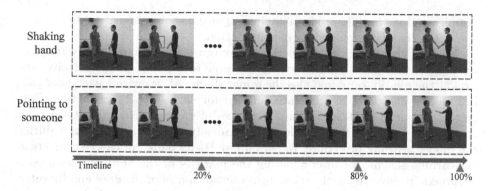

Fig. 1. Illustration of two example activities from NTU RGB+D dataset [32]. Though sufficient discrimination information can be used to distinguish these two activities when their full sequences are observed, at the early stages (e.g., when only 20% is observed), these two activities are quite similar (with only subtle discrimination information contained, as labelled by the red boxes). This makes early prediction hard. (Color figure online)

activity prediction, only the beginning segments of the sequences are observed. This makes early activity prediction much more challenging than recognition. More specifically, when performing early activity prediction, the observed beginning segments of many activities can be very similar, i.e., there may be only subtle discrepancies among them for discrimination. Thus due to the lack of significant discrimination information, these partially observed segments can be easily "mis-predicted" into other categories. For example, in Fig. 1, at the early stage (20% observation ratio), the "pointing to someone" sample can be easily mis-predicted into the "shaking hand" class, since there are only very subtle differences between them. Here we call the easily mis-predicted segments as the *hard instances*, and the classes that they are easily mis-predicted into as their *interference classes*. We also call the pair containing a *hard instance* and its corresponding *interference class* as a *hard pair*.

To deal with the challenging task of 3D early activity prediction, some existing works [13,40] focus on inferring or distilling information from the full activity sequences that contain more sufficient discrimination information, to assist activity prediction from the partial sequences. Though remarkable progress has been achieved by the previous methods [10,11,42], most of them do not explicitly consider the *hard pair* discrimination issue, i.e., specifically investigating the relationships within each *hard pair* in order to exploit their minor discrepancies for better early activity prediction.

As mentioned above, the high similarities of the partially-observed activity sequences between the *hard instance* and its corresponding *interference class* make 3D early activity prediction challenging. Thus to achieve reliable prediction performance, a desired prediction model should be discriminative and powerful enough in comprehending the relationships within the confusing *hard pair* samples and meanwhile prudentially investigating their inherent subtle discrepancies that can be exploited for discrimination.

Inspired by this, in this paper, we propose a novel Hardness-AwaRe Discrimination Network (HARD-Net), that is able to explicitly mine, perceive, and exploit the relationships and also the minor discrepancies within each *hard pair*, in order to achieve a discriminative model for early activity prediction. Concretely, in our HARD-Net, a Hard Instance-Interference Class (HI-IC) bank is specifically designed, that is able to dynamically record the *hard pairs* during the model learning procedure. Based on our HI-IC bank, an effective adversarial learning scheme for discriminating the features of the *hard pair* samples is proposed. To investigate the relationship between a *hard instance* and its *interference class*, a feature generator is designed, which produces confusing yet plausible *hard instance* features by conditioning on the similarities of this instance to the corresponding *interference class*. Meanwhile, to obtain the ability of mining subtle discrimination information within the features of *hard pair* samples, a class discriminator is further designed that pushes the prediction model to distinguish the confusing features of the *hard instance* from its *interference classes*. Therefore, with the adversarial learning going on, the generated features of the *hard instance* become more confusing with regard to its *interference class*, which in turn promote the capability of the class discriminator in mining the subtle differences that exist within the features of the *hard pair* samples for class discrimination. As a result, the proposed HARD-Net with the class discriminator as the classifier becomes very powerful in handling *hard pairs* that are often very hard to be discriminated by the early activity prediction models.

2   Related Work

3D Human Activity Recognition. Some of the existing methods [1,12,24, 26,32,46] used RNN/LSTM-based methods for 3D human activity recognition. Besides the RNN/LSTM models, 2D convolutional neural networks (CNNs) have also been investigated in this domain [14]. More recently, graph convolutional networks (GCNs) become prevalent for handling 3D activity recognition [21, 33,34,37,38,44]. Yan *et al.* [44] proposed to use spatial-temporal GCN for 3D activity recognition. Shi *et al.* [34] proposed an adaptive graph convolutional network to adaptively learn the topology of the graph for various layers, and employed the second-order information of the raw skeleton data as an extra input stream to boost the performance.

Early Human Activity Prediction. Unlike the activity recognition that is able to observe the full-length activity sequences which often contain sufficient discrimination information, in early activity prediction, only the segments from the beginning parts of the activity sequences can be used. Due to the drop of discrimination information, early activity prediction becomes much more challenging than activity recognition. Different approaches [7,10,13,16–19,23,40,42,43] have been proposed for early activity prediction. Ke *et al.* [13] proposed to learn latent global information from full-length sequences and local information from partial-length sequences. Wang *et al.* [40] introduced a teacher-student learning architecture to transfer knowledge from the long-term sequences to the shorter-term sequences.

Overall, the aforementioned works on 3D early activity prediction do not focus on improving the discrimination ability of the prediction model by specifically handling the very similar *hard pair* samples, though the discrimination ability for the *hard pairs* is often one of the bottlenecks in early activity prediction. Different from these works, we construct an HI-IC bank to explicitly and dynamically record the *hard pair* samples, and propose a novel HARD-Net with adversarial learning to push the prediction model to be able to specifically discriminate *hard pair* samples by exploiting their relationships and mining their subtle discrimination information.

Hard Example Learning. Explicitly learning from hard-to-predict examples has been shown to be very helpful for a wide range of computer vision and machine learning tasks [6,9,22,27,28,36,41]. For example, Shrivastava *et al.* [36] proposed a hard example mining scheme to automatically select hard data to improve the object classification performance. Felzenszwalb *et al.* [6] proposed a margin-sensitive method for handling hard negative examples with a latent SVM to iteratively fix the latent values for positive examples and optimize the latent SVM objective function.

Unlike these methods on hard example learning, we focus on improving the ability of mining subtle discrimination information within the *hard pairs*, by explicitly pairing the easily mis-predicted early activity segments with their corresponding *interference classes* via an adversarial learning scheme. An HI-IC bank is also introduced to specifically store the *hard instances* and their *interference classes*, in order to facilitate the comprehending of the relationships and minor differences within the pairs. This thus boosts the discrimination capability of the early activity prediction model.

3 Method

3.1 Problem Formulation

Given a full-length activity sequence $S = \{s_t\}_{t=1}^{T}$, where s_t denotes the t_{th} frame, and T represents the sequence length, following existing works [10,13], the full-length sequence S is first divided into N segments, i.e., each segment contains $\frac{T}{N}$ frames. Thus a partial sequence can be denoted as $P = \{s_t\}_{t=1}^{\tau}$, where $\tau = i \cdot \frac{T}{N}$ and $i \leq N$. The task of early activity prediction is to identify the activity category $c \in \mathbb{C} = \{1, 2, ..., C\}$ that the partial activity sequence P belongs to, based on various observation ratios.

3.2 Hardness-AwaRe Discrimination Network

3.2.1 Overview

The overall architecture of our end-to-end Hardness-AwaRe Discrimination Network (HARD-Net) is shown in Fig. 2. As mentioned above, certain activities can be quite similar at their early stages. Thus 3D early activity prediction often suffers from lack of sufficient discrimination information when at low observation ratios. Here we introduce a new method that is able to explicitly record the hard pairs, that lack sufficient discrimination information, using an HI-IC bank, and investigate the relationship between the *hard instances* and their corresponding *interference classes* via an adversarial learning scheme, which thus enhances the capability of our prediction model in mining the minor discrimination information within the *hard samples* in feature space, for better activity prediction.

3.2.2   Hard Instance-Interference Class (HI-IC) Bank

We design an HI-IC bank to record *hard pairs*, where each pair contains a *hard instance* as well as its corresponding *interference class*, as shown in the top part of Fig. 2. This thus enables our model to get aware of the specific categories that a hard partial activity sequence can be easily mis-predicted into.

As shown in Fig. 2, the base structure of our network includes a feature encoder $\mathbb{E}$ that learns features for the partial activity sequence, and a classifier (denoted as class discriminator in Fig. 2) for class prediction. At each training iteration, each partial activity sequence (P) is fed to the encoder to obtain the original features f^{ori}, which are then further fed to the class discriminator to produce the prediction scores $\hat{y}$. If the partial sequence instance (P) is wrongly predicted by the class discriminator, given the prediction scores $\hat{y}$, the activity class c_{r_1} that has the rank-1 score in $\hat{y}$ is considered as the *interference class* (c^I) of P (i.e., $c^I = c_{r_1}$), as c_{r_1} has the most ambiguous information regarding to P.

We regard the wrongly predicted partial sequence (P) that does not have sufficient discrimination information, as the *hard instance* (P^H). We can then pack the pair of P^H and c^I as a *hard pair* that is then stored into the HI-IC bank, as illustrated in Fig. 2.

3.2.3   Adversarial Hardness-AwaRe Discrimination Learning Scheme

To investigate the relationship within the *hard pair*, we design a feature generator $(\mathbb{G})$ conditioning on the *hard instance* and its corresponding *interference class* from the pair, in order to derive latent features that are confusing yet plausible for representing the original *hard instance*. Meanwhile, to enhance the capability of our network in mining subtle discrimination information, we design a class discriminator $(\mathbb{D}^{cls})$ by granting the prediction model with the power of distinguishing the generated latent features of the *hard instance* from its *interference class*. With such an adversarial learning scheme, the generated latent features of the *hard instance* become more and more confusing with regard to its *interference class*, which in turn boosts the power of the class discriminator in mining the subtle discrimination information to distinguish the confusing *hard instance* from its *interference class*. As a result, the overall discrimination capability of the prediction model is strengthened during the adversarial learning procedure.

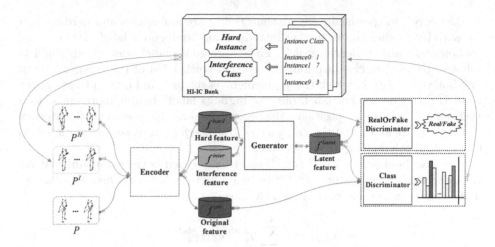

Fig. 2. Illustration of our end-to-end HARD-Net. Our network is constructed on a replaceable feature encoder (e.g., CNN skeleton encoder [13] or GCN skeleton encoder [34]) that encodes features for partial sequences. Following the red arrows representing both backbone training phase and inference phase, partial sequences (P) are fed to the encoder followed by the classifier to obtain classification scores that serve as the criterion for storing *hard pairs* into the HI-IC bank. Then following the green arrows representing adversarial learning phase, we randomly select a *hard pair* including a *hard instance* (P^H) and an *interference class* sample (P^I) from the HI-IC bank for feature encoding. The obtained feature pairs, f^{hard} and f^{inter}, are further taken into account for adversarial learning, in order to improve the capability of our prediction model in mining subtle discrepancies within each *hard pair* in the feature space. (Color figure online)

Feature Generator. We design a generator ($\mathbb{G}$) that exploits the relationship between the *hard instance* and the corresponding *interference class*, in order to produce latent features that are confusing and hard to predict, yet are still plausible and retain inherent information representing the original *hard instance*.

Concretely, to exploit the relationship between the *hard instances* and the corresponding *interference classes*, for a *hard instance* (P^H), we refer to the HI-IC bank and identify its *interference class*. We then randomly sample an interference instance (P^I) from the *interference class*. Thus we get the paired hard samples (P^H and P^I). These two samples are fed to the feature encoder to obtain the feature pair (f^{hard} and f^{inter}) representing these two samples.

In our design, we aim to generate latent features (f^{latent}) of the *hard instance* that are very confusing and ambiguous w.r.t its *interference class*. Thus beside feeding the original features (f^{hard}) of the *hard instance* to the generator ($\mathbb{G}$), the *interference class* features (f^{inter}) are used as the interference information and are also fed to $\mathbb{G}$, as shown in Fig. 2. Therefore, conditioning on the aggregated features of P^H and P^I, the produced latent features (f^{latent}) from $\mathbb{G}$ become very confusing and hard for discriminating between these two classes.

Moreover, to specifically ensure that f^{latent} are ambiguous and hard enough for activity prediction, we further introduce an "ambiguous label" for the *hard instance* to assist the learning of $\mathbb{G}$. This "ambiguous label" can be explained as follows. Usually, in classification, the ground-truth label of the category j is a one-hot vector (y), in which the j_{th} element is set to 1 and other places are set to 0. Unlike this one-hot label, our "ambiguous label" here is represented as a vector y^{amb}, where two positions, that correspond to the ground-truth category of the *hard instance* and the category of its *interference class*, are both set to 0.5, and all other elements are set to 0.

Such an "ambiguous label" (y^{amb}) can then be used as the constraint to drive the generated latent features f^{latent} to be ambiguous between these two classes. This constraint can be formulated as follows:

$$\mathcal{L}_{amb}^{\mathbb{G}} = -\sum_{k=1}^{K} y_k^{amb} \cdot \log \hat{y}_k^{latent} \tag{1}$$

where K denotes the total number of the activity classes, and $\hat{y}^{latent}$ is the output vector of the class discriminator that performs classification based on the generated latent features (f^{latent}).

The constraint in Eq. (1) ensures that the generated latent features (f^{latent}) are ambiguous enough. However, as mentioned before, f^{latent} still needs to be plausible and retain inherent information for representing the original *hard instance*. To achieve this, we apply a real-or-fake constraint on f^{latent} to make it plausible, as well as a mean absolute error constraint to drive f^{latent} to be closer to the features (f^{hard}) of *hard instance*.

The mean absolute error constraint $(\mathcal{L}_{con}^{\mathbb{G}})$, for narrowing the distance between f^{latent} and f^{hard}, is formulated in Eq. (2). The real-or-fake constraint $(\mathcal{L}_{rof}^{\mathbb{G}})$, brought by the RealOrFake Discriminator $(\mathbb{D}^{rof})$ for ensuring that the generated features (f^{latent}) and the original features (f^{hard}) still stay in the same feature domain, is formulated in Eq. (3).

$$\mathcal{L}_{con}^{\mathbb{G}} = ||f^{latent} - f^{hard}||_1 \tag{2}$$

$$\mathcal{L}_{rof}^{\mathbb{G}} = E[\log \mathbb{D}^{rof}(f^{hard})] + E[\log[1 - \mathbb{D}^{rof}(f^{latent})]] \tag{3}$$

The overall objective function for the generator $(\mathbb{G})$ can thus be formulated as:

$$\mathcal{L}^{\mathbb{G}} = \mathcal{L}_{con}^{\mathbb{G}} + \lambda_1 \mathcal{L}_{rof}^{\mathbb{G}} + \lambda_2 \mathcal{L}_{amb}^{\mathbb{G}} \tag{4}$$

With the above objective function, $\mathbb{G}$ is thus able to generate latent features (f^{latent}) that are very confusing with regard to the *interference class*, yet still retain inherent information for representing the input *hard instance*.

Class Discriminator. To obtain strong discrimination power, we design a class discriminator $(\mathbb{D}^{cls})$ that is able to distinguish the generated latent features (f^{latent}) of each *hard instance* from its corresponding *interference class*. To learn our class discriminator, a classification constraint ($\mathcal{L}^{\mathbb{D}^{cls}}$) is applied on $\mathbb{D}^{cls}$ that pushes it to predict the accurate label (y) of the original *hard instance* based on the confusing latent features (f^{latent}):

$$\mathcal{L}^{\mathbb{D}^{cls}} = -\sum_{k=1}^{K} y_k \cdot \log \hat{y}_k^{latent} \tag{5}$$

Therefore, with the adversarial learning going on, the generated latent features (f^{latent}) for representing the *hard instance* become more and more confusing with regard to its *interference class* (i.e., contain less and less discrimination information for $\mathbb{D}^{cls}$ to do class distinguishing). This, however, in turn boosts the power of $\mathbb{D}^{cls}$ in comprehending the remaining subtle discriminative information in f^{latent} for distinguishing it from its *interference class*, i.e., $\mathbb{D}^{cls}$ thus becomes more and more powerful in mining the very minor discrimination information for class distinguishing.

Note that during adversarial learning, beside feeding in the generated latent features (f^{latent}) to train $\mathbb{D}^{cls}$, the original features (f^{ori}) encoded from the original samples are also fed to $\mathbb{D}^{cls}$ during training, as shown in Fig. 2. Therefore the below objective function is also applied when learning $\mathbb{D}^{cls}$:

$$\mathcal{L}_{ori}^{\mathbb{D}^{cls}} = -\sum_{k=1}^{K} y_k \cdot \log \hat{y}_k^{ori} \tag{6}$$

As mentioned before, in our adversarial learning scheme, the original features and the generated features are kept in the same domain. Thus such a training scheme (combining Eq. (5) and (6)) is able to stabilize the overall network learning, which further yields a powerful $\mathbb{D}^{cls}$ for mining subtle discrimination information contained in both the latent features and the original features for class distinguishing. Therefore the obtained class discriminator $\mathbb{D}^{cls}$, that has very strong power in mining subtle discrimination information for distinguishing the *hard instances* from the *interference classes*, can act as the final classifier for activity prediction.

3.2.4 Training and Testing

Each training iteration of our HARD-Net is comprised of two phases, namely backbone training with HI-IC bank populating, and adversarial learning.

Backbone Training and HI-IC Bank Filling. The backbone of our network mainly consists of an encoder $\mathbb{E}$ and a class discriminator $\mathbb{D}^{cls}$, as shown in Fig. 2. This backbone can be trained based on Eq. (6). To fill the HI-IC bank, a mini-batch of original partial sequences P with batch size B is first fed to the encoder $\mathbb{E}$ to extract features f^{ori}. Then based on f^{ori}, the class discriminator produces predicted scores, which serve as a criterion for storing *hard pairs* into our HI-IC bank, i.e., if a sample is mis-predicted, this sample and its mis-predicted class are packed as a *hard pair*, which will be stored into the HI-IC bank.

Algorithm 1: Learning procedure of our HARD-Net.

Input: Partial skeleton sequences (P) and ground-truth labels (c^τ)

while *not converge* **do**

    Backbone learning and HI-IC Bank Filling

        Calculate f^{ori} by $\mathbb{E}$;

        Calculate $\hat{y}^{ori}$ by $\mathbb{D}^{cls}$;

        Calculate $\mathcal{L}_{ori}^{\mathbb{D}^{cls}}$ with Eq. (6);

        Update $\mathbb{E}$ and $\mathbb{D}^{cls}$;

        if *rank-1($\hat{y}^{ori}$)* $!= c^\tau$ **then**

            $P^H \leftarrow P$;

            $c^I \leftarrow$ *rank-1($\hat{y}$)*;

            HI-IC Bank $\leftarrow \{P^H; c^I\}$;

        end

    end

    Adversarial HARD-Net Learning

        Freeze $\mathbb{E}$;

        Select and sample P^H and P^I from HI-IC Bank;

        Calculate f^{hard} and f^{inter} by $\mathbb{E}$;

        Calculate f^{latent} by $\mathbb{G}$;

        Calculate $\mathcal{L}^{\mathbb{D}^{cls}}$ and $\mathcal{L}^{\mathbb{D}^{rof}}$;

        Freeze $\mathbb{G}$; Update $\mathbb{D}^{rof}$ and $\mathbb{D}^{cls}$;

        Calculate $\mathcal{L}^{\mathbb{G}}$;

        Freeze $\mathbb{D}^{rof}$ and $\mathbb{D}^{cls}$; Update $\mathbb{G}$;

    end

end

Adversarial Learning Scheme. During adversarial learning, the parameters of the encoder $\mathbb{E}$ are first frozen. We then sample rB *hard pairs* from the HI-IC bank, where $0 < r \leq 1$. If there are not enough pairs in the bank (i.e., at the early stage of training process), all pairs in the bank are selected and repeated to reach rB. Otherwise, we follow the first-in-first-out strategy to select the rB *hard pairs*. Based on the *interference class* from each sampled *hard pair*, we sample an instance of it from the dataset as P^I. After that, P^H and P^I are fed into the encoder $\mathbb{E}$ for feature encoding, and then the encoded features f^{hard} and f^{inter} are fed into the generator $\mathbb{G}$ to attain f^{latent}. The generated latent features f^{latent} are fed into the RealOrFake discriminator $\mathbb{D}^{rof}$ and the class discriminator $\mathbb{D}^{cls}$ with ground-truth category to update $\mathbb{D}^{rof}$ and $\mathbb{D}^{cls}$. Finally, the f^{latent} are used to update $\mathbb{G}$ with the ambiguous label. This training procedure is detailed in Algorithm 1.

Testing. As shown by red arrows in Fig. 2, at the inference phase, we input a partial skeleton sequence to the encoder to obtain features, which are then fed to $\mathbb{D}^{cls}$ for activity prediction. As $\mathbb{D}^{cls}$ has strong capabilities in mining subtle discrimination information for distinguishing hard samples from their similar interference classes, our network becomes powerful in early activity prediction.

4   Experiments

We test the proposed method for 3D early action prediction on the NTU RGB+D dataset [32], and 3D early gesture prediction on the First Person Hand Action (FPHA) dataset [8]. We conduct extensive experiments on these two datasets as below.

NTU RGB+D dataset [32] is a large dataset that has been widely used for 3D action recognition and 3D early action prediction. It contains more than 56 thousands videos and over 4 million frames from 60 activity categories. Each human skeleton in the dataset possesses 25 human body joints represented by 3D coordinates. This dataset is very challenging for 3D early action prediction, as it contains a large number of samples that are confusing at the beginning of the activity sequences. There are two standard evaluation protocols provided by the dataset. The first protocol is the Cross Subject (CS) protocol, where 20 subjects are employed for training and the remaining 20 subjects are left for testing. The second protocol is the Cross View (CV) protocol, where two viewpoints are employed for training, and the third one is for testing.

First Person Hand Action (FPHA) dataset [8] is a challenging 3D hand gesture dataset. The samples in this dataset are the first-person hand activities interacting with 3D objects recorded by six subjects. It contains over 100K frames of 45 different hand activity categories. Each hand skeleton attains 21 hand joints interpreted by 3D coordinates. We test our method on FPHA by following the standard evaluation protocol as [8], where 600 activity sequences are employed for training and the remaining 575 activity sequences are for testing.

Evaluated Models. To test the efficacy of our method, we test two different models, namely "w/o HARD-Net" and "w/ HARD-Net". (1) "w/o HARD-Net": This is actually the backbone model of our network, that contains the feature encoder and the classifier. (2) "w/ HARD-Net": This is our proposed activity prediction model (HARD-Net) that has strong capabilities in discriminating *hard pair* samples via Hardness-AwaRe Discrimination adversary leaning.

4.1   Implementation Details

To comprehensively evaluate the efficacy of our HARD-Net, we specifically construct our method above two state-of-the-art baseline encoders, namely the CNN encoder [20] and the GCN encoder [34], as shown in Table 4. The details of these two baseline encoders can be found in the corresponding papers [20,34]. We also design our generator and real-or-fake discriminator by following Radford *et al.* [31], and implement the class discriminator based on multi-layer perceptron. The weights λ_1 and λ_2 in Eq. (4) are both set to 1.

All experiments are performed based on the Pytorch framework. Adam [15] algorithm is used to train our end-to-end network. The batch size B, learning rate, betas and weight decay are set to 128, 2×10^{-4}, (0.9, 0.999), and 1×10^{-5}, respectively. We set the size of HI-IC bank to be 5000 for the very large NTU RGB+D dataset and 100 for the much smaller FPHA dataset. In each training iteration, the proportion r between the original instances and the *hard pair* instances used for network learning is set to 4 : 1.

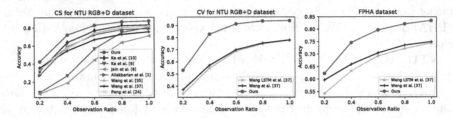

Fig. 3. Comparison of 3D early activity prediction performance on NTU RGB+D and FPHA datasets. Our method outperforms state-of-the-arts by a large margin.

Table 1. Performance comparison (%) on NTU RGB+D (cross-subject protocol). Our method outperforms the backbone model ("w/o HARD-Net") significantly. It also outperforms the state-of-the-art 3D early activity prediction methods by a large margin.

Methods	Observation ratios					
	20%	40%	60%	80%	100%	AUC
Ke *et al.* [14]	8.34	26.97	56.78	75.13	80.43	45.63
Jain *et al.* [12]	7.07	18.98	44.55	63.84	71.09	37.38
Aliakbarian *et al.* [1]	27.41	59.26	72.43	78.10	79.09	59.98
Wang *et al.* [40]	35.85	58.45	73.86	80.06	82.01	60.97
Pang *et al.* [30]	33.30	56.94	74.50	80.51	81.54	61.07
Weng *et al.* [42]	35.56	54.63	67.08	72.91	75.53	57.51
Ke *et al.* [13]	32.12	63.82	77.02	82.45	83.19	64.22
w/o HARD-Net	37.82	67.87	79.22	83.39	84.52	66.91
w/ HARD-Net	**42.39**	**72.24**	**82.99**	**86.75**	**87.54**	**70.56**

4.2 Experiments on 3D Early Action Prediction

We compare the proposed HARD-Net with the state-of-the-art approaches on NTU RGB+D. The comparison results with different observation ratios are shown in Table 1 (cross subject protocol), and Table 2 (cross view protocol).

Results on Cross Subject Protocol. Comparison results on cross subject protocol are shown in Table 1 and Fig. 3 (left). As shown in Table 1, our proposed

Table 2. Performance comparison (%) on NTU RGB+D (cross-view protocol). We observe only [42] has reported early action prediction results on the cross-view protocol.

Methods	Observation ratios					
	20%	40%	60%	80%	100%	AUC
LSTM [42]	33.86	54.70	68.85	74.86	77.84	57.93
Weng *et al.* [42]	37.22	57.18	69.92	75.41	77.99	59.71
w/o HARD-Net	47.71	78.95	88.49	91.51	91.79	75.50
w/ HARD-Net	**53.15**	**82.87**	**91.34**	**93.71**	**94.03**	**78.84**

HARD-Net achieves the best performance over all observation ratios, which indicates the efficacy of our proposed HARD-Net. Compared to the state-of-the-art works and the backbone model, our method outperforms them significant, especially when the observation ratio is very low. The significant improvements demonstrate that our proposed approach can mine minor yet significant discrepancies for discrimination.

Moreover, following [1,30,42], we also use the area under curve metric, denoted as AUC, which is used to illustrate the average precision over all observation ratios to investigate the overall efficacy of our proposed HARD-Net. As shown in Table 1, our approach achieves the highest AUC of 70.56%, compared to the existing methods and also the backbone model ("w/o HARD-Net"). Note that our HARD-Net outperforms the backbone model by 3.65%, which further demonstrates that the proposed adversarial learning scheme can well-perceive and comprehend the subtle differences within *hard classes* and facilitate the discrimination capabilities of the class discriminator.

Ablation study on different loss weights in Eq. 4 are also conducted. Our method achieves AUC 70.6% under full losses ($\lambda_1 = \lambda_2 = 1$). Below we analyze the impact of each loss: 1) When removing ambiguous loss (setting its weight to 0) and keeping other two losses, the AUC drops to 67.9%. 2) When removing reconstruction loss and keeping other two losses, AUC drops to 67.7%. 3) When removing real/fake loss and keeping other two losses, AUC drops to 68.0%.

Results on Cross View Protocol. We also evaluate our HARD-Net on cross view protocol as in [42] and the comparison results are shown in Table 2 and Fig. 3 (middle). As shown in Table 2, our proposed HARD-Net model outperforms the existing works by a large margin over all observations ratios, which demonstrates the efficacy of our approach.

It is worth noting that the average accuracy score AUC of the HARD-Net exceeds the previous work [42] by 19.13% and exceeds baseline encoder by 3.34% which indicates our class discriminator can benefit from adversarial learning scheme and obtain more discrimination abilities for 3D early activity prediction.

4.3 Experiments on 3D Early Gesture Prediction

To demonstrate the efficacy of our proposed HARD-Net on 3D gesture dataset, extensive experiments are conducted on a publicly available 3D hand gesture dataset, namely FPHA. As illustrated in Table 3 and Fig. 3 (right), our proposed HARD-Net achieves better performance consistently over all observation ratios compared to Weng *et al.* [42].

Compared to the baseline model, at the early stages when the observation ratio is very low that lack sufficient discrimination information, since our HARD-Net is powerful in mining minor discrepancies, it achieves the most significant performance gain by 9.57% at the 20% observation ratio.

Table 3. Quantitative results (%) comparison on FPHA with state-of-the-arts.

Methods	Observation ratios					
	20%	40%	60%	80%	100%	AUC
LSTM [42]	54.26	63.30	69.22	72.17	74.43	64.11
Weng *et al.* [42]	59.65	65.91	70.43	73.57	74.96	66.66
w/o HARD-Net	62.26	74.61	79.65	82.09	83.48	72.17
w/ HARD-Net	**71.83**	**82.78**	**86.09**	**87.13**	**87.30**	**78.56**

4.4 Ablation Study

In this section, extensive ablation experiments are conducted based on the NTU-RGB+D dataset (cross-subject protocol), that is widely used by existing works [13,30,40,42] in early activity prediction community.

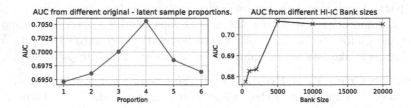

Fig. 4. Left: Evaluation of the impact of using different proportions of original samples and *hard pair* samples for network training. When proportion between original sample number and *hard pair* sample number is 4:1, our model achieves the highest prediction accuracy. Right: Evaluation of the impact of different HI-IC bank sizes.

Impact of Bank Size. We evaluate the performance of the HI-IC bank in different bank sizes. The result is shown in Fig. 4 (right). The AUC reflecting average precision over all observation ratios increases rapidly from a smaller bank size to a larger one, and then remains stable when the bank size is large enough (e.g., size 5000). This can be explained by the number of *hard pairs* in a dataset, and when the intrinsic threshold is reached, the further performance gain is limited.

Impact of Proportions Between Original Features and Latent Features for Training. Our experimental results in Fig. 4 (left) show that the optimal proportion of original features and latent features used for network training is 4:1. This can be explained as: if we use too much original features for network training, then less useful discrimination information will be mined via our adversarial learning scheme. However, if too much latent features are employed for training, it may lead to performance drops over the original samples. Moreover, small performance differences achieved by different ratios (1:1 to 6:1) indicate that our HARD-Net is not sensitive to ratios. In Fig. 4 (left), all achieved AUCs of the HARD-Net are in a small range (69.5% to 70.6%), which shows robustness of our method against ratios. Besides, these AUCs achieved all outperform the baseline (66.9%) by a large margin, validating the efficacy of our HARD-Net.

Table 4. Performance gain (%) brought by our HARD-Net with different backbones.

Backbone	Methods	Observation ratios				
		20%	40%	60%	80%	100%
CNN backbone [13]	w/o HARD-Net	34.01	63.16	75.87	81.39	82.24
	w/ HARD-Net	**35.86**	**64.97**	**77.12**	**82.22**	**82.98**
	Δ	**+1.85**	**+1.81**	**+1.25**	**+0.83**	**+0.74**
GCN backbone [34]	w/o HARD-Net	37.82	67.87	79.22	83.39	84.52
	w/ HARD-Net	**42.39**	**72.24**	**82.99**	**86.75**	**87.54**
	Δ	**+4.57**	**+4.37**	**+3.77**	**+3.36**	**+3.02**

Impact of Backbone Encoder. We extensively test our algorithm on a CNN backbone and a GCN backbone, and show the efficacy of the proposed method. As shown in Table 4, our HARD-Net boosts early prediction performance on both backbone models obviously, especially at the very low observation ratios. This indicates that our HARD-Net is powerful in mining subtle discrimination information for early activity prediction.

5   Conclusion

In this paper, we propose a novel Hardness-AwaRe Discrimination Network (HARD-Net) for 3D early activity prediction. The proposed HARD-Net is able to explicitly investigate the relationship between an easily mis-predicted instance, named *hard instance*, and the particular category that it is mis-predicted into, named *interference class*. An adversarial learning scheme is proposed to mine subtle discrepancies between this *hard instance - interference class* pair by generating ambiguous and less discriminative latent features conditioned on that particular pair to represent original *hard instances*. We further design a class discriminator to distinguish the derived latent features from the corresponding *interference classes*. With such a network design, our proposed HARD-Net achieves state-of-the-art performance on two challenging datasets.

Acknowledgement. This work is supported by SUTD Project PIE-SGP-AI-2020-02, SUTD Project SRG-ISTD-2020-153, the National Natural Science Foundation of China under Grant 61991411, and Grant U1913204, the National Key Research and Development Plan of China under Grant 2017YFB1300205, and the Shandong Major Scientific and Technological Innovation Project (MSTIP) under Grant 2018CXGC1503.

References

1. Aliakbarian, M., Saleh, F., Salzmann, M., Fernando, B., Petersson, L., Andersson, L.: Encouraging LSTMs to anticipate actions very early (2017)
2. Cai, Y., Ge, L., Cai, J., Magnenat-Thalmann, N., Yuan, J.: 3D hand pose estimation using synthetic data and weakly labeled RGB images. IEEE Trans. Pattern Anal. Mach. Intell. (2020)
3. Cai, Y., Ge, L., Cai, J., Yuan, J.: Weakly-supervised 3D hand pose estimation from monocular RGB images. In: Proceedings of the European Conference on Computer Vision, pp. 666–682 (2018)
4. Cai, Y., et al.: Exploiting spatial-temporal relationships for 3D pose estimation via graph convolutional networks. In: Proceedings of the IEEE International Conference on Computer Vision, pp. 2272–2281 (2019)
5. Cai, Y., Huang, L., Wang, Y., et al.: Learning progressive joint propagation for human motion prediction. In: Proceedings of the European Conference on Computer Vision (2020)
6. Felzenszwalb, P., Girshick, R., Mcallester, D., Ramanan, D.: Object detection with discriminatively trained part-based models. IEEE Trans. Pattern Anal. Mach. Intell. **32**, 1627–45 (2010)
7. Gammulle, H., Denman, S., Sridharan, S., Fookes, C.: Predicting the future: a jointly learnt model for action anticipation. In: Proceedings of the IEEE International Conference on Computer Vision, pp. 5561–5570 (2019)
8. Garcia-Hernando, G., Yuan, S., Baek, S., Kim, T.K.: First-person hand action benchmark with RGB-D videos and 3D hand pose annotations. In: Proceedings of Computer Vision and Pattern Recognition (CVPR) (2018)
9. Girshick, R.: Fast r-cnn. In: 2015 IEEE International Conference on Computer Vision (ICCV), pp. 1440–1448 (December 2015)
10. Hu, J.-F., Zheng, W.-S., Ma, L., Wang, G., Lai, J.: Real-time RGB-D activity prediction by soft regression. In: Leibe, B., Matas, J., Sebe, N., Welling, M. (eds.) ECCV 2016. LNCS, vol. 9905, pp. 280–296. Springer, Cham (2016). https://doi.org/10.1007/978-3-319-46448-0_17
11. Hu, J.F., Zheng, W.S., Ma, L., Wang, G., Lai, J., Zhang, J.: Early action prediction by soft regression. IEEE Trans. Pattern Anal. Mach. Intell. **41**(11), 2568–2583 (2018)

12. Jain, A., Singh, A., Koppula, H., Soh, S., Saxena, A.: Recurrent neural networks for driver activity anticipation via sensory-fusion architecture. Arxiv (2015)
13. Ke, Q., Bennamoun, M., Rahmani, H., An, S., Sohel, F., Boussaid, F.: Learning latent global network for skeleton-based action prediction. IEEE Trans. Image Process. **29**, 959–970 (2020)
14. Ke, Q., Bennamoun, M., An, S., Sohel, F.A., Boussaïd, F.: A new representation of skeleton sequences for 3D action recognition. In: IEEE Conference on Computer Vision and Pattern Recognition, pp. 4570–4579 (2017)
15. Kingma, D.P., Ba, J.: Adam: a method for stochastic optimization. In: ICLR (2015)
16. Kong, Y., Tao, Z., Fu, Y.: Adversarial action prediction networks. IEEE Trans. Pattern Anal. Mach. Intell. **42**(3), 539–553 (2020)
17. Kong, Y., Gao, S., Sun, B., Fu, Y.: Action prediction from videos via memorizing hard-to-predict samples. In: AAAI (2018)
18. Kong, Y., Kit, D., Fu, Y.: A discriminative model with multiple temporal scales for action prediction. In: Fleet, D., Pajdla, T., Schiele, B., Tuytelaars, T. (eds.) ECCV 2014. LNCS, vol. 8693, pp. 596–611. Springer, Cham (2014). https://doi.org/10.1007/978-3-319-10602-1_39
19. Kong, Y., Tao, Z., Fu, Y.: Deep sequential context networks for action prediction. In: IEEE Conference on Computer Vision and Pattern Recognition, pp. 3662–3670 (2017)
20. Li, C., Zhong, Q., Xie, D., Pu, S.: Co-occurrence feature learning from skeleton data for action recognition and detection with hierarchical aggregation. In: IJCAI, pp. 786–792 (2018)
21. Li, M., Chen, S., Chen, X., Zhang, Y., Wang, Y., Tian, Q.: Actional-structural graph convolutional networks for skeleton-based action recognition. In: The IEEE Conference on Computer Vision and Pattern Recognition (CVPR) (2019)
22. Lin, T., Goyal, P., Girshick, R., He, K., Dollár, P.: Focal loss for dense object detection. IEEE Trans. Pattern Anal. Mach. Intell. **42**(2), 318–327 (2020)
23. Liu, J., Shahroudy, A., Wang, G., Duan, L., Kot, A.C.: Skeleton-based online action prediction using scale selection network. IEEE Trans. Pattern Anal. Mach. Intell. **42**(6), 1453–1467 (2020)
24. Liu, J., Wang, G., Duan, L., Abdiyeva, K., Kot, A.C.: Skeleton-based human action recognition with global context-aware attention LSTM networks. IEEE Trans. Image Process. **27**(4), 1586–1599 (2018)
25. Liu, J., et al.: Feature boosting network for 3D pose estimation. IEEE Trans. Pattern Anal. Mach. Intell. **42**(2), 494–501 (2020)
26. Liu, J., Shahroudy, A., Xu, D., Wang, G.: Spatio-temporal LSTM with trust gates for 3D human action recognition. In: Leibe, B., Matas, J., Sebe, N., Welling, M. (eds.) ECCV 2016. LNCS, vol. 9907, pp. 816–833. Springer, Cham (2016). https://doi.org/10.1007/978-3-319-46487-9_50
27. Loshchilov, I., Hutter, F.: Online batch selection for faster training of neural networks (2015)
28. Lou, Y., Bai, Y., Liu, J., Wang, S., Duan, L.: Veri-wild: a large dataset and a new method for vehicle re-identification in the wild. In: Proceedings of the IEEE Conference on Computer Vision and Pattern Recognition, pp. 3235–3243 (2019)
29. Moreno-Noguer, F.: 3D human pose estimation from a single image via distance matrix regression. In: Proceedings of the IEEE Conference on Computer Vision and Pattern Recognition, pp. 2823–2832 (2017)
30. Pang, G., Wang, X., Hu, J.F., Zhang, Q., Zheng, W.S.: DBDNet: learning bidirectional dynamics for early action prediction. In: IJCAI, pp. 897–903 (2019)

31. Radford, A., Metz, L., Chintala, S.: Unsupervised representation learning with deep convolutional generative adversarial networks. In: Bengio, Y., LeCun, Y. (eds.) 4th International Conference on Learning Representations, ICLR 2016, San Juan, Puerto Rico, May 2–4, 2016, Conference Track Proceedings (2016). http://arxiv.org/abs/1511.06434

32. Shahroudy, A., Liu, J., Ng, T.T., Wang, G.: NTU RGB+D: a large scale dataset for 3D human activity analysis. In: Proceedings of the IEEE Conference on Computer Vision and Pattern Recognition, pp. 1010–1019 (2016)

33. Shi, L., Zhang, Y., Cheng, J., Lu, H.: Skeleton-based action recognition with directed graph neural networks. In: The IEEE Conference on Computer Vision and Pattern Recognition (CVPR) (2019)

34. Shi, L., Zhang, Y., Cheng, J., Lu, H.: Two-stream adaptive graph convolutional networks for skeleton-based action recognition. In: CVPR (2019)

35. Shotton, J., et al.: Real-time human pose recognition in parts from single depth images. In: Proceedings of the IEEE Conference on Computer Vision and Pattern Recognition, pp. 1297–1304 (2011)

36. Shrivastava, A., Mulam, H., Girshick, R.: Training region-based object detectors with online hard example mining (2016)

37. Si, C., Chen, W., Wang, W., Wang, L., Tan, T.: An attention enhanced graph convolutional LSTM network for skeleton-based action recognition. In: The IEEE Conference on Computer Vision and Pattern Recognition (CVPR) (2019)

38. Tang, Y., Tian, Y., Lu, J., Li, P., Zhou, J.: Deep progressive reinforcement learning for skeleton-based action recognition. In: The IEEE Conference on Computer Vision and Pattern Recognition (CVPR) (June 2018)

39. Wang, J., Liu, Z., Wu, Y., Yuan, J.: Learning actionlet ensemble for 3D human action recognition. IEEE Trans. Pattern Anal. Mach. Intell. **36**(5), 914–927 (2013)

40. Wang, X., Hu, J., Lai, J., Zhang, J., Zheng, W.: Progressive teacher-student learning for early action prediction. In: IEEE Conference on Computer Vision and Pattern Recognition, pp. 3551–3560 (2019)

41. Wang, X., Shrivastava, A., Gupta, A.: A-fast-rcnn: hard positive generation via adversary for object detection. In: The IEEE Conference on Computer Vision and Pattern Recognition (CVPR) (2017)

42. Weng, J., Jiang, X., Zheng, W., Yuan, J.: Early action recognition with category exclusion using policy-based reinforcement learning. IEEE Trans. Circuits Syst. Video Technol. 1 (2020)

43. Xu, W., Yu, J., Miao, Z., Wan, L., Ji, Q.: Prediction-CGAN: human action prediction with conditional generative adversarial networks. In: Proceedings of the ACM International Conference on Multimedia (2019)

44. Yan, S., Xiong, Y., Lin, D.: Spatial temporal graph convolutional networks for skeleton-based action recognition. In: AAAI (2018)

45. Yang, S., Liu, J., Lu, S., Er, M.H., Kot, A.C.: Collaborative learning of gesture recognition and 3D hand pose estimation with multi-order feature analysis. In: Proceedings of the European Conference on Computer Vision (2020)

46. Zhu, W., et al.: Co-occurrence feature learning for skeleton based action recognition using regularized deep LSTM networks. In: AAAI (2016)

Pseudo RGB-D for Self-improving Monocular SLAM and Depth Prediction

Lokender Tiwari[1]([✉]) [ID], Pan Ji[2] [ID], Quoc-Huy Tran[2] [ID], Bingbing Zhuang[2] [ID], Saket Anand[1] [ID], and Manmohan Chandraker[2,3] [ID]

[1] IIIT-Delhi, Delhi, India
lokendert@iiitd.ac.in
[2] NEC Labs America, Princeton, USA
[3] UCSD, San Diego, USA

Abstract. Classical monocular Simultaneous Localization And Mapping (SLAM) and the recently emerging convolutional neural networks (CNNs) for monocular depth prediction represent two largely disjoint approaches towards building a 3D map of the surrounding environment. In this paper, we demonstrate that the coupling of these two by leveraging the strengths of each mitigates the other's shortcomings. Specifically, we propose a joint narrow and wide baseline based self-improving framework, where on the one hand the CNN-predicted depth is leveraged to perform *pseudo RGB-D* feature-based SLAM, leading to better accuracy and robustness than the monocular RGB SLAM baseline. On the other hand, the bundle-adjusted 3D scene structures and camera poses from the more principled geometric SLAM are injected back into the depth network through novel wide baseline losses proposed for improving the depth prediction network, which then continues to contribute towards better pose and 3D structure estimation in the next iteration. We emphasize that our framework only requires *unlabeled monocular videos* in both training and inference stages, and yet is able to outperform state-of-the-art *self-supervised monocular* and *stereo* depth prediction networks (*e.g.*, Monodepth2) and feature-based monocular SLAM system (*i.e.*, ORB-SLAM). Extensive experiments on KITTI and TUM RGB-D datasets verify the superiority of our self-improving geometry-CNN framework.

Keywords: Self-supervised learning · Self-improving · Monocular depth prediction · Monocular SLAM

1 Introduction

One of the most reliable cues towards 3D perception from a monocular camera arises from camera motion that induces multiple-view geometric constraints [20]

Electronic supplementary material The online version of this chapter (https://doi.org/10.1007/978-3-030-58621-8_26) contains supplementary material, which is available to authorized users.

A. Vedaldi et al. (Eds.): ECCV 2020, LNCS 12356, pp. 437–455, 2020.
https://doi.org/10.1007/978-3-030-58621-8_26

wherein the 3D scene structure is encoded. Over the years, Simultaneous Localization And Mapping (SLAM) [6,21,32] has been long studied to simultaneously recover the 3D scene structure of the surrounding and estimate the ego-motion of the agent. With the advent of Convolutional Neural Networks (CNNs), unsupervised learning of single-view depth estimation [13,16,60] has emerged as a promising alternative to the traditional geometric approaches. Such methods rely on CNNs to extract meaningful depth cues (*e.g.*, shading, texture, and semantics) from a single image, yielding very promising results.

Despite the general maturity of monocular geometric SLAM [9,10,30] and the rapid advances in unsupervised monocular depth prediction approaches [2,15,29,38,44,57], they both still have their own limitations.

Monocular SLAM. Traditional monocular SLAM has well-known limitations in robustness and accuracy as compared to those leveraging active depth sensors, *e.g.*, RGB-D SLAM [31]. This performance issue is due to the inherent scale ambiguity of depth recovery from monocular cameras, which causes the so-called scale drift in both the camera trajectory and 3D scene depth, and thus lowers robustness and accuracy of conventional monocular SLAM. In addition, the triangulation-based depth estimation employed by traditional SLAM methods is degenerate under pure rotational camera motion [20].

Unsupervised Monocular Depth Prediction. Most of the unsupervised and self-supervised methods [2,15,16,60] formulate single image depth estimation as a novel-view synthesis problem, with appearance based photometric losses being central to the training strategy. Usually, these models train two networks, one each for *pose* and *depth*. As photometric losses largely rely on the brightness consistency assumption, nearly all existing self-supervised approaches operate in a narrow-baseline setting optimizing the loss over a snippet of 2–5 consecutive frames. Consequently, models like MondoDepth2 [15], work very well for close range points, but generate inaccurate depth estimates for points that are farther away (*e.g.*, see 0th iteration in Fig. 6). While it is well known that a wide-baseline yields better depth estimates for points at larger depth, a straightforward extension of existing CNN based approaches is inadequate for the following two reasons. A wide baseline in a video sequence implies a larger temporal window, which in most practical scenarios will violate the brightness consistency assumption, rendering the photometric loss ineffective. Secondly, larger temporal windows (wider baselines) would also imply more occluded regions that behave as outliers. Unless these aspects are effectively handled, training of CNN based depth and pose networks in the wide baseline setting will lead to inaccuracies and biases.

In view of the limitations in both monocular geometric SLAM and unsupervised monocular depth estimation approaches, a particularly interesting question to ask is whether these two approaches can complement each other (see Sect. 5) and mitigate the issues discussed above. Our work makes contributions towards answering this question. Specifically, we propose a *self-supervised, self-improving* framework of these two tasks, which is shown to improve the robustness and accuracy on each of them.

While the performance gap between geometric SLAM and self-supervised learning-based SLAM methods is still large, incorporating depth information drastically improves the robustness of geometric SLAM methods (*e.g.*, see RGB-D SLAM vs. RGB SLAM on the KITTI Odometry leaderboard [14]). Inspired by this success of RGB-D SLAM, we postulate the use of an unsupervised CNN-based depth estimation model as a *pseudo depth sensor*, which allows us to design our self-supervised approach, pseudo RGB-D SLAM (pRGBD-SLAM) that only uses monocular cameras and yet achieves significant improvements in robustness and accuracy as compared to RGB SLAM.

Our fusion of geometric SLAM and CNN-based monocular depth estimation turns out to be symbiotic and this complementary nature sets the basis of our self-improving framework. To improve the depth predictions, we make use of two main modifications in the training strategy. First, we eschew the learning based pose estimates in favor of geometric SLAM based estimates (an illustrative motivation is shown in Fig. 1). Second, we make use of common tracked keypoints from neighboring *keyframes* and impose a symmetric depth transfer and a depth consistency loss on the CNN model. These adaptations are based on the observation that both pose estimates and sparse 3D feature point estimates from geometric SLAM are robust, as most techniques typically apply multiple bundle adjustment iterations over wide baseline depth estimates of common keypoints. This simple observation and the subsequent modification is key to our self-improving framework, which can leverage any unsupervised CNN-based depth estimation model and a modern monocular SLAM method. In this paper, we test our framework, with ORBSLAM [31] as the geometric SLAM method and MonoDepth2 [15] as the CNN-based model. We show that our self-improving framework outperforms previously proposed self-supervised approaches that utilizes monocular, stereo, and monocular-plus-stereo cues for self-supervision (see Table 1) and a strong feature based RGB-SLAM baseline (see Table 5). The framework runs in a simple alternating update fashion: first, we use depth maps from the CNN-based depth network and run pRGBD-SLAM; second, we inject the outputs of pRGBD-SLAM, *i.e.*, the relative camera poses and common tracked keypoints and keyframes to fine-tune the depth network parameters to improve the depth prediction; then, we repeat the process until we see no improvement. Our specific contributions are summarized here:

- We propose a self-improving strategy to inject into depth prediction networks the supervision from SLAM outputs, which stem from more generally applicable geometric principles.
- We introduce two wide baseline losses, *i.e.*, the symmetric depth transfer loss and the depth consistency loss on common tracked points, and propose a joint narrow and wide baseline based depth prediction learning setup, where appearance based losses are computed on narrow baselines and purely geometric losses on wide baselines (non-consecutive temporally distant keyframes).
- Through extensive experiments on KITTI [14] and TUM RGB-D [40], our framework is shown to outperform both monocular SLAM system (*i.e.*, ORB-

SLAM [30]) and the state-of-the-art unsupervised single-view depth prediction network (*i.e.*, Monodepth2 [15]).

2   Related Work

Monocular SLAM. Visual SLAM has a long history of research in the computer vision community. Due to its well-understood underlying geometry, various geometric approaches have been proposed in the literature, ranging from the classical MonoSLAM [6], PTAM [21], DTAM [32] to the more recent LSD-SLAM [10], ORB-SLAM [30] and DSO [9]. More recently, in view of the successful application of deep learning in a wide variety of areas, researchers have also started to exploit deep learning approaches for SLAM, in the hope that it can improve certain components of geometric approaches or even serve as a complete alternative. Our work makes further contributions along this line of research.

Monocular Depth Prediction. Inspired by the pioneering work by Eigen et al. [8] on learning single-view depth estimation, a vast amount of learning methods [3,12,26] emerge along this line of research. The earlier works often require ground truth depths for fully-supervised training. However, per-pixel depth ground truth is generally hard or prohibitively costly to obtain. Therefore, many self-supervised methods that make use of geometric constraints as supervision signals are proposed. Godard et al. [16], relies on the photo-consistency between the left-right cameras of a calibrated stereo. Zhou et al. [60] learn monocular depth prediction as well as ego-motion estimation, thereby permitting unsupervised learning with only a monocular camera. This pipeline has inspired a large amount of follow-up works that utilize various additional heuristics, including 3D geometric constraints on point clouds [29], direct visual odometry [44], joint learning with optical flow [57], scale consistency [2], and others [5,15,38,63].

Using Depth to Improve Monocular SLAM. Approaches [27,41,53,56] leveraging CNN-based depth estimates to tackle issues in monocular SLAM have been proposed. CNN-SLAM [41] uses learned depth maps to initialize keyframes' depth maps in LSD-SLAM [10] and refines them via a filtering framework. Yin et al. [56] use a combination of CNNs and conditional random fields to recover scale from the depth predictions and iteratively refine ego-motion and depth estimates. DVSO [53] trains a single CNN to predict both the left and right disparity maps, forming a virtual stereo pair. The CNN is trained with photoconsistency between stereo images and consistency with depths estimated by Stereo DSO [46]. More recently, CNN-SVO [27] uses depths learned from stereo images to initialize depths of keypoints and reduce their corresponding uncertainties in SVO [11]. In contrast to our self-supervised approach, [41,56] use *ground truth* depths for training depth networks while [27,53] need *stereo* images.

Using SLAM to Improve Monocular Depth Prediction. Depth estimates from geometric SLAM have been leveraged for training monocular depth estimation networks in recent works [1,22]. In [1], sparse depth maps by Stereo ORB-SLAM [31] are first converted into dense ones via an auto-encoder, which are then integrated into geometric constraints for training the depth network. [22] employ depths and poses by ORB-SLAM [30] as supervision signals for training the depth and pose networks respectively. This approach only considers five consecutive frames, thus restricting its operation in the narrow-baseline setting.

3 Method: A Self-improving Framework

Our self-improving framework leverages the strengths of each, the unsupervised single-image depth estimation and the geometric SLAM approaches, to mitigate the other's shortcomings. On one hand, the depth network typically generates reliable depth estimates for nearby points, which assist in improving the geometric SLAM estimates of poses and sparse 3D points (Sect. 3.1). On the other hand, geometric SLAM methods rely on a more holistic view of the scene to generate robust pose estimates as well as identify *persistent* 3D points that are visible across many frames, thus providing an opportunity to perform wide-baseline and reliable sparse depth estimation. Our framework leverages these sparse, but robust estimates to improve the noisier depth estimates of the farther scene points by minimizing a blend of the symmetric transfer and depth consistency losses (Sect. 3.2) and the commonly used appearance based loss. In the following iteration, this improved depth estimate further enhances the capability of geometric SLAM and the cycle continues until the improvements become negligible. Even in the absence of ground truth, our self-improving framework continues to produce better pose and depth estimates.

An overview of the proposed self-improving framework is shown in Fig. 2, which iterates between improving poses and improving depths. Our pose refinement and depth refinement steps are then detailed in Sect. 3.1 and 3.2 respectively. An overview of narrow and wide baseline losses we use for improving the depth network is shown in Fig. 3 and details are provided in Sect. 3.2.

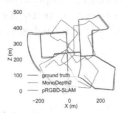

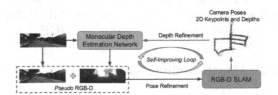

Fig. 1. MonoDepth2 [15] pose network camera poses vs. pseudo RGBD-SLAM camera poses where depth(D) is from CNN. The pose network from [15] leads to significant drift.

Fig. 2. Overview of Our Self-Improving Framework. It alternates between pose refinement (blue arrows; Sect. 3.1) and depth refinement (red arrows; Sect. 3.2). (Color figure online)

3.1   Pose Refinement

Pseudo RGB-D for Improving Monocular SLAM. We employ a well explored and widely used geometry-based SLAM system, i.e., the RGB-D version of ORB-SLAM [31], to process the pseudo RGB-D data, yielding camera poses as well as 3D map points and the associated 2D keypoints. Any other geometric SLAM system that provides these output estimates can also be used in place of ORB-SLAM. A trivial direct use of pseudo RGB-D data to run RGB-D ORB-SLAM is not possible, because CNN might predict depth at a very different scale compared to depth measurements from real active sensors, e.g., LiDAR. Keeping the above difference in mind, we discuss an important adaptation in order for RGB-D ORB-SLAM to work well in our setting. We first note that RGB-D ORB-SLAM transforms the depth data into disparity on a virtual stereo to reuse the framework of stereo ORB-SLAM. Specifically, considering a keypoint with 2D coordinates (u_l, v_l) (i.e., u_l and v_l denote the horizontal and vertical coordinates respectively) and a CNN-predicted depth d_l, the corresponding 2D keypoint coordinates (u_r, v_r) on the virtual rectified right view are $u_r = u_l - \frac{f_x b}{d_l}$, $v_r = v_l$, where f_x is the horizontal focal length and b is the virtual stereo baseline.

Adaptation. In order to have a reasonable range of disparity, we mimic the setup of the KITTI dataset [14] by making the baseline adaptive, $b = \frac{b^{\text{KITTI}}}{d_{max}^{\text{KITTI}}} *$ d_{max}, where d_{max} represents the maximum CNN-predicted depth of the input sequence, and $b^{\text{KITTI}} = 0.54$ and $d_{max}^{\text{KITTI}} = 80$ (both in meters) are respectively the actual stereo baseline and empirical maximum depth value of the KITTI dataset.

We also summarize the overall pipeline of RGB-D ORB-SLAM here. The 3D map is initialized at the very first frame of the sequence due to the availability of depth. After that, the following main tasks are performed: i) track the camera by matching 2D keypoints against the local map, ii) enhance the local map via local bundle adjustment, and iii) detect and close loops for pose-graph optimization and full bundle adjustment to improve camera poses and scene depths. As we will show in Sect. 4.4, using pseudo RGB-D data leads to better robustness and accuracy as compared to using only RGB data.

3.2   Depth Refinement

We start from the pre-trained depth network of Monodepth2 [15], a state-of-the-art monocular depth estimation network, and fine-tune its network parameters with the camera poses, 3D map points and the associated 2D keypoints produced by the above pseudo RGB-D ORB-SLAM (pRGBD-SLAM). In contrast to Monodepth2, which relies only on the narrow baseline photometric reconstruction loss between adjacent frames for short-term consistencies, we propose wide baseline symmetric depth transfer and sparse depth consistency losses to introduce long-term consistencies. Our final loss (Eq. (4)) consists of both narrow and wide baseline losses. The narrow baseline losses, i.e., photometric and

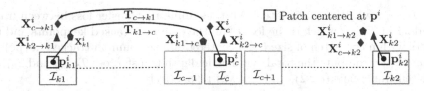

Fig. 3. Narrow and Wide Baseline Losses. Narrow baseline photometric and smoothness losses involve keyframe $\mathcal{I}_c$ and temporally *adjacent* frames $\mathcal{I}_{c-1}$ and $\mathcal{I}_{c+1}$, and wide baseline symmetric depth transfer and depth consistency losses involve keyframe $\mathcal{I}_c$ and temporally *farther* keyframes $\mathcal{I}_{k1}$ and $\mathcal{I}_{k2}$. Refer to the text below.

smoothness losses, involve the current keyframe $\mathcal{I}_c$ and its temporally adjacent frames $\mathcal{I}_{c-1}$ and $\mathcal{I}_{c+1}$, while wide baseline losses are computed on the current keyframe $\mathcal{I}_c$ and the two neighboring keyframes $\mathcal{I}_{k1}$ and $\mathcal{I}_{k2}$ that are temporally farther than $\mathcal{I}_{c-1}$ and $\mathcal{I}_{c+1}$ (see Fig. 3). Next, we introduce the notation and describe the losses.

Notation. Let $\mathcal{X}$ represent the set of common tracked keypoints visible in all the three keyframes $\mathcal{I}_{k1}$, $\mathcal{I}_c$ and $\mathcal{I}_{k2}$ obtained from pRGBD-SLAM. Note that $k1$ and $k2$ are two neighboring keyframes of the current frame c (*i.e.*, $k1 < c < k2$) in which keypoints are visible. Let $\mathbf{p}_{k1}^i = [p_{k1}^{i1}, p_{k1}^{i2}]$, $\mathbf{p}_c^i = [p_c^{i1}, p_c^{i2}]$ and $\mathbf{p}_{k2}^i = [p_{k2}^{i1}, p_{k2}^{i2}]$ be the 2D coordinates of the ith common tracked keypoint in the keyframes $\mathcal{I}_{k1}$, $\mathcal{I}_c$ and $\mathcal{I}_{k2}$ respectively, and the associated depth values obtained from pRGBD-SLAM are represented by $d_{k1}^i(\text{SLAM})$, $d_c^i(\text{SLAM})$, and $d_{k2}^i(\text{SLAM})$ respectively. The depth values corresponding to the keypoints $\mathbf{p}_{k1}^i$, $\mathbf{p}_c^i$ and $\mathbf{p}_{k2}^i$ can also be obtained from the depth network and are represented by $d_{k1}^i(\mathbf{w})$, $d_c^i(\mathbf{w})$, and $d_{k2}^i(\mathbf{w})$ respectively, where $\mathbf{w}$ stands for the depth network parameters.

Symmetric Depth Transfer Loss. Given the camera intrinsic matrix $\mathbf{K}$, and the depth value $d_c^i(\mathbf{w})$ of the ith keypoint $\mathbf{p}_c^i$, the 2D coordinates of the keypoint $\mathbf{p}_c^i$ can be back-projected to its corresponding 3D coordinates as: $\mathbf{X}_c^i(\mathbf{w}) = \mathbf{K}^{-1}[\mathbf{p}_c^i, 1]^T d_c^i(\mathbf{w})$. Let $\mathbf{T}_{c \to k1}^{\text{SLAM}}$ represent the relative camera pose of frame $k1$ w.r.t. frame c obtained from pRGBD-SLAM. Using $\mathbf{T}_{c \to k1}^{\text{SLAM}}$, we can transfer the 3D point $\mathbf{X}_c^i(\mathbf{w})$ from frame c to $k1$ as: $\mathbf{X}_{c \to k1}^i(\mathbf{w}) = \mathbf{T}_{c \to k1}^{\text{SLAM}} \mathbf{X}_c^i(\mathbf{w}) = [x_{c \to k1}^i(\mathbf{w}), y_{c \to k1}^i(\mathbf{w}), d_{c \to k1}^i(\mathbf{w})]^T$. Here, $d_{c \to k1}^i(\mathbf{w})$ is the transferred depth of the ith keypoint from frame c to frame $k1$. Following the above procedure, we can obtain the transferred depth $d_{k1 \to c}^i(\mathbf{w})$ of the same i^{th} keypoint from frame $k1$ to frame c. The symmetric depth transfer loss of the keypoint $\mathbf{p}_c^i$ between frame pair c and $k1$, is the sum of absolute errors (ℓ_1 distance) between the transferred network-predicted depth $d_{c \to k1}^i(\mathbf{w})$ and existing network-predicted depth $d_{k1}^i(\mathbf{w})$ in the target keyframe $k1$, and vice-versa. It can be written as:

$$\mathcal{T}_{c \leftrightarrow k1}^i(\mathbf{w}) = |d_{c \to k1}^i(\mathbf{w}) - d_{k1}^i(\mathbf{w})| + |d_{k1 \to c}^i(\mathbf{w}) - d_c^i(\mathbf{w})|. \tag{1}$$

Similarly, we can compute the symmetric depth transfer loss of the same i^{th} keypoint between frame pair c and $k2$, *i.e.*, $\mathcal{T}_{c \leftrightarrow k2}^i(\mathbf{w})$, and between $k1$ and $k2$, *i.e.*, $\mathcal{T}_{k1 \leftrightarrow k2}^{i}(\mathbf{w})$. We accumulate the total symmetric transfer loss between frame c and $k1$ in $\mathcal{T}_{c \leftrightarrow k1}$, which is the loss of all the common tracked keypoints and the points within the patch of size 5×5 centered at the common tracked keypoints. Similarly, we compute the total symmetric depth transfer loss $\mathcal{T}_{c \leftrightarrow k2}$ and $\mathcal{T}_{k1 \leftrightarrow k2}$ between frame pair $(c, k2)$, and $(k1, k2)$ respectively.

Depth Consistency Loss. The role of the depth consistency loss is to make depth network's prediction consistent with the refined depth values obtained from the pRGBD-SLAM. Note that depth values from pRGBD-SLAM undergo multiple optimization over wide baselines, hence are more accurate and capture long-term consistencies. We inject these long-term consistent depths from pRGBD-SLAM to depth network through the depth consistency loss. The loss for the frame c can be written as follows:

$$\mathcal{D}_c = \frac{\sum_{i \in \mathcal{X}} |d_c^i(\mathbf{w}) - d_c^i(\text{SLAM})|}{|\mathcal{X}|}. \tag{2}$$

Photometric Reconstruction Loss. Denote the relative camera pose of frame $\mathcal{I}_{c\text{-}1}$ and $\mathcal{I}_{c+1}$ w.r.t. current keyframe $\mathcal{I}_c$ obtained from pRGBD-SLAM by $\mathbf{T}_{c\text{-}1 \rightarrow c}^{\text{SLAM}}$ and $\mathbf{T}_{c+1 \rightarrow c}^{\text{SLAM}}$ respectively. Using frame $\mathcal{I}_{c+1}$, $\mathbf{T}_{c+1 \rightarrow c}^{\text{SLAM}}$, network-predicted depth map $d_c(\mathbf{w})$ of the keyframe $\mathcal{I}_c$, and the camera intrinsic $\mathbf{K}$, we can synthesize the current frame $\mathcal{I}_c$ [15,16]. Let the synthesized frame be represented in the functional form as: $\mathcal{I}_{c+1 \rightarrow c}(d_c(\mathbf{w}), \mathbf{T}_{c+1 \rightarrow c}^{\text{SLAM}}, \mathbf{K})$. Similarly we can synthesize $\mathcal{I}_{c\text{-}1 \rightarrow c}(d_c(\mathbf{w}), \mathbf{T}_{c\text{-}1 \rightarrow c}^{\text{SLAM}}, \mathbf{K})$ using frame $\mathcal{I}_{c\text{-}1}$. The photometric reconstruction error between the synthesized and the original current frame [13,16,60] is then computed as:

$$\mathcal{P}_c = pe(\mathcal{I}_{c+1 \rightarrow c}(d_c(\mathbf{w}), \mathbf{T}_{c+1 \rightarrow c}^{\text{SLAM}}, \mathbf{K}), \mathcal{I}_c) + pe(\mathcal{I}_{c\text{-}1 \rightarrow c}(d_c(\mathbf{w}), \mathbf{T}_{c\text{-}1 \rightarrow c}^{\text{SLAM}}, \mathbf{K}), \mathcal{I}_c), \tag{3}$$

where we follow [15,16] to construct the photometric reconstruction error function $pe(\cdot, \cdot)$. Additionally, we adopt the more robust per-pixel minimum error, multi-scale strategy, auto-masking, and depth smoothness loss $\mathcal{S}_c$ from [15]. Our final loss for fine-tuning the depth network at the depth refinement step is the weighted sum of narrow baseline losses (*i.e.*, photometric ($\mathcal{P}_c$) and smoothness loss ($\mathcal{S}_c$)), and wide baseline losses (*i.e.*, symmetric depth transfer ($\mathcal{T}_{c \leftrightarrow k1}, \mathcal{T}_{c \leftrightarrow k2}, \mathcal{T}_{k1 \leftrightarrow k2}$) and depth consistency loss ($\mathcal{D}_c$)):

$$\mathcal{L} = \alpha \mathcal{P}_c + \beta \mathcal{S}_c + \gamma \mathcal{D}_c + \mu(\mathcal{T}_{c \leftrightarrow k1} + \mathcal{T}_{c \leftrightarrow k2} + \mathcal{T}_{k1 \leftrightarrow k2}). \tag{4}$$

4   Experiments

We conduct experiments to evaluate depth refinement and pose refinement steps of our self-improving framework with the state-of-the-arts in self-supervised depth estimation and RGB-SLAM based pose estimation respectively.

4.1 Datasets and Evaluation Metrics

KITTI Dataset. Our experiments are mostly performed on the KITTI dataset [14], which contains outdoor driving sequences for road scene understanding [7,39]. We further split KITTI experiments into two parts: one focused on depth refinement evaluation and the other on pose refinement. For depth refinement evaluation we train/fine-tune the depth network using the Eigen train split [8] which contains 28 training sequences and evaluate depth prediction on the Eigen test split [8] following the baselines [4,28,29,35,44,54,55,57,60,64]. For pose refinement evaluation, we train/fine-tune the depth network using KITTI odometry sequences 00–08 and test on sequences 09–10 and 11–21. Note, for evaluation on sequences 09–10 we use the ground-truth trajectories provided by [14], while for evaluation on sequences 11–21, since the ground-truth is not available we use the pseudo ground-truth trajectories obtained by running stereo version of ORB-SLAM on these sequences.

TUM RGB-D Dataset. For completeness and to demonstrate the capability of our self-improving framework on indoor scenes, we evaluate on the TUM RGB-D dataset [40], which consists of indoor sequences captained by a hand-held camera. We choose *freiburg3* sequences because only they have *undistorted* RGB images and ground truth available to train/fine-tune and evaluate respectively. We use 6 of 8 *freiburg3* sequences for training/fine-tune and the remaining 2 for evaluation.

Metrics for Pose Evaluation. For quantitative pose evaluation, we compute the Root Mean Square Error (*RMSE*), Relative Translation (*Rel Tr*) error, and Relative Rotation (*Rel Rot*) error of the predicted camera trajectory. Since monocular SLAM systems can only recover camera poses up to a global scale, we align the camera trajectory estimated by each method with the ground truth one using the EVO toolbox [18]. We then use the official evaluation code from the KITTI Odometry benchmark to compute the *Rel Tr* and *Rel Rot* errors for all sub-trajectories with length in $\{100, \ldots, 800\}$ meters.

Metrics for Depth Evaluation. For quantitative depth evaluation, we use the standard metrics, including the Absolute Relative (*Abs Rel*) error, Squared Relative (*Sq Rel*) error, *RMSE*, *RMSE log*, $\delta < 1.25$ (namely *a1*), $\delta < 1.25^2$ (namely *a2*), and $\delta < 1.25^3$ (namely *a3*) as defined in [8]. Again, since the depths from monocular images can only be estimated up to scale, we align the predicted depth map with the ground truth one using their median depth values. Following [8] and other baselines, we also clip the depths to 80 m.

Note. In all tables, best performances are in **bold** and second bests are underlined.

4.2 Implementation Details

We implement our framework based on Monodepth2 [15] and ORB-SLAM [31], *i.e.*, we use the depth network of Monodepth2 and the RGB-D version of ORB-SLAM for depth refinement and pose refinement respectively. We would like to emphasize, that our self-improving strategy is not specific to MonoDepth2 or ORB-SLAM. Any other depth network that allows to incorporate SLAM outputs and any SLAM system that can provide the desired SLAM outputs can be put into the self-improving framework. We set the weight of the smoothness loss term of the final loss (Eq. (4)) $\beta = 0.001$ similar as in [15] and α, γ, and μ to 1. The ablation study results on disabling different loss terms can be found in Table 3. A single self-improving loop takes 0.6 h on a NVIDIA TITAN Xp 8GB GPU.

KITTI Eigen Split/Odometry Experiments. We pre-train MonoDepth2 using monocular videos of the KITTI Eigen split training set with the hyper-parameters as suggested in MonoDepth2 [15]. We use an input/output resolution of 640×192 for training/fine-tuning and scale it up to the original resolution while running pRGBD-SLAM. We use same hyperparameters as for KITTI Eigen split to train/fine-tune the depth model on KITTI Odometry train sequences mentioned in Sect. 4.1. During a self-improving loop, we *discard* pose network of MonoDepth2 and instead use camera poses from pRGBD-SLAM.

Outlier Removal. Before running a depth refinement step, we run an outlier removal step on the SLAM outputs. Specifically, we filter out outlier 3D map points and the associated 2D keypoints that satisfy at least one of the following conditions: i) it is observed in less than 3 keyframes, ii) its reprojection error in the current keyframe $\mathcal{I}_c$ is larger than 3 pixels.

Camera Intrinsics. Monodepth2 computes the average camera intrinsics for the KITTI dataset and uses it for the training. However, for our fine-tuning of the depth network, using the average camera intrinsics leads to inferior performance, because we use the camera poses from pRGBD-SLAM, which runs with different camera intrinsics. Therefore, we use different camera intrinsics for different sequences when fine-tuning the depth network.

For fine-tuning the depth network pre-trained on KITTI Eigen split training sequences, we run pRGBD-SLAM on all the training sequences, and extract camera poses, 2D keypoints and the associated depths from keyframes. For pRGBD-SLAM(RGB-D ORB-SLAM), we use the default setting of ORB-SLAM, except for the adjusted b described in Sect. 3.1. The same above procedure is followed for depth model pre-trained on KITTI Odometry training sequences. The average number of keyframes used in a self-improving loop is $\sim 9K$ and $\sim 10K$ for KITTI Eigen split and KITTI Odometry experiments respectively. At each depth refinement step, we fine-tune the depth network parameters with 1 epoch only, using learning rate 1e−6, keeping all the other hyperparameters the same as pre-training. For both KITTI Eigen split and KITTI Odometry experiments we report results after *5 self-improving loops*.

TUM RGB-D Experiments. For TUM RGB-D, we pre-train/fine-tune the depth network on 6 *freiburg3* sequences, and test on 2 *freiburg3* sequences. The average number of keyframes in a self-improving loop is $\sim 3.5K$. We use an input/output resolution of 480×320 for pre-training/fine-tuning and scale it up to the original resolution while running pRGBD-SLAM. We report results after *3 self-improving loops*. Other details can be found in the supplementary material.

4.3 Monocular Depth/Depth Refinement Evaluation

In the following, we evaluate the performance of our depth estimation on the KITTI Raw Eigen split test set and TUM RGB-D *frieburg3* sequences.

Results on KITTI Eigen Split Test Set. We show the depth evaluation results on the Eigen split test set in Table 1. From the table, it is evident that our refined depth model (pRGBD-Refined) outperforms all the competing monocular (M) unsupervised methods by non-trivial margins, including MonoDepth2-M retrained depth model, and even surpasses the unsupervised methods with stereo (S) training, *i.e.*, Monodepth2-S, and combined monocular-stereo (MS) training, *i.e.*, MonoDepth2-MS, in most metrics. Our method also outperforms several ground-truth depth supervised methods [8,26]. The reason is probably that the aggregated cues from multiple views with wide baseline losses (*e.g.*, our symmetric depth transfer, depth consistency losses) lead to more well-posed depth recovery, and hence even higher accuracy than learning with the pre-calibrated stereo rig with smaller baselines. Further analysis is provided in Sect. 5. Figure 4 shows some qualitative results, where our method (pRGBD-Refined) shows visible improvements. Refer supplementary material for more qualitative results.

Results on TUM RGB-D Sequences. The depth evaluation results on the two TUM *frieburg3* RGB-D sequences is shown in Table 2. Our refined depth model (pRGBD-Refined) outperforms pRGBD-Initial/Monodepth2-M in both sequences and all metrics. Refer supplementary material for qualitative results.

4.4 Monocular SLAM/Pose Refinement Evaluation

In this section, we evaluate pose estimation/refinement on the KITTI Odometry sequences 09 and 10, KITTI Odometry test set sequences 11–21, and two TUM *frieburg3* RGB-D sequences.

Results on KITTI Odometry Sequences 09 and 10. We show the quantitative results on seqs 09 and 10 in Table 4. It can be seen that our pRGBD-Initial outperforms RGB ORB-SLAM [30] both in terms of RSME and Rel Tr. Our pRGBD-Refined further improves pRGBD-Initial in all metrics, which verifies the effectiveness of our self-improving mechanism in terms of pose estimation. The higher Rel Rot errors of our method compared to RGB ORB-SLAM could be due to the high uncertainty of CNN-predicted depths for far-away points, which affects our rotation estimation [20]. In addition, our methods outperform all the competing supervised and self-supervised methods by a large margin, except for the supervised method of [51] with lower Rel Tr than ours on sequence 10. Note that we evaluate the camera poses produced by the pose network of Monodepth2-M [15] in Table 4, yielding much higher errors than ours. Figure 5(a) shows the camera trajectories estimated for sequence 09 by RGB ORB-SLAM, our pRGBD-Initial, and pRGBD-Refined. It is evident that, although all the methods perform loop closure successfully, our methods generate camera trajectories that align better with the ground truth.

Table 1. Depth evaluation result on KITTI Eigen split test set. M: self-supervised monocular supervision, and S: self-supervised stereo supervision, D: depth supervision. '-' means the result is not available from the paper. pRGBD-Refined outperforms all the self-supervised monocular methods and several stereo only and combined monocular and stereo methods. Our results are after *5 self-improving loop*s.

	Method	Train	Lower is better				Higher is better		
			Abs Rel	Sq Rel	RMSE	RMSE log	a1	a2	a3
self-supervised	Yang[55]	M	0.182	1.481	6.501	0.267	0.725	0.906	0.963
	Mahjourian[29]	M	0.163	1.240	6.220	0.250	0.762	0.916	0.968
	Klodt[22]	M	0.166	1.490	5.998	-	0.778	0.919	0.966
	DDVO[44]	M	0.151	1.257	5.583	0.228	0.810	0.936	0.974
	GeoNet[57]	M	0.149	1.060	5.567	0.226	0.796	0.935	0.975
	DF-Net[64]	M	0.150	1.124	5.507	0.223	0.806	0.933	0.973
	Ranjan[35]	M	0.148	1.149	5.464	0.226	0.815	0.935	0.973
	EPC++[28]	M	0.141	1.029	5.350	0.216	0.816	0.941	0.976
	Struct2depth(M)[4]	M	0.141	1.026	5.291	0.215	0.816	0.945	0.979
	WBAF [59]	M	0.135	0.992	5.288	0.211	0.831	0.942	0.976
	MonoDepth2-M (re-train) [15]	M	0.117	0.941	4.889	0.194	0.873	0.957	0.980
	MonoDepth2-M (original) [15]	M	<u>0.115</u>	<u>0.903</u>	<u>4.863</u>	<u>0.193</u>	**0.877**	<u>0.959</u>	<u>0.981</u>
	pRGBD-Refined	M	**0.113**	**0.793**	**4.655**	**0.188**	<u>0.874</u>	**0.960**	**0.983**
	Garg[13]	S	0.152	1.226	5.849	0.246	0.784	0.921	0.967
	3Net (R50)[34]	S	0.129	0.996	5.281	0.223	0.831	0.939	0.974
	Monodepth2-S[15]	S	0.109	0.873	4.960	0.209	0.864	0.948	0.975
	SuperDepth [33]	S	0.112	0.875	4.958	0.207	0.852	0.947	0.977
	monoResMatch [43]	S	0.111	0.867	4.714	0.199	0.864	<u>0.954</u>	<u>0.979</u>
	DepthHints [49]	S	<u>0.106</u>	<u>0.780</u>	<u>4.695</u>	<u>0.193</u>	<u>0.875</u>	**0.958**	**0.980**
	DVSO[53]	S	**0.097**	**0.734**	**4.442**	**0.187**	**0.888**	**0.958**	**0.980**
	UnDeepVO [24]	MS	0.183	1.730	6.570	0.268	-	-	-
	EPC++ [28]	MS	<u>0.128</u>	<u>0.935</u>	<u>5.011</u>	<u>0.209</u>	<u>0.831</u>	<u>0.945</u>	**0.979**
	Monodepth2-MS[15]	MS	**0.106**	**0.818**	**4.750**	**0.196**	**0.874**	**0.957**	**0.979**
	Eigen[8]	D	0.203	1.548	6.307	0.282	0.702	0.890	0.890
	Liu[26]	D	0.201	1.584	6.471	0.273	0.680	0.898	0.967
	Kuznietsov[23]	DS	0.113	0.741	4.621	0.189	0.862	0.960	0.986
	SVSM FT[28]	DS	<u>0.094</u>	<u>0.626</u>	4.252	0.177	0.891	0.965	0.984
	Guo[19]	DS	0.096	0.641	<u>4.095</u>	<u>0.168</u>	<u>0.892</u>	<u>0.967</u>	<u>0.986</u>
	DORN[12]	D	**0.072**	**0.307**	**2.727**	**0.120**	**0.932**	**0.984**	**0.994**

Table 2. Depth evaluation results on two *frieburg3* RGB-D sequences. pRGBD-R results are after *3 self-improving loops*.

	TUM RGBD Sequences						
	Lower is better				Higher is better		
Method	AbRel	SqRel	RMSE	RMSElog	a1	a2	a3
pRGBD-I	0.397	0.848	1.090	0.719	0.483	0.722	0.862
pRGBD-R	**0.307**	**0.341**	**0.743**	**0.655**	**0.522**	**0.766**	**0.873**

Table 3. Ablation study on 1st self-improving loop using KITTI Eigen split. The best performance is in **bold**.

Loss	\multicolumn{4}{Lower is better}				\multicolumn{3}{Higher is better}		
	AbsRel	SqRel	RMSE	RMSElog	a1	a2	a3
w/o $\mathcal{D}_c$	**0.117**	0.958	4.956	0.194	0.862	0.955	0.980
w/o $\mathcal{T}_c$	0.118	0.955	4.867	0.194	0.872	0.957	0.980
w/o $\mathcal{P}_c$	**0.117**	0.942	4.855	0.194	**0.873**	**0.958**	0.980
all losses	**0.117**	**0.931**	**4.809**	**0.192**	**0.873**	**0.958**	**0.981**

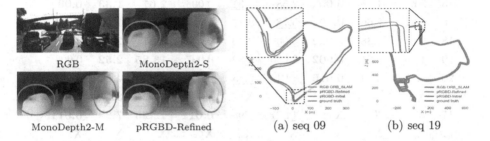

RGB   MonoDepth2-S

MonoDepth2-M   pRGBD-Refined   (a) seq 09   (b) seq 19

Fig. 4. Qualitative depth evaluation on KITTI Raw Eigen's split test set.

Fig. 5. Qualitative pose evaluation results on KITTI sequences.

Table 4. Quantitative pose evaluation results on KITTI Odometry Sequences 09 and 10. '-' means the result is not available from the paper.

	Method	Seq. 09			Seq. 10		
		RMSE	RelTr	RelRot	RMSE	RelTr	RelRot
Supervised	DeepVO [47]	–	–	–	–	8.11	0.088
	ESP-VO [48]	–	–	-	–	9.77	0.102
	GFS-VO [50]	–	–	–	–	6.32	0.023
	GFS-VO-RNN [50]	–	–	–	–	7.44	0.032
	BeyondTracking [51]	–	–	–	–	**3.94**	**0.017**
	DeepV2D [42]	79.06	8.71	0.037	48.49	12.81	0.083
Self-Supervised	SfMLearner [60]	**24.31**	8.28	0.031	20.87	12.20	**0.030**
	GeoNet [57]	158.45	28.72	0.098	43.04	23.90	0.090
	Depth-VO [58]	–	11.93	0.039	–	12.45	0.035
	vid2depth [29]	–	–	–	–	21.54	0.125
	UnDeepVO [24]	–	7.01	0.036	-	10.63	0.046
	Wang et al. [45]	–	9.88	0.034	-	12.24	0.052
	CC[35]	29.00	**6.92**	**0.018**	**13.77**	7.97	0.031
	DeepMatchVO [37]	27.08	9.91	0.038	24.44	12.18	0.059
	Li et al. [25]	–	8.10	0.028	–	12.90	0.032
	Monodepth2-M [15]	55.47	11.47	0.032	20.46	**7.73**	0.034
	SC-SfMLearer [2]	–	11.2	0.034	–	10.1	0.050
	RGB ORB-SLAM	18.34	7.42	**0.004**	8.90	5.85	**0.004**
	pRGBD-Initial	12.21	4.26	0.011	8.30	5.55	0.017
	pRGBD-Refined	**11.97**	**4.20**	0.010	**6.35**	4.40	0.016

Results on KITTI Odometry Test Set. The KITTI Odometry leaderboard requires complete camera trajectories of all frames of all the sequences. Since

Table 5. Pose evaluation results on KITTI Odometry test set. Since the ground truth for the KITTI Odometry test set is not available we run Stereo ORB-SLAM [31] to get the complete camera trajectories and use them as the pseudo ground truth to evaluate. 'X' denotes tracking failure.

Seq	RGB ORB-SLAM			pRGBD-Initial			pRGBD-Refined		
	RMSE	RelTr	RelRot	RMSE	RelTr	RelRot	RMSE	RelTr	RelRot
11	14.83	7.69	**0.003**	6.68	3.28	0.016	**3.64**	**2.96**	0.015
13	6.58	2.39	**0.006**	6.83	2.52	0.008	**6.43**	**2.31**	0.007
14	4.81	5.19	**0.004**	4.30	4.14	0.014	**2.15**	**3.06**	0.014
15	3.67	1.78	**0.004**	2.58	1.61	0.005	**2.07**	**1.33**	0.004
16	6.21	2.66	**0.002**	5.78	2.14	0.006	**4.65**	**1.90**	0.004
18	6.63	2.38	**0.002**	5.50	2.30	0.008	**4.37**	**2.21**	0.006
19	18.68	4.91	**0.002**	23.96	2.82	0.007	**13.85**	**2.52**	0.006
20	9.19	6.74	**0.016**	8.94	5.43	0.027	**7.03**	**4.50**	0.022
12	X	X	X	X	X	X	**94.2**	**32.94**	0.026
17	X	X	X	14.71	8.98	**0.011**	12.23	7.23	0.011
21	X	X	X	X	X	X	X	X	X

Table 6. Pose evaluation results on two *frieburg3* RGB-D sequences, walking_xyz (W) and large_cabinet_validation (L).

	RGB SLAM			pRGBD-Initial			pRGBD-Refined		
	RMSE	RlTr	RlRot	RMSE	RlTr	RlRot	RMSE	RlTr	RlRot
W	X	X	X	0.23	0.02	0.52	**0.09**	**0.01**	**0.30**
L	1.72	0.02	**0.32**	1.40	**0.01**	0.34	**0.39**	**0.01**	0.33

we keep the default setting from ORB-SLAM, causing tracking failures in a few sequences, to facilitate quantitative evaluation on this test set (*i.e.*, sequences 11–21), we use pseudo-ground-truth computed as mentioned in Sect. 4.1 to evaluate all the competing methods in Table 5. From the results, RGB ORB-SLAM fails on three challenging sequences due to tracking failures, whereas our pRGBD-Initial fails on two sequences and our pRGBD-Refined fails only on one sequence. Among the sequences where all the competing methods succeed, our pRGBD-Initial reduces the RMSEs of RGB ORB-SLAM by a considerable margin for all sequences except for sequence 19. After our self-improving mechanism, our pRGBD-Refined further boosts the performance, reaching the best results both in terms of RMSE and Rel Tr. Figure 5(b) shows qualitative comparisons on sequence 19.

Results on TUM RGB-D Sequences. Performance of pose refinement step on the two TUM RGB-D sequences is shown in Table 6. The result shows increased robustness and accuracy by pRGBD-Refined. In particular, RGB

ORB-SLAM fails on walking_xyz, while pRGBD-Refined succeeds and achieves the best performance on both sequences. Refer supplementary material for qualitative results.

5   Analysis of Self-improving Loops

In this section, we analyze the behaviour of three different evaluation metrics for depth estimation: Squared Relative (Sq Rel) error, RMSE error and accuracy metric a2, as defined in Sect. 4. The pose estimation is evaluated using the absolute trajectory pose error. In Fig. 6, we use the KITTI Eigen split dataset and report these metrics for each iteration of the self-improving loop. The evaluation metrics corresponding to the 0th self-improving loop are of the pre-trained MonoDepth2-M. We summarize the findings from the plots in Fig. 6 as below:

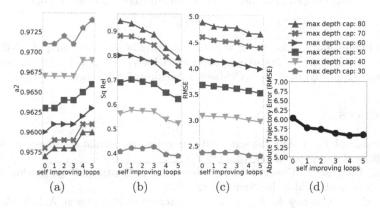

Fig. 6. Depth/Pose evaluation metric w.r.t. self-improving loops. Depth evaluation metrics in (a–c) are computed at different max depth caps ranging from 30–80 m.

- A comparison of evaluation metrics of farther scene points (*e.g.*max depth 80) with nearby points (*e.g.*max depth 30) at the 0th self-improving loop shows that the pre-trained MonoDepth2 performs poorly for farther scene points compared to nearby points.
- In the subsequent self-improving loops, we can see the rate of reduction in the Sq Rel and RMSE error is significant for farther away points compared to nearby points, *e.g.*, slope of error curves in Fig. 6(b–c) corresponding to max depth 80 is steeper than that of max depth 30. This validates our hypothesis of including wider baseline losses that help the depth network predict more accurate depth values for farther points. Overall, our joint narrow and wide baseline based learning setup helps improve the depth prediction of both the nearby and farther away points, and outperforms MonoDepth2 [15].

- The error plot in Fig. 6(d) shows a decrease in pose error with self-improving loops and complements the improvement in depth evaluation metrics as shown in Fig.6(a)–(c). We terminate the self-improvement loop once there is no further improvement, *i.e.*, at the 5th iteration.

6   Conclusion

In this work, we propose a self-improving framework to couple geometrical and learning based methods for 3D perception. A win-win situation is achieved—both the monocular SLAM and depth prediction are improved by a significant margin without any additional active depth sensor or ground truth label. Currently, our self-improving framework only works in an off-line mode, so developing an online real-time self-improving system remains one of our future works. Another avenue for our future works is to move towards more challenging settings [52], e.g., rolling shutter cameras [36,61] or uncalibrated cameras [17,62].

Acknowledgements. This work was part of L. Tiwari's internship at NEC Labs America, in San Jose. L. Tiwari was supported by Visvesvarya Ph.D. Fellowship. S. Anand was supported by Infosys Center for Artificial Intelligence, IIIT-Delhi.

References

1. Andraghetti, L., et al.: Enhancing self-supervised monocular depth estimation with traditional visual odometry. In: 3DV (2019)
2. Bian, J.W., et al.: Unsupervised scale-consistent depth and ego-motion learning from monocular video. In: NeurIPS (2019)
3. Bloesch, M., Czarnowski, J., Clark, R., Leutenegger, S., Davison, A.J.: CodeSLAM-learning a compact, optimisable representation for dense visual slam. In: CVPR (2018)
4. Casser, V., Pirk, S., Mahjourian, R., Angelova, A.: Depth prediction without the sensors: leveraging structure for unsupervised learning from monocular videos. In: AAAI (2019)
5. Chen, Y., Schmid, C., Sminchisescu, C.: Self-supervised learning with geometric constraints in monocular video: connecting flow, depth, and camera. In: ICCV (2019)
6. Davison, A.J., Reid, I.D., Molton, N.D., Stasse, O.: Monoslam: real-time single camera slam. IEEE Trans. Pattern Anal. Mach. Intell. **29**(6), 1052–1067 (2007)
7. Dhiman, V., Tran, Q.H., Corso, J.J., Chandraker, M.: A continuous occlusion model for road scene understanding. In: CVPR (2016)
8. Eigen, D., Puhrsch, C., Fergus, R.: Depth map prediction from a single image using a multi-scale deep network. In: NeurIPS (2014)
9. Engel, J., Koltun, V., Cremers, D.: Direct sparse odometry. IEEE Trans. Pattern Anal. Mach. Intell. **40**(3), 611–625 (2017)
10. Engel, J., Schöps, T., Cremers, D.: LSD-SLAM: large-scale direct monocular SLAM. In: Fleet, D., Pajdla, T., Schiele, B., Tuytelaars, T. (eds.) ECCV 2014. LNCS, vol. 8690, pp. 834–849. Springer, Cham (2014). https://doi.org/10.1007/978-3-319-10605-2_54

11. Forster, C., Pizzoli, M., Scaramuzza, D.: SVO: fast semi-direct monocular visual odometry. In: ICRA (2014)
12. Fu, H., Gong, M., Wang, C., Batmanghelich, K., Tao, D.: Deep ordinal regression network for monocular depth estimation. In: CVPR (2018)
13. Garg, R., Bg, V.K., Carneiro, G., Reid, I.: Unsupervised CNN for single view depth estimation: geometry to the rescue. In: Leibe, B., Matas, J., Sebe, N., Welling, M. (eds.) ECCV 2016. LNCS, vol. 9912, pp. 740–756. Springer, Cham (2016). https://doi.org/10.1007/978-3-319-46484-8_45
14. Geiger, A., Lenz, P., Urtasun, R.: Are we ready for autonomous driving? The KITTI vision benchmark suite. In: CVPR (2012)
15. Godard, C., Aodha, O.M., Firman, M., Brostow, G.J.: Digging into self-supervised monocular depth estimation. In: ICCV (2019)
16. Godard, C., Mac Aodha, O., Brostow, G.J.: Unsupervised monocular depth estimation with left-right consistency. In: CVPR (2017)
17. Gordon, A., Li, H., Jonschkowski, R., Angelova, A.: Depth from videos in the wild: unsupervised monocular depth learning from unknown cameras. In: CVPR (2019)
18. Grupp, M.: EVO: python package for the evaluation of odometry and slam (2017). https://github.com/MichaelGrupp/evo
19. Guo, X., Li, H., Yi, S., Ren, J., Wang, X.: Learning monocular depth by distilling cross-domain stereo networks. In: Ferrari, V., Hebert, M., Sminchisescu, C., Weiss, Y. (eds.) ECCV 2018. LNCS, vol. 11215, pp. 506–523. Springer, Cham (2018). https://doi.org/10.1007/978-3-030-01252-6_30
20. Hartley, R., Zisserman, A.: Multiple View Geometry in Computer Vision. Cambridge University Press, Cambridge (2003)
21. Klein, G., Murray, D.: Parallel tracking and mapping for small AR workspaces. In: ISMAR (2007)
22. Klodt, M., Vedaldi, A.: Supervising the new with the old: learning SFM from SFM. In: Ferrari, V., Hebert, M., Sminchisescu, C., Weiss, Y. (eds.) ECCV 2018. LNCS, vol. 11214, pp. 713–728. Springer, Cham (2018). https://doi.org/10.1007/978-3-030-01249-6_43
23. Kuznietsov, Y., Stuckler, J., Leibe, B.: Semi-supervised deep learning for monocular depth map prediction. In: CVPR (2017)
24. Li, R., Wang, S., Long, Z., Gu, D.: UnDeepVO: monocular visual odometry through unsupervised deep learning. In: ICRA (2018)
25. Li, Y., Ushiku, Y., Harada, T.: Pose graph optimization for unsupervised monocular visual odometry. In: ICRA (2019)
26. Liu, F., Shen, C., Lin, G., Reid, I.: Learning depth from single monocular images using deep convolutional neural fields. IEEE Trans. Pattern Anal. Mach. Intell. **38**(10), 2024–2039 (2015)
27. Loo, S.Y., Amiri, A.J., Mashohor, S., Tang, S.H., Zhang, H.: CNN-SVO: improving the mapping in semi-direct visual odometry using single-image depth prediction. In: ICRA (2019)
28. Luo, C., et al.: Every pixel counts++: joint learning of geometry and motion with 3D holistic understanding. arXiv preprint arXiv:1810.06125 (2018)
29. Mahjourian, R., Wicke, M., Angelova, A.: Unsupervised learning of depth and ego-motion from monocular video using 3D geometric constraints. In: CVPR (2018)
30. Mur-Artal, R., Montiel, J.M.M., Tardos, J.D.: ORB-SLAM: a versatile and accurate monocular slam system. IEEE Trans. Robot. **31**(5), 1147–1163 (2015)
31. Mur-Artal, R., Tardós, J.D.: ORB-SLAM2: an open-source slam system for monocular, stereo, and RGB-D cameras. IEEE Trans. Robot. **33**(5), 1255–1262 (2017)

32. Newcombe, R.A., Lovegrove, S.J., Davison, A.J.: DTAM: dense tracking and mapping in real-time. In: ICCV (2011)
33. Pillai, S., Ambruş, R., Gaidon, A.: Superdepth: self-supervised, super-resolved monocular depth estimation. In: ICRA (2019)
34. Poggi, M., Tosi, F., Mattoccia, S.: Learning monocular depth estimation with unsupervised trinocular assumptions. In: 3DV (2018)
35. Ranjan, A., et al.: Competitive collaboration: joint unsupervised learning of depth, camera motion, optical flow and motion segmentation. In: CVPR (2019)
36. Schubert, D., Demmel, N., Usenko, V., Stuckler, J., Cremers, D.: Direct sparse odometry with rolling shutter. In: ECCV (2018)
37. Shen, T., et al.: Beyond photometric loss for self-supervised ego-motion estimation. In: ICRA (2019)
38. Sheng, L., Xu, D., Ouyang, W., Wang, X.: Unsupervised collaborative learning of keyframe detection and visual odometry towards monocular deep slam. In: ICCV (2019)
39. Song, S., Chandraker, M.: Robust scale estimation in real-time monocular SFM for autonomous driving. In: CVPR (2014)
40. Sturm, J., Engelhard, N., Endres, F., Burgard, W., Cremers, D.: A benchmark for the evaluation of RGB-D slam systems. In: IROS (2012)
41. Tateno, K., Tombari, F., Laina, I., Navab, N.: CNN-SLAM: real-time dense monocular slam with learned depth prediction. In: CVPR (2017)
42. Teed, Z., Deng, J.: DeepV2D: video to depth with differentiable structure from motion. arXiv preprint arXiv:1812.04605 (2018)
43. Tosi, F., Aleotti, F., Poggi, M., Mattoccia, S.: Learning monocular depth estimation infusing traditional stereo knowledge. In: CVPR (2019)
44. Wang, C., Miguel Buenaposada, J., Zhu, R., Lucey, S.: Learning depth from monocular videos using direct methods. In: CVPR (2018)
45. Wang, R., Pizer, S.M., Frahm, J.M.: Recurrent neural network for (un-) supervised learning of monocular video visual odometry and depth. In: CVPR (2019)
46. Wang, R., Schworer, M., Cremers, D.: Stereo DSO: large-scale direct sparse visual odometry with stereo cameras. In: ICCV (2017)
47. Wang, S., Clark, R., Wen, H., Trigoni, N.: DeepVO: towards end-to-end visual odometry with deep recurrent convolutional neural networks. In: ICRA (2017)
48. Wang, S., Clark, R., Wen, H., Trigoni, N.: End-to-end, sequence-to-sequence probabilistic visual odometry through deep neural networks. Int. J. Robot. Res. 37(4–5), 513–542 (2018)
49. Watson, J., Firman, M., Brostow, G.J., Turmukhambetov, D.: Self-supervised monocular depth hints. In: ICCV (2019)
50. Xue, F., Wang, Q., Wang, X., Dong, W., Wang, J., Zha, H.: Guided feature selection for deep visual odometry. In: Jawahar, C.V., Li, H., Mori, G., Schindler, K. (eds.) ACCV 2018. LNCS, vol. 11366, pp. 293–308. Springer, Cham (2019). https://doi.org/10.1007/978-3-030-20876-9_19
51. Xue, F., Wang, X., Li, S., Wang, Q., Wang, J., Zha, H.: Beyond tracking: selecting memory and refining poses for deep visual odometry. In: CVPR (2019)
52. Yang, N., Wang, R., Gao, X., Cremers, D.: Challenges in monocular visual odometry: photometric calibration, motion bias, and rolling shutter effect. IEEE Robot. Autom. Lett. 3(4), 2878–2885 (2018)
53. Yang, N., Wang, R., Stückler, J., Cremers, D.: Deep virtual stereo odometry: leveraging deep depth prediction for monocular direct sparse odometry. In: Ferrari, V., Hebert, M., Sminchisescu, C., Weiss, Y. (eds.) ECCV 2018. LNCS, vol. 11212, pp. 835–852. Springer, Cham (2018). https://doi.org/10.1007/978-3-030-01237-3_50

54. Yang, Z., Wang, P., Wang, Y., Xu, W., Nevatia, R.: Lego: learning edge with geometry all at once by watching videos. In: CVPR (2018)
55. Yang, Z., Wang, P., Xu, W., Zhao, L., Nevatia, R.: Unsupervised learning of geometry with edge-aware depth-normal consistency. In: AAAI (2018)
56. Yin, X., Wang, X., Du, X., Chen, Q.: Scale recovery for monocular visual odometry using depth estimated with deep convolutional neural fields. In: ICCV (2017)
57. Yin, Z., Shi, J.: GeoNet: unsupervised learning of dense depth, optical flow and camera pose. In: CVPR (2018)
58. Zhan, H., Garg, R., Saroj Weerasekera, C., Li, K., Agarwal, H., Reid, I.: Unsupervised learning of monocular depth estimation and visual odometry with deep feature reconstruction. In: CVPR (2018)
59. Zhou, L., Kaess, M.: Windowed bundle adjustment framework for unsupervised learning of monocular depth estimation with u-net extension and clip loss. IEEE Robot. Autom. Lett. 5(2), 3283–3290 (2020)
60. Zhou, T., Brown, M., Snavely, N., Lowe, D.G.: Unsupervised learning of depth and ego-motion from video. In: CVPR (2017)
61. Zhuang, B., Tran, Q.H.: Image stitching and rectification for hand-held cameras. In: ECCV (2020)
62. Zhuang, B., Tran, Q.H., Lee, G.H., Cheong, L.F., Chandraker, M.: Degeneracy in self-calibration revisited and a deep learning solution for uncalibrated slam. In: IROS (2019)
63. Zou, Y., Ji, P., Tran, Q.H., Huang, J.B., Chandraker, M.: Learning monocular visual odometry via self-supervised long-term modeling. In: ECCV (2020)
64. Zou, Y., Luo, Z., Huang, J.-B.: DF-Net: unsupervised joint learning of depth and flow using cross-task consistency. In: Ferrari, V., Hebert, M., Sminchisescu, C., Weiss, Y. (eds.) ECCV 2018. LNCS, vol. 11209, pp. 38–55. Springer, Cham (2018). https://doi.org/10.1007/978-3-030-01228-1_3

Interpretable and Generalizable Person Re-identification with Query-Adaptive Convolution and Temporal Lifting

Shengcai Liao[1(✉)] and Ling Shao[1,2]

[1] Inception Institute of Artificial Intelligence (IIAI), Abu Dhabi, UAE
{scliao,ling.shao}@ieee.org
[2] Mohamed bin Zayed University of Artificial Intelligence, Abu Dhabi, UAE

Abstract. For person re-identification, existing deep networks often focus on representation learning. However, without transfer learning, the learned model is fixed as is, which is not adaptable for handling various unseen scenarios. In this paper, beyond representation learning, we consider how to formulate person image matching directly in deep feature maps. We treat image matching as finding local correspondences in feature maps, and construct query-adaptive convolution kernels on the fly to achieve local matching. In this way, the matching process and results are interpretable, and this explicit matching is more generalizable than representation features to unseen scenarios, such as unknown misalignments, pose or viewpoint changes. To facilitate end-to-end training of this architecture, we further build a class memory module to cache feature maps of the most recent samples of each class, so as to compute image matching losses for metric learning. Through direct cross-dataset evaluation, the proposed Query-Adaptive Convolution (QAConv) method gains large improvements over popular learning methods (about 10%+ mAP), and achieves comparable results to many transfer learning methods. Besides, a model-free temporal cooccurrence based score weighting method called TLift is proposed, which improves the performance to a further extent, achieving state-of-the-art results in cross-dataset person re-identification. Code is available at https://github.com/ShengcaiLiao/QAConv.

1 Introduction

Person re-identification is an active research topic in computer vision. It aims at finding the same person as the query image from a large volume of gallery images. With the progress in deep learning, person re-identification has been largely advanced in recent years. However, when generalization ability becomes an important concern, required by practical applications, existing methods usually lack satisfactory performance, evidenced by direct cross-dataset evaluation

Electronic supplementary material The online version of this chapter (https://doi.org/10.1007/978-3-030-58621-8_27) contains supplementary material, which is available to authorized users.

[10,53]. To address this, many transfer learning, domain adaptation, and unsupervised learning methods, performed on the target domain, have been proposed. However, these methods require heavy computations in deployment, limiting their application in practical scenarios where the deployment machine may have limited resources to support deep learning and users may cannot wait for a time-consuming adaptation stage. Therefore, improving the baseline model's generalization ability to support ready usage is still of urgent importance.

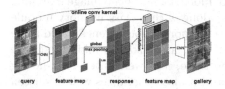

Fig. 1. QAConv constructs adaptive convolution kernels on the fly from query feature maps, and perform convolutions and max pooling on gallery feature maps to find the best local correspondences.

Fig. 2. Examples of the interpreted local correspondences from the outputs of the QAConv.

Most existing person re-identification methods compute a fixed representation vector, also known as a feature vector, for each image, and employ a typical distance or similarity metric (e.g. Euclidean distance or cosine similarity) for image matching. Without domain adaptation or transfer learning, the learned model is fixed as is, which is not adaptable for handling various unseen scenarios. Therefore, when generalization ability is a concern, it is expected to have an adaptive ability for the given model architecture.

In this paper, we focus on generalizable and ready-to-use person re-identification, through direct cross-dataset evaluation. Beyond representation learning, we consider how to formulate query-adaptive image matching directly in deep feature maps. Specifically, we treat image matching as finding local correspondences in feature maps, and construct query-adaptive convolution kernels on the fly to achieve local matching (see Fig. 1). In this way, the learned model benefits from adaptive convolution kernels in the final layer, specific to each image, and the matching process and result are interpretable (see Fig. 2), similar to traditional feature correspondence approaches [2,26]. Probably because finding local correspondences through query-adaptive convolution is a common process among different domains, this explicit matching is more generalizable than representation features to unseen scenarios, such as unknown misalignments, pose or viewpoint changes. We call this Query-Adaptive Convolution QAConv. To facilitate end-to-end training of this architecture, we further build a class memory module to cache feature maps of the most recent samples of each class, so as to compute image matching losses for metric learning.

Through direct cross-dataset evaluation without further transfer learning, the proposed method achieves comparable results to many transfer learning methods for person re-identification. Besides, to explore the prior spatial-temporal structure of a camera network, a model-free temporal cooccurrence based score weighting method is proposed, named Temporal Lifting (TLift). This is also computed on the fly for each query image, without statistical learning of a transition time model in advance. As a result, TLift improves person re-identification to a further extent, resulting in state-of-the-art results in cross-dataset evaluations.

To summarize, the novelty of this work include (i) a new deep image matching approach with query-adaptive convolutions, along with a class memory module for end-to-end training, and (ii) a model-free temporal cooccurrence based score weighting method. The advantages of this work are also two-fold. First, the proposed image matching method is interpretable, it is well-suited in handling misalignments, pose or viewpoint changes, and it also generalizes well in unseen domains. Second, both QAConv and TLift can be computed on the fly, and they are complementary to many other methods. For example, QAConv can serve as a better pre-trained model for transfer learning, and TLift can be readily applied by most person re-identification algorithms as a post-processing step.

2   Related Works

Deep learning approaches have largely advanced person re-identification in recent years [52]. However, due to limited labeled data and a big diversity in real-world surveillance, these methods usually have poor generalization ability in unseen scenarios. To address this, many unsupervised domain adaption (UDA) methods have been proposed [3,7,15,16,30,43,51,55,64], which show improved cross-dataset results than traditional methods, though requiring further training on the target domain. QAConv is orthogonal to transfer learning methods as it can provide a better baseline model for them (see Sect. 5.4 and Table 3).

There are many representation learning methods proposed to deal with viewpoint changes and misalignments in person re-identification, such as part-aligned feature representations [38,39,42,58], pose-adapted feature representations [33,57], human parsing based representations [14], local neighborhood matching [1,17], and attentional networks [18,24,25,31,35,49,50]. While these methods present high accuracy when trained and tested on the same dataset, their generalization ability to other datasets is mostly unknown. Besides, beyond representation learning, QAConv focuses on image matching via local correspondences.

Generalizable person re-identification was first studied in our previous works [10,53], where direct cross-dataset evaluation was proposed. More recently, Song et al. [36] proposed a domain-invariant mapping network by meta-learning, and Jia et al. [12] applied the IBN-Net [29] to improve generalizability, while QAConv is preliminarily reported in [19]. QAConv is orthogonal to methods of network design, for example, it can also be applied on the IBN-Net for improvements.

For deep feature matching, Kronecker-Product Matching (KPM) [34] computes a cosine similarity map by outer product for softly aligned element-wise subtraction. Besides, Bilinear Pooling [22,37,40] and Non-local Neural Networks [44] also apply the outer product for part-aligned or self-attended representation learning. Different to the above methods, QAConv is a convolutional matching method but not simply outer product especially when its kernel size $s > 1$. It is explicitly designed for local correspondence matching, interpretation, and generalization, in a straightforward way without other branches.

For post-processing, re-ranking is a technique of refining matching scores, which further improves person re-identification [23,33,56,62]. Besides, temporal information is also a useful cue to facilitate cross-camera person re-identification [27,41]. While existing methods model transition times across different cameras but encounter difficulties in complex transition time distributions, the proposed TLift method applies cooccurrence constraint within each camera to avoid estimating transition times, and it is model-free and can be computed on the fly.

For memory based loss, ECN [64] proposed an exemplar memory which caches feature vectors of every instance for UDA. This makes the instance-level label inference convenient but limits its scalability. In contrast, class memory is independently designed [19], which is more efficient working in class level.

3   Query-Adaptive Convolution

3.1   Query-Adaptive Convolutional Matching

For face recognition and person re-identification, most existing methods do not explicitly consider the relationship between two input images under matching, but instead, like classification, they treat each image independently and apply the learned model to extract a fixed feature representation. Then, image matching is simply a distance measure between two representation vectors, regardless of the direct relationship between the actual contents of the two images.

In this paper, we consider the relationship between two images, and try to formulate adaptive image matching directly in deep feature maps. Specifically, we treat image matching as finding local correspondences in feature maps, and construct query-adaptive convolution kernels on the fly to achieve local matching. As shown in Fig. 1 and Fig. 3, to match two images, each image is firstly fed forward into a backbone CNN, resulting in a final feature map of size $[1, d, h, w]$, where d is the number of output channels, and h and w are the height and width of the feature map, respectively. Then, the channel dimension of both feature maps is normalized by the $\ell2$-norm. After that, local patches of size $[s, s]$ at every location of the query feature map are extracted, and then reorganized into $[hw, d, s, s]$ as a convolution kernel, with input channels d, output channels hw, and kernel size $[s, s]$. This acts as a query-adaptive convolution kernel, with parameters constructed on the fly from the input, in contrast to fixed convolution kernels in the learned model. Upon this, the adaptive kernel can be used to perform a convolution on another feature map, resulting in $[1, hw, h, w]$ similarities.

Since feature channels are $\ell2$-normalized, when $s = 1$, the convolution in fact measures the cosine similarity at every location of the two feature maps. Besides, since the convolution kernel is adaptively constructed from the image content, these similarity values exactly reflect the local matching results between the two input images. Therefore, an additional global max pooling (GMP) operation will output the best local matches, and the maximum indices found by GMP indicate the best locations of local correspondences, which can be further used to interpret the matching result, as shown in Fig. 2. Note that GMP can also be done along the hw axis of the $[1, hw, h, w]$ similarity map. That is, seeking the best matches can be carried out from both sides of the images. Concatenating the output will result in a similarity vector of size $2hw$ for each pair of images.

3.2 Network Architecture

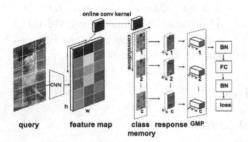

Fig. 3. Architecture of the QAConv. GMP: global max pooling. BN: batch normalization. FC: fully connection.

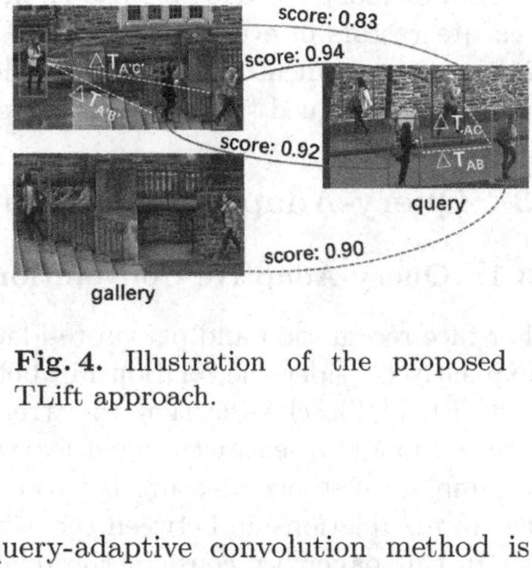

Fig. 4. Illustration of the proposed TLift approach.

The architecture of the proposed query-adaptive convolution method is shown in Fig. 3, which consists of a backbone CNN, the QAConv layer for local matching, a class memory layer for training, a global max pooling layer, a BN-FC-BN block, and, finally, a similarity output by a sigmoid function for evaluation in the test phase or loss computation in the training phase. The output size of the FC layer is 1, which acts as a binary classifier or a similarity metric, indicating whether or not one pair of images belongs to the same class. The two BN (batch normalization [11]) layers are all one-dimensional. They are used to normalize the similarity output and stabilize the gradient during training.

3.3 Class Memory and Update

We propose a class memory module to facilitate the end-to-end training of the QAConv network. Specifically, a $[c, d, h, w]$ tensor buffer is registered, where c is the number of classes. For each mini batch of size b, the $[b, d, h, w]$ feature map tensor of the mini batch will be updated into the memory buffer. We use a direct assignment update strategy, that is, each $[1, d, h, w]$ sample of class i from the mini batch will be assigned into location i of the $[c, d, h, w]$ memory buffer.

An exponential moving average update can also be used here. However, in our experience this is inferior to the direct replacement update. There might be two reasons for this. First, the replacement update caches feature maps of the most recent samples of each class, so as to reflect the most up-to-date state of the current model for loss computation. Second, since our task is to carry out image matching with local details in feature maps for correspondences, exponential moving average may smooth the local details of samples from the same class.

3.4 Loss Function

With a mini batch of size $[b, d, h, w]$ and class memory of size $[c, d, h, w]$, $b \times c$ pairs of similarity values will be computed by QAConv after the BN-FC-BN block. We use a sigmoid function to map the similarity values into $[0, 1]$, and compute the binary cross entropy loss. Since there are far more negative than positive pairs, to balance them and enable online hard example mining, we apply the focal loss [21] to weight the binary cross entropy. That is,

$$\ell(\theta) = -\frac{1}{b} \sum_{i=1}^{b} \sum_{j=1}^{c} (1 - \hat{p}_{ij}(\theta))^{\gamma} log(\hat{p}_{ij}(\theta)), \tag{1}$$

where θ is the network parameter, $\gamma = 2$ is the focusing parameter [21], and

$$\hat{p}_{ij} = \begin{cases} p_{ij} & \text{if } y_{ij} = 1, \\ 1 - p_{ij} & \text{otherwise,} \end{cases} \tag{2}$$

where $y_{ij} = 1$ indicates a positive pair, while a negative pair otherwise, and $p_{ij} \in [0, 1]$ is the sigmoid probability.

4   Temporal Lifting

For person re-identification, to explore the prior spatial-temporal structure of a camera network, usually a transition time model is learned to measure the transition probability. However, for a complex camera network and various person transition patterns, it is not easy to learn a robust transition time distribution. In contrast, in this paper a model-free temporal cooccurrence based score weighting method is proposed, which is called Temporal Lifting (TLift). TLift does not model cross-camera transition times which could be variable and complex. Instead, TLift makes use of a group of nearby persons in each single camera, and find similarities between them.

Figure 4 illustrates the idea. A basic assumption is that people nearby in one camera are likely still nearby in another camera. Therefore, their corresponding matches in other cameras can serve as pivots to enhance the weights of other nearby persons. In Fig. 4, A is the query person. E is more similar than A' to A in another camera. With nearby persons B and C, and their top retrievals B' and C' acting as pivots, the matching score of A' can be temporally lifted since it is a nearby person of B' and C', while the matching score of E will be reduced since there is no such pivot.

Formally, suppose A is the query person in camera Q, then, the set of nearby persons to A in camera Q is defined as $R = \{B|\Delta T_{AB} < \tau, \forall B \in Q\}$, where ΔT_{AB} is the within-camera time difference between persons A and B, and τ is a threshold on ΔT to define nearby persons. Then, for each person in R, cross-camera person retrieval will be performed on a gallery camera G by QAConv or other methods, and the overall top K retrievals for R are defined as the pivot set P. Next, each person in P acts as an ensemble point for 1D kernel density estimation on within-camera time differences in G, and the temporal matching probability between A and any person X in camera G will be computed as

$$p_t(A, X) = \frac{1}{|P|} \sum_{B \in P} e^{-\frac{\Delta T_{BX}^2}{\sigma^2}}, \tag{3}$$

where σ is the sensitivity parameter of the time difference. Then, this temporal probability is used to weight the similarity score of appearance models using a multiplication fusion as $p(A, X) = (p_t(A, X) + \alpha)p_a(A, X)$, where $p_a(A, X)$ is the appearance based matching probability (e.g. by QAConv), and α is a regularizer.

This way, true positives near pivots will be lifted, while hard negatives far from pivots will be suppressed. Note that this is also computed on the fly for each query image, without learning a transition time model in advance. Therefore, it does not require training data, and can be readily applied by many other person re-identification methods.

5   Experiments

5.1   Implementation Details

The proposed method is implemented in PyTorch, based upon an adapted version [61] of the open source person re-identification library (open-reid)[1]. Person images are resized to 384×128. The backbone network is the ResNet-152 [8] pre-trained on ImageNet, unless otherwise stated. The layer3 feature map of the backbone network is used, since the size of the layer4 feature map is too small. A 1×1 convolution with 128 channels is further appended to reduce the final feature map size. The batch size of samples for training is 32. The SGD optimizer is applied, with a learning rate of 0.001 for the backbone network, and 0.01 for newly added layers. They are decayed by 0.1 after 40 epochs, and the training stops at 60 epochs. The whole QAConv is end-to-end jointly trained, while class memory is updated only after the loss computation. Considering the memory consumption and the efficiency, the kernel size of QAConv is set to $s = 1$. Parameters for TLift are $\tau = 100$, $\sigma = 200$, $K = 10$, and $\alpha = 0.2$. They are not sensitive in a broad range, as analyzed in the Appendix.

A random occlusion module is implemented for data augmentation, which is similar to the random erasing [63] and cutout [5] methods (see Appendix for

---

[1] https://cysu.github.io/open-reid/.

comparisons). Specifically, a square area is generated with the size randomly sampled at most $0.8 \times width$ of the image. Then this square area is filled with white pixels. It is useful for QAConv because random occlusion forces QAConv to learn various local correspondences, instead of only saliency but easy ones. Beyond this, only a random horizontal flipping is used for data augmentation.

5.2 Datasets

Experiments were conducted on four large person re-identification datasets, Market-1501 [59], DukeMTMC-reID [6,60], CUHK03 [17], and MSMT17 [45]. The Market-1501 dataset contains 32,668 images of 1501 identities captured from 6 cameras. There are 12,936 images from 751 identities for training, and 19,732 images from 750 identities for testing . The DukeMTMC-reID is a subset of the multi-target and multi-camera pedestrian tracking dataset DukeMTMC [6]. It includes 1,812 identities and 36,411 images, where 16,522 images of 702 identities are used for training, and the remainings for test. The CUHK03 dataset includes 13,164 images of 1,360 pedestrians. We adopted the CUHK03-NP protocol provided in [62], where images of 767 identities were used for training, and other images of 700 identities were used for test. Besides, we used the detected subset for evaluation, which is more challenging. The MSMT17 dataset is the largest person re-identification dataset to date, which contains 4,101 identities and 126,441 images captured from 15 cameras. It is divided into a training set of 32,621 images from 1,041 identities, and a test set with the remaining images from 3,010 identities.

Cross-dataset evaluation was performed in these datasets, by training on the training subset of one dataset (except that in MSMT17 we used all images for training following [51,55]), and evaluating on the test subset of another dataset. The cumulative matching characteristic (CMC) and mean Average Precision (mAP) were used as the performance evaluation metrics. All evaluations followed the single-query evaluation protocol.

The Market-1501 and DukeMTMC-reID datasets are with frame numbers available, so that it is able to evaluate the proposed TLift method. The DukeMTMC-reID dataset has a good global and continuous record of frame numbers, and it is synchronized by providing offset times. In contrast, the Market-1501 dataset has only independent frame numbers for each session of videos from each camera. Accordingly we simply made a cumulative frame record by assuming continuous video sessions. After that, frame numbers were converted to seconds in time by dividing the Frames Per Second (FPS) in video records, where $FPS = 59.94$ for the DukeMTMC-reID dataset and $FPS = 25$ for the Market-1501 dataset.

5.3 Ablation Study

Some ablation studies have been conducted to understand the proposed method, in the context of direct cross-dataset evaluation between the Market-1501 and DukeMTMC-reID datasets. First, to understand the QAConv loss, several other loss functions, including the classical softmax based cross entropy loss, the center loss [13, 46], the Arc loss (derived from the ArcFace method [4] which is effective for face recognition), and the proposed class memory based loss, are evaluated for comparison. For these compared loss functions, the global average pooling of layer4 (better than layer3) of the ResNet-152 is used for feature representation, and the cosine similarity measure is adopted instead of the QAConv similarity. For the class memory loss, feature vectors are cached in memory instead of learnable parameters, and the same BN layer and Eq. (1) are applied after calculating the cosine similarity values between mini-batch features and memory features.

From results shown in Table 1, it is obvious that QAConv improves existing loss functions by a large margin, with 13.7%-19.5% improvements in Rank-1, and 9.6%-11.1% in mAP. Interestingly, large margin classifiers improves the softmax cross-entropy baseline when trained on the Market-1501 dataset, but do not have such improvements when trained on DukeMTMC-reID. This is probably due to many ambiguously labeled or closely walking persons in DukeMTMC-reID (see Sect. 5.5), which may confuse the strict large margin training. Note that the class memory based loss only performs comparable to other existing losses, indicating that the large improvement of QAConv is mainly due to the new matching mechanism, rather than the class memory based loss function. Besides, the Arc loss published recently is one of the best face recognition method, but it does not seem to be powerful when applied in person re-identification[2]. In our experience, the choice of loss functions does not largely influence person re-identification performance. Similar as in face recognition, existing studies [4, 46] show that new loss functions do have improvements, but cannot be regarded as significant ones over the softmax cross entropy baseline. Therefore, we may conclude that the large improvement observed here is due to the new matching scheme, instead of different loss configurations (see Appendix for more analyses).

Table 1. Role of loss functions (%).

Method	Market→Duke		Duke→Market	
	Rank-1	mAP	Rank-1	mAP
Softmax cross-entropy	34.9	18.4	48.5	21.4
Arc loss [4]	35.3	17.1	48.9	21.4
Center loss [13, 46]	38.9	22.1	48.8	22.0
Class memory loss	40.7	21.8	47.8	20.5
QAConv	**54.4**	**33.6**	**62.8**	**31.6**

Next, to understand the role of re-ranking (RR), the k-reciprocal encoding method [62] is applied upon QAConv. From results shown in Table 2, it

---

[2] We have tried different hyper parameters and reported the best results. The best margin values were found to be 0.5 on Market-1501 and 0.2 on DukeMTMC-reID.

can be seen that enabling re-ranking do improve the performance a lot, especially with mAP, which is increased by 18.8% under Market→Duke, and 19.6% under Duke→Market. This improvement is much more significant based on QAConv than that based on other methods as reported in [62]. This is probably because the new QAConv matching scheme better measures the similarity between images, which benefits the reverse neighbor based re-ranking method.

Furthermore, based on QAConv and re-ranking, the contribution of TLift is evaluated, compared to a recent method called TFusion (TF) [28], which is originally designed to iteratively improve transfer learning. From results shown in Table 2, it can be observed that employing TLift to explore temporal information further improves the results, with Rank-1 improved by 8.2%-10.2%, and mAP by 7.0%-8.8%. This improvement is complementary to re-ranking, so they can be combined. As for the existing method TFusion, it appears to be not stable, as a large improvement can be observed under Market→Duke, but little improvement can be obtained under Duke→Market, or even the mAP is clearly decreased[3]. This may be because TFusion is based on learning transition time distributions across cameras, which is not easy to deal with complex camera networks and person transitions as in the Market-1501 (various repeated presences per person in one camera). In contrast, the TLift method only depends on single-camera temporal information which is relatively more easy to handle. Note that TLift can also be generally applied to other methods for improvements, as shown in the Appendix. Besides, as shown in Table 2, directly applying TLift to QAConv without re-ranking also improves the performance a lot.

Table 2. Performance (%) of different post-processing methods.

Method	Market→Duke		Duke→Market	
	Rank-1	mAP	Rank-1	mAP
QAConv	54.4	33.6	62.8	31.6
QAConv + TLift	62.7	45.3	61.5	40.6
QAConv + RR [62]	61.8	52.4	68.5	51.2
QAConv + RR + TF [28]	**70.7**	**61.9**	68.6	47.2
QAConv + RR + TLift	70.0	61.2	**78.7**	**58.2**

Finally, to understand the effect of the backbone network, the QAConv results with the ResNet-50 as backbone are also reported in Tables 3 and 4, compared to the default ResNet-152 (denoted as $QAConv_{50}$ and $QAConv_{152}$, respectively). As can be observed, a larger network ResNet-152 does have a better performance due to its larger learning capability. It can improve the Rank-1 accuracy over the $QAConv_{50}$ by 1.3%-7.3%, and the mAP by 0.8%-5.5%. Besides, there are also consistent improvements in case of combining re-ranking and TLift. Hence, it seems

---

[3] Note that TFusion parameters were optimized on each dataset to get the best results, but for TLift we used fixed parameters for all datasets (see Appendix for analysis).

that this larger network, which contains more learnable parameters, does not have the overfitting problem when equipped with QAConv. Note that, though ResNet-152 is a very large network requiring heavy computation, in practice, it can be efficiently reduced by knowledge distillation [9].

5.4 Comparison to the State of the Arts

There are a great number of person re-identification methods since this is a very active research area. Here we only list recent results for comparison due to limited space. The cross-dataset evaluation results on the four datasets are listed in Tables 3 and 4. Considering that many person re-identification methods employ the ResNet-50 network, for a fair comparison, the following analysis is based on the QAConv$_{50}$ results. Note that this paper mainly focuses on cross-dataset evaluation. Therefore, some recent methods performing unsupervised learning on the target dataset are not compared here, such as the TAUDL [15], UTAL [16], and UGA [48], and also partially due to the fact that they use single-camera target identity labels for training. There are mainly two groups of methods listed in Table 3, namely unsupervised transfer learning based methods, and direct cross-dataset evaluation based methods. The first group of methods require images from the target dataset for unsupervised learning, which are not directly comparable to the second one that directly evaluates on the target

Table 3. Comparison of the state-of-the-art cross-dataset evaluation results (%) with DukeMTMC-reID and Market-1501 as the target datasets.

Method	Training Source	Target	Test: Duke R1	mAP	Training Source	Target	Test: Market R1	mAP
PUL, TOMM18 [7]	Market	Duke	30.4	16.4	Duke	Market	44.7	20.1
TJ-AIDL, CVPR18 [43]	Market	Duke	44.3	23.0	Duke	Market	58.2	26.5
MMFA, BMVC18 [20]	Market	Duke	45.3	24.7	Duke	Market	56.7	27.4
CFSM, AAAI19 [3]	Market	Duke	49.8	27.3	Duke	Market	61.2	28.3
DECAMEL, TPAMI19 [54]	-	-	-	-	Multi	Market	60.2	32.4
PAUL, CVPR19 [51]	Market	Duke	56.1	35.7	Duke	Market	66.7	36.8
ECN, CVPR19 [64]	Market	Duke	63.3	40.4	Duke	Market	75.1	43.0
CDS, ICME19 [47]	Market	Duke	67.2	42.7	Duke	Market	71.6	39.9
ECN baseline, CVPR19 [64]	Market		28.9	14.8	Duke		43.1	17.7
PN-GAN, ECCV18 [32]	Market		29.9	15.8	-		-	-
QAConv$_{50}$	Market		48.8	28.7	Duke		58.6	27.2
QAConv$_{152}$	Market		54.4	33.6	Duke		62.8	31.6
QAConv$_{50}$ + RR + TLift	Market		64.5	55.1	Duke		74.6	51.5
QAConv$_{152}$ + RR + TLift	Market		70.0	61.2	Duke		78.7	58.2
MAR, CVPR19 [55]	MSMT	Duke	67.1	48.0	MSMT	Market	67.7	40.0
PAUL, CVPR19 [51]	MSMT	Duke	72.0	53.2	MSMT	Market	68.5	40.1
MAR baseline, CVPR19 [55]	MSMT		43.1	28.8	MSMT		46.2	24.6
PAUL baseline, CVPR19 [51]	MSMT		65.7	45.6	MSMT		59.3	31.0
QAConv$_{50}$	MSMT		69.4	52.6	MSMT		72.6	43.1
QAConv$_{152}$	MSMT		72.2	53.4	MSMT		73.9	46.6
QAConv$_{50}$ + RR + TLift	MSMT		80.3	77.2	MSMT		86.5	72.2
QAConv$_{152}$ + RR + TLift	MSMT		82.2	78.4	MSMT		88.4	76.0

dataset in consideration of real applications. The proposed QAConv method belongs to the second group. There are very few existing results of the same setting for the second group, except some baselines of other recent methods and the PN-GAN [32] which aims at augmenting source training data by GAN. For the comparison to the transfer learning methods, we consider that QAConv can serve as a better pre-trained model for them, and computing the RR + TLift on the fly is also more efficient than training on target dataset.

DukeMTMC-reID Dataset. As can be observed from Table 3, when trained on the Market-1501 dataset, QAConv achieves the best performance in the direct evaluation group with a large margin. When compared to transfer learning methods, QAConv also outperforms many of them except some very recent methods, indicating that QAConv enables the network to learn how to match two person images, and the learned model generalizes well in unseen domains without transfer learning. Besides, by enabling re-ranking and TLift, the proposed method achieves the best result among all except Rank-1 of CDS. Note that the re-ranking and TLift methods can also be incorporated into other methods, though. Therefore, we list their results separately. However, both of these are calculated on the fly without learning in advance, so together with QAConv, it appears that a ready-to-use method with good generalization ability can also be achieved even without further UDA, which is a nice solution considering that UDA requires heavy computation for deep learning in deployment phase.

When trained on MSMT17, QAConv itself beats all other methods except the transfer learning method PAUL. This is also the second best result among all existing methods taking DukeMTMC-reID as the target dataset, regardless of the training source. This clearly indicates QAConv's superiority in learning from large-scale data. It is preferred in practice in the sense that, when trained with large-scale data, there may be no need to adapt the learned model in deployment.

Market-1501 Dataset. With Market-1501 as the target dataset as shown in Table 3, similarly, when trained with MSMT17, QAConv itself also achieves the best performance among others except Rank-1 of ECN. This can be considered a large advancement in cross-dataset evaluation, which is a better evaluation strategy for understanding the generalization ability of algorithms. Besides, when equipped with RR+TLift, the proposed method achieves the state of the art, with Rank-1 accuracy of 86.5% and mAP of 72.2%. Note that this comparison is not in a sense of *fair*. We would like to share that beyond many recent efforts in UDA, enlarging the training data and exploiting on-the-fly computations in re-ranking and temporal fusion may also lead to good performance in unknown domain, with the advantage of no cost in training deep models everywhere.

CUHK03 Dataset. The CUHK03 and MSMT17 datasets present large domain gaps to others. For CUHK03, it can be observed from Table 4 that, with either Market-1501 or DukeMTMC-reID dataset as training set, QAConv without UDA performs better than a UDA method PUL [7], and fairly comparable to another recent transfer learning method CDS [47]. However, all methods perform not well

Table 4. Comparison of the state-of-the-art cross-dataset evaluation results (%) with CUHK03-NP (detected) and MSMT17 as the target datasets.

Method	Training		Test: CUHK03		Training		Test: MSMT	
	Source	Target	R1	mAP	Source	Target	R1	mAP
PUL, TOMM18 [7]	Market	CUHK03	7.6	7.3	-	-	-	-
CDS, ICME19 [47]	Market	CUHK03	9.1	8.7	-	-	-	-
PTGAN [45], CVPR18	-	-	-	-	Market	MSMT	10.2	2.9
ECN, CVPR19 [64]	-	-	-	-	Market	MSMT	25.3	8.5
QAConv$_{50}$	Market		9.9	8.6	Market		22.6	7.0
QAConv$_{152}$	Market		14.1	11.8	Market		25.6	8.2
PUL, TOMM18 [7]	Duke	CUHK03	5.6	5.2	-	-	-	-
CDS, ICME19 [47]	Duke	CUHK03	8.1	7.1	-	-	-	-
PTGAN [45], CVPR18	-	-	-	-	Duke	MSMT	11.8	3.3
ECN, CVPR19 [64]	-	-	-	-	Duke	MSMT	30.2	10.2
QAConv$_{50}$	Duke		7.9	6.8	Duke		29.0	8.9
QAConv$_{152}$	Duke		11.0	9.4	Duke		32.7	10.4
QAConv$_{50}$	MSMT		25.3	22.6	-	-	-	-
QAConv$_{152}$	MSMT		32.6	28.1	-	-	-	-

on the CUHK03 dataset. Only with the large MSMT17 data set as the source training data, the proposed method performs relatively better.

MSMT17 Dataset. With the MSMT17 as target, only QAConv does not require adaptation in Table 4. However, it performs better than PTGAN [45] and in part comparable to ECN [64]. This further confirms the generalizability of QAConv under large domain gaps, since without UDA it is already in part comparable to the state-of-the-art UDA methods. Note that TLift is not applicable on CUHK03 and MSMT17 due to no temporal information provided.

5.5 Qualitative Analysis and Discussion

A unique characteristic of the proposed QAConv method is its interpretation ability of the matching. Therefore, we show some qualitative matching results in Fig. 5 for a better understanding of the proposed method. The model used here is trained on the MSMT17 dataset, and the evaluations are done on the query subsets of the Market-1501 and DukeMTMC-reID datasets. Results of both positive pairs and hard negative pairs are shown. Note that only reliable correspondences with matching scores over 0.5 are shown, and the local positions are coarse due to the 24×8 size of the feature map. As can be observed from Fig. 5, the proposed method is able to find correct local correspondences for positive pairs of images, even if there are notable misalignments in both scale and position, pose/viewpoint changes, occlusions, and mix up of other persons, thanks to the local matching mechanism of QAConv instead of global feature representations. Besides, for hard negative pairs, the matching of QAConv still appears to be mostly reasonable, by linking visually similar parts or even the same person (may be ambiguously labeled or walking closely to other persons).

Note that the QAConv method gains the matching capability by automatic learning, from supervision of only class labels but not local correspondence labels.

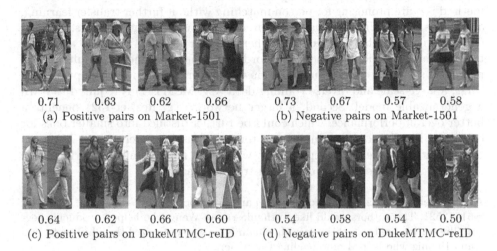

0.71   0.63   0.62   0.66   0.73   0.67   0.57   0.58
(a) Positive pairs on Market-1501   (b) Negative pairs on Market-1501

0.64   0.62   0.66   0.60   0.54   0.58   0.54   0.50
(c) Positive pairs on DukeMTMC-reID   (d) Negative pairs on DukeMTMC-reID

Fig. 5. Examples of qualitative matching results by the proposed QAConv method using the model trained on the MSMT17 dataset. Numbers represent similarity scores.

The QAConv network was trained on an NVIDIA DGX-1 server, with two V100 GPU cards. With the backbone network ResNet-50, the training time of QAConv on the DukeMTMC-reID dataset was 1.22 h. In contrast, the most efficient softmax baseline took 0.72 h for training. For deployment, the ECN [64] reported 1 h of transfer learning time with DukeMTMC-reID as target, while MAR [55] and DECAMEL [54] reported 10 and 35.2 h of total learning time, respectively, compared to the ready-to-use QAConv. For inference, with the DukeMTMC-reID dataset as target, QAConv took 26 s for feature extraction and 26 s for similarity computation. In contrast, the softmax baseline took 26 s for feature extraction and 0.2 s for similarity computation. Besides, the proposed method took 303 s for reranking, and 67 s for TLift. This is still efficient, especially for RR+TLift, compared to transfer learning for deployment. Therefore, the overall solution of QAConv+RR+TLift is promising in practical applications.

For further analysis on memory usage, please see the Appendix. As for the TLift, it can only be applied on datasets with good time records. Though this information is easy to obtain in real surveillance, most existing person re-identification datasets do not contain it. Another drawback of TLift is that it cannot be applied to arbitrary query images beyond a camera network, though once an initial match is found, it can be used to refine the search. Besides, it cannot help when there is no nearby person with the query.

6   Conclusion

In this paper, through extensive experiments we show that the proposed QAConv method is quite promising for person matching without further transfer learning, and it has a much better generalization ability than existing baselines. Though QAConv can also be plugged into other transfer learning methods as a better pre-trained model, in practice, according to the experimental results of this paper, we suggest a ready-to-use solution which works in the following principles. First, a large-scale and diverse training data (e.g. MSMT17) is required to learn a generalizable model. Second, a larger network (e.g. ResNet-152) benefits a better overall performance, which could be further distilled into smaller ones for efficiency. Finally, score re-ranking and temporal fusion model such as TLift can be computed on the fly in deployment, which can largely improve performance and they are more efficient to use than transfer learning.

Acknowledgements. This work was partly supported by the NSFC Project #61672521. The authors would like to thank Yanan Wang who helped producing several illustration figures in this paper, Jinchuan Xiao who optimized the TLift code, and Anna Hennig who helped proofreading the paper.

References

1. Ahmed, E., Jones, M., Marks, T.K.: An improved deep learning architecture for person re-identification. In: Proceedings of the IEEE Conference on Computer Vision and Pattern Recognition, pp. 3908–3916 (2015)
2. Bay, H., Tuytelaars, T., Van Gool, L.: Speeded-up robust features (SURF). Comput. Vis. Image Underst. **110**(3), 346–359 (2008)
3. Chang, X., Yang, Y., Xiang, T., Hospedales, T.M.: Disjoint label space transfer learning with common factorised space. Proc. AAAI Conf. Artif. Intell. **33**, 3288–3295 (2019)
4. Deng, J., Guo, J., Xue, N., Zafeiriou, S.: Arcface: additive angular margin loss for deep face recognition. In: Proceedings of the IEEE Conference on Computer Vision and Pattern Recognition, pp. 4690–4699 (2019)
5. DeVries, T., Taylor, G.W.: Improved regularization of convolutional neural networks with cutout. arXiv preprint arXiv:1708.04552 (2017)
6. Ergys, R., Francesco, S., Roger, Z., Rita, C., Carlo, T.: Performance measures and a data set for multi-target, multi-camera tracking. In: ECCV workshop on Benchmarking Multi-Target Tracking (2016)
7. Fan, H., Zheng, L., Yan, C., Yang, Y.: Unsupervised person re-identification: clustering and fine-tuning. TOMM **14**(4), 83 (2018)
8. He, K., Zhang, X., Ren, S., Sun, J.: Deep residual learning for image recognition. In: Proceedings of IEEE Computer Society Conference on Computer Vision and Pattern Recognition, pp. 770–778 (2016)
9. Hinton, G., Vinyals, O., Dean, J.: Distilling the knowledge in a neural network. arXiv preprint arXiv:1503.02531 (2015)
10. Hu, Y., Yi, D., Liao, S., Lei, Z., Li, S.Z.: Cross dataset person Re-identification. In: ACCV Workshop on Human Identification for Surveillance (HIS), pp. 650–664 (2014)

11. Ioffe, S., Szegedy, C.: Batch normalization: accelerating deep network training by reducing internal covariate shift. In: International Conference on Machine Learning, pp. 448–456 (2015)

12. Jia, J., Ruan, Q., Hospedales, T.M.: Frustratingly easy person re-identification: Generalizing person re-id in practice. In: British Machine Vision Conference (2019)

13. Jin, H., Wang, X., Liao, S., Li, S.Z.: Deep person re-identification with improved embedding and efficient training. In: 2017 IEEE International Joint Conference on Biometrics (IJCB), pp. 261–267. IEEE (2017)

14. Kalayeh, M.M., Emrah, B., Gökmen, M., Kamasak, M.E., Shah, M.: Human semantic parsing for person re-identification. In: Proceedings of IEEE Computer Society Conference on Computer Vision and Pattern Recognition, pp. 1062–1071 (2018)

15. Li, M., Zhu, X., Gong, S.: Unsupervised person re-identification by deep learning tracklet association. In: Proceedings of the European Conference on Computer Vision (ECCV), pp. 737–753 (2018)

16. Li, M., Zhu, X., Gong, S.: Unsupervised tracklet person re-identification. TPAMI **42**(7), 1770–1782 (2019)

17. Li, W., Zhao, R., Xiao, T., Wang, X.: DeepReID: deep filter pairing neural network for person re-identification. In: IEEE Conference on Computer Vision and Pattern Recognition (2014)

18. Li, W., Zhu, X., Gong, S.: Harmonious attention network for person re-identification. In: Proceedings of the IEEE Conference on Computer Vision and Pattern Recognition, pp. 2285–2294 (2018)

19. Liao, S., Shao, L.: Interpretable and generalizable deep image matching with adaptive convolutions. CoRR abs/1904.10424v1 (23, April 2019), http://arxiv.org/abs/1904.10424v1

20. Lin, S., Li, H., Li, C.T., Kot, A.C.: Multi-task mid-level feature alignment network for unsupervised cross-dataset person re-identification. In: The British Machine Vision Conference (BMVC) (2018)

21. Lin, T.Y., Goyal, P., Girshick, R., He, K., Dollár, P.: Focal loss for dense object detection. In: Proceedings of the IEEE International Conference on Computer Vision, pp. 2980–2988 (2017)

22. Lin, T.Y., Roychowdhury, A., Maji, S.: Bilinear CNN models for fine-grained visual recognition. In: IEEE International Conference on Computer Vision (2015)

23. Liu, C., Loy, C.C., Gong, S., Wang, G.: Pop: person re-identification post-rank optimisation. In: International Conference on Computer Vision (2013)

24. Liu, H., Feng, J., Qi, M., Jiang, J., Yan, S.: End-to-end comparative attention networks for person re-identification. IEEE Trans. Image Process. **26**(7), 3492–3506 (2017)

25. Liu, X., et al.: Hydraplus-net: attentive deep features for pedestrian analysis. In: Proceedings of the IEEE International Conference on Computer Vision, pp. 350–359 (2017)

26. Lowe, D.G.: Distinctive image features from scale-invariant keypoints. Int. J. Comput. Vis. **60**(2), 91–110 (2004)

27. Lv, J., Chen, W., Li, Q., Yang, C.: Unsupervised cross-dataset person re-identification by transfer learning of spatial-temporal patterns. In: Proceedings of IEEE Computer Society Conference on Computer Vision and Pattern Recognition, pp. 7948–7956 (2018)

28. Lv, J., Chen, W., Li, Q., Yang, C.: Unsupervised cross-dataset person re-identification by transfer learning of spatial-temporal patterns. In: 2018 IEEE Conference on Computer Vision and Pattern Recognition, CVPR 2018, Salt Lake City, UT, USA, 18–22 June 2018, pp. 7948–7956 (2018)
29. Pan, X., Luo, P., Shi, J., Tang, X.: Two at once: Enhancing learning and generalization capacities via ibn-net. In: Proceedings of the European Conference on Computer Vision (ECCV), pp. 464–479 (2018)
30. Peng, P., Xiang, T., Wang, Y., Pontil, M., Gong, S., Huang, T., Tian, Y.: Unsupervised cross-dataset transfer learning for person re-identification. In: Proceedings of the IEEE Conference on Computer Vision and Pattern Recognition, pp. 1306–1315 (2016)
31. Qian, X., Fu, Y., Jiang, Y.G., Xiang, T., Xue, X.: Multi-scale deep learning architectures for person re-identification. In: Proceedings of the IEEE International Conference on Computer Vision, pp. 5399–5408 (2017)
32. Qian, X., et al.: Pose-normalized image generation for person re-identification. In: Proceedings of the European Conference on Computer Vision (ECCV), pp. 650–667 (2018)
33. Saquib Sarfraz, M., Schumann, A., Eberle, A., Stiefelhagen, R.: A pose-sensitive embedding for person re-identification with expanded cross neighborhood re-ranking. In: Proceedings of the IEEE Conference on Computer Vision and Pattern Recognition, pp. 420–429 (2018)
34. Shen, Y., Xiao, T., Li, H., Yi, S., Wang, X.: End-to-end deep kronecker-product matching for person re-identification. In: Proceedings of the IEEE Conference on Computer Vision and Pattern Recognition, pp. 6886–6895 (2018)
35. Si, J., et al.: Dual attention matching network for context-aware feature sequence based person re-identification. In: Proceedings of the IEEE Conference on Computer Vision and Pattern Recognition, pp. 5363–5372 (2018)
36. Song, J., Yang, Y., Song, Y.Z., Xiang, T., Hospedales, T.M.: Generalizable person re-identification by domain-invariant mapping network. In: Proceedings of the IEEE Conference on Computer Vision and Pattern Recognition, pp. 719–728 (2019)
37. Suh, Y., Wang, J., Tang, S., Mei, T., Lee, K.M.: Part-aligned bilinear representations for person re-identification. In: Proceedings of the European Conference on Computer Vision (ECCV), pp. 402–419 (2018)
38. Suh, Y., Wang, J., Tang, S., Mei, T., Mu Lee, K.: Part-aligned bilinear representations for person re-identification. In: Proceedings of the European Conference on Computer Vision (ECCV), pp. 402–419 (2018)
39. Sun, Y., Zheng, L., Yang, Y., Tian, Q., Wang, S.: Beyond part models: person retrieval with refined part pooling (and a strong convolutional baseline). In: Proceedings of the European Conference on Computer Vision (ECCV) (2018)
40. Ustinova, E., Ganin, Y., Lempitsky, V.: Multi-Region bilinear convolutional neural networks for person re-identification. In: IEEE International Conference on Advanced Video and Signal Based Surveillance (2017)
41. Wang, G., Lai, J., Huang, P., Xie, X.: Spatial-temporal person re-identification. In: AAAI Conference on Artificial Intelligence (2019)
42. Wang, G., Yuan, Y., Chen, X., Li, J., Zhou, X.: Learning discriminative features with multiple granularities for person re-identification. In: 2018 ACM Multimedia Conference on Multimedia Conference, pp. 274–282. ACM (2018)
43. Wang, J., Zhu, X., Gong, S., Li, W.: Transferable joint attribute-identity deep learning for unsupervised person re-identification. In: Proceedings of the IEEE Conference on Computer Vision and Pattern Recognition, pp. 2275–2284 (2018)

44. Wang, X., Girshick, R., Gupta, A., He, K.: Non-local neural networks. In: Proceedings of the IEEE Conference on Computer Vision and Pattern Recognition, pp. 7794–7803 (2018)
45. Wei, L., Zhang, S., Gao, W., Tian, Q.: Person transfer gan to bridge domain gap for person re-identification. In: Proceedings of the IEEE Conference on Computer Vision and Pattern Recognition, pp. 79–88 (2018)
46. Wen, Yandong., Zhang, Kaipeng., Li, Zhifeng, Qiao, Yu.: A discriminative feature learning approach for deep face recognition. In: Leibe, Bastian, Matas, Jiri, Sebe, Nicu, Welling, Max (eds.) ECCV 2016. LNCS, vol. 9911, pp. 499–515. Springer, Cham (2016). https://doi.org/10.1007/978-3-319-46478-7_31
47. Wu, J., Liao, S., Wang, X., Yang, Y., Li, S.Z., et al.: Clustering and dynamic sampling based unsupervised domain adaptation for person re-identification. In: 2019 IEEE International Conference on Multimedia and Expo (ICME), pp. 886–891. IEEE (2019)
48. Wu, J., Yang, Y., Liu, H., Liao, S., Lei, Z., Li, S.Z.: Unsupervised graph association for person re-identification. In: Proceedings of the IEEE International Conference on Computer Vision, pp. 8321–8330 (2019)
49. Xu, J., Zhao, R., Zhu, F., Wang, H., Ouyang, W.: Attention-aware compositional network for person re-identification. In: Proceedings of the IEEE Conference on Computer Vision and Pattern Recognition, pp. 2119–2128 (2018)
50. Xu, S., Cheng, Y., Gu, K., Yang, Y., Chang, S., Zhou, P.: Jointly attentive spatial-temporal pooling networks for video-based person re-identification. In: Proceedings of the IEEE International Conference on Computer Vision, pp. 4733–4742 (2017)
51. Yang, Q., Yu, H.X., Wu, A., Zheng, W.S.: Patch-based discriminative feature learning for unsupervised person re-identification. In: Proceedings of the IEEE Conference on Computer Vision and Pattern Recognition, pp. 3633–3642 (2019)
52. Ye, M., Shen, J., Lin, G., Xiang, T., Shao, L., Hoi, S.C.H.: Deep Learning for Person Re-identification: A Survey and Outlook. arXiv preprint arXiv:2001.04193 (2020)
53. Yi, D., Lei, Z., Liao, S., Li, S.Z.: Deep metric learning for person re-identification. In: International Conference on Pattern Recognition, pp. 34–39 (December 2014)
54. Yu, H.X., Wu, A., Zheng, W.S.: Unsupervised person re-identification by deep asymmetric metric embedding. In: IEEE Transactions on Pattern Analysis and Machine intelligence (2019)
55. Yu, H.X., Zheng, W.S., Wu, A., Guo, X., Gong, S., Lai, J.H.: Unsupervised person re-identification by soft multilabel learning. In: Proceedings of the IEEE Conference on Computer Vision and Pattern Recognition, pp. 2148–2157 (2019)
56. Yu, R., Zhou, Z., Bai, S., Bai, X.: Divide and fuse: a re-ranking approach for person re-identification. In: The British Machine Vision Conference (BMVC) (2017)
57. Zhao, H., et al.: Spindle net: Person re-identification with human body region guided feature decomposition and fusion. In: Proceedings of the IEEE Conference on Computer Vision and Pattern Recognition, pp. 1077–1085 (2017)
58. Zhao, L., Li, X., Zhuang, Y., Wang, J.: Deeply-learned part-aligned representations for person re-identification. In: Proceedings of the IEEE International Conference on Computer Vision, pp. 3219–3228 (2017)
59. Zheng, L., Shen, L., Tian, L., Wang, S., Wang, J., Tian, Q.: Scalable person re-identification: a benchmark. In: Proceedings of IEEE International Conference on Computer Vision (2015)
60. Zheng, Z., Zheng, L., Yang, Y.: Unlabeled samples generated by GAN improve the person re-identification baseline in vitro. In: International Conference on Computer Vision, pp. 3774–3782 (2017)

61. Zhong, Z., Zheng, L., Zheng, Z., Li, S., Yang, Y.: Camstyle: A novel data augmentation method for person re-identification. In: IEEE Transactions on Image Processing (2018)
62. Zhong, Z., Zheng, L., Cao, D., Li, S.: Re-ranking person re-identification with k-reciprocal encoding. In: Proceedings of the IEEE Conference on Computer Vision and Pattern Recognition, pp. 1318–1327 (2017)
63. Zhong, Z., Zheng, L., Kang, G., Li, S., Yang, Y.: Random erasing data augmentation. In: Proceedings of the AAAI Conference on Artificial Intelligence (AAAI) (2020)
64. Zhong, Z., Zheng, L., Luo, Z., Li, S., Yang, Y.: Invariance matters: Exemplar memory for domain adaptive person re-identification. In: Proceedings of the IEEE Conference on Computer Vision and Pattern Recognition, pp. 598–607 (2019)

Self-supervised Bayesian Deep Learning for Image Recovery with Applications to Compressive Sensing

Tongyao Pang[1], Yuhui Quan[2], and Hui Ji[1(✉)]

[1] Department of Mathematics, National University of Singapore,
Singapore 119076, Singapore
{matpt,matjh}@nus.edu.sg
[2] School of Computer Science and Engineering,
South China University of Technology, Guangzhou 510006, China
csyhquan@scut.edu.cn

Abstract. In recent years, deep learning emerges as one promising technique for solving many ill-posed inverse problems in image recovery, and most deep-learning-based solutions are based on supervised learning. Motivated by the practical value of reducing the cost and complexity of constructing labeled training datasets, this paper proposed a self-supervised deep learning approach for image recovery, which is dataset-free. Built upon Bayesian deep network, the proposed method trains a network with random weights that predicts the target image for recovery with uncertainty. Such uncertainty enables the prediction of the target image with small mean squared error by averaging multiple predictions. The proposed method is applied for image reconstruction in compressive sensing (CS), i.e., reconstructing an image from few measurements. The experiments showed that the proposed dataset-free deep learning method not only significantly outperforms traditional non-learning methods, but also is very competitive to the state-of-the-art supervised deep learning methods, especially when the measurements are few and noisy.

Keywords: Self-supervised learning · Bayesian neural network · Compressive sensing · Image recovery

1 Introduction

Image recovery is about recovering an image of high quality from its related measurement. Many image recovery tasks are to solve a linear inverse problem:

$$y = Ax + n, \tag{1}$$

Electronic supplementary material The online version of this chapter (https://doi.org/10.1007/978-3-030-58621-8_28) contains supplementary material, which is available to authorized users.

where $\boldsymbol{y}$ denotes the available measurement, $\boldsymbol{x}$ denotes the latent image to recover, $\boldsymbol{n}$ denotes the measurement noise which is often modeled by i.i.d. random variables componently. The operator $\boldsymbol{A}$ denotes a linear degradation/measuring process on the latent image, which is usually non-invertible/ill-conditioned. When $\boldsymbol{A}$ is non-invertible such that the number of unknowns is larger than the number of independent equations, the solution is not unique. How to resolve such solution ambiguity is the main challenge when solving (1). When $\boldsymbol{A}$ is invertible but ill-conditioned, how to suppress the magnification of measurement noise $\boldsymbol{n}$ during the recovery becomes the main concern. In the past, the most prominent approach is the regularization method, which imposes certain image prior on the solution for resolving solution ambiguities and suppressing noise amplification.

Recently, deep learning emerges as a powerful tool for solving many image recovery problems ; see $e.g.$ [11, 25, 26, 32–34, 38, 44–46]. These deep-learning-based solutions are all using supervised learning, $i.e.$, a DNN (deep neural network) is trained on a labeled dataset which contains a large amount of the pairs of measurement and truth image. In supervised learning, the performance of the model will be significantly impacted by the characters of training dataset, including the amount of training samples and the correlation between the dataset and target image. In many scenarios, it is often very costly or infeasible to build a large-scale high-quality dataset closely related to the data for processing, $e.g.$ magnetic resonance imaging (MRI) and computed tomography (CT) scanning for medical imaging. Certainly, there is a need to develop deep-learning methods for image recovery that provide state-of-the-art performance, while not requesting any additional training data.

In comparison to active ongoing studies on supervised learning methods, there are few works on unsupervised deep learning for solving ill-posed linear systems arising from imaging systems. Recently, image denoising, often served as one sub-module in most image recovery tasks, saw rapid progresses along this line. For example, deep image prior (DIP) [41], SURE-Net[39], and Self2Self denoising [35]. Nevertheless, image denoising is very different from general image recovery tasks. With $\boldsymbol{A} = \mathbf{I}$, image denoising is not an ill-posed problem, and its focus is on noise suppression. In contrast, most image recovery problems require solving an ill-posed linear inverse system. In addition to noise suppression, how to resolve solution ambiguity is another main concern. It is non-trivial to generalize these denoisers to solving ill-posed image recovery problems. One might use these denoisers for post-processing to refine the estimate from an image recovery method. However, this straightforward way does not work well in practice, as the artifacts in estimates are rather different from measurement noises.

There is great practical value of a deep learning method for solving image recovery problems without the need of constructing any training dataset. Thus, this paper aims at developing a self-supervised deep learning method for image recovery and applying it for image reconstruction in CS.

1.1 Main Idea

In our setting, there is no training dataset available for unsupervised learning, and only the sensed measurement y and the sensing matrix A are given. It is shown in DIP [41] that, if an early stop is adopted to avoid overfitting, a convolutional neural network (CNN) tends to predict structured results even on random input. Similar to DIP, we also learn a deep neural network (DNN) $\mathcal{F}_\theta$, parameterized by weights θ, to predict x, by taking a random initialization ϵ_0 as the DNN input:

$$x = \mathcal{F}_\theta(\epsilon_0). \tag{2}$$

As x, y are related by (2), we have then

$$y = A\mathcal{F}_\theta(\epsilon_0) + n. \tag{3}$$

The maximum likelihood estimate (MLE) for x is given by

$$\min_\theta \text{dist}(A\mathcal{F}_\theta(\epsilon_0), y), \tag{4}$$

where $\text{dist}(\cdot, \cdot)$ is determined by the statistical model of the measurement noise, e.g. $\| \cdot \|_2^2$ for Gaussian white noise. As there is significant redundancy in the parameters θ, MLE is vulnerable to overfitting. The maximum A posterior (MAP) estimation addresses such an issue by imposing prior knowledge on θ. Let $p(\theta)$ denote the prior probability distribution of θ and consider Gaussian white noise with noise variance $\tilde{\sigma}^2$. Then, an MAP estimator is given by

$$\min_\theta \frac{1}{2\tilde{\sigma}^2}\|A\mathcal{F}_\theta(\epsilon_0) - y\|_2^2 - \log p(\theta). \tag{5}$$

It can be seen that an MAP estimator is the MLE estimator regularized by the term relating to the prior of θ. Indeed, the early stopping used in DIP for avoiding overfitting can be interpreted as adding regularization on the MLE with some implicit prior. Other explicit priors are used as well in image classification, e.g. the sum-of-squares-based weight decay regularization which assumes the network weights to follow i.i.d. normal distribution [17]. While an early-stopping-based regularization is used in DIP to avoid overfitting in the case $A = I$, the matrix A in many image recovery problems is non-invertible, i.e., there are more unknowns than independent linear equations. In such a case, an MAP estimator with the form of (5) is sometimes not efficient enough to resolve the solution ambiguity, arising from the non-trivial null space of A:

$$\text{null}(A) := \{x \neq 0 : Ax = 0\}. \tag{6}$$

Aiming at addressing such ineffectiveness of the MLE/MAP estimator, this paper proposed a self-supervised deep learning method for image recovery, which is built on the framework of Bayesian Neural Network (BNN). Briefly, the weights of a BNN are not deterministic, but random variables following certain probability distributions. Instead of learning deterministic weights, we learn the parameters of the probability distributions of these random weights. The motivation of our approach to tackle ill-posed image recovery problems comes from

(1) Model uncertainty (*i.e.* weight uncertainty) is helpful to correct the prediction bias caused by the network architecture; see *e.g.* [3,5,15,20].
(2) An ensemble of multiple realizations of a Bayesian model provides more accurate inferences than a single deterministic model; see *e.g.* [2,22].

In a nutshell, from the perspective of Bayesian approximation, the proposed BNN-based approach is about learning an approximation to the minimum mean square error (MMSE) estimate defined by

$$\widehat{x} := \arg\min_{u} \mathbb{E}_{(x|y)} \|u - x\|_2^2 = \mathbb{E}_{(x|y)}(x|y) = \int x p(x|y) dx = \int \mathcal{F}_\theta(\epsilon_0) p(\theta|y) d\theta, \tag{7}$$

where $p(\theta|y)$ is the posterior probability distribution function of θ. Considering the number of weights and the non-linear structure of the network, the computation of $p(\theta|y)$ is intractable, and thus we use the joint distribution of independent normal distributions $q(\theta|\mu, \sigma)$ to approximate $p(\theta|y)$:

$$q(\theta|\mu, \sigma): \quad \theta_i \sim \mathcal{N}(\mu_i, \sigma_i^2), \quad \theta = \{\theta_i\}, \quad \mu = \{\mu_i\}, \quad \sigma = \{\sigma_i\}. \tag{8}$$

The cost is defined by the KL divergence between $q(\theta|\mu, \sigma)$ and $p(\theta|y)$:

$$(\mu^*, \sigma^*) = \arg\min_{\mu, \sigma} D_{\mathrm{KL}}(q(\theta|\mu, \sigma)\|p(\theta|y)). \tag{9}$$

Once the model is trained with learned distribution parameters μ^* and σ^*, we have a prediction that approximates the MMSE estimate:

$$x^* = \int \mathcal{F}_\theta(\epsilon_0) q(\theta|\mu^*, \sigma^*) d\theta. \tag{10}$$

See Sect. 3 for a more detailed discussion.

1.2   Main Contributions

Built on BNN, this paper proposed a self-supervised learning method for image recovery and applied to solve image reconstruction in CS. The proposed method is dataset-free without requiring any external training sample. The experiments showed that the proposed approach not only significantly outperformed representative traditional non-learning methods, but also is very competitive to supervised deep learning methods with state-of-the-art performance. Indeed, the proposed method has its advantages when the measurement is noisy.

The results of this paper have significance in both theoretical research and practical applications. In the dataset-free setting, existing works showed that a DNN can effectively learn meaningful image structures from a noisy image by avoiding overfitting. However, for solving an ill-posed linear problem, the solution ambiguity requires new techniques to avoid more likely overfitting (with perturbations from null space), when training a DNN using only the measurement itself. This paper is the first work that showed the weight uncertainty induced by BNN is effective to handle overfitting, and learning ensemble can lead to accurate prediction. These results showed great potential of BNN in solving ill-posed linear inverse problems arising from imaging systems.

CS is one powerful sensing modality for designing imaging systems with faster sampling and lower energy consumption. It is about reconstructing signals/images, which are sparse in certain transform domain, using the measurements much less than that traditional uniform sampling (see e.g. [7,13]). It has received great attention in a wide range of applications, including medical imaging [8,16,27] and computational photography [1,14]. Image reconstruction is one key module in CS-based imaging systems. Existing supervised learning methods requires a large amount of training samples to provide state-of-the-art performance, which often is a challenging task in practice. For instance, it takes a long time to collect fully-sampled true images in MRI. The performance of the proposed self-supervised method is very competitive to state-of-the-art supervised learning method, while there is no any prerequisite on training samples. Such a dataset-free setting makes the proposed method very appealing in practice.

2 Related Work

As this paper is about deep learning for image recovery, we only give a very brief review on those non-learning methods and focus more on deep learning methods.

2.1 Regularization Methods with Pre-defined Image Prior

Regularization method is one widely-used technique for image recovery, which imposes certain image prior on the image to resolve solution ambiguities. In most regularization methods, the problem of image recovery is re-formulated to some optimization problem, and its minimizer is defined as the estimate of the truth. Many regularization methods have been proposed for solving various image recovery problems, including CS image reconstruction. The ℓ_1-norm relating regularization methods (e.g. [6,23,24,27,40]) assume the image gradients are sparse and use ℓ_1-norm relating regularizations for image recovery. The non-local methods exploit the recurrence prior of local image patches for image reconstruction. For instance, the low-rank regularization method [12] assumes that the stack of matched patches is of low rank; the BM3D-based regularization method [10,30] employs the BM3D denoiser [9] for regularizing patch stacks. Non-local wavelet frame method [36] regularizes the image in non-local wavelet tight frame when recovering the image.

2.2 Supervised Deep Learning Methods

Recently, deep learning has became one powerful technique for solving ill-inverse problems in imaging. Most existing such solutions are based on supervised learning, i.e., the DNN is trained on a dataset with the pairs of measurement/image. Earlier works take an end-to-end approach that learns the direct map between the measurement and the image, and the main difference among them is NN architecture, *e.g.* image recovery [37,43] and CS image reconstruction [21,32]. As imaging physics encoded in A is not utilized in an end-to-end approach, their performance is not significantly better than traditional regularization methods.

A more promising approach is the optimization unrolling with learnable prior. The idea is unrolling the iterative scheme of a regularization method (e.g. ℓ_1-norm relating method) and replacing the operations related to image prior by a CNN. For image deconvolution, Meinhardt *et al.* [29] unrolled the primal-dual hybrid gradient method, Zhang *et al.* [46] unrolled a half-quadratic splitting method, and Nan *et al.* [33] unrolled a VEM-based iterative scheme. Proximal forward backward splitting scheme and Douglas-Rachford iteration are unrolled in [11] and [26] for medical image reconstruction. ADMM-Net [44] unrolled the alternating direction method of multipliers (ADMM) for MRI image reconstruction, and Liu *et al.* [26] proposed another scheme of the ADMM method with different variable splitting scheme. For CS image reconstruction, ISTA-Net [45] unrolled the iterative shrinkage-thresholding algorithm (ISTA). A scalable Laplacian pyramid reconstructive adversarial network (LAPRAN) [42] was proposed for CS image reconstruction which is adaptive to different CS ratios. In SCSNet [38], sensing and reconstruction of CS is integrated in one network.

2.3   Unsupervised Deep Learning Method for Image Denoising

There are few existing works on unsupervised deep learning methods for image recovery problems. Most existing unsupervised deep learning methods focus on image denoising, one sub-module often seen in image recovery. Based on the observation that regular image structures appear before random patterns during the training, Ulyanov *et al.* [41] proposed the deep image prior (DIP) method for image denoising by using early stopping to avoid overfitting. By simplifying a deep decoder, Heckel and Hand [19] proposed to use an under-parameterized NN for avoiding overfitting. Based on Stein's unbiased risk estimator (SURE), Soltanayeva and Chun [39] proposed a regularization on NN weights to train a denoising NN without training data. Quan *et al.* [35] proposed to use dropout in both training and testing for learning a denoising NN without training data.

Image denoising does not need to consider the ill-posedness of the matrix A. Thus, these denoising methods cannot be easily generalized to solve ill-posed image recovery problems with good performance. Using the SURE denoiser, Zhussip *et al.* [47] developed a SURE-AMP method for CS image reconstruction based on the denoiser approximate message passing (AMP) framework. However, its performance is not competitive to other supervised learning methods.

3   Main Body

In this section, we give a detailed discussion on the proposed self-supervised BNN for CS image reconstruction. Let y denote the measurement, x denote the image to predict, n denote the measurement noise, and they are related by

$$y = Ax + n. \tag{11}$$

Let $\mathcal{F}_\theta$ denote the BNN whose weights $\theta = \{\theta_i\}$ are random variables. Consider a random initialization ϵ_0. As shown in (7), provided the posterior probability distribution function $p(\theta|y)$, the MMSE estimate of the truth x is given by

$$\widehat{x} = \int \mathcal{F}_\theta(\epsilon_0)p(\theta|y)d\theta. \tag{12}$$

The amount of weights and complexity of a DNN makes $p(\theta|y)$ computationally intractable. Thus, we take a Bayesian approximation approach to approximate $p(\theta|y)$ by using the following independent joint normal distribution $q(\theta|\mu,\sigma)$:

$$\theta_i \sim \mathcal{N}(\mu_i, \sigma_i), \tag{13}$$

where $\mu = \{\mu_i\}, \sigma = \{\sigma_i\}$, mean and s.t.d., are distribution parameters.

3.1   Training

The training of the BNN is done by minimizing the distance between the prediction of the NN and the MMSE estimate in (12). As we use $q(\theta|\mu,\sigma)$ defined by (13) to approximate $p(\theta|y)$ in the MMSE estimate, we proposed to train the BNN by minimizing the KL divergence between $q(\theta|\mu,\sigma)$ and $p(\theta|y)$:

$$\min_{\mu,\sigma} \mathrm{KL}(q(\theta|\mu,\sigma)\|p(\theta|y)). \tag{14}$$

The minimum of the KL-divergence is difficult to find for general distributions. Thus, we further simplify the optimization problem by assuming that the prior distribution $p(\theta)$ can be well approximated by the joint distribution of i.i.d. normal distribution with zero mean and standard deviation $\bar{\sigma}$. Then, we have

Proposition 1. *Suppose that the measurement noise n is Gaussian white noise such that $p(n) \sim \prod_i \exp(\frac{-n_i^2}{2\tilde{\sigma}^2})$ and $p(\theta) \sim \prod_i \exp(\frac{-\theta_i^2}{2\bar{\sigma}^2})$. Then, we have*

$$\begin{aligned}
&\min_{\mu,\sigma} \mathrm{KL}(q(\theta|\mu,\sigma)\|p(\theta|y)) \\
&\Leftrightarrow \min_{\mu,\sigma} \mathbb{E}_{\theta \sim q(\theta|\mu,\sigma)} \|A\mathcal{F}_\theta(\epsilon_0) - y\|_2^2 + \lambda_1(\|\mu\|_2^2 + \|\sigma\|_2^2) - \lambda_2 \sum_i \log \sigma_i,
\end{aligned} \tag{15}$$

where $\lambda_1 = \tilde{\sigma}^2/\bar{\sigma}^2$ and $\lambda_2 = 2\tilde{\sigma}^2$.

Proof. See the supplementary for the detailed derivation.

For the data-dependent term $\mathbb{E}_{\theta \sim q(\theta|\mu,\sigma)}\|A\mathcal{F}_\theta(\epsilon_0) - y\|_2^2$ in (15), we only sample one instance of θ from the distribution $q(\theta|\mu,\sigma)$ to approximate the expectation at each iteration for computational efficiency, which can be interpreted as a variation of the stochastic gradient descent (SGD) algorithm. It is noted that each σ_i denotes the standard deviation, which should be always positive. Thus, we adopt the same re-parameterization trick as [5] that re-expresses σ_i by

$$\sigma_i = \log(1 + \exp(\rho_i)). \tag{16}$$

Then, at every iteration of BNN training, we first randomly draw sample ϵ from standard normal distribution $\mathcal{N}(0,1)$ and then generate the network weights by

$$\theta_i = \mu_i + \log(1 + \exp(\rho_i)) \cdot \epsilon. \tag{17}$$

More details on back-propagation for BNN can be found in the related materials.

3.2   Testing

Once the BNN is trained by (15) with estimated distribution parameters $\boldsymbol{\mu}^*$ and $\boldsymbol{\sigma}^*$, we have now an approximation to the posterior probability distribution $p(\boldsymbol{\theta}|\boldsymbol{y})$, *i.e.* $q(\boldsymbol{\theta}|\boldsymbol{\mu}^*, \boldsymbol{\sigma}^*)$. The approximate MMSE estimate is then given by

$$\boldsymbol{x}^* = \int \mathcal{F}_{\boldsymbol{\theta}}(\boldsymbol{\epsilon}_0) q(\boldsymbol{\theta}|\boldsymbol{\mu}^*, \boldsymbol{\sigma}^*) d\boldsymbol{\theta}. \tag{18}$$

Although both $\mathcal{F}_{\boldsymbol{\theta}}(\boldsymbol{\epsilon}_0)$ and $q(\boldsymbol{\theta}|\boldsymbol{\mu}^*, \boldsymbol{\sigma}^*)$ have explicit forms, the above integration is still intractable. Instead, we use Monte Carlo (MC) integration in practice

$$\boldsymbol{x}^* \approx \frac{1}{T} \sum_{j=1}^{T} \mathcal{F}_{\boldsymbol{\theta}^j}(\boldsymbol{\epsilon}_0), \tag{19}$$

where $\{\boldsymbol{\theta}^j\}$ are the realizations of random variable $\boldsymbol{\theta}$ from the distribution $q(\boldsymbol{\theta}|\boldsymbol{\mu}^*, \boldsymbol{\sigma}^*)$ and T is the total sampling number.

3.3   Network Structure

We adopt the decoder part of a plain encoder-decoder NN as our NN with the following motivations. (a) Different from classic encoder-decoder network which takes images as inputs, our network input is a random vector which is very similar to the "code" generated by the encoder. Thus, the encoder does not see its function in our setting. (b) Owing to the need to learn both $\boldsymbol{\mu}$ and $\boldsymbol{\sigma}$, the parameter number of the BNN is twice as that of its deterministic counterpart. Using only the decoder results in the reduction of model size, which is helpful for avoiding overfitting and reducing computational cost.

The NN architecture is illustrated in Fig. 1. To recover an image of size $H \times W \times C$, the size of our NN input is $H/32 \times W/32 \times 128$. The input is forwarded into our decoder NN which contains five decoder blocks. Each of the first four decoder blocks sequentially connects an upsampling layer with a scaling factor of 2, and two Bayesian convolution layers both of which have 128 channels and are equipped with the leaky ReLU. The last decoder block contains an upsampling layer with a scaling factor of 2, and three aforementioned Bayesian convolution layers whose numbers of output channels are 64, 32, C with two LReLUs and one Sigmoid layer followed respectively. A Bayesian convolution layer is a convolution layer such that its weights are generated by (17) during training and testing.

4   Image Reconstruction in CS

In this section, we apply the proposed self-supervised image recovery method to solve image reconstruction problem in CS. Mathematically, CS image reconstruction can be formulated as solving an under-determined linear system:

$$\boldsymbol{y} = \boldsymbol{A}\boldsymbol{x} + \boldsymbol{n}, \tag{20}$$

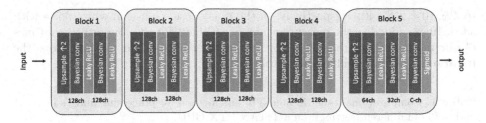

Fig. 1. Structure of the BNN used in the proposed method.

where $A \in \mathbb{R}^{M \times N}$ (or $\mathbb{C}^{M \times N}$) denotes the sensing matrix with $M \ll N$, y denotes the measurements collected by sensors, and n denotes noise. The experiments are conducted on two settings of CS: one is the block-wise random Gaussian CS problem in natural image acquisition, and the other is the random Fourier downsampling CS problem in magnetic resonance imaging (MRI).

4.1 Implementation Details

Our method is implemented using Pytorch. For convolution layers, the kernel size is 3×3 and both the stride and padding number is 1. The bi-linear interpolation is used for upsampling layers. For leaky ReLUs, the negative slope is fixed to 0.01. The BNN parameter μ is initialized using the normal distribution as [18]. The initial value of ρ is drawn from the uniform distribution on $[-5, -4]$. The model is trained by the Adam optimizer with fixed learning rate 10^{-4}. The parameter λ_1 and λ_2 in (15) are updated as follows:

$$\lambda_1 = \gamma_1 (\tilde{\sigma} + 10^{-3})^2, \quad \lambda_2 = \gamma_2 (\tilde{\sigma} + 10^{-3})^2, \tag{21}$$

with $\gamma_1 = 0.05$ and $\gamma_2 = 0.25$. The training procedure is stopped either the maximum iteration number 10^5 is reached or the residual $\|A\mathcal{F}_\theta(\epsilon_0) - y\|_2^2 / M$ is less than $(\tilde{\sigma} + 10^{-3})^2$. The sampling number T used in the MC approximation during prediction is set to 100. For comparison to other methods, we cite the results directly from the literature if possible; otherwise, we run the codes from the authors with the effort on the tuning-up of parameters to reproduce. If none is available, we leave it blank in the table.

4.2 CS on Natural Image Acquisition

For the CS-based reconstruction on natural images, we follow the setting of one recent deep-learning-based method, *i.e.* ISTA-Net [45]. Two datasets are used for testing. One is "Set11" [45] with 11 images and the other is "BSD68" [28] with 68 images. These images are cropped into non-overlapped blocks of size 33×33 to generate the measurements. The sensing matrix A of size $M \times N$ ($N = 1089$) is first sampled from independent standard normal distribution entry-wisely and then orthogonalized row-wisely. The CS ratio, *i.e.* M/N, is set to $4\%, 10\%, 25\%, 40\%$ respectively. In the noisy case, Gaussian white noise with s.t.d. 10 are added to the measurements. See Table 1 for the computational times of our method. The time varies for different settings as our method stops the iteration when the residual meets tolerance.

Table 1. Computational time (in hours) of our method for processing images in Set11 and Set68 in different settings, on a TITAN RTX GPU.

Dataset	$\tilde{\sigma}$	40%	25%	10%	4%	Dataset	$\tilde{\sigma}$	40%	25%	10%	4%
Set11	0	4.7	4.9	5.1	5.8	Set68	0	25.0	24.9	25.3	40.9
	10	1.2	1.2	0.7	0.4		10	11.3	10.7	7.3	5.3

Table 2. Average PSNR(dB)/SSIM results of different methods on Set11 [45] and BSD68 [28] in noiseless CS-based natural image reconstruction.

Dataset	Method	40%	25%	10%	4%
Set11	TVAL3	30.52/0.90	26.44/0.80	21.35/0.59	17.45/0.41
	DAMP	33.49/0.93	28.21/0.85	21.16/0.60	15.69/0.35
	ReconNet	-/-	25.54/0.76	22.68/0.64	19.98/0.53
	ISTA	**35.97/0.96**	**32.59/0.93**	26.64/0.81	21.59/0.62
	DIP	33.28/0.92	31.33/0.91	27.40/**0.83**	23.15/0.69
	Ours	35.71/0.95	32.30/0.92	**27.49/0.83**	**23.26/0.70**
BSD68	TVAL3	29.39/0.86	26.48/0.77	22.49/0.58	19.10/0.42
	DAMP	28.03/0.79	25.57/0.70	21.92/0.52	17.11/0.33
	ReconNet	-/-	25.31/0.71	23.16/0.60	21.28/0.50
	ISTA	**32.17/0.92**	**29.36/0.85**	**25.32**/0.70	22.40/0.56
	DIP	30.10/0.87	27.78/0.80	24.82/0.69	22.51/**0.58**
	Ours	31.28/0.90	28.63/0.84	25.24/**0.71**	**22.52/0.58**

Four methods are included in the comparison, *i.e.* TVAL3 [23], D-AMP [31], ReconNet [21] and ISTA [45]. The first two are regularization methods while the last two are supervised deep learning methods. In addition, we also include the DIP method [41], which is a recent unsupervised learning technique for image recovery. There is no work that directly extends the original DIP for CS image reconstruction. Thus, following the DIP for image super-resolution [41], we implement a DIP-based CS reconstruction method by using the cost function

$$\min_{\theta} \| A\mathcal{F}_{\theta}(\epsilon_0) - y \|_2^2. \tag{22}$$

Table 3. Average PSNR(dB)/SSIM results of different methods on Set11 [45] and BSD68 [28] in CS-based natural image reconstruction with noise level $\tilde{\sigma} = 10$.

Dataset	Method	40%	25%	10%	4%
Set11	TVAL3	26.66/0.72	24.75/0.67	21.02/0.54	17.28/0.39
	DAMP	29.25/0.86	26.35/0.80	20.84/0.58	15.56/0.35
	ReconNet	-/-	24.36/0.66	22.00/0.57	19.62/0.49
	ISTA	27.98/0.75	27.26/0.75	24.55/0.70	20.79/0.56
	DIP	28.87/0.83	27.36/0.79	24.19/0.68	21.27/0.55
	Ours	**30.39/0.88**	**28.67/0.84**	**25.23/0.76**	**21.91/0.64**
BSD68	TVAL3	26.15/0.68	24.80/0.63	22.03/0.52	18.93/0.39
	DAMP	26.55/0.72	24.87/0.65	21.70/0.51	16.96/0.33
	ReconNet	-/-	24.12/0.61	22.36/0.53	20.77/0.46
	ISTA	26.68/0.70	25.84/0.68	**23.86**/0.60	**21.64**/0.50
	DIP	25.24/0.64	24.07/0.59	22.46/0.51	21.13/0.45
	Ours	**28.13/0.81**	**26.47/0.75**	23.79/**0.64**	21.54/**0.53**

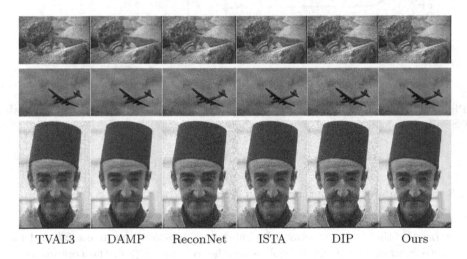

  TVAL3   DAMP   ReconNet   ISTA   DIP   Ours

Fig. 2. Results of Gaussian CS image reconstruction using noisy input with ratio 25%.

The NN used for CS image reconstruction is the same as that for super-resolution, *i.e.* an encoder-decoder NN with skip-connections whose model size is comparable to ours. The NN stops if it hits the maximum iteration number 2×10^4 or the residual $\|A\mathcal{F}_\theta(\epsilon_0) - y\|_2^2/M$ reaches the same tolerance $(\tilde{\sigma} + 10^{-3})^2$ as ours.

See Table 2 and Table 3 for the quantitative results in both noiseless and noisy cases. See Fig. 2 for visual comparison of some examples. Generally, our method outperformed two traditional non-learning methods (TVAL3 and DAMP) by a large margin and the unsupervised deep-learning method DIP. Even compared to the state-of-the-art (SOTA) supervised learning methods, our method remains very competitive. The SOTA supervised learning methods (*e.g.* ISTA) have small advantages when the measurements are noise-free and ours has noticeable advantages when the measurements are noisy.

Additionally, we compared the behavior of the BNN method to the DIP method over iterations. We run 10^5 steps for both without early stopping on a sample image. See Fig. 3 for the trace of PSNR and residual value over iteration. It can be seen that our method is more stable to the iteration number than the DIP method in terms of PSNR value. Moreover, the residual of DIP decreases to zero eventually even in the presence of noise, which causes overfitting, while the residual of our method does not vanish. This indicates the advantages of our BNN over the DIP method when processing noisy measurements.

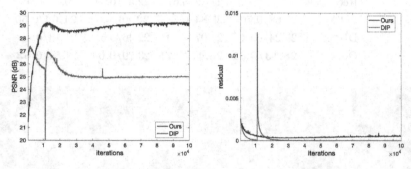

Fig. 3. PSNR (left) and residual(right) over iterations of DIP and our method with CS ratio 25% and noise level $\tilde{\sigma} = 10$ on the natural image "boats".

4.3   CS Image Reconstruction in MRI

For CS in MRI, we use the down-sampled data in the k-space. The sensing matrix A is the dot production of a random down-sampling mask M and the discrete Fourier transform F. Following the setting in [26], the contaminated measurements are generated by $y = M \odot F(x + n_1 + in_2)$, where the entries of n_1 and n_2 follow i.i.d. normal distribution of mean zero and s.t.d. $\tilde{\sigma}$. Then the noise n in (20) takes the form of $n = M \odot F(n_1 + in_2)$, which is complex and also follows i.i.d. Gaussian distribution entry-wisely. The dataset is the same as [26] with 21 MRI images from ADNI (Alzheimer's Disease Neuroimaging Initiative). We test three types of down-sampling masks of sampling ratio 25%, namely, 1D Gaussian mask, 2D Gaussian mask, and radial mask shown in Fig. 4. In the noisy case, Gaussian white noise with s.t.d. $\tilde{\sigma}$ as 10% of the maximum pixel value of the MRI image are added to the down-sampled k-space measurements. We compare the performance of our method with the simple zero-filling method (ZF) [4], TV-regularization-based method [27], ADMM-Net [44], the plug-in methods in [26] with three different networks: SCAE, SNLAE, and GAN, and DIP [41]. See

Table 4. Average PSNR(dB)/SSIM of the results of of different methods for CS-based MRI reconstruction.

Method	1D Gaussian		2D Gaussian		radial	
$\hat{\sigma}$	0	10%	0	10%	0	10%
ZF	23.06/0.62	20.37/0.26	25.30/0.50	22.38/0.36	25.45/0.51	22.38/0.36
TV	25.77/0.76	22.25/0.37	32.79/0.90	24.92/0.49	32.32/0.90	25.16/0.49
ADMM-Net	28.99/0.87	22.98/0.44	34.97/0.94	25.84/0.60	33.67/0.93	25.96/0.61
SCAE	29.37/0.88	22.72/0.63	35.61/0.95	26.06/0.74	33.94/0.94	26.13/0.70
SNLAE	29.06/0.86	24.39/0.56	32.85/0.86	26.15/0.67	32.53/0.88	26.38/0.66
GAN	27.47/0.82	23.32/0.69	32.94/0.91	26.31/0.75	32.26/0.90	25.53/0.74
DIP	**31.80/0.92**	23.38/0.68	35.63/0.95	24.41/0.72	33.81/0.94	24.54/0.73
Ours	31.38/0.91	**25.65/0.76**	**36.10/0.96**	**27.12/0.82**	**34.08/0.95**	**27.07/0.82**

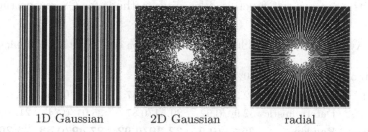

     1D Gaussian   2D Gaussian   radial

Fig. 4. Three different types of sampling masks of sample ratio 25%.

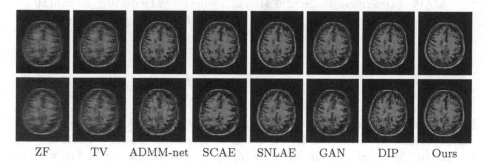

  ZF   TV   ADMM-net   SCAE   SNLAE   GAN   DIP   Ours

Fig. 5. MRI reconstruction results with 1D Gaussian mask of sampling ratio 25%; the first row corresponds to the noiseless case and the second row noisy case.

Table 4 for the quantitative comparison of different method and Fig. 5 for visual comparison on some sample images. It can be seen that our method outperformed all the other methods in all settings, except that DIP performs best in the noiseless case with 1D Gaussian mask.

4.4 Ablation Study

Ablation study is conducted on CS reconstruction for image acquisition on the dataset Set11 to show how much performance improvement weight uncertainty of BNN can bring in. Two deterministic versions of the BNN are used for comparison. One is the MLE estimator which trains the NN with deterministic weights: $\min_\theta \|A\mathcal{F}_\theta(\epsilon_0) - y\|_2^2$. The other is the MAP estimator which trains the NN with deterministic weights using (5) and a Gaussian prior on the weights:

$$\min_\theta \|A\mathcal{F}_\theta(\epsilon_0) - y\|_2^2 - 2\tilde{\sigma}^2 \log(p(\theta)) = \|A\mathcal{F}_\theta(\epsilon_0) - y\|_2^2 + \gamma\tilde{\sigma}^2\|\theta\|_2^2, \qquad (23)$$

where γ is set to 0.05 after tuning-up and $\tilde{\sigma}^2$ replaced with a small perturbation $(\tilde{\sigma} + 10^{-3})^2$ as ours. All these two versions and ours use the same architecture and stopping criteria. See Table 5 for the comparison. Clearly, the BNN with random weights significantly outperformed the other two deterministic versions. This clearly indicates the effectiveness of weight uncertainty in BNN on handling the overfitting in our self-supervised learning methods for CS reconstruction.

Table 5. Average PSNR(dB)/SSIM results of ablation studies on natural image Set11.

$\tilde{\sigma}$	Method	Weights	40%	25%	10%	4%
0	MLE	Deterministic	32.34/0.92	29.43/0.87	25.13/0.75	21.13/0.61
	MAP	Deterministic	32.90/0.92	29.51/0.87	25.01/0.74	21.08/0.60
	Ours	Random	**35.71/0.95**	**32.30/0.92**	**27.49/0.83**	**23.26/0.70**
10	MLE	Deterministic	28.87/0.84	27.24/0.80	23.92/0.70	20.35/0.56
	MAP	Deterministic	28.82/0.84	27.18/0.80	23.90/0.70	20.25/0.56
	Ours	Random	**30.39/0.88**	**28.67/0.84**	**25.23/0.76**	**21.91/0.64**

5   Conclusion

Built Bayesian neural network with random weights, this paper proposed a self-supervised framework of deep learning with state-of-the-art performance for reconstructing an image from fewer and noisy measurements in CS. The work in this paper not only has its value in the applications of CS-based imaging systems, but also provides a new insight for developing dataset-free un-supervised/self-supervised deep learning methods for other image recovery problems.

Acknowledgment. Tongyao Pang and Hui Ji would like to acknowledge the support from Singapore MOE Academic Research Fund (AcRF) Tier 2 research project (MOE2017-T2-2-156), and Yuhui Quan would like to acknowledge the support of National Natural Science Foundation of China (61872151, U1611461).

References

1. Arce, G., Brady, D., Carin, L., Arguello, H., Kittle, D.: Compressive coded aperture spectral imaging: an introduction. IEEE Signal Process. Mag. **31**(1), 105–115 (2013)

2. Baldi, P., Sadowski, P.J.: Understanding dropout. In: NeurIPS, pp. 2814–2822 (2013)
3. Barber, D., Bishop, C.M.: Ensemble learning in Bayesian neural networks. Nato ASI Ser. F Comput. Syst. Sci. **168**, 215–238 (1998)
4. Bernstein, M.A., Fain, S.B., Riederer, S.J.: Effect of windowing and zero-filled reconstruction of MRI data on spatial resolution and acquisition strategy. J. Magn. Reson. Imaging **14**(3), 270–280 (2001)
5. Blundell, C., Cornebise, J., Kavukcuoglu, K., Wierstra, D.: Weight uncertainty in neural network. In: ICML, pp. 1613–1622 (2015)
6. Cai, J., Ji, H., Liu, C., Shen, Z.: Blind motion deblurring from a single image using sparse approximation. In: CVPR, pp. 104–111 (2009)
7. Candes, E.J., Tao, T.: Near-optimal signal recovery from random projections: universal encoding strategies? IEEE Trans. Inf. Theor. **52**(12), 5406–5425 (2006)
8. Chen, G., Tang, J., Leng, S.: Prior image constrained compressed sensing (PICCS): a method to accurately reconstruct dynamic CT images from highly undersampled projection data sets. Med. Phys. **35**(2), 660–663 (2008)
9. Dabov, K., Foi, A., Katkovnik, V., Egiazarian, K.: Image denoising by sparse 3-d transform-domain collaborative filtering. IEEE Trans. Image Process. **16**(8), 2080–2095 (2007)
10. Danielyan, A., Katkovnik, V., Egiazarian, K.: Bm3d frames and variational image deblurring. IEEE Trans. Image Process. **21**(4), 1715–1728 (2011)
11. Ding, Q., Chen, G., Zhang, X., Huang, Q., Ji, H., Gao, H.: Low-dose CT with deep learning regularization via proximal forward backward splitting. Phys. Med. Biol. **65**(12), 125009 (2020)
12. Dong, W., Shi, G., Li, X., Ma, Y., Huang, F.: Compressive sensing via nonlocal low-rank regularization. IEEE Trans. Image Process. **23**(8), 3618–3632 (2014)
13. Donoho, D.L.: Compressed sensing. IEEE Trans. Inf. Theory **52**(4), 1289–1306 (2006)
14. Duarte, M., et al.: Single-pixel imaging via compressive sampling. IEEE Signal Process. Mag. **25**(2), 83–91 (2008)
15. Gal, Y., Ghahramani, Z.: Dropout as a Bayesian approximation: Representing model uncertainty in deep learning. In: ICML, pp. 1050–1059 (2016)
16. Gamper, U., Boesiger, P., Kozerke, S.: Compressed sensing in dynamic MRI. Magn. Reson. Med. **59**(2), 365–373 (2008)
17. Goodfellow, I., Bengio, Y., Courville, A.: Deep Learning. MIT Press, Cambridge (2016)
18. He, K., Zhang, X., Ren, S., Sun, J.: Delving deep into rectifiers: Surpassing human-level performance on imagenet classification. In: ICCV, pp. 1026–1034 (2015)
19. Heckel, R., Hand, P.: Deep decoder: Concise image representations from untrained non-convolutional networks. arXiv preprint arXiv:1810.03982 (2018)
20. Kendall, A., Gal, Y.: What uncertainties do we need in Bayesian deep learning for computer vision? In: NeurIPS, pp. 5574–5584 (2017)
21. Kulkarni, K., Lohit, S., Turaga, P., Kerviche, R., Ashok, A.: Reconnet: Non-iterative reconstruction of images from compressively sensed measurements. In: CVPR, pp. 449–458 (2016)
22. Lakshminarayanan, B., Pritzel, A., Blundell, C.: Simple and scalable predictive uncertainty estimation using deep ensembles. In: NeurIPS, pp. 6402–6413 (2017)
23. Li, C., Yin, W., Jiang, H., Zhang, Y.: An efficient augmented lagrangian method with applications to total variation minimization. Comput. Optim. Appl. **56**(3), 507–530 (2013)

24. Li, M., Fan, Z., Ji, H., Shen, Z.: Wavelet frame based algorithm for 3d reconstruction in electron microscopy. SIAM J. Sci. Comput. **36**(1), B45–B69 (2014)
25. Liu, J., Chen, N., Ji, H.: Learnable Douglas-Rachford iteration and its applications in dot imaging. Inverse Prob. Imaging **14**(4), 683 (2020)
26. Liu, J., Kuang, T., Zhang, X.: Image reconstruction by splitting deep learning regularization from iterative inversion. In: Frangi, A.F., Schnabel, J.A., Davatzikos, C., Alberola-López, C., Fichtinger, G. (eds.) MICCAI 2018. LNCS, vol. 11070, pp. 224–231. Springer, Cham (2018). https://doi.org/10.1007/978-3-030-00928-1_26
27. Lustig, M., Donoho, D., Pauly, J.M.: Sparse MRI: The application of compressed sensing for rapid MR imaging. Magn. Reson. Med. **58**(6), 1182–1195 (2007)
28. Martin, D., Fowlkes, C., Tal, D., Malik, J.: A database of human segmented natural images and its application to evaluating segmentation algorithms and measuring ecological statistics. In: ICCV, vol. 2, pp. 416–423. IEEE (2001)
29. Meinhardt, T., Moller, M., Hazirbas, C., Cremers, D.: Learning proximal operators: Using denoising networks for regularizing inverse imaging problems. In: ICCV, pp. 1781–1790 (2017)
30. Metzler, C.A., Maleki, A., Baraniuk, R.: Bm3d-amp: a new image recovery algorithm based on bm3d denoising. In: ICIP, pp. 3116–3120. IEEE (2015)
31. Metzler, C.A., Maleki, A., Baraniuk, R.: From denoising to compressed sensing. IEEE Trans. Inf. Theory **62**(9), 5117–5144 (2016)
32. Mousavi, A., Patel, A., Baraniuk, R.: A deep learning approach to structured signal recovery. In: Allerton, pp. 1336–1343. IEEE (2015)
33. Nan, Y., Quan, Y., Ji, H.: Variational-EM-based deep learning for noise-blind image deblurring. In: CVPR, pp. 3626–3635 (June 2020)
34. Nan, Y., Ji, H.: Deep learning for handling kernel/model uncertainty in image deconvolution. In: CVPR, pp. 2388–2397 (June 2020)
35. Quan, Y., Chen, M., Pang, T., Ji, H.: Self2self with dropout: Learning self-supervised denoising from single image. In: CVPR, pp. 1890–1898 (2020)
36. Quan, Y., Ji, H., Shen, Z.: Data-driven multi-scale non-local wavelet frame construction and image recovery. J. Sci. Comput. **63**(2), 307–329 (2015). https://doi.org/10.1007/s10915-014-9893-2
37. Schuler, C., B., C., Harmeling, S., Scholkopf, B.: A machine learning approach for non-blind image deconvolution. In: CVPR, pp. 1067–1074 (2013)
38. Shi, W., Jiang, F., Liu, S., Zhao, D.: Scalable convolutional neural network for image compressed sensing. In: CVPR, pp. 12290–12299 (2019)
39. Soltanayev, S., Chun, S.: Training deep learning based denoisers without ground truth data. In: NeurIPS, pp. 3257–3267 (2018)
40. Tang, J., Nett, B.E., Chen, G.: Performance comparison between total variation (TV)-based compressed sensing and statistical iterative reconstruction algorithms. Phys. Med. Biol. **54**(19), 5781 (2009)
41. Ulyanov, D., Vedaldi, A., Lempitsky, V.: Deep image prior. In: CVPR, pp. 9446–9454 (2018)
42. Xu, K., Zhang, Z., Ren, F.: Lapran: A scalable laplacian pyramid reconstructive adversarial network for flexible compressive sensing reconstruction. In: ECCV, pp. 485–500 (2018)
43. Xu, L., Ren, J.S., Liu, C., Jia, J.: Deep convolutional neural network for image deconvolution. In: NIPS, pp. 1790–1798 (2014)
44. Yang, Y., Sun, J., Li, H., Xu, Z.: Deep ADMM-Net for compressive sensing MRI. In: NeurIPS, pp. 10–18 (2016)
45. Zhang, J., Ghanem, B.: Ista-net: Interpretable optimization-inspired deep network for image compressive sensing. In: CVPR, pp. 1828–1837 (2018)

46. Zhang, J., Pan, J., Lai, W.S., Lau, R.W., Yang, M.H.: Learning fully convolutional networks for iterative non-blind deconvolution. In: CVPR, pp. 3817–3825 (2017)
47. Zhussip, M., Soltanayev, S., Chun, S.: Training deep learning based image denoisers from undersampled measurements without ground truth and without image prior. In: CVPR, pp. 10255–10264 (2019)

Graph-PCNN: Two Stage Human Pose Estimation with Graph Pose Refinement

Jian Wang$^{(\boxtimes)}$, Xiang Long, Yuan Gao, Errui Ding, and Shilei Wen

Department of Computer Vision Technology(VIS), Baidu Inc., Beijing, China
{wangjian33,longxiang,gaoyuan18,dingerrui,wenshilei}@baidu.com

Abstract. Recently, most of the state-of-the-art human pose estimation methods are based on heatmap regression. The final coordinates of keypoints are obtained by decoding heatmap directly. In this paper, we aim to find a better approach to get more accurate localization results. We mainly put forward two suggestions for improvement: 1) different features and methods should be applied for rough and accurate localization, 2) relationship between keypoints should be considered. Specifically, we propose a two-stage graph-based and model-agnostic framework, called Graph-PCNN, with a localization subnet and a graph pose refinement module added onto the original heatmap regression network. In the first stage, heatmap regression network is applied to obtain a rough localization result, and a set of proposal keypoints, called guided points, are sampled. In the second stage, for each guided point, different visual feature is extracted by the localization subnet. The relationship between guided points is explored by the graph pose refinement module to get more accurate localization results. Experiments show that Graph-PCNN can be used in various backbones to boost the performance by a large margin. Without bells and whistles, our best model can achieve a new state-of-the-art 76.8% AP on COCO `test-dev` split.

Keywords: Human pose estimation · Keypoint localization · Two stage · Graph pose refinement

1 Introduction

Human pose estimation [1] is a fundamental yet challenging computer vision problem, that aims to localize keypoints (human body joints or parts). It is the basis of other related tasks and various downstream vision applications, including video pose estimation [43], tracking [10,42] and human action recognition [22,39,44]. This paper is interested in 2D pose estimation to detect the spatial location (i.e. 2D coordinate) of keypoints for persons in a top-down manner. Keypoint localization is a very challenging task, even for humans. It is really difficult to locate the keypoint coordinates precisely, since the variation of clothing,

---

Both authors contributed equally to this work.

© Springer Nature Switzerland AG 2020
A. Vedaldi et al. (Eds.): ECCV 2020, LNCS 12356, pp. 492–508, 2020.
https://doi.org/10.1007/978-3-030-58621-8_29

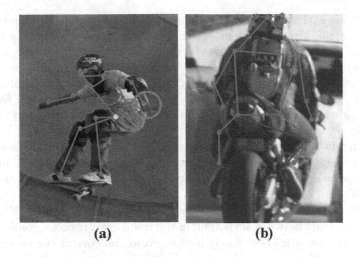

(a)   (b)

Fig. 1. Example of 2D pose estimation. The green points and lines indicate keypoints and their connections that are correctly predicted, while the red ones indicate incorrect predictions. We have observed two important characteristics of keypoint localization: 1) different features and processes are preferred for rough and accurate localization, 2) relationship between keypoints should be considered. (Color figure online)

the occlusion between the limbs, the deformation of human joints under different poses and the complex unconstrained background, will affect the keypoint recognition and localization [49].

Most existing state-of-the-art methods use CNNs to get the heatmap of each keypoint [7,8,11,12,16,20,25,29,32–34,36,40–42,45,46]. Then the heatmap will be directly decoded to keypoint coordinates. However, these approaches do not take into account two important characteristics of human pose estimation: 1) different features and processes are preferred for rough and accurate localization, 2) relationship between keypoints should be considered.

First, humans perform keypoint localization in a two-step manner [2,4,15, 26]. For example, for the blue point in Fig. 1(a), we will first perform a rough localization based on the context information, including fingers and arms shown in the blue circle, to determine whether there is a wrist keypoint in a nearby area. This step can be treated as a proposal process. After rough localization, we will further observe the detail structure of the wrist itself to determine the accurate location of wrist keypoint, which can be seen as a refinement process.

We get inspiration from object detection. In object detection methods, proposal and refinement are performed based on two different feature map achieved by two separate subnets. We suggest that the proposal and refinement processes in keypoint localization should also be based on different feature maps. Therefore, we apply two different subnets to get feature maps for proposal and refinement respectively. Besides, two-stage method is very common in object detection, and can achieve excellent results in terms of both effectiveness and performance.

A natural idea is that we apply the design of the two-stage to keypoint localization task, let the first stage focus on the proposal process, improving the recall of keypoint, and let the second stage focus on the refinement process, improving the localization accuracy. Therefore, we introduce the concept of Guided Point. First, we select the guided points based on the heatmap as rough proposals in the first stage. Then in the second stage, based on the corresponding features of selected guided points, we perform coordinate refinement for accurate keypoint regression.

Secondly, in the case of complicated clothing and occlusion, the relationship between the keypoints is very important to judge its location. For example, the yellow keypoint in Fig. 1(a), due to the occlusion, we cannot see its location directly. We can only infer it from the location of other related keypoints. In addition, due to the structural limitations of the human body, there exist obvious mutual constraints between keypoints. In the refinement process, considering the relationship between keypoints may help to avoid and correct the misprediction. For example, in Fig. 1(b), the keypoints in red are the wrong predictions of the left leg. By considering the connection between them and other keypoints, we can find out and correct these errors more easily.

However, in the traditional heatmap based method, we cannot know the location of keypoints before decoding the heatmap to coordinates. This makes it difficult for us to build a pose graph that connects keypoint features at different locations. After introducing guided points, we can know the rough locations of keypoints, such that we can build a pose graph between keypoints easily. Therefore, we propose a graph pose refinement (GPR) module, which is an extension of graph convolutional network, to improve the accuracy of keypoint localization.

The main contributions of this paper include:

- This paper proposes a model-agnostic two-stage keypoint localization framework, Graph-PCNN, which can be used in any heatmap based keypoint localization method to bring significant improvement.
- A graph pose refinement module is proposed to consider the relationship between keypoints at different locations, and further improve the localization accuracy.
- Our method set a new stage-of-the-art on COCO `test-dev` split.

2   Related Work

The classical approach of human pose estimation is using the pictorial structures framework with a pre-defined pose or part templates not depending on image data, which limit the expressiveness of the model [31,47].

Convolution Neural Networks (CNNs) have dramatically changed the direction of pose estimation methods. Since the introduction of "DeepPose" [38] by Toshev et al., most recent pose estimation systems have generally adopted CNNs as their backbone. There are mainly two kinds of methods to get the locations of keypoints: directly regress coordinates and estimate the heatmaps of the keypoints first, and then decode to coordinates.

Coordinate Based Methods. Only a few methods regress coordinates of key-points directly. DeepPose [38] formulate pose estimation as a CNN-based regression problem directly towards body joints in a holistic fashion. Fan et al., [13] propose to integrate both the body part appearance and the holistic view of each local part for more accurate regression. A few other methods [6,35] further improve performance, but there is still a gap between with heatmap based methods.

Heatmap Based Methods. The heatmap representation is first introduce by Tompson et al. [37], and then quickly becomes the most popular solution in state-of-the-art methods. A lot of research works improve the network architectures to improve the effectiveness of heatmap regression [3,7,8,11,12,16,20,23,25,28, 29,32–34,36,40–42,45,46]. For example, Hourglass [28] and its follow-ups [9, 12,45] consist of blocks of several pooling and upsampling layers, which looks like an hourglass, to capture information at every scale. SimpleBaseline [42] adds several deconvolutional layers to enlarge the resolution of output feature maps, which is quite simple but performs better. The HRNet [33] model has outperformed all existing methods on public dataset by maintaining a high-resolution representation through the whole process.

Hybrid Methods. Some works speculate that heatmap will introduce a statistical error and try to combine heatmap estimation with coordinate offset regression for better localization accuracy [18,30]. But in these methods, heatmap estimation and coordinate regression are performed at the same time on the same feature map, without refinement process to gradually improve accuracy.

Refinement Methods. Many works focus on coordinate refinement to improve the accuracy of keypoints localization [4,6,15,26]. Instead of predicting absolute joint locations, Carreira et al. refine pose estimation by predicting error feedback at each iteration [6], Bulat et al. design a cascaded architecture for mining part relationships and spatial context [4]. Some other works use a human pose refinement network to exploit dependencies between input and output spaces [15,26]. However, they can not effectively combine heatmap estimation and coordinate regression, and the relationship between different keypoints is not considered during refinement. Our method will introduce the relationship between keypoints for more effective refinement. Zhang et al. [50] builds a pose graph directly on heatmaps and uses Graph Neural Network for refinement. However, it essentially only considers the relationship between heatmap weights at the same location, while the visual information of keypoints is completely ignored. In our framework, pose graph is built on the visual features at the position of corresponding keypoints, which is more conducive to subsequent refinement.

3   Two Stage Pose Estimation Framework

In the top-down manner pose estimation methods, single person pose estimator aims to locate K keypoints $\mathbf{P} = \{\mathbf{p}_1, \mathbf{p}_2, ..., \mathbf{p}_k\}$ from an image $\mathbf{I}$ of size $W \times H \times 3$, where $\mathbf{p}_k$ is a 2D-coordinates. Heatmap based methods transform

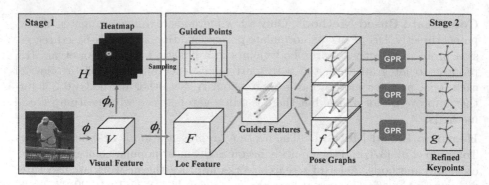

Fig. 2. Overall architecture of two stage pose estimation framework. In the first stage, heatmap regressor is applied to obtain a rough localization heatmap, and a set of guided points are sampled. In the second stage, guided points with corresponding localization features are constructed as pose graphs and then feed into a graph pose refinement (GPR) module to get refined results.

this problem to estimating K heatmaps $\{\mathbf{H}_1, \mathbf{H}_2, ..., \mathbf{H}_k\}$ of size $W' \times H' \times K$, where each heatmap $\mathbf{H}_k$ will be decoded to the corresponding coordinates $\mathbf{p}_k$ during the test phase.

Our method simply follows the popular methods to generate the heamap in the first stage. A common pipeline is first to use a deep convolutional network ϕ to extract visual features $\mathbf{V}$ from image $\mathbf{I}$,

$$\mathbf{V} = \phi(\mathbf{I}). \tag{1}$$

A heapmap regressor ϕ_h, typically ended with a 1×1 convolutional layer, is applied to estimating the heatmaps,

$$\{\mathbf{H}_1, \mathbf{H}_2, ..., \mathbf{H}_k\} = \phi_h(\mathbf{V}). \tag{2}$$

The refinement network is added after the heatmap regression, without any changes to the existing network architecture in the first stage. Therefore, our method can be applied to any heatmap based models easily. The overall architecture of our method is shown in Fig. 2. At first, we apply a localization subnet ϕ_l to transform the visual feature to the same spacial scale as heatmaps,

$$\mathbf{F} = \phi_l(\mathbf{V}), \tag{3}$$

where the size of $\mathbf{F}$ is $W' \times H' \times C$. During training, N guided points $\{\mathbf{s}_k^1, \mathbf{s}_k^2, ..., \mathbf{s}_k^N\}$ are sampled for each heatmap $\mathbf{H}_k$, while the best guided points $\mathbf{s}_k^*$ is selected for heapmap $\mathbf{H}_k$ during testing. For sake of simplification, we omit the superscript in the following formula. For any guided point $\mathbf{s}_k$, guided feature $\mathbf{f}_k = \mathbf{F}[\mathbf{s}_k]$ at the corresponding location and its confidence score $h_k = \mathbf{H_k}[\mathbf{s}_k]$ can be extracted.

Subsequently, we can build N pose graph for $N \times K$ guided features, and introduce a graph pose refinement (GPR) module to refine the visual features by considering the relationship between keypoints.

(a)   (b)   (c)

Fig. 3. Illustration of sampling region. Taking right wrist as example, (a), (b), (c) show the three kinds of guided points respectively, points which are close to the ground truth keypoint, points which are far away from the ground truth keypoint and points which have high heat response, where the yellow circle points indicate sampled guided points and the yellow star points indicate ground truth of right wrist. (Color figure online)

$$\{\mathbf{g}_1, \mathbf{g}_2, ..., \mathbf{g}_K\} = \text{GPR}(\{\mathbf{f}_1, \mathbf{f}_2, ..., \mathbf{f}_K\}, \{h_1, h_2, ..., h_K\}). \tag{4}$$

Finally, the refined classification result $\mathbf{c}_k$ and offset regression result $\mathbf{r}_k$ are achieved based on the refined feature $\mathbf{g}_k$. The refined coordinate of keypoint is

$$\mathbf{p}_k = \mathbf{s}_k + \mathbf{r}_k. \tag{5}$$

In the following, we first describe the guided point sampling strategy in Sect. 3.1. Second, we show the detail structure of graph pose refinement module in Sect. 3.2. Third, we introduce the loss used for training in Sect. 3.3. Finally, we show how to integrate our framework to existing backbones and elaborate the details of training and testing in Sect. 3.4.

3.1   Guided Point Sampling

Locating human joint based on the peak of heatmap is frequently-used in modern human pose estimators, and they modeled the target of heatmap by generating gaussian distribution around the ground truth. But due to the complex image context and human action, the joint heat may not be satisfy gaussian distribution strictly which, together with quantisation affect of image resolution downsampling, leads to an insufficient precision of this localization method. However, the peak of heatmap is always close to the true location of joint, which make it adequate to regress the true location.

To achieve the goal of obtaining refined coordinate based on the peak of heatmap, we sample several guided points and train coordinate refinement in stage2. Concretely, we equally sample three kinds of guided points for training: (a) points which are close to the ground truth keypoint, (b) points which are far away from the ground truth keypoint, and (c) points which have high heat response. And the kth ground truth keypoint is denoted as $\mathbf{t}_k$. As exhibited in Fig. 3, (a) and (b) are randomly sampled within the red region and blue region, respectively, and the red region which centered at ground truth has a radius of 3σ, where σ is same with the standard deviation for generating gaussian heatmap target. (c) is randomly sampled from the top N highest response points at the heatmap.

Due to the different characterization of different keypoints, we sample guided points for each keypoint individually, and the total amount of the three kinds of guided points for each keypoint is set equally to N.

After the N guided points $\{\mathbf{s}_k^1, \mathbf{s}_k^2, ..., \mathbf{s}_k^N\}$ are sampled, we divide them into two sets, positive set and negative set, denoted as

$$
\begin{aligned}
\mathcal{S}_k^+ &= \{\mathbf{s}_k \mid \|\mathbf{s}_k - \mathbf{t}_k\| < 3\sigma\} \\
\mathcal{S}_k^- &= \{\mathbf{s}_k \mid \|\mathbf{s}_k - \mathbf{t}_k\| \geq 3\sigma\}
\end{aligned}
\tag{6}
$$

and $N_k^+ = |\mathcal{S}_k^+|$, $N_k^- = |\mathcal{S}_k^-|$. Then all of the corresponding guided feature extracted from $\mathbf{F}$ by means of bilinear interpolation are feeded into stage2 for refinement while only the guided points from positive set contributed to the coordinate regression.

According to the above label assignment manner, (a) and (b) are definite positive and negative samples, and the influence of proportion between them will be explored in Sect. 4.3. While (c) is almost negative samples during the beginning stage of training and turns to positive samples as the training schedule goes on. We suppose that (c) can not only accelerate the feature learning at the beginning of training, because (c) are hard negative samples for classification at this stage, but also contribute to the learning of regression when the classification status of feature is relatively stable, as (c) are almost positive samples at this period. Further more, (c) is not necessarily positive when the model converges roughly because of some prediction error caused by hard situation. In this circumstances, (c) can also be regarded as hard negative samples for helping the model to be trained better.

3.2   Graph Pose Refinement

In most of previous works, many fields have been well studied for human pose estimation, such as network structure, data preprocessing and postprocessing, post refinement, etc. However, in these works, the localization of human keypoints are conducted independently for each keypoint while the relationship between different keypoints is ignored all along. Intuitively, the human keypoints construct a salient graph structure base on the pattern of human body, and they have clear adjacent relation with each other. So we consider that the localization

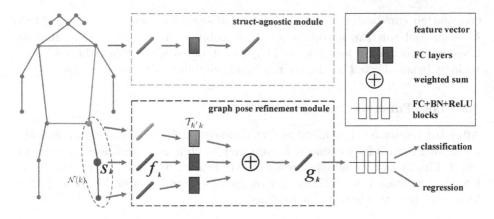

Fig. 4. The structure of graph pose refinement module. The relationship between keypoints is taken into account in contrast to the struct-agnostic module.

of keypoints can be infered better with the help of the information hinted by this relationship. For instance, in our framework, if we know that a guided point is left elbow, then the positive guided points of left wrist should tend to have higher response on left wrist a priori, as left wrist is adjacent to left elbow. So that more supervision can be imposed upon the feature of these keypoints than treating them independently.

To take advantage of the information implicit in the graph structure mentioned above, we propose a graph pose refinement module to model it, and then refine the feature of these keypoints. As shown in Fig. 4, we build a graph and conduct graph convolution for each keypoint. The output embedding feature can be computed by

$$
\mathbf{g}_k = \frac{1}{Z_k} \sum_{\mathbf{s}_{k'} \in \mathcal{N}(k)} \omega_{k'} \mathcal{T}_{k' k}(\mathbf{f}_{k'})
$$

$$
\omega_{k'} =
\begin{cases}
h_{k'} \mathbb{1}(R_{k'}), & k' \neq k \\
1, & k' = k
\end{cases}
\tag{7}
$$

where $\mathcal{N}(k)$ represents for a point set containing the guided point $\mathbf{s}_k$ and its neighbours, $\mathcal{T}_{k' k}$ for the linear transformation from guided point $\mathbf{s}_{k'}$ to $\mathbf{s}_k$, and $\mathbb{1}$ for the indicator function. $Z_k = \sum_{\mathbf{s}_{k'} \in \mathcal{N}(k)} \omega_{k'}$ is used for normalization. $R_{k'}$ is a boolean type parameter encoding the reliability of a guided point which works for filtering out points of low quality, and its definition will be explained in detail in Sect. 3.4.

Specially, as defined in (7), this graph convolution is an extension of traditional graph convolution, it is designed by considering the characteristic of pose estimation problem. Firstly, we add a weight for each message generated from $\mathbf{s}_k'$ to $\mathbf{s}_k$, which can control the contribution of each message according to the intensity and reliability of $\mathbf{s}_k'$. With the constraint of these weights, the graph convolution can be trained more stably. Further more, we set $\omega_{k'} = 1$ when $k' = k$. And this can make the graph convolution degrading to a traditional linear transformation for s_k when $\mathbb{1}(R_{k'}) = 0$ for all $s_{k'} \in \mathcal{N}(k)$ where $k' \neq k$, without being affected by the intensity and reliability of s_k itself.

3.3 Loss Function

After the refinement module above, the embedded feature is sent to a module containing several fully connected layers and batch norm layers, as illustrated in Fig. 4. Finally two predictions are outputed, denoted as c_k and r_k, for classification and regression respectively. Giving ground truth keypoint location t_k, the losses for these two branches are defined as

$$L_k^{cls} = \frac{1}{2} \left[\frac{1}{N_k^+} \sum_{s_k^i \in \mathcal{S}_k^+} \alpha_k^i \mathcal{L}_{cls}(c_k^i, 1) + \frac{1}{N_k^-} \sum_{s_k^i \in \mathcal{S}_k^-} \mathcal{L}_{cls}(c_k^i, 0) \right] \tag{8}$$

$$\alpha_k^i = exp(-\frac{(s_k^i - t_k)^2}{2\sigma^2})$$

and

$$L_k^{reg} = \frac{1}{N_k^+} \sum_{s_k^i \in \mathcal{S}_k^+} \mathcal{L}_{reg}(r_k^i, t_k - s_k^i), \tag{9}$$

where $\mathcal{L}^{cls}$ and $\mathcal{L}^{reg}$ are softmax cross-entropy loss and L1 loss. The total loss of the stage2, can be expressed as

$$L^{s_2} = \frac{\sum_k \gamma_k (L_k^{cls} + \lambda L_k^{reg})}{\sum_k \gamma_k}, \tag{10}$$

where γ_k is the target weight of keypoint k. And λ is a loss weight which is set to 16 constantly. And the total loss of Graph-PCNN is

$$L = L^{s_1} + L^{s_2}, \tag{11}$$

where L^{s_1} is the traditional heatmap regression loss for stage1.

3.4 Network Architecture

Network Architecture. In previous works such as [18,30], there is also a coordinate refinement after heatmap decoding, and their coordinate refinement branch share the same feature map with heatmap prediction branch. However, the rough and accurate localization always need different embedding feature, further more, it is hard to conduct particular feature refinement for either of these two branches. In order to alleviate the above problems, we copy the last stage of the backbone network to produce two different feature maps with the same size followed by heatmap regression convolution and graph pose refinement module respectively. By means of this modification, the network can learn more particular feature for two different branches, and easily conduct guided points sampling for further feature refinement.

Training and Testing. For the proposed two stage pose estimation framework, several operations are specific in the training and testing phase.

Firstly, in order to make the stage2 be trained sufficiently, we sample multiple guided points for each keypoint following the strategy described in Sect. 3.1 during training, and the amount of guided points N is various according to the input size. While during testing, only one guided point is generated by decoding the predicted heatmap, and the output score of it is gathered as the corresponding heat response score from stage1. Following most of previous works [33,42], a quarter offset in the direction from the highest response to the second highest response is added to the position of heatmap peak for higher precision, when decoding this guided point from heatmap.

Secondly, the definition of guided point reliability metric $R_{k'}$ is different for training and testing, which is represented as

$$R_{k'} = \begin{cases} ||\mathbf{s}_{k'} - \mathbf{t}_{k'}|| < \delta & \text{in training phase} \\ h_{k'} > \xi & \text{in testing phase} \end{cases} \quad (12)$$

At the training phase, the ground truth is available for measuring this reliability, and the guided points which are close to their corresponding ground truth can be regarded reliable. δ is a distance threshold controling the close degree which equals to 2σ. While at the testing phase the ground truth is unknown, so for insurance, the guided points which heat responses are high enough are qualified to pass message to their neighbour points. And ξ is a threshold for gating the heat response, which is set to 0.85 constantly.

Finally, during training, we shuffle the guided points of one keypoint after the guided point sampling in order to create more various situation of graph combination, which can make the graph pose refinement module more generalized.

4   Experiments

4.1   Dataset

In this paper, we use the most popular human pose estimation dataset, COCO. The COCO keypoint dataset [24] presents challenging images with multi-person pose of various body scales and occlusion patterns in unconstrained environments. It contains 200,000 images and 250,000 person samples. Each person instance is labelled with 17 joints. We train our models on train2017(includes 57 K images and 150 K person instances) with no extra data, and conduct ablation study on val2017. Then we test our models on test-dev for comparison with the state-of-the-art methods. In evaluation, we use the metric of Object Keypoint Similarity (OKS) for COCO to report the model performance.

4.2   Implementation Details

For fair comparison, we follow the same training configuration as [42] and [33] for ResNet and HRNet respectively.

To construct the localization subnet, we copy the conv5 stage, which spatial size is 1/32 to the input size, and the last three deconvolution layers for ResNet series networks, while copying the stage4, which has three high resolution modules, for HRNet series networks. For ablation study, we also add 128×96 input size in our experiment following [49]. And we set N as 48, 192 and 432 corresponding to the three input sizes of 128×96, 256×192 and 384×288 during all of our experiment except the ablation study of N. During inference, we use person detectors of AP 56.4 and 60.9 for COCO `val2017` and `test-dev` respectively, while for pose estimation, we evaluate single model and only use flipping test strategy for testing argumentation.

Table 1. Ablation study on COCOval2017

Method	Size	stage1 AP	stage2 AP
SBN	128×96	59.3	-
Graph-PCNN	128×96	**61.1**	**64.6**
SBN	256×192	70.4	-
Graph-PCNN	256×192	**71.3**	**72.6**
SBN	384×288	72.2	-
Graph-PCNN	384×288	**72.7**	**73.6**

4.3   Ablation Studies

We use ResNet-50 backbone to perform ablation study on COCO `val2017`.

Two Stage Pose Estimation Framework. Firstly, we evaluate the effectiveness of our proposed two stage pose estimation framework. As Table 1 shows, the stage2 of Graph-PCNN gives 5.3%, 2.2%, 1.4% AP gain comparing to original simple baseline network(SBN) at the three input sizes, which demonstrates that our regression based two stage framework is more effective than decoding joint location from heatmap. Further more, we test the stage1 of Graph-PCNN which shares the same network architecture with SBN. It should be noted that training with Graph-PCNN can also boost the performance of heatmap, and 1.8%, 0.9%, 0.5% AP gain are got as shown. That means we can also get considerable performance boosting without any extra computing cost during inference if we only use the stage1 of Graph-PCNN.

Sampling Strategy. Secondly, we study the influence of the proportion of different kinds of guided points and the total amount of guided points N based on ResNet-50 with 128×96 input size. In order to avoid exploring the proportion

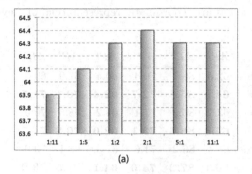
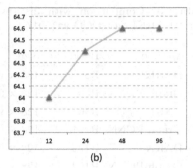

Fig. 5. Influence of the proportion and total amount of guided points in sampling. (a) is the results on different proportions, while the x-axis represents the proportions between positive guided points and negative guided points. (b) is the results on different values of the total amount, while the x-axis represents the values of N.

Table 2. Effectiveness of the graph pose refinement(GPR) module.

Method	Size	stage1 AP	stage2 AP
struct-agnostic	128×96	60.7	63.8
GPR-va	128×96	61.2	52.1
GPR-vb	128×96	61.1	64.5
GPR-vc	128×96	60.8	64.3
GPR	128×96	61.1	**64.6**

among all the three kinds of guided points, we simplify the proportion study by using only definite positive points and negative points, and then we set different proportion between them with N unchanged. From the results shown in Fig. 5 (a), we can come to that the proportion ranging from 1:2 to 2:1 is already appropriate, and the sampling strategy proposed in Sect. 3.1 can fit this proportion range in any situation. In addition, we try different N based on the strategy in Sect. 3.1 and finally select 48 as the value of N according to the results reported in Fig. 5 (b).

Graph Pose Refinement Module. Finally, we evaluate the contribution of the proposed graph pose refinement(GPR) module. In this study, we compare proposed GPR with a struct-agnostic baseline module and several variants of GPR(GPR-va, GPR-vb, GPR-vc). GPR-va set $\omega_{k'} = 1$ for all $\{k'|\mathbf{s}_{k'} \in \mathcal{N}(k)\}$ in (7), GPR-vb set $\omega_{k'} = \mathbb{1}(R_{k'})$ for $\{k'|\mathbf{s}_{k'} \in \mathcal{N}(k), k' \neq k\}$ with the heat response factor dropped, and GPR-vc dropped the guided points shuffling operation mentioned in Sect. 3.4. The comparison results are displayed in Table 2. We can see that GPR boosts the stage1 AP and stage2 AP by 0.4% and 0.8% respectively, comparing to the struct-agnostic baseline. And the performance of GPR is better than all of its other variants, which reveals the importance of parameter $\omega_{k'}$ and the guided points shuffling operation. Especially, the reliabil-

Table 3. Comparison with distribution-aware coordinate representation of keypoint(DARK) on COCO `val2017`.

Method	Backbone	Size	AP	AP^{50}	AP^{75}	AP^M	AP^L	AR
DARK	R50	128 × 96	62.6	86.1	70.4	60.4	67.9	69.5
Graph-PCNN	R50	128 × 96	**64.6**	**86.4**	**72.7**	**62.4**	**70.1**	**71.5**
DARK	R101	128 × 96	63.2	86.2	71.1	61.2	68.5	70.0
Graph-PCNN	R101	128 × 96	**64.8**	**86.6**	**73.1**	**62.6**	**70.3**	**71.7**
DARK	R152	128 × 96	63.1	86.2	71.6	61.3	68.1	70.0
Graph-PCNN	R152	128 × 96	**66.1**	**87.2**	**74.6**	**64.1**	**71.5**	**73.0**
DARK	HR32	128 × 96	70.7	88.9	78.4	67.9	76.6	76.7
Graph-PCNN	HR32	128 × 96	**71.5**	**89.0**	**79.0**	**68.4**	**77.6**	**77.3**
DARK	HR48	128 × 96	71.9	89.1	79.6	69.2	78.0	77.9
Graph-PCNN	HR48	128 × 96	**72.8**	**89.2**	**80.1**	**69.9**	**79.0**	**78.6**
DARK	HR32	256 × 192	75.6	90.5	82.1	71.8	82.8	80.8
Graph-PCNN	HR32	256 × 192	**76.2**	**90.3**	**82.6**	**72.5**	**83.2**	**81.2**
DARK	HR32	384 × 288	76.6	90.7	82.8	72.7	83.9	81.5
Graph-PCNN	HR32	384 × 288	**77.2**	**90.7**	**83.6**	**73.5**	**84.0**	**82.1**

Table 4. Comparison with model-agnostic human pose refinement network(PoseFix) on COCO `val2017`.

Method	Backbone	Size	AP	AP^{50}	AP^{75}	AP^M	AP^L	AR
PoseFix	R50	256 × 192	72.1	88.5	78.3	68.6	78.2	-
Graph-PCNN	R50	256 × 192	**72.6**	**89.1**	**79.3**	**69.1**	**79.7**	78.1

ity factor $\mathbb{1}(R_{k'})$ affects the performance greatly. Thus, we believe that GPR can refine the feature of a guided point by taking advantage of the supervision signal of its neighbour keypoint which is good located, as we supposed in Sect. 3.2.

4.4   Comparison with Other Methods with Coordinate Refinement

DARK [49] is a state-of-the-art method which improved traditional decoding by a more precise refinement based on Taylor-expansion. We follow the training settings of DARK and compare our refinement results with it. From Table 3 we can observe that our Graph-PCNN generally outperforms DARK over different network architecture and input size. This suggests that regression based refinement predicts coordinate more precise than analyzing the distribution of response signal from heatmap, as the response signal itself may not satisfy gaussian distribution strictly because of complex human pose and image context while regression is regardless of these drawback.

PoseFix [26] is a model-agnostic method which refines a existing pose result from any other method by a independent model. A coarse-to-fine coordinate

Table 5. Comparison with the state-of-the-arts methods on COCO `test-dev`.

Method	Backbone	Size	AP	AP^{50}	AP^{75}	AP^M	AP^L	AR
CMU-Pose [5]	-	-	61.8	84.9	67.5	57.1	68.2	66.5
Mask-RCNN [17]	R50-FPN	-	63.1	87.3	68.7	57.8	71.4	-
G-RMI [30]	R101	353 × 257	64.9	85.5	71.3	62.3	70.0	69.7
AE [27]	-	512 × 512	65.5	86.8	72.3	60.6	72.6	70.2
Integral Pose [35]	R101	256 × 256	67.8	88.2	74.8	63.9	74.0	-
CPN [8]	ResNet-Inception	384 × 288	72.1	91.4	80.0	68.7	77.2	78.5
RMPE [14]	PyraNet [45]	320 × 256	72.3	89.2	79.1	68.0	78.6	-
CFN [19]	-	-	72.6	86.1	69.7	78.3	64.1	-
CPN(ensemble)[8]	ResNet-Inception	384 × 288	73.0	91.7	80.9	69.5	78.1	79.0
Posefix [26]	R152+R152	384 × 288	73.6	90.8	81.0	70.3	79.8	79.0
CSM+SCARB [32]	R152	384 × 288	74.3	91.8	81.9	70.7	80.2	80.5
CSANet [48]	R152	384 × 288	74.5	91.7	82.1	71.2	80.2	80.7
MSPN [21]	MSPN	384 × 288	76.1	93.4	83.8	72.3	81.5	81.6
Simple Base [42]	R152	384 × 288	73.7	91.9	81.1	70.3	80.0	79.0
UDP [18]	R152	384 × 288	74.7	91.8	82.1	71.5	80.8	80.0
Graph-PCNN	R152	384 × 288	**75.1**	91.8	**82.3**	**71.6**	**81.4**	**80.2**
HRNet [33]	HR32	384 × 288	74.9	92.5	82.8	71.3	80.9	80.1
UDP [18]	HR32	384 × 288	76.1	92.5	83.5	72.8	82.0	81.3
Graph-PCNN	HR32	384 × 288	**76.4**	**92.5**	**83.8**	**72.9**	**82.4**	**81.3**
HRNet [33]	HR48	384 × 288	75.5	92.5	83.3	71.9	81.5	80.5
DARK [49]	HR48	384 × 288	76.2	92.5	83.6	72.5	82.4	81.1
UDP [18]	HR48	384 × 288	76.5	**92.7**	84.0	73.0	82.4	81.6
PoseFix [26]	HR48+R152	384 × 288	76.7	92.6	84.1	73.1	82.6	81.5
Graph-PCNN	HR48	384 × 288	**76.8**	92.6	**84.3**	**73.3**	**82.7**	**81.6**

estimation schedule ended by coordinate calculation following integral loss [35] is used to enhance the precision. We conduct comparison with PoseFix by using same backbone and input size with its model from refinement stage and the performance of human detectors for these two methods are comparable, AP 55.3 vs 56.4 for PoseFix(using CPN) and our Graph-PCNN respectively. As illustraed in Table 4, we achieve a competable result with PoseFix, but PoseFix included input from CPN which need an extra R50 network while our method only need an extra R50 conv5 stage as refinement branch.

4.5 Comparison to State of the Art

We compare our Graph-PCNN with other top-performed methods on COCO `test-dev`. As Table 5 reports, our method with HR48 backbone at the input size of 384 × 288 achieves the best AP(76.8), and improves HR48 with the same input size(75.5) by a large margin(+1.3). Mean while, It also outperforms other competitors with same backbone and input size settings, such as DARK(76.2), UDP(76.5) and PoseFix(76.7), which illustrates the advantages of our method.

5   Conclusions

In this paper, we propose a two stage human pose estimator for the top-down pose estimation network, which improves the overall localization performance by introducing different features for rough and accurate localization. Meanwhile, a graph pose refinement module is proposed to refine the feature for pose regression by taking the relationship between keypoints into account, which boosts the performance of our two stage pose estimator further. Our proposed method is model-agnostic and can be added on most of the mainstream backbone. Even better, more improvement can be explored by drawing on the successful experience of the two stage detection framework in the future.

References

1. Andriluka, M., Pishchulin, L., Gehler, P., Schiele, B.: 2D human pose estimation: new benchmark and state of the art analysis. In: CVPR (2014)
2. Belagiannis, V., Rupprecht, C., Carneiro, G., Navab, N.: Robust optimization for deep regression. In: Proceedings of the IEEE International Conference on Computer Vision, pp. 2830–2838 (2015)
3. Belagiannis, V., Zisserman, A.: Recurrent human pose estimation. In: FG (2017)
4. Bulat, A., Tzimiropoulos, G.: Human pose estimation via convolutional part heatmap regression. In: Leibe, B., Matas, J., Sebe, N., Welling, M. (eds.) ECCV 2016. LNCS, vol. 9911, pp. 717–732. Springer, Cham (2016). https://doi.org/10.1007/978-3-319-46478-7_44
5. Cao, Z., Simon, T., Wei, S.E., Sheikh, Y.: Realtime multi-person 2D pose estimation using part affinity fields. In: CVPR (2017)
6. Carreira, J., Agrawal, P., Fragkiadaki, K., Malik, J.: Human pose estimation with iterative error feedback. In: CVPR (2016)
7. Chen, X., Yuille, A.L.: Articulated pose estimation by a graphical model with image dependent pairwise relations. In: NeurIPS (2014)
8. Chen, Y., Wang, Z., Peng, Y., Zhang, Z., Yu, G., Sun, J.: Cascaded pyramid network for multi-person pose estimation. In: CVPR (2018)
9. Chen, Y., Shen, C., Wei, X.S., Liu, L., Yang, J.: Adversarial posenet: a structure-aware convolutional network for human pose estimation. In: ICCV (2017)
10. Cho, N.G., Yuille, A.L., Lee, S.W.: Adaptive occlusion state estimation for human pose tracking under self-occlusions. Pattern Recogn. **46**(3), 649–661 (2013)
11. Chu, X., Ouyang, W., Li, H., Wang, X.: Structured feature learning for pose estimation. In: CVPR (2016)
12. Chu, X., Yang, W., Ouyang, W., Ma, C., Yuille, A.L., Wang, X.: Multi-context attention for human pose estimation. In: CVPR (2017)
13. Fan, X., Zheng, K., Lin, Y., Wang, S.: Combining local appearance and holistic view: dual-source deep neural networks for human pose estimation. In: CVPR (2015)
14. Fang, H.S., Xie, S., Tai, Y.W., Lu, C.: RMPE: regional multi-person pose estimation. In: ICCV (2017)
15. Fieraru, M., Khoreva, A., Pishchulin, L., Schiele, B.: Learning to refine human pose estimation. In: CVPR (2018)

16. Gkioxari, G., Toshev, A., Jaitly, N.: Chained predictions using convolutional neural networks. In: Leibe, B., Matas, J., Sebe, N., Welling, M. (eds.) ECCV 2016. LNCS, vol. 9908, pp. 728–743. Springer, Cham (2016). https://doi.org/10.1007/978-3-319-46493-0_44

17. He, K., Gkioxari, G., Dollár, P., Girshick, R.: Mask R-CNN. In: ICCV (2017)

18. Huang, J., Zhu, Z., Guo, F., Huang, G.: The devil is in the details: delving into unbiased data processing for human pose estimation. arXiv preprint arXiv:1911.07524 (2019)

19. Huang, S., Gong, M., Tao, D.: A coarse-fine network for keypoint localization. In: ICCV (2017)

20. Ke, L., Chang, M.C., Qi, H., Lyu, S.: Multi-scale for human pose estimation. In: ECCV (2018)

21. Li, W., et al.: Rethinking on multi-stage networks for human pose estimation. arXiv preprint arXiv:1901.00148 (2019)

22. Liang, Z., Wang, X., Huang, R., Lin, L.: An expressive deep model for human action parsing from a single image. In: ICME. IEEE (2014)

23. Lifshitz, I., Fetaya, E., Ullman, S.: Human pose estimation using deep consensus voting. In: Leibe, B., Matas, J., Sebe, N., Welling, M. (eds.) ECCV 2016. LNCS, vol. 9906, pp. 246–260. Springer, Cham (2016). https://doi.org/10.1007/978-3-319-46475-6_16

24. Lin, T.-Y., et al.: Microsoft COCO: common objects in context. In: Fleet, D., Pajdla, T., Schiele, B., Tuytelaars, T. (eds.) ECCV 2014. LNCS, vol. 8693, pp. 740–755. Springer, Cham (2014). https://doi.org/10.1007/978-3-319-10602-1_48

25. Liu, W., Chen, J., Li, C., Qian, C., Chu, X., Hu, X.: A cascaded inception of inception network with attention modulated feature fusion for human pose estimation. In: AAAI (2018)

26. Moon, G., Chang, J.Y., Lee, K.M.: Posefix: model-agnostic general human pose refinement network. In: CVPR (2019)

27. Newell, A., Huang, Z., Deng, J.: Associative embedding: End-to-end learning for joint detection and grouping. In: NeurIPS (2017)

28. Newell, A., Yang, K., Deng, J.: Stacked hourglass networks for human pose estimation. In: Leibe, B., Matas, J., Sebe, N., Welling, M. (eds.) ECCV 2016. LNCS, vol. 9912, pp. 483–499. Springer, Cham (2016). https://doi.org/10.1007/978-3-319-46484-8_29

29. Ning, G., Zhang, Z., He, Z.: Knowledge-guided deep fractal neural networks for human pose estimation. IEEE Trans. Multimedia **20**, 1246–1259 (2017)

30. Papandreou, G., et al.: Towards accurate multi-person pose estimation in the wild. In: CVPR (2017)

31. Pishchulin, L., Andriluka, M., Gehler, P., Schiele, B.: Poselet conditioned pictorial structures. In: CVPR (2013)

32. Su, K., Yu, D., Xu, Z., Geng, X., Wang, C.: Multi-person pose estimation with enhanced channel-wise and spatial information. In: CVPR (2019)

33. Sun, K., Xiao, B., Liu, D., Wang, J.: Deep high-resolution representation learning for human pose estimation. In: CVPR (2019)

34. Sun, X., Shang, J., Liang, S., Wei, Y.: Compositional human pose regression. In: ICCV (2017)

35. Sun, X., Xiao, B., Wei, F., Liang, S., Wei, Y.: Integral human pose regression. In: ECCV, September 2018

36. Tang, W., Yu, P., Wu, Y.: Deeply learned compositional models for human pose estimation. In: ECCV (2018)

37. Tompson, J.J., Jain, A., LeCun, Y., Bregler, C.: Joint training of a convolutional network and a graphical model for human pose estimation. In: NeurIPS (2014)
38. Toshev, A., Szegedy, C.: Deeppose: Human pose estimation via deep neural networks. In: CVPR (2014)
39. Wang, C., Wang, Y., Yuille, A.L.: An approach to pose-based action recognition. In: CVPR (2013)
40. Wei, S.E., Ramakrishna, V., Kanade, T., Sheikh, Y.: Convolutional pose machines. In: CVPR (2016)
41. Xiao, Ouyang, W., Wang, X., et al.: CRF-CNN: modeling structured information in human pose estimation. In: NeurIPS (2016)
42. Xiao, B., Wu, H., Wei, Y.: Simple baselines for human pose estimation and tracking. In: ECCV (2018)
43. Xiaohan Nie, B., Xiong, C., Zhu, S.C.: Joint action recognition and pose estimation from video. In: CVPR (2015)
44. Yan, S., Xiong, Y., Lin, D.: Spatial temporal graph convolutional networks for skeleton-based action recognition. In: AAAI (2018)
45. Yang, W., Li, S., Ouyang, W., Li, H., Wang, X.: Learning feature pyramids for human pose estimation. In: ICCV (2017)
46. Yang, W., Ouyang, W., Li, H., Wang, X.: End-to-end learning of deformable mixture of parts and deep convolutional neural networks for human pose estimation. In: CVPR (2016)
47. Yang, Y., Ramanan, D.: Articulated human detection with flexible mixtures of parts. IEEE Trans. Pattern Anal. Mach. Intell. **35**(12), 2878–2890 (2012)
48. Yu, D., Su, K., Geng, X., Wang, C.: A context-and-spatial aware network for multi-person pose estimation. arXiv preprint arXiv:1905.05355 (2019)
49. Zhang, F., Zhu, X., Dai, H., Ye, M., Zhu, C.: Distribution-aware coordinate representation for human pose estimation. arXiv preprint arXiv:1910.06278 (2019)
50. Zhang, H., Ouyang, H., Liu, S., Qi, X., Shen, X., Yang, R., Jia, J.: Human pose estimation with spatial contextual information. arXiv preprint arXiv:1901.01760 (2019)

Semi-supervised Learning with a Teacher-Student Network for Generalized Attribute Prediction

Minchul Shin[✉][iD]

Search Solutions Inc., Gyeonggi-do, Republic of Korea
min.stellastra@gmail.com

Abstract. This paper presents a study on semi-supervised learning to solve the visual attribute prediction problem. In many applications of vision algorithms, the precise recognition of visual attributes of objects is important but still challenging. This is because defining a class hierarchy of attributes is ambiguous, so training data inevitably suffer from class imbalance and label sparsity, leading to a lack of effective annotations. An intuitive solution is to find a method to effectively learn image representations by utilizing unlabeled images. With that in mind, we propose a multi-teacher-single-student (MTSS) approach inspired by the multi-task learning and the distillation of semi-supervised learning. Our MTSS learns task-specific domain experts called teacher networks using the label embedding technique and learns a unified model called a student network by forcing a model to mimic the distributions learned by domain experts. Our experiments demonstrate that our method not only achieves competitive performance on various benchmarks for fashion attribute prediction, but also improves robustness and cross-domain adaptability for unseen domains.

Keywords: Semi-supervised learning · Unlabeled data · Visual attributes

1 Introduction

Visual attributes are qualities of objects that facilitate the cognitive processes of human beings. Therefore, predicting the attributes of an object accurately has many useful applications in the real world. For example, a search engine can use predictions to screen products with undesirable attributes instead of using noisy metadata provided by anonymous sellers [2]. Attribute prediction is essentially a multi-label classification problem that aims to determine if an image contains certain attributes (*e.g., colors and patterns*). However, attribute prediction is known as a very challenging task based on the expense of annotation, difficulty

Electronic supplementary material The online version of this chapter (https://doi.org/10.1007/978-3-030-58621-8_30) contains supplementary material, which is available to authorized users.

© Springer Nature Switzerland AG 2020
A. Vedaldi et al. (Eds.): ECCV 2020, LNCS 12356, pp. 509–525, 2020.
https://doi.org/10.1007/978-3-030-58621-8_30

GT: Person, Dog, Car,Skateboard   GT: Flower, Rose, Skirt, Pleated

(a) Multi-Instance, Single-Attribute   (b) Single-Instance, Multi-Attribute

Fig. 1. Two different types of visual attribute prediction tasks. Image (a) was sampled from MSCOCO [21] and image (b) was sampled from DeepFashion [23].

in defining a class hierarchy, and simultaneous appearance of multiple attributes in objects. Although recent works have shown competitive results on various benchmark datasets [7,12], we identified several additional conditions that must be satisfied to solve the aforementioned issues for such methods to be useful in real-world applications: domain-agnostic training, the use of unlabeled data, and robustness/generalization.

Use of Unlabeled Data. The human perception of an attribute is intrinsically subjective, leading to sparse and ambiguous annotations in datasets. Additionally, attributes have very-long-tailed distributions, leading to severe class imbalance (*e.g., pattern: solid versus zebra*). Therefore, even if the total number of annotated images is large, models can suffer from extreme shortages of usable training data following class balancing in many cases [9]. An intuitive solution is to use unlabeled images for training in a semi-supervised manner to enhance the ability of a model to represent the visual information included in images. The teacher-student paradigm proposed in this paper facilitates this process with no preprocessing tricks, such as obtaining pseudo-labels based on predictions [5,40,41].

Robustness/Generalization. Robustness measures how stable and reliable a network is when making decisions in the presence of unexpected perturbations in inputs [43]. Generalization measures how well a trained model performs with a domain shift or if a model can be generalized sufficiently with insufficient training data [17,26]. In addition to high accuracy on benchmarks, these aspects must also be considered as crucial factors for measuring model quality.

Domain-agnostic Training. To achieve state-of-the-art performance on benchmarks, recent works [4,22,23,42] have relied on domain-specific auxiliary information during training. However, such methods have obvious limitations in terms of expansion to additional domains, which is an important feature for real applications. Therefore, it is preferable to avoid using any domain-specific information during training. We refer to this principle as domain-agnostic training.

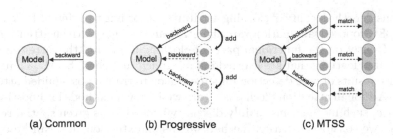

(a) Common   (b) Progressive   (c) MTSS

Fig. 2. Conceptual summary of previous methods and the proposed method denoted, as common [4,22,23,42], progressive [2], and MTSS.

Main Contributions. In this paper, we propose a multi-teacher-single-student (MTSS) method for generalized visual attribute prediction that aims to train a single unified model to predict all different attributes in a single forward operation. This paper makes the following main contributions. First, we introduce an MTSS method that learns from multiple domain experts in a semi-supervised manner utilizing unlabeled images. We also show the advantages of label embedding for training teacher networks, which are used as domain experts. Second, we demonstrate that the MTSS method can enhance model quality in terms of robustness and generalization without sacrificing benchmark performance by learning from images in distinct domains. Finally, our approach is fully domain agnostic, meaning it requires no domain-specific auxiliary supervision, such as landmarks, pose detection, or text description, for training. By learning using only attribute labels, our model not only outperforms previous methods under the same conditions [6,8,16], but also achieves competitive results relative to existing domain-specific methods [22,23,32,38] that use additional auxiliary labels for supervision (Figs. 2,3,4).

2 Related Work

Visual Attributes. Visual attribute prediction (VAP) is a multi-label classification task that has been widely studied [2,4,22,23,30,32,38,42]. There are two main types of VAP tasks, as illustrated in Fig. 1. We are particularly interested in type (b), which is frequently observed in the fashion domain. This is because our interests lie in the ability of a model to predict multiple simultaneously appearing attributes, rather than localizing interest regions, which is a matter of detection. The attributes appearing in type (b) are not necessarily related to visual similarity. For example, they may only be related to low-level characteristics, such as color, which makes the task more interesting. In the fashion domain, some works [4,22,23,42] have used landmarks for clothing items, pose detection, or textual item descriptions to improve overall accuracy. However, such strong requirements regarding auxiliary information limit such methods to domain-specific solutions. In contrast, some works have focused on attention mechanisms [32,38,42]. Zhang et al. [42] proposed a task-aware attention mechanism that considers the locality of clothing attributes according to different tasks. Wang *et al.* [38] proposed landmark-aware attention and category-driven attention. The former focuses on the functional parts of clothing items and the latter enhances task-related features that are learned in an implicit manner. VSAM [32] uses the inference results of pose detection for supervision to train a pose-guided attention model. Although attention mechanisms are well-known methods for boosting performance, such mechanisms mainly discuss where to focus given spatial regions. Because we consider attention mechanisms as an extension that can be applied to any existing method with slight modifications, such mechanisms are given little attention in this work. One of main challenges in VAP is solving multi-task learning problems in the presence of label sparsity and class imbalances in training data [1]. A single unified model is preferred for saving on the cost of inference and is expected to provide robust predictions for all targets (*e.g.*, *style, texture, and patterns in fashion*). The most commonly used method is to train multiple binary classifiers [8, 23]. However, because the numbers of effective annotations for each task differ significantly based on label sparsity, models can easily encounter overfitting on tasks with many annotations and underfitting on tasks with few annotations over the same number of iterations. Balancing such bias during training is important issue that must be solved for VAP. Adhikari *et al.* [2] proposed a progressive learning approach that add branches for individual models for attributes progressively as training proceeds. Lu *et al.* [24] proposed a tree-like deep architecture that automatically widens the network for multi-task learning in a greedy manner. However, the implementation of such methods is tricky and requires considerable engineering work.

Label Embedding. Label embedding (LE) refers to an important family of multi-label classification algorithms that have been introduced in various studies [3, 28, 36]. Such approaches jointly learn mapping functions φ and λ that project embeddings of an image x and label y in a common intermediate space Z. Compared to direct attribute prediction, which requires training a single binary classifier for each attribute, LE has the following advantages. First, attributes are not required to be independent because one can simply move the embedding of x closer to the correct label y than to any incorrect labels y' based on ranking loss [15, 37]. Second, prediction classes can be readily expanded because classes can be predicted by measuring the shortest distance to the center point of a feature cluster assigned to an unseen class [18, 19, 33, 35]. We found a number of studies examining LE in terms of zero-shot learning. In this paper, we highlight the usage of LE to train domain experts for an MTSS model for semi-supervised learning.

Semi-supervised Learning and Distillation. Recent works [5, 40, 41] have investigated semi-supervised learning (SSL) methods [10] that use large-scale unlabeled images to improve supervision. Yalniz *et. al.* [41] proposed a self-training method representing a special form of SSL and achieved state-of-the-art accuracy by leveraging a large amount of unannotated data. Distillation can also be considered as a form of SSL in that a teacher model makes predictions for unlabeled data and the results are used for supervision to train a student model [11, 29, 31]. Such strategies have yielded impressive results on many vision tasks. Inspired by these methods, we propose a training method that takes advantage of SSL for VAP. We found that a teacher-student paradigm of distillation is very effective for performing VAP because relevant visual attributes can be grouped into particular attribute types (*e.g.* {*stripe, dot*} ∈ *pattern*), meaning a student model can effectively learn from multiple teachers that are specialists for each attribute type.

3 Methodology

3.1 Overview

The main concept of the MTSS approach is to integrate multiple teachers (MT) that are domain experts for each attribute type into a single student (SS) to construct a unified model that can predict multiple attributes in a single forward operation. Given an image x and attribute type $\alpha_k \in \{\alpha_1, \alpha_2, ..., \alpha_\kappa\}$, our goal is to predict the attribute classes $c_p \in \{c_1^{\alpha_k}, c_2^{\alpha_k}, ..., c_P^{\alpha_k}\}$, where κ and P are the number of target attribute types and final predictions to be outputted, respectively. Suppose α_k is a *pattern* and *color*. Then, the corresponding c_p could be *dot, stripe* or *red*. P may differ depending on which confidence score is used for the predicted result to reach a final decision regarding each attribute type α_k. Our assumption is that attributes existing in the real world have a conceptual hierarchy, implying that relevant visual attributes are grouped into a particular attribute type α_k. For example, >1000 attributes in DeepFashion are grouped into 6 attribute types which means only 6 teachers are required for training student. The training procedure can be divided into two stages for the teacher and student. The design details and advantages of our two-stage training method are discussed in the following subsections.

3.2 Teacher Models for Individual Attributes

Given pairs of an image x and ground-truth label y in a training set $\zeta_{\alpha_k} = \{(x_n, y_n)\}, n = 1, ..., N\}$ for an attribute type α_k with $x_n \in X$ and $y_n \in Y$, our goal is to train a teacher model $\varphi_{\alpha_k}^T : X \to Y$. Our teacher model uses the pairwise ranking loss of metric learning to learn image representations. Specifically, a group of randomly initialized label embeddings obtained from an attribute dictionary $d_{\alpha_k}^n = \lambda_{\alpha_k}(y_n)$, where $d_{\alpha_k}^n \in R^D$ and D denotes the dimension of the image embedding, is defined and $d_{\alpha_k}^n$ is learned to represent the center of the feature cluster of the n-th class for attribute type α_k. Given a label embedding $d_{\alpha_k}^n$ that corresponds to the most representative feature point of the n-th class of α_k, our goal is to locate an embedding of the positive image $e^+ = \varphi_{\alpha_k}^T(I^+)$ to be close to $d_{\alpha_k}^n$ and an embedding of the negative image $e^- = \varphi_{\alpha_k}^T(I^-)$ far away. The positive image is sampled from the same attribute class as $d_{\alpha_k}^n$ and the negative image is sampled from a randomly selected class that is different from the class of the positive image. Although various sampling strategies can be adopted to enhance overall performance [39], such optimization was omitted

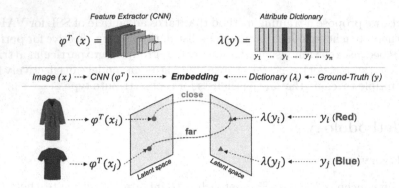

Fig. 3. Training of a teacher model. The feature extractor φ^T and mapping λ are learned to place the image embedding (left) and label embedding (right) close to each other in latent space.

in this work because it lies outside of our focus. The general form of an objective can be written as (1).

$$\ell'(d^n_{\alpha_k}, e^+, e^-) = max\{0, (1 - \|d^n_{\alpha_k}\|_2 \cdot \|e^+\|_2 + \|d^n_{\alpha_k}\|_2 \cdot \|e^-\|_2)\}, \qquad (1)$$

where $\|\ \|_2$ represents L2 normalization. Lin *et al.* [20] introduced the concept of focal loss, which can help a model focus on difficult and misclassified examples by multiplying a factor $(1 - p_t)^\gamma$ by the standard cross-entropy loss, where p_t is the estimated probability for the target class. As a modification for metric learning, we applied focal loss to our method by defining the probability p_t as the cosine similarity between d and e, which is bounded in the range of $[0, 1]$. The final objective is defined as (2).

$$p_t = 0.5 \times max\{0, (1 + \|d^n_{\alpha_k}\|_2 \cdot \|e^+\|_2 - \|d^n_{\alpha_k}\|_2 \cdot \|e^-\|_2)\},$$
$$\ell^{Teacher}(p_t) = -(1 - p_t)^\gamma log(p_t), \qquad (2)$$

where γ is the hyperparameter to be found. Note that γ was set to 1.0 in our experiments unless stated otherwise. We found that if γ is adjusted carefully, adopting the modified focal loss term significantly boosts overall scores, which will be discussed in the experimental section. During training, an image can be duplicated in more than one attribute class if the multiple labels exist in the ground truth. We expect the number of experts (i.e., trained teacher models $\varphi^T \in \{\varphi^T_{\alpha_1}, \varphi^T_{\alpha_2}, ..., \varphi^T_{\alpha_\kappa}\}$) to be the same as the number of attribute types κ.

3.3   Unified Student Model for All Attributes

The goal of the student stage is to integrate the trained teacher models $\varphi^T \in \{\varphi^T_{\alpha_1}, \varphi^T_{\alpha_2}, ..., \varphi^T_{\alpha_k}\}$ into a single unified network φ^S and boost overall performance by utilizing unlabeled images $\mathcal{U}$ in a semi-supervised manner. A distillation method that aims to transfer knowledge acquired in one model, namely a

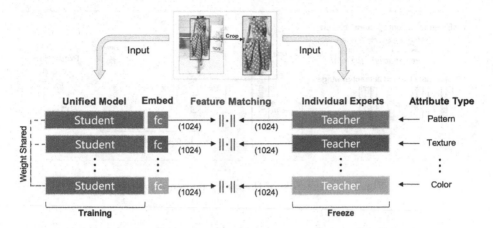

Fig. 4. Training of a student model. The teachers, which represent individual experts for each attribute type, are independent models, unlike the unified student model that shares weights before its fully connected layers.

teacher, to another model, namely a student, was adopted for this purpose. The core idea of distillation is very simple. A student model is trained to mimic the feature distribution of φ^T. Learning distributions from trained models can be achieved simply by matching the image embedding of a teacher $e_{\alpha_k}^T = \varphi_{\alpha_k}^T(x)$ to the embedding of a student $e_{\alpha_k}^S = \varphi^S(x; \alpha_k)$ according to a target attribute type α_k. It is assumed that if φ^S is able to reproduce the same feature distribution as φ^T, then prediction can be performed by simply measuring the distance between $e_{\alpha_k}^S$ and a learned dictionary $d_{\alpha_k} \in \{d_{\alpha_k}^1, d_{\alpha_k}^2, ..., d_{\alpha_k}^N\}$ that was already obtained in the teacher stage. A student φ^S consists of S branches of fully-connected layers following the last pooling layer of the backbone, where each branch is in charge of projecting pooled descriptors into an attribute-specific embedding space $Z_{\alpha_k} \in R^D$. The weights before the fully connected branches are shared. Given an attribute type α_k, we maximize the cosine-similarity between $e_{\alpha_k}^S$ and $e_{\alpha_k}^T$. Therefore objective is formulated as

$$\ell'' = \Sigma_{k=1}^{\kappa} \{1 - \|e_{\alpha_k}^S\|_2 \cdot \|e_{\alpha_k}^T\|_2\}, \qquad (3)$$

where $\|\|_2$ is L2 normalization. Because φ^S learns from φ^T, training can be unstable with large lr values if φ^T frequently produces outlier points based on a lack of generalization or excessive input noise. To alleviate such unwanted effects, we assign additional weight to an objective if the distance from the embedding of teacher $e_{\alpha_k}^T$ to the closest label embedding $\hat{d}_{\alpha_k}$ is small because a smaller distance indicates a more certain prediction. A large distance to $\hat{d}_{\alpha_k}$ could indicate the presence of outliers, so such signals are suppressed during gradient updating. The final objective is formulated as

$$\ell^{Student} = \Sigma_{k=1}^{\kappa} \{(\|\hat{d}_{\alpha_k}\|_2 \cdot \|e_{\alpha_k}^T\|_2)^{\beta} \times (1 - \|e_{\alpha_k}^S\|_2 \cdot \|e_{\alpha_k}^T\|_2)\}, \qquad (4)$$

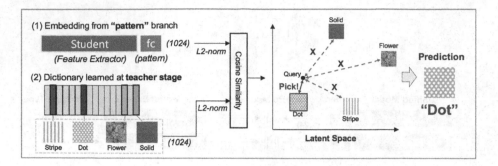

Fig. 5. An example of the query inference process.

where β is the hyperparameter to be found. β is set to 1.0 unless stated otherwise. Although we observed that β has a very small effect on final accuracy, this setting enables to use a large lr value at the very beginning of training phase by suppressing noise.

3.4   Query Inference

Given a query image x and ground-truth label y of an attribute type α_k, the attribute-specific embedding of α_k can be calculated as $e_{\alpha_k}^S = \varphi_{\alpha_k}^S(x; \alpha_k)$. Let the dictionary learned at the teacher stage $d_{\alpha_k} = \lambda_{\alpha_k}(y)$, where $d_{\alpha_k} \in \{d_{\alpha_k}^1, ..., d_{\alpha_k}^N\}$ and N is the number of classes included in α_k. Because d_{α_k} represents the center points of the feature cluster for each ground-truth class y and $\varphi_{\alpha_k}^S$ can reproduce the same distribution with $\varphi_{\alpha_k}^T$, a prediction can be obtained by finding the $\hat{d}_{\alpha_k}$ that maximizes the cosine similarity with $e_{\alpha_k}^S$. An example of the query inference process for predicting a *pattern* is presented in Fig. 5.

4   Experiment

4.1   Experimental Setup

Datasets. All models were evaluated on several fashion-related benchmarks, namely three public datasets called iMaterialist-Fashion-2018 (iMatFashion) [13], Deepfashion Category and Attribute Prediction Benchmark (DeepFashion) [23], and DARN [16], and one private dataset called FiccY. We cropped the images using a CenterNet [44] based fashion object detector unless ground-truth bounding boxes were not provided for a dataset. We omit the detailed statistics of the public datasets because they have been well described in many related works [8,13,34,38]. FiccY is a private fashion dataset collected from our database and annotated by human experts. It contains 520 K high-resolution images labeled for four types of attributes (*Category, Color, Pattern, and Texture*) produced by both sellers and wearers. We intentionally include the result on FiccY in our experiments because it cab reflect a real service environment. However, the majority of experiments were performed on the DeepFashion benchmark for reproducibility.

4.2 Performance on Benchmarks

Multi-label Classification. Previous studies [6, 8, 16, 22, 23, 32, 38] on attribute prediction in the fashion domain have reported R@3 scores according to attribute types using DeepFashion, so we used the same metric for fair comparison. The results are listed in Table 1. The R@3 scores of both the teacher and student are listed for various experimental settings of γ and lr. The results demonstrate that with the default hyperparameter of $\gamma = 1$, our teacher model outperforms previous methods [6, 8, 16] that use no auxiliary supervision. We determined that the optimized values of γ and lr differ depending on the task, meaning these hyperparameters must be selected carefully according to the target attribute type. When training a student, we observed higher gains when a greater lr value was adopted. Adjusting the β value enables the use of greater lr values while avoiding instability at the beginning of training. The student trained by the tuned teachers yields very similar results compared to state-of-the-art domain-specific methods [22, 23, 32, 38].

Table 1. Top-k recall for attribute prediction on the DeepFashion [23] dataset. *-marked methods use additional labels for training, such as landmark, pose, or text descriptions. $\gamma = tune$ indicates that the best-performing γ was selected according to the attribute types. Recall was measured when the model achieved the best F1@1 score. *Style* is not compared to the other methods because style scores were not reproducible using publicly released code. Overall $1^{st}/2^{nd}$ best in **blue**/green. (Color figure online)

Method	Category		Texture		Fabric		Shape		Part		Style		All	
	top-3	top-5	top-3	top-5	top-3	top-5	top-3	top-5	top-3	top-5	top-3	top-5	top-3	top-5
WTBI [6]	43.73	66.26	24.21	32.65	25.38	36.06	23.39	31.26	26.31	33.24	-	-	27.46	35.37
DARN [16]	59.48	79.58	36.15	48.15	36.64	48.52	35.89	46.93	39.17	50.14	-	-	42.35	51.95
* FashionNet [23]	82.58	90.17	37.46	49.52	39.30	49.84	39.47	48.59	44.13	54.02	-	-	45.52	54.61
Corbiere et al. [8]	86.30	92.80	53.60	63.20	39.10	48.80	50.10	59.50	38.80	48.90	-	-	23.10	30.40
* Wang et al. [38]	90.99	95.78	50.31	65.48	40.31	48.23	53.32	61.05	40.65	56.32	-	-	51.53	60.95
* VSAM + FL [32]	-	-	56.28	65.45	41.73	52.01	55.69	65.40	43.20	53.95	-	-	-	-
* Liu et al. [22]	**91.16**	96.12	56.17	**65.83**	43.20	**53.52**	**58.28**	67.80	46.97	57.42	-	-	-	-
Ours(T)($\gamma = 0$)	88.45	91.21	55.91	62.02	43.90	50.54	55.92	61.46	45.40	51.55	30.13	35.44	-	-
Ours(T)($\gamma = 1$)	89.32	92.92	57.89	64.71	44.51	51.57	56.22	62.34	46.09	52.73	30.88	36.19	-	-
Ours(T)($\gamma = tune$)	90.01	93.74	58.14	65.29	44.03	52.18	56.74	62.66	46.83	53.66	31.31	37.13	-	-
Ours(S)($\gamma = 1, \beta = 1$)	89.65	93.29	**58.33**	65.17	**45.05**	52.26	56.91	63.17	46.08	54.30	32.30	38.41	57.85	63.91
Ours(S)($\gamma = tune, \beta = 1$)	90.17	93.98	57.58	64.66	44.87	52.47	57.34	63.69	**47.36**	54.99	**32.64**	**39.24**	**58.02**	**64.35**

LE versus Binary Cross-Entropy.

In this section, we analyze the strength of LE compared to the most commonly used method, namely binary cross-entropy (BCE). Table 2 compares our implementation to the existing method. The student model is trained by the teacher networks, which are trained using LE. The

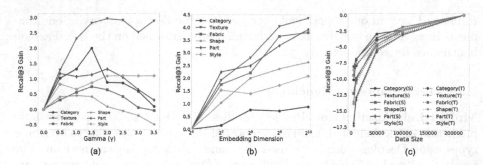

Fig. 6. Effects of training hyperparameters evaluated on DeepFashion: (a) gamma γ, (b) embedding dimension D, and (c) training data size. These plots represent analysis of the changes in R@3 (gain or degradation) compared to the baseline trained with (a) $\gamma = 0$, (b) $D = 256$, and (c) a data size equal to the size of the entire training set. Part (c) presents results for both the student (S) and teacher (T).

state-of-the-art method based on DeepFashion uses BCE for training. One can see that LE consistently yields better results when k=1, implying that prediction is more accurate with a small number of trials. We directly compared the LE and BCE methods by replacing the objective of our teacher model with BCE. The results in Table 3 reveal consistent improvements in terms of recall at k=1 for all evaluation datasets.

Hyperparameter tuning. The optimal value of γ for training a teacher network is analyzed in Fig. 6 (a). While the use of focal loss consistently improves the results by up to 3% in terms of R@3, we found that the optimal value differs depending on the attribute types. Therefore, it is highly recommended to tune γ carefully for each attribute type. The feature dimension is also an important engineering factor. Figure 6 (b) presents the effects of the embedding dimension D. Better results can be observed for higher dimension, but processing high-dimensional features may require more resources and inference time. The result in Table 1 demonstrate that tuning the hyperparameters of not only γ and D, but also lr, significantly improves performance. The combination of optimal values is different for each attribute type.

Training Depending on Data Size. Figure 6 (c) presents the performance degradation when the number of usable training images decreases. Overall R@3 scores are presented for both the teacher and student networks. The number of training images was intentionally limited to observe how performance degrades with a shortage of training data. As expected, a lack of training data significantly

Table 2. The top-k F1 scores and recall values evaluated on DeepFashion.

Method	F1@1	R@1	F1@3	R@3	F1@5	R@5
Liu et al. [22]	31.35	37.17	**22.50**	57.30	**16.82**	**65.66**
Ours(S)($\gamma = tune, \beta = 1$)	**33.53**	**40.11**	22.41	**58.02**	16.16	64.35

Table 3. Comparison between LE and BCE. The F1@1 scores were measured on FiccY.

Dataset	Category	Pattern	Color	Texture	Avg. Δ
BCE	85.12	67.46	65.81	57.43	–
LE	**85.52**	**67.71**	**66.31**	**57.98**	+0.62%

degrades the overall results. It is noteworthy that the student always outperforms the teachers slightly in terms of gain, even though the teachers are very weak. However, this also indicates that the final performance of the student is strongly bounded by that of teachers. We found no relationship between training data size and the amount of performance gain for a student compared to a teacher.

4.3   Effectiveness of Distillation

Comparison Between a Teacher and Student. In Table 4, we compare classification performances between a teacher and student on the DeepFashion, iMatFashion, and FiccY datasets. Results were evaluated for all three datasets to determine if the proposed method exhibits consistent experimental trends in different domains. The results illustrate that the student achieves a better score than the teachers in all cases. Our interpretation is that the use of unlabeled images can induce improvements because the negative effects caused by missed annotations are successfully suppressed during training by matching features from teacher and student for supervision. As a proof, only slight improvements in terms of F1 score are observed for *categories* which has relatively dense annotations than the others. Our MTSS approach is a semi-supervised form of multitask learning, so we analyze the benefits of learning representations simultaneously from multiple tasks in Table 5. Given four different attribute types for FiccY, we compare students trained with the supervision of either single or multiple domain experts. We found consistent improvements in both cases. However, higher improvement is achieved in general with supervision from all four teachers than a single teacher. We can conclude that the proposed model benefits from knowledge transfer between different tasks, which were represented

Table 4. Performance comparison between a teacher (T) and student (S) model. F1@1 scores are measured on the iMatFashion, DeepFashion, and FiccY datasets.

	iMatFashion [13]							
	Category	Gender	Material	Pattern	Style	Neckline	Sleeve	Color
T	**42.50**	88.50	49.27	37.84	27.70	40.11	76.72	47.75
S	42.43	**90.46**	**50.45**	**38.04**	**28.60**	**41.93**	**77.66**	**49.69**

	DeepFashion [23]						FiccY			
	Category	Texture	Fabric	Shape	Part	Style		Category	Pattern	Color	Texture
T	74.45	25.61	24.97	30.45	21.12	12.63	T	85.10	67.69	66.15	57.74
S	**74.85**	**26.08**	**25.14**	**30.81**	**21.40**	**13.29**	S	**85.78**	**68.31**	67.60	**59.19**

Table 5. R@3 gain of a student model compared to teachers with either single (S-Single) or multiple (S-Multi) teachers evaluated on FiccY. Δ_{single} and Δ_{multi} denote the results for S-Single and S-Multi, respectively, as percentages.

Attribute	T	S-Single	Δ_{single}	S-Multi	Δ_{multi}	Δ_{multi} - Δ_{single}
Category	96.70	97.30	+0.62%	**97.30**	+0.62%	+0.00%
Pattern	90.11	90.73	+0.69%	**91.14**	+1.14%	+0.45%
Color	88.97	89.54	+0.64%	**89.98**	+1.14%	+0.50%
Texture	92.06	93.57	+1.64%	**94.22**	+2.35%	+0.71%

by different attribute types in this experiment. Figure 8 analyzes the differences in R@3 scores between teachers and students for each class with $\alpha_\kappa = shape$. Although the result by class is generally improved based on distillation, performance degradation occurs in some classes. Such results are found when the recall of a teacher is poor meaning that poorly generalized teacher networks can negatively affect optimization for minor classes, while having a small effect on overall performance. Another possibility is that certain rare attributes lead to underfitting for certain classes, even with a large number of training epochs. Regardless, the overall results with distillation for all classes exhibit consistent improvement.

t-SNE Visualization of Distributions. The distribution of teacher and student embeddings is visualized in Fig. 7. The result illustrates that the student reproduces almost the same distribution as the teacher. We omitted visualizations for attribute types other than *patterns* because similar results were observed for all attribute types. The feature clusters for each prediction class are clearly formed for both the teacher and student.

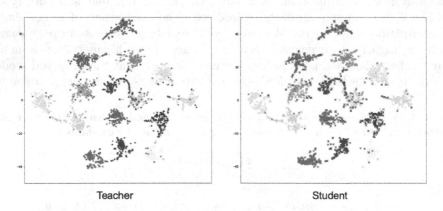

Teacher   Student

Fig. 7. T-distributed stochastic neighbor embedding (t-SNE) [25] visualization of pattern attribute types for FiccY. The left and right parts present the distributions of image embeddings extracted from the teacher and student models, respectively.

Table 6. Cross-domain adaptability. F1@1 scores of five commonly appearing classes were measured on DARN [16]. The symbol † indicates a model trained with both original training set images and external large-scale unlabeled images.

Model	Train set	Test set	Flower	Stripe	Dot	Check	Leopard	All
T	DeepFashion	DARN	70.46	66.11	67.47	51.00	40.90	64.31
S	DeepFashion	DARN	71.16	67.81	68.97	52.02	43.97	65.15
$S^\dagger$	DeepFashion+ DARN	DARN	**71.84**	**68.70**	**70.03**	**52.62**	**44.09**	**65.49**
T	iMatFashion	DARN	74.73	75.33	74.25	68.53	51.84	72.34
S	iMatFashion	DARN	75.37	76.74	74.90	69.77	55.72	73.10
$S^\dagger$	iMatFashion+ DARN	DARN	**75.81**	**76.78**	**75.20**	**70.34**	**57.61**	**73.22**

4.4 Robustness and Generalization

Cross-domain Adaptability. In this subsection, we examine the capability of SSL to transfer knowledge for adaptation to a new target domain [27]. Validating the cross-domain adaptability of VAP is difficult because the classes provided in each dataset differ. For fair comparison, we carefully selected five commonly appearing classes (*Flower, Stripe, Dot, Check, and Leopard*) for three datasets (*DeepFashion, iMatFashion, and DARN*) and created subsets containing only images from one of the selected classes. The results in Table 6 reveal that training a student is effective for cross-domain classification. We observe clear improvements for all five classes relative to the teacher models. We also examined the strength of SSL useful for deploying a model in a realistic environment. Because the training process for a student requires no annotations, the training images for two distinct domains (DARN and DeepFashion/iMatFashion) were mixed for training. This combination provided a significant performance boost for the testing set. This could be an useful feature in real production environments because a model can be easily fine-tuned for a service domain after being trained in a constrained environment.

Table 7. Results for robustness to corruption and perturbation in inputs. Recall was measured on both clean and corrupted versions of the DeepFashion dataset. The symbol $\dagger$ indicates a model trained with both clean images and 1M, 10M, or 15M corrupted images.

Model	Mixing	Clean R@3	Δ	Corrupted R@3	Δ
Baseline(S)	No	57.47	–	42.96	–
$S^\dagger$-1M	Yes	57.75	+0.49%	**43.96**	+2.33%
$S^\dagger$-10M	Yes	**57.80**	+0.57%	43.88	+2.14%
$S^\dagger$-15M	Yes	57.74	+0.47%	43.84	+2.05%

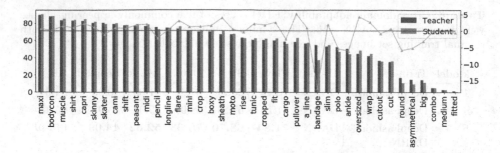

Fig. 8. R@3 scores for each class of attribute type *shape* for the DeepFashion dataset. The yellow line represents the changes in performance before and after training the student. The green line is a baseline representing a change of zero percent. (Color figure online)

Robustness to Corruption and Perturbation. Hendrycks *et al.* [14] introduced a benchmark for image classifier robustness to corruption and perturbation. Images were transformed using 15 types of algorithmically generated corruptions with five levels of severity, resulting in a total of 15×5 corrupted images. By following this same procedure, we created a corrupted version of the DeepFashion dataset and measured the drop in R@3 compared to the clean version. The result are listed in Table 7. Testing on the corrupted images degrades the overall R@3 score by an average of approximately 25%. In Table 6, the benefits of mixing two distinct domains in terms of domain adaptation were explored. Following the same strategy, we mixed clean images with 1M, 10M, and 15M randomly sampled corrupted images and trained a student model to improve robustness to corruption. It should be noted that the mixed images were only used for training the student, meaning the teachers never saw the corrupted images. The results reveal that a student trained with mixed images achieves an increase of up to 2.33% in terms of R@3 score compared to a baseline student trained using only clean images. Additionally, performing fine-tuning using corrupted images does not degrade the model's performance on clean images. In fact, such fine-tuning slightly improves performance on clean images. The greatest performance gain can be observed when including only 1M corrupted images. The optimal ratio between clean and corrupted images should be identified because the noisy signals generated by corrupted images can make optimization less effective if they dominate the clean images.

5   Conclusion

In this paper, we proposed an MTSS approach to solving the VAP problem. Our method trains a unified model according to multiple domain experts, which enables it to predict multiple attributes that appear simultaneously in objects using a single forward operation. The core idea of MTSS is to transfer knowledge by forcing a student to reproduce the feature distributions learned by teachers.

We demonstrated that such a strategy is highly effective for VAP, which suffers from a lack of effective annotations. Furthermore, our method can achieve competitive results on benchmarks using attribute labels alone, and improve robustness and domain adaptability without sacrificing accuracy fine-tuning with unlabeled images.

References

1. Abdulnabi, A.H., Wang, G., Lu, J., Jia, K.: Multi-task CNN model for attribute prediction. IEEE Trans. Multimedia **17**(11), 1949–1959 (2015)
2. Adhikari, S.S., Singh, S., Rajagopal, A., Rajan, A.: Progressive fashion attribute extraction. arXiv preprint arXiv:1907.00157 (2019)
3. Akata, Z., Perronnin, F., Harchaoui, Z., Schmid, C.: Label-embedding for image classification. IEEE Trans. Pattern Anal. Mach. Intell. **38**(7), 1425–1438 (2015)
4. Arslan, H.S., Sirts, K., Fishel, M., Anbarjafari, G.: Multimodal sequential fashion attribute prediction. Information **10**(10), 308 (2019)
5. Cevikalp, H., Benligiray, B., Gerek, O.N.: Semi-supervised robust deep neural networks for multi-label image classification. Pattern Recogn. **100**, 107164 (2019)
6. Chen, H., Gallagher, A., Girod, B.: Describing clothing by semantic attributes. In: Fitzgibbon, A., Lazebnik, S., Perona, P., Sato, Y., Schmid, C. (eds.) ECCV 2012. LNCS, vol. 7574, pp. 609–623. Springer, Heidelberg (2012). https://doi.org/10.1007/978-3-642-33712-3_44
7. Chen, Z.M., Wei, X.S., Wang, P., Guo, Y.: Multi-label image recognition with graph convolutional networks. In: Proceedings of the IEEE Conference on Computer Vision and Pattern Recognition, pp. 5177–5186 (2019)
8. Corbiere, C., Ben-Younes, H., Ramé, A., Ollion, C.: Leveraging weakly annotated data for fashion image retrieval and label prediction. In: Proceedings of the IEEE International Conference on Computer Vision, pp. 2268–2274 (2017)
9. Dal Pozzolo, A., Caelen, O., Bontempi, G.: When is undersampling effective in unbalanced classification tasks? In: Appice, A., Rodrigues, P.P., Santos Costa, V., Soares, C., Gama, J., Jorge, A. (eds.) ECML PKDD 2015. LNCS (LNAI), vol. 9284, pp. 200–215. Springer, Cham (2015). https://doi.org/10.1007/978-3-319-23528-8_13
10. van Engelen, J.E., Hoos, H.H.: A survey on semi-supervised learning. Mach. Learn. **109**(2), 373–440 (2019). https://doi.org/10.1007/s10994-019-05855-6
11. Gong, C., Chang, X., Fang, M., Yang, J.: Teaching semi-supervised classifier via generalized distillation. In: IJCAI, pp. 2156–2162 (2018)
12. Guo, H., Zheng, K., Fan, X., Yu, H., Wang, S.: Visual attention consistency under image transforms for multi-label image classification. In: Proceedings of the IEEE Conference on Computer Vision and Pattern Recognition, pp. 729–739 (2019)
13. Guo, S., et al.: The imaterialist fashion attribute dataset. arXiv preprint arXiv:1906.05750 (2019)
14. Hendrycks, D., Dietterich, T.: Benchmarking neural network robustness to common corruptions and perturbations. arXiv preprint arXiv:1903.12261 (2019)
15. Hoffer, E., Ailon, N.: Deep metric learning using triplet network. In: Feragen, A., Pelillo, M., Loog, M. (eds.) SIMBAD 2015. LNCS, vol. 9370, pp. 84–92. Springer, Cham (2015). https://doi.org/10.1007/978-3-319-24261-3_7
16. Huang, J., Feris, R.S., Chen, Q., Yan, S.: Cross-domain image retrieval with a dual attribute-aware ranking network. In: Proceedings of the IEEE International Conference on Computer Vision, pp. 1062–1070 (2015)

17. Kawaguchi, K., Kaelbling, L.P., Bengio, Y.: Generalization in deep learning. arXiv preprint arXiv:1710.05468 (2017)
18. Khodadadeh, S., Boloni, L., Shah, M.: Unsupervised meta-learning for few-shot image classification. In: Advances in Neural Information Processing Systems, pp. 10132–10142 (2019)
19. Kim, J., Kim, T., Kim, S., Yoo, C.D.: Edge-labeling graph neural network for few-shot learning. In: Proceedings of the IEEE Conference on Computer Vision and Pattern Recognition, pp. 11–20 (2019)
20. Lin, T.Y., Goyal, P., Girshick, R., He, K., Dollár, P.: Focal loss for dense object detection. In: Proceedings of the IEEE International Conference on Computer Vision, pp. 2980–2988 (2017)
21. Lin, T.-Y., et al.: Microsoft COCO: common objects in context. In: Fleet, D., Pajdla, T., Schiele, B., Tuytelaars, T. (eds.) ECCV 2014. LNCS, vol. 8693, pp. 740–755. Springer, Cham (2014). https://doi.org/10.1007/978-3-319-10602-1_48
22. Liu, J., Lu, H.: Deep fashion analysis with feature map upsampling and landmark-driven attention. In: Proceedings of the European Conference on Computer Vision (ECCV), pp. 0–0 (2018)
23. Liu, Z., Luo, P., Qiu, S., Wang, X., Tang, X.: Deepfashion: powering robust clothes recognition and retrieval with rich annotations. In: Proceedings of the IEEE Conference on Computer Vision and Pattern Recognition, pp. 1096–1104 (2016)
24. Lu, Y., Kumar, A., Zhai, S., Cheng, Y., Javidi, T., Feris, R.: Fully-adaptive feature sharing in multi-task networks with applications in person attribute classification. In: Proceedings of the IEEE Conference on Computer Vision and Pattern Recognition, pp. 5334–5343 (2017)
25. Maaten, L.V.D., Hinton, G.: Visualizing data using t-SNE. J. Mach. Learn. Res. 9(Nov), 2579–2605 (2008)
26. Neyshabur, B., Bhojanapalli, S., McAllester, D., Srebro, N.: Exploring generalization in deep learning. In: Advances in Neural Information Processing Systems, pp. 5947–5956 (2017)
27. Orbes-Arteainst, M., et al.: Knowledge distillation for semi-supervised domain adaptation. In: Zhou, L., Sarikaya, D., Kia, S.M., Speidel, S., Malpani, A., Hashimoto, D., Habes, M., Löfstedt, T., Ritter, K., Wang, H. (eds.) OR 2.0/MLCN -2019. LNCS, vol. 11796, pp. 68–76. Springer, Cham (2019). https://doi.org/10.1007/978-3-030-32695-1_8
28. Palatucci, M., Pomerleau, D., Hinton, G.E., Mitchell, T.M.: Zero-shot learning with semantic output codes. In: Advances in Neural Information Processing Systems, pp. 1410–1418 (2009)
29. Papernot, N., Abadi, M., Erlingsson, U., Goodfellow, I., Talwar, K.: Semi-supervised knowledge transfer for deep learning from private training data. arXiv preprint arXiv:1610.05755 (2016)
30. Park, S., Shin, M., Ham, S., Choe, S., Kang, Y.: Study on fashion image retrieval methods for efficient fashion visual search. In: Proceedings of the IEEE Conference on Computer Vision and Pattern Recognition Workshops (2019)
31. Park, W., Kim, D., Lu, Y., Cho, M.: Relational knowledge distillation. In: Proceedings of the IEEE Conference on Computer Vision and Pattern Recognition, pp. 3967–3976 (2019)
32. Quintino Ferreira, B., Costeira, J.P., Sousa, R.G., Gui, L.Y., Gomes, J.P.: Pose guided attention for multi-label fashion image classification. In: Proceedings of the IEEE International Conference on Computer Vision Workshops (2019)
33. Ren, M., et al.: Meta-learning for semi-supervised few-shot classification. arXiv preprint arXiv:1803.00676 (2018)

34. Shin, M., Park, S., Kim, T.: Semi-supervised feature-level attribute manipulation for fashion image retrieval. In: Proceedings of the British Machine Vision Conference (2019)
35. Snell, J., Swersky, K., Zemel, R.: Prototypical networks for few-shot learning. In: Advances in Neural Information Processing Systems, pp. 4077–4087 (2017)
36. Socher, R., Ganjoo, M., Manning, C.D., Ng, A.: Zero-shot learning through cross-modal transfer. In: Advances in Neural Information Processing Systems, pp. 935–943 (2013)
37. Sohn, K.: Improved deep metric learning with multi-class n-pair loss objective. In: Advances in Neural Information Processing Systems, pp. 1857–1865 (2016)
38. Wang, W., Xu, Y., Shen, J., Zhu, S.C.: Attentive fashion grammar network for fashion landmark detection and clothing category classification. In: Proceedings of the IEEE Conference on Computer Vision and Pattern Recognition, pp. 4271–4280 (2018)
39. Wu, C.Y., Manmatha, R., Smola, A.J., Krahenbuhl, P.: Sampling matters in deep embedding learning. In: Proceedings of the IEEE International Conference on Computer Vision, pp. 2840–2848 (2017)
40. Xie, Q., Hovy, E., Luong, M.T., Le, Q.V.: Self-training with noisy student improves imagenet classification. arXiv preprint arXiv:1911.04252 (2019)
41. Yalniz, I.Z., Jégou, H., Chen, K., Paluri, M., Mahajan, D.: Billion-scale semi-supervised learning for image classification. arXiv preprint arXiv:1905.00546 (2019)
42. Zhang, S., Song, Z., Cao, X., Zhang, H., Zhou, J.: Task-aware attention model for clothing attribute prediction. IEEE Trans. Circuits Syst. Video Technol. **30**(4), 1051–1064 (2019)
43. Zheng, S., Song, Y., Leung, T., Goodfellow, I.: Improving the robustness of deep neural networks via stability training. In: Proceedings of the IEEE Conference on Computer Vision and Pattern Recognition, pp. 4480–4488 (2016)
44. Zhou, X., Wang, D., Krähenbühl, P.: Objects as points. arXiv preprint arXiv:1904.07850 (2019)

Unsupervised Domain Adaptation with Noise Resistible Mutual-Training for Person Re-identification

Fang Zhao[1], Shengcai Liao[1(✉)], Guo-Sen Xie[1], Jian Zhao[2], Kaihao Zhang[3], and Ling Shao[1,4]

[1] Inception Institute of Artificial Intelligence, Abu Dhabi, UAE
{fang.zhao,shengcai.liao,guosen.xie,ling.shao}@inceptioniai.org
[2] Institute of North Electronic Equipment, Beijing, China
zhaojian90@u.nus.edu
[3] Tencent AI Lab, Shenzhen, China
super.khzhang@gmail.com
[4] Mohamed bin Zayed University of Artificial Intelligence, Abu Dhabi, UAE

Abstract. Unsupervised domain adaptation (UDA) in the task of person re-identification (re-ID) is highly challenging due to large domain divergence and no class overlap between domains. Pseudo-label based self-training is one of the representative techniques to address UDA. However, label noise caused by unsupervised clustering is always a trouble to self-training methods. To depress noises in pseudo-labels, this paper proposes a Noise Resistible Mutual-Training (NRMT) method, which maintains two networks during training to perform collaborative clustering and mutual instance selection. On one hand, collaborative clustering eases the fitting to noisy instances by allowing the two networks to use pseudo-labels provided by each other as an additional supervision. On the other hand, mutual instance selection further selects reliable and informative instances for training according to the peer-confidence and relationship disagreement of the networks. Extensive experiments demonstrate that the proposed method outperforms the state-of-the-art UDA methods for person re-ID.

Keywords: Unsupervised domain adaptation · Person re-identification · Collaborative clustering · Mutual instance selection

1 Introduction

Person re-identification (re-ID), which aims at retrieving images of the same person from the database given a person image, has advanced considerably relying on the power of deep learning technology in recent years [19,29,32,34,35,48,50, 51,53,58]. However, due to the problem of domain shift [17], the performance of a deep re-ID model that performs well in a source domain may drop significantly when applied to a target domain. Besides, it is usually not easy to obtain labels

© Springer Nature Switzerland AG 2020
A. Vedaldi et al. (Eds.): ECCV 2020, LNCS 12356, pp. 526–544, 2020.
https://doi.org/10.1007/978-3-030-58621-8_31

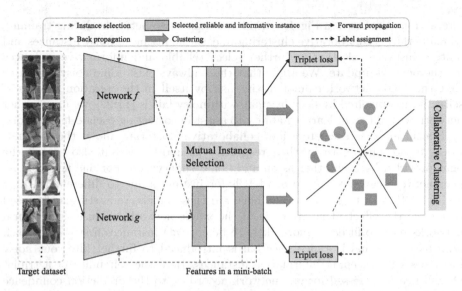

Fig. 1. Overview of the proposed Noise Resistible Mutual-Training (NRMT). NRMT maintains two networks during training, which performs collaborative clustering to ease the fitting to noisy instances and mutual instance selection to further select reliable and informative instances for the network update.

of target data in practice, which hinders supervised fine-tuning of the deep model on the target data.

To learn a deep re-ID model which generalizes well in the target domain without using labels from this domain, unsupervised domain adaptation (UDA) approaches are proposed given labeled source data and unlabeled target data [5, 21,24,45,56,57]. Different from the traditional setting of UDA which assumes that the source and target domains share the same classes, UDA in person re-ID is under an open-set scenario, *i.e.*, the two domains have totally different person identities (classes). Thus, it is a more challenging task.

Self-training is an effective strategy for UDA in person re-ID [8,11,31,49], which performs clustering with the pre-trained source model to assign pseudo-labels to samples of the target dataset, then alternately updates the model with the pseudo-labels on target data and re-assigns the labels with the updated model to make the model adapt to the target data progressively. In the early stage of training, pseudo-labels assigned by clustering usually contain lots of noises due to the divergence between the source and target domains. The model can correct some of them by learning from clean labels. However, as the number of training iteration increases, some noisy instances are fitted by the model and cannot be corrected anymore. These noises eventually harm the self-training model performance on the target data.

In order to address the problem mentioned above, we propose *Noise Resistible Mutual-Training* (NRMT) to effectively reduce the impact of noisy instances throughout the training process by leveraging dual networks with information interaction. As shown in Fig. 1, NRMT maintains two networks during training, which performs collaborative clustering to ease the fitting to noisy instances and mutual instance selection to further select reliable and informative instances for the network update. We argue that there always exist some noisy instances that the single network cannot distinguish by itself in the iteration process of self-training. Inspired by deep learning with noisy labels [14,22], we use another network with different learning ability to assist in correcting pseudo-label errors.

Specifically, for each iteration, collaborative clustering allows the two networks to not only learn by their respective pseudo-labels but also exploit the ones provided by each other as an additional supervision. For one network, its peer network can provide various labels for instances due to different learning ability. Although there also exists noises in these labels, they still can be used to reduce the effect of label errors of the single network because deep neural networks tend to fit easy (more likely to be correct) instances first [1]. For each mini-batch, mutual instance selection is introduced to further filter out noisy instances while keeping informative instances. Here the reliability of a triplet of instances is assessed for one network according to the prediction confidence of its peer network on this triplet. Informative instances are also important for improving the network performance. Thus, we further measure the amount of information of the triplet by the relationship disagreement of the predictions across the networks. Combining collaborative clustering at each iteration and mutual instance selection within each mini-batch, the proposed NRMT can effectively depress noises in pseudo-labels and improve the performance of both the two networks.

Our main contributions can be summarized as follows: 1) We present a novel noise resistible mutual-training method for unsupervised domain adaptation in person re-ID, which exploits dual network interaction to depress noises in pseudo-labels of unsupervised iterative training on the target data. 2) We introduce a collaborative clustering to ease the fitting to noisy instances by the memorization effects of deep networks. 3) We propose a mutual instance selection based on the peer-confidence and relationship disagreement of networks on triplets of instances to select reliable and informative instances in a mini-batch.

2   Related Work

Unsupervised Domain Adaptation. Our work is related to unsupervised domain adaptation (UDA) [3,28,36,37]. Some methods are proposed to match distributions between the source and target domains [20,33]. Long et al. [20] embed features of task-specific layers in a reproducing kernel Hilbert space to explicitly match the mean embeddings of different domain distributions. Sun et al. [33] propose to learn a linear transformation that aligns the second-order statistics of feature distributions between the two domains. There are also several works that learn domain-invariant features [12,37]. Ganin et al. [12] introduce a gradient reversal layer to learn features invariant to domain via an adversarial loss. The aforementioned methods only consider the closed-set scenario.

Recently, some works are introduced to address the problem of open set domain adaptation [10, 23, 27], where several classes are unknown in the two domains (or in the target domain). However, classes of the two domains are entirely different for UDA in person re-ID, which presents a greater challenge.

UDA for Person re-ID. There are many research works that have been proposed for unsupervised cross-domain person re-ID [5, 24, 25, 30, 31, 38, 40–42, 44, 46, 56, 57]. Some of them focus on image-level domain invariance. Wei *et al.* [39] propose a person transfer generative adversarial network to bridge the domain gap, which considers the style transfer and person identity keeping. Deng *et al.* [7] generate target image samples through the coordination between a CycleGAN and an Siamese network. Several works also try to improve the model generalization from the view of feature learning. Wang *et al.* [38] establish an identity-discriminative and attribute-sensitive feature representation space transferable to any new (unseen) target domain. Qi *et al.* [25] develop a camera-aware domain adaptation to reduce the discrepancy across sub-domains in cameras and utilize the temporal continuity in each camera to provide discriminative information.

Recently, some methods are developed based on the self-training framework. Fu *et al.* [11] present a self-similarity grouping to explore the potential similarities by both global and local appearance cues. Zhang *et al.* [49] propose a self-training method with progressive augmentation framework to offer complementary data information by different learning strategies for self-training. In contrast, our method provides complementary information through dual network interaction. Ge *et al.* [13] present a mutual mean-teaching framework to softly refine the pseudo-labels in the target domain. Note that our method and [13] are complementary and can be combined.

Deep Learning with Noisy Labels. There exist several works that aim at improving the training of deep models with noisy labels. Decoupling [22] trains two networks simultaneously, and then updates models only using the instances that have different predictions from these two networks. Co-teaching [14] proposes to select small-loss instances of each network as the useful knowledge and transfer such useful instances to its peer network for the further training. Yu *et al.* [47] combine the disagreement strategy with Co-teaching, which trains two deep neural networks with the disagreement-update step (data update) and the cross-update step (parameters update). These methods mainly focus on the classification problem, which cannot be directly applied to the metric learning problem in our task.

3   Our Method

Given a labeled training dataset $\{\mathbf{X}^s, \mathbf{Y}^s\}$ from the source domain and an unlabeled training dataset $\mathbf{X}^t$ from the target domain where identities of persons are different from the ones in the source domain, we aim to learn discriminative feature representations for target testing dataset. In this section, we present the proposed Noise Resistible Mutual-Training (NRMT) method, which incorporates the interaction of dual networks to depress noises in pseudo-labels produced by unsupervised clustering in a self-training process. Now, we proceed to explain each component of our NRMT in details.

3.1 Self-training with Clustering

Since the ground truth labels of the target person images are not available, one way to fine-tune the target model is to consider the target labels as latent variables that can be inferred in the learning process. Thus, a typical self-training framework for unsupervised domain adaptation aims to minimize the following loss function:

$$\min_{\hat{\mathbf{Y}}^t, \mathbf{W}} \mathcal{L}(\hat{\mathbf{Y}}^t, f(\mathbf{X}^t; \mathbf{W})), \tag{1}$$

where $\hat{\mathbf{Y}}^t$ denotes the estimated target labels, $\mathbf{X}^t$ is the set of target images and f denotes the target model parameterized by $\mathbf{W}$.

In the case of person re-ID, source and target domains do not share the common label space. Thus, one cannot directly apply the classifier trained on the source dataset to estimate the target identities. Similar with [8,31], we perform clustering on CNN features to assign pseudo-labels to instances with the most confident predictions and assume that they are mostly correct. Once the target model is updated with these pseudo-labels, the remaining instances with less confidence are continuously explored by the model adapted better to the target domain. Therefore, to minimize the loss function in Eq. (1), we firstly initialize the model parameters $\mathbf{W}$ on the source data $\{\mathbf{X}^s, \mathbf{Y}^s\}$ and then apply an alternating block coordinate descent algorithm: 1) Fix $\mathbf{W}$ and minimize the loss w.r.t $\hat{\mathbf{Y}}^t$ through clustering. 2) Fix $\hat{\mathbf{Y}}^t$ and optimize the loss w.r.t $\mathbf{W}$ by stochastic gradient descent.

3.2 Mutual-Training with Collaborative Clustering

The problem of self-training based models [8,31] is that the quality (correctness) of pseudo-labels generated by unsupervised clustering on the target data heavily affects the model performance. Although the deep learning model in self-training can avoid fitting noisy instances in the early stage of training due to the memorization effects of deep neural networks [1] and improve the performance progressively as more and more instances with high confidence are explored, there inevitably exist some label errors that cannot be corrected and would be overfitted as the training proceeds. These accumulated errors eventually impede the performance growth.

In order to reduce the label error accumulation throughout the training process, the proposed NRMT maintains two neural networks f parameterized by $\mathbf{W}_f$ and g parameterized by $\mathbf{W}_g$ simultaneously during training, and allows them to share clustering information by collaborative clustering at each iteration to reduce the effect of their respective label errors.

To make f and g have different learning abilities, we use different random seeds to pre-train f and g on the source dataset $\mathbf{X}^s$ with labels $\mathbf{Y}^s$ by the triplet loss and the Softmax loss [31]. Here f and g have the same network architecture to facilitate the deployment. Because deep neural networks are highly non-convex models, different initializations can still lead to different local optima even with the same architecture and optimization algorithm [14]. Then, we use the pre-trained f and g to extract features on the target dataset $\mathbf{X}^t$ and obtain two sets of pseudo-labels $\hat{\mathbf{Y}}^t_f$ and $\hat{\mathbf{Y}}^t_g$ through applying clustering to the features. Since the target domain has classes different from the source domain, we drop the Softmax loss and fine-tune the networks on the target data only using the triplet loss with the pseudo-labels. To share clustering information, f and g consider both their own pseudo-labels and the ones of their peer networks. Thus, we have a joint loss function for each network:

$$\mathcal{L}_f = \mathcal{L}_{tri}(\hat{\mathbf{Y}}^t_f, f(\mathbf{X}^t; \mathbf{W}_f)) + \mathcal{L}_{tri}(\hat{\mathbf{Y}}^t_g, f(\mathbf{X}^t; \mathbf{W}_f)), \qquad (2)$$

$$\mathcal{L}_g = \mathcal{L}_{tri}(\hat{\mathbf{Y}}^t_g, g(\mathbf{X}^t; \mathbf{W}_g)) + \mathcal{L}_{tri}(\hat{\mathbf{Y}}^t_f, g(\mathbf{X}^t; \mathbf{W}_g)), \qquad (3)$$

where $\mathcal{L}_{tri}$ is the batch-sampling triplet loss [16].

Different from self-training where the network assigns new pseudo-labels to the training instances at each iteration only according to its own parameter update, in NRMT, to make the learning more robust, the two networks f and g collaboratively assign pseudo-labels, $i.e.$, each instance has two pseudo-labels from f and g, respectively. The study on memorization in deep networks [1] suggests that deep networks tend to prioritize learning easy patterns. Usually noisy instances caused by clustering are relatively hard examples, thus if one instance is assigned two labels, the networks will fit the clean (easy) one first to become robust and the error may be eliminated at the next iteration. The joint loss functions in Eq. (2) and Eq. (3) are similar to Co-training [2] where classifiers are trained on two views (two independent sets of features). However, here we have two networks but only have a single view, and we utilize the memorization effect of deep networks to handle the error in labels.

3.3   Mutual Instance Selection

Although collaborative clustering across networks is able to ease the fitting to noisy instances for each iteration, these noisy instances still have impact on the network training in a mini-batch, especially in the advanced stage of training. To further select reliable and informative instances in a mini-batch, we introduce a mutual instance selection strategy by considering both the peer-confidence and relationship disagreement of the two networks.

Reliable Instance Selection by Peer-Confidence. In order to select reliable instances for training, we consider using the prediction confidence of the peer network to measure the reliability of instances for one network. We argue that in the metric learning, the relationship of one pair of instances with other pairs in the feature space can provide more information about the network prediction than the distance between one instance and another one. Thus, we compute the prediction confidence based on the relationship of a triplet of instances.

Given an instance x, its corresponding positive instance x_p and negative instance x_n from a mini-batch, we encode the relationship of the triplet $\{x, x_p, x_n\}$ by the difference between the Euclidean distances of the positive and negative pairs in the feature space:

$$\mathcal{D}(x, x_p, x_n; f) = ||f(x) - f(x_p)||_2 - ||f(x) - f(x_n)||_2, \tag{4}$$

$$\mathcal{D}(x, x_p, x_n; g) = ||g(x) - g(x_p)||_2 - ||g(x) - g(x_n)||_2, \tag{5}$$

where $f(x)$ and $g(x)$ is the features extracted by the networks f and g, respectively. The smaller the difference is, the higher the confidence is. If the difference of the peer network g (resp. f) of f (resp. g) for the triplet $\{x, x_p, x_n\}$ is smaller than a threshold T_c:

$$\mathcal{D}(x, x_p, x_n; g) < T_c, \tag{6}$$

$$\text{resp. } \mathcal{D}(x, x_p, x_n; f) < T_c, \tag{7}$$

we call $\{x, x_p, x_n\}$ as a peer-confident triplet of instances for f (resp. g) and use this peer-confident triplet to update f (resp. g). Because the two networks have different learning abilities, we expect that they can filter out various noisy instances [14] to make up for each other's mistakes.

Informative Instance Selection by Relationship Disagreement. The peer-confidence of the network can pick up reliable (clean) instances in a mini-batch, but these instances usually contain lots of easy instances which provide limited information for the network performance improvement. To further select more informative instances, we propose to use the relationship disagreement between one network and its peer network to measure the amount of information on the basis of the peer-confidence.

Similar to the peer-confidence, we compute the relationship disagreement on a triplet of instances. We first define the prediction inconsistency of the two networks f and g combined with Eq. (4) and Eq. (5) as:

$$\mathcal{I}(x, x_p, x_n; f, g) = \mathcal{D}(x, x_p, x_n; f) - \mathcal{D}(x, x_p, x_n; g). \tag{8}$$

Larger absolute value of the inconsistency indicates that the triplet of instances has larger amount of information. It can be considered that there is the relationship disagreement between the predictions of two networks for the triplet $\{x, x_p, x_n\}$ if the absolute value of the prediction inconsistency is larger than a threshold T_d:

$$|\mathcal{I}(x, x_p, x_n; f, g)| > T_d \tag{9}$$

The networks are only updated on the mini-batch data with the relationship disagreement. Furthermore, when combined with the peer-confidence, Eq. (9)

can be rewritten with the absolute symbol removed:

$$\mathcal{I}(x, x_p, x_n; f, g) > T_d, \tag{10}$$

$$\mathcal{I}(x, x_p, x_n; g, f) > T_d. \tag{11}$$

The intuition is that, for the item within the absolute symbol in Eq. (9) which is smaller than $-T_d$, because T_d is not less than zero and $\{x, x_p, x_n\}$ meets the peer-confidence condition in Eq. (6) or Eq. (7), we have

$$\mathcal{D}(x, x_p, x_n; f) < \mathcal{D}(x, x_p, x_n; g) - T_d < \mathcal{D}(x, x_p, x_n; g) < T_c, \tag{12}$$

$$\text{or } \mathcal{D}(x, x_p, x_n; g) < \mathcal{D}(x, x_p, x_n; f) - T_d < \mathcal{D}(x, x_p, x_n; f) < T_c. \tag{13}$$

As a result, when T_c is set to a proper small value, for the network f or g, the triplet $\{x, x_p, x_n\}$ is actually an easy instance that can be ignored during training. Figure 2 illustrates three types of triplets of instances obtained by the proposed mutual instance selection strategy, where we consider instances selection for the network f according to the prediction of the network g.

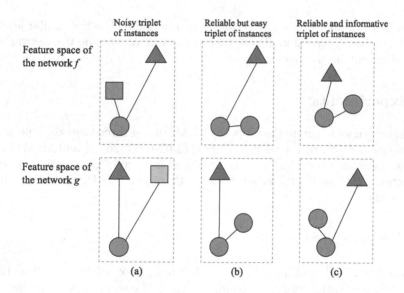

Fig. 2. Three types of triplets of instances obtained by the proposed mutual instance selection strategy. Different shapes (circle, triangle and square) denote different ground truth class labels and different colors (blue, green and yellow) denote different pseudo-labels. (a) Noisy triplet of instances obtained by $\mathcal{D}(x, x_p, x_n; g) \geq T_c$; (b) Reliable but easy triplet of instance obtained by $\mathcal{D}(x, x_p, x_n; g) < T_c$ but $\mathcal{I}(x, x_p, x_n; f, g) \leq T_d$; (c) Reliable and informative triplet of instances obtained by $\mathcal{D}(x, x_p, x_n; g) < T_c$ and $\mathcal{I}(x, x_p, x_n; f, g) > T_d$. (Best viewed in color). (Color figure online)

For the clarity, the training process of NRMT is summarized in Algorithm 1. It is worth noting that we only maintain two networks in the stage of training

---

Algorithm 1. Noise Resistible Mutual-Training (NRMT)

Input: Deep networks f and g, labeled source training dataset $\{\mathbf{X}^s, \mathbf{Y}^s\}$, unlabeled
　　target training dataset $\mathbf{X}^t$, maximal number of updates N_{max}, maximal number of
　　iterations I_{max}.
Output: f and g.
1: Pre-train f and g on $\{\mathbf{X}^s, \mathbf{Y}^s\}$ with different random seeds, respectively;
2: **for** $I = 1$ **to** I_{max} **do**
3:　　Extract features $f(x)$ and $g(x)$ on $\mathbf{X}^t$;
4:　　Perform clustering on $f(x)$ and $g(x)$ to generate pseudo-labels $\hat{\mathbf{Y}}_f^t$ and $\hat{\mathbf{Y}}_g^t$;
5:　　**for** $N = 1$ **to** N_{max} **do**
6:　　　　Sample mini-batches $\mathcal{M}(\hat{\mathbf{Y}}_f^t)$ and $\mathcal{M}(\hat{\mathbf{Y}}_g^t)$ from $\mathbf{X}^t$ with $\hat{\mathbf{Y}}_f^t$ and $\hat{\mathbf{Y}}_g^t$;
7:　　　　Obtain $\mathcal{M}_f(\hat{\mathbf{Y}}_f^t)$ and $\mathcal{M}_f(\hat{\mathbf{Y}}_g^t)$ by Eq. (6) and Eq. (10);
8:　　　　Obtain $\mathcal{M}_g(\hat{\mathbf{Y}}_f^t)$ and $\mathcal{M}_g(\hat{\mathbf{Y}}_g^t)$ by Eq. (7) and Eq. (11);
9:　　　　Update f with both $\mathcal{M}_f(\hat{\mathbf{Y}}_f^t)$ and $\mathcal{M}_f(\hat{\mathbf{Y}}_g^t)$ as in Eq. (2);
10:　　　　Update g with both $\mathcal{M}_g(\hat{\mathbf{Y}}_f^t)$ and $\mathcal{M}_g(\hat{\mathbf{Y}}_g^t)$ as in Eq. (3);
11:　　**end for**
12: **end for**

---

and the performance of the two networks can be aligned to the similar level via
the information interaction. Thus, we can use any one of the two networks for
the deployment in practice.

4　Experiments

In this section, we evaluate the proposed NRMT using three large-scale person re-
ID datasets, $i.e.$, Market-1501 [52], DukeMTMC-reID [26,54] and MSMT17 [39]
and the performance evaluations are presented in term of Cumulative Matching
Characteristic (CMC) and mean Average Precision (mAP) under the single-
query setting.

4.1　Datasets

Market-1501 [52] contains 32,668 labeled images of 1,501 identities. 12,936
images of 751 identities form the training set. 3,368 query images from the other
750 identities and 19,732 gallery images (with 2,793 distractors) are used as the
test set. The bounding boxes of persons are generated by Deformable Part Model
(DPM) [9]. **DukeMTMC-reID** [26,54] includes 36,411 labeled images of 1,404
identities. 702 identities are randomly selected for training and the rest is used
for testing. There are 16,522 training images, 2,228 query images and 17,661
gallery images. **MSMT17** [39] is the largest re-ID dataset consisting of 126,441
bounding boxes of 4,101 identities taken by 12 outdoor and 3 indoor cameras.
32,621 images of 1,041 identities are used for training.

4.2   Implementation Details

We adopt ResNet-50 [15] as the architectures of the two networks and initialize them with the parameters pre-trained on ImageNet [6]. All images are resized to 256×128. Random horizontal flipping and random erasing [55] are employed for training data augmentation. We use the Softmax and triplet losses to pre-train the two networks on the source dataset with different random seeds, respectively. The margin m in the triplet loss is 0.5. For each mini-batch, we randomly sample 32 identities and 4 images per identity. The SGD optimizer with a momentum of 0.9 is used to train the networks and the learning rate is 6e-5.

The peer-confidence threshold T_c is set to 1.0 and the relationship disagreement threshold T_d is set to 0.5. The HDBSCAN clustering algorithm [4] is adopted to produce pseudo-labels for each iteration, which does not require the number of clusters as prior parameter. The number of minimum samples for each cluster is set to 8. The maximal number of iterations is 30. At the first half of the iterative process, we train the networks only using collaborative clustering. Then we add mutual instance selection to further select clean and informative data in mini-batches for the network update.

Table 1. Evaluation on different values of the threshold T_c. Results of the two networks f and g are reported, respectively.

T_c	Duke → Market		Market → Duke	
	mAP	R1	mAP	R1
0	71.0/70.2	86.9/86.5	61.1/60.9	76.9/76.6
0.5	71.5/70.6	87.3/87.0	61.7/61.4	77.5/77.0
1.0	**72.2/71.1**	**88.0/87.5**	**62.3/62.0**	**78.1/77.5**
1.5	72.0/71.0	87.7/87.3	62.0/61.8	78.0/77.2
2.0	71.7/70.7	87.4/87.0	61.7/61.5	77.7/77.0

Table 2. Evaluation on different values of the threshold T_d. Results of the two networks f and g are reported, respectively.

T_d	Duke → Market		Market → Duke	
	mAP	R1	mAP	R1
0.3	71.2/70.4	87.3/86.8	61.3/61.0	77.0/76.7
0.4	71.6/70.7	87.6/87.2	61.8/61.5	77.5/77.1
0.5	**72.2/71.1**	**88.0/87.5**	**62.3/62.0**	**78.1/77.5**
0.6	72.0/70.8	87.7/87.3	62.1/61.7	77.8/77.3
0.7	71.5/70.4	87.3/86.9	61.6/61.3	77.2/76.8

Table 3. Evaluation on different numbers of the minimum samples for each cluster in HDBSCAN. Results of the two networks f and g are reported, respectively.

Min. Samp.	Duke → Market		Market → Duke	
	mAP	R1	mAP	R1
6	71.5/70.7	86.8/86.5	61.7/61.6	77.1/77.0
8	**72.2/71.1**	**88.0/87.5**	**62.3/62.0**	**78.1/77.5**
10	71.9/71.1	87.5/87.1	61.8/61.4	77.7/ 76.9

4.3   Parameter Analysis

We first study impacts of some important parameter settings in the proposed NRMT, including the peer-confidence threshold T_c, the relationship disagreement threshold T_d and the number of minimum samples in the HDBSCAN clustering algorithm.

Peer-Confidence Threshold T_c. To analyze the impact of T_c in Eq. (6) and Eq. (7), we fix the relationship disagreement threshold $T_d = 0.5$ in all experiments. The results are listed in Table 1. We can observe that a proper value of T_c is important for NRMT to filter out noisy instances, which provides a reasonable assessment of the noise confidence. The best performance is achieved when T_c is set to 1.0.

Relationship Disagreement Threshold T_d. We also conduct experiments to investigate the impact of T_d in Eq. (10) and Eq. (11). In all experiments, we fix the peer-confidence threshold $T_c = 1.0$. As reported in Table 2, when $T_d = 0.5$, we can obtain the best results. When T_d is set to a larger value, fewer instances are selected for update, which is likely to discard instances that are actually informative. Too small values of T_d will allow most of the instances to be involved in update, which may contain too many easy instance and thus cannot provide effective information for improving the network.

Number of Minimum Samples. To evaluate the influence of the number of minimum samples in HDBSCAN, we report the results of {6, 8, 10} minimum samples in Table 3. As we can see, the number 8 yields the superior performance. Note that our NRMT is not very sensitive to this prior clustering parameter.

4.4   Ablation Study

We further validate the effectiveness of each component in the proposed NRMT, including collaborative clustering, instance selection by the peer-confidence and relationship disagreement on Market-1501 and DukeMTMC-reID. The results are shown in Table 4. As we can see, by sharing clustering information between two networks on the whole dataset, "Ours w/ CC" improves the performance of both the two networks compared with "Separate Training". This demonstrates that the collaborative clustering is able to ease the fitting to noisy instances

Table 4. Performance evaluation of components in the proposed NRMT on Market-1501 and DukeMTMC-reID. **Separate Training:** Train the two networks separately. **CC:** Collaborative clustering. **SC:** Instance selection by the peer-confidence. **SD:** Instance selection by the relationship disagreement. Results of the two networks f and g are reported, respectively.

Methods	DukeMTMC-reID → Market-1501			
	mAP	R1	R5	R10
Direct Transfer	33.0/32.3	63.3/62.3	77.2/76.5	82.3/81.9
Separate Training	54.2/53.0	76.4/75.6	88.3/87.8	92.2/91.8
Ours w/ CC	68.9/68.2	85.9/85.5	94.0/94.1	96.2/96.1
Ours w/ CC+SC	70.9/70.1	86.8/86.3	94.3/94.2	96.3/96.1
Ours w/ CC+SC+SD	**72.2/71.1**	**88.0/87.5**	**94.7/94.5**	**96.5/96.4**
Methods	Market-1501 → DukeMTMC-reID			
	mAP	R1	R5	R10
Direct Transfer	30.2/30.2	47.3/46.5	61.9/61.8	68.2/68.1
Separate Training	48.7/48.2	67.2/66.5	80.3/80.0	84.3/84.1
Ours w/ CC	59.1/58.7	75.8/75.4	85.5/85.2	88.2/88.3
Ours w/ CC+SC	60.6/60.2	76.6/76.3	86.1/85.9	88.9/88.7
Ours w/ CC+SC+SD	**62.3/62.0**	**78.1/77.5**	**87.0/86.8**	**89.7/89.2**

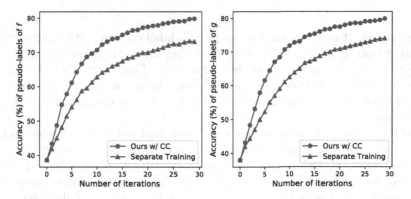

Fig. 3. Comparison on the accuracy of pseudo-labels in the iteration process for DukeMTMC-reID → Market-1501.

caused by unsupervised clustering by exploiting different learning abilities of two networks and the memorization effect of deep networks. "Ours w/ CC+SC" and "Ours w/ CC+SC+SD" further obtain better results by prediction information interaction between the networks in mini-batches, which can pick up clean and informative instances to update the networks.

To explore the ability of correcting label errors of collaborative clustering, Fig. 3 illustrates the accuracy of pseudo-labels generated by clustering in the iter-

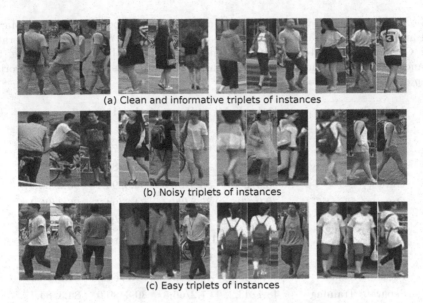

(a) Clean and informative triplets of instances

(b) Noisy triplets of instances

(c) Easy triplets of instances

Fig. 4. Examples of (a) clean and informative, (b) noisy and (c) easy triplets of instances obtained by the proposed mutual instance selection strategy in a mini-batch. Only the clean and informative triplets are used for the network update. For each triplet, the first two ones are positive examples and the last one is negative example.

ation process. It can be seen that the pseudo-label accuracy of the two networks f and g trained with collaborative clustering are both improved significantly compared with the networks trained separately. This shows that sharing clustering information between two networks on the whole dataset can effectively correct label errors at each iteration and reduce the accumulation of noises during training.

In Fig. 4, we show some examples of clean and informative, noisy and easy triplets of instances obtained by the proposed mutual instance selection strategy. We can observe that the clean and informative triplets selected by our strategy contains negative examples with similar appearances and positive examples with large variations. Meanwhile, our strategy can filter out not only noisy triplets but also easy triplets. This indicates that our strategy is able to act as a robust online hard example mining for the triplet loss in training with noisy labels.

4.5   Comparison with State-of-the-art Methods

In this section, we compare the proposed NRMT with the state-of-the-art unsupervised person re-ID methods on the transfers between DukeMTMC-reID and Market-1501 and the transfers from DukeMTMC-reID/Market-1501 to MSMT17. Here we reports the averaged performance of the two networks f and g in NRMT.

Table 5. Comparison with the state-of-the-art UDA methods on Market-1501 and DukeMTMC-reID. The averaged performance of the two networks f and g is reported.

Methods	Market-1501				DukeMTMC-reID			
	mAP	R1	R5	R10	mAP	R1	R5	R10
LOMO [18]	8.0	27.2	41.6	49.1	4.8	12.3	21.3	26.6
BoW [52]	14.8	35.8	52.4	60.3	8.3	17.1	28.8	34.9
UMDL [24]	12.4	34.5	52.6	59.6	7.3	18.5	31.4	37.6
PUL [8]	20.5	45.5	60.7	66.7	16.4	30.0	43.4	48.5
DECAMEL [45]	32.4	60.2	–	–	–	–	–	–
PTGAN [39]	–	38.6	–	66.1	–	27.4	–	50.7
SPGAN+LMP [7]	26.7	57.7	75.8	82.4	26.2	46.4	62.3	68.0
TJ-AIDL [38]	26.5	58.2	74.8	81.1	23.0	44.3	59.6	65.0
HHL [56]	31.4	62.2	78.8	84.0	27.2	46.9	61.0	66.7
ARN [17]	39.4	70.3	80.4	86.3	33.4	60.2	73.9	79.5
UDAP [31]	53.7	75.8	89.5	93.2	49.0	68.4	80.1	83.5
MAR [46]	40.0	67.7	81.9	–	48.0	67.1	79.8	–
ECN [57]	43.0	75.1	87.6	91.6	40.4	63.3	75.8	80.4
CR-GAN+LMP [5]	33.2	64.5	79.8	85.0	33.3	56.0	70.5	74.6
PCB-R-PAST [49]	54.6	78.4	–	–	54.3	72.4	–	–
SSG [11]	58.3	80.0	90.0	92.4	53.4	73.0	80.6	83.2
ACT [43]	60.6	80.5	–	–	54.5	72.4	–	–
NRMT	**71.7**	**87.8**	**94.6**	**96.5**	**62.2**	**77.8**	**86.9**	**89.5**

Table 6. Comparison with the state-of-the-arts on transfers from DukeMTMC-reID and Market-1501 to MSMT17.

Methods	DukeMTMC-reID $\rightarrow$ MSMT17				Market-1501 $\rightarrow$ MSMT17			
	mAP	R1	R5	R10	mAP	R1	R5	R10
PTGAN [39]	3.3	11.8	–	27.4	2.9	10.2	–	24.4
ECN [57]	10.2	30.2	41.5	46.8	8.5	25.3	36.3	42.1
SSG [11]	13.3	32.2	–	51.2	13.2	31.6	–	49.6
NRMT	**20.6**	**45.2**	**57.8**	**63.3**	**19.8**	**43.7**	**56.5**	**62.2**

Table 5 shows the results on the transfers between DukeMTMC-reID and Market-1501. We first compare the proposed NRMT with two hand-crafted features, *i.e.*, LOMO [18] and Bag-of-Words (BoW) [52]. We can see that deep learning features can significantly improve the performance. Three unsupervised methods including UMDL [24], PUL [8] and DECAMEL [45] are compared. Our method surpasses these methods by a large margin by adapting to the target data from the source data progressively. We also compare with the unsupervised domain adaptation methods, including UDAP [31], MAR [46], ECN [57],

PCB-R-PAST [49], SSG [11], ACT [43], etc. our method still achieves the best performance. Especially, our NRMT outperforms PCB-R-PAST [49], which also focuses on the improvement of label quality, by 17.1%/9.4% on mAP/Rank-1 accuracy for DukeMTMC-reID → Market-1501 and by 7.9%/5.4% for Market-1501 → DukeMTMC-reID. This demonstrates the effectiveness of information interactions between dual networks for noise reduction. Moreover, our NRMT also exceeds the second best method ACT [43] by clear margins.

We also evaluate our NRMT on transfers from DukeMTMC-reID and Market-1501 to MSMT17 in Table 6. The results obtained by NRMT are 20.6%/45.2% on mAP/R1 accuracy for DukeMTMC-reID → MSMT17 and 19.8%/43.7% for Market-1501 → MSMT17, which all exceed the second best method, *i.e.*, SSG [11]. This further demonstrates the superiority of our NRMT on the large-scale dataset.

5 Conclusions

This paper proposed a noise resistible mutual-training method (NRMT) for unsupervised domain adaptation (UDA) in person re-ID to effectively depress label noises in a self-training process. In NRMT, two networks are maintained during training. For each iteration, these two networks share clustering information to ease the fitting to noisy instances. For each mini-batch update, the networks also exchange prediction information to further select both reliable and informative instances. Extensive experimental results demonstrate that the proposed NRMT achieves the state-of-the-art performance for UDA in person re-ID.

Acknowledgments. This work was supported by the National Natural Science Foundation of China under Grant 61702163.

References

1. Arpit, D., et al.: A closer look at memorization in deep networks. In: International Conference on Machine Learning (ICML) (2017)
2. Blum, A., Mitchell, T.: Combining labeled and unlabeled data with co-training. In: Proceedings of the Eleventh Annual Conference on Computational Learning Theory (1998)
3. Bousmalis, K., Trigeorgis, G., Silberman, N., Krishnan, D., Erhan, D.: Domain separation networks. In: Advances in Neural Information Processing Systems (NeurIPS) (2016)
4. Campello, R.J.G.B., Moulavi, D., Sander, J.: Density-based clustering based on hierarchical density estimates. In: Pei, J., Tseng, V.S., Cao, L., Motoda, H., Xu, G. (eds.) PAKDD 2013. LNCS (LNAI), vol. 7819, pp. 160–172. Springer, Heidelberg (2013). https://doi.org/10.1007/978-3-642-37456-2_14
5. Chen, Y., Zhu, X., Gong, S.: Instance-guided context rendering for cross-domain person re-identification. In: Proceedings of the IEEE International Conference on Computer Vision (ICCV) (2019)

6. Deng, J., Dong, W., Socher, R., Li, L.J., Li, K., Fei-Fei, L.: Imagenet: a large-scale hierarchical image database. In: Proceedings of the IEEE Conference on Computer Vision and Pattern Recognition (CVPR) (2009)

7. Deng, W., Zheng, L., Ye, Q., Kang, G., Yang, Y., Jiao, J.: Image-image domain adaptation with preserved self-similarity and domain-dissimilarity for person re-identification. In: Proceedings of the IEEE Conference on Computer Vision and Pattern Recognition (CVPR) (2018)

8. Fan, H., Zheng, L., Yan, C., Yang, Y.: Unsupervised person re-identification: clustering and fine-tuning. ACM Trans. Multimed. Comput. Commun. Appl. (TOMCCAP) **14**(4), 1–18 (2018)

9. Felzenszwalb, P., McAllester, D., Ramanan, D.: A discriminatively trained, multiscale, deformable part model (2008)

10. Feng, Q., Kang, G., Fan, H., Yang, Y.: Attract or distract: exploit the margin of open set. In: Proceedings of the IEEE International Conference on Computer Vision (ICCV) (2019)

11. Fu, Y., Wei, Y., Wang, G., Zhou, Y., Shi, H., Huang, T.S.: Self-similarity grouping: a simple unsupervised cross domain adaptation approach for person re-identification. In: Proceedings of the IEEE International Conference on Computer Vision (ICCV) (2019)

12. Ganin, Y., et al.: Domain-adversarial training of neural networks. J. Mach. Learn. Res. (JMLR) **17**(1), 2096–2130 (2016)

13. Ge, Y., Chen, D., Li, H.: Mutual mean-teaching: pseudo label refinery for unsupervised domain adaptation on person re-identification. In: International Conference on Learning Representations (ICLR) (2020)

14. Han, B., et al.: Co-teaching: robust training of deep neural networks with extremely noisy labels. In: Advances in Neural Information Processing Systems (NeurIPS) (2018)

15. He, K., Zhang, X., Ren, S., Sun, J.: Deep residual learning for image recognition. In: Proceedings of the IEEE Conference on Computer Vision and Pattern Recognition (CVPR) (2016)

16. Hermans, A., Beyer, L., Leibe, B.: In defense of the triplet loss for person re-identification. arXiv preprint arXiv:1703.07737 (2017)

17. Li, Y.J., Yang, F.E., Liu, Y.C., Yeh, Y.Y., Du, X., Frank Wang, Y.C.: Adaptation and re-identification network: an unsupervised deep transfer learning approach to person re-identification. In: Proceedings of the IEEE Conference on Computer Vision and Pattern Recognition Workshops (CVPRW) (2018)

18. Liao, S., Hu, Y., Zhu, X., Li, S.Z.: Person re-identification by local maximal occurrence representation and metric learning. In: Proceedings of the IEEE Conference on Computer Vision and Pattern Recognition (CVPR) (2015)

19. Liu, Z., Wang, J., Gong, S., Lu, H., Tao, D.: Deep reinforcement active learning for human-in-the-loop person re-identification. In: Proceedings of the IEEE International Conference on Computer Vision (ICCV) (2019)

20. Long, M., Cao, Y., Wang, J., Jordan, M.: Learning transferable features with deep adaptation networks. In: International Conference on Machine Learning (ICML) (2015)

21. Lv, J., Chen, W., Li, Q., Yang, C.: Unsupervised cross-dataset person re-identification by transfer learning of spatial-temporal patterns. In: Proceedings of the IEEE Conference on Computer Vision and Pattern Recognition (2018)

22. Malach, E., Shalev-Shwartz, S.: Decoupling "when to update" from "how to update". In: Advances in Neural Information Processing Systems (NeurIPS) (2017)

23. Panareda Busto, P., Gall, J.: Open set domain adaptation. In: Proceedings of the IEEE International Conference on Computer Vision (ICCV) (2017)
24. Peng, P., et al.: Unsupervised cross-dataset transfer learning for person re-identification. In: Proceedings of the IEEE Conference on Computer Vision and Pattern Recognition (CVPR) (2016)
25. Qi, L., Wang, L., Huo, J., Zhou, L., Shi, Y., Gao, Y.: A novel unsupervised camera-aware domain adaptation framework for person re-identification. In: Proceedings of the IEEE International Conference on Computer Vision (ICCV) (2019)
26. Ristani, E., Solera, F., Zou, R., Cucchiara, R., Tomasi, C.: Performance measures and a data set for multi-target, multi-camera tracking. In: Hua, G., Jégou, H. (eds.) ECCV 2016. LNCS, vol. 9914, pp. 17–35. Springer, Cham (2016). https://doi.org/10.1007/978-3-319-48881-3_2
27. Saito, K., Yamamoto, S., Ushiku, Y., Harada, T.: Open set domain adaptation by backpropagation. In: Ferrari, V., Hebert, M., Sminchisescu, C., Weiss, Y. (eds.) ECCV 2018. LNCS, vol. 11209, pp. 156–171. Springer, Cham (2018). https://doi.org/10.1007/978-3-030-01228-1_10
28. Shu, R., Bui, H.H., Narui, H., Ermon, S.: A dirt-t approach to unsupervised domain adaptation. In: Proceedings of the International Conference on Learning Representations (ICLR) (2018)
29. Song, C., Huang, Y., Ouyang, W., Wang, L.: Mask-guided contrastive attention model for person re-identification. In: Proceedings of the IEEE Conference on Computer Vision and Pattern Recognition (CVPR) (2018)
30. Song, J., Yang, Y., Song, Y.Z., Xiang, T., Hospedales, T.M.: Generalizable person re-identification by domain-invariant mapping network. In: Proceedings of the IEEE Conference on Computer Vision and Pattern Recognition (CVPR) (2019)
31. Song, L., et al.: Unsupervised domain adaptive re-identification: theory and practice. arXiv preprint arXiv:1807.11334 (2018)
32. Suh, Y., Wang, J., Tang, S., Mei, T., Lee, K.M.: Part-aligned bilinear representations for person re-identification. In: Ferrari, V., Hebert, M., Sminchisescu, C., Weiss, Y. (eds.) Computer Vision – ECCV 2018. LNCS, vol. 11218, pp. 418–437. Springer, Cham (2018). https://doi.org/10.1007/978-3-030-01264-9_25
33. Sun, B., Feng, J., Saenko, K.: Return of frustratingly easy domain adaptation. In: Thirtieth AAAI Conference on Artificial Intelligence (2016)
34. Sun, Y., Zheng, L., Deng, W., Wang, S.: SVDNet for pedestrian retrieval. In: Proceedings of the IEEE International Conference on Computer Vision (ICCV) (2017)
35. Sun, Y., Zheng, L., Yang, Y., Tian, Q., Wang, S.: Beyond part models: person retrieval with refined part pooling (and a strong convolutional baseline). In: Ferrari, V., Hebert, M., Sminchisescu, C., Weiss, Y. (eds.) ECCV 2018. LNCS, vol. 11208, pp. 501–518. Springer, Cham (2018). https://doi.org/10.1007/978-3-030-01225-0_30
36. Tzeng, E., Hoffman, J., Darrell, T., Saenko, K.: Simultaneous deep transfer across domains and tasks. In: Proceedings of the IEEE International Conference on Computer Vision (ICCV) (2015)
37. Tzeng, E., Hoffman, J., Saenko, K., Darrell, T.: Adversarial discriminative domain adaptation. In: Proceedings of the IEEE Conference on Computer Vision and Pattern Recognition (CVPR) (2017)
38. Wang, J., Zhu, X., Gong, S., Li, W.: Transferable joint attribute-identity deep learning for unsupervised person re-identification. In: Proceedings of the IEEE Conference on Computer Vision and Pattern Recognition (CVPR) (2018)

39. Wei, L., Zhang, S., Gao, W., Tian, Q.: Person transfer Gan to bridge domain gap for person re-identification. In: Proceedings of the IEEE Conference on Computer Vision and Pattern Recognition (CVPR) (2018)
40. Wu, A., Zheng, W.S., Lai, J.H.: Unsupervised person re-identification by camera-aware similarity consistency learning. In: Proceedings of the IEEE International Conference on Computer Vision (ICCV) (2019)
41. Xie, G.S., et al.: Attentive region embedding network for zero-shot learning. In: Proceedings of the IEEE Conference on Computer Vision and Pattern Recognition (CVPR) (2019)
42. Xie, G.S., et al.: Region graph embedding network for zero-shot learning. In: Vedaldi, A., Bischof, H., Brox, T., Frahm, J.M. (eds.) ECCV 2020. LNCS, vol. 12349, pp. 562–580. Springer, Cham (2020). https://doi.org/10.1007/978-3-030-58548-8_33
43. Yang, F., et al.: Asymmetric co-teaching for unsupervised cross-domain person re-identification. In: Thirtieth AAAI Conference on Artificial Intelligence (AAAI) (2020)
44. Yang, Q., Yu, H.X., Wu, A., Zheng, W.S.: Patch-based discriminative feature learning for unsupervised person re-identification. In: Proceedings of the IEEE Conference on Computer Vision and Pattern Recognition (CVPR) (2019)
45. Yu, H.X., Wu, A., Zheng, W.S.: Unsupervised person re-identification by deep asymmetric metric embedding. IEEE Trans. Pattern Anal. Mach. Intell. (TPAMI) **42**, 956–973 (2018)
46. Yu, H.X., Zheng, W.S., Wu, A., Guo, X., Gong, S., Lai, J.H.: Unsupervised person re-identification by soft multilabel learning. In: Proceedings of the IEEE Conference on Computer Vision and Pattern Recognition (CVPR) (2019)
47. Yu, X., Han, B., Yao, J., Niu, G., Tsang, I., Sugiyama, M.: How does disagreement help generalization against label corruption? In: International Conference on Machine Learning (ICML) (2019)
48. Zhang, K., Luo, W., Ma, L., Liu, W., Li, H.: Learning joint gait representation via quintuplet loss minimization. In: Proceedings of the IEEE Conference on Computer Vision and Pattern Recognition (CVPR) (2019)
49. Zhang, X., Cao, J., Shen, C., You, M.: Self-training with progressive augmentation for unsupervised cross-domain person re-identification. In: Proceedings of the IEEE International Conference on Computer Vision (ICCV) (2019)
50. Zhao, H., et al.: Spindle net: person re-identification with human body region guided feature decomposition and fusion. In: Proceedings of the IEEE Conference on Computer Vision and Pattern Recognition (CVPR) (2017)
51. Zhao, L., Li, X., Zhuang, Y., Wang, J.: Deeply-learned part-aligned representations for person re-identification. In: Proceedings of the IEEE International Conference on Computer Vision (ICCV) (2017)
52. Zheng, L., Shen, L., Tian, L., Wang, S., Wang, J., Tian, Q.: Scalable person re-identification: a benchmark. In: Proceedings of the IEEE International Conference on Computer Vision (ICCV) (2015)
53. Zheng, Z., Yang, X., Yu, Z., Zheng, L., Yang, Y., Kautz, J.: Joint discriminative and generative learning for person re-identification. In: Proceedings of the IEEE Conference on Computer Vision and Pattern Recognition (CVPR) (2019)
54. Zheng, Z., Zheng, L., Yang, Y.: Unlabeled samples generated by gan improve the person re-identification baseline in vitro. In: Proceedings of the IEEE International Conference on Computer Vision (ICCV) (2017)
55. Zhong, Z., Zheng, L., Kang, G., Li, S., Yang, Y.: Random erasing data augmentation. arXiv preprint arXiv:1708.04896 (2017)

56. Zhong, Z., Zheng, L., Li, S., Yang, Y.: Generalizing a Person retrieval model hetero- and homogeneously. In: Ferrari, V., Hebert, M., Sminchisescu, C., Weiss, Y. (eds.) ECCV 2018. LNCS, vol. 11217, pp. 176–192. Springer, Cham (2018). https://doi.org/10.1007/978-3-030-01261-8_11
57. Zhong, Z., Zheng, L., Luo, Z., Li, S., Yang, Y.: Invariance matters: exemplar memory for domain adaptive person re-identification. In: Proceedings of the IEEE Conference on Computer Vision and Pattern Recognition (CVPR) (2019)
58. Zhou, S., Wang, J., Wang, J., Gong, Y., Zheng, N.: Point to set similarity based deep feature learning for person re-identification. In: Proceedings of the IEEE Conference on Computer Vision and Pattern Recognition (CVPR) (2017)

DPDist: Comparing Point Clouds Using Deep Point Cloud Distance

Dahlia Urbach[1]([✉]), Yizhak Ben-Shabat[2], and Michael Lindenbaum[1]

[1] Technion IIT, Haifa, Israel
dahliau@technion.ac.il, mic@cs.technion.ac.il
[2] Australian National University, Australian Centre for Robotic Vision,
Canberra, Australia
yizhak.benshabat@anu.edu.au
https://github.com/dahliau/DPDist

Abstract. We introduce a new deep learning method for point cloud comparison. Our approach, named Deep Point Cloud Distance (DPDist), measures the distance between the points in one cloud and the estimated surface from which the other point cloud is sampled. The surface is estimated locally using the 3D modified Fisher vector representation. The local representation reduces the complexity of the surface, enabling effective learning, which generalizes well between object categories. We test the proposed distance in challenging tasks, such as similar object comparison and registration, and show that it provides significant improvements over commonly used distances such as Chamfer distance, Earth mover's distance, and others.

Keywords: 3D point clouds · 3D computer vision · 3D deep learning · Distance · Registration

1 Introduction

Recent advancements in 3D sensor technology have led to the integration of 3D sensors into many application domains, such as virtual and augmented reality, robotic vision, and autonomous systems. These sensors supply a set of 3D points, sampled on surfaces in the scene, known as a point cloud. As raw, memory-efficient outputs of 3D sensors, point clouds are a common 3D data representation. However, in contrast to more traditional data (e.g., images), point clouds are unstructured, unordered, and may have a varying number of points. Therefore, unlike traditional signals, they may not be represented as values on some regular grid, and are difficult to process using common signal processing tools.

In particular, many applications, such as registration, retrieval, autoencoding etc., require comparisons between two or more point clouds. Comparing point

Electronic supplementary material The online version of this chapter (https://doi.org/10.1007/978-3-030-58621-8_32) contains supplementary material, which is available to authorized users.

© Springer Nature Switzerland AG 2020
A. Vedaldi et al. (Eds.): ECCV 2020, LNCS 12356, pp. 545–560, 2020.
https://doi.org/10.1007/978-3-030-58621-8_32

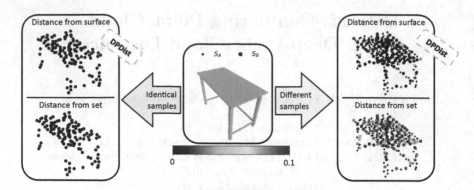

Fig. 1. Distance to surface vs. distance to sampled point cloud. Every figure contains two samples of the same object (a table). The samples can be identical (left) or different (right). The distance of every point from one point cloud (disks) to the surface estimated from the other point cloud (+) is given as a colored code in the top two figure. One can see that even for the different sampling most of the distances are still small. The distance from every point in one point cloud (disks) to the nearest point in the other point cloud (+) is presented as a color code in the bottom figure. We can see that different sampling produce significantly larger distances. (Color figure online)

clouds is difficult for two main reasons. First, because they are not a function on a grid, point clouds cannot be compared using a common metric (such as Euclidean metric). Second, when comparing point clouds sampled from a 3D surface, we usually want to compare their underlying surfaces and not the given sample points. Common methods for point cloud comparisons, such as Chamfer distance and Earth mover's distance (see more in Sect. 2) compare the given clouds directly, rely on the unstable correspondence process, and are sensitive to sampling.

We propose a method for comparing point clouds that measures the distance between the surfaces that they were sampled on (see Fig. 1). We use a deep learning network to estimate the distance function from a point to an underlying continuous surface. By using a vector representation of point clouds, the 3D modified Fisher vector (3DmFV) [3], the network can work directly with point clouds, despite their irregular structure. An additional advantage of the particular 3DmFV representation is its grid structure, which enables the network to process a local surface representation rather than calculating a representation at every location. Working on local representation reduces the complexity of the surface, enabling efficient and effective learning, which generalizes well between object categories.

We test the robustness of the proposed approach to sampling, on challenging tasks such as detecting small transformation and comparing similar objects. We incorporate it into the learning process of a registration network. We compare it to several, more traditional, distance measures and show its advantage, which is significant, especially when the task is harder and the point cloud is sparser.

The main contributions of this paper are as follow:

- DPDist: A new distance measure between point clouds, that operates directly on raw unstructured points, but measures distance to the underlying continuous surface.
- An algorithm that implements the DPDist using local implicit representation, in an effective and efficient way.
- An improved variant of the PCRNet registration algorithm [20] that applies the proposed DPDist in a training loss function.

2   Related Work

2.1   Point Cloud Distance

Point clouds may represent raw data, but in the context of 3D sensors, they represent the surfaces on which they are sampled. As such, the distance between the point clouds, should preferably refers to the distance between the sampled continuous surfaces and be robust to sampling and noise.

Most distances between points sets are based on an underlying metric, assigning a distance value $\|a-b\|$ to every two points a, b. Like the majority of previous work, we use a Euclidean metric.

Consider two point clouds $S_A, S_B \subseteq \mathbb{R}^3$, with N_A, N_B points in each cloud, respectively. An early method for comparing point sets is the Hausdorff distance $(\mathcal{D}_H())$, which builds on the minimum distance from a point to a set

$$d(x, y) = \|x - y\|_2 \tag{1}$$

$$D(x, S) = \min_{y \in S} d(x, y) \tag{2}$$

and calculates a symmetric max min distance [8].

$$\mathcal{D}_H(S_A, S_B) = \max\{\max_{a \in S_A} D(a, S_B), \max_{b \in S_B} D(b, S_A)\} \tag{3}$$

The Chamfer distance $(\mathcal{D}_{CD}())$ (Eq. 4), and partial Hausdorff $(\mathcal{D}_{PH(f)}())$ (Eq. 5) distances are two of its variants, based on averaging (instead of taking the maximum) and on robustly ignoring a fraction $(1 - f)$ of the points that are far from the other object, respectively

$$\mathcal{D}_{CD}(S_A, S_B) = \frac{1}{N_A} \sum_{a \in S_A} \min_{y \in S_B} d(a, y)^2 + \frac{1}{N_B} \sum_{b \in S_B} \min_{y \in S_A} d(b, y)^2 \tag{4}$$

$$\mathcal{D}_{PH(f)}(S_A, S_B) = \max\{T | \frac{|\{a|D(a, S_B) \leq T\}|}{|S_A|} < f, \frac{|\{b|D(b, S_A) \leq T\}|}{|S_B|} < f\} \tag{5}$$

where $a \in S_A, b \in S_B$.

Unlike the Hausdorff distance and its variant, which rely on finding the nearest neighbor to every point, the Earth mover's distance ($\mathcal{D}_{EMD}$) (Eq. 6), also known as the Wasserstein distance, is based on finding the 1-1 correspondence (or bijection, ξ) between the two point sets, so that the sum of distances between corresponding points is minimal:

$$\mathcal{D}_{EMD}(S_A, S_B) = \min_{\xi: S_A \to S_B} \sum_{a \in S_A} \|a - \xi(a)\|_2 \tag{6}$$

Clearly, this distance measure is limited to point clouds with the same point count. Other measures includes Geodesic distances, which provide deformation insensitive measures [4], and measures relying on the distance between point cloud vector descriptors, such as PointNet [16], VoxNet [13], or 3DmFV [3].

Here we focus on the direct distances, and specifically on CD and EMD like measures. These are the leading approaches for assessing subtle differences in point clouds, and are used as evaluation distances and as loss functions for the training of neural networks [1,6,7,11,20,24,25]. They highly depend on point correspondence, which makes them sensitive to sampling and noise. In this work, we propose a novel method for comparing 3D point clouds where the measured distance is between the points from the first point cloud to the surface represented by the second (without constructing a mesh) and the other way round. As a distance, it can replace the aforementioned distances as a loss function for various neural networks that require point cloud comparison.

2.2   Deep Learning on 3D Point Clouds

Processing a point cloud with deep learning is challenging because point cloud representation is unordered, unstructured, and has unknown number of points, rendering it an unnatural input for deep networks. Several methods were proposed to overcome these challenges. One approach quantizes the points into a voxels grid. This approach allows to directly use 3D convolutional neural networks (CNNs) [13] but induces quantization. Another approach encodes the point cloud with a kd-tree and uses it to learn shared weights for nodes in the tree [10]. A recent popular network for processing point clouds is PointNet [16]. It computes features separately for each input point, and then extracts a global feature using a permutation-independent (symmetric) function (e.g., max/avg). An effective variant, PointNet++ [17], applies the PointNet encoding locally and hierarchically.

The recent 3D modified Fisher Vector (3DmFV) approach [3] builds on the well known Fisher Vectors [19] which, in turn, are based on the Fisher Kernel (FK) principles [9]. 3DmFV represents each point's deviations from a mixture of Gaussians, lying on a regular grid, and then applies symmetric functions to get a global generalized Fisher Vector representation [19] that integrates into a 3D CNN. The proposed method uses the regular grid structure of the 3DmFV representation, to extract multiple local patches without any recalculation.

2.3 Deep Implicit Function

Two recent works that are the closest to the proposed method rely on continuous implicit surface representations. Occupancy Networks [14] learn an indicator function between the inner part and the outer part of a model, where the boundary between these parts is an implicit function representing the continuous boundary. DeepSDF [15] learns a signed distance function to the surface, and estimates the boundary as the zero surface. Both methods rely on a global presentation of the point cloud and the estimated surface. Occupancy Networks [14] rely on PointNet, and DeepSDF specifies each model's surface by an decoder training process, which provides both the weights and the latent layer values (without using an encoder).

Unlike DeepSDF, our proposed algorithm is able to transform a point cloud into underlying surface representation quickly and can therefore use the distance-to-surface principle to calculate the distance between point clouds. Unlike Occupancy Networks, our proposed method calculates the distance to the surface and can therefore estimate whether two point clouds, sampled from surfaces, are indeed sampled from the same surface. In addition, unlike the two aforementioned methods, our approach can extract and use effective local representation, which makes both the learning and the estimate, accurate and effective.

2.4 Registration

Registration between point clouds is a fundamental task. A popular, classic (non-learnable) algorithm is Iterative Closest Point (ICP) [18], which relies on finding closest point pairs. To overcome ICP's convergence to local minima, recent ICP variants [23] start by finding a good initialization at some computational expense.

Recently, deep-learning-based registration methods have been introduced. Some methods [2, 21] learn to regress the transformation parameters. The PCR-Net [20], however, learns by comparing point clouds. The method aims to find the transformation that converts the measured point cloud (source) to the model (template). At every iteration, it maintains a temporary transformed source, calculated with a currently available transformation. A pose estimation network gets PointNet representations of this temporary cloud and of the template cloud, and provides a new, improved, transformation (represented as translation parameters and quaternion rotation parameters). The network is trained by a loss based on the Chamfer distance or the EMD between the temporary cloud and the template cloud. In this paper, we test the proposed new distance DPDist, as a loss function for training the PCRNet.

3 The Deep Distance

We now present the main contribution of this work: a new method for fine comparison of point clouds that we call the Deep Point Cloud Distance (DPDist).

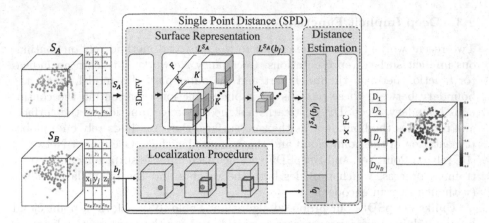

Fig. 2. SPD illustration. Two point clouds (S_A, S_B) are input to a network that estimates the distance between each query point from S_B to the surface represented by S_A. SPD uses 3DmFV to extract surface representation L^{S_A} (4D tensor) from S_A. It then applies a localization procedure for each query point b_j to find its closest grid point (the grid is distributed uniformly: K^3). Given the grid point, it extracts the corresponding 4D sub tensor $L^{S_A}(b_j)$. The SPD processes each query point b_j and its matched local representation to produce its distance from the surface $D_j = D_{SPD}(b_j, S_A)$.

The DPDist method is based on estimating the distances of points from one cloud to the underlying continuous surface corresponding to the other point cloud.

Let $A, B \in \mathbb{R}^3$ be two continuous surfaces, and $S_A = \{a_i\}_{i=1}^{N_A}, S_B = \{b_j\}_{j=1}^{N_B}$ two sets of points sampled from them. Our aim is to find the distance from each point $b_j \in S_B$ to the closest point in A (and vice versa). Formally, this distance is

$$D(b, A) = \min_{y \in A} d(b, y), \tag{7}$$

where $D()$ is the distance from a point to a surface and $d()$ is any distance between points in $\mathbb{R}^3$ (here we shall use the Euclidean distance). The surface A is not available, and hence we propose to use the approximation

$$\hat{D}(b, A) = \phi(b, S_A), \tag{8}$$

where ϕ is a learned distance function depending on the samples of A.

In principle, we can design a network that accepts the coordinates of the point b, and some representation of the cloud of points S_A (e.g., PointNet [16] or 3DmFV [3]). This network should learn to provide the distance. We found, however that due to the large variation of objects, learning the distance function is hard. In particular, even with a complex network and a lot of examples, the results we obtained were not very good. See [14,15] where a similar task of estimating an implicit function by using at least 7 fully connected layers to express the complicated geometric features.

3.1 Local Representation

Because of the difficulty of learning the distance function, we prefer to model the surface A by parts, so that every part is less geometrically complex. In principle, we could derive such a piecewise model by training a different distance estimator for every spatial region. Instead, we choose the more elegant approach of using the 3D modified Fisher Vector (3DmFV) representation.

The 3DmFV representation is preferable for two reasons. First, it provides better performance than some other representations (and in particular, it is better than the PointNet; see [3]). In addition, it uses a grid structure therefore allowing us to specify a local, partial representation by choosing a subgrid.

First, we calculate the global representation from the cloud S_A. To that end, we decide on the representation grid size K, and specify the 3D grid of size $K \times K \times K$. This grid specifies a mixture of Gaussians, one for each grid point. The representation itself results from calculating the derivative of the Gaussian with respect to its parameters, and taking maximum, minimum, and average statistics over these derivatives; see [3,19]. Overall, we calculate F statistics for each Gaussian and concatenate them to a global representation $L^{S_A} \in \mathbb{R}^{K \times K \times K \times F}$. This representation is calculated once for every point cloud and the partial representations are simply cut from it.

We extract the local representation from this global representation as follows: for each query point b_j from the point cloud S_B, find its nearest grid point and specify a subgrid of size k, centered at this grid point (k is always odd). Then, the entries of L^{S_A} corresponding to the subgrid are extracted and concatenated to give the local representation, denoted $L^{S_A}(b_j) \in \mathbb{R}^{k \times k \times k \times F}$. Note that the local representation is a representation of the point cloud S_A, in a region determined by b_j.

3.2 Point-to-Implicit-Surface Distance Estimation

To estimate the distance from a point to the underlying surface associated with a point cloud, we use a learned function. We chose a multilayer, fully connected neural network (FC). The input to this network is the coordinate vector specifying the point b_j, concatenated to the local representation $L^{S_A}(b_j)$. The neural network, denoted ψ, learns to estimate the distance $\psi(b_j, L^{S_A}(b_j))$ between the point and the corresponding local representation. The full process, which gets the point b_j and the point cloud S_A (as input), is carried out as follows. The process first extracts the local representation corresponding to b_j from the 3DmFV vector; concatenates it to the point coordinates; and processes them using the network ψ. We denote it: SPD (Single point distance). That is

$$\hat{D}(b_j, A) = D_{SPD}(b_j, S_A) = \psi(b_j, L^{S_A}(b_j)) \tag{9}$$

The detailed description of the SPD neural network is presented in Fig. 2. We found indeed that learning the distance function of a local representation is faster, more accurate, and can generalize better for unseen objects. Furthermore,

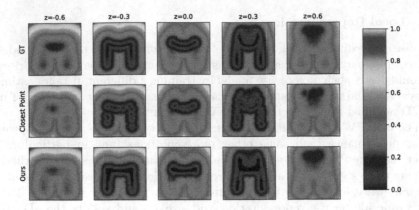

Fig. 3. Three distance map to the "Chair" CAD model: The ground truth distance from the CAD model (upper row), the distance to the 'closest point' in a 128 point cloud, sampled from the CAD model (middle row), and the proposed SPD network distance from the surface specified by the same point cloud (bottom row). For each 3D distance map, several XY slices are shown. Clearly, the SPD is smoother than the distance to the point cloud, and approximates the ground truth better.

representing a surface A directly from a set of points S_A using the proposed approach yields sampling invariance (sparse or non-uniform point clouds may represent the same surface). Figure 3, illustrates the distance maps of SPD and compares it to the distance to the cloud, for a dense set of query points that lie on a regular XY grid in different spatial depths ($z = \{-0.6, -0.3, 0, 0.3, 0.6\}$). The advantage of the SPD for this low sampling data ($N = 128$) is clear.

3.3   Estimating the Distance Between Point Clouds

The point to surface distance estimates, developed above may be easily applied to measuring the distance between two point clouds. The average distance between all points in one cloud, S_B, to the underlying surface corresponding to the other point cloud is a straightforward choice. Here we often use a symmetrized version to yield the DPDist distance between two point clouds.

$$\mathcal{D}_{DPDist}(A, B) = \frac{1}{N_A} \sum_{i=1}^{N_A} D_{SPD}(a_i, S_B) + \frac{1}{N_B} \sum_{j=1}^{N_B} D_{SPD}(b_j, S_A) \qquad (10)$$

3.4   DPDist as Loss Function for Training Neural Networks

After DPDist is fully trained, it can be used as a distance estimator building block that can be connected as a loss function to various tasks. Generally, its weights can either be frozen or adjusted to the desired task, but as a loss function, the weights remain constant. It takes two point clouds S_A, S_B as input and outputs the estimated distance between the underlying represented surfaces.

DPDist uses a neural network, and is therefore differentiable and can be easily integrated as a loss function for training in different tasks.

4   Experiments

We start by conducting a thorough analysis of the method's robustness to sampling. We would like to emphasize that for tasks involving coarse differences between point clouds, such as categorization and coarse registration, distances that are based directly on the point clouds (e.g., Chamfer distance) or even distances based on vector representations (e.g., distance between 3DmFVs) should suffice. The distance suggested here can detect small, subtle differences. Therefore, we chose to experiment with the following challenging tasks: object instance identification, detecting small translations, and detecting small rotations. We compare our method with the following methods: Hausdorff (H), Chamfer distance (CD), partial Hausdorff (PHx), Earth mover's distance (EMD), and 3DmFV representation (we have shortened PH(f) to PHx, where f = 0.x). We then evaluate our method's effectiveness as a loss function in training a registration network. We provide additional point cloud generation experiments in the supplemental material. They show that DPDist's main advantage of sampling invariance can introduces significant challenges for this task. Additionally, we report in the supplemental results on real-world data that align with our findings on CAD models.

4.1   Setup Details

Dataset. The experiments were conducted on the ModelNet40 dataset [22]. It contains 12311 CAD models from 40 object categories. We normalize each shape to 80% of the unit sphere. We split the data into train/test as in [16].

Training Data. For training the SPD network, we sample points from the CAD models, keeping their distance from the surface. At each training step, we sample two point clouds. The first, representing the surface, contains N points sampled from the surface. The second set is the query set. It contains $0.5N$ points sampled from the surface, $0.25N$ points sampled uniformly in the region closer than 0.1 to the surface, and $0.25N$ points uniformly sampled from the unit cube. Adding the second type of query points increases the density of the query points near the surface and enhances the learning of small geometric details; see [15]).

Training Details. Figure 2 describes the overall architecture of SPD: We set the 3DmFV layer parameters to $K = 8$, $k = 5$, and the Gaussian's sigma to 0.125. Additional ablation study of the influence of 3DmFV parameters is provided in the supplemental material. To evaluate the parameters we used marching cubes [12] for surface reconstruction as done in [14,15]. The neural network is composed of three fully connected layers with a size of 1024, each processed with

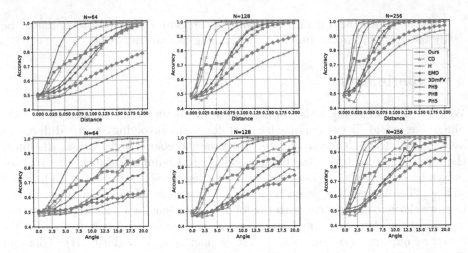

Fig. 4. Detecting translation (top) and rotations (bottom) using DPDist and other point cloud distances, for different point cloud densities $N = 64, 128, 256$.

RELU activation. During training, we minimized the mean L_1 distance between the network output (our prediction) and the real distance (GT) associated with each query point, i.e. $Loss = \frac{1}{|S_B|} \sum_{x_i \in S_B} ||D_{SPD}(x_i, S_A) - GT(x_i)||_1$.

We train separate networks for each point cloud input size $N = 32, 64, 128, 256, 512, 1024$. Our early experiments showed that we could use a network trained for an input size of $N = 512$ for all different size inputs; however, training each network for a specific input size yields higher accuracy.

For the following experiments, we trained the SPD using the train-set of the "Chair" category, which contains 889 CAD models. For specific training parameters and further details, please refer to the supplemental material.

Evaluation Data Sampling. To obtain the data for evaluation, we first sample N_C points uniformly from each CAD model using Farthest Point Sampling (FPS) [5]. We then sample two disjoint sets S_A, S_B with N_A, N_B. In most experiments, $N_A = N_B = N_C/2 = N$.

4.2 Robustness to Sampling

Discriminating between very different objects from dense samples of their surfaces is easy, and performance is highly independent of the actual sampling. Harder tasks and sparser sampling are more challenging. We consider here three discrimination tasks associated with similar objects, perform them using different distances between their point cloud samples, and examine their robustness to sampling.

We conducted the experiments on the "Chair" category test-set from Modelnet40 with a total of 100 chairs.

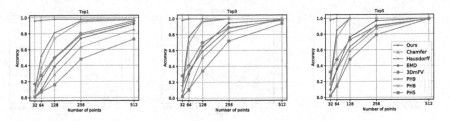

Fig. 5. Instance identification test - comparing top 1, 3, and 5 accuracy in nearest neighbor identification robustness test. The DPDist is the clear winner.

Translation Detection Test: The task here is to discriminate between a stationary object and a moving object. Thus, in every test we consider three point clouds: S_A - a sampling of the object in its original position, S_B - a sampling of the object after it was translated (see below), and S_C - another sampling of the same object without movement. We can discriminate between the stationary object and the moving object, when the following distance inequality $\mathcal{D}(S_C, S_A) < \mathcal{D}(S_B, S_A)$ holds true. We consider translation along the 26 directions $(\theta_1, \theta_2, \theta_3), \theta_{1,2,3} \in \{0, 1, -1\}$, with magnitude uniformly sampled in $[0, 0.2]$. The accuracy is defined separately for each translation magnitude, as a fraction of successful tests. The results are shown in Fig. 4 (top). As expected, for large translations and high sampling density, almost all methods succeed. For small densities and translations, some distances succeed better and the proposed method is best by a large margin. Remarkably, the partial Hausdorff variants succeed better than CD and EMD.

Rotation Detection Test: This time, the task is to discriminate between a stationary object and a rotated object. The rotation angle magnitude is uniformly sampled in $[0, 20]$ (deg) around the 26 directions specified in the last section. Figure 4 (bottom) shows that DPDist performs significantly better than the other distances in the presence of small rotations and densities.

Identification Test: We evaluate the sampling robustness of different point cloud distances by testing their ability to discriminate between objects from the same category. Given two samples of the same CAD object S_A and S'_A, we test if the two samples of the same object are closer to each other than to samples of any other objects from the same category. For each point cloud S_A, we sort its distances from S'_A and from the point cloud of all other objects. The identification is considered "Top m" successful if S'_A is in the "m" closest point clouds to S_A (out of 100). The success rate is the fraction of objects for which the identification is "Top m" successful. Figure 5 shows that the proposed DPDist method is robust and works much better than other methods. Remarkably, the partial Hausdorff distance is the next best method, and it easily overcomes the more standard CD and EMD.

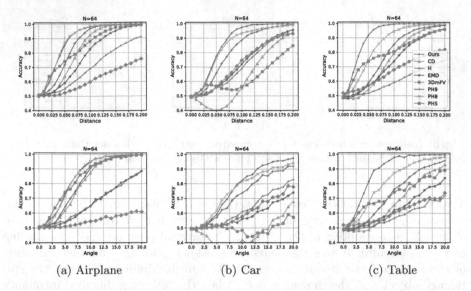

Fig. 6. Translation (top) and rotation (bottom) detection experiments evaluated on (a) "Airplane" (b) "Car", and (c) "Table" as before. Remarkably, DPDist is still the best method, although the SPD was trained over another category ("Chair") and not over these categories. The generalization ability is a result of using local surface modelling.

4.3 Local Representation Learning Generalization

A major advantage of the proposed method is its ability to learn local representations, which makes the implicit surface estimation effective and computationally efficient.

Objects are different from each other but often share local parts. That is, many objects contain planes, corners, curved surfaces, etc. We therefore hypothesize that training the SPD network over local representations of just one reasonably rich object category (e.g., "Chair") makes the learning universal, because the category includes enough variations of small patches to generalize for unseen categories. Therefore, although we train our SPD only on the "Chair" category, we expect it to generalize to other categories.

We test our hypothesis by performing the translation and rotation tests over "Airplane", "Car", and "Table" categories on a network that was trained only on "Chair". Figure 6 shows that the DPDist is still the most discriminative method and more robust to sampling than the other tested methods.

4.4 Applying Point Cloud Distances to Registration Learning

In this section, we use an existing registration network, and show its performance improvement when using DPDist as its loss function. We choose the recent PCR-Net [20] (see also Sect. 2.4) since it compares point clouds in its training loss function. Figure 7 illustrates the challenges of using a correspondence-based distance vs. the proposed distance. We compare DPDist to CD and EMD distances as the loss function for training the iterative PCRNet.

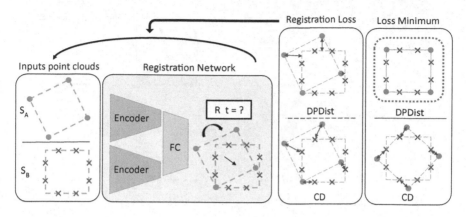

Fig. 7. Learning registration using DPDist vs. CD distances as loss function. Two point cloud (S_A, S_B) are input to a network that regresses the transformation between them. The chosen loss highly effects the output transformation as the DPDist loss minimum is consistent with zero distance, while Chamfer loss results in nonzero distance for different sampling of the same object.

We used the PCRNet as described in [20], but found that limiting the output rotation angle to $45°$ improves the stability when training on small point clouds. In the context of iterative registration, this modification does not limit the registration range. Using the "Chair" category and following [20], we randomly generate 5070 different transformations for training and other 5070 transformations for testing. The transformations include a rotation between $[-45, 45]°$ about one random direction and a translation between $[-0.1, 0.1]$ in another random direction. The evaluation metric at test time is the "success ratio" [20]: the percentage of point clouds with a transformation error under a given 'max error' threshold.

To demonstrate the DPDist robustness, we conduct the following experiment: Given a CAD model input, we generate two point clouds by sampling $2N$ points using FPS, and randomly divide it into two disjoint clouds S'_A and S_B of size N. We transform S'_A to S_A and store the transformation matrix as ground truth. We then train PCRNet using different distance losses. Figure 8 shows the results for training PCRNet using CD, EMD, and the proposed DPDist for different sample sizes. The significant advantage of DPDist is clear.

The relative inaccuracy of the CD and EMD distances is due to the different samples. To demonstrate that this is indeed the reason, we added another success ratio curve with 128 points, but this time, $S'_A = S_B$. That is, the two point clouds (before transformation) are identical. We can see in Fig. 8d that this time, the performance with CD and EMD is nearly perfect.

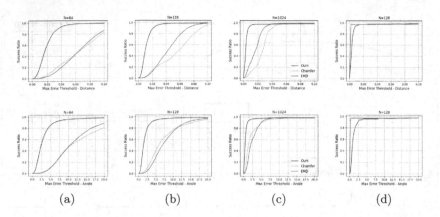

Fig. 8. Results for training PCRNet using CD, EMD, and the proposed DPDist over a single category ("Chair"). We show the success ratio vs. max error threshold for translation (top) and rotation (bottom) of the following input point cloud sizes: $N = 64, 128, 1024$ (a, b, c). DPDist consistently achieves high performance for various densities. (d) shows the case of two identical samples $S'_A = S_B$, where CD and EMD gets nearly perfect results thanks to the high correspondence between the clouds' samples.

5   Conclusions

Comparing between point clouds is a fundamental data analysis task. For point clouds, obtained by sampling some underlying continuous object surfaces, the actual hidden goal is to compare between these objects. Most methods, however, rely on distances between the raw points and are therefore sensitive to the uncertainties involved in the sampling process.

The method proposed here, DPDist, estimates the distances of points from one cloud to the underlying continuous surface corresponding to the other point cloud and the other way around. DPDist is fast and effective. It is more accurate than the commonly used Chamfer distance and Earth mover's distance methods. Its advantage is significant especially for difficult tasks, such as discriminating between similar objects, and for sparse point clouds. For example, using a DPDist dependent loss function, we were able to train a registration network so that it provides good results with point clouds of size 64.

Unlike other methods, we provide a fast process for generating the surface representation. This representation is local but is created efficiently as part of a global representation. Being local enables us to represent fine surface details with a moderately deep network, and makes the learned network universal: learning the local descriptions associated with one category is general enough to represent local descriptions of objects from other categories.

Acknowledgments. This research was supported by the Israel Science Foundation, and by the Israeli Ministry of Science and Technology.

References

1. Achlioptas, P., Diamanti, O., Mitliagkas, I., Guibas, L.: Learning representations and generative models for 3D point clouds. arXiv preprint arXiv:1707.02392 (2017)
2. Aoki, Y., Goforth, H., Srivatsan, R.A., Lucey, S.: PointNetLK: robust & efficient point cloud registration using PointNet. In: Proceedings of the IEEE Conference on Computer Vision and Pattern Recognition, pp. 7163–7172 (2019)
3. Ben-Shabat, Y., Lindenbaum, M., Fischer, A.: 3DmFV: three-dimensional point cloud classification in real-time using convolutional neural networks. IEEE Robot. Autom. Lett. **3**(4), 3145–3152 (2018)
4. Bronstein, A.M., Bronstein, M.M., Guibas, L.J., Ovsjanikov, M.: Shape google: geometric words and expressions for invariant shape retrieval. ACM Trans. Graph. (TOG) **30**(1), 1–20 (2011)
5. Eldar, Y., Lindenbaum, M., Porat, M., Zeevi, Y.Y.: The farthest point strategy for progressive image sampling. IEEE Trans. Image Process. **6**(9), 1305–1315 (1997)
6. Fan, H., Su, H., Guibas, L.J.: A point set generation network for 3D object reconstruction from a single image. In: Proceedings of the IEEE Conference on Computer Vision and Pattern Recognition, pp. 605–613 (2017)
7. Groueix, T., Fisher, M., Kim, V., Russell, B., Aubry, M.: AtlasNet: a Papier-Mâché approach to learning 3D surface generation. In: CVPR 2018 (2018)
8. Huttenlocher, D.P., Klanderman, G.A., Rucklidge, W.J.: Comparing images using the Hausdorff distance. IEEE Trans. Pattern Anal. Mach. Intell. **15**(9), 850–863 (1993)
9. Jaakkola, T., Haussler, D.: Exploiting generative models in discriminative classifiers. In: Advances in Neural Information Processing Systems, pp. 487–493 (1999)
10. Klokov, R., Lempitsky, V.: Escape from cells: deep kd-networks for the recognition of 3D point cloud models. In: 2017 IEEE International Conference on Computer Vision (ICCV), pp. 863–872. IEEE (2017)
11. Li, J., Chen, B.M., Hee Lee, G.: SO-Net: self-organizing network for point cloud analysis. In: Proceedings of the IEEE Conference on Computer Vision and Pattern Recognition (CVPR), June 2018
12. Lorensen, W.E., Cline, H.E.: Marching cubes: a high resolution 3D surface construction algorithm. ACM SIGGRAPH Comput. Graph. **21**(4), 163–169 (1987)
13. Maturana, D., Scherer, S.: VoxNet: a 3D convolutional neural network for real-time object recognition. In: 2015 IEEE/RSJ International Conference on Intelligent Robots and Systems (IROS), pp. 922–928. IEEE (2015)
14. Mescheder, L., Oechsle, M., Niemeyer, M., Nowozin, S., Geiger, A.: Occupancy networks: learning 3D reconstruction in function space. In: Proceedings of the IEEE Conference on Computer Vision and Pattern Recognition, pp. 4460–4470 (2019)
15. Park, J.J., Florence, P., Straub, J., Newcombe, R., Lovegrove, S.: DeepSDF: learning continuous signed distance functions for shape representation. In: Proceedings of the IEEE Conference on Computer Vision and Pattern Recognition, pp. 165–174 (2019)
16. Qi, C.R., Su, H., Mo, K., Guibas, L.J.: PointNet: deep learning on point sets for 3D classification and segmentation. In: Proceedings of the IEEE Conference on Computer Vision and Pattern Recognition, pp. 652–660 (2017)
17. Qi, C.R., Yi, L., Su, H., Guibas, L.J.: PointNet++: deep hierarchical feature learning on point sets in a metric space. In: Advances in Neural Information Processing Systems, pp. 5099–5108 (2017)

18. Rusinkiewicz, S., Levoy, M.: Efficient variants of the ICP algorithm. In: Proceedings Third International Conference on 3-D Digital Imaging and Modeling, pp. 145–152. IEEE (2001)
19. Sánchez, J., Perronnin, F., Mensink, T., Verbeek, J.: Image classification with the fisher vector: theory and practice. Int. J. Comput. Vis. **105**(3), 222–245 (2013). https://doi.org/10.1007/s11263-013-0636-x
20. Sarode, V., et al.: PCRNet: point cloud registration network using PointNet encoding. arXiv preprint arXiv:1908.07906 (2019)
21. Wang, Y., Solomon, J.M.: Deep closest point: learning representations for point cloud registration. In: Proceedings of the IEEE International Conference on Computer Vision, pp. 3523–3532 (2019)
22. Wu, Z., et al.: 3D ShapeNets: a deep representation for volumetric shapes. In: The IEEE Conference on Computer Vision and Pattern Recognition (CVPR), June 2015
23. Yang, J., Li, H., Campbell, D., Jia, Y.: Go-ICP: a globally optimal solution to 3D ICP point-set registration. IEEE Trans. Pattern Anal. Mach. Intell. **38**(11), 2241–2254 (2015)
24. Yang, Y., Feng, C., Shen, Y., Tian, D.: FoldingNet: point cloud auto-encoder via deep grid deformation. In: 31st Meeting of the IEEE/CVF Conference on Computer Vision and Pattern Recognition, CVPR 2018, pp. 206–215. IEEE Computer Society (2018)
25. Zhao, Y., Birdal, T., Deng, H., Tombari, F.: 3D point capsule networks. In: Proceedings of the IEEE/CVF Conference on Computer Vision and Pattern Recognition (CVPR), June 2019

Bi-directional Cross-Modality Feature Propagation with Separation-and-Aggregation Gate for RGB-D Semantic Segmentation

Xiaokang Chen[1]([✉]) [iD], Kwan-Yee Lin[2] [iD], Jingbo Wang[3] [iD], Wayne Wu[2] [iD],
Chen Qian[2] [iD], Hongsheng Li[3] [iD], and Gang Zeng[1] [iD]

[1] Key Laboratory of Machine Perception (MOE), School of EECS,
Peking University, Beijing, China
pkucxk@pku.edu.cn
[2] SenseTime Research, Tai Po, Hong Kong
[3] The Chinese University of Hong Kong, Sha Tin, Hong Kong

Abstract. Depth information has proven to be a useful cue in the semantic segmentation of RGB-D images for providing a geometric counterpart to the RGB representation. Most existing works simply assume that depth measurements are accurate and well-aligned with the RGB pixels and models the problem as a cross-modal feature fusion to obtain better feature representations to achieve more accurate segmentation. This, however, may not lead to satisfactory results as actual depth data are generally noisy, which might worsen the accuracy as the networks go deeper.

In this paper, we propose a unified and efficient Cross-modality Guided Encoder to not only effectively recalibrate RGB feature responses, but also to distill accurate depth information via multiple stages and aggregate the two recalibrated representations alternatively. The key of the proposed architecture is a novel Separation-and-Aggregation Gating operation that jointly filters and recalibrates both representations before cross-modality aggregation. Meanwhile, a Bi-direction Multi-step Propagation strategy is introduced, on the one hand, to help to propagate and fuse information between the two modalities, and on the other hand, to preserve their specificity along the long-term propagation process. Besides, our proposed encoder can be easily injected into the previous encoder-decoder structures to boost their performance on RGB-D semantic segmentation. Our model outperforms state-of-the-arts consistently on both in-door and out-door challenging datasets (Code of this work is available at https://charlescxk.github.io/).

Keywords: RGB-D semantic segmentation · Cross-modality feature propagation

---

Electronic supplementary material The online version of this chapter (https://doi.org/10.1007/978-3-030-58621-8_33) contains supplementary material, which is available to authorized users.

A. Vedaldi et al. (Eds.): ECCV 2020, LNCS 12356, pp. 561–577, 2020.
https://doi.org/10.1007/978-3-030-58621-8_33

1   Introduction

Semantic segmentation, which aims at assigning each pixel with different semantic labels, is a long-standing task. Besides exploiting various contextual information from the visual cues [5,11,12,14,24,41], depth data have recently been utilized as supplementary information to RGB data to achieve improved segmentation accuracy [4,6,15,19,22,25,31,42]. Depth data naturally complements RGB signals by providing the 3D geometry to 2D visual information, which is robust to illumination changes and helps better distinguishing various objects.

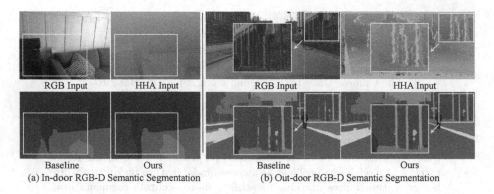

(a) In-door RGB-D Semantic Segmentation   (b) Out-door RGB-D Semantic Segmentation

Fig. 1. (a) RGB-D baseline, which is designed with a habitual cross-modality fusion schema, results in inaccurate classification on the area that exists substantial variations between RGB and Depth modalities. (b) The depth measurements in out-door environments are noisy. Without proposed modules, the results will degrade dramatically

Although significant advances have been achieved in RGB semantic segmentation, directly feeding the complementary depth data into existing RGB semantic segmentation frameworks [24] or simply ensemble results of two modalities [6] might lead to inferior performance. The key challenges lie in two aspects. (1) *The substantial variations between RGB and Depth modalities.* RGB and depth data show different characteristics. How to effectively identify their differences and unify the two types of information into an efficient representation for semantic segmentation is still an open problem. (2) *The uncertainty of depth measurements.* Depth data provided with existing benchmarks are mainly captured by Time-of-Flight or structured light cameras, such as Kinect, AsusXtion and RealSense *etc.* The depth measurements are generally noisy due to different object materials and limited distance measurement range. The noise is more apparent for out-door scenes and results in undesirable segmentation, as shown in Fig. 1.

· Most existing RGB-D based methods mainly focus on tackling the first challenge. Standard practice is to use the depth data[1] as another input and adopt Fully Convolutional Network (FCN)-like architectures with feature fusion schemas, *e.g.*, convolution and modality-based affinity *etc.*, to fuse the features of two modalities [6,17,25,36]. The fused feature is then used to recalibrate the subsequent RGB feature responses or predicted results. Although these methods provide plausible solutions to unify the two types of information, the assumption of the input depth data being accurate and well-aligned with RGB signals might not be true, making these methods sensitive to in-the-wild samples. Moreover, how to ensure that the network fully utilizes information from both modalities remains an open problem. Recently, some works [37,42] attempt to tackle the second challenge by diminishing the network's sensitivity to the quality of depth measurements. Instead of utilizing depth data as an extra input, they propose to distill the depth features via multi-task learning and regard depth data as extra supervision for training. Specifically, [37] introduces a two-stage framework, which first predicts several intermediate tasks including depth estimation and then uses the outputs of these intermediate tasks as the multi-modal input to final tasks. [42] proposes a pattern-affinitive propagation with jointly predicting depth, surface normal and semantic segmentation to capture correlative information between modalities. We argue that there exists an inherent inefficacy in such design, *i.e.* the interaction and correlation of RGB and depth information are only implicitly modeled. The complementarity of the two types of data for semantic segmentation was not well studied in this way.

Motivated by the above observations, we propose to tackle both two challenges in a simple yet effective framework by introducing a novel cross-modality guided encoder to FCN-like RGB-D semantic segmentation backbones. The key idea of the proposed framework is to leverage both channel-wise and spatial-wise correlation of the two modalities to firstly squeeze the exceptional feature responses of depth, which effectively suppresses feature responses from the low-quality depth measurements, and then use the suppressed depth representations to refine RGB features. In practice, we devise the steps bi-directionally due to the in-door RGB sources also contain noisy features. In contrast to depth data, the RGB noisy features are usually caused by similar appearance of different neighboring objects. We denote the above process as *depth-feature recalibration* and *RGB-feature recalibration*, respectively. We therefore introduce a new gate unit, namely the *Separation-and-Aggregation Gate (SA-Gate)*, to improve the quality of the multi-modality representation by encouraging the network to recalibrate and spotlight the modality-specific feature of each modality first, and then selectively aggregate the informative features from both modalities for the final segmentation. To effectively take advantage of the differences of features between the two modalities, we further introduce the *Bi-direction Multi-step*

---

[1] Raw depth map or its encoded representation–HHA map, which includes horizontal disparity, height above ground and norm angle. For more detail about HHA, please refer to [13].

Propagation (BMP) that encourages the two streams to better preserve their specificity during the information interaction process in the encoder stage.

Our contributions can be summarized into three-fold:

- We propose a novel bi-directional cross-modality guided encoder for RGB-D semantic segmentation. With the proposed *SA-Gate* and *BMP* modules, we could effectively diminish the influence of noisy depth measurements, and also allow incorporating sufficiently complementary information to form discriminative representations for segmentation.
- Comprehensive evaluation on the NYUD V2 dataset shows significant improvements by our approach when integrated into state-of-the-art RGB semantic segmentation networks, which demonstrate the generalization of our encoder as a plug-and-play module.
- The proposed method achieves state-of-the-art performances on both in-door and challenging out-door semantic segmentation datasets.

2   Related Work

2.1   RGB-D Semantic Segmentation

With the development of depth sensors, recently there is a surge of interest in leveraging depth data as a geometry augmentation for RGB semantic segmentation task, dubbed as RGB-D semantic segmentation [3,20,23,25,31,42]. According to specific functionality of depth information suited in different architectures, current RGB-D based methods could be roughly divided into two categories.

Most of the works treat depth data as an additional input source to recalibrate the RGB feature responses either implicitly or explicitly. Long *et al.* [24] shows simply averaging final score maps of RGB and D modalities helps enforce the inter-object discrimination in the in-door setting. Li *et al.* [22] utilize the LSTM layers to selectively fuse the feature from the two modalities input. With a similar target, [6] proposes locality-sensitive deconvolution networks along with a gated fusion module. Several recent works [9,17,30] extend the RGB feature recalibration process from the final outputs of a dual-path network to different stages of the backbone, encouraging better recalibration with multi-level cross-modality feature fusion. To guide the recalibration with explicit cross-modality interaction modeling, some works [20,26,31,35] tailor general 2D operations to 2.5D behaviors with depth guidance. For example, [31] proposes depth-aware convolution and pooling operations to help recalibrating RGB feature responses in depth-consistent regions. [20] proposes a depth-aware gate module that adaptively selects the pooling field size in a CNN according to object scale. 3DGNN [26] introduces a 3D graph neural network to model accurate context with geometry cues provided by depth. Alternatively, some approaches regard the depth data as an extra supervised signal to recalibrate the RGB counterpart in a multi-task learning manner. For example, [42] proposes a pattern affinity propagation network to regularize and boost complementary tasks. [37] introduces a multi-modal distillation model to pass the valid messages from depth to RGB features.

Different from previous works that hold the ideal assumption of depth source's quality and mainly focus on in-door setting, we try to extend the task to the in-the-wild environment, *e.g.*, CityScapes dataset. The out-door setting is more challenging due to the inevitable noisy signals contained in the depth data. In this work, we try to recalibrate RGB feature responses from a filtered depth representation and vice versa, which effectively enhance the strength of representations for both modalities.

2.2 Attention Mechanism

Attention mechanisms have been widely utilized in kinds of computer vision tasks, serving as the tools to spotlight the most representative and informative regions of input signals [11,16,21,29,32,33]. For example, to improve the performance of the image/video classification task, SENet [16] introduces a self recalibrate gating mechanism by model importance among different channels of feature maps. Based on similar spirits, SKNet [21] designs a channel-wise attention module to select kernel sizes to adaptively adjust its receptive field size based on multiple scales of input information. [32] introduces a non-local operation which explores the similarity of each pair of points in space. For the segmentation task, a well-designed attention module could encourage the network to learn helpful context information effectively. For instance, DFN [39] introduces a channel attention block to select the more discriminative features from multi-level feature maps to get more accurate semantic information. DANet [11] proposes two types of attention modules to model the semantic inter-dependencies in spatial and channel dimensions respectively.

However, the main challenge of RGB-D semantic segmentation task is how to make full use of cross-modality data under the substantial variations and noisy signals between modalities. The proposed SA-Gate is the first to focus on the noisy features of cross-modalities by tailoring the attention mechanisms. The SA-Gate module is specialized for suppressing the exceptional noisy feature of depth data and recalibrate its counterpart RGB feature responses in a unified manner at first, and then fuses the cross-modality information with a softmax gating that is guided by the recalibrated features, achieving effective and efficient cross-modality feature aggregation.

3 Method

RGB-D semantic segmentation needs to aggregate features from both RGB and depth modalities. However, both modalities have inevitably noisy information. Specifically, depth measurements are inaccurate due to the characteristics of depth sensors and RGB features might generate confusing results due to the high appearance similarity between the objects. An effective cross-modality aggregation scheme should be able to identify their strengths from each feature as well as unify the most informative cross-modality features into an efficient representation. To this end, we put forward a novel cross-modality guided encoder.

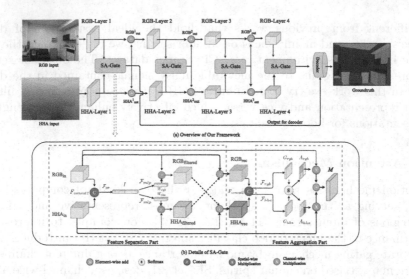

(a) Overview of Our Framework

(b) Details of SA-Gate

Feature Separation Part   Feature Aggregation Part

Fig. 2. (a) The overview of our network. We employ an encoder-decoder architecture. The input of the network is a pair of RGB-HHA images. During training, each pair of feature maps (*e.g.*, outputs of RGB-Layer1 and HHA-Layer1) are fused by a SA-Gate and propagated to the next stage of the encoder for further feature transformation. Fusion results of the first and the last SA-Gates would be propagated to the segmentation decoder (DeepLab V3+). (b) The architecture of the SA-Gate, which contains two parts, Feature Separation (FS) and Feature Aggregation (FA)

The overall framework of the proposed approach is depicted in Fig. 2 (a), which consists of a cross-modality guided encoder and a segmentation decoder. Given RGB-D data as inputs[2], our encoder recalibrates and fuses the complementary information from the two modalities via the SA-Gate unit, and then propagates the fused multi-modal features along with modality-specific features via the Bi-direction Multi-step Propagation (BMP) module. The information is then decoded by a segmentation decoder network to generate the segmentation map. We will detail each component in the remaining parts of this section.

3.1 Bi-direction Guided Encoder

Separation-and-Aggregation (SA) Gate. To ensure informative feature propagation between modalities, the SA-Gate is designed with two operations. One is feature recalibration on each single modality, and the other is cross-modality feature aggregation. The operations are in terms of Feature Separation (FS) and Feature Aggregation (FA) parts, as illustrated in Fig. 2 (b).

Feature Separation (FS) . We take depth stream for example. Due to physical characteristics of depth sensors, noisy signals in depth modality frequently show

---

[2] Note that we use HHA map to encode the depth measurements.

up in regions close to object's boundaries or partial surfaces outside the scope of depth sensors, as shown in the second column of Fig. 3. Hence, the network is expected to first filter noisy signals surrounding these local regions to avoid misleading information propagation on the process of recalibrating complementary RGB modality and aggregating cross-modality features. In practice, we exploit high confident activations in RGB stream to filter out exceptional depth activations at the same level. To do so, global spatial information of both modalities should be embedded and squeezed to obtain a cross-modality attention vector first. We achieve this by a global average pooling along the channel-wise dimensions of two modalities, which is followed by concatenation and a MLP operation to obtain attention vector. Suppose we have two input feature maps denoted as $\text{RGB}_{in} \in \mathbb{R}^{C \times H \times W}$ and $\text{HHA}_{in} \in \mathbb{R}^{C \times H \times W}$, above operations could be formulated as

$$I = \mathcal{F}_{gp}(\text{RGB}_{in} \parallel \text{HHA}_{in}), \tag{1}$$

where $\parallel$ denotes the concatenation of feature maps from two modalities, $\mathcal{F}_{gp}$ refers to global average pooling, $I = (I_1, \ldots, I_k, \ldots, I_{2C})$ is the cross-modality global descriptor for collecting expressive statistics for the whole inputs. Then, the cross-modality attention vector for the depth input is learned by

$$W_{hha} = \sigma(\mathcal{F}_{mlp}(I)), \quad W_{hha} \in \mathbb{R}^C, \tag{2}$$

where $\mathcal{F}_{mlp}$ denotes MLP network, σ denotes sigmoid function scaling the weight value into $(0, 1)$. By doing so, the network can take advantage of the most informative visual appearance and geometry features, and thus tends to effectively suppress the importance of noisy features in depth stream. Then, we could obtain a less noisy depth representation, namely Filtered HHA, through a channel-wise multiplication $\circledast$ between input depth feature maps and the cross-modality gate:

$$\text{HHA}_{filtered} = \text{HHA}_{in} \circledast W_{hha}. \tag{3}$$

With a filtered depth representation counterpart, the RGB feature responses could be recalibrated with more accurate depth information. We devise the recalibration operation as the summation of the two modalities:

$$\text{RGB}_{rec} = \text{HHA}_{filtered} + \text{RGB}_{in}, \tag{4}$$

where RGB_{rec} denotes recalibrated RGB feature maps. The general idea behind the formula is that, instead of directly using element-wise product to reweight RGB feature with regarding depth features as recalibrate coefficients, the proposed operation using summation could be viewed as some kind of offset to refine RGB feature responses at corresponding positions, as demonstrated in Table 2.

In practice, we implement *recalibration step* in a symmetric and bi-directional manner, such that low confident activations in RGB stream could also be suppressed in the same manner and filtered RGB information $\text{RGB}_{filtered}$ could inversely recalibrate the depth feature responses to form a more robust depth representation HHA_{rec}. We visualize feature maps of HHA before and after Feature Separation Part in Fig. 3. The RGB counterpart is shown in the supplementation.

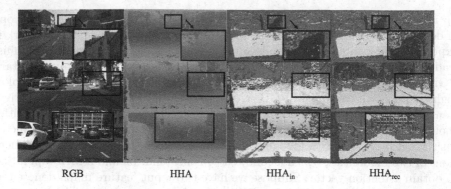

RGB   HHA   HHA$_{in}$ HHA$_{rec}$

Fig. 3. Visualization of depth features before and after FSP on CityScapes validation set. We can observe that objects have more precise shapes after FSP and invalid partial surfaces are completed. More explanation is illustrated in the supplemental material

Feature Aggregation (FA) . RGB and D features are strongly complementary to each other. To make full use of their complementarity, we need to complementarily aggregate the cross-modality features at a certain position in space according to their characterization capabilities. To achieve this, we consider both characteristics of these two modalities and generate spatial-wise gates for both RGB$_{in}$ and HHA$_{in}$ to control information flow of each modality feature map with soft attention mechanism, which is visualized in Fig. 2 (b) and marked by the second red frame. To make the gate more precise, we use recalibrated RGB and HHA feature maps from *FS* part, *i.e.*, RGB$_{rec} \in \mathbb{R}^{C \times H \times W}$ and HHA$_{rec} \in \mathbb{R}^{C \times H \times W}$, to generate the gate. We first concatenate these two feature maps to combine their features at a certain position in space. Then we define two mapping functions to map high-dimensional feature to two different spatial-wise gates:

$$\mathcal{F}_{rgb} : F_{concat2} \rightarrow G_{rgb} \in \mathbb{R}^{1 \times H \times W}, \tag{5}$$

$$\mathcal{F}_{hha} : F_{concat2} \rightarrow G_{hha} \in \mathbb{R}^{1 \times H \times W}, \tag{6}$$

where $F_{concat2} \in \mathbb{R}^{2C \times H \times W}$ is the concatenated feature, G_{rgb} is the spatial-wise gate for RGB feature map, and G_{hha} is the spatial-wise gate for HHA feature map. In practice, we use a 1×1 convolution to implement this mapping function. A softmax function is applied on these two gates:

$$A_{rgb}^{(i,j)} = \frac{e^{G_{rgb}^{(i,j)}}}{e^{G_{rgb}^{(i,j)}} + e^{G_{hha}^{(i,j)}}}, \ A_{hha}^{(i,j)} = \frac{e^{G_{hha}^{(i,j)}}}{e^{G_{rgb}^{(i,j)}} + e^{G_{hha}^{(i,j)}}} \tag{7}$$

where $A_{rgb}, A_{hha} \in R^{1 \times H \times W}$ and $A_{rgb}^{(i,j)} + A_{hha}^{(i,j)} = 1$. $G_{rgb}^{(i,j)}$ is the weight assigned to each position in the RGB feature map and $G_{hha}^{(i,j)}$ is the weight assigned to each position in the HHA feature map. The final merged feature M can be obtained by weighting the RGB and HHA maps:

$$M_{i,j} = \text{RGB}_{in}^{(i,j)} \cdot A_{rgb}^{(i,j)} + \text{HHA}_{in}^{(i,j)} \cdot A_{hha}^{(i,j)}. \tag{8}$$

So far, we have added gated RGB and HHA feature maps to obtain the fused feature maps M. Since SA-Gate is injected into the encoder stage, we then average the fused features and the original input to obtain RGB_{out} and HHA_{out} respectively, which share similar spirits with residual learning.

Bi-directional Multi-step Propagation (BMP). By normalizing the sum of two weights at each position to 1, the numerical scale of the weighted feature will not significantly differ from the input RGB or HHA. Therefore, it has no negative influence on the learning of the encoder or the loading of the pre-trained parameters. For each layer l, we use the output M^l generated by the l-th SA-Gate to refine the raw output of the l-th layer in the encoder: $\text{RGB}_{out}^l = (\text{RGB}_{in}^l + M^l)/2$, $\text{HHA}_{out}^l = (\text{HHA}_{in}^l + M^l)/2$. This is a bi-directional propagation process and the refined results will be propagated to the next layer in the encoder for more accurate and efficient encoding of the two modalities.

3.2 Segmentation Decoder

The decoder can adopt almost any design of decoder from SOTA RGB-based segmentation networks, since SA-Gate is a plug-and-play module and can make good use of complementary information of cross-modality on encoder stage. We show results of combining our encoder with different decoders in Table 6. We choose DeepLabV3+ [2] as our decoder for it achieves the best performance.

4 Experiments

We conduct comprehensive experiments on in-door NYU Depth V2 and out-door CityScapes datasets in terms of two metrics: mean Intersection-over-Union ($mIoU$) and pixel accuracy (pixel acc.). We also evaluate our model on SUN-RGBD dataset (Please refer to the supplemental material for more details).

4.1 Datasets

NYU Depth V2. [27] contains 1449 RGB-D images with 40-class labels, in which 795 images are used for training and the rest 654 images are for testing.

CityScapes. [8] contains images from 27 cities. There are 2975 images for training, 500 for validation and 1525 for testing. Each image has a resolution of 2048×1024 and is fine-annotated with pixel-level labels of 19 semantic classes. **We do not use additional coarse annotations in our experiments**.

4.2 Implementation Details

We use PyTorch framework. For data augmentation, we use random horizontal flipping and scaling with scales [0.5,1.75]. When comparing with SOTA methods, we adopt flipping and multi-scale inference strategies as a test-time augmentation to boost the performance. More details are shown in the supplemental material.

Table 1. Comparison of efficiency on NYUDV2 test set. We use ResNet-50 as backbone and DeepLab V3 + [2] as decoder. FLOPS are estimated for input of $3 \times 480 \times 480$

Methods	Params/M	FLOPs/G	mIoU (%)
RGB-D baseline	78.2	269.6	46.7
Ours	**63.4**	**204.9**	**50.4**

4.3 Efficiency Analysis

To verify whether the proposed cross-modality feature propagation helps and is efficient, we compare the final model with the RGB-D baseline. We average predictions of two parallel DeepLab V3+ as RGB-D baseline. As shown in Table 1, the proposed method achieves better performance with significantly less memory requirement and computational cost when compared with baseline. The results indicate that aimlessly adding parameters to a multi-modality network will not bring extra representational power to better recognize objects. In contrast, a well-design cross-modality mechanism, like proposed cross-modality feature propagation, helps to learn more powerful representations to improve performance more efficiently.

Table 2. Ablation study on *feature separation (FS)* part on NYU Depth V2 test set. No decoder is used here

Backbone	Concat	Self-global	Cross-global	Product	Proposed	mIoU(%)
Res50	✓					47.8
Res50		✓				47.5
Res50			✓			47.8
Res50				✓		47.5
Res50					✓	**48.6**

4.4 Ablation Study

We perform ablation studies on our design choices under same hyperparameters.

Feature Separation. We employ the FS operation before the feature aggregation in SA-Gate, to filter out noisy features for bi-directional recalibration step. To verify effectiveness of this operation, we ablate each design of FS in Table 2. Note that we ablate four different architectures and replace all FS parts in the network for comparison. 'Concat' represents we concatenate RGB_{in} and HHA_{in} feature maps and directly pass them to feature aggregation part. 'Self-global' represents we filter single modality features with its own global information. 'Cross-global' represents the filtered RGB is added to input RGB and vice versa.

The filtering guidance comes from cross-modality global information. 'Product' means we multiply RGB_{in} by $HHA_{filtered}$ and vice versa. We see that from column 2 to 4, not using cross-modality information to filter noisy feature or refine features without explicit cross-modality recalibration lead to about 1% drop. On the other hand, the last two columns indicate the cross-modality guidance (Eq. 4) is more appropriate and effective than cross-modality re-weighting when doing cross-modality recalibration. Overall, these results show that proposed FS operator effectively filters incorrect messages and recalibrates feature responses, achieving the best performance among all compared designs.

Feature Aggregation. We employ the SA-Gating mechanism to adaptively select the feature from the cross-modal data, according to their different characteristics at each spatial location. This gate can effectively control information flow of multimodal data. To evaluate the validity of the design, we perform ablation study on feature aggregation, as shown in Table 3. The experiment setting is kept the same as above. 'Addition' represents directly adding the recalibrated RGB and HHA feature maps. 'Conv' represents conducting convolution on the concatenated feature map. 'Proposed' represents the FA operator. We see that FA operator leads to the best result, since it considers the spatial-wise relationship between two modalities and can better explore the complementary information.

Table 3. Ablation study on *feature aggregation (FA)* part on NYU Depth V2 test set. No decoder is used here

Backbone	Addition	Conv	Proposed	mIoU (%)
Res50	✓			47.8
Res50		✓		48.0
Res50			✓	**48.6**

Table 4. Ablation study on encoder design on NYU Depth V2 test set. '*' means we average two outputs of RGB and HHA to get final output. No decoder is used here

Backbone	Block1	Block2	Block3	Block4	mIoU (%)
Res50*					45.9
Res50*	✓				47.8
Res50*		✓			47.5
Res50*			✓		46.8
Res50*				✓	44.3
Res50*	✓	✓			47.9
Res50*	✓	✓	✓		48.3
Res50*	✓	✓	✓	✓	48.0
Res50	✓	✓	✓	✓	**48.6**

Table 5. Ablation study for BMP and SA-Gate. No decoder is used here

Method	mIoU (%)
Res50 (Average of Dual Path)	45.9
Res50 + SA-Gate	47.4 (1.5% ↑)
Res50 + BMP	47.8 (1.9% ↑)
Res50 + BMP + SA-Gate	**48.6** (2.7% ↑)

Table 6. The plug-and-play property evaluation of the proposed model on NYU Depth V2 test set. **Method** indicates different decoders, SA-Gate indicates the proposed fusion module. **RGB**: RGB image as inputs; **RGB-D**: the simple method which only average final score maps of RGB path and HHA path. Note that we reproduce these methods using official open-source code and all experiments use the same setting as our method

Method	RGB (%$mIoU$)	RGB-D (%$mIoU$)	RGB-D w SA-Gate (%$mIoU$)
DeepLab V3 [1]	44.7	46.5	**49.1** (2.6 ↑)
PSPNet [43]	43.1	46.2	**48.2** (2.0 ↑)
DenseASPP [38]	42.3	45.7	**47.8** (2.1 ↑)
OCNet [40]	44.5	47.6	**49.1** (1.5 ↑)
DeepLab V3+ [2]	44.3	46.7	**50.4** (3.7 ↑)
DANet [11]	43.0	45.5	**48.6** (3.1 ↑)
FastFCN [34]	45.4	47.6	**50.1** (2.5 ↑)

Design of Encoder. We verify and analyze the effectiveness of proposed BMP to our encoder, and how it functions with the SA-Gate. Toward this end, we conduct two ablation studies as shown in Table 4 and 5. We use ResNet-50 as our backbone here and directly upsampling the final score map by a factor of 16, without using a segmentation decoder. The first row in Table 4 and 5 is the baseline that averages score maps generated by two ResNet-50 (RGB & D).

For the first ablation, we gradually embed SA-Gate unit behind different layers of ResNet50. Note that we generate score maps for both two sides and average them as final segmentation result. This setting is different from those above, because last block of ResNet may not be equipped with a SA-Gate in this part, i.e., no fused feature is generated from last block. From Table 4, we observe that if SA-Gate is embedded into a higher stage, it will lead to relatively worse performance. Besides, when stacking SA-Gate stage by stage, the additional gain continuously reduces. These two phenomena show that features of different modalities are more different in lower stage and an early fusing will achieve better performance. Table 5 shows results of second experiment. We observe that both SA-Gate and BMP can boost performance. Meanwhile, they complement each other and performs better in the presence of the other component. Moreover, when associating Table 5 and 2, we see that SA-Gate helps BMP better propagate valid information than other gate mechanisms. It demonstrates effectiveness and importance of a more accurate representation to the feature propagation.

The Plug-and-Play Property of Proposed Encoder. We conduct ablation study to validate the flexibility and effectiveness of our method for different types of decoders. Following recent RGB-based semantic segmentation algorithms, we splice their decoders with our model to form modified RGB-D versions (*i.e.,* RGB-D w SA-Gate), as shown in Table 6. We see that in the column 2 and 4, our method consistently helps achieving significant improvements against original RGB versions. Besides, comparing with naive RGB-D modifications, our method also boosts the performance at least 1.5% *mIoU*. Especially, with the decoders in Deeplab V3+ [2], our method achieves 3.7% *mIoU* improvements. The results verify both the flexibility and effectiveness of our method for various decoders.

4.5   Visualization of SA-Gate

We visualize first SA-Gate in our model to see what it has learned, as shown in Fig. 4. *Note that the black region in GT represents ignored pixels when calculating IoU. We reproduce RDFNet-101 [25]in PyTorch with 48.7% mIoU on NYU Depth V2, which is close to the result in the original paper (49.1%).* Red represents a higher weight assigned to RGB and blue represents a higher weight assigned to HHA. From column 4, we can see that RGB has a stronger response at boundary and HHA responds well in glare and dark areas. The phenomenon is reasonable since RGB feature has more details in high contrast areas and HHA feature is not affected by lighting conditions. From row 1, details inside yellow boxes are lost in HHA while obvious in RGB. Our method successfully identifies chair legs and distinguishes table that looks similar to chair. In row 2, glare blurs the border of the photo frame. Since our model focuses more on HHA in this area, it predicts the photo frame more completely than RDFNet. Besides, our model captures more details than RDFNet on clothes stand. In row 3, cabinet in dark red is hard to recognize in RGB but with identifiable features in HHA. Improper fusion of RGB and HHA leads to erroneous semantics for this area (column 3). While our model pays more attention to HHA in this area to achieve more precise results.

4.6   Comparing with State-of-the-arts

NYU Depth V2. Results are shown in Table 7. Our model achieves leading performance. On the consideration of a fair comparison to [17,37,42] that utilize ResNet-50 as backbone, we also use same backbone and achieve 51.3% *mIoU*, which is still better than these methods. Specifically, [17,25] try to use channel-wise attention or vanilla convolution to extract complementary feature, which are more implicit than our model in selecting valid feature from complementary information. Besides, we can see that utilizing depth data as extra supervision

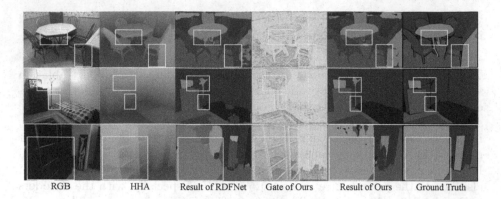

RGB   HHA   Result of RDFNet   Gate of Ours   Result of Ours   Ground Truth

Fig. 4. Visualization of feature selection through SA-Gate on NYUD V2 test set. For each row, we show (1) RGB, (2) HHA, (3) results of RDFNet-101, (4) visualization of SA-Gate, (5) results of ours, (6) GT. Red represents a higher weight assigned to RGB and blue represents a higher weight assigned to HHA. Best viewed in color (Color figure online)

Table 7. State-of-the-art comparison experiments on NYU Depth V2 test set

Method	mIoU (%)	Pixel Acc. (%)
3DGNN [26]	43.1	–
Kong *et al.* [20]	44.5	72.1
CFN [23]	47.7	–
RDF-101 [25]	49.1	75.6
PADNet [37]	50.2	75.2
ACNet [17]	48.3	–
PAP [42]	50.4	76.2
Ours	**52.4**	**77.9**

(such as [37,42]) could make network more robust than general RGB-D methods that take both RGB and depth as input sources [6,25,26]. However, our results demonstrate that once the input RGB-D information could be effectively recalibrated and aggregated, higher performance could be obtained.

CityScapes. We achieve 81.7% $mIoU$ on validation set and 82.8% $mIoU$ on test set, which are both leading performances. Table 8 shows results on test set. We observe that due to serious noise of depth measurements in this dataset, most of previous RGB-D based methods even worse than RGB-based methods. However, our method effectively distills depth feature and extracts valid information in it and boosts the performance. Note that [7] is a contemporary work and we outperform them by 0.7%. We exclude the results of GSCNN [28] for fair comparison, since it uses a stronger backbone WideResNet instead of

Table 8. Cityscapes test set accuracies. '*' means RGB-D based methods

Method	roa.	sid.	bui.	wal.	fen.	pol.	lig.	sig.	veg.	ter.	sky	per.	rid.	car	tru.	bus	tra.	mot.	bic.	mIoU
CCNet [18]	–	–	–	–	–	–	–	–	–	–	–	–	–	–	–	–	–	–	–	81.4
BFP [10]	98.7	87.0	93.5	59.8	63.4	68.9	76.8	80.9	93.7	72.8	95.5	87.0	72.1	96.0	77.6	89.0	86.9	69.2	77.6	81.4
DANet [11]	98.6	86.1	93.5	56.1	63.3	69.7	77.3	81.3	93.9	72.9	95.7	87.3	72.9	96.2	76.8	89.4	86.5	72.2	78.2	81.5
ACFNet [41]	98.7	87.1	93.9	60.2	63.9	71.1	78.6	81.5	94.0	72.9	95.9	88.1	74.1	96.5	76.6	89.3	81.5	72.1	79.2	81.8
LDFNet* [19]	–	–	–	–	–	–	–	–	–	–	–	–	–	–	–	–	–	–	–	71.3
Shu Kong et al.* [20]	–	–	–	–	–	–	–	–	–	–	–	–	–	–	–	–	–	–	–	78.2
PADNet * [37]	–	–	–	–	–	–	–	–	–	–	–	–	–	–	–	–	–	–	–	80.3
Choi et al.* [7]	98.8	88.0	93.9	60.5	63.3	71.3	78.1	81.3	94.0	72.9	96.1	87.9	74.5	96.5	77.0	88.0	85.9	72.7	79.0	82.1
RGB baseline (Deeplab V3+ [2])	98.7	87.1	93.9	61.0	63.8	71.5	78.6	82.6	93.9	72.6	95.9	88.3	74.8	96.5	68.9	86.1	86.4	73.6	79.1	81.8
RGB-D baseline*	98.7	86.7	93.7	57.8	61.8	70.0	77.3	81.8	93.9	72.2	95.9	87.9	74.1	96.3	70.7	87.9	80.3	72.2	78.6	80.9
Ours*	98.7	87.3	93.9	63.8	62.7	70.8	77.9	82.2	93.9	72.8	95.9	88.2	75.2	96.5	80.4	91.6	89.0	73.2	78.9	**82.8**

ResNet-101. However, we still outperform GSCNN by 0.9% mIoU on the validation set and achieve the same performance as it on test set.

5 Conclusion

In this work, we propose a cross-modality guided encoder along with SA-Gate and BMP modules to address two key challenges in RGB-D semantic segmentation, *i.e.,* the effective unified representation for different modalities and the robustness to low-quality depth source. Meanwhile, our proposed encoder can act as a plug-and-play module, which can be easily injected to current state-of-the-art RGB semantic segmentation frameworks to boost their performances.

Acknowledgments. This work is supported by the National Key Research and Development Program of China (2017YFB1002601, 2016QY02D0304), National Natural Science Foundation of China (61375022, 61403005, 61632003), Beijing Advanced Innovation Center for Intelligent Robots and Systems (2018IRS11), and PEK-SenseTime Joint Laboratory of Machine Vision.

References

1. Chen, L.C., Papandreou, G., Schroff, F., Adam, H.: Rethinking atrous convolution for semantic image segmentation. arXiv preprint arXiv:1706.05587 (2017)
2. Chen, L.-C., Zhu, Y., Papandreou, G., Schroff, F., Adam, H.: Encoder-decoder with atrous separable convolution for semantic image segmentation. In: Ferrari, V., Hebert, M., Sminchisescu, C., Weiss, Y. (eds.) ECCV 2018. LNCS, vol. 11211, pp. 833–851. Springer, Cham (2018). https://doi.org/10.1007/978-3-030-01234-2_49
3. Chen, X., Lin, K.Y., Qian, C., Zeng, G., Li, H.: 3D sketch-aware semantic scene completion via semi-supervised structure prior. In: CVPR (2020)
4. Chen, Y., Mensink, T., Gavves, E.: 3D neighborhood convolution: learning depth-aware features for RGB-D and RGB semantic segmentation. In: 3DV. IEEE (2019)

5. Cheng, B., et al.: SPGNet: semantic prediction guidance for scene parsing. In: ICCV (2019)

6. Cheng, Y., Cai, R., Li, Z., Zhao, X., Huang, K.: Locality-sensitive deconvolution networks with gated fusion for RGB-D indoor semantic segmentation. In: CVPR (2017)

7. Choi, S., Kim, J.T., Choo, J.: Cars can't fly up in the sky: improving urban-scene segmentation via height-driven attention networks. In: Proceedings of the IEEE/CVF Conference on Computer Vision and Pattern Recognition (2020)

8. Cordts, M., et al.: The cityscapes dataset for semantic urban scene understanding. In: CVPR (2016)

9. Deng, L., Yang, M., Li, T., He, Y., Wang, C.: RFBNet: deep multimodal networks with residual fusion blocks for RGB-D semantic segmentation. arXiv preprint arXiv:1907.00135 (2019)

10. Ding, H., Jiang, X., Liu, A., Thalmann, N.M., Wang, G.: Boundary-aware feature propagation for scene segmentation. In: ICCV (2019)

11. Fu, J., et al.: Dual attention network for scene segmentation. In: CVPR (2019)

12. Fu, J., et al.: Adaptive context network for scene parsing. In: ICCV (2019)

13. Gupta, S., Girshick, R., Arbeláez, P., Malik, J.: Learning rich features from RGB-D images for object detection and segmentation. In: Fleet, D., Pajdla, T., Schiele, B., Tuytelaars, T. (eds.) ECCV 2014. LNCS, vol. 8695, pp. 345–360. Springer, Cham (2014). https://doi.org/10.1007/978-3-319-10584-0_23

14. He, J., Deng, Z., Qiao, Y.: Dynamic multi-scale filters for semantic segmentation. In: CVPR (2019)

15. He, Y., Chiu, W.C., Keuper, M., Fritz, M.: STD2P: RGBD semantic segmentation using spatio-temporal data-driven pooling. In: ICCV (2017)

16. Hu, J., Shen, L., Sun, G.: Squeeze-and-excitation networks. In: CVPR (2018)

17. Hu, X., Yang, K., Fei, L., Wang, K.: ACNet: attention based network to exploit complementary features for RGBD semantic segmentation. arXiv preprint arXiv:1905.10089 (2019)

18. Huang, Z., Wang, X., Huang, L., Huang, C., Wei, Y., Liu, W.: CCNet: criss-cross attention for semantic segmentation. In: ICCV (2019)

19. Hung, S.W., Lo, S.Y., Hang, H.M.: Incorporating luminance, depth and color information by a fusion-based network for semantic segmentation. In: ICIP. IEEE (2019)

20. Kong, S., Fowlkes, C.C.: Recurrent scene parsing with perspective understanding in the loop. In: CVPR (2018)

21. Li, X., Wang, W., Hu, X., Yang, J.: Selective kernel networks. In: CVPR (2019)

22. Li, Z., Gan, Y., Liang, X., Yu, Y., Cheng, H., Lin, L.: LSTM-CF: unifying context modeling and fusion with LSTMs for RGB-D scene labeling. In: Leibe, B., Matas, J., Sebe, N., Welling, M. (eds.) ECCV 2016. LNCS, vol. 9906, pp. 541–557. Springer, Cham (2016). https://doi.org/10.1007/978-3-319-46475-6_34

23. Lin, D., Chen, G., Cohen-Or, D., Heng, P.A., Huang, H.: Cascaded feature network for semantic segmentation of RGB-D images. In: ICCV (2017)

24. Long, J., Shelhamer, E., Darrell, T.: Fully convolutional networks for semantic segmentation. In: CVPR (2015)

25. Park, S.J., Hong, K.S., Lee, S.: RDFNet: RGB-D multi-level residual feature fusion for indoor semantic segmentation. In: ICCV (2017)

26. Qi, X., Liao, R., Jia, J., Fidler, S., Urtasun, R.: 3D graph neural networks for RGBD semantic segmentation. In: ICCV (2017)

27. Silberman, N., Hoiem, D., Kohli, P., Fergus, R.: Indoor segmentation and support inference from RGBD images. In: Fitzgibbon, A., Lazebnik, S., Perona, P., Sato, Y., Schmid, C. (eds.) ECCV 2012. LNCS, vol. 7576, pp. 746–760. Springer, Heidelberg (2012). https://doi.org/10.1007/978-3-642-33715-4_54

28. Takikawa, T., Acuna, D., Jampani, V., Fidler, S.: Gated-SCNN: gated shape CNNs for semantic segmentation (2019)

29. Wang, F., et al.: Residual attention network for image classification. In: CVPR (2017)

30. Wang, J., Wang, Z., Tao, D., See, S., Wang, G.: Learning common and specific features for RGB-D semantic segmentation with deconvolutional networks. In: Leibe, B., Matas, J., Sebe, N., Welling, M. (eds.) ECCV 2016. LNCS, vol. 9909, pp. 664–679. Springer, Cham (2016). https://doi.org/10.1007/978-3-319-46454-1_40

31. Wang, W., Neumann, U.: Depth-aware CNN for RGB-D segmentation. In: Ferrari, V., Hebert, M., Sminchisescu, C., Weiss, Y. (eds.) ECCV 2018. LNCS, vol. 11215, pp. 144–161. Springer, Cham (2018). https://doi.org/10.1007/978-3-030-01252-6_9

32. Wang, X., Girshick, R., Gupta, A., He, K.: Non-local neural networks. In: CVPR (2018)

33. Woo, S., Park, J., Lee, J.-Y., Kweon, I.S.: CBAM: convolutional block attention module. In: Ferrari, V., Hebert, M., Sminchisescu, C., Weiss, Y. (eds.) ECCV 2018. LNCS, vol. 11211, pp. 3–19. Springer, Cham (2018). https://doi.org/10.1007/978-3-030-01234-2_1

34. Wu, H., Zhang, J., Huang, K., Liang, K., Yu, Y.: FastFCN: rethinking dilated convolution in the backbone for semantic segmentation. arXiv preprint arXiv:1903.11816 (2019)

35. Xing, Y., Wang, J., Chen, X., Zeng, G.: 2.5D convolution for RGB-D semantic segmentation. In: ICIP. IEEE (2019)

36. Xing, Y., Wang, J., Chen, X., Zeng, G.: Coupling two-stream RGB-D semantic segmentation network by idempotent mappings. In: ICIP. IEEE (2019)

37. Xu, D., Ouyang, W., Wang, X., Sebe, N.: PAD-Net: multi-tasks guided prediction-and-distillation network for simultaneous depth estimation and scene parsing. In: CVPR (2018)

38. Yang, M., Yu, K., Zhang, C., Li, Z., Yang, K.: DenseASPP for semantic segmentation in street scenes. In: CVPR (2018)

39. Yu, C., Wang, J., Peng, C., Gao, C., Yu, G., Sang, N.: Learning a discriminative feature network for semantic segmentation. In: CVPR (2018)

40. Yuan, Y., Wang, J.: OCNet: object context network for scene parsing. arXiv preprint arXiv:1809.00916 (2018)

41. Zhang, F., et al.: ACFNet: attentional class feature network for semantic segmentation. In: ICCV (2019)

42. Zhang, Z., Cui, Z., Xu, C., Yan, Y., Sebe, N., Yang, J.: Pattern-affinitive propagation across depth, surface normal and semantic segmentation. In: CVPR (2019)

43. Zhao, H., Shi, J., Qi, X., Wang, X., Jia, J.: Pyramid scene parsing network. In: CVPR (2017)

DataMix: Efficient Privacy-Preserving Edge-Cloud Inference

Zhijian Liu[1], Zhanghao Wu[1,2], Chuang Gan[3], Ligeng Zhu[1], and Song Han[1(✉)]

[1] Massachusetts Institute of Technology, Cambridge, MA, USA
songhan@mit.edu
[2] Shanghai Jiao Tong University, Shanghai, China
[3] MIT-IBM Watson AI Lab, Cambridge, MA, USA

Abstract. Deep neural networks are widely deployed on edge devices (*e.g.*, for computer vision and speech recognition). Users either perform the inference locally (*i.e.*, edge-based) or send the data to the cloud and run inference remotely (*i.e.*, cloud-based). However, both solutions have their limitations: edge devices are heavily constrained by insufficient hardware resources and cannot afford to run large models; cloud servers, if not trustworthy, will raise serious privacy issues. In this paper, we mediate between the resource-constrained edge devices and the privacy-invasive cloud servers by introducing a novel *privacy-preserving edge-cloud inference* framework, DataMix. We off-load the majority of the computations to the cloud and leverage a pair of mixing and de-mixing operation, inspired by *mixup*, to protect the privacy of the data transmitted to the cloud. Our framework has three advantages. First, it is *privacy-preserving* as the mixing cannot be inverted without the user's private mixing coefficients. Second, our framework is *accuracy-preserving* because our framework takes advantage of the space spanned by images, and we train the model in a mixing-aware manner to maintain accuracy. Third, our solution is *efficient* on the edge since the majority of the workload is delegated to the cloud, and our mixing and de-mixing processes introduce very few extra computations. Also, our framework introduces small communication overhead and maintains high hardware utilization on the cloud. Extensive experiments on multiple computer vision and speech recognition datasets demonstrate that our framework can greatly reduce the local computations on the edge (to fewer than 20% of FLOPs) with negligible loss of accuracy and no leakages of private information.

1 Introduction

The high performance and superior accuracy of deep neural networks always comes at the expense of larger model size and more computations. Meanwhile,

---

Z. Liu and Z. Wu—Indicates equal contributions; order determined by a coin toss.

Electronic supplementary material The online version of this chapter (https://doi.org/10.1007/978-3-030-58621-8_34) contains supplementary material, which is available to authorized users.

A. Vedaldi et al. (Eds.): ECCV 2020, LNCS 12356, pp. 578–595, 2020.
https://doi.org/10.1007/978-3-030-58621-8_34

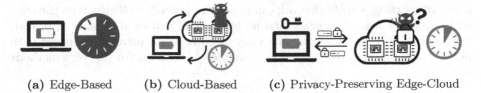

(a) Edge-Based   (b) Cloud-Based   (c) Privacy-Preserving Edge-Cloud

Fig. 1. Edge devices are *resource-limited*, and cloud servers are *privacy-invasive*. Our proposed framework takes advantage of both, providing *low-latency, privacy-preserving* model inference.

large models are difficult to be deployed on resource-constrained edge devices (such as mobile phones, self-driving cars and smart speakers): mobile applications interact with the users and require low latency, while edge devices have limited hardware resources and tight power budgets. To address these challenges, researchers have proposed to either directly design the compact models [23,47] or accelerate the existing models by compression [17,27,52]. However, a bottleneck is the ceiling of accuracy that small models can achieve. To our best knowledge, it is rather challenging to achieve high accuracy with very compact models: *e.g.*, with roughly 200M FLOPs of computations, the state-of-the-art mobile models [22] can only achieve 75% of top-1 accuracy on ImageNet, which is still 10% lower than the best performances [59].

In contrast, cloud servers have much more computation resources and power budgets than edge devices. With the next generation wireless network (*i.e.*, 5G) approaching, the high bandwidth and low latency of the technique will lead to a fundamental change of the way we process information both on the edge devices and cloud servers, which will affect the paradigm of AI computing. The communication latency will be significantly reduced, and the cloud servers can then handle the computation for the edge devices without sacrificing the real-time experience. Taking advantage of the computation power on both cloud and edge will offer new opportunities for efficient AI computing. However, cloud-based solutions raise privacy issues as the cloud servers might be malicious. User's data, in many cases, is very privacy-sensitive: *i.e.*, users may not want to disclose their personal information to the cloud (such as identity, age, and health status). Therefore, privacy-preserving inference is of critical importance.

This paper presents a novel perspective to tackle this challenge, in Fig. 1. We introduce the *privacy-preserving edge-cloud inference* framework, DataMix, to bring the best of the privacy-preserving edge devices and resource-abundant cloud servers together. We delegate the majority of the computations to the cloud, therefore reducing the local resource requirements. Inspired by *mixup* [60], we design the mixing and de-mixing operation for the privacy of the data transmitted to the cloud and train the model in an mixing-aware manner to maintain the accuracy. Our framework is a general method for cloud-edge inference and can be applied to multiple modalities. We evaluate our proposed framework on multiple tasks on two modalities including facial attribute classification and keyword spotting. Our framework can greatly improve the efficiency on the edge with negligible loss of accuracy and no leakages of private information, providing a superior trade-off among efficiency, accuracy and privacy compared with previous approaches. We provide an example of attacking for different methods in Fig. 2.

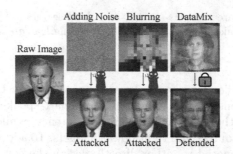

Fig. 2. Adding noise and blurring are not secure: the original image can be recovered by the GAN-based attack model. Our proposed DataMix preserves the privacy better.

2   Related Work

2.1   Efficient Inference

Considerable efforts have been made to design efficient models under tight resource constraints on the edge devices while maintaining the high performances at the same time, such as SqueezeNets [10,24], MobileNets [23,47] ShuffleNets [37,61], TSM [34] and modifications of Transformers [50,57]. Another approach to achieve the efficient inference is to compress and accelerate the existing large models. For instance, some have proposed to prune the separate neurons [17,18] or the entire channels [20,21,35]; others have proposed to quantize the network [8,30,55,62] to accelerate the model inference.

However, it is very challenging to achieve the state-of-the-art performance with these compact models. In this paper, we provide a new solution to the efficient inference, that is to make use of the computing power on the cloud without compromising privacy.

2.2   Privacy-Preserving Inference

There have been extensive investigations on the problem of privacy in the machine learning. Osia *et al.* [41] summarized the previous works into mainly three categories: dataset publishing [2,3,26], model sharing [1,9,48] and private inference. Our paper falls into the last category, which is to perform the inference on the cloud without leaking any private information.

Table 1. Our framework achieves high *efficiency* on the edge, introduces small *network communication overhead*, attains full *resource utilization* on the cloud, and protects both *input and output privacy*. As for our framework, we send the output activation of the first or second convolution layer to the cloud (the last two columns). In this table, the red entries are unsatisfactory. *Numbers are adopted from the authors' oral presentation at ICLR'19.

	Baseline (cloud)	Osia *et al.* [41, 42]	Tramer *et al.* [53]	DataMix (Ours)	
Computation (on edge)	0%	93%	~1%*	1%	13%
Transmission size	0.6 MB	0.4 MB	86.5 MB	12.3 MB	3.1 MB
Transmission time	0.4 s	0.3 s	33.1 s	6.5 s	1.7 s
GPU utilization (cloud)	100%	100%	~10%*	100%	100%
Input privacy	✗	✓	✓	✓	✓
Output privacy	✗	✗	✓	✓	✓

Researchers proposed different approaches of private inference on specific tasks: *e.g.*, performing activity recognition on extremely low-resolution videos [6, 7, 45, 46]. As for the face-related tasks, researchers have introduced various face de-identification methods to help protect the privacy [4, 28, 33, 38, 39]. The k-anonymity [12–15, 51] methods are also proposed to protect the information in the data by averaging k closest samples, but it does not take model inference into consideration. However, most of them either require much computation or compromise to the model accuracy degradation.

Inspired by the generative adversarial networks (GANs) [11], researchers proposed to train one neural network to obfuscate the input data and train another neural network to recover the original data in an adversarial manner [5, 29, 31, 40, 43, 44, 58]. However, these obfuscators are dedicated to one particular adversarial attacker and might not be able to generalize to other attack methods. Furthermore, these works do not take the efficiency into consideration, and some obfuscators are very computationally expensive: *e.g.*, 30× more FLOPs than ResNet-18 [44].

2.3 Hybrid Edge-Cloud Inference

There are several preliminary attempts [32, 41, 42, 54] to leverage both the edge and the cloud during the model inference. However, in order to maintain the accuracy, these frameworks need to send very deep layers to the cloud (*e.g.*, after conv5-1 for VGG-16), which means that more than 90% of the computation still needs to be performed on the edge. Besides, they only consider the input privacy; however, the outputs (*i.e.*, prediction result) might also be very sensitive: *e.g.*, for facial attribute classification, the prediction of the user's age can lead to some ageist behaviors, and it therefore should be equally important as the input.

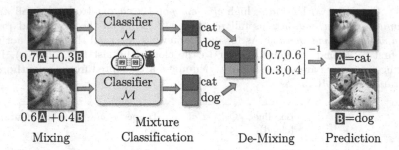

Fig. 3. A motivating example for privacy-preserving inference with mixing. The classifier trained with *mixup* takes mixtures of images and outputs mixed predictions. We can then solve the correct labels from the outputs. Though having the mixtures and the parameter weights of the classifier, the cloud cannot recover the private information without the private coefficients owned by users for mixing.

Recently, Tramer *et al.* [53] proposed to delegate the executions of all linear layers in a DNN from a trusted processor (TEEs) to another untrusted processor. However, it requires two processors to be co-located as it requires transmitting the activations of *all* the linear layers, which brings a large communication overhead. Therefore, the approach is not suitable for the edge-cloud inference setting where the transmission cost between the edge (trusted processor) and the cloud (untrusted processor) is expensive. We highlight the differences (w.r.t. efficiency, privacy and communication overhead) between these previous frameworks and our proposed framework in Table 1. Specifically, all benchmarks in the table are conducted on VGG-16 [49] with input image size of 224×224, and the transmissions are over the 4G LTE network with upload speed of 15 Mbps, download speed of 30 Mbps, and network delay of 25 ms. The transmission time can be calculated by Eq. 1.

$$T = \frac{\text{input feature}}{\text{upload bandwidth}} + \frac{\text{output feature}}{\text{download bandwidth}} + 2 \times \text{network latency} \quad (1)$$

3   A Motivating Example

Mixup [60] is a general training technique for neural networks. It encourages the model to behave linearly on the mix of training examples to smooth the decision boundaries among classes for generalization. That is to say, neural networks can not only learn from the raw images but also the space spanned by two random images from the distribution of images.

As shown in Fig. 3, the two raw images of a cat A and a dog B are mixed with a pair of coefficients $c = [0.7, 0.3]^T$. We denote the mixture as:

$$m(c) = [A, B] \cdot c = 0.7A + 0.3B.$$

When fed with the mixed images, a neural network trained for the dataset with *mixup* can output the mixed probabilities $\tilde{y}_{m(c)}$ for the classes with the same coefficients c, *i.e.*

$$\tilde{y}_{m(c)} = [\tilde{y}_A, \tilde{y}_B] \cdot c.$$

The case will be more interesting, when we introduce another mixture of the same two images with a different pair of coefficients $c' = [0.6, 0.4]^T$. The new mixture is denoted by $m(c')$. The model output for the $m(c')$ provides another mixed probabilities $\tilde{y}_{m(c')}$ with the coefficients c'. Aware of the exact values of the two pairs of coefficients $C = [c; c']$, we can solve (or de-mix) the correct prediction from the $[\tilde{y}_{m(c)}, \tilde{y}_{m(c')}]$ for the each raw inputs, where

$$[\tilde{y}_A, \tilde{y}_B] = [\tilde{y}_{m(c)}, \tilde{y}_{m(c')}] \cdot C^{-1}.$$

The mixing and de-mixing operation lead to an effective and efficient protocol for privacy-preserving inference on the cloud. Using the operation, we can offload the model trained with *mixup* to the cloud and only transmit mixed inputs and outputs for model inference. Since both the attackers on the network and cloud cannot access the mixing coefficients, they are not able to recover the original private data from the mixed inputs and outputs transmitted. That makes the method an excellent privacy-preserving operation. On the other hand, the operation only contains a small amount of computation, *i.e.* a few additions, making the method efficient enough for the resource-constrained edge devices. That enables computational-intensive neural network inference for privacy-sensitive situations on mobiles and IoTs.

However, though mixing and de-mixing operation protect the original data from being recovered, some of the private information can still be recognized, *e.g.* the color of the cat in Fig. 3. We extend the method to our DataMix for a general privacy-preserving framework.

4 Method

In this section, we will extend the privacy-preserving inference method in Sect. 3. We first describe the problem setting; we then extend the mixing and de-mixing for DataMix with larger group size and intermediate features processing; we finally analyze the design choice and provide some techniques for practice.

4.1 Problem Setting

In this paper, we focus on the *model inference on the edge devices*. As these edge devices are tightly resource-constrained, it is beneficial for them to offload the computations to the cloud for fast and efficient inference. We assume that *the users' data is privacy-sensitive* and *the cloud is malicious*. In our setting, the cloud has the weights of the neural network model, and both the attackers on the internet and cloud want to recover users' private information from each request. We also assume that the attackers and the cloud do not relate multiple requests. For the users, they have multiple inputs to be inferred, *e.g.* image classification in smart albums, and automatic medical analysis in hospitals with many patients.

In another situation, the users can have a pool of images, from which they can sample some random images to protect privacy of the real image for inference. Our goal is to develop an effective and efficient method so that attackers cannot recover users' data and the inferred results from users' requests to the cloud.

4.2   DataMix

The motivating example provides a prototype for privacy-preserving inference on the cloud. With the classifier trained by *mixup*, users can send the privacy-preserving mixtures to the cloud for inference. Practically, we further extend it as DataMix for more secure and better performance.

Larger Group for Mixing. Mixing the images with random (private) coefficients can hide information in the data. That is because the entropy of the mixed data increases when pixels from the two images are combined. If we enlarge the group of images for mixing, the entropy will become larger as more information is combined, and therefore better preserve the privacy. It is also the case for the output predictions.

We examine the intuition on figures of human faces from CelebA dataset. We trained attack models for the personal identity for the raw images and mixtures with different group sizes, separately. In Fig. 4, we observe that by mixing two images, the attack success rate will be reduced by 43.7%. When the group size S_G of image mixing increases, the success rate further decreases, *i.e.* better preserves the privacy. In our DataMix, rather than merely using a group size of 2 for privacy protection, we apply a larger group for mixing, *e.g.* $S_G = 8$.

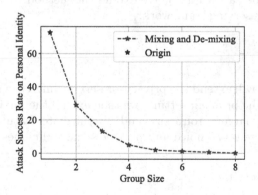

Fig. 4. The influence of the group size of mixing for data privacy on CelebA dataset. The personal identity accuracy indicates the how much information can an attacker get from the data transmitted to the cloud. When the group size increases, the privacy of the data are better preserved.

Intermediate Features Mixing. We also extend the mixing and de-mixing operation to the intermediate features, *e.g.* the outputs of the first convolution layer.

In the motivating example (Sect. 3), we apply mixing and de-mixing directly to the raw images and outputs of the classifier so that all the attackers on the network and the cloud can only access to the mixed data. The *mixup* training technique can be regarded as a regularization term for the classifier that encourages the model to perform well on the space spanned by two images from the dataset. After we enlarge the group size S_G for mixing, the space becomes more complicated. To improve the performance of the classifier on the mixtures, instead of applying mixing to the raw data, we adopt the operation on the intermediate features. The model can then learn to leverage the first several layers to re-project raw data to another space that provides better mixtures for the suffix layers to model the patterns.

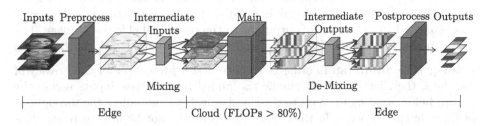

Inputs Preprocess   Intermediate   Main   Intermediate Postprocess Outputs
                          Inputs   Outputs

                          Mixing   De-Mixing

| Edge | Cloud (FLOPs > 80%) | Edge |

Fig. 5. Partitioning the neural network into three parts, our framework leverages the mixing and de-mixing operation to protect the privacy of data transmitted to the cloud.

Formalization. To formalize our method, we illustrate our method in Fig. 5. We partition the neural network model $\mathcal{M}$ sequentially into three parts:

$$\mathcal{M} = \mathcal{M}_{\text{post}}^{\text{E}} \circ \mathcal{M}_{\text{main}}^{\text{C}} \circ \mathcal{M}_{\text{pre}}^{\text{E}}, \tag{2}$$

where $\mathcal{M}_{\text{pre}}^{\text{E}}$, $\mathcal{M}_{\text{main}}^{\text{C}}$ and $\mathcal{M}_{\text{post}}^{\text{E}}$ represent the *preprocess*, *main* and *postprocess* models, respectively.

During the inference, the edge first runs the preprocess model on the *raw inputs* and sends its output (*i.e.*, *intermediate input*) to the cloud; then, the cloud runs the main model and transmits its output (*i.e.*, *intermediate output*) back to the edge; finally, the edge executes the postprocess model to obtain the *final output*. Throughout the whole procedure, the attackers can only access to the intermediate input instead of the raw input data.

However, without mixing the transmissions, the framework is not yet secure for both input and output privacy. For the input data, it is possible to train a neural network to approximate the inverse function of the preprocess model if it is relatively shallow, and the cloud can then recover the input by running the inference of the *inverse* preprocess model on intermediate inputs. For the output data, the cloud can simply rerun the inference of the postprocess model on intermediate outputs since the cloud has access to the weights.

To solve these issues, as previously mentioned, we apply the mixing after the preprocess model and de-mixing before the postprocess model on the intermediate features (inputs and outputs). As mentioned above, the preprocess model is used as the projection of the raw image to the space that is more mixing-friendly, and the postprocess model works similarly for the outputs. The mixing and de-mixing operation are both computed on the edge so that only the mixed data are exposed to public, protecting the privacy of original data.

Mixing-Aware Training. With intermediate feature mixing, we design a mixing-aware training process to improve the model performance on the mixture. While training, a batch of training data is firstly fed into the preprocess model; we then mix images in each group with coefficients randomly sampled from orthogonal matrix group; the main model takes the mixtures and generates intermediate outputs; after de-mixing, we feed the outputs to the postprocess model and calculate the loss of outputs with the correct label for each image. The whole model, including $\mathcal{M}_{\text{pre}}^{\text{E}}$, $\mathcal{M}_{\text{main}}^{\text{C}}$ and $\mathcal{M}_{\text{post}}^{\text{E}}$, is updated with the gradients.

Inter-group Shuffling. In order to achieve better protection for the privacy of the data, the clients can also shuffle the mixed intermediate inputs across the groups before sending the data to the cloud, which will not affect the throughput of the inference process. In that case, the cloud will not be able to relate data from the same group, which further increases the ambiguity.

4.3   Design Analysis

We analyze several properties of our DataMix to be a good privacy-preserving method for model inference on the cloud.

Non-Invertibility. The private-preserving method for the data transmitted to the cloud should be *non-invertible* without the private keys. Otherwise, the cloud can very easily recover the raw input x_{raw} from the transmitted input x_{in}:

$$x_{\text{raw}} = (\mathcal{M}_{\text{pre}}^{\text{E}})^{-1}(\mathcal{E}^{-1}(x_{\text{in}})), \tag{3}$$

where $\mathcal{E}^{-1}$ is the inverse function of the mixing for the transmitted data, and $(\mathcal{M}_{\text{pre}}^{\text{E}})^{-1}$ is the inverse preprocess model (approximated by neural networks). The output y can be recovered by rerunning the inference $\mathcal{M}(x)$.

In our DataMix, the coefficients are randomly generated by the users and kept private. For mixed images, one of the possible attacking methods is to average a large set of mixtures that contain the same raw image (not easy to collect without knowledge of what and how the images are mixed). In that way, when the images are i.i.d. and have a mean of 0, the average of the mixtures will be equal to the same raw image in expectation. However, the method is applicable only for training, where one image will appear in many mixtures with different groups of images. Since our DataMix focuses on inference, an image is only mixed with a same group of images, and the average of these mixtures is still a mixture with unknown coefficients (not close to zero) for the attackers.

Without the coefficients, our DataMix is non-invertible because the attackers do not have access to the data being mixed and do not know how they are mixed.

Compatibility. Let us consider an extreme case where we leverage a complicated cryptographic hash function (such as MD5) as our encryption for the transmitted data. It is indeed secure as it is empirically not invertible; however, it breaks the *continuity* and *locality* of the input data, which are the foundations for the convolution to be effective. Our DataMix encourages the model to make inference on the space spanned by inputs. As shown in *mixup* [60], the space is compatible with neural networks, when $S_G = 2$. We will provide extensive experiments for larger group size S_G in the next section, where our mixing and de-mixing operation also only causes negligible accuracy degradation for larger S_G.

Efficiency. The privacy-preserving inference on the cloud should also be very efficient to compute locally on the edge while protecting the privacy of the data sent to the cloud. The mixing and de-mixing operation is only composed of a few additions and very efficient. The computation on the edge is still small, even when we place some of the layers for the preprocess and postprocess model on the edge.

5 Experiments

We conduct experiments on two modalities, facial attribute classification and keyword spotting, to demonstrate the consistent effectiveness of our framework.

Table 2. Privacy-preserving facial attribute classification on CelebA. All of our DataMix have $S_G = 8$. The red entries are unsatisfactory (efficiency: the fewer FLOPs the better; privacy: the lower attack success rate the better). We require fewer computations on the edge, while maintaining higher accuracy and lower attack success rate.

	Efficiency($\downarrow$better)		Accuracy($\uparrow$better)		Privacy($\downarrow$better)	
	Params	FLOPs	Valid	Test	ID	Attrs.
Baseline (all on Edge)	11.21M	1.50G	91.6%	91.0%	0.1%	50.0%
Baseline (all on Cloud)	0	0	91.6%	91.0%	85.5%	79.3%
Adding Noise $\mathcal{N}(0,4)$ [44]	0	0	89.2%	88.6%	46.5%	73.1%
Adding Noise $\mathcal{N}(0,8)$ [44]	0	0	88.5%	87.9%	35.3%	70.8%
Blurring (16×16) [44,46]	0	0	89.6%	89.0%	52.2%	73.1%
Blurring (8×8) [44,46]	0	0	87.9%	87.3%	25.6%	68.7%
Face Anonymizer [44]	11.38M	47.13G	90.5%	89.8%	62.6%	76.3%
DataMix (Ours) ($N_{pre}=1$)	**0.05M**	**0.09G**	91.2%	90.7%	**0.6%**	**51.5%**
DataMix (Ours) ($N_{pre}=2$)	0.12M	0.28G	91.2%	90.7%	**0.6%**	**51.6%**
DataMix (Ours) ($N_{pre}=3$)	0.20M	0.46G	**91.4%**	**91.0%**	**0.6%**	**51.5%**

5.1   Computer Vision: Facial Attribute Classification

We test our framework on two large-scale facial attribute classification benchmarks, CelebA [36] and LFWA [36], and design three attack methods to evaluate the preservation of privacy.

Setups. CelebA contains more than 200,000 celebrity images of more than 10,000 identities. LFWA has more than 10,000 images of more than 5,000 identities. For the two datasets, each image is annotated with 40 attributes, some of which are privacy-sensitive (*e.g.*, age). With Han *et al.* [16] as baseline and ResNet-18 [19] as backbone, we train the models for 100 epochs for CelebA and 600 epochs for LFWA using SGD with weight decay of 10^{-4}. We also decay the learning rate with cosine annealing when training. We evaluate the average classification accuracy over 40 attributes on both datasets. We run all experiment for four times and report the average results.

Metrics. Previous work [31] uses the pixel-wise reconstruction error (*e.g.*, PSNR) as a measurement for privacy: lower PSNR means worse reconstruction and better privacy. However, reconstruction error is not directly correlated to privacy: *e.g.*, a small distortion will lead to a large reconstruction error, but we can still identify the person from the image with small distortion. Instead, we propose to train an attack model and use the attack success rate as the evaluation metrics for privacy: a lower success rate indicates better privacy. We consider both the *person identity* and the *facial attributes* as the private information under attack. Concretely, we train two separate attack models for each privacy-preserving methods (with similar architecture as the baseline model)

Table 3. Privacy-preserving facial attribute classification on LFWA. *Bal.* denotes the balanced accuracy on the test set and *Recon.* represents the inverse mean square error of the reconstructed images (GAN-based) with the raw inputs. All of our DataMix have $S_G = 8$. *The GAN-based attack model is applied on the encrypted input image without the *preprocessing* model for fair comparison.

	Efficiency(↓better)		Accuracy(↑better)		Privacy(↓better)	
	Params	FLOPs	Test	Bal.	Recon.	Attrs.
Baseline (all on Edge)	11.21M	1.50G	91.1%	87.1%	−0.56	50.0%
Baseline (all on Cloud)	0	0	91.1%	87.1%	−0.00	87.1%
Adding Noise $\mathcal{N}(0,4)$ [44]	0	0	88.5%	82.5%	−0.03	82.7%
Adding Noise $\mathcal{N}(0,8)$ [44]	0	0	87.7%	81.3%	−0.02	81.4%
Blurring (16×16) [44,46]	0	0	88.8%	83.4%	−0.03	83.6%
Blurring (8×8) [44,46]	0	0	87.0%	80.6%	−0.07	77.8%
DataMix (Ours) ($N_{pre} = 1$)	**0.05M**	**0.09G**	90.5%	86.8%	**−0.37***	**50.6%**
DataMix (Ours) ($N_{pre} = 2$)	0.12M	0.28G	90.7%	86.9%	**−0.37***	50.7%
DataMix (Ours) ($N_{pre} = 3$)	0.20M	0.46G	**90.7%**	**87.1%**	**−0.37***	50.6%

that takes the mixed intermediate input $\tilde{x}_k^C$ and predicts the person identity and facial attributes corresponding to the input data x_k. We report the class-balanced attack success rate for the facial attributes (lower the better).

Baselines and Model Settings. We compare our framework with two hand-crafted approaches (*i.e.*, adding noise and blurring) and one learning-based adversarial obfuscator (*i.e.*, face anonymizer) [44,45]. As for our framework, we investigate different group sizes S_G (*i.e.*, number of images to be mixed) and different model partitions (how many computations to be offloaded to the cloud): the preprocess model contains N_{pre} convolution blocks, and the postprocess model contains the final layer only.

Recovering Face. We designed a GAN-based attack model to recovers the raw images from the mixed inputs. Since the cloud cannot relate the data from the same group, as mentioned in Sect. 4.2, the attacker has to reconstruct all the faces x_k from the transmitted mixed intermediate input $\tilde{x}_k^C$:

$$\tilde{x}_k^C = \mathcal{M}_{pre}^E(X) \cdot c. \tag{4}$$

We use Pix2Pix [25] as our attack model to recover the raw input image. We train the model to recover all inputs x_k's from the mixed intermediate input due to the ambiguity: *i.e.*, the model does not know which x_k corresponds to the desired image. During training, we use the Chamfer distance as the optimization objective, since the ordering of the outputs does not matter:

$$\mathcal{L}(x, y) = \sum_k \min_i \|x_k - y_i\|_1 + \sum_k \min_i \|y_k - x_i\|_1,$$

Fig. 6. Qualitative results of defending methods on the CelebA dataset. The images represent the ones accessible to the clouds and attackers with different defending methods. Personal identity and some private attributes like hair style are still recognizable with strong noise. Instead, our DataMix provides much better privacy.

where x and y are the original input images and the model's reconstructions, respectively. We train an attack model for each of the privacy-preserving methods, including adding noise, blurring, and our DataMix.

Results. In Fig. 6, we show the images accessible to the cloud and attackers with different defending methods. Our DataMix provides the best protection for personal identity and private facial attributes like hair style. As in Table 2, compared with hand-crafted on CelebA dataset, our framework achieves 3% higher accuracy and much better privacy (more than 20× lower attack success rate on person ID). Another interesting observation is that adding large Gaussian noise is not secure for protecting the person identity, which also indicates that the pixel-wise error is indeed not a good privacy metrics as large noise will lead to large pixel-wise error. Compared with adversarial obfuscator [44], our framework achieves higher accuracy and significantly better privacy with two orders of magnitude fewer FLOPs on the edge. This is not surprising, since these obfuscators usually use the encoder-decoder framework and are rather computationally expensive. Apart from the personal identity, our framework can also protect the output privacy, which is quantified by the attack success rate on facial attributes (including personal information such as age). However, previous approaches do not take this into consideration.

Similar conclusions can be drawn from the experiments on the LFWA dataset. As in Table 3, our framework outperforms the hand-crafted approaches on both the accuracy and the privacy, including the error of the reconstruction and the output privacy. Specifically, the reconstruction error is the mean square error between the reconstructed images and the raw inputs. To calculate the reconstruction error for DataMix, we first match the reconstructed images and the raw images in the same group so that the sum of the mean square error between each pair of images is the minimum. We take the average inverse mean square error of these pairs of images as the reconstruction error. The mixing operation increases the ambiguity of the transmitted data, which prevents the attacker from recovering the original images, giving a much lower inverse mean square error for our DataMix.

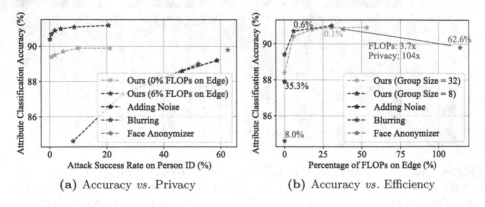

(a) Accuracy *vs.* Privacy   (b) Accuracy *vs.* Efficiency

Fig. 7. Our framework provides much better trade-offs among efficiency, accuracy and privacy. In 7b, the number next to each framework represents the attack success rate (the lower the better).

Trade-offs. In Fig. 7, we present the trade-offs between accuracy *vs.* privacy (by changing the group size) and accuracy *vs.* efficiency (by changing the number of layers to execute on the edge). In Fig. 7a, the space spanned by images becomes more complicated as the group size increases, leading to the accuracy degradation. At the same time, our framework achieves better privacy since the combination of a larger group introduces more ambiguities to the mixed data for the attacker. In Fig. 7b, when more convolution blocks are executed on the edge, more local computation will be required, making the fast and efficient inference more challenging. On the other hand, more layers on the edge means higher capacity for the projection from the raw images to the mixture space that is more friendly for classifier training, leading to a better performance of the main model for the mixing and de-mixing operation.

5.2   Speech: Keyword Spotting

Speech data also contains personal information such as speaker identity and sensitive content. We conduct experiments on Speech Commands [56] to show the generalization of our framework on different modalities.

Setups. The Speech Commands dataset has more than 100,000 utterances from 35 classes. For each utterance, we extract the normalized spectrogram from the waveform at a sampling rate of 16 kHz. We then leverage ResNet-18 [19] as baseline, which takes the spectrogram as input and classifies which class each utterance belongs to. We train all models for 100 epochs using SGD with cosine annealing for the learning rate decay.

Metrics. Similar to the previous task, we evaluate how the *speaker identity* and the *output content* are protected using two separate attack models. We then consider the attack success rates of these models as indicators of privacy.

Results. We present the quantitative results in Table 4. Adding noise only improves the privacy a little bit, but the accuracy degradation is significant.

Table 4. Privacy-preserving keyword spotting on Speech Commands. *ID.* represents the speaker ID on the test set and *Key.* denotes the accuracy of keyword spotting.

	Efficiency($\downarrow$better)		Accuracy($\uparrow$better)		Privacy($\downarrow$better)	
	Params	FLOPs	Val.	Test	ID.	Key.
Baseline (all on Edge)	11.18M	1.27G	96.6%	96.5%	1.2%	3.8%
Baseline (all on Cloud)	0	0	96.6%	96.5%	99.8%	96.5%
Adding Noise $\mathcal{N}(0,4)$ [44]	0	0	92.8%	91.5%	94.4%	91.5%
Adding Noise $\mathcal{N}(0,8)$ [44]	0	0	90.3%	89.1%	89.9%	89.1%
DataMix (Ours) ($N_{\mathrm{pre}} = 1$)	**0.02M**	**0.03G**	92.5%	92.2%	**19.4%**	**18.8%**
DataMix (Ours) ($N_{\mathrm{pre}} = 2$)	0.10M	0.18G	93.5%	93.3%	**15.2%**	**19.4%**
DataMix (Ours) ($N_{\mathrm{pre}} = 3$)	0.17M	0.34G	**94.4%**	**94.6%**	**12.7%**	**19.4%**

This is because large Gaussian noise will also weaken the capability of the model to extract effective features from the input and classify the keywords in the noised utterance. In contrast, our framework mixes the examples with different personal identities and contents with private random coefficient, making the data transmitted to the cloud ambiguous and non-invertible.

6   Conclusion

In this paper, we introduce the *privacy-preserving edge-cloud inference* framework, DataMix, to bring the best of the resource-hungry edge devices and the privacy-invasive cloud servers together for the model inference. We propose to delegate most of the model computations to the cloud and carefully design a mixing and de-mixing operation to protect the privacy of the data transmitted to the cloud. Our framework is efficient, accurate and privacy-preserving: extensive experiments on two computer vision datasets and a speech recognition dataset demonstrate that DataMix can greatly reduce the local computations on the edge with negligible loss of accuracy and no leakages of private information.

Acknowledgements. We thank MIT-IBM Watson AI Lab, MIT Quest for Intelligence, Samsung and Facebook for supporting this research. We thank AWS Machine Learning Research Awards for providing the computation resource.

References

1. Abadi, M., et al.: Deep learning with differential privacy. In: CCS (2016)
2. Agrawal, D., Aggarwal, C.C.: On the design and quantification of privacy preserving data mining algorithms. In: PODS (2001)
3. Agrawal, R., Srikant, R.: Privacy-Preserving Data Mining. In: SIGMOD (2000)
4. Bitouk, D., Kumar, N., Dhillon, S., Belhumeur, P.N., Nayar, S.K.: Face Swapping: Automatically Replacing Faces in Photographs. In: SIGGRAPH (2008)
5. Chen, J., Konrad, J., Ishwar, P.: VGAN-based image representation learning for privacy-preserving facial expression recognition. In: CVPRW (2018)
6. Chen, J., Wu, J., Konrad, J., Ishwar, P.: Semi-coupled two-stream fusion ConvNets for action recognition at extremely low resolutions. In: WACV (2017)
7. Chou, E., et al.: Privacy-preserving action recognition for smart hospitals using low-resolution depth images. In: NeurIPS Workshop (2018)
8. Courbariaux, M., Hubara, I., Soudry, D., El-Yaniv, R., Bengio, Y.: Binarized neural networks: training deep neural networks with weights and activations constrained to +1 or -1. arXiv (2016)
9. Dwork, C.: Differential privacy: a survey of results. In: TAMC (2008)
10. Gholami, A., et al.: SqueezeNext: hardware-aware neural network design. In: CVPR Workshop (2018)
11. Goodfellow, I., et al.: Generative adversarial nets. In: NIPS (2014)
12. Gross, R., Airoldi, E., Malin, B., Sweeney, L.: Integrating utility into face de-identification. In: Privacy Enhancing Technologies (2005)
13. Gross, R., Sweeney, L., Cohn, J., De la Torre, F., Baker, S.: Face de-identification. In: Senior, A. (ed.) Protecting Privacy in Video Surveillance. Springer, London (2009). https://doi.org/10.1007/978-1-84882-301-3_8

14. Gross, R., Sweeney, L., De La Torre, F., Baker, S.: Semi-supervised learning of multi-factor models for face de-identification. In: CVPR (2008)
15. Gross, R., Sweeney, L., De la Torre, F., Baker, S.: Model-based face de-identification. In: CVPRW (2006)
16. Han, H., Jain, A., Wang, F., Shan, S., Chen, X.: Heterogeneous face attribute estimation: a deep multi-task learning approach. In: TPAMI (2018)
17. Han, S., Mao, H., Dally, W.: Deep compression: compressing deep neural networks with pruning, trained quantization and Huffman coding. In: ICLR (2016)
18. Han, S., Pool, J., Tran, J., Dally, W.: Learning both weights and connections for efficient neural networks. In: NIPS (2015)
19. He, K., Zhang, X., Ren, S., Sun, J.: Deep residual learning for image recognition. In: CVPR (2016)
20. He, Y., Lin, J., Liu, Z., Wang, H., Li, L.-J., Han, S.: AMC: AutoML for model compression and acceleration on mobile devices. In: Ferrari, V., Hebert, M., Sminchisescu, C., Weiss, Y. (eds.) ECCV 2018. LNCS, vol. 11211, pp. 815–832. Springer, Cham (2018). https://doi.org/10.1007/978-3-030-01234-2_48
21. He, Y., Zhang, X., Sun, J.: Channel pruning for accelerating very deep neural networks. In: ICCV (2017)
22. Howard, A., et al.: Searching for MobileNetV3. arXiv (2019)
23. Howard, A.G., et al.: MobileNets: efficient convolutional neural networks for mobile vision applications. arXiv (2017)
24. Iandola, F.N., Han, S., Moskewicz, M.W., Ashraf, K., Dally, W., Keutzer, K.: SqueezeNet: AlexNet-Level Accuracy with 50x Fewer Parameters and ¡0.5MB Model Size. arXiv (2016)
25. Isola, P., Zhu, J.Y., Zhou, T., Efros, A.: Image-to-image translation with conditional adversarial networks. In: CVPR (2017)
26. Iyengar, V.S.: Transforming data to satisfy privacy constraints. In: KDD (2002)
27. Jaderberg, M., Vedaldi, A., Zisserman, A.: Speeding up convolutional neural networks with low rank expansions. In: BMVC (2014)
28. Jourabloo, A., Yin, X., Liu, X.: Attribute preserved face de-identification. In: ICB (2015)
29. Kim, T.H., Kang, D., Pulli, K., Choi, J.: Training with the invisibles: obfuscating images to share safely for learning visual recognition models. arXiv (2019)
30. Krishnamoorthi, R.: Quantizing deep convolutional networks for efficient inference: a whitepaper. arXiv (2018)
31. Leroux, S., Verbelen, T., Simoens, P., Dhoedt, B.: Privacy aware offloading of deep neural networks. In: ICML Workshop (2018)
32. Li, M., Lai, L., Suda, N., Chandra, V., Pan, D.Z.: PrivyNet: a flexible framework for privacy-preserving deep neural network training. arXiv (2017)
33. Li, T., Lin, L.: AnonymousNet: natural face de-identification with measurable privacy. In: CVPR Workshop (2019)
34. Lin, J., Gan, C., Han, S.: TSM: temporal shift module for efficient video understanding. In: ICCV, pp. 7083–7093 (2019)
35. Liu, Z., Li, J., Shen, Z., Huang, G., Yan, S., Zhang, C.: Learning efficient convolutional networks through network slimming. In: ICCV (2017)
36. Liu, Z., Luo, P., Wang, X., Tang, X.: Deep learning face attributes in the wild. In: ICCV (2015)
37. Ma, N., Zhang, X., Zheng, H.-T., Sun, J.: ShuffleNet V2: practical guidelines for efficient CNN architecture design. In: Ferrari, V., Hebert, M., Sminchisescu, C., Weiss, Y. (eds.) Computer Vision – ECCV 2018. LNCS, vol. 11218, pp. 122–138. Springer, Cham (2018). https://doi.org/10.1007/978-3-030-01264-9_8

38. Newton, E.M., Sweeney, L., Malin, B.: Preserving privacy by de-identifying face images. In: TKDE (2005)
39. Oh, S.J., Benenson, R., Fritz, M., Schiele, B.: Faceless person recognition: privacy implications in social media. In: Leibe, B., Matas, J., Sebe, N., Welling, M. (eds.) ECCV 2016. LNCS, vol. 9907, pp. 19–35. Springer, Cham (2016). https://doi.org/10.1007/978-3-319-46487-9_2
40. Oh, S.J., Fritz, M., Schiele, B.: Adversarial image perturbation for privacy protection: a game theory perspective. In: ICCV (2017)
41. Osia, S.A., et al.: A hybrid deep learning architecture for privacy-preserving mobile analytics. In: TKDD (2017)
42. Osia, S.A., Taheri, A., Shamsabadi, A.S., Katevas, K., Haddadi, H., Rabiee, H.R.: Deep Private-feature extraction. In: TKDE (2018)
43. Raval, N., Machanavajjhala, A., Cox, L.P.: Protecting visual secrets using adversarial nets. In: CVPRW (2017)
44. Ren, Z., Lee, Y.J., Ryoo, M.S.: Learning to anonymize faces for privacy preserving action detection. In: Ferrari, V., Hebert, M., Sminchisescu, C., Weiss, Y. (eds.) ECCV 2018. LNCS, vol. 11205, pp. 639–655. Springer, Cham (2018). https://doi.org/10.1007/978-3-030-01246-5_38
45. Ryoo, M.S., Kim, K., Yang, H.J.: Extreme low resolution activity recognition with multi-siamese embedding learning. In: AAAI (2018)
46. Ryoo, M.S., Rothrock, B., Fleming, C., Yang, H.J.: Privacy-preserving human activity recognition from extreme low resolution. In: AAAI (2017)
47. Sandler, M., Howard, A., Zhu, M., Zhmoginov, A., Chen, L.C.: MobileNetV2: inverted residuals and linear bottlenecks. In: CVPR (2018)
48. Shokri, R., Shmatikov, V.: Privacy-preserving deep learning. In: CCS (2015)
49. Simonyan, K., Zisserman, A.: Very deep convolutional networks for large-scale image recognition. In: ICLR (2015)
50. So, D., Le, Q., Liang, C.: The evolved transformer. In: ICML (2019)
51. Sweeney, L.: K-anonymity: a model for protecting privacy. Int. J. Uncertain. Fuzziness Knowl.-Based Syst. (2002)
52. Tian, Y., Krishnan, D., Isola, P.: Contrastive representation distillation. ICLR (2020)
53. Tramèr, F., Boneh, D.: Slalom: fast, verifiable and private execution of neural networks in trusted hardware. In: ICLR (2019)
54. Wang, J., Zhang, J., Bao, W., Zhu, X., Cao, B., Yu, P.S.: Not just privacy: improving performance of private deep learning in mobile cloud. In: KDD (2018)
55. Wang, K., Liu, Z., Lin, Y., Lin, J., Han, S.: HAQ: hardware-aware automated quantization with mixed precision. In: CVPR (2019)
56. Warden, P.: Speech commands: a dataset for limited-vocabulary speech recognition. arXiv (2018)
57. Wu, Z., Liu, Z., Lin, J., Lin, Y., Han, S.: Lite transformer with long-short range attention. In: ICLR (2020)
58. Wu, Z., Wang, Z., Wang, Z., Jin, H.: Towards privacy-preserving visual recognition via adversarial training: a pilot study. In: Ferrari, V., Hebert, M., Sminchisescu, C., Weiss, Y. (eds.) ECCV 2018. LNCS, vol. 11220, pp. 627–645. Springer, Cham (2018). https://doi.org/10.1007/978-3-030-01270-0_37
59. Xie, Q., Hovy, E., Luong, M.T., Le, Q.V.: Self-training with noisy student improves ImageNet classification. In: arXiv (2019)
60. Zhang, H., Cisse, M., Dauphin, Y.N., Lopez-Paz, D.: MIXUP: beyond empirical risk minimization. In: ICLR (2018)

61. Zhang, X., Zhou, X., Lin, M., Sun, J.: ShuffleNet: an extremely efficient convolutional neural network for mobile devices. In: CVPR (2018)
62. Zhu, C., Han, S., Mao, H., Dally, W.: Trained ternary quantization. In: ICLR (2017)

Neural Re-rendering of Humans
from a Single Image

Kripasindhu Sarkar[1]([⊠]), Dushyant Mehta[1], Weipeng Xu[2], Vladislav Golyanik[1],
and Christian Theobalt[1]

[1] MPI for Informatics, SIC, Saarbrücken, Germany
ksarkar@mpi-inf.mpg.de
[2] Facebook Reality Labs, Pittsburgh, USA

Abstract. Human re-rendering from a single image is a starkly under-constrained problem, and state-of-the-art algorithms often exhibit undesired artefacts, such as over-smoothing, unrealistic distortions of the body parts and garments, or implausible changes of the texture. To address these challenges, we propose a new method for neural re-rendering of a human under a novel user-defined pose and viewpoint, given one input image. Our algorithm represents body pose and shape as a parametric mesh which can be reconstructed from a single image and easily reposed. Instead of a colour-based UV texture map, our approach further employs a learned high-dimensional UV feature map to encode appearance. This rich implicit representation captures detailed appearance variation across poses, viewpoints, person identities and clothing styles better than learned colour texture maps. The body model with the rendered feature maps is fed through a neural image-translation network that creates the final rendered colour image. The above components are combined in an end-to-end-trained neural network architecture that takes as input a source person image, and images of the parametric body model in the source pose and desired target pose. Experimental evaluation demonstrates that our approach produces higher quality single-image re-rendering results than existing methods.

Keywords: Neural rendering · Pose transfer · Novel view synthesis

1 Introduction

Algorithms to realistically render dressed humans under controllable poses and viewpoints are essential for character animation, 3D video, or virtual and augmented reality, to name a few. Over the past decades, computer graphics

---

Project webpage: http://gvv.mpi-inf.mpg.de/projects/NHRR/.

Electronic supplementary material The online version of this chapter (https://doi.org/10.1007/978-3-030-58621-8_35) contains supplementary material, which is available to authorized users.

© Springer Nature Switzerland AG 2020
A. Vedaldi et al. (Eds.): ECCV 2020, LNCS 12356, pp. 596–613, 2020.
https://doi.org/10.1007/978-3-030-58621-8_35

and vision have developed impressive methods for high-fidelity artist-driven and reconstruction-based human modelling, high-quality animation, and photo-realistic rendering. However, these often require sophisticated multi-camera setups, and deep expertise in animation and rendering, and are thus costly, time-consuming and difficult to use. Recent advances in monocular human reconstruction and neural network-based image synthesis open up a radically different approach to the problem, neural re-rendering of humans from a single image. Given a single reference image of a person, the goal is to synthesise a photo-real image of this person in, for instance, a user-controlled new pose, modified body proportions, the same or different garments, or a combination of these.

Fig. 1. Given an image of a person, our neural re-rendering approach allows synthesis of images of the person in different poses, or with different clothing obtained from another reference image.

There has been tremendous progress in monocular human capture and re-rendering [2,6,12,19,21,23,25,33,45] towards this goal. However, owing to the starkly underconstrained nature of the problem, true photo-realism under all possible conditions has not yet been achieved. Methods frequently exhibit unwanted over-smoothing and a lack of details in the rendered image, unrealistic distortions of body parts and garments, or implausible texture alterations.

We, therefore, propose a new algorithm for monocular neural re-rendering of a dressed human under a novel user-defined pose and viewpoint, which has starkly improved visual quality, see Figs. 1, 3, 6, 7. We take inspiration from recent work on neural rendering of general scenes with a continuous [48] or a multi-dimensional feature representation with implicit [50] or explicit [47] occlusion handling that are learned from multi-view images or videos.

Our algorithm represents body pose and shape with the SMPL parametric human surface model [29], which can be easily reposed. Instead of modelling appearance as explicit colour maps, *e.g..*, learned colour-based UV texture maps on the body surface [12,33], we employ a learned high-dimensional UV feature map to encode appearance. This rich implicit representation learns the detailed appearance variation across poses, viewpoints, person identities and clothing styles. Given a single image of a person, we predict pixel correspondences to the SMPL [29] mesh using DensePose [37]. We then extract partial UV texture maps based on the observed body regions and use a neural network to convert it to a complete UV feature map, with a d-dimensional feature per texel. The UV feature map is then rendered in the desired target pose and passed through a neural image translation network that creates the final rendered image. These components are combined in an end-to-end trained neural architecture. In quantitative experiments and a user study to judge the qualitative results, we show that the visual quality of our results improves over the current state of the art.

To summarise, our **contributions** are as follows:

- A new end-to-end trainable method that combines monocular parametric 3D body modelling, a learned detail-preserving neural-feature based body appearance representation, and a neural network based image-synthesis network to enable highly realistic human re-rendering from a single image;
- state-of-the-art results on the DeepFashion dataset [27] which are confirmed with quantitative metrics, and qualitatively with a user study.

2   Related Work

While our proposed approach relates to many sub-fields of visual computing, for brevity we only elaborate on the immediately relevant work on human body re-enactment and neural rendering methods for object and scene rendering.

2.1   Classical Methods for Novel View Synthesis

Earlier methods for image-based 3D reconstruction and novel view synthesis rely on traditional concepts of multi-view geometry, explicit 3D shape and appearance reconstruction, and classical computer graphics or image-based rendering. Methods based on light fields use ray space representations or coarse multi-view geometry models for novel view synthesis [4,11,22]. To achieve high quality, dense camera arrays are required, which is impractical. Other algorithms capture and operate on dense depth maps [60], layered depth images [41], 3D point clouds [1,26,40], meshes [32,52], or surfels [5,36,55] for dynamic scenes. Multi-view stereo can be combined with fusion algorithms operating with implicit geometry and achieving more temporally consistent reconstructions over short time windows [9,13,34]. Dynamic scene capture and novel view synthesis were also shown with a low number of RGB or RGB-D cameras [15,49,57,58]. While reconstruction is fast and feasible with fewer cameras, the coarse approximate geometry often leads to rendering artefacts.

2.2   Neural Rendering of Scenes and Objects

Recently, neural rendering approaches have shown promising results for scenes and objects. Image-based rendering (IBR) methods reconstruct scene geometry with classical techniques and use it to render novel views [7,8]. Lack of observations can cause high uncertainty in novel views. On the other hand, neural rendering approaches [47,48,51,66] can generate higher-quality results by leveraging collections of training data. Many applications of neural rendering have been recently shown, ranging from synthesising view-dependent effects [51,66] to learning the shape and appearance priors from sparse data [39,56].

Only a few works on neural scene representation and rendering can handle dynamic scenes [19,28]. Some methods combine explicit dynamic scene reconstruction and traditional graphics rendering with neural re-rendering [18,19,31,51]. Thies *et al.* [50] combine *neural textures* with the classical graphics pipeline for novel view synthesis of static objects and monocular video re-rendering. Their technique requires a scene-specific geometric proxy which has to be reconstructed before the training. Instead of more complex joint reasoning of the geometry and appearance needed from the intermediate representation by neural rendering approaches such as that of Sitzmann *et al.* [48], for our human-specific application scenario the coarse geometry is handled by the posable SMPL mesh, with a feature map similar to the Thies *et al.* [50] capturing clothing appearance, which includes fine-scaled geometry, and clothing textures.

Several approaches address related problems such as generating images of humans in new poses [3,30,33,35,62], or body re-enactment from monocular videos [6], which are discussed next.

2.3   Human Re-enactment and Novel View Rendering

Recent work on photo-realistic human body re-enactment and novel view rendering can be sub-classified along various dimensions.

Object-agnostic approaches [44,45] model deformable objects directly in the image space. Siarohin *et al.* [45] learn keypoints in a self-supervised manner and capture deformations in the vicinity of the keypoints using affine transforms. Features extracted from the source image are deformed to the target using the predicted transformations and passed on to a generator. Additional predictions of dis-occluded regions indicate to the generator the regions which have to be rendered based on the context. Zhu et al.. [65] leverage geometric constraints and optical flow for synthesising novel views of humans from a single image.

Object-specific techniques have the same core components as above, *i.e.*, colour or feature transformation from source to target, occlusion reasoning or inpainting, and photo-realistic image generation from the warped feature or colour image. The key difference is that the feature transformation, occlusion reasoning, and inpainting are guided by an underlying object model, which, in our case, is a parametric human body mesh. Kim *et al.* [19] achieve full control over the head pose and facial expressions in photo-realistic renderings of a target actor by an adversarial training with a performance of the target actor. Dense-Pose Transfer [33] uses direct texture transfer from the input image to the SMPL model, inpaints the occluded regions of the texture and renders it in a new pose. This image is blended with the image resulting from direct conditional generation from the input image, input Densepose, and target Densepose. Zablotskaia *et al.* [59] generate subsequent video frames of human motion and use a direct warping guided by the reference frame, previously generated frame, and Dense-Pose representations of the past and future frames. Their method does not rely on an explicit UV texture map. ClothFlow [14] implicitly captures the geometric transformation between the source and target image by estimating dense flow. Chan*et al.* [6] learn a subject-specific puppeteering system using a video of the subject such that all parts of the subject's body are seen in advance. The GAN-based rendering is driven by 2D pose extracted from the target subject. Zhou *et al.* [63] also learn a personalised model using piecewise affine transforms of the part-segmented source image for modelling pose changes, generating the person image in front of a clean background plate, with a second stage fusing a given background image with the generated person's image. In contrast to Liu *et al.* [25], we transfer appearance from source to target image using a UV feature map. Instead of directly predicting missing regions of the UV texture map, coordinate-based inpainting [12] predicts correspondence between all regions on the UV texture map and the input image pixels. This results in more texture details in body regions that become dis-occluded when re-posing. As shown in Sect. 4, our UV feature map based approach yields results of much higher quality in comparisons. Shysheya *et al.* [42] explicitly model the body texture and implicitly handle the shape. In contrast, while we explicitly handle the coarse shape, we use a UV feature map to model the fine-scaled shape and clothing texture implicitly. Lazowa *et al.* [21] propose an approach for reconstruction of textured 3D human models from a single image. Similar to our approach, it extracts a partial UV texture map using DensePose but inpaints the UV texture map using a GAN based supervision directly applied to the texture map. Additionally—and similar to Alldieck *et al.* [2]—it predicts a displacement map on top of the SMPL mesh to capture clothing details not present in the SMPL model. Our approach does not explicitly model clothing details.

In contrast to existing methods, we propose a new end-to-end trainable method that combines monocular parametric 3D body modeling [12,33], a learned neural detail-preserving surface feature representation [50], and a neural image-synthesis network for highly realistic human re-rendering from a single image.

3   Method

Given an image I_s of a person, we synthesise a new image of the person in a different target body pose. Our approach comprises of four distinct steps. The first step uses DensePose [37] to predict dense correspondences between the input image I_s and the SMPL model. This allows a *UV texture map* T_s to be extracted for the visible regions. The second step uses a U-Net [38] based network, which we term *FeatureNet*, to construct the full *UV feature map* F_s from the partial RGB UV texture map T_s. F_s contains a d-dimensional feature representation for all texels, both visible and occluded in the source image. The third step takes a target pose as input, and 'renders' the UV feature map F_s to produce a d dimensional *Feature image* $R_{s \to t}$. The fourth step uses a generator network based on Pix2PixHD [54], which we term *RenderNet*, to generate a photorealistic image $I_{s \to t}$ of the reposed person, from the input Feature image. The overview of our pipeline is shown in Fig. 2.

3.1 Extracting a Partial UV Texture Map from the Input Image

The pixels of the input image are transformed into UV space through matches predicted with DensePose. We use the ResNet-101 based variant of DensePose for predicting the correspondences for the body regions visible in the image. The network is pre-trained on COCO-DensePose dataset and provides 24 body segments and their part-specific U, V coordinates of SMPL model. For easier mapping, the 24 part-specific UV maps are combined to form a single UV Texture map T_s in the format provided in SURREAL dataset [53] through a pre-computed lookup table. Note that one could putatively use monocular 3D pose estimation methods (*e.g..*, [17]) to compute SMPL parameters, and subsequently, the DensePose of the input image. However, frequent misalignments of the predictions with the end-effector positions in the image lead to significant artefacts in the UV texture map for hands and feet in that case and thus such an approach is not advised [2].

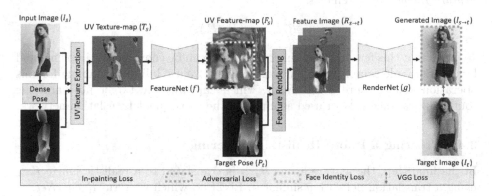

Fig. 2. Pipeline Overview: Given a source image I_s, we extract the UV texture map T_s of an underlying parametric body mesh model for the body regions visible in the image. FeatureNet converts the partial UV texture map to a full UV feature map, which encodes a richer 16D representation at each texel. Given a new pose P_t, the parametric body mesh can be re-posed and textured with the UV feature map to produce an intermediate Feature Image $R_{s\rightarrow t}$. RenderNet converts the intermediate 16-channel feature image to a realistic image.

3.2 Generating the Full UV Feature Map

The partial (on account of occlusion) texture map T_s is converted to a full UV feature map F_s using a U-Net-like convolutional network f, which we term *FeatureNet*. That is,

$$F_s = f(T_s).$$

RenderNet comprises of four down-sampling blocks followed by four up-sampling blocks. Therefore, a partial input texture of the spatial dimension of 256×256 is transformed into a spatial dimension of 16×16 in the middle-most layer. Each downsampling block consists of two convolutions followed by maxpool operation.

For up-sampling blocks, we use bilinear upsampling followed by two convolutions. The final convolutional layer produces a d-dimensional (channel) *UV feature map* which is used subsequently for rendering a feature image. The first three channels of the UV feature map can be supervised to in-paint the input partial UV texture map T_s, thus having a small subset of the feature channels resemble the classical colour texture map. Our experiments use 16 feature channels.

3.3   Intermediate Feature Image Rendering

The SMPL mesh can be reposed using a target pose P_t, which can be extracted from a target image I_t, or obtained from a different source. In our case, when given a target image I_t, we directly obtain the DensePose output, which is equivalent to the reposed SMPL model. Given the source feature map F_s, we render the SMPL mesh through the DensePose output P_t to produce a d-dimensional *Feature Image* $R_{s \to t}$. That is,

$$R_{s \to t} = r(F_s, P_t).$$

Note that this feature rendering operation r can be conveniently implemented by differentiable sampling. In our experiments, we use bilinear sampling for this operation. The feature image $R_{s \to t}$, which captures the target pose and the source appearance is then used as input to the subsequent translation network.

3.4   Creating a Photo-Realistic Rendering

In the final step, the feature image $R_{s \to t}$ is translated to a realistic image $I_{s \to t}$ using a translation network g similar to Pix2Pix, which we term *RenderNet*:

$$I_{s \to t} = g(R_{s \to t}).$$

RenderNet comprises of (a) 3 down-sampling blocks, (b) 6 residual blocks, (c) 3 up-sampling blocks and finally (d) a convolution layer with Tanh activation that gives the final output. The discriminator for adversarial training of RenderNet al.so uses the multiscale design of Pix2PixHD [54]. In our experiments, we use a three scale discriminator network for adversarial training.

3.5   Training Details and Loss Functions

During training, our system takes pairs of images (I_s, I_t) of the same person (but in different poses) as input. Partial texture T_s extracted from the source image I_s is passed through the above-mentioned operations to produce the generated output $I_{s \to t}$. That is,

$$I_{s \to t} = g \circ r \circ f(T_s, P_t).$$

Note that all operations g, r and f are differentiable. We train the entire system *end-to-end* and optimise the parameters of FeatureNet ($g(\cdot)$) and RenderNet ($f(\cdot)$). We use the combination of the following loss functions in our system:

- **Perceptual Loss**: We use a perceptual loss based on the VGG Network [16]—the difference between the activations on different layers of the pre-trained VGG network [46] applied on the generated image $I_{s \to t}$ and ground truth image target image I_t.

$$L_p = \sum \frac{1}{N^j} |p^j(I_{s \to t}) - p^j(I_t)|.$$

Here, p^j is the activation and N^j the number of elements of the j-th layer in the ImageNet pre-trained VGG network.

- **Adversarial Loss**: We use a multiscale discriminator D of Pix2PixHD [54] for enforcing adversarial loss L_{adv} in our system. The multiscale discriminator D is conditioned on both the generated image and rendered feature image.

- **Face Identity Loss**: We use a pre-trained network to ensure that the extracted UV feature map and RenderNet preserve the *face identity* on the cropped face of the generated and the ground truth image.

$$L_{face} = |N_{face}(I_{s \to t}) - N_{face}(I_t)|.$$

Here, N_{face} is the pre-trained SphereFaceNet [24]

- **Intermediate in-Painting Loss**: To mimic classical colour texture map, we enforce a loss L_{tex} on the first three channels of the output of the in-painting network. This loss is set to the sum of 1) l_1 distance of the visible part of the partial source texture and generated texture and 2) l_1 distance of the visible part of the partial target texture and generated texture.

The final loss on the generator is then

$$L_G = \lambda_{VGG} L_p + \lambda_{face} L_{face} + \lambda_{tex} L_{tex} + \lambda_{GAN} L_{adv}.$$

The conditional discriminator D is updated every step enforcing binary cross-entropy loss on real and fake images. We train the networks end-to-end using Adam optimiser [20] with an initial learning rate of 2×10^4, β_1 as 0.5 and no weight decay. The loss weights are set empirically to $\lambda_{GAN} = 1, \lambda_{VGG} = 10, \lambda_{face} = 5, \lambda_{tex} = 1$. For speed, we pre-compute DensePose on the images and directly read them as input.

During testing, the system takes as input a single image of a person and a target Densepose. The target pose can be extracted by DensePose RCNN on the image of the source person in a different pose (used in the experiments on DeepFashion dataset), or alternatively it can be obtained by reposing the SMPL mesh of the source body. In many cases, the actor can be a completely different person (see Fig. 5 and 7). The neural texture is then rendered using the given target Densepose which is followed by the translation network to generate a realistic image of the source person in the target pose.

4   Experimental Results

4.1   Experimental Setup

Datasets We use the In-shop Clothes Retrieval Benchmark of DeepFashion dataset [27] for our main experiments. The dataset comprises of 52,712 images of fashion models with 13,029 different clothing items in different poses. For training and testing, we consider the split provided by Siarohin *et al.* [43], which is also used by other related works [12,33]. We also show qualitative results of our method with Fashion dataset [59]. Fashion dataset has 500 training and 100 test videos, each containing roughly 350 frames. The videos are single person sequences, containing different people catwalking in different clothes.

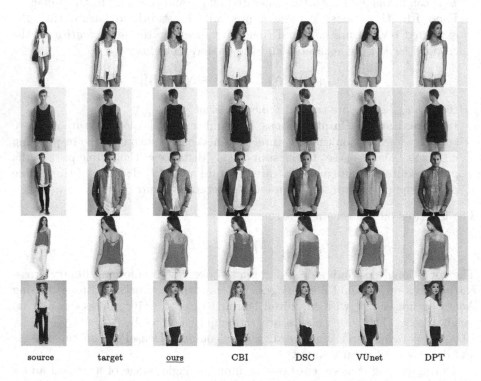

source   target   ours   CBI   DSC   VUnet   DPT

Fig. 3. Results of our method, CBI [12], DSC [43], VUnet [10] and DPT [33]. Our approach produces higher quality renderings than the competing methods.

4.2   Comparison with SOTA

We compare our results with four state-of-the-art methods, *i.e.,* Coordinate Based Inpainting (CBI) [12], Deformable GAN (DSC) [43], Variational U-Net (VUnet) [10] and Dense Pose Transfer (DPT) [33]. The qualitative results are shown in Fig. 3. It can be observed that our results show higher realism and better preserve identity and garment details compared to the other methods.

The quantitative results are provided in Table 1. Due to inconsistent reporting (or unavailability) of the metrics for the existing approaches, we computed them ourselves. To this end, we collected the results of 176 testing pairs for each state-of-the-art method (the testing pairs and results were kindly provided by the authors of Coordinate Based In-painting [12]) and used them for this report. We use the following two metrics for comparison - 1) Structural Similarity Index (SSIM) [64] 2) Learned Perceptual Image Patch Similarity (LPIPS) [61]. SSIM is a structure preservation metric widely used in the existing literature. Though it is an excellent metric for assessment of image degradation quality, it often does not reflect human perception [61]. On the other hand, the recently introduced LPIPS claims to capture human judgment better than existing hand-designed metrics. In terms of SSIM, we perform as well as the existing methods, whereas we significantly outperform them on LPIPS metric. Please note that similar to other learning-based methods, our approach will struggle with poses that are far from those seen in the training set. However, our method performs well in such scenarios for many cases. Qualitative results on some target poses outside of training dataset distribution are shown in Fig. 5.

Table 1. Comparison with state-of-the-art methods, using various perceptual metrics, Structural Similarity Index (SSIM) [64] and Learned Perceptual Image Patch Similarity (LPIPS) [61]. ↑ (↓) means higher (lower) is better.

	SSIM ↑	LPIPS ↓
CBI [12]	0.766	0.178
DSC [43]	0.750	0.214
VUnet [10]	0.739	0.202
DPT [33]	0.759	0.206
Ours	**0.768**	**0.164**
GT	1.0	0.0

4.3   User Study

To assess the qualitative impact of our method, we perform an extensive user study which compares it with two other state-of-the-art pose transfer methods – Coordinate Base Inpainting (CBI) [12] and DensePose Transfer (DPT) [33]. We train on the DeepFashion dataset [27] and generate renderings on the test split. The user study follows several criteria. First, it covers as large a variety of source and target poses. Second, the ratio between the male and female samples reflects the same ratio of the dataset. It also contains failure cases as those shown in Fig. 8 with difficult decisions. In total, we prepare 26 samples containing the source image (explicitly marked as such) and three novel views generated by CBI, DPT and our method (labeled as view A, B or C in randomised order). For each sample, two questions are asked: 1) Which view looks the most like the person in the source image? and 2) Which view looks the most realistic?

The user study was performed with a browser interface, the order of questions is randomised, and 46 anonymous participants submitted their answers. The results are as follows. The first question has been answered in 46.2% of the cases in favour of CBI, and in 53.8% of the cases in favour of our method. In all cases, DPT has always been the last choice. The second question has been answered by 30.8% of the participants in favour of CBI, and by 69.2% of the participants in favour of our approach. Again, DPT was preferred in no case.

The user study shows that our method achieves state-of-the-art quality in preserving the identity, and significantly outperforms the baselines in the realism of the generated images. In 23% of the overall cases, the participants have preferred CBI as the best identity-preserving method and, at the same time, our method as those producing the most realistic renderings. In contrast, there was only one case (3.8%) when our method had been voted as the best identity-preserving and, at the same time, CBI was chosen as the approach producing most realistic renderings.

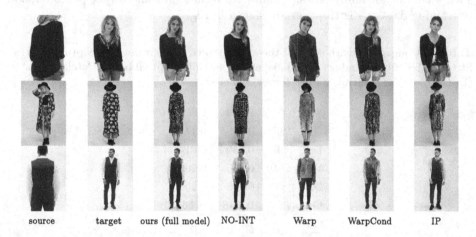

| source | target | ours (full model) | NO-INT | Warp | WarpCond | IP |

Fig. 4. Results of different baselines and our full model. *No-Int* has no intermediate loss, *Warp* and *WarpCond* perform translation on warped partial texture and *IP* inpaints full colour-texture followed by translation. Under extreme poses and strong occlusions, our method outperforms all the baselines (see Sect. 4.4).

| Source | Target | Pose Transfer | Source | Target | Pose Transfer | Source | Target | Pose Transfer |

Fig. 5. Generalisation of our method to new body poses. The images of the target pose are obtained from the internet.

4.4   Ablation Study

To study the advantage of the learned neural texture over other natural choices of texture-based human re-rendering, we created the following three baselines.

IP - This baseline involves two stages. First, we train an inpainting network to generate the full UV texture map from the partial UV texture map extracted from the input image. We use the same in-painting loss function as described in Sect. 3.5 for training this network. After the convergence, we fix and use this network to generate full colour texture from a partial input texture. This full 3-channel UV texture map is then rendered into an intermediate image, and translated through a trained RenderNet g.

Warp - In this experiment, we warp the incomplete partial UV texture map to the target pose. The reposed incomplete 3-channel intermediate image is then fed to a trained RenderNet g to produce realistic output.

WarpCond - In this experiment, we warp the partial source texture to the target pose P_T similar to the previous experiment. In addition to the reposed incomplete texture, we also condition the generator network g with the target DensePose image. The target DensePose image acts as a cue to the generator when the texture information is missing.

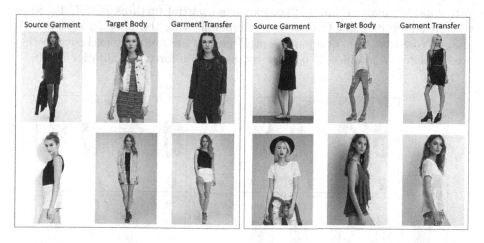

Fig. 6. Garment Transfer: Our approach can also be used to render garments from a source image onto the person in a target image.

In all these baselines, the architecture of RenderNet g and the losses on the generated image are the same as ours. The only difference is in the number of input channels to RenderNet. Besides, we perform an ablation experiment with an identical pipeline as ours, except we do not enforce any intermediate texture loss – **No-Int**. The qualitative results of all the networks are shown in Fig. 4.

It can be seen that our methods using a richer intermediate representation (*full* and *No-Int*) produce more realistic images than the other baselines. *Baseline-IP* performs well but produces smooth output compared to the other methods. Because of the lack of details, *Baseline-Warp* often produces non-realistic output in both face and garment regions. When the incomplete texture information is supervised with additional DesnsePose image (as in *Baseline-WarpCond*), the output is of higher quality. In the presence of strong occlusions, the method fails, as the translation network g is incapable of performing both inpainting and realistic rendering at the same time. In contrast, our methods performed well in all the scenarios. Adding the intermediate texture loss (to mimic real texture) to the part of neural texture helps our network to converge faster. However, over a large number of iterations, the quality of the final result without such intermediate loss (*No-Int*) is similar to that with intermediate loss.

4.5   Garment Transfer

Our method can be naturally extended to perform garment transfer without any further training. Given an image of a person with the source body, we extract the partial texture T_s' of the 'body regions' (*e.g.*, face, hands and regions with garments which remain unchanged). We use part indices provided by DensePose to extract the partial texture of the required body parts. Next, we extract the partial texture T_t' of the 'garment regions' of an image with the desired target garments. We make a union of the extracted partial textures $T_s' \cup T_t'$ based on their texel regions and feed it to our pipeline with the pose P_s of the body image. Note that texel occupancies of T_s' and T_t' are mutually exclusive as they are extracted from different body parts. See Fig. 6 for the qualitative results.

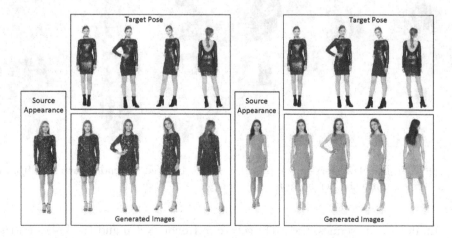

Fig. 7. Motion transfer on the Fashion dataset [59]. Our approach also can generate realistic renderings for a sequence of poses given a single source image.

4.6   Motion Transfer

Even though we did not train specifically for generating videos, our method can be applied to each frame of a driving video to create motion transfer. To this end, we keep the source image of the imitator fixed and use the pose from the actor of the driving video (for each frame) in our system to create image animation. We perform the experiment on Fashion Dataset [59] and show our results in Fig. 7.

Please refer to the supplemental document and the accompanying video for more results.

5   Discussion

Limitations. Even though we produce high-quality novel views which are preserving the identity and look very realistic, there remain certain limitations for future work to address. Figure 8 visualises two representative examples which are difficult for our as well as competing methods. In the first row, the head in the source image is only partially visible so that the methods have high uncertainty in the frontal facial view and change the gender to female (*e.g.,* hallucinate long hair). In the second case, the source texture is too fine-grained for the methods so that some hallucinate repeating patterns and the other ones generate patterns reminiscent of noise. In this case, our method generates a texture which is neither repetitive nor looks like noise, and which is still far from the reference.

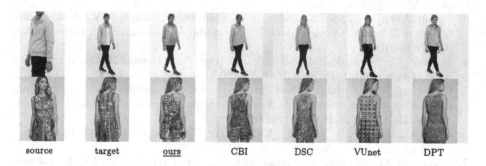

| source | target | ours | CBI | DSC | VUnet | DPT |

Fig. 8. Limitations: Even though our method produces better quality results than all competing approaches, it nevertheless has some limitations, which are also shared by all competing methods. The top row highlights failures arising out of biases in the training set, while the bottom row highlights failures owing to fine scaled textures which are not effectively captured by any approach.

Future Extension of Our UV Feature Maps. Instead of sampling RGB textures from the input image to construct a partial UV texture map, learned CNN based features could be used to construct a more informative partial UV feature map, which putatively captures off-geometry details not modelled by the SMPL mesh. Then FeatureNet would convert this partial UV feature map to a full UV feature map. Another alternative would be to use displacement map prediction similar to prior work [2, 21] to capture off-geometry details.

6 Conclusion

In this work, we present an approach for human image synthesis, which allows us to change the camera view and the pose and garments of the subject in the source image. Our approach uses a high-dimensional UV feature map to encode appearance as an intermediate representation, which is then re-posed and translated using a generator network to achieve realistic rendering of the input subject in a novel pose. We qualitatively and quantitatively demonstrate the efficacy of our proposed approach at better preserving identity and garment details compared to the other competing methods. Our system, trained once for pose guided image synthesis, can be directly used for other tasks such as garment transfer and motion transfer.

Acknowledgements. This work was supported by the ERC Consolidator Grant 4DReply (770784).

References

1. Agarwal, S., Furukawa, Y., Snavely, N., Simon, I., Curless, B., Seitz, S.M., Szeliski, R.: Building Rome in a day. Commun. ACM **54**(10), 105–112 (2011)
2. Alldieck, T., Pons-Moll, G., Theobalt, C., Magnor, M.: Tex2shape: detailed full human body geometry from a single image. In: International Conference on Computer Vision (ICCV) (2019)
3. Balakrishnan, G., Zhao, A., Dalca, A.V., Durand, F., Guttag, J.V.: Synthesizing images of humans in unseen poses. In: Computer Vision and Pattern Recognition (CVPR) (2018)
4. Buehler, C., Bosse, M., McMillan, L., Gortler, S.J., Cohen, M.F.: Unstructured lumigraph rendering. In: SIGGRAPH (2001)
5. Carceroni, R.L., Kutulakos, K.N.: Multi-view scene capture by surfel sampling: from video streams to non-rigid 3d motion, shape and reflectance. Int. J. Comput. Vision (IJCV) **49**(2), 175–214 (2002)
6. Chan, C., Ginosar, S., Zhou, T., Efros, A.A.: Everybody dance now. In: International Conference on Computer Vision (ICCV) (2019)
7. Chaurasia, G., Duchêne, S., Sorkine-Hornung, O., Drettakis, G.: Depth synthesis and local warps for plausible image-based navigation. ACM Trans. Graphics **32**, 1–13 (2013)
8. Debevec, P., Yu, Y., Borshukov, G.: Efficient view-dependent image-based rendering with projective texture-mapping. In: Eurographics Workshop on Rendering (1998)
9. Dou, M., et al.: Fusion4d: real-time performance capture of challenging scenes. ACM Trans. Graph. **35**(4), 1–13 (2016)
10. Esser, P., Sutter, E., Ommer, B.: A variational u-net for conditional appearance and shape generation. In: Computer Vision and Pattern Recognition (CVPR), pp. 8857–8866 (2018)
11. Gortler, S.J., Grzeszczuk, R., Szeliski, R., Cohen, M.F.: The lumigraph. In: SIGGRAPH, pp. 43–54 (1996)
12. Grigor'ev, A.K., Sevastopolsky, A., Vakhitov, A., Lempitsky, V.S.: Coordinate-based texture inpainting for pose-guided human image generation. In: Computer Vision and Pattern Recognition (CVPR), pp. 12127–12136 (2019)

13. Guo, K., Xu, F., Yu, T., Liu, X., Dai, Q., Liu, Y.: Real-time geometry, albedo, and motion reconstruction using a single RGB-D camera. ACM Trans. Graph. 36(4) (2017)

14. Han, X., Hu, X., Huang, W., Scott, M.R.: Clothflow: a flow-based model for clothed person generation. In: Proceedings of the IEEE/CVF International Conference on Computer Vision (ICCV), October 2019

15. Huang, Z.: Deep volumetric video from very sparse multi-view performance capture. In: Ferrari, V., Hebert, M., Sminchisescu, C., Weiss, Y. (eds.) ECCV 2018. LNCS, vol. 11220, pp. 351–369. Springer, Cham (2018). https://doi.org/10.1007/978-3-030-01270-0_21

16. Johnson, J., Alahi, A., Fei-Fei, L.: Perceptual losses for real-time style transfer and super-resolution. In: Leibe, B., Matas, J., Sebe, N., Welling, M. (eds.) ECCV 2016. LNCS, vol. 9906, pp. 694–711. Springer, Cham (2016). https://doi.org/10.1007/978-3-319-46475-6_43

17. Kanazawa, A., Black, M.J., Jacobs, D.W., Malik, J.: End-to-end recovery of human shape and pose. In: Computer Vision and Pattern Regognition (CVPR) (2018)

18. Kim, H., et al.: Neural style-preserving visual dubbing. ACM Trans. Graphics (TOG) 38(6), 178:1–178:13 (2019)

19. Kim, H., et al.: Deep videoportraits. ACM Trans. Graphics (TOG) 37 (2018)

20. Kingma, D.P., Ba, J.: Adam: a method for stochastic optimization. In: International Conference on Learning Representations (ICLR) (2015)

21. Lazova, V., Insafutdinov, E., Pons-Moll, G.: 360-degree textures of people in clothing from a single image. In: International Conference on 3D Vision (3DV), pp. 643–653 (2019)

22. Levoy, M., Hanrahan, P.: Light field rendering. In: SIGGRAPH, p. 31–42 (1996)

23. Liu, L., et al.: Neural rendering and reenactment of human actor videos. ACM Trans. Graphics (TOG) (2019)

24. Liu, W., Wen, Y., Yu, Z., Li, M., Raj, B., Song, L.: Sphereface: deep hypersphere embedding for face recognition. In: Computer Vision and Pattern Recognition (CVPR), pp. 212–220 (2017)

25. Liu, W., Piao, Z., Jie, M., Luo, W., Ma, L., Gao, S.: Liquid warping GAN: a unified framework for human motion imitation, appearance transfer and novel view synthesis. In: International Conference on Computer Vision (ICCV) (2019)

26. Liu, Y., Dai, Q., Xu, W.: A point-cloud-based multiview stereo algorithm for free-viewpoint video. IEEE Trans. Vis. Comput. Graphics (TVCG) 16(3), 407–418 (2010)

27. Liu, Z., Luo, P., Qiu, S., Wang, X., Tang, X.: Deepfashion: powering robust clothes recognition and retrieval with rich annotations. In: Computer Vision and Pattern Recognition (CVPR), pp. 1096–1104 (2016)

28. Lombardi, S., Simon, T., Saragih, J., Schwartz, G., Lehrmann, A., Sheikh, Y.: Neural volumes: learning dynamic renderable volumes from images. ACM Trans. Graph. (SIGGRAPH) 38(4) (2019)

29. Loper, M., Mahmood, N., Romero, J., Pons-Moll, G., Black, M.J.: SMPL: a skinned multi-person linear model. ACM Trans. Graphics (Proc. SIGGRAPH Asia) 34(6), 248:1–248:16 (2015)

30. Ma, L., Sun, Q., Georgoulis, S., van Gool, L., Schiele, B., Fritz, M.: Disentangled person image generation. In: Computer Vision and Pattern Recognition (CVPR) (2018)

31. Martin Brualla, R., et al.: Lookingood: enhancing performance capture with real-time neural re-rendering. ACM Trans. Graphics (TOG) 37 (2018)

32. Matsuyama, T., Xiaojun Wu, Takai, T., Wada, T.: Real-time dynamic 3-D object shape reconstruction and high-fidelity texture mapping for 3-D video. IEEE Trans. Circuits Syst. Video Technol. **14**(3), 357–369 (2004)
33. Neverova, N., Alp Güler, R., Kokkinos, I.: Dense pose transfer. In: Ferrari, V., Hebert, M., Sminchisescu, C., Weiss, Y. (eds.) ECCV 2018. LNCS, vol. 11207, pp. 128–143. Springer, Cham (2018). https://doi.org/10.1007/978-3-030-01219-9_8
34. Orts-Escolano, S., et al.: Holoportation: virtual 3D teleportation in real-time. In: Annual Symposium on User Interface Software and Technology, pp. 741–754 (2016)
35. Pandey, R., et al.: Volumetric capture of humans with a single RGBD camera via semi-parametric learning. In: Computer Vision and Pattern Recognition (CVPR) (2019)
36. Pfister, H., Zwicker, M., van Baar, J., Gross, M.: Surfels: surface elements as rendering primitives. In: SIGGRAPH, pp. 335–342 (2000)
37. Gueler, R.A., Neverova, N., Kokkinos, I.: Densepose: dense human pose estimation in the wild. In: Computer Vision and Pattern Recognition (CVPR) (2018)
38. Ronneberger, O., Fischer, P., Brox, T.: U-Net: convolutional networks for biomedical image segmentation. In: Navab, N., Hornegger, J., Wells, W.M., Frangi, A.F. (eds.) MICCAI 2015. LNCS, vol. 9351, pp. 234–241. Springer, Cham (2015). https://doi.org/10.1007/978-3-319-24574-4_28
39. Saito, S., Huang, Z., Natsume, R., Morishima, S., Kanazawa, A., Li, H.: PIFU: pixel-aligned implicit function for high-resolution clothed human digitization. In: International Conference on Computer Vision (ICCV) (2019)
40. Schonberger, J.L., Frahm, J.M.: Structure-from-motion revisited. In: Computer Vision and Pattern Recognition (CVPR), pp. 4104–4113 (2016)
41. Shade, J., Gortler, S., He, L.W., Szeliski, R.: Layered depth images. In: SIGGRAPH, pp. 231–242 (1998)
42. Shysheya, A., et al.: Textured neural avatars. In: Computer Vision and Pattern Recognition (CVPR) (2019)
43. Siarohin, A., Lathuilière, S., Sangineto, E., Sebe, N.: Appearance and pose-conditioned human image generation using deformable GANs. Trans. Pattern Anal. Mach. Intell. (TPAMI) (2019)
44. Siarohin, A., Lathuilière, S., Tulyakov, S., Ricci, E., Sebe, N.: Animating arbitrary objects via deep motion transfer. In: Computer Vision and Pattern Recognition (CVPR) (2019)
45. Siarohin, A., Lathuilière, S., Tulyakov, S., Ricci, E., Sebe, N.: First order motion model for image animation. In: Conference on Neural Information Processing Systems (NeurIPS) (2019)
46. Simonyan, K., Zisserman, A.: Very deep convolutional networks for large-scale image recognition. arXiv preprint arXiv:1409.1556 (2014)
47. Sitzmann, V., Thies, J., Heide, F., Nießner, M., Wetzstein, G., Zollhöfer, M.: Deepvoxels: Learning persistent 3D feature embeddings. In: Computer Vision and Pattern Recognition (CVPR) (2019)
48. Sitzmann, V., Zollhöfer, M., Wetzstein, G.: Scene representation networks: continuous 3D-structure-aware neural scene representations. In: Advances in Neural Information Processing Systems (NeurIPS) (2019)
49. Tao, Y., et al.: Doublefusion: real-time capture of human performance with inner body shape from a depth sensor. In: Computer Vision and Pattern Recognition (CVPR) (2018)
50. Thies, J., Zollhöfer, M., Nießner, M.: Deferred neural rendering: image synthesis using neural textures. ACM Trans. Graphics (TOG) **38** (2019)

51. Thies, J., Zollhöfer, M., Theobalt, C., Stamminger, M., Nießner, M.: Image-guided neural object rendering. In: International Conference on Learning Representations (ICLR) (2020)
52. Tung, T., Nobuhara, S., Matsuyama, T.: Complete multi-view reconstruction of dynamic scenes from probabilistic fusion of narrow and wide baseline stereo. In: International Conference on Computer Vision (ICCV). pp. 1709–1716 (2009)
53. Varol, G., et al.: Learning from synthetic humans. In: Computer Vision and Pattern Recognition (CVPR) (2017)
54. Wang, T.C., Liu, M.Y., Zhu, J.Y., Tao, A., Kautz, J., Catanzaro, B.: High-resolution image synthesis and semantic manipulation with conditional GANs. In: Computer Vision and Pattern Recognition (CVPR) (2018)
55. Waschbüsch, M., Würmlin, S., Cotting, D., Sadlo, F., Gross, M.: Scalable 3D video of dynamic scenes. Visual Comput. **21**(8), 629–638 (2005)
56. Xu, Z., Bi, S., Sunkavalli, K., Hadap, S., Su, H., Ramamoorthi, R.: Deep view synthesis from sparse photometric images. ACM Trans. Graph. **38**(4), 76:1–76:13 (2019)
57. Yu, T., et al.: Bodyfusion: real-time capture of human motion and surface geometry using a single depth camera. In: International Conference on Computer Vision (ICCV), pp. 910–919 (2017)
58. Yu, T., et al: Simulcap: single-view human performance capture with cloth simulation. In: Computer Vision and Pattern Recognition (CVPR) (2019)
59. Zablotskaia, P., Siarohin, A., Sigal, L., Zhao, B.: DwNet: dense warp-based network for pose-guided human video generation. In: British Machine Vision Conference (BMVC) (2019)
60. Zhang, L., Curless, B., Seitz, S.M.: Spacetime stereo: shape recovery for dynamic scenes. In: Computer Vision and Pattern Recognition (CVPR) (2003)
61. Zhang, R., Isola, P., Efros, A.A., Shechtman, E., Wang, O.: The unreasonable effectiveness of deep features as a perceptual metric. In: Computer Vision and Pattern Recognition (CVPR) (2018)
62. Zhao, B., Wu, X., Cheng, Z.Q., Liu, H., Jie, Z., Feng, J.: Multi-view image generation from a single-view. In: ACM International Conference on Multimedia, pp. 383–391 (2018)
63. Zhou, Y., Wang, Z., Fang, C., Bui, T., Berg, T.L.: Dance dance generation: motion transfer for internet videos. In: International Conference on Computer Vision Workshops (ICCVW) (2019)
64. Wang, Z., Bovik, A.C., Sheikh, H.R., Simoncelli, E.P.: Image quality assessment: from error visibility to structural similarity. IEEE Trans. Image Process. **13**(4), 600–612 (2004)
65. Zhu, H., Su, H., Wang, P., Cao, X., Yang, R.: View extrapolation of human body from a single image. In: Computer Vision and Pattern Recognition (CVPR) (2018)
66. Zhu, J.Y., et al.: Visual object networks: image generation with disentangled 3D representations. In: Conference on Neural Information Processing Systems (NeurIPS), pp. 118–129 (2018)

Reversing the Cycle: Self-supervised Deep Stereo Through Enhanced Monocular Distillation

Filippo Aleotti[1], Fabio Tosi[1], Li Zhang[2], Matteo Poggi[1(✉)],
and Stefano Mattoccia[1]

[1] University of Bologna, Viale del Risorgimento 2, Bologna, Italy
m.poggi@unibo.it
[2] China Agricultural University, Beijing, China

Abstract. In many fields, self-supervised learning solutions are rapidly evolving and filling the gap with supervised approaches. This fact occurs for depth estimation based on either monocular or stereo, with the latter often providing a valid source of self-supervision for the former. In contrast, to soften typical stereo artefacts, we propose a novel self-supervised paradigm reversing the link between the two. Purposely, in order to train deep stereo networks, we distill knowledge through a monocular completion network. This architecture exploits single-image clues and few sparse points, sourced by traditional stereo algorithms, to estimate dense yet accurate disparity maps by means of a consensus mechanism over multiple estimations. We thoroughly evaluate with popular stereo datasets the impact of different supervisory signals showing how stereo networks trained with our paradigm outperform existing self-supervised frameworks. Finally, our proposal achieves notable generalization capabilities dealing with domain shift issues. Code available at https://github.com/FilippoAleotti/Reversing.

Keywords: Stereo matching · Self-supervised learning · Distillation

1 Introduction

Among techniques to infer depth, stereo is an effective and well-established strategy to accomplish this task deploying two cameras. Stereo methods, at first, compute disparity by matching corresponding points across the two images and then recover depth through triangulation, determining the parameters of the stereo rig

---

F. Aleotti and F. Tosi—Joint first authorship
L. Zhang—Work done while at University of Bologna.

Electronic supplementary material The online version of this chapter (https://doi.org/10.1007/978-3-030-58621-8_36) contains supplementary material, which is available to authorized users.

© Springer Nature Switzerland AG 2020
A. Vedaldi et al. (Eds.): ECCV 2020, LNCS 12356, pp. 614–632, 2020.
https://doi.org/10.1007/978-3-030-58621-8_36

beforehand with calibration. Nowadays, deep learning architectures have outperformed traditional methods by a large margin in terms of accuracy on standard benchmarks. Nonetheless, state-of-the-art solutions require a large amount of data and ground-truth labels to learn how to perform *matching*, i.e. find in the other view corresponding pixels. The advent of self-supervised solutions based on image reprojection overcomes this limitation at the cost of weak performance in presence of occluded and texture-less regions, i.e. where the matching does not occur.

In recent years, single-image depth estimation methods, in general up to a scale factor, gained ever-increasing attention. In this field, despite the *ill-posed* nature of the problem, deep learning architectures achieved outstanding results as reported in the literature. By construction, a monocular method does not infer depth by matching points between different views of the same scene. Therefore, compared to stereo approaches, monocular ones infer depth relying on different cues and thus potentially not affected by some inherent issues of stereo, such as occlusions. Even if supervision is sourced from stereo images [9], a set of practices suited for the specific monocular task allow networks to avoid undesired artifacts in correspondence of occlusions [10]. Starting from these observations, we argue that a single image method could potentially strengthen a stereo one, especially in occluded areas, but it would suffer the inherent scale factor issue. Purposely, in this paper we prove that traditional stereo methods and monocular cues can be effectively deployed jointly in a *monocular completion network* able to alleviate both problems, and thus beneficial to obtain accurate and robust depth predictions.

Our contributions can be summarized as follows: i) A new general-purpose methodology to source accurate disparity annotations in a self-supervised manner given a stereo dataset without additional data from active sensors. To the best of our knowledge, our proposal is the first leveraging at training time a novel self-supervised monocular completion network aimed specifically at ameliorate annotations in critical regions such as occluded areas. ii) In order to reduce as much as possible inconsistent disparity annotations, we propose a novel consensus mechanism over multiple predictions exploiting input randomness of the monocular completion network. iii) The generated proxies are dense and accurate even if we do not rely on any active depth sensor (e.g. LiDAR). iv) Our proxies allow for training heterogeneous deep stereo networks outperforming self-supervised state-of-the-art strategies on KITTI. Moreover, the networks trained with our method show higher generalization to unseen environments.

2   Related Work

In this section, we review the literature relevant to our work.

Traditional and Deep Stereo. Depth from stereo images has a longstanding history in computer vision and several hand-designed methods based on some of the steps outlined in [34] have been proposed. For instance, a fast yet noisy solution can be obtained by simply matching pixels according to a robust function [56] over a fixed window (Block Matching), while a better accuracy-speed trade-off is obtained by running Semi-Global Matching (SGM) [13]. Recently, deep learning proved unpaired performance at tackling stereo correspondence. Starting from matching cost computation [3,26,57], deep networks at first replaced single steps in the pipeline [34], moving then to optimization [36], disparity selection [37] and refinement [8]. The first end-to-end model was proposed by Mayer et al. [28], deploying a 1D correlation layer to encode pixel similarities and feed them to a 2D network. In alternative, Kendall et al. [17] stacked features to build a cost volume, processed by 3D convolutions to obtain disparity values through a differentiable *argmin* operation. These two pioneering works paved the way for more complex and effective 2D [15,23,30] and 3D [1,55,58] architectures. Finally, multi-task frameworks combining stereo with semantic segmentation [5,53] and edge detection [39,40] proved to be effective as well. On the other hand, deep learning stereo methods able to learn directly from images largely alleviate the need for ground-truth labels. These have been used either for domain adaptation or for training from scratch a deep stereo network. In the former case, Tonioni et al. [41,42] leveraged traditional algorithms and confidence measures, in [44] developed a modular architecture able to be updated in real time leveraging image reprojection and in [43] made use of meta-learning for the same purpose. In the latter, an iterative schedule to train an unsupervised stereo CNN has been proposed in [61], Godard et al. [9] trained a naive stereo network using image reprojection. Zhong et al. first showed the fast convergence of 3D networks when trained with image reprojection [59], then adopted a RNN LSTM network using stereo video sequences [60]. Wang et al. [49] improved their stereo network thanks to a rigid-aware direct visual odometry module, while in [20] the authors exploited the relationship between optical flow and stereo. Joung et al. [16] trained a network from scratch selecting good matches obtained by a pretrained model. Finally, in [38] a semi-supervised framework leveraging raw LiDAR and image reprojection has been proposed.

Monocular Depth. Single image depth estimation is attractive, yet an *ill-posed* problem. Nonetheless, modern deep learning strategies showed impressive performance up to a scale factor. The first successful attempt in this field followed a supervised paradigm [6,21,24]. Seminal works switching to self-supervision are [9] and [62], respectively requiring stereo pairs and monocular video sequences in place of ground-truth depth labels. Both methods paved the way for self-supervised monocular methods [10,32]. In recent works [45,50], *proxies* labels have been distilled from traditional stereo algorithms [13] in order to strengthen the supervision from stereo pairs.

In parallel to our work, Watson et al. [51] used monocular depth networks to train stereo models from single images through view synthesis.

Depth Completion. Finally, we mention methods that aim at filling a sparse or low resolution depth map, traditionally output of a LiDAR, to obtain dense estimates. Two main categories exist, respectively based on depth only [7,19,25] or guided by images [2,4,14,27,48,54]. Although inspired by these works, our strategy is not comparable with them since processing very different input cues and deployed for other purposes.

3 Method

This section describes our strategy in detail, that allows us to distill highly accurate disparity annotations for a stereo dataset made up of raw rectified images only and then use them to supervise deep stereo networks. It is worth noting that, by abuse of notation, we use depth and disparity interchangeably although our proxy extraction method works entirely in the disparity domain. For our purposes, we rely on two main stages, as depicted in Fig. 1: 1) we train a monocular completion network (MCN) from sparse disparity points sourced by traditional stereo methods and 2) we train deep stereo networks using highly reliable points from MCN, selected by a novel consensus mechanism.

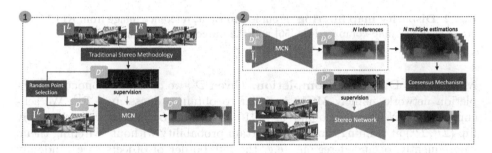

Fig. 1. Overview of our methodology. ① Sparse disparity points from a traditional stereo method are given as input to a monocular completion network (MCN). Then, in ② we leverage MCN to distill accurate proxies through the proposed consensus mechanism. Such labels guide the training of a deep stereo network.

3.1 Monocular Completion Network (MCN)

Stereo algorithms struggle on occluded regions due to the difficulties to find correspondences between images. On the contrary, monocular methods do not rely on matching and thus, they are potentially not affected by this problem. In this stage, our goal is to obtain a strong guidance even on occluded areas relying on a monocular depth network. However, monocular estimates intrinsically suffer the scale factor ambiguity due to the lack of geometric constraints. Therefore, since stereo pairs are always available in our setup, we also leverage on reliable sparse disparity input points from traditional stereo algorithms in addition to the reference image. Thanks to this combination, MCN is able to predict dense depth maps preserving geometrical information.

Reliable Disparity Points Extraction. At first, we rely on a traditional stereo matcher $\mathcal{S}$ (e.g. [56]) to obtain an initial disparity map $\mathcal{D}$ from a given stereo pair $(\mathcal{I}^L, \mathcal{I}^R)$ as

$$\mathcal{D} = \mathcal{S}(\mathcal{I}^L, \mathcal{I}^R) \tag{1}$$

However, since such raw disparity map contains several outliers, especially on ill-posed regions such as occlusions or texture-less areas as it can be noticed in Fig. 2, a filtering strategy $\mathcal{F}$ (e.g. [46]) is applied to discard spurious points

$$\mathcal{D}' = \mathcal{F}(\mathcal{S}(\mathcal{I}^L, \mathcal{I}^R)) \tag{2}$$

By doing so, only a subset $\mathcal{D}'$ of highly reliable points is preserved from $\mathcal{D}$ at the cost of a sparser disparity map. However, most of them do not belong to occluded regions thus not enabling supervision on such areas. This can be clearly perceived observing the outcome of a filtering strategy in Fig. 2.

Fig. 2. Disparity map filtering. From left to right, reference image from KITTI, the noisy disparity map computed by [56] and the outcome of filtering [46].

Monocular Disparity Completion. Given $\mathcal{D}'$, we deploy a monocular completion network, namely MCN, in order to obtain a dense map $\mathcal{D}^\mathbb{O}$. We self-supervise MCN from stereo and, as in [10], to handle occlusions we horizontally flip $(\mathcal{I}^L, \mathcal{I}^R)$ at training time with a certain probability without switching them. Consequently, occluded regions (e.g. the left border of objects) are randomly swapped with not-occluded areas (e.g. the right borders), preventing to always expect high error on left and low error on right borders, thus forcing the network to handle both. This strategy turns out ineffective in case of self-supervised stereo, since after horizontal flip the stereo pair have to be switched in order to keep the same search direction along the epipolar line, thus making occlusions occur in the same regions (see the supplementary material for details). Even if this technique helps to alleviate errors in occluded regions, a pure monocular network struggles compared to a stereo method at determining the correct depth. This is well-known in the literature and shown in our experiments as well. Thus, we adopt a completion approach leveraging sparse reliable points provided by a traditional stereo method constraining the predictions to be properly scaled. Given the set of filtered points, only a small subset $\mathcal{D}''$, with $||\mathcal{D}''|| \ll ||\mathcal{D}'||$, is randomly selected and used as input, while $\mathcal{D}'$ itself is used for supervision purposes. The output of MCN is defined as

$$\mathcal{D}^\mathbb{O} = \mathrm{MCN}(\mathcal{I}^L, \mathcal{D}'' \xleftarrow{p \ll 1} (\mathcal{D}')) \tag{3}$$

with $x \xleftarrow{p} (y)$ a random uniform sampling function extracting x values out of y per-pixel values with probability p. This sampling is crucial to both improve MCN accuracy, as shown in our experiments, as well as for the final distillation step discussed in the remainder. Once trained, MCN is able to infer scaled dense disparity maps $\mathcal{D}^\mathbb{O}$, as can be perceived in Fig. 3. Looking at the rightmost and central disparity maps, we can notice how the augmentation protocol enables to alleviate occlusion artifacts. Moreover, our overall completion strategy, compared to the output of the monocular network without disparity seeds (leftmost and center disparity maps), achieves much higher accuracy as well as correctly handles occlusions. Therefore, we effectively combine stereo from non-occluded regions and monocular prediction in occluded areas. Finally, we point out that we aim at specializing MCN on the training set to generate labels on it since its purpose is limited to distillation.

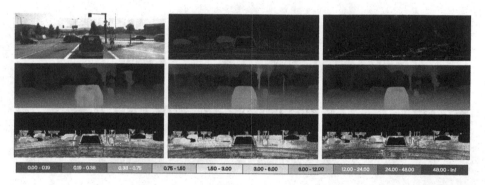

Fig. 3. Occlusion handling and scale recovery. The first row depicts the reference image from KITTI, the ground-truth and the disparity map by [56] filtered with [46]. In the middle, from left to right the output of monocular depth network [45] trained without occlusion augmentation, the same network using the occlusion augmentation and our MCN. Last row shows the corresponding error maps. Best viewed with colors. (Color figure online)

3.2   Proxy Distillation for Deep Stereo

Eventually, we leverage the trained MCN to distill offline proxy labels beneficial to supervise stereo networks. However, such data might still contain some inconsistent predictions, as can be perceived in the rightmost disparity map of Fig. 3. Therefore, our goal is to discard them, keeping trustworthy reliable depth estimates to train deep stereo networks.

Consensus Mechanism and Distillation. To this aim, given an RGB image $\mathcal{I}$ and the relative $\mathcal{D}'$, we perform N inferences of MCN by feeding it with $\mathcal{D}''_i$ and $\tilde{\mathcal{I}}_i$, with $i \in [1, N]$. Respectively, $\mathcal{D}''_i$ is sampled from $\mathcal{D}'$ according to the strategy introduced in Sect. 3.1 and $\tilde{\mathcal{I}}_i$ is obtained through random augmentation (explained later) applied to $\mathcal{I}$. This way, we exploit consistencies and contradictions among multiple $\mathcal{D}_i^{\mathbb{O}}$ to obtain reliable proxy labels $\mathcal{D}^{\mathbb{P}}$, defined as

$$\mathcal{D}^{\mathbb{P}} \xleftarrow{\sigma^2(\{\mathcal{D}_i^{\mathbb{O}}\}_{i=1}^N)<\gamma} \mu(\{\mathcal{D}_i^{\mathbb{O}}\}_{i=1}^N) \tag{4}$$

where $x \xleftarrow{\sigma^2(y)<\gamma} \mu(y)$ is a function that, given N values y for the same pixel, samples the mean value $\mu(y)$ only if the variance $\sigma^2(y)$ is smaller γ. Being distillation performed offline, this step does not need to be differentiable.

Figure 4 shows that such a strategy allows us to largely regularize $\mathcal{D}^{\mathbb{P}}$ compared to $\mathcal{D}^{\mathbb{O}}$, preserving thin structures, e.g. the poles on the right side, yet achieving high density. It also infers significant portions of occluded regions compared to proxies sourced from traditional methods (e.g. SGM).

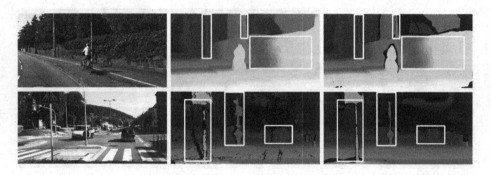

Fig. 4. Proxy distillation. The first row depicts, from left to right, the reference image, the disparity map computed by a single inference of MCN and the one filtered and regularized using our consensus mechanism. The second row shows the reference image, the disparity map generated by SGM [13] filtered using the left-right consistency check strategy and our disparity map. Images from KITTI.

Deep Stereo Training. Once highly accurate proxy labels $\mathcal{D}^{\mathbb{P}}$ are available on the same training set, we exploit them to train deep stereo networks in a self-supervised manner. In particular, a regression loss is used to minimize the difference between stereo predictions and $\mathcal{D}^{\mathbb{P}}$.

4   Experiments

In this section, we first introduce the datasets used in this work, then we thoroughly evaluate our proposal, proving that sourcing labels with a monocular completion approach is beneficial to train deep stereo networks.

4.1   Datasets

KITTI. The KITTI (K) dataset [29] contains 61 scenes (about 42,382 stereo pairs), with a typical image size of 1242×375, captured using a stereo rig mounted on a moving car equipped with a LiDAR sensor. We conducted experiments using all of the raw KITTI images for training excluding scenes from the KITTI 2015 training set containing 200 ground-truth images used for testing purposes. This results in a training split 29K rectified stereo images.

DrivingStereo. DrivingStereo [52] (DS) is a recent large-scale dataset depicting autonomous driving scenarios in various weather conditions, containing more than 180k stereo pairs with high-quality disparity annotations generated by means of a model-guided filtering method from multi-frame LiDAR points. For our purposes, we split the dataset into a training set and a testing set consisting of 97681 and 1k images respectively.

Middlebury v3. The Middlebury v3 dataset [33] provides 15 stereo pairs depicting indoor scenes, with high precision and dense ground-truth disparities obtained using structured light. We rely on this dataset for generalization purposes, using images and the ground-truth disparity maps at quarter resolution.

ETH3D. The ETH3D high-resolution dataset [35] depicts heterogeneous scenes consisting of 27 grayscale stereo pairs with ground-truth depth values. As for Middlebury v3, we run generalization experiments on it.

4.2   Implementation Details

Traditional Stereo Methods. We consider two main non-learning based solutions, characterized by different peculiarities, to generate accurate sparse disparity points from a rectified stereo pair. In particular, we use the popular semi-global matching algorithm SGM [13], exploiting the left-right consistency check (LRC) to remove wrong disparity assignments, and the WILD strategy proposed in [46] that selects highly reliable values from the maps computed by the local algorithm Block-Matching (BM) [56] exploiting traditional confidence measures. We refer to these methods (i.e. stereo method followed by a filtering strategy) as SGM/L and BM/W, respectively.

Monocular Completion Network. We adopt the publicly available self-supervised monocular architecture monoResMatch [45] trained with the supervision of disparity proxy labels specifically suited for our purposes. We modify the network to exploit accurate sparse annotations as input by concatenating them with the RGB image. We set the random sampling probability in Eq. 3 as $p = \frac{1}{1000}$. In our experiments, we train from scratch the MCN network following the same training protocol defined in [45] except for the augmentation procedure which includes the flipping strategy (with 0.25 probability) aimed at handling occlusion artifacts [10]. Instead, we empirically found out that generating $\mathcal{D}^O$ using a larger set of points helps to achieve more accurate predictions at inference time. In particular, we fix $p = \frac{1}{20}$ and $p = \frac{1}{200}$ for BM/W and SGM/L respectively. Finally, for the consensus mechanism, we fix $N = 50$, the threshold $\gamma = 3$ and apply for each $\mathcal{I}_i$ color augmentation and random horizontal flip (with 0.5 probability). Please see the supplementary material for more details about hyper-parameters.

Stereo Networks. We considered both 2D and 3D deep stereo architectures, ensuring a comprehensive validation of our proposal. In particular, we designed a baseline architecture, namely Stereodepth, by extending [10] to process stacked left and right images, and iResNet [23] as examples of the former case, while PSMNet [1] and GWCNet [11] as 3D architectures. At training time, the models predict disparities $\mathcal{D}^S$ at multiple scales in which each intermediate prediction is upsampled at the input resolution. A weighted smooth L1 loss function (the lower the scale, the lower the weight) minimizes the difference between $\mathcal{D}^S$ and the disparity provided by the proxy $\mathcal{D}^{\mathbb{P}}$ considering only valid pixels, using Adam [18] as optimizer ($\beta_1 = 0.9$ and $\beta_2 = 0.999$). We adopt the original PyTorch [31] implementation of the networks if available. Moreover, all the models have been trained to fit a single Titan X GPU. More details are provided in the supplementary material.

4.3 Evaluation of Proxy Label Generators

At first, we first evaluate the accuracy of proxies produced by our self-supervised approach with respect to traditional methods. We consider both D1 and EPE, computed on disparities, as error metrics on both non-occluded (*Noc*) and all regions (*All*). In particular, D1 represents the percentage of pixels for which the estimation error is ≥ 3 px and $\geq 5\%$ of its ground-truth value, while EPE is obtained by averaging the absolute difference between predictions and ground-truths. In addition, the density and the overlap with the ground-truth are reported to take into account filtering strategies. Table 1 reports a thorough evaluation of different methodologies and filtering techniques. It can be noticed how BM and SGM have different performances due to their complementarity (local vs semi-global), but containing several errors. Filtering strategies help to remove outliers, at the cost of sparser maps. Notice that restoring the full density through *hole-filling* [13] slightly improves the results of SGM/L, but it is not meaningful for BM/W since filtered maps are too sparse. Unsurprisingly, even if the depth maps produced by the vanilla monoResMatch are fully-dense, they are not accurate due to its inherent monocular nature. On the contrary, our monocular strategy MCN produces dense yet accurate maps thanks to the initial disparity guesses, regardless the sourcing stereo algorithm. Moreover, by applying Augmentation techniques (A) on the RGB image or selecting Random

Table 1. Evaluation of proxy generators. We tested proxies generated by different strategies on the KITTI 2015 training set.

Method	Configuration					Statistics		Noc		All	
	Source	Filter	A	R	C	Density(%)	Overlap(%)	D1(%)	EPE	D1(%)	EPE
MONO	monoResMatch	–	–	–	–	100.0	100.0	26.63	2.96	27.00	2.99
BM	BM	–	–	–	–	100.0	100.0	34.48	16.14	35.46	16.41
SGM	SGM	–	–	–	–	100.0	100.0	6.65	1.67	8.12	2.16
BM/L	BM	LRC	–	–	–	57.89	62.09	16.09	6.42	16.22	6.46
SGM/L	SGM	LRC	–	–	–	86.47	92.28	3.99	1.00	4.01	1.00
SGM/L(*hole-filling*)	SGM	LRC	–	–	–	100.0	100.0	6.56	1.34	7.68	1.57
BM/W	BM	WILD	–	–	–	12.33	10.43	1.33	0.81	1.35	0.81
MCN-SGM/L	SGM	LRC	–	–	–	100.0	100.0	6.36	1.27	7.80	1.50
MCN-SGM/L-R	SGM	LRC	–	✓	–	100.0	100.0	5.28	1.13	5.73	1.21
MCN-SGM/L-AC	SGM	LRC	✓	–	✓	95.36	97.36	5.58	1.17	5.58	1.15
MCN-SGM/L-RC	SGM	LRC	–	✓	✓	93.50	96.32	2.95	0.86	3.14	0.89
MCN-SGM/L-ARC	SGM	LRC	✓	✓	✓	92.53	95.76	2.78	0.84	2.92	0.86
MCN-BM/W	BM	WILD	–	–	–	100.0	100.0	11.86	1.93	12.50	2.03
MCN-BM/W-R	BM	WILD	–	✓	–	100.0	100.0	6.79	1.40	7.11	1.45
MCN-BM/W-AC	BM	WILD	✓	–	✓	91.45	94.76	8.36	1.53	8.64	1.57
MCN-BM/W-RC	BM	WILD	–	✓	✓	91.12	95.28	3.79	0.95	4.03	1.0
MCN-BM/W-ARC	BM	WILD	✓	✓	✓	86.82	93.56	3.16	0.90	3.27	0.92

input points (R), allow to increase variance and to exploit our Consensus mechanism (C) to filter out unreliable values, thus achieving even better results. In fact, the consensus mechanism is able to discard wrong predictions preserving high density, reaching best performances when A and R are both applied. It is worth noting that if R is not performed, the network is fed with all the available guesses both at training and testing time, with remarkably worse results compared to configuration using random sampling.

Disparity Completion Comparison. We validate our MCN combined with the consensus mechanism comparing it to GuideNet [52], a supervised architecture designed to generate high-quality disparity annotations exploiting multiframe LiDAR points and stereo pairs as input. Following [52], we measure the valid pixels, correct pixels, and accuracy (i.e. 100.0 - D1) on 142 images of the KITTI 2015 training set. Table 2 clearly shows how MCN trained in a self-supervised manner achieves comparable accuracy with respect to GuideNet-LiDAR by exploiting sparse disparity estimates from both [46] and [13] but with a significantly higher number of points, even on foreground regions (Obj). Notice that LiDAR indicates that the network is fed with LiDAR points filtered according to [48]. To further demonstrate the generalization capability of MCN to produce highly accurate proxies relying on points from heterogeneous sources, we feed MCN-BM/W-R with raw LiDAR measurements. By doing so, our network notably outperforms GuideNet in this configuration, despite it leverages a single RGB image and has not been trained on LiDAR points.

Table 2. Model-guided comparison. Comparison between our self-supervised MCN model and the supervised GuideNet stereo architecture [52] using 142 ground-truth images of the KITTI 2015 training set. † indicates that the network requires LiDAR points at training time. Accuracy is defined as 100-D1.

Model	All			Obj		
	Valid	Correct	Accuracy (%)	Valid	Correct	Accuracy (%)
MCN-BM/W-ARC	11,551,461	11,247,966	97.37	1,718,267	1,642,872	95.61
MCN-SGM/L-ARC	12,201,763	11,860,923	97.20	1,788,154	1,672,222	93.52
MCN-LiDAR	11,773,897	11,636,787	98.83	1,507,222	1,459,726	96.84
GuideNet-LiDAR [52]	†2,973,882	2,915,110	98.02	221,828	210,912	95.07

4.4 Ablation Study

In this subsection, we support the statement that a completion approach provides a better supervision compared to traditional stereo algorithms. We first run experiments on KITTI and then use our best configuration on DrivingStereo as well, showing that it is effective on multiple large stereo datasets.

KITTI. For the ablation study, reported in Table 3, we consider both 3D (PSM-Net) and 2D (Stereodepth) networks featuring different computational complexity. First, we train the baseline configuration of the networks, i.e. relying image reconstruction loss functions (PHOTO) only as in [10] (see supplementary material for more details). Then, we leverage disparity values sourced by traditional stereo algorithms in which outliers have been removed by the filtering strategies adopted. Such labels provide a useful guidance for stereo networks and allow to obtain more accurate models w.r.t. the baselines. Nonetheless, proxies produced by MCN prove to be much more effective than traditional ones, improving both D1 and EPE by a notable margin regardless the stereo algorithm used to extract the input guesses. Moreover, it can be perceived that best results are obtained when the complete consensus mechanism is enabled.

Finally, we rely also on filtered LiDAR measurements from [48] in order to show differences with respect to supervision from active sensors. Noteworthy, models trained using proxies distilled by ARC configuration of MCN prove to be comparable or even better than using LiDAR with PSMNet and Stereodepth. This behaviour can be explained due to a more representative and accurate supervision on occluded areas than traditional stereo and filtered LiDAR, thus making the deep networks more robust even there, as clearly shown in Fig. 5.

DrivingStereo. We validate the proposed strategy also on DrivingStereo, proving that our distillation approach is able to largely improve the performances of stereo networks also on different datasets. In particular, in Table 4 we compare Stereodepth and PSMNet errors when trained using MCN-BM/W-ARC method (i.e. the best configuration on KITTI) with LiDAR and BM/W. Again, our proposal outperforms BM/W, and reduces the gap with high quality LiDAR supervision. Moreover, to verify generalization capabilities, we test on KITTI

Table 3. Ablation study. We trained Stereodepth and PSMNet on KITTI using supervision signals from different proxy generators and tested on KITTI 2015.

Backbone	Supervision	Noc		All	
		D1(%)	EPE	D1(%)	EPE
Stereodepth	PHOTO	**6.50**	**1.30**	7.12	**1.40**
PSMNet	PHOTO	6.62	**1.30**	7.67	1.50
Stereodepth	SGM-L	5.22	**1.13**	5.43	**1.15**
Stereodepth	SGM/L(*hole-filling*)	6.06	1.16	6.38	1.21
Stereodepth	BM/W	**5.19**	1.16	**5.37**	1.18
PSMNet	SGM/L	5.46	1.19	5.61	1.21
PSMNet	SGM/L(*hole-filling*)	6.06	1.23	6.32	1.26
PSMNet	BM/W	6.89	1.59	7.03	1.60
Stereodepth	MCN-SGM/L-R	5.11	1.11	5.37	1.14
Stereodepth	MCN-BM/W-R	4.75	1.05	4.96	1.07
Stereodepth	MCN-SGM/L-ARC	4.56	1.08	4.77	1.11
Stereodepth	MCN-BM/W-ARC	4.21	1.06	4.39	1.07
PSMNet	MCN-SGM/L-R	4.39	1.05	4.60	1.07
PSMNet	MCN-BM/W-R	4.30	1.06	4.49	1.08
PSMNet	MCN-SGM/L-ARC	4.02	1.05	4.20	1.07
PSMNet	MCN-BM/W-ARC	<u>**3.68**</u>	<u>**0.99**</u>	<u>**3.85**</u>	<u>**1.01**</u>
Stereodepth	LiDAR/SGM [48]	3.95	1.07	4.10	1.09
PSMNet	LiDAR/SGM [48]	**3.93**	**1.05**	**4.07**	**1.07**

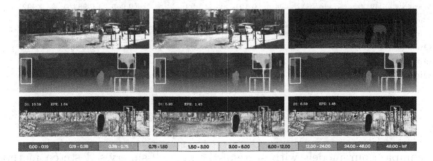

Fig. 5. Impact of proxies. From top, input stereo pair and ground-truth disparity map, predictions by Stereodepth trained with SGM/L (left), LiDAR (center) and our MCN-BM/W-ARC (right), error maps. Best viewed with colors. (Color figure online)

also correspondent models trained on DrivingStereo, without performing any fine-tuning (DS → K), and vice versa (K → DS). It can be noticed that the gap between KITTI models (see Table 3) and those trained on DrivingStereo gets smaller, proving that the networks are able to perform matching correctly even in cross-validation scenario. We want to point out that this is due to our proxies, as can be clearly perceived by looking at rows 1–2 vs 3–4 in Table 4.

Table 4. Cross-validation analysis. We tested on the Target dataset models trained on the Source one, leveraging different proxies. Notice that no fine-tuning on the target dataset is performed in case of cross-validation.

Backbone	Supervision	Source → Target					
		DS → DS		K → DS		DS → K	
		D1(%)	EPE	D1(%)	EPE	D1(%)	EPE
Stereodepth	BM/W	**4.46**	**1.20**	**4.67**	**1.10**	**6.35**	**1.36**
PSMNet	BM/W	8.81	1.94	5.06	1.30	7.07	1.65
Stereodepth	MCN-BM/W-ARC	2.47	0.94	2.97	0.96	5.64	1.22
PSMNet	MCN-BM/W-ARC	**1.87**	**0.86**	<u>**2.32**</u>	<u>**0.88**</u>	5.16	<u>**1.17**</u>
Stereodepth	LiDAR [52]	1.20	0.69	3.60	1.23	4.57	<u>**1.17**</u>
PSMNet	LiDAR [52]	<u>**0.59**</u>	<u>**0.54**</u>	2.64	1.03	<u>**4.52**</u>	1.26

Table 5. Comparison with state-of-the-art. Results of different self-supervised stereo networks on the KITTI 2015 training set with max depth set to 80m. **Ours** indicates networks trained using MCN-BM/W-ARC labels. * indicates networks trained on the same KITTI 2015 data, therefore not directly comparable with other methods.

Method	RMSE	RMSE log	D1 (%)	EPE	$\delta < 1.25$	$\delta < 1.25^2$	$\delta < 1.25^3$
Godard et al.[9] (stereo)	5.742	0.202	10.80	–	0.928	0.966	0.980
Lai et al.[20]	4.186	0.157	8.62	1.46	0.950	0.979	0.990
Wang et al.[49] (stereo only)	4.187	0.135	7.07	–	0.955	0.981	0.990
Zhong et al.[59]	4.857	0.165	6.42	–	0.956	0.976	0.985
Wang et al.[49] (stereo videos)	3.404	0.121	5.94	–	0.965	0.984	0.992
Zhong et al.[60]*	(3.176)	(0.125)	(5.14)	–	(0.967)	–	–
Ours (Stereodepth)	3.882	0.117	4.39	1.07	0.971	0.988	<u>**0.993**</u>
Ours (GWCNet)	3.614	0.111	3.93	1.04	0.974	<u>**0.989**</u>	<u>**0.993**</u>
Ours (iResNet)	3.464	<u>**0.108**</u>	3.88	1.02	<u>**0.975**</u>	0.988	<u>**0.993**</u>
Ours (PSMNet)	3.764	0.115	<u>**3.85**</u>	<u>**1.01**</u>	0.974	0.988	<u>**0.993**</u>

4.5   Comparison with State-of-the-Art

We compare our models with state-of-the-art self-supervised stereo methods. Table 5 reports, in addition to D1 and EPE, also RMSE and RMSE log as depth error measurements and $\delta < 1.25, \delta < 1.25^2, \delta < 1.25^3$ accuracy metrics according to [49,61]. Notice that some of these methods exploit additional information, such as stereo videos [49] or adaptation strategies [60]. Proxies distilled by MCN-BM/W-ARC can be successfully exploited using both 2D and 3D architectures, enabling even the simplest 2D network Stereodepth to outperform all the competitors. Our strategy is effective, allowing all the adopted backbones to improve depth estimation by a notable margin on 6 metrics out of 7. Furthermore, we test our PSMNet trained using MCN-BM/W-ARC proxies on the KITTI 2015 online benchmark, reporting the results in Table 6. Our model not only outperforms [12] and self-supervised competitors, as can be also perceived in Fig. 6, but also supervised strategies [28,44] on both non-occluded and all areas.

Table 6. KITTI 2015 online benchmark. We submitted PSMNet, trained on MCN-BM/W-ARC labels, on the KITTI 2015 online stereo benchmark. In blue self-supervised methods, while in red supervised strategies. We indicate with E2E architectures trained in an end-to-end manner, while SF on the SceneFlow dataset [28].

Models	Dataset	E2E	D1-bg (%)	D1-fg (%)	D1-All (%)	D1-Noc (%)
Zbontar and LeCun (acrt) [57]	K	–	2.89	8.88	3.89	3.33
Tonioni et al. [44]	SF+K	✓	3.75	9.20	4.66	4.27
Mayer et al. [28]	SF+K	✓	4.32	4.41	4.34	4.05
Chang and Chen [1] (PSMNet)	SF+K	✓	1.86	4.62	2.32	2.14
Guo et al. [11] (GWCNet)	SF+K	✓	1.74	3.93	2.11	1.92
Zhang et al. [58]	SF+K	✓	**1.48**	**3.46**	**1.81**	**1.63**
Hirschmuller [12]	–	–	8.92	20.59	10.86	9.47
Zhou et al. [61]	K	✓	–	–	9.91	–
Li and Yuan [22]	K	✓	6.89	19.42	8.98	7.39
Tulyakov et al. [47]	K	–	3.78	10.93	4.97	4.11
Joung et al. [16]	K	–	–	–	4.47	–
Ours(PSMNet)	K	✓	**3.13**	**8.70**	**4.06**	**3.86**

Fig. 6. KITTI 2015 online benchmark qualitatives. From left to right, the reference images, and the disparity maps computed by [12], [22] and our PSMNet trained on MCN-BM/W-ARC labels.

4.6 Generalization

Finally, we show experiments supporting that supervision from our MCN-BM/W-ARC labels achieves good generalization to different domains. To this aim, we run our KITTI networks on Middlebury v3 and ETH3D, framing completely different environments.

Table 7 shows the outcome of this evaluation. We report, on top, the performance of fully supervised methods trained on SceneFlow [28] and fine-tuned on KITTI for comparison. On bottom, we report self-supervised frameworks trained on the KITTI split from the previous experiments. All networks are transferred without fine-tuning. Compared to existing self-supervised strategies (rows 4–6),

Table 7. Generalization test on Middlebury v3 and ETH3D. We evaluate networks trained in self-supervised (blue) or supervised (red) fashion on KITTI (K) and SceneFlow dataset (SF) [28].

Method	Training Dataset	Middlebury v3 [33]		ETH3D [35]	
		BAD2 (%)	EPE	BAD2 (%)	EPE
Zhang et al.[58]	SF+K	**18.90**	**3.44**	**3.43**	0.91
Chang and Chen [1] (PSMNet)	SF+K	20.04	3.01	13.07	1.35
Guo et al.[11](GWCNet)	SF+K	21.36	3.29	19.96	1.88
Wang et al.[49](stereo only)	K	30.55	4.77	11.17	1.47
Wang et al.[49](stereo videos)	K	31.63	5.23	19.59	1.97
Lai et al.[20](stereo videos)	K	45.18	6.42	10.15	1.01
Ours(Stereodepth)	K	27.43	3.72	6.94	1.31
Ours(iResNet)	K	25.08	3.85	6.29	0.81
Ours(GWCNet)	K	20.75	3.17	**3.50**	**0.48**
Ours(PSMNet)	K	**19.56**	**2.99**	4.00	0.51

Reference   GT   Wang [49]   Lai [20]   **Ours** [1]   Zhang [58]

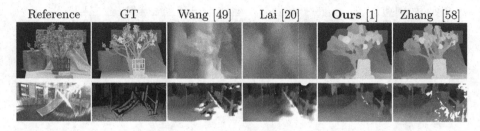

Fig. 7. Examples of generalization. First row shows disparity maps obtained on a stereo pair from the Middlebury v3 dataset, while second from ETH3D. Methods in **blue** are self-supervised, while in red are supervised with ground-truth. (Color figure online)

networks trained with our proxies achieve much better generalization on both the datasets, performing comparable (or even better) with ground-truth supervised networks. Figure 7 shows few examples from the two datasets, where the structure of the scene is much better recovered when trained on our proxies.

5   Conclusion

This paper proposed a novel strategy to source reliable disparity proxy labels in order to train deep stereo networks in a self-supervised manner leveraging a monocular completion paradigm. Well-known stereo artefacts are soften by learning on such labels, that can be obtained from large RGB stereo datasets in which no additional depth information (e.g. LiDAR or active sensors) is available. Through an extensive ablation study on two popular stereo datasets, we proved that our approach is able to infer accurate yet dense maps starting from points sourced by (potentially) any traditional stereo algorithm, and that such labels provide a strong supervision for both 2D and 3D stereo networks with different complexity. We showed that these networks outperform state-of-the-art self-supervised methods on KITTI by a large margin and are, in terms of generalization on Middlebury v3 and ETH3D, comparable or even better than ground-truth supervised stereo networks.

Acknowledgments.. We gratefully acknowledge the support of NVIDIA Corporation with the donation of the Titan Xp GPU used for this research.

References

1. Chang, J.R., Chen, Y.S.: Pyramid stereo matching network. In: The IEEE Conference on Computer Vision and Pattern Recognition (CVPR). IEEE (2018)
2. Chen, Y., Yang, B., Liang, M., Urtasun, R.: Learning joint 2D–3D representations for depth completion. In: IEEE International Conference on Computer Vision (ICCV), pp. 10023–10032. IEEE (2019)
3. Chen, Z., Sun, X., Wang, L., Yu, Y., Huang, C.: A deep visual correspondence embedding model for stereo matching costs. In: The IEEE International Conference on Computer Vision (ICCV). IEEE (2015)
4. Cheng, X., Wang, P., Yang, R.: Depth estimation via affinity learned with convolutional spatial propagation network. In: European Conference on Computer Vision (ECCV), pp. 103–119. Springer, Heidlelberg (2018)
5. Dovesi, P.L., et al.: Real-time semantic stereo matching. In: IEEE International Conference on Robotics and Automation (ICRA). IEEE (2020)
6. Eigen, D., Puhrsch, C., Fergus, R.: Depth map prediction from a single image using a multi-scale deep network. In: Advances in Neural Information Processing Systems, pp. 2366–2374. MIT Press (2014)
7. Eldesokey, A., Felsberg, M., Khan, F.S.: Propagating confidences through cnns for sparse data regression. arXiv preprint arXiv:1805.11913 (2018)
8. Gidaris, S., Komodakis, N.: Detect, replace, refine: deep structured prediction for pixel wise labeling. In: The IEEE Conference on Computer Vision and Pattern Recognition (CVPR). IEEE (2017)
9. Godard, C., Mac Aodha, O., Brostow, G.J.: Unsupervised monocular depth estimation with left-right consistency. In: IEEE Conference on Computer Vision and Pattern Recognition (CVPR). IEEE (2017)
10. Godard, C., Mac Aodha, O., Brostow, G.J.: Digging into self-supervised monocular depth estimation. In: IEEE International Conference on Computer Vision (ICCV). IEEE (2019)
11. Guo, X., Yang, K., Yang, W., Wang, X., Li, H.: Group-wise correlation stereo network. In: IEEE Conference on Computer Vision and Pattern Recognition, pp. 3273–3282. IEEE (2019)
12. Hirschmuller, H.: Accurate and efficient stereo processing by semi-global matching and mutual information. In: IEEE Computer Society Conference on Computer Vision and Pattern Recognition, 2005. CVPR 2005, vol. 2, pp. 807–814. IEEE (2005)
13. Hirschmuller, H.: Stereo processing by semiglobal matching and mutual information. IEEE TPAMI **30**(2), 328–341 (2008)
14. Huang, Z., Fan, J., Cheng, S., Yi, S., Wang, X., Li, H.: Hms-net: hierarchicalmulti-scale sparsity-invariant network for sparse depth completion. IEEE Trans. Image Process. **29**, 3429–3441 (2019)

15. Ilg, E., Saikia, T., Keuper, M., Brox, T.: Occlusions, motion and depth boundaries with a generic network for disparity, optical flow or scene flow estimation. In: Ferrari, V., Hebert, M., Sminchisescu, C., Weiss, Y. (eds.) ECCV 2018. LNCS, vol. 11216, pp. 626–643. Springer, Cham (2018). https://doi.org/10.1007/978-3-030-01258-8_38

16. Joung, S., Kim, S., Park, K., Sohn, K.: Unsupervised stereo matching usingconfidential correspondence consistency. IEEE Trans. Intell. Transp. Syst. **21**, 2190–2203 (2019)

17. Kendall, A., et al.: End-to-end learning of geometry and context for deep stereo regression. In: The IEEE International Conference on Computer Vision (ICCV). IEEE (2017)

18. Kingma, D., Ba, J.: Adam: A method for stochastic optimization. arXiv preprint arXiv:1412.6980 (2014)

19. Ku, J., Harakeh, A., Waslander, S.L.: In defense of classical image processing: fast depth completion on the cpu. In: 2018 15th Conference on Computer and Robot Vision (CRV), pp. 16–22. IEEE (2018)

20. Lai, H.Y., Tsai, Y.H., Chiu, W.C.: Bridging stereo matching and optical flow via spatiotemporal correspondence. In: IEEE Conference on Computer Vision and Pattern Recognition (CVPR). IEEE (2019)

21. Laina, I., Rupprecht, C., Belagiannis, V., Tombari, F., Navab, N.: Deeper depth prediction with fully convolutional residual networks. In: 3DV. IEEE (2016)

22. Li, A., Yuan, Z.: Occlusion aware stereo matching via cooperative unsupervised learning. In: Jawahar, C.V., Li, H., Mori, G., Schindler, K. (eds.) ACCV 2018. LNCS, vol. 11366, pp. 197–213. Springer, Cham (2019). https://doi.org/10.1007/978-3-030-20876-9_13

23. Liang, Z., et al.: Learning for disparity estimation through feature constancy. In: The IEEE Conference on Computer Vision and Pattern Recognition (CVPR). IEEE (2018)

24. Liu, F., Shen, C., Lin, G., Reid, I.: Learning depth from single monocular images using deep convolutional neural fields. IEEE Trans Pattern Anal. Mach. Intell. **38**(10), 2024–2039 (2016)

25. Liu, L.K., Chan, S.H., Nguyen, T.Q.: Depth reconstruction from sparse samples: representation, algorithm, and sampling. IEEE Trans. Image Process. **24**(6), 1983–1996 (2015)

26. Luo, W., Schwing, A.G., Urtasun, R.: Efficient deep learning for stereo matching. In: IEEE Conference on Computer Vision and Pattern Recognition, pp. 5695–5703. IEEE (2016)

27. Ma, F., Cavalheiro, G.V., Karaman, S.: Self-supervised sparse-to-dense: self-supervised depth completion from lidar and monocular camera. In: 2019 International Conference on Robotics and Automation (ICRA), pp. 3288–3295. IEEE (2019)

28. Mayer, N., et al.: A large dataset to train convolutional networks for disparity, optical flow, and scene flow estimation. In: The IEEE Conference on Computer Vision and Pattern Recognition (CVPR). IEEE (2016)

29. Menze, M., Geiger, A.: Object scene flow for autonomous vehicles. In: Conference on Computer Vision and Pattern Recognition (CVPR). IEEE (2015)

30. Pang, J., Sun, W., Ren, J.S., Yang, C., Yan, Q.: Cascade residual learning: a two-stage convolutional neural network for stereo matching. In: The IEEE International Conference on Computer Vision (ICCV) Workshops. IEEE (2017)

31. Paszke, A., et al.: Pytorch: an imperative style, high-performance deep learning library. In: Advances in Neural Information Processing Systems, pp. 8024–8035. MIT Press (2019)

32. Poggi, M., Tosi, F., Mattoccia, S.: Learning monocular depth estimation with unsupervised trinocular assumptions. In: 6th International Conference on 3D Vision (3DV). IEEE (2018)

33. Scharstein, D., et al.: High-resolution stereo datasets with subpixel-accurate ground truth. In: Jiang, X., Hornegger, J., Koch, R. (eds.) GCPR 2014. LNCS, vol. 8753, pp. 31–42. Springer, Cham (2014). https://doi.org/10.1007/978-3-319-11752-2_3

34. Scharstein, D., Szeliski, R.: A taxonomy and evaluation of dense two-frame stereo correspondence algorithms. Int. J. Comput. Vis. **47**(1–3), 7–42 (2002)

35. Schops, T., et al.: A multi-view stereo benchmark with high-resolution images and multi-camera videos. In: IEEE Conference on Computer Vision and Pattern Recognition, pp. 3260–3269. IEEE (2017)

36. Seki, A., Pollefeys, M.: Patch based confidence prediction for dense disparity map. In: BMVC. BMVA (2016)

37. Shaked, A., Wolf, L.: Improved stereo matching with constant highway networks and reflective confidence learning. In: The IEEE Conference on Computer Vision and Pattern Recognition (CVPR). IEEE (2017)

38. Smolyanskiy, N., Kamenev, A., Birchfield, S.: On the importance of stereo for accurate depth estimation: an efficient semi-supervised deep neural network approach. In: IEEE Conference on Computer Vision and Pattern Recognition (CVPR) Workshops. IEEE (2018)

39. Song, X., Zhao, X., Fang, L., Hu, H., Yu, Y.: Edgestereo: an effective multi-task learning network for stereo matching and edge detection. Int. J. Comput. Vis. **128**, 1–21 (2020)

40. Song, X., Zhao, X., Hu, H., Fang, L.: EdgeStereo: a context integrated residual pyramid network for stereo matching. In: Jawahar, C.V., Li, H., Mori, G., Schindler, K. (eds.) ACCV 2018. LNCS, vol. 11365, pp. 20–35. Springer, Cham (2019). https://doi.org/10.1007/978-3-030-20873-8_2

41. Tonioni, A., Poggi, M., Mattoccia, S., Di Stefano, L.: Unsupervised adaptation for deep stereo. In: The IEEE International Conference on Computer Vision (ICCV). IEEE (2017)

42. Tonioni, A., Poggi, M., Mattoccia, S., Di Stefano, L.: Unsupervised domain adaptation for depth prediction from images. IEEE Trans. Pattern Anal. Mach. Intell. **42**, 2396–2409 (2019)

43. Tonioni, A., Rahnama, O., Joy, T., Di Stefano, L., Thalaiyasingam, A., Torr, P.: Learning to adapt for stereo. In: The IEEE Conference on Computer Vision and Pattern Recognition (CVPR). IEEE (2019)

44. Tonioni, A., Tosi, F., Poggi, M., Mattoccia, S., Stefano, L.D.: Real-time self-adaptive deep stereo. In: The IEEE Conference on Computer Vision and Pattern Recognition (CVPR). IEEE (2019)

45. Tosi, F., Aleotti, F., Poggi, M., Mattoccia, S.: Learning monocular depth estimation infusing traditional stereo knowledge. In: The IEEE Conference on Computer Vision and Pattern Recognition (CVPR). IEEE (2019)

46. Tosi, F., Poggi, M., Tonioni, A., Di Stefano, L., Mattoccia, S.: Learning confidence measures in the wild. In: BMVC. BMVA (2017)

47. Tulyakov, S., Ivanov, A., Fleuret, F.: Weakly supervised learning of deep metrics for stereo reconstruction. In: IEEE Conference on Computer Vision and Pattern Recognition, pp. 1339–1348. IEEE (2017)

48. Uhrig, J., Schneider, N., Schneider, L., Franke, U., Brox, T., Geiger, A.: Sparsity invariant CNNs. In: International Conference on 3D Vision (3DV). IEEE (2017)
49. Wang, Y., Wang, P., Yang, Z., Luo, C., Yang, Y., Xu, W.: Unos: unified unsupervised optical-flow and stereo-depth estimation by watching videos. In: IEEE Conference on Computer Vision and Pattern Recognition, pp. 8071–8081. IEEE (2019)
50. Watson, J., Firman, M., Brostow, G.J., Turmukhambetov, D.: Self-supervised monocular depth hints. In: IEEE International Conference on Computer Vision (ICCV). IEEE (2019)
51. Watson, J., Mac Aodha, O., Turmukhambetov, D., Brostow, G.J., Firman, M.: Learning stereo from single images. In: European Conference on Computer Vision (ECCV). Springer, Heidelberg (2020)
52. Yang, G., Song, X., Huang, C., Deng, Z., Shi, J., Zhou, B.: Drivingstereo: a large-scale dataset for stereo matching in autonomous driving scenarios. In: IEEE Conference on Computer Vision and Pattern Recognition (CVPR). IEEE (2019)
53. Yang, G., Zhao, H., Shi, J., Deng, Z., Jia, J.: SegStereo: exploiting semantic information for disparity estimation. In: Ferrari, V., Hebert, M., Sminchisescu, C., Weiss, Y. (eds.) ECCV 2018. LNCS, vol. 11211, pp. 660–676. Springer, Cham (2018). https://doi.org/10.1007/978-3-030-01234-2_39
54. Yang, Q., Yang, R., Davis, J., Nistér, D.: Spatial-depth super resolution for range images. In: 2007 IEEE Conference on Computer Vision and Pattern Recognition, pp. 1–8. IEEE (2007)
55. Yu, L., Wang, Y., Wu, Y., Jia, Y.: Deep stereo matching with explicit cost aggregation sub-architecture. In: Thirty-Second AAAI Conference on Artificial Intelligence. AAAI Press (2018)
56. Zabih, R., Woodfill, J.: Non-parametric local transforms for computing visual correspondence. In: Eklundh, J.-O. (ed.) ECCV 1994. LNCS, vol. 801, pp. 151–158. Springer, Heidelberg (1994). https://doi.org/10.1007/BFb0028345
57. Zbontar, J., LeCun, Y.: Stereo matching by training a convolutional neural network to compare image patches. J. Mach. Learn. Res. **17**(1–32), 2 (2016)
58. Zhang, F., Prisacariu, V., Yang, R., Torr, P.H.: Ga-net: guided aggregation net for end-to-end stereo matching. In: IEEE Conference on Computer Vision and Pattern Recognition, pp. 185–194. IEEE (2019)
59. Zhong, Y., Li, H., Dai, Y.: Self-supervised learning for stereo matching with self-improving ability. arXiv preprint arXiv:1709.00930 (2017)
60. Zhong, Y., Li, H., Dai, Y.: Open-world stereo video matching with deep RNN. In: Ferrari, V., Hebert, M., Sminchisescu, C., Weiss, Y. (eds.) ECCV 2018. LNCS, vol. 11206, pp. 104–119. Springer, Cham (2018). https://doi.org/10.1007/978-3-030-01216-8_7
61. Zhou, C., Zhang, H., Shen, X., Jia, J.: Unsupervised learning of stereo matching. In: The IEEE International Conference on Computer Vision (ICCV). IEEE (2017)
62. Zhou, T., Brown, M., Snavely, N., Lowe, D.G.: Unsupervised learning of depth and ego-motion from video. In: The IEEE Conference on Computer Vision and Pattern Recognition (CVPR). IEEE (2017)

PIPAL: A Large-Scale Image Quality Assessment Dataset for Perceptual Image Restoration

Gu Jinjin[1(✉)], Cai Haoming[1,2], Chen Haoyu[1], Ye Xiaoxing[1], Jimmy S. Ren[3], and Dong Chao[2,4]

[1] The School of Data Science, The Chinese University of Hong Kong, Shenzhen, China
{jinjingu,haomingcai,haoyuchen,xiaoxingye}@link.cuhk.edu.cn
[2] ShenZhen Key Lab of Computer Vision and Pattern Recognition, SIAT-SenseTime Joint Lab, Shenzhen Institutes of Advanced Technology, Chinese Academy of Sciences, Shenzhen, China
chao.dong@siat.ac.cn
[3] SenseTime Research, Science Park, Hong Kong
rensijie@sensetime.com
[4] SIAT Branch, Shenzhen Institute of Artificial Intelligence and Robotics for Society, Shenzhen, China

Abstract. Image quality assessment (IQA) is the key factor for the fast development of image restoration (IR) algorithms. The most recent IR methods based on Generative Adversarial Networks (GANs) have achieved significant improvement in visual performance, but also presented great challenges for quantitative evaluation. Notably, we observe an increasing inconsistency between perceptual quality and the evaluation results. Then we raise two questions: (1) Can existing IQA methods objectively evaluate recent IR algorithms? (2) When focus on beating current benchmarks, are we getting better IR algorithms? To answer these questions and promote the development of IQA methods, we contribute a large-scale IQA dataset, called Perceptual Image Processing Algorithms (PIPAL) dataset. Especially, this dataset includes the results of GAN-based methods, which are missing in previous datasets. We collect more than 1.13 million human judgments to assign subjective scores for PIPAL images using the more reliable "Elo system". Based on PIPAL, we present new benchmarks for both IQA and super-resolution methods. Our results indicate that existing IQA methods cannot fairly evaluate GAN-based IR algorithms. While using appropriate evaluation methods is important, IQA methods should also be updated along with the development of IR algorithms. At last, we improve the performance of IQA networks on GAN-based distortions by introducing anti-aliasing pooling. Experiments show the effectiveness of the proposed method.

Keywords: Perceptual image restoration · Image quality assessment · Generative adversarial network · Perceptual super-resolution

© Springer Nature Switzerland AG 2020
A. Vedaldi et al. (Eds.): ECCV 2020, LNCS 12356, pp. 633–651, 2020.
https://doi.org/10.1007/978-3-030-58621-8_37

1   Introduction

Image restoration (IR) is a classic low-level vision problem that aims to reconstruct high-quality images from distorted low-quality inputs. Typical IR tasks include image super-resolution (SR), denoising, enhancement, etc. The whirlwind of deep-learning progress has produced a steady stream of promising IR algorithms that could generate less-distorted or perceptual-friendly images. Nevertheless, one of the key bottlenecks that restrict IR methods' future development is the "evaluation mechanism". Although it is nearly effortless for human eyes to distinguish perceptually better images, it is challenging for an algorithm to measure visual quality fairly. In this work, we will focus on the analysis of existing evaluation methods, and introduce a new image quality assessment (IQA) dataset, which not only includes the most recent IR methods but also has the largest scale/diversity. The motivation will be first stated as follows.

IR methods are generally evaluated by measuring the similarity between the reconstructed images and ground-truth images via IQA metrics, such as PSNR [18] and SSIM [49]. Recently, some non-reference IQA methods, such as Ma [31] and Perceptual Index (PI) [4], are introduced to evaluate the recent perceptual-oriented algorithms. To some extent, these IQA methods are the chief reason for the considerable progress of the IR field. However, while new algorithms have been continuously improving IR performance, we notice an increasing inconsistency between quantitative results and perceptual quality. For example, literature [4] reveals that the superiority of PSNR values does not always accord with better visual quality. Although Blua *et al.* suggest that PI is more relevant to human judgment, algorithms with high PI scores (e.g., ESRGAN [47] and RankSRGAN [61]) could still produce images with obvious unrealistic artifacts. These conflicts lead us to rethink the evaluation methods for IR tasks.

An important reason for this situation is the invention of Generative Adversarial Networks (GANs) [17] and GAN-based IR methods [20,47], bringing completely new characteristics to the output images. In general, these methods often fabricate seemingly realistic yet fake details and textures. This presents a great challenge for existing IQA methods, which cannot distinguish the GAN-generated textures from noises and real details. We naturally raise two questions: (1) Can existing IQA methods objectively evaluate current IR methods, especially GAN-based methods? (2) With the focus on beating benchmarks on the flawed IQA methods, are we getting better IR algorithms? A few works have made early attempt to answer these questions by proposing new benchmarks for IR and IQA methods. Yang *et al.* [51] conduct a comprehensive evaluation of traditional SR algorithms. Blau *et al.* [4] analyze the perception-distortion trade-off phenomenon and suggest the use of multiple IQA methods. However, these prior studies usually apply unreliable human ratings of image quality, and are generally insufficient in the number of IR/IQA methods. Especially, the results of GAN-based methods are missing in the above works.

To touch the heart of this problem, we need to have a better understanding of the new challenges brought by GAN. The first issue is to build a new IQA dataset with GAN-based algorithms. An IQA dataset includes a lot of distorted images with visual quality levels annotated by humans. It can be used to measure the consistency of the prediction of IQA method and human judgment. In this work, we contribute a novel IQA dataset, namely Perceptual Image Processing ALgorithms dataset (PIPAL). The proposed dataset distinguishes from previous datasets in three aspects: (1) In addition to traditional distortion types (e.g., Gaussian noise/blur), PIPAL contains the outputs of several kinds of IR algorithms, including traditional algorithms, deep-learning-based algorithms and GAN-based algorithms. In particular, this is the first time for the results of GAN-based algorithms to appear in an IQA dataset. (2) We employ the Elo rating system [14] to assign subjective scores, involving more than 1.13 million human judgments. Comparing with existing rating systems (e.g., five gradations [42] and Swiss system [36]), the Elo rating system provides much more reliable probability-based rating results. Furthermore, it has good extensibility, allowing users to update the dataset by directly adding new distortion types. (3) The proposed dataset contains 29k images in total, including 250 high-quality reference images, and each of which has 116 distortions. To date, PIPAL is the largest IQA dataset with complete subjective scoring.

With the PIPAL dataset, we are able to answer the questions above. (1) We build a benchmark using the proposed PIPAL dataset for existing IQA methods. Experiments indicate that PIPAL poses challenges for these IQA methods. Evaluating IR algorithms only using existing metrics is not appropriate. Our research also shows that compared with the widely-used metrics (e.g., PSNR and PI), PieAPP [37] and LPIPS [60] are more suitable for evaluating IR algorithms, especially GAN-based algorithms. (2) We then review the development of SR algorithms in recent years. The results show that the recent SR algorithms achieve great progress in the average subjective image quality scores. However, we find that none of the existing IQA methods is always effective in evaluating SR algorithms. With the invention of new IR technologies, the corresponding evaluation methods also need to be adjusted to continuously promote the development of the IR field. (3) We also study the characteristics of GAN-based distortion by comparing them with some well-studied traditional distortions. Based on the results, we argue that existing IQA methods' low tolerance toward spatial misalignment may be one of the key reasons for their performance drop. By introducing anti-aliasing pooling to the existing IQA networks, we are able to improve their performance on GAN-based distortions.

2   Related Work

Image Restoration. As a fundamental computer vision problem, IR aims at recovering a high-quality image from its degraded observations. In past decades, plenty of IR algorithms have been proposed to continuously improve the performance. The early algorithms use hand-craft features [9,53] or exploit image priors [45,46] in optimization problems to reconstruct images. Since the pioneer work of using Convolution Neural Networks (CNNs) to learn the IR mappings [12,21], the deep-learning-based algorithms have dominated IR research due to their remarkable performance and usability [16,19]. Recently, with the invention of GAN [17], GAN-based IR methods [39,61] are not limited to getting a higher PSNR performance but trying to have better perceptual effect. However, these IR algorithms are not perfect. The results of those algorithms also include various image defects, and they are different from the traditional distortions that are often discussed in previous IQA researches. With the development of IR algorithms and the emergence of new technologies, evaluating the results of these algorithms becomes more and more challenging. In this paper, we mainly focus on the restoration of low-resolution images, noisy images, and images degraded by both resolution reduction and noise.

Table 1. Comparison with the previous datasets. We include the outputs of GAN-based algorithms as a novel distortion type. Note that BAPPS [60] and PieAPP are perceptual similarity dataset (as opposed to IQA datasets), and are marked with "*"

Dataset	# Ref. images	Image types	Distortion types	# Distort. types	# Distort. images	# Human judgments	Judgment type
LIVE [42]	29	Image	Traditional	5	0.8k	25k	MOS (Five gradations)
CSIQ [27]	30	Image	Traditional	6	0.8k	5k	MOS (Direct ranking)
TID2008 [36]	25	Image	Traditional	17	1.7k	256k	MOS (Swiss system)
TID2013 [35]	25	Image	Traditional	24	3.0k	524k	MOS (Swiss system)
BAPPS* [60]	187.7k	Patch (256 × 256)	Trad. + alg. outputs	425	375.4k	484.3k	Prob. of Preference
PieAPP* [37]	200	Patch (256 × 256)	Trad. + alg. outputs	75	20.3k	2.3 m	Prob. of Preference
PIPAL (Ours)	250	patch (288 × 288)	Trad. + alg. outputs *including* GAN	40	29 k	1.13 m	MOS (Elo rating system)

Image Quality Assessment. The IQA methods were developed to measure the perceptual quality of images after degradation or post-processing operation. According to different usage scenarios, IQA methods can be divided into full-reference methods (FR-IQA) and no-reference methods (NR-IQA). FR-IQA methods measure the similarity between two images from the perspective of information or perceptual feature similarity, and have been widely used in the evaluation of image/video coding, restoration and communication quality. Beyond the most widely-used PSNR, FR-IQA methods follow a long line of works that can trace back to SSIM [49], which first introduces structural information in measuring image similarity. After that, various FR-IQA methods have been proposed to bridge the gap between the results of IQA methods and human judgments. Similar to other computer vision problems, advanced data-driven methods have also motivated the investigation of applications of IQA [37,60]. In addition to the above FR-IQA methods, NR-IQA methods are proposed to assess image quality without a reference image. Some popular NR-IQA methods include NIQE [34], Ma *et al.* [31], BRISQUE [33], and PI [4]. In some recent works, NR-IQA and FR-IQA methods are combined to measure IR algorithms [4]. Despite of the progress of IQA methods, only a few IQA methods (e.g., PSNR, SSIM and PI) are frequently used to evaluate IR methods.

Image Quality Assessment Datasets. In order to evaluate and develop IQA methods, many datasets have been proposed, such as LIVE [42], CSIQ [27], TID2008 and TID2013 [35,36]. There are also some perceptual similarity datasets such as PieAPP [37], and BAPPS [60]. These datasets provide both distorted images and the corresponding subjective scores, and they have served as baselines for evaluation of IQA methods. IQA datasets are mainly distinguished from each other in three aspects: (1) the collecting of the reference images, (2) the number of distortions included and their types and (3) the collecting strategy of subjective score. A quick comparison of these datasets can be found in Table 1.

Table 2. Our distortions types. In addition to the existing distortions, we include 19 different GAN-based algorithms distortions

Sub-type	Distortion types
Traditional	Gaussian blur, motion blur, image compression, Gaussian noise, spatial warping, bilateral filter, comfort noise.
Super-Resolution	interpolation method, traditional methods, SR with kernel mismatch, PSNR-oriented methods, GAN-based methods.
Denoising	mean filtering, traditional methods, deep-learning-based methods.
Mixture restoration	SR of noisy images, SR after denoising, SR after compression noise removal

3   Perceptual Image Processing ALgorithms Dataset

We then describe the peculiarities of the proposed dataset from the aforementioned aspects of (1) the collecting of the reference images, (2) the number of distortions and their types, and (3) the collecting of subjective score, respectively.

Collection of Reference Images. In the proposed dataset, we select 250 image patches from two high-quality image datasets – DIV2K [1] and Flickr2K [44].

We mainly focus on the area that is relatively hard to restore, such as high-frequency textures. Thus, we crop patches of the representative texture areas from the selected images. The selected reference images are representative of a wide variety of real-world textures, including but not limited to buildings, trees and grasses, animal fur, human faces, text, and artificial textures. The size of these images is 288×288, which could meet the requirements of most IQA methods.

Image Distortions. In our dataset, we have 40 distortion types and these distortions can be divided into four sub-types. An overview of these distortion types is shown in Table 2. The first sub-type includes some traditional distortions (e.g., blur, noise, and compression), which are usually performed by basic low-level image editing operations. In some datasets, these distortions can be very severe; however, in our dataset, we constrain the situation of severe distortions as we want these distortions to be comparable to the IR results, which are not likely to be very low-quality. The second sub-type includes the SR results from a lot of real algorithms. Although some recent datasets [37,60] have covered some of the SR results, they contain results that are inferior in algorithms number and types to our dataset. We divided the used SR algorithms into three categories – traditional algorithms, PSNR-oriented algorithms, and GAN-based algorithms. The results of traditional algorithms can be understood to some extent as loss of detail. The PSNR-oriented algorithms are usually based on deep-learning technology. Comparing with the traditional algorithms, their outputs tend to have sharper edges and higher PSNR performances. The outputs of GAN-based algorithms are more complicated and challenging for IQA methods. They do not quite match the quality of detail loss, as they usually contain texture-like noises, or the quality of noise, as their texture-like noise is similar to the ground truth to some extent, just not accurate. An example of GAN-based distortions is shown in Fig. 1. Measure the similarity of incorrect yet similar features are of great importance to the development of perceptual SR. The third sub-type includes the outputs of several denoising algorithms. Similar to image SR, the used denoising

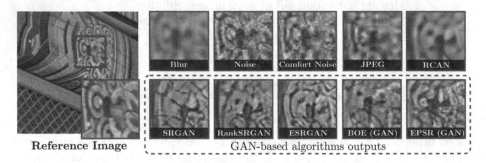

Fig. 1. Visualizing different distortions. Unlike the distortions in the upper row, which do not follow the natural image distribution. The GAN-based outputs are actually similar to natural images. However, their details are wrong

algorithms contain both model-based algorithms and deep learning-based algorithms. In addition to Gaussian noise, we also include JPEG compression noise removal results. At last, we include the restoration results of the mixture image degradation. As revealed in [38,54], performing denoising and SR sequentially or jointly will bring new artifacts or different blur effects that barely occur in other IR tasks.

In summary, we have 40 different distortion types and 116 different distortion levels, totally 29 k distortion images. Note that although the number of distortion types is less than some of the existing datasets, we contain a lot of new distortion types and, especially, a large number of real algorithms results and GAN results. This allows our proposed dataset to provide a more objective benchmark for not only IQA methods but also IR methods.

Elo Rating for Mean Opinion Score. Having distorted images, Mean Opinion Score (MOS) is to be provided for each distortion image. There are several methodologies used to assess the visual quality of an image [35,37,42,60]. Early datasets [42] use "five-gradations rating" method where images are assigned into five categories directly. Using this method will result in a huge bias when the user has a little experience. In recent years, datasets usually collect MOS through a large number of pairwise selections using the Swiss rating system [35,36]. However, as revealed in [37], the way this pairwise MOS is calculated makes it dependent on specific set, which means the MOS scores of two distorted images can change considerably when they are included in two different datasets. In order to eliminate this set-dependence effect, Prashnani *et al.* [37] propose to build dataset only based on the probability of pairwise preference. This method can provide a more accurate propensity probability. However, it not only requires a large number of human judgments, but also can not provide the MOS for distortion types, which is important for building benchmarks. In the proposed dataset, we employ Elo rating system [14] to bring pairwise preference probability and rating system together. The use of Elo system not only provides reliable human ratings but also reduces the number of required human judgments.

The Elo rating system is a statistic-based rating method and is first proposed for assessing chess player levels. We assume that the user preference between two images I_A and I_B follows a Logistic distribution parameterized by their Elo Scores [15]. Given their Elo scores R_A and R_B, the expected probability of preference is given by:

$$P_{A>B} = \frac{1}{1 + 10^{(R_B - R_A)/M}}, P_{B>A} = \frac{1}{1 + 10^{(R_A - R_B)/M}}, \quad (1)$$

where $P_{A>B}$ indicates the probability that one user would prefer I_A than I_B, and M is the parameter of the distribution. In our dataset we use $M = 400$. Once the user makes a choice, we then update the Elo score for both I_A and I_B use the following rule

$$R'_A = R_A + K \times (S_A - P_{A>B}), R'_B = R_B + K \times (S_B - P_{B>A}), \quad (2)$$

where K is the change step in one judgment and is set to 16. S_A indicates whether I_A is chosen: $S_A = 1$ if I_A wins and $S_A = 0$ if I_A fails. With thousands of human

Table 3. The SRCC results with respect to different distortion sub-types. ↑ means the higher the better, while ↓ means the lower the better. Higher coefficients matche perceptual scores better. The values with top 3 performance are marked in **blod**

Method	Traditional distortion	Denoising SR	SR full SR	Traditional SR	PSNR. SR	*GAN-based SR*
PSNR ↑	0.3589	0.4542	0.4099	0.4782	0.5462	0.2839
NQM ↑	0.2561	0.5650	0.4742	0.5374	0.6462	0.3410
UQI ↑	0.3455	0.6246	0.5257	0.6087	0.7060	0.3385
SSIM ↑	0.3910	0.6684	0.5209	0.5856	0.6897	0.3388
MS-SSIM ↑	0.3967	0.6942	0.5596	0.6527	0.7528	0.3823
IFC ↑	0.3708	**0.7440**	0.5651	**0.7062**	**0.8244**	0.3217
VIF ↑	0.4516	**0.7282**	0.5917	0.6927	**0.7864**	0.3857
VSNR-FR ↑	0.4030	0.5938	0.5086	0.6146	0.7076	0.3128
RFSIM ↑	0.3450	0.4520	0.4232	0.4593	0.5525	0.2951
GSM ↑	0.5645	0.6076	0.5361	0.6074	0.6904	0.3523
SR-SIM ↑	0.6036	0.6727	0.6094	0.6561	0.7476	0.4631
FSIM ↑	0.5760	0.6882	0.5896	0.6515	0.7381	0.4090
FSIM$_C$ ↑	0.5724	0.6866	0.5872	0.6509	0.7374	0.4058
VSI ↑	0.4993	0.5745	0.5475	0.6086	0.6938	0.3706
MAD ↓	0.3769	0.7005	0.5424	0.6720	0.7575	0.3494
LPIPS-Alex ↓	0.5935	0.6688	0.5614	0.5487	0.6782	0.4882
LPIPS-VGG ↓	0.4087	0.7197	0.6119	0.6077	0.7329	0.4816
PieAPP ↓	**0.6893**	**0.7435**	**0.7172**	**0.7352**	**0.8097**	**0.5530**
WaDIQaM ↑	**0.6127**	0.7157	**0.6621**	**0.6944**	0.7628	**0.5343**
DISTS ↓	**0.6213**	0.7190	**0.6544**	0.6685	0.7733	**0.5527**
NIQE ↓	0.1107	-0.0059	0.0320	0.0599	0.1521	0.0155
Ma *et al.* ↑	0.4526	0.4963	0.3676	0.6176	0.7124	0.0545
PI ↓	0.3631	0.3107	0.1953	0.4833	0.5710	0.0187

judgments, the Elo scores for each distorted images will converge. The average of the Elo scores in the last few steps will be assigned as the MOS subjective score. The averaging operation aims at reducing the randomness of Elo changes.

An example might help to understand the Elo system. Assume that $R_A = 1500$ and $R_b = 1600$, then we have $P_{A>B} \approx 0.36$ and $P_{B>A} \approx 0.64$. In this situation, if I_A is chosen, the updated Elo score for I_A will be $R'_A = 1500 + 16 \times (1 - 0.36) \approx 1510$ and the new score for I_B is $R'_B = 1600 + 16 \times (0 - 0.64) \approx 1594$; if I_B is chosen, the new score will be $R'_A \approx 1494$ and $R'_B \approx 1605$. Note that as the expected probability for different images being chosen are different, the value change of the Elo scores will also be different. This also indicates that when the quality is too different, the winner will not get a lot from winning the bad image. According to Eq. (1), a 200 of score difference indicates 76% chance to win, and 400 indicates the chance more than 90%. At first, we assign an Elo score of 1400 for each distortion image. After numerous human judgment (in our dataset, we have 1.13 million human judgments), the Elo score for each image are collected.

Another superiority of employing Elo system is that our dataset could be dynamic, and can be extended in the future. The Elo system has been widely used to evaluate the relative level of players in electronic games, where the players are constantly changing and the Elo system can provide ratings for new players in a few gameplays. Recall that one of the chief reasons that "these IQA methods are facing challenges" is the invention of GAN and GAN-based IR methods. What if other novel image generation technologies are proposed in the future? Do people need to build a new dataset to include those new algorithms? With the extendable characteristic of Elo system, one can easily add new distortion types into this dataset and follow the same rating process. Elo system will automatically adjust the Elo score for all the distortions without re-rating for the old distortions.

4   Results

In this section, we conduct a comprehensive study using the proposed PIPAL dataset. We first build a benchmark for IQA methods. Through this benchmark, we can answer the question that "can existing IQA methods objectively evaluate recent IR algorithms?" We then build a benchmark for some recent SR algorithms to explore the relationship between the development of IQA methods and IR research. We can get the answer of "are we getting better IR algorithms by beating benchmarks on these IQA methods?" At last, we study the characteristics of GAN-based distortion by comparing them with some existing distortion types. We also improve the performance of IQA networks on GAN-based distortions by introducing anti-aliased pooling layers.

4.1   Evaluations on IQA Methods

We select a set of commonly-used IQA methods to build the benchmark. For the FR-IQA methods, we include: PSNR [18], NQM [10], UQI [48], SSIM [49], MS-SSIM [50], IFC [41], VIF [40], VSNR-FR [7], RFSIM [57], GSM [30], SR-SIM [55], FSIM and $FSIM_C$ [58], SFF [8], VSI [56], SCQI [2], LPIPS-Alex and -VGG [60], PieAPP [37], WaDIQaM [5] and DISTS [11]. We also include some popular NR-IQA methods: NIQE [34], Ma [31], and PI [4]. All these methods are calculated using the official implementation released by the authors. As in many previous works [42], we evaluate IQA methods mainly using Spearman rank order correlation coefficients (SRCC) and Kendall rank order correlation coefficients (KRCC) [23]. These two indexes evaluate the monotonicity of methods: whether the scores of high-quality images are higher (or lower) than low-quality images. We first evaluate the IQA methods using all types of distortions in PIPAL dataset. A clear exhibition for both SRCC and KRCC rank coefficients is shown in Fig. 2. The first conclusion is that even the best IQA method (i.e., PieAPP) provides only 0.71 SRCC score, which is much lower than their performance in TID2013 dataset (about 0.90). This indicates that the proposed PIPAL dataset is challenging for existing IQA methods and there is a large room for future improvement. Moreover, a high overall correlation performance does not necessarily indicate the high performance on each sub-type of distortions. As the focus of this paper, we want to analyze the performance of IQA using IR results, especially the outputs of GAN-based algorithms. Specifically, we take SR sub-type as an example and show the performance of IQA methods in evaluating SR algorithms. In Table 3, we show the SRCC results with respect to different distortion sub-types, including traditional distortions, denoising outputs, all SR outputs, and the outputs of traditional SR, PSNR-oriented SR and GAN-based SR algorithms. Analysis of Table 3 leads to the following conclusions. First, although performing well in evaluating traditional and PSNR-oriented SR algorithms, almost all IQA methods suffer from severe performance drop when evaluating GAN-based algorithms. This confirms the conclusion of Blua *et al.* [4] that less distortion (e.g., higher PSNR values) may be related to lower perceptual performance for GAN-based IR algorithms. Second, despite of the severe performance drop, several IQA methods still outperform the others on GAN-based algorithms. Coincidentally, they are all recent works and based on deep networks.

We next present the analysis of IQA methods as IR evaluation metrics. In Fig. 3, we show the scatter plots of subjective scores vs. the average values of some commonly-used image quality metrics for 23 SR algorithms. Among them, PSNR and SSIM are the most common measures, IFC is suggested by Yang *et al.* [51], NIQE and PI are suggested in recent works [4,61] for their good performance on GAN-based SR algorithms. LPIPS [60] and PieAPP [37] are selected according to our benchmark. As can be seen that, although widely used, PSNR, SSIM and IFC are anti-correlated with the subjective scores, thus are inappropriate for evaluating GAN-based algorithms. It is worth noting that IFC shows good performance on denoising, traditional SR and PSNR-oriented SR according to Table 3, but drops severely on GAN-based distortions. NIQE and PI show moderate performance on evaluating IR algorithms, and LPIPS and PieAPP are the most correlated. Note that different from the work of Blau *et al.* [4] where they collect perceptual quality only based on whether the image looks real, we collect subjective scores based on the perceptual similarity with the ground truth. Therefore, in evaluating the performance of the IR algorithms from the perspective of reconstructing ground truth, the suggestions given by our work are more appropriate.

4.2   Evaluations on IR Methods

One of the most important applications of IQA technology is to evaluate IR algorithms. IQA methods have been the chief reason for the progress in the IR field as a means of comparing the performance. However, evaluating IR methods only with specific IQA methods also narrows the focus of IR research and converts it to competitions only on the quantitative numbers (e.g., PSNR competitions [6,44] and PI competition [3]). As stated above, existing IQA methods may be inadequate in evaluating IR algorithms. We wonder that with the focus on beating benchmarks on the flawed IQA methods, are we getting better IR

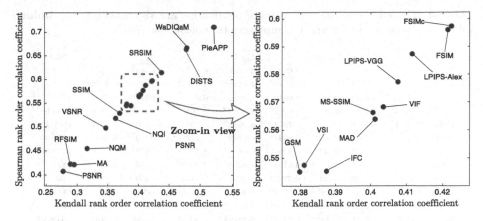

Fig. 2. Quantitative comparison of IQA methods. The right fidgure is the zoom-in view. Higher coefficient matches perceptual score better

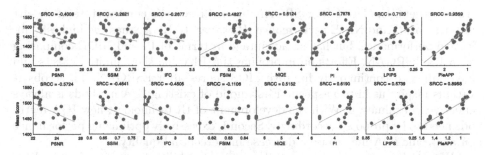

Fig. 3. Analysis of IQA methods in evaluating IR methods. The first row shows the scatter plots of MOS score vs. IQA methods for all SR algorithms. The second row gives scatter plots for GAN-based SR algorithms

algorithms? To answer this question, we take SR task as a representative and select 12 SR algorithms to build a benchmark. These are all representative algorithms and selected from the pre-deep-learning era (since 2013) to the present. The results are shown in Table 4. One can observe that before 2017 (when GAN was applied to SR) the PSNR performance improves continuously. Especially, the deep-learning-based algorithms improve PSNR by about 1.4 dB. These efforts do improve the subjective performance – the average MOS values increases by about 90 in 4 years. After SRGAN was proposed, the PSNR decreased by about 2.6 dB compared to the state-of-the-art PSNR performance at that time (EDSR), but the MOS value increased by about 50 suddenly. In contrast, RCAN was proposed to defeat EDSR in terms of PSNR. Its PSNR performance is a little higher than EDSR but its MOS score is even lower than EDSR. When noting that the mainstream metrics (PSNR and SSIM) had conflicted with the subjective performance, PI was proposed to evaluate perceptual SR algorithms [4]. After that, ESRGAN and RankSRGAN have been continuously improving PI performance.

Table 4. The ×4 SR results. The years of publication are also provided. The **bolded** values are the top 2 values and the superscripts indicate the ranking

Method	Year	PSNR ↑	SSIM ↑	Ma ↑	NIQE ↓	PI ↓	LPIPS ↓	MOS ↑
YY [52]	2013	23.35^8	0.6897^7	4.5486^{10}	6.4174^8	5.9344^7	0.3574^{12}	1367.71^8
TSG [45]	2013	23.55^7	0.6775^8	4.1298^{12}	6.4163^7	6.1433^{10}	0.3570^{11}	1387.24^7
A+ [46]	2014	23.82^6	0.6919^6	4.3852^{11}	6.3645^5	5.9897^9	0.3491^{10}	1354.52^{12}
SRCNN [12]	2014	23.93^5	0.6966^5	4.6094^9	6.5657^{10}	5.9781^8	0.3316^8	1363.68^{11}
FSRCNN [13]	2016	24.07^4	0.7013^3	4.6686^8	6.9985^{11}	6.1649^{11}	0.3281^7	1367.49^9
VDSR [24]	2016	24.13^3	0.6984^4	4.7799^7	7.4436^{12}	6.3319^{12}	0.3484^9	1364.90^{10}
EDSR [29]	2017	$\mathbf{25.17}^2$	$\mathbf{0.7541}^2$	5.7634^6	6.4560^9	5.3463^6	0.3016^6	1447.44^6
SRGAN [28]	2017	22.57^{10}	0.6494^{11}	8.4215^3	3.9527^3	2.7656^3	$\mathbf{0.2687}^2$	1494.14^3
RCAN [62]	2018	$\mathbf{25.21}^1$	$\mathbf{0.7569}^1$	5.9260^5	6.4121^6	5.2430^5	0.2992^5	1455.31^5
BOE [32]	2018	22.68^9	0.6582^9	$\mathbf{8.5209}^2$	$\mathbf{3.7945}^1$	$\mathbf{2.6368}^2$	0.2933^4	1481.51^4
ESRGAN [47]	2018	22.51^{11}	0.6566^{10}	8.3424^4	4.7821^4	3.2198^4	$\mathbf{0.2517}^1$	$\mathbf{1534.25}^1$
RankSRGAN [61]	2019	22.11^{12}	0.6392^{12}	$\mathbf{8.6882}^1$	$\mathbf{3.8155}^2$	$\mathbf{2.5636}^1$	0.2755^3	$\mathbf{1518.29}^2$

Among them, the latest RankSRGAN has achieved the current state-of-the-art in terms of PI and NIQE performance. However, ESRGAN has the highest subjective performance, but has no advantage in terms of PI and NIQE comparing with RankSRGAN. Efforts on improving the PI value show limited effects and have failed to continuously improve MOS scores after ESRGAN. These observations inspire us in two aspects. First, none of existing IQA methods is always effective in evaluation. With the development of IR technology, new IQA methods need to be proposed accordingly. Second, excessively optimizing performance on a specific IQA may cause a decrease in perceptual quality.

We conduct experiments to explore this possibility by performing gradient ascend/descend of certain IQA methods. According to Blau et al. [4], distortion and perceptual quality are at odds with each other. In order to simulate the situation where there is a perception-distortion trade-off, we constrain the PSNR value to be equal to that of the initial distorted image during optimization. We use the output of ESRGAN as the initial image, and the results are shown in Fig. 4. We can see that some images show superior numerical performance when evaluated using certain IQA methods, but may not be dominant in other metrics. Their best-cases also show different visual effects. Even for some IQA methods (LPIPS and DISTS) with good performance on GAN-based distortion, their best-cases still contain serious artifacts. This indicates that evaluating and developing new IQA methods plays an important role in future research.

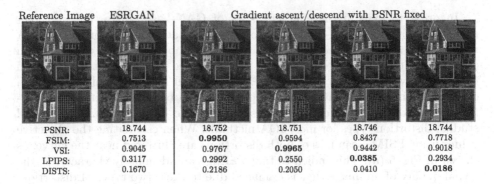

	Reference Image	ESRGAN	Gradient ascent/descend with PSNR fixed			
PSNR:		18.744	18.752	18.751	18.746	18.744
FSIM:		0.7513	**0.9950**	0.9594	0.8437	0.7718
VSI:		0.9045	0.9767	**0.9965**	0.9442	0.9018
LPIPS:		0.3117	0.2992	0.2550	**0.0385**	0.2934
DISTS:		0.1670	0.2186	0.2050	0.0410	**0.0186**

Fig. 4. Best-case images with respect to different IQA methods, with identical PSNR. These are computed by gradient ascent/descent optimization on certain IQA methods

4.3 Discussion of GAN-based Distortion

Recall that LPIPS, PieAPP and DISTS perform relatively better in evaluating GAN-based distortion. The effectiveness of these methods may be attributed to the following reasons. Compared with other IQA methods, deep-learning-based IQA methods can extract image features more effectively. For traditional distortion types, such as blur, compression and noise, the distorted images usually disobey the prior distribution of natural images. Early IQA methods can assess these images by measuring low-level statistic image features such as image gradient and structural information. These strategies are also effective for the outputs of traditional and PSNR-oriented algorithms. However, most of these strategies fail in the case of GAN-based distortion, as the way that GAN-based distortions differ from the reference images is less apparent. They may have similar image statistic features with the reference image. In this case, deep networks are able to capture these unapparent features and distinguish such distortions to some extent.

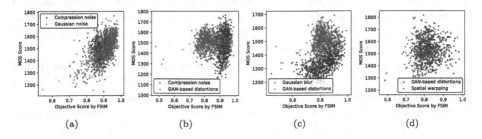

    (a)   (b)   (c)   (d)

Fig. 5. Examples of scatter plots for pairs of distortion types. For distortion types that are easy to measure, samples are well clustered along the fitted curve. For others that are difficult for IQA method, the samples will not be well clustered. The samples of distortion types which have similar behavior will overlap with each other

In order to explore the characteristics of GAN-based distortion, we compare them with some well-studied distortions. As stated in [35], for a good IQA method, the subjective scores in the scatter plot should increase coincide with objective values, and the samples are well clustered along the fitted curve. In the case of two distortion types, if the IQA method behaves similarly for both of them, their samples on the scatter plot will also be well clustered and overlaped. For example, the additive Gaussian noise and lossy compression are well studied distortion types for most IQA methods. When calculating the objective values using FSIM$_C$, samples of both distortions are clustered near the curve, as shown in Fig. 5 (a). This indicates that FSIM$_C$ can adequately characterize the visual quality of an image that is damaged due to these two types of distortion. Then we study GAN-based distortion by comparing it with some existing distortion types using FSIM$_C$: Fig. 5 (b) shows the result of GAN-based distortion and compression noise, and Fig. 5 (c) shows the result of GAN-based distortion and Gaussian blur. It can be seen that the samples of compression noise and Gaussian blur barely intersect with GAN-based samples. FSIM$_C$ largely underestimates the visual quality of GAN-based distortion. In Fig. 5 (d), we show the result of GAN-based distortion and spatial warping distortion. As can be seen, these two distortion types behave unexpectedly similar. FSIM$_C$ cannot handle them and presents the same random and diffused state. The quantitative results also verify this phenomenon. For spatial warping distortion type, the SRCC of FSIM$_C$ is 0.31, and it is close to the performance of GAN-based distortion, which is 0.41. Thus we argue that the spatial warping distortion and GAN-based distortion pose similar challenges to FSIM$_C$.

As revealed in experimental psychology [22,25], the mutual interference between visual information may cause the Visual Masking effects. According to this theory, some key reasons that IQA methods tend to underestimate both GAN-based distortion and spatial warping distortion are as follows. Firstly, for the edges with strong intensity change, the human visual system (HVS) is sensitive to the contour and shape, but not sensitive to the error and misalignment of the edges. Secondly, the ability of HVS to distinguish texture decreases in the region with dense textures. When the extracted features of the texture are similar, the HVS will ignore part of the subtle differences and misalignment of textures. However, most of the traditional and deep-learning-based IQA methods require good alignment for the inputs. This partially causes the drop of performance of these IQA methods on GAN-based distortion.

This finding provides us an insight that if we explicitly consider the spatial misalignment, we may improve the performance of IQA methods on GAN-based distortion. We explore this possibility by introducing anti-aliasing pooling layer to IQA networks. IQA networks extract features by cascaded convolution operations. If we want the IQA networks to be robust to small misalignment, the extracted features should at least be invariant to this misalignment/shift. CNN should have been shift-invariant, as the standard convolution operations are shift-invariant. However, IQA networks are usually not shift-invariant, as some commonly used downsampling layers, such as max/average pooling and stridedconvolution ignore the sampling theorem [59]. These operations are employed in VGG [43] and Alex [26] networks, which are popular backbone architectures for feature extraction (e.g., in LPIPS and DISTS). As suggested by Zhang [59], one can fix this by introducing anti-aliasing pooling. We conduct this experiment based on LPIPS-Alex and introduce l2 pooling [11] and BlurPool [59] layers to replace its max pooling layers. l2 pooling fixes this problem by low-pass filtering before downsampling, and BlurPool further improves the performance by low-pass filtering between dense max operation and subsampling. The results are shown in Table 5. We observe increased correlation in both PIPAL full set and GAN-based distortion subset. The results demonstrate the effectiveness of improving anti-aliasing pooling and indicate that the lack of robustness to small misalignment is one of the reasons for the decline in the performance of GAN-based distortion.

Table 5. The SRCC and KRCC performance of LPIPS with different pooling layers. The anti-aliasing pooling layers (l2 pooling [11] and BlurPool [59]) improve the performance both on the PIPAL full set and GAN-based distortion subset

Test set	LPIPS baseline		LPIPS + l2 pooling		LPIPS + BlurPool	
	SRCC	KRCC	SRCC	KRCC	SRCC	KRCC
PIPAL	0.5604	0.3910	0.5816	0.4080	**0.5918**	**0.4160**
PIPAL GAN distort	0.4862	0.3339	0.4942	0.3394	**0.5135**	**0.3549**

5   Conclusion

In this paper, we construct a novel IQA dataset, namely PIPAL and establish benchmarks for both IQA methods and IR algorithms. Our results indicate that existing IQA methods face challenges in evaluating perceptual IR algorithms, especially GAN-based algorithms. We also shed light on improving IQA networks by introducing anti-aliasing pooling layers. Experiments demonstrate the effectiveness of the proposed strategy.

Acknowledgement. This work is partially supported by SenseTime Group Limited, the National Natural Science Foundation of China (61906184), Science and Technology Service Network Initiative of Chinese Academy of Sciences (KFJ-STS-QYZX-092), Shenzhen Basic Research Program (JSGG20180507182100698, CXB201104220032A), the Joint Lab of CAS-HK, Shenzhen Institute of Artificial Intelligence and Robotics for Society. The corresponding author is Chao Dong.

References

1. Agustsson, E., Timofte, R.: Ntire 2017 challenge on single image super-resolution: Dataset and study. In: Proceedings of the IEEE Conference on Computer Vision and Pattern Recognition Workshops, pp. 126–135 (2017)

2. Bae, S.H., Kim, M.: A novel image quality assessment with globally and locally consilient visual quality perception. IEEE Trans. Image Process. **25**(5), 2392–2406 (2016)
3. Blau, Y., Mechrez, R., Timofte, R., Michaeli, T., Zelnik-Manor, L.: The 2018 PIRM challenge on perceptual image super-resolution. In: Leal-Taixé, L., Roth, S. (eds.) ECCV 2018. LNCS, vol. 11133, pp. 334–355. Springer, Cham (2019). https://doi.org/10.1007/978-3-030-11021-5_21
4. Blau, Y., Michaeli, T.: The perception-distortion tradeoff. In: Proceedings of the IEEE Conference on Computer Vision and Pattern Recognition, pp. 6228–6237 (2018)
5. Bosse, S., Maniry, D., Müller, K.R., Wiegand, T., Samek, W.: Deep neural networks for no-reference and full-reference image quality assessment. IEEE Trans. Image Process. **27**(1), 206–219 (2017)
6. Cai, J., et al.: Ntire 2019 challenge on real image super-resolution: Methods and results. In: 2019 IEEE/CVF Conference on Computer Vision and Pattern Recognition Workshops (CVPRW), pp. 2211–2223. IEEE (2019)
7. Chandler, D.M., Hemami, S.S.: VSNR: a wavelet-based visual signal-to-noise ratio for natural images. IEEE Trans. Image Process. **16**(9), 2284–2298 (2007)
8. Chang, H.W., Yang, H., Gan, Y., Wang, M.H.: Sparse feature fidelity for perceptual image quality assessment. IEEE Trans. Image Process. **22**(10), 4007–4018 (2013)
9. Dabov, K., Foi, A., Katkovnik, V., Egiazarian, K.: Image denoising by sparse 3-d transform-domain collaborative filtering. IEEE Trans. Image Process. **16**(8), 2080–2095 (2007)
10. Damera-Venkata, N., Kite, T.D., Geisler, W.S., Evans, B.L., Bovik, A.C.: Image quality assessment based on a degradation model. IEEE Trans. Image Process. **9**(4), 636–650 (2000)
11. Ding, K., Ma, K., Wang, S., Simoncelli, E.P.: Image quality assessment: Unifying structure and texture similarity. arXiv preprint arXiv:2004.07728 (2020)
12. Dong, C., Loy, C.C., He, K., Tang, X.: Image super-resolution using deep convolutional networks. IEEE Trans. Pattern Anal. Mach. Intell. **38**(2), 295–307 (2015)
13. Dong, Chao., Loy, Chen Change, Tang, Xiaoou: Accelerating the super-resolution convolutional neural network. In: Leibe, Bastian, Matas, Jiri, Sebe, Nicu, Welling, Max (eds.) ECCV 2016. LNCS, vol. 9906, pp. 391–407. Springer, Cham (2016). https://doi.org/10.1007/978-3-319-46475-6_25
14. Elo, A.E.: The Rating of Chessplayers, Past and Present. Arco publishing, New York (1978)
15. Elo, A.E.: Logistic Probability as a Rating Basis. The Rating of Chessplayers, Past&Present. ISHI Press International, New York p. 10453 (2008)
16. Feng, R., Gu, J., Qiao, Y., Dong, C.: Suppressing model overfitting for image super-resolution networks. In: 2019 IEEE/CVF Conference on Computer Vision and Pattern Recognition Workshops (CVPRW), pp. 1964–1973. IEEE (2019)
17. Goodfellow, I., et al.: Generative adversarial nets. In: Advances in Neural Information Processing Systems, pp. 2672–2680 (2014)
18. Group, V.Q.E., et al.: Final report from the video quality experts group on the validation of objective models of video quality assessment. In: VQEG meeting, Ottawa, Canada, March 2000 (2000)
19. Gu, J., Lu, H., Zuo, W., Dong, C.: Blind super-resolution with iterative kernel correction. In: Proceedings of the IEEE Conference on Computer Vision and Pattern Recognition, pp. 1604–1613 (2019)

20. Gu, J., Shen, Y., Zhou, B.: Image processing using multi-code GAN prior. In: Proceedings of the IEEE/CVF Conference on Computer Vision and Pattern Recognition, pp. 3012–3021 (2020)
21. Jain, V., Seung, S.: Natural image denoising with convolutional networks. In: Advances in Neural Information Processing Systems, pp. 769–776 (2009)
22. Kahneman, D.: Method, findings, and theory in studies of visual masking. Psychol. Bull. **70**(6p1), 404 (1968)
23. Kendall, M., Stuart, A.: The advanced theory of statistics; charles griffin & co. Ltd. (London) **83**, 62013 (1977)
24. Kim, J., Kwon Lee, J., Mu Lee, K.: Accurate image super-resolution using very deep convolutional networks. In: Proceedings of the IEEE Conference on Computer Vision and Pattern Recognition, pp. 1646–1654 (2016)
25. Kolers, P.A.: Intensity and contour effects in visual masking. Vis. Res. **2**(9–10), 277-IN4 (1962)
26. Krizhevsky, A., Sutskever, I., Hinton, G.E.: ImageNet classification with deep convolutional neural networks. In: Advances in Neural Information Processing Systems, pp. 1097–1105 (2012)
27. Larson, E.C., Chandler, D.M.: Most apparent distortion: full-reference image quality assessment and the role of strategy. J. Electron. Imaging **19**(1), 011006 (2010)
28. Ledig, C., et al.: Photo-realistic single image super-resolution using a generative adversarial network. In: Proceedings of the IEEE Conference on Computer Vision and Pattern Recognition, pp. 4681–4690 (2017)
29. Lim, B., Son, S., Kim, H., Nah, S., Mu Lee, K.: Enhanced deep residual networks for single image super-resolution. In: Proceedings of the IEEE Conference on Computer Vision and Pattern Recognition Workshops, pp. 136–144 (2017)
30. Liu, A., Lin, W., Narwaria, M.: Image quality assessment based on gradient similarity. IEEE Trans. Image Process. **21**(4), 1500–1512 (2011)
31. Ma, C., Yang, C.Y., Yang, X., Yang, M.H.: Learning a no-reference quality metric for single-image super-resolution. Comput. Vis. Image Underst. **158**, 1–16 (2017)
32. Navarrete Michelini, Pablo., Zhu, Dan, Liu, Hanwen: Multi–scale recursive and perception–distortion controllable image super–resolution. In: Leal-Taixé, Laura, Roth, Stefan (eds.) ECCV 2018. LNCS, vol. 11133, pp. 3–19. Springer, Cham (2019). https://doi.org/10.1007/978-3-030-11021-5_1
33. Mittal, A., Moorthy, A.K., Bovik, A.C.: No-reference image quality assessment in the spatial domain. IEEE Trans. Image Process. **21**(12), 4695–4708 (2012)
34. Mittal, A., Soundararajan, R., Bovik, A.C.: Making a "Completely Blind" image quality analyzer. IEEE Signal Process. Lett. **20**(3), 209–212 (2012)
35. Ponomarenko, N., et al.: Image database tid2013: peculiarities, results and perspectives. Signal Process Image Commun. **30**, 57–77 (2015)
36. Ponomarenko, N., Lukin, V., Zelensky, A., Egiazarian, K., Carli, M., Battisti, F.: TID 2008-a database for evaluation of full-reference visual quality assessment metrics. Adv. Mod. Radioelectronics **10**(4), 30–45 (2009)
37. Prashnani, E., Cai, H., Mostofi, Y., Sen, P.: Pieapp: Perceptual image-error assessment through pairwise preference. In: Proceedings of the IEEE Conference on Computer Vision and Pattern Recognition, pp. 1808–1817 (2018)
38. Qian, G., Gu, J., Ren, J.S., Dong, C., Zhao, F., Lin, J.: Trinity of pixel enhancement: a joint solution for demosaicking, denoising and super-resolution. arXiv preprint arXiv:1905.02538 (2019)
39. Sajjadi, M.S., Scholkopf, B., Hirsch, M.: Enhancenet: single image super-resolution through automated texture synthesis. In: Proceedings of the IEEE International Conference on Computer Vision, pp. 4491–4500 (2017)

40. Sheikh, H.R., Bovik, A.C.: Image information and visual quality. IEEE Trans. Image Process. **15**(2), 430–444 (2006)
41. Sheikh, H.R., Bovik, A.C., De Veciana, G.: An information fidelity criterion for image quality assessment using natural scene statistics. IEEE Trans. Image Process. **14**(12), 2117–2128 (2005)
42. Sheikh, H.R., Sabir, M.F., Bovik, A.C.: A statistical evaluation of recent full reference image quality assessment algorithms. IEEE Trans. Image Process. **15**(11), 3440–3451 (2006)
43. Simonyan, K., Zisserman, A.: Very deep convolutional networks for large-scale image recognition. arXiv preprint arXiv:1409.1556 (2014)
44. Timofte, R., Agustsson, E., Van Gool, L., Yang, M.H., Zhang, L.: Ntire 2017 challenge on single image super-resolution: Methods and results. In: Proceedings of the IEEE Conference on Computer Vision and Pattern Recognition Workshops, pp. 114–125 (2017)
45. Timofte, R., De Smet, V., Van Gool, L.: Anchored neighborhood regression for fast example-based super-resolution. In: Proceedings of the IEEE International Conference on Computer Vision, pp. 1920–1927 (2013)
46. Timofte, Radu., De Smet, Vincent, Van Gool, Luc: A+: adjusted anchored neighborhood regression for fast super-resolution. In: Cremers, Daniel, Reid, Ian, Saito, Hideo, Yang, Ming-Hsuan (eds.) ACCV 2014. LNCS, vol. 9006, pp. 111–126. Springer, Cham (2015). https://doi.org/10.1007/978-3-319-16817-3_8
47. Wang, Xintao: ESRGAN: enhanced super-resolution generative adversarial networks. In: Leal-Taixé, Laura, Roth, Stefan (eds.) ECCV 2018. LNCS, vol. 11133, pp. 63–79. Springer, Cham (2019). https://doi.org/10.1007/978-3-030-11021-5_5
48. Wang, Z., Bovik, A.C.: A universal image quality index. IEEE Signal Process. Lett. **9**(3), 81–84 (2002)
49. Wang, Z., Bovik, A.C., Sheikh, H.R., Simoncelli, E.P., et al.: Image quality assessment: from error visibility to structural similarity. IEEE Trans. Image Process. **13**(4), 600–612 (2004)
50. Wang, Z., Simoncelli, E.P., Bovik, A.C.: Multiscale structural similarity for image quality assessment. In: The Thrity-Seventh Asilomar Conference on Signals, Systems and Computers, 2003. vol. 2, pp. 1398–1402 (2003)
51. Yang, Chih-Yuan., Ma, Chao, Yang, Ming-Hsuan: Single-image super-resolution: a benchmark. In: Fleet, David, Pajdla, Tomas, Schiele, Bernt, Tuytelaars, Tinne (eds.) ECCV 2014. LNCS, vol. 8692, pp. 372–386. Springer, Cham (2014). https://doi.org/10.1007/978-3-319-10593-2_25
52. Yang, C.Y., Yang, M.H.: Fast direct super-resolution by simple functions. In: Proceedings of the IEEE International Conference on Computer Vision, pp. 561–568 (2013)
53. Yang, J., Wright, J., Huang, T.S., Ma, Y.: Image super-resolution via sparse representation. IEEE Trans. Image Process. **19**(11), 2861–2873 (2010)
54. Zhang, K., Zuo, W., Zhang, L.: Learning a single convolutional super-resolution network for multiple degradations. In: Proceedings of the IEEE Conference on Computer Vision and Pattern Recognition, pp. 3262–3271 (2018)
55. Zhang, L., Li, H.: Sr-sim: A fast and high performance IQA index based on spectral residual. In: 2012 19th IEEE International Conference on Image Processing, pp. 1473–1476. IEEE (2012)
56. Zhang, L., Shen, Y., Li, H.: VSI: a visual saliency-induced index for perceptual image quality assessment. IEEE Trans. Image Process. **23**(10), 4270–4281 (2014)

57. Zhang, L., Zhang, L., Mou, X.: RFSIM: a feature based image quality assessment metric using riesz transforms. In: 2010 IEEE International Conference on Image Processing. pp. 321–324. IEEE (2010)
58. Zhang, L., Zhang, L., Mou, X., Zhang, D.: FSIM: a feature similarity index for image quality assessment. IEEE Trans. Image Process. **20**(8), 2378–2386 (2011)
59. Zhang, R.: Making convolutional networks shift-invariant again. In: International Conference on Machine Learning, pp. 7324–7334 (2019)
60. Zhang, R., Isola, P., Efros, A.A., Shechtman, E., Wang, O.: The unreasonable effectiveness of deep features as a perceptual metric. In: Proceedings of the IEEE Conference on Computer Vision and Pattern Recognition, pp. 586–595 (2018)
61. Zhang, W., Liu, Y., Dong, C., Qiao, Y.: Ranksrgan: Generative adversarial networks with ranker for image super-resolution. In: Proceedings of the IEEE International Conference on Computer Vision, pp. 3096–3105 (2019)
62. Zhang, Y., Li, K., Li, K., Wang, L., Zhong, B., Fu, Y.: Image super-resolution using very deep residual channel attention networks. In: Proceedings of the European Conference on Computer Vision (ECCV), pp. 286–301 (2018)

Why Do These Match? Explaining the Behavior of Image Similarity Models

Bryan A. Plummer[1]([envelope]), Mariya I. Vasileva[2], Vitali Petsiuk[1], Kate Saenko[1,3], and David Forsyth[2]

[1] Boston University, Boston, MA 02215, USA
{bplum,vpetsiuk,saenko}@bu.edu
[2] University of Illinois at Urbana-Champaign, Urbana, IL 61801, USA
{mvasile2,daf}@illinois.edu
[3] MIT-IBM Watson AI Lab, Cambridge, MA 02142, USA

Abstract. Explaining a deep learning model can help users understand its behavior and allow researchers to discern its shortcomings. Recent work has primarily focused on explaining models for tasks like image classification or visual question answering. In this paper, we introduce Salient Attributes for Network Explanation (SANE) to explain image similarity models, where a model's output is a score measuring the similarity of two inputs rather than a classification score. In this task, an explanation depends on both of the input images, so standard methods do not apply. Our SANE explanations pairs a saliency map identifying important image regions with an attribute that best explains the match. We find that our explanations provide additional information not typically captured by saliency maps alone, and can also improve performance on the classic task of attribute recognition. Our approach's ability to generalize is demonstrated on two datasets from diverse domains, Polyvore Outfits and Animals with Attributes 2. Code available at: https://github. com/VisionLearningGroup/SANE.

Keywords: Explainable AI · Image similarity models · Fashion compatibility · Image retrieval

1 Introduction

Many problems in artificial intelligence that require reasoning about complex relationships can be solved by learning a feature embedding to measure similarity between images and/or other modalities such as text. Examples of these tasks include image retrieval [18,27,38], zero-shot recognition [2,23,35] or scoring

---

B. A. Plummer and M. I. Vasileva—Equal contribution

Electronic supplementary material The online version of this chapter (https:// doi.org/10.1007/978-3-030-58621-8_38) contains supplementary material, which is available to authorized users.

A. Vedaldi et al. (Eds.): ECCV 2020, LNCS 12356, pp. 652–669, 2020.
https://doi.org/10.1007/978-3-030-58621-8_38

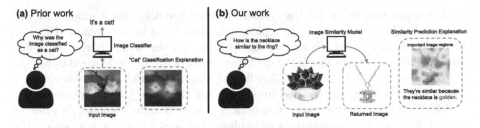

Fig. 1. Existing explanation methods focus on image classification problems (left), whereas we explore explanations for image similarity models (right). We pair a saliency map, which identifies important image regions but often provides little interpretable information, with an attribute (*e.g.*, golden), which is more human-interpretable and, thus, a more useful explanation than saliency alone

fashion compatibility [11,15,31,33]. Reasoning about the behavior of similarity models can aid researchers in identifying potential improvements, show where two images differ for anomaly detection, promote diversity in fashion recommendation by ensuring different traits are most prominent in the top results, or simply help users understand the model's predictions which can build trust [32]. However, prior work on producing explanations for neural networks has primarily focused on explaining classification models [7,25,26,28,29,40] and does not directly apply to similarity models. Given a *single* input image, such methods produce a saliency map which identifies pixels that played a significant role towards a particular class prediction (see Fig. 1a for an example). On the other hand, a similarity model requires at least *two* images to produce a score. The interaction between both images defines which features are more important, so replacing just one of the images can result in identifying different salient traits.

Another limitation of existing work is that saliency alone may be insufficient as an explanation of (dis)similarity. When similarity is determined by the presence or absence of an object, a saliency map may be enough to understand model behavior. However, for the image pair in Fig. 1b, highlighting the necklace as the region that contributes most to the similarity score is reasonable, but uninformative given that there are no other objects in the image. Instead, what is important is that the necklace shares a similar color with the ring. Whether these attributes or salient parts are a better fit can vary depending on the image pairs and the domain they come from. For example, an image can be matched as formal-wear because of a shirt's collar (salient part), while two images of animals can match because both have stripes (attribute).

Guided by this intuition, we introduce *Salient Attributes for Network Explanation (SANE)*. Our approach generates a saliency map to explain a model's similarity score, paired with an attribute explanation that identifies important image properties. SANE is a "black box" method, meaning it can explain any network architecture and only needs to measure changes to a similarity score with different inputs. Unlike a standard classifier, which simply predicts the most likely attributes for a given image, our explanation method predicts which attributes are important for the similarity score predicted by a model. Predictions are made for each image in a pair, and allowed to be non-symmetric: *e.g.*, the explanation for why the ring in Fig. 1b matches the necklace may be that it contains "black", even though the explanation for why the necklace matches the ring could be that it is "golden." A different similarity model may also result in different attributes being deemed important for the same pair of images.

SANE combines three major components: an attribute predictor, a prior on the suitability of each attribute as an explanation, and a saliency map generator. Our underlying assumption is that at least one of the attributes present in each image should be able to explain the similarity score assigned to the pair. Given an input image, the attribute predictor provides a confidence score and activation map for each attribute, while the saliency map generator produces regions important for the match. During training, SANE encourages overlap between the similarity saliency and attribute activation. At test time, we rank attributes as explanations for an image pair based on a weighted sum of this attribute-saliency map matching score, the explanation suitability prior of the attribute, and the likelihood that the attribute is present in the image. Although we only evaluate the top-ranked attribute, in practice multiple attributes could be used to explain a similarity score. We find that using saliency maps as supervision for the attribute activation maps during training not only improves the attribute-saliency matching, resulting in better attribute explanations, but also boosts attribute recognition performance using standard metrics like average precision.

We evaluate several candidate saliency map generation methods which are primarily adaptations of "black box" approaches that do not rely on a particular model architecture or require access to network parameters to produce a saliency map [7,26,28,40]. These methods generally identify important regions by measuring a change in the output class score resulting from a perturbation of the input image. Similarity models, however, often rely on a learned embedding space to reason about relationships between images, where proximity between points or the lack thereof indicates some degree of correspondence. An explanation system for embedding models must therefore consider how distances between embedded points, and thus their similarity, change based on perturbing one or both input images. We explore two strategies for adapting these approaches to our task. First, we manipulate just a single image (the one we wish to produce an explanation for) while keeping the other image fixed. Second, we manipulate both images to allow for more complex interactions between the pair. See Subsect. 3.2 for details and a discussion on the ramifications of this choice.

Our paper makes the following contributions: 1) we provide the first quantitative study of explaining the behavior of image similarity models; 2) we propose a novel explanation approach that combines saliency maps and attributes; 3) we validate our method with a user study combined with metrics designed to link our explanations to model performance, and find that it produces more informative explanations than adaptations of prior work to this task, and further improves attribute recognition performance.

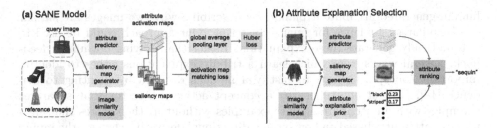

Fig. 2. Approach Overview. (a) During training we use the saliency map generator (Subsect. 3.2) to the important regions when compared to many reference images. Then, we encourage at least one ground truth attribute's activation maps to match each saliency map (details in Subsect. 3.1). (b) At test time, we rank attribute explanations by how well the saliency and attribute activation maps match, along with the likelihood of the attribute and its explanation suitability prior (details in Subsect. 3.3). We assume the image similarity model has been pretrained and is kept fixed in all our experiments

2   Related Work

Saliency-Based Explanations. Saliency methods can generally be split into "white box" and "black box" approaches. "White box" methods assume access to internal components of a neural network, either in the form of gradients or activations of specific layers [4,5,25,29,30,39,41,43]. Most of these methods produce a saliency map by using some version of backpropagation from class probability to an input image. In contrast, "black box" methods require no knowledge of model internals (*e.g.* weights or gradients). They obtain saliency maps perturbing the input in a predefined way and measuring the effect of that perturbation on the model output, such as class score. We adapt and compare three "black box" and one "white box" methods for our saliency map generator in Fig. 2. "Black box" approaches include a Sliding Window [40], which masks image regions sequentially, and Randomized Input Sampling for Explanations (RISE) [26], which masks random sets of regions. Both measure the effect removing these regions has on the class score. LIME [28] first obtains a super-pixel representation of an image. Super-pixel regions are randomly deleted, and their importance is estimated using Lasso. "White box" Mask [7] learns a saliency map directly by using different perturbation operators and propagating the error to a low resolution mask. Although there exists limited work that adapts certain saliency methods to the image similarity setting [8], they present qualitative results only, *i.e.*, these methods are not evaluated quantitatively on their explanation accuracy as done in our work.

Natural Language and Attribute-Based Explanations. Instead of producing saliency maps, which can sometimes be difficult to interpret, researchers have explored methods of producing text-based explanations. These include methods which justify a model's answer in the visual question answering task [17,22], rationalize the behavior of a self-driving vehicle [20], or describe why a category was selected in fine-grained object classification [13]. Lad *et al.* [21] used human-generated attribute explanations describing why two images are similar or dissimilar as guidance for image clustering. Our approach could be used to automatically generate these explanations rather than relying on human feedback. Several works exist which learn attribute explanations either to identify important concepts [3,6] or to justify a model's decision by pointing to evidence [14]. Kim *et al.* [19] learns a concept activation vector that separates examples with an attribute against examples without it, then scores the sensitivity of attributes based on how often a directional derivative changes the inputs towards the concept. However, all these methods were designed to explain categorical predictions rather than similarity models. To the best of our knowledge, ours is the first work which uses attribute explanations in the image similarity setting.

Interpretable Image Similarity. Attributes are often used to provide a sense of interpretability to for image similarity tasks [9,24,37] or aid in the retrieval of similar images using attribute information [1,42]. However, these methods typically require that a model be trained with a particular architecture in order to provide an interpretabile output (*i.e.*, they cannot be directly applied to any pretrained model). In contrast, SANE is able to explain the predictions of any image similarity model regardless of architecture.

3   Salient Attributes for Network Explanations (SANE)

We are given a fixed model that predicts the similarity between two images, and must explain why a query image is similar to a reference image. While typical models for predicting similarity are learned from data, *e.g.*, with an embedding method and triplet loss, our approach is agnostic as to how the model being explained is built. SANE consists of three components: the attribute explanation model (Subsect. 3.1), the saliency map generator (Subsect. 3.2), and an attribute explanation suitability prior (Subsect. 3.3). Although we train a CNN to produce attribute predictions, the image similarity model we wish to explain is kept fixed. At test time, one recovers a saliency map for the match from the query image in a pair, then uses the attribute explanation model and attribute suitability prior to rank each attribute's ability to explain the image similarity model. See Fig. 2 for an overview of our approach.

3.1   Attribute Explanation Model

Suppose we have access to pairs of images (I_r, I_q), where I_r denotes a reference image, and I_q a query image. We wish to obtain an explanation for the match between I_r and I_q. Associated with each pair is a saliency map $\mathbf{m}_q$ produced by a saliency map generator (described in Subsect. 3.2). To compute a saliency map for $\mathbf{m}_r$ instead, we need simply to swap the query and reference images, which would likely result in a different saliency map than $\mathbf{m}_q$. Finally, assume we have access to binary attribute annotations a_i, $i = 1, \ldots, A$, and let $\mathbf{a}_{gt} \in \{0, 1\}^A$ be the set of ground truth attribute annotations for a given query image.

If no attribute annotations are provided, an attribute discovery method could be employed (*e.g.*, [10,34]). We explore an attribute discovery method in the supplementary.

Our attribute explanation model produces confidence scores $\hat{\mathbf{a}} \in \mathbb{R}^A$ for I_q. Unlike a standard attribute classifier, however, our goal is not just to predict the most likely attributes in I_q, but rather to identify which attributes contribute the most to the similarity score $s(I_r, I_q)$ produced by the similarity model we wish to obtain explanations for. To accomplish this, the layer activations for attribute a_i before the global average pooling layer are defined as an attribute activation map $\mathbf{n}_i$. This attribute activation map represents a downsampled mask of an image that identifies prominent regions in I_q for that attribute. We encourage at least one ground truth attribute's activation map for image I_q to match saliency map $\mathbf{m}_q$ as a form of regularization. Our underlying assumption, which we validate empirically, is that at least one of the ground truth attributes of I_q should be able to explain why I_q is similar to I_r. Thus, at least one of the attribute activation maps $\mathbf{n}_i$ should closely resemble the saliency map for the match, $\mathbf{m}_q$.

Each attribute confidence score is obtained using a global average pooling layer on its attribute activation map, followed by a softmax. A traditional loss function for multi-label classification would be binary cross-entropy, which makes independent (*i.e.*, noncompetitive) predictions reflecting the likelihood of each attribute in an image. However, this typically results in a model where attribute scores are not comparable. For example, a confidence score of 0.6 may be great for attribute A, but a horrible score for attribute B. Thus, such a loss function would be ill-suited for our purposes since we need a ranked list of attributes for each image. Instead, our attribute explanation model is trained using a Huber loss [16], sometimes referred to as a smooth ℓ_1 loss, which helps encourage sparsity in predictions. This provides a competitive loss across attributes and thus can help ensure calibrated attribute confidence scores that can be used to rank attribute prevalence in an image. More formally, given a set of confidence scores $\hat{\mathbf{a}}$ and attribute labels $\mathbf{a}_{gt}$, our loss is,

$$L_{Huber}(\hat{\mathbf{a}}, \mathbf{a}_{gt}) = \begin{cases} \frac{1}{2}(\mathbf{a}_{gt} - \hat{\mathbf{a}})^2 & \text{for } |\mathbf{a}_{gt} - \hat{\mathbf{a}}| \leq 1 \\ |\mathbf{a}_{gt} - \hat{\mathbf{a}}| & \text{otherwise.} \end{cases} \tag{1}$$

Note that multiple attributes can be present in the image; and that this loss operates on attributes, not attribute activation maps. Since the confidence scores sum to one (due to the softmax), we scale a binary label vector by the number of ground truth attributes A_{gt} (*e.g.*, if there are four attributes for an image, its label would be 0.25 for each ground truth attribute, and zero for all others).

Leveraging Saliency Maps During Training. We explicitly encourage our model to identify attributes that are useful in explaining the predictions of an image similarity model by finding which attributes best describe the regions of high importance to similarity predictions. To accomplish this, we first find a set of regions that may be important to the decisions of an image similarity model by generating a set of K saliency maps $\mathcal{M}_q$ for up to K reference images similar to the query. For the query image under consideration, we also construct a set of attribute activation maps $\mathcal{N}_{gt}$ corresponding to each ground truth attribute. Then, for each saliency map in $\mathcal{M}_q$, we find its best match in $\mathcal{N}_{gt}$. We match saliency maps to attributes rather than the other way around since not all annotated attributes are necessarily relevant to the explanation of $s(I_r, I_q)$. We use an ℓ_2 loss between the selected attribute activation map and saliency map, $i.e.$,

$$L_{hm} = \frac{1}{K} \sum_{\forall m \in \mathcal{M}_q} \min_{\forall n \in \mathcal{N}_{gt}} \|m - n\|_2 . \tag{2}$$

Combined with the attribute classification loss, our model's complete loss is:

$$L_{total} = L_{Huber} + \lambda L_{hm}, \tag{3}$$

where λ is a scalar parameter.

3.2 Saliency Map Generator

Most "black box" methods produce a saliency map by measuring the effect manipulating the input image ($e.g.$, by removing image regions) has on a model's similarity score. If a large drop in similarity is measured, then the region must be significant. If almost no change is measured, then the model considers the image region irrelevant. The saliency map is generated by averaging the similarity scores for each pixel over all instances where it was altered. The challenge is determining the best way of manipulating the input image to discover these important regions. A key benefit of "black box" methods is that they do not require having access to underlying model parameters. We compare three black box methods: a simple Sliding Window baseline [40], LIME [28], which determines how much super-pixel regions affect the model predictions, and RISE [26], an efficient high-performing method that constructs a saliency map using random masking. We also compare to "white box" learned Mask [7], which was selected due to its high performance and tendency to produce compact saliency maps. We now describe how we adapt these models for our task; see supplementary for additional details on each method.

Computing Similarity Scores. Each saliency method we compare is designed to operate on a single image, and measures the effect manipulating the image has on the prediction of a specific object class. However, an image similarity model's predictions are arrived at using two or more input images. Let us consider the case where we are comparing only two images – a query image ($i.e.$ the image we want to produce an explanation for) and a single reference image, although our approach extends to consider multiple reference images. Even though we do not have access to a class label, we can measure the effect manipulating an image has on the similarity score between the query and reference images. Two approaches are possible: manipulating both images, or only the query image.

Manipulating both images would result in NM forward passes through the image similarity model (for N query and M reference image manipulations), which is prohibitively expensive unless $M << N$. But we only need an accurate saliency map for the query image, so we set $M << N$ in our experiments. There is another danger: for example, consider two images of clothing items that are similar if either they both contain or do not contain a special button. Masking out the button in one image and not the other would cause a drop in similarity score, but masking out the button in both images would result in high image similarity. These conflicting results could make accurately identifying the correct image regions contributing to a score difficult.

The alternative is to **manipulate the query image** alone, and use a fixed reference. We evaluate saliency maps produced by both methods in Subsect. 4.1.

3.3 Selecting Informative Attributes

At test time, given a similarity model and a pair of inputs, SANE generates a saliency map and selects an attribute to show to the user. We suspect that some attributes are not useful for explaining a given image similarity model. Thus, we take into account each attribute's usefulness by learning concept activation vectors (CAVs) [19] over the final image similarity embedding. These CAVs identify which attributes are useful in explaining a layer's activations by looking at whether an attribute positively affects the model's predictions. CAVs are defined as the vectors that are orthogonal to the classification boundary of a linear classifier trained to recognize an attribute over a layer's activations. Then, the sensitivity of each concept to an image similarity model's predictions (the TCAV score) is obtained by finding the fraction of features that are positively influenced by the concept using directional derivatives computed via triplet loss with a margin of machine epsilon. Note that this creates a single attribute ranking over the entire image similarity embedding (*i.e.*, it is agnostic to the image pair being explained), which we use as an attribute explanation suitability prior. Finally, attributes are ranked as explanations using a weighted combination of the TCAV scores, the attribute confidence score $\hat{a}$, and how well the attribute activation map $\mathbf{n}$ matches the generated saliency map $\mathbf{m}_q$. *I.e.*,

$$e(\mathbf{m}_q, \hat{a}, \mathbf{n}, \mathrm{TCAV}) = \phi_1\hat{a} + \phi_2\,d_{\cos}(\mathbf{m}_q, \mathbf{n}) + \phi_3\mathrm{TCAV}, \qquad (4)$$

where $d_{\cos}$ denotes cosine similarity, and ϕ_{1-3} are scalars estimated via grid search on held out data. See the supplementary for additional details.

4 Experiments

Datasets. We evaluate our approach using two datasets from different domains to demonstrate its ability to generalize. The Polyvore Outfits dataset [33] consists of 365,054 fashion product images annotated with 205 attributes and composed into 53,306/10,000/5,000 train/test/validation outfits. Animals with Attributes 2 (AwA) [36] consists of 37,322 natural images of 50 animal classes annotated with 85 attributes, and is split into 40 animal classes for training, and 10 used at test time. To evaluate our explanations, we randomly sample 20,000 (query, reference) pairs of images for each dataset from the test set, where 50% of the pairs are annotated as similar images.

Image Similarity Models. For Polyvore Outfits we use the type-aware embedding model released by Vasileva *et al.* [33]. This model captures item compatibility (*i.e.* how well two pieces of clothing go together) using a set of learned projections on top of a general embedding, each of which compares a specific pair of item types (*i.e.* a different projection is used when comparing a top-bottom pair than when comparing a top-shoe pair). For AwA we train a feature representation using a 18-layer ResNet [12] with a triplet loss function that encourages animals of the same type to embed nearby each other. For each dataset/model, cosine similarity is used to compare an image pair's feature representations.

4.1   Saliency Map Evaluation

Metrics. Following Petsiuk *et al.* [26], we evaluate the generated saliency maps using insertion and deletion metrics which measure the change in performance of the model being explained as pixels are inserted into a blank image, or deleted from the original image. For our task, we generate saliency maps for all query images, and insert or delete pixels in that image only. If a saliency map correctly captures the most important image regions, we should expect a sharp drop in performance as pixels are deleted (or a sharp increase as they are inserted). We report the area under the curve (AUC) created as we insert/delete pixels at a rate of 1% per step for both metrics. We normalize the similarity scores for each image pair across these thresholds so they fall in a [0–1] interval.

Results. Table 1 compares the different saliency map generation methods on the insertion and deletion tasks. RISE performs best on most metrics, with the exception of LIME doing better on the deletion metric on AwA. This is not surprising, since LIME learns which super-pixels contribute to a similarity score. For AwA this means that parts of the animals could be segmented out and deleted or inserted in their entirety before moving onto the next super-pixel. On Polyvore Outfits, however, the important components may be along the boundaries of objects (*e.g.* the cut of a dress), something not well represented by super-pixel segmentation. Although Mask does not perform as well as other approaches, it tends to produce the most compact regions of salient pixels as it searches for a saliency map with minimal support (see the supplementary for examples). Notably, we generally obtain better performance when the reference image is kept fixed and only the query image is manipulated. This may be due to issues stemming from noisy similarity scores as discussed in Subsect. 3.2, and suggests extra care must be taken when manipulating both images.

4.2   Attribute Prediction Evaluation

Metrics. For the standard task of attribute recognition we use mean average precision (mAP) computed over predictions of all images in the test set. Two additional metrics are used to evaluate our attribute explanations using the

Table 1. Comparison of candidate saliency map generator methods described in Subsect. 3.2. We report AUC for the insertion and deletion metrics described in Subsect. 4.1

Method	Fixed Reference?	Polyvore outfits		Animals with attributes 2	
		Insertion (↑)	Deletion (↓)	Insertion (↑)	Deletion (↓)
Sliding Window	Y	57.1	50.6	76.9	76.9
LIME	Y	55.6	52.7	76.9	**71.7**
Mask	Y	56.1	51.8	72.4	75.9
RISE	Y	**61.2**	**46.8**	**77.8**	74.9
Sliding Window	N	56.6	51.1	77.6	76.5
Mask	N	55.6	52.6	72.9	76.6
RISE	N	58.5	50.6	77.7	73.8

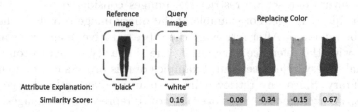

Fig. 3. Attribute replacement example. First, SANE explains why the similarity model predicted that the leggings (reference) and dress (query) have a compatibility score 0.16 – because the dress is "white" and leggings are "black." Then, we artificially change the color of the dress and re-compute similarity. Since our explanation was a good one, this lowers compatibility most of the time. However, this can be noisy as compatibility did improve for the black dress, but still useful as we show in Sect. 4.3

(query, reference) image pairs from the saliency map experiments, which are similar to the evaluation of saliency maps. Given the set of attributes we know exist in the image, we select which attribute among them best explains the similarity score using Eq. (4), and then see the effect *deleting* that attribute from the image has on the similarity score. Analogically, we select the attribute which best explains the similarity score from those which are *not* present in the image, and measure the effect *inserting* that attribute has on the similarity score. Intuitively, if an attribute is critical for an explanation, the similarity score should shift more than if a different attribute is selected. Scores for these metrics are expressed in terms of relative change. When inserting important missing attributes, we expect that the similarity score to improve, and vice versa: when deleting important attributes, we would expect the similarity score to drop.

We provide an example of attribute removal and its effect on similarity in Fig. 3. Note that, because the attribute explanation is a color in this example, we can easily remove the attribute by replacing with another color. We see that when we modify the white dress to be a different color, the similarity score drops significantly. The only exception is when we make the dress the same color (black) as the attribute explanation of the pants it is being compared to. This demonstrates in a causal way how our predicted explanation attributes can play a significant role in the similarity scores.

Since most attributes are not as easily replaced as colors, in order to insert or remove a particular attribute, we find a *representative image* in the test set which is most similar to the query image in all other attributes. For example, let us consider a case where we want to remove the attribute "striped" from a query image. We would search through the database for images which are most similar in terms of non-striped attributes, but which have not been labeled as being "striped". We rank images using average confidence for each attribute computed over the three attribute models we compare in Table 2. After obtaining these average attribute confidence scores, we use cosine similarity between the non-explanatory attributes to score candidate representative images. On the Polyvore Outfits dataset we restrict the images considered to be of the same type as the query image. For example, if the query image is a shoe, then only a shoe can be retrieved. After retrieving the representative image, we compute its similarity with the reference image using the similarity model to compare with the original (query, reference) pair. Examples of this process can be found in the supplementary. Since any retrieved image may inadvertently change multiple attributes, we average the scores over the top-k representative images.

Compared Methods. We provide three baseline approaches: a random baseline, a sample attribute classifier (*i.e.* no attribute activation maps), and a modified version of FashionSearchNet [1] – an attribute recognition model which also creates a weakly-supervised attribute activation map, for comparison. To validate our model choices, we also compare using a binary cross-entropy loss L_{BCE} to the Huber loss for training our attribute predictors. Additional details on these models can be found in the supplementary.

Results. Table 2 shows the performance of the compared attribute models for our metrics. Our attribute explanation metrics demonstrate the effectiveness of our attribute explanations, with our model, which matches saliency maps and includes TCAV scores, obtaining best performance on both datasets. This shows that "inserting" or "deleting" the attribute predicted by SANE from the query image affects the similarity model's predicted score more than inserting or deleting the attribute suggested by baselines. Notably, our approach outperforms FashionSearchNet + Map Matching (MM), which can be considered a weakly-supervised version of SANE trained for attribute recognition. The fifth line of Table 2 reports that TCAV consistently outperforms many methods on insertion, but has mixed results on deletion. This is partly due to the fact that other models, including SANE, are trained to reason about attributes that actually exist in an image, whereas for insertion the goal is to predict which attribute *not* present in the image would affect the similarity score most significantly. Thus, using a bias towards globally informative attributes (*i.e.*, TCAV scores) is more consistently useful for insertion. Finally, training an attribute classifier using

Table 2. Comparison of how attribute recognition (mAP) and attribute explanation (insertion, deletion) metrics described in Subsect. 4.2 are affected for different approaches. We use fixed-reference RISE as our saliency map generator for both datasets

Method	Polyvore outfits			Animals with attributes 2		
	mAP	Insert ($\uparrow$)	Delete ($\downarrow$)	mAP	Insert ($\uparrow$)	Delete ($\downarrow$)
Random	–	25.3	–6.3	–	2.1	–8.5
Attribute classifier - L_{BCE}	53.2	25.7	–5.8	65.1	–2.5	–2.3
Attribute classifier - $L_{BCE} + L_{hm}$	**56.0**	25.8	–5.9	**67.2**	–1.7	–2.0
FashionSearchNet - L_{BCE} [1]	54.7	25.5	–5.2	65.9	–1.6	–2.6
TCAV [19]	–	28.0	–8.5	–	3.4	–22.0
Attribute classifier - L_{Huber}	–	25.9	–8.1	–	–0.8	–2.8
FashionSearchNet - L_{Huber} [1]	–	26.1	–7.6	–	–0.3	–3.5
FashionSearchNet - L_{Huber} + MM	–	26.6	–10.1	–	1.2	–6.8
SANE	–	26.3	–9.8	–	0.1	–3.0
SANE + MM	–	27.1	–10.9	–	6.0	–10.7
SANE + MM + TCAV (Full)	–	**31.5**	**–11.8**	–	**6.2**	**–24.1**

saliency maps for supervision (L_{hm}) leads to a 3% improvement on the standard attribute recognition task measured in mAP over a simple attribute classifier while using the same number of parameters. This also outperforms FashionSearchNet, which treats localizing important image regions as a latent variable rather than using saliency maps for supervision. While L_{BCE} does perform well for attribute recognition, it does poorly as an explanation as shown in Table 2.

We provide qualitative examples of our explanations in Fig. 4. Examples demonstrate that our explanations pass important sanity checks. Notice that "golden", "striped" and "printed" in the first two columns of Fig. 4 are sensibly localized, and are also reasonable explanations for the match, while a more abstract explanation like "fashionable" is linked to the high heel, the curve of the sole, and the straps of the shoe. Note further that the explanations are nontrivial: they more often than not differ from the most likely attribute in the query image, as predicted by a standard attribute classifier. In other words, our explanation model is utilizing information from each pair of images and the saliency map characterizing the match to produce a sensible, interpretable explanation.

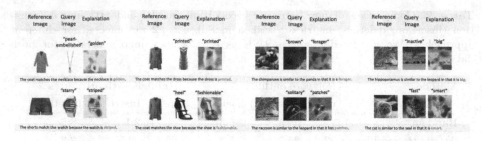

Fig. 4. Qualitative results of our attribute explanations for pairs of examples on the Polyvore Outfits and the AwA datasets. The attribute predicted as explanation for each reference-query match is shown above the saliency map. The most likely attribute for the query image as predicted by our attribute classifier is shown directly above it

4.3   User Study

A key component of evaluating whether our explanations are sensible and interpretable is conducting a user study. This is not a straightforward task, as one has to carefully formulate the prompts asked in a manner suitable for answering questions like, *Are our explanations useful? Do they provide insight that helps users understand what the similarity model is doing? Are they consistent?*

We formulate our study as a "guessing game" whereby a unique image triplet (A, B, C) is presented in each question that asks users to select whether the image similarity model predicted B or C as a better match for A. Images B and C are selected such that the pairs (A, B) and (A, C) have sufficiently different similarity scores. We adopt an A/B testing methodology and present seven different versions of the study: (a) a control case whereby no explanations are presented and users have to guess whether (A, B) or (A, C) are more similar based on their own intuition; (b) image pairs (A, B) and (A, C) are presented along with a random saliency map for images B and C; (c) image pairs (A, B) and (A, C) are presented along with the corresponding saliency maps for B and C generated using RISE [26]; (d) (A, B) and (A, C) are presented along with a sentence explanation containing the most likely predicted attribute made by our attribute predictor for B and C; (e) image pairs (A, B) and (A, C) are presented along with a sentence explanation containing the SANE explanation attribute for each pair; (f) image pairs (A, B) and (A, C) are presented along with *both* randomly generated saliency maps for B and C and the SANE attributes for each pair; and (g) image pairs (A, B) and (A, C) are presented along with *both* the corresponding RISE saliency maps for B and C and the SANE attributes for each pair. Examples of each question type are provided in the supplementary.

Our underlying hypothesis is that, if our model produces sensible and human-interpretable explanations, there must be a correlation between the resulting explanations and the similarity score for an image pair, and that relationship should be discernible by a human user. Thus, user accuracy on guessing which pair the model predicted is a better match based on the provided explanations or lack thereof should be a good indicator as to the reliability of our explanations.

Table 3. User study results. We report user's accuracy given different information in guessing which image pair the image similarity model thought was a better match and the portion of users who felt the explanations were helpful

Explanation type	Polyvore outfits		Animals with attributes 2	
	Accuracy	Helpful?	Accuracy	Helpful?
(a) Control case (no explanations)	66.0	–	42.0	–
(b) Random maps	66.0	20.0	50.0	18.2
(c) Saliency maps	56.7	**100.0**	53.8	**87.5**
(d) Predicted attr's	62.5	66.7	53.1	69.2
(e) SANE attr's	68.9	66.7	**61.3**	**87.5**
(f) Random maps + SANE Attr's	65.8	75.0	59.2	66.7
(g) Saliency maps + SANE Attr's	**70.0**	55.6	52.5	62.5

Using a web form, we present a total of 50 unique image triplets to 59 subjects in the age range 14–50, each triplet forming a question as described above. We keep question order the same for each subject pool and question type. Each user sees 10 unique questions, and subject pools across different study versions are kept disjoint. Each question type received 1–4 unique responses.

Study Results. Table 3 reports the results of our study. In addition to reporting the user's ability to correctly identify the image pair the similarity model thought was a better match, we also report the percent of subjects for each study type that reported finding the provided explanations helpful in answering the questions. Comparing lines 1 and 2 of Table 3, we see that, as expected, users find random saliency maps to be confusing, as 20% or fewer ever report them as useful. Comparing lines 1 and 3 of Table 3, we see that accuracy goes down on the Polyvore Outfits dataset when users are shown saliency maps, suggesting that users may have misinterpreted why the particular image regions were highlighted; yet all users thought the maps were helpful in answering the questions. However, on AwA, accuracy increases significantly with providing saliency in addition to image pairs, along with 87.5% of users finding them helpful.

Comparing the most likely attribute vs. our SANE explanation attributes in lines 4 and 5 of Table 3, respectively, we see that users demonstrate improved understanding of the image similarity model's behavior and also find the attribute explanations to be helpful. On Polyvore Outfits, although users consider the most likely attributes as explanations about as helpful as SANE explanations, the 6.4% difference in accuracy suggests that SANE explanations do indeed provide valuable information about the match. On AwA, SANE attributes result in the highest user accuracy, with 87.5% of users reporting them useful vs. only 69.2% reporting most likely attributes used as explanations useful. Notably, SANE attributes reports improving user accuracy by 3–7.5% on both datasets over the control case or using saliency maps alone. Line 7 of Table 3 shows that on Polyvore Outfits, users do best if they are provided both saliency maps and SANE attributes as explanations, even though they did not find this type of explanation most useful. We suspect this could be due to natural human bias, *i.e.* users' intuition disagreeing with the image similarity model's predictions.

Overall, the study results suggest that (1) users find our explanations helpful; (2) our explanations consistently outperform baselines; (3) the type of explanation that proves most helpful depends on the dataset: users like having saliency maps as a guide, although their performance is best using both saliency maps and explanation attributes on a fashion dataset, while on a natural image dataset, having attribute explanations for each image pair helps the most.

5   Conclusion

In this paper we introduced SANE, a method of explaining an image similarity model's behavior by identifying attributes that are important to the similarity score paired with saliency maps indicating significant image regions. We confirm that our SANE explanations improve a person's understanding of a similarity model's behavior through a user study to supplement automatic metrics. In future work, we believe closely integrating the saliency generator and attribute explanation model, enabling each component to take advantage of the predictions of the other, would help improve performance.

Acknowledgements.. This work is funded in part by a DARPA XAI grant, NSF Grant No. 1718221, and ONR MURI Award N00014-16-1-2007.

References

1. Ak, K.E., Kassim, A.A., Lim, J.H., Tham, J.Y.: Learning attribute representations with localization for flexible fashion search. In: The IEEE Conference on Computer Vision and Pattern Recognition (CVPR) (2018)
2. Bansal, A., Sikka, K., Sharma, G., Chellappa, R., Divakaran, A.: Zero-shot object detection. In: The European Conference on Computer Vision (ECCV) (2018)
3. Bau, D., Zhou, B., Khosla, A., Oliva, A., Torralba, A.: Network dissection: Quantifying interpretability of deep visual representations. In: The IEEE Conference on Computer Vision and Pattern Recognition (CVPR) (2017)
4. Cao, C., et al.: Look and think twice: Capturing top-down visual attention with feedback convolutional neural networks. In: The IEEE International Conference on Computer Vision (ICCV) (2015)
5. Chang, C.H., Creager, E., Goldenberg, A., Duvenaud, D.: Explaining image classifiers by counterfactual generation. In: The International Conference on Learning Representations (2019)
6. Fong, R., Vedaldi, A.: Net2vec: Quantifying and explaining how concepts are encoded by filters in deep neural networks. In: The IEEE Conference on Computer Vision and Pattern Recognition (CVPR) (2018)

7. Fong, R.C., Vedaldi, A.: Interpretable explanations of black boxes by meaningful perturbation. In: The IEEE International Conference on Computer Vision (ICCV) (2017)

8. Gordo, A., Larlus, D.: Beyond instance-level image retrieval: Leveraging captions to learn a global visual representation for semantic retrieval. In: The IEEE Conference on Computer Vision and Pattern Recognition (CVPR) (2017)

9. Han, X., Song, X., Yin, J., Wang, Y., Nie, L.: Prototype-guided attribute-wise interpretable scheme for clothing matching. In: Proceedings of the 42nd International ACM SIGIR Conference on Research and Development in Information Retrieval (2019)

10. Han, X., et al.: Automatic spatially-aware fashion concept discovery. In: The IEEE International Conference on Computer Vision (ICCV) (2017)

11. Han, X., Wu, Z., Jiang, Y.G., Davis, L.S.: Learning fashion compatibility with bidirectional LSTMs. In: ACM International Conference on Multimedia (2017)

12. He, K., Zhang, X., Ren, S., Sun, J.: Deep residual learning for image recognition. In: The IEEE Conference on Computer Vision and Pattern Recognition (CVPR) (2016)

13. Hendricks, L.A., Akata, Z., Rohrbach, M., Donahue, J., Schiele, B., Darrell, T.: Generating visual explanations. In: The European Conference on Computer Vision (ECCV) (2016)

14. Hendricks, L.A., Hu, R., Darrell, T., Akata, Z.: Grounding visual explanations. In: The European Conference on Computer Vision (ECCV) (2018)

15. Hsiao, W.L., Grauman, K.: Creating capsule wardrobes from fashion images. In: The IEEE Conference on Computer Vision and Pattern Recognition (CVPR) (2018)

16. Huber, P.J.: Robust estimation of a location parameter. Ann. Stat. **53**(1), 73–101 (1964)

17. Huk Park, D., et al.: Multimodal explanations: Justifying decisions and pointing to the evidence. In: The IEEE Conference on Computer Vision and Pattern Recognition (CVPR) (2018)

18. Kiapour, M.H., Han, X., Lazebnik, S., Berg, A.C., Berg, T.L.: Where to buy it: matching street clothing photos to online shops. In: The IEEE International Conference on Computer Vision (ICCV) (2015)

19. Kim, B., et al.: Interpretability beyond feature attribution: Quantitative testing with concept activation vectors (TCAV). In: The International Conference on Machine Learning (ICML) (2018)

20. Kim, J., Rohrbach, A., Darrell, T., Canny, J., Akata, Z.: Textual explanations for self-driving vehicles. In: The European Conference on Computer Vision (ECCV) (2018)

21. Lad, S., Parikh, D.: Interactively guiding semi-supervised clustering via attribute-based explanations. In: The European Conference on Computer Vision (ECCV) (2014)

22. Li, Q., Tao, Q., Joty, S., Cai, J., Luo, J.: VQA-E explaining, elaborating, and enhancing your answers for visual questions. In: The European Conference on Computer Vision (ECCV) (2018)

23. Li, Y., Zhang, J., Zhang, J., Huang, K.: Discriminative learning of latent features for zero-shot recognition. In: The IEEE Conference on Computer Vision and Pattern Recognition (CVPR) (2018)

24. Liao, L., He, X., Zhao, B., Ngo, C.W., Chua, T.S.: Interpretable multimodal retrieval for fashion products. In: Proceedings of the 26th ACM International Conference on Multimedia (2018)

25. Nguyen, A., Dosovitskiy, A., Yosinski, J., Brox, T., Clune, J.: Synthesizing the preferred inputs for neurons in neural networks via deep generator networks. In: Advances in Neural Information Processing Systems (NIPS) (2016)

26. Petsiuk, V., Das, A., Saenko, K.: Rise: Randomized input sampling for explanation of black-box models. In: British Machine Vision Conference (BMVC) (2018)

27. Radenović, F., Iscen, A., Tolias, G., Avrithis, Y., Chum, O.: Revisiting oxford and Paris: large-scale image retrieval benchmarking. In: The IEEE Conference on Computer Vision and Pattern Recognition (CVPR) (2018)

28. Ribeiro, M.T., Singh, S., Guestrin, C.: "why should i trust you?": explaining the predictions of any classifier. In: Proceedings of the 22nd ACM SIGKDD International Conference on Knowledge Discovery and Data Mining (2016)

29. Selvaraju, R.R., Cogswell, M., Das, A., Vedantam, R., Parikh, D., Batra, D.: Grad-cam: visual explanations from deep networks via gradient-based localization. In: The IEEE International Conference on Computer Vision (ICCV) (2017)

30. Simonyan, K., Vedaldi, A., Zisserman, A.: Deep inside convolutional networks: visualising image classification models and saliency maps. In: ICLR Workshop (2014)

31. Tan, R., Vasileva, M.I., Saenko, K., Plummer, B.A.: Learning similarity conditions without explicit supervision. In: The IEEE International Conference on Computer Vision (ICCV) (2019)

32. Teach, R.L., Shortliffe, E.H.: An analysis of physician attitudes regarding computer-based clinical consultation systems. Comput. Biomed. Res. **14**(6), 542–558 (1981)

33. Vasileva, M.I., Plummer, B.A., Dusad, K., Rajpal, S., Kumar, R., Forsyth, D.: Learning type-aware embeddings for fashion compatibility. In: The European Conference on Computer Vision (ECCV) (2018)

34. Vittayakorn, S., Umeda, T., Murasaki, K., Sudo, K., Okatani, T., Yamaguchi, K.: Automatic attribute discovery with neural activations. In: The European Conference on Computer Vision (ECCV) (2016)

35. Wang, X., Ye, Y., Gupta, A.: Zero-shot recognition via semantic embeddings and knowledge graphs. In: The IEEE Conference on Computer Vision and Pattern Recognition (CVPR) (2018)

36. Xian, Y., Lampert, C.H., Schiele, B., Akata, Z.: Zero-shot learning—a comprehensive evaluation of the good, the bad and the ugly. IEEE Trans. Pattern Anal. Mach. Intell. **41**(9), 2251–2265 (2019). https://doi.org/10.1109/TPAMI.2018.2857768

37. Yang, X., et al.: Interpretable fashion matching with rich attributes. In: Proceedings of the 42nd International ACM SIGIR Conference on Research and Development in Information Retrieval (2019)

38. Yelamarthi, S.K., Reddy, S.K., Mishra, A., Mittal, A.: A zero-shot framework for sketch based image retrieval. In: The European Conference on Computer Vision (ECCV) (2018)

39. Yosinski, J., Clune, J., Nguyen, A., Fuchs, T., Lipson, H.: Understanding neural networks through deep visualization. In: Deep Learning Workshop, International Conference on Machine Learning (ICML) (2015)

40. Zeiler, M.D., Fergus, R.: Visualizing and understanding convolutional networks. In: The European Conference on Computer Vision (ECCV) (2014)

41. Zhang, J., Lin, Z., Brandt, J., Shen, X., Sclaroff, S.: Top-down neural attention by excitation backprop. In: The European Conference on Computer Vision (ECCV) (2016)

42. Zhao, B., Feng, J., Wu, X., Yan, S.: Memory-augmented attribute manipulation networks for interactive fashion search. In: The IEEE Conference on Computer Vision and Pattern Recognition (CVPR) (2017)
43. Zhou, B., Khosla, A., A., L., Oliva, A., Torralba, A.: Learning deep features for discriminative localization. In: The IEEE Conference on Computer Vision and Pattern Recognition (CVPR) (2016)

CooGAN: A Memory-Efficient Framework for High-Resolution Facial Attribute Editing

Xuanhong Chen[1,2], Bingbing Ni[1,2]([✉]), Naiyuan Liu[1], Ziang Liu[1], Yiliu Jiang[1], Loc Truong[1], and Qi Tian[3]

[1] Shanghai Jiao Tong University, Shanghai, China
{chen19910528,nibingbing,acemenethil,jiangyiliu}@sjtu.edu.cn,
liunaiyuan27@gmail.com,ttanloc@gmail.com
[2] Huawei HiSilicon, Shenzhen, China
{chenxuanhong,nibingbing}@hisilicon.com
[3] Huawei, China
tian.qi1@huawei.com

Abstract. In contrast to great success of memory-consuming face editing methods at a low resolution, to manipulate high-resolution (HR) facial images, *i.e.*, typically larger than 768^2 pixels, with very limited memory is still challenging. This is due to the reasons of 1) intractable huge demand of memory; 2) inefficient multi-scale features fusion. To address these issues, we propose a NOVEL pixel translation framework called *Cooperative GAN*(CooGAN) for HR facial image editing. This framework features a local path for fine-grained local facial patch generation (*i.e.*, patch-level HR, LOW memory) and a global path for global low-resolution (LR) facial structure monitoring (*i.e.*, image-level LR, LOW memory), which largely reduce memory requirements. Both paths work in a cooperative manner under a local-to-global consistency objective (*i.e.*, for smooth stitching). In addition, we propose a lighter selective transfer unit for more efficient multi-scale features fusion, yielding higher fidelity facial attributes manipulation. Extensive experiments on CelebA-HQ well demonstrate the memory efficiency as well as the high image generation quality of the proposed framework.

Keywords: Generative adversarial networks · Conditional GANs · Face attributes editing

---

X. Chen—Work done during an internship at Huawei HiSilicon.
N. Liu—Contributed to the work while he was a research assistant at Shanghai Jiao Tong University.

Electronic supplementary material The online version of this chapter (https://doi.org/10.1007/978-3-030-58621-8_39) contains supplementary material, which is available to authorized users.

A. Vedaldi et al. (Eds.): ECCV 2020, LNCS 12356, pp. 670–686, 2020.
https://doi.org/10.1007/978-3-030-58621-8_39

1 Introduction

Recent development in the field of deep learning has led to a rising interest in Facial Attribute Editing (FAE) [4,5,8,18,20,25,26,30,33], *e.g.*, manipulating structural, emotional, semantic attributes (*e.g.*, gender, hair color, age etc.) of a given facial image. These functionalities are highly demanded in emerging mobile applications. However, most of recent facial editing approaches [18,20,26] are based on deep image-to-image translation model, which can ONLY deal with low resolution (LR) facial images due to excessive deep model (*e.g.*, GAN [20,31]) size. Concretely, as the image size increases, the memory consumption of deep models also increases dramatically, even surpasses the limitation of device. Such limitation forbids many promising applications on mobile platforms (*e.g.*, ZAO [1] processing images in the cloud service). Although upgrading hardware can alleviate these issues, it is a non-cheap solution and still hard to process higher resolution images (*e.g.*, 4K). In this work, we aim at developing a memory-efficient deep framework to effectively handle high resolution facial attribute editing on resource-constrained devices (Fig. 1).

There are two major challenges of deep model based HR facial attribute editing: 1. *Constrained Computational and Memory Resource.* In some mobile scenarios (*e.g.*, smartphone, AR/VR glasses) with only limited computational and memory resources, it is infeasible to use popular image editing models [18,30] which require sophisticated networks. To address this issue, methods based on model pruning and operator simplifying [13,23,28,34,35] have been proposed to reduce the inference computational complexity. However, the metric-based way to reduce the model size will do harm to model perceptual representation ability, the output facial image quality is usually largely sacrificed. 2. *Inefficient Multi-scale Features Fusion.* In order to achieve high-level semantic manipulation while maintaining local details during image generation, multi-scale features fusion is widely adopted. It is a common practice to utilize skip connection, *e.g.*, U-Net [27].

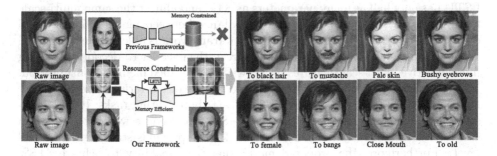

Fig. 1. For the purpose of translating the high-resolution images in the memory-constrained platform, we develop a memory-efficient image translation framework. In the figure, we show some high resolution (768 × 768) editing results and its manipulation labels, more results can be found in the suppl. From the results, it can be seen that our framework can effectively edit both regional and global attributes. (Color figure online)

However, fixed schemes, such as skip connection, usually result in infeasible or even self-contradicting fusion output (*e.g.*, during style transfer, content is well-preserved but failed to change the image style), and flexible schemes such as GRU [9] lead to additional computational burden (*e.g.*, applying GRU-based unit [20] directly for multi-scale features fusion can achieve excellent fusion effects, but will increase network parameters by more than four times).

In this work, a novel image translation framework for high resolution facial attribute editing, dubbed *CooGAN*, is proposed to explicitly address above issues. It adopts a divide-and-combine strategy to break the **intractable** whole HR image generation task down to several sub-tasks for reducing memory cost greatly. More concretely, the pipeline of our framework consists of a series of local HR patch generation sub-tasks and a global LR image generation sub-task. To handle these two types of sub-tasks, our framework is also composed of two paths, *i.e.*, local network path and global network path. Namely, the local sub-task is to generate an HR patch with edited attributes and fine-grained details. And a global sub-task is to generate an LR whole facial snapshot with structural coordinates to guide the local workers properly recognize the correct patch semantics. As only tiny size patch (*e.g.*, 64 × 64) generation sub-task is involved in the pipeline, the proposed framework avoids processing large size feature maps. As a result, this framework is very light-weighted and suited for resource constrained scenarios. In addition, a local-to-global consistency objective is proposed to enforce the cooperation of sub modules, which guarantees between-patch contextual consistency and appearance smoothness in fine-grained scale.

Moreover, we design a variant of SRU [19], *Light Selective Transfer Unit (LSTU)*, for multi-scale features fusion. The GRU-based STU [20] has similar functions, but needs two states (one state obtained from encoder, another from the higher level hidden state) to inference the selected skip feature. As a result, it has to face a heavy computing burden and is not friendly to GPU acceleration. Unlike STU, our SRU-based LSTU just need a single hidden state to get the gating signal, which greatly reduces the complexity of the unit. Actually, the LSTU has only half as many parameters as STU and almost the same multi-scale features fusion effect, achieving a good balance between model efficiency and output image fidelity. Under this design, the framework is able to selectively and efficiently transfer the shallow semantics from the encoder to decoder, enabling more effective multi-scale features fusion with constrained memory consumption.

We extensively experiment the proposed framework in terms of both qualitative and quantitative evaluations on facial attribute editing tasks. It is demonstrated that our framework can process 768 × 768 images well with less than 1 GB GPU memory rather than over 6.3 GB (the previous state-of-the-art [20] level). Furthermore, the proposed model has the capability to achieve 84.8% facial attribute editing accuracy on CelebA-HQ.

2 Related Works

Image-to-Image Translation. Image-to-image translation aims at learning cross-domain mapping in supervised or unsupervised settings [20]. In the early stage, methods like pix2pix use image-conditional GANs to train paired images in the supervised manner. Later, unpaired image-to-image translation frameworks, which work in the unsupervised manner, are also proposed in [21, 36]. For instance, CycleGAN [36] learns the mapping between one image distribution and another and introduces a cycle consistency loss to preserve key features during translation. However, all those mentioned frameworks can only perform coarse-grain translation from one image distribution to another. They cannot modify specific attributes of an image, which is more important in real world applications. To address such problem, StarGAN [8] is proposed to learn multiple domain translation using only a single model.

Facial attribute editing is a typical task of image-to-image translation that focuses on editing specific attributes within image. Early methods [6, 15, 29] are proposed to learn only a single attribute editing model like expression transfer [29], age progression [15], and so on. To achieve arbitrary attribute editing, IcGAN [25] uses a conditional GAN to process encoded features. To strengthen the relations between high-feature space and attributes, Lample G. et al. [17] applies adversarial constraints. StarGAN [8] and AttGAN [12] use target attribute vector input and achieve great success. Based on them, the STGAN [20] is further proposed for simultaneously enhancing the attribute manipulation ability and image quality. Especially, Ran Yi et al. [32] propose a hierarchical facial editing method to modify the face via facial features patches. In our work, we analyze the limitation of StarGAN, AttGAN, [32] and STGAN and then further develop a patch-based model to make facial attribute editing suitable for higher resolution images.

3 Methodology

3.1 Overview

The proposed CooGAN framework for conditional facial generation presents two innovative modules, the global module and the local module. The global module is designed to generate LR translated facial image, and the local module aims at generating HR facial image patches and stitching them together. A cooperation mechanism is introduced to make these two modules work together, so that the global module provides the local module with a global-to-local spatial consistency constraint. In addition, to guarantee the performance and edit-ability of the generated images, we propose a well-designed unit, LSTU, to filter the features from latent space and infuse them with detail information inside the naive skip connection.

3.2   The Cascaded Global-to-Local Face Translation Architecture

The *CooGAN* consists of two interdependent generation modules. We depict the framework architecture in Fig. 2.

Global Module. The global module is designed for generating the translated snapshot X_t' which carries the whole image spatial coordinate information. As is shown in Fig. 2, this module contains two components: the global-aware generator $\mathcal{G}_g$ and the global-aware discriminator $\mathcal{D}_g$. $\mathcal{G}_g$ has a conventional U-Net [27] structure strengthening by LSTU. It takes in X_i', the down-sampled image of X_i, and generates the LR snapshot X_t'. Following the ACGAN-like [24] fashion, the discriminator $\mathcal{D}_g$ has two headers sharing the same extracted features. Note that the main purpose of global module is to guarantee the global semantic consistency of the final result.

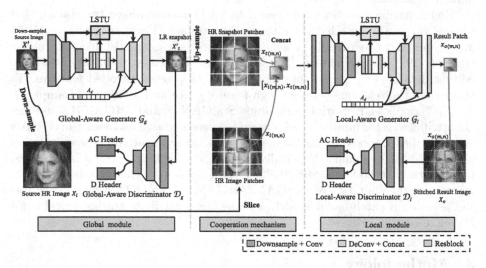

Fig. 2. An overview of our framework. Our framework has two modules–the global image translation module on the left side and the local patch refinement module on the right side. Either module contains one generator and one discriminator, as indicated in the graph. The two modules cooperate with each other through the cooperation mechanism in the middle. All specific operations are elaborated in the graph.

Local Module. The local module is designed for processing HR image patches under the guidance of LR translated patches with limited computational resources. As is shown in Fig. 2, the local module contains two components: the local-aware generator $\mathcal{G}_l$ and the local-aware discriminator $\mathcal{D}_l$. We concatenate the original HR image patch and up-sampled snapshot patch, *i.e.*, $[x_i(m, n), x_t(m, n)]$, where (m, n) represents the coordinate of the patch in the whole image. Then the local-aware generator $\mathcal{G}_l$ processes the concatenated data sequentially to generate translated HR patches. The generated patches can be stitched together *without overlap* to synthesize the final output X_o. To avoid the inconsistency between generated patches, the local-aware discriminator $\mathcal{D}_l$ is introduced to make the final stitched output smooth and seamless. Note that the discriminator is just introduced temporarily for training our model, so it will not add any memory cost during the inference stage.

Cooperation Mechanism. In the previous sections, we discuss the functions of the local module and the global module. The global module requires more computational resources while the size of image increases, which seriously limits its application for generating HR images. The local module decomposes the full-size image into patches and translates the patches into HR ones, which is suitable for processing HR image without adding any memory cost. But the local module alone cannot correctly perceive the patch semantics and is inapplicable for editing local attributes, as can be seen in Fig. 6. To this end, we introduce a valid cooperation mechanism to encourage the two generators to cooperate well for achieving satisfying generation performance. As shown in the middle of Fig. 2, the global snapshot X'_t is up-sampled to the same size as HR image X_i. Then, we decompose X'_t and X_i into a series of patches and concatenate the patches with the same coordinate. The global patch carries the global spatial coordinate information while the HR image contains detailed textural information. By this way, we combine the global spatial information and detailed textural information together. Namely, the global spatial information is used to make the generated local patch smoother and globally consistent, and the detailed textural information is used to preserve the quality of the generation.

3.3 Objective Function

Our framework contains two sets of adversarial training in global and local module, respectively. For the sake of clarity, we first describe the inference stage of either module. In the global module, we get the snapshot X'_t by the generator $\mathcal{G}_g$ and then feed X'_t into the discriminator $\mathcal{D}_g$. Following [20], we use A_d to facilitate the adversarial training, where A_d denotes the difference attribute vector between the target A_t and the source A_i. The training process can be defined as:

$$X'_t = \mathcal{G}_g \left(X'_i, A_d \right). \tag{1}$$

In the local module, $x_{i(i,j)}$ denotes a patch of HR image X_i and $x_{t(i,j)}$ denotes a patch of up-sampled snapshot image X_t. We use $[\cdot, \cdot]$ to represent the channel-wise patch concatenation,

$$x_{o(i,j)} = \mathcal{G}_l \left(\left[x_{i(i,j)}, x_{t(i,j)} \right], A_d \right). \tag{2}$$

Image Reconstruction Loss. We apply the reconstruction loss in our framework to improve image quality and avoid as many editing miscues as possible. This loss is achieved in a self-supervision manner, *i.e.*, we feed the framework with zero condition input, which can be depicted as:

$$\mathcal{L}_{rec} = ||x - \mathcal{G}(x, 0)||_1. \tag{3}$$

Adversarial Loss. For FAE task, there is no ground-truth for training. Adversarial loss is usually applied due to its domain-to-domain translation nature. To alleviate the instability of adversarial loss in training process, gradient penalty [10] is introduced:

$$\max_{\mathcal{D}_{adv}} \mathcal{L}_{\mathcal{D}_{adv}} = \mathbb{E}_x \big[\mathcal{D}_{adv}(x) \big] - \mathbb{E}_{\hat{y}} \big[\mathcal{D}_{adv}(\hat{y}) \big] + $$
$$\lambda \mathbb{E}_{\hat{x}} \big[(\| \nabla_{\hat{x}} \mathcal{D}_{adv}(\hat{x}) \|_2 - 1)^2 \big], \tag{4}$$

$$\max_{\mathcal{G}} \mathcal{L}_{\mathcal{G}_{adv}} = \mathbb{E}_{x, A_d} \big[\mathcal{D}_{adv}(\mathcal{G}(x, A_d)) \big], \tag{5}$$

where $\mathcal{D}_{adv}$ denotes the adversarial header of the discriminator. x is the input real image, $\hat{y}$ is the generated image and $\hat{x}$ is the sampled point along straight lines between the real image distribution and generated image distribution.

Attribute Editing Loss. As discussed above, attribute editing has no ground truth. Following the ideology of ACGAN [24], we add a classification header sharing features with adversarial header. Such a design effectively realizes adversarial training along with attribute learning. The attribute editing loss below supervises the two modules of our framework individually,

$$\mathcal{L}_{\mathcal{D}_a} = - \sum_{i=1}^{N_c} \Big[a_t^{(i)} \log \mathcal{D}_a^{(i)}(x) + (1 - a_t^{(i)}) \log (1 - \mathcal{D}_a^{(i)}(x)) \Big], \tag{6}$$

where $a_t^{(i)}$ represents the i-th target attribute and $\mathcal{D}_a$ represents the classification header. We use the combination of the above three types of loss to train our model.

3.4   Light Selective Transfer Unit

As mentioned above, we proposed a high-resolution facial image editing framework which can process HR images with limited computational resources. In order to further improve the generation performance, we improve the manner of multi-scale features fusion.

The most popular method for multi-scale features fusion in image-to-image translation is the *Skip Connection* structure. It helps the network balance the contradiction between the pursuit of larger receptive field and loss of more

Table 1. PSNR/SSIM performance and average attribute editing accuracy of models with different Skip Connection (SC) numbers, SCi represents model with i skip connections.

SC type	SC0	SC1	SC2	SC3	SC4
PSNR	20.2	24.29	28.0	29.9	34.8
SSIM	0.643	0.777	0.877	0.916	0.989
Average Acc.	89.5%	82.4%	75.1%	69.1%	63.9%

details. One of its classic deployments is U-Net. However, there is a fatal draw-back of the original skip connection: it will degrade the function of deeper parts and further damage the effectiveness of condition injection. From Table 1, it is obvious that PSNR increases but attribute editing accuracy decreases when the skip connection number multiplies. A detailed graph showing the editing accu-racy of each specific attribute is given in suppl. STGAN [20] tries to alleviate the problem with the STU, a variant of GRU [7,9], which uses the latent fea-ture to control the information transfer in the skip connection through the unit. This feature carries the conditional information added through the concatena-tion. Unfortunately, such a unit omits the spatial and temporal complexity and it is a time-consuming process for the underlying feature to bubble up from the bottleneck to drive the STU of each layer.

To explicitly address the mentioned problem, we present our framework to employ *Light Selective Transfer Unit (LSTU)* to efficiently and selectively trans-fer the encoder feature. LSTU is an SRU-based unit with totally different infor-mation flow. Compare to STU, our LSTU discards the dependence on the two states when calculating the gating signal, which greatly reduces our parameters but the unit is still efficient. The detailed structure of LSTU is shown in Fig. 3. Without loss of generality, we choose the l-th LSTU as an analysis example. The l-th layer feature coming from the encoder side denotes as x^l. h^{l+1} denotes the feature in the adjacent deeper layer. It contains the filtered latent state infor-mation of that layer. h^{l+1} is firstly concatenated with attribute difference A_d to obtain up-sampled hidden state $\hat{h}^{l+1}$. Then $\hat{h}^{l+1}$ is used to independently calculate the masks f^l, r^l for the forget-gate and reset-gate. $\mathbf{W_T}$, $\mathbf{W_{1 \times 1}}$, $\mathbf{W_f}$ and $\mathbf{W_r}$ represent parameter matrix of transpose convolution, linear transform, forget gate and update gate. The further process is similar to SRU. The equation of gates is shown on the right side of Fig. 3.

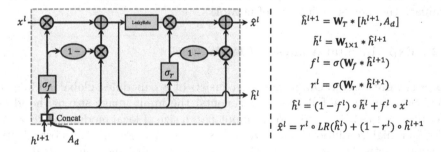

$$\hat{h}^{l+1} = \mathbf{W}_T * [h^{l+1}, A_d]$$

$$\bar{h}^l = \mathbf{W}_{1 \times 1} * \hat{h}^{l+1}$$

$$f^l = \sigma(\mathbf{W}_f * \hat{h}^{l+1})$$

$$r^l = \sigma(\mathbf{W}_r * \hat{h}^{l+1})$$

$$\bar{h}^l = (1 - f^l) \circ \bar{h}^l + f^l \circ x^l$$

$$\hat{x}^l = r^l \circ LR(\hat{h}^l) + (1 - r^l) \circ \hat{h}^{l+1}$$

Fig. 3. The structure of proposed LSTU. The design of LSTU is inspired by SRU, which makes LSTU more lightweight than STU and suitable for GPU parallel acceleration. The right side shows the mathematical expression of the inference process of our LSTU. LR is short for LeakyRelu.

4   Experiments

In this section, we take the following steps to evaluate the performance of proposed CooGAN: firstly, we discuss the facial attribute editing performance and the memory usage of our overall framework. Then, we verify the effectiveness of the LSTU in multi-scale features fusion.

4.1   Implement Details

Dataset. We train our model on the CelebA [22] dataset (for the global module) and CelebA-HQ dataset (for the local module). CelebA is a collection of more than 200,000 human facial images and labels of 40 attributes for each of the images. CelebA-HQ is an artificial high resolution dataset translated from CelebA utilizing the method from [14]. Considering the distinguishability of attributes, in the experiments we mainly use 13 attributes including *Bald, Bangs, Black Hair, Blond Hair, Brown Hair, Bushy Eyebrows, Eyeglasses, Male, Mouth Slightly Open, Mustache, No Beard, Pale Skin and Young.*

Optimizer. The optimizer used in our training process is Adam [16] with hyperparameter $\beta = (0.5, 0.99)$ and learning rate 0.0002. And we apply the learning rate decay strategy with the decay rate 0.1 and decay epoch 100.

Quantitative Metric. We design and train an attribute-classifier to quantify attribute editing effects in form of attribute classes accuracy. It takes form of the classifying part in our discriminator and achieves a 96.0% average accuracy on CelebA. And for the texture sharpness of reconstruction results, Peak Signal to Noise Ratio(PSNR) and Structural Similarity(SSIM) are used. They are two common metrics to measure the generated image fidelity. In addition, we use the floating-point operations per second (FLOPS) and model parameter size to evaluate the memory efficiency of networks.

4.2   Experimental Analysis of Framework

Experiment Setup. Our framework consists of two modules: global module and local module. In the following experiments, the input image size of the global module is set as 256×256 and the input patch size of local module is 128×128. The final high resolution image size is 768×768. We use the two most effective previous methods (AttGAN, STGAN) and our global module as comparison

Table 2. Reconstruction PSNR and SSIM (CelebA 128×128) of different facial attribute editing frameworks.

Method	StarGAN	AttGAN	STGAN	Ours
PSNR/SSIM	22.86/0.848	24.1/0.821	31.6/0.934	**32.1/0.938**

objects. Since these models are not originally designed for HR image processing, we make proper adjustments of increasing their layers (AttGAN-HR: 7 layers, STGAN-HR: 7 layers, LSTU-net-HR: 7 layers) to expand their perceptual fields. We expand the models based on their official codes and train them with their original loss functions. Those adjusted approaches can be denoted as: AttGAN-HR (AttGAN), STGAN-HR (STGAN) and LSTU-net-HR(global module). The complete model structure setting of our framework, AttGAN-HR and STGAN-HR can be found in suppl. Models are evaluated in the CelebA-HQ dataset.

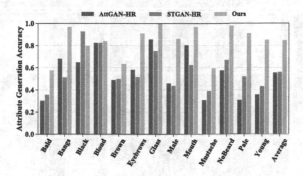

Fig. 4. Our CooGAN against other approaches in attribute translation accuracy. Our framework transcend others in most attributes.

Table 3. GPU memory consumption and average inference time (200 images), measured on a 2080ti, for different framework with batch size 1, input images at the 768 × 768 pixels.

Method	AttGAN-HR	STGAN-HR	Ours
GPU memory	3692 MB	6306 MB	**985** MB
Inference time	0.0414 s	0.0502 s	**0.0645** s

Quantitative Evaluation. We report the attribute editing accuracy of our framework and other methods in Fig. 4. Our framework can achieve 84.8% in classification accuracy, 31.3 in PSNR and 0.923 in SSIM. In terms of memory usage, the largest memory consumption occurs in the global module for the reason that the local module only processes image patches with considerably small sizes. As is shown in Table 3, while processing 768 × 768 images the maximum consumption of our framework is below all compared models. This is on the grounds that the peak consumption in our framework happens when the global module processes 256 × 256 images, while other models process 768 × 768 images directly. Besides, our framework reduces large memory consumption at the cost of only a little inference time increase, as there is *no overlap* when our framework

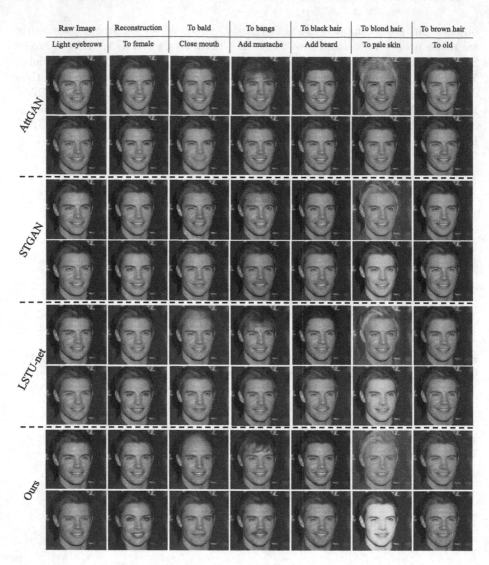

Fig. 5. High resolution (768 × 768) facial attribute editing results of AttGAN-HR, STGAN-HR, LSTU-net-HR and our framework. The textural details can be observed clearly through zooming in. More HR editing results can be found in suppl.

processes patches, i.e., our advantage. Therefore, the prominent image processing efficiency of our framework is verified.

Qualitative Analysis. The visual effect comparison are shown in Fig. 5. In general, our CooGAN can effectively generate high-resolution images with smooth patch boundaries. As is shown, images generated by our CooGAN possess more prominent characteristics as well as better textural quality. For local attributes like bald and mustache, our framework performs much more accurate and significant modification. And in global attributes like gender and age, our CooGAN translates the image more thoroughly and successfully. It is worth mentioning that HR translated images have better effect in some attributes like age and gender. This is because these attribute can not be properly manifested in LR images, *e.g.*, the wrinkles of aging are obvious only in high resolution.

Fig. 6. 768 × 768 results generated by our framework without and with global module.

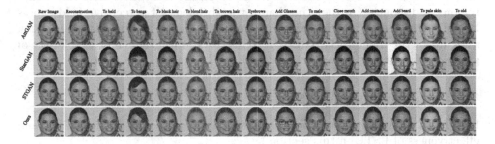

Fig. 7. Low resolution (128 × 128) facial attribute editing results of StarGAN, AttGAN, STGAN and our LSTU-Net. (Color figure online)

User Study. To further validate the subjective performance of our framework, it is important to conduct survey on a crowdsourcing platform to acquire people's general opinions of our HR edited images and compare our CooGAN with the modified HR versions of previous state-of-the-art models including STGAN-HR and AttGAN-HR. In the survey, 50 questions of 14 attributes/quality are involved, *i.e.*, 13 previous mentioned attributes and the smoothness of the image. And in each question, a randomly selected image and its edited versions by the three models are presented in random sequences. And people will determine whether an edited image has correct and realistic attributes and is generally smooth and consistent. The user study results are given in Table 4. In comparison, our CooGAN shows its excellent performance in attributes editing. Because our framework generates image by patches, it is slightly inferior to the other two methods in terms of smoothness.

Table 4. User study results of our framework, STGAN-HR and AttGAN-HR.

Method	Bald	Bangs	Black	Blond	Brown	Eyebrows	Glasses
AttGAN-HR	25.7%	66.2%	61.2%	89.4%	47.9%	51.7%	86.7%
STGAN-HR	38.8%	50.5%	78.2%	92.8%	53.5%	65.4%	71.5%
Ours	**61.5%**	**93.3%**	**90.6%**	**94.9%**	**73.6%**	**84.9%**	**94.8%**
Method	Male	Mouth	Mustache	No Beard	Pale	Young	Smooth
AttGAN-HR	36.8%	83.2%	29.8%	51.4%	34.5%	41.4%	99.8%
STGAN-HR	51.3%	67.7%	49.1%	67.8%	61.4%	49.9%	99.9%
Ours	**87.9%**	**91.7%**	**61.9%**	**93.8%**	**98.1%**	**64.7%**	91.5%

Ablation Study for Cooperative Mechanism. In this part, our focus is to study the importance and effectiveness of global module in the framework. We train our framework with and without the global module. The results show that without global module the attribute editing will be impractical, as is shown in Fig. 6. On the one hand, without the spatial information given by global snapshot, the local module can not well recognize local attributes like bangs regions and mouth. On the other hand, the shrinkage of perceptive field causes loss of global semantics, rendering certain features (like gender characteristics), difficult for the framework to discern. Thus, we confirm the irreplaceable role of global module in our framework.

4.3   Ablation Study for LSTU

LSTU is a vital component of our framework. We will verify the feasibility and effectiveness of LSTU in this section.

Experiment Setup. Our LSTU is a lightweight and equally effective replacement unit for STU [20] (previous state-of-the-art multi-scale features fusion unit proposed in STGAN). To validate its effect and efficiency, we replicate the network structure of STGAN and replace its STU with our LSTU. We name this new network LSTU-Net to be distinguished from the original STGAN. Besides, we also compare LSTU-Net with StarGAN and AttGAN since they are the most relative and effective facial attribute editing methods that can be found. However, these compared models are not designed for HR image generation tasks. They are mostly purposed to process 128×128 images. To be fair, all models are tested on 128×128 CelebA dataset.

In addition, the official codes and pretrained models of those three compared models are used in our experiments for the sake of impartiality.

Quantitative Evaluation. The attribute editing accuracy and image fidelity comparisons are shown in Fig. 8 and Table 2. For attribute accuracy, our LSTU-Net outperforms StarGAN, AttGAN and ties with STGAN. This circumstance remains the same for image fidelity manifested through PSNR and SSIM. These results show that LSTU has the same level of effect as STU, whether in attribute

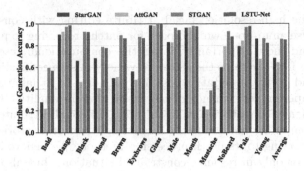

Fig. 8. Our LSTU-Net against other approaches in attribute translation accuracy.

Table 5. The total number of parameters and FLOPS in the skip connections within a five-layer network with different input channels for the LSTU and STU. G denotes GFLOPS.

Channel	16 in-channel		32 in-channel		64 in-channel	
metric	param	FLOPS	param	FLOPS	param	FLOPS
STU	1.63M	5.06G	6.40M	19.99G	25.37M	79.50G
LSTU	0.81M	2.02G	3.20M	7.85G	12.68M	30.80G

editing accuracy or image fidelity. The complexity and computational cost comparisons of different units are demonstrated in Table 5. Compared to STU, our LSTU reduces parameter size by half and FLOPS by over 60%. These comparisons prove the effectiveness and efficiency of implementing LSTU.

Qualitative Analysis. The visual effect comparison of 128×128 images is given in Fig. 7. The editing result of our LSTU is better than StarGAN, AttnGAN, and achieves similar visual effect with STGAN. In almost all attributes, our model not only performs thorough and accurate translation, but also suffer the least from textural artifacts. In contrast, the results of StarGAN have distinct color difference due to the lack of skip connection. AttGAN's results contain fake artificial textures and half-changed features in attributes like bald and hair color. Such inconsistency is caused by the direct use of skip connection. STGAN is more successful than AttGAN and StarGAN. And its edited features and ours are equally effective. These comparison results present the state-of-the-art level attribute editing ability and tremendous efficiency advantage of our LSTU. More results generated by the LSTU-Net are given in the suppl.

5   Conclusion

In this paper, we study the problem of high-resolution facial attribute editing in resource-constrained conditions. By proposing the patch-based local-global framework *CooGAN* along with the multi-scale features fusion method LSTU, we attained the high-resolution facial attribute editing with constrained GPU memory. We use an up-to-down approach for patch processing to preserve global consistency and most importantly, lower the computational resource requirement. And the LSTU retains the attribute variety in skip connection with fewer parameters and lower computation cost. Experiments on facial attribute editing exhibit the superior performance of our framework in quality, attribute accuracy and efficiency to state-of-arts in the scope of facial attribute editing with constrained computational resources. Theoretically, our framework has no image size limitation owing to its patch processing method. We believe it has promising prospect not only in resource-constrained situations, but also in extremely high-resolution image processing tasks.

Acknowledgements. This work was supported by National Science Foundation of China (61976137, U1611461, U19B2035) and STCSM(18DZ1112300).

References

1. ZAO. https://baike.baidu.com/item/ZAO/23721314
2. IEEE Conference on Computer Vision and Pattern Recognition, CVPR 2018, Salt Lake City, UT, USA, 18–22 June 2018. IEEE Computer Society (2018)
3. IEEE Conference on Computer Vision and Pattern Recognition, CVPR 2019, Long Beach, CA, USA, 16–20 June 2019. Computer Vision Foundation/IEEE (2018)
4. Awiszus, M., Ackermann, H., Rosenhahn, B.: Learning disentangled representations via independent subspaces. CoRR abs/1908.08989 (2019)
5. Cao, J., Huang, H., Li, Y., Liu, J., He, R., Sun, Z.: Biphasic learning of GANs for high-resolution image-to-image translation. CoRR abs/1904.06624 (2019)
6. Chen, Y., et al.: Facelet-bank for fast portrait manipulation. In: 2018 IEEE Conference on Computer Vision and Pattern Recognition, CVPR 2018, Salt Lake City, UT, USA, 18–22 June 2018 [2], pp. 3541–3549 (2018). https://doi.org/10.1109/CVPR.2018.00373
7. Cho, K., et al.: Learning phrase representations using RNN encoder-decoder for statistical machine translation. In: Moschitti, A., Pang, B., Daelemans, W. (eds.) Proceedings of the 2014 Conference on Empirical Methods in Natural Language Processing, EMNLP 2014, 25–29 October 2014, Doha, Qatar, A Meeting of SIGDAT, A Special Interest Group of the ACL, pp. 1724–1734. ACL (2014). https://www.aclweb.org/anthology/volumes/D14-1/
8. Choi, Y., Choi, M., Kim, M., Ha, J., Kim, S., Choo, J.: StarGAN: unified generative adversarial networks for multi-domain image-to-image translation. In: 2018 IEEE Conference on Computer Vision and Pattern Recognition, CVPR 2018, Salt Lake City, UT, USA, 18–22 June 2018 [2], pp. 8789–8797 (2018). https://doi.org/10.1109/CVPR.2018.00916
9. Chung, J., Gülçehre, Ç., Cho, K., Bengio, Y.: Empirical evaluation of gated recurrent neural networks on sequence modeling. CoRR abs/1412.3555 (2014)
10. Gulrajani, I., Ahmed, F., Arjovsky, M., Dumoulin, V., Courville, A.C.: Improved training of Wasserstein GANs. CoRR abs/1704.00028 (2017)
11. Guyon, I., et al. (eds.): Advances in Neural Information Processing Systems 30: Annual Conference on Neural Information Processing Systems 2017, Long Beach, CA, USA, 4–9 December 2017 (2017)

12. He, Z., Zuo, W., Kan, M., Shan, S., Chen, X.: AttGAN: facial attribute editing by only changing what you want. IEEE Trans. Image Process. **28**(11), 5464–5478 (2019). https://doi.org/10.1109/TIP.2019.2916751
13. Howard, A.G., et al.: MobileNets: efficient convolutional neural networks for mobile vision applications. CoRR abs/1704.04861 (2017)
14. Karras, T., Aila, T., Laine, S., Lehtinen, J.: Progressive growing of GANs for improved quality, stability, and variation. In: 6th International Conference on Learning Representations, ICLR 2018, Vancouver, BC, Canada, 30 April–3 May 2018, Conference Track Proceedings. OpenReview.net (2018). https://openreview.net/group?id=ICLR.cc/2018/Conference
15. Kemelmacher-Shlizerman, I., Suwajanakorn, S., Seitz, S.M.: Illumination-aware age progression. In: 2014 IEEE Conference on Computer Vision and Pattern Recognition, CVPR 2014, Columbus, OH, USA, 23–28 June 2014, pp. 3334–3341. IEEE Computer Society (2014). https://doi.org/10.1109/CVPR.2014.426. https://ieeexplore.ieee.org/xpl/conhome/6909096/proceeding
16. Kingma, D.P., Ba, J.: Adam: a method for stochastic optimization. In: Bengio, Y., LeCun, Y. (eds.) 3rd International Conference on Learning Representations, ICLR 2015, San Diego, CA, USA, 7–9 May 2015, Conference Track Proceedings (2015). https://iclr.cc/archive/www/doku.php%3Fid=iclr2015:accepted-main.html
17. Lample, G., Zeghidour, N., Usunier, N., Bordes, A., Denoyer, L., Ranzato, M.: Fader networks: manipulating images by sliding attributes. In: Guyon et al. [11], pp. 5967–5976
18. Lee, C., Liu, Z., Wu, L., Luo, P.: MaskGAN: towards diverse and interactive facial image manipulation. CoRR abs/1907.11922 (2019)
19. Lei, T., Zhang, Y., Wang, S.I., Dai, H., Artzi, Y.: Simple recurrent units for highly parallelizable recurrence. In: Riloff, E., Chiang, D., Hockenmaier, J., Tsujii, J. (eds.) Proceedings of the 2018 Conference on Empirical Methods in Natural Language Processing, Brussels, Belgium, 31 October–4 November 2018, pp. 4470–4481. Association for Computational Linguistics (2018). https://doi.org/10.18653/v1/d18-1477
20. Liu, M., et al.: STGAN: a unified selective transfer network for arbitrary image attribute editing. In: IEEE Conference on Computer Vision and Pattern Recognition, CVPR 2019, Long Beach, CA, USA, 16–20 June 2019 [3], pp. 3673–3682
21. Liu, M., Breuel, T., Kautz, J.: Unsupervised image-to-image translation networks. In: Guyon et al. [11], pp. 700–708
22. Liu, Z., Luo, P., Wang, X., Tang, X.: Deep learning face attributes in the wild. In: Proceedings of International Conference on Computer Vision (ICCV), December 2015
23. Ma, N., Zhang, X., Zheng, H.-T., Sun, J.: ShuffleNet V2: practical guidelines for efficient CNN architecture design. In: Ferrari, V., Hebert, M., Sminchisescu, C., Weiss, Y. (eds.) Computer Vision – ECCV 2018. LNCS, vol. 11218, pp. 122–138. Springer, Cham (2018). https://doi.org/10.1007/978-3-030-01264-9_8
24. Odena, A., Olah, C., Shlens, J.: Conditional image synthesis with auxiliary classifier GANs. In: Precup, D., Teh, Y.W. (eds.) Proceedings of the 34th International Conference on Machine Learning, ICML 2017, Sydney, NSW, Australia, 6–11 August 2017. Proceedings of Machine Learning Research, vol. 70, pp. 2642–2651. PMLR (2017). http://proceedings.mlr.press/v70/
25. Perarnau, G., van de Weijer, J., Raducanu, B., Álvarez, J.M.: Invertible conditional GANs for image editing. CoRR abs/1611.06355 (2016)
26. Qian, S., et al.: Make a face: towards arbitrary high fidelity face manipulation. CoRR abs/1908.07191 (2019)

27. Ronneberger, O., Fischer, P., Brox, T.: U-Net: convolutional networks for biomedical image segmentation. In: Navab, N., Hornegger, J., Wells, W.M., Frangi, A.F. (eds.) MICCAI 2015. LNCS, vol. 9351, pp. 234–241. Springer, Cham (2015). https://doi.org/10.1007/978-3-319-24574-4_28

28. Sandler, M., Howard, A.G., Zhu, M., Zhmoginov, A., Chen, L.: Inverted residuals and linear bottlenecks: mobile networks for classification, detection and segmentation. CoRR abs/1801.04381 (2018)

29. Thies, J., Zollhöfer, M., Nießner, M., Valgaerts, L., Stamminger, M., Theobalt, C.: Real-time expression transfer for facial reenactment. ACM Trans. Graph. **34**(6), 183:1–183:14 (2015). https://doi.org/10.1145/2816795.2818056

30. Xu, S., Huang, H., Hu, S., Liu, W.: FaceShapeGene: a disentangled shape representation for flexible face image editing. CoRR abs/1905.01920 (2019)

31. Xu, T., et al.: AttnGAN: fine-grained text to image generation with attentional generative adversarial networks. In: 2018 IEEE Conference on Computer Vision and Pattern Recognition, CVPR 2018, Salt Lake City, UT, USA, 18–22 June 2018 [2], pp. 1316–1324. https://doi.org/10.1109/CVPR.2018.00143

32. Yi, R., Liu, Y., Lai, Y., Rosin, P.L.: APDrawingGAN: generating artistic portrait drawings from face photos with hierarchical GANs. In: IEEE Conference on Computer Vision and Pattern Recognition, CVPR 2019, Long Beach, CA, USA, 16–20 June 2019, pp. 10743–10752. Computer Vision Foundation/IEEE (2019). https://doi.org/10.1109/CVPR.2019.01100

33. Yin, W., Liu, Z., Loy, C.C.: Instance-level facial attributes transfer with geometry-aware flow. In: The Thirty-Third AAAI Conference on Artificial Intelligence, AAAI 2019, The Thirty-First Innovative Applications of Artificial Intelligence Conference, IAAI 2019, The Ninth AAAI Symposium on Educational Advances in Artificial Intelligence, EAAI 2019, Honolulu, Hawaii, USA, 27 January–1 February 2019, pp. 9111–9118. AAAI Press (2019). https://doi.org/10.1609/aaai.v33i01.33019111. https://www.aaai.org/Library/AAAI/aaai19contents.php

34. Zhang, X., Zhou, X., Lin, M., Sun, J.: ShuffleNet: an extremely efficient convolutional neural network for mobile devices. In: 2018 IEEE Conference on Computer Vision and Pattern Recognition, CVPR 2018, Salt Lake City, UT, USA, 18–22 June 2018 [2], pp. 6848–6856

35. Zhao, C., Ni, B., Zhang, J., Zhao, Q., Zhang, W., Tian, Q.: Variational convolutional neural network pruning. In: IEEE Conference on Computer Vision and Pattern Recognition, CVPR 2019, Long Beach, CA, USA, 16–20 June 2019 [3], pp. 2780–2789

36. Zhu, J., Park, T., Isola, P., Efros, A.A.: Unpaired image-to-image translation using cycle-consistent adversarial networks. In: IEEE International Conference on Computer Vision, ICCV 2017, Venice, Italy, 22–29 October 2017, pp. 2242–2251. IEEE Computer Society (2017). https://doi.org/10.1109/ICCV.2017.244. https://ieeexplore.ieee.org/xpl/conhome/8234942/proceeding

Progressive Transformers for End-to-End Sign Language Production

Ben Saunders[✉], Necati Cihan Camgoz, and Richard Bowden

University of Surrey, Guildford, England
{b.saunders,n.camgoz,r.bowden}@surrey.ac.uk

Abstract. The goal of automatic Sign Language Production (SLP) is to translate spoken language to a continuous stream of sign language video at a level comparable to a human translator. If this was achievable, then it would revolutionise Deaf hearing communications. Previous work on predominantly isolated SLP has shown the need for architectures that are better suited to the continuous domain of full sign sequences.

In this paper, we propose Progressive Transformers, the first SLP model to translate from discrete spoken language sentences to continuous 3D sign pose sequences in an end-to-end manner. A novel counter decoding technique is introduced, that enables continuous sequence generation at training and inference. We present two model configurations, an end-to-end network that produces sign direct from text and a stacked network that utilises a gloss intermediary. We also provide several data augmentation processes to overcome the problem of drift and drastically improve the performance of SLP models.

We propose a back translation evaluation mechanism for SLP, presenting benchmark quantitative results on the challenging RWTH-PHOENIXWeather-2014T (PHOENIX14T) dataset and setting baselines for future research. Code available at https://github.com/BenSaunders27/ProgressiveTransformersSLP.

Keywords: Sign language production · Continuous sequence synthesis · Transformers · Sequence-to-sequence · Human pose generation

1 Introduction

Sign language is the language of communication for the Deaf community, a rich visual language with complex grammatical structures. As it is their native language, most Deaf people prefer using sign as their main medium of communication, as opposed to a written form of spoken language. Sign Language Production (SLP), converting spoken language to continuous sign sequences, is

Electronic supplementary material The online version of this chapter (https://doi.org/10.1007/978-3-030-58621-8_40) contains supplementary material, which is available to authorized users.

© Springer Nature Switzerland AG 2020
A. Vedaldi et al. (Eds.): ECCV 2020, LNCS 12356, pp. 687–705, 2020.
https://doi.org/10.1007/978-3-030-58621-8_40

therefore essential in involving the Deaf in the predominantly spoken language of the wider world. Previous work has been limited to the production of concatenated isolated signs [53,64], highlighting the need for improved architectures to properly address the full remit of continuous sign language.

In this paper, we propose *Progressive Transformers*, the first SLP model to translate from text to continuous 3D sign pose sequences in an end-to-end manner. Our novelties include an alternative formulation of transformer decoding for continuous variable sequences, where there is no pre-defined vocabulary. We introduce a counter decoding technique to predict continuous sequences of variable lengths by tracking the production progress, hence the name *Progressive Transformers*. This approach also enables the driving of timing at inference, producing stable sign pose outputs. We also propose several data augmentation methods that assist in reducing drift in model production.

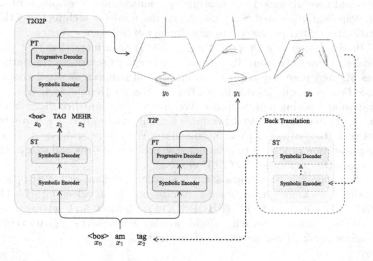

Fig. 1. Overview of the progressive transformer architecture, showing text to Gloss to Pose (T2G2P) and Text to Pose (T2P) model configurations. (PT: progressive transformer, ST: symbolic transformer)

An overview of our approach is shown in Fig. 1. We evaluate two different model configurations, first translating from spoken language to sign pose via gloss[1] intermediary (T2G2P), as this has been shown to increase translation performance [7]. In the second configuration we go direct, translating end-to-end from spoken language to sign (T2P).

To evaluate performance, we propose a back translation evaluation method for SLP, using a Sign Language Translation (SLT) model to translate back to spoken language (dashed lines in Fig. 1). We evaluate on the challenging RWTH-PHOENIX-Weather-2014T (PHOENIX14T) dataset, presenting several benchmark results to underpin future research. We also share qualitative results to

---

[1] Glosses are a written representation of sign, defined as minimal lexical items.

give further insight of the models performance to the reader, producing accurate sign pose sequences of an unseen text sentence.

The rest of this paper is organised as follows: In Sect. 2, we go over the previous research on SLT and SLP. In Sect. 3, we introduce our Progressive Transformer SLP model. Section 4 outlines the evaluation protocol and presents quantitative results, whilst Sect. 5 showcases qualitative examples. Finally, we conclude the paper in Sect. 6 by discussing our findings and possible future work.

2   Related Work

Sign Language Recognition and Translation: Sign language has been a focus of computer vision researchers for over 30 years [4,52,57], primarily on isolated Sign Language Recognition (SLR) [46,56] and, relatively recently, the more demanding task of Continuous Sign Language Recognition (CSLR) [6,33]. However, the majority of work has relied on manual feature representations [11] and statistical temporal modelling [60]. The availability of larger datasets, such as RWTH-PHOENIX-Weather-2014 (PHOENIX14) [17], have enabled the application of deep learning approaches such as Convolutional Neural Networks (CNNs) [32,34,36] and Recurrent Neural Networks (RNNs) [12,35].

Distinct to SLR, the task of SLT was recently introduced by Camgoz et al. [7], aiming to directly translate sign videos to spoken language sentences [15,31,45,62]. SLT is more challenging than CSLR due to the differences in grammar and ordering between sign and spoken language. Transformer based models are the current state-of-the-art in SLT, jointly learning the recognition and translation tasks [8].

Sign Language Production: Previous approaches to SLP have extensively used animated avatars [20,27,42] that can generate realistic sign production, but rely on phrase lookup and pre-generated sequences. Statistical Machine Translation (SMT) has also been applied to SLP [28,37], relying on static rule-based processing that can be difficult to encode.

Recently, deep learning approaches have been applied to the task of SLP [15,53,61]. Stoll et al. present an initial SLP model using a combination of Neural Machine Translation (NMT) and Generative Adversarial Networks (GANs) [54]. The authors break the problem into three separate processes that are trained independently, producing a concatenation of isolated 2D skeleton poses [16] mapped from sign glosses via a look-up table. Contrary to Stoll et al., our paper focuses on automatic sign production and learning the mapping between text and skeleton pose sequences directly, instead of providing this a priori.

The closest work to this paper is that of Zelinka et al., who build a neural-network-based translator between text and synthesised skeletal pose [63]. The authors produce a single sign for each source word with a set size of 7 frames, generating sequences with a fixed length and ordering. In contrast, our model allows a dynamic length of output sign sequence, learning the correct length and ordering of each word from the data, whilst using counter decoding to determine the end of sequence generation. Unlike [63], who work on a proprietary dataset, we produce results on the publicly PHOENIX14T, providing a benchmark for future SLP research

Neural Machine Translation: NMT aims to learn a mapping between language sequences, generating a target sequence from a source sequence of another language. RNNs were first proposed to solve the sequence-to-sequence problem, with Kalchbrenner et al. [26] introducing a single RNN that iteratively applied a hidden state computation. Further models were later developed [10,55] that introduced encoder-decoder architectures, mapping both sequences to an intermediate embedding space. Bahdanau et al. [3] overcame the bottleneck problem by adding an attention mechanism that facilitated a soft-search over the source sentence for the context most useful to the target word prediction.

Transformer networks [58], a recent NMT breakthrough, are based solely on attention mechanisms, generating a representation of the entire source sequence with global dependencies. Multi-Headed Attention (MHA) is used to model different weighted combinations of an input sequence, improving the representation power of the model. Transformers have achieved impressive results in many classic Natural Language Processing (NLP) tasks such as language modelling [13,65] and sentence representation [14] alongside other domains including image captioning [40,66] and action recognition [19]. Related to this work, transformer networks have previously been applied to continuous output tasks such as speech synthesis [41,50,59], music production [24] and image generation [47].

Applying NMT methods to continuous output tasks is a relatively underresearched problem. Encoder-decoder models and RNNs have been used to map text to a human action sequence [1,49] whilst adversarial discriminators have enabled the production of realistic pose [18,39]. In order to determine sequence length of continuous outputs, previous works have used a fixed output size that limits the models flexibility [63], a binary end-of-sequence (EOS) flag [22] or a continuous representation of an EOS token [44].

3   Progressive Transformers

In this section, we introduce *Progressive Transformers*, an SLP model which learns to translate spoken language sentences to continuous sign pose sequences. Our objective is to learn the conditional probability $p(Y|X)$ of producing a sequence of signs $Y = (y_1, ..., y_U)$ with U time steps, given a spoken language sentence $X = (x_1, ..., x_T)$ with T words. Gloss can also be used as intermediary supervision for the network, formulated as $Z = (z_1, ..., z_N)$ with N glosses, where the objective is then to learn the conditional probabilities $p(Z|X)$ and $p(Y|Z)$.

Producing a target sign sequence from a reference text sequence poses several challenges. Firstly, the sequences have drastically varying length, with the number of frames much larger than the number of words ($U >> T$). The sequences

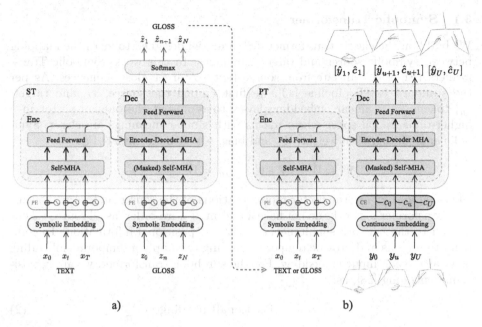

Fig. 2. Architecture details of (a) symbolic and (b) progressive Transformers. (ST: symbolic transformer, PT: progressive transformer, PE: positional encoding, CE: counter embedding, MHA: multi-head attention)

also have a non-monotonic relationship due to the different vocabulary and grammar used in sign and spoken languages. Finally, the target signs inhabit a continuous vector space requiring a differing representation to the discrete space of text.

To address the production of continuous sign sequences, we propose a progressive transformer-based architecture that allows translation from a symbolic to a continuous sequence domain. We first formalise a Symbolic Transformer architecture, converting an input to a symbolic target feature space, as detailed in Fig. 2a. This is used in our Text to Gloss to Pose (T2G2P) model to convert from spoken language to gloss representation as an intermediary step before pose production, as seen in Fig. 1.

We then describe the Progressive Transformer architecture, translating from a symbolic input to a continuous output representation, as shown in Fig. 2b. We use this model for the production of realistic and understandable sign language sequences, either via gloss supervision in the T2G2P model or direct from spoken language in our end-to-end Text to Pose (T2P) model. To enable sequence length prediction of a continuous output, we introduce a counter decoding that allows the model to track the progress of sequence generation. In the remainder of this section we describe each component of the architecture in detail.

3.1   Symbolic Transformer

We build on the classic transformer [58], a model designed to learn the mapping between symbolic source and target languages. In this work, Symbolic Transformers (Fig. 2a) translate from source text to target gloss sequences. As per the standard NMT pipeline [43], we first embed the source, x_t, and target, z_n, tokens via a linear embedding layer, to represent the one-hot-vector in a higher-dimensional space where tokens with similar meanings are closer. Symbolic embedding, with weight, W, and bias, b, can be formulated as:

$$w_t = W^x \cdot x_t + b^x, \quad g_n = W^z \cdot z_n + b^z \tag{1}$$

where w_t and g_n are the vector representations of the source and target tokens.

Transformer networks do not have a notion of word order, as all source tokens are fed to the network simultaneously without positional information. To compensate for this and provide temporal ordering, we apply a temporal embedding layer after each input embedding. For the symbolic transformer, we apply positional encoding [58], as:

$$\hat{w}_t = w_t + \text{PositionalEncoding}(t) \tag{2}$$

$$\hat{g}_n = g_n + \text{PositionalEncoding}(n) \tag{3}$$

where PositionalEncoding is a predefined sinusoidal function conditioned on the relative sequence position t or n.

Our symbolic transformer model consists of an encoder-decoder architecture. The encoder first learns the contextual representation of the source sequence through self-attention mechanisms, understanding each input token in relation to the full sequence. The decoder then determines the mapping between the source and target sequences, aligning the representation sub-spaces and generating target predictions in an auto-regressive manner.

The symbolic encoder (E_S) consists of a stack of L identical layers, each containing 2 sub-layers. Given the temporally encoded source embeddings, $\hat{w}_t$, a MHA mechanism first generates a weighted contextual representation, performing multiple projections of scaled dot-product attention. This aims to learn the relationship between each token of the sequence and how relevant each time step is in the context of the full sequence. Formally, scaled dot-product attention outputs a vector combination of values, V, weighted by the relevant queries, Q, keys, K, and dimensionality, d_k:

$$\text{Attention}(Q, K, V) = \text{softmax}(\frac{QK^T}{\sqrt{d_k}})V \tag{4}$$

MHA stacks parallel attention mechanisms in h different mappings of the same queries, keys and values, each with varied learnt parameters. This allows different representations of the input to be generated, learning complementary information in different sub-spaces. The outputs of each head are then concatenated together and projected forward via a final linear layer, as:

$$\text{MHA}(Q, K, V) = [head_1, ..., head_h] \cdot W^O,$$
$$\text{where } head_i = \text{Attention}(QW_i^Q, KW_i^K, VW_i^V) \tag{5}$$

and W^O, W_i^Q, W_i^K and W_i^V are weights related to each input variable.

The outputs of MHA are then fed into the second sub-layer of a non-linear feed-forward projection. A residual connection [23] and subsequent layer norm [2] is employed around each of the sub-layers, to aid training. The final symbolic encoder output can be formulated as:

$$h_t = E_S(\hat{w}_t | \hat{w}_{1:T}) \tag{6}$$

where h_t is the contextual representation of the source sequence.

The symbolic decoder (D_S) is an auto-regressive architecture that produces a single token at each time-step. The positionally embedded target sequences, $\hat{g}_n$, are passed through an initial MHA self-attention layer similar to the encoder, with an extra masking operation. Alongside the fact that the targets are offset from the inputs by one position, the masking of future frames prevents the model from attending to subsequent time steps in the sequence.

A further MHA sub-layer is then applied, which combines encoder and decoder representations and learns the alignment between the source and target sequences. The final sub-layer is a feed forward layer, as in the encoder. After all decoder layers are processed, a final non-linear feed forward layer is applied, with a softmax operation to generate the most likely output token at each time step. The output of the symbolic decoder can be formulated as:

$$z_{n+1} = \underset{i}{\text{argmax}} \, D_S(\hat{g}_n | \hat{g}_{1:n-1}, h_{1:T}) \tag{7}$$

where z_{n+1} is the output at time $n + 1$, from a target vocabulary of size i.

3.2   Progressive Transformer

We now adapt our symbolic transformer architecture to cope with continuous outputs, in order to convert source sequences to a continuous target domain. In this work, Progressive Transformers (Fig. 2b) translate from the symbolic domains of gloss or text to continuous sign pose sequences that represent the motion of a signer producing a sentence of sign language. The model must produce skeleton pose outputs that can both express an accurate translation of the given input sequence and a realistic sign pose sequence.

We represent each sign pose frame, y_u, as a continuous vector of the 3D joint positions of the signer. These joint values are first passed through a linear embedding layer, allowing sign poses of similar content to be closely represented in the dense space. The continuous embedding layer can be formulated as:

$$j_u = W^y \cdot y_u + b^y \tag{8}$$

where j_u is the embedded 3D joint coordinates of each frame, y_u.

We next apply a counter embedding layer to the sign poses as temporal embedding (CE in Fig. 2). The counter, c, holds a value between 0 and 1, representing the frame position relative to the total sequence length. The joint embeddings, j_u, are concatenated with the respective counter value, c_u, formulated as:

$$\hat{j}_u = [j_u, \text{CounterEmbedding}(u)] \tag{9}$$

where CounterEmbedding is a linear projection of the counter value for frame u.

At each time-step, counter values are predicted alongside the skeleton pose, as shown in Fig. 3, with sequence generation concluded once the counter reaches 1. We call this process Counter Decoding, determining the progress of sequence generation and providing a way to predict the end of sequence without the use of a tokenised vocabulary.

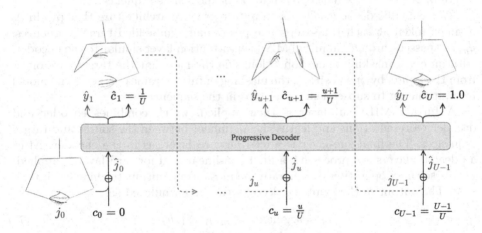

Fig. 3. Counter decoding example, showing the simultaneous prediction of sign pose, $\hat{y}_u$, and counter, $\hat{c}_u \in \{0 : 1\}$, with $\hat{c} = 1.0$ denoting end of sequence

The counter provides the model with information relating to the length and speed of each sign pose sequence, determining the sign duration. At inference, we drive the sequence generation by replacing the predicted counter value, $\hat{c}$, with the ground truth timing information, c^*, to produce a stable output sequence.

The Progressive Transformer also consists of an encoder-decoder architecture. Due to the input coming from a symbolic source, the encoder has a similar setup to the symbolic transformer, learning a contextual representation of the input sequence. As the representation will ultimately be used for the end goal of SLP, these representations must also contain sufficient context to fully and accurately reproduce sign. Taking as input the temporally embedded source embeddings, $\hat{w}_t$, the encoder can be formulated as:

$$r_t = E_S(\hat{w}_t | \hat{w}_{1:T}) \tag{10}$$

where E_S is the symbolic encoder and r_t is the encoded source representation.

The progressive decoder (D_P) is an auto-regressive model that produces a sign pose frame at each time-step, alongside the counter value described above. Distinct from symbolic transformers, the progressive decoder produces continuous sequences that hold a sparse representation in a large continuous sub-space. The counter-concatenated joint embeddings, $\hat{j}_u$, are extracted as target input, representing the sign information of each frame.

A self-attention MHA sub-layer is first applied, with target masking to avoid attending to future positions. A further MHA mechanism is then used to map the symbolic representations from the encoder to the continuous domain of the decoder, learning the important alignment between spoken and sign languages.

A final feed forward sub-layer follows, with each sub-layer followed by a residual connection and layer normalisation as before. No softmax layer is used as the skeleton joint coordinates can be regressed directly and do not require stochastic prediction. The progressive decoder output can be formulated as:

$$[\hat{y}_{u+1}, \hat{c}_{u+1}] = D_P(\hat{j}_u | \hat{j}_{1:u-1}, r_{1:T})\tag{11}$$

where $\hat{y}_{u+1}$ corresponds to the 3D joint positions representing the produced sign pose of frame $u+1$ and $\hat{c}_{u+1}$ is the respective counter value. The decoder learns to generate one frame at a time until the predicted counter value reaches 1, determining the end of sequence. Once the full sign pose sequence is produced, the model is trained end-to-end using the Mean Squared Error (MSE) loss between the predicted sequence, $\hat{y}_{1:U}$, and the ground truth, $y^*_{1:U}$:

$$L_{MSE} = \frac{1}{U} \sum_{i=1}^{u} (y^*_{1:U} - \hat{y}_{1:U})^2\tag{12}$$

The progressive transformer outputs, $\hat{y}_{1:U}$, represent the 3D skeleton joint positions of each frame of a produced sign sequence. To ease the visual comparison with reference sequences, we apply Dynamic Time Warping (DTW) [5] to align the produced sign pose sequences. Animating a video from this sequence is then a trivial task, plotting the joints and connecting the relevant bones, with timing information provided from the counter. These 3D joints could subsequently be used to animate an avatar [30,42] or condition a GAN [25,67].

4 Quantitative Experiments

In this section, we share our SLP experimental setup and report experimental results. We first provide dataset and evaluation details, outlining back translation. We then evaluate both symbolic and progressive transformer models, demonstrating results of data augmentation and model configuration.

4.1 Sign Language Production Dataset

Forster et al. released PHOENIX14 [17] as a large video-based corpus containing parallel sequences of German Sign Language - Deutsche Gebärdensprache

(DGS) and spoken text extracted from German weather forecast recordings. This dataset is ideal for computational sign language research due to the provision of gloss level annotations, becoming the primary benchmark for both SLR and CSLR.

In this work, we use the publicly available PHOENIX14T dataset introduced by Camgoz et al. [7], a continuous SLT extension of the original PHOENIX14. This corpus includes parallel sign videos and German translation sequences with redefined segmentation boundaries generated using the forced alignment approach of [36]. 8257 videos of 9 different signers are provided, with a vocabulary of 2887 German words and 1066 different sign glosses from a combined 835,356 frames.

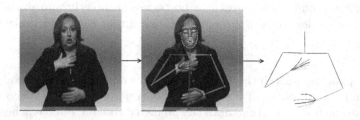

Fig. 4. Skeleton pose extraction, using OpenPose [9] and 2D to 3D mapping [63]

We train our SLP network to generate sequences of 3D skeleton pose. 2D joint positions are first extracted from each video using OpenPose [9]. We then utilise the skeletal model estimation improvements presented in [63] to lift the 2D joint positions to 3D. An iterative inverse kinematics approach is applied to minimise 3D pose whilst maintaining consistent bone length and correcting misplaced joints. Finally, we apply skeleton normalisation similar to [53] and represent 3D joints as x, y and z coordinates. An example is shown in Fig. 4.

4.2   Evaluation Details

In this work, we propose back-translation as a means of SLP evaluation, translating back from produced sign to spoken language. This provides a measure of how understandable the productions are, and how much translation content is preserved. Evaluation of a generative model is often difficult but we find a close correspondence between back translation score and the visual production quality. We liken it to the wide use of the inception score for generative models [51], using a pre-trained classifier. Similarly, SLP work used an SLR discriminator to evaluate isolated skeletons [61], but did not measure the translation performance.

We utilise the state-of-the-art SLT [8] as our back translation model, modified to take sign pose sequences as input. This is again trained on the PHOENIX14T dataset, ensuring a robust translation from sign to text. We generate spoken language translations of the produced sign pose sequences and compute BLEU and ROUGE scores. We provide BLEU n-grams from 1 to 4 for completeness.

In the following experiments, our symbolic and progressive transformer models are each built with 2 layers, 8 heads and embedding size of 256. All parts of our network are trained with Xavier initialisation [21], Adam optimization [29] with default parameters and a learning rate of 10^{-3}. Our code is based on Kreutzer et al.'s NMT toolkit, JoeyNMT [38], and implemented using PyTorch [48].

4.3 Symbolic Transformer: Text to Gloss

Our first experiment measures the performance of the symbolic transformer architecture for sign language understanding. We train our symbolic transformer to predict gloss representations from source spoken language sentences. Table 1 shows our model achieves state-of-the-art results, significantly outperforming that of Stoll et al. [53], who use an encoder-decoder network with 4 layers of 1000 Gated Recurrent Units (GRUs). This supports our use of the proposed transformer architecture for sign language understanding.

Table 1. Symbolic transformer results for Text to Gloss translation

Approach:	DEV SET					TEST SET				
	BLEU-4	BLEU-3	BLEU-2	BLEU-1	ROUGE	BLEU-4	BLEU-3	BLEU-2	BLEU-1	ROUGE
Stoll et al. [53]	16.34	22.30	32.47	50.15	48.42	15.26	21.54	32.25	50.67	48.10
Ours	20.23	27.36	38.21	55.65	55.41	19.10	26.24	37.10	55.18	54.55

Table 2. Progressive Transformer results for Gloss to Sign Pose production, with multiple data augmentation techniques. FP: future prediction, GN: Gaussian Noise

Approach:	DEV SET					TEST SET				
	BLEU-4	BLEU-3	BLEU-2	BLEU-1	ROUGE	BLEU-4	BLEU-3	BLEU-2	BLEU-1	ROUGE
Base	7.04	9.10	13.12	24.20	25.53	5.03	6.89	10.81	23.03	23.31
Future Prediction	9.96	12.71	17.83	30.03	31.03	8.38	11.04	16.41	28.94	29.73
Just Counter	11.04	13.86	19.05	31.16	32.45	9.16	11.96	17.41	30.08	30.41
Gaussian Noise	11.88	15.07	20.61	32.53	34.19	10.02	12.96	18.58	31.11	31.83
FP & GN	11.93	15.08	20.50	32.40	34.01	10.43	13.51	19.19	31.80	32.02

4.4 Progressive Transformer: Gloss to Pose

In our next set of experiments, we evaluate our progressive transformer and its capability to produce a continuous sign pose sequence from a given symbolic input. As a baseline, we train a progressive transformer model to translate from gloss to sign pose without augmentation, with results shown in Table 2 (Base).

We believe our base progressive model suffers from prediction drift, with erroneous predictions accumulating over time. As transformer models are trained to predict the next time-step of all ground truth inputs, they are often not robust to noise in target inputs. At inference time, with predictions based off previous outputs, errors are propagated throughout the full sequence generation, quickly leading to poor quality production. The impact of drift is heightened due to the continuous distribution of the target skeleton poses. As neighbouring frames differ little in content, a model learns to just copy the previous ground truth input and receive a small loss penalty. We thus experiment with various data augmentation approaches in order to overcome drift and improve performance.

Future Prediction. Our first data augmentation method is conditional future prediction, requiring the model to predict more than just the next frame in the sequence. Experimentally, we find the best performance comes from a prediction of all of the next 10 frames from the current time step. As can be seen in Table 2, prediction of future time steps increases performance from the base architecture. We believe this is because the model now cannot rely on just copying the previous frame, as there are more considerable changes to the skeleton positions in 10 frames time. The underlying structure and movement of sign has to be learnt, encoding how each gloss is represented and reproduced in the training data.

Just Counter. Inspired by the memorisation capabilities of transformer models, we next experiment with a pure memorisation approach. Only the counter values are provided as target input to the model, as opposed to the usual full 3D skeleton joint positions. We show a further performance increase with this approach, considerably increasing the BLEU-4 score as shown in Table 2.

Table 3. Results of the Text2Pose (T2P) and Text2Gloss2Pose (T2G2P) network configurations for Text to Sign Pose production

Configuration:	DEV SET					TEST SET				
	BLEU-4	BLEU-3	BLEU-2	BLEU-1	ROUGE	BLEU-4	BLEU-3	BLEU-2	BLEU-1	ROUGE
T2P	11.82	14.80	19.97	31.41	33.18	10.51	13.54	19.04	31.36	32.46
T2G2P	11.43	14.71	20.71	33.12	34.05	9.68	12.53	17.62	29.74	31.07

We believe the just counter model setup helps to allay the effect of drift, as the model now must learn to decode the target sign pose solely from the counter position, without relying on the ground truth joint embeddings it previously had access to. This setup is now identical at both training and inference, with the model having to generalise only to new data rather than new prediction inputs.

Gaussian Noise. Our final augmentation experiment examines the effect of applying noise to the skeleton pose sequences during training, increasing the variety of data to train a more robust model. For each joint, statistics on the positional distribution of the previous epoch are collected, with randomly sampled noise applied to the inputs of the next epoch. Applied noise is multiplied by a noise factor, r_n, with empirical validation suggesting $r_n = 5$ gives the best performance. An increase of Gaussian noise causes the model to become more robust to prediction inputs, as it must learn to correct the augmented inputs back to the target outputs.

Table 2 (FP & GN) shows that the best BLEU-4 performance comes from a combination of future prediction and Gaussian noise augmentation. The model must learn to cope with both multi-frame prediction and a noisy input, building a firm robustness to drift. We continue with this setup for further experiments.

4.5   Text2Pose V Text2Gloss2Pose

Our final experiment evaluates the two network configurations outlined in Fig. 1, sign production either direct from text or via a gloss intermediary. Text to Pose (T2P) consists of a single progressive transformer model with spoken language input, learning to jointly translate from the domain of spoken language to sign and subsequently produce meaningful sign representations. Text to Gloss to Pose (T2G2P) uses an initial symbolic transformer to convert to gloss, which is then input into a further progressive transformer to produce sign pose sequences.

As can be seen from Table 3, the T2P model outperforms that of T2G2P. This is surprising, as a large body of previous work has suggested that using gloss as intermediary helps networks learn [7]. However, we believe this is because there is more information available within spoken language compared to a gloss representation, with more tokens per sequence to predict from. Predicting gloss sequences as an intermediary can act as a bottleneck, as all information required for production needs to be present in the gloss. Therefore, any contextual information present in the source text can be lost.

The success of the T2P network shows that our progressive transformer model is powerful enough to complete two sub-tasks; firstly mapping spoken language sequences to a sign representation, then producing an accurate sign pose recreation. This is important for future scaling and application of the SLP model architecture, as many sign language domains do not have gloss availability.

Furthermore, our final BLEU-4 scores outperform similar end-to-end Sign to Text methods which do not utilize gloss information [7] (9.94 BLEU-4). Note that this is an unfair direct comparison, but it does provide an indication of model performance and the quality of the produced sign pose sequences.

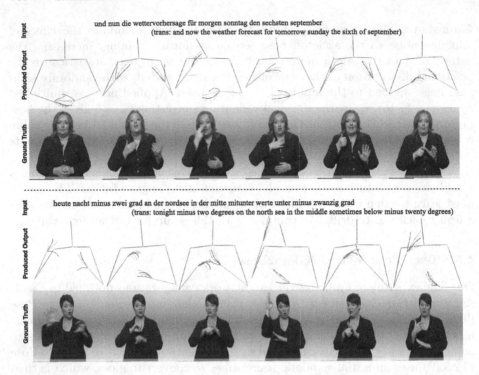

Fig. 5. Examples of produced sign pose sequences. The top row shows the spoken language input from the unseen validation set alongside English translation. The middle row presents our produced sign pose sequence from this text input, with the bottom row displaying the ground truth video for comparison.

5   Qualitative Experiments

In this section we report qualitative results for our progressive transformer model. We share snapshot examples of produced sign pose sequences in Fig. 5, with more examples provided in supplementary material. The unseen spoken language sequence is shown as input alongside the sign pose sequence produced by our Progressive Transformer model, with ground truth video for comparison.

As can be seen from the provided examples, our SLP model produces visually pleasing and realistic looking sign with a close correspondence to the ground truth video. Body motion is smooth and accurate, whilst hand shapes are meaningful if a little under-expressed. We find that the most difficult production occurs with proper nouns and specific entities, due to the lack of grammatical context and examples in the training data.

These examples show that regressing continuous sequences can be successfully achieved using an attention-based mechanism. The predicted joint locations for neighbouring frames are closely positioned, showing that the model has learnt the subtle movement of the signer. Smooth transitions between signs are produced, highlighting a difference from the discrete generation of spoken language.

6   Conclusion

Sign Language Production (SLP) is an important task to improve communication between the Deaf and hearing. Previous work has focused on producing concatenated isolated signs instead of full continuous sign language sequences. In this paper, we proposed Progressive Transformers, a novel transformer architecture that can translate from discrete spoken language to continuous sign pose sequences. We introduced a counter decoding that enables continuous sequence generation without the need for an explicit end of sequence token. Two model configurations were presented, an end-to-end network that produces sign direct from text and a stacked network that utilises a gloss intermediary.

We evaluated our approach on the challenging PHOENIX14T dataset, setting baselines for future research with a back translation evaluation mechanism. Our experiments showed the importance of several data augmentation techniques to reduce model drift and improve SLP performance. Furthermore, we have shown that a direct text to pose translation configuration can outperform a gloss intermediary model, meaning SLP models are not limited to only training on data where expensive gloss annotation is available.

As future work, we would like to expand our network to multi-channel sign production, focusing on non-manual aspects of sign language such as body pose, facial expressions and mouthings. It would be interesting to condition a GAN to produce sign videos, learning a prior for each sign represented in the data.

Acknowledgements. This work received funding from the SNSF Sinergia project 'SMILE' (CRSII2 160811), the European Union's Horizon2020 research and innovation programme under grant agreement no. 762021 'Content4All' and the EPSRC project 'ExTOL' (EP/R03298X/1). This work reflects only the authors view and the Commission is not responsible for any use that may be made of the information it contains. We would also like to thank NVIDIA Corporation for their GPU grant.

References

1. Ahn, H., Ha, T., Choi, Y., Yoo, H., Oh, S.: Text2Action: generative adversarial synthesis from Language to action. In: International Conference on Robotics and Automation (ICRA) (2018)
2. Ba, J.L., Kiros, J.R., Hinton, G.E.: Layer Normalization. arXiv preprint arXiv:1607.06450 (2016)
3. Bahdanau, D., Cho, K., Bengio, Y.: Neural machine translation by jointly learning to align and translate. arXiv:1409.0473 (2014)
4. Bauer, B., Hienz, H., Kraiss, K.F.: Video-based continuous sign language recognition using statistical methods. In: Proceedings of 15th International Conference on Pattern Recognition (ICPR) (2000)
5. Berndt, D.J., Clifford, J.: Using dynamic time warping to find patterns in time series. In: AAA1 Workshop on Knowledge Discovery in Databases (KDD) (1994)
6. Camgoz, N.C., Hadfield, S., Koller, O., Bowden, R.: SubUNets: end-to-end hand shape and continuous sign language recognition. In: Proceedings of the IEEE International Conference on Computer Vision (ICCV) (2017)

702   S. Ben et al.

7. Camgoz, N.C., Hadfield, S., Koller, O., Ney, H., Bowden, R.: Neural sign language translation. In: Proceedings of the IEEE Conference on Computer Vision and Pattern Recognition (CVPR) (2018)
8. Camgoz, N.C., Koller, O., Hadfield, S., Bowden, R.: Sign language transformers: joint end-to-end sign language recognition and translation. In: Proceedings of the IEEE Conference on Computer Vision and Pattern Recognition (CVPR) (2020)
9. Cao, Z., Hidalgo, G., Simon, T., Wei, S.E., Sheikh, Y.: OpenPose: realtime multi-person 2D pose estimation using part affinity fields. In: Proceedings of the IEEE Conference on Computer Vision and Pattern Recognition (CVPR) (2017)
10. Cho, K., van Merriënboer, B., Bahdanau, D., Bengio, Y.: On the properties of neural machine translation: encoder-decoder approaches. In: Proceedings of the Syntax, Semantics and Structure in Statistical Translation (SSST) (2014)
11. Cooper, H., Ong, E.J., Pugeault, N., Bowden, R.: Sign Language Recognition using Sub-units. J. Mach. Learn. Res. (JMLR) **13**, 2205–2231 (2012)
12. Cui, R., Liu, H., Zhang, C.: Recurrent convolutional neural networks for continuous sign language recognition by staged optimization. In: Proceedings of the IEEE Conference on Computer Vision and Pattern Recognition (CVPR) (2017)
13. Dai, Z., Yang, Z., Yang, Y., Carbonell, J., Le, Q.V., Salakhutdinov, R.: Transformer-XL: attentive language models beyond a fixed-length context. In: International Conference on Learning Representations (ICLR) (2019)
14. Devlin, J., Chang, M.W., Lee, K., Toutanova, K.: BERT: pre-training of deep bidirectional transformers for language understanding. In: Proceedings of the Conference of the North American Chapter of the Association for Computational Linguistics (ACL) (2018)
15. Duarte, A.C.: Cross-modal neural sign language translation. In: Proceedings of the ACM International Conference on Multimedia (ICME) (2019)
16. Ebling, S., et al.: SMILE: swiss German sign language dataset. In: Proceedings of the International Conference on Language Resources and Evaluation (LREC) (2018)
17. Forster, J., Schmidt, C., Koller, O., Bellgardt, M., Ney, H.: Extensions of the sign language recognition and translation corpus RWTH-PHOENIX-weather. In: Proceedings of the International Conference on Language Resources and Evaluation (LREC) (2014)
18. Ginosar, S., Bar, A., Kohavi, G., Chan, C., Owens, A., Malik, J.: Learning individual styles of conversational gesture. In: Proceedings of the IEEE Conference on Computer Vision and Pattern Recognition (CVPR) (2019)
19. Girdhar, R., Carreira, J., Doersch, C., Zisserman, A.: Video action transformer network. In: Proceedings of the IEEE Conference on Computer Vision and Pattern Recognition (CVPR) (2019)
20. Glauert, J., Elliott, R., Cox, S., Tryggvason, J., Sheard, M.: VANESSA: a system for communication between deaf and hearing people. Technology and Disability (2006)
21. Glorot, X., Bengio, Y.: Understanding the difficulty of training deep feedforward neural networks. In: Proceedings of the International Conference on Artificial Intelligence and Statistics (AISTATS) (2010)
22. Graves, A.: Generating sequences with recurrent neural networks. arXiv preprint arXiv:1308.0850 (2013)
23. He, K., Zhang, X., Ren, S., Sun, J.: Deep residual learning for image recognition. In: Proceedings of the IEEE Conference on Computer Vision and Pattern Recognition (CVPR) (2016)

24. Huang, C.Z.A., et al.: Music transformer. In: International Conference on Learning Representations (ICLR) (2018)
25. Isola, P., Zhu, J.Y., Zhou, T., Efros, A.A.: Image-to-image translation with conditional adversarial networks. In: Proceedings of the IEEE Conference on Computer Vision and Pattern Recognition (CVPR) (2017)
26. Kalchbrenner, N., Blunsom, P.: Recurrent continuous translation models. In: Proceedings of the Conference on Empirical Methods in Natural Language Processing (EMNLP) (2013)
27. Karpouzis, K., Caridakis, G., Fotinea, S.E., Efthimiou, E.: Educational resources and implementation of a Greek sign language synthesis architecture. Comput. Educ. **49**(1), 54–74 (2007)
28. Kayahan, D., Güngör, T.: A hybrid translation system from Turkish spoken language to Turkish sign language. In: IEEE International Symposium on Innovations in Intelligent Systems and Applications (INISTA) (2019)
29. Kingma, D.P., Ba, J.: Adam: a method for stochastic optimization. In: Proceedings of the International Conference on Learning Representations (ICLR) (2014)
30. Kipp, M., Heloir, A., Nguyen, Q.: Sign language avatars: animation and comprehensibility. In: International Workshop on Intelligent Virtual Agents (IVA) (2011)
31. Ko, S.K., Kim, C.J., Jung, H., Cho, C.: Neural sign language translation based on human keypoint estimation. Appl. Sci. **9**(13), 2683 (2019)
32. Koller, O., Camgoz, N.C., Bowden, R., Ney, H.: Weakly supervised learning with multi-stream CNN-LSTM-HMMs to discover sequential parallelism in sign language videos. IEEE Transactions on Pattern Analysis and Machine Intelligence (TPAMI) (2019)
33. Koller, O., Forster, J., Ney, H.: Continuous sign language recognition: towards large vocabulary statistical recognition systems handling multiple signers. Computer Vision and Image Understanding (CVIU) (2015)
34. Koller, O., Ney, H., Bowden, R.: Deep hand: how to train a cnn on 1 million hand images when your data is continuous and weakly labelled. In: Proceedings of the IEEE Conference on Computer Vision and Pattern Recognition (CVPR) (2016)
35. Koller, O., Zargaran, S., Ney, H.: Re-sign: re-aligned end-to-end sequence modelling with deep recurrent CNN-HMMs. In: Proceedings of the IEEE Conference on Computer Vision and Pattern Recognition (CVPR) (2017)
36. Koller, O., Zargaran, S., Ney, H., Bowden, R.: Deep sign: hybrid CNN-HMM for continuous sign language recognition. In: Proceedings of the British Machine Vision Conference (BMVC) (2016)
37. Kouremenos, D., Ntalianis, K.S., Siolas, G., Stafylopatis, A.: Statistical machine translation for Greek to Greek sign language using parallel corpora produced via rule-based machine translation. In: IEEE 31st International Conference on Tools with Artificial Intelligence (ICTAI) (2018)
38. Kreutzer, J., Bastings, J., Riezler, S.: Joey NMT: a minimalist NMT toolkit for novices. In: Proceedings of the Conference on Empirical Methods in Natural Language Processing (EMNLP) (2019)
39. Lee, H.Y., et al.: Dancing to music. In: Advances in Neural Information Processing Systems (NIPS) (2019)
40. Li, G., Zhu, L., Liu, P., Yang, Y.: Entangled transformer for image captioning. In: Proceedings of the IEEE International Conference on Computer Vision (CVPR) (2019)
41. Li, N., Liu, S., Liu, Y., Zhao, S., Liu, M.: Neural speech synthesis with transformer network. In: Proceedings of the AAAI Conference on Artificial Intelligence (2019)

42. McDonald, J., et al.: Automated technique for real-time production of lifelike animations of American sign language. Universal Access in the Information Society (UAIS) (2016)
43. Mikolov, T., Sutskever, I., Chen, K., Corrado, G.S., Dean, J.: Distributed representations of words and phrases and their compositionality. In: Advances in Neural Information Processing Systems (NIPS) (2013)
44. Mukherjee, S., Ghosh, S., Ghosh, S., Kumar, P., Roy, P.P.: Predicting video-frames using encoder-convlstm combination. In: IEEE International Conference on Acoustics, Speech and Signal Processing (ICASSP) (2019)
45. Orbay, A., Akarun, L.: Neural sign language translation by learning tokenization. arXiv preprint arXiv:2002.00479 (2020)
46. Özdemir, O., Camgöz, N.C., Akarun, L.: Isolated sign language recognition using improved dense trajectories. In: Proceedings of the Signal Processing and Communication Application Conference (SIU) (2016)
47. Parmar, N., et al.: Image transformer. In: International Conference on Machine Learning (ICML) (2018)
48. Paszke, A., et al.: Automatic differentiation in pyTorch. In: NIPS Autodiff Workshop (2017)
49. Plappert, M., Mandery, C., Asfour, T.: Learning a bidirectional mapping between human whole-body motion and natural language using deep recurrent neural networks. Rob. Auton. Syst. **109**, 13–26 (2018)
50. Ren, Y., et al.: Fastspeech: fast, robust and controllable text to speech. In: Advances in Neural Information Processing Systems (NIPS) (2019)
51. Salimans, T., Goodfellow, I., Zaremba, W., Cheung, V., Radford, A., Chen, X.: Improved techniques for training GANs. In: Advances in Neural Information Processing Systems (NIPS) (2016)
52. Starner, T., Pentland, A.: Real-time American sign language recognition from video using hidden markov models. In: Shah, M., Jain, R. (eds.) Motion-Based Recognition. Computational Imaging and Vision, vol. 9, pp. 227–243. Springer, Dordrecht (1997). https://doi.org/10.1007/978-94-015-8935-2_10
53. Stoll, S., Camgoz, N.C., Hadfield, S., Bowden, R.: Sign language production using neural machine translation and generative adversarial networks. In: Proceedings of the British Machine Vision Conference (BMVC) (2018)
54. Stoll, S., Camgoz, N.C., Hadfield, S., Bowden, R.: Text2Sign: towards sign language production using neural machine translation and generative adversarial networks. International Journal of Computer Vision (IJCV) (2020)
55. Sutskever, I., Vinyals, O., Le, Q.V.: Sequence to sequence learning with neural networks. In: Proceedings of the Advances in Neural Information Processing Systems (NIPS) (2014)
56. Süzgün, M., et al.: Hospisign: an interactive sign language platform for hearing impaired. J. Naval Sci. Eng. **11**(3), 75–92 (2015)
57. Tamura, S., Kawasaki, S.: Recognition of sign language motion images. Pattern Recogn. **21**(4), 343–353 (1988)
58. Vaswani, A., et al.: Attention is all you need. In: Advances in Neural Information Processing Systems (NIPS) (2017)
59. Vila, L.C., Escolano, C., Fonollosa, J.A., Costa-jussà, M.R.: End-to-end speech translation with the transformer. In: Advances in Speech and Language Technologies for Iberian Languages (IberSPEECH) (2018)
60. Vogler, C., Metaxas, D.: Parallel midden Markov models for American sign language recognition. In: Proceedings of the IEEE International Conference on Computer Vision (ICCV) (1999)

61. Xiao, Q., Qin, M., Yin, Y.: Skeleton-based Chinese sign language recognition and generation for bidirectional communication between deaf and hearing people. In: Neural Networks (2020)
62. Yin, K.: Sign Language translation with transformers. arXiv preprint arXiv:2004.00588 (2020)
63. Zelinka, J., Kanis, J.: Neural sign language synthesis: words are our glosses. In: The IEEE Winter Conference on Applications of Computer Vision (WACV) (2020)
64. Zelinka, J., Kanis, J., Salajka, P.: NN-based Czech sign language synthesis. In: International Conference on Speech and Computer (SPECOM) (2019)
65. Zhang, Z., Han, X., Liu, Z., Jiang, X., Sun, M., Liu, Q.: ERNIE: enhanced language representation with informative entities. In: 57th Annual Meeting of the Association for Computational Linguistics (ACL) (2019)
66. Zhou, L., Zhou, Y., Corso, J.J., Socher, R., Xiong, C.: End-to-end dense video captioning with masked transformer. In: Proceedings of the IEEE Conference on Computer Vision and Pattern Recognition (CVPR) (2018)
67. Zhu, J.Y., Park, T., Isola, P., Efros, A.A.: Unpaired image-to-image translation using cycle-consistent adversarial networks. In: Proceedings of the IEEE Conference on Computer Vision and Pattern Recognition (CVPR) (2017)

Mask TextSpotter v3: Segmentation Proposal Network for Robust Scene Text Spotting

Minghui Liao[1], Guan Pang[2], Jing Huang[2], Tal Hassner[2], and Xiang Bai[1]($\boxtimes$)

[1] Huazhong University of Science and Technology, Wuhan, China
{mhliao,xbai}@hust.edu.cn
[2] Facebook AI, Menlo Park, USA
{gpang,jinghuang,thassner}@fb.com

Abstract. Recent end-to-end trainable methods for scene text spotting, integrating detection and recognition, showed much progress. However, most of the current arbitrary-shape scene text spotters use region proposal networks (RPN) to produce proposals. RPN relies heavily on manually designed anchors and its proposals are represented with axis-aligned rectangles. The former presents difficulties in handling text instances of extreme aspect ratios or irregular shapes, and the latter often includes multiple neighboring instances into a single proposal, in cases of densely oriented text. To tackle these problems, we propose Mask TextSpotter v3, an end-to-end trainable scene text spotter that adopts a Segmentation Proposal Network (SPN) instead of an RPN. Our SPN is anchor-free and gives accurate representations of arbitrary-shape proposals. It is therefore superior to RPN in detecting text instances of extreme aspect ratios or irregular shapes. Furthermore, the accurate proposals produced by SPN allow masked RoI features to be used for decoupling neighboring text instances. As a result, our Mask TextSpotter v3 can handle text instances of extreme aspect ratios or irregular shapes, and its recognition accuracy won't be affected by nearby text or background noise. Specifically, we outperform state-of-the-art methods by **21.9%** on the Rotated ICDAR 2013 dataset (rotation robustness), **5.9%** on the Total-Text dataset (shape robustness), and achieve state-of-the-art performance on the MSRA-TD500 dataset (aspect ratio robustness). Code is available at: https://github.com/MhLiao/MaskTextSpotterV3

Keywords: Scene text · Detection · Recognition

M. Liao—Work done while an intern at Facebook.

Electronic supplementary material The online version of this chapter (https://doi.org/10.1007/978-3-030-58621-8_41) contains supplementary material, which is available to authorized users.

A. Vedaldi et al. (Eds.): ECCV 2020, LNCS 12356, pp. 706–722, 2020.
https://doi.org/10.1007/978-3-030-58621-8_41

1   Introduction

Reading text in the wild is of great importance, with abundant real-world applications, including Photo OCR [2], reading menus, and geo-location. Systems designed for this task generally consist of text detection and recognition components, where the goal of text detection is localizing the text instances with their bounding boxes whereas text recognition aims to recognize the detected text regions by converting them into a sequence of character labels. Scene text spotting/end-to-end recognition is a task that combines the two tasks, requiring both detection and recognition.

Proposals   RoI Features   Results   Proposals   RoI Features   Results

(a) RPN-based   (b) SPN-based

Fig. 1. Comparisons between RPN and SPN. Left: the state-of-the-art, RPN-based text spotter (Mask TextSpotter v2 [21]); Right: our SPN-based text spotter (Mask TextSpotter v3). Although RPN proposals are localized well with the axis-aligned rectangles, its RoI features contain multiple text instances, resulting in inaccurate detection/recognition. By comparison, the proposals of our SPN are more accurate, thereby producing only a single text instance for each RoI feature and leading to accurate detection/recognition results. RoIs are shown with image regions

The challenges of scene text reading mainly lie in the varying orientations, extreme aspect ratios, and diverse shapes of scene text instances, which bring difficulties to both text detection and recognition. Thus, *rotation robustness, aspect ratio robustness*, and *shape robustness* are necessary for accurate scene text spotters. Rotation robustness is important in scene text images, where text cannot be assumed to be well aligned with the image axes. Aspect ratio robustness is especially important for non-Latin scripts where the text is often organized in long text lines rather than words. Shape robustness is necessary for handling text of irregular shapes, which frequently appears in logos.

A recent popular trend is to perform scene text spotting by integrating both text detection and recognition into a unified model [3,20], as the two tasks are naturally closely related. Some such scene text spotters are designed to detect and recognize multi-oriented text instances, such as Liu et al. [27] and He et al. [15]. Mask TextSpotter v1 [30], Qin et al. [34], and Mask TextSpotter v2 [21] can further handle text instances of arbitrary shapes. Mask TextSpotter series adopt Region Proposal Network (RPN) [35] to generate proposals and extract RoI features of the proposals for detection and recognition. Qin et al. [34] directly apply Mask R-CNN [11] for detection, which also uses RPN to produce proposals. These methods made great progress towards rotation robustness and shape robustness. The architectures of these methods, however, were not designed to be fully robust to rotations, aspect ratios, and shapes. Although these methods can deal with the scattered text instances of various orientations and diverse shapes, they can fail on densely oriented text instances or text lines of extreme aspect ratios due to the limitations of RPN.

The limitations of RPN mainly lie in two aspects: (1) The manually pre-designed anchors are defined using axis-aligned rectangles which cannot easily match text instances of extreme aspect ratios. (2) The generated axis-aligned rectangular proposals can contain multiple neighboring text instances when text instances are densely positioned. As evident in Fig. 1, the proposals produced by Mask TextSpotter v2 [21] are overlapped with each other and its RoI features therefore include multiple neighboring text instances, causing errors for detection and recognition. As shown in Fig. 1, the errors can be one or several characters, which may not be embodied in the performance if a strong lexicon is given. Thus, the evaluation without lexicon or with a generic lexicon is more persuasive.

In this paper, we propose a Segmentation Proposal Network (SPN), designed to address the limitations of RPN-based methods. Our SPN is anchor-free and gives accurate polygonal representations of the proposals. Without restrictions by pre-designed anchors, SPN can handle text instances of extreme aspect ratios or irregular shapes. Its accurate proposals can then be fully utilized by applying our proposed hard RoI masking into the RoI features, which can suppress neighboring text instances or background noise. This is beneficial in cases of densely oriented or irregularly shaped texts, as shown in Fig. 1. Consequently, Mask TextSpotter v3 is proposed by adopting SPN into Mask TextSpotter v2.

Our experiments show that Mask TextSpotter v3 significantly improves robustness to rotations, aspect ratios, and shapes. On the Rotated ICDAR 2013 dataset where the images are rotated with various angles, our method surpasses the state-of-the-art on both detection and end-to-end recognition by more than **21.9%**. On the Total-Text dataset [4] containing text instances of various shapes, our method outperforms the state-of-the-art by **5.9%** on the end-to-end recognition task. Our method also achieves state-of-the-art performance on the MSRA-TD500 dataset [45] labeled with text lines of extreme aspect ratios, as well as the ICDAR 2015 dataset that includes many low-resolution small text instances with a generic lexicon. To summarize, our contributions are three-fold:

1. We describe **Segmentation Proposal Network (SPN)**, for an accurate representation of arbitrary-shape proposals. The anchor-free SPN overcomes the limitations of RPN in handling text of extreme aspect ratios or irregular shapes, and provides more accurate proposals to improve recognition robustness. To our knowledge, it is the first arbitrary-shape proposal generator for end-to-end trainable text spotting.
2. We propose **hard RoI masking** to apply polygonal proposals to RoI features, effectively suppressing background noise or neighboring text instances.
3. Our proposed **Mask TextSpotter v3** significantly improves robustness to rotations, aspect ratios, and shapes, beating/achieving state-of-the-art results on several challenging scene text benchmarks.

2   Related Work

Current text spotting methods can be roughly classified into two categories: (1) *two-stage scene text spotting* methods, whose detector and recognizer are trained separately; (2) *end-to-end trainable scene text spotting* methods, which integrate detection and recognition into a unified model.

Two-Stage Scene Text Spotting. Two-stage scene text spotting methods use two separate networks for detection and recognition. Wang et al. [41] tried to detect and classify characters with CNNs. Jaderberg et al. [17] proposed a scene text spotting method consisting of a proposal generation module, a random forest classifier to filter proposals, a CNN-based regression module for refining the proposals, and a CNN-based word classifier for recognition. TextBoxes [23] and TextBoxes++ [22] combined their proposed scene text detectors with CRNN [37] and re-calculated the confidence score by integrating the detection confidence and the recognition confidence. Zhan et al. [46] proposed to apply multi-modal spatial learning into the scene text detection and recognition system.

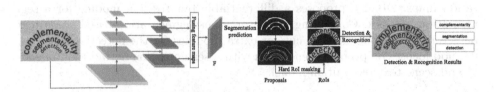

Fig. 2. Overview of Mask TextSpotter v3. "F": fused feature map for segmentation. We use the original image regions to represent RoIs for better visualization

End-to-End Trainable Scene Text Spotting. Recently, end-to-end trainable scene text spotting methods have dominated this area, benefiting from the complementarity of text detection and recognition. Li et al. [20] integrated a horizontal text detector and a sequence-to-sequence text recognizer into a unified network. Meanwhile, Bušta et al. [3] used a similar architecture while its detector can deal with multi-oriented text instances. After that, Liu et al. [27] and He et al. [15] further improved performance by adopting better detection and recognition methods, respectively.

Mask TextSpotter v1 [30] is the first end-to-end trainable arbitrary-shape scene text spotter, consisting of a detection module based on Mask R-CNN [11] and a character segmentation module for recognition. Following Mask TextSpotter v1 [30], several arbitrary-shape scene text spotters appeared concurrently. Mask TextSpotter v2 [21] further extends Mask TextSpotter v1 by applying a spatial attentional module for recognition, which alleviated the problem of character-level annotations and improved the performance significantly. Qin et al. [34] also combine a Mask R-CNN detector and an attention-based recognizer to deal with arbitrary-shape text instances. Xing et al. [43] propose to simultaneously detect/recognize the characters and the text instances, using the text instance detection results to group the characters. TextDragon [7] detects and recognizes text instances by grouping and decoding a series of local regions along with their centerline.

Qin et al. [34] use the mask maps from a Mask R-CNN detector to perform RoI masking on the RoI features, which is beneficial to recognition. However, the detector that adopts RPN to produce proposals may produce inaccurate mask maps, causing further recognition errors. Different from Qin et al. [34], our Mask TextSpotter v3 obtains accurate proposals and applies our hard RoI masking on the RoI features for both detection and recognition modules. Thus, it can detect and recognize densely oriented/curved text instances accurately.

Segmentation-Based Scene Text Detectors. Zhang et al [47] first use FCN to obtain the salient map of the text region, then estimate the text line hypotheses by combining the salient map and character components (using MSER). Finally, another FCN predicts the centroid of each character to remove the false hypotheses. He et al. [13] propose Cascaded Convolutional Text Networks (CCTN) for text center lines and text regions. PSENet [42] adopts a progressive scale expansion algorithm to get the bounding boxes from multi-scale segmentation maps. DB [24] proposes a differentiable binarization module for a segmentation network. Comparing to the previous segmentation-based scene text detectors that adopt multiple cues or extra modules for the detection task, our method focuses on proposal generation with a segmentation network for an end-to-end scene text recognition model.

3   Methodology

Mask TextSpotter v3 consists of a ResNet-50 [12] backbone, a Segmentation Proposal Network (SPN) for proposal generation, a Fast R-CNN module [8] for refining proposals, a text instance segmentation module for accurate detection, a character segmentation module and a spatial attentional module for recognition. The pipeline of Mask TextSpotter v3 is illustrated in Fig. 2. It provides polygonal representations for the proposals and eliminates added noise for the RoI features, thus achieving accurate detection and recognition results.

3.1   Segmentation Proposal Network

As shown in Fig. 2, our proposed SPN adopts a U-Net [36] structure to make it robust to scales. Unlike the FPN-based RPN [26,35], which produces proposals of different scales from multiple stages, SPN generates proposals from segmentation masks, predicted from a fused feature map F that concatenates feature maps of various receptive fields. F is of size $\frac{H}{4} \times \frac{W}{4}$, where H and W are the height and width of the input image respectively. The configuration of the segmentation prediction module for F is shown in the supplementary. The predicted text segmentation map S is of size $1 \times H \times W$, whose values are in the range of $[0, 1]$.

Segmentation Label Generation. To separate the neighboring text instances, it is common for segmentation-based scene text detectors to shrink the text regions [42, 49]. Inspired by Wang et al. [42] and DB [24], we adopt the Vatti clipping algorithm [39] to shrink the text regions by clipping d pixels. The offset pixels d can be determined as $d = A(1 - r^2)/L$, where A and L are the area and perimeter of the polygon that represents the text region, and r is the shrink ratio, which we empirically set to 0.4. An example of the label generation is shown in Fig. 3.

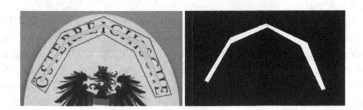

Fig. 3. Illustration of the segmentation label generation. Left: Red and green polygons are the original annotation and the shrunk region respectively. Right: segmentation label; black and white indicate the values of 0 and 1 respectively (Color figure online)

Proposal Generation. Given a text segmentation map, S, whose values are in the range of $[0, 1]$, we first binarize S into a binary map B:

$$B_{i,j} = \begin{cases} 1 & \text{if } S_{i,j} >= t, \\ 0 & \text{otherwise.} \end{cases} \tag{1}$$

Here, i and j are the indices of the segmentation or binary map and t is set to 0.5. Note that B is of the same size as S and the input image.

We then group the connected regions in the binary map B. These connected regions can be considered as shrunk text regions since the text segmentation labels are shrunk, as described above. Thus, we dilate them by un-clipping $\hat{d}$ pixels using the Vatti clipping algorithm, where $\hat{d}$ is calculated as $\hat{d} = \hat{A} \times \hat{r}/\hat{L}$. Here, $\hat{A}$ and $\hat{L}$ are the area and perimeter of the predicted shrunk text regions. $\hat{r}$ is set to 3.0 according to the value of the shrink ratio r.

As explained above, the proposals produced by SPN can be accurately represented as polygons, which are the contours of text regions. Thus, SPN generates suitable proposals for text lines with extreme aspect ratios and densely oriented/irregularly shaped text instances.

3.2 Hard RoI Masking

Since the custom RoI Align operator only supports the axis-aligned rectangular bounding boxes, we use the minimum, axis-aligned, rectangular bounding boxes of the polygon proposals to generate the RoI features to keep the RoI Align operator simple.

Qin et al. [34] proposed RoI masking which multiplies the mask probability map and the RoI feature, where the mask probability map is generated by a Mask R-CNN detection module. However, the mask probability maps may be inaccurate since they are predicted by the proposals from RPN. For example, it may contain multiple neighboring text instances for densely oriented text. In our case, accurate polygonal representations are designed for the proposals, thus we can directly apply the proposals to the RoI features through our proposed hard RoI masking.

Hard RoI masking multiplies binary polygon masks with the RoI features to suppress background noise or neighboring text instances, where a polygon mask M indicates an axis-aligned rectangular binary map with all 1 values in the polygon region and all 0 values outside the polygon region. Assuming that R_0 is the RoI feature and M is the polygon mask, which is of size 32×32, the masked RoI feature R can be calculated as $R = R_0 * M$, where $*$ indicates element-wise multiplication. M can be easily generated by filling the polygon proposal region with 1 while setting the values outside the polygon to 0. We report an ablation study on the hard RoI masking in Sect. 4.7, where we compare the proposed hard RoI masking with other operators including the RoI masking in Qin et al. [34].

After applying hard RoI masking, the background regions or neighboring text instances are suppressed in our masked RoI features, which significantly reduce the difficulties and errors in the detection and recognition modules.

3.3   Detection and Recognition

We follow the main design of its text detection and recognition modules of Mask TextSpotter v2 [21] for the following reasons: (1) Mask TextSpotter v2 is the current state-of-the-art with competitive detection and recognition modules. (2) Since Mask TextSpotter v2 is a representative method in the RPN-based scene text spotters, we can fairly compare our method with it to verify the effectiveness and robustness of our proposed SPN.

For detection, the masked RoI features generated by the hard RoI masking are fed into the Fast R-CNN module for further refining the localizations and the text instance segmentation module for precise segmentation. The character segmentation module and spatial attentional module are adopted for recognition.

3.4   Optimization

The loss function L is defined as below:

$$L = L_s + \alpha_1 L_{rcnn} + \alpha_2 L_{mask}. \tag{2}$$

L_{rcnn} and L_{mask} are defined in Fast R-CNN [8] and Mask TextSpotter v2 [21] respectively. L_{mask} consists of a text instance segmentation loss, a character segmentation loss, and a spatial attentional decoder loss. L_s indicates the SPN loss. Finally, following Mask TextSpotter v2 [21], we set α_1 and α_2 to 1.0.

We adopt dice loss [32] for SPN. Assuming that S and G are the segmentation map and the target map, the segmentation loss L_s can be calculated as:

$$I = \sum (S * G); \quad U = \sum S + \sum G; \quad L_s = 1 - \frac{2.0 \times I}{U}, \quad (3)$$

where I and U indicate the intersection and union of the two maps, and $*$ represents element-wise multiplication.

4 Experiments

We evaluate our method, testing robustness to four types of variations: rotations, aspect ratios, shapes, and small text instances, on different standard scene text benchmarks. We further provide an ablation study of our hard RoI masking.

4.1 Datasets

SynthText [9] is a synthetic dataset containing 800k text images. It provides annotations for word/character bounding boxes and text sequences.

Rotated ICDAR 2013 Dataset (RoIC13) is generated from the ICDAR 2013 dataset [19], whose images are focused around the text content of interest. The text instances are in the horizontal direction and labeled by axis-aligned rectangular boxes. Character-level segmentation annotations are given and so we can get character-level bounding boxes. The dataset contains 229 training and 233 testing images. To test rotation robustness, we create the Rotated ICDAR 2013 dataset by rotating the images and annotations in the test set of the ICDAR 2013 benchmark with some specific angles, including 15°, 30°, 45°, 60°, 75°, and 90°. Since all text instances in the ICDAR 2013 dataset are horizontally oriented, we can easily control the orientations of the text instances and find the relations between performances and text orientations. We use the evaluation protocols in the ICDAR 2015 dataset, because the ones in ICDAR 2013 only support axis-aligned bounding boxes.

MSRA-TD500 Dataset. [45] is a multi-language scene text detection benchmark that contains English and Chinese text, including 300 training images and 200 testing images. Text instances are annotated in the text-line level, thus there are many text instances of extreme aspect ratios. This dataset does not contain recognition annotations.

Total-Text Dataset. [4,5] includes 1,255 training and 300 testing images. It offers text instances of various shapes, including horizontal, oriented, and curved shapes, which are annotated with polygonal bounding boxes and transcriptions. Note that although character-level annotations are provided in the Total-Text dataset, we do not use them for fair comparisons with previous methods [21,31].

ICDAR 2015 Dataset (IC15). [18] consists of 1,000 training images and 500 testing images, which are annotated with quadrilateral bounding boxes. Most of the images are of low resolution and contain small text instances.

4.2   Implementation Details

For a fair comparison with Mask TextSpotter v2 [21], we use the same training data and training settings described below. Data augmentation follows the official implementation of Mask TextSpotter v2[1], including multi-scale training and pixel-level augmentations. Since our proposed SPN can deal with text instances of arbitrary shapes and orientations without conflicts, we adopt a more radical rotation data augmentation. The input images are randomly rotated with an angle range of $[-90°, 90°]$ while the original Mask TextSpotter v2 uses an angle range of $[-30°, 30°]$. Note that the Mask TextSpotter v2 is trained with the same rotation augmentation as ours for the experiments on the RoIC13 dataset.

The model is optimized using SGD with a weight decay of 0.001 and a momentum of 0.9. It is first pre-trained with SynthText and then fine-tuned with a mixture of SynthText, the ICDAR 2013 dataset, the ICDAR 2015 dataset, the SCUT dataset [48], and the Total-Text dataset for 250k iterations. The sampling ratio among these datasets is set to $2 : 2 : 2 : 1 : 1$ for each mini-batch of eight.

During pre-training, the learning rate is initialized with 0.01 and then decreased to a tenth at 100k iterations and 200k iterations respectively. During fine-tuning, we adopt the same training scheme while using 0.001 as the initial learning rate. We choose the model weights of 250k iterations for both pre-training and fine-tuning. In the inference period, the short sides of the input images are resized to 1000 on the RoIC13 dataset and the Total-Text dataset, 1440 on the IC15 dataset, keeping the aspect ratios.

Fig. 4. Qualitative results on the RoIC13 dataset. Top: Mask TextSpotter v2; Bottom: Mask TextSpotter v3. More results in the supplementary

---

[1] https://github.com/MhLiao/MaskTextSpotter.

Fig. 5. Qualitative results on the MSRA-TD500 dataset. Top: Mask TextSpotter v2; Bottom: Mask TextSpotter v3

Fig. 6. Qualitative results on the Total-Text dataset. Top: Mask TextSpotter v2; Bottom: Mask TextSpotter v3. The yellow text with red background are some inaccurate recognition results. Only inaccurate recognition results are visualized (Color figure online)

4.3 Rotation Robustness

We test for rotation robustness by conducting experiments on the RoIC13 dataset. We compare the proposed Mask TextSpotter v3 with two state-of-the-art methods Mask TextSpotter v2[3] and CharNet[2], with their official implementations. For a fair comparison, Mask TextSpotter v2 is trained with the same data and data augmentation as ours. Some qualitative comparisons on the RoIC13 dataset are shown in Fig. 4. We can see that Mask TextSpotter v2 fails on detecting and recognizing the densely oriented text instances while Mask TextSpotter v3 can successfully handle such cases.

We use the pre-trained model with a large backbone (Hourglass-88 [33]) for CharNet since the official implementation does not provide the ResNet-50 backbone. Note that the official pre-trained model of CharNet is trained with different training data. Thus, it is not suitable to directly compare the performance with Mask TextSpotter v3. However, we can observe the performance variations under

---

[2] https://github.com/MalongTech/research-charnet.

different rotation angles. The detection and end-to-end recognition performance of CharNet drop dramatically when the rotation angle is large.

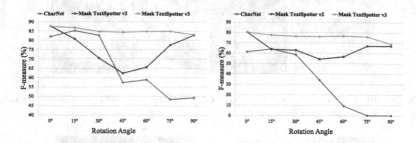

Fig. 7. Detection (left) and end-to-end recognition (right) results on the RoIC13 dataset with different rotation angles. The recognition results are evaluated without lexicon. Mask TextSpotter v2 is trained with the same rotation augmentation as Mask TextSpotter v3. CharNet is tested with the official released pre-trained model, with a backbone of Hourglass-88 [33]

Detection Task. As shown in Fig. 7, the detection performance of Mask TextSpotter v2 drops dramatically when the rotation angles are 30°, 45°, and 60°. In contrast, the detection results of Mask TextSpotter v3 are much more stable with various rotation angles. The maximum performance gap between Mask TextSpotter v3 and Mask TextSpotter v2 occurs when the rotation angle is 45°. As shown in Table 1, Mask TextSpotter v3 outperforms Mask TextSpotter v2 by **26.8%**, **18.0%**, and **22.0%** in terms of Precision, Recall, and F-measure, with a rotation angle of 45°. Note that it is reasonable that the two methods achieve almost the same results with 0° and 90°, since 0° indicates without rotation and the bounding boxes are also in the shape of axis-aligned rectangles when the rotation angle is 90°.

Table 1. Quantitative results on the RoIC13 dataset. The evaluation protocol is the same as the one in the IC15 dataset. The end-to-end recognition task is evaluated without lexicon. *CharNet is tested with the officially released pre-trained model; Mask TextSpotter v2 (MTS v2) is trained with the same rotation augmentation as Mask TextSpotter v3 (MTS v3). "P", "R", and "F" indicate precision, recall and F-measure. "E2E" is short for end-to-end recognition. More results are in the supplementary

Method	RoIC13 dataset (Rotation Angle: 45°)						RoIC13 dataset (Rotation Angle: 60°)					
	Detection			E2E			Detection			E2E		
	P	R	F	P	R	F	P	R	F	P	R	F
CharNet* [43]	57.8	56.6	57.2	34.2	33.5	33.9	65.5	53.3	58.8	10.3	8.4	9.3
MTS v2* [21]	64.8	59.9	62.2	66.4	45.8	54.2	70.5	61.2	65.5	68.2	48.3	56.6
MTS v3	**91.6**	**77.9**	**84.2**	**88.5**	**66.8**	**76.1**	**90.7**	**79.4**	**84.7**	**88.5**	**67.6**	**76.6**

End-to-End Recognition Task. The trend of the end-to-end recognition results is similar to the detection results, as shown in Fig. 7. The performance gaps between Mask TextSpotter v2 and Mask TextSpotter v3 are especially large when the rotation angles are 30°, 45°, and 60°. Mask TextSpotter v3 surpasses Mask TextSpotter v2 by more than **19.2%** in terms of F-measure with the rotation angle of 45° and 60°. The detailed results of 45° rotation angle are listed in Table 1, where Mask TextSpotter v3 achieves **22.1**, **21.0**, and **21.9** performance gain compared to the previous state-of-the-art method Mask TextSpotter v2.

The qualitative and quantitative results on the detection task and end-to-end recognition task prove the rotation robustness of Mask TextSpotter v3. The reason is the RPN used in Mask TextSpotter v2 would result in errors in both detection and recognition when dealing with densely oriented text instances. In contrast, the proposed SPN can generate accurate proposals and exclude the neighboring text instances by hard RoI masking in such cases. More qualitative and quantitative results are provided in the supplementary.

4.4 Aspect Ratio Robustness

Aspect ratio robustness is verified by our experimental results on the MSRA-TD500 dataset, which contains many text lines of extreme aspect ratios. Since there are no recognition annotations, we disable our recognition module and evaluate only on the detection task. Our qualitative and quantitative results are shown in Fig. 5 and Table 2.

Although Mask TextSpotter v2 is the existing state-of-the-art, end-to-end recognition method, it fails to detect long text lines due to the limitation of RPN. Compared with Mask TextSpotter v2, Mask TextSpotter v3 achieves a **9.3%** performance gain, which proves its superiority in handling text lines of extreme aspect ratios. Moreover, Mask TextSpotter v3 even outperforms state-of-the-art methods designed for text line detection [1,29,38], further showing its robustness to aspect ratio variations.

Table 2. Quantitative detection results on the MSRA-TD500 dataset

Method	P	R	F
He et al. [14]	71	61	69
DeepReg [16]	77	70	74
RRD [25]	87	73	79
PixelLink [6]	83.0	73.2	77.8
Xue et al. [44]	83.0	77.4	80.1
CRAFT [1]	88.2	78.2	82.9
Tian et al. [38]	84.2	**81.7**	82.9
MSR [38]	87.4	76.7	81.7
DB (without DCN) [24]	86.6	77.7	81.9
Mask TextSpotter v2 [21]	80.8	68.6	74.2
Mask TextSpotter v3	**90.7**	77.5	**83.5**

4.5 Shape Robustness

Robustness to shape variations is evaluated with end-to-end recognition performance on the Total-Text dataset, which contains text instances of various shapes, including horizontal, oriented, and curved shapes. Some qualitative results are shown in Fig. 6, where we can see that our method obtains more accurate detection and recognition results compared with Mask TextSpotter v2, especially on text instances with irregular shapes or with large spaces between neighboring characters. The quantitative results listed in Table 3 show that our method outperforms Mask TextSpotter v2 by **5.9%** in terms of F-measure when no lexicon is provided. Both the qualitative and quantitative results demonstrate the superior robustness to shape variations offered by our method.

Table 3. Quantitative end-to-end recognition results on the Total-Text dataset. "None" means recognition without any lexicon. "Full" lexicon contains all words in the test set. The values in the table are the F-measure. The evaluation protocols are the same as those in Mask TextSpotter v2

Method	None	Full
Mask TextSpotter v1 [30]	52.9	71.8
CharNet [43] Hourglass-57	63.6	–
Qin et al. [34] Inc-Res	63.9	–
Boundary TextSpotter [40]	65.0	76.1
ABCNet [28]	64.2	75.7
Mask TextSpotter v2 [21]	65.3	77.4
Mask TextSpotter v3	**71.2**	**78.4**

4.6 Small Text Instance Robustness

The challenges in the IC15 dataset mainly lie in the low-resolution and small text instances. As shown in Table 4, Mask TextSpotter v3 outperforms Mask TextSpotter v2 on all tasks with different lexicons, demonstrating the superiority of our method on handling small text instances in low-resolution images.

Although TextDragon [7] achieves better results on some tasks with the strong/weak lexicon, our method outperforms it by large margins, 7.1% and 9.0%, with the generic lexicon. We argue that there are no such strong/weak lexicons with only 100/1000+ words in most real-world applications, thus performance with a generic lexicon of 90k words is more meaningful and more challenging. Regardless, the reason for the different behaviors with different lexicons is that the attention-based recognizer in our method can learn the language knowledge while the CTC-based recognizer in TextDragon is more independent for the character prediction. Mask TextSpotter v3 relies less on the correction of the strong lexicon, which is also one of the advantages.

Table 4. Quantitative results on the IC15 dataset in terms of F-measure. "S", "W" and "G" mean recognition with strong, weak, and generic lexicon respectively. The values in the bracket (such as 1,600 and 1,400) indicate the short side of the input images. Note that in most real-world applications there are no such strong/weak lexicons with only 100/1000+ words. Thus, performance with the generic lexicon of 90k words is more meaningful

Method	Word Spotting			E2E Recognition			FPS
	S	W	G	S	W	G	
TextBoxes++ [22]	76.5	69.0	54.4	73.3	65.9	51.9	-
He *et al.* [15]	85.0	80.0	65.0	82.0	77.0	63.0	–
Mask TextSpotter v1 [30] (1600)	79.3	74.5	64.2	79.3	73.0	62.4	2.6
TextDragon [7]	**86.2**	**81.6**	68.0	82.5	**78.3**	65.2	2.6
CharNet [43] R-50	–	–	–	80.1	74.5	62.2	–
Boundary TextSpotter [40]	–	–	–	79.7	75.2	64.1	–
Mask TextSpotter v2 [21] (1600)	82.4	78.1	73.6	83.0	77.7	73.5	2.0
Mask TextSpotter v3 (1440)	83.1	79.1	**75.1**	**83.3**	78.1	**74.2**	2.5

4.7 Ablation Study

It is important to apply polygon-based proposals to the RoI features. There are two attributions for such an operator: "direct/indirect" and "soft/hard". "direct/indirect" means using the segmentation/binary map directly or through additional layers; "soft/hard" indicates a soft probability mask map whose values are from $[0, 1]$ or a binary polygon mask map whose values are 0 or 1. We conduct experiments on four types of combinations and the results show that our proposed hard RoI masking (Direct-hard) is simple yet achieves the best performance. Results and discussions are in the supplementary.

4.8 Limitations

Although Mask TextSpotter v3 is far more robust to rotated text variations than the existing state-of-the-art scene text spotters, it still suffers minor performance disturbance with some extreme rotation angles, e.g. 90°, as shown in Fig. 7, since it is hard for the recognizer to judge the direction of the text sequence. In the future, we plan to make the recognizer more robust to such rotations.

5 Conclusion

We propose Mask TextSpotter v3, an end-to-end trainable arbitrary-shape scene text spotter. It introduces SPN to generate proposals, represented with accurate polygons. Thanks to the more accurate proposals, Mask TextSpotter v3 is much more robust on detecting and recognizing text instances with rotations or irregular shapes than previous arbitrary-shape scene text spotters that use RPN for proposal generation. Our experiment results on the Rotated ICDAR 2013 dataset with different rotation angles, the MSRA-TD500 dataset with long text lines, and the Total-Text dataset with various text shapes demonstrate the robustness to rotations, aspect ratios, and shape variations of Mask TextSpotter v3. Moreover, results on the IC15 dataset show that the proposed Mask TextSpotter v3 is also robust in detecting and recognizing small text instances. We hope the proposed SPN could extend the application of OCR to other challenging domains [10] and offer insights to proposal generators used in other object detection/instance segmentation tasks.

References

1. Baek, Y., Lee, B., Han, D., Yun, S., Lee, H.: Character region awareness for text detection. In: Proceeding Conference Computer Vision Pattern Recognition, pp. 9365–9374 (2019)
2. Bissacco, A., Cummins, M., Netzer, Y., Neven, H.: PhotoOCR: reading text in uncontrolled conditions. In: Proceeding International Conference Computer Vision, pp. 785–792 (2013)
3. Busta, M., Neumann, L., Matas, J.: Deep textspotter: an end-to-end trainable scene text localization and recognition framework. In: Proceeding International Conference Computer Vision, pp. 2223–2231 (2017)
4. Ch'ng, C.K., Chan, C.S.: Total-text: a comprehensive dataset for scene text detection and recognition. In: Proceeding International Conference on Document Analysis and Recognition, pp. 935–942 (2017)
5. Ch'ng, C.-K., Chan, C.S., Liu, C.-L.: Total-Text: toward orientation robustness in scene text detection. Int. J. Doc. Anal. Recogn. (IJDAR) 23(1), 31–52 (2019). https://doi.org/10.1007/s10032-019-00334-z
6. Deng, D., Liu, H., Li, X., Cai, D.: Pixellink: detecting scene text via instance segmentation. In: AAAI Conference on Artificial Intelligence (2018)
7. Feng, W., He, W., Yin, F., Zhang, X.Y., Liu, C.L.: TextDragon: an end-to-end framework for arbitrary shaped text spotting. In: Proceeding International Conference Computer Vision (2019)
8. Girshick, R.B.: Fast R-CNN. In: Proceeding International Conference Computer Vision, pp. 1440–1448 (2015)
9. Gupta, A., Vedaldi, A., Zisserman, A.: Synthetic data for text localisation in natural images. In: Proceeding Conference on Computer Vision and Pattern Recognition (2016)
10. Hassner, T., Rehbein, M., Stokes, P.A., Wolf, L.: Computation and palaeography: potentials and limits. Dagstuhl Rep. 2(9), 184–199 (2012)
11. He, K., Gkioxari, G., Dollár, P., Girshick, R.: Mask r-cnn. In: Proceeding Conference Computer Vision Pattern Recognition, pp. 2961–2969 (2017)
12. He, K., Zhang, X., Ren, S., Sun, J.: Deep residual learning for image recognition. In: Proceeding Conference Computer Vision Pattern Recognition, pp. 770–778 (2016)
13. He, T., Huang, W., Qiao, Y., Yao, J.: Accurate text localization in natural image with cascaded convolutional text network. CoRR abs/1603.09423 (2016)
14. He, T., Huang, W., Qiao, Y., Yao, J.: Text-attentional convolutional neural network for scene text detection. Trans. Image Process. 25(6), 2529–2541 (2016)

15. He, T., Tian, Z., Huang, W., Shen, C., Qiao, Y., Sun, C.: An end-to-end textspotter with explicit alignment and attention. In: Proceeding Conference Computer Vision Pattern Recognition, pp. 5020–5029 (2018)
16. He, W., Zhang, X., Yin, F., Liu, C.: Deep direct regression for multi-oriented scene text detection. In: Proceeding Conference Computer Vision Pattern Recognition (2017)
17. Jaderberg, M., Simonyan, K., Vedaldi, A., Zisserman, A.: Reading text in the wild with convolutional neural networks. Int. J. Comput. Vision **116**(1), 1–20 (2016). https://doi.org/10.1007/s11263-015-0823-z
18. Karatzas, D., et al.: ICDAR 2015 competition on robust reading. In: Proceedings International Conference on Document Analysis and Recognition, pp. 1156–1160 (2015)
19. Karatzas, D., et al.: ICDAR 2013 robust reading competition. In: Proceedings of the International Conference on Document Analysis and Recognition, pp. 1484–1493 (2013)
20. Li, H., Wang, P., Shen, C.: Towards end-to-end text spotting with convolutional recurrent neural networks. In: Proceedings International Conference Computer Vision, pp. 5248–5256 (2017)
21. Liao, M., Lyu, P., He, M., Yao, C., Wu, W., Bai, X.: Mask TextSpotter: An end-to-end trainable neural network for spotting text with arbitrary shapes. Trans. Pattern Anal. Mach. Intell., 1–1 (2019)
22. Liao, M., Shi, B., Bai, X.: TextBoxes++: a single-shot oriented scene text detector. Trans. Image Processing **27**(8), 3676–3690 (2018)
23. Liao, M., Shi, B., Bai, X., Wang, X., Liu, W.: TextBoxes: A fast text detector with a single deep neural network. In: AAAI Conference on Artificial Intelligence (2017)
24. Liao, M., Wan, Z., Yao, C., Chen, K., Bai, X.: Real-time scene text detection with differentiable binarization. In: AAAI Conference on Artificial Intelligence, pp. 11474–11481 (2020)
25. Liao, M., Zhu, Z., Shi, B., Xia, G.S., Bai, X.: Rotation-sensitive regression for oriented scene text detection. In: Proceedings Conference Computer Vision Pattern Recognition, pp. 5909–5918 (2018)
26. Lin, T.Y., Dollár, P., Girshick, R., He, K., Hariharan, B., Belongie, S.: Feature pyramid networks for object detection. In: Proceeding Conference Computer Vision Pattern Recognition, pp. 2117–2125 (2017)
27. Liu, X., Liang, D., Yan, S., Chen, D., Qiao, Y., Yan, J.: FOTS: Fast oriented text spotting with a unified network. In: Proceeding Conference Computer Vision Pattern Recognition, pp. 5676–5685 (2018)
28. Liu, Y., Chen, H., Shen, C., He, T., Jin, L., Wang, L.: Abcnet: Real-time scene text spotting with adaptive bezier-curve network. In: Proceeding Conference Computer Vision Pattern Recognition, pp. 9809–9818 (2020)
29. Liu, Z., Lin, G., Yang, S., Feng, J., Lin, W., Goh, W.L.: Learning markov clustering networks for scene text detection. In: Proceeding Conference Computer Vision Pattern Recognition, pp. 6936–6944 (2018)
30. Lyu, P., Liao, M., Yao, C., Wu, W., Bai, X.: Mask TextSpotter: An end-to-end trainable neural network for spotting text with arbitrary shapes. In: European Conference Computer Vision, pp. 71–88 (2018)
31. Lyu, P., Yao, C., Wu, W., Yan, S., Bai, X.: Multi-oriented scene text detection via corner localization and region segmentation. In: Proceedings Conference Computer Vision Pattern Recognition, pp. 7553–7563 (2018)

32. Milletari, F., Navab, N., Ahmadi, S.A.: V-net: Fully convolutional neural networks for volumetric medical image segmentation. In: International Conference on 3D Vision, pp. 565–571 (2016)

33. Newell, A., Yang, K., Deng, J.: Stacked hourglass networks for human pose estimation. In: European Conference Computer Vision, pp. 483–499 (2016)

34. Qin, S., Bissacco, A., Raptis, M., Fujii, Y., Xiao, Y.: Towards unconstrained end-to-end text spotting. In: Proceedings of the International Conference Computer Vision (2019)

35. Ren, S., He, K., Girshick, R., Sun, J.: Faster r-cnn: Towards real-time object detection with region proposal networks. In: Neural Information Processing System, 91–99 (2015)

36. Ronneberger, O., Fischer, P., Brox, T.: U-net: Convolutional networks for biomedical image segmentation. In: Int. Conf. on Medical image computing and computer-assisted intervention. pp. 234–241. Springer (2015)

37. Shi, B., Bai, X., Yao, C.: An end-to-end trainable neural network for image-based sequence recognition and its application to scene text recognition. Trans. Pattern Anal. Mach. Intell. **39**(11), 2298–2304 (2017)

38. Tian, Z., Shu, M., Lyu, P., Li, R., Zhou, C., Shen, X., Jia, J.: Learning shape-aware embedding for scene text detection. In: Proceeding Conference Computer Vision Pattern Recognition, pp. 4234–4243 (2019)

39. Vatti, B.R.: A generic solution to polygon clipping. Commun. ACM **35**(7), 56–64 (1992)

40. Wang, H., et al.: All you need is boundary: toward arbitrary-shaped text spotting. In: AAAI Conference on Artificial Intelligence, pp. 12160–12167 (2020)

41. Wang, T., Wu, D.J., Coates, A., Ng, A.Y.: End-to-end text recognition with convolutional neural networks. In: International Conference Pattern Recognition (2012)

42. Wang, W., Xie, E., Li, X., Hou, W., Lu, T., Yu, G., Shao, S.: Shape robust text detection with progressive scale expansion network. In: Proceeding Conference Computer Vision Pattern Recognition, pp. 9336–9345 (2019)

43. Xing, L., Tian, Z., Huang, W., Scott, M.R.: Convolutional character networks. In: Proceeding Conference Computer Vision Pattern Recognition (2019)

44. Xue, C., Lu, S., Zhan, F.: Accurate scene text detection through border semantics awareness and bootstrapping. In: European Conference Computer Vision, pp. 355–372 (2018)

45. Yao, C., Bai, X., Liu, W., Ma, Y., Tu, Z.: Detecting texts of arbitrary orientations in natural images. In: Proceeding Conference Computer Vision Pattern Recognition (2012)

46. Zhan, F., Xue, C., Lu, S.: GA-DAN: geometry-aware domain adaptation network for scene text detection and recognition. In: Proceeding Conference Computer Vision Pattern Recognition (2019)

47. Zhang, Z., Zhang, C., Shen, W., Yao, C., Liu, W., Bai, X.: Multi-oriented text detection with fully convolutional networks. In: Proceeding Conference Computer Vision Pattern Recognition, pp. 4159–4167 (2016)

48. Zhong, Z., Jin, L., Zhang, S., Feng, Z.: DeepText: A unified framework for text proposal generation and text detection in natural images. CoRR abs/1605.07314 (2016)

49. Zhou, X., Yao, C., Wen, H., Wang, Y., Zhou, S., He, W., Liang, J.: EAST: an efficient and accurate scene text detector. In: Proceeding Conference Computer Vision Pattern Recognition, pp. 2642–2651 (2017)

Making Affine Correspondences Work in Camera Geometry Computation

Daniel Barath[1,2]($\boxtimes$), Michal Polic[3], Wolfgang Förstner[5], Torsten Sattler[3,4], Tomas Pajdla[3], and Zuzana Kukelova[2]

[1] Machine Perception Research Laboratory, SZTAKI in Budapest, Budapest, Hungary
barath.daniel@sztaki.mta.hu
[2] VRG, Faculty of Electrical Engineering, Czech Technical University in Prague, Prague, Czech Republic
[3] Czech Institute of Informatics, Robotics and Cybernetics, CTU in Prague, Prague, Czech Republic
[4] Chalmers University of Technology in Gothenburg, Gothenburg, Sweden
[5] Institute of Geodesy and Geoinformation, University of Bonn, Bonn, Germany

Abstract. Local features *e.g.* SIFT and its affine and learned variants provide region-to-region rather than point-to-point correspondences. This has recently been exploited to create new minimal solvers for classical problems such as homography, essential and fundamental matrix estimation. The main advantage of such solvers is that their sample size is smaller, *e.g.*, only two instead of four matches are required to estimate a homography. Works proposing such solvers often claim a significant improvement in run-time thanks to fewer RANSAC iterations. We show that this argument is not valid in practice if the solvers are used naively. To overcome this, we propose guidelines for effective use of region-to-region matches in the course of a full model estimation pipeline. We propose a method for refining the local feature geometries by symmetric intensity-based matching, combine uncertainty propagation inside RANSAC with preemptive model verification, show a general scheme for computing uncertainty of minimal solvers results, and adapt the sample cheirality check for homography estimation. Our experiments show that affine solvers can achieve accuracy comparable to point-based solvers at faster run-times when following our guidelines. We make code available at https://github.com/danini/affine-correspondences-for-camera-geometry.

1 Introduction

Estimating the geometric relationship between two images, such as homography or the epipolar geometry, is a fundamental step in computer vision approaches such as Structure-from-Motion [50], Multi-View Stereo [51], and SLAM [41].

Electronic supplementary material The online version of this chapter (https://doi.org/10.1007/978-3-030-58621-8_42) contains supplementary material, which is available to authorized users.

Traditionally, geometric relations are estimated from point correspondences (PCs) between the two images [27,42]. This ignores that correspondences are often rather established between image regions than between individual points to which the descriptors are finally often assigned. These regions, *i.e.* patches extracted around keypoints found by detectors such as DoG [31] or MSER [34], are oriented and have a specified size. Thus, they provide an affine transformation mapping feature regions to each other for each match [31,44].

Many works [3–5,8,16,24,25,29,43,45,48] used this additional information provided by affine correspondences (ACs), *i.e.*, region-to-region matches, to design minimal solvers for camera geometry estimation. As each correspondence carries more information, such solvers require fewer matches than their traditional point-based counterparts. For example, only three affine correspondences are required to estimate the fundamental matrix [8] compared to seven point correspondences [27]. This increases the probability of drawing an all-inlier sample, thus decreasing the required number of RANSAC [17] iterations. Also, ACs are known to be more robust against view changes than point correspondences [38].

In terms of noise on the measurements, affine solvers are affected differently than their point-based counterparts. If the points are well-spread in the images, the amount of noise is small compared to the distances between the points. In this case, the influence of the noise on the solution computed by a minimal solver is relatively small. In contrast, the comparatively small regions around the keypoints that define the affine features are much more affected by the same level of noise. As such, we observe that affine correspondence-based solvers are significantly less accurate than point correspondence-based ones if used naively. Yet, when explicitly modelling the impact of noise, we observe that affine solvers can achieve a similar level of accuracy as classical approaches while offering faster run-times. Based on our observations, we provide a practical guide for making affine correspondences work well in camera geometry computation.

Contribution. (1) We demonstrate how to use affine solvers to obtain accurate results at faster RANSAC run-times than achieved by pure point-based solvers. (2) We present strategies for all parts of the camera geometry estimation pipeline designed to improve the performance of affine solvers. This includes the refinement of the affinities defined by the features, rejection of samples based on cheirality checks, uncertainty propagation to detect and reject models that are too uncertain, and the importance of local optimization. (3) Through detailed experiments, we evaluate the impact of each strategy on the overall performance of affine solvers, both in terms of accuracy and run-time, showing that affine solvers can achieve a similar or higher accuracy than point-based approaches at faster run-times. These experiments validate our guidelines. (4) We make various technical contributions, such as a novel method for refining affine correspondence based on image intensity; a new minimal solver for fundamental matrix estimation; a strategy for combining the SPRT [11] test with uncertainty propagation for rejecting models early; the adaptation of the sample cheirality test, which is often used for point-based homography estimation, to affine features, and a general scheme for deriving covariance matrices for minimal solvers.

2 Related Work

Our guide to best use affine correspondences for camera geometry estimation problems analyzes the individual stages of the pipeline leading from matches to transformation estimates. The following reviews prior work for each stage.

Affine features are described by a point correspondence and a 2×2 linear transformation. For obtaining them, one can apply one of the traditional affine-covariant feature detectors, thoroughly surveyed in [37], such as MSER, Hessian-Affine, or Harris-Affine detectors. An alternative way to acquiring affine features is via view-synthesizing, as done, *e.g.*, by Affine-SIFT [40], and MODS [39] or by learning-based approaches, *e.g.*, Hes-Aff-Net [38], which obtains affine regions by running CNN-based shape regression on Hessian keypoints.

Matching Affine Regions. Given the noise in the parameters of the regions around affine matches, a natural approach to good estimation is to use high precision least squares matching (LSM) for refining affine correspondences [1,18,33]. Similar to template matching via cross correlation, a small patch from one image is located within a larger region in a second image. While arbitrary geometric and radiometric models are possible, practical approaches mostly consider *affine transformations*. As a maximum likelihood estimator, LSM provides the *covariance matrix of the estimated parameters*, the Cramer-Rao bound, reaching standard deviations for parallaxes down to below 1/100 pixel [26]. Intensity-based refinement has been used [16,48] for pose estimation. Yet, no analysis on the accuracy of the derived uncertainty is known and a symmetric formulation of the problem is missing so far. In this paper, we close this gap by providing both.

Affine solvers use ACs for geometric model estimation. Bentolila and Francos [8] proposed a method for estimating the epipolar geometry between two views using three ACs by interpreting the problem via conic constraints. Perdoch et al. [43] proposed two techniques for approximating the pose based on two and three matches by converting each AC to three PCs and applying standard estimation techniques. Raposo et al. [48] proposed a solution for essential matrix estimation from two ACs. Baráth et al. [3,5] showed that the relationship between ACs and epipolar geometry is linear and geometrically interpretable. Eichhardt et al. [16] proposed a method that uses two ACs for relative pose estimation based on general central-projective views. Similarly, [24,25] proposed minimal solutions for relative pose from a single affine correspondence when the camera is mounted to a moving vehicle. Homographies can also be estimated from two ACs as first shown by Köser [4,29]. Pritts et al. [45] used affine features for simultaneous estimation of affine image rectification and lens distortion.

Uncertainty analysis of image data provides several approaches useful for our task. Variances or covariance matrices are often used to model the *uncertainty of the input parameters*, *i.e.*, image intensities, coordinates of keypoints, and affine parameters. Assuming an ideal camera with a linear transfer function, the variance of the image noise increases linearly with the intensity [15,54]. In practice, the complexity of the internal camera processing requires an estimate of the variance function $\sigma_n^2(I)$ from the given images [19]. The accuracy of keypoint coordinates usually lies in the order of the rounding error, *i.e.*, $1/\sqrt{12} \approx 0.3$ pixels. We exploit here the uncertainty of the image intensities and keypoints for deriving realistic covariance matrices of the affine correspondences.

Algorithm 1. Robust model estimation pipeline with ACs

Require: I_1, I_2 – images
1:    $\mathcal{A} \leftarrow$ DetectACs(I_1, I_2)▷ Sect. 4.1, *default*: SIFT desc. [32], DoG shape adapt. [31]
2:    $\widehat{\mathcal{A}} \leftarrow$ RefineACs$(\mathcal{A})$ ▷ Sec. 3.2, *default*: symmetric LSM refinement
3:    $\theta^*, q^* \leftarrow 0, 0$ ▷ Best model and its quality
4:    **while** ¬Terminate **do**   ▷ Robust estimation, *default*: GC-RANSAC [6]
5:       $S \leftarrow$ Sample$(\widehat{\mathcal{A}})$ ▷ *default*: PROSAC sampler [10]
6:       **if** ¬ TestSample(S) **then**   ▷ Sample degeneracy and cheirality tests, Sec. 3.4
7:         continue
8:       $\theta \leftarrow$ ModelEstimation(S) ▷ *default*: F – Sec. 3.3, E – [5], H – [4]
9:       **if** ¬ TestModel(θ) **then**   ▷ *default*: tests from USAC [46]
10:      continue
11:      **if** ¬ Preemption(θ) **then**   ▷ Sec. 3.5, *default*: SPRT [11] + uncertainty test
12:        continue
13:      $q \leftarrow$ Validate$(\theta, \widehat{\mathcal{A}})$ ▷ Model quality calculation, *default*: MSAC score [55]
14:      **if** $q > q^*$ **then**
15:        $q^*, \theta^* \leftarrow q, \theta$
16:        $\theta' \leftarrow$ LocalOptimization$(\theta, \widehat{\mathcal{A}})$ ▷ *note*: only PCs are used from the ACs
17:        **if** ¬ TestModel(θ') **then**   ▷ *default*: tests from USAC [46]
18:          continue
19:        $q' \leftarrow$ Validate$(\theta', \widehat{\mathcal{A}})$▷ Model quality calculation, *default*: MSAC score [55]
20:        **if** $q' > q^*$ **then**
21:          $q^*, \theta^* \leftarrow q', \theta'$

Propagation of input uncertainty to model parameters through the estimation depends on the model being estimated. The uncertainty of a homography estimated from four or more points has been based on the SVD [13,47] and Lie groups [7]. The uncertainty of an estimated fundamental matrix has been also based on the SVD [53], but also on minimal representations [14]. Finally, the uncertainty for essential matrices has been derived using a minimal representation [21]. As far as we know, the propagation for affine solvers has not been presented before, and there was no general scheme for deriving covariance matrices for the solutions of minimal solvers. In this paper, we provide an efficient and general scheme and the uncertainty propagation for all minimal solvers used.

3   A Practical Guide to Using Affine Features

As argued in Sect. 1, and shown experimentally in Sect. 4.1, using ACs instead of point correspondences (PCs) leads to minimal solvers with smaller sample sizes but also to less accurate results. In the following, we analyze the individual stages of a classical matching pipeline and discuss how state-of-the-art results can be obtained with ACs. The pipeline is summarized in Algorithm 1.

3.1 Definition of Affine Correspondences

Affine correspondences are defined in this paper as

$$\text{AC}: \quad \{\boldsymbol{y}_0, \boldsymbol{z}_0, {}^{h}\text{A}\} \quad \text{with} \quad {}^{h}\text{A} = \begin{bmatrix} \text{A} & \boldsymbol{c} \\ \boldsymbol{0}^{\mathsf{T}} & 1 \end{bmatrix} \tag{1}$$

where the keypoint coordinates in the left and the right image are $\boldsymbol{y}_0$, and $\boldsymbol{z}_0$, and the local affinity is $\text{A} = \text{A}_2 \text{A}_1^{-1}$, *e.g.* derived from the affine frames A_1 and A_2 representing the affine shape of the underlying image region. The matching and refinement of the affine region correspondences, presented in the next section, refers to the homogeneous matrix ${}^{h}\text{A}$, specifically the translation vector $\boldsymbol{c}$, initially $\boldsymbol{0}$, and the affinity matrix A in (1).

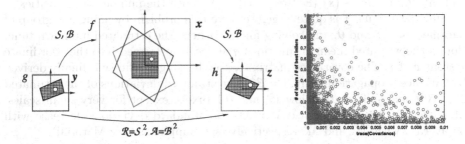

Fig. 1. Left: Relations between two image patches $g(\boldsymbol{y})$ (blue) and $h(\boldsymbol{h})$ (green) and the mean patch $f(\boldsymbol{x})$ (which is the black within the red region). The two patches g and h are related by geometric $\mathcal{B}$ and a radiometric $\mathcal{S}$ affinities. The correspondence is established by patch f, lying in the middle between g and h. We choose the maximum square (black), *i.e.* all the pixels in g and h which map into the black square of the reference image f. **Right**: The inlier ratio (vertical) of 100k homographies from real scenes as a function of the trace of their covariance matrices (horizontal). This shows that uncertain models (on right) generate small numbers of inliers. We use this to reject uncertain models.

3.2 Matching and Refining Affine Correspondences

For refining the ACs, we propose an intensity-based matching procedure which (i) is symmetric and provides (ii) a statistic for the coherence between the data and the model and (iii) a covariance matrix for the estimates of the parameters. Let the two image windows in the two images be $g(\boldsymbol{y})$ and $h(\boldsymbol{z})$. We assume, both windows are noisy observations of an unknown underlying signal $f(\boldsymbol{x})$, with individual geometric distortion, brightness, and contrast. We want to find the geometric distortion $\boldsymbol{z} = \mathcal{A}(\boldsymbol{y})$ and the radiometric distortion $h = \mathcal{R}(g) = pg + q$. Classical matching methods assume the geometric and radiometric distortion of one of the two windows is zero, *e.g.* $g(\boldsymbol{y}) = f(\boldsymbol{x})$, with $\boldsymbol{y} = \boldsymbol{x}$. We break this asymmetry by placing the unknown signal $f(\boldsymbol{x})$ in the middle between the observed signals g and h: $g(\boldsymbol{y}) \xrightarrow{\mathcal{S,B}} f(\boldsymbol{x}) \xrightarrow{\mathcal{S,B}} h(\boldsymbol{z})$ leading to $\mathcal{R} = \mathcal{S}^2$ and $\mathcal{A} = \mathcal{B}^2$. Assuming affinities for the geometric and radiometric distortions, the model is shown in Fig. 1(left).

The geometric and the radiometric models are

$$\boldsymbol{x} = \mathtt{B}\boldsymbol{y} + \boldsymbol{b}, \quad \boldsymbol{z} = \mathtt{B}\boldsymbol{x} + \boldsymbol{b} \quad \text{and} \quad f = sg + t, \quad h = sf + t. \tag{2}$$

In the following, we collect the eight unknown parameters of the two affinities in a single vector $\boldsymbol{\theta} = [b_{11}, b_{21}, b_{12}, b_{22}, b_1, b_2, s, t]^{\mathsf{T}}$.

Now, we assume the intensities g and h to be noisy with variances $\sigma_n(g)$ and $\sigma_m(h)$ depending on g and h. Hence, in an ML-approach, we minimize the weighted sum $\Omega(\boldsymbol{\theta}, f) = \sum_j n_j^2(\boldsymbol{\theta}, f)/\sigma_{n_j}^2 + \sum_k m_k^2(\boldsymbol{\theta}, f)/\sigma_{m_k}^2$ w.r.t. the unknown parameters $\boldsymbol{\theta}$ and f, where the residuals are $n_j(\boldsymbol{\theta}, f) = g_j - s^{-1}\left(f\left(\mathtt{B}\boldsymbol{y}_j + \boldsymbol{b}\right) - t\right)$ and $m_k(\boldsymbol{\theta}, f) = h_k - \left(sf\left(\mathtt{B}^{-1}(\boldsymbol{z}_k - \boldsymbol{b})\right) + t\right)$. Since the number of intensities in the unknown signal f is quite large, we solve this problem by fixing one group of parameters of f and $\boldsymbol{\theta}$, and solving for the other. The estimated unknown function is the weighted mean of functions g and h transformed into the coordinate system $\boldsymbol{x}$ of f. The covariance matrix $\widehat{\Sigma}_{\widehat{a}\widehat{a}}$ of the sought affinity is finally derived by variance propagation from $\mathcal{A} = \mathcal{B}^2$. The standard deviations of the estimated affinity $\widehat{\mathtt{A}}$ and shift $\widehat{\boldsymbol{c}}$ are below 1% and 0.1 pixels, except for very small scales. Moreover, for the window size $M \times M$, the standard deviations decrease with on average with M^2 and M, respectively (see Supplementary Material).

3.3   Solvers Using Affine Correspondences

In this paper we consider three important camera geometry problems: estimating planar homography, and two cases of estimating relative pose of two cameras: uncalibrated and calibrated cameras. We also include the semi-calibrated case, *i.e.* unknown focal length, in the supplementary material.

Homography from 1AC + 1PC: The problem of estimating a planar homography $\mathtt{H} \in \mathbb{R}^{3\times3}$ is well-studied with simple linear solutions from point and/or affine correspondences. The homography $\mathtt{H}$ has eight degrees of freedom. Since each PC gives two linear constraints on $\mathtt{H}$ and each AC gives six linear constraints on $\mathtt{H}$, the minimal number of correspondences necessary to estimate the unknown homography is either 4PC or 1AC+1PC. Both the well-known 4PC [27] and the 1AC+1PC [4,29] solvers are solving a system of eight linear equations in nine unknowns and are therefore equivalent in terms of efficiency.

Fundamental Matrix from 2AC + 1PC: The problem of estimating the relative pose of two uncalibrated cameras, *i.e.*, estimating the fundamental matrix $\mathtt{F} \in \mathbb{R}^{3\times3}$, has a well-known 7PC solver [27]. The fundamental matrix $\mathtt{F}$ has seven degrees of freedom, since it is a 3×3 singular matrix, *i.e.*, $\det(\mathtt{F}) = 0$. The well-known epipolar constraint gives one linear constraint on $\mathtt{F}$ for each PC. We propose a solver for estimating the unknown $\mathtt{F}$ using the linear constraints proposed in [3]. Here, we briefly describe this new 2AC+1PC solver.

Each AC gives three linear constraints on the epipolar geometry [3]. Therefore, the minimal number of correspondences necessary to estimate the unknown F is 2AC+1PC. The solver first uses seven linear constraints, $i.e.$, six from two ACs and one from a PC, rewritten in a matrix form as $\mathtt{M}\mathbf{f} = \mathbf{0}$, where $\mathbf{f} = \mathrm{vec}(\mathtt{F})$, to find a 2-dimensional null-space of the matrix $\mathtt{M}$. The unknown fundamental matrix is parameterized as $\mathtt{F} = x\mathtt{F}_1 + \mathtt{F}_2$, where $\mathtt{F}_1$ and $\mathtt{F}_2$ are matrices created from the 2-dimensional null-space of $\mathtt{M}$ and x is a new unknown. This parameterization is substituted into the constraint $\det(\mathtt{F}) = 0$, resulting in a polynomial of degree three in one unknown. The final 2AC+1PC solver is performing the same operations as the 7PC solver, $i.e.$, the computation of the null-space of a 7×9 matrix and finding the roots of a univariate polynomial of degree three.

Essential Matrix from 2ACs: The problem of estimating the relative pose of two calibrated cameras, $i.e.$, estimating the unknown essential matrix $\mathtt{E} \in \mathbb{R}^{3\times3}$, has five degrees of freedom (three for rotation and two for translation) and there exists the well-known 5PC solver [42]. This problem has recently been solved from two ACs [5,16]. Each AC gives three linear constraints on $\mathtt{E}$ [5]. Thus, two ACs provide more constraints than degrees of freedom. One approach to solve for $\mathtt{E}$ from two ACs is to use just five out of six constraints, which results in the same operations as the well-known 5PC solver [42] does. Another one is to solve an over-constrained system as suggested in [5]. In the experiments, we used the solver of [5] since it has lower computational complexity and similar accuracy.

3.4 Sample Rejection via Cheirality Checks

It is well-known for homography fitting that some minimal samples can be rejected without estimating the implied homography as they would lead to impossible configurations. Such a configuration occurs when the plane flips between the two views, $i.e.$, the second camera sees it from the back. This cheirality constraint is implemented in the most popular robust approaches, $e.g.$, USAC [46] and OpenCV's RANSAC. We thus adapt this constraint via a simple strategy by converting each AC to three PCs. Given affine correspondence $\{\boldsymbol{y}_0, \boldsymbol{z}_0, \mathtt{A}\}$, where $\boldsymbol{y}_0 = [y_{01}, y_{02}]^\mathsf{T}$ and $\boldsymbol{z}_0 = [z_{01}, z_{02}]^\mathsf{T}$ are the keypoint coordinates in the left, respectively the right images, the generated point correspondences are $(\boldsymbol{y}_0 + [1,0]^\mathsf{T}, \boldsymbol{z}_0 + \mathtt{A}[1,0]^\mathsf{T})$ and $(\boldsymbol{y}_0 + [0,1]^\mathsf{T}, \boldsymbol{z}_0 + \mathtt{A}[0,1]^\mathsf{T})$. When estimating the homography using the 1AC+1PC solver, the affine matrix is converted to these point correspondences and the cheirality check is applied to the four PCs.

Note that any direct conversion of ACs to (non-colinear) PCs is theoretically incorrect since the AC is a local approximation of the underlying homography [4]. However, it is a sufficiently good approximation for the cheirality check.

3.5 Uncertainty-Based Model Rejection

Before evaluating the consensus of a model, it is reasonable to check its quality, especially to eliminate configurations close to a singularity, see [22]. We can use the covariance matrix $\Sigma_{\theta\theta}$ of each solution to decide on its further suitability. To do so, we propose a new general way of deriving the $\Sigma_{\theta\theta}$ for minimal problems.

All problems we address here are based on a set of constraints $g(y, \theta) = 0$ on some observations y, and parameters θ, $e.g.$ the F estimation constraint $g_i(y_i, \theta)$ is of the form $x_i^T F y_i = 0$, hence $(y, \theta) = ([x; y], \text{vec}(F))$. We want to use classical variance propagation for implicit functions [21], Sect. 2.7.5. If Σ_{yy} is given, the determination of $\Sigma_{\theta\theta}$ is based on linearizing g at a point (y, θ) and using the Jacobians $A = \partial g / \partial y$ and $B = \partial g / \partial \theta$ leading to covariance matrix $\Sigma_{\theta\theta} = B^{-1} A \Sigma_{yy} A^T B^{-T}$, provided B can be inverted. Using constraints g we derive Jacobians A, B algebraically. Further, given the kth solution of a minimal problem (a system of equations defining a minimal problem has, in general, more than one solution), $i.e.$ a pair (y, θ_k), we can compute $\Sigma_{\theta_k \theta_k}$ without needing to know how the problem was solved and how this specific solution has been selected.

However, the number of constraints in g in most of the minimal problems is smaller than the number of parameters θ, $e.g.$ 7 constraints vs. 9 elements of F. Then the matrix B cannot be inverted. We propose to append the minimum number of constraints $h(\theta) = 0$ (between the parameters only, $e.g.$ det(F) and $||F|| = 1$) such that the number of all constraints $(g; h)$ is identical to the number of parameters. This leads to a regular matrix B, except for critical configurations.

Such algebraic derivations should be checked to ensure the equivalence of the algebraic and numerical solution, best by Monte Carlo simulations. However, the empirically obtained covariance matrix $\widehat{\Sigma}_{\theta\theta}$ is regular and cannot be directly compared to the algebraically derived $\Sigma_{\theta\theta}$ if it is singular ($e.g.$ for $\theta = \text{vec}(F)$). We propose to project both matrices on the tangent space of the manifold of θ, leading to regular covariance matrices as follows: let $J(\Sigma)$ be an orthonormal base of the column space of some covariance matrix Σ, then $\Sigma_r = J^T \Sigma J$ is regular; $e.g.$ using $J = \text{null}(\text{null}(\Sigma)^T)$ where $\text{null}(\Sigma)$ is the nullspace of Σ. Hence, with $J = J(\Sigma_{\theta\theta})$ the two covariance matrices $\Sigma_{\theta\theta,r} = J^T \Sigma_{\theta\theta} J$ and $\widehat{\Sigma}_{\theta\theta,r} = J^T \widehat{\Sigma}_{\theta\theta} J$ are regular. They can be compared and an identity test can be performed checking the hypothesis $\mathbb{E}(\widehat{\Sigma}_{\theta\theta,r}) = \Sigma_{\theta\theta,r}$, see [21], p. 71. The used constraints and detailed discussion for all listed minimal problems are in the supplementary material.

The covariance matrix $\Sigma_{\theta\theta}$ can be used in following way. Since we do not have a reference configuration, we eliminate models where the condition number $c = \text{cond}(\Sigma_{\theta\theta})$ is too large, since configurations close to singularity show large condition numbers in practice. For reasons of speed, it is useful to compute an approximation for the condition number, $e.g.$ $c_s = \text{tr}(\Sigma_{\theta\theta}) \text{tr}(\Sigma_{\theta\theta}^{-1})$, if the inverse covariance matrix can be obtained efficiently, which is the case in our context since $\Sigma_{\theta\theta}^{-1} = B(A\Sigma_{yy}A^T)^{-1}B^T$.[1] A weaker measure is $\text{tr}(\Sigma_{\theta\theta})$, which is more efficient to calculate than the previous ones. It essentially measures the average variance of the parameters θ. Thus, it can identify configurations where parameters are very uncertain. We use this measure in the following for deriving a prior for preemptive model verification by the Sequential Probability Ratio Test [11].

---

[1] For $\Sigma = \text{Diag}([a,b])$, with $a > b$, the condition number is $c = a/b$, while the approximation is $c_s = (a + b)^2 / (ab)$, which for $a \gg b$ converges to the condition number.

We experimentally found, for each problem, the parameters of exponential (for points solvers) and log-normal (for affine solvers) distributions for the trace. These parameters are used to measure the likelihood of the model being too uncertain to lead to a large number of inliers. In our experiments, for the sake of simplicity, we model the trace values by normal distributions for all solvers. Thus, we used the a-priori determined parameters to initialize the mean and standard deviation from all tested image pairs. Finally, we get a probability for each model being acceptable or not. Note that the selection of the provably correct probabilistic kernel for a particular problem and scene is a direction for future research. However, it is not a crucial issue due to being used only for rejecting too uncertain models early to avoid unnecessary calculations.

As a final step, we feed the determined probability to the Sequential Probability Ratio Test (SPRT) [11,35] as a prior knowledge about the model to be verified. This is done by initializing the model probability, which is sequentially updated in SPRT, to the a priori estimated one.

3.6 Local Optimization

Minimal solvers do not take noise in their input measurements into account during the estimation process. However, noise affects the estimated models. As such, not every all-inlier sample leads to the best possible transformation [12]. As shown in [20,49], starting from the algebraic solution and performing only a single iteration of ML estimation is often sufficient to obtain a significantly better estimate. They show that this strategy approaches the optimal result with an error below 10%–40% of the parameters' standard deviations while only increasing the computation time by a factor of ∼2. A strongly recommended approach is thus to use local optimization [6,12,30] inside RANSAC: every time a new best model is found, ML-based refinement on its inliers is used to account for noise in the input parameters. While this adds a computational overhead, it can be shown that this overhead is small and is compensated by the observation that local optimization (LO) typically helps RANSAC to terminate early. Moreover LO is usually applied rarely [12]. As we show in Sect. 4.5, local optimization is crucial to obtain accurate geometry estimates when using ACs.

4 Experiments

In this section, different algorithmic choices are tested on homography, fundamental and essential matrix fitting problems to provide a pipeline which leads to results superior to point-based methods.

Experimental Setup. Tests for epipolar geometry estimation were performed on the benchmark of [9]. The used datasets are the TUM dataset [52] consisting of videos, of resolution 640×480, of indoor scenes. The KITTI dataset [23] consists of consecutive frames of a camera mounted to a vehicle. The images are of resolution 1226×370. Both KITTI and TUM have image pairs with short baselines. The Tanks and Temples dataset [28] provides images of real-world objects for image-based reconstruction and, thus, contains mostly wide-baseline pairs. The images are of sizes between 1080×1920 and 1080×2048. The benchmark provides 1 000 image pairs for each dataset with ground truth epipolar geometry. Homography estimation was tested on the scenes of the HPatches dataset [2]. RANSAC's inlier-outlier threshold is set to 1.0 px (F), 1.0 px (E) and 5 px (H).

When evaluating F and E matrices, we calculate the normalized symmetric geometry errors (NSGD). The symmetric geometry error (SGD) was proposed in [58]. It generates virtual correspondences using the ground-truth F and computes the epipolar distance to the estimated F. It then reverts their roles to compute symmetric distance. The SGD error (in pixels) causes comparability issues for images of different resolutions. Therefore, it is normalized into the range of $[0, 1]$ by regularizing by factor $f = \frac{1}{\sqrt{h^2 + w^2}}$, where h and w are the height and width of the image, respectively.

The error of the estimated homographies is measured by, first, projecting the first image to the second one and back to get the commonly visible area. Each pixel in the visible area is projected by the ground truth and estimated homographies and the error is calculated as the L_2 distance of the projected points. Finally, the error is averaged over all pixels of the visible area.

4.1 Matchers and Descriptors

To estimate ACs in real images, we applied the VLFeat library [56] since it is available for multiple programming languages and, thus, we considered it a practical choice. VLFeat provides several options either for the feature descriptor or the affine shape adaptation technique. To select the best-performing combination, first, we detected ACs using all of the possible combinations. Note that we excluded the multi-scale versions of Harris-Laplace and Hessian-Laplace [36] affine shape adaptations since they were computationally expensive. Correspondences are filtered by the standard second nearest neighbor ratio test [32]. Next, we estimated fundamental matrices using affine and point-based solvers and *vanilla* RANSAC [17]. Figure 2 reports the cumulative distribution function (CDF) of the NSGD errors calculated from the estimated fundamental matrices.

Curves showing affine solvers have circles as markers. We applied PC-based methods (crosses) considering only the locations of the correspondences and ignoring the affinities. The line style denotes the feature descriptor: straight line – SIFT [32], dotted – LIOP [57]. Affine shape adaption techniques (DoG, Hessian [37], Harris- and Hessian-Leplace [36]) are shown by color. Applying VLFeat with any affine shape adaptation increases the extraction time by $\approx 10\%$.

The first and most dominant trend visible from the plots is that methods exploiting ACs are significantly less accurate then point-based approaches when the naive approach is used: vanilla RANSAC. The SIFT descriptor [32] and DoG affine shape adaptation lead to the most accurate results. Consequently, we use this combination in the experiments. In the next sections we will show ways of making the affine solvers similarly or more accurate than point-based methods.

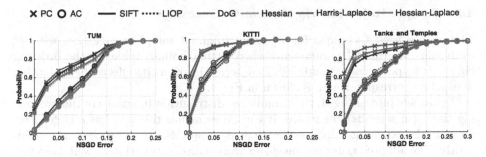

Fig. 2. Fundamental matrix estimation on datasets TUM, KITTI and Tanks and Temples (from benchmark [9]; 1000 image pairs each) using ACs detected by different descriptors and detectors. The CDFs of NSGD errors are shown. Vanilla RANSAC was applied followed by a LS fitting on all inliers.

4.2 Match Refinement

We demonstrate how the proposed refinement affects the accuracy of the affine matches by analysing the fulfillment of the constraints in the case of calibrated cameras. For each AC, the three constraints consist of the epipolar constraint $c_p = c_p^\mathsf{T} e = 0$ for the image coordinates, and the two constraints $c_a = C_a^\mathsf{T} e = 0$ for the affinity **A**. Assuming the pose, $i.\,e.$ the essential matrix, is known we determine a test statistics for the residuals c_p and c_a as $d_p = ||c_p||_{\sigma_{c_p}^2}$ and $d_a = ||c_a||_{\Sigma_{c_a c_a}}$. For the ACs of five image pairs, we used the LOWE keypoint coordinates and the scale and direction differences for deriving approximate affinities and refined them using the proposed LSM refinement technique (see Sect. 3.2).

Figure 3, left shows the CDF of the improvement caused by the proposed technique in the point coordinates (r_p) and in the affine parameters (r_a); a bigger value is better. The method improves both the affine parameters and point coordinates significantly. In Fig. 3, right the inconsistency with the epipolar constraints are shown; smaller values are better. The refined ACs are better, in terms of fulfilling the epipolar constraints, than the input ACs.

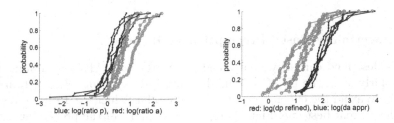

Fig. 3. Left: Improvement of the points (r_p) and affine parameters (r_a) after the proposed refinement, larger values are better. **Right:** Inconsistency with the epipolar constraints before (blue) and after (red) the refinement, smaller values are better. The CDFs are calculated from five images. (Color figure online)

4.3   Sample Rejection via Cheirality Checks

The widely used technique for rejecting minimal samples early (*i.e.* without estimating the model parameters) when fitting homographies is the ordering check of the point correspondences as described earlier. Its effect when adapting it to affine correspondences is shown in Fig. 4.

In the left plot of Fig. 4, the cumulative distribution functions of the processing times (in seconds) are shown. It can be seen that this test has a huge effect on point-based homography estimation as it speeds up the procedure significantly. The adapted criterion speeds up affine-based estimation as well, however, not that dramatically. Note, that affine-based estimation is already an order of magnitude faster than point-based methods and for AC-based homography estimation the cumulative distribution curve of the processing time is already very steep, Fig. 4 (left). This means that affine-based estimators do not perform many iterations and skipping model verification for even the half of the cases would not affect the time curve significantly. The avg. processing time of affine-based estimation is dropped from 8.6 to 7.7 ms. The right plot shows the $\log_{10}$ iteration numbers of the methods. The test does not affect the iteration number significantly. It sometimes leads to more iterations due to not checking samples of impossible configurations, however, this is expected and does not negatively affect the time.

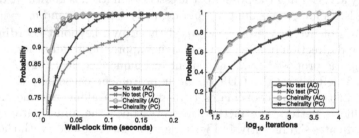

Fig. 4. Effect of cheirality test on point and affine-based homography estimators.

4.4   Uncertainty-Based Preemptive Verification

As it is described earlier, we combined the SPRT test [11,35] (parameters are set similarly as in the USAC [46] implementation) with model uncertainty calculation to avoid the expensive validation of models which are likely to be worse than the current best one. Figure 5 reports the CDFs of the processing time for (a) homography, (b) fundamental and (c) essential matrix fitting. Note that we excluded uncertainty-based verification for essential matrices since the solvers became too complex and, thus, the uncertainty calculation was slow for being applied to every estimated model.

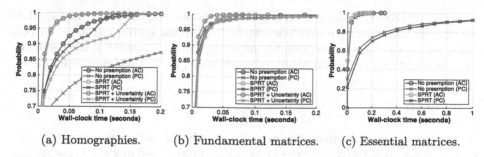

(a) Homographies. (b) Fundamental matrices. (c) Essential matrices.

Fig. 5. Evaluating pre-emptive model verification strategies for affine (AC) and point-based (PC) robust estimation. The CDFs of the processing times (in seconds) are shown. Being fast is interpreted as a curve close to the top-left corner.

It can clearly be seen that the proposed combination of the SPRT and the uncertainty check leads to the fastest robust estimation both for H and F fitting. For E estimation from ACs, using the SPRT test is also important, leading to faster termination. Most importantly, using affine correspondences, compared to point-based solvers, leads to a significant speed-up for all problems.

4.5 The Importance of Local Optimization

While the speed-up caused by using ACs has clearly been demonstrated, the other most important aspect is to get accurate results. As it is shown in Fig. 2, including AC-based solvers in *vanilla* RANSAC leads to significantly less accurate results than using PCs. A way of making the estimation by ACs accurate is to use a locally optimized RANSAC, where the initial model is estimated by an AC-based minimal solver and the local optimization performs the model polishing solely on the inlier PCs. We tested state-of-the-art local optimization techniques, *i.e.*, LO-RANSAC [12], LO'-RANSAC [30], GC-RANSAC [6]. The results are reported in Fig. 6. It can be seen that affine-based estimation

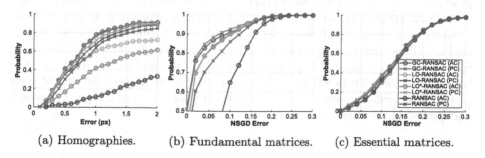

(a) Homographies. (b) Fundamental matrices. (c) Essential matrices.

Fig. 6. Evaluating local optimization techniques for affine (AC) and point-based (PC) robust model estimation. The CDFs of the geometric errors are shown. Being accurate is interpreted as a curve close to the top-left corner.

with GC-RANSAC is always among the top-performing methods. In (a), it is marginally more accurate than considering only point correspondences. In (b), using point correspondences is slighly more accurate. Compared to the results of *vanilla* RANSAC, the results of affine-based robust estimation improved notably.

5   Discussion

In summary of the investigated approaches, the best practices to accurately and efficiently use affine solvers are the following. Affinity-based model estimation has an accuracy similar or better to point-based solvers if (i) the detector and affine shape refining method is carefully selected; and (ii), most importantly, if a locally optimized RANSAC is applied to polish every new so-far-the-best model using *only* the point locations from the inlier correspondences. Consequently, affine features are used for estimating models from minimal samples, while their point counterparts are used to obtain accurate results. Efficiency is achieved by (iii) adapting strategies well-established for point correspondences, *e.g.*, cheirality check for homography estimation. (iv) Also, uncertainty-based model rejection and other preemptive verification techniques have a significant impact when speeding up the robust estimation procedure.

6   Conclusions

In this paper, we have considered the problem of using affine correspondences (ACs) for camera geometry estimation, *i.e.*, homography and epipolar geometry computation. Compared to classical approaches based on point correspondences (PCs), minimal solvers based on ACs offer the advantage of smaller sample sizes and, thus, the possibility to significantly accelerate RANSAC-based randomized robust estimation. However, noise has a larger negative impact on affine solvers as their input measurements typically originate from a smaller image region compared to point solvers. As we have shown, this significantly decreases the accuracy of the affine solvers. In this work, we have thus collected a set of "best practices", including novel contributions such as refining the local feature geometry and uncertainty-based model rejection techniques, for using ACs in practice. Through extensive experiments, we have shown that following our guidelines enables affine solvers to be used effectively, resulting in similar accuracy but faster run-times compared to point-based solvers. We believe that our guide will be valuable for both practitioners aiming to improve the performance of their pipelines as well as researchers working on ACs as it covers a topic previously unexplored in the literature.

Acknowledgements. This research was supported by project Exploring the Mathematical Foundations of Artificial Intelligence (2018-1.2.1-NKP-00008), the Research Center for Informatics project CZ.02.1.01/0.0/0.0/16 019/0000765, the MSMT LL1901 ERC-CZ grant, the Swedish Foundation for Strategic Research (Semantic Mapping and Visual Navigation for Smart Robots), the Chalmers AI Research Centre (CHAIR) (VisLo-cLearn), the European Regional Development Fund under IMPACT No. CZ.02.1.01/0.0/0.0/15 003/0000468, EU H2020 ARtwin No. 856994, and EU H2020 SPRING No. 871245 Projects.

References

1. Ackermann, F.: Digital image correlation: performance and potential application in photogrammetry. Photogrammetric Rec. **11**(64), 429–439 (1984)
2. Balntas, V., Lenc, K., Vedaldi, A., Mikolajczyk, K.: HPatches: a benchmark and evaluation of handcrafted and learned local descriptors. In: Proceedings of the IEEE Conference on Computer Vision and Pattern Recognition, pp. 5173–5182 (2017)
3. Baráth, D., Tóth, T., Hajder, L.: A minimal solution for two-view focal-length estimation using two affine correspondences. In: IEEE Conference on Computer Vision and Pattern Recognition (2017)
4. Barath, D., Hajder, L.: A theory of point-wise homography estimation. Pattern Recogn. Lett. **94**, 7–14 (2017)
5. Barath, D., Hajder, L.: Efficient recovery of essential matrix from two affine correspondences. IEEE Trans. Image Process. **27**(11), 5328–5337 (2018)
6. Barath, D., Matas, J.: Graph-cut RANSAC. In: Proceedings of the IEEE Conference on Computer Vision and Pattern Recognition, pp. 6733–6741 (2018)
7. Begelfor, E., Werman, M.: How to put probabilities on homographies. IEEE Trans. Pattern Anal. Mach. Intell. **27**(10), 1666–1670 (2005), http://dx.doi.org/10.1109/TPAMI.2005.200
8. Bentolila, J., Francos, J.M.: Conic epipolar constraints from affine correspondences. Comput. Vis. Image Understand. **122**, 105–114 (2014)
9. Bian, J.W., Wu, Y.H., Zhao, J., Liu, Y., Zhang, L., Cheng, M.M., Reid, I.: An evaluation of feature matchers forfundamental matrix estimation. arXiv preprint arXiv:1908.09474 (2019), https://jwbian.net/fm-bench
10. Chum, O., Matas, J.: Matching with PROSAC-progressive sample consensus. In: Proceedings of the IEEE Conference on Computer Vision and Pattern Recognition. vol. 1, pp. 220–226. IEEE (2005)
11. Chum, O., Matas, J.: Optimal randomized RANSAC. IEEE Trans. Pattern Anal. Mach. Intell. **30**(8), 1472–1482 (2008)
12. Chum, O., Matas, J., Kittler, J.: Locally optimized RANSAC. In: Michaelis, B., Krell, G. (eds.) DAGM 2003. LNCS, vol. 2781, pp. 236–243. Springer, Heidelberg (2003). https://doi.org/10.1007/978-3-540-45243-0_31
13. Criminisi, A.: Accurate Visual Metrology from Single and Multiple Uncalibrated Images. Springer, London (2001). https://doi.org/10.1007/978-0-85729-327-5
14. Csurka, G., Zeller, C., Zhang, Z., Faugeras, O.D.: Characterizing the uncertainty of the fundamental matrix. Comput. Vis. Image Understand. **68**(1), 18–36 (1997)
15. Dainty, J.C., Shaw, R.: Image Science. Academic Press (1974)
16. Eichhardt, I., Chetverikov, D.: Affine correspondences between central cameras for rapid relative pose estimation. In: Proceedings of the European Conference on Computer Vision, pp. 482–497 (2018)

17. Fischler, M., Bolles, R.: Random sampling consensus: a paradigm for model fitting with application to image analysis and automated cartography. Commun. ACM **24**, 381–395 (1981)
18. Förstner, W.: On the geometric precision of digital correlation. In: Hakkarainen, J., Kilpelä, E., Savolainen, A. (eds.) International Archives of Photogrammetry. vol. XXIV-3, pp. 176–189. ISPRS Symposium, Communication III, Helsinki, June 1982, http://www.ipb.uni-bonn.de/pdfs/Forstner1982Geometric.pdf
19. Förstner, W.: Image preprocessing for feature extraction in digital intensity, color and range images. In: Geomatic Methods for the Analysis of Data in Earth Sciences, Lecture Notes in Earth Sciences, vol. 95/2000, pp. 165–189. Springer, Heidelberg (2000), http://www.ipb.uni-bonn.de/pdfs/Forstner2000Image.pdf
20. Förstner, W., Khoshelham, K.: Efficient and accurate registration of point clouds with plane to plane correspondences. In: 3rd International Workshop on Recovering 6D Object Pose (2017), http://www.ipb.uni-bonn.de/pdfs/foerstner17efficient.pdf
21. Förstner, W., Wrobel, B.P.: Photogrammetric Computer Vision. Springer, Heidelberg (2016), http://www.ipb.uni-bonn.de/book-pcv/
22. Frahm, J.M., Pollefeys, M.: RANSAC for (quasi-)degenerate data (QDEGSAC). vol. 1, pp. 453–460 (2006)
23. Geiger, A., Lenz, P., Urtasun, R.: Are we ready for autonomous driving? The KITTI vision benchmark suite. In: 2012 IEEE Conference on Computer Vision and Pattern Recognition, pp. 3354–3361. IEEE (2012)
24. Guan, B., Zhao, J., Li, Z., Sun, F., Fraundorfer, F.: Minimal solutions for relative pose with a single affine correspondence (2020)
25. Hajder, L., Baráth, D.: Relative planar motion for vehicle-mounted cameras from a single affine correspondence. In: IEEE International Conference on Robotics and Automation (2020)
26. Haralick, R., Shapiro, L.G.: Computer and Robot Vision, vol. II. Addison-Wesley, Reading, MA (1992), http://www.ipb.uni-bonn.de/pdfs/Forstner1993Image.pdf
27. Hartley, R., Zisserman, A.: Multiple View Geometry in Computer Vision. Cambridge University Press (2003)
28. Knapitsch, A., Park, J., Zhou, Q.Y., Koltun, V.: Tanks and temples: benchmarking large-scale scene reconstruction. ACM Trans. Graph. **36**(4), 78 (2017)
29. Köser, K.: Geometric Estimation with Local Affine Frames and Free-form Surfaces. Shaker (2009)
30. Lebeda, K., Matas, J., Chum, O.: Fixing the locally optimized RANSAC. In: British Machine Vision Conference, pp. 1–11. Citeseer (2012)
31. Lowe, D.: Distinctive image features from scale-invariant keypoints. Int. J. Comput. Vis. **60**(2) (2004)
32. Lowe, D.G.: Object recognition from local scale-invariant features. In: International Conference on Computer Vision. IEEE (1999)
33. Lucas, B.D., Kanade, T.: An iterative image registration technique with an application to stereo vision. In: IJCAI 1981. pp. 674–679 (1981)
34. Matas, J., Chum, O., Urban, M., Pajdla, T.: Robust wide-baseline stereo from maximally stable extremal regions. Image Vis. Comput. **22**(10), 761–767 (2004), http://www.sciencedirect.com/science/article/pii/S0262885604000435
35. Matas, J., Chum, O.: Randomized RANSAC with sequential probability ratio test. In: Tenth IEEE International Conference on Computer Vision (ICCV 2005), vol. 1. vol. 2, pp. 1727–1732. IEEE (2005)
36. Mikolajczyk, K., Schmid, C.: Indexing based on scale invariant interest points. In: Proceedings Eighth IEEE International Conference on Computer Vision, vol. 1, pp. 525–531. IEEE (2001)

37. Mikolajczyk, K., Schmid, C.: An affine invariant interest point detector. In: Heyden, A., Sparr, G., Nielsen, M., Johansen, P. (eds.) ECCV 2002. LNCS, vol. 2350, pp. 128–142. Springer, Heidelberg (2002). https://doi.org/10.1007/3-540-47969-4_9
38. Mishkin, D., Radenovic, F., Matas, J.: Repeatability is not enough: learning affine regions via discriminability. In: Proceedings of the European Conference on Computer Vision, pp. 284–300 (2018)
39. Mishkin, D., Matas, J., Perdoch, M.: Mods: fast and robust method for two-view matching. Comput. Vis. Image Understand. **141**, 81–93 (2015)
40. Morel, J.M., Yu, G.: ASIFT: a new framework for fully affine invariant image comparison. SIAM J. Imag. Sci. **2**(2), 438–469 (2009)
41. Mur-Artal, R., Tardós, J.D.: ORB-SLAM2: an open-source SLAM system for monocular. Stereo and RGB-D Cameras. TRO **33**(5), 1255–1262 (2017)
42. Nistér, D.: An efficient solution to the five-point relative pose problem. In: IEEE TPAMI, pp. 756–770 (2004)
43. Perdoch, M., Matas, J., Chum, O.: Epipolar geometry from two correspondences. In: International Conference on Computer Vision (2006)
44. Philbin, J., Chum, O., Isard, M., Sivic, J., Zisserman, A.: Object retrieval with large vocabularies and fast spatial matching. In: Conference on Computer Vision and Pattern Recognition (2007)
45. Pritts, J.B., Kukelova, Z., Larsson, V., Lochman, Y., Chum, O.: Minimal solvers for rectifying from radially-distorted conjugate translations. In: IEEE Transactions on Pattern Analysis and Machine Intelligence (2020)
46. Raguram, R., Chum, O., Pollefeys, M., Matas, J., Frahm, J.M.: USAC: a universal framework for random sample consensus. IEEE Trans. Pattern Anal. Mach. Intell. **35**(8), 2022–2038 (2012)
47. Raguram, R., Frahm, J.M., Pollefeys, M.: Exploiting uncertainty in random sample consensus. In: International Conference on Computer Vision, pp. 2074–2081. IEEE (2009)
48. Raposo, C., Barreto, J.P.: Theory and practice of structure-from-motion using affine correspondences. In: IEEE Conference on Computer Vision and Pattern Recognition, pp. 5470–5478 (2016)
49. Schneider, J., Stachniss, C., Förstner, W.: On the quality and efficiency of approximate solutions to bundle adjustment with epipolar and trifocal constraints. In: ISPRS Annals of Photogrammetry, Remote Sensing and Spatial Information Sciences. vol. IV-2/W3, pp. 81–88 (2017). https://www.isprs-ann-photogramm-remote-sens-spatial-inf-sci.net/IV-2-W3/81/2017/isprs-annals-IV-2-W3-81-2017.pdf
50. Schönberger, J.L., Frahm, J.M.: Structure-from-motion revisited. In: CVPR, June 2016
51. Schönberger, J.L., Zheng, E., Pollefeys, M., Frahm, J.M.: Pixelwise view selection for unstructured multi-view stereo. In: European Conference on Computer Vision (ECCV) (2016)
52. Sturm, J., Engelhard, N., Endres, F., Burgard, W., Cremers, D.: A benchmark for the evaluation of RGB-D SLAM systems. In: 2012 IEEE/RSJ International Conference on Intelligent Robots and Systems, pp. 573–580. IEEE (2012)
53. Sur, F., Noury, N., Berger, M.O.: Computing the uncertainty of the 8 point algorithm for fundamental matrix estimation (2008)
54. Szeliski, R.: Computer Vision: Algorithms and Applications. Springer, London (2010). https://doi.org/10.1007/978-1-84882-935-0
55. Torr, P.H.S.: Bayesian model estimation and selection for epipolar geometry and generic manifold fitting. In: IJCV (2002)

56. Vedaldi, A., Fulkerson, B.: VLFeat: An open and portable library of computer vision algorithms. In: Proceedings of the 18th ACM International Conference on Multimedia, pp. 1469–1472 (2010)
57. Wang, Z., Fan, B., Wu, F.: Local intensity order pattern for feature description. In: 2011 International Conference on Computer Vision, pp. 603–610. IEEE (2011)
58. Zhang, Z.: Determining the epipolar geometry and its uncertainty: a review. Int. J. Comput. Vis. 27(2), 161–195 (1998)

Sub-center ArcFace: Boosting Face Recognition by Large-Scale Noisy Web Faces

Jiankang Deng[1]([✉]), Jia Guo[2], Tongliang Liu[3], Mingming Gong[4],
and Stefanos Zafeiriou[1]

[1] Imperial College, London, UK
{j.deng16,s.zafeiriou}@imperial.ac.uk
[2] InsightFace, London, UK
guojia@gmail.com
[3] University of Sydney, Sydney, Australia
tongliang.liu@sydney.edu.au
[4] University of Melbourne, Melbourne, Australia
mingming.gong@unimelb.edu.au

Abstract. Margin-based deep face recognition methods (e.g. Sphere-Face, CosFace, and ArcFace) have achieved remarkable success in unconstrained face recognition. However, these methods are susceptible to the massive label noise in the training data and thus require laborious human effort to clean the datasets. In this paper, we relax the intra-class constraint of ArcFace to improve the robustness to label noise. More specifically, we design K sub-centers for each class and the training sample only needs to be close to any of the K positive sub-centers instead of the only one positive center. The proposed sub-center ArcFace encourages one dominant sub-class that contains the majority of clean faces and non-dominant sub-classes that include hard or noisy faces. Extensive experiments confirm the robustness of sub-center ArcFace under massive real-world noise. After the model achieves enough discriminative power, we directly drop non-dominant sub-centers and high-confident noisy samples, which helps recapture intra-compactness, decrease the influence from noise, and achieve comparable performance compared to ArcFace trained on the manually cleaned dataset. By taking advantage of the large-scale raw web faces (Celeb500K), sub-center Arcface achieves state-of-the-art performance on IJB-B, IJB-C, MegaFace, and FRVT.

Keywords: Face recognition · Sub-class · Large-scale · Noisy data

---

J. Deng and J. Guo—Equal contributions.

Electronic supplementary material The online version of this chapter (https://doi.org/10.1007/978-3-030-58621-8_43) contains supplementary material, which is available to authorized users.

A. Vedaldi et al. (Eds.): ECCV 2020, LNCS 12356, pp. 741–757, 2020.
https://doi.org/10.1007/978-3-030-58621-8_43

1 Introduction

Face representation using Deep Convolutional Neural Network (DCNN) embedding with margin penalty [5,15,26,32] to simultaneously achieve intra-class compactness and inter-class discrepancy is the method of choice for state-of-the-art face recognition. To avoid the sampling problem in the Triplet loss [26], margin-based softmax methods [5,15,31,32] focused on incorporating margin penalty into a more feasible framework, the softmax loss, which has global sample-to-class comparisons within the multiplication step between the embedding feature and the linear transformation matrix. Naturally, each column of the linear transformation matrix is viewed as a class center representing a certain class [5].

Even though remarkable advances have been achieved by the margin-based softmax methods [5,8,15,31,32,39], they all need to be trained on well-annotated clean datasets [5,30], which require intensive human efforts. Wang et al. [30] found that faces with label noise significantly degenerate the recognition accuracy and manually built a high-quality dataset including 1.7M images of 59 K celebrities. However, it took 50 annotators to work continuously for one month to clean the dataset, which further demonstrates the difficulty of obtaining a large-scale clean dataset for face recognition.

Since accurate manual annotations can be expensive [30], learning with massive noisy data[1] has recently attracted much attention [4,11,14,33,41]. However, computing time-varying weights for samples [11] or designing piece-wise loss functions [41] based on the current model's predictions can only alleviate the influence from noisy data to some extent as the robustness and improvement depend on the initial performance of the model. Besides, the co-mining

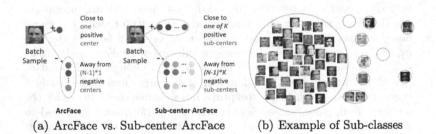

(a) ArcFace vs. Sub-center ArcFace   (b) Example of Sub-classes

Fig. 1. (a) Difference between ArcFace and the proposed sub-center ArcFace. In this paper, we introduce sub-class into ArcFace to relax the intra-class constraint, which can effectively improve robustness under noise. (b) The sub-classes of one identity from the CASIA dataset [40] after using the sub-center ArcFace loss ($K = 10$). Noisy samples and hard samples (e.g. profile and occluded faces) are automatically separated from the majority of clean samples.

---

[1] Generally, there are two types of label noise in face recognition [11,30,33,41]: one is open-set label noise, i.e., faces whose true labels are out of the training label set but are wrongly labeled to be within the set; and the other one is close-set label noise, i.e., faces whose true labels are in the training label set but are wrongly labeled.

method [33] requires to train twin networks together thus it is less practical for training large models on large-scale datasets.

As shown in Fig. 1(a), the objective of ArcFace [5] has two parts: (1) intra-class compactness: pushing the sample close to the corresponding positive center; and (2) inter-class discrepancy: pushing the sample away from all other negative centers. If a face is a noisy sample, it does not belong to the corresponding positive class. In ArcFace, this noisy sample generates a large wrong loss value, which impairs the model training. In this paper, we relax the intra-class constraint of forcing all samples close to the corresponding positive center by introducing sub-classes into ArcFace. For each class, we design K sub-centers and the training sample only needs to be close to any of the K positive sub-centers instead of the only one positive center. As illustrated in Fig. 1(b), the proposed sub-center ArcFace will encourage one dominant sub-class that contains the majority clean faces and multiple non-dominant sub-classes that include hard or noisy faces. This happens because the intra-class constraint of sub-center ArcFace enforces the training sample to be close to one of the multiple positive sub-classes but not all of them. The noise is likely to form a non-dominant sub-class and will not be enforced into the dominant sub-class. Therefore, sub-center ArcFace is more robust to noise. Extensive experimental results in this paper indicate that the proposed sub-center ArcFace is more robust than ArcFace [5] under massive real-world noises.

Although the proposed sub-center ArcFace can effectively separate clean data from noisy data. However, hard samples are also kept away. The existing of sub-centers can improve the robustness but also undermine the intra-class compactness, which is important for face recognition [34]. As the devil of face recognition is in the noise [30], we directly drop non-dominant sub-centers and high-confident noisy samples after the model achieves enough discriminative power. By pushing hard samples close to the dominant sub-center, we gradually recapture intra-class compactness and further improve the accuracy.

To summarise, our key contributions are as follows:

- We introduce sub-class into ArcFace to improve its robustness on noisy training data. The proposed sub-center ArcFace consistently outperforms ArcFace under massive real-world noises.
- By dropping non-dominant sub-centers and high-confident noisy samples, our method can achieve comparable performance compared to ArcFace trained on the manually cleaned dataset.
- Sub-center Arcface can be easily implemented by using the parallel toolkit and thus enjoys scalability to large-scale datasets. By taking advantage of the large-scale raw web faces (e.g. Celeb500K [1]), the proposed sub-center Arcface achieves state-of-the-art performance on IJB-B, IJB-C, MegaFace, and FRVT 1:1 Verification.

2   Related Work

Face Recognition with Margin Penalty. The pioneering work [26] uses the Triplet loss to exploit triplet data such that faces from the same class are closer than faces from different classes by a clear Euclidean distance margin. Even though the Triplet loss makes perfect sense for face recognition, the sample-to-sample comparisons are local within mini-batch and the training procedure for the Triplet loss is very challenging as there is a combinatorial explosion in the number of triplets especially for large-scale datasets, requiring effective sampling strategies to select informative mini-batch [25,26] and choose representative triplets within the mini-batch [21,28,36]. Some works tried to reduce the total number of triplets with proxies [19,23], i.e., sample-to-sample comparison is changed into sample-to-proxy comparison. However, sampling and proxy methods only optimise the embedding of partial classes instead of all classes in one iteration step.

Margin-based softmax methods [5,8,15,31,32] focused on incorporating margin penalty into a more feasible framework, softmax loss, which has extensive sample-to-class comparisons. Compared to deep metric learning methods (e.g., Triplet [26], Tuplet [21,28]), margin-based softmax methods conduct global comparisons at the cost of memory consumption on holding the center of each class. Sample-to-class comparison is more efficient and stable than sample-to-sample comparison as (1) the class number is much smaller than sample number, and (2) each class can be represented by a smoothed center vector which can be updated during training.

Face Recognition Under Noise. Most of the face recognition datasets [1,2, 9,40] are downloaded from the Internet by searching a pre-defined celebrity list, and the original labels are likely to be ambiguous and inaccurate [30]. Learning with massive noisy data has recently drawn much attention in face recognition [11,33,37,41] as accurate manual annotations can be expensive [30] or even unavailable.

Wu et al. [37] proposed a semantic bootstrap strategy, which re-labels the noisy samples according to the probabilities of the softmax function. However, automatic cleaning by the bootstrapping rule requires time-consuming iterations (e.g. twice refinement steps are used in [37]) and the labelling quality is affected by the capacity of the original model. Hu et al. [11] found that the cleanness possibility of a sample can be dynamically reflected by its position in the target logit distribution and presented a noise-tolerant end-to-end paradigm by employing the idea of weighting training samples. Zhong et al. [41] devised a noise-resistant loss by introducing a hypothetical training label, which is a convex combination of the original label with probability ρ and the predicted label by the current model with probability $1-\rho$. However, computing time-varying fusion weight [11] and designing piece-wise loss [41] contain many hand-designed hyper-parameters. Besides, re-weighting methods are susceptible to the performance of the initial model. Wang et al. [33] proposed a co-mining strategy which uses the loss values as the cue to simultaneously detect noisy labels, exchange the high-confidence clean faces to alleviate the error accumulation caused by the sampling bias, and re-weight the predicted clean faces to make them dominate the discriminative model training. However, the co-mining method requires training twin networks simultaneously and it is challenging to train large networks (e.g. ResNet100 [10]) on a large-scale dataset (e.g. MS1M [9] and Celeb500K [1]).

Face Recognition with Sub-classes. Practices and theories that lead to "sub-class" have been studied for a long time [42,43]. The concept of "sub-class" applied in face recognition was first introduced in [42,43], where a mixture of Gaussians was used to approximate the underlying distribution of each class. For instance, a person's face images may be frontal view or side view, resulting in different modalities when all images are represented in the same data space. In [42,43], experimental results showed that subclass divisions can be used to effectively adapt to different face modalities thus improve the performance of face recognition. Wan et al. [29] further proposed a separability criterion to divide every class into sub-classes, which have much less overlaps. The new within-class scatter can represent multi-modality information, therefore optimising this within-class scatter will separate different modalities more clearly and further increase the accuracy of face recognition. However, these work [29,42,43] only employed hand-designed feature descriptor on tiny under-controlled datasets.

Concurrent with our work, Softtriple [22] presents a multi-center softmax loss with class-wise regularizer. These multi-centers can capture the hidden distribution of the data better [20] due to the fact that they can capture the complex geometry of the original data and help reduce the intra-class variance. On the fine-grained visual retrieval problem, the Softtriple [22] loss achieves better performance than the softmax loss as capturing local clusters is essential for this task. Even though the concept of "sub-class" has been employed in face recognition [29,42,43] and fine-grained visual retrieval [22], none of these work has considered the large-scale (e.g. 0.5 million classes) face recognition problem under massive noise (e.g. around 50% noisy samples within the training data).

3 The Proposed Approach

3.1 ArcFace

ArcFace [5] introduced an additive angular margin penalty into the softmax loss,

$$\ell_{\mathrm{ArcFace}} = -\log \frac{e^{s\cos(\theta_{y_i}+m)}}{e^{s\cos(\theta_{y_i}+m)} + \sum_{j=1,j\neq y_i}^{N} e^{s\cos\theta_j}}, \tag{1}$$

where θ_j is the angle between the embedding feature $\mathbf{x}_i \in \mathbb{R}^{512\times1}$ of the i-th face sample and the j-th class center $W_j \in \mathbb{R}^{512\times1}$. Given that the corresponding class label of $\mathbf{x}_i$ is y_i, θ_{y_i} represents the angle between $\mathbf{x}_i$ and the ground-truth center W_{y_i}. $m = 0.5$ is the angular margin parameter, $s = 64$ is the feature re-scale parameter, and N is the total class number. As there is a ℓ_2 normalisation step on both $\mathbf{x}_i$ and W_j, $\theta_j = arccos\left(W_j^T\mathbf{x}_i\right)$.

Taking advantage of parallel acceleration on both $\mathbf{x}_i$ and W_j, the implementation of ArcFace[2] can efficiently handle million-level identities on a single server with 8 GPUs (11 GB 1080ti). This straightforward solution has changed the ingrained belief that large-scale global comparison with all classes is usually not attainable due to the bottleneck of GPU memory [26,28].

---

[2] https://github.com/deepinsight/insightface/tree/master/recognition.

3.2   Sub-center ArcFace

Even though ArcFace [5] has shown its power in efficient and effective face feature embedding, this method assumes that training data are clean [5,30]. However, this is not true especially when the dataset is in large scale. How to enable ArcFace to be robust to noise is one of the main challenges that impeding the development of face representation and recognition [30]. In this paper, we address this problem by proposing the idea of using sub-classes for each identity, which can be directly adopted by ArcFace and will significantly increase its robustness.

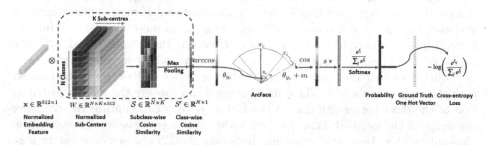

Fig. 2. Training the deep face recognition model by minimizing the proposed sub-center ArcFace loss. The main contribution in this paper is highlighted by the blue dashed box. Based on a ℓ_2 normalisation step on both embedding feature $\mathbf{x}_i \in \mathbb{R}^{512 \times 1}$ and all sub-centers $W \in \mathbb{R}^{N \times K \times 512}$, we get the subclass-wise similarity score $\mathcal{S} \in \mathbb{R}^{N \times K}$ by a matrix multiplication $W^T \mathbf{x}_i$. After a max pooling step, we can easily get the class-wise similarity score $\mathcal{S}' \in \mathbb{R}^{N \times 1}$. The following steps are same as ArcFace [5]. (Color figure online)

Table 1. The strictness and robustness analysis of different comparison strategies. In the angular space, "Min" is closest and "Max" is farest. "intra" refers to comparison between the training sample and the positive sub-centers (K). "inter" refers to comparison between the training sample and all negative sub-centers ($(N-1) \times K$). "outlier" denotes the open-set noise and "label flip" denotes the close-set noise.

Constraints	Sub-center?	Strictness?	Robustness to outlier?	Robustness to label flip?
(1) Min(inter) - Min(intra) $\geq$ m	$\checkmark$	+++	++	+
(2) Max(inter) - Min(intra) $\geq$ m	$\checkmark$	+	++	++
(3) Min(inter) - Max(intra) $\geq$ m		++++		
(4) Max(inter) - Max(intra) $\geq$ m		++		+

Foster Sub-classes. As illustrated in Fig. 2, we set a sufficiently large K for each identity. Based on a ℓ_2 normalisation step on both embedding feature $\mathbf{x}_i \in \mathbb{R}^{512 \times 1}$ and all sub-centers $W \in \mathbb{R}^{N \times K \times 512}$, we get the subclass-wise similarity

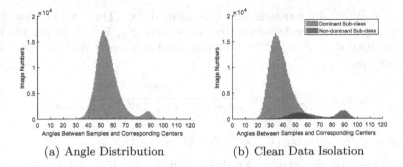

(a) Angle Distribution   (b) Clean Data Isolation

Fig. 3. (a) Angle distribution of samples to their corresponding centers predicted by the pre-trained ArcFace model [5]. Noise exists in the CASIA dataset [30,40]. (b) Angle distribution of samples from the dominant and non-dominant sub-classes. Clean data are automatically isolated by sub-center ArcFace (K=10).

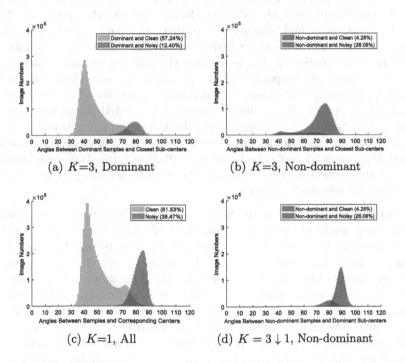

(a) K=3, Dominant   (b) K=3, Non-dominant

(c) K=1, All   (d) $K = 3 \downarrow 1$, Non-dominant

Fig. 4. Data distribution of ArcFace (K=1) and the proposed sub-center ArcFace (K=3) before and after dropping non-dominant sub-centers. MS1MV0 [9] is used here. $K = 3 \downarrow 1$ denotes sub-center ArcFace with non-dominant sub-centers dropping.

scores $\mathcal{S} \in \mathbb{R}^{N \times K}$ by a matrix multiplication $W^T \mathbf{x}_i$. Then, we employ a max pooling step on the subclass-wise similarity score $\mathcal{S} \in \mathbb{R}^{N \times K}$ to get the class-wise similarity score $\mathcal{S}' \in \mathbb{R}^{N \times 1}$. The proposed sub-center ArcFace loss can be formulated as:

$$\ell_{\mathrm{ArcFace_{subcenter}}} = -\log \frac{e^{s \cos(\theta_{i,y_i}+m)}}{e^{s \cos(\theta_{i,y_i}+m)} + \sum_{j=1, j \neq y_i}^{N} e^{s \cos \theta_{i,j}}}, \tag{2}$$

where $\theta_{i,j} = arccos\left(\max_k \left(W_{j_k}^T \mathbf{x}_i\right)\right)$, $k \in \{1, \cdots, K\}$.

Robustness and Strictness Analysis. Given a large K, sub-classes are able to capture the complex distribution of the whole training data. Except for applying max pooling on the subclass-wise cosine similarity score, we can also consider other different comparison strategies. In Table 1, we give the strictness and robustness analysis of four comparison strategies. (1) adds angular margin between the closest inter-class sub-center and the closest intra-class sub-center. For intra-class comparison, choosing the closest positive sub-center can relax the intra-class constraint and improve the robustness under noise. For inter-class comparison, choosing the closest negative sub-center will enhance the inter-class constraint as sub-centers can better capture the complex geometric distributions of the whole data set compared to a single center for each class. However, the enhanced inter-class comparison is less robust under the close-set noise. The training procedure of (2) can not converge as the initial status between inter-classes is orthogonal and relaxing both of the inter-class and intra-class comparisons will disorient the training, as there is no loss from inter-class comparisons. (3) and (4) can not foster sub-classes as stiffening intra-class comparison will compress sub-centers into one point in the high-dimension feature space thus undermine the robustness to noise.

Dominant and Non-dominant Sub-classes. In Fig. 1(b), we have visualised the clustering results of one identity from the CASIA dataset [40] after employing the sub-center ArcFace loss ($K = 10$) for training. It is obvious that the proposed sub-center ArcFace loss can automatically cluster faces such that hard samples and noisy samples are separated away from the dominant clean samples. Note that some sub-classes are empty as $K = 10$ is too large for a particular identity. In Fig. 3(a) and Fig. 3(b), we show the angle distribution on the CASIA dataset [40]. We use the pre-trained ArcFace model [5] to predict the feature center of each identity and then calculate the angle between the sample and its corresponding feature center. As we can see from Fig. 3(a), most of the samples are close to their centers, however, there are some noisy samples which are far away from their centers. This observation on the CASIA dataset matches the noise percentage estimation $(9.3\% - 13.0\%)$ in [30]. To automatically obtain a clean training dataset, the noisy tail is usually removed by a hard threshold (e.g. angle $\geq 77°$ or cosine ≤ 0.225). Since sub-center ArcFace can automatically divide the training samples into dominant sub-classes and non-dominant sub-classes, we visualise these two different kinds of samples in Fig. 3(b). As we can see from the two histograms, sub-center ArcFace can automatically separate clean samples from hard and noisy samples. More specifically, the majority of clean faces (85.6%) go to the dominant sub-class, while the rest hard and noisy faces go to the non-dominant sub-classes.

Drop Non-dominant Sub-centers and High-confident Noises. Even though using sub-classes can improve the robustness under noise, it undermines the intra-class compactness as hard samples are also kept away as shown in Fig. 3(b). In [9], MS1MV0 (around 10M images of 100 K identities) is released with the estimated noise percentage around 47.1% − 54.4% [30]. In [6], MS1MV0 is refined by a semi-automatic approach into a clean dataset named MS1MV3 (around 5.1M images of 93 K identities). Based on these two datasets, we can get clean and noisy labels on MS1MV0. In Fig. 4(a) and Fig. 4(b), we show the angle distributions of samples to their closest sub-centers (training settings: [MS1MV0, ResNet-50, Sub-center ArcFace $K=3$]). In general, there are four categories of samples: (1) easy clean samples belonging to dominant sub-classes (57.24%), (2) hard noisy samples belonging to dominant sub-classes (12.40%), (3) hard clean samples belonging to non-dominant sub-classes (4.28%), and (4) easy noisy samples belonging to non-dominant sub-classes (26.08%). In Fig. 4(c), we show the angle distribution of samples to their corresponding centers from the ArcFace model (training settings: [MS1MV0, ResNet50, ArcFace $K=1$]). By comparing the percentages of noisy sample in Fig. 4(a) and Fig. 4(c), we find that sub-center ArcFace can significantly decrease the noise rate to around one third (from 38.47% to 12.40%) and this is the reason why sub-center Arc-Face is more robust under noise. During the training of sub-center ArcFace, samples belonging to non-dominant sub-classes are pushed to be close to these non-dominant sub-centers as shown in Fig. 4(b). Since we have not set any constraint on sub-centers, the sub-centers of each identity can be quite different and even orthogonal. In Fig. 4(d), we show the angle distributions of non-dominant samples to their dominant sub-centers. By combining Fig. 4(a) and Fig. 4(d), we find that even though the clean and noisy data have some overlaps, a constant angle threshold (between 70° and 80°) can be easily searched to drop most of high-confident noisy samples.

Based on the above observations, we propose a straightforward approach to recapture intra-class compactness. We directly drop non-dominant sub-centers after the network has enough discriminative power. Meanwhile, we introduce a constant angle threshold to drop high-confident noisy data. After that, we retrain the model from scratch on the automatically cleaned dataset.

Comparison with Re-weighting Methods. The main difference between the proposed sub-center ArcFace and re-weighting methods [11,41] is that sub-center ArcFace is less affected by the noisy data from the beginning of the model training. By contrast, the discriminative power of the initial model is important for both NT [11] and NR [41] methods as their adaptive weights are predicted from the model.

Our sub-center ArcFace achieves high accuracy in face recognition while keeps extreme simplicity, only adding two hyper-parameters: the sub-center number and the constant threshold to drop high-confident noisy data.

4   Experiments

4.1   Experimental Settings

Datasets. Our training datasets include MS1MV0 (~10M images of 100 K identities) [9], MS1MV3 (~5.1M faces of 91 K identities) [6], and Celeb500K [1]. MS1MV0 is a raw data with the estimated noise percentage around 47.1%–54.4% [30]. MS1MV3 is cleaned from MS1MV0 by a semi-automatic approach [6]. Celeb500K [1] is collected as MS1MV0 [9], using half of the MS1M name list [9] to search identities from Google and download the top-ranked face images. Our testing datasets consist of IJB-B [35], IJB-C [17], MegaFace [13], and Face Recognition Vendor Test (FRVT). Besides, we also report our final results on widely used verification datasets (e.g. LFW [12], CFP-FP [27], and AgeDB-30 [18]).

Implementation Details. For data pre-possessing, we follow ArcFace [5] to generate the normalised face crops (112 × 112) by utilising five facial points predicted by RetinaFace [7]. We employ ResNet-50 and ResNet-100 [5,10] to get the 512-D face embedding feature. Following [5], the feature scale s is set to 64 and the angular margin m is set to 0.5. All experiments in this paper are implemented by MXNet [3]. We set the batch size for back-propagation as 512 and train models on 8 NVIDIA Tesla P40 (24 GB) GPUs. We set momentum to 0.9 and weight decay to 5e–4. For the training of ArcFace on MS1MV0 and MS1MV3, the learning rate starts from 0.1 and is divided by 10 at the 100K, 160K, and 220 K iteration steps. We finish the training process at 240 K steps. For the training of the proposed sub-center ArcFace, we also employ the same learning rate schedule to train the first round of model ($K = 3$). Then, we drop non-dominant sub-centers ($K = 3 \downarrow 1$) and high-confident noisy data ($> 75°$) by using the first round model through an off-line way. Finally, we retrain the model from scratch using the automatically cleaned data.

4.2   Ablation Study

To facilitate comparisons, we abbreviate different settings by the experiment number (E*) in the table and only focus on the TAR@FAR=1e-4 of IJB-C, which is more objective and less affected by the noise within the test data [38].

Real-world Noise. In Table 2, we conduct extensive experiments to investigate the proposed Sub-center ArcFace. We train ResNet-50 networks on different datasets (MS1MV0, MS1MV3 and Celeb500K) with different settings. From Table 2, we have the following observations: **(a)** ArcFace has an obvious performance drop (from E14 96.50% to E1 90.27%) when the training data is changed from the clean MS1MV3 to the noisy MS1MV0. By contrast, sub-center ArcFace is more robust (E2 93.72%) under massive noise. **(b)** Too many sub-centers (too large K) can obviously undermine the intra-class compactness and decrease the accuracy (from E2 93.72% to E5 67.94%). This observation indicates that robustness and strictness should be balanced during training, thus we select K=3 in this paper. **(c)** The nearest sub-center assignment by the max pooling

Table 2. Ablation experiments of different settings on MS1MV0, MS1MV3 and Celeb500K. The 1:1 verification accuracy (TAR@FAR) is reported on the IJB-B and IJB-C datasets. ResNet-50 is used for training.

Settings	IJB-B					IJB-C				
	1e−6	1e−5	**1e-4**	1e−3	1e−2	1e−6	1e−5	**1e-4**	1e−3	1e−2
(1) MS1MV0,K=1	34.14	74.74	87.87	93.27	96.40	67.08	81.11	90.27	94.59	97.08
(2) MS1MV0,K=3	40.89	85.62	91.70	94.88	96.93	86.18	90.59	93.72	95.98	97.60
(3) MS1MV0,K=3, softmax pooling [22]	38.4	85.49	91.53	94.76	96.83	85.43	90.40	93.55	95.87	97.36
(4) MS1MV0,K=5	39.24	85.48	91.47	94.68	96.96	85.49	90.38	93.62	95.88	97.59
(5) MS1MV0,K=10	19.81	49.03	63.84	76.09	87.73	45.98	55.74	67.94	79.44	89.29
(6) MS1MV0, $K = 3 \downarrow 1$, drop > 70°	47.61	90.60	94.44	96.44	97.71	90.40	94.05	95.91	97.42	98.42
(7) MS1MV0, $K = 3 \downarrow 1$, drop > 75°	46.78	89.40	94.56	96.49	97.83	89.17	94.03	95.92	97.40	98.41
(8) MS1MV0, $K = 3 \downarrow 1$, drop > 80°	38.05	88.26	94.04	96.19	97.64	86.16	93.09	95.74	97.19	98.33
(9) MS1MV0, $K = 3 \downarrow 1$, drop > 85°	42.89	87.06	93.33	96.05	97.59	81.53	92.01	95.10	97.01	98.24
(10) MS1MV0, K=3, regularizer [22]	39.92	85.51	91.53	94.77	96.92	85.44	90.41	93.64	95.85	97.40
(11) MS1MV0,Co-mining [33]	40.96	85.57	91.80	94.99	97.10	86.31	90.71	93.82	95.95	97.63
(12) MS1MV0,NT [11]	40.84	85.56	91.57	94.79	96.83	86.14	90.48	93.65	95.86	97.54
(13) MS1MV0,NR [41]	40.86	85.53	91.58	94.77	96.80	86.07	90.41	93.60	95.88	97.44
(14) MS1MV3, K=1	35.86	91.52	95.13	96.61	97.65	90.16	94.75	96.50	97.61	98.40
(15) MS1MV3, K=3	40.16	91.30	94.84	96.66	97.74	90.64	94.68	96.35	97.66	98.48
(16) MS1MV3, $K = 3 \downarrow 1$	40.18	91.32	94.87	96.70	97.81	90.67	94.74	96.43	97.66	98.47
(17) Celeb500K, K=1	42.42	88.18	90.96	92.19	93.00	88.18	90.87	92.15	95.47	97.64
(18) Celeb500K, K=3	43.84	90.91	93.76	95.12	96.00	90.92	93.66	94.90	96.21	98.02
(19) Celeb500K, $K = 3 \downarrow 1$	44.64	92.71	95.65	96.94	97.89	92.73	95.52	96.91	97.87	98.42

is slightly better than the softmax pooling [22] (E2 93.72% vs. E3 93.55%). Thus, we choose the more efficient max pooling operator in the following experiments. **(d)** Dropping non-dominant sub-centers and high-confident noisy samples can achieve better performance than adding regularization [22] to enforce compactness between sub-centers (E7 95.92% vs. E10 93.64%). Besides, The performance of our method is not very sensitive to the constant threshold (E6 95.91%, E7 95.92% and E8 95.74%), and we select 75° as the threshold for dropping high-confident noisy samples in the following experiments. **(e)** Co-mining [33] and re-weighting methods [11,41] can also improve the robustness under massive noise, but sub-center ArcFace can do better through automatic clean and noisy data isolation during training (E7 95.92% vs. E11 93.82%, E12 93.65% and E13 93.60%). **(f)** On the clean dataset (MS1MV3), sub-center ArcFace achieves similar performance as ArcFace (E16 96.43% vs. E14 96.50%). **(g)** The proposed sub-center ArcFace trained on noisy MS1MV0 can achieve comparable performance compared to ArcFace trained on manually cleaned MS1MV3 (E7 95.92% vs. E14 96.50%). **(h)** By enlarging the training data, sub-center ArcFace can easily achieve better performance even though it is trained from noisy web faces (E19 96.91% vs. E13 96.50%).

Table 3. Ablation experiments of different settings under synthetic open-set and close-set noise. The 1:1 verification accuracy (TAR@FAR) is reported on the IJB-B and IJB-C datasets. ResNet-50 is used for training.

Settings	IJB-B					IJB-C				
	1e−6	1e−5	**1e-4**	1e−3	1e−2	1e−6	1e−5	**1e-4**	1e−3	1e−2
Synthetic Open-set Noise										
(1) 75%CleanID,K=1	37.49	90.02	94.48	96.48	97.72	90.10	94.18	96.00	97.45	98.38
(2) 75%CleanID+ 25%NoisyID,K=1	37.80	86.68	92.96	94.72	95.80	86.19	92.03	94.52	95.89	97.29
(3) 75%CleanID+ 25%NoisyID,K=3	38.31	87.87	94.17	95.83	97.15	87.23	93.01	95.57	96.95	97.75
(4) 75%CleanID+ 25%NoisyID,$K = 3 \downarrow 1$	38.36	88.14	94.20	96.15	97.94	87.51	93.27	95.89	97.29	98.43
(5) 50%CleanID, K=1	34.43	89.36	93.97	96.26	97.63	88.35	93.49	95.65	97.28	98.35
(6) 50%CleanID+ 50%NoisyID,K=1	35.96	81.45	90.77	92.69	94.56	80.97	88.49	92.25	93.84	95.10
(7) 50%CleanID+ 50%NoisyID,K=3	34.15	85.13	92.62	94.98	96.77	84.43	91.00	94.50	95.79	97.33
(8) 50%CleanID+ 50%NoisyID,$K = 3 \downarrow 1$	34.55	86.43	93.85	96.13	97.37	85.22	91.82	95.50	96.73	98.16
Synthetic Close-set Noise										
(9) 75%CleanIM,K=1	38.44	89.41	94.76	96.42	97.71	89.31	94.19	96.19	97.39	98.43
(10) 75%CleanIM+ 25%NoisyIM,K=1	36.16	83.46	92.29	94.85	95.61	82.20	91.24	94.28	95.58	97.58
(11) 75%CleanIM+ 25%NoisyIM,K=3	36.09	83.16	91.45	94.33	95.23	81.28	90.02	93.57	94.96	96.32
(12) 75%CleanIM+ 25%NoisyIM,$K = 3 \downarrow 1$	37.79	85.50	94.03	95.53	97.42	84.09	93.17	95.13	96.85	97.61
(13) 50%CleanIM,K=1	36.85	90.50	94.59	96.49	97.65	90.46	94.32	96.08	97.44	98.33
(14) 50%CleanIM+ 50%NoisyIM,K=1	17.54	43.10	71.76	82.08	93.38	28.40	55.46	75.80	88.22	94.68
(15) 50%CleanIM+ 50%NoisyIM,K=3	17.47	41.63	66.42	78.70	91.37	26.03	54.23	72.04	86.36	94.19
(16) 50%CleanIM+ 50%NoisyIM,$K = 3 \downarrow 1$	22.19	68.11	85.86	88.13	95.08	44.34	69.25	78.12	90.51	96.16

Synthetic Noise. In Table 3, we investigate the robustness of the proposed sub-center ArcFace under synthetic open-set and close-set noise. We train ResNet-50 networks on MS1MV3 with different noise types and levels. To simulate the training data with controlled open-set noise, we randomly select 75% and 50% identities from MS1MV3 [6] and the face images of the rest identities are assigned with random labels of selected identities. To simulate the training data with controlled close-set noise, we use all identities ($\sim$100K) from MS1MV3 [6] but randomly select 25% and 50% face images of each identity and assign random labels to these face images.

From Table 3, we have the following observations: **(a)** Performance drops as the ratio of synthetic noise increases, especially for the close-set noise (E2 94.52% vs. E6 92.25% and E10 94.28% vs. E14 75.80%). In fact, close-set noise is also found to be more harmful than open-set noise in [30]. **(b)** Under the open-set noise, the proposed sub-center can effectively enhance the robustness of ArcFace (E3 95.57% vs. E2 94.52% and E7 94.50% vs. E6 92.25%). By dropping non-dominant sub-centers and high-confident noisy samples, the performance of sub-center arcface can even approach Arcface trained on the clean dataset (E4 95.89% vs. E1 96.00% and E8 95.50% vs. E5 95.65%). (c) Under the close-set noise, the performance of sub-center Arcface is worse than ArcFace (E11 93.57% vs. E10 94.28% and E15 72.04% vs. E14 75.80%), as the inter-class constraint of sub-center Arcface is more strict than ArcFace. By dropping non-dominant sub-centers and high-confident noisy samples, the performance of sub-center Arcface outperforms ArcFace (E12 95.13% vs. E10 94.28% and E16 78.12% vs. E14 75.80%) but still lags behind ArcFace trained on the clean dataset (E12 95.13% vs. E9 96.19% and E16 78.12% vs. E13 96.08%), which indicates the capacity of the network to drop noisy samples depends on its initial discriminative power. Sub-center ArcFace trained on 50% close-set noise is far from accurate (E15 72.04%) and the step of dropping noisy samples is also not accurate. Therefore, it is hard to catch up with ArcFace trained on the clean dataset. However, in the real-world data, close-set noise is not dominant, much less than 50% (e.g. only a small part of celebrities frequently appear in others' album).

4.3   Benchmark Results

Results on IJB-B [35] **and IJB-C** [35]. We employ the face detection scores and the feature norms to re-weigh faces within templates [16,24]. In Table 4, we compare the TAR (@FAR=1e-4) of ArcFace and the proposed sub-center Arc-Face trained on noisy data (e.g. MS1MV0 and Celeb500K). The performance of ArcFace significantly drops from 96.61% to 90.42% on the IJB-C dataset when the training data is changed from the manually cleaned data (MS1MV3) to the raw noisy data (MS1MV0). By contrast, the proposed sub-center ArcFace is robust to massive noise and can achieve similar results compared with Arc-Face trained on the clean data (96.28% vs. 96.61%). When we apply sub-center ArcFace on large-scale training data (Celeb500K), we further improve the TAR (@FAR=1e-4) to 95.75% and 96.96% on IJB-B and IJB-C, respectively.

Results on MegaFace [13]. We adopt the refined version of MegaFace [5] to give a fair evaluation. As shown in Table 4, the identification accuracy of ArcFace obviously drops from 98.40% to 96.52% when the training data is changed from MS1MV3 to MS1MV0, while the proposed sub-center ArcFace is more robust under massive noise within MS1MV0, achieving the identification accuracy of 98.16%. ArcFace trained on MS1MV3 only slightly outperforms our method trained on MS1MV0 under both verification and identification protocols. Finally, the sub-center ArcFace model trained on the large-scale Celeb500K dataset achieves state-of-the-art identification accuracy of 98.78% on the MegaFace dataset.

Table 4. Column 2–3: 1:1 verification TAR (@FAR=1e-4) on the IJB-B and IJB-C dataset. Column 4–5: Face identification and verification evaluation on MegaFace Challenge1 using FaceScrub as the probe set. "Id" refers to the rank-1 face identification accuracy with 1M distractors, and "Ver" refers to the face verification TAR at 10^{-6} FAR. Column 6-8: The 1:1 verification accuracy on the LFW, CFP-FP and AgeDB-30 datasets. ResNet-100 is used for training.

Settings	IJB		MegaFace		Quick Verification Datasets		
	IJB-B	IJB-C	Id	Ver	LFW	CFP-FP	AgeDB-30
MS1MV0, K=1	87.91	90.42	96.52	96.75	99.75	97.17	97.26
MS1MV0, $K = 3 \downarrow 1$	94.94	96.28	98.16	98.36	99.80	98.80	98.31
MS1MV3, K=1 [5,6]	95.25	96.61	98.40	98.51	99.83	98.80	**98.45**
Celeb500K, $K = 3 \downarrow 1$	**95.75**	**96.96**	**98.78**	**98.69**	**99.86**	**99.11**	98.35

Results on LFW [12],**CFP-FP** [27], **and AgeDB-30** [18]. We follow the *unrestricted with labelled outside data* protocol to report the verification performance. As reported in Table 4, sub-center ArcFace trained on noisy MS1MV0 achieves comparable performance compared to ArcFace trained on clean MS1MV3. Moreover, our method trained on the noisy Celeb500K outperforms ArcFace [5], achieving the verification accuracy of 99.86%, 99.11%, 98.35% on LFW, CFP-FP and AgeDB-30, respectively.

Results on FRVT. The Face Recognition Vendor Test (FRVT) is the most strict industry-level face recognition test, and the participants need to submit the whole face recognition system (e.g. face detection, alignment and feature embedding) to the organiser. No test image has been released for hyper-parameter searching and the submission interval is no less than three months. Besides, the submitted face recognition system should complete face detection and face feature embedding within 1000 ms on Intel Xeon CPU (E5-2630 v4 @ 2.20 GHz processors) by using the single-thread inference. We build our face recognition system by RetinaFace (ResNet-50) [7] and sub-center ArcFace (ResNet-100), and accelerate the inference by the openVINO toolkit. In Table 5, we show the top-performing 1:1 algorithms measured on false non-match rate (FNMR) across several different tracks. As we can see from the results, the proposed sub-center ArcFace trained on the Celeb500K dataset achieves state-of-the-art performance on the wild track (0.0303, rank 3rd). Considering several hundred of industry submissions to FRVT, the overall performance of our single model is very impressive.

Table 5. FRVT 1:1 verification results. Sub-center ArcFace ($K = 3 \downarrow 1$) employs ResNet-100 and is trained on the Celeb500K dataset. FNMR is the proportion of mated comparisons below a threshold set to achieve the false match rate (FMR) specified. FMR is the proportion of impostor comparisons at or above that threshold.

Rank	Submissions	WILD FNMR @FMR $\leq$ 1e–5	VISA FNMR @FMR $\leq$ 1e–6	VISA FNMR @FMR $\leq$ 1e–4	MUGSHOT FNMR @FMR $\leq$ 1e–5	MUGSHOT FNMR @FMR $\leq$ 1e–5 DT=14 YRS	VISABORDER FNMR @FMR $\leq$ 1e–6
1	deepglint-002	0.0301	0.0027	0.0004	0.0032	0.0041	0.0043
2	everai-paravision-003	0.0302	0.0050	0.0011	0.0036	0.0053	0.0092
3	Sub-center ArcFace	0.0303	0.0081	0.0027	0.0055	0.0087	0.0083
4	dahua-004	0.0304	0.0058	0.0019	0.0036	0.0051	0.0051
5	xforwardai-000	0.0305	0.0072	0.0018	0.0036	0.0051	0.0074
6	visionlabs-008	0.0308	0.0036	0.0007	0.0031	0.0044	0.0045
7	didiglobalface-001	0.0308	0.0092	0.0016	0.0030	0.0048	0.0088
8	vocord-008	0.0310	0.0038	0.0008	0.0042	0.0054	0.0045
9	paravision-004	0.0311	0.0046	0.0012	0.0030	0.0041	0.0091
10	ntechlab-008	0.0312	0.0061	0.0011	0.0056	0.0106	0.0042
11	tevian-005	0.0325	0.0062	0.0020	0.0057	0.0081	0.0070
12	sensetime-003	0.0355	0.0027	0.0005	0.0027	0.0033	0.0051
13	yitu-003	0.0360	0.0026	0.0003	0.0066	0.0083	0.0064

5 Conclusion

In this paper, we have proposed sub-center ArcFace which first enforces sub-classes by nearest sub-center selection and then only keeps the dominant sub-center to achieve intra-class compactness. As we relax the intra-class compactness from beginning, the proposed sub-center ArcFace is robust under massive label noise and can easily train face recognition models from raw downloaded data. Extensive experimental results show that our method consistently outperforms ArcFace on real-world noisy datasets and achieve comparable performance compared to using manually refined data.

Acknowledgements. Jiankang Deng acknowledges the Imperial President's PhD Scholarship. Tongliang Liu acknowledges support from the Australian Research Council Project DE-190101473. Stefanos Zafeiriou acknowledges support from the Google Faculty Fellowship, EPSRC DEFORM (EP/S010203/1) and FACER2VM (EP/N007743/1). We are thankful to Nvidia for the GPU donations.

References

1. Cao, J., Li, Y., Zhang, Z.: Celeb-500k: A large training dataset for face recognition. In: ICIP (2018)
2. Cao, Q., Shen, L., Xie, W., Parkhi, O.M., Zisserman, A.: Vggface2: A dataset for recognising faces across pose and age. In: FG (2018)
3. Chen, T., Li, M., Li, Y., Lin, M., Wang, N., Wang, M., Xiao, T., Xu, B., Zhang, C., Zhang, Z.: Mxnet: A flexible and efficient machine learning library for heterogeneous distributed systems. arXiv:1512.01274 (2015)

4. Cheng, J., Liu, T., Ramamohanarao, K., Tao, D.: Learning with bounded instance- and label-dependent label noise. ICML (2020)
5. Deng, J., Guo, J., Xue, N., Zafeiriou, S.: Arcface: Additive angular margin loss for deep face recognition. In: CVPR (2019)
6. Deng, J., Guo, J., Zhang, D., Deng, Y., Lu, X., Shi, S.: Lightweight face recognition challenge. In: ICCV Workshops (2019)
7. Deng, J., Guo, J., Zhou, Y., Yu, J., Kotsia, I., Zafeiriou, S.: Retinaface: Single-stage dense face localisation in the wild. arXiv:1905.00641 (2019)
8. Deng, J., Zhou, Y., Zafeiriou, S.: Marginal loss for deep face recognition. In: CVPR Workshops (2017)
9. Guo, Y., Zhang, L., Hu, Y., He, X., Gao, J.: Ms-celeb-1m: A dataset and benchmark for large-scale face recognition. In: ECCV (2016)
10. He, K., Zhang, X., Ren, S., Sun, J.: Deep residual learning for image recognition. In: CVPR (2016)
11. Hu, W., Huang, Y., Zhang, F., Li, R.: Noise-tolerant paradigm for training face recognition cnns. In: CVPR (2019)
12. Huang, G.B., Ramesh, M., Berg, T., Learned-Miller, E.: Labeled faces in the wild: A database for studying face recognition in unconstrained environments. Tech. rep. (2007)
13. Kemelmacher-Shlizerman, I., Seitz, S.M., Miller, D., Brossard, E.: The megaface benchmark: 1 million faces for recognition at scale. In: CVPR (2016)
14. Liu, T., Tao, D.: Classification with noisy labels by importance reweighting. TPAMI (2015)
15. Liu, W., Wen, Y., Yu, Z., Li, M., Raj, B., Song, L.: Sphereface: Deep hypersphere embedding for face recognition. In: CVPR (2017)
16. Masi, I., Tran, A.T., Hassner, T., Sahin, G., Medioni, G.: Face-specific data augmentation for unconstrained face recognition. IJCV (2019)
17. Maze, B., Adams, J., Duncan, J.A., Kalka, N., Miller, T., Otto, C., Jain, A.K., Niggel, W.T., Anderson, J., Cheney, J.: Iarpa janus benchmark-c: Face dataset and protocol. In: ICB (2018)
18. Moschoglou, S., Papaioannou, A., Sagonas, C., Deng, J., Kotsia, I., Zafeiriou, S.: Agedb: The first manually collected in-the-wild age database. In: CVPR Workshops (2017)
19. Movshovitz-Attias, Y., Toshev, A., Leung, T.K., Ioffe, S., Singh, S.: No fuss distance metric learning using proxies. In: ICCV (2017)
20. Müller, R., Kornblith, S., Hinton, G.: Subclass distillation. arXiv:2002.03936 (2020)
21. Oh Song, H., Xiang, Y., Jegelka, S., Savarese, S.: Deep metric learning via lifted structured feature embedding. In: CVPR (2016)
22. Qian, Q., Shang, L., Sun, B., Hu, J., Li, H., Jin, R.: Softtriple loss: Deep metric learning without triplet sampling. In: ICCV (2019)
23. Qian, Q., Tang, J., Li, H., Zhu, S., Jin, R.: Large-scale distance metric learning with uncertainty. In: CVPR (2018)
24. Ranjan, R., Bansal, A., Xu, H., Sankaranarayanan, S., Chen, J.C., Castillo, C.D., Chellappa, R.: Crystal loss and quality pooling for unconstrained face verification and recognition. arXiv:1804.01159 (2018)
25. Rippel, O., Paluri, M., Dollar, P., Bourdev, L.: Metric learning with adaptive density discrimination. In: ICLR (2016)
26. Schroff, F., Kalenichenko, D., Philbin, J.: Facenet: A unified embedding for face recognition and clustering. In: CVPR (2015)
27. Sengupta, S., Chen, J.C., Castillo, C., Patel, V.M., Chellappa, R., Jacobs, D.W.: Frontal to profile face verification in the wild. In: WACV (2016)

28. Sohn, K.: Improved deep metric learning with multi-class n-pair loss objective. In: NeurIPS (2016)
29. Wan, H., Wang, H., Guo, G., Wei, X.: Separability-oriented subclass discriminant analysis. TPAMI (2017)
30. Wang, F., Chen, L., Li, C., Huang, S., Chen, Y., Qian, C., Loy, C.C.: The devil of face recognition is in the noise. In: ECCV (2018)
31. Wang, F., Cheng, J., Liu, W., Liu, H.: Additive margin softmax for face verification. SPL (2018)
32. Wang, H., Wang, Y., Zhou, Z., Ji, X., Gong, D., Zhou, J., Li, Z., Liu, W.: Cosface: Large margin cosine loss for deep face recognition. In: CVPR (2018)
33. Wang, X., Wang, S., Wang, J., Shi, H., Mei, T.: Co-mining: Deep face recognition with noisy labels. In: ICCV (2019)
34. Wen, Y., Zhang, K., Li, Z., Qiao, Y.: A discriminative feature learning approach for deep face recognition. In: ECCV (2016)
35. Whitelam, C., Taborsky, E., Blanton, A., Maze, B., Adams, J.C., Miller, T., Kalka, N.D., Jain, A.K., Duncan, J.A., Allen, K.: Iarpa janus benchmark-b face dataset. In: CVPR Workshops (2017)
36. Wu, C.Y., Manmatha, R., Smola, A.J., Krahenbuhl, P.: Sampling matters in deep embedding learning. In: ICCV (2017)
37. Wu, X., He, R., Sun, Z., Tan, T.: A light cnn for deep face representation with noisy labels. TIFS (2018)
38. Xie, W., Li, S., Zisserman, A.: Comparator networks. In: ECCV (2018)
39. Yang, J., Bulat, A., Tzimiropoulos, G.: Fan-face: a simple orthogonal improvement to deep face recognition. In: AAAI (2020)
40. Yi, D., Lei, Z., Liao, S., Li, S.Z.: Learning face representation from scratch. arXiv:1411.7923 (2014)
41. Zhong, Y., Deng, W., Wang, M., Hu, J., Peng, J., Tao, X., Huang, Y.: Unequal-training for deep face recognition with long-tailed noisy data. In: CVPR (2019)
42. Zhu, M., Martínez, A.M.: Optimal subclass discovery for discriminant analysis. In: CVPR Workshops (2004)
43. Zhu, M., Martinez, A.M.: Subclass discriminant analysis. TPAMI (2006)

Foley Music: Learning to Generate Music from Videos

Chuang Gan[1,2(✉)], Deng Huang[2], Peihao Chen[2], Joshua B. Tenenbaum[1],
and Antonio Torralba[1]

[1] Massachusetts Institute of Technology, Cambridge, USA
[2] MIT-IBM Watson AI Lab, Cambridge, USA
ganchuang1990@gmail.com,
http://foley-music.csail.mit.edu

Abstract. In this paper, we introduce *Foley Music*, a system that can synthesize plausible music for a silent video clip about people playing musical instruments. We first identify two key intermediate representations for a successful video to music generator: body keypoints from videos and MIDI events from audio recordings. We then formulate music generation from videos as a motion-to-MIDI translation problem. We present a *Graph−Transformer* framework that can accurately predict MIDI event sequences in accordance with the body movements. The MIDI event can then be converted to realistic music using an off-the-shelf music synthesizer tool. We demonstrate the effectiveness of our models on videos containing a variety of music performances. Experimental results show that our model outperforms several existing systems in generating music that is pleasant to listen to. More importantly, the MIDI representations are fully interpretable and transparent, thus enabling us to perform music editing flexibly. We encourage the readers to watch the supplementary video with audio turned on to experience the results.

Keywords: Audio-visual · Sound generation · Pose · Foley

1 Introduction

Date Back to 1951, British computer scientist, Alan Turing was the first to record computer-generated music that took up almost an entire floor of the laboratory. Since then, computer music has become an active research field. Recently, the emergence of deep neural networks facilitates the success of generating expressive music by training from large-scale music transcriptions datasets [11,28,40,46,62]. Nevertheless, music is often accompanied by the players interacting with the instruments. Body and instrument interact with nuanced gestures to produce unique music [23]. Studies from cognitive psychology suggest that humans, including young children, are remarkably capable of integrating the correspondences between acoustic and visual signals to perceive the world around them. For example, the McGurk effect [37] indicates that the visual signals people receive from seeing a person speak can influence the sound they hear.

© Springer Nature Switzerland AG 2020
A. Vedaldi et al. (Eds.): ECCV 2020, LNCS 12356, pp. 758–775, 2020.
https://doi.org/10.1007/978-3-030-58621-8_44

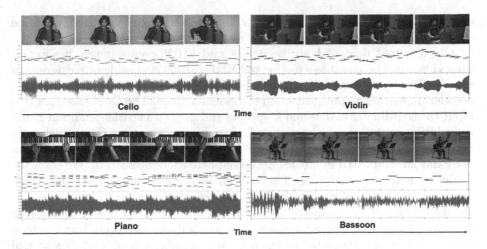

Fig. 1. Given a video of people playing instrument, our system can predict the corresponding MIDI events, and generate plausible musics.

An interesting question then arises: given a silent video clip of a musician playing an instrument, could we develop a computational model to automatically generate a piece of plausible music in accordance with the body movements of that musician? Such capability serves as the foundations for a variety of applications, such as adding sound effects to videos automatically to avoid tedious manual efforts; or creating auditory immersive experiences in virtual reality;

In this paper, we seek to build a system that can learn to generate music by seeing and listening to a large-scale music performance videos (See Fig. 1). However, it is an extremely challenging computation problem to learn a mapping between audio and visual signals from unlabeled video in practice. First, we need a visual perception module to recognize the physical interactions between the musical instrument and the player's body from videos; Second, we need an audio representation that not only respects the major musical rules about structure and dynamics but also easy to predict from visual signals. Finally, we need to build a model that is able to associate these two modalities and accurately predict music from videos.

To address these challenges, we identify two key elements for a successful video to music generator. For the visual perception part, we extract key points of the human body and hand fingers from video frames as intermediate visual representations, and thus can explicitly model the body parts and hand movements. For the music, we propose to use Musical Instrument Digital Interface (MIDI), a symbolic musical representation, that encodes timing and loudness information for each note event, such as note-on and note-off. Using MIDI musical representations offers several unique advantages: 1) MIDI events capture the expressive timing and dynamics information contained in music; 2) MIDI is a sequence of symbolic representation, thus relatively easy to fit into machine learning models; 3) MIDI representation is fully interpretable and flexible; 4) MIDI could be easily converted to realistic music with a standard audio synthesizer.

Given paired data of body keypoints and MIDI events, music generation from videos can be posed as a motion to MIDI translation problem. We develop a *Graph–Transformer* module, which consists of a GCN encoder and a Transfomer decoder, to learn a mapping function to associate them. The GCN encoder takes input the coordinates of detected keypoints and applies a spatial-temporal graph convolution strategy to produce latent feature vectors over time. The transformer decoder can then effectively capture the long-term relationships between human body motion and MIDI events using the self-attention mechanism. We train the model to generate music clips of accordion, bass, bassoon, cello, guitar, piano, tuba, ukulele, and violin, using large-scale music performance videos. To evaluate the quality of our predicted sounds, we conduct listener study experiments measured by correctness, least noise, synchronization, and overall preferences. We show the music generated by our approach significantly outperforms several strong baselines. In summary, our work makes the following contributions:

- We present a new model to generate synchronized and expressive music from videos.
- This paper proposes body keypoint and MIDI as an intermediate representation for transferring knowledge across two modalities, and we empirically demonstrate that such representations are key to success.
- Our system outperforms previous state-of-the-art systems on music generation from videos by a large margin.
- We additionally demonstrate that MIDI musical representations facilitate new applications on generating different styles of music, which seems impossible before.

2   Related Work

2.1   Audio-Visual Learning

Cross-modal learning from vision and audio has attracted increasing interest in recent years [2,4,32,44,57]. The natural synchronization between vision and sound has been leveraged for learning diverse tasks. Given unlabeled training videos, Owens et al. [44] used sound clusters as supervision to learn visual feature representation, and Aytar et al. [4] utilized the scene to learn the audio representations. Follow up works [2,33] further investigated to jointly learn the visual and audio representation using a visual-audio correspondence task. Instead of learning feature representations, recent works have also explored to localize sound source in images or videos [3,26,29,48,64], biometric matching [39], visual-guided sound source separation [15,19,60,64], auditory vehicle tracking [18], multi-modal action recognition [21,35,36], audio inpainting [66], emotion recognition [1], audio-visual event localization [56], multi-modal physical scene understanding [16], audio-visual co-segmentation [47], aerial scene recognition [27] and audio-visual embodied navigation [17].

2.2 Motion and Sound

Several works have demonstrated the strong correlations between sound and motion. For example, the associations between speech and facial movements can be used for facial animations from speech [31,55], generating high-quality talking face from audio [30,54], separate mixed speech signals of multiple speakers [14,42], and even lip-reading from raw videos [12]. Zhao *et al.* [63] and Zhou *et al.* [68] have demonstrated to use optical flow like motion representations to improve the quality of visual sound separations and sound generations. There are also some recent works to explore the correlations between body motion and sound by predicting gestures from speech [22], body dynamics from music [50], or identifying a melody through body language [15]. Different from them, we mainly focus on generating music from videos according to body motions.

2.3 Music Generation

Generating music has been an active research area for decades. As opposed to handcrafted models, a large number of deep neural network models have been proposed for music generation [8,11,24,28,40,46,59,62,65]. For example, MelodyRNN [59] and DeepBach [24] can generate realistic melodies and bach chorales. WaveNet [40] showed very promising results in generating realistic speech and music. Song from PI [11] used a hierarchical RNN model to simultaneously generate melody, drums, and chords, thus leading to a pop song. Huang *et al.* [28] proposed a music transformer model to generate expressive piano music from MIDI event. Hawlhorne *et al.* [25] created a new MAESTRO Dataset to factorize piano music modeling and generation. A detailed survey on deep learning for music generation can be found at [5]. However, there is little work on exploring the problem of generating expressive music from videos.

2.4 Sound Generation from Videos

Back in the 1920s, Jack Foley invented *Foley*, a technique that can create convincing sound effects to movies. Recently, a number of works have explored the ideas of training neural networks to automate Foley. Owens *et al.* [43] investigated the task of predicting the sound emitted by interacting objects with a drumstick. They first used a neural network to predict sound features and then performed an exemplar-based retrieval algorithm instead of directly generating the sound. Chen *et al.* [10] proposed to use the conditional generative adversarial networks for cross-modal generation on lab-collected music performance videos. Zhou *et al.* [68] introduced a SampleRNN-based method to directly predict a generate waveform from an unconstraint video dataset that contains 10 types of sound recorded in the wild. Chen *et al.* [9] proposed a perceptual loss to improve the audio-visual semantic alignment. Chen *et al.* [45] introduced an information bottleneck to generate visually aligned sound. Recent works [20,38,67] also attempt to generate 360/stereo sound from videos. However, these works all use appearances or optical flow for visual representations, and spectrograms or waveform for audio representations. Concurrent to our work, [32,52] also study using MIDI for music transcription and generation.

3 Approach

In this section, we describe our framework of generating music from videos. We first introduce the visual and audio representations used in our system (Sect. 3.1). Then we present a new Graph–Tansformer model for MIDI events prediction from body pose features (Sect. 3.2). Finally, we introduce the training objective and inference procedures (Sect. 3.3). The pipeline of our system is illustrated in Fig. 2.

3.1 Visual and Audio Representations

Visual Representations. Existing work on video to sound generation either use the appearances [43,68] or optical flow [68] as the visual representations. Though remarkable results have achieved, they exhibit limited abilities to applications that require the capture of the fine-grained level correlations between motion and sound. Inspired by previous success on associating vision with audio signals through the explicit movement of the human body parts and hand fingers [22,50], we use the human pose features to capture the body motion cues. This is achieved by first detecting the human body and hand keypoints from each video frame and then stacking their 2D coordinates over time as structured visual representations. In practice, we use the open-source OpenPose toolbox [6] to extract the 2D coordinates of human body joints and adopt a pre-trained hand detection model and the OpenPose [6] hand API [51] to predict the coordinates of hand keypoints. In total, we obtain 25 keypoints for the human body parts and 21 keypoints for each hand.

Audio Representations. Choosing the correct audio representations is very important for the success of generating expressive music. We have explored several audio representations and network architectures. For example, we have explored to directly generate raw waveform using RNN [43,68] or predict sound spectrograms using GAN [10]. However, none of these models work well on generating realistic music from videos. These results are not surprising since music is highly compositional and contains many structured events. It is extremely hard for a machine learning model to discover these rules contained in the music.

We choose the Musical Instrument Digital Interface (MIDI) as the audio representations. MIDI is composed of timing information note-on and note-off events. Each event also defines note pitch. There is also additional velocity information contained in note-on events that indicates how hard the note was played. We first use a music transaction software[1] to automatically detect MIDI events from the audio track of the videos. For a 6-s video clip, it typically contains around 500 MIDI

---

[1] https://www.lunaverus.com/.

events, although the length might vary for different music. To generate expressive timing information for music modeling, we adopt similar music performance encoding proposed by Oore *et al.* [41], which consists of a vocabulary of 88 note-on events, 88 note-off events, 32 velocity bins and 32 time-shift events. These MIDI events could be easily imported into a standard synthesizer to generate the waveforms of music.

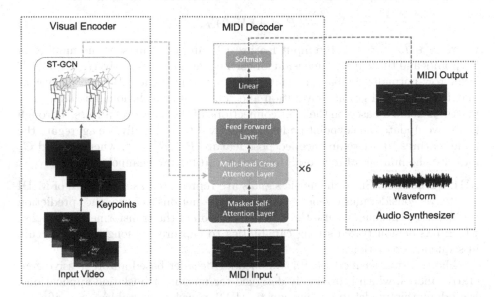

Fig. 2. An overview of our model architecture. It consists of three components: a visual encoder, a MIDI decoder, and an audio synthesizer. The visual encoder takes video frames to extract keypoint coordinates, use GCN to capture the body dynamic and produce a latent representation over time. The MIDI decoder take the video sequence representation to generate a sequence of MIDI event. Finally the MIDI event is converted to the waveform with a standard audio synthesizer.

3.2   Body Motions to MIDI Predictions

We build a *Graph–Tansformer* module to model the correlations between the human body parts and hand movements with the MIDI events. In particular, we first adopt a spatial-temporal graph convolutional network on body keypoint coordinates over time to capture body motions and then feed the encoded pose features to a music transformer decoder to generate a sequence of the MIDI events.

Visual Encoder. Given the 2D keypoints coordinates are extracted from the raw videos, we adopt a Graph CNN to explicitly model the spatial-temporal relationships among different keypoints on the body and hands. Similar to [61], we first represent human skeleton sequence as an undirected spatial-temporal graph $G = (V, E)$, where the node $v_i \in \{V\}$ corresponds to a key point of the human body and edges reflect the natural connectivity of body keypoints.

The input for each node are 2D coordinates of a detected human body keypoint over time T. To model the spatial-temporal body dynamics, we first perform a spatial GCN to encode the pose features at each frame independently and then a standard temporal convolution is applied to the resulting tensor to aggregate the temporal cues. The encoded pose feature P is defined as:

$$P = AXW_SW_T, \tag{1}$$

where $X \in R^{V \times T \times C_n}$ is the input features; V and C_n represent the number of keypoints and the feature dimension for each input node, respectively; $A \in R^{V \times V}$ is the row-normalized adjacency matrix of the graph; W_S and W_T are the weight matrices of spatial graph convolution and temporal convolution. The adjacency matrix is defined based on the joint connections of the body and fingers. Through GCN, we update the keypoint node features over time. Finally, we aggregate the node features to arrive an encoded pose feature $P \in R^{T_v \times C_v}$, where T_v and C_v indicate the number of temporal dimension and feature channels.

MIDI Decoder. Since the music signals are represented as a sequence of MIDI events, we consider music generation from body motions as a sequence prediction problem. To this end, we use the decoder portion of the transformer model [28], which has demonstrated strong capabilities to capture the long-term structure in sequence predictions.

The transformer model [58] is an encoder-decoder based autoregressive generative model, which is originally designed for machine translation applications. We adapt this model to our motion to MIDI translation problem. Specifically, given a visual representation $P \in R^{T_v \times C_v}$, the decoder of transformers is responsible for predicting a sequence of MIDI events $M \in R^{T_m \times L}$, where T_m and L denote a total number of MIDI events contained in a video clip and the vocabulary size of MIDI events. At each time step, the decoder takes the previously generated feature encoding over the MIDI event vocabulary and visual pose features as input and predicts the next MIDI event.

The core mechanism used in the Transformer is the *scale dot-product self-attention* module. This self-attention layer first transforms a sequence of vectors into query Q, key K, and values V, and then output a weighted sum of value$V$, where the weight is calculated by dot products of the key K and query Q. Mathematical:

$$\text{Attention}(Q, K, V) = \text{softmax}(\frac{QK^t}{\sqrt{D_k}})V \tag{2}$$

Instead of performing single attention function, *multi-head attention* is a common used strategy, which allows the model to integrate information from different independent representations.

Different from the vanilla Transformer model, which only uses positional sinusoids to represent timing information, we adopt relative position representations [49] to allow attention to explicitly know the distance between two tokens in a sequence. This is s critically important for modeling music application [28], since music has rich polyphonic sound, and the relative difference matter significantly to timing and pitch. To address this issue, we follow the strategy used in [28] to jointly learn an ordered relative position embedding R for each possible pairwise distance among pairs of query and key on each head as:

$$\text{RelativeAttention}(Q, K, V) = \text{softmax}(\frac{QK^t + R}{\sqrt{D_k}})V \qquad (3)$$

For our MIDI decoder, we first use a masked self-attention module with relative position embedding to encode input MIDI events, where queries, keys, and values are all from the same feature encoding and only depend only on the current and previous positions to maintain the auto-aggressive property. The output of masked self-attention module $M \in R^{T_m \times C_m}$ and pose features $P \in R^{T_v \times C_v}$ are then passed into a multi-head attention module, computed as:

$$\text{CrossAttention}(M, P) = \text{softmax}(\frac{MW^M(PW^P)^t}{\sqrt{D_k}})(PW^V) \qquad (4)$$

The pointwise feed-forward layer takes the input from cross multi-head attention layer, and further transforms it through two fully connected layers with ReLU activation as:

$$\text{FeedFoward} = \max(0, xW_1 + b_1)W_2 + b_2 \qquad (5)$$

The output of feed-forward layers is passed into a softmax layer to produce probability distributions of the next token over the vocabulary.

Music Synthesizer. MIDI can get rendered into a music waveform using a standard synthesizer. It is also possible to train a neural synthesizer [25] for the audio rendering. We leave it to future work.

3.3   Training and Inference

Our graph−transformer model is fully differentiable, thus can be trained in an end-to-end fashion. During training, we take input 2D coordinates of the human skeleton and predict a sequence of MIDI events. At each generation process, the MIDI decoder takes visual encoder features over time, previous and current MIDI event tokens as input and predict the next MIDI event. The training objective is to minimize the cross-entropy loss given a source target sequence of MIDI events. Given the testing video, our model generates MIDI events by performing a beam-search with a beam size of 5.

4   Experiments

In this section, we introduce the experimental setup, comparisons with state-of-the-arts, and ablation studies on each model component.

4.1 Experimental Setup

Datasets: We conduct experiments on three video datasets of music performances, namely URMP [34], AtinPiano and MUSIC [64]. URMP is a high-quality multi-instrument video dataset recorded in a studio and provides MIDI file for each recorded video. AtinPiano is a YouTube channel, including piano video recordings with camera looking down on the keyboard and hands. We use [53] to extract the hands from the videos. MUSIC is an untrimmed video dataset downloaded by querying keywords from Youtube. It contains around 1000 music performance videos belonging to 11 categories. In the paper, we MUSIC and AtinPiano datasets for comparisons with state-of-the-arts, and URMP dataset for ablated study.

Implementation Details: We implement our framework using Pytorch. We first extract the coordinates of body and hand keypoints for each frame using OpenPose [6]. Our GCN encoder consists of 10-layers with residual connections. When training the graph CNN network, we use a batch normalization layer for input 2D coordinates to keep the scale of the input the same. During training, we also perform random affine transformations on the skeleton sequences of all frames as data augmentation to avoid overfitting. The MIDI decoder consists of 6 identical decoder blocks. For each block, the dimension of the attention layer and feed-forward layer are set to 512 and 1024, respectively. The number of attention head is set to 8. For the audio data pre-processing, we first use the toolbox to extract MIDI events from audio recordings. During training, we randomly take a 6-s video clip from the dataset. A software synthesizer[2] is applied to obtain the final generated music waveforms.

   We train our model using Adam optimizer with $\beta_1 = 0.9$, $\beta_2 = 0.98$ and $\epsilon = 10^{-9}$. We schedule the learning rate during training with a warm-up period. Specifically, the learning rate is linearly increased to 0.0007 for the first 4000 training steps, and then decreased proportionally to the inverse square root of the step number.

4.2 Comparisons with State-of-the-arts

We use 9 instruments from MUSIC and AtinPiano dataset to compare against previous systems, including accordion, bass, bassoon, cello, guitar, piano, tuba, ukulele, and violin.

Baseline: we consider 3 state-of-the-art systems to compare against. For fair comparisons, we use the same pose feature representations extracted from GCN for all these baselines.

- **SampleRNN:** We follow the sequence-to-sequence pipeline used in [68]. Specifically, we used the pose features to initial the coarsest tier RNN of the SampleRNN, which serves as a sound generator.

---

[2] https://github.com/FluidSynth/fluidsynth.

- **WaveNet:** We take a conditional WaveNet as our sound generator. To consider the video content during sound generation, we use pose features as the local condition. All other settings are the same as [40].
- **GAN-based Model:** We adopt the framework proposed in [10]. Specifically, taking the pose feature as input, an encoder-decoder is adopted to generate a spectrogram. A discriminator is designed to determine whether the spectrogram is real or fake, conditional on the input pose feature. We transform the spectrogram to waveform by iSTFT.

Table 1. Human evaluation on model comparisons.

Method	GAN-based	SampleRNN	WaveNet	Ours
Accordion	12%	16%	8%	**64%**
Bass	8%	8%	12%	**72%**
Bassoon	10%	14%	6%	**70%**
Cello	8%	14%	12%	**66%**
Guitar	12%	26%	6%	**56%**
Piano	14%	10%	10%	**66%**
Tuba	8%	20%	10%	**62%**
Ukulele	10%	14%	14%	**62%**
Violin	10%	18%	14%	**58%**

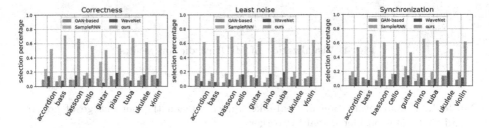

Fig. 3. Human evaluation results of forced-choice experiments in term of correctness, least noise, and synchronization.

Qualitative Evaluation with Human Study: Similar to the task of image or video generation, the quality of the generated sound can be very subjective. For instance, it could be possible to generate music not similar to the ground truth by applying distance metrics, but still sound like a reasonable match to the video content. We carried out a listening study to qualitatively compare the perceived quality of generated music on the Amazon Mechanical Turk (AMT).

We first conduct a forced-choice evaluation [68] to directly compare the proposed method against three baselines. Specifically, we show the four videos with the same video content but different sounds synthesized from our proposed method and three baselines to AMT turkers. They are instructed to choose the best video-sound pair. We use four criteria proposed in [68]:

- **Correctness:** which music recording is most relevant to video content;
- **Least noise:** which music recording has least noise;
- **Synchronization:** which music recording temporally aligns with the video content best;
- **Overall:** which sound they prefer to listen to overall.

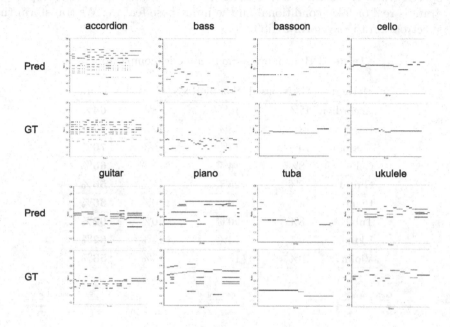

Fig. 4. Visualization of MIDI prediction results.

Table 2. Human evaluation on real-fake. Success mean the percentage of generate sound that were considered real by worker.

Method	Sample RNN	WaveNet	GAN	Ours	Oracle
Success	12%	8%	12%	**38%**	50%

For each instrument category, we choose 50 video clips for evaluation. There are 450 video clips in total. Every question for each test video has been labeled by three independent turkers and the results are reported by majority voting. Table 1 shows overall preference rate for all categories. We find that our method beat the baseline systems for all the instrument categories. To in-depth understand the benefit of our approach, we further analyze the correctness, least noise and synchronization in Fig. 3. We can observe that our approach also consistently outperform baseline systems across all the evaluation criteria by a large-margin. These results further support our claims that the MIDI event representations help improve sound quality, semantic alignment, and temporal synchronization for music generation from videos.

Visualizations: In Fig. 4, we first show the MIDI prediction and ground truth. We can observe that our predicted MIDI event are reasonable similar to the ground truth. We also visualize the sound spectrogram generated by different approaches in Fig. 5. We can find that our model does generate more structured harmonic components than other baselines.

Qualitative Evaluation with Real or Fake Study. In this task, we would like to assess whether the generated audios can fool people into thinking that they are real. We provide two videos with real (originally belonging to this video) and fake (generated by computers) audio to the AMT turkers. The turkers are required to choose the video that they think is real. The criteria for being fake can be bad synchronization, artifacts, or containing noise. We evaluated the ranking of 3 AMT turkers, each was given 100 video pairs. To be noted, an oracle score of 50% would indicate perfect confusions between real and fake. The results in Table 2 demonstrate that, our generated music was hard to distinguish from the real audio recordings than other systems.

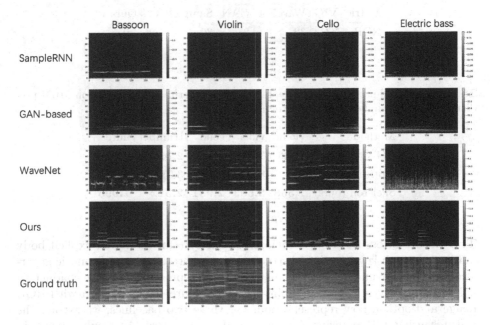

Fig. 5. Qualitative comparison results on sound spectrogram generated by different methods. We report the fraction of generated images

Quantitative Evaluation with Automatic Metrics. We adopt the Number of Statistically-Different Bins (NDB) [13] as automatic metrics to evaluate the diversity of generated sound. Specifically, we first transform sound to log-spectrogram. Then, we cluster the spectrogram in the training set into $k = 50$ Voronoi cells by k-means algorithm. Each generated sound in the testing set is assigned to the nearest cell. NDB indicated the number of cells in which the training samples are significantly different from the number of testing examples. Except for the baselines mentioned above, we also compare with VIG baseline [8] which uses perception loss. The results are listed in Table 3. Our method achieve significantly lower NDB, demonstrating that we can generate more diverse sound.

4.3 Ablated Study

In this section, we perform in-depth ablation studies to assess the impact of each component of our model. We use 5 instruments from URMP dataset for quantitative evaluations, including violin, viola, cello, trumpet, and flute. Since this dataset provides the ground-truth MIDI file, we use negative log-likelihood (NLL) of MIDI event prediction on the validation set as an evaluation metric.

Table 3. Automatic metrics for different models. For NDB, lower is better.

Metric	VIG	WaveNet	GAN	SampleRNN	Ours
NDB	33	32	25	30	**20**

Table 4. Ablated study on visual representation in term of NELL loss on MIDI prediction. **Lower** number means **Better** results.

Method	violin	viola	cello	trumpet	flute
RGB image	1.586	3.772	3.077	2.748	2.219
Optical flow	1.581	3.859	3.178	3.013	2.046
Skeleton (ours)	**1.558**	**3.603**	**2.981**	**2.512**	**1.995**

The effectiveness of Body Motions. In our system, we exploit explicit body motions through keypoint-based structure representations to guide music generation. To further understand the ability of these representations, we conduct an ablated study by replacing keypoint-based structure representation with RGB image and optical flow representation. For these two baselines, we extract the features using I3D network [7] pre-trained on Kinetics. As results shown in Table 4, keypoint-based representation achieve better MIDI prediction accuracy than other options. We hope our findings could inspire more works using the keypoints-based visual representations to solve more challenging audio-visual scene analysis tasks.

The effectiveness of Music Transformers. We adopt a music transformers framework for the sequence predictions. To verify its efficacy, we replace this module with GRU, and keep the other parts of the pipeline the same. The comparison results are shown in Table 5. We can find that the music transformer module improves NEL loss over the GRU baseline. These results demonstrated the benefits of our designed choices using the transformer to capture the long-term dependencies in music.

4.4   Music Editing with MIDI

Since MIDI representation is fully interpretable and transparent, we can easily perform the music editing by manipulating the MIDI file. To demonstrate the flexibility of MIDI representations, we show an example in Fig. 6. Here, we simply manipulate the key of the predicted MIDI, showing its capability to generate music with different styles. These result validate that the MIDI events are flexible and interpretable, thus enabling new applications on controllable music generation, which seem impossible for previous systems which use the waveform or spectrogram as the audio representations.

Table 5. Ablated study on sequence prediction model in term of NELL loss on MIDI prediction. **Lower** number means **Better** results.

Method	violin	viola	cello	trumpet	flute
GRU	1.631	3.747	3.06	2.631	2.101
Transformers w/o hands (ours)	1.565	3.632	3.014	2.805	2.259
Transformers w hands (ours)	**1.558**	**3.603**	**2.981**	**2.512**	**1.995**

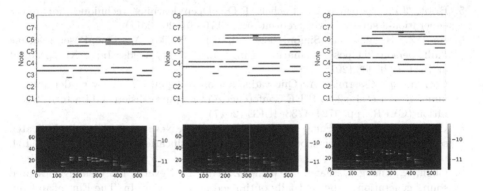

Fig. 6. Music key editing results by manipulating MIDI.

5   Conclusions and Future Work

In this paper, we introduce a *foley music* system to generate expressive music from videos. Our model takes video as input, detects human skeletons, recognizes interactions with musical instruments over time and then predicts the corresponding MIDI files. We evaluated the quality of our approach using human evaluation, showing that the performance of our algorithm was significantly better than baselines. The results demonstrated that the correlations between visual and music signals can be well established through body keypoints and MIDI representations. We additionally show our framework can be easily extended to generate music with different styles through the MIDI representations.

In the future, we plan to train a WaveNet [40] like neural music synthesizer that can generate waveform from MIDI events. Therefore, the whole system can be end-to-end trainable. We envision that our work will open up future research on studying the connections between video and music using intermediate body keypoints and MIDI event representations.

Acknowledgement. This work is supported by ONR MURI N00014-16-1-2007, the Center for Brain, Minds, and Machines (CBMM, NSF STC award CCF-1231216), and IBM Research.

References

1. Albanie, S., Nagrani, A., Vedaldi, A., Zisserman, A.: Emotion recognition in speech using cross-modal transfer in the wild. In: ACM Multimedia (2018)
2. Arandjelovic, R., Zisserman, A.: Look, listen and learn. In: 2017 IEEE International Conference on Computer Vision (ICCV), pp. 609–617. IEEE (2017)
3. Arandjelović, R., Zisserman, A.: Objects that sound. arXiv preprint arXiv:1712.06651 (2017)
4. Aytar, Y., Vondrick, C., Torralba, A.: SoundNet: learning sound representations from unlabeled video. In: Advances in Neural Information Processing Systems, pp. 892–900 (2016)
5. Briot, J.P., Hadjeres, G., Pachet, F.D.: Deep learning techniques for music generation-a survey. arXiv preprint arXiv:1709.01620 (2017)
6. Cao, Z., Hidalgo, G., Simon, T., Wei, S.E., Sheikh, Y.: OpenPose: realtime multi-person 2D pose estimation using part affinity fields. In: arXiv preprint arXiv:1812.08008 (2018)
7. Carreira, J., Zisserman, A.: Quo vadis, action recognition? A new model and the kinetics dataset. In: 2017 IEEE Conference on Computer Vision and Pattern Recognition (CVPR), pp. 4724–4733. IEEE (2017)
8. Chen, K., Zhang, C., Fang, C., Wang, Z., Bui, T., Nevatia, R.: Visually indicated sound generation by perceptually optimized classification. In: ECCV, vol. 11134, pp. 560–574 (2018)
9. Chen, K., Zhang, C., Fang, C., Wang, Z., Bui, T., Nevatia, R.: Visually indicated sound generation by perceptually optimized classification. In: The European Conference on Computer Vision, pp. 560–574 (2018)
10. Chen, L., Srivastava, S., Duan, Z., Xu, C.: Deep cross-modal audio-visual generation. In: ACM Multimedia 2017, pp. 349–357 (2017)
11. Chu, H., Urtasun, R., Fidler, S.: Song from pi: a musically plausible network for pop music generation. In: ICLR (2017)
12. Chung, J.S., Senior, A.W., Vinyals, O., Zisserman, A.: Lip reading sentences in the wild. In: CVPR, pp. 3444–3453 (2017)
13. Engel, J.H., Agrawal, K.K., Chen, S., Gulrajani, I., Donahue, C., Roberts, A.: GANSynth: adversarial neural audio synthesis. In: ICLR (2019)

14. Ephrat, A., et al.: Looking to listen at the cocktail party: a speaker-independent audio-visual model for speech separation. ACM Trans. Graph. (TOG) **37**(4), 112 (2018)

15. Gan, C., Huang, D., Zhao, H., Tenenbaum, J.B., Torralba, A.: Music gesture for visual sound separation. In: CVPR, pp. 10478–10487 (2020)

16. Gan, C., et al.: ThreeDWorld: a platform for interactive multi-modal physical simulation. arXiv preprint arXiv:2007.04954 (2020)

17. Gan, C., Zhang, Y., Wu, J., Gong, B., Tenenbaum, J.B.: Look, listen, and act: towards audio-visual embodied navigation. In: ICRA (2020)

18. Gan, C., Zhao, H., Chen, P., Cox, D., Torralba, A.: Self-supervised moving vehicle tracking with stereo sound. In: ICCV, pp. 7053–7062 (2019)

19. Gao, R., Feris, R., Grauman, K.: Learning to separate object sounds by watching unlabeled video. In: ECCV, pp. 35–53 (2018)

20. Gao, R., Grauman, K.: 2.5 d visual sound. arXiv preprint arXiv:1812.04204 (2018)

21. Gao, R., Oh, T.H., Grauman, K., Torresani, L.: Listen to look: action recognition by previewing audio. In: CVPR, pp. 10457–10467 (2020)

22. Ginosar, S., Bar, A., Kohavi, G., Chan, C., Owens, A., Malik, J.: Learning individual styles of conversational gesture. In: CVPR, pp. 3497–3506 (2019)

23. Godøy, R.I., Leman, M.: Musical Gestures: Sound, Movement, and Meaning. Routledge, Abingdon (2010)

24. Hadjeres, G., Pachet, F., Nielsen, F.: DeepBach: a steerable model for bach chorales generation. In: ICML, pp. 1362–1371 (2017)

25. Hawthorne, C., et al.: Enabling factorized piano music modeling and generation with the maestro dataset. In: ICLR (2019)

26. Hershey, J.R., Movellan, J.R.: Audio vision: using audio-visual synchrony to locate sounds. In: Solla, S.A., Leen, T.K., Müller, K. (eds.) Advances in Neural Information Processing Systems, vol. 12, pp. 813–819 (2000)

27. Hu, D., et al.: Cross-task transfer for multimodal aerial scene recognition. In: ECCV (2020)

28. Huang, C.Z.A., et al.: Music transformer: generating music with long-term structure (2018)

29. Izadinia, H., Saleemi, I., Shah, M.: Multimodal analysis for identification and segmentation of moving-sounding objects. IEEE Trans. Multimed. **15**(2), 378–390 (2013)

30. Jamaludin, A., Chung, J.S., Zisserman, A.: You said that?: Synthesising talking faces from audio. Int. J. Comput. Vis. **127**, 1–13 (2019)

31. Karras, T., Aila, T., Laine, S., Herva, A., Lehtinen, J.: Audio-driven facial animation by joint end-to-end learning of pose and emotion. ACM Trans. Graph. (TOG) **36**(4), 94 (2017)

32. Koepke, A.S., Wiles, O., Moses, Y., Zisserman, A.: Sight to sound: an end-to-end approach for visual piano transcription. In: ICASSP, pp. 1838–1842 (2020)

33. Korbar, B., Tran, D., Torresani, L.: Co-training of audio and video representations from self-supervised temporal synchronization. arXiv preprint arXiv:1807.00230 (2018)

34. Li, B., Liu, X., Dinesh, K., Duan, Z., Sharma, G.: Creating a multitrack classical music performance dataset for multimodal music analysis: challenges, insights, and applications. IEEE Trans. Multimed. **21**(2), 522–535 (2018)

35. Long, X., et al.: Multimodal keyless attention fusion for video classification. In: AAAI (2018)

36. Long, X., Gan, C., de Melo, G., Wu, J., Liu, X., Wen, S.: Attention clusters: purely attention based local feature integration for video classification. In: CVPR (2018)

37. McGurk, H., MacDonald, J.: Hearing lips and seeing voices. Nature **264**(5588), 746–748 (1976)
38. Morgado, P., Nvasconcelos, N., Langlois, T., Wang, O.: Self-supervised generation of spatial audio for 360 video. In: NIPS (2018)
39. Nagrani, A., Albanie, S., Zisserman, A.: Seeing voices and hearing faces: cross-modal biometric matching. arXiv preprint arXiv:1804.00326 (2018)
40. Oord, A.V.D., et al.: WaveNet: a generative model for raw audio. In: ICLR (2017)
41. Oore, S., Simon, I., Dieleman, S., Eck, D., Simonyan, K.: This time with feeling: learning expressive musical performance. Neural Comput. Appl. **32**, 1–13 (2018)
42. Owens, A., Efros, A.A.: Audio-visual scene analysis with self-supervised multisensory features. In: ECCV (2018)
43. Owens, A., Isola, P., McDermott, J., Torralba, A., Adelson, E.H., Freeman, W.T.: Visually indicated sounds. In: Proceedings of the IEEE Conference on Computer Vision and Pattern Recognition, pp. 2405–2413 (2016)
44. Owens, A., Wu, J., McDermott, J.H., Freeman, W.T., Torralba, A.: Ambient sound provides supervision for visual learning. In: Leibe, B., Matas, J., Sebe, N., Welling, M. (eds.) ECCV 2016. LNCS, vol. 9905, pp. 801–816. Springer, Cham (2016). https://doi.org/10.1007/978-3-319-46448-0_48
45. Peihao, C., Yang, Z., Mingkui, T., Hongdong, X., Deng, H., Chuang, G.: Generating visually aligned sound from videos. IEEE Trans. Image Process. **29**, 8292–8302 (2020)
46. Roberts, A., Engel, J., Raffel, C., Hawthorne, C., Eck, D.: A hierarchical latent vector model for learning long-term structure in music. arXiv preprint arXiv:1803.05428 (2018)
47. Rouditchenko, A., Zhao, H., Gan, C., McDermott, J., Torralba, A.: Self-supervised audio-visual co-segmentation. In: ICASSP 2019–2019 IEEE International Conference on Acoustics, Speech and Signal Processing (ICASSP), pp. 2357–2361. IEEE (2019)
48. Senocak, A., Oh, T.H., Kim, J., Yang, M.H., Kweon, I.S.: Learning to localize sound source in visual scenes. arXiv preprint arXiv:1803.03849 (2018)
49. Shaw, P., Uszkoreit, J., Vaswani, A.: Self-attention with relative position representations. arXiv preprint arXiv:1803.02155 (2018)
50. Shlizerman, E., Dery, L., Schoen, H., Kemelmacher-Shlizerman, I.: Audio to body dynamics. In: CVPR, pp. 7574–7583 (2018)
51. Simon, T., Joo, H., Matthews, I., Sheikh, Y.: Hand keypoint detection in single images using multiview bootstrapping. In: CVPR (2017)
52. Su, K., Liu, X., Shlizerman, E.: Audeo: audio generation for a silent performance video. arXiv preprint arXiv:2006.14348 (2020)
53. Submission, A.: At your fingertips: automatic piano fingering detection. In: ICLR (2020)
54. Suwajanakorn, S., Seitz, S.M., Kemelmacher-Shlizerman, I.: Synthesizing Obama: learning lip sync from audio. ACM Trans. Graph. (TOG) **36**(4), 95 (2017)
55. Taylor, S., et al.: A deep learning approach for generalized speech animation. ACM Trans. Graph. (TOG) **36**(4), 93 (2017)
56. Tian, Y., Shi, J., Li, B., Duan, Z., Xu, C.: Audio-visual event localization in unconstrained videos. In: Ferrari, V., Hebert, M., Sminchisescu, C., Weiss, Y. (eds.) ECCV 2018. LNCS, vol. 11206, pp. 252–268. Springer, Cham (2018). https://doi.org/10.1007/978-3-030-01216-8_16
57. Tian, Y., Krishnan, D., Isola, P.: Contrastive multiview coding. In: ECCV (2020)
58. Vaswani, A., et al.: Attention is all you need. In: NIPS, pp. 5998–6008 (2017)

59. Waite, E., et al.: Generating long-term structure in songs and stories. Webblog Post. Magenta, vol. 15 (2016)

60. Xu, X., Dai, B., Lin, D.: Recursive visual sound separation using minus-plus net. In: Proceedings of the IEEE International Conference on Computer Vision, pp. 882–891 (2019)

61. Yan, S., Xiong, Y., Lin, D.: Spatial temporal graph convolutional networks for skeleton-based action recognition. In: AAAI (2018)

62. Yang, L.C., Chou, S.Y., Yang, Y.H.: MidiNet: a convolutional generative adversarial network for symbolic-domain music generation. arXiv preprint arXiv:1703.10847 (2017)

63. Zhao, H., Gan, C., Ma, W.C., Torralba, A.: The sound of motions. In: ICCV (2019)

64. Zhao, H., Gan, C., Rouditchenko, A., Vondrick, C., McDermott, J., Torralba, A.: The sound of pixels. In: Ferrari, V., Hebert, M., Sminchisescu, C., Weiss, Y. (eds.) ECCV 2018. LNCS, vol. 11205, pp. 587–604. Springer, Cham (2018). https://doi.org/10.1007/978-3-030-01246-5_35

65. Zhao, K., Li, S., Cai, J., Wang, H., Wang, J.: An emotional symbolic music generation system based on LSTM networks. In: 2019 IEEE 3rd Information Technology, Networking, Electronic and Automation Control Conference (ITNEC), pp. 2039–2043 (2019)

66. Zhou, H., Liu, Z., Xu, X., Luo, P., Wang, X.: Vision-infused deep audio inpainting. In: ICCV, pp. 283–292 (2019)

67. Zhou, H., Xu, X., Lin, D., Wang, X., Liu, Z.: Sep-stereo: visually guided stereophonic audio generation by associating source separation. In: ECCV (2020)

68. Zhou, Y., Wang, Z., Fang, C., Bui, T., Berg, T.L.: Visual to sound: generating natural sound for videos in the wild. In: CVPR (2018)

Contrastive Multiview Coding

Yonglong Tian[1]($\boxtimes$), Dilip Krishnan[2], and Phillip Isola[1]

[1] MIT CSAIL, Cambridge, USA
yonglong@mit.edu
[2] Google Research, Cambridge, USA

Abstract. Humans view the world through many sensory channels, e.g., the long-wavelength light channel, viewed by the left eye, or the high-frequency vibrations channel, heard by the right ear. Each view is noisy and incomplete, but important factors, such as physics, geometry, and semantics, tend to be shared between all views (e.g., a "dog" can be seen, heard, and felt). We investigate the classic hypothesis that a powerful representation is one that models view-invariant factors. We study this hypothesis under the framework of multiview contrastive learning, where we learn a representation that aims to maximize mutual information between different views of the same scene but is otherwise compact. Our approach scales to any number of views, and is view-agnostic. We analyze key properties of the approach that make it work, finding that the contrastive loss outperforms a popular alternative based on cross-view prediction, and that the more views we learn from, the better the resulting representation captures underlying scene semantics. Code is available at: http://github.com/HobbitLong/CMC/.

1 Introduction

A foundational idea in coding theory is to learn compressed representations that nonetheless can be used to reconstruct the raw data. This idea shows up in contemporary representation learning in the form of autoencoders [64] and generative models [23,39], which try to represent a data point or distribution as losslessly as possible. Yet lossless representation might not be what we really want, and indeed it is trivial to achieve – the raw data itself is a lossless representation. What we might instead prefer is to keep the "good" information (signal) and throw away the rest (noise). How can we identify what information is signal and what is noise?

To an autoencoder, or a max likelihood generative model, a bit is a bit. No one bit is better than any other. Our conjecture in this paper is that some bits *are* in fact better than others. Some bits code important properties like semantics, physics, and geometry, while others code attributes that we might

Electronic supplementary material The online version of this chapter (https://doi.org/10.1007/978-3-030-58621-8_45) contains supplementary material, which is available to authorized users.

A. Vedaldi et al. (Eds.): ECCV 2020, LNCS 12356, pp. 776–794, 2020.
https://doi.org/10.1007/978-3-030-58621-8_45

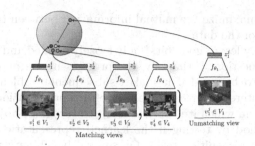

Fig. 1. Given a set of sensory views, a deep representation is learnt by bringing views of the *same* scene together in embedding space, while pushing views of *different* scenes apart. Here we show and example of a 4-view dataset (NYU RGBD [52] and its learned representation. The encodings for each view may be concatenated to form the full representation of a scene.

consider less important, like incidental lighting conditions or thermal noise in a camera's sensor.

We revisit the classic hypothesis that the good bits are the ones that are shared between multiple *views* of the world, for example between multiple sensory modalities like vision, sound, and touch [69]. Under this perspective "presence of dog" is good information, since dogs can be seen, heard, and felt, but "camera pose" is bad information, since a camera's pose has little or no effect on the acoustic and tactile properties of the imaged scene. This hypothesis corresponds to the inductive bias that the way you view a scene should not affect its semantics. There is significant evidence in the cognitive science and neuroscience literature that such view-invariant representations are encoded by the brain (e.g., [14, 31, 69]. In this paper, we specifically study the setting where the different views are different image channels, such as luminance, chrominance, depth, and optical flow. The fundamental supervisory signal we exploit is the *co-occurrence*, in natural data, of multiple views of the same scene. For example, we consider an image in Lab color space to be a paired example of the co-occurrence of two views of the scene, the L view and the ab view: $\{L, ab\}$.

Our goal is therefore to learn representations that capture information shared between multiple sensory channels but that are otherwise compact (i.e. discard channel-specific nuisance factors). To do so, we employ contrastive learning, where we learn a feature embedding such that views of the same scene map to nearby points (measured with Euclidean distance in representation space) while views of different scenes map to far apart points. In particular, we adapt the recently proposed method of Contrastive Predictive Coding (CPC) [56], except we simplify it – removing the recurrent network – and generalize it – showing how to apply it to arbitrary collections of image channels, rather than just to temporal or spatial predictions. In reference to CPC, we term our method *Contrastive Multiview Coding* (CMC), although we note that our formulation is arguably equally related to Instance Discrimination [79]. The contrastive objective in our formulation, as in CPC and Instance Discrimination, can be understood as attempting to maximize the mutual information between the representations of multiple views of the data.

We intentionally leave "good bits" only loosely defined and treat its definition as an empirical question. Ultimately, the proof is in the pudding: we consider a representation to be good if it makes subsequent problem solving easy, on tasks of human interest. For example, a useful representation of images might be a feature space in which it is easy to learn to recognize objects. We therefore evaluate our method by testing if the learned representations transfer well to standard semantic recognition tasks. On several benchmark tasks, our method achieves results competitive with the state of the art, compared to other methods for self-supervised representation learning. We additionally find that the quality of the representation improves as a function of the number of views used for training. Finally, we compare the contrastive formulation of multiview learning to the recently popular approach of cross-view prediction, and find that in head-to-head comparisons, the contrastive approach learns stronger representations.

The core ideas that we build on: contrastive learning, mutual information maximization, and deep representation learning, are not new and have been explored in the literature on representation and multiview learning for decades [3,44,63,80]. Our main contribution is to set up a framework to extend these ideas to *any number of views*, and to empirically study the factors that lead to success in this framework. A review of the related literature is given in Sect. 2; and Fig. 1 gives a pictorial overview of our framework. Our main contributions are:

- We apply contrastive learning to the multiview setting, attempting to maximize mutual information between representations of different views of the same scene (in particular, between different image channels).
- We extend the framework to learn from *more than two* views, and show that the quality of the learned representation improves as number of views increase. Ours is the first work to explicitly show the benefits of multiple views on representation quality.
- We conduct controlled experiments to measure the effect of mutual information estimates on representation quality. Our experiments show that the relationship between mutual information and views is a subtle one.
- Our representations rival state of the art on popular benchmarks.
- We demonstrate that the contrastive objective is superior to cross-view prediction.

2   Related Work

Unsupervised representation learning is about learning transformations of the data that make subsequent problem solving easier [7]. This field has a long history, starting with classical methods with well established algorithms, such as principal components analysis (PCA [36]) and independent components analysis

(ICA [32]). These methods tend to learn representations that focus on low-level variations in the data, which are not very useful from the perspective of downstream tasks such as object recognition.

Representations better suited to such tasks have been learnt using deep neural networks, starting with seminal techniques such as Boltzmann machines [64,70], autoencoders [29], variational autoencoders [39], generative adversarial networks [23] and autoregressive models [55]. Numerous other works exist, for a review see [7]. A powerful family of models for unsupervised representations are collected under the umbrella of "self-supervised" learning [34,59,63,78,83–85]. In these models, an input X to the model is transformed into an output $\hat{X}$, which is supposed to be close to another signal Y (usually in Euclidean space), which itself is related to X in some meaningful way. Examples of such X/Y pairs are: luminance and chrominance color channels of an image [85], patches from a single image [56], modalities such as vision and sound [57] or the frames of a video [78]. Clearly, such examples are numerous in the world, and provides us with nearly infinite amounts of training data: this is one of the appeals of this paradigm. Time contrastive networks [67] use a triplet loss framework to learn representations from aligned video sequences of the same scene, taken by different video cameras. Closely related to self-supervised learning is the idea of multi-view learning, which is a general term involving many different approaches such as co-training [8], multi-kernel learning [12] and metric learning [6,87]; for comprehensive surveys please see [44,80]. Nearly all existing works have dealt with one or two views such as video or image/sound. However, in many situations, many more views are available to provide training signals for any representation.

The objective functions used to train deep learning based representations in many of the above methods are either reconstruction-based loss functions such as Euclidean losses in different norms e.g. [33], adversarial loss functions [23] that learn the loss in addition to the representation, or contrastive losses e.g. [24,25,35] that take advantage of the co-occurence of multiple views.

Some of the prior works most similar to our own (and inspirational to us) are Contrastive Predictive Coding (CPC) [56], Deep InfoMax [30], and Instance Discrimination [79]. These methods, like ours, learn representations by contrasting between congruent and incongruent representations of a scene. CPC learns from two views – the past and future – and is applicable to sequential data, either in space or in time. Deep Infomax [30] considers the two views to be the input to a neural network and its output. Instance Discrimination learns to match two sub-crops of the same image. CPC and Deep InfoMax have recently been extended in [28] and [4] respectively. These methods all share similar mathematical objectives, but differ in the definition of the views. Our method differs from these works in the following ways: we extend the objective to the case of *more than two* views, and we explore a different set of view definitions, architectures, and application settings. In addition, we contribute a unique empirical investigation of this paradigm of representation learning. The idea of contrastive learning has started to spread over many other tasks in various other domains [37,47,60,73,74,76,81,86].

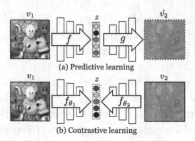

Fig. 2. Predictive learning vs contrastive learning. Cross-view prediction (**Top**) learns latent representations that predict one view from another, with loss measured in the *output* space. Common prediction losses, such as the $\mathcal{L}_1$ and $\mathcal{L}_2$ norms, are *unstructured*, in the sense that they penalize each output dimension independently, perhaps leading to representations that do not capture all the shared information between the views. In contrastive learning (**Bottom**), representations are learnt by contrasting congruent and incongruent views, with loss measured in *representation* space. The red dotted outlines show where the loss function is applied. (Color figure online)

3   Method

Our goal is to learn representations that capture information shared between multiple sensory views without human supervision. We start by reviewing previous predictive learning (or reconstruction-based learning) methods, and then elaborate on contrastive learning within two views. We show connections to mutual information maximization and extend it to scenarios including more than two views. We consider a collection of M views of the data, denoted as $V_1, \ldots, V_M$. For each view V_i, we denote v_i as a random variable representing samples following $v_i \sim \mathcal{P}(V_i)$.

3.1   Predictive Learning

Let V_1 and V_2 represent two views of a dataset. For instance, V_1 might be the luminance of a particular image and V_2 the chrominance. We define the *predictive learning* setup as a deep nonlinear transformation from v_1 to v_2 through latent variables z, as shown in Fig. 2. Formally, $z = f(v_1)$ and $\hat{v}_2 = g(z)$, where f and g represent the encoder and decoder respectively and $\hat{v}_2$ is the prediction of v_2 given v_1. The parameters of the encoder and decoder models are then trained using an objective function that tries to bring $\hat{v}_2$ "close to" v_2. Simple examples of such an objective include the $\mathcal{L}_1$ or $\mathcal{L}_2$ loss functions. Note that these objectives assume independence between each pixel or element of v_2 given v_1, i.e., $p(v_2|v_1) = \Pi_i p(v_{2i}|v_1)$, thereby reducing their ability to model correlations or complex structure. The predictive approach has been extensively used in representation learning, for example, colorization [84,85] and predicting sound from vision [57].

3.2 Contrastive Learning with Two Views

The idea behind contrastive learning is to learn an embedding that separates (contrasts) samples from two different distributions. Given a dataset of V_1 and V_2 that consists of a collection of samples $\{v_1^i, v_2^i\}_{i=1}^N$, we consider contrasting congruent and incongruent pairs, i.e. samples from the joint distribution $x \sim p(v_1, v_2)$ or $x = \{v_1^i, v_2^i\}$, which we call *positives*, versus samples from the product of marginals, $y \sim p(v_1)p(v_2)$ or $y = \{v_1^i, v_2^j\}$, which we call *negatives*.

We learn a "critic" (a discriminating function) $h_\theta(\cdot)$ which is trained to achieve a high value for positive pairs and low for negative pairs. Similar to recent setups for contrastive learning [24,50,56], we train this function to correctly select a single positive sample x out of a set $S = \{x, y_1, y_2, ..., y_k\}$ that contains k negative samples:

$$\mathcal{L}_{contrast} = -\mathop{\mathbb{E}}_{S}\left[\log \frac{h_\theta(x)}{h_\theta(x) + \sum_{i=1}^k h_\theta(y_i)}\right] \tag{1}$$

To construct S, we simply fix one view and enumerate positives and negatives from the other view, allowing us to rewrite the objective as:

$$\mathcal{L}_{contrast}^{V_1,V_2} = -\mathop{\mathbb{E}}_{\{v_1^1,v_2^1,...,v_2^{k+1}\}}\left[\log \frac{h_\theta(\{v_1^1, v_2^1\})}{\sum_{j=1}^{k+1} h_\theta(\{v_1^1, v_2^j\})}\right] \tag{2}$$

where k is the number of negative samples v_2^j for a given sample v_1^1. In practice, k can be extremely large (e.g., 1.2 million in ImageNet), and so directly minimizing Eq. 2 is infeasible. In Sect. 3.4, we show two approximations that allow for tractable computation.

Implementing the Critic. We implement the critic $h_\theta(\cdot)$ as a neural network. To extract compact latent representations of v_1 and v_2, we employ two encoders $f_{\theta_1}(\cdot)$ and $f_{\theta_2}(\cdot)$ with parameters θ_1 and θ_2 respectively. The latent representations are extracted as $z_1 = f_{\theta_1}(v_1)$, $z_2 = f_{\theta_2}(v_2)$. We compute their cosine similarity as score and adjust its dynamic range by a hyper-parameter τ:

$$h_\theta(\{v_1, v_2\}) = \exp\left(\frac{f_{\theta_1}(v_1) \cdot f_{\theta_2}(v_2)}{\|f_{\theta_1}(v_1)\| \cdot \|f_{\theta_2}(v_2)\|} \cdot \frac{1}{\tau}\right) \tag{3}$$

Loss $\mathcal{L}_{contrast}^{V_1,V_2}$ in Eq. 2 treats view V_1 as anchor and enumerates over V_2. Symmetrically, we can get $\mathcal{L}_{contrast}^{V_2,V_1}$ by anchoring at V_2. We add them up as our two-view loss:

$$\mathcal{L}(V_1, V_2) = \mathcal{L}_{contrast}^{V_1,V_2} + \mathcal{L}_{contrast}^{V_2,V_1} \tag{4}$$

After the contrastive learning phase, we use the representation z_1, z_2, or the concatenation of both, $[z_1, z_2]$, depending on our paradigm. This process is visualized in Fig. 1.

Connecting to Mutual Information. The optimal critic h_θ^* is proportional to the density ratio between the joint distribution $p(z_1, z_2)$ and the product of marginals $p(z_1)p(z_2)$ (proof provided in supplementary material):

$$h_\theta^*(\{v_1, v_2\}) \propto \frac{p(z_1, z_2)}{p(z_1)p(z_2)} \propto \frac{p(z_1|z_2)}{p(z_1)} \qquad (5)$$

This quantity is the pointwise mutual information, and its expectation, in Eq. 2, yields an estimator related to mutual information. A formal proof is given by [56,61], which we recapitulate in supplement, showing that:

$$I(z_i; z_j) \geq \log(k) - \mathcal{L}_{contrast} \qquad (6)$$

where, as above, k is the number of negative pairs in sample set S. Hence minimizing the objective $\mathcal{L}$ maximizes the lower bound on the mutual information $I(z_i; z_j)$, which is bounded above by $I(v_i; v_j)$ by the data processing inequality. The dependency on k also suggests that using more negative samples can lead to an improved representation; we show that this is indeed the case (see supplement). We note that recent work [46] shows that the bound in Eq. 6 can be very weak; and finding better estimators of mutual information is an important open problem.

3.3 Contrastive Learning with More Than Two Views

We present more general formulations of Eq. 2 that can handle any number of views. We call them the "core view" and "full graph" paradigms, which offer different tradeoffs between efficiency and effectiveness. These formulations are visualized in Fig. 3.

Suppose we have a collection of M views $V_1, \ldots, V_M$. The "core view" formulation sets apart one view that we want to optimize over, say V_1, and builds pair-wise representations between V_1 and each other view $V_j, j > 1$, by optimizing the sum of a set of pair-wise objectives:

$$\mathcal{L}_C = \sum_{j=2}^{M} \mathcal{L}(V_1, V_j) \qquad (7)$$

A second, more general formulation is the "full grap" where we consider all pairs $(i, j), i \neq j$, and build $\binom{n}{2}$ relationships in all. By involving all pairs, the objective function that we optimize is:

$$\mathcal{L}_F = \sum_{1 \leq i < j \leq M} \mathcal{L}(V_i, V_j) \qquad (8)$$

Both these formulations have the effect that information is prioritized in proportion to the number of views that share that information. This can be seen in the information diagrams visualized in Fig. 3. The number in each partition of the diagram indicates how many of the pairwise objectives, $\mathcal{L}(V_i, V_j)$, that partition

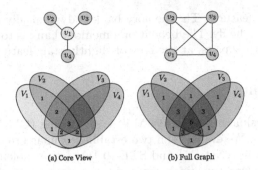

(a) Core View   (b) Full Graph

Fig. 3. Graphical models and information diagrams [1] associated with the core view and full graph paradigms, for the case of 4 views, which gives a total of 6 learning objectives. The numbers within the regions show how much "weight" the total loss places on each partition of information (i.e. how many of the 6 objectives that partition contributes to). A region with no number corresponds to 0 weight. For example, in the full graph case, the mutual information between all 4 views is considered in all 6 objectives, and hence is marked with the number 6.

contributes to. Under both the core view and full graph objectives, a factor, like "presence of dog", that is common to all views will be preferred over a factor that affects fewer views, such as "depth sensor noise".

The computational cost of the bivariate score function in the full graph formulation is combinatorial in the number of views. However, it is clear from Fig. 3 that this enables the full graph formulation to capture more information between different views, which may prove useful for downstream tasks. For example, the mutual information between V_2 and V_3 or V_2 and V_4 is completely ignored in the core view paradigm (as shown by a 0 count in the information diagram). Another benefit of the full graph formulation is that it can handle missing information (e.g. missing views) in a natural manner.

3.4   Implementing the Contrastive Loss

Better representations using $\mathcal{L}_{contrast}^{V_1, V_2}$ in Eq. 2 are learnt by using many negative samples. In the extreme case, we include every data sample in the denominator for a given dataset. However, computing the full softmax loss is prohibitively expensive for large dataset such as ImageNet. One way to approximate this full softmax distribution, as well as alleviate the computational load, is to use Noise-Contrastive Estimation [24,79] (see supplement). Another solution, which we also adopt here, is to randomly sample m negatives and do a simple $(m+1)$-way softmax classification. This strategy is also used in concurrent work [4,28] and dates back to [71].

Memory Bank. Following [79], we maintain a memory bank to store latent features for each training sample. Therefore, we can efficiently retrieve m negative samples from the memory buffer to pair with each positive sample without recomputing their features. The memory bank is dynamically updated with features computed on the fly. The benefit of a memory bank is to allow contrasting against more negative pairs, at the cost of slightly stale features.

4   Experiments

We extensively evaluate Contrastive Multiview Coding (CMC) on a number of datasets and tasks. We evaluate on two established image representation learning benchmarks: ImageNet [15] and STL-10 [11] (see supplement). We further validate our framework on video representation learning tasks, where we use image and optical flow modalities, as the two views that are jointly learned. The last set of experiments extends our CMC framework to more than two views and provides empirical evidence of its effectiveness.

4.1   Benchmarking CMC on ImageNet

Following [84], we evaluate task generalization of the learned representation by training 1000-way *linear* classifiers on top of different layers. This is a standard benchmark that has been adopted by many papers in the literature.

Setup. Given a dataset of RGB images, we convert them to the *Lab* image color space, and split each image into L and ab channels, as originally proposed in SplitBrain autoencoders [85]. During contrastive learning, L and ab from the same image are treated as the positive pair, and ab channels from other randomly selected images are treated as a negative pair (for a given L). Each split represents a view of the orginal image and is passed through a separate encoder. As in SplitBrain, we design these two encoders by evenly splitting a given deep network, such as AlexNet [42], into sub-networks across the channel dimension. By concatenating representations layer-wise from these two encoders, we achieve the final representation of an input image. As proposed by previous literature [3,30,56,79,87], the quality of such a representation is evaluated by freezing the weights of encoder and training linear classifier on top of each layer.

Implementation. Unless otherwise specified, we use PyTorch [58] default data augmentation. Following [79], we set the temperature τ as 0.07 and use a momentum 0.5 for memory update. We use 16384 negatives. The supplementary material provides more details on our hyperparameter settings.

CMC with ResNets. We verify the effectiveness of CMC with larger networks such as ResNets [27]. We experiment on learning from luminance and chrominance views in two colorspaces, $\{L, ab\}$ and $\{Y, DbDr\}$ (see Sect. 4.6 for validation of this choice), and we vary the width of the ResNet encoder for each view. We use the feature after the global pooling layer to train the linear classifier, and the results are shown in Table 1. $\{L, ab\}$ achieves 68.3% top-1 single crop accuracy with ResNet50x2 for each view, and switching to $\{Y, DbDr\}$ further brings about 0.7% improvement. On top of it, strengthening data augmentation with

Table 1. Top-1/5 *Single* crop classification accuracy (%) on ImageNet with a supervised logistic regression classifier. We evaluate CMC using ResNet50 with different width as encoder for *each* of the two views (e.g., L and ab). "RA" stands for RandAugment [13].

Setting	ResNet-50 x0.5	ResNet-50 x1	ResNet-50 x2
$\{L, ab\}$	57.5/80.3	64.0/85.5	68.3/88.2
$\{Y, DbDr\}$	58.4/81.2	64.8/86.1	69.0/88.9
$\{Y, DbDr\}$ + RA	**60.0/82.3**	**66.2/87.0**	**70.6/89.7**

RandAugment [13] yields better or comparable results to other state-of-the-art methods [4, 18, 26, 28, 40, 48, 79, 87].

CMC with AlexNet. As many previous unsupervised methods are evaluated with AlexNet [42] on ImageNet [10, 15–17, 21, 41, 53, 54, 83–85], we also include the the results of CMC using this network in supplementary material.

4.2 CMC on Videos

We apply CMC on videos by drawing insight from the two-streams hypothesis [22, 66], which posits that human visual cortex consists of two distinct processing streams: the ventral stream, which performs object recognition, and the dorsal stream, which processes motion. In our formulation, given an image i_t that is a frame centered at time t, the ventral stream associates it with a neighbouring frame i_{t+k}, while the dorsal stream connects it to optical flow f_t centered at t. Therefore, we extract i_t, i_{t+k} and f_t from two modalities as three views of a video; for optical flow we use the TV-L1 algorithm [82]. Two separate contrastive learning objectives are built within the ventral stream (i_t, i_{t+k}) and within the dorsal stream (i_t, f_t). For the ventral stream, the negative sample for i_t is chosen as a random frame from another randomly chosen video; for the dorsal stream, the negative sample for i_t is chosen as the flow corresponding to a random frame in another randomly chosen video.

Pre-training. We train CMC on UCF101 [72] and use two CaffeNets [42] for extracting features from images and optical flows, respectively. In our implementation, f_t represents 10 continuous flow frames centered at t. We use batch size of 128 and contrast each positive pair with 127 negative pairs.

Action recognition. We apply the learnt representation to the task of action recognition. The spatial network from [68] is a well-established paradigm for evaluating pre-trained RGB network on action recognition task. We follow the same spirit and evaluate the transferability of our RGB CaffeNet on UCF101 and HMDB51 datasets. We initialize the action recognition CaffeNet up to conv5 using the weights from the pre-trained RGB CaffeNet. The averaged accuracy over three splits is present in Table 2. Unifying both ventral and dorsal streams during pre-training produces higher accuracy for downstream recognition than using only single stream. Increasing the number of views of the data from 2 to 3 (using both streams instead of one) provides a boost for UCF-101.

Table 2. Test accuracy (%) on UCF-101 which evaluates *task* transferability and on HMDB-51 which evaluates *task* and *dataset* transferability. Most methods either use single RGB view or additional optical flow view, while VGAN explores sound as the second view. * indicates different network architecture.

Method	# of Views	UCF-101	HMDB-51
Random	–	48.2	19.5
ImageNet	–	67.7	28.0
VGAN* [77]	2	52.1	–
LT-Motion* [45]	2	53.0	–
TempCoh [51]	1	45.4	15.9
Shuffle and Learn [49]	1	50.2	18.1
Geometry [20]	2	55.1	23.3
OPN [43]	1	56.3	22.1
ST Order [9]	1	58.6	25.0
Cross and Learn [65]	2	58.7	**27.2**
CMC (V)	2	55.3	–
CMC (D)	2	57.1	–
CMC (V + D)	3	**59.1**	26.7

4.3   Does Representation Quality Improve as Number of Views Increases?

We further extend our CMC learning framework to multiview scenarios. We experiment on the NYU-Depth-V2 [52] dataset which consists of 1449 labeled images. We focus on a deeper understanding of the behavior and effectiveness of CMC. The views we consider are: luminance (L channel), chrominance (ab channel), depth, surface normal [19], and semantic labels.

Setup. To extract features from each view, we use a neural network with 5 convolutional layers, and 2 fully connected layers. As the size of the dataset is relatively small, we adopt the sub-patch based contrastive objective (see supplement) to increase the number of negative pairs. Patches with a size of 128×128 are randomly cropped from the original images for contrastive learning (from images of size 480×640). For downstream tasks, we discard the fully connected layers and evaluate using the convolutional layers as a representation.

To measure the quality of the learned representation, we consider the task of predicting semantic labels from the representation of L. We follow the *core view paradigm* and use L as the core view, thus learning a set of representations by contrasting different views with L. A UNet style architecture [62] is utilized to perform the segmentation task. Contrastive training is performed on the above architecture that is equivalent of the UNet's encoder. After contrastive training is completed, we initialize the encoder weights of the UNet from the L encoder

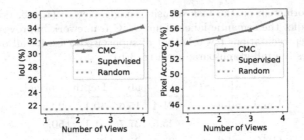

Fig. 4. We show the Intersection over Union (IoU) (left) and pixel accuracy (right) for the NYU-Depth-V2 dataset, as CMC is trained with increasingly more views from 1 to 4. As more views are added, both these metrics steadily increase. The views are (in order of inclusion): L, ab, depth and surface normals.

Table 3. Results on the task of predicting semantic labels from **L channel** representation which is learnt using the patch-based contrastive loss and all 4 views. We compare CMC with *Random* and *Supervised* baselines, which serve as lower and upper bounds respectively. Th core-view paradigm refers to Fig. 3(a), and full-view Fig. 3(b).

	Pixel	
	Accuracy (%)	mIoU (%)
Random	45.5	21.4
CMC (core-view)	57.1	34.1
CMC (full-graph)	57.0	34.4
Supervised	**57.8**	**35.9**

(which are equivalent architectures) and keep them frozen. Only the decoder is trained during this finetuning stage.

Since we use the patch-based contrastive loss, in the 1 view setting case, CMC coincides with DIM [30]. The 2–4 view cases contrast L with ab, and then sequentially add depth and surface normals. The semantic labeling results are measured by mean IoU over all classes and pixel accuracy, shown in Fig. 4. We see that the performance steadily improves as new views are added. We have tested different orders of adding the views, and they all follow a similar pattern.

We also compare CMC with two baselines. First, we randomly initialize and freeze the encoder, and we call this the *Random* baseline; it serves as a lower bound on the quality since the representation is just a random projection. Rather than freezing the randomly initialized encoder, we could train it jointly with the decoder. This end-to-end *Supervised* baseline serves as an upper bound. The results are presented in Table 3, which shows our CMC produces high quality feature maps even though it's unaware of the downstream task.

Table 4. Performance on the task of using single view v (L, ab, depth or surface normal) to predict the semantic labels. Our CMC framework improves the quality of unsupervised representations towards that of supervised ones, for all of views investigated. This uses the full-graph paradigm Fig. 3(b).

	Metric (%)	L	ab	Depth	Normal
Rand.	mIoU	21.4	15.6	30.1	29.5
	pix. acc.	45.5	37.7	51.1	50.5
CMC	mIoU	34.4	26.1	39.2	37.8
	pix. acc.	57.0	49.6	**59.4**	57.8
Sup.	mIoU	**35.9**	**29.6**	**41.0**	**41.5**
	pix. acc.	**57.8**	**52.6**	59.1	**59.6**

4.4 Is CMC Improving All Views?

A desirable unsupervised representation learning algorithm operating on multiple views or modalities should improve the quality of representations for all views. We therefore investigate our CMC framework beyond L channel. To treat all views fairly, we train these encoders following the *full graph paradigm*, where each view is contrasted with all other views.

We evaluate the representation of each view v by predicting the semantic labels from only the representation of v, where v is L, ab, depth or surface normals. This uses the full-graph paradigm. As in the previous section, we compare CMC with *Random* and *Supervised* baselines. As shown in Table 4, the performance of the representations learned by CMC using full-graph significantly outperforms that of randomly projected representations, and approaches the performance of the fully supervised representations. Furthermore, the full-graph representation provides a good representation learnt for all views, showing the importance of capturing different types of mutual information across views.

4.5 Predictive Learning vs. Contrastive Learning

While experiments in Sect. 4.1 show that contrastive learning outperforms predictive learning [85] in the context of Lab color space, it's unclear whether such an advantage is due to the natural inductive bias of the task itself. To further understand this, we go beyond chrominance (ab), and try to answer this question when geometry or semantic labels are present.

We consider three view pairs on the NYU-Depth dataset: (1) L and depth, (2) L and surface normals, and (3) L and segmentation map. For each of them, we train two identical encoders for L, one using contrastive learning and the other with predictive learning. We then evaluate the representation quality by training a linear classifier on top of these encoders on the STL-10 dataset.

The comparison results are shown in Table 5, which shows that contrastive learning consistently outperforms predictive learning in this scenario where both

Table 5. We compare predictive learning with contrastive learning by evaluating the learned encoder on unseen dataset and task. The contrastive learning framework consistently outperforms predictive learning.

	Accuracy on STL-10 (%)	
Views	Predictive	Contrastive
L, depth	55.5	**58.3**
L, normal	58.4	**60.1**
L, Seg. map	57.7	**59.2**
Random	25.2	
Supervised	65.1	

the task and the dataset are unknown. We also include "random" and "supervised" baselines similar to that in previous sections. Though in the unsupervised stage we only use 1.3K images from a dataset much different from the target dataset STL-10, the object recognition accuracy is close to the supervised method, which uses an end-to-end deep network directly trained on STL-10.

Given two views V_1 and V_2 of the data, the predictive learning approach approximately models $p(v_2|v_1)$. Furthermore, losses used typically for predictive learning, such as pixel-wise reconstruction losses usually impose an independence assumption on the modeling: $p(v_2|v_1) \approx \Pi_i p(v_{2i}|v_1)$. On the other hand, the contrastive learning approach by construction does not assume conditional independence across dimensions of v_2. In addition, the use of random jittering and cropping between views allows the contrastive learning approach to benefit from spatial co-occurrence (contrasting in space) in addition to contrasting across views. We conjecture that these are two reasons for the superior performance of contrastive learning approaches over predictive learning.

4.6 How Does Mutual Information Affect Representation Quality?

Given a fixed set of views, CMC aims to maximize the mutual information between representations of these views. We have found that maximizing information in this way indeed results in strong representations, but it would be incorrect to infer that information maximization (infomax) is the key to good representation learning. In fact, this paper argues for precisely the opposite idea: that cross-view representation learning is effective because it results in a kind of information minimization, *discarding* nuisance factors that are not shared between the views.

The resolution to this apparent dilemma is that we want to maximize the "good" information – the *signal* – in our representations, while minimizing the "bad" information – the *noise*. The idea behind CMC is that this can be achieved by doing infomax learning on two views that share signal but have independent noise. This suggests a "Goldilocks principle" [38]: a good collection of views is one that shares some information but not too much. Here we test this hypothesis

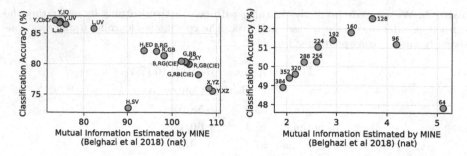

Fig. 5. How does mutual information between views relate to representation quality? (Left) classification accuracy against estimated MI between channels of different color spaces; (Right) classification accuracy vs estimated MI between patches at different distances in pixels. MI estimated using MINE [5].

on two domains: learning representations on images with different colorspaces forming the two views; and learning representations on pairs of patches extracted from an image, separated by varying spatial distance.

In patch experiments we randomly crop two RGB patches of size 64×64 from the same image, and use these patches as the two views. Their relative position is fixed. Namely, the two patches always starts at position (x, y) and $(x + d, y + d)$ with (x, y) being randomly sampled. While varying the distance d, we start from 64 to avoid overlapping. There is a possible bias that with an image of relatively small size (e.g., 512×512), a large d (e.g., 384) will always push these two patches around boundary. To minimize this bias, we use high resolution images (e.g. $2k$) from DIV2K [2] dataset.

Figure 5 shows the results of these experiments. The left plot shows the result of learning representations on different colorspaces (splitting each colorspace into two views, such as (L, ab), (R, GB) etc.). We then use the MINE estimator [5] to estimate the mutual information between the views. We measure representation quality by training a linear classifier on the learned representations on the STL-10 dataset [11]. The plots clearly show that using colorspaces with minimal mutual information give the best downstream accuracy (For the outlier HSV in this plot, we conjecture the representation quality is harmed by the periodicity of H. Note that the H in HED is not periodic.). On the other hand, the story is more nuanced for representations learned between patches at different offsets from each other (Fig. 5, right). Here we see that views with too little or too much MI perform worse; a sweet spot in the middle exists which gives the best representation. That there exists such a sweet spot should be expected. If two views share *no* information, then, in principle, there is no incentive for CMC to learn anything. If two views share all their information, no nuisances are discarded and we arrive back at something akin to an autoencoder or generative model, that simply tries to represent all the bits in the multiview data.

These experiments demonstrate that the relationship between mutual information and representation quality is meaningful but not direct. Selecting optimal views, which just share relevant signal, has been further discussed in a follow-up work [75] of CMC, and may be a fruitful direction for future research.

5 Conclusion

We have presented a contrastive learning framework which enables the learning of unsupervised representations from multiple views or modalities of a dataset. The principle of maximization of mutual information enables the learning of powerful representations. A number of empirical results show that our framework performs well compared to predictive learning and scales with the number of views.

References

1. Information Diagram - Wikipedia. https://en.wikipedia.org/wiki/Information_diagram
2. Agustsson, E., Timofte, R.: Ntire 2017 challenge on single image super-resolution: dataset and study. In: CVPR (2017)
3. Arora, S., Khandeparkar, H., Khodak, M., Plevrakis, O., Saunshi, N.: A theoretical analysis of contrastive unsupervised representation learning. In: ICML (2019)
4. Bachman, P., Hjelm, R.D., Buchwalter, W.: Learning representations by maximizing mutual information across views. arXiv preprint arXiv:1906.00910 (2019)
5. Belghazi, M.I., et al.: Mine: mutual information neural estimation. arXiv preprint arXiv:1801.04062 (2018)
6. Bellet, A., Habrard, A., Sebban, M.: Similarity learning for provably accurate sparse linear classification. arXiv preprint arXiv:1206.6476 (2012)
7. Bengio, Y., Courville, A., Vincent, P.: Representation learning: a review and new perspectives. TPAMI 35, 1798–1828 (2013)
8. Blum, A., Mitchell, T.: Combining labeled and unlabeled data with co-training. In: COLT. ACM (1998)
9. Buchler, U., Brattoli, B., Ommer, B.: Improving spatiotemporal self-supervision by deep reinforcement learning. In: ECCV (2018)
10. Caron, M., Bojanowski, P., Joulin, A., Douze, M.: Deep clustering for unsupervised learning of visual features. In: Ferrari, V., Hebert, M., Sminchisescu, C., Weiss, Y. (eds.) Computer Vision – ECCV 2018. LNCS, vol. 11218, pp. 139–156. Springer, Cham (2018). https://doi.org/10.1007/978-3-030-01264-9_9
11. Coates, A., Ng, A., Lee, H.: An analysis of single-layer networks in unsupervised feature learning. In: AISTATS (2011)
12. Cortes, C., Mohri, M., Rostamizadeh, A.: Learning non-linear combinations of kernels. In: NIPS (2009)
13. Cubuk, E.D., Zoph, B., Shlens, J., Le, Q.V.: Randaugment: practical data augmentation with no separate search. arXiv preprint arXiv:1909.13719 (2019)
14. Den Ouden, H.E., Kok, P., De Lange, F.P.: How prediction errors shape perception, attention, and motivation. Front. Psychol. 3, 548 (2012)
15. Deng, J., Dong, W., Socher, R., Li, L.J., Li, K., Fei-Fei, L.: ImageNet: a large-scale hierarchical image database. In: CVPR (2009)
16. Doersch, C., Gupta, A., Efros, A.A.: Unsupervised visual representation learning by context prediction. In: CVPR (2015)

17. Donahue, J., Krähenbühl, P., Darrell, T.: Adversarial feature learning. In: ICLR (2017)
18. Donahue, J., Simonyan, K.: Large scale adversarial representation learning. In: NIPS (2019)
19. Eigen, D., Fergus, R.: Predicting depth, surface normals and semantic labels with a common multi-scale convolutional architecture. In: ICCV (2015)
20. Gan, C., Gong, B., Liu, K., Su, H., Guibas, L.J.: Geometry guided convolutional neural networks for self-supervised video representation learning. In: CVPR (2018)
21. Gidaris, S., Singh, P., Komodakis, N.: Unsupervised representation learning by predicting image rotations. In: ICLR (2018)
22. Goodale, M.A., Milner, A.D.: Separate visual pathways for perception and action. Trends Neurosci. **15**, 2025Grave (1992)
23. Goodfellow, I., et al.: Generative adversarial nets. In: NIPS (2014)
24. Gutmann, M., Hyvärinen, A.: Noise-contrastive estimation: a new estimation principle for unnormalized statistical models. In: AISTATS (2010)
25. Hadsell, R., Chopra, S., LeCun, Y.: Dimensionality reduction by learning an invariant mapping. In: CVPR (2006)
26. He, K., Fan, H., Wu, Y., Xie, S., Girshick, R.: Momentum contrast for unsupervised visual representation learning. arXiv preprint arXiv:1911.05722 (2019)
27. He, K., Zhang, X., Ren, S., Sun, J.: Deep residual learning for image recognition. In: CVPR (2016)
28. Hénaff, O.J., Razavi, A., Doersch, C., Eslami, S., Oord, A.V.D.: Data-efficient image recognition with contrastive predictive coding. arXiv preprint arXiv:1905.09272 (2019)
29. Hinton, G.E., Salakhutdinov, R.R.: Reducing the dimensionality of data with neural networks. Science **313**, 504–507 (2006)
30. Hjelm, R.D., Fedorov, A., Lavoie-Marchildon, S., Grewal, K., Trischler, A., Bengio, Y.: Learning deep representations by mutual information estimation and maximization. In: ICLR (2019)
31. Hohwy, J.: The Predictive Mind. Oxford University Press, Oxford (2013)
32. Hyvärinen, A., Karhunen, J., Oja, E.: Independent Component Analysis, vol. 46. Wiley, Hoboken (2004)
33. Isola, P., Zhu, J.Y., Zhou, T., Efros, A.A.: Image-to-image translation with conditional adversarial networks. In: CVPR (2017)
34. Isola, P., Zoran, D., Krishnan, D., Adelson, E.H.: Learning visual groups from co-occurrences in space and time. arXiv preprint arXiv:1511.06811 (2015)
35. Ji, X., Henriques, J.F., Vedaldi, A.: Invariant information clustering for unsupervised image classification and segmentation. In: ICCV (2019)
36. Jolliffe, I.: Principal Component Analysis. Springer, Heidelberg (2011). https://doi.org/10.1007/b98835
37. Kawakami, K., Wang, L., Dyer, C., Blunsom, P., Oord, A.V.D.: Learning robust and multilingual speech representations. arXiv preprint arXiv:2001.11128 (2020)
38. Kidd, C., Piantadosi, S.T., Aslin, R.N.: The goldilocks effect: human infants allocate attention to visual sequences that are neither too simple nor too complex. PloS One **7**, e36399 (2012)
39. Kingma, D.P., Welling, M.: Auto-encoding variational Bayes. arXiv preprint arXiv:1312.6114 (2013)
40. Kolesnikov, A., Zhai, X., Beyer, L.: Revisiting self-supervised visual representation learning. In: CVPR (2019)
41. Krähenbühl, P., Doersch, C., Donahue, J., Darrell, T.: Data-dependent initializations of convolutional neural networks. arXiv preprint arXiv:1511.06856 (2015)

42. Krizhevsky, A., Sutskever, I., Hinton, G.E.: ImageNet classification with deep convolutional neural networks. In: NIPS (2012)
43. Lee, H.Y., Huang, J.B., Singh, M., Yang, M.H.: Unsupervised representation learning by sorting sequences. In: ICCV (2017)
44. Li, Y., Yang, M., Zhang, Z.M.: A survey of multi-view representation learning. TKDE **31**, 1863–1883 (2018)
45. Luo, Z., Peng, B., Huang, D.A., Alahi, A., Fei-Fei, L.: Unsupervised learning of long-term motion dynamics for videos. In: CVPR (2017)
46. McAllester, D., Statos, K.: Formal limitations on the measurement of mutual information. arXiv preprint arXiv:1811.04251 (2018)
47. Miech, A., Alayrac, J.B., Smaira, L., Laptev, I., Sivic, J., Zisserman, A.: End-to-end learning of visual representations from uncurated instructional videos. In: CVPR (2020)
48. Misra, I., van der Maaten, L.: Self-supervised learning of pretext-invariant representations. arXiv preprint arXiv:1912.01991 (2019)
49. Misra, I., Zitnick, C.L., Hebert, M.: Shuffle and learn: unsupervised learning using temporal order verification. In: Leibe, B., Matas, J., Sebe, N., Welling, M. (eds.) ECCV 2016. LNCS, vol. 9905, pp. 527–544. Springer, Cham (2016). https://doi.org/10.1007/978-3-319-46448-0_32
50. Mnih, A., Kavukcuoglu, K.: Learning word embeddings efficiently with noise-contrastive estimation. In: NIPS (2013)
51. Mobahi, H., Collobert, R., Weston, J.: Deep learning from temporal coherence in video. In: ICML (2009)
52. Silberman, N., Hoiem, D., Kohli, P., Fergus, R.: Indoor segmentation and support inference from RGBD images. In: Fitzgibbon, A., Lazebnik, S., Perona, P., Sato, Y., Schmid, C. (eds.) ECCV 2012. LNCS, vol. 7576, pp. 746–760. Springer, Heidelberg (2012). https://doi.org/10.1007/978-3-642-33715-4_54
53. Noroozi, M., Favaro, P.: Unsupervised learning of visual representations by solving jigsaw puzzles. In: Leibe, B., Matas, J., Sebe, N., Welling, M. (eds.) ECCV 2016. LNCS, vol. 9910, pp. 69–84. Springer, Cham (2016). https://doi.org/10.1007/978-3-319-46466-4_5
54. Noroozi, M., Pirsiavash, H., Favaro, P.: Representation learning by learning to count. In: ICCV (2017)
55. Oord, A.V.D., Kalchbrenner, N., Kavukcuoglu, K.: Pixel recurrent neural networks. arXiv preprint arXiv:1601.06759 (2016)
56. Oord, A.V.D., Li, Y., Vinyals, O.: Representation learning with contrastive predictive coding. arXiv preprint arXiv:1807.03748 (2018)
57. Owens, A., Isola, P., McDermott, J., Torralba, A., Adelson, E.H., Freeman, W.T.: Visually indicated sounds. In: CVPR (2016)
58. Paszke, A., et al.: Pytorch: an imperative style, high-performance deep learning library. In: NIPS (2019)
59. Pathak, D., Krahenbuhl, P., Donahue, J., Darrell, T., Efros, A.A.: Context encoders: feature learning by inpainting. In: CVPR (2016)
60. Piergiovanni, A., Angelova, A., Ryoo, M.S.: Evolving losses for unlabeled video representation learning. In: CVPR (2020)
61. Poole, B., Ozair, S., Oord, A.V.D., Alemi, A.A., Tucker, G.: On variational bounds of mutual information. In: ICML (2019)
62. Ronneberger, O., Fischer, P., Brox, T.: U-Net: convolutional networks for biomedical image segmentation. In: Navab, N., Hornegger, J., Wells, W.M., Frangi, A.F. (eds.) MICCAI 2015. LNCS, vol. 9351, pp. 234–241. Springer, Cham (2015). https://doi.org/10.1007/978-3-319-24574-4_28

63. Sa, V.: Sensory modality segregation. In: NIPS (2004)
64. Salakhutdinov, R., Hinton, G.: Deep Boltzmann machines. In: AISTATS (2009)
65. Sayed, N., Brattoli, B., Ommer, B.: Cross and learn: cross-modal self-supervision. arXiv preprint arXiv:1811.03879 (2018)
66. Schneider, G.E.: Two visual systems. Science **163**, 895–902 (1969)
67. Sermanet, P., et al.: Time-contrastive networks: self-supervised learning from video. In: ICRA (2018)
68. Simonyan, K., Zisserman, A.: Two-stream convolutional networks for action recognition in videos. In: NIPS (2014)
69. Smith, L., Gasser, M.: The development of embodied cognition: six lessons from babies. Artif. Life **11**, 13–29 (2005)
70. Smolensky, P.: Information processing in dynamical systems: foundations of harmony theory. Tech. rep., Colorado University at Boulder Department of Computer Science (1986)
71. Sohn, K.: Improved deep metric learning with multi-class n-pair loss objective. In: NIPS (2016)
72. Soomro, K., Zamir, A.R., Shah, M.: Ucf101: a dataset of 101 human actions classes from videos in the wild. arXiv preprint arXiv:1212.0402 (2012)
73. Sun, C., Baradel, F., Murphy, K., Schmid, C.: Contrastive bidirectional transformer for temporal representation learning. arXiv preprint arXiv:1906.05743 (2019)
74. Tian, Y., Krishnan, D., Isola, P.: Contrastive representation distillation. In: ICLR (2020)
75. Tian, Y., Sun, C., Poole, B., Krishnan, D., Schmid, C., Isola, P.: What makes for good views for contrastive learning. arXiv preprint arXiv:2005.10243 (2020)
76. Tschannen, M., et al.: Self-supervised learning of video-induced visual invariances. arXiv preprint arXiv:1912.02783 (2019)
77. Vondrick, C., Pirsiavash, H., Torralba, A.: Generating videos with scene dynamics. In: NIPS (2016)
78. Wang, X., Gupta, A.: Unsupervised learning of visual representations using videos. In: ICCV (2015)
79. Wu, Z., Xiong, Y., Yu, S.X., Lin, D.: Unsupervised feature learning via nonparametric instance discrimination. In: CVPR (2018)
80. Xu, C., Tao, D., Xu, C.: A survey on multi-view learning. arXiv preprint arXiv:1304.5634 (2013)
81. Ye, M., Zhang, X., Yuen, P.C., Chang, S.F.: Unsupervised embedding learning via invariant and spreading instance feature. In: CVPR (2019)
82. Zach, C., Pock, T., Bischof, H.: A duality based approach for realtime TV-L^1 optical flow. In: Hamprecht, F.A., Schnörr, C., Jähne, B. (eds.) DAGM 2007. LNCS, vol. 4713, pp. 214–223. Springer, Heidelberg (2007). https://doi.org/10.1007/978-3-540-74936-3_22
83. Zhang, L., Qi, G.J., Wang, L., Luo, J.: AET vs. AED: unsupervised representation learning by auto-encoding transformations rather than data. In: CVPR (2019)
84. Zhang, R., Isola, P., Efros, A.A.: Colorful image colorization. In: Leibe, B., Matas, J., Sebe, N., Welling, M. (eds.) ECCV 2016. LNCS, vol. 9907, pp. 649–666. Springer, Cham (2016). https://doi.org/10.1007/978-3-319-46487-9_40
85. Zhang, R., Isola, P., Efros, A.A.: Split-brain autoencoders: unsupervised learning by cross-channel prediction. In: CVPR (2017)
86. Zhuang, C., Andonian, A., Yamins, D.: Unsupervised learning from video with deep neural embeddings. arXiv preprint arXiv:1905.11954 (2019)
87. Zhuang, C., Zhai, A.L., Yamins, D.: Local aggregation for unsupervised learning of visual embeddings. arXiv preprint arXiv:1903.12355 (2019)

Regional Homogeneity: Towards Learning Transferable Universal Adversarial Perturbations Against Defenses

Yingwei Li[1]([✉]), Song Bai[2], Cihang Xie[1], Zhenyu Liao[3], Xiaohui Shen[4], and Alan Yuille[1]

[1] Johns Hopkins University, Baltimore, USA
yingwei.li@jhu.edu
[2] University of Oxford, Oxford, UK
[3] Kuaishou Technology, Palo Alto, USA
[4] ByteDance Research, Mountain View, USA

Abstract. This paper focuses on learning transferable adversarial examples specifically against defense models (models to defense adversarial attacks). In particular, we show that a simple universal perturbation can fool a series of state-of-the-art defenses.

Adversarial examples generated by existing attacks are generally hard to transfer to defense models. We observe the property of regional homogeneity in adversarial perturbations and suggest that the defenses are less robust to regionally homogeneous perturbations. Therefore, we propose an effective transforming paradigm and a customized gradient transformer module to transform existing perturbations into regionally homogeneous ones. Without explicitly forcing the perturbations to be universal, we observe that a well-trained gradient transformer module tends to output input-independent gradients (hence universal) benefiting from the under-fitting phenomenon. Thorough experiments demonstrate that our work significantly outperforms the prior art attacking algorithms (either image-dependent or universal ones) by an average improvement of 14.0% when attacking 9 defenses in the transfer-based attack setting. In addition to the cross-model transferability, we also verify that regionally homogeneous perturbations can well transfer across different vision tasks (attacking with the semantic segmentation task and testing on the object detection task). The code is available here: https://github.com/LiYingwei/Regional-Homogeneity.

Keywords: Transferable adversarial example · Universal attack

1 Introduction

Deep neural networks are demonstrated vulnerable to adversarial examples [66], crafted by adding imperceptible perturbations to clean images. The variants of

Electronic supplementary material The online version of this chapter (https://doi.org/10.1007/978-3-030-58621-8_46) contains supplementary material, which is available to authorized users.

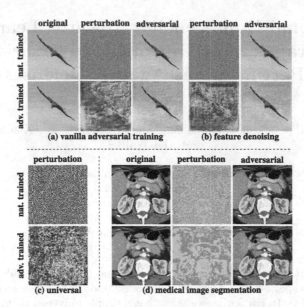

Fig. 1. Illustration of region homogeneity property of adversarial perturbations by white-box attacking naturally trained models (top row) and adversarially trained models (bottom row). The adversarially trained models are acquired by (a) vanilla adversarial training [48,75], (b) adversarial training with feature denoising [75], (c) universal adversarial training [60], and (d) adversarial training for medical image segmentation [41]

adversarial attacks [3,5,9,12,20,21,30,32,33,55,62,63,79] cast a security threat when deploying machine learning systems. To mitigate this, large efforts have been devoted to adversarial defense [7,34,47,71], via adversarial training [42,48, 67,68,73,75,76], randomized transformation [14,26,45,74] *etc.*.

The focus of this work is to attack defense models, especially in the transfer-based attack setting where models' architectures and parameters remain unknown to attackers. In this case, the adversarial examples generated for one model, which possess the property of "transferability", may also be misclassified by other models. To the best of our knowledge, learning transferable adversarial examples for attacking defense models is still an open problem.

Our work stems from the observation of *regional homogeneity* on adversarial perturbations in the white-box setting. As Fig. 1(a) shows, we plot the adversarial perturbations generated by attacking a naturally trained Resnet-152 [28] model (top) and an representative defense one (*i.e.*, an adversarially trained model [48,75]). It suggests that the patterns of two kinds of perturbations are visually different. Concretely, the perturbations of defense models reveal a coarser level of granularity, and are more locally correlated and more structured than that of the naturally trained model. The observation also holds when attacking different defense models (*e.g.*, adversarial training with feature denoising [75], Fig. 1(b)), generating different types of adversarial examples (image-dependent or universal perturbations [60], Fig. 1(c)), or tested on different data domains (CT scans [57], Fig. 1(d)).

Motivated by this observation, we suggest that *regionally homogeneous perturbations* are strong in attacking defense models, which is especially helpful to learn transferable adversarial examples in the transfer-based attack setting. Hence, we propose to transform the existing perturbations (those derive from differentiating naturally trained models) to the regionally homogeneous ones. To this end, we develop a novel transforming paradigm (Fig. 2) to craft regionally homogeneous perturbations, and accordingly a gradient transformer module (Fig. 3), to encourage local correlations within the pre-defined regions.

The proposed gradient transformer module is quite light-weight, with only $12 + 2K$ trainable parameters in total, where a 3×3 convolutional layer (bias enabled) incurs 12 parameters and K is the number of region partitions. According to our experiments, it leads to under-fitting (large bias and small variance) if the module is trained with a large number of images. In general vision tasks, an under-fitting model is undesirable. However in our case, once the gradient transformer module becomes quasi-input-independent (*i.e.*, aforementioned large bias and small variance), it will output a nearly fixed pattern whatever the input is. Then, our work is endowed with a desirable property, *i.e.*, seemingly training to generate image-dependent perturbations, yet get the universal ones. We note our mechanism is different from other universal adversarial generations [50,54] as we do not explicitly force the perturbation to be universal.

Comprehensive experiments are conducted to verify the effectiveness of the proposed regionally homogeneous perturbation (RHP). Under the transfer-based attack setting, RHP successfully attacks 9 latest defenses [26,35,42,48,68,74,75] and improves the top-1 error rates by 21.6% in average, where three of them are the top submissions in the NeurIPS 2017 defense competition [39] and the Competition on Adversarial Attacks and Defenses 2018. Compared with the state-of-the-art attack methods, RHP not only outperforms universal adversarial perturbations (*e.g.*, UAP [50] by 19.2% and GAP [54] by 15.6%), but also outperforms image-dependent perturbations (FGSM [23] by 12.9%, MIM [16] by 12.6% and DIM [16,77] by 9.58%). The achievement over image-dependent perturbations is especially valuable as it is known that image-dependent perturbations generally perform better as they utilized information from the original images. Since it is universal, RHP is more general (natural noises are not related to the target image), more efficient (without additional computational power), and more flexible (*e.g.*, without knowing the target image, people can stick a pattern on the lens to attack artificial intelligence surveillance cameras).

Moreover, we also evaluate the cross-task transferability of RHP and demonstrate that RHP generalizes well in cross-task attack, *i.e.*, attacking with the semantic segmentation task and testing on the object detection task.

2   Related Work

Transfer-Based Attacks. Practically, attackers cannot easily access the internal information of target models (including its architecture, parameters and outputs). A typical solution is to generate adversarial examples with strong transferability. Szegedy *et al.* [66] first discuss the transferability of adversarial examples that the same input can successfully attack different models. Liu *et al.* [46] then develop a stronger attack to successfully circumvent an online image classification system with ensemble attacks, which is later analysed by [44]. Based on one of the most well-known attack methods, Fast Gradient Sign Method (FGSM) [23] and its iteration-based version (I-FGSM) [38], many follow-ups are then proposed to further improve the transferability by adopting momentum term [16], smoothing perturbation [80], constructing diverse inputs [77], augmenting ghost models [40] and smoothing gradient [17], respectively. Recent works [4,49,52–54,72] also suggest to train generative models for creating adversarial examples. Besides transfer-based attacks, query-based [5,8,11,25,78] attacks are also very popular black-box attack settings.

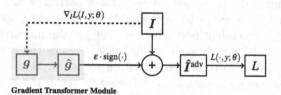

Gradient Transformer Module

Fig. 2. Illustration of the transforming paradigm, where I is an original image with the corresponding label y, and the gradient g is computed from the naturally trained model θ. Our work learns a mapping to transform gradient from g to $\hat{g}$

Universal Adversarial Perturbations. Above are all image-dependent perturbation attacks. Moosavi-Dezfooli *et al.* [50] craft universal perturbations which can be directly added to any test images to fool the classifier with a high success rate. Poursaeed *et al.* [54] propose to train a neural network for generating adversarial examples by explicitly feeding random noise to the network during training. After obtaining a well-trained model, they use a fixed input to generate universal adversarial perturbations. Researchers also explore to produce universal adversarial perturbations by different methods [36,51] or on different tasks [29,54]. All these methods construct universal adversarial perturbations explicitly or data-independently. Unlike them, we provide an implicit data-driven alternative to generate universal adversarial perturbations.

Defense Methods. Xie *et al.* [74] and Guo *et al.* [26] break transferability by applying input transformation such as random padding/resizing [74], JPEG compression [18], and total variance minimization [59]. Injecting adversarial examples during training improves the robustness of deep neural network, termed as adversarial training. These adversarial examples can be pre-generated [42,68] or generated on-the-fly during training [35,48,75]. Adversarial training is also applied to universal adversarial perturbations [1,60].

Normalization. To induce regionally homogeneous perturbations, our work resorts to a new normalization strategy. This strategy appears similar to some normalization techniques, such as batch normalization [31], layer normalization [2], instance normalization [69], group normalization [70], *etc.*. While these techniques aim to help the model converge faster and speed up the learning procedure for different tasks, the goal of our proposed region norm is to explicitly enforce the region structure and build homogeneity within regions.

3 Regionally Homogeneous Perturbations

As shown in Sect. 1, regionally homogeneous perturbations appear to be strong in attacking defense models. To acquire regionally homogeneous adversarial examples, we propose a gradient transformer module to generate regionally homogeneous perturbations from existing regionally non-homogeneous perturbations (*e.g.*, perturbations in the top row of Fig. 1). In the following, we detail the transforming paradigm in Sect. 3.1 and the core component called gradient transformer module in Sect. 3.2, respectively. In Sect. 3.3, we observe an underfitting phenomenon and illustrate that the proposed gradient transformer module becomes quasi-input-independent, which benefits crafting universal adversarial perturbations.

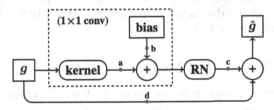

Fig. 3. Structure of the gradient transformer module, which has a newly proposed Region Norm (RN) layer, 1×1 convolutional layer (bias enabled) and identity mapping. We insert four probes (a, b,c and d) to assist analysis in Sect. 3.3 and Sect. 4.2

3.1 Transforming Paradigm

To learn regionally homogeneous adversarial perturbations, we propose to use a shallow network T, which we call gradient transformer module, to transform the gradients that are generated by attacking naturally trained models.

Concretely, we consider Fast Gradient Sign Method (FGSM) [23] which generates adversarial examples by

$$I^{\text{adv}} = I + \epsilon \cdot \text{sign}\left(\nabla_I L(I, y; \theta)\right), \tag{1}$$

where L is the loss function of the model θ, and sign($\cdot$) denotes the sign function. y is the ground-truth label of the original image I. FGSM ensures that the generated adversarial example I^{adv} is within the ϵ-ball of I in the L_∞ space.

Based on FGSM, we build pixel-wise connections via the additional gradient transformer module T, so that we may have regionally homogeneous perturbations. Therefore, Eq. (1) becomes

$$I^{\text{adv}} = I + \epsilon \cdot \text{sign}\left(T(\nabla_I L(I, y; \theta); \theta_T)\right), \tag{2}$$

where θ_T is trainable parameter of gradient transformer module T, and we omit θ_T where possible for simplification. The challenge we are facing now is how to train the gradient transformer module $T(\cdot)$ with the limited supervision. We address this by proposing a new transforming paradigm illustrated in Fig. 2. It consists of four steps, as we 1) compute the gradient $g = \nabla_I L(I, y; \theta)$ by attacking the naturally trained model θ; 2) get the transformed gradient $\hat{g} = T(g; \theta_T)$ via the gradient transformer module; 3) construct the adversarial image $\hat{I}$ by adding the transformed perturbation to the clean image I, forward $\hat{I}$ to the same model θ, and obtain the classification loss $L(\hat{I}^{\text{adv}}, y; \theta)$; and 4) freeze the clean image I and the model θ, and update the parameters θ_T of $T(\cdot; \theta_T)$ by **maximizing** $L(\hat{I}^{\text{adv}}, y; \theta)$. The last step is implemented via stochastic gradient ascent (*e.g.*, we use the Adam optimizer [37] in our experiments).

With the new transforming paradigm, one can potentially embed desirable properties via using the gradient transformer module $T(\cdot)$, and in the meantime, keep a high error rate on the model θ. As we will show below, $T(\cdot)$ is customized to generate regionally homogeneous perturbations specially against defense models. Meanwhile, since we freeze the most part of the computation graph and leave a limited number of parameters (that is θ_T) to optimize, the learning procedure is very fast.

3.2   Gradient Transformer Module

With the transforming paradigm aforementioned, we introduce the architecture of the core module, termed as gradient transformer module. The gradient transformer module aims at increasing the correlation of pixels in the same region, therefore inducing regionally homogeneous perturbations. As shown in Fig. 3, given a loss gradient g as the input, the gradient transformer module $T(\cdot)$ is

$$\hat{g} = T(g; \theta_T) = \text{RN}\left(\text{conv}(g)\right) + g, \tag{3}$$

where conv($\cdot$) is a 1×1 convolutional layer and RN($\cdot$) is the newly proposed region norm layer. θ_T is the module parameters, which goes to the region norm layer (γ and β below) and the convolutional layer. A residual connection [28] is also incorporated. Since RN($\cdot$) is initialized as zero [24], the residual connection allows us to insert the gradient transformer module into any gradient-based attack methods without breaking its initial behavior (*i.e.*, the transformed gradient $\hat{g}$ initially equals to g). Since the initial gradient g is able to craft stronger adversarial example (compared with random noises), the gradient transformer module has a proper initialization. The region norm layer consists of two parts, including a region split function and a region norm operator.

Region Split Function splits an image (or equivalently, a convolutional feature map) into K regions. Let $r(\cdot,\cdot)$ denote the region split function. The input of $r(\cdot,\cdot)$ is a pixel coordinate while the output is an index of the region which the pixel belongs to. With a region split function, we can get a partition $\{P_1, P_2, ..., P_K\}$ of an image, where $P_k = \{(h,w) \mid r(h,w) = k, 1 \le k \le K\}$.

In Fig. 4, we show 4 representatives of region split functions on a toy 6×6 image, including 1) vertical partition $(h,w) = w$, 2) horizontal partition $r(h,w) = h$, 3) grid partition $r(h,w) = \lfloor h/3 \rfloor + 2\lfloor w/3 \rfloor$, and 4) slash partition (parallel to an increasing line with the slope equal to 0.5).

Region Norm Operator links pixels within the same region P_k, defined as

$$y_i = \gamma_k \bar{x}_i + \beta_k, \quad \bar{x}_i = \frac{1}{\sigma_k}(x_i - \mu_k), \tag{4}$$

where x_i and y_i are the i-th input and output, respectively. And $i = (n, c, h, w)$ is a 4D vector indexing the features in (N, C, H, W) order, where N is the batch axis, C is the channel axis, and H and W are the spatial height and width axes. We define S_k as a set of pixels that belong to the region P_k, that is, $r(h,w) = k$.

μ_k and σ_k in Eq. (4) are the mean and standard deviation (std) of the kth region, computed by

$$\mu_k = \frac{1}{m_k} \sum_{j \in S_k} x_j,$$

$$\sigma_k = \sqrt{\frac{1}{m_k} \sum_{j \in S_k} (x_j - \mu_k)^2 + \text{const},} \tag{5}$$

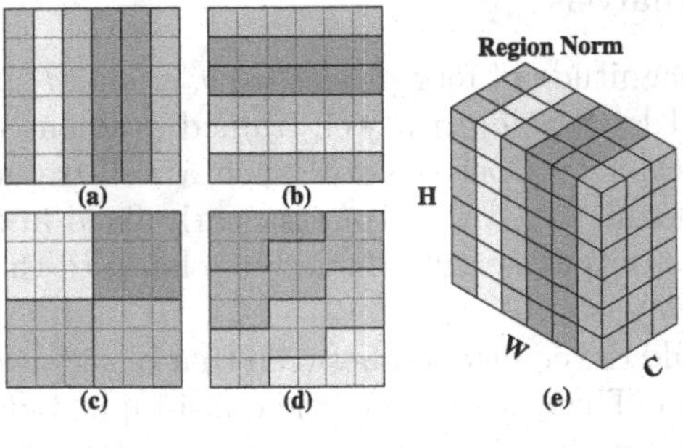

Fig. 4. Toy examples of region split functions, including (a) vertical partition, (b) horizontal partition, (c) grid partition, and (d) slash partition. (e) illustrates the region norm operator with the region split function (a), where C is the channel axis, H and W are the spatial axes. Each pixel indicates an N-dimensional vector, where N is the batch size

where const is a small constant for numerical stability. m_k is the size of S_k. Here $m_k = NC|P_k|$ and $|\cdot|$ is the cardinality of a given set. In the testing phase, the moving mean and moving std during training are used instead. Since we split the image to regions, the trainable scale γ and shift β in Eq. (4) are also learned per-region.

We illustrate the region norm operator in Fig. 4(e). To analyze the benefit, we compute the derivatives as

$$\frac{\partial L}{\partial \beta_k} = \sum_{j \in S_k} \frac{\partial L}{\partial y_j}, \quad \frac{\partial L}{\partial \gamma_k} = \sum_{j \in S_k} \frac{\partial L}{\partial y_j} \bar{x}_j,$$

$$\frac{\partial L}{\partial x_i} = \frac{1}{m_k \sigma_k}(m_k \frac{\partial L}{\partial \bar{x}_i} - \sum_{j \in S_k} \frac{\partial L}{\partial \bar{x}_j} - \bar{x}_i \sum_{j \in S_k} \frac{\partial L}{\partial \bar{x}_j} \bar{x}_j),$$

(6)

where L is the loss to optimize, and $\frac{\partial L}{\partial \bar{x}_i} = \frac{\partial L}{\partial y_i} \cdot \gamma_k$. It is not surprising that the gradient of γ or β is computed by all pixels in the related region. However, the gradient of a pixel with an index i is also computed by all pixels in the same region. More significantly in Eq. (6), the second term, $\sum_{j \in S_k} \frac{\partial L}{\partial \bar{x}_j}$, and the third term, $\bar{x}_i \sum_{j \in S_k} \frac{\partial L}{\partial \bar{x}_j} \bar{x}_j$, are shared by all pixels in the same region. Therefore, the pixel-wise connections within the same region are much denser after inserting the region norm layer.

Comparison with Other Normalizations. Compared with existing normalizations (*e.g.*, Batch Norm [31], Layer Norm [2], Instance Norm [69] and Group Norm [70]), which aims to speed up learning, there are two main difference: 1) the goal of Region Norm is to generate regionally homogeneous perturbations, while existing methods mainly aim to stabilize and speed up training; 2) the formulation of Region Norm is splitting an image to regions and normalize each region individually, while other methods do not split along spatial dimension.

3.3 Universal Analysis

By analyzing the magnitude of four probes (a, b, c, and d) in Fig. 3, we observe that $|b| >> |a|$ and $|c| >> |d|$ in a well-trained gradient transformer module (more results in Sect. 4.2). Consequently, such a well-trained module becomes quasi-input-independent, *i.e.*, the output is nearly fixed and less related to the input. Note that the output is still a little bit related to the input which is the reason why we use "quasi-".

Here, we first build the connection between that observation and under-fitting to explain the reason. Then, we convert the quasi-input-independent module to an input-independent module for generating universal adversarial perturbations.

Under-Fitting and the Quasi-Input-Independent Module. People figure out the trade-off between bias and variance of a model, *i.e.*, the price for achieving a small bias is a large variance, and vice versa [6,27]. Under-fitting occurs when the model shows low variance (but inevitable bias). An extremely low variance function gives a nearly fixed output whatever the input, which we term as quasi-input-independent. Although in the most machine learning situation people do not expect this case, the quasi-input-independent function is desirable for generating universal adversarial perturbation.

Therefore, to encourage under-fitting, we go to the opposite direction of preventing under-fitting suggestions in [22]. On the one hand, to minimize the model capacity, our gradient transformer module only has $(12+2K)$ parameters, where a 3×3 convolutional layer (bias enabled) incurs 12 parameters and K is the number of region partitions. On the other hand, we use a large training data set $\mathcal{D}$ (5k images or more) so that the model capacity is relatively small. We then will have a quasi-input-independent module.

From Quasi-Input-Independent to Input-Independent. According to the analysis above, we already have a quasi-input-independent module. To generate a universal adversarial perturbation, following the post-process strategy of Poursaeed *et al.* [54], we use a fixed vector as input of the module. Then following FGSM [23], the final universal perturbation will be $u = \epsilon \cdot \text{sign}(T(z))$, where z is a fixed input. Recall that $\text{sign}(\cdot)$ denotes the sign function, and $T(\cdot)$ denotes the gradient transformer module.

4 Experiments

In this section, we demonstrate the effectiveness of the proposed regionally homogeneous perturbation (RHP) by attacking a series of defense models. The code is made publicly available.

4.1 Experimental Setup

Dataset and Evaluation Metric. Without loss of generality, we randomly select 5000 images from the ILSVRC 2012 [15] validation set to access the transferability of attack methods. For the evaluation metric, we use the improvement of top-1 error rate after attacking, *i.e.*, the difference between the error rate of adversarial images and that of clean images.

Table 1. The error rates (%) of defense methods on our dataset which contains 5000 randomly selected ILSVRC 2012 validation images

Defenses	TVM	HGD	R& P	Inc_{ens3}	Inc_{ens4}	IncRes_{ens}	PGD	ALP	FD
Error Rate	37.4	18.6	19.9	25.0	24.5	21.3	40.9	48.6	35.1

Attack Methods. For performance comparison, we reproduce five representative attack methods, including fast gradient sign method (FGSM) [23], momentum iterative fast gradient sign method (MIM) [16], momentum diverse inputs iterative fast gradient sign method (DIM) [16,77], universal adversarial perturbations (UAP) [50], and the universal version of generative adversarial perturbations (GAP) [54]. If not specified otherwise, we follow the default parameter setup in each method respectively.

To keep the perturbation quasi-imperceptible, we generate adversarial examples in the ϵ-ball of original images in the L_∞ space. The maximum perturbation ϵ is set as 16 or 32. The adversarial examples are generated by attacking a naturally trained network, Inception v3 (IncV3) [65], Inception v4 (IncV4) or Inception Resnet v2 (IncRes) [64]. We use IncV3 and $\epsilon = 16$ in default.

Defense Methods. As our method is to attack defense models, we reproduce nine defense methods for performance evaluation, including input transformation [26] through total variance minimization (TVM), high-level representation guided denoiser (HGD) [42], input transformation through random resizing and padding (R&P) [74], three ensemble adversarially trained models (Inc$_{ens3}$, Inc$_{ens4}$ and IncRes$_{ens}$) [68], adversarial training with project gradient descent white-box attacker (PGD) [48,75], adversarial logits pairing (ALP) [35], and feature denoising adversarially trained ResNeXt-101 (FD) [75].

Among them, HGD [42] and R&P [74] are the *rank-1 submission* and *rank-2 submission* in the NeurIPS 2017 defense competition [39], respectively. FD [75] is the *rank-1 submission* in the Competition on Adversarial Attacks and Defenses 2018. The top-1 error rates of these methods on our dataset are shown in Table 1.

Implementation Details. To train the gradient transformer module, we randomly select another 5000 images from the validation set of ILSVRC 2012 [15] as the training set. Note that the training set and the testing set are disjoint.

For the region split function, we choose $r(h, w) = w$ as default, and will discuss different region split functions in Sect. 4.4. We train the gradient transformer module for 50 epochs. When testing, we use a zero array as the input of the gradient transformer module to get universal adversarial perturbations, *i.e.* the fixed input $z = 0$.

4.2   Under-Fitting and Universal

To verify the connections between under-fitting and universal adversarial perturbations, we change the number of training images so that the models are supposed to be under-fitting (due to the model capacity becomes low compared to large dataset) or not. Specifically, we select 4, 5k or 45k images from the validation set of ILSVRC 2012 as the training set. We insert four probes a, b, c, and d in the gradient transformer module as shown in Fig. 3 and compare their values in Fig. 5 with respect to the training iterations.

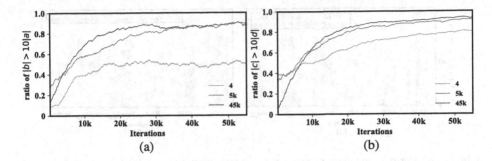

Fig. 5. Universal analysis of RHP. In (a), we plot the ratio of the number of variables in probe pairs (a, b) satisfying that $|b| > 10|a|$ to the total number of variables when training with 4, 5k or 45k images. In (b), we plot the case of $|c| > 10|d|$

When the gradient transformer module is well trained with 5k or 45k images, we observe that: 1) c overwhelms d, indicating the residual learning branch dominates the final output, *i.e.*, $\hat{g} \approx c$; and 2) b overwhelms a, indicating the output of the convolutional layer is less related to the input gradient g. Based on the two observations, we conclude that the gradient transformer module is quasi-input-independent when the module is under-fitted by a large number of training images in this case. Such a property is beneficial to generate universal adversarial perturbations (see Sect. 3.3).

When the number of training images is limited (say 4 images), we observe that b does not overwhelm a, indicating the output of the conv layer is related to the input gradient g, since a small training set cannot lead to under-fitting.

This conclusion is further supported by Fig. 6(a): when training with 4 images, the performance gap between universal inference (use a fixed zero as the input of the gradient transformer module) and image dependent inference (use the loss gradient as the input) is quite large. The gap is reduced when using more data for training.

To provide a better understanding of our implicit universal adversarial perturbation generating mechanism, we present an ablation study by comparing our method with other 3 strategies of generating universal adversarial perturbation with the same region split function. The compared includes 1) RP: Randomly assigns the Perturbation as $+\epsilon$ and $-\epsilon$ for each region; 2) OP: iteratively Optimizes the Perturbation to maximize classification loss on the naturally trained model (the idea of [50]); 3) TU: explicitly Trains a Universal adversarial perturbations. The only difference between TU and our proposed RHP is that random noises take the place of the loss gradient **g** in Fig. 2 (following [54]) and are fed to the gradient transformer module. RHP is our proposed implicitly method, and the gradient transformer module becomes quasi-input-independent without taking random noise as the training input.

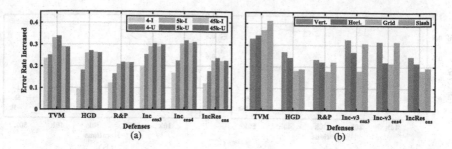

Fig. 6. (a) Performance comparison of universal (denoted by -U) inference and image dependent inference (denoted by -I) by varying the number of training images (4, 5k or 45k). (b) Performance comparison among four split functions, including vertical partition, horizontal partition, grid partition and slash partition

We evaluate above four settings on $IncRes_{ens}$, the error rates increase by 14.0%, 19.4%, 19.3%, and 24.6% for RP, OP, TU, and RHP respectively. Since our implicit method has a proper initialization (Sect. 3.2), we observe that our implicit method constructs stronger universal adversarial perturbations.

4.3   Transferability Toward Defenses

We first conduct the comparison in Table 2 when the maximum perturbation $\epsilon = 16$ and 32, respectively.

A first glance shows that compared with other representatives, the proposed RHP provides much stronger attack toward defenses. For example, when attacking HGD [42] with $\epsilon = 16$, RHP outperforms FGSM [23] by 24.0%, MIM [16] by 19.5%, DIM [16,77] by 14.9%, UAP [50] by 24.9%, and GAP [54] by 25.5%, respectively. Second, universal methods generally perform worse than image-dependent methods as the latter can access and utilize the information from the clean images. Nevertheless, RHP, as a universal method, still beats those image-dependent methods by a large margin. At last, we observe that our method gains more when the maximum perturbation ϵ becomes larger.

The performance comparison is also done when generating adversarial examples by attacking IncV4 or IncRes. Here we do not report the performance of GAP, because the official code does not support generating adversarial examples with IncV4 or IncRes. As shown in Table 3 and Table 4, RHP still keeps strong against defense models. Meanwhile, it should be mentioned that when the model for generating adversarial perturbations is changed, RHP still generates universal adversarial examples. The only difference is that the gradients used in the training phase are changed, which then leads to a different set of parameters in the gradient transformer module.

Table 2. The increase of error rates (%) after attacking. The adversarial examples are generated with IncV3. In each cell, we show the results when the maximum perturbation $\epsilon = 16/32$, respectively. The left 3 columns (FGSM, MIM and DIM) are image-dependent methods while the right 3 columns are (UAP, GAP and RHP) are universal methods

Methods	FGSM [23]	MIM [16]	DIM [16,77]	UAP [50]	GAP [54]	RHP (ours)
TVM	21.9/45.3	18.2/37.1	21.9/41.0	4.78/12.1	18.5/50.1	**33.0/56.9**
HGD	2.84/20.7	7.30/18.7	11.9/32.1	1.94/11.3	1.34/37.9	**26.8/57.5**
R& P	6.80/13.9	7.52/13.7	12.0/21.9	2.42/6.66	3.52/26.9	**23.3/56.1**
Inc$_{ens3}$	10.0/17.9	11.4/17.3	16.7/26.1	1.00/7.82	5.48/33.3	**32.5/60.8**
Inc$_{ens4}$	9.34/15.9	10.9/16.5	16.2/25.0	1.80/8.34	4.14/29.4	**31.6/58.7**
IncRes$_{ens}$	6.86/13.3	7.76/13.6	10.8/19.6	1.88/5.60	3.76/22.5	**24.6/57.0**
PGD	1.90/12.8	1.36/6.86	1.84/7.70	0.04/1.04	1.28/10.2	**2.40/25.8**
ALP	17.0/32.3	15.3/24.4	15.5/24.7	7.98/11.5	15.6/30.0	**17.8/39.4**
FD	1.62/13.3	1.00/7.48	1.34/8.22	-0.1/0.40	0.56/11.1	**2.38/24.5**

Table 3. The increase of error rates (%) after attacking. The adversarial examples are generated with IncV4. In each cell, we show the results when the maximum perturbation $\epsilon = 16/32$, respectively. The left 3 columns (FGSM, MIM and DIM) are image-dependent methods while the right 2 columns are (UAP and RHP) are universal methods

Methods	FGSM [23]	MIM [16]	DIM [16,77]	UAP [50]	RHP (ours)
TVM	22.4/46.3	20.1/40.4	22.7/42.9	6.28/18.2	**37.1/58.4**
HGD	4.00/21.1	10.0/23.9	16.3/37.1	1.42/9.94	**23.4/59.8**
R& P	8.68/15.1	10.2/17.4	14.7/25.0	2.42/6.52	**20.2/57.6**
Inc$_{ens3}$	10.1/18.3	13.4/20.3	18.7/28.6	2.08/7.68	**27.5/60.3**
Inc$_{ens4}$	9.72/17.4	13.1/19.0	17.9/26.5	1.94/6.92	**26.7/62.5**
IncRes$_{ens}$	7.58/14.7	9.96/16.6	13.6/22.1	2.34/6.78	**21.2/58.5**
PGD	2.02/12.8	1.50/7.54	1.82/8.02	0.28/2.12	**2.20/29.7**
ALP	17.3/32.1	14.8/25.1	15.2/24.8	10.1/15.9	**20.3/42.1**
FD	1.42/13.4	1.24/8.18	1.62/8.74	0.16/1.18	**1.90/31.8**

4.4   Region Split Functions

In this section, we discuss the choice of region split functions, *i.e.*, vertical partition, horizontal partition, grid partition and slash partition (parallel to an increasing line with the slope equal to 0.5). Figure 6(b) shows the transferability to the defenses, which demonstrates that different region split functions are almost equivalently effective and all are stronger than our strongest baseline (DIM). Moreover, we observe an interesting phenomenon as presented in Fig. 7.

Table 4. The increase of error rates (%) after attacking. The adversarial examples are generated with IncRes. In each cell, we show the results when the maximum perturbation $\epsilon = 16/32$, respectively. The left 3 columns (FGSM, MIM and DIM) are image-dependent methods while the right 2 columns are (UAP and RHP) are universal methods

Methods	FGSM [23]	MIM [16]	DIM [16,77]	UAP [50]	RHP (**ours**)
TVM	20.6/44.1	20.3/39.4	24.6/44.0	7.10/24.7	**37.1/57.4**
HGD	5.34/22.3	15.0/28.1	23.7/44.1	2.14/10.6	**26.9/62.1**
R& P	10.1/15.8	13.4/22.1	22.5/34.5	2.50/8.36	**25.1/61.4**
Inc$_{ens3}$	11.7/19.4	17.4/24.6	25.8/37.1	1.88/8.28	**29.7/62.3**
Inc$_{ens4}$	10.5/17.2	15.1/22.5	22.4/33.7	1.74/7.22	**29.8/63.3**
IncRes$_{ens}$	10.4/16.3	13.6/22.6	20.2/32.5	1.96/8.18	**26.8/62.8**
PGD	2.06/13.8	1.84/8.80	**2.36/9.26**	0.40/3.78	2.20/**28.3**
ALP	17.5/32.6	12.3/25.9	12.6/25.9	7.12/17.0	**22.8/43.5**
FD	1.72/14.7	1.62/9.48	1.78/10.1	-0.1/3.06	**2.20/32.2**

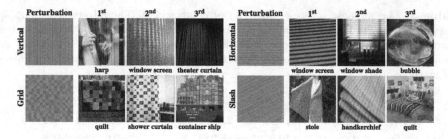

Fig. 7. Four universal adversarial perturbations generated by different region split functions, and the corresponding top-3 target categories

In each row of Fig. 7, we exhibit the universal adversarial perturbation generated by a certain kind of region split functions, followed by the top-3 categories to which the generated adversarial examples are most likely to be misclassified. For each category, we show a clean image as an exemplar. Note that our experiments are about the non-targeted attack, indicating the target class is undetermined and solely relies on the region split function.

As can be seen, the regionally homogeneous perturbations with different region split functions seem to be targeting at different categories, with an inherent connection between the low-level cues (*e.g.*, texture, shape) they share. For example, when using grid partition, the top-3 target categories are quilt, shower curtain, and container ship, respectively, and one can observe that images in the three categories generally have grid-structured patterns.

Motivated by these qualitative results, we have a preliminary hypothesis that the regionally homogeneous perturbations tend to attack the low-level part of a model. The claim is not supported by a theoretical proof, however, it inspires us

Table 5. Comparison of cross-task transferability. We attack segmentation model and test on the detection model Faster R-CNN, and report the value of mAP (lower is better for attacking methods). "−" denotes the baseline performance without attacks

Attacks	−	FGSM	MIM	DIM	RHP (ours)
mAP	69.2	43.1	41.6	36.2	**31.6**

to test the cross-task transferability of RHP. As it is a common strategy to share the low-level CNN architecture/information in multi-task learning systems [58], we conjecture that RHP can well transfer between different tasks (see below).

4.5   Cross-Task Transferability

To demonstrate the cross-task transferability of RHP, we attack with the semantic segmentation task and test on the object detection task.

In more detail, we attack a semantic segmentation model (an Xception-65 [13] based deeplab-v3+ [10]) on the Pascal VOC 2012 segmentation val [19], and obtain the adversarial examples. Then, we take a VGG16 [61] based Faster-RCNN model [56], trained on MS COCO [43] and VOC2007 trainval, as the testing model. To avoid testing images occurred in the training set of detection model, the testing set is the union of VOC2012 segmentation val and VOC2012 detection trainval, then we remove the images in VOC2007 dataset. The baseline performance of the clean images is mAP 69.2. Here mAP score is the average of the precisions at different recall values.

As shown in Table 5, RHP reports the lowest mAP with object detection, which demonstrates the stronger cross-task transferability than the baseline image-dependent perturbations, i.e., FGSM [23], MIM [16], and DIM [16,77].

5   Conclusion

By white-box attacking naturally trained models and defense models, we observe the regional homogeneity of adversarial perturbations. Motivated by this observation, we propose a transforming paradigm and a gradient transformer module to generate the regionally homogeneous perturbation (RHP) specifically for attacking defenses. RHP possesses three merits, including 1) transferability: we demonstrate that RHP well transfers across different models (i.e., transfer-based attack) and different tasks; 2) universal: taking advantage of the under-fitting of the gradient transformer module, RHP generates universal adversarial examples without explicitly enforcing the learning procedure towards it; 3) strong: RHP successfully attacks 9 representative defenses and outperforms the state-of-the-art attacking methods by a large margin.

Recent studies [42,75] show that the mechanism of some defense models can be interpreted as a "denoising" procedure. Since RHP is less like noise compared with other perturbations, it would be interesting to reveal the property of RHP from a denoising perspective in future works. Meanwhile, although evaluated with the non-targeted attack, RHP is supposed to be strong targeted attack as well, which requires further exploration and validation.

Acknowledgements. We thank Yuyin Zhou and Zhishuai Zhang for their insightful comments and suggestions. This work was partially supported by the Johns Hopkins University Institute for Assured Autonomy with grant IAA 80052272.

References

1. Akhtar, N., Liu, J., Mian, A.: Defense against universal adversarial perturbations. In: CVPR (2018)
2. Ba, J.L., Kiros, J.R., Hinton, G.E.: Layer normalization. arXiv preprint arXiv:1607.06450 (2016)
3. Bai, S., Li, Y., Zhou, Y., Li, Q., Torr, P.H.: Metric attack and defense for person re-identification. arXiv preprint arXiv:1901.10650 (2019)
4. Baluja, S., Fischer, I.: Learning to attack: adversarial transformation networks. In: AAAI (2018)
5. Bhagoji, A.N., He, W., Li, B., Song, D.: Practical black-box attacks on deep neural networks using efficient query mechanisms. In: Ferrari, V., Hebert, M., Sminchisescu, C., Weiss, Y. (eds.) ECCV 2018. LNCS, vol. 11216, pp. 158–174. Springer, Cham (2018). https://doi.org/10.1007/978-3-030-01258-8_10
6. Bishop, C.M.: The bias-variance decomposition. In: Pattern Recognition and Machine Learning, pp. 147–152. Springer, Heidelberg (2006)
7. Borkar, T., Heide, F., Karam, L.: Defending against universal attacks through selective feature regeneration. In: CVPR (2020)
8. Brendel, W., Rauber, J., Bethge, M.: Decision-based adversarial attacks: reliable attacks against black-box machine learning models. In: ICLR (2018)
9. Cao, Y., et al.: Adversarial sensor attack on lidar-based perception in autonomous driving. In: ACM SIGSAC CCS (2019)
10. Chen, L.-C., Zhu, Y., Papandreou, G., Schroff, F., Adam, H.: Encoder-decoder with Atrous separable convolution for semantic image segmentation. In: Ferrari, V., Hebert, M., Sminchisescu, C., Weiss, Y. (eds.) ECCV 2018. LNCS, vol. 11211, pp. 833–851. Springer, Cham (2018). https://doi.org/10.1007/978-3-030-01234-2_49
11. Chen, P.Y., Zhang, H., Sharma, Y., Yi, J., Hsieh, C.J.: ZOO: zeroth order optimization based black-box attacks to deep neural networks without training substitute models. In: Proceedings of the 10th ACM Workshop on Artificial Intelligence and Security (2017)
12. Chen, X., Liu, C., Li, B., Lu, K., Song, D.: Targeted backdoor attacks on deep learning systems using data poisoning. arXiv preprint arXiv:1712.05526 (2017)
13. Chollet, F.: Xception: deep learning with depthwise separable convolutions. In: ICCV (2017)
14. Das, N., et al.: SHIELD: fast, practical defense and vaccination for deep learning using JPEG compression. In: KDD. ACM (2018)
15. Deng, J., Dong, W., Socher, R., Li, L.J., Li, K., Fei-Fei, L.: ImageNet: a large-scale hierarchical image database. In: CVPR (2009)
16. Dong, Y., et al.: Boosting adversarial attacks with momentum. In: CVPR (2018)
17. Dong, Y., Pang, T., Su, H., Zhu, J.: Evading defenses to transferable adversarial examples by translation-invariant attacks. In: CVPR (2019)

18. Dziugaite, G.K., Ghahramani, Z., Roy, D.M.: A study of the effect of JPG compression on adversarial images. arXiv preprint arXiv:1608.00853 (2016)
19. Everingham, M., Eslami, S.A., Van Gool, L., Williams, C.K., Winn, J., Zisserman, A.: The pascal visual object classes challenge: a retrospective. IJCV **111**(1), 98–136 (2015)
20. Eykholt, K., et al.: Robust physical-world attacks on deep learning visual classification. In: CVPR (2018)
21. Gao, L., Zhang, Q., Song, J., Liu, X., Shen, H.T.: Patch-wise attack for fooling deep neural network. arXiv preprint arXiv:2007.06765 (2020)
22. Goodfellow, I., Bengio, Y., Courville, A.: Deep Learning. MIT Press, Cambridge (2016)
23. Goodfellow, I.J., Shlens, J., Szegedy, C.: Explaining and harnessing adversarial examples. In: ICLR (2015)
24. Goyal, P., et al.: Accurate, large minibatch SGD: training ImageNet in 1 hour. arXiv preprint arXiv:1706.02677 (2017)
25. Guo, C., Frank, J.S., Weinberger, K.Q.: Low frequency adversarial perturbation. arXiv preprint arXiv:1809.08758 (2018)
26. Guo, C., Rana, M., Cissé, M., van der Maaten, L.: Countering adversarial images using input transformations. In: ICLR (2018)
27. Haykin, S.S.: Finite sample-size considerations. In: Neural Networks and Learning Machines, vol. 3, pp. 82–86. Pearson Upper Saddle River (2009)
28. He, K., Zhang, X., Ren, S., Sun, J.: Deep residual learning for image recognition. In: CVPR (2016)
29. Hendrik Metzen, J., Chaithanya Kumar, M., Brox, T., Fischer, V.: Universal adversarial perturbations against semantic image segmentation. In: ICCV (2017)
30. Huang, L., et al.: Universal physical camouflage attacks on object detectors. In: CVPR (2020)
31. Ioffe, S., Szegedy, C.: Batch normalization: Accelerating deep network training by reducing internal covariate shift. In: ICML (2015)
32. Jia, R., Konstantakopoulos, I.C., Li, B., Spanos, C.: Poisoning attacks on data-driven utility learning in games. In: ACC (2018)
33. Jin, W., Li, Y., Xu, H., Wang, Y., Tang, J.: Adversarial attacks and defenses on graphs: a review and empirical study. arXiv preprint arXiv:2003.00653 (2020)
34. Jin, W., Ma, Y., Liu, X., Tang, X., Wang, S., Tang, J.: Graph structure learning for robust graph neural networks. arXiv preprint arXiv:2005.10203 (2020)
35. Kannan, H., Kurakin, A., Goodfellow, I.: Adversarial logit pairing. arXiv preprint arXiv:1803.06373 (2018)
36. Khrulkov, V., Oseledets, I.: Art of singular vectors and universal adversarial perturbations. In: CVPR (2018)
37. Kingma, D.P., Ba, J.: Adam: a method for stochastic optimization. arXiv preprint arXiv:1412.6980 (2014)
38. Kurakin, A., Goodfellow, I., Bengio, S.: Adversarial examples in the physical world. In: ICLR Workshop (2017)
39. Kurakin, A., et al.: Adversarial attacks and defences competition. arXiv preprint arXiv:1804.00097 (2018)
40. Li, Y., Bai, S., Zhou, Y., Xie, C., Zhang, Z., Yuille, A.: Learning transferable adversarial examples via ghost networks. In: AAAI (2020)

41. Li, Y., et al.: Volumetric medical image segmentation: a 3D deep coarse-to-fine framework and its adversarial examples. In: Lu, L., Wang, X., Carneiro, G., Yang, L. (eds.) Deep Learning and Convolutional Neural Networks for Medical Imaging and Clinical Informatics. ACVPR, pp. 69–91. Springer, Cham (2019). https://doi.org/10.1007/978-3-030-13969-8_4

42. Liao, F., Liang, M., Dong, Y., Pang, T., Hu, X., Zhu, J.: Defense against adversarial attacks using high-level representation guided denoiser. In: CVPR (2018)

43. Lin, T.-Y., et al.: Microsoft COCO: common objects in context. In: Fleet, D., Pajdla, T., Schiele, B., Tuytelaars, T. (eds.) ECCV 2014. LNCS, vol. 8693, pp. 740–755. Springer, Cham (2014). https://doi.org/10.1007/978-3-319-10602-1_48

44. Liu, L., et al.: Deep neural network ensembles against deception: ensemble diversity, accuracy and robustness. In: 2019 IEEE 16th International Conference on Mobile Ad Hoc and Sensor Systems (MASS), pp. 274–282. IEEE (2019)

45. Liu, X., Cheng, M., Zhang, H., Hsieh, C.-J.: Towards robust neural networks via random self-ensemble. In: Ferrari, V., Hebert, M., Sminchisescu, C., Weiss, Y. (eds.) ECCV 2018. LNCS, vol. 11211, pp. 381–397. Springer, Cham (2018). https://doi.org/10.1007/978-3-030-01234-2_23

46. Liu, Y., Chen, X., Liu, C., Song, D.: Delving into transferable adversarial examples and black-box attacks. In: ICLR (2017)

47. Ma, X., et al.: Characterizing adversarial subspaces using local intrinsic dimensionality. In: ICLR (2018)

48. Madry, A., Makelov, A., Schmidt, L., Tsipras, D., Vladu, A.: Towards deep learning models resistant to adversarial attacks. In: ICLR (2018)

49. Mao, X., Chen, Y., Li, Y., He, Y., Xue, H.: GAP++: learning to generate target-conditioned adversarial examples. arXiv preprint arXiv:2006.05097 (2020)

50. Moosavi-Dezfooli, S.M., Fawzi, A., Fawzi, O., Frossard, P.: Universal adversarial perturbations. In: CVPR (2017)

51. Mopuri, K.R., Garg, U., Babu, R.V.: Fast feature fool: a data independent approach to universal adversarial perturbations. In: BMVC (2017)

52. Naseer, M.M., Khan, S.H., Khan, M.H., Khan, F.S., Porikli, F.: Cross-domain transferability of adversarial perturbations. In: Advances in Neural Information Processing Systems, pp. 12905–12915 (2019)

53. Poursaeed, O., Jiang, T., Yang, H., Belongie, S., Lim, S.N.: Fine-grained synthesis of unrestricted adversarial examples. arXiv preprint arXiv:1911.09058 (2019)

54. Poursaeed, O., Katsman, I., Gao, B., Belongie, S.: Generative adversarial perturbations. In: CVPR (2017)

55. Qiu, H., Xiao, C., Yang, L., Yan, X., Lee, H., Li, B.: SemanticAdv: generating adversarial examples via attribute-conditional image editing. arXiv preprint arXiv:1906.07927 (2019)

56. Ren, S., He, K., Girshick, R., Sun, J.: Faster R-CNN: towards real-time object detection with region proposal networks. In: NeurIPS (2015)

57. Roth, H.R., et al.: DeepOrgan: multi-level deep convolutional networks for automated pancreas segmentation. In: Navab, N., Hornegger, J., Wells, W.M., Frangi, A.F. (eds.) MICCAI 2015. LNCS, vol. 9349, pp. 556–564. Springer, Cham (2015). https://doi.org/10.1007/978-3-319-24553-9_68

58. Ruder, S.: An overview of multi-task learning in deep neural networks. arXiv preprint arXiv:1706.05098 (2017)

59. Rudin, L.I., Osher, S., Fatemi, E.: Nonlinear total variation based noise removal algorithms. Physica D Nonlinear Phenomena **60**(1–4), 259–268 (1992)

60. Shafahi, A., Najibi, M., Xu, Z., Dickerson, J., Davis, L.S., Goldstein, T.: Universal adversarial training. In: AAAI (2020)

61. Simonyan, K., Zisserman, A.: Very deep convolutional networks for large-scale image recognition. In: ICLR (2015)

62. Sun, M., et al.: Data poisoning attack against unsupervised node embedding methods. arXiv preprint arXiv:1810.12881 (2018)

63. Sun, Y., Wang, S., Tang, X., Hsieh, T.Y., Honavar, V.: Adversarial attacks on graph neural networks via node injections: a hierarchical reinforcement learning approach. In: Proceedings of the Web Conference 2020, pp. 673–683 (2020)

64. Szegedy, C., Ioffe, S., Vanhoucke, V., Alemi, A.A.: Inception-v4, Inception-ResNet and the impact of residual connections on learning. In: AAAI (2017)

65. Szegedy, C., Vanhoucke, V., Ioffe, S., Shlens, J., Wojna, Z.: Rethinking the inception architecture for computer vision. In: CVPR (2016)

66. Szegedy, C., et al.: Intriguing properties of neural networks. In: ICLR (2014)

67. Tang, X., Li, Y., Sun, Y., Yao, H., Mitra, P., Wang, S.: Transferring robustness for graph neural network against poisoning attacks. In: Proceedings of the 13th International Conference on Web Search and Data Mining, pp. 600–608 (2020)

68. Tramèr, F., Kurakin, A., Papernot, N., Boneh, D., McDaniel, P.: Ensemble adversarial training: attacks and defenses. In: ICLR (2018)

69. Ulyanov, D., Vedaldi, A., Lempitsky, V.: Instance normalization: the missing ingredient for fast stylization. arXiv preprint arXiv:1607.08022 (2016)

70. Wu, Y., He, K.: Group normalization. In: Ferrari, V., Hebert, M., Sminchisescu, C., Weiss, Y. (eds.) ECCV 2018. LNCS, vol. 11217, pp. 3–19. Springer, Cham (2018). https://doi.org/10.1007/978-3-030-01261-8_1

71. Xiao, C., Zhong, P., Zheng, C.: Enhancing adversarial defense by k-winners-take-all. In: ICLR (2020)

72. Xiao, C., Li, B., Zhu, J.Y., He, W., Liu, M., Song, D.: Generating adversarial examples with adversarial networks. In: IJCAI (2018)

73. Xie, C., Tan, M., Gong, B., Yuille, A., Le, Q.V.: Smooth adversarial training. arXiv preprint arXiv:2006.14536 (2020)

74. Xie, C., Wang, J., Zhang, Z., Ren, Z., Yuille, A.: Mitigating adversarial effects through randomization. In: ICLR (2018)

75. Xie, C., Wu, Y., Maaten, L.v.d., Yuille, A.L., He, K.: Feature denoising for improving adversarial robustness. In: CVPR (2019)

76. Xie, C., Yuille, A.: Intriguing properties of adversarial training at scale. In: ICLR (2020)

77. Xie, C., Zhang, Z., Zhou, Y., Bai, S., Wang, J., Ren, Z., Yuille, A.L.: Improving transferability of adversarial examples with input diversity. In: CVPR (2019)

78. Yang, C., Kortylewski, A., Xie, C., Cao, Y., Yuille, A.: PatchAttack: a black-box texture-based attack with reinforcement learning. arXiv preprint arXiv:2004.05682 (2020)

79. Zhang, Z., Zhu, X., Li, Y., Chen, X., Guo, Y.: Adversarial attacks on monocular depth estimation. arXiv preprint arXiv:2003.10315 (2020)

80. Zhou, W., et al.: Transferable adversarial perturbations. In: Ferrari, V., Hebert, M., Sminchisescu, C., Weiss, Y. (eds.) Computer Vision – ECCV 2018. LNCS, vol. 11218, pp. 471–486. Springer, Cham (2018). https://doi.org/10.1007/978-3-030-01264-9_28

Author Index